Elizabeth-Jane Smith.   15/2/90.
HND FOOD TECHNOLOGY.
THE WEST OF SCOTLAND
        AGRICULTURAL COLLEGE.
        AUCHINCRUIVE.

As additional learning tools, McGraw-Hill also publishes a laboratory manual and a study guide to supplement your understanding of these textbooks. Here is the information your bookstore manager will need to order them for you:
49235-2   LABORATORY MANUAL TO ACCOMPANY MICROBIOLOGY, Fifth Edition; 49236-0   STUDY GUIDE TO ACCOMPANY MICROBIOLOGY, Fifth Edition.

Microbiology

MICHAEL J. PELCZAR, JR.

Professor Emeritus
University of Maryland

E. C. S. CHAN

Associate Professor of Microbiology
McGill University

NOEL R. KRIEG

Alumni Distinguished Professor of Microbiology
Virginia Polytechnic Institute and State University

With the assistance of
MERNA FOSS PELCZAR

FIFTH EDITION

McGRAW-HILL BOOK COMPANY

New York Auckland Bogotá Guatemala Hamburg Lisbon
London Madrid Mexico New Delhi Panama Paris San Juan São Paulo
Singapore Sydney Tokyo Toronto

MICROBIOLOGY, FIFTH EDITION
INTERNATIONAL EDITION

3 4 5 6 7 8 9 0 FSP 8 9 4 3 2 1 0 9

This book was set in Melior by Ruttle, Shaw & Wetherill, Inc.
The editors were Mary Jane Martin, Kathleen M. Civetta, and James W.
Bradley.
The designer was Hermann Strohbach.
The production supervisor was Charles Hess.

Library of Congress Cataloging-in-Publication Data

Pelczar, Michael Joseph, date
  Microbiology

  Includes bibliographies and indexes.
  1. Microbiology.   I. Chan, Eddie Chin Sun, date
II. Krieg, Noel R.   III. Title
QR 41.2. P4   1986   576   84-23932
ISBN 0–07–049234–4

Cover photograph: Sporangiophores of a zygomycete, Phycomces sp. (J. Robert
Waaland, University of Washington.)

When ordering this title use ISBN 0–07–Y66494–3

Contents

Preface

"Messieurs, c'est les microbes qui auront le dernier mot." *Pasteur*

For the major part of the twentieth century the physical sciences have dominated science and engineering. This situation was due to a large degree to the development of the atomic bomb, and the achievements of the Soviet Union in outer space. The successful launching of the first satellite into space (Sputnik) in 1957 by the Soviet Union accelerated physical science research and development programs in the United States by the government, by universities, and by industry. We became engaged in a race for leadership in science and technology.

We are now experiencing a rapid shift of national priorities in research and development. As we approach the twenty-first century, we see biology emerging as one of the top priorities in the field of science, and among the biological sciences microbiology has gained new stature. Microorganisms and their activities are increasingly central to many of the concerns of society both nationally and internationally. The problems of the global environment, the recognition of the need to recycle natural resources, the discovery of recombinant DNA and the resulting high technology of genetic engineering—these and other developments have placed microbiology in the limelight.

Microbiology is emerging as the key biological science. Microorganisms provide the models used in molecular biology for research. This research at the molecular level has provided, and continues to provide, the answers to numerous fundamental questions in genetics, metabolism, and cell forms and functions. Microorganisms also provide model systems for studying the relationships between species in mixed populations.

There is growing recognition of the potential of microorganisms in many applied areas. The ability of microorganisms to decompose materials such as herbicides, pesticides, and oils in oil spills; the potential of microorganisms as food supplements; the exploitation of microbial activity to produce energy such as methane gas for rural consumption; and the potential of new therapeutic substances produced by microorganisms—these and other uses of microorganisms are becoming increasingly attractive.

Recombinant DNA technology, commonly referred to as genetic engineering, is one of the principal thrusts of the emerging high technologies in the biological sciences. Recombinant DNA technology makes it feasible to consider genetically manipulated (engineered) microorganisms for commercial production of new and valuable products for a variety of purposes, e.g., medicinals, fuel, and food.

This fifth edition of MICROBIOLOGY retains many of the features that have proved successful in the first four editions, particularly the balance between *fundamental* or basic microbiology and *applied* microbiology. This approach emphasizes the importance of integrating new knowledge gained through basic research with applied research and development programs. A strong continuum of research and development, from the basic to the applied, facilitates the development of benefits for society.

One of the new features of this edition is a presentation of the classification of bacteria in a totally new format following the scheme introduced in the first volume of the recently published *Bergey's Manual of Systematic Bacteriology.* (One of us, Noel R. Krieg, served as editor of the first volume.)

We have also expanded and revised the material on metabolism, bacterial genetics, and genetic engineering and reorganized the section on microorganisms and disease. Careful attention has been given to updating of information in all aspects of the discipline. Many new summary tables have been developed, and new illustrations selected. New review questions, and updated references, follow each chapter.

The subject material is presented in eight parts. As a new feature, each part now opens with an essay providing added insight into the material that follows. Each chapter begins with a chapter outline and an introduction. Many chapters now contain boxed essays highlighting important discoveries and developments in microbiology. As in the past, the order of arrangement of chapters lends itself to adjustments in any sequence desired by the instructor.

A considerable amount of the artwork has been drawn by Dr. Erwin F. Lessel (a microbiologist in his own right). We have found this to be a distinct asset in terms of improving the pedagogical value of illustrations.

Three valuable supplementary publications are available to accompany this new edition: an INSTRUCTOR'S MANUAL, a STUDENT'S GUIDE, and LABORATORY EXPERIMENTS IN MICROBIOLOGY. Each has been revised to conform with the subject matter in the fifth edition of MICROBIOLOGY. We have provided extensive cross-referencing among these four publications. The INSTRUCTOR'S MANUAL includes suggested lecture and laboratory schedules, chapter summaries, sources of audiovisual aids, sources of laboratory equipment and reagents, as well as sample test questions. The STUDENT'S GUIDE has been developed to assist the student in his or her efforts to comprehend the subject matter. It provides for each chapter a concise statement of the content (an overview), a comprehensive topical outline, and a series of self-study questions of several types.

The writing of a textbook on a subject as comprehensive as microbiology requires considerable assistance from a large number of professional colleagues. Among these, we wish to acknowledge the following persons who were generous in their assistance, particularly in commenting upon drafts of various chapters: Phillip M. Achey, University of Florida; Ronald L. Crawford, University of Minnesota; Loretta C. Ellias, Florida State University; Louis R. Fina, Kansas State University; Thomas R. Jewell, University of Wisconsin-Eau Claire; Ted R. Johnson, St. Olaf College; Robert J. Janssen, University of Arizona; David Kafkewitz, Rutgers University; Joseph S. Layne, Memphis State University; Haideh Lightfoot, Eastern Washington University; David Pramer, Rutgers University; Ramond J. Seidler, Oregon State University; Robert Todd, South Dakota State University; Anne H. Williams, Evergreen Valley College; and Fred D. Williams, Iowa State University.

Special thanks are due Malcolm G. Baines, McGill University; John O. Corliss, University of Maryland; A. C. Dornbush, Medical Research Division, American

Cyanamid Co.; Jerome J. Motta, University of Maryland; and Robert C. Bates, Virginia Polytechnic Institute and State University, who provided extra help with certain chapters.

We are grateful to our colleagues at the McGraw-Hill Book Company, Kathleen Civetta, Editor; James W. Bradley, Editing Supervisor; and Charles Hess, Production Supervisor, for their pleasant cooperation and assistance in the task of preparing and publishing this book. Thanks are due also to Karen Jacques and Edna Khalil for their skillful assistance in the preparation of manuscript.

In the writing of this text, each chapter has been the primary responsibility of one author. However, each of us has read and critiqued all the chapters. As previously mentioned, we have had the benefit of reviews of each chapter from several of our professional associates. In the end we take collective responsibility for the complete content of this text.

Michael J. Pelczar, Jr.
E. C. S. Chan
Noel R. Krieg

PART ONE
INTRODUCTION TO MICROBIOLOGY

The introduction of agar for pure culture study of microorganisms

The discovery of the world of microorganisms came about as investigators developed microscopes and used them to examine droplets of natural fluids. Menageries of microbes were revealed from a variety of specimens. Initially these observations were a source of great curiosity. During the period from 1600 to 1800, considerable information accumulated about the occurrence of these microscopic forms of life. Great debates emerged as to the origin of these microbes. The controversy centered on the question of whether they arose from nonliving materials, i.e., spontaneous generation.

Simultaneous with the development of evidence over hundreds of years to disprove spontaneous generation was the growing acceptance of the concept that these microorganisms were the cause of many conditions that occur in everyday life, ranging from food spoilage to diseases of humans, other animals, and plants. In the latter part of the nineteenth century a major problem confronted investigators searching for evidence to prove that a specific kind of microorganism was responsible for a disease or spoilage. How could they isolate in pure culture the microorganism suspected of causing the change and prove that it was the causal agent? What the researchers needed was a solid nutrient substance upon which a specimen could be spread so that individual microbial cells would be distant from each other. Upon incubation, each cell would reproduce, resulting in a mass of identical cells (a colony). A small portion of the colony could then be transferred to a fresh medium and be maintained as a pure culture for subsequent experiments, i.e., the pure culture technique.

Robert Koch (1843–1885) was particularly concerned with the need to develop a technique for the isolation of microorganisms in pure culture in order to establish the causative agent of a disease. He experimented with slices of sterile potatoes as the solid surface upon which to grow colonies of bacteria. This proved unsatisfactory for a variety of reasons. He tried gelatin as a solidifying agent. This had the desirable feature of being a transparent gel, but it had the serious disadvantage of becoming liquid above 25°C, which is below the optimum temperature for the growth of human disease-producing bacteria.

The solution to this problem was provided in 1883 by a German housewife, Fannie E. Hesse, who spent part of her time working in the laboratory of her husband, Walther Hesse. Hesse, a physician, was a former student of Robert Koch. Frau Hesse suggested to her husband that he use agar, a polysaccharide of algal origin, as a substitute for gelatin in microbiological media. She had gained experience with the characteristics of agar in the process of making jelly; the agar increased the consistency. His experiments with agar as a substitute for gelatin were dramatically successful. This observation was of such significance that Hesse promptly reported the experiments with agar to Robert Koch. Koch immediately recognized the great value of agar as a solidifying agent for microbiological media and adopted its use. Agar goes into solution (1.5 percent) at 100°C and solidifies at 45°C. Upon jelling at 45°C, it remains solid at elevated temperatures—temperatures just below 100°C. This feature makes it possible to incubate the inoculated media at almost any desired temperature and still have the medium remain solid.

It is a remarkable fact that agar, introduced for use as a solidifying agent in microbiological media just about 100 years ago, has not been replaced. Agar is as important now as it was then. The manner in which it was discovered adds credence to one of Louis Pasteur's observations: "Chance favors the prepared mind."

Preceding page. A compound microscope made by John Marshall of London about 1700 after a design by Robert Hooke. A condensing lens on a jointed arm allowed the instrument to be tilted on a ball-and-socket joint. (*Courtesy of the Armed Forces Institute of Pathology, Washington, D.C.*)

Chapter 1   The Scope of Microbiology

Microbiology is the study of living organisms of microscopic size, which include bacteria, fungi, algae, protozoa, and the infectious agents at the borderline of life that are called viruses. It is concerned with their form, structure, reproduction, physiology, metabolism, and classification. It includes the study of their distribution in nature, their relationship to each other and to other living organisms, their effects on human beings and on other animals and plants, their abilities to make physical and chemical changes in our environment, and their reactions to physical and chemical agents.

Microorganisms are closely associated with the health and welfare of human beings; some microorganisms are beneficial and others are detrimental. For example, microorganisms are involved in the making of yogurt, cheese, and wine; in the production of penicillin, interferon, and alcohol; and in the processing of domestic and industrial wastes. Microorganisms can cause disease, spoil food, and deteriorate materials like iron pipes, glass lenses, and wood pilings.

Most microorganisms are unicellular. In unicellular organisms all the life processes are performed by a single cell. In the so-called higher forms of life, organisms are composed of many cells that are arranged in tissues and organs to perform specific functions. Regardless of the complexity of an organism, the cell is the basic structural unit of life. All living cells are fundamentally similar.

The word *cell* was first used more than two centuries ago by an Englishman, Robert Hooke (1635–1703), in his descriptions (1665) of the fine structure of cork and other plant materials. The honeycomblike structure he observed in a thin slice of cork (see Fig. 1-1) was due to the cell walls of cells that were once

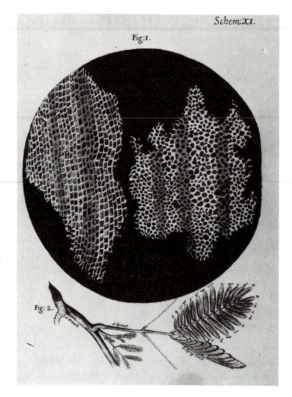

Figure 1-1. Robert Hooke's drawing of a thin slice of cork as he observed it under the microscope. This drawing was included in a report made to the Royal Society (London) in 1665. He is credited as being the first person to use the word *cell*. *(Courtesy of National Library of Medicine.)*

living. But the concept of the cell as the structural unit of life—the **cell theory**—is credited to two Germans, Matthias Schleiden and Theodor Schwann, who in 1838–1839 described cells as the basic structural and functional units of all organisms. Schleiden and Schwann recognized that all cells, no matter what the organism, are very similar in structure. As the concept of the cell as the basic unit of life gained acceptance, investigators speculated on the nature of the substance contained within the cell. **Protoplasm** (Greek *proto*, "first"; *plasm*, "formed substance", introduced to characterize the living material of a cell), is a colloidal organic complex consisting largely of protein, lipids, and nucleic acids. These substances are enclosed by membranes or cell walls; and the protoplasm always contains nuclei or an equivalent nuclear substance (see Fig. 1-2). Developments in electron-microscope techniques have made it possible to reveal the complex intricacies of intracellular organization (see Fig. 1-3).

All biological systems have the following characteristics in common: (1) the ability to reproduce; (2) the ability to ingest or assimilate food substances and metabolize them for energy and growth; (3) the ability to excrete waste products; (4) the ability to react to changes in their environment—sometimes called **irritability**; and (5) susceptibility to mutation. In the study of microbiology, we encounter "organisms" which may represent the borderline of life. These are the viruses, which are simpler in structure and composition than single cells. Viruses provide an exciting challenge and an opportunity to gain a better un-

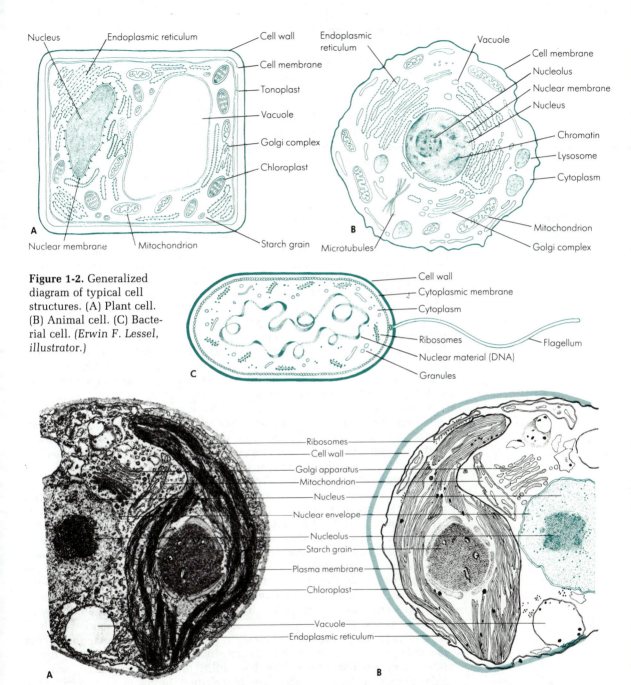

Figure 1-2. Generalized diagram of typical cell structures. (A) Plant cell. (B) Animal cell. (C) Bacterial cell. *(Erwin F. Lessel, illustrator.)*

Figure 1-3. (A) Electron micrograph of the alga *Chlamydomonas reinhardii* (X15,000), a eucaryotic cell. *(Courtesy of George E. Palade, The Rockefeller University, by permission of Holt, New York, publishers of Ariel G. Leowy and Philip Seikovitz, Cell Structure and Function, 1969.)* (B) Schematic representation of (A). *(Erwin F. Lessel, illustrator.)*

derstanding of the nature of complex organic substances that may bridge the gap between the living and the nonliving worlds.

Viruses are obligate parasites; that is, they are obligated to grow within an appropriate host cell—plant, animal, or microbe. They cannot multiply outside a host cell. However, when a virus enters an appropriate living cell, it is able to direct the synthesis of hundreds of identical viruses, using the cell's energy and biochemical machinery. A virus is made up of substances unique to life: **nucleic acids** (chemicals that make up genetic material) and **proteins** (complex nitrogenous substances found in various forms in animals and plants).

MICROBIOLOGY AS A FIELD OF BIOLOGY

Biologists are known for their differing opinions as to how the huge field of biology can best be subdivided. Historically, the divisions followed the major groups of life, as in zoology (animals), botany (plants), entomology (insects), and microbiology (microorganisms). Another manner of subdividing the subject matter of biology is based on the level at which the study is conducted: for example, studies at the level of molecular constituents of the cell (molecular biology); studies at the level of the cell (cell biology); studies at the level of the intact organism (organismal biology); and studies of groups of organisms (population biology). Still another approach is to establish divisions on the basis of form and function, as in morphology or anatomy, physiology, metabolism, and genetics. In some colleges and universities, the study of microbiology is carried out in a department of biology; in others, it is in a department of microbiology or a department of molecular biology.

Irrespective of where microbiology is placed in the broad field of biology, microorganisms have some characteristics which make them ideal specimens for the study of numerous fundamental life processes. This is possible because, at the cellular level, many life processes are performed in the same manner whether they be in microbe, mouse, or human.

Microorganisms are exceptionally attractive models for studying fundamental life processes. They can be grown conveniently in test tubes or flasks, thus requiring less space and maintenance than larger plants and animals. They grow rapidly and reproduce at an unusually high rate; some species of bacteria undergo almost 100 generations in a 24-h period. The metabolic processes of microorganisms follow patterns that occur among higher plants and animals. For example, yeasts utilize glucose in essentially the same manner as cells of mammalian tissue; the same system of enzymes is present in these diverse organisms. The energy liberated during the breakdown of glucose is "trapped" and made available for the work to be performed by the cells, whether they be bacteria, yeasts, protozoa, or muscle cells. In fact, the mechanism by which organisms (or their cells) utilize energy is fundamentally the same throughout the biological world. The source of energy does, of course, vary among organisms. Plants are characterized by their ability to use radiant energy, whereas animals require chemical substances as their fuel. In this respect some microorganisms are like plants, others like animals; and some have the unique ability of using either radiant energy or chemical energy and thus are like both plants and animals. Furthermore, some microorganisms, the bacteria in particular, are able to utilize a great variety of chemical substances as their energy source—ranging from simple inorganic substances to complex organic substances.

In microbiology we can study organisms in great detail and observe their life processes while they are actively metabolizing, growing, reproducing, aging, and dying. By modifying their environment we can alter metabolic activities, regulate growth, and even change some details of their genetic pattern—all without destroying the organisms.

For example, **bacteriophages**, which are viruses that infect and reproduce in bacteria, demonstrate the complete sequence of host-parasite reactions and provide a model by which virus-host cell reactions can be postulated for infections in higher plants and animals. Bacteriophages have been of inestimable value in elucidating many biological phenomena, including those concerned with genetics.

Microorganisms have a wider range of physiological and biochemical potentialities than all other organisms combined. For example, some bacteria are able to utilize atmospheric nitrogen for the synthesis of proteins and other complex organic nitrogenous compounds. Other species require inorganic or organic nitrogen compounds as the initial building blocks for their nitrogenous constituents. Some microorganisms synthesize all their vitamins, while others need to be furnished vitamins. By reviewing the nutritional requirements of a large collection of microorganisms, it is possible to arrange them from those with the simplest to those with the most complex requirements. The increasing complexity of nutritional requirements in such an arrangement is also a reflection of the decreasing synthetic capacity of the organisms so arranged. In addition, this kind of arrangement provides information about the steps in the synthesis of various metabolites, e.g., from atmospheric oxygen to inorganic nitrogen salts to amino acids. The biochemist has used microorganisms having varying degrees of synthetic ability to investigate pathways of synthesis.

In his presidential address to the Society of American Bacteriologists (now The American Society for Microbiology) in 1942, the late Selman A. Waksman (Fig. 1-4) observed:

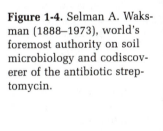

Figure 1-4. Selman A. Waksman (1888–1973), world's foremost authority on soil microbiology and codiscoverer of the antibiotic streptomycin.

There is no field of human endeavor, whether it be in industry or agriculture, or in the preparation of food or in connection with problems of shelter or clothing, or in the conservation of human or animal health and the combating of disease, where the microbe does not play an important and often dominant role.

Waksman, longtime professor of microbiology at Rutgers University, in 1952 was awarded the Nobel prize in physiology or medicine for the part he played in the discovery of the antibiotic streptomycin, which is produced by a soil bacterium.

THE PLACE OF MICROORGANISMS IN THE LIVING WORLD

In biology as in any other field, classification means the orderly arrangement of units under study into groups of larger units. Present-day classification in biology was established by the work of Carolus Linnaeus (1707–1778), a Swedish botanist. His books on the classification of plants and animals are considered to be the beginning of modern botanical and zoological nomenclature, a system of naming plants and animals. Nomenclature in microbiology, which came much later, was based on the principles established for the plant and animal kingdoms.

Until the eighteenth century, the classification of living organisms placed all organisms into one of two kingdoms, plant and animal. As previously stated, in microbiology we study some organisms that are predominantly plantlike, others that are animallike, and some that share characteristics common to both plants and animals. Since there are organisms that do not fall naturally into either the plant or the animal kingdom, it was proposed that new kingdoms be established to include those organisms which typically are neither plants nor animals.

Haeckel's Kingdom Protista

One of the earliest of these proposals was made in 1866 by a German zoologist, E. H. Haeckel. He suggested that a third kingdom, *Protista*, be formed to include those unicellular microorganisms that are typically neither plants nor animals. These organisms, the protists, include bacteria, algae, fungi, and protozoa. (Viruses are not cellular organisms and therefore are not classified as protists.) Bacteria are referred to as lower protists; the others—algae, fungi, and protozoa— are called higher protists.

Procaryotic and Eucaryotic Protists

Haeckel's kingdom *Protista* left some questions unanswered. For example, what criteria could be used to distinguish a bacterium from a yeast or certain microscopic algae? Satisfactory criteria were unavailable until late in the 1940s when more definitive observation of internal cell structure was made possible with the aid of the powerful magnification provided by electron microscopy. It was discovered that in some cells, for example typical bacteria, the nuclear substance was not enclosed by a nuclear membrane. In other cells, such as typical algae and fungi, the nucleus was enclosed in a membrane. This discovery—the absence of membrane-bound internal structures in one group of protists (bacteria) and the presence of membrane-bound structures in all the others (fungi, algae, and protozoa)—was a discovery of fundamental significance. Further research has revealed additional differences in the internal structure of these cells.

These two cell types, as characterized in Table 1-1, have been designated

Table 1-1. Features Distinguishing Procaryotic from Eucaryotic Cells

Feature	Procaryotic Cells	Eucaryotic Cells
Groups where found as unit of structure	Bacteria	Algae, fungi, protozoa, plants, and animals
Size range of organism	1–2 by 1–4 μm or less	Greater than 5 μm in width or diameter
Genetic system		
Location	Nucleoid, chromatin body, or nuclear material	Nucleus, mitochondria, chloroplasts
Structure of nucleus	Not bounded by nuclear membrane; one circular chromosome	Bounded by nuclear membrane; more than one chromosome
	Chromosome does not contain histones; no mitotic division	Chromosomes have histones; mitotic nuclear division
	Nucleolus absent; functionally related genes may be clustered	Nucleolus present; functionally related genes not clustered
Sexuality	Zygote nature is merozygotic (partial diploid)	Zygote is diploid
Cytoplasmic nature and structures		
Cytoplasmic streaming	Absent	Present
Pinocytosis	Absent	Present
Gas vacuoles	Can be present	Absent
Mesosome	Present	Absent
Ribosomes	70S,* distributed in the cytoplasm	80S arrayed on membranes as in endoplasmic reticulum; 70S in mitochondria and chloroplasts
Mitochondria	Absent	Present
Chloroplasts	Absent	May be present
Golgi structures	Absent	Present
Endoplasmic reticulum	Absent	Present
Membrane-bound (true) vacuoles	Absent	Present
Outer cell structures		
Cytoplasmic membranes	Generally do not contain sterols; contain part of respiratory and, in some, photosynthetic machinery	Sterols present; do not carry out respiration and photosynthesis
Cell wall	Peptidoglycan (murein or mucopeptide) as component	Absence of peptidoglycan
Locomotor organelles	Simple fibril	Multifibrilled with "9 + 2" microtubules

Table 1-1. (continued)

Feature	Procaryotic Cells	Eucaryotic Cells
Pseudopodia	Absent	Present in some
Metabolic mechanisms	Wide variety, particularly that of anaerobic energy-yielding reactions; some fix nitrogen gas; some accumulate poly-β-hydroxybutyrate as reserve material	Glycolysis is pathway for anaerobic energy-yielding mechanism
DNA base ratios as moles % of guanine + cytosine (G + C %)	28 to 73	About 40

* S refers to the Svedberg unit, the sedimentation coefficient of a particle in the ultracentrifuge.
NOTE: Definitions of technical words are provided in the glossary at the back of the book.

procaryotic and **eucaryotic;** organisms of each cell type are called procaryotes and eucaryotes, respectively.

Bacteria are procaryotic microorganisms. The eucaryotic microorganisms in-

Figure 1-5. The bacterium *Escherichia coli*, a typical procaryotic cell. Note the absence of any discrete intracellular organelle structures. The light area represents nuclear material; the dark area is ribosomal material. (*Courtesy I. D. J. Burdett and R. G. E. Murray, J. Bacteriol* **119:***1039, 1974.*) Inset: *E. coli* cells as seen by light microscopy.

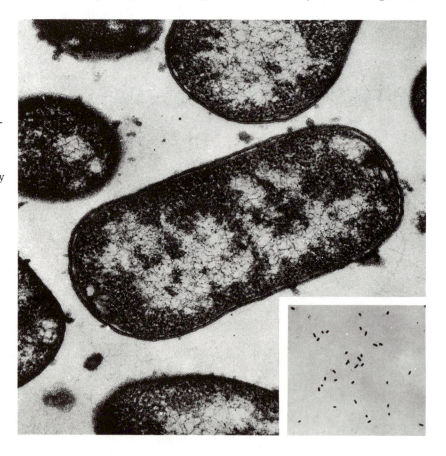

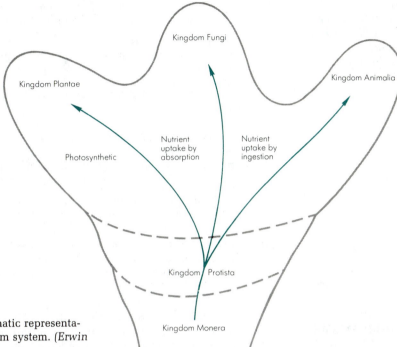

Kingdom Fungi

Kingdom Plantae

Kingdom Animalia

Nutrient uptake by absorption

Nutrient uptake by ingestion

Photosynthetic

Kingdom Protista

Kingdom Monera

Figure 1-6. A simplified schematic representation of Whittaker's five-kingdom system. *(Erwin F. Lessel, illustrator.)*

clude the protozoa, fungi, and algae. (Plant and animal cells are also eucaryotic.) Viruses are left out of this scheme of classification. Examples of typical procaryotic and eucaryotic cells are shown in Figs. 1-2, 1-3, and 1-5.

Whittaker's Five-Kingdom Concept

A more recent and comprehensive system of classification, the five-kingdom system, was proposed by R. H. Whittaker (1969). This system of classification, shown in Fig. 1-6, is based on three levels of cellular organization which evolved to accommodate three principal modes of nutrition: photosynthesis, absorption, and ingestion. The procaryotes are included in the kingdom *Monera*; they lack the ingestive mode of nutrition. Unicellular eucaryotic microorganisms are placed in the kingdom *Protista*; all three nutritional types are represented here. In fact, as shown in Fig. 1-6, the nutritional modes are continuous: the mode of nutrition of the microalgae is photosynthetic; the mode of nutrition of the protozoa is ingestive; and the mode of nutrition in some other protists is absorptive, with some overlap to the photosynthetic and ingestive modes. The multicellular and multinucleate eucaryotic organisms are found in the kingdoms *Plantae* (multicellular green plants and higher algae), *Animalia* (multicellular animals), and *Fungi* (multinucleate higher fungi). Their diversified nutritional modes lead to a more diversified cellular organization. Microorganisms are found in three of the five kingdoms: *Monera* (bacteria and cyanobacteria), *Protista* (microalgae and protozoa), and *Fungi* (yeasts and molds).

Kingdom *Procaryotae* after *Bergey's Manual of Systematic Bacteriology*

Bergey's Manual of Systematic Bacteriology places all bacteria in the kingdom *Procaryotae* which in turn is divided into 4 divisions as follows:

Division 1 Gracilicutes
   Procaryotes with a complex cell-wall structure characteristic of Gram-negative bacteria
Division 2 Firmicutes
   Procaryotes with a cell-wall structure characteristic of Gram-positive bacteria
Division 3 Tenericutes
   Procaryotes that lack a cell wall
Division 4 Mendosicutes
   Procaryotes that show evidence of an earlier phylogenetic origin than those bacteria included in Divisions 1 and 2 (above)

Bergey's Manual is the international standard for bacterial taxonomy. More detailed descriptions of groups of bacteria are given in Part Four of this book.

A comparable manual of classification does not exist for fungi, algae, or protozoa. There are, however, schemes of classification for each group that have wide acceptance and usage. An international system for classification and nomenclature of viruses is in the process of development.

GROUPS OF MICROORGANISMS

The major groups of protists are briefly described below. Although viruses are not protists or cellular organisms, they are included for two reasons: (1) the techniques used to study viruses are microbiological in nature; and (2) viruses are causative agents of diseases, hence, diagnostic procedures for their identification are employed in the clinical microbiological laboratory as well as the plant pathology laboratory.

Algae are relatively simple organisms. The most primitive types are unicellular. Others are aggregations of similar cells with little or no differentiation in structure or function. Still other algae, such as the large brown kelp, have a complex structure with cell types specialized for particular functions. Regardless of size or complexity, all algal cells contain chlorophyll and are capable of photosynthesis. Algae are found most commonly in aquatic environments or in damp soil.

Viruses are very small noncellular parasites or pathogens of plants, animals, and bacteria as well as other protists. They are so small that they can be visualized only by the electron microscope. Viruses can be cultivated only in living cells.

Bacteria are unicellular procaryotic organisms or simple associations of similar cells. Cell multiplication is usually by binary fission.

Protozoa are unicellular eucaryotic organisms. They are differentiated on the basis of morphological, nutritional, and physiological characteristics. Their role in nature is varied, but the best-known protozoa are the few that cause disease in human beings and animals.

Fungi are eucaryotic lower plants devoid of chlorophyll. They are usually multicellular but are not differentiated into roots, stems, and leaves. They range in size and shape from single-celled microscopic yeasts to giant multicellular mushrooms and puffballs. We are particularly interested in those organisms commonly called molds, the mildews, the yeasts, and the plant pathogens known as rusts. True fungi are composed of filaments and masses of cells which make up the body of the organism, known as a mycelium. Fungi reproduce by fission, by budding, or by means of spores borne on fruiting structures that are quite distinctive for certain species.

Some morphological and characteristic features of these various microbial groups are shown in Fig. 1-7 and Table 1-2.

Microbiologists may specialize in the study of certain groups of microorganisms. Strictly speaking, bacteriology is the study of bacteria, but the term is

Figure 1-7. Morphological features of various groups of microorganisms. (Note that this illustration is only intended to convey the impression of morphological diversity. No size relationship between groups can be obtained from it. The wide range in microbial sizes does not permit both constancy in magnification and showing of meaningful morphological details at the same time.) (A) *Escherichia coli* (X1,000). (B) Tobacco mosaic virus (X100,000). *(Hitachi, Ltd., Tokyo.)* (C) *Rickettsia tsutsugamushi* in cytoplasm of infected cell (X940). *(N.J. Kramis and The Rocky Mountain Laboratory, U.S. Public Health Service.)* (D) *Candida utilis* (X2,000 approx.). *(Courtesy of G. Svihla, J. L. Dainko, and F. Schlenk, J Bacteriol, 85:399, 1963.)* (E) *Aspergillus* sp. *(Courtesy of Douglas F. Lawson.)* (F) Amoeba. *(Carolina Biological Supply Co.)* (G) *Chlorella infusionum* (X1,000). *(Courtesy of Robert W. Krauss.)*

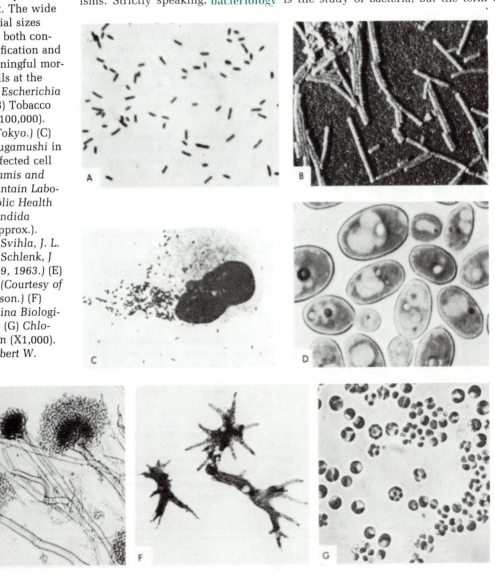

Table 1-2. Some Characteristics of Major Groups of Microorganisms *(Erwin F. Lessel, illustrator)*

Group	Morphology	Size	Important Characteristics	Practical Significance
Bacteria	5 μm	Typical: 0.5–1.5 μm by 1.0–3.0 μm Range: 0.2 by 100 μm	Procaryotic; unicellular, simple internal structure; grow on artificial laboratory media; reproduction asexual, characteristically by simple cell division	Some cause disease; some perform important role in natural cycling of elements which contributes to soil fertility; useful in industry for manufacture of valuable compounds; some spoil foods and some make foods
Viruses	100 nm	Range: 0.015–0.2 μm	Do not grow on artificial laboratory media—require living cells within which they are reproduced; all are obligate parasites; electron microscopy required to see viruses	Cause diseases in humans, other animals, and plants; also infect microorganisms
Fungi: Yeasts	5 μm	Range: 5.0–10.0 μm	Eucaryotic; unicellular; laboratory cultivation much like that of bacteria; reproduction by asexual cell division, budding, or sexual processes	Production of alcoholic beverages; also used as food supplement; some cause disease
Fungi: Molds	20 μm	Range: 2.0–10.0 μm by several mm	Eucaryotic; multicellular, with many distinctive structural features; cultivated in laboratory much like bacteria; reproduction by asexual and sexual processes	Responsible for decomposition (deterioration) of many materials; useful for industrial production of many chemicals, including penicillin; cause diseases of humans, other animals, and plants
Protozoa	50 μm	Range: 2.0–200 μm	Eucaryotic; unicellular; some cultivated in laboratory much like bacteria; some are intracellular parasites; reproduction by asexual and sexual processes	Food for aquatic animals; some cause disease
Algae	10 μm	Range: 1.0 μm to many feet	Eucaryotic; unicellular and multicellular; most occur in aquatic environments; contain chlorophyll and are photosynthetic; reproduction by asexual and sexual processes	Important to the production of food in aquatic environments; used as food supplement and in pharmaceutical preparations; source of agar for microbiological media; some produce toxic substances

often used as a synonym for microbiology. Protozoology is the study of protozoa; a special branch of protozoology called parasitology deals exclusively with the parasitic or disease-producing protozoa and other parasitic micro- and macroorganisms. Mycology is the study of fungi such as yeasts and molds. Virology is the science that deals with viruses. Phycology is the study of algae. Further specialization in some aspect of the biology of a particular group of organisms is not uncommon; e.g., bacterial genetics, algal physiology, and bacterial cytology.

DISTRIBUTION OF MICROORGANISMS IN NATURE

Microorganisms occur nearly everywhere in nature. They are carried by air currents from the earth's surface to the upper atmosphere. Even those indigenous to the ocean may be found many miles away on mountaintops. Microbes are found in the bottom of the ocean at its greatest depths. Fertile soil teems with them. They are carried by streams and rivers into lakes and other large bodies of water; and if human wastes containing harmful bacteria are discharged into streams, diseases may be spread from one place to another. Microorganisms occur most abundantly where they find food, moisture, and a temperature suitable for their growth and multiplication. Since the conditions that favor the survival and growth of many microorganisms are those under which people normally live, it is inevitable that we live among a multitude of microbes. They are in the air we breathe and the food we eat. They are on the surfaces of our bodies, in our alimentary tracts, and in our mouths, noses, and other body orifices. Fortunately most microorganisms are harmless to us; and we have means of resisting invasion by those that are potentially harmful.

APPLIED AREAS OF MICROBIOLOGY

Microorganisms affect the well-being of people in a great many ways. As we have already stated, they occur in large numbers in most natural environments and bring about many changes, some desirable and others undesirable. The diversity of their activities ranges from causing diseases in humans, other animals, and plants to the production and deposition of minerals, the formation of coal, and the enhancement of soil fertility.

There are many more species of microorganisms that perform important roles in nature than there are disease-producing species.

A summary of the major fields of applied microbiology appears in Table 1-3.

MICROBIOLOGY AND THE ORIGIN OF LIFE

Many explanations have been offered for the origin of life on earth. One of the more acceptable of these proposals suggests that life originated in the sea following millions of years of a chemical evolutionary process. According to this hypothesis the inorganic compounds of the atmosphere, under the influence of ultraviolet light, electrical discharges, and/or high temperatures, interacted to form organic compounds which precipitated into the sea, where they accumulated. These organic compounds, subjected to additional physical effects of the environment, combined to form amino acids. The amino acids interacted to form peptides, polypeptides, and other more complex organic substances which served as the precursors of the first form of life.

Table 1-3. Major Fields of Applied Microbiology

Field	Some Applied Areas
Medical microbiology	Causative agents of disease; diagnostic procedures; diagnostic procedures for identification of causative agents; preventive measures
Aquatic microbiology	Water purification; microbiological examination; biological degradation of waste; ecology
Aeromicrobiology	Contamination and spoilage; dissemination of diseases
Food microbiology	Food preservation and preparation; foodborne diseases and their prevention
Agricultural microbiology	Soil fertility; plant and animal diseases
Industrial microbiology	Production of medicinal products such as antibiotics and vaccines; fermented beverages, industrial chemicals; production of proteins and hormones by genetically engineered microorganisms
Exomicrobiology	Exploration for life in outer space
Geochemical microbiology	Coal, mineral and gas formation; prospecting for deposits of coal, oil, and gas; recovery of minerals from low-grade ores

The time scale of chemical evolution, biological evolution, and the emergence of microbial life is shown in Fig. 1-8.

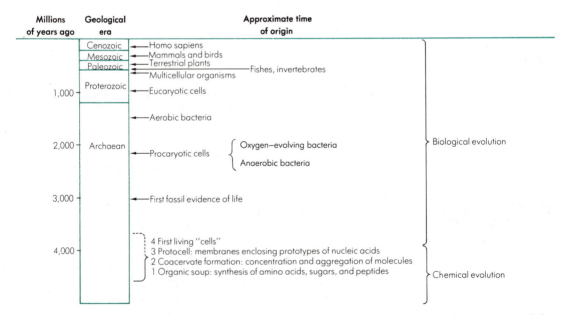

Figure 1-8. Time scale of the chemical evolution, the biological evolution, and the occurrence of microbial life.

QUESTIONS

1 List the characteristics common to all biological systems.
2 Why are microorganisms useful as subjects for research in the field of biology?
3 Explain why a knowledge of microbiology is useful in understanding life processes in higher plants and animals.
4 How did the term *protists* arise? What organisms do we refer to by use of this term? What is the difference between lower protists and higher protists?
5 Discuss the differences between procaryotic and eucaryotic cells.
6 How do viruses differ from other microorganisms?
7 What is the basis of the five-kingdom classification scheme according to Whittaker? Give a reason why it is so widely accepted in the biological community.
8 Discuss the place of microorganisms in Whittaker's five-kingdom classification scheme.
9 Why is *Bergey's Manual of Systematic Bacteriology* so important to bacteriologists?
10 Where are microorganisms found in nature? How may they be transferred from place to place?
11 Name several applied areas of microbiology. Describe the importance of microorganisms in each of these applied fields.

REFERENCES

Delauney, A., and H. Erni (eds.): *The World of Microbes*, vol. 4, *Encyclopedia of the Life Sciences*, Doubleday, Garden City, N.Y., 1965. *A beautifully illustrated and well-summarized account of microbes in relation to human beings. It is an inspiring introductory book for the new student.*

Edmonds, P.: *Microbiology, An Environmental Perspective*, Macmillan, New York, 1978. *This volume serves to introduce nonspecialists to the important functions carried out by microorganisms in nature.*

Jennings, R. K., and R. F. Acker: *The Protistan Kingdom*, Van Nostrand Reinhold, New York, 1970. *This small book dealing with the protists and viruses is both informative and entertaining. It may provide the stimulus for further study by students and nonstudents alike.*

National Academy of Sciences: *Microbial Processes: Promising Technologies for Developing Countries*, National Academy of Sciences, Washington, D.C., 1981. *A report of scientific and technological developments in microbiology now available that might be applicable to less-developed countries for improvement of their environment and their economy.*

Postgate, J.: *Microbes and Man*, Penguin, Baltimore, 1975. *A very good introduction to microbes. It gives an account of how microbes keep our terrestrial biochemistry moving and influence our food supply at every stage. Other aspects of applied microbiology are also covered. Additional topics include microbes in evolution and in the future.*

Rossmore, H. W.: *The Microbes, Our Unseen Friends*, Wayne State, Detroit, 1976. *An elementary discussion of the many useful activities of microorganisms.*

Whittaker, R. H.: "New Concepts of Kingdoms of Organisms," *Science*, **163**:150–160, 1969. *A now classic paper stating the case that evolutionary relations are better represented by new classifications than by the traditional two kingdoms of plants and animals. A new scheme of classification consists of five kingdoms.*

Chapter 2   The History of Microbiology

History is the story of the achievements of men and women, but it records relatively few outstanding names and events. Many important contributions were made by people whose names have been forgotten and whose accomplishments have been lost in the longer and deeper shadows cast by those who caught the fancy of the chroniclers. It has been said that in science the credit goes to the one who convinces the world, not to the one who first had the idea. So, in the development of microbiology, the outstanding names are often of those who convinced the world—who developed a technique, a tool, or a concept that was generally adopted, or who explained their findings so clearly or dramatically that the science grew and prospered.

Antony van Leeuwenhoek's lucid reports on the ubiquity of microbes enabled Louis Pasteur 200 years later to discover the involvement of these creatures in fermentation reactions and allowed Robert Koch, Theobald Smith, Pasteur, and many others to discover the association of microbes with disease. Koch is remembered for his isolation of the bacteria that cause anthrax and tuberculosis and for the rigid criteria he demanded before a specific bacterium be held as the cause of a disease. His important contributions to the creation of the science of microbiology won him the 1905 Nobel prize.

The building of the Panama Canal dramatized Walter Reed's studies of the epidemiology of yellow fever, but historians remember that Theobald Smith's work on transmission of Texas fever pointed the way for Walter Reed's subsequent work.

18

In diagnosis by laboratory methods, G. F. I. Widal and August von Wasserman presented those who followed them with tools and ideas with which to work. Paul Ehrlich's discovery of a chemical compound that would destroy the syphilis spirochete in the human body without injury to tissue cells paved the way for future developments in the use of chemicals in treating disease. For this he shared the Nobel prize in 1908 with Elie Metchnikoff, who discovered a system in the human body that combated infection.

Though of relatively short duration, the history of microbiology is filled with thrilling achievements. We have won many battles with microorganisms and have learned not only to make them work for us but also to control some of those that work against us.

THE MICROSCOPE

Microbiology began when people learned to grind lenses from pieces of glass and combine them to produce magnifications great enough to enable microbes to be seen. During the thirteenth century Roger Bacon (1220–1292) postulated that disease is produced by invisible living creatures. This suggestion was made again by Girolamo Fracastoro of Verona (1483–1553) and Anton von Plenciz in 1762, but these people had no proof. As early as 1658, a monk named Athanasius Kircher (1601–1680) referred to "worms" invisible to the naked eye in decaying bodies, meat, milk, and diarrheal secretions. Although his description lacked accuracy, Kircher was the first person to recognize the significance of bacteria and other microbes in disease. In 1665 Robert Hooke's description of cells in a piece of cork established the fact that the bodies of "animals and plants, complex as they may appear, are yet composed of a few elementary parts frequently repeated"—a quotation not from Hooke but from Aristotle's description of the cellular structure of living things back in the fourth century B.C.

Although he was probably not the first to see bacteria and protozoa, Antony van Leeuwenhoek, who lived in Delft, Holland, from 1632 to 1723, was the first to report his observations with accurate descriptions and drawings (Fig. 2-1). Leeuwenhoek had the means and opportunity to pursue his hobby of lens grinding and microscope making. During his lifetime he made more than 250 microscopes consisting of home-ground lenses mounted in brass and silver, the most powerful of which would magnify about 200 to 300 times (Fig. 2-2). These microscopes bear little resemblance to the compound light microscope of today, which is capable of magnifications of 1,000 to 3,000 times. However, the lenses of Leeuwenhoek's microscopes were well made and Leeuwenhoek had the openness of mind that is so very important in an investigator. His descriptions of protozoa were so accurate that many of the forms he described are easily recognized today.

Leeuwenhoek carefully recorded his observations in a series of letters to the British Royal Society. In one of the first letters, dated September 7, 1674, addressed to Henry Oldenburg, Secretary of the Royal Society, he described the "very little animalcules" which we recognize as free-living protozoa. On October 9, 1676, he wrote:

In the year 1675, I discovered living creatures in rain water which had stood but a few days in a new earthen pot, glazed blue within. This invited me to view this

Figure 2-1. Antony van Leeuwenhoek (1632–1723), a Dutch student of natural history whose hobby was making microscopes, is shown here with one of the more than 250 microscopes that he made. His best lenses were capable of magnifications up to ×270, and he was the first person to report descriptions of microorganisms in detail. *(Courtesy of Lambert-Hudnut, Division Warner-Lambert Pharmaceutical Company.)*

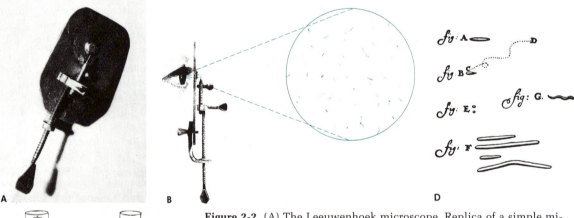

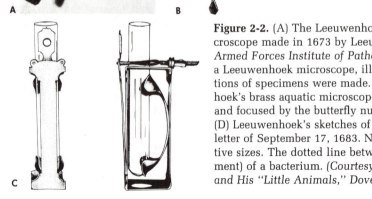

Figure 2-2. (A) The Leeuwenhoek microscope. Replica of a simple microscope made in 1673 by Leeuwenhoek. *(From the collection of the Armed Forces Institute of Pathology, Washington, D.C.)* (B) Side view of a Leeuwenhoek microscope, illustrating the manner in which observations of specimens were made. (C) Front and side views of Leeuwenhoek's brass aquatic microscope. Lenses could be slotted in at the front and focused by the butterfly nut at the rear. *(Erwin F. Lessel, illustrator.)* (D) Leeuwenhoek's sketches of bacteria from the human mouth, from letter of September 17, 1683. Note particularly shapes of cells and relative sizes. The dotted line between **C** and **D** indicates motility (movement) of a bacterium. *(Courtesy of C. Dobell, Antony van Leeuwenhoek and His "Little Animals," Dover, New York, 1960.)*

water with great attention, especially those little animals appearing to me ten thousand times less than those . . . which may be perceived in the water with the naked eye.

He described his little animals in great detail, leaving little doubt that he saw bacteria, fungi, and many forms of protozoa. For example, he reported that on June 16, 1675, while examining well water into which he had put a whole pepper the day before:

I discovered, in a tiny drop of water, incredibly many very little animalcules, and these of divers sorts and sizes. They moved with bendings, as an eel always swims with its head in front, and never tail first, yet these animalcules swam as well backwards as forwards, though their motion was very slow.

His enthusiastic letters were read with interest by the British scientists, but the importance of his discoveries evidently went unappreciated. The talents and astuteness of this remarkable man can best be appreciated by reading Dobell's biography of Leeuwenhoek.

Before the time of Pasteur, microorganisms were studied mainly to satisfy curiosity concerning their characteristics and their relationships to higher living forms, without awareness of their importance in fermentation and disease.

SPONTANEOUS GENERATION VERSUS BIOGENESIS

The discovery of microbes spurred interest in the origin of living things, and argument and speculation grew. As far as human beings were concerned, the Greek explanation that the goddess Gaea was able to create people from stones and other inanimate objects had been largely discarded. But even the astute Aristotle (384–322 B.C.) taught that animals might originate spontaneously from the soil, plants, or other unlike animals, and his influence was still strongly felt in the seventeenth century. About 40 B.C., Virgil (70–19 B.C.) gave directions for the artificial propagation of bees. This was but one of many fanciful tales of a similar nature that persisted into the seventeenth century. For example, it was accepted as a fact that maggots could be produced by exposing meat to warmth and air, but Francesco Redi (1626–1697) doubted this. Proof that his skepticism was well founded came from an experiment in which he placed meat in a jar covered with gauze. Attracted by the odor of the meat, flies laid eggs on the covering, and from the eggs maggots developed. Hence the experiment established the fact that the origin of the maggots was the flies and not the meat. This experiment and others involving mice and scorpions appear to have settled the matter so far as these forms of life were concerned. But microbes were another matter; surely such minute creatures needed no parents!

There appeared champions for and challengers of the theory that living things can originate spontaneously, each with a new and sometimes fantastic explanation or bit of experimental evidence. In 1749, while experimenting with meat exposed to hot ashes, John Needham (1713–1781) observed the appearance of organisms not present at the start of the experiment and concluded that the bacteria originated from the meat. About the same time, Lazaro Spallanzani (1729–1799) boiled beef broth for an hour and then sealed the flasks. No microbes appeared following incubation. But his results, confirmed in repeated

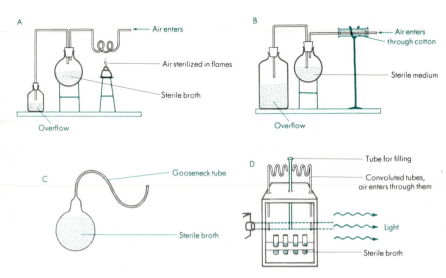

Figure 2-3. The theory of spontaneous generation was disproved with the devices illustrated here, all of which eliminated airborne bacteria. Schwann heat-sterilized the air which flowed through the glass tube to his culture flask (A). Schröder and von Dusch filtered the air entering the culture flask through cotton (B). Simple goose-necked flasks (C) were devised by Pasteur. Tyndall constructed a dust-free incubation chamber (D).

experiments, failed to convince Needham, who insisted that air was essential to the spontaneous production of microscopic beings and that it had been excluded from the flasks by sealing them. This argument was answered some 60 or 70 years later independently by two other investigators, Franz Schulze (1815–1873) and Theodor Schwann (1810–1882). Schulze passed air through strong acid solutions into boiled infusions, whereas Schwann passed air into his flasks through red-hot tubes (Fig. 2-3A). In neither case did microbes appear. But the die-hard advocates of spontaneous generation were still not convinced. Acid and heat altered the air so that it would not support growth, they said. About 1850, H. Schröder and T. von Dusch performed a more convincing experiment by passing air through cotton into flasks containing heated broth (Fig. 2-3B). Thus the microbes were filtered out of the air by the cotton fibers so that growth did not occur, and a basic technique of plugging bacterial culture tubes with cotton stoppers was initiated.

The concept of spontaneous generation was revived for the last time by Felix-Archimede Pouchet (1800–1872), who published in 1859 an extensive report "proving" its occurrence. But Pouchet reckoned without the ingenious, tireless, and stubborn Pasteur (1822–1895). Irritated by Pouchet's logic and data, Pasteur performed experiments that ended the argument for all time. He prepared a flask with a long, narrow gooseneck opening (Fig. 2-3C). The nutrient solutions were heated in the flask, and air—untreated and unfiltered—could pass in or out; but the germs settled in the gooseneck, and no microbes appeared in the solution.

Pasteur reported his results with a great flourish at the Sorbonne in Paris on April 7, 1864. His flasks would yield no sign of life, he said:

For I have kept from them, and am still keeping from them, that one thing which is above the power of man to make; I have kept from them the germs that float in the air. I have kept from them life.

In his exuberance, Pasteur sent a few darts at those he disagreed with:

There is no condition known today in which you can affirm that microscopic beings come into the world without germs, without parents like themselves. They who allege it have been the sport of illusions, of ill-made experiments, vitiated by errors which they have not been able to perceive and have not known how to avoid.

Finally, John Tyndall (1820–1893) conducted experiments in a specially designed box to prove that dust carried the germs (Fig. 2-3D). He demonstrated that if no dust was present, sterile broth remained free of microbial growth for indefinite periods.

FERMENTATION

Louis Pasteur (Fig. 2-4) began his brilliant career as professor of chemistry at the University of Lille, France. A principal industry of France being the manufacture of wines and beer, Pasteur studied the methods and processes involved in order to help his neighbors produce a consistently good product. He found that fermentation of fruits and grains, resulting in alcohol, was brought about by microbes. By examining many batches of "ferment," he found microbes of different sorts. In good lots one type predominated, and in the poor products another kind was present. By proper selection of the microbe, the manufacturer might be assured of a consistently good and uniform product. Pasteur suggested that the undesirable types of microbes might be removed by heating—not enough to hurt the flavor of the fruit juice, but enough to destroy a very high percentage of the microbial population. He found that holding the juices at a temperature of 62.8°C (145°F) for half an hour did the job. Today pasteurization is widely used in fermentation industries, but we are most familiar with it in the dairy industry.

Figure 2-4. Louis Pasteur in his laboratory. *(Courtesy of Institut Pasteur, Paris.)*

THE GERM THEORY OF DISEASE

Even before Pasteur had proved by experiment that bacteria are the cause of some diseases, many observant students had expressed strong arguments for the germ theory of disease. Fracastoro of Verona suggested that diseases might be due to invisible organisms transmitted from one person to another. In 1762 von Plenciz not only stated that living agents are the cause of disease but suspected that different germs were responsible for different diseases. That the concept of parasitism was becoming quite general is reflected in the following bit of doggerel written by Jonathan Swift (1667–1745) early in the eighteenth century:

> So naturalists observe, a flea
>   Hath smaller fleas that on him prey;
> And these have smaller fleas to bit 'em;
>   And so proceed ad infinitum.

This is better known in the colloquial version:

> Big bugs have little bugs,
>   Upon their backs to bit 'em;
> And little bugs have smaller ones,
>   And so ad infinitum.

Oliver Wendell Holmes (1809–1894), a successful physician as well as a scholar, insisted that puerperal fever, a disease of childbirth, was contagious and that it was probably caused by a germ carried from one mother to another by midwives and physicians. He wrote *The Contagiousness of Puerperal Fever* in 1842. At approximately the same time, the Hungarian physician Ignaz Philipp Semmelweis (1818–1865) was pioneering in the use of antiseptics in obstetrical practice. Deaths due to infections associated with childbirth were reduced in the cases handled according to his instructions, which minimized chances for infection. As part of his crusade he published *The Cause, Concept and Prophylaxis of Childbed Fever* in 1861. Still, most physicians ignored his advice, and it was not until about 1890, when the work of Joseph Lister in England had become known, that the importance of antisepsis was fully appreciated by the medical profession.

Pasteur's success in solving the problem of fermentation led the French goverment to request that he investigate pebrine, a silkworm disease that was ruining an important French industry. For several years Pasteur struggled with this problem, heartaches and disappointments following one after another. Eventually he isolated the parasite causing the disease. He also showed that silkworm farmers could eliminate the disease by using only healthy, disease-free caterpillars for breeding stock.

Turning from silk to wool, Pasteur next tackled the problem of anthrax, a disease of cattle, sheep, and sometimes human beings. He grew the microbes in laboratory flasks after isolating them from the blood of animals that had died of the disease. Meanwhile Robert Koch (1843–1910) was busy with the anthrax problem in Germany. Koch, a quiet, meticulous physician, sometimes neglected his medical practice to play with the fascinating new science of bacteriology. It was he who discovered the typical bacilli with squarish ends in the blood of cattle that had died of anthrax. He grew these bacteria in cultures in his laboratory, examined them microscopically to be sure he had only one kind present, and then injected them into other animals to see if these became infected and developed clinical symptoms of anthrax. From these experimentally infected animals he isolated microbes like those he had originally found in sheep that died of anthrax. This was the first time a bacterium had been proved to be the cause of an animal disease. (Pebrine is caused by a protozoan rather than by a bacterium.) This series of observations led to the establishment of **Koch's postulates,** which provided guidelines to identify the causative agent of an infectious disease. Koch's postulates are: (1) A specific organism can always be found in association with a given disease. (2) The organism can be isolated and grown in pure culture in the laboratory. (3) The pure culture will produce the disease when inoculated into a susceptible animal. (4) It is possible to recover the organism in pure culture from the experimentally infected animal.

LABORATORY TECHNIQUES AND PURE CULTURES

As we have previously stated, microorganisms occur in nature in extremely large populations made up of many different species. In order to study the characteristics of a particular species it is first necessary to separate it from all other species. Laboratory procedures have been developed that make it possible to isolate microorganisms representing each species and to grow (cultivate) each of the species separately. The growth of a mass of cells of the same species in a laboratory vessel (such as a test tube) is called a **pure culture.**

Pure cultures of bacteria were first obtained by Joseph Lister in 1878 using serial dilutions in liquid media. With a specially constructed syringe he diluted a fluid (probably milk) containing a mixture of bacteria until a single organism was delivered into a container of sterile milk. After incubation, bacteria in this container were of a single kind, identical to the parent cell. Lister named the organism *Bacterium lactis.*

Meanwhile Koch was carefully refining methods for the study of bacteria. He found that by smearing bacteria on a glass slide and adding certain dyes to them, individual cells could be seen more clearly with the microscope. He added gelatin and other solidifying materials such as agar to media in order to obtain isolated growths of organisms known as **colonies,** each of which contained millions of individual bacterial cells packed tightly together. From these colonies, pure cultures could be transferred to other media. The development of a liquefiable solid-culture medium was of fundamental importance.

Using techniques he had devised, Koch studied with painstaking care material taken from patients with pulmonary tuberculosis. After performing a series of rigid tests, as he had done with the anthrax bacillus, he announced the discovery of the microorganism that causes tuberculosis.

The importance of pure cultures to the development of the science of microbiology cannot be overestimated, since by using pure-culture techniques the microorganisms responsible for many infections, certain fermentations, nitrogen fixation in soil, and other activities were isolated and identified. However, strict adherence to pure-culture techniques and Koch's postulates sometimes led investigators up dead-end streets. Early investigators did not know about viruses, nor did they know about the cooperation of two or more microorganisms in causing disease or in bringing about a desirable fermentation such as we find in the ripening of cheese. Today we are as much interested in mixed microbial populations and the effects they produce as we are in pure cultures. Further advances in marine microbiology, rumen microbiology, microbiology of the intestinal tract, and many other systems will depend upon understanding first the physiology of individual microorganisms in pure culture and then upon the ecological relationships of the total microbial populations in a given environment.

PROTECTION AGAINST INFECTION: IMMUNITY

Pasteur continued to make discoveries concerning the cause and prevention of infectious diseases. About 1880 he isolated the bacterium responsible for chicken cholera and grew it in pure culture. Here again, the practical Pasteur made use of the fundamental techniques devised by the more theoretical Koch. To prove that he really had isolated the organisms responsible for chicken cholera, Pasteur arranged for a public demonstration where he repeated an experiment (Fig. 2-5) that had been successful in many previous trials. He inoculated healthy chickens with his pure cultures, but to his dismay, the chickens failed to get sick and die! Reviewing each step of the experiment, Pasteur found that he had accidentally used cultures several weeks old instead of the fresh ones grown especially for the demonstration. Some weeks later he repeated the experiment, using two groups of chickens. One of these groups had been inoculated at the first demonstration with the old cultures that had proved ineffective, and the second had not been previously exposed. Both groups received bacteria from fresh young cultures. This time the chickens in the second group got sick and died, but those in the first group remained hale and hearty. This puzzled Pasteur, but he soon found the explanation. In some way bacteria could lose their ability to produce disease, i.e., their **virulence,** after standing and growing old. But these **attenuated** (having decreased virulence) bacteria still retained their capacity for stimulating the host to produce substances, i.e. **antibodies,** that protect against subsequent exposure to virulent organisms.

Figure 2-5. The principle of immunization was demonstrated by Pasteur when he inoculated chickens with cultures of chicken cholera bacteria several weeks old and the chickens remained healthy. They did not become sick when inoculated with a fresh culture several weeks later although this fresh culture killed chickens that had not received the attenuated (old) culture.

This demonstration explained the principle involved in Edward Jenner's successful use of cowpox virus, in 1798, to immunize people against smallpox (Fig. 2-6). Pasteur next applied this principle to the prevention of anthrax, and again

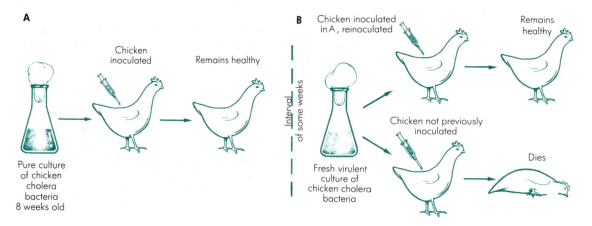

A

Chicken inoculated

Remains healthy

Pure culture of chicken cholera bacteria 8 weeks old

B

Chicken inoculated in A, reinoculated

Remains healthy

Interval of some weeks

Chicken not previously inoculated

Fresh virulent culture of chicken cholera bacteria

Dies

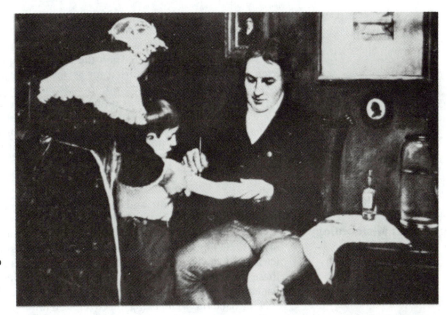

Figure 2-6. Edward Jenner vaccinating (inoculating) James Phipps with cowpox material, which resulted in development of resistance to smallpox infection. *(Courtesy of Culver Pictures, New York.)*

it worked. He called the attenuated cultures **vaccines,** a term derived from the Latin *vacca,* meaning "cow." Pasteur was honoring Jenner when he applied the term *vaccination* to immunization with attenuated cultures of bacteria that had no connection with cows.

Pasteur's fame was by now well established throughout France, and the belief became prevalent that he could work miracles with bacteria and the control of infections. It was not surprising, then, that he was given an even greater challenge: he was asked to work on a disease affecting human beings. As he was a chemist and not a physician, studying a human disease might prove risky. But Pasteur again accepted the challenge to be of service to humanity and set out to make a vaccine for hydrophobia, or rabies, a disease transmitted to people by bites of dogs, cats, and other animals. Because it was invariably fatal, when a boy named Joseph Meister was bitten by a mad wolf, his family did not hesitate to take the one chance in thousands that Pasteur could make a vaccine that would save him.

Figure 2-7. Rabies vaccine is made by inoculating a rabbit with saliva from a rabid dog. Virus in the extract of the rabbit's spinal cord is attenuated before injection into a patient.

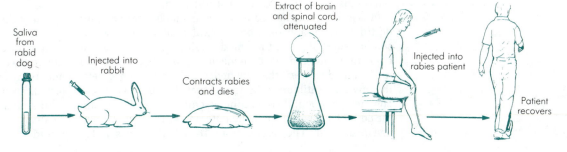

Saliva from rabid dog

Injected into rabbit

Contracts rabies and dies

Extract of brain and spinal cord, attenuated

Injected into rabies patient

Patient recovers

It had been established that the rabies virus was too minute to be seen even with a microscope; it had never been grown in laboratory culture, and it was not a bacterium. The disease could be produced in rabbits by inoculating them with saliva from mad dogs. Then the brain and spinal cord could be removed from the infected rabbits, dried for several days, pulverized, and mixed with glycerin. Injecting this mixture into dogs protected them against rabies (Fig. 2-7). But vaccinating dogs was quite different from treating a sick boy. Perhaps the worried Pasteur was as surprised as anyone else when after the crucial trial, which took several weeks, Joseph Meister did not die. As with Jenner's vaccination for smallpox, the principles of the preventive treatment of rabies have not changed.

WIDENING HORIZONS

Medical Microbiology

The success of Pasteur and Koch brought honors and accolades from their appreciative countrymen. Koch became Professor of Hygiene and Director of the Institute for Infective Diseases, which was founded for him at the University of Berlin. France showed its gratitude by establishing the Pasteur Institute in Paris in 1888. To each of these men came scholars from all over the world, and these students later carried the spirit and knowledge of Koch and Pasteur to America and throughout Europe.

New bacteria were being discovered with increasing frequency, and their disease-producing capacities were proved by Koch's postulates. After Edwin Klebs in 1883 and Frederick Loeffler in 1884 discovered the diphtheria bacillus and demonstrated that it produced its poisons in a laboratory flask, Emil von Behring and Shibasaburo Kitasato devised a method of producing immunity to infections caused by these organisms by injecting their **toxins** (poisons) into animals so that an **antitoxin** (a substance that neutralizes toxin) would develop. Similarly, Kitasato cultivated *Clostridium tetani*, the cause of tetanus (lockjaw), and with von Behring made antitoxin for the prevention and treatment of this disease. For his work on serum therapy, von Behring was awarded the Nobel prize in physiology or medicine in 1901. During this time, D. E. Salmon and Theobald Smith demonstrated that immunity to many infections could be produced by inoculation with killed cultures of microorganisms.

Working in Pasteur's laboratory, Elie Metchnikoff (Fig. 2-8A), a Russian, described how certain leukocytes (white blood cells) could ingest (eat) disease-producing bacteria in the body. He called these special defenders against infection **phagocytes** ("eating cells") and the process **phagocytosis** (Fig. 2-8B). Metchnikoff formulated the theory that the phagocytes were the body's first and most important line of defense against infection.

In Germany, one of Koch's students had a different concept of how the body destroys bacteria. Paul Ehrlich explained immunity on the basis of certain soluble substances in the blood. Today we know that there is much merit in both arguments and that both mechanisms play their parts. Ehrlich made another important discovery that opened the door to future developments in chemotherapy and antibiotics: he found that the 606th compound tested, an organic chemical compound containing arsenic, would destroy the syphilis microbe in the body. This was the first chemotherapeutic substance scientifically discovered and evaluated. The period from 1880 to1900 was indeed a golden time for

Figure 2-8. (A) Elie Metchnikoff was the first person to recognize the role of phagocytes in combating bacterial infections. *(Courtesy of René Dubos.)* (B) The process of phagocytosis, the ingestion of particulate matter by certain cells, shown in three steps. Phagocytosis is a natural defense mechanism against disease. *(Erwin F. Lessel, illustrator.)*

1 μm

microbiology because the infant science grew to adolescence during those years. As shown in Table 2-1, most of the causative agents of bacterial diseases that had plagued the world for centuries were isolated and identified.

The students of Koch and Pasteur continued to discover the causative agents of diseases and new methods for diagnosis; e.g., the Widal test for typhoid fever and the Wassermann test for syphilis made diagnosis of these diseases accurate and quick.

In the 1860s, while all these things were happening on the continent of Europe, an English surgeon, Joseph Lister, was trying to combat the microbes that caused postoperative and wound infections. Deaths from these infections were frequent in the nineteenth century. Disinfectants as such were unknown, but since carbolic acid would kill bacteria, Lister used a dilute solution of this acid to soak surgical dressings. Wounds protected in this way did not become infected, and healing took place rapidly. So remarkable was Lister's success that the technique was quickly accepted, and this antiseptic surgical practice established the principles of present-day aseptic ("without infection") techniques.

Table 2-1. Discovery of Causative Agents of Microbial Diseases Following Establishment of Germ Theory of Disease

Date	Disease of Infection	Causative Agent*	Discoverer†
1876	Anthrax	*Bacillus anthracis*	Koch
1879	Gonorrhea	*Neisseria gonorrhoeae*	Neisser
1880	Typhoid fever	*Salmonella typhi*	Eberth
1880	Malaria	*Plasmodium spp.*	Laveran (C. Alphonse)
1881	Wound infections	*Staphylococcus aureus*	Ogston
1882	Tuberculosis	*Mycobacterium tuberculosis*	Koch
1882	Glanders	*Pseudomonas mallei*	Loeffler and Schütz
1883	Cholera	*Vibrio cholerae*	Koch
1883	Diphtheria	*Corynebacterium diphtheriae*	Klebs
1884	Diphtheria	*Corynebacterium diphtheriae*	Loeffler
1885	Swine erysipelas	*Erysipelothrix rhusiopathiae*	Loeffler
1885	Tetanus	*Clostridium tetani*	Nicolaier
1886	Bacterial pneumonia	*Streptococcus pneumoniae*	Fraenkel
1887	Meningitis	*Neisseria meningitidis*	Weichselbaum
1887	Malta fever	*Brucella spp.*	Bruce
1888	Equine strangles	*Streptococcus spp.*	Schütz
1889	Chancroid	*Hemophilus ducreyi*	Ducrey
1892	Gas gangrene	*Clostridium perfringens*	Welch and Nuttall
1894	Plague	*Yersinia pestis*	Kitasato and Yersin
1895	Fowl typhoid	*Salmonella gallinarum*	Moore
1896	Botulism (food poisoning)	*Clostridium botulinum*	Van Ermengem
1897	Bang's disease (bovine abortion)	*Brucella abortus*	Bang
1898	Dysentery	*Shigella dysenteriae*	Shiga
1898	Pleuropneumonia of cattle	*Mycoplasma mycoides*	Nocard and Roux
1905	Syphilis	*Treponema pallidum*	Schaudin and Hoffmann
1906	Whooping cough	*Bordetella pertussis*	Bordet and Gengou
1909	Rocky Mountain spotted fever	*Rickettsia rickettsii*	Ricketts
1912	Tularemia	*Francisella tularensis*	McCoy and Chapin

* Present name of causative agent; original name, in many instances, was different.

† In some instances the individual simply observed the causative agent; in other instances the investigator isolated the agent in pure culture.

Microbiologists began to appear in America at the turn of the century. Some had studied under Koch, and others had visited the Pasteur Institute or watched Lister operate in a mist of carbolic acid vapors or a spray of bichloride of mercury. Whether they actually received their training in the "laboratories of the masters," first- and second-generation bacteriologists in Europe and America were greatly influenced by them. These Americans who actually studied under Pasteur and Koch—William Henry Welch at Johns Hopkins, Harold Clarence Ernst at Harvard, F. G. Novy at Michigan, and H. L. Russell at Wisconsin—were, as P. F. Clark, longtime professor at the University of Wisconsin, said, "responsible directly or indirectly for the development of bacteriology in the United States." Their students and their students' students constitute the who's who of those who built the science of microbiology in America.

Agricultural, Industrial, and Food Microbiology

In sketching the history of bacteriology, it is natural to emphasize developments that have strikingly affected the health of people. But discoveries like those of Pasteur and Koch were almost immediately applied to the fields of agriculture and industry. Here we can describe only briefly a few of these applications and the scientists who originated them.

The field of soil microbiology was opened in the late 1800s by the Russian Sergei Winogradsky, who showed the importance of bacteria in taking nitrogen from the atmosphere, combining it with other elements, and making it available as plant food and hence as animal food. In 1888, the mutually beneficial, or symbiotic, relationship between bacteria and the leguminous plants, such as clover and alfalfa, was shown by H. Hellriegel and H. Wilfarth. In 1901, Martinus Willem Beijerinck (1851–1931), a famous Dutch microbiologist, found the free-living nitrogen-fixing bacterium *Azotobacter* and described its usefulness in promoting soil fertility (see Chap. 25).

Emil Christian Hansen (1842–1909), a Dane, opened the way to the study of industrial fermentations. He developed the pure-culture study of yeasts and bacteria used in vinegar manufacture, and pure cultures known as starters were soon used to encourage the study of fermentation processes. For example, L. Adametz (1889), an Austrian, used pure cultures in cheese manufacturing, and H. W. Conn in Connecticut and H. Weigmann in Germany developed pure-culture starters for butter production (1890–1897).

Late in the nineteenth century, T. J. Burrill, working in Illinois, found that in pears a disease known as fire blight was caused by a bacterium. This discovery opened a new area for microbiology, namely plant pathology—the study of plant diseases. Additional discoveries followed. In 1886 A. E. Mayer described a mottling disease of the tobacco plant and transmitted it to healthy plants by transferring sap from an infected plant. About the same time, Erwin F. Smith of the U.S. Department of Agriculture transmitted the disease peach yellows from diseased to healthy plants by the process of budding. With such leads as these to work on, Dmitrii Iwanowski demonstrated the viral nature of the infective agents of these plant diseases. Transmission of virus diseases of plants by insect was suggested by an observant Japanese farmer named Hashimoto in 1894 and independently in 1907 by the American workers A. B. Ball, A. Adams, and J. C. Shaw. Proof that insects could harbor viruses and transmit them from diseased to healthy plants was not provided until 1915, by E. Smith and P. A. Bonquet. The tobacco mosaic virus was isolated in crystalline form in 1935 by Wendell M. Stanley and John H. Northrup. For their valuable contributions to knowledge of the nature of viruses and the crystallization of virus protein, they were awarded the Nobel prize in chemistry in 1946.

From these few examples we can see that the science of microbiology grew up in less than a quarter of a century. Its early years were exciting, and by the beginning of the twentieth century, people in the street and on the farm knew of bacteria and were learning what these organisms could do and how they could be controlled.

The historical events leading to the establishment of microbiology as a science are summarized in Table 2-2.

Table 2-2. A Chronological Arrangement of Events Important in the History of Microbiology

Era	Investigator	Contribution
1500–1600	Girolamo Fracastoro (1483–1553)	Theory that invisible living seeds caused disease
1600–1700	Francesco Redi (1626–1697)	Performed experiments to disprove spontaneous generation
	Antony van Leeuwenhoek (1632–1723)	First to observe and accurately record and report microorganisms
1700–1800	John Needham (1713–1781)	Performed experiments, results supported concept of spontaneous generation
	Lazaro Spallanzani (1729–1799)	Did experiments, results disproved spon'aneous generation
	Edward Jenner (1749–1823)	Discovered vaccination for smallpox using cowpox vaccine
1800–1900	Theodor Schwann (1810–1882)	Performed experiments, results disproved spontaneous generation
	Franz Schultze (1815–1873)	Performed experiments, results disproved spontaneous generation
	Justus von Liebig (1803–1873)	Supported concept of physicochemical theory of fermentation
	Jacob Henle (1809–1885)	Established principles for germ theory of disease
	Oliver Wendell Holmes (1809–1894)	Stressed contagiousness of puerperal fever; that agent was carried from one mother to another by doctors
	Ignaz Philipp Semmelweis (1818–1865)	Introduced use of antiseptics
	Louis Pasteur (1822–1895)	Established germ theory of fermentation and germ theory of disease, developed immunization techniques
	Florence Nightingale (1820–1910)	Organized hospitals which minimized cross-infection
	Joseph Lister (1827–1912)	Developed aseptic techniques; isolated bacteria in pure culture
	Thomas J. Burrill (1839–1916)	Discovered bacterial disease of plants
	John Tyndall (1820–1893)	Developed fractional sterilization to kill spores (Tyndallization)
	Fanny Hesse (1850–1934)	Suggested use of agar as a solidifying material for micro-biological media
	Robert Koch (1843–1910)	Developed pure culture technique and Koch's postulates; discovered causative agents of anthrax and tuberculosis
	Paul Ehrlich (1854–1915)	Developed modern concept of chemotherapy and chemotherapeutic agents
	Elie Metchnikoff (1845–1916)	Discovered phagocytosis
	Hans Christian Gram (1853–1933)	Developed important procedure for differential staining of bacteria, the Gram stain
	Sergei N. Winogradsky (1856–1953)	Discovered nitrogen-fixing bacteria in soil
	William Henry Welch (1850–1934)	One of first great American microbiologists; discovered relation of clostridia to gas gangrene
	Theobald Smith (1859–1934)	Early American microbiologist; discovered transmission of Texas fever by cattle tick
1900–1910	Walter Reed (1851–1902)	Reported transmission of yellow fever by mosquito
	Jules Bordet (1870–1961) and Octave Gengou (1875–1957)	Discovered complement-fixation reaction

Table 2-2. (*continued*)

Era	Investigator	Contribution
	August von Wassermann (1866–1925)	Introduced complement-fixation test for syphilis
	Martinus Willem Beijerinck (1851–1931)	Utilized principle of enrichment cultures; confirmed finding of first virus
	Frederick W. Twort (1877–1950) Felix H. d'Herelle (1873–1949)	{ Independently discovered bacteriophages, viruses that destroy bacteria
	Howard T. Ricketts (1871–1910)	Reported Rocky Mountain spotted fever transmitted by wood tick and Mexican typhus transmitted by body louse

MICROBIOLOGY AND MODERN BIOLOGY: MOLECULAR BIOLOGY

As new laboratory techniques and experimental procedures were developed, our knowledge of the characteristics of microorganisms accumulated rapidly. Extensive information about the biochemical activities of microorganisms became available. An analysis of the data suggested that there was much in common among different microorganisms—the differences were likely to be variations on a major central biochemical pathway. At about the same time there was a growing recognition of the unity of the biochemical life processes in microorganisms and higher forms of life, including human beings. Consequently it became attractive to use microorganisms as a tool to explore fundamental life processes. Microorganisms offer numerous advantages for this kind of research: they reproduce (grow) very rapidly, they can be cultured (grown) in small or vast quantities conveniently and rapidly, their growth can be manipulated easily by chemical or physical means, and their cells can be broken apart and the contents separated into fractions of various particle sizes. These characteristics, as well as others, make microorganisms a very convenient research model for determining exactly how various life processes take place in terms of specific chemical reactions and the specific structures involved. Scientists from many disciplines recognized the usefulness of microorganisms as experimental models. Thus it was not surprising that physicists, geneticists, chemists, and biologists joined with microbiologists in what is now known as **molecular biology**. Salvador E. Luria (Fig. 2-9), professor of biology at the Massachusetts Institute

Figure 2-9. Salvador E. Luria, Professor of Biology at the Massachusetts Institute of Technology, was awarded the Nobel prize in 1969 for his research in molecular biology.

of Technology and one of the major contributors to this field, defines molecular biology as follows:

It is the program of interpreting the specific structures and functions of organisms in terms of molecular structure.

The results and rewards from this field of research have been spectacular. The contributions include elucidation of enzyme structure and mode of action, cellular regulatory mechanisms, energy metabolism, protein synthesis, structure of viruses, function of membranes, and the structure and function of nucleic acids (including DNA). Most of the basic knowledge about DNA and genetic processes at the molecular level has been obtained through research with bacteria and bacteriophages (viruses that infect bacteria). The significance of these discoveries in molecular biology to all of biology is underscored by the fact that numerous Nobel prizes have been awarded to researchers for their work in this field.

MICROBIOLOGY AND SOCIETY

From the foregoing account it is clear that microbiology has become increasingly important to our society, and microbiology has emerged as one of the most important branches of the life sciences. Microbiologists have made significant contributions to basic biological sciences as well as in the applied areas of public health and medical sciences, agriculture, industry, and environmental sciences.

The most dramatic current development in applied microbiology is the ability to alter an organism's genetic makeup, commonly referred to as **genetic engineering.** The detailed knowledge that has been obtained about the structure and function of DNA, together with the discovery of enzymes that "cut, unzip, or rebuild" the molecule, has made it possible to alter the DNA structure of microorganisms. New pieces of DNA can be inserted into a DNA molecule in a process called **recombination.** Thus a microorganism can be engineered, through modification of its DNA, to produce new substances, such as human proteins. Bacteria have been genetically modified to produce human insulin and interferon, for example. Genetically engineered microorganisms hold great potential for the production of drugs and vaccines, for improvement of agricultural crops, and for other products and processes.

QUESTIONS

1 Why was the belief in spontaneous generation an obstacle to the development of the science of microbiology? What is the relationship between the germ theory of fermentation and the germ theory of disease?
2 What is meant by the pure-culture concept? Does it exist in natural environments? Explain.
3 How did Leeuwenhoek's work influence the contributions of Pasteur and Koch?
4 In what way are the contributions of Ehrlich and Fleming related?
5 In what way did Koch's postulates influence the development of microbiology?

6 Why is the period 1880–1900 significant for the emergence of microbiology as a science?

7 Why was the introduction of the use of agar important to microbiology?

8 For what contribution to microbiology is each of the following remembered: (a) H. W. Conn, (b) Erwin F. Smith, (c) Emil Christian Hansen, and (d) T. J. Burrill?

9 Name an important contribution to microbiology made by each of the following: (a) Jenner, (b) Metchnikoff, (c) Lister, and (d) Welch.

10 Describe the process of phagocytosis.

11 How did Ehrlich's concept of disease differ from the phenomenon of phagocytosis?

12 Many researchers in the biological sciences use microorganisms as a model system to explore life processes. Explain why this is so.

13 For what contributions are the following microbiologists remembered: (a) Winogradsky, (b) Hellriegel and Wilfarth, and (c) Beijerinck?

14 List some of the applied fields of microbiology and make a general statement of the importance of microorganisms in each field.

15 What, in your judgment, were the five most important major discoveries between 1800 and 1900 which contributed to the establishment of microbiology as a science?

16 Prior to the 1930s universities and colleges had departments of bacteriology. Since that time there has been a shift toward a change in name to departments of microbiology. What is the explanation for this change?

REFERENCES

Brock, Thomas (ed.): *Milestones in Microbiology*, Prentice-Hall, Englewood Cliffs, N.J., 1961. *A compilation of historically important papers which are useful supplementary material. Reading these papers will give an insight into how current theories have developed out of the past and will help develop an understanding of experimental design.*

Bulloch, W.: *The History of Bacteriology*. Oxford, London, 1938. *The most complete and authoritative history of the development of bacteriology, it includes an extensive bibliography and a long list of biographical notices of some of the early workers in bacteriology.*

Clark, Paul F.: *Pioneer Microbiologists of America*, Wisconsin, Madison, 1961. *The people who have made microbiology a science in the United States. The author's entertaining style makes this an enjoyable as well as informative book, and his acquaintance with many of those he writes about adds a personal touch.*

Dobell, Clifford: *Antony van Leeuwenhoek and His "Little Animals,"* Dover, New York, 1960. *A collection of the writings of the founder of microbiology, with a historical background to set the environment in which he lived.*

Doetsch, Raymond N.: *Microbiology: Historical Contributions from 1776 to 1908*, Rutgers, New Brunswick, N.J., 1960. *Illustrates, in the field of general microbiology, how science has been and is concerned with ideas as they arise from the conditions of time and circumstance.*

Dowling, H. F.: *Fighting Infection, Conquests of the Twentieth Century*, Harvard, Cambridge, Mass. 1977. *One of the great success stories of the twentieth century is retold in this interesting history of the diagnosis, prevention, and treatment of infectious diseases.*

Dubos, René J.: *Louis Pasteur: Free Lance of Science*, Little, Brown, Boston, 1950. *An interesting account of the life and contributions of Louis Pasteur, written in a pleasing style.*

Lechevalier, H., and M. Solotorovsky: *Three Centuries of Microbiology*, McGraw-Hill, New York, 1965. *A definitive history of microbiology, covering the important discoveries and theories from the period of the invention of the microscope to our own time.*

van Iterson, G., L. den Dooren de Jong, A. J. Kluyver, C. B. Van Neil, and T. D. Brock: *Martinus Beijerinck, His Life and Work*, Science Tech, Madison, Wis., 1984. *Beijerinck's contributions to the founding of general microbiology were numerous and highly significant. This biography provides an excellent story of his life and professional accomplishments.*

Chapter 3

The Characterization, Classification, and Identification of Microorganisms

Characterization, classification, and identification are major objectives in all branches of the biological sciences.

Classification is a means of bringing order to the bewildering variety of organisms in nature. Once we learn the characteristics of an organism we can compare it with other organisms to discover similarities and differences. The human mind tends to arrange similar things together in groups and to distinguish these groups from one another.

In order to identify and classify microorganisms, we must first learn their **characteristics.** It is usually not feasible to study the characteristics of a single microorganism, because of its small size; therefore, we study the characteristics of a **culture**—a population of microorganisms. If we study the characteristics of a culture containing many microorganisms (usually millions or billions of cells of only one kind), it is as if we are studying the characteristics of a single organism.

As stated earlier, a culture that consists of a single kind of microorganism (one living species), regardless of the number of individuals, in an environment free of other living organisms is called a **pure culture**. (Strictly speaking, this is an **axenic culture**; however, microbiologists customarily refer to such a culture as a pure culture. In the strict, technical sense a pure culture is one grown from a single cell.)

Determining the characteristics of microorganisms is not only prerequisite for classification but is done for other reasons as well. As we have already pointed out, microorganisms play many important, indeed essential, roles in nature. It is therefore desirable to determine the characteristics of species that enable these activities to occur.

MAJOR CHARACTERISTICS OF MICROORGANISMS

The major characteristics of microorganisms fall into the following categories:

1 *Morphological characteristics.* Cell shape, size, and structure; cell arrangement; occurrence of special structures and developmental forms; staining reactions; and motility and flagellar arrangement

2 *Chemical composition.* The various chemical constituents of the cells

3 *Cultural characteristics.* Nutritional requirements and physical conditions required for growth, and the manner in which growth occurs

4 *Metabolic characteristics.* The way in which cells obtain and use their energy, carry out chemical reactions, and regulate these reactions

5 *Antigenic characteristics.* Special large chemical components (antigens) of the cell, distinctive for certain kinds of microorganisms

6 *Genetic characteristics.* Characteristics of the hereditary material of the cell (deoxyribonucleic acid, or DNA); and occurrence and function of other kinds of DNA that may be present, such as plasmids

7 *Pathogenicity.* The ability to cause disease in various plants or animals or even other microorganisms

8 *Ecological characteristics.* Habitat and the distribution of the organism in nature and the interactions between and among species in natural environments

Morphological Characteristics

Unlike other kinds of microbial characteristics, determination of morphological features usually requires studying individual cells of a pure culture. Microorganisms are very small and their size is usually expressed in micrometers (μm). One μm is equivalent to 0.001 millimeter (mm) or about 0.00004 in; consequently, routine examination of microbial cells requires the use of a high-power microscope, usually at a magnification of about 1,000 diameters.

The use of electron microscopy provides magnification of thousands of diameters and makes it possible to see fine details of cell structure. Numerous techniques are available for the microscopic examination of microorganisms. The technique selected depends upon the information which is being sought. Some of the techniques are described in Chap. 4.

Chemical Characteristics

Microbial cells consist of a wide variety of organic compounds. When cells are broken apart and their components subjected to chemical analysis, each kind of microorganism is found to have a characteristic chemical composition. Both qualitative and quantitative differences in composition occur among various species. For example, the occurrence of lipopolysaccharide in cell walls is characteristic of Gram-negative bacteria but not Gram-positive bacteria; on the other hand, many Gram-positive bacteria have cell walls that contain teichoic acids, compounds not found in Gram-negative bacteria. Fungal and algal cell walls are very different in composition from those of bacteria. A major distinction among viruses is made on the basis of the kind of nucleic acid they possess, namely ribonucleic acid (RNA) or deoxyribonucleic acid (DNA).

Cultural Characteristics

Each kind of microorganism has specific growth requirements. Many microorganisms can be grown in or on a culture medium (a mixture of nutrients used in the laboratory to support growth and multiplication of microorganisms). Some microorganisms can grow in a medium containing only inorganic compounds, whereas others require a medium containing organic compounds (amino acids, sugars, purines or pyrimidines, vitamins, or coenzymes). Some require complex natural substances (peptone, yeast autolysate, blood cells, or blood serum), and some cannot as yet be grown in an artificial laboratory medium and can be propagated only in a living host or living cells. For example, rickettsias require a host in which to grow, such as an animal, a fertilized chicken egg (chick embryo), an arthropod, or a culture of mammalian tissue cells. The host serves as a very complex "medium" for such nutritionally demanding microorganisms.

In addition to specific nutrients, each kind of organism also requires specific physical conditions for growth. For example, some bacteria grow best at high temperatures and cannot grow below 40°C; others grow best in the cold and cannot grow above 20°C; still others, such as bacteria pathogenic to humans, require a temperature close to that of the human body (i.e., 37°C). The gaseous atmosphere required for growth is also important; for instance, some bacteria require oxygen for growth; oxygen is lethal to others and they can grow only in its absence. Light may be another important physical condition: certain bacteria, such as cyanobacteria, require light as a source of energy, whereas others may be indifferent to light or may even find it deleterious to their growth.

Each kind of microorganism grows in a characteristic manner. For example, growth in a liquid medium may be abundant or sparse; it may be evenly dispersed throughout the medium, or it may occur only as a sediment at the bottom or only as a thin film or **pellicle** at the top. On solid media, microbes grow as **colonies**—distinct, compact masses of cells that are macroscopically visible. Colonies are characterized by their size, shape, texture, consistency, color, and other notable features.

Metabolic Characteristics

The life processes of the microbial cell are a complex integrated series of chemical reactions collectively referred to as **metabolism.** The variety of these reactions affords many opportunities to characterize and differentiate various groups of microorganisms. For instance, some organisms may obtain energy by absorbing light, others by oxidizing various organic or inorganic compounds, and others by redistributing the atoms within certain molecules so that the molecules become less stable. Organisms also differ in the ways in which they synthesize their cell components during growth. The various chemical reactions of an organism are catalyzed by proteins called **enzymes,** and the complement of enzymes possessed by one kind of organism, as well as the ways in which those enzymes are regulated, can differ significantly from that of other organisms.

Antigenic Characteristics

Certain chemical compounds of microbial cells are called **antigens.** Antigenic characterization of a microorganism has great practical importance. If microbial cells enter the animal body, the animal responds to their antigens by forming specific blood serum proteins called **antibodies,** which bind to the antigens. Antibodies are highly specific for the antigens that induce their formation. Because different kinds of microorganisms have different types of antigens, antibodies are widely used as tools for the rapid identification of particular kinds of microorganisms.

The numerous applications of antigen-antibody reactions will be discussed in detail in Chap. 34. For the present this might be explained as a "lock and key system." Because of the highly specific nature of the reaction, if we know the identity of one part of the system (antigen or antibody) we can identify the other. For example, if we take typhoid bacterium antibody and mix it with a suspension of unknown bacterial cells, and a positive reaction occurs, we can conclude that the cells are those of the typhoid organism. If no reaction occurs, then these bacterial cells are of some species other than the typhoid bacterium.

Genetic Characteristics

The double-stranded chromosomal DNA of each kind of microorganism has certain features that are constant and characteristic for that organism and useful for its classification:

1 ***DNA base composition.*** It is important to note that the DNA molecule is made up of base pairs: guanine-cytosine and adenine-thymine. Of the total number of nucleotide bases present in the DNA, that percentage represented by guanine plus cytosine is termed the moles % G + C value (or more briefly, mol% G + C). Values for various organisms range from 23 to 75. Some examples are listed in Table 3-1.

2 ***The sequence of nucleotide bases in the DNA.*** This sequence is unique for each kind of organism and is the most fundamental of all the characteristics of an organism; consequently, it has great significance for microbial classification.

In addition to chromosomal DNA, plasmid DNA may sometimes be present in microbial cells. Plasmids are circular DNA molecules that are capable of autonomous replication within bacterial cells, and their presence can confer special characteristics on the cells that contain them, such as the ability to make toxins (toxigenicity), to become resistant to various antibiotics, or to use unusual chemical compounds as nutrients.

Pathogenicity

The ability to cause disease, or pathogenicity, of some microorganisms is certainly a dramatic characteristic and it stimulated much of the early work with microorganisms. Although we now know that relatively few species of microorganisms cause disease, certain microorganisms are pathogenic for animals or plants, and some microorganisms may cause disease in other microorganisms.

Table 3-1. Some Examples of the DNA Base Composition of Bacteria

Species	Moles % G + C Content of DNA
Azospirillum brasilense	70–71
A. lipoferum	69–70
Campylobacter fetus	32–35
C. jejuni	31
Klebsiella pneumoniae	56–58
K. terrigena	57
Neisseria gonorrhoeae	50–53
N. elongata	53–54
Pseudomonas aeruginosa	67
P. cichorii	59
Wolinella recta	42–46
W. succinogenes	46–49

SOURCE: Krieg, N. R. (ed.), *Bergey's Manual of Systematic Bacteriology*, vol. 1, Williams & Wilkins, Baltimore, 1984.

For example, bacteria known as bdellovibrios are predatory on other bacteria, and viruses called bacteriophages can infect and destroy bacterial cells.

Ecological Characteristics

The habitat of a microorganism is important in characterizing that organism. For example, microorganisms normally found in marine environments generally differ from those in freshwater environments. The microbial population of the oral cavity differs from that of the intestinal tract. Some kinds of microorganisms are widely distributed in nature, but others may be restricted to a particular environment. The relation of an organism to its environment is often complex and may involve special characteristics of the organism that are not yet known.

MICROBIAL CLASSIFICATION, NOMENCLATURE, AND IDENTIFICATION

Once the characteristics of microorganisms have been determined and appropriately catalogued, the process of classification can begin.

Classification

In microbiology, taxa are initially constructed from strains. A strain is made up of all the descendents of a pure culture; it is usually a succession of cultures derived from an initial colony. Each strain has a specific history and designation.

Taxonomic Groups
(Taxa)

For example, strain ATCC 19554 is a strain of spirilla isolated originally from pond water in Blacksburg, Virginia in 1965 by Wells and Krieg, and cultures of this strain are maintained at the American Type Culture Collection (ATCC), Rockville, Maryland. Cultures of the same species that were isolated from other sources would be considered different strains.

The basic taxonomic group (taxon) is the species, i.e., a collection of strains having similar characteristics. Bacterial species consist of a special strain called the type strain together with all other strains that are considered sufficiently similar to the type strain as to warrant inclusion in the species. The type strain is the strain that is designated to be the permanent reference specimen for the species. Unfortunately, it is not always the strain that is most typical of all the strains included in the species, but it is the strain to which all other strains must be compared to see if they resemble it closely enough to belong to the species. Therefore, type strains are particularly important and special attention is given to their maintenance and preservation, particularly by national reference collections such as the ATCC in the United States or the National Collection of Type Cultures in England. Many other culture collections are maintained throughout the world.

In the definition just given for a bacterial species, the phrase "considered sufficiently similar to the type strain" indicates that the definition contains an element of subjectivity. In other words, the criteria which one taxonomist believes to constitute "sufficient similarity" may be quite different from those used by another taxonomist. At present there are no specific criteria for a bacterial species that are universally accepted. However, certain criteria based on DNA homology experiments (described later in this chapter) are probably more widely accepted today than any others and eventually may lead to a unifying concept for defining a species.

Just as a bacterial species is composed of a collection of similar strains, a bacterial **genus** is composed of a collection of similar species. One of the species is designated the **type species,** and this serves as the permanent example of the genus; that is, other species must be judged to be sufficiently similar to the type species to be included with it in the genus. Unfortunately, there is even less agreement about the criteria for a bacterial genus than there is for a bacterial species.

Taxonomic groups of higher rank than genus are listed below, and the same considerations about subjectivity apply here as well:

Family	A group of similar genera
Order	A group of similar families
Class	A group of similar orders
Division	A group of similar classes
Kingdom	A group of similar divisions

The Goals of Classification

Taxonomists strive to make classifications that have the following two qualities:

1 *Stability.* Classifications that are subject to frequent, radical changes lead to confusion. Every attempt should be made to devise classifications that need only minor changes as new information becomes available.
2 *Predictability.* By knowing the characteristics of one member of a taxonomic group, it should be possible to assume that the other members of the same group probably have similar characteristics. If this cannot be done, the classification has little value.

General Methods of Classifying Bacteria

Three methods are used for arranging bacteria into taxa:

The Intuitive Method. A microbiologist who is thoroughly familiar with the properties of the organisms he or she has been studying for several years may decide that the organisms represent one or more species or genera. The trouble with this method is that the characteristics of an organism that seem important to one person may not be so important to another, and different taxonomists may arrive at very different groupings. However, some classification schemes based on the intuitive method have proved to be quite useful.

Numerical Taxonomy. In an effort to be more objective about grouping bacteria, a scientist may determine many characteristics (usually 100 to 200) for each strain studied, giving each characteristic equal weight. Then using a computer he or she calculates the **% similarity (%S)** of each strain to every other strain. For any two strains, this is:

$$\%S = \frac{NS}{NS + ND}$$

where NS is the number of characteristics that are the same (positive or negative) for the two strains, and ND is the number of characteristics that are different. (The method is sometimes made more rigorous by making NS equal to the number of *positive* characteristics that are the same for the two strains, since what organisms can do may be more important than what they cannot do.) Those strains having a high %S to each other are placed into groups; those groups having a high %S to each other are in turn placed into larger groups, and so on (see Fig. 3-1). The degree of similarity needed to rank a group as a species, genus, or other taxon is a matter of judgment on the part of the taxonomist. This method of classification has great practical usefulness as well as being relatively unbiased in its approach; it also yields classifications that have a high degree of stability and predictability.

Genetic Relatedness. The third and most reliable method of classification is

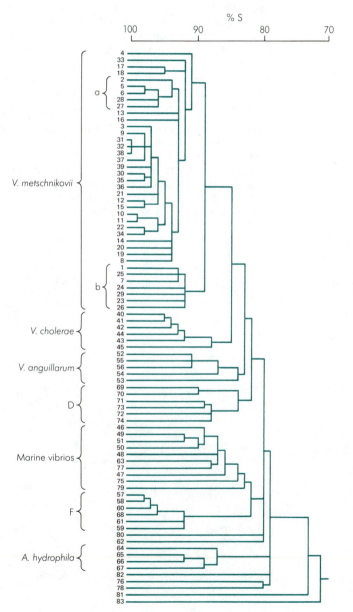

Figure 3-1. Diagram showing the arrangement of 83 strains of oxidase-positive, vibriolike bacteria according to a numerical taxonomic study. Some of the resulting groups represent species (e.g., *Vibrio metschnikovii*, *V. cholerae*, etc.); others are designated only by vernacular names (Marine vibrios, Group D, etc.). *Courtesy of J. V. Lee, T. J. Donovan, and A. L. Furniss, Int J Syst Bacteriol 28:99–111, 1978.)*

based on the degree of genetic relatedness between organisms. This method is the most objective of all and is based on the most fundamental aspect of organisms, their hereditary material (DNA). In the 1960s the development of that branch of science known as molecular biology provided techniques by which the DNA of one organism could be compared with that of other organisms. At first only crude comparisons could be made, based on mol% G + C values. It is true that two organisms of the same or similar species that are very closely related will have very similar mol% G + C values, and it is also true that two organisms having quite different mol% G + C values are not very closely related. *However, it is important to realize that organisms that are completely unrelated may have similar mol% G + C values.* Therefore, much more precise methods of comparison were needed—namely, methods by which the DNA molecules from various organisms could be compared with respect to the *sequence of their component nucleotides.* This sequence is the most fundamental characteristic of an organism. Modern techniques have now made it possible to make such a comparison. The basic principles can be described briefly as follows:

1 ***DNA homology experiments.*** The double-stranded DNA molecules from two organisms are heated to convert them to single strands. The single strands from one organism are then mixed with those from the other organism and allowed to cool. If the two organisms are closely related, heteroduplexes will form. In other words, a strand from one organism will pair with a strand from the other organism (see Fig. 3-2). If the two organisms are not closely related, no heteroduplexes will form. This method is most useful at the species level of classification.

Figure 3-2. Schematic diagram illustrating the basic principle behind DNA homology experiments.

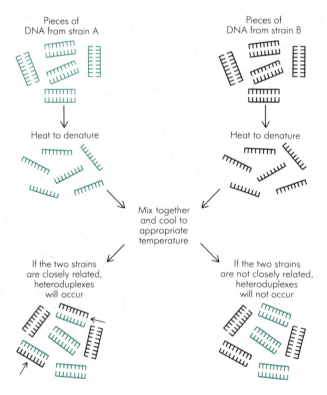

2 *Ribosomal RNA homology experiments and ribosomal RNA oligonucleotide cataloging.* Two organisms may not be so closely related as to give a high level of DNA homology, yet they may still have some degree of relatedness. Ribosomes, the small granular-appearing structures within the cell which manufacture proteins, are composed of proteins and RNA. The ribosomal RNA (rRNA) is coded for by only a small fraction of the DNA molecule, the rRNA cistrons. In all bacteria so far studied, the nucleotide sequence of these rRNA genes has been found to be highly conserved; that is, during evolution, the nucleotide sequence has changed more slowly than that of the bulk of the DNA molecule. This means that even if two organisms are only distantly related and show no significant DNA homology, there still may be considerable similarity in the nucleotide sequences of their rRNA cistrons. The degree of similarity that exists can therefore be used as a measure of relatedness between organisms, but at a level beyond that of species (at the level of genus, family, order, etc.). RNA homology experiments and RNA oligonucleotide cataloging are two modern methods used to determine the degree of similarity between the rRNA cistrons of different organisms. The techniques are complex and are being used by only a few laboratories.

Classifications based on genetic relatedness come the closest to achieving the taxonomic goals of stability and predictability. Moreover, the data obtained for such classifications allow microbiologists to infer the way in which bacteria have evolved, so that the present-day bacterial genera and species can be arranged in a hierarchy that reflects their ancestral relationships, i.e., in a phylogenetic classification. Much of the work is still fragmentary, but some of the results, especially those obtained by Dr. C. R. Woese of the University of Illinois and his colleagues, have already revolutionized current thinking about how bacteria have evolved and how they are related to one another. In fact, it is now apparent that present-day bacteria evolved by at least two very different major routes from an early ancestral form and that they now comprise two very large groups: the eubacteria (which are the traditional, familiar ones that have received the most study) and the archaeobacteria (consisting of methane-producers, extreme halophiles, and thermoacidophiles). It has been proposed that these two groups be considered as two separate kingdoms of life, and, indeed, they do seem as distantly related to each other as they are to eucaryotic organisms. Although the kingdom question is still debatable, data obtained from rRNA oligonucleotide cataloging nevertheless make it clear that the archaeobacteria are separated from other bacteria by a great phylogenetic gulf (see Fig. 3-3).

Nomenclature

Each species of microorganism has only one officially accepted name, by international agreement. This system provides for precise communication. If an organism were to be called *Escherichia coli* in one country and *Coprobacterium intestinale* in another, chaos would result. It would be difficult to know that the same organism was being studied.

The name of a species is merely a convenient label. It is not necessarily even descriptive, although some names are. For example, *Micrococcus luteus* means "yellow berry" in Latin, and *Proteus vulgaris* is Latin for "common organism of many shapes." Some species are named after persons: for example, *Escherichia coli*—the organism of the colon, named after Theodor Escherich (a German bacteriologist); or *Clostridium barkeri*—the spindle-shaped organism,

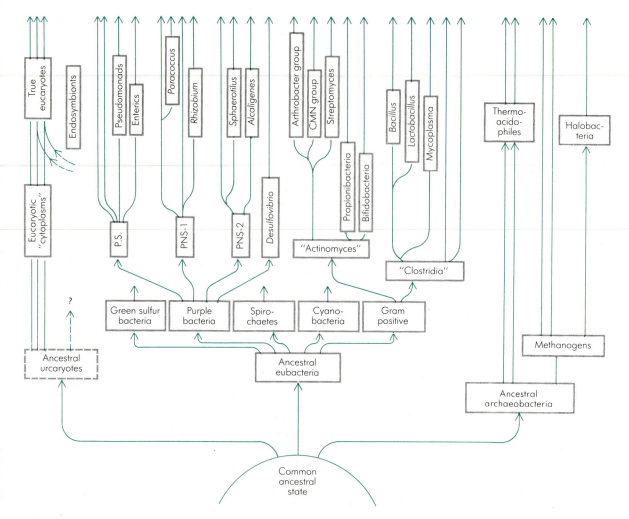

Figure 3-3. Schematic representation of the major lines of procaryotic descent, based on rRNA oligonucleotide cataloging. The archaeobacteria arose from an early ancestral form according to the pathway at the center. (A third line of descent which led to eucaryotic organisms is depicted at the left.) (*Courtesy of G. E. Fox et al., Science* **209**:457–463, 1980.)

named after H. A. Barker (an American biochemist). Some names are even nonsensical [e.g., *Runella slithyformis*—"the organism whose shape resembles runes (characters of an ancient alphabet) and which is slithy," the latter term being taken from Lewis Carroll's poem "Jabberwocky" from *Alice in Wonderland*]. The important point is that names are only convenient designations. For example, instead of referring to "the rod-shaped, acid-fast bacterium that is slow-growing, is stimulated by glycerol, causes pulmonary tuberculosis in humans, is spread mainly by airborne droplets, forms buff-colored colonies, synthesizes niacin, reduces nitrate to nitrite, and is pathogenic for guinea pigs," it is much more convenient simply to say "*Mycobacterium tuberculosis.*"

Although it might seem that microbial names could be constructed almost at random, the fact is that certain rules must be followed. Bacteria, for example, are named according to rules set down in the *International Code of Nomenclature of Bacteria*; other codes govern the naming of algae, fungi, and viruses.

One rule in bacteriological nomenclature is that a name must be written as a Latin or latinized **binomial** (two words) and must follow certain rules of Latin grammar. The first word in the binomial is the genus name and is always capitalized. The second word is the specific epithet and is never capitalized. Both the genus name and specific epithet are given in italics (or underlined, which means "italics" to a printer). Bacteria are sometimes referred to by **common** or **colloquial** names, which have no official standing in nomenclature and are never italicized (for example, the "colon bacillus," which is *E. coli*, or the "tubercle bacillus," which is *M. tuberculosis*). Such names do not lead to precise communication; for instance, many bacteria occur in the colon besides *E. coli*, and other organisms besides *M. tuberculosis* can cause tuberculosis.

Those bacterial names which have official standing in microbiology were published in the *Approved Lists of Bacterial Names* in January, 1980. Any new or additional names must be published in the *International Journal of Systematic Bacteriology* in order to achieve official recognition.

The International Code of Nomenclature of Bacteria was developed with reference to the much earlier established International Codes of Zoological and Botanical Nomenclature. All of these codes incorporate certain common principles as listed below.

1 Each distinct kind of organism is designated as a species.
2 The species is designated by a Latin binomial to provide a characteristic international label (binomial system of nomenclature).
3 Regulation is established for the application of names.
4 A law of priority ensures the use of the oldest available legitimate name.
5 Designation of categories is required for classification of organisms.
6 Requirements are given for effective publication of new specific names, as well as guidance in coining new names.

Identification

An organism must be classified before it can be **identified**—that is, given a name. This is true even if the classification is merely the recognition that the organism is different from any known organism. (For example, this occurred with the Legionnaires' disease agent, which caused the famous pneumonia epidemic in 1976 in Philadelphia; this organism was unlike bacteria of any established species; it has now been classified in a new bacterial genus, *Legionella*, and has been assigned the species name *L. pneumophila*.) Once an organism is classified, a few of its characteristics are selected by which it can be identified by other microbiologists. In order to be useful for identification, *the combination of characteristics chosen must occur only in that particular kind of organism* and in no other. The characteristics chosen should also be ones that are *easy to determine*, such as shape, staining reactions, and sugar fermentations. For example, DNA homology experiments, while very useful for classifying an organism, would be quite unsatisfactory for the routine identification of an organism because of the complexity of the procedure.

Many identification schemes are in the form of **keys,** which give identifying characteristics arranged in a logical fashion. **Identification tables** are also useful and generally contain more characteristics than do keys, with the information arranged in an easy-to-read, summarized form.

The Past and Present State of Bacterial Taxonomy

The first classification scheme for bacteria was published in 1773, and many more have appeared since. The early schemes were based only on morphological characteristics, but as the science of microbiology developed, other kinds of characteristics became increasingly important for classifications. Each successive classification scheme reflected the level of knowledge available at the time, and this continues to be true. Even present arrangements of bacteria are only provisional, subject to modification or replacement as new information appears.

Many classification schemes presently exist, but most cover only one or a few groups of bacteria. One classification scheme is unique, however, because of its broad scope and wide acceptance: *Bergey's Manual of Determinative Bacteriology*. This international reference work not only provides descriptions of all established genera and species of bacteria, but it also provides a practical arrangement of these taxa that is useful for their identification, together with appropriate keys and tables. Eight editions of *Bergey's Manual* have appeared since 1923, and a new edition is now in preparation, part of which has already been published. The title has been changed to *Bergey's Manual of Systematic Bacteriology* to reflect an increased coverage of bacterial characterization, classification, and taxonomic problems, in addition to the identification aspects. *Bergey's Manual* is written by hundreds of authors from around the world, each an authority on a particular bacterial group.

The arrangement of bacterial taxa in the new edition of *Bergey's Manual* is mainly along traditional, practical lines. Each volume is divided into a number of sections, each bearing a vernacular name such as "The Spirochetes" or "Gram-Negative Anaerobic Cocci" rather than a formal taxonomic name. The emphasis is largely on genera and species and, because of the present incomplete and fragmentary understanding of the real relationships that exist among bacteria, no attempt is made to adhere to any comprehensive, formal taxonomic hierarchy. The present classification scheme has considerable practical value, but the editorial board of *Bergey's Manual* regards it only as an interim arrangement that must eventually give way to a new, general, comprehensive classification scheme based on genetic relatedness. This is expected to provide greater stability and predictability, to lead to improved identification schemes, and to aid our understanding of the origin of present-day genera and species.

QUESTIONS

1 Define the following terms:

Classification	Taxon
Nomenclature	Taxa
Identification	Bacterial strain
Pure culture	Bacterial species
Mol% G + C	%S
Phylogenetic	*Bergey's Manual*

2 Why is it essential to classify microorganisms and what must be done before they can be classified?

3 List and describe briefly the major kinds of microbial characteristics and indicate those that must be determined by examination of individual cells.

4 Why is the type strain the most important strain in a bacterial species?

5 Explain the subjectivity that exists in the definition of a bacterial species.

6 (a) If two microorganisms have an identical mol% G + C value for their DNA, are they necessarily related? Explain.

(b) If two microorganisms have very different mol% G + C values for their DNA, are they necessarily unrelated? Explain.

7 In DNA homology experiments, we directly compare the entire genome (all the DNA) of one organism with that of another organism. What are we comparing when we do rRNA homology experiments or rRNA oligonucleotide cataloging?

8 What advantages do rRNA homology experiments and rRNA oligonucleotide cataloging offer compared to DNA homology experiments?

9 What is the reason that each taxon has only one officially recognized name?

10 What function does the name of a bacterial species serve?

11 Give an example of a bacterial name and write it in its proper form.

12 What makes *Bergey's Manual* unique among microbiological publications?

13 What is the present philosophy of the editorial board of *Bergey's Manual* toward bacterial classification and what sorts of changes may occur in future editions?

REFERENCES

Fox, G. E., E. Stackebrandt, R. B. Hespell, J. Gibson, J. Maniloff, T. A. Dyer, R. S. Wolfe, W. E. Balch, R. S. Tanner, L. J. Magrum, L. B. Zablen, R. Blakemore, R. Gupta, L. Bonen, B. J. Lewis, D. A. Stahl, K. R. Luehrsen, K. N. Chen, and C. R. Woese: "The Phylogeny of Prokaryotes," *Science,* **209**:457–463, 1980. *A summary of the results of rRNA oligonucleotide cataloging that led to the recognition of two major lines of evolutionary descent for bacteria.*

Gerhardt, P., R. G. E. Murray, R. N. Costilow, E. W. Nester, W. A. Wood, N. R. Krieg, and G. B. Phillips (eds.): *Manual of Methods for General Bacteriology,* American Society for Microbiology, Washington, D.C., 1981. *Section V of this book describes methods for bacterial characterization, numerical taxonomy, determining DNA base composition, and experiments in DNA and RNA homology.*

Krieg, N. R. (ed.): *Bergey's Manual of Systematic Bacteriology,* vol. 1, Williams & Wilkins, Baltimore, 1984. *This is the first of the four subvolumes of the new edition of Bergey's Manual and covers mainly the Gram-negative bacteria of medical and industrial importance. Introductory chapters provide an overview of bacterial classification, classification methods, nomenclature, and identification.*

Lapage, S. P., P. H. A. Sneath, E. F. Lessel, V. B. D. Skerman, H. P. R. Seeliger, and W. A. Clark (eds.): *International Code for Nomenclature of Bacteria* (1975 revision), American Society for Microbiology, Washington, D.C., 1975. *Presents the general considerations, principles, and rules for the naming of bacteria.*

Sneath, P. H. A.: "Classification of Microorganisms," in J. R. Norris and M. H. Richmond (eds.): *Essays in Microbiology,* John Wiley, Chichester, United Kingdom, 1978. *An excellent general introduction to microbial classification.*

Trüper, H. G., and J. Kramer: "Principles of Characterization and Identification of Prokaryotes," in M. P. Starr, H. Stolp, H. G. Trüper, A. Balows, and H. G. Schlegel (eds.): *The Prokaryotes: A Handbook on Habitats, Isolation, and Identification of Bacteria,* Springer-Verlag, New York, 1981. *Provides a brief overview of systematic bacteriology, including developments and trends in taxonomy.*

Chapter 4

The Microscopic Examination of Microorganisms

The **microscope** is the instrument most characteristic of the microbiology laboratory. The **magnification** it provides enables us to see microorganisms and their structures otherwise invisible to the naked eye. The magnifications attainable by microscopes range from X100 to X400,000. In addition, several different kinds of microscopy are available, and many techniques have been developed by which specimens of microorganisms can be prepared for examination. Each type of microscopy and each method of preparing specimens for examination offers advantages for demonstration of specific morphological features. In this chapter we shall describe some of the microbiologists' methods for observing the morphological characteristics of microorganisms. The techniques used to make these examinations are provided in the laboratory manual.

MICROSCOPES AND MICROSCOPY

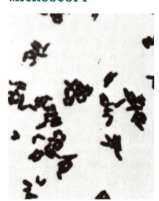

Microscopes are of two categories, **light** (or **optical**) and **electron,** depending upon the principle on which magnification is based. **Light microscopy,** in which magnification is obtained by a system of optical lenses using light waves, includes: (1) **bright-field,** (2) **dark-field,** (3) **fluorescence,** and (4) **phase-contrast** microscopy. The **electron microscope,** as the name suggests, uses a beam of electrons in place of light waves to produce the image. Specimens can be examined by either **transmission** or **scanning** electron microscopy.

In a first microbiology course, students perform most of their examinations, if not all, with the bright-field microscope. This is the most widely used instrument for routine microscopic work. The other types of microscopy are used for special purposes or research investigations. However, students should be acquainted with their applications, since each has some unique feature that is useful for demonstrating particular structures of the cell.

Figure 4-1. A stained preparation of bacteria as seen by bright-field microscopy.

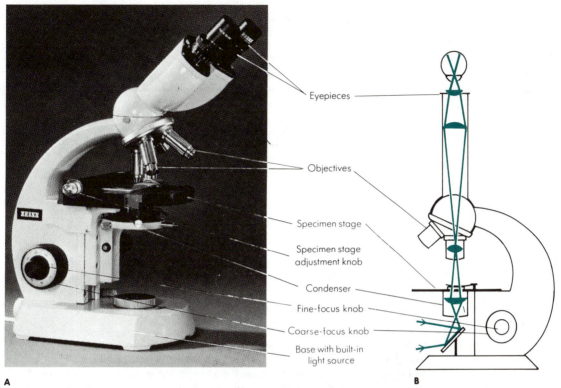

Labels in figure:
Eyepieces
Objectives
Specimen stage
Specimen stage adjustment knob
Condenser
Fine-focus knob
Coarse-focus knob
Base with built-in light source

A

B

Figure 4-2. The student microscope. (A) Identification of parts. (B) Cutaway sketch of student microscope showing optimal parts and path of light. *(Courtesy of Carl Zeiss, New York.)*

Bright-Field Microscopy

In bright-field microscopy, the microscopic field (the area observed) is brightly lighted and the microorganisms appear dark because they absorb some of the light. Ordinarily, microorganisms do not absorb much light, but staining them with a dye greatly increases their light-absorbing ability (Fig. 4-1), resulting in greater contrast and color differentiation. The optical parts of a typical bright-field microscope and the path the light rays follow to produce enlargement, or magnification, of the object are shown in Fig. 4-2. Generally microscopes of this type produce a useful magnification of about X1,000 to X2,000. At magnifications greater than X2,000 the image becomes fuzzy for reasons we will explain now.

Resolving Power

The basic limitation of the bright-field microscope is one not of magnification but of **resolving power,** the ability to distinguish two adjacent points as distinct and separate. Mere increase in size (greater magnification) without the ability to distinguish structural details (greater resolution) is not beneficial. To state it differently, the largest magnification produced by a microscope may not be the most useful because the image obtained may be unclear or fuzzy. The more lines

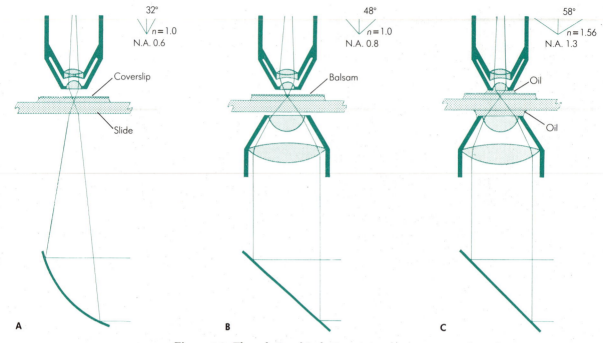

Figure 4-3. The relationship between angular aperture and resolution. (A) A narrow cone of light enters the low-power objective; the total angle is 64°, θ is 32°, and the numerical aperture (N.A.) is 0.6. (B) A substage condenser increases the size of the cone of light to 96°; θ is 48°, N.A. is 0.8. (C) With the oil-immersion objective, the size of the cone of light is increased to 116°; θ is 58°, N.A. is 1.3. The refractive index (n) for air is 1.0; for oil, 1.3. The resolution of the lens system, as described in the text, increases as the numerical aperture increases. *(From P. Gray,* Handbook of Basic Microtechnique, *McGraw-Hill, New York, 1964.)*

or dots per unit area that can be seen distinctly as separate lines or dots, the greater is the resolving power of the microscope system. The resolving power of a microscope is a function of the wavelength of light used and the numerical aperture (NA) of the lens system.

Numerical Aperture

The angle θ subtended by the optical axis and the outermost rays still covered by the objective is the measure of the **aperture** of the objective; it is the **half-aperture angle** (Fig. 4-3). The magnitude of this angle is expressed as a sine value. The sine value of the half-aperture angle multiplied by the refractive index n of the medium filling the space between the front lens and the cover slip gives the numerical aperture (NA): NA $= n \sin \theta$

With dry objectives the value of n is 1, since 1 is the refractive index of air. When immersion oil is used (Fig. 4-3) as the medium, n is 1.56, and if θ is 58°, then

$$NA = n \sin \theta = 1.56 \times \sin 58° = 1.56 \times 0.85 = 1.33$$

The degree to which microscope objectives can be altered to increase the NA is limited: the maximum NA for a dry objective is less than 1.0, and oil-immersion objectives have an NA value of slightly greater than 1.0 (1.2 to 1.4). The wavelength of light used in optical microscopes is also limited; the visible light range is between 400 nm (blue light) and 700 nm (red light), or 0.4 μm to 0.7 μm. (The abbreviation nm stands for nanometer and is equal to 0.001 μm, or 10^{-9}m.)

Thus it is apparent that the resolving power of the optical microscope is restricted by the limiting values of the NA and the wavelength of visible light.

Limit of Resolution

The limit of resolution is the smallest distance by which two objects can be separated and still be distinguishable as two separate objects. The greatest resolution in light microscopy is obtained with the shortest wavelength of visible light and an objective with the maximum NA. The relationship between NA and resolution can be expressed as follows:

$$d = \frac{\lambda}{2\text{NA}}$$

where d = resolution and λ = wavelength of light. Using the values 1.3 for NA and 0.55 μm, the wavelength of green light, for λ, resolution can be calculated as

$$d = \frac{0.55}{2 \times 1.30} = 0.21 \text{ μm}$$

From these calculations we may conclude that the smallest details that can be seen by the typical light microscope are those having dimensions of approximately 0.2 μm.

Magnification

Magnification beyond the resolving power is of no value since the larger image will be less distinct in detail and fuzzy in appearance. The situation is analogous to that of a movie screen: if we move closer to the screen the image is larger but is also less sharp than when viewed at a distance.

Most laboratory microscopes are equipped with three objectives, each capable of a different degree of magnification. These are referred to as the oil-immersion, high-dry, and low-power objectives. The primary magnification provided by each objective is engraved on its barrel. The total magnification of the system is determined by multiplying the magnifying power of the objective by that of the eyepiece. Generally, an eyepiece having a magnification of X10 is used, although eyepieces of higher or lower magnifications are available.

Dark-Field Microscopy

The effect produced by the dark-field technique is that of a dark background against which objects are brilliantly illuminated. This is accomplished by equipping the light microscope with a special kind of condenser that transmits a hollow cone of light from the source of illumination, as shown in Fig. 4-4. Most of the light directed through the condenser does not enter the objective; the field is essentially dark. However, some of the light rays will be scattered (diffracted) if the transparent medium contains objects such as microbial cells. This diffracted light will enter the objective and reach the eye; thus the object or microbial cell, in this case, will appear bright in an otherwise dark micro-

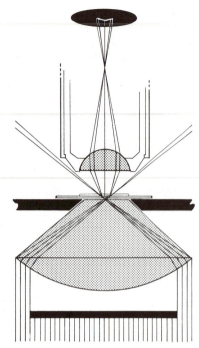

Figure 4-4. Path of light through a dark-field microscope system. Note that only those light waves which strike an object in the microscopic field are "bent" toward the observer's eye. *(Erwin F. Lessel, illustrator.)*

Figure 4-5. Dark-field and bright-field microscopy. The appearance of a white blood cell (eosinophil) surrounded by red blood cells, as viewed by (A) dark-field and (B) bright-field microscopy. *(From Scope, courtesy of The Upjohn Company.)*

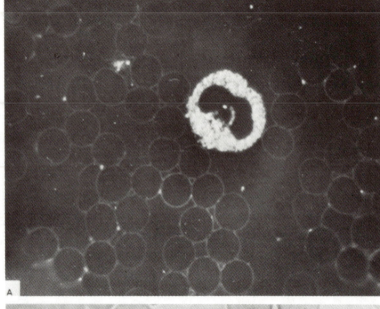

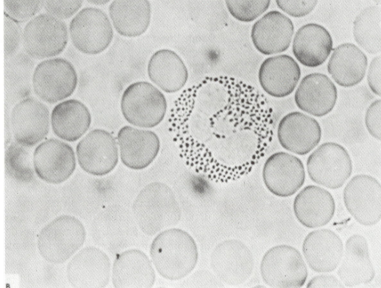

scopic field (Fig. 4-5). Dark-field microscopy is particularly valuable for the examination of *unstained microorganisms* suspended in fluid—wet-mount and hanging-drop preparations.

Fluorescence Microscopy

Many chemical substances absorb light. After absorbing light of a particular wavelength and energy, some substances will then emit light of a longer wave-

Microscopic preparations:

a The yeast *Candida albicans* stained by the Gram method; the blue color indicates that it is a Gram-positive organism.

b Gram stain of *Streptococcus pyrogenes* shows that this species is Gram-positive.

c Gram stain of specimen combining *Neisseria gonorrhoeae*, the causative agent of gonorrhea. This bacterium is Gram-negative; the cells appear red. In this specimen the gonococci appear as small diplococci within a white blood cell.

d Flagella stain of *Pseudomonas aeruginosa*. A special staining technique is required to demonstrate with light microscopy the presence of flagella on bacteria.

e Capsule stain of *Klebsiella pneumoniae*, the bacterial cells appear dark within a clear envelope.

f *Thiospirillum jenense* with sulfur globules (X1100).

g *Trypanosoma cruzi*, causative agent of Chagas' disease as seen in a stained blood film.

h *Plasmodium vivax*, causative agent of malaria. Stain of a blood film showing trophozoite in red blood cell.

(Figures a to e courtesy of Microbiology Service, Clinical Pathology Department, National Institutes of Health; Fig. f courtesy of R. L. Gherna, American Type Culture Collection; Figs. g and h courtesy of National Medical Audiovisual Center, Centers for Disease Control, Atlanta, Georgia.)

stained with fluorescein-labeled Group A specific antiserum. The unstained cells in the background are Group C streptococci; they do not react with the labeled Group A antiserum.

b *Neisseria meningitidis* in cerebrospinal fluid.

c Poliomyelitis virus, Mahoney strain, in HeLa cell culture, stained with fluorescein-labeled poliomyelitis antibody. Note the intracytoplasmic accumulation of virus in the cell near the center of the field.

d Negri bodies or rabies virus in mouse brain tissue stained orange. This preparation is stained with rhodamine-labeled antiserum. This technique permits detection of virus aggregates smaller than the Negri bodies observable by light microscopy.

e A mixed smear of *Salmonella typhosa* and *Salmonella virginia* stained with fluorescein-labeled *S. typhosa* antiserum and counterstained with Flazo Orange. The fluorescein-labeled antiserum combines only with the typhoid bacteria.

f A preparation from a pure culture of *Salmonella virginia* stained in the same manner as described in (e). They are stained by Flazo Orange counterstain, which is nonspecific.

g Enteropathogenic *Escherichia coli* in a fecal smear from a case of infant diarrhea.

h *Blastomyces dermatitidis* as seen by direct fluorescent antibody stain.

(Figures a, c, and d courtesy of Baltimore Biological Laboratory; Figs. e for Disease Control, DHEW, Atlanta, Georgia; Fig. b courtesy of National

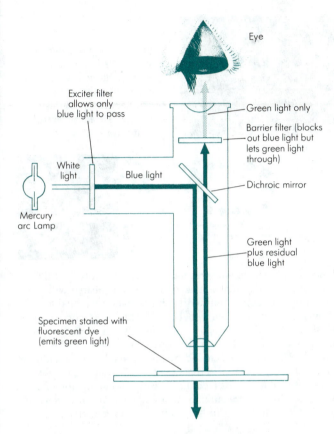

Eye

Exciter filter
allows only
blue light to pass

Green light only

Barrier filter (blocks
out blue light but
lets green light
through)

White
light

Blue light

Dichroic mirror

Mercury
arc Lamp

Green light
plus residual
blue light

Specimen stained with
fluorescent dye
(emits green light)

Figure 4-6. The special features of fluorescence microscopy. A high-intensity mercury lamp is used as the light source and emits white light. The exciter filter transmits only blue light to the specimen and blocks out all other colors. The blue light is reflected downward to the specimen by a dichroic mirror (which reflects light of certain colors but transmits light of other colors). The specimen is stained with a fluorescent dye: certain portions of the specimen retain the dye, others do not. The stained portions absorb blue light and emit green light, which passes upward, penetrates the dichroic mirror, and reaches the barrier filter. This filter allows the green light to pass to the eye; however, it blocks out any residual blue light from the specimen which may not have been completely deflected by the dichroic mirror. Thus the eye perceives the stained portions of the specimen as glowing green against a jet black background, whereas the unstained portions of the specimen are invisible. (*Erwin F. Lessel, illustrator.*)

length and a lesser energy content. Such substances are called **fluorescent** and the phenomenon is termed **fluorescence.** Application of this phenomenon is the basis of fluorescence microscopy. In practice, microorganisms are stained with a fluorescent dye and then illuminated with blue light; the blue light is absorbed and green light emitted by the dye.

The special features of fluorescence microscopy with the respect to illumination of the specimen are shown in Fig. 4-6. The function of the **exciter filter** is to remove all but the blue light; the **barrier filter** blocks out blue light and allows green light (or other light emitted by the fluorescing specimen) to pass through and reach the eye. Barrier filters are selected on the basis of the dye used.

An example of direct staining of bacteria with a fluorescent dye is shown in Fig. 4-7.

The Fluorescent Antibody Technique—Immunofluorescence

It is possible to chemically combine fluorescent dyes with antibodies, i.e., substances that combine with specific microorganisms. Antibodies to which a fluorescent dye is attached are referred to as **labeled antibodies**. Thus labeled antibodies can be mixed with a suspension of bacteria and then the preparation examined by fluorescent microscopy. The bacterial cells that have combined

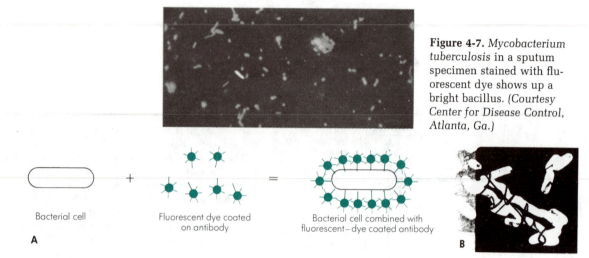

Figure 4-7. *Mycobacterium tuberculosis* in a sputum specimen stained with fluorescent dye shows up a bright bacillus. *(Courtesy Center for Disease Control, Atlanta, Ga.)*

Bacterial cell

Fluorescent dye coated on antibody

Bacterial cell combined with fluorescent–dye coated antibody

A

B

Figure 4-8. Fluorescence staining technique and microscopy. (A) The direct fluorescent antibody staining technique. When a bacterial cell is incubated with specific antibody that is conjugated (combined) with a fluorescent dye, the dye–antibody conjugate will cover the surface of the cell. The technique is performed on a glass slide, the excess fluorescent dye–antibody conjugate is washed off, and the preparation is examined by ultraviolet light microscopy. The bacterial cell will glow brilliantly as a result of fluorescence caused by the ultraviolet illumination of the dye-coated bacterial cell. Any bacterial cells not covered by the dye do not fluoresce and hence are not visible by this technique. (B) Photomicrograph of a fluorescent-stained *Proteus mirabilis* preparation as described above. *(Courtesy of Judith Hoeniger, F. M. Clinits, and E. A. Clinits, J. Bacteriol,* **98**:226, 1969.)

with the labeled antibody will be visible in the microscopic preparation (see Fig. 4-8). This procedure is known as the **fluorescent antibody technique;** the phenomenon is termed **immunofluorescence.** Theoretically, it is possible to identify a **single microbial cell** by this procedure. The application of this test in diagnostic procedures is discussed in Chap. 34.

Phase-Contrast Microscopy

Phase-contrast microscopy is extremely valuable for studying living unstained cells and is widely used in applied and theoretical biological studies. It uses a conventional light microscope fitted with a phase-contrast objective and a phase-contrast condenser. This special optical system makes it possible to distinguish unstained structures within a cell which differ only slightly in their refractive indices or thicknesses.

In principle, this technique is based on the fact that light passing through one material and into another material of a slightly different refractive index and/or thickness will undergo a change in **phase.** These differences in phase, or wavefront irregularities, are translated into variations in brightness of the structures and hence are detectable by the eye.

With phase-contrast microscopy it is possible to reveal differences in cells and their structures not discernible by other microscopic methods. A compari-

A

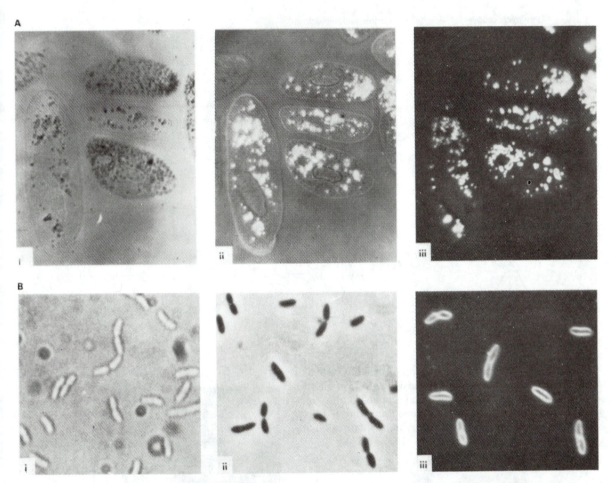

Figure 4-9. (A) Phase-contrast microscopy compared with bright-field and dark-field microscopy. The same specimen of a protozoan as seen by each method: (i) bright-field; (ii) phase-contrast; (iii) dark-field. *(Courtesy of O. W. Richards, Research Department, American Optical Company.)* (B) Photomicrographs of living, unstained, rod-shaped cells of *Pseudomonas fluorescens*. The bacilli are 0.7 to 0.8 µm in width. They can be seen only indistinctly by ordinary bright-field microscopy (i) but are readily visible by phase-contrast (ii) or dark-field microscopy (iii). *(Courtesy N. R. Krieg.)*

son of a specimen viewed by bright-field, dark-field, and phase-contrast microscopy is shown in Fig. 4-9.

Transmission Electron Microscopy

Electron microscopy differs markedly and in many respects from the optical microscopic techniques. The electron microscope provides tremendous useful magnification, because of the much higher resolution obtainable with the ex-

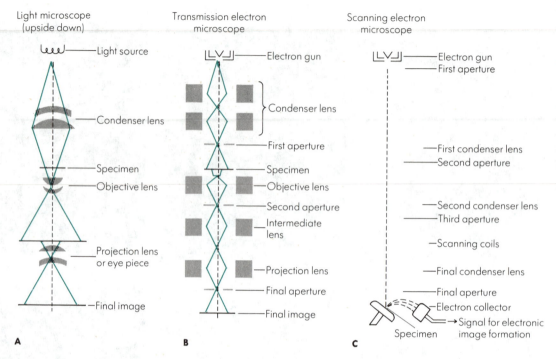

Light microscope (upside down)

- Light source
- Condenser lens
- Specimen
- Objective lens
- Projection lens or eye piece
- Final image

A

Transmission electron microscope

- Electron gun
- Condenser lens
- First aperture
- Specimen
- Objective lens
- Second aperture
- Intermediate lens
- Projection lens
- Final aperture
- Final image

B

Scanning electron microscope

- Electron gun
- First aperture
- First condenser lens
- Second aperture
- Second condenser lens
- Third aperture
- Scanning coils
- Final condenser lens
- Final aperture
- Electron collector
- Specimen
- Signal for electronic image formation

C

Figure 4-10. Diagrammatic comparison of imaging systems in (A) optical microscope, (B) transmission electron microscope, and (C) scanning electron microscope. *(Courtesy of L. A. Bulla, Jr., G. St. Julian, C. W. Hesseltine, and F. L. Baker, Scanning Electron Microscopy, in Methods in Microbiology, vol. 8, Academic, New York, 1973.)*

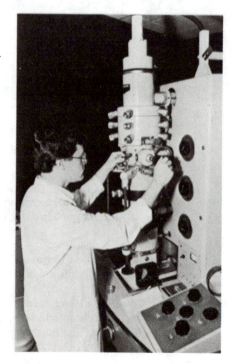

Figure 4-11. A high-resolution electron microscope. *(Courtesy of George Hatjygeorge, Fordham University.)*

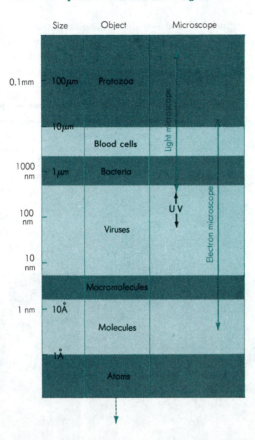

Figure 4-12. Relative size of microbes, molecules, and atoms is depicted here, together with an indication of the useful range of different types of microscopes. *(Courtesy of A. J. Rhodes and C. E. van Rooyen, Textbook of Virology, Williams & Wilkins, Baltimore, 1968.)*

tremely short wavelength of the electron beam used to magnify the specimen. As shown in Fig. 4-10, the electron microscope uses electron beams and magnetic fields to produce the image, whereas the light microscope uses light waves and glass lenses.

With an electron microscope (Fig. 4-11) employing 60- to 80-kV electrons the wavelength is only 0.05 Å. Å is the abbreviation for **angstrom;** 1 Å equals 1/100,000,000 (10^{-8}) cm or 1/10,000 (10^{-4}) μm. (Compare this electron wavelength with the light wavelengths used for optical microscopes.) It is possible to resolve objects as small as 10 Å (Fig. 4-12). The resolving power of the electron microscope is more than *100 times* that of the light microscope, and it produces useful magnification up to ×400,000.

For electron microscopy, the specimen to be examined is prepared as an extremely thin dry film on small screens and is introduced into the instrument at a point between the magnetic condenser and the magnetic objective; this point is comparable to the stage of the light microscope. The magnified image may be viewed on a fluorescent screen through an airtight "window" or recorded on a photographic plate by a camera built into the instrument.

Numerous techniques are available for use with electron microscopy which extend its usefulness in characterizing cellular structure. Some of these are described below.

Shadow-casting

This technique involves depositing an extremely thin layer of metal (e.g., platinum) at an oblique angle on the organism so that the organism produces a shadow on the uncoated side. The shadowing technique produces a topographical representation of the surface of the specimen (see Fig. 4-13).

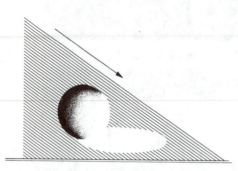

Figure 4-13. Shadow-casting technique. The specimen is dried on a special grid which is placed in a vacuum jar. Atoms of a heavy metal such as platinum are projected (from a highly heated filament) at an angle that produces a "shadow" behind the particles being examined. Examination of the shadowed image provides information as to the shape of the specimen particles. (*Erwin F. Lessel, illustrator.*)

Figure 4-14. Electron micrographs of tobacco rattle virus as seen in three different preparations (A, B, and C). This virus characteristically appears as particles of two different sizes; the larger particle measures 184 by 25 nm and the smaller particle measures 74 by 25 nm. (A) Shadow-cast preparation using chromium. (B) Negative-stain preparation using potassium phosphotungstate. (C) Ultrathin section of infected leaf showing intracellular virus crystals, stained with uranyl acetate and lead citrate. (*Courtesy of M. Kenneth Corbett.*)

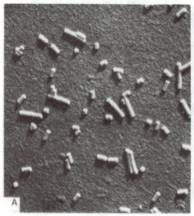
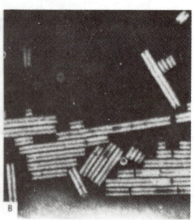
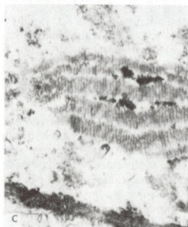

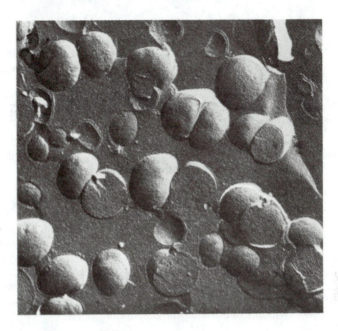

Figure 4-15. Electron micrograph of freeze-etched preparation of *Neisseria gonorrheae*. This bacterium measures approximately 0.6 to 1.0 μm in diameter. *(Courtesy of Ivan L. Roth.)*

Negative Staining

An electron-dense material such as phosphotungstic acid can be used as a "stain" to outline the object. The electron-opaque phosphotungstate does not penetrate structures but forms thick deposits in crevices (see Fig. 4-14). Fine detail of objects such as viruses or bacterial flagella can be seen by this technique.

Ultrathin Sectioning

In order to make observations of intracellular structures, the material for examination must be extremely thin. An intact microbial cell is too thick to allow distinct visualization of its internal fine structure by electron microscopy. However, techniques are available for sectioning (slicing) a bacterial cell; for example, bacterial cells can be embedded in a plastic material and then this "block" can be cut into ultrathin slices, as thin as 60 nm. These slices are then prepared for microscopic examination. As you might expect, the slices will reveal cells sliced at different levels and at different angles. Improvement in contrast of structures is possible through use of special electron-microscope stains such as uranium and lanthanum salts.

Freeze-Etching

Freeze-etching was developed to prepare sections of the specimen without resorting to the chemical treatment of the fixation process, which can produce artifacts. The specimen is sectioned while contained in a frozen block. Carbon replicas of these exposed surfaces are then prepared which reveal internal structures of the cell (see Fig. 4-15).

Localization of Cell Constituents

Special techniques have been developed making it possible to locate chemical constituents of the cell. For example, thin sections of a cell can be treated with ferritin-labeled antibody. Ferritin is an iron-containing substance of high density that markedly affects passage of the electron beam. The combination of this ferritin-labeled antibody with antigen in the cell produces a complex which manifests a higher contrast in the electron-microscope image.

Localization of Enzymes in Thin Sections

Electron-microscope techniques have been developed to locate the position of enzymes within cells. The intracellular localization of the enzyme isocitrate dehydrogenase of *Escherichia coli* is shown in Fig. 4-16. This was accomplished by first preparing ultrathin sections of the bacterial cells, followed by an immunochemical technique by which colloidal gold is affixed specifically to the isocitrate dehydrogenase enzyme.

Autoradiography

Autoradiography is a cytochemical method in which the location of a particular chemical constituent in a specimen is determined by observing the site at which radioactive material becomes positioned. The cells are first exposed to the radioactive substance to permit its uptake. In practice, the specimen prepared for microscopic examination is covered with a layer of photographic emulsion and stored in the dark for a period of time. The ionizing radiation emitted during the decay of the radioactive substance produces latent images in the emulsion, and, after photographic processing, the developed image is seen as grains of silver in the preparation.

Scanning Electron Microscopy

In scanning electron microscopy the specimen is subjected to a narrow electron beam which rapidly moves over (scans) the surface of the specimen. This causes the release of a shower of secondary electrons and other types of radiation from

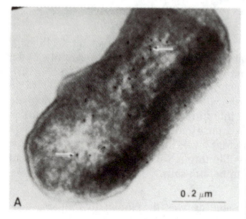
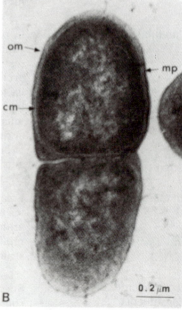

Figure 4-16. Intracellular localization of isocitrate dehydrogenase in *Escherichia coli* as seen by electron microscopy. (A) Ultrathin section of *E. coli* treated with a specific fraction of antiserum to the enzyme and then with a protein–gold particle complex which reacts with the antiserum. Location of gold particles and hence the enzyme is shown by arrows. (B) Control section treated with preimmune serum and protein–gold, showing outer membrane (om), peptidoglycan (mp), and cytoplasmic membrane (cm). *(Courtesy of J. R. Swafford et al., Science* **121:**295, 1983.)

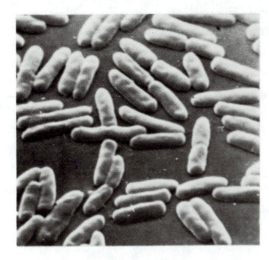

Figure 4-17. Scanning electron micrograph of cells of *Pseudomonas aeruginosa*. The average size of this bacterium is 0.5 to 1.0 μm by 1.5 to 4 μm. (*Courtesy of David Greenwood.*)

the specimen surface. The intensity of these secondary electrons depends on the shape and the chemical composition of the irradiated object. The secondary electrons are collected by a detector which generates an electronic signal. These signals are then scanned in the manner of a television system to produce an image on a cathode ray tube. The image system for the scanning electron microscope is shown in Fig. 4-10.

The scanning electron microscope lacks the resolving power obtainable with the transmission electron microscope but has the advantage of revealing a striking three-dimensional picture. The surface topography of a specimen can be revealed with a clarity and a depth of field not possible by any other method. An example of a scanning electron microscope picture (micrograph) is shown in Fig. 4-17.

LIMITATIONS OF ELECTRON MICROSCOPY

Despite the great advantage of tremendous resolution and magnification, there are several limitations to electron microscopy. For example, the specimen being examined is in a chamber that is under a very high vacuum. Thus cells cannot be examined in a living state. In addition, the drying process may alter some morphological characteristics. Another limitation of the technique is the low penetration power of the electron beam, necessitating the use of thin sections to reveal the internal structures of the cell.

The real problem confronting the researcher who attempts to unravel the fine intracellular structure of the microbial cell is identification of intracellular material. Frequently it is necessary to correlate results obtained with the same organism viewed by different microscopic techniques, e.g., phase-contrast, bright-field (stained preparations), and electron microscopy. Each method contributes different kinds of information. Interpretation of this information, particularly comparison of what is revealed by each technique, makes it possible to identify cellular structures. But considerable experience in microscopy is required before a researcher can correctly interpret the results.

Table 4-1. A Comparison of different Types of Microscopy

Type of Microscopy	Maximum Useful Magnification	Appearance of Specimen	Useful Applications
Bright-field	1,000–2,000	Specimens stained or unstained; bacteria generally stained and appear color of stain	For gross morphological features of bacteria, yeasts, molds, algae, and protozoa
Dark-field	1,000–2,000	Generally unstained; appears bright or "lighted" in an otherwise dark field	For microorganisms that exhibit some characteristic morphological feature in the living state and in fluid suspension, e.g., spirochetes
Fluorescence	1,000–2,000	Bright and colored; color of the fluorescent dye	Diagnostic techniques where fluorescent dye fixed to organism reveals the organism's identity
Phase-contrast	1,000–2,000	Varying degrees of "darkness"	For examination of cellular structures in living cells of the larger microorganisms, e.g., yeasts, algae, protozoa, and some bacteria
Electron	200,000–400,000	Viewed on fluorescent screen	Examination of viruses and the ultrastructure of microbial cells

Some of the major features of the several kinds of microscopy are summarized in Table 4-1.

PREPARATIONS FOR LIGHT-MICROSCOPE EXAMINATIONS

Two general techniques are used to prepare specimens for light-microscope examination. One is to suspend organisms in a liquid (the wet-mount or the hanging-drop techniques), and the other is to dry, fix, and stain films or smears of the specimen.

The Wet-Mount and Hanging-Drop Techniques

Wet preparations permit examination of organisms in a normal living condition. A wet mount is made by placing a drop of fluid containing the organisms onto a glass slide and covering the drop with a cover slip. To reduce the rate of evaporation and exclude the effect of air currents, the drop may be ringed with petroleum jelly or a similar material to provide a seal between the slide and cover slip. A special slide with a circular concave depression is sometimes used for examination of wet preparations. A suspension of microbial specimen is placed on a cover slip, then inverted over the concave depression to produce a "hanging drop" of the specimen.

Examination of microorganisms in wet preparation is desirable in the following instances:

1 The morphology of spiral bacteria is greatly distorted when these bacteria are dried and stained; they should be examined in living condition. For example, in the examination of serous exudates suspected of containing the spirochete that causes syphilis, the wet preparations are examined by dark-field microscopy. This provides a sharp contrast between the organisms and the dark background. The normal arrangement of cells can also be better determined in a wet preparation.
2 The observation of bacteria to determine whether or not they are motile obviously requires that they be suspended in a liquid medium, free to move about.
3 To observe cytological changes occurring during cell division and to determine the rate at which the division occurs, the organisms must be examined in the living state (i.e., wet mount). Spore formation and germination must also be observed in living cells.
4 Some cell inclusion bodies, e.g., vacuoles and lipid material, can be observed readily by this method.

When wet preparations are examined by bright-field microscopy, it is extremely important to control the light source. The reason is that the lack of a stain makes the cells less distinctly visible; adjustment of the intensity of the light source can enhance their visibility. Partially closing the substage condenser diaphragm helps to increase contrast; however, some resolving power is lost. Dark-field and phase-contrast microscopy offer the distinct advantage of providing both high contrast and high resolving power for examination of unstained preparations.

Fixed, Stained Smears

Fixed, stained preparations are most frequently used for the observation of the morphological characteristics of bacteria. The advantages of this procedure are that (1) the cells are made more clearly visible after they are colored, and (2) differences between cells of different species and within the same species can be demonstrated by use of appropriate staining solutions (differential or selective staining).

The essential steps in the preparation of a fixed, stained smear are (1) preparation of the film or smear, (2) fixation, and (3) application of one or more staining solutions.

Microbiological Stains

A large number of colored organic compounds (dyes) are available for staining microorganisms. These compounds are generally rather complex in terms of molecular structure. On this basis they may be classified into groups such as **triphenylmethane dyes, oxazine dyes,** and **thiazine dyes.**

A more practical classification for the cytologist is one based on the chemical behavior of the dye; namely, **acid, basic,** or **neutral.** An acid (or anionic) dye is one in which the charge on the dye ion is negative; a basic (or cationic) dye is one in which the charge carried by the dye ion is positive. A neutral dye is a complex salt of a dye acid with a dye base, e.g., eosinate of methylene blue. Acid dyes generally stain basic cell components, and basic dyes generally stain acidic cell components.

The process of staining may involve ion-exchange reactions between the stain and active sites at the surface of or within the cell. For example, the colored ions of the dye may replace other ions on cellular components. Certain chemical groupings of cell proteins or nucleic acids may be involved in salt formation with positively charged ions such as Na^+ or K^+. Thus we might view these peripheral areas of the cell as carrying a negative charge in combination with positively charged ions; for example,

$$(\text{Bacterial cell}^-)(Na^+)$$

In a basic dye like methylene blue, the colored ion is positively charged (a cation), and if we represent this ion by the symbol MB, the dye, which is actually methylene blue chloride, may be represented as

$$MB^+Cl^-$$

The ionic exchange which takes place during staining can be represented by the following equation, in which the MB^+ cation replaces the Na^+ cation in the cell:

$$(\text{Bacterial cell}^-)(Na^+) + (MB^+)(Cl^-) \rightarrow (\text{bacterial cell}^-)(MB^+)\ ^+ (Na^+Cl^-)$$

Simple Staining. The coloration of bacteria by applying a single solution of stain to a fixed smear is termed **simple staining.** The fixed smear is flooded with a dye solution for a specified period of time, after which this solution is washed off with water and the slide blotted dry. The cells usually stain uniformly. However, with some organisms, particularly when methylene blue is used, some granules in the interior of the cell may appear more deeply stained than the rest of the cell, indicating a different type of chemical substance.

Differential Staining. Staining procedures that make visible the differences between bacterial cells or parts of a bacterial cell are termed **differential staining** techniques. They are slightly more elaborate than the simple staining technique in that the cells may be exposed to more than one dye solution or staining reagent.

Gram Staining. One of the most important and widely used differential staining techniques in microbiology is **Gram staining.** This technique was introduced by Christian Gram in 1884. In this process the fixed bacterial smear is subjected to the following staining reagents in the order listed: crystal violet, iodine solution, alcohol (decolorizing agent), and safranin or some other suitable counterstain. Bacteria stained by the Gram method fall into two groups: **Gram-positive** bacteria, which retain the crystal violet and hence appear deep violet in color; and **Gram-negative** bacteria, which lose the crystal violet, are counterstained by the safranin, and hence appear red in color. Why does this procedure stain some bacteria purple-violet and others red?

The most plausible explanations for this phenomenon are associated with the structure and composition of the cell wall. (See Chap. 5 for a discussion of the relative differences between the cell walls of Gram-negative and Gram-positive bacteria.) Differences in the thickness of cell walls between these two groups may be important; the cell walls of Gram-negative bacteria are generally thinner than those of Gram-positive bacteria. Gram-negative bacteria contain a higher percentage of lipid than do Gram-positive bacteria. Experimental evidence suggests that during staining of Gram-negative bacteria the alcohol treatment extracts the lipid, which results in increased porosity or permeability of the cell wall. Thus the crystal violet–iodine (CV-I) complex can be extracted and the Gram-negative organism is decolorized. These cells subsequently take on the color of the safranin counterstain. The cell walls of Gram-positive bacteria, because of their different composition (lower lipid content), become dehydrated during treatment with alcohol. The pore size decreases, permeability is reduced, and the CV-I complex cannot be extracted. Therefore these cells remain purple-violet.

Another explanation, somewhat similar, is also based on permeability differences between the two groups of bacteria. In Gram-positive bacteria, the CV-I complex is trapped in the wall following ethanol treatment, which presumably causes a diminution in the diameter of the pores in the cell-wall peptidoglycan. Walls of Gram-negative bacteria have a very much smaller amount of peptidoglycan, which is less extensively cross-linked than that in the walls of Gram-positive bacteria. The pores in the peptidoglycan of Gram-negative bacteria remain sufficiently large even after ethanol treatment to allow the CV-I complex to be extracted. These two explanations are not mutually exclusive, and it is likely that both may contribute to the explanation of the mechanism of the Gram stain. Furthermore, if Gram-positive cells are treated with lysozyme (an enzyme) to remove the cell wall, the resulting structures, called **protoplasts** (cells lacking walls), will be stained by the CV-I complex. However, they are easily decolorized by alcohol. All this evidence points to the cell-wall structure of Gram-positive bacteria as the site of retention of the primary stain.

Although Gram-negative organisms consistently fail to retain the primary crystal violet stain, Gram-positive organisms may sometimes show variations in this respect, i.e., a **Gram-variable reaction.** For example, old cultures of Gram-positive bacteria lose the ability to retain the crystal violet and hence will be stained by the safranin. A similar effect may sometimes be due to changes in the environment of the organism or a slight modification in staining technique.

Within some groups of bacteria, such as the archaeobacteria (see Chap. 5), some are Gram-positive and others Gram-negative; yet the cell wall structure and chemical composition of these bacteria is very different from that of other groups of Gram-positive and Gram-negative bacteria.

Gram-positive bacteria differ from Gram-negative bacteria in other characteristics besides staining reaction. Gram-positive bacteria are usually more susceptible to penicillin and less susceptible to disintegration by mechanical treatment or exposure to some enzymes than Gram-negative bacteria. Gram-negative bacteria as a group are more susceptible to other antibiotics such as streptomycin. There are other differences between these two groups of bacteria.

The Gram stain has its greatest use in characterizing bacteria. This staining technique is not generally applicable for other groups of microorganisms such as protozoa and fungi; however, yeasts consistently stain Gram-positive.

Other Differential Stains. There are numerous other staining techniques designed to identify some particular feature of cell structure or composition. These techniques are summarized here. Detailed descriptions of these procedures appear in the laboratory manual.

NAME OF STAINING TECHNIQUE	APPLICATION
Acid-fast stain	Distinguishes acid-fast bacteria such as *Mycobacterium* spp. from non-acid-fast bacteria
Endospore stain	Demonstrates spore structure in bacteria as well as free spores
Capsule stain	Demonstrates presence of capsules surrounding cells
Flagella stain	Demonstrates presence and arrangement of flagella
Cytoplasmic inclusion stains	Identifies intracellular deposits of starch, glycogen, polyphosphates, hydroxybutyrate, and other substances
Giemsa stain	Particularly applicable for staining rickettsias and some protozoa

QUESTIONS

1 Define the following terms:

Resolving power   Fluorescence
Limit of resolution, d Autoradiography
NA (numerical aperture)   Anionic dye
Angle θ Cationic dye

2 What are the usual magnifications obtainable with light microscopy? What determines its useful limit?

3 Assume that a yeast cell is examined by (a) bright-field, (b) phase-contrast, and (c) dark-field microscopy. Describe the likely differences in the appearance of the cell when viewed by these methods.

4 Why are microorganisms stained?

5 What is the function of oil when used with the oil-immersion objective?

6 Name several different staining techniques and describe their particular applications.

7 Compare the kind of image obtained with scanning electron microscopy with that obtained using transmission electron microscopy.

8 Compare the resolving power of the electron microscope with that of the light microscope.

9 Name two limitations of electron microscopy.

10 What are some major differences between Gram-positive and Gram-negative bacteria?

11 Why is the Gram stain one of the most important and widely used stains in bacteriology?

12 Compare the appearance of microorganisms as seen by dark-field and by phase-contrast microscopy.

13 Describe two special applications of fluorescence microscopy.

REFERENCES

Bradbury, S.: *The Microscope, Past and Present*, Pergamon, New York, 1968. *A 272-page paperback describing developments in microscopy from the first magnification devices to the electron microscope.*

Clark, G.: *Staining Procedures*, 3d ed., Williams & Wilkins, Baltimore, 1973. *This volume is organized in three main sections: animal histology, botanical sciences, and microbiology. A section on fluorescence microscopy is included.*

Gerhardt, P., R. G. E. Murray, R. N. Costilow, E. W. Nester, W. A. Wood, N. R. Krieg, and G. B. Phillips: *Manual of Methods for General Bacteriology*, American Society for Microbiology, Washington, D.C., 1981. *The first four units of this manual cover the topics of light microscopy, specimen preparation for light microscopy, staining procedures, and electron microscopy. An excellent supplement to the usual laboratory manuals.*

Gray, P.: *Handbook of Basic Microscopic Technique*, 3d ed., McGraw-Hill, New York, 1964. *The principles of microscopy are clearly presented in Part I, "The Microscope." Other sections describe preparation of microscopic slides and slide-making techniques.*

Ohnsorge, J., and R. Holm: *Scanning Electron Microscopy: An Introduction for Physicians and Biologists*, Publishing Science Group, Acton, Mass., 1973. *A guide to the main principles of the techniques for instrument operation and specimen preparation.*

Spencer, Michael: *Fundamentals of Light Microscopy*, IUPAB Biophysics Series, Cambridge, New York, 1982. *The basics of light microscopy are presented in a succinct manner in this paperback.*

Weakley, Brenda S.: *A Beginner's Handbook in Biological Electron Microscopy*, Williams & Wilkins, Baltimore, 1972. *A very good introduction to the techniques employed in electron microscopy, including fixation, embedding, audioradiography, and freeze-etching. The author discusses the pros and cons of various methods.*

PART TWO
MICROORGANISMS– BACTERIA

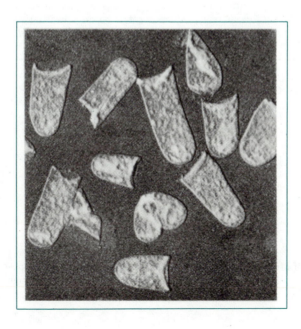

Some elegant experiments with bacterial flagella

It had been known for many years that bacterial flagella—the helically shaped filaments projecting from bacterial cells—are somehow responsible for the ability of bacterial cells to swim. Bacteria lacking flagella, and mutants that have lost the ability to make flagella, are unable to swim; moreover, mutants that make only straight flagella are unable to swim, indicating that a helical flagellar shape is required for swimming. Prior to 1973, however, it was not clear just how flagella operated. One theory was that if the flagella could rotate as on a bearing, this could cause the flagella to screw through the medium, just as a rotating corkscrew can penetrate a piece of cork. The discovery of disk-shaped structures at the base of bacterial flagella suggested that there might indeed be a "flagellar motor" that could cause flagella to spin. Unfortunately, individual bacterial flagella were so thin that they could not be observed microscopically "in action" on a living bacterium.

In 1973, Michael Silverman and Melvin Simon at the University of California, San Diego, performed experiments that indicated unequivocally that bacterial flagella do actually rotate. They realized that a bacterial flagellum might be analogous to the shaft of an electric motor: if the motor housing is bolted to a table, then the shaft will rotate. However, if the motor is not bolted to a table but rather is grasped by the shaft and held up in the air, *then the shaft will be stationary and the motor housing will rotate.* Using this analogy, Silverman and Simon devised experiments whereby they could prevent a bacterial flagellum from rotating. The

consequence of this would be that the *bacterial cell would rotate instead*—and this was something that could be easily seen with an ordinary microscope.

Silverman and Simon chose a mutant bacterium that had a single straight flagellum on one side of the cell. This mutant could not swim because the flagellum was not helical; nevertheless, if the rotational hypothesis were correct, the flagellum should still be able to spin. They prepared antibodies against the flagellum and then added a mixture of the bacteria and the antibodies to a slide. The slide became coated with the antibodies, and the flagella adhered to the antibodies. Thus, each cell became "tethered" by its flagellum to the slide. In this condition the flagella of these cells could not rotate; however, each tethered cell began to spin like a pinwheel!

It was difficult to explain how such behavior could occur unless, in a free-swimming cell, the flagella spun—probably by means of a rotary motor at their base. Additional experiments on untethered cells also supported the idea of flagellar rotation. Silverman and Simon found that small latex beads could be attached to a bacterial flagellum by means of antibodies; such beads, which were easily visible with a microscope, were observed to rotate rapidly about an invisible axis (the straight flagellum)!

It is now generally accepted that bacterial flagella do rotate as on a bearing—a type of motion that may be unique among living organisms.

Preceding page. Cell walls of *Mycobacterium tuberculosis* obtained from cells exposed to extremely high pressures under special conditions. Note the shape maintained by the "hollow" cell fragments which is indicative of the rigid structure of the cell wall. ×41,500 *(Courtesy of E. Ribi.)*

Chapter 5　The Morphology and Fine Structure of Bacteria

Among the major characteristics of bacterial cells are their size, shape, structure, and arrangement. These characteristics constitute the **morphology** of the cell. Depending on the species, individual cells are spherical, rodlike, or helical, although many variations of these three basic shapes occur. Furthermore, in certain species of bacteria the cells are arranged in groups, the most common of which are pairs, clusters, chains, trichomes, and filaments. It is important to recognize these patterns of shape and arrangement, since they are often characteristic of a taxonomic group, e.g., a genus. Some bacteria also possess appendages, which can be made visible by special staining techniques or by electron microscopy. All of these morphological features are regared as the gross morphological characteristics of bacterial cells.

The bacterial cell possesses a detailed internal structure. The discovery of this internal structure was made possible by the development of electron-microscope techniques and of instruments for slicing a bacterial cell into extremely thin sections. The terms **microbial cytology** and **bacterial anatomy** have become commonplace in microbiological literature.

The various structures of a bacterial cell differ from one another not only in their physical features but also in their chemical characteristics and in their functions. Thus biologists today seek to integrate the structural, chemical, and functional properties of the bacterial cell. This area of research studied by biologists is sometimes referred to as **biochemical cytology.**

THE SIZE, SHAPE, AND ARRANGEMENT OF BACTERIAL CELLS

Size

Bacteria are very small, most being approximately 0.5 to 1.0 μm in diameter. An important consequence of the small size of microorganisms is that the **surface area/volume ratio** of bacteria is exceedingly high compared to the same ratio for larger organisms of similar shape (Table 5-1). A relatively large surface through which nutrients can enter (or waste products leave) compared to a small volume of cell substance to be nourished accounts for the unusually high rate of growth and metabolism of bacteria. Moreover, because of the high surface area/volume ratio, the mass of cell substance to be nourished is very close to the surface; therefore, no circulatory mechanism is needed to distribute the nutrients that are taken in, and there is thought to be little or no cytoplasmic movement within a bacterial cell. Despite these advantages, a high surface area/ volume ratio limits the size of bacteria to microscopic dimensions.

Shape and Arrangement

The shape of a bacterium is governed by its rigid cell wall; however, exactly what attribute of this rigid material determines that a cell will have a particular shape is not yet understood. Typical bacterial cells are spherical (**cocci;** singular, **coccus**); straight rods (**bacilli;** singular, **bacillus**); or rods that are helically curved (**spirilla;** singular, **spirillum**) as illustrated in Fig. 5-1. Although most bacterial species have cells that are of a fairly constant and characteristic shape, some have cells that are **pleomorphic,** i.e., that can exhibit a variety of shapes (Fig. 5-2).

Table 5-1. Comparison of the Surface Area/Volume Ratio of Spherical Organisms of Different Sizes*

Diameter of Sphere, μm	Surface Area, μm² $(4\pi r^2)$	Volume, μm³ $(4\pi r^3/3)$	Surface Area / Volume, μm⁻¹ $(3/r)$
1 μm	3.1	0.52	6
1,000 μm	3.1×10^6	5.2×10^8	0.006
1,000,000 μm	3.1×10^{12}	5.2×10^{17}	0.000006

* For a given volume, the geometrical shape that has the smallest surface area/volume ratio is a sphere; i.e., if two organisms have the same volume, one being spherical and the other cylindrical, the cylindrical organism has the greater surface area/volume ratio.

Figure 5-1. Bacteria are generally either (A) spherical (cocci); (B) rodlike (rods or bacilli); or (C) helical (spirilla). However, there are many modifications of these three basic forms. (Erwin F. Lessel, illustrator.)

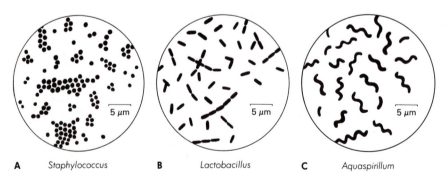

| A   Staphylococcus | B   Lactobacillus | C   Aquaspirillum |

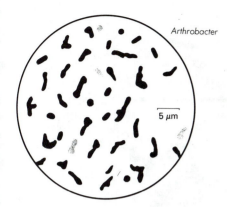

Arthrobacter

5 μm

Figure 5-2. Drawing of pleomorphic cells of the genus *Arthrobacter*. *(Erwin F. Lessel, illustrator.)*

A Diplococci:

B Streptococci:

C Tetrads:

D Staphylococci:

E Sarcinae:

Figure 5-3. Characteristic arrangements of cocci, with schematic illustrations of patterns of multiplication. (A) Diplococci: cells divide in one plane and remain attached predominantly in pairs. (B) Streptococci: cells divide in one plane and remain attached to form chains. (C) Tetracocci: cells divide in two planes and characteristically form groups of four cells. (D) Staphylococci: cells divide in three planes, in an irregular pattern, producing "bunches" of cocci. (E) Sarcinae: cells divide in three planes, in a regular pattern, producing a cuboidal arrangement of cells.

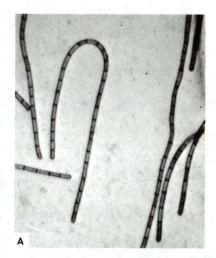

A

Figure 5-4. Photomicrograph of the trichomes of *Saprospira grandis*, composed of individual cylindrical cells that are 1 to 5 μm long and closely attached to one another (X1,650). *(Courtesy of G. J. Hageage, Jr.)*

Bacterial cells are usually arranged in a manner characteristic of their particular species. Although it is rare that all the cells of a species are arranged in the same manner, it is the predominant arrangement that is the important feature.

Cocci appear in several characteristic arrangements, depending on the plane of cellular division and whether the daughter cells stay together following division (Fig. 5-3). Bacilli are not arranged in patterns as complex as those of cocci, and most occur singly or in pairs **(diplobacilli).** But some species, such as *Bacillus subtilis*, form chains **(streptobacilli);** others, such as *Beggiatoa* and *Saprospira* species, form **trichomes,** which are similar to chains but have a much larger area of contact between the adjacent cells (Fig. 5-4). In other bacillus species, such as *Corynebacterium diphtheriae*, the cells are lined side by side like matchsticks **(palisade arrangement)** and at angles to one another (Fig. 5-5).

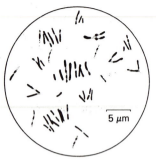

Figure 5-5. Drawing of the cells of *Corynebacterium diphtheriae* showing palisade arrangements. *(Erwin F. Lessel, illustrator.)*

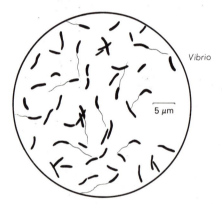

Figure 5-7. Drawing of cells of the genus *Vibrio*, showing the characteristic curved shape and the polar flagella. The flagella are not visible by ordinary staining procedures. *(Erwin F. Lessel, illustrator.)*

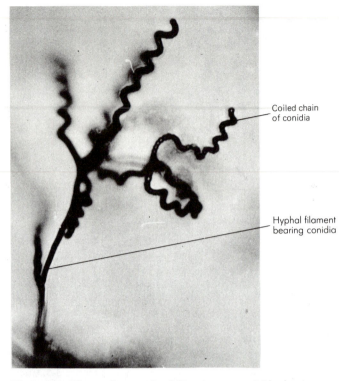

Coiled chain of conidia

Hyphal filament bearing conidia

Vibrio

Figure 5-6. Photomicrograph of *Streptomyces viridochromogenes.* This bacterium produces coiled chains of spores (called conidia) which develop at the ends of vegetative filaments called hyphae. *(Courtesy of Mary P. Lechevalier.)*

Still others, such as *Streptomyces* species, form long, branched, multinucleate filaments called **hyphae** (singular, **hypha**) which collectively form a **mycelium** (Fig. 5-6). (Note that the terms *hyphae* and *mycelium* are also commonly applied to the filaments formed by fungi, described in Chap. 17).

Curved bacteria are usually curved with a twist. Bacteria with less than one complete twist or turn have a **vibrioid** shape (Fig. 5-7), whereas those with one or more complete turns have a **helical** shape. Spirilla are rigid helical bacteria, whereas **spirochetes** are highly flexible (Fig. 5-8).

In addition to the common bacterial shapes, many others also occur: pear-shaped cells (e.g., *Pasteuria*); lobed spheres (e.g., *Sulfolobus*); rods with squared rather than the usual hemispherical ends (e.g., *Bacillus anthracis*); disks arranged like stacks of coins (e.g., *Caryophanon*); rods with helically sculptured surfaces (e.g., *Seliberia*); and many others.

BACTERIAL STRUCTURES

Examination of a bacterial cell reveals various component structures. Some of these are external to the cell wall (Fig. 5-9); others are internal to the cell wall

(Fig. 5-10). Some structures are present in only certain species; some are more characteristic of certain species than of others; and still other cellular parts, such as the cell wall, are naturally common to almost all bacteria. The following are brief descriptions of the readily evident structures of bacteria.

Figure 5-8. Drawings of spirochetes (A and B) and spirilla (C). Spirochetes are flexible and can twist and contort their shape, whereas spirilla are relatively rigid. *(Erwin F. Lessel, illustrator.)*

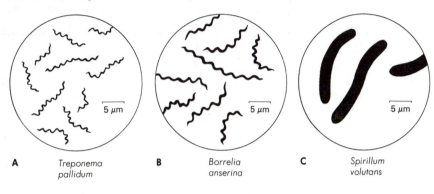

A   *Treponema pallidum*

B   *Borrelia anserina*

C   *Spirillum volutans*

Figure 5-9. Drawing of the major structures external to the bacterial cell wall. Certain structures, e.g., capsules, flagella, and pili, are not common to all bacterial cells. *(Erwin F. Lessel, illustrator.)*

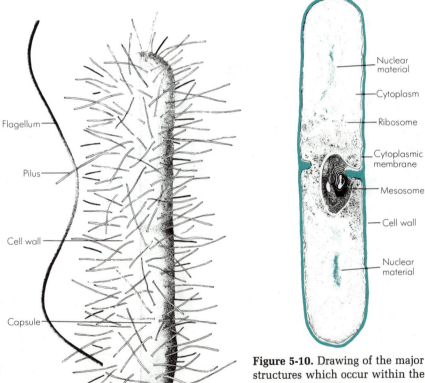

Flagellum

Pilus

Cell wall

Capsule

Nuclear material

Cytoplasm

Ribosome

Cytoplasmic membrane

Mesosome

Cell wall

Nuclear material

Figure 5-10. Drawing of the major structures which occur within the bacterial cell wall. Certain structures, e.g., mesosomes, are not common to all bacterial cells. *(Erwin F. Lessel, illustrator.)*

STRUCTURES EXTERNAL TO THE CELL WALL

Flagella and Motility

Bacterial **flagella** (singular, **flagellum**) are hairlike, helical appendages that protrude through the cell wall and are responsible for swimming motility. They are much thinner than the flagella or cilia of eucaryotes, being 0.01 to 0.02 μm in diameter, and they are also much simpler in structure. Their location on the cell varies depending on the bacterial species and may be **polar** (at one or both ends of the bacterium) or **lateral** (along the sides of the bacterium). Some arrangements of bacterial flagella are shown in Fig. 5-11. A flagellum is composed of three parts (Fig. 5-12): a **basal body** associated with the cytoplasmic membrane and cell wall, a short **hook,** and a helical **filament** which is usually several times as long as the cell. Some Gram-negative bacteria have a sheath surrounding the flagellum; this sheath is continuous with the outer membrane of the Gram-negative cell wall. The chemical composition of the basal body is unknown, but the hook and filament are composed of protein subunits (monomers) arranged in a helical fashion. The protein of the filament is known as **flagellin.**

Unlike a hair, a flagellum grows at its tip rather than at the base. Flagellin monomers synthesized within the cell are believed to pass along the hollow center of the flagellum and are added to the distal end of the filament.

Figure 5-11. Drawings of various arrangements of bacterial flagella. (A) Monotrichous; a single polar flagellum. (B) Lophotrichous; a cluster of polar flagella. (C) Amphitrichous; flagella, either single or clusters, at both cell poles. (D) Peritrichous; surrounded by lateral flagella. *(Erwin F. Lessel, illustrator.)*

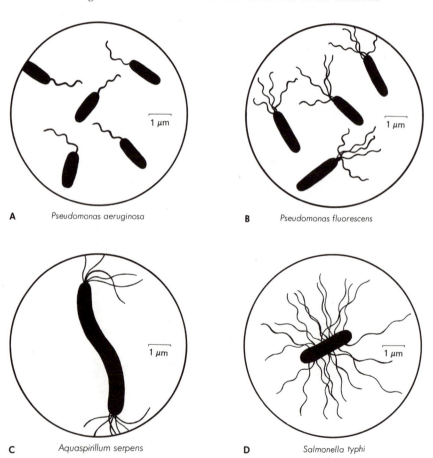

A   *Pseudomonas aeruginosa*

B   *Pseudomonas fluorescens*

C   *Aquaspirillum serpens*

D   *Salmonella typhi*

Hydrodynamics of Flagella

Large motile bodies such as boats and fish make use of the inertia of water for their propulsion. When pushed against with, for example, an oar, a propeller blade, or fins, the water temporarily acts as a solid, thereby enabling the boat or fish to generate a forward propulsive force. However, the small size of bacteria prohibits their use of the inertia of water to gain propulsive force, because the drag forces due to the viscosity of water become thousands of times greater than any forces that can be generated from inertia. The difficulty would be similar to what we would encounter if we attempted to row a boat on a lake filled with thick molasses. However, bacteria can swim many times their own length per second under analogous conditions!

Bacteria propel themselves by rotating their helical flagella. The principle involved can be illustrated by imagining the penetration of a piece of cork by a corkscrew. If one tries to ram the corkscrew directly through the cork, great force will probably be needed. On the other hand if one merely rotates the corkscrew, the cork can be easily penetrated. In the case of bacteria, the cork is analogous to the viscous medium and the corkscrew to the helical flagellum. It is apparent from this analogy that a mutant bacterium having straight rather than helical flagella would be unable to swim. The nature of the rotary motor that spins each corkscrew-shaped flagellum is still not understood, but the rings found in the basal body (Fig. 5-12) are probably involved. It is known that the flagellar motor is driven by the protonmotive force, i.e., the force derived from the electrical potential and the hydrogen-ion gradient across the cytoplasmic membrane (see Chap. 10). Moreover, recent studies suggest that the concentration of cGMP (guanosine 3',5'-cyclic phosphoric acid) within the cell governs the direction in which the rotation occurs.

Bacteria having polar flagella swim in a back-and-forth fashion; they reverse their direction of swimming by reversing the direction of flagellar rotation. Bacteria having lateral flagella swim in a more complicated manner. Their flagella operate in synchrony to form a bundle that extends behind the cell (Fig.

Figure 5-12. The mechanism of attachment of flagella to a Gram-negative bacterial cell (*Pseudomonas aeruginosa*). (A) Prior to electron-microscope examination, the cells were partially lysed and then negatively stained to make the point of flagellar attachment (basal body) more visible (X80,000 approx.). (B) Isolated flagella showing basal body at one end. (C) Model of basal body illustrating its structure and attachment to a Gram-negative bacterium. The flagella of Gram-positive bacteria have only two basal rings. (*Courtesy of T. Iino, University of Tokyo.*)

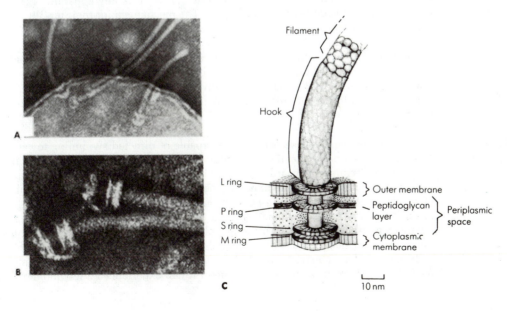

Filament

Hook

L ring
P ring
S ring
M ring

Outer membrane
Peptidoglycan layer
Cytoplasmic membrane

Periplasmic space

10 nm

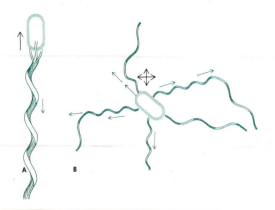

Figure 5-13. Diagram of the configuration and arrangement of peritrichous flagella during swimming and tumbling. The small arrows indicate the direction of propagation of helical waves along the flagella. (A) During swimming the flagella are in the form of left-handed helices and rotate counterclockwise in synchrony to form a bundle. The large arrow indicates the direction of swimming. (B) During tumbling the flagella reverse their rotation, portions of the flagella acquire a short wavelength and right-handed configuration, and the bundle flies apart. The cell cannot swim under these conditions and instead exhibits a chaotic motion, as symbolized by the large crossed arrows. *(Courtesy of R. M. MacNab and M. K. Ornston, J Mol Biol **112**:1, 1977.)*

5-13). However, when the flagellar motors reverse, conformational changes occur along the flagella, the bundle flies apart, and the cell tumbles wildly. Finally, the flagellar motors resume their normal direction, the flagellar bundle again forms, and the cell begins to swim—but now in a different direction. This sequence of events occurs repeatedly, so that the motility becomes a series of periods of swimming (runs) punctuated by periods of tumbling (twiddles), with a change in direction after each tumble.

Swimming Motility Without Flagella

Certain helical bacteria (spirochetes) exhibit swimming motility, particularly in highly viscous media, yet they lack external flagella. However, they have flagellalike structures located within the cell, just beneath the outer cell envelope (see Fig. 13-1). These are called **periplasmic flagella;** they have also been termed **axial fibrils** or **endoflagella.** They are responsible for the motility of spirochetes, but how they accomplish this is not yet clear. Other helical bacteria called **spiroplasmas** are able to swim in viscous media, yet lack any apparent organelles for motility, even periplasmic flagella. The mechanism for their motility is completely unknown.

Gliding Motility

Some bacteria, e.g., *Cytophaga* species, are motile only when they are in contact with a solid surface. As they glide they exhibit a sinuous, flexing motion. This kind of movement is comparatively slow, only a few μm per second. The mechanism of gliding motility is unknown; no organelles responsible for motility have been observed.

Bacterial Chemotaxis

Many, perhaps most, motile bacteria are capable of directed swimming toward or away from various chemical compounds—a phenomenon called **bacterial chemotaxis.** Swimming toward a chemical is termed **positive chemotaxis;** swimming away is **negative chemotaxis.** Although chemicals may act as attractants or repellents, the stimulus is in fact not the chemical itself but rather a change in the concentration of the chemical with time, i.e., a **temporal gradient.** Such gradients are sensed by means of protein **chemoreceptors** which are located on the cytoplasmic membrane and are specific for various attractants and repellents.

By means of its chemoreceptors, a bacterium continually compares its immediate environment with the environment it had experienced a few moments earlier. To illustrate this, suppose we are observing the behavior of a bacterium that has peritrichous flagella and for which glucose is an attractant. If the cell is placed in a homogeneous glucose broth, the glucose concentration remains constant regardless of the direction of the bacterium's swimming, and the glucose-specific chemoreceptors can sense no change in glucose concentration. Consequently, the cell exhibits a normal swimming pattern—periods of swimming with intermittent periods of tumbling. Suppose that the cell is now placed in a long capillary tube with a higher concentration of glucose at one end than at the other. If the cell happens to swim toward the higher concentration of glucose (i.e., in the "right" direction), the chemoreceptors sense that the glucose concentration is increasing with time. This results in suppression of normal tumbling, causing the cell to swim smoothly ahead for a long period before it tumbles. On the other hand, if the cell happens to swim toward the end of the tube where there is less glucose (i.e., in the "wrong" direction), the chemoreceptors sense that the glucose concentration is decreasing with time, and no suppression of tumbling occurs. Therefore, the cell soon tumbles, changes direction, and tries again until finally the "right" direction is achieved. (In a gradient of a repellent compound, the right direction would be *down* the gradient, i.e., toward a *decreasing* concentration, and the wrong direction would be up the gradient.)

Tactic responses are not limited to chemical gradients. For instance, phototrophic bacteria exhibit positive **phototaxis** toward increasing light intensities and are repelled by decreasing light intensities. Still another type of taxis is exhibited by *Aquaspirillum magnetotacticum*; this organism exhibits directed swimming in response to the earth's magnetic field or to local magnetic fields (magnets placed near the culture). This is attributed to a chain of magnetite inclusions **(magnetosomes)** within the cell, which allows the cell to become oriented as a magnetic dipole (Fig. 5-14). Because of the downward inclination of the Earth's magnetic field in the regions where these bacteria have been found, **magnetotaxis** may serve to direct the cells downward in aquatic environments toward oxygen-deficient areas more favorable for growth.

Figure 5-14. Negatively stained cell of *Aquaspirillum magnetotacticum* showing a particle chain (PC) of highly electron-dense magnetite inclusions (magnetosomes) within the cell. The bar represents 1 μm. (*Courtesy of D. L. Balkwill, D. Maratea and R. P. Blakemore, J. Bacteriol* **141**:1399, *1980.*)

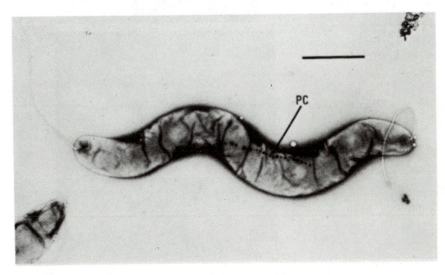

Pili (Fimbriae)

Pili (singular, **pilus**) are hollow, nonhelical, filamentous appendages that are thinner, shorter, and more numerous than flagella (Fig. 5-15). They do not function in motility, since they are found on nonmotile as well as motile species. There are, however, several functions associated with different types of pili. One type, known as the F pilus (or sex pilus), serves as the port of entry of genetic material during bacterial mating (see Chap. 12). Some pili play a major role in human infection by allowing pathogenic bacteria to attach to epithelial cells lining the respiratory, intestinal, or genitourinary tracts. This attachment prevents the bacteria from being washed away by the flow of mucous or body fluids and permits the infection to be established.

Capsules

Some bacterial cells are surrounded by a viscous substance forming a covering layer or envelope around the cell wall. If this layer can be visualized by light microscopy using special staining methods, it is termed a **capsule.** If the layer is too thin to be seen by light microscopy it is termed a **microcapsule;** if it is so abundant that many cells are embedded in a common matrix, the material is called **slime.**

By light microscopy, capsules appear to be amorphous gelatinous areas surrounding a cell (Fig. 5-16A); however, special techniques designed to preserve delicate structures for observation by electron microscopy have revealed that capsules consist of a mesh or network of fine strands (Fig. 5-16B).

In many instances capsular material is not highly water-soluble and therefore does not readily diffuse away from the cells that produce it. In other instances

Figure 5-15. Fimbriated bacteria. (A) *Shigella flexneri:* dividing bacilli with numerous fimbriae surrounding the cells (X20,000). (B) *Salmonella typhi:* dividing bacilli with numerous fimbriae and a few flagella (the very long appendages) (X12,500). *(Courtesy of J. P. Duguid and J. F. Wilkinson and The Society of General Microbiology: Symposium XI. 1961.)*

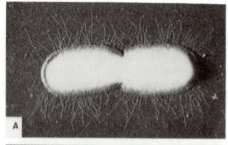
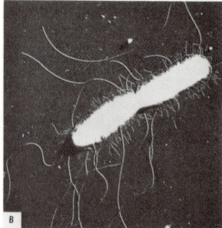

Figure 5-16. Bacterial capsules as seen by light microscopy (A) and electron microscopy (B). (A) India-ink preparation of a capsulated bacterium isolated from a paper-mill operation. The particles of carbon in the ink cannot penetrate the capsules (white areas around the cells). *Courtesy of P. M. Borick, Wallace and Tiernan, Inc.)* (B) Freeze-etch preparation of Gram-positive rod-shaped bacteria isolated from acid mine water, showing a fibrillar polymer network surrounding the cells. The freeze-fracture process has also revealed various internal and surface structures of the cells. *(Courtesy of P. R. Dugan, C. B. MacMillan, and R. M. Pfister, J. Bacteriol **101**:982, 1970.)*

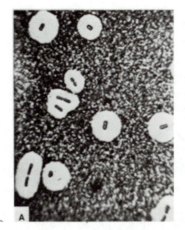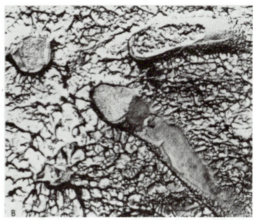

the material is highly water-soluble and dissolves in the medium, sometimes dramatically increasing the viscosity of the broth in which the organisms are cultured.

Capsules can serve a number of functions, depending on the bacterial species. (1) They may provide protection against temporary drying by binding water molecules. (2) They may block attachment of bacteriophages. (3) They may be antiphagocytic; i.e., they inhibit the engulfment of pathogenic bacteria by white blood cells and thus contribute to invasive or infective ability (virulence). (4) They may promote attachment of bacteria to surfaces; for example, *Streptococcus mutans*, a bacterium associated with producing dental caries, firmly adheres to the smooth surfaces of teeth because of its secretion of a water-insoluble capsular glucan. (5) If capsules are composed of compounds having an electrical charge, such as sugar–uronic acids, they may promote the stability of bacterial suspension by preventing the cells from aggregating and settling out, because cells bearing similarly charged surfaces tend to repel one another.

Most bacterial capsules are composed of polysaccharides. Capsules composed of a single kind of sugar are termed **homopolysaccharides;**   are usually synthesized outside the cell from disaccharides by exocellular enzymes. The synthesis of glucan (a polymer of glucose) from sucrose by *S. mutans* is an example. Other capsules are composed of several kinds of sugars and are termed **heteropolysaccharides;** these are usually synthesized from sugar precursors that are activated (energized) within the cell, attached to a lipid carrier molecule, transported across the cytoplasmic membrane, and polymerized outside the cell. The capsule of *Klebsiella pneumoniae* is an example.

A few capsules are polypeptides. For example, the capsule of the anthrax organism, *B. anthracis*, is composed entirely of a polymer of glutamic acid. Moreover, this peptide is an unusual one because the glutamic acid is the rare D optical isomer rather than the usual L isomer commonly found in nature.

Sheaths

Some species of bacteria, particularly those from freshwater and marine environments, form chains or trichomes that are enclosed by a hollow tube called a **sheath.** This structure is most readily visualized when some of the cells have

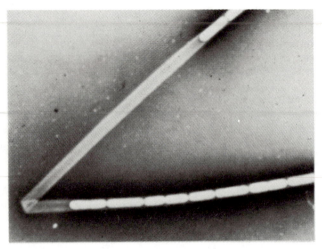

Figure 5-17. Sheathed bacteria. Sheath and cells of *Sphaerotilus natans* stained with nigrosin. Dimensions of individual cells are 1 μm by 2 to 6 μm, and the sheaths may reach a length of several millimeters. *(Courtesy of J. L. Stokes, J Bacteriol* **67**:279, 1954.)

migrated from it (Fig. 5-17). Sheaths may sometimes become impregnated with ferric or manganese hydroxides, which strengthen them.

Prosthecae and Stalks

Prosthecae (singular, prostheca) are semirigid extensions of the cell wall and cytoplasmic membrane and have a diameter that is always less than that of the cell. They are characteristic of a number of aerobic bacteria from freshwater and marine environments. Some bacterial genera such as *Caulobacter* have a single prostheca; others such as *Stella* and *Ancalomicrobium* have several (Fig. 5-18). Prosthecae increase the surface area of the cells for nutrient absorption, which is advantageous in dilute environments. Some prosthecate bacteria may form a new cell (bud) at the end of a prostheca; others have an adhesive substance at the end of a prostheca that aids in attachment to surfaces.

Although the term stalk is sometimes used interchangeably with the terms *prostheca* or *hypha*, it is perhaps better to restrict its use to certain nonliving ribbonlike or tubular appendages that are excreted by the cell, such as those found in the genera *Gallionella* or *Planctomyces* (see Chap. 15). These stalks aid in attachment of the cells to surfaces.

THE CELL WALL

Beneath such external structures as capsules, sheaths, and flagella and external to the cytoplasmic membrane is the cell wall, a very rigid structure that gives shape to the cell. Its main function is to prevent the cell from expanding and eventually bursting because of uptake of water, since most bacteria live in hypotonic environments (i.e., environments having a lower osmotic pressure than exists within the bacterial cells). The rigidity of the wall can be readily demonstrated by subjecting bacteria to very high pressures or other severe physical conditions: most bacterial cells retain their original shapes during and after such treatments. To obtain isolated cell walls for analysis, bacteria usually must be mechanically disintegrated by drastic means, as by sonic or ultrasonic treatment or by exposure to extremely high pressures with subsequent sudden

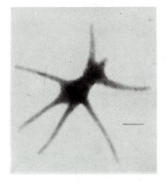

Figure 5-18. *Ancalomicrobium adetum,* a budding bacterium with several prosthecae per cell. Electron micrograph of whole cell, negatively stained. The bar represents 1.0 μm. *(Courtesy of J. T. Staley, J Bacteriol 95:1921, 1968.)*

release of pressure. The broken cell walls are then separated from the rest of the components of the disintegrated cells by differential centrifugation. Isolated cell walls, devoid of other cellular constituents, retain the original contour of the cells from which they were derived.

Among the ordinary or typical bacteria (which are sometimes called eubacteria to distinguish them from the phylogenetically distinct group known as the archaeobacteria, discussed in Chap. 3), the walls of Gram-negative species are generally thinner (10 to 15 nm) than those of Gram-positive species (20 to 25 nm). The walls of Gram-negative archaeobacteria are also thinner than those of Gram-positive archaeobacteria. Since the chemical composition of the walls of archaeobacteria is quite different from that of eubacteria, wall thickness rather than chemical composition may be the major factor in the Gram reaction.

The cell wall constitutes a significant portion of the dry weight of the cell; depending on the species and culture conditions, it may account for as much as 10 to 40 percent. Bacterial cell walls are usually essential for bacterial growth and division. Cells whose walls have been completely removed (i.e., protoplasts) are incapable of normal growth and division.

Structure and Chemical Composition

Peptidoglycan

For eubacteria, the shape-determining part of the cell wall is largely **peptidoglycan** (sometimes called **murein**), an insoluble, porous, cross-linked polymer of enormous strength and rigidity. Peptidoglycan is found only in procaryotes; it occurs in the form of a "bag-shaped macromolecule" surrounding the cytoplasmic membrane. Peptidoglycan differs somewhat in composition and structure from one species to another, but it is basically a polymer of N-acetylglucosamine, N-acetylmuramic acid, L-alanine, D-alanine, D-glutamate, and a diamino acid (LL- or meso-diaminopimelic acid, L-lysine, L-ornithine, or L-diaminobutyric acid). The structure of this polymer is depicted in Figs. 11-6 and 11-7. It is important to realize that as tough as peptidoglycan is, it is also in a dynamic state. That is, in order for the cell to grow and divide, portions of the peptidoglycan must continually be degraded by wall-associated hydrolytic enzymes so that new polymer can be added.

Walls of Archaeobacteria

Although most archaeobacteria possess cell walls, these do not contain peptidoglycan, and their cell-wall fine structure and chemical composition is very different from that of eubacteria. Their walls are usually composed of **proteins, glycoproteins,** or **polysaccharides.** A few genera, such as *Methanobacterium,* have walls composed of **pseudomurein,** a polymer whose structure superficially resembles eubacterial peptidoglycan but which differs markedly in chemical composition (see Chap. 15).

Walls of Gram-Positive Eubacteria

Gram-positive bacteria usually have a much greater amount of peptidoglycan in their cell walls than do Gram-negative bacteria; it may account for 50 percent or more of the dry weight of the wall of some Gram-positive species, but only

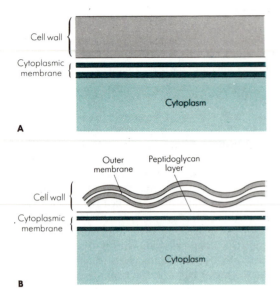

Figure 5-19. Schematic interpretation of cell walls of eubacteria from electron-microscope observations. (A) Gram-positive bacteria, showing thick wall consisting mainly of peptidoglycan. Although the wall is often homogeneous in appearance, in some bacteria it may consist of several layers. (B) Gram-negative bacteria, showing outer membrane and thin peptidoglycan layer. *(Courtesy of A. I. Laskin and H. A. Lechevalier (eds.), Handbook of Microbiology, CRC Press, Inc., Cleveland, 1974.)*

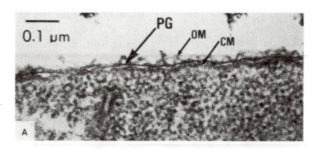
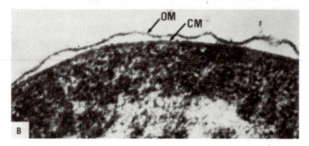

Figure 5-20. (A) Thin section of *Aquaspirillum serpens* showing the wavy outer membrane (OM), the peptidoglycan layer (PG), and the cytoplasmic membrane (CM). (B) Companion preparation of a spheroplast formed by treatment of the cells with a chelating agent and lysozyme. The peptidoglycan layer is missing. *(From R. G. E. Murray, P. Steed and H. E. Elson, Can J Microbiol* **11:**547, 1965.)

about 10 percent of the wall of Gram-negative bacteria. Other substances may occur in addition to peptidoglycan. For instance, the walls of *Streptococcus pyogenes* contain **polysaccharides** that are covalently linked to the peptidoglycan and which can be extracted with hot dilute hydrochloric acid. The walls of *Staphylococcus aureus* and *Streptococcus faecalis* contain **teichoic acids**—acidic polymers of ribitol phosphate or glycerol phosphate—which are covalently linked to peptidoglycan and which can be extracted with cold dilute acid. Teichoic acids bind magnesium ions, and there is some evidence that they help to protect bacteria from thermal injury by providing an accessible pool of these cations for stabilization of the cytoplasmic membrane. The walls of most Gram-positive bacteria contain very little lipid, but those of *Mycobacterium*, *Corynebacterium*, and certain other genera are exceptions, being rich in lipids called **mycolic acids.** These compounds have the following general structure:

$$R_1-CH-CH-COOH$$
$$\quad\ \ |\quad\ \ |$$
$$\quad\ \ OH\ \ R_2$$

where R_1 and R_2 are long hydrocarbon chains. The ability of mycobacteria to exhibit acid-fast staining (i.e., when stained, the cells cannot be decolorized easily despite treatment with dilute acids) is correlated with the presence of cell wall mycolic acids. A mycolic acid derivative called **cord factor** (trehalose dimycolate) is toxic and plays an important role in the diseases caused by *C. diphtheriae* and *M. tuberculosis*, described in Chap. 36.

Walls of Gram-Negative Eubacteria

The walls of Gram-negative bacteria are more complex than those of Gram-positive bacteria. The most interesting difference is the presence of an **outer membrane** that surrounds a thin underlying layer of peptidoglycan (Figs. 5-19 and 5-20). Because of this membrane, the walls of Gram-negative bacteria are rich in lipids (11 to 22 percent of the dry weight of the wall), in contrast to those of Gram-positive bacteria. This outer membrane serves as an impermeable barrier to prevent the escape of important enzymes, such as those involved in cell wall growth, from the space between the cytoplasmic membrane and the outer membrane **(periplasmic space).** The outer membrane also serves as a barrier to various external chemicals and enzymes that could damage the cell. For example, the walls of many Gram-positive bacteria can be easily destroyed by treatment with an enzyme called **lysozyme,** which selectively dissolves peptidoglycan; however, Gram-negative bacteria are refractory to this enzyme because large protein molecules cannot penetrate the outer membrane. Only if the outer membrane is first damaged, as by removal of stabilizing magnesium ions by a chelating agent, can the enzyme penetrate and attack the underlying peptidoglycan layer (see Fig. 5-20B).

The outer membrane of the Gram-negative cell wall is anchored to the underlying peptidoglycan by means of **Braun's lipoprotein** (Fig. 5-21). The membrane is a bilayered structure consisting mainly of **phospholipids, proteins,** and **lipopolysaccharide (LPS).** The LPS has toxic properties and is also known as **endotoxin.** It occurs only in the outer layer of the membrane (Fig. 5-21) and is composed of three covalently linked parts: (1) **Lipid A,** firmly embedded in the membrane; (2) **core polysaccharide,** located at the membrane surface; and (3) polysaccharide **O antigens,** which extend like whiskers from the membrane surface into the surrounding medium (Fig. 5-21). Many of the serological prop-

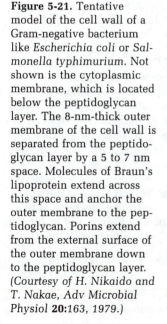

Figure 5-21. Tentative model of the cell wall of a Gram-negative bacterium like *Escherichia coli* or *Salmonella typhimurium*. Not shown is the cytoplasmic membrane, which is located below the peptidoglycan layer. The 8-nm-thick outer membrane of the cell wall is separated from the peptidoglycan layer by a 5 to 7 nm space. Molecules of Braun's lipoprotein extend across this space and anchor the outer membrane to the peptidoglycan. Porins extend from the external surface of the outer membrane down to the peptidoglycan layer. *(Courtesy of H. Nikaido and T. Nakae, Adv Microbial Physiol* **20**:163, 1979.)

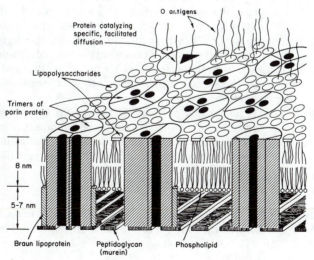

Figure 5-22. Thin section of an *Escherichia coli* cell that was plasmolyzed in a 20% sucrose solution, causing the protoplast to contract. Numerous adhesions are evident between the cytoplasmic membrane and the outer membrane of the cell wall. The light, fibrillar area in the center of the cell is the nuclear material. The bar represents 0.1 μm. *(Courtesy of M. E. Bayer, J Gen Microbiol* **53:***395, 1968.)*

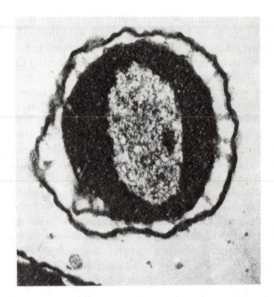

erties of Gram-negative bacteria are attributable to O antigens; they can also serve as receptors for bacteriophage attachment.

Although impermeable to large molecules such as proteins, the outer membrane can allow smaller molecules, such as nucleosides, oligosaccharides, monosaccharides, peptides, and amino acids, to pass across. This is accomplished by means of channels in special proteins called **porins,** which span the membrane (Fig. 5-21). The various porins are specific for different kinds or classes of small molecules, and some can even allow certain essential large molecules to penetrate, such as vitamin B_{12}. Many porins also serve as receptors for attachment of bacteriophages and bacteriocins.

One of the questions posed by the structure of Gram-negative cell walls is: How can water-insoluble, lipophilic substances such as LPS pass from their place of synthesis within the cytoplasm and cytoplasmic membrane across a watery periplasmic space to be inserted into the outer membrane? A likely explanation has been provided by the discovery of numerous **adhesions,** or points of direct contact between the two membranes (Fig. 5-22). These adhesions seem to be the export sites for newly synthesized LPS and porins, and they are also the sites at which pili and flagella are made.

Macromolecular Surface Arrays

The cell walls of some bacteria, both Gram-negative and Gram-positive, are covered by a mosaic layer of protein subunits (Fig. 5-23). The functions of these mosaic layers are not well understood, but at least one function is to protect Gram-negative bacteria against attack and penetration by other small, predatory bacteria known as bdellovibrios.

STRUCTURES INTERNAL TO THE CELL WALL

Immediately beneath the cell wall is the **cytoplasmic membrane.** This structure is approximately 7.5 nm (0.0075 μm) thick and is composed primarily of phos-

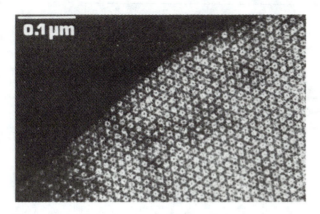

Figure 5-23. Electron micrograph showing a macromolecular surface array of protein subunits of the outer surface of a cell-wall fragment from *Aquaspirillum serpens*. *(Courtesy of R. G. E. Murray. From N. R. Krieg, Bacteriol Rev* **40**:55, 1976.)

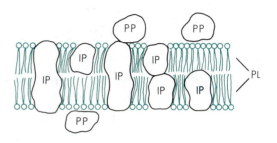

Figure 5-24. Schematic interpretation of the structure of the cytoplasmic membrane. Phospholipids (PL) are arranged in a bilayer such that the polar portions (circles) face outward and the nonpolar portions (filaments) face inward. IP = integral protein; PP = peripheral protein. Note that some integral proteins, such as transport proteins, are believed to span the membrane.

The Cytoplasmic Membrane

pholipids (about 20 to 30 percent) and proteins (about 60 to 70 percent). The phospholipids form a bilayer in which most of the proteins are tenaciously held (integral proteins) (Fig. 5-24); these proteins can be removed only by destruction of the membrane, as with treatment by detergents. Other proteins are only loosely attached (peripheral proteins) and can be removed by mild treatments such as osmotic shock. The lipid matrix of the membrane has fluidity, allowing the components to move around laterally. This fluidity appears to be essential for various membrane functions and is dependent on factors such as temperature and on the proportion of unsaturated fatty acids to saturated fatty acids present in the phospholipids.

A significant difference exists between the phospholipids of eubacteria and those of archaeobacteria. In eubacteria the phospholipids are phosphoglycerides, in which straight-chain fatty acids are ester-linked to glycerol (Fig. 5-25). In archaeobacteria, the lipids are polyisoprenoid branched-chain lipids, in which long-chain branched alcohols (phytanols) are ether-linked to glycerol (Fig. 5-25).

The cytoplasmic membrane is a hydrophobic barrier to penetration by most water-soluble molecules. However, specific proteins in the membrane allow, indeed facilitate, the passage of small molecules (i.e., nutrients and waste products) across the membrane; these transport systems are discussed in Chap. 11. The cytoplasmic membrane also contains various enzymes involved in respiratory metabolism and in synthesis of capsular and cell-wall components; moreover, because of its impermeability to protons (hydrogen ions), the cytoplasmic membrane is the site of generation of the protonmotive force—the force that drives ATP synthesis in many organisms, certain nutrient transport systems, and flagellar motility (see Chaps. 10 and 11). Consequently the cytoplasmic

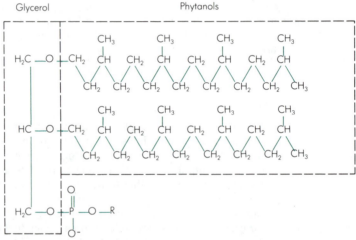

Figure 5-25. (A) Example of a eubacterial phospholipid, showing two unbranched, long-chain fatty acids ester-linked to glycerol. (B) Example of an archaeobacterial phospholipid, showing two branched phytanol chains that are ether-linked to glycerol. (R is any of several compounds such as ethanolamine, choline, serine, inositol, or glycerol.)

membrane is an extremely important functional structure, and damage to it by physical or chemical agents can result in the death of the cell.

Proteins are synthesized within the cell, but some can pass across the cytoplasmic membrane barrier to the outside; examples of such exported molecules are the protein components of cell walls (e.g., porins or lipoproteins) or the exocellular enzymes that are secreted by many bacteria into their culture medium, such as penicillinases, proteinases and amylases. Other proteins made within the cell may pass into the cytoplasmic membrane and remain there (e.g., enzymes such as cytochromes and membrane-bound dehydrogenases). The mechanism by which transport of these proteins occurs into or across the cytoplasmic membrane is unknown. A related question is: How does a cell "know" which of the many kinds of proteins within the cell to transport out of the cell? This question has been partially answered: The genes that code for these proteins carry a message that results in the addition of a sequence of about 20 extra amino acids (the **signal peptide**) to the proteins during their synthesis within the cell. Unlike ordinary proteins, proteins carrying a signal peptide are destined to be transported into or across the cytoplasmic membrane. According to one hypothesis, special membrane proteins might bind the signal peptide at the inner surface of the cytoplasmic membrane and form a channel by which the protein can traverse the membrane. Whatever its function, the signal peptide is subsequently removed by a proteolytic enzyme and does not appear in the final, transported protein.

Protoplasts, Spheroplasts

Protoplasts

A **protoplast** is that portion of a bacterial cell consisting of the cytoplasmic membrane and the cell material bounded by it. Protoplasts can be prepared from Gram-positive bacteria by treating the cells with an enzyme such as lysozyme, which selectively dissolves the cell wall, or by culturing the bacteria in the presence of an antibiotic such as penicillin, which prevents the formation of the cell wall. In either case, the osmotic pressure of the medium must be sufficiently high to protect the organisms from bursting. Bacteria normally occur in hypotonic environments (i.e., environments having a lower osmotic pressure than that within the bacterial cells) and they continually take up water by osmosis; thus, they tend to expand, pressing the cytoplasmic membrane tightly against the rigid cell wall. In the absence of a rigid cell wall, there is nothing to prevent the continued expansion and eventual bursting of a protoplast. This bursting can be prevented by preparing protoplasts in an **isotonic** medium, i.e., in a medium that has an osmotic pressure similar to that of the protoplast. Such osmotically protected protoplasts are soft and fragile and are spherical, regardless of the original shape of the cell.

Spheroplasts

Round, osmotically fragile forms of Gram-negative bacteria can be prepared by procedures similar to those used for the protoplasts of Gram-positive bacteria. However, the cell walls of Gram-negative bacteria differ from those of Gram-positive bacteria by possessing an outer membrane. Although the peptidoglycan of the cell wall may be destroyed by lysozyme or its synthesis inhibited by antibiotics, the flexible outer membrane of the cell wall remains (Fig. 5-20B). Because the treated cell has two membranes, the cytoplasmic membrane of the protoplast plus the outer membrane of the cell wall, the cell is called a **spheroplast** rather than a protoplast.

Some bacteria, the mycoplasmas, never have cell walls and are bounded by only a cytoplasmic membrane; therefore, they have many of the properties of protoplasts, yet they manage to thrive nonetheless. Most mycoplasmas are parasites of animals, plants, or arthropods, and therefore live in osmotically favorable or isotonic environments. Some are able to attain a degree of rigidity by incorporating cholesterol into their cytoplasmic membranes. Most mycoplasmas have a more or less spherical shape, but one genus, *Spiroplasma*, consists of helical cells. How such cells are able to maintain this shape in the absence of a cell wall is unknown.

Membranous Intrusions and Intracellular Membrane Systems

Bacterial cells do not contain membrane-enclosed organelles corresponding to the mitochondria and chloroplasts of eucaryotic cells. However, bacteria may have specialized invaginations of the cytoplasmic membrane that can increase their surface area for certain functions.

Figure 5-26. Thin section of the Gram-positive bacterium *Streptococcus faecalis,* showing the beginning stages of cell division occurring beneath a thickened equatorial ridge of the cell wall (arrows). A central mesosome (m) is present in each cell and is seen to be a complex invagination of the cytoplasmic membrane. Nuclear material (n) appears as a light, fibrillar area. *(Courtesy of J. M. Garland, A. R. Archibald and J. M. Baddiley, J Gen Microbiol **89:**73, 1975.)*

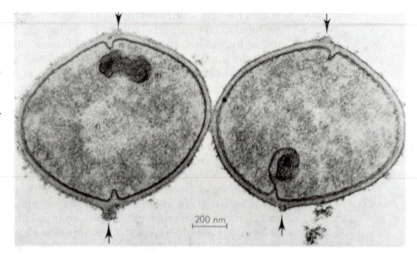

Many bacteria, especially Gram-positive bacteria, possess membrane invaginations in the form of systems of convoluted tubules and vesicles termed **mesosomes.** Those known as **central mesosomes** penetrate deeply into the cytoplasm, are located near the middle of the cell, and seem to be attached to the cell's nuclear material; they are thought to be involved in DNA replication and cell division (Fig. 5-26). In contrast, **peripheral mesosomes** show only a shallow penetration into the cytoplasm, are not restricted to a central location, and are not associated with nuclear material; they seem to be involved in export of exocellular enzymes such as penicillinase.

Extensive intracellular membrane systems occur in methane-oxidizing bacteria, in certain chemoautotrophic bacteria (Fig. 5-27), and in nearly all phototrophic bacteria. They serve to increase surface area for various metabolic activities. For example, in phototrophic bacteria they are the site of the photosynthetic apparatus of the cell; the infoldings provide a large surface area to accommodate a high content of light-absorbing pigments. In the phototrophs known as cyanobacteria, special intracellular membranes (thylakoids) occur that seem to be separate from the cytoplasmic membrane.

The Cytoplasm

The cell material bounded by the cytoplasmic membrane may be divided into (1) the cytoplasmic area, granular in appearance and rich in the macromolecular RNA-protein bodies known as **ribosomes,** on which proteins are synthesized; (2) the chromatinic area, rich in DNA; and (3) the fluid portion with dissolved substances. Unlike animal or plant cells, there is no endoplasmic reticulum to which ribosomes are bound; some ribosomes are free in the cytoplasm, and others, especially those involved in the synthesis of proteins to be transported out of the cell, are associated with the inner surface of the cytoplasmic membrane. When the ribosomes of procaryotes undergo sedimentation in a centrifuge, they have a sedimentation coefficient of 70 Svedberg units (70S) and are composed of two subunits, a 50S and a 30S subunit. This is in contrast to the ribosomes of eucaryotic organisms, which have a sedimentation coefficient of 80S and are composed of a 60S and a 40S subunit.

Cytoplasmic Inclusions and Vacuoles

Concentrated deposits of certain substances are detectable in the cytoplasm of some bacteria. Volutin granules, also known as metachromatic granules, are composed of polyphosphate. They stain an intense reddish-purple color with dilute methylene blue and can be observed by light microscopy. By electron microscopy they appear as round, dark areas (Fig. 5-28). Volutin serves as a reserve source of phosphate. Another polymer often found in aerobic bacteria, especially under high-carbon, low-nitrogen culture conditions, is a chloroform-soluble, lipidlike material, poly-β-hydroxybutyrate, (PHB), which can serve as a reserve carbon and energy source. PHB granules can be stained with lipid-soluble dyes such as Nile blue. By electron microscopy they appear as clear round areas (Fig. 5-28). Polysaccharide granules, i.e., glycogen, can be stained brown with iodine. By electron microscopy they appear as dark granules (Fig. 5-28). Another type of inclusion is represented by the intracellular globules of elemental sulfur that may accumulate in certain bacteria growing in environments rich in hydrogen sulfide.

Some bacteria that live in aquatic habitats form gas vacuoles that provide buoyancy. By light microscopy these are bright, refractile bodies; by electron microscopy they are seen to have a regular shape: hollow, rigid cylinders with more or less conical ends and having a striated protein boundary. This boundary is impermeable to water, but the various dissolved gases in the culture medium can penetrate it to fill the cavity. The identifying feature of gas vacuoles is that they can be made to collapse under pressure and thereby lose their refractility.

Nuclear Material

In contrast to eucaryotic cells, bacterial cells contain neither a distinct membrane-enclosed nucleus nor a mitotic apparatus. However, they do contain an

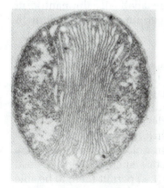

Figure 5-27. Electron micrograph of a thin section of a chemoautotrophic bacterium, *Nitrosococcus oceanus,* showing an extensive intracellular membrane system. *(Courtesy of S. W. Watson.)*

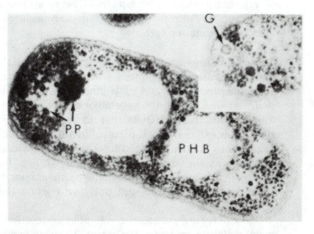

Figure 5-28. Thin section of *Pseudomonas pseudoflava* showing polyphosphate (volutin) granules (PP), poly-β-hydroxybutyrate granules (PHB), and glycogenlike granules (G). *(Courtesy of G. Auling, M. Reh and H. G. Schlegel, Int J Syst bacteriol **28:**82, 1978.)*

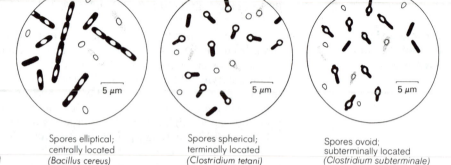

Figure 5-29. Drawings showing the location, size, and shape of endospores in cells of various species of *Bacillus* and *Clostridium*. (Erwin F. Lessel, illustrator.)

Spores elliptical; centrally located (*Bacillus cereus*)

Spores spherical; terminally located (*Clostridium tetani*)

Spores ovoid; subterminally located (*Clostridium subterminale*)

area near the center of the cell that is regarded as a nuclear structure, and the DNA of the cell is confined to this area. Because it is not a discrete nucleus, this nebulous structure has been designated by such terms as the **nucleoid;** the **chromatin body;** the **nuclear equivalent;** and even the **bacterial chromosome,** since it consists of a single, circular DNA molecule in which all the genes are linked. The nucleoid can be made visible under the light microscope by Feulgen staining, which is specific for DNA. By electron microscopy it appears as a light area with a delicate fibrillar structure (for example, see Figs. 5-22 and 5-26). The behavior of the nucleoid in growing, dividing bacteria has been observed by use of phase-contrast microscopy with a medium having a high refractive index.

SPORES AND CYSTS

Certain species of bacteria produce spores, either within the cell **(endospores)** or external to the cell **(exospores).** The spore is a metabolically dormant form which, under appropriate conditions, can undergo germination and outgrowth to form a vegetative cell.

Endospores

These structures are unique to bacteria. They are thick-walled, highly refractile bodies that are produced (one per cell) by *Bacillus, Clostridium, Sporosarcina, Thermoactinomyces,* and a few other genera. The shapes of endospores and also their location within the vegetative cell vary depending on the species (Fig. 5-29). The structural changes that occur during the development of endospores have been extensively studied in *Bacillus* and *Clostridium* species (Fig. 5-30). Endospores are usually produced by cells growing in rich media but which are approaching the end of active growth. Various factors such as aging or heat treatment are needed to activate the dormant spores (i.e., permit them to be able to undergo germination and outgrowth when they are placed in a suitable medium).

Endospores are extremely resistant to desiccation, staining, disinfecting chemicals, radiation, and heat. For example, the endospores of *Clostridium botulinum* type A have been reported to resist boiling for several hours. The degree of heat resistance of endospores varies with the bacterial species, but most can resist treatment at 80°C for at least 10 minutes. What causes this heat resistance has been a subject of intense study, but the explanation is still not clear. During sporulation, a dehydration process occurs in which most of the water in the developing spore is expelled; the resulting dehydrated state may be an important factor for heat resistance.

All endospores contain large amounts of **dipicolinic acid (DPA),** a unique compound that is undetectable in the vegetative cells yet can account for 10 to 15 percent of the spore's dry weight. It occurs in combination with large amounts of calcium and is probably located in the core, i.e., in the central part of the spore. The **calcium-DPA** complex may possibly play a role in the heat resistance of endospores. Synthesis of DPA and the uptake of calcium occur during advanced stages of sporulation.

Figure 5-30. Structural changes in the bacterial cell during sporulation. *(Erwin F. Lessel, illustrator; redrawn, with modifications, from L. E. Hawker and A. H. Linton, Microorganisms— Function, Form and Environment, 2d ed., University Park Press, Baltimore, 1979.)*

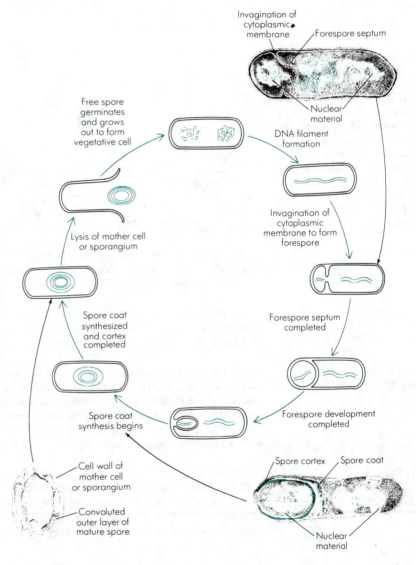

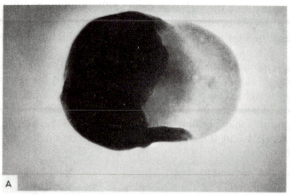

Figure 5-31. Outgrowth of spores from cultures of *Bacillus mycoides:* (A) grown 2 h at 35°C (X44,000), (B) grown 1¾ h at 35°C (X46,000). The two halves of the severed spore coat appear at the ends of the vegetative cell. *(SAB photos LS 203 and 204 courtesy of G. Knaysi, R. F. Baker, and J. Hillier, J Bacteriol,* **53:**525, *1947.)*

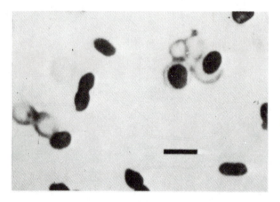

Figure 5-32. Outgrowth of cysts from cultures of an *Azotobacter* strain. Vegetative cells are also evident. *[Courtesy of Y.-T. Tchan and P. B. New, from N. R. Krieg (ed.), Bergey's Manual of Systematic Bacteriology, vol. 1, Williams & Wilkins, Baltimore, 1984.]*

During **germination,** endospores lose their resistance to heat and staining. Subsequent **outgrowth** occurs, characterized by synthesis of new cell material and development of the organism into a growing cell (Fig. 5-31).

Exospores

Cells of the methane-oxidizing genus *Methylosinus* form exospores, i.e., spores external to the vegetative cell, by budding at one end of the cell. These are desiccation- and heat-resistant, but unlike endospores they do not contain DPA.

Conidiospores and Sporangiospores

The large group of bacteria known as the actinomycetes form branching hyphae; spores develop, singly or in chains, from the tips of these hyphae by crosswall formation (septation). If the spores are contained in an enclosing sac **(sporangium),** they are termed **sporangiospores;** if not, they are called **conidiospores** (or **conidia**) (Fig. 5-6). The spores do not have the high heat resistance of endospores, but they can survive long periods of drying.

Cysts

Cysts are dormant, thick-walled, desiccation-resistant forms that develop by differentiation of a vegetative cell and which can later germinate under suitable conditions (Fig. 5-32). In some ways cysts resemble endospores; however, their structure and chemical composition are different and they do not have the high heat resistance of endospores. The classic example of a cyst is the structurally

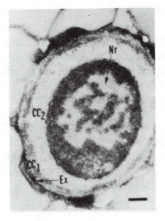

Figure 5-33. Fine structure of an *Azotobacter* cyst. The exosporium (Ex) and the two layers of exine (CC₁ and CC₂) are visible. A nuclear region (Nr) and a cytoplasmic region containing ribosomes are observable within the central body. [*Courtesy of Y.-T. Tchan and P. B. New, from N. R. Krieg (ed.), Bergey's Manual of Systematic Bacteriology, vol. 1, Williams & Wilkins, Baltimore, 1984.*]

complex type produced by the genus *Azotobacter* (Fig. 5-33). Several other bacteria can differentiate into cystlike forms, but these seem to lack the degree of structural complexity characteristic of *Azotobacter* cysts.

QUESTIONS

1 How does the cell's surface area/volume ratio compare with that of larger organisms? What advantages does a high surface area/volume ratio offer? What constraints does it place on a cell?

2 What bacterial cell structures may help to increase the cell's surface area/ volume ratio?

3 If you performed a microscopic examination of an appropriately stained preparation of *Staphylococcus aureus*, would you expect all the cells to be arranged in clusters? Explain.

4 Explain why some species of cocci appear as chains but others appear in a cuboidal arrangement.

5 Draw a typical bacterial cell and identify all parts.

6 Contrast propulsion by a bacterial flagellum with that by a screw propeller on a submarine.

7 What functions might chemotaxis, phototaxis, and magnetotaxis have for bacteria in their natural habitats?

8 What problems associated with the shape and motility of spiroplasmas still remain to be solved?

9 What function might a capsule serve for the following bacteria?
(*a*) a pathogenic bacterium
(*b*) a soil bacterium where the soil is periodically subjected to drought conditions
(*c*) a bacterium living in a flowing stream

10 Why are Gram-negative eubacteria usually much easier to disrupt by sonic oscillation than Gram-positive eubacteria?

11 Compare the structure and chemistry of the cell walls of Gram-positive eubacteria versus those of Gram-negative eubacteria. List some major differences between the cell walls of archaeobacteria versus those of eubacteria.

12 What function do the porins of the outer membrane of a Gram-negative eubacterial cell wall serve? What functions do cytoplasmic membrane/outer membrane adhesions serve?

13 In what kinds of bacteria and in what kinds of bacterial cell structures would we be most likely to find the following compounds: (a) peptidoglycan, (b) teichoic acids, (c) calcium dipicolinate, (d) cholesterol, (e) lipopolysaccharide, (f) phytanols ether-linked to glycerol?

14 Is spore formation in bacteria a method of reproduction or a means of multiplication? Explain.

15 What are the similarities and differences between protoplasts and spheroplasts?

16 Is it proper to refer to bacterial cells as containing a typical nucleus? Explain.

17 Name several cytoplasmic inclusions or substances. What function might be associated with each of these?

REFERENCES

Adler, J.: "The Sensing of Chemicals by Bacteria," *Sci Am*, **234**:40–47, 1976. *A concise and well-illustrated discussion of bacterial chemotaxis.*

Berg, H. C.: "How Bacteria Swim," *Sci Am*, **233**:36–44, 1975. *A clear exposition of the swimming behavior of bacteria and the action of their flagella.*

Ghosh, B. K. (ed.): *Organization of Prokaryotic Cell Membranes*, CRC Press, Boca Raton, Fl., 1981. *A series of reviews of the structure and function of bacterial membranes, gas vacuoles, mesosomes, and other structures.*

Ingraham, J. L., O. Maaløe, and F. C. Neidhardt: *Growth of the Bacterial Cell*, Sinauer, Sunderland, Mass., 1983. *Chapter 1 provides a succinct summary of the composition, organization, and structure of the bacterial cell.*

Kandler, O.: "Cell Wall Structures and Their Phylogenetic Implications," *Syst Appl Microbiol*, **3**:149–160, 1982. *A concise overview of the various kinds of bacterial walls and their chemical composition, including the archaeobacteria.*

Levinson, H., A. I. Sonenshein, and D. J. Tipper (eds.): *Sporulation and Germination*, American Society for Microbiology, Washington, D.C., 1981. *A collection of review articles and research papers dealing with all aspects of bacterial resting forms.*

Starr, M. P., H. Stolp, H. G. Trüper, A. Balows, and H. G. Schlegel (eds.): *The Prokaryotes: A Handbook on Habitats, Isolation, and Identification of Bacteria*, Springer-Verlag, New York, 1981. *The introductory chapters provide a useful overview of bacterial morphology as well as a guide to the pertinent literature. Illustrations and discussions of the morphology of nearly all types of bacteria are presented throughout the two volumes of this monumental work.*

Chapter 6   The Cultivation of Bacteria

Except for certain ecological studies where bacterial populations are examined in their natural habitats, bacteria are usually cultivated and studied under laboratory conditions. Numerous **media** (singular, **medium**) have been developed for bacterial cultivation. Because the nutritional requirements of bacteria vary widely, there are great differences in the chemical compositions of the media used in the laboratory. Bacteria also exhibit wide differences with respect to the physical conditions favoring their growth, such as temperature, pH, and gaseous environment. The successful cultivation of bacteria requires an awareness of all of these factors.

NUTRITIONAL REQUIREMENTS

All forms of life, from microorganisms to human beings, share certain nutritional requirements for growth and normal functioning. The following observations substantiate this statement and also illustrate the great diversity of nutritional types found among bacteria.

1 All organisms require a **source of energy.** Some rely on chemical compounds for their energy and are designated as **chemotrophs.** Others can utilize radiant energy (light) and are called **phototrophs.** Both chemotrophs and phototrophs exist among bacteria (see Table 6-1 for examples).

2 All organisms require a **source of electrons** for their metabolism. Some organisms can use reduced inorganic compounds as electron donors and are termed **litho-trophs** (some may be **chemolithotrophs,** others **photolithotrophs**). Other organisms use organic compounds as electron donors and are called **organotrophs** (some are **chemoorganotrophs,** others **photoorganotrophs**). Examples appear in Table 6-1.

3 All organisms require **carbon** in some form for use in synthesizing cell compo-

99

Table 6-1. Nutritional Characterization of Bacteria

Bacteria	Energy		Electron Donor		Carbon for Assimilation	
	Phototrophic	Chemotrophic	Lithotrophic	Organotrophic	Autotrophic	Heterotrophic
Chromatium okenii	+		+		+	
Rhodospirillum rubrum						
(anaerobic conditions)	+			+		+
(aerobic conditions)		+		+		+
Nitrosomonas europaea		+	+		+	
Desulfovibrio desulfuricans		+	+			+
Pseudomonas pseudoflava						
(H₂ supplied)		+	+		+	
(no H₂ supplied)		+		+		+
Escherichia coli		+		+		+

nents. All organisms require at least small amounts of CO_2. However, some can use CO_2 as their major, or even sole, source of carbon; such organisms are termed **autotrophs.** Others require organic compounds as their carbon source and are termed **heterotrophs** (Table 6-1).

4 All organisms require **nitrogen** in some form for cell components. Bacteria are extremely versatile in this respect. Unlike eucaryotes, some bacteria can use atmospheric nitrogen. Others thrive on inorganic nitrogen compounds such as nitrates, nitrites, or ammonium salts, and still others derive nitrogen from organic compounds such as amino acids.

5 All organisms require **oxygen, sulfur** and **phosphorus** for cell components. Oxygen is provided in various forms, such as water; component atoms of various nutrients; or molecular oxygen. Sulfur is needed for synthesis of certain amino acids (cysteine, cystine, and methionine). Some bacteria require organic sulfur compounds, some are capable of utilizing inorganic sulfur compounds, and some can even use elemental sulfur. Phosphorus, usually supplied in the form of phosphate, is an essential component of nucleotides, nucleic acids, phospholipids, teichoic acids, and other compounds.

6 All living organisms require *metal ions*, such as K^+, Ca^{2+}, Mg^{2+}, and Fe^{2+} for normal growth. Other metal ions are also needed but usually only at very low concentrations, such as Zn^{2+}, Cu^{2+}, Mn^{2+}, Mo^{6+}, Ni^{2+}, B^{3+}, and Co^{2+}; these are often termed **trace elements** and often occur as contaminants of other components of culture media in amounts sufficient to support bacterial growth.

Not all the biological functions of metal ions are known, but Fe^{2+}, Mg^{2+}, Zn^{2+}, Mo^{6+}, Mn^{2+}, and Cu^{2+}, are known to be cofactors for various enzymes (see Chap. 9).

Most bacteria do not require Na^+, but certain marine bacteria, cyanobacteria, and photosynthetic bacteria do require it. For those members of the archaeobacteria known as the "red extreme halophiles," the requirement is astonishing: they cannot grow with less than 12 to 15 percent NaCl! They require this high level of NaCl for maintenance of the integrity of their cell walls and for the stability and activity of certain of their enzymes.

7 All living organisms contain **vitamins** and **vitaminlike compounds.** These function either as coenzymes for several enzymes (see Chap. 9) or as the building blocks for coenzymes. Some bacteria are capable of synthesizing their entire requirement of vitamins from other compounds in the culture medium, but others cannot do so and will not grow unless the required vitamins are supplied pre-formed to them in the medium (see Table 6-2). Research in bacterial nutrition led to the discovery of some of the vitamins required by humans, and metabolic studies with bacteria contributed to our understanding of how these vitamins are synthesized and how they function.

8 All living organisms require water, and in the case of bacteria all nutrients must be in aqueous solution before they can enter the cells. Water is a highly polar compound that is unequaled in its ability to dissolve or disperse cellular components and to provide a suitable milieu for the various metabolic reactions of a cell. Moreover, the high specific heat of water provides resistance to sudden, transient temperature changes in the environment. Water is also a chemical reactant, being required for the many hydrolytic reactions carried out by a cell.

NUTRITIONAL TYPES OF BACTERIA

From the generalizations in the preceding paragraphs, it is apparent that bacteria can be divided into many groups on the basis of their nutritional requirements. The major separation is into two groups, phototrophs and chemotrophs.

Phototrophs

Among the phototrophic bacteria are species that use inorganic compounds as their source of electrons (i.e., photolithotrophs). For example, *Chromatium okenii* uses H_2S as its electron donor, oxidizing it to elemental sulfur:

$$H_2S \rightarrow S + 2e^- + 2H^+$$

Some phototrophic bacteria use organic compounds such as fatty acids and alcohols as electron donors and are therefore photoorganotrophs. For example, *Rhodospirillum rubrum* can use succinate as an electron donor:

$$\text{Succinate} \rightarrow \text{fumarate} + 2e^- + 2H^+$$

Certain phototrophic bacteria are not restricted to being phototrophic. As indicated before, chemotrophs rely on chemical compounds rather than light for their energy, and under some circumstances a phototrophic bacterium can grow as a chemotroph. For example, in the absence of O_2 (i.e., under anaerobic conditions) *R. rubrum* is dependent on light as its source of energy and lives as a photoorganotroph; however, in the presence of O_2 it can grow in the dark as a chemoorganotroph.

Chemotrophs

Among the chemotrophic bacteria are species that use inorganic compounds as

Table 6-2. Vitamin Requirements for Some Bacteria

Vitamin	Species Exhibiting Requirement (or Growth Stimulation)
Thiamine (B_1)	*Bacillus anthracis*
Riboflavin	*Clostridium tetani*
Niacin (nicotinic acid)	*Brucella abortus*
Pyridoxine (B_6)	*Lactobacillus* spp.
Biotin	*Leuconostoc mesenteroides*
Pantothenic acid	*Morganella morganii*
Folic acid	*Leuconostoc dextranicum*
Cobalamin (B_{12})	*Lactobacillus* spp.
Vitamin K	*Bacteroides melaninogenicus*

their source of electrons (i.e., chemolithotrophs). For example, bacteria of the genus *Nitrosomonas* use ammonia as their electron source, obtaining energy by oxidizing ammonia to nitrite:

$$NH_4^+ + \tfrac{3}{2} O_2 + H_2O \rightarrow NO_2^- + 2H_3O^+$$

This reaction involves a net transfer of 6 electrons, causing a valence change of the nitrogen atom from -3 to $+3$.

Many other chemotrophic bacteria use organic compounds, such as sugars and amino acids, as electron donors and are therefore chemoorganotrophs.

Certain bacteria can grow as either chemolithotrophs or chemoorganotrophs. For example, *Pseudomonas pseudoflava* can use either the organic compound glucose or the inorganic compound H_2 as its source of electrons:

$$\underset{\text{glucose}}{C_6H_{12}O_6} + 6H_2O \rightarrow 6CO_2 + 24H^+ + 24e^-$$

$$H_2 \rightarrow 2H^+ + 2e^-$$

Autotrophs and Heterotrophs

Autotrophs

As indicated before, the chemolithotrophic bacteria of the genus *Nitrosomonas* are able to oxidize ammonia to nitrite, thereby obtaining sufficient energy to assimilate the carbon of CO_2 into cell components (CO_2 fixation):

$$CO_2 + 4e^- + 4H^+ \rightarrow (CH_2O) + H_2O$$

where (CH_2O) represents carbohydrate. Organisms that can use CO_2 as their sole source of carbon for assimilation are termed autotrophs.

Until recently it was thought that all chemolithotrophic bacteria were autotrophs. Although this is true for most chemolithotrophs, a few are now recognized as being chemolithotrophic heterotrophs **(mixotrophs);** i.e., they obtain energy by utilizing inorganic electron donors, but obtain most of their carbon from organic compounds. One such organism is *Desulfovibrio desulfuricans*, which uses electrons from H_2 for the reduction of sulfate, yet derives most of its carbon from organic compounds in the culture medium.

Some autotrophs are facultative autotrophs; i.e., they can either live as autotrophs, deriving their carbon from CO_2, or they can live as heterotrophs, deriving their carbon from organic compounds. For example, *P. pseudoflava* can live as a heterotroph, using glucose as a source of carbon for assimilation (and also as its source of electrons, as mentioned above); however, if H_2 is provided as the electron source, then it can use CO_2 as its sole carbon source and can grow as an autotroph.

Cultivation of Autotrophs. In terms of chemical complexity of nutrient substances required for growth, the autotrophic bacteria exhibit the simplest requirements. For example, a medium of the composition shown in Table 6-3 supports the growth of *Nitrosomonas europaea*. (Because this medium is composed of known chemical compounds, it is called a **chemically defined** or **synthetic medium.**) The fact that an organism can grow and reproduce in such a mixture of inorganic compounds indicates that it has an elaborate capacity for synthesis. That is, the organism can transform these compounds into the carbohydrates, proteins, nucleic acids, lipids, vitamins, and other complex organic substances that constitute the living cell.

Table 6-3. Medium for *Nitrosomonas europaea*

Ingredient	Amount
NH₄Cl*	0.8 g
MgSO₄·7H₂O	0.2 g
K₂HPO₄	0.016 g
CaCl₂·2H₂O	0.02 g
Chelated iron	0.001 g
MnCl₂·4H₂O	0.0002 g
Na₂MoO₄·2H₂O	0.0001 g
ZnSO₄·7H₂O	0.0001 g
CuSO₄·5H₂O	0.00002 g
CoCl₂·6H₂O	0.000002 g
Distilled water	1,000 ml
Atmospheric CO₂†	

* The ammonium salt serves not only as the nitrogen source for this organism but also as the electron donor. The organism obtains energy by oxidizing ammonium ions to nitrite ions.

† Carbon dioxide is the sole carbon source.

Table 6-4. Minimal Nutritional Requirements of Some Heterotrophic Bacteria

Bacteria	Inorganic Salts	Organic Carbon	Atmo- spheric N_2	Inorganic Nitrogen	One Amino Acid	Two or More Amino Acids	One Vitamin	Two or More Vitamins
Azospirillum brasilense	+	+	+					
Escherichia coli	+	+		+				
Salmonella typhi	+	+		+	+			
Proteus vulgaris	+	+		+	+		+	
Staphylococcus aureus	+	+		+		+	+	
Lactobacillus acidophilus	+	+		+		+		+

Heterotrophs

Heterotrophic bacteria have been studied more extensively than the autotrophs because heterotrophs, in a sense, are of more immediate concern to us: it is here that we find all the species that cause diseases of human beings, other animals, and plants, as well as those that constitute the greater part of the microbial population in our immediate environment. However, we need to emphasize this does not mean that autotrophs are less important. On the contrary, they are of utmost importance in less conspicuous but indispensable processes in nature such as the cycling of elements through biological systems.

Cultivation of Heterotrophs. The heterotrophic bacteria, although they constitute one major nutritional group, vary considerably in the specific nutrients required for growth, particularly with respect to their organic carbon sources, nitrogen sources, and vitamin requirements. As indicated in Table 6-4, the requirements may be relatively simple or complex, depending on the species. This is shown more specifically in Table 6-5, where chemically defined media for the growth of *Escherichia coli* and lactobacilli are compared. From this table it is evident that *E. coli* has much simpler nutritional requirements than lactobacilli. Organisms such as lactobacilli that have elaborate requiremnts for specific nutrients, i.e., vitamins and other growth-promoting substances, are designated **fastidious heterotrophs.**

Obligate Parasites

Some bacteria have not yet been successfully cultivated on an artificial medium, and their nutritional and physical requirements are not understood. At present, such bacteria can be propagated only in association with a living host which, in a sense, serves as the medium. One example is the bacterium that causes leprosy, *Mycobacterium leprae*, which can be cultivated by infecting mice or armadillos. Other examples include the rickettsias, the chlamydias, and the spirochete that causes syphilis, *Treponema pallidum*.

BACTERIOLOGICAL MEDIA

Chemically defined media are needed for the cultivation of autotrophs and are also useful for defining the nutritional requirements of heterotrophs. However, for the routine cultivation of heterotrophs, chemically defined media are not generally used. Instead, certain complex raw materials such as **peptones, meat**

Table 6-5. Composition of Media Supporting Growth of Lactobacilli and *Escherichia coli* (Heterotrophic Bacteria)

MEDIUM FOR CULTIVATION OF LACTOBACILLI*

Casein hydrolysate	5 g
Glucose	10 g
Solution A	10 ml
Solution B	5 ml
L-Asparagine	250 ml
L-Tryptophan	50 mg
L-Cystine	100 mg
DL-Methionine	100 mg
Cysteine	100 mg
Ammonium citrate	2 g
Sodium acetate (anhydrous)	6 g
Adenine, guanine, xanthine, and uracil, each	10 mg
Riboflavin, thiamine, panththenate, and niacin, each	500 μg
Pyridoxamine	200 μg
Pyridoxal	100 μg
Pyridoxine	200 μg
Inositol and choline, each	10 μg
p-Aminobenzoic acid	200 μg
Biotin	5 μg
Folic acid (synthetic)	3 μg

Make up to 1 liter with distilled water.
Solution A: K_2HPO_4 and KH_2PO_4, each 25 g, into distilled water to a volume of 250 ml
Solution B: $FeSO_4 \cdot 7H_2O$, 0.5 g; $MnSO_4 \cdot 2H_2O$, 2.0 g; NaCl, 0.5 g; and $MgSO_4 \cdot 7H_2O$, 10g
Dissolve in distilled water to a volume of 250 ml

MEDIUM FOR CULTIVATION OF *E. COLI*

$NH_4H_2PO_4$	1 g
Glucose	5 g
NaCl	5 g
$MgSO_4 \cdot 7H_2O$	0.2 g
K_2HPO_4	1 g
H_2O	1,000 ml

* SOURCE: M. Rogosa et al., *J Bacteriol*, **54**:13, 1947.

extract, and **yeast extract** are used, and the resulting media support the growth of a wide variety of heterotrophic bacteria. **Agar** is included as a nonnutritive solidifying agent when a solid medium is desired. A description of these raw materials is given in Table 6-6. Examples of relatively simple liquid and solid media that support the growth of many common heterotrophs are **nutrient broth** and **nutrient agar** (Table 6-7). The addition of yeast extract to each of these formulas improves the nutrient quality, since yeast extract contains several of the B vitamins and other growth-promoting substances. Other complex supplements such as bovine rumen fluid, animal blood, blood serum, or extracts of plant and animal tissues may be required for the cultivation of certain fastidious heterotrophs.

Types of Media

Many special-purpose media are needed to facilitate recognition, enumeration, and isolation of certain types of bacteria. To meet these needs, the microbiologist

Table 6-6. Characteristics of Several Complex Materials Used as Ingredients of Media

Raw Material	Characteristic	Nutritional Value
Beef extract	An aqueous extract of lean beef tissue concentrated to a paste	Contains the water-soluble substances of animal tissue, which include carbohydrates, organic nitrogen compounds, water-soluble vitamins, and salts
Peptone	The product resulting from the digestion of proteinaceous materials, e.g., meat, casein, and gelatin; digestion of the protein material is accomplished with acids or enzymes; many different peptones (depending upon the protein used and the method of digestion) are available for use in bacteriological media; peptones differ in their ability to support growth of bacteria	Principal source of organic nitrogen; may also contain some vitamins and sometimes carbohydrates, depending upon the kind of proteinaceous material digested
Agar	A complex carbohydrate obtained from certain marine algae; processed to remove extraneous substances	Used as a solidification agent for media; agar, dissolved in aqueous solutions, gels when the temperature is reduced below 45°C; agar not considered a source of nutrient to the bacteria
Yeast extract	An aqueous extract of yeast cells, commercially available as a powder	A very rich source of the B vitamins; also contains organic nitrogen and carbon compounds

Table 6-7. Composition of Nutrient Broth and Nutrient Agar

Nutrient broth	
Beef extract	3 g
Peptone	5 g
Water	1,000 ml
Nutrient agar	
Beef extract	3 g
Peptone	5 g
Agar	15 g
Water	1,000 ml

has available numerous media which, on the basis of their application or function, may be classified as follows.

Selective Media

These media provide nutrients that enhance the growth and predominance of a particular type of bacterium and do not enhance (and may even inhibit) other types of organisms that may be present. For instance, a medium in which cellulose is the only carbon source will specifically select for or enrich the growth of cellulose-utilizing organisms when it is inoculated with a soil sample containing many kinds of bacteria. As an example of a different type of selective medium, the isolation of the gonorrhea-causing organism, *Neisseria gonorrhoeae*, from a clinical specimen is facilitated by the use of media containing certain antibiotics; these antibiotics do not affect *N. gonorrhoeae* but do inhibit the growth of contaminating bacteria.

Differential Media

Certain reagents or supplements, when incorporated into culture media, may allow differentiation of various kinds of bacteria. For example, if a mixture of bacteria is inoculated onto a blood-containing agar medium (blood agar), some of the bacteria may hemolyze (destroy) the red blood cells; others do not. Thus one can distinguish between **hemolytic** and **nonhemolytic** bacteria on the same medium.

Assay Media

Media of prescribed compositions are used for the assay of vitamins, amino acids, and antibiotics. Media of special composition are also available for testing disinfectants.

Media for Enumeration of Bacteria

Specific kinds of media are used for determining the bacterial content of such materials as milk and water. Their composition must adhere to prescribed specifications.

Media for Characterization of Bacteria

A wide variety of media are conventionally used to determine the type of growth produced by bacteria, as well as to determine their ability to produce certain chemical changes.

Maintenance Media

Satisfactory maintenance of the viability and physiological characteristics of a culture over time may require a medium different from that which is optimum for growth. Prolific, rapid growth may also be associated with rapid death of the cells at the end of the growth phase. For example, glucose in a medium frequently enhances growth, but acid harmful to the cells is likely to be produced. Therefore, omission of the glucose is preferable in a maintenance medium.

Solid and Semisolid Media

In addition to liquid media, solid and semisolid media are widely used for cultivation of bacteria. Solid media are useful for isolating bacteria or for determining the characteristics of colonies. The solidifying agent is usually **agar,** which at concentrations of 1.5 to 2.0 percent forms firm, transparent gels that are not degraded by most bacteria. **Silica gel** is sometimes used as an inorganic solidifying agent for autotrophic bacteria.

Semisolid media, prepared with agar at concentrations of 0.5 percent or less, have a soft, custardlike consistency and are useful for the cultivation of microaerophilic bacteria (see Gaseous Requirements later in this chapter) or for determination of bacterial motility.

Preparation of Media

Some naturally occurring substances are used for the cultivation of bacteria. Notable among these is milk, usually skimmed rather than whole. Such natural materials are merely dispensed into tubes or flasks and sterilized before use. Media of the nutrient broth or nutrient agar type are prepared by compounding the required individual ingredients or, more conveniently, by adding water to a dehydrated product which contains all the ingredients. Practically all media are available commercially in powdered form.

The preparation of bacteriological media usually involves the following steps:

1 Each ingredient, or the complete dehydrated medium, is dissolved in the appropriate volume of distilled water.
2 The pH of the fluid medium is determined with a pH meter and adjusted if necessary.
3 If a solid medium is desired, agar is added and the medium is boiled to dissolve the agar.
4 The medium is dispensed into tubes or flasks.
5 The medium is sterilized, generally by autoclaving. Some media (or specific ingredients) that are heat-labile are sterilized by filtration.

PHYSICAL CONDITIONS REQUIRED FOR GROWTH

In addition to knowing the proper nutrients for the cultivation of bacteria, it is also necessary to know the physical environment in which the organisms will grow best. Just as bacteria vary greatly in their nutritional requirements, so do they exhibit diverse responses to physical conditions such as temperature, gaseous conditions, and pH.

Temperature

Since all processes of growth are dependent on chemical reactions and since the rates of these reactions are influenced by temperature, the pattern of bacterial growth can be profoundly influenced by this condition. The temperature that allows for most rapid growth during a short period of time (12 to 24 h) is known as the **optimum growth temperature.** (It should be noted, however, that the optimum growth temperature thus defined may not necessarily be optimum for other cellular activities.)

Table 6-8 shows the optimum temperture for several bacteria and also the range of temperatures within which they will grow. It can be seen that the maximum temperature at which growth occurs is usually quite close to the optimum temperature, whereas the minimum temperature for growth is usually much lower than the optimum. On the basis of their temperature relationships, bacteria are divided into three main groups:

1 **Psychrophiles** are able to grow at 0°C or lower, though they grow best at higher temperatures. Many microbiologists restrict the term *psychrophile* to organisms that can grow at 0°C but have an optimum temperature of 15°C or lower and a maximum temperature of about 20°C; the term *psychrotroph* or *facultative psychrophile* is used for those organisms able to grow at 0°C but which grow best at temperatures in the range of about 20 to 30°C (e.g., see Fig. 6-1).

During isolation of strict psychrophiles it is usually necessary to maintain the source samples (for example, Antarctic soil samples) at cold temperatures from the time they are collected and also to chill all media before attempting isolation. This is because strict psychrophiles usually die if they are even temporarily exposed to room temperature. Even at optimum growth temperatures, it often takes two or three weeks for colonies of psychrophiles to develop.

The physiological factors responsible for the low temperature maxima for strict psychrophiles are not entirely clear, but some factors that have been implicated are heat instability of ribosomes and various enzymes, increased leakage of cell

Table 6-8. Characteristics of Several Species of Bacteria with Regard to Temperatures at Which They Grow

	Temperature of Growth, °C		
	Minimum	Optimum	Maximum
Vibrio marinus strain MP-1	-1	15	20
Vibrio psychroerythrus	0	15	19
Pseudomonas fluorescens	4	25–30	40
Staphylococcus aureus	6.5	30–37	46
Corynebacterium diphtheriae	15	37	40
Neisseria gonorrhoeae	30	35–36	38.5
Streptococcus thermophilus	20	40–45	50
Thermoactinomyces vulgaris	27–30	60	65–70
Thermus aquaticus	40	70–72	79

SOURCE: Data from R. Y. Morita, *Bacteriol Rev*, **39**:144, 1975, and from *Bergey's Manual of Determinative Bacteriology*, 8th ed, Williams & Wilkins, Baltimore, 1974.

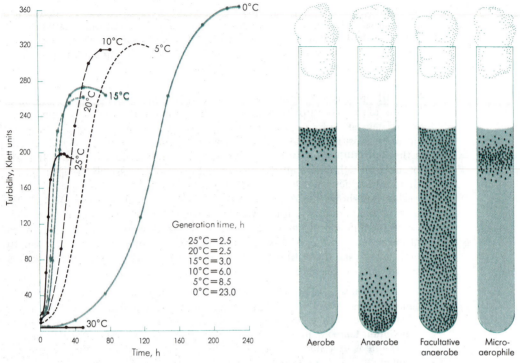

Figure 6-1. Effect of temperature on the growth of a psychrotrophic *Bacillus* sp. Note that rate of growth (measured turbidimetrically in Klett units) is more rapid at 25°C than at 0°C, although the total quantity of cells at the termination of growth is greater at the lower temperature. (*Courtesy of J. L. Stokes in Low Temperature Biology of Food Stuffs, Pergamon, New York, 1968.*)

Figure 6-2. Schematic illustration of the growth of bacteria in deep agar tubes, showing differences in response to atmospheric oxygen.

components, and impaired transport of nutrients above the maximum temperature.

2 **Mesophiles** grow best within a temperature range of approximately 25 to 40°C. For example, all bacteria that are pathogenic for humans and warm-blooded animals are mesophiles, most growing best at about body temperature (37°C).

3 **Thermophiles** grow best at temperatures above 45°C. The growth range of many thermophiles extends into the mesophilic region; these species are designated **facultative thermophiles.** Other thermophiles cannot grow in the mesophilic range; these are termed **true thermophiles, obligate thermophiles,** or **stenothermophiles.**

Factors that have been implicated in the ability to grow at high temperatures are an increased thermal stability of ribosomes, membranes, and various enzymes. Loss of the fluidity that exists within the lipid bilayer of the cytoplasmic membrane may be a factor governing the minimum temperature.

It is important to note that a bacterial species may not manifest the same characteristics in every detail when grown at different temperatures. For example, *Serratia marcescens* forms a blood-red to orange pigment when cultured at 25°C but produces little or no pigment when cultured at 37°C. Similarly, *Lactobacillus plantarum* does not require the amino acid phenylalanine for growth when cultured at 25°C but does require it at 37°C.

Gaseous Requirements

The principal gases that affect bacterial growth are oxygen and carbon dioxide. Bacteria display such a wide variety of responses to free oxygen that it is convenient to divide them into four groups on the following bases:

1 **Aerobic bacteria** require oxygen for growth and can grow when incubated in an air atmosphere (i.e., 21 percent oxygen).
2 **Anaerobic bacteria** do not use oxygen to obtain energy; moreover, oxygen is toxic for them and they cannot grow when incubated in an air atmosphere. Some can tolerate low levels of oxygen (**nonstringent** or **tolerant anaerobes**), but others (**stringent** or **strict anaerobes**) cannot tolerate even low levels and may die upon brief exposure to air.
3 **Facultatively anaerobic bacteria** do not require oxygen for growth, although they may use it for energy production if it is available. They are not inhibited by oxygen and usually grow as well under an air atmosphere as they do in the absence of oxygen.
4 **Microaerophilic bacteria** require low levels of oxygen for growth but cannot tolerate the level of oxygen present in an air atmosphere.

Figure 6-2 shows diagrammatically how these four classes can be distinguished by their patterns of growth in tubes in deep agar media where the diffusion of oxygen into the medium is a controlling factor.

Oxygen Toxicity

Oxygen is both beneficial and poisonous to living organisms. It is beneficial because its strong oxidizing ability makes it an excellent terminal electron acceptor for the energy-yielding process known as respiration. However, oxygen is also a toxic substance. Aerobic and facultative organisms have developed protective mechanisms that greatly mitigate this toxicity, but microaerophiles and anaerobes are deficient in these mechanisms and are restricted to habitats where little or no oxygen is present. The following factors are among those that have been implicated in oxygen toxicity.

1 *Oxygen inactivation of enzymes.* Molecular oxygen can directly oxidize certain essential reduced groups, such as thiol (-SH) groups, or enzymes, resulting in enzyme inactivation. For instance, the enzyme complex known as nitrogenase, responsible for nitrogen fixation, is irreversibly destroyed by even small amounts of oxygen.
2 *Damage due to toxic derivatives of oxygen.* Various cellular enzymes catalyze chemical reactions involving molecular oxygen; some of these reactions can result in addition of a single electron to an oxygen molecule, thereby forming a **superoxide radical** ($O_2^{-\cdot}$):

$$O_2 + e^- \rightarrow O_2^{-\cdot} \tag{1}$$

Superoxide radicals can inactivate vital cell components. However, recent studies suggest that their greatest detrimental action is through production of even more toxic substances such as **hydrogen peroxide** (H_2O_2) and **hydroxyl radicals** (OH·) by means of the following reactions:

$$2O_2^{-\cdot} + 2H^+ \rightarrow O_2 + H_2O_2 \tag{2}$$

$$O_2^{-\cdot} + H_2O_2 \xrightarrow{\text{chelated iron}} O_2 + OH^- + OH\cdot \tag{3}$$

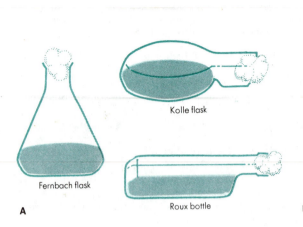

Kolle flask

Fernbach flask

Roux bottle

A   **B**

Figure 6-3. Methods for providing increased aeration during incubation. (A) Culture vessels of several designs that provide a large surface area for a shallow layer of medium. (B) An example of an incubator-shaker. The environmental chamber provides controlled conditions of temperature, humidity, and illumination. Within the chamber, flasks are fixed firmly on a platform which rotates in a circular manner, thus agitating the fluid medium constantly during incubation and exposing more culture surface to the gas phase. *(Courtesy of New Brunswick Scientific Company.)*

Hydroxyl radicals are among the most reactive free radicals known to organic chemistry and can damage almost every kind of molecule found in living cells. Hydrogen peroxide is not a free radical, but it is a powerful oxidizing agent that is highly toxic to many kinds of cells. Another toxic derivative of oxygen is an energized form known as **singlet oxygen,** $(^1\Delta_g)O_2$, which is produced in biological systems by certain photochemical reactions.

Aerobic and facultative organisms have developed various protective mechanisms against the toxic forms of oxygen. One is the enzyme known as **superoxide dismutase,** which eliminates superoxide radicals by greatly increasing the rate of reaction 2 above. The hydrogen peroxide produced by this reaction can in turn be dissipated by **catalase** and **peroxidase** enzymes:

$$2H_2O_2 \xrightarrow{\text{catalase}} 2H_2O + O_2 \tag{4}$$

$$H_2O_2 + \text{reduced substrate} \xrightarrow{\text{peroxidase}} 2H_2O + \text{oxidized substrate} \tag{5}$$

Note that elimination of either superoxide radicals or hydrogen peroxide can prevent the formation of the highly dangerous hydroxyl radicals, since *both reactants* are required for reaction (3).

In general, anaerobic bacteria have either no superoxide dismutase or only relatively low levels compared to aerobes. Many anaerobes are also deficient in catalase and/or peroxidase. This may help to explain, at least in part, their sensitivity to oxygen, although other factors are probably involved as well.

Cultivation of Aerobic Bacteria. To grow aerobic or facultative bacteria in tubes or small flasks, incubation of the medium under normal atmospheric conditions is generally satisfactory. However, when aerobic organisms are to be grown in large quantities, it is advantageous to increase the exposure of the medium to the atmosphere. This can be accomplished by dispensing the medium in shallow layers, for which special containers are available. Aeration can also be increased by constantly shaking the inoculated liquid cultures (Fig. 6-3).

Cultivation of Anaerobic Bacteria. Stringent anaerobes can be grown only by taking special precautions to exclude all atmospheric oxygen from the medium. Such an environment can be established by using one of the following methods:

1 *Prereduced media.* During preparation, the culture medium is boiled for several minutes to drive off most of the dissolved oxygen. A reducing agent, e.g., cysteine, is added to further lower the oxygen content. Oxygen-free N_2 is bubbled through the medium to keep it anaerobic. The medium is then dispensed into tubes which are being flushed with oxygen-free N_2, stoppered tightly, and sterilized by autoclaving. Such tubes can be stored for many months before being used. During inoculation, the tubes are continuously flushed with oxygen-free CO_2 by means of a cannula (Fig. 6-4), restoppered, and incubated.

2 *Anaerobic chamber.* This refers to a plastic anaerobic glove box (Fig. 6-5) that

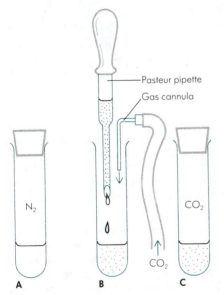

Figure 6-4. Use of prereduced media for cultivation of stringent anaerobes. (A) Tube of prereduced medium containing an atmosphere of oxygen-free N_2. (B) To inoculate, the stopper is removed and a gas cannula inserted to flush the tube continuously with oxygen-free CO_2 and maintain anaerobic conditions. The medium is inoculated with a few drops of culture by means of a Pasteur pipette. (C) After inoculation the tube is restoppered and incubated.

Figure 6-5. (A) Schematic diagram of the various parts of an anaerobic chamber (top view). (a) Glove ports and rubber gloves that allow the operator to perform manipulations within the chamber. (b) Air lock with inner and outer doors. Media are placed within the air lock with the inner door remaining sealed; air is removed by a vacuum pump connection (c) and replaced with N_2 through (d). The inner door is opened and the media are placed within the main chamber, which contains an atmosphere of $H_2 + CO_2 + N_2$. A circulator (e) circulates the gas atmosphere through pellets of palladium catalyst (f), causing any residual oxygen in the media to be used up by reaction with H_2. After media have become completely anaerobic they can be inoculated and placed in an incubator (g) located within the chamber. (B) Photograph of an anaerobic chamber. (*Courtesy of The Germfree Laboratories, Inc.*)

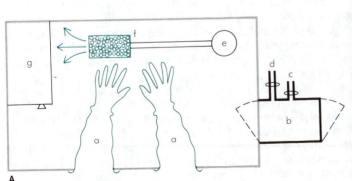

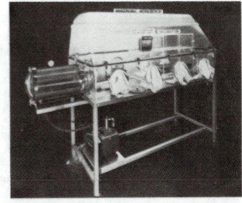

Figure 6-6. Anaerobic jar: GasPak system. (A) Media are inoculated and then placed in the jar. Water is added to the GasPak generator envelope, causing the evolution of H_2 and CO_2. The H_2 reacts with O_2 on the surface of the palladium catalyst, forming water and establishing anaerobic conditions. The CO_2 aids growth of fastidious anaerobes which sometimes fail to grow, or grow only poorly, in its absence. An anaerobic indicator strip (a pad saturated with methylene blue solution) changes from blue to colorless in the absence of oxygen. (B) The GasPak Anaerobic System with inoculated Petri dishes, the GasPak generator envelope, and the anaerobic indicator strip. *(Courtesy of BBL Microbiology Systems.)*

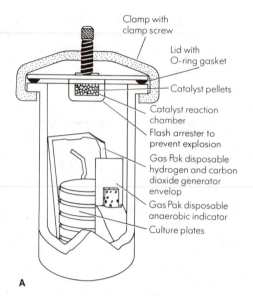

Clamp with clamp screw
Lid with O-ring gasket
Catalyst pellets
Catalyst reaction chamber
Flash arrester to prevent explosion
Gas Pak disposable hydrogen and carbon dioxide generator envelop
Gas Pak disposable anaerobic indicator
Culture plates

A   B

contains an atmosphere of H_2, CO_2, and N_2. Culture media are placed within the chamber by means of an air lock which can be evacuated and refilled with N_2. From the air lock the media are placed within the main chamber. Any O_2 in the media is slowly removed by reaction with the H_2, forming water; this reaction is aided by a palladium catalyst. After being rendered oxygen-free, the media are inoculated within the chamber (by means of the glove ports) and incubated (also within the chamber).

Nonstringent anaerobes can be cultured within an anaerobic jar such as that depicted in Fig. 6-6. Inoculated media are placed in the jar along with an H_2 + CO_2 generating system. After the jar is sealed, the oxygen present in the atmosphere within the jar, as well as that dissolved in the culture medium, is gradually used up through reaction with the hydrogen in the presence of a catalyst.

Acidity or Alkalinity (pH)

For most bacteria the optimum pH for growth lies between 6.5 and 7.5, and the limits generally lie somewhere between 5 and 9. However, a few bacteria prefer more extreme pH values for growth. For example, *Thiobacillus thiooxidans* has an optimum pH of 2.0 to 3.5 and can grow in a range between pH 0.5 and 6.0. On the other hand, an unclassified bacterium isolated from an alkaline spring in California was found to grow best at a pH of 9.0 to 9.5 and could grow within a range from 8.0 to 11.4.

When bacteria are cultivated in a medium originally adjusted to a given pH, for example, 7.0, it is very likely that this pH will change as a result of the chemical activities of the organism. If a carbohydrate is present it may be fermented or oxidized to organic acids, thus decreasing the pH of the medium. If the salt of an organic acid is supplied as a carbon source (e.g., sodium malate), its oxidation by bacteria will cause an increase in pH. Such shifts in pH may be so great that further growth of the organism is eventually inhibited.

Radical shifts in pH can be prevented by incorporating a buffer (i.e., a substance that resists change in pH) into the medium. A buffer is a mixture of a weak acid and its conjugate base [e.g., acetic acid (CH_3COOH) and acetate (CH_3COO^-)]. Such mixtures have maximum buffering capacity at the pH where the concentration of the acid equals that of its conjugate base. This pH value is called the pK_a and is the negative logarithm of the dissociation constant of the acid. Phosphate buffer, i.e., a combination of $H_2PO_4^-$ and HPO_4^{2-} having a pK_a of 6.8, is widely used in bacteriological media. Some of the nutritional ingredients of the medium, such as peptones, also possess some buffering capacity because the component amino acids provide weak acid/conjugate base systems (e.g., —COOH/—COO$^-$, —NH$_3^+$/—NH$_2$, —NH$_2^+$/—NH). The extent to which a medium should or may be buffered depends on its intended purpose and is limited by the buffering capacity of the compounds used. Some large fermentation apparatuses are equipped with automatic controls that maintain a constant pH.

Miscellaneous Physical Requirements

Temperature, the gaseous environment, and pH are the major physical factors to be taken into consideration in establishing the optimum conditions for the growth of most species of bacteria. However, some bacteria have additional requirements. For example, phototrophic bacteria must be exposed to a source of *illumination*, since light is their source of energy. Bacterial growth may also be influenced by *hydrostatic pressure*. Bacteria have been isolated from the deepest ocean trenches where the pressure is measured in tons per square inch, and many of these organisms will not grow in the laboratory unless the medium is subjected to a similar pressure.

CHOICE OF MEDIA AND CONDITIONS OF INCUBATION

From this brief excursion into the nutritional requirements of bacteria, it is apparent that to grow bacteria successfully the laboratory worker must provide the proper kind of medium and an appropriate set of physical conditions. A great deal of information is available to a microbiologist with respect to choice of media, preparation of media, and the physical and nutritional conditions required for the cultivation of the various genera and species of bacteria. Much of this information can be obtained by consulting the three references listed at the end of this chapter.

QUESTIONS

1 What nutritional requirements in terms of chemicals are needed by all forms of life for growth and cellular maintenance?

2 Distinguish between (a) phototrophs and chemotrophs, (b) lithotrophs and organotrophs, and (c) autotrophs and heterotrophs.

3 Suppose you want to determine the occurrence of the vitamin biotin in a batch of milk. *Leuconostoc mesenteroides* requires biotin for growth. Using this bacterium and a chemically defined medium to which a small sample of sterilized milk can be added, devise an experiment that could indicate whether biotin is present in milk. What substance should be omitted during preparation of the chemically defined medium?

4 What conditions of cultivation would allow you to grow selectively:
(a) *Thiobacillus thiooxidans* from a mixed culture of bacteria
(b) *Neisseria gonorrhoeae* from a clinical specimen

(c) The stenothermophile *Clostridium thermosaccharolyticum* from a can of spoiled corn

(d) an extreme halophile from a sample of sea salt

(e) a nitrogen-fixing bacterium from a soil sample

5 Most medically important anaerobic bacteria are nonstringent anaerobes, but some are the stringent type. Describe two methods recommended for cultivating the latter organisms.

6 Is nutrient broth a "universal" medium (that is, can it support the growth of every kind of bacteria)? Explain.

7 If you wanted to buffer a culture medium to keep the pH at 5.5, you would try to use a buffer having what pK_a? Why?

8 List some of the ways whereby you could enhance the growth of an aerobic bacterium.

9 Under what circumstances might you wish to use silica gel as a solidifying agent for a culture medium?

10 Indicate the various toxic derivatives of oxygen and explain how aerobic organisms might protect themselves against these derivatives.

REFERENCES

Gerhardt, P., R. G. E. Murray, R. N. Costilow, E. W. Nester, W. A. Wood, N. R. Krieg, and G. B. Phillips (eds): *Manual of Methods for General Bacteriology*, American Society for Microbiology, Washington, D.C., 1981. *A "how-to-do-it" manual that deals with the choice and preparation of media, the biophysical conditions required for growth, and the selective isolation and cultivation of many kinds of bacteria.*

Krieg, N. R. (ed.): *Bergey's Manual of Systematic Bacteriology*, vol. 1, Williams & Wilkins, Baltimore, 1984. *This together with subsequent volumes provides detailed descriptions of the genera and species of bacteria, including the nutritional and physical conditions required for isolation and cultivation.*

Starr, M. P., H. Stolp, H. G. Trüper, A. Balows, and H. G. Schlegel (eds.): *The Prokaryotes: A Handbook on Habitats, Isolation, and Identification of Bacteria*, Springer-Verlag, New York, 1981. *The introductory chapters of this book contain summaries of the various nutritional and physical factors that affect bacteria; other chapters provide specific information about the habitat, isolation, and cultivation of each bacterial group.*

Chapter 7   Reproduction and Growth

When bacteria are inoculated into a suitable medium and incubated under
appropriate conditions, a tremendous increase in the number of cells occurs
within a relatively short time. With some species the maximum population is
reached within 24 h, but others require a much longer period of incubation to
reach maximum growth. The term *growth* as commonly applied to bacteria and
other microorganisms usually refers to *changes in the total population* rather
than an increase in the size or mass of an individual organism. More frequently
than not, the inoculum contains thousands of organisms; growth denotes the
increase in number beyond that present in the original inoculum. Therefore,
determination of growth requires quantitative measurement of the total population of cells or cell crops at the time of inoculation and again after incubation.
In this chapter we will discuss how bacteria reproduce and how their growth
can be measured.

REPRODUCTION

Modes of Cell Division

The most common, and no doubt the most important, mode of cell division in
the usual growth cycle of bacterial populations is **transverse binary fission,** in
which a single cell divides after developing a transverse septum (crosswall)
(Fig. 7-1 A, B, C). Transverse binary fission is an asexual reproductive process.
(Infrequently, in some species, binary fission may be preceded by a mating or
conjugation of cells; this sexual process is discussed in Chap. 12.)

Binary Fission

115

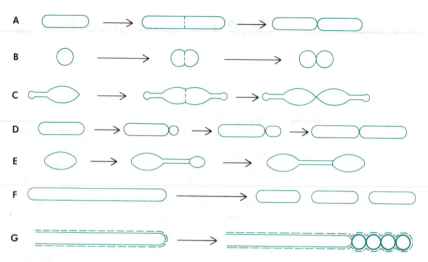

Figure 7-1. Schematic drawing of modes of cell division in various bacteria. Transverse binary fission occurs in *Bacillus subtilis* (A), *Streptococcus faecalis* (B), and the prosthecate bacterium *Prosthecobacter fusiformis* (C); in the latter species the small round area at the tip of each prostheca is a holdfast—a means of attachment to surfaces. Budding occurs in *Rhodopseudomonas acidophila* (D) and *Hyphomicrobium vulgare* (E); in the latter species the mother cell produces a prostheca on which a terminal bud forms; this bud develops into a daughter cell. (F) Fragmentation occurs in the filamentous cells of a *Nocardia* species. (G) Formation of conidiospores by a *Streptomyces* species. A hypha that gives rise to spores is covered by a sheath (represented here by a dashed line); septation occurs at the hyphal tip to produce a chain of conidiospores still enclosed by the sheath.

Budding

Some bacteria, such as *Rhodopseudomonas acidophila*, reproduce by budding, a process in which a small protuberance (*bud*) develops at one end of the cell; this enlarges and eventually develops into a new cell which separates from the parent (Fig. 7-1D). In some budding bacteria, such as *Hyphomicrobium* species, the bud may develop at the end of a prostheca (Fig. 7-1E).

Fragmentation

Bacteria that produce extensive filamentous growth, such as *Nocardia* species, reproduce by fragmentation of the filaments into small bacillary or coccoid cells, each of which gives rise to new growth (Fig. 7-1F).

Formation of Conidiospores or Sporangiospores

Species of the genus *Streptomyces* and related bacteria produce many spores per organism by developing crosswalls (septation) at the hyphal tips; each spore gives rise to a new organism (Fig. 7-1G).

New Cell Formation (Macromolecular Synthesis)

A bacterial cell inoculated into a fresh medium selectively takes up nutrients from its environment. Many biochemical syntheses then take place. The nutrients are converted into cell substance—RNA, DNA, proteins, enzymes, and other macromolecules. Cell mass and cell size increase, and new cell wall building blocks are synthesized. Subsequently, the process of binary fission is initiated, ultimately resulting in the formation of two new cells.

Septum Formation

In transverse binary fission, septum formation does not begin until the chromosome content of the cell has been doubled; i.e., cell division is triggered by completion of DNA replication (discussed in Chap. 11). The first step is an inward growth of the cytoplasmic membrane at the middle of the cell; a mesosome is usually attached to the cytoplasmic membrane at this location, particularly in Gram-positive cells, and may have a role in the synthesis of new membrane material. The next step is the inward growth of the cell wall to form a septum that ultimately splits to allow separation of the two daughter cells.

For example, during growth of the Gram-positive coccus *Streptococcus faecalis*, all of the new wall material formed by the dividing cell is made during synthesis of the septum. Septum formation begins beneath an **equatorial ridge** in the cell wall (see Figs. 7-2 and 5-26). New cell-wall material is synthesized in this region and, as the septum forms, this material becomes one half of the wall of each daughter cell. Some plasticity must be present in order for the new wall to achieve its final, more or less hemispherical shape; this is believed to be due to two factors: (1) the turgor pressure of the protoplast against the newly synthesized wall and (2) a certain amount of reorganization of the peptidoglycan due to breakage of some of the chemical bonds by hydrolytic enzymes and subsequent formation of new bonds at a different location.

In Gram-positive rods such as *Bacillus subtilis*, the transverse septum is formed in a manner similar to that for *S. faecalis*, although no ridge is present at the middle of the cell. Moreover, only about 15 percent of the new wall of a daughter cell is derived from formation of the septum. The remainder is synthesized along the cylindrical part of the cell, since a bacillus grows mainly by *elongation* rather than just by septum formation as does a coccus. Perhaps there are just a few discrete regions where new wall is synthesized and inserted into old wall, or perhaps new wall is made and inserted into old wall all along the length of the cell, but this is a question yet unanswered. However, there is strong evidence that the youngest portion of the cylindrical wall is that layer which is immediately adjacent to the cytoplasmic membrane. During growth and extension of the wall, the older, outer layers of the wall become more thinly spread out. They are eventually destroyed by degradative enzymes located in the cell wall and are replaced from below by the newer wall material. Thus the wall is not static like a plastic coating; rather, it is in a dynamic state, with old, outer material continually being destroyed and new, inner material continually being added as the cell elongates.

The stages of septum formation in Gram-negative bacilli such as *Escherichia coli* are depicted in Fig. 7-3.

A number of basic questions remain to be answered concerning transverse binary fission:

1 What ensures that each daughter cell will receive a complete genome? That is, during septum formation in any bacterial cell, it is essential that the DNA be precisely distributed to the daughter cells so that each receives a complete genome.

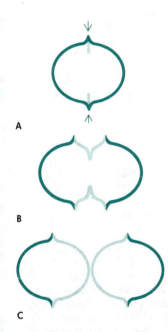

Figure 7-2. Schematic diagram illustrating septum formation in Gram-positive cocci such as *Streptococcus faecalis*. (A) Synthesis of new wall begins at the equatorial ridge (arrows). (B) From this site new wall grows peripherally, pushing apart the hemispheres of old wall. Half of the entire wall of each daughter cell is derived from the septum.

Figure 7-3. Septum formation in the Gram-negative bacterium *Escherichia coli*. In stages 3 to 6, a bleb, or fold, of the outer membrane occurs at the site where the septum will be formed; it is not evident in the final stages of septum formation (stage 7). The outer membrane does not invaginate until the final stages of septum formation (stage 7); however, the cytoplasmic membrane and the peptidoglycan layer grow inward in the early stages (stages 4 and 5). A mesosome, linked to the cytoplasmic membrane, is present during the early stages of septum formation (stage 4). *(Courtesy of I. D. J. Burdett and R. G. E. Murray, J Bacteriol* **119**:*1039, 1974.)*

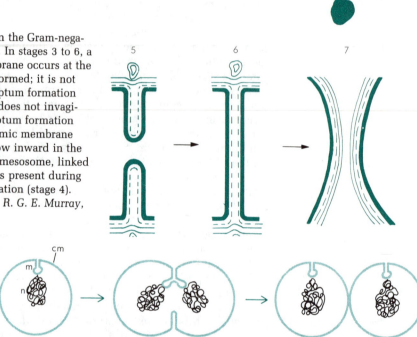

Figure 7-4. Illustration of the hypothesized role of the central mesosome in segregation of DNA into daughter bacterial cells. (A) Cell before binary fission, showing cytoplasmic membrane (cm), central mesosome (m), and the nucleoid (n). (The cell wall is not shown.) In this model the nucleoid is shown attached to the mesosome. Evidence for such attachment has been obtained by electron microscopy of thin sections of bacteria. (B) During binary fission both the DNA of the nucleoid and its attachment site to the mesosome are duplicated. The mesosome begins to divide because of synthesis of membrane between the DNA-mesosome attachment sites. (C) During binary fission each mesosome is "pushed" to the middle of a daughter cell because of synthesis of cytoplasmic membrane between the mesosomes. Because a nucleoid is attached to each mesosome, the nucleoids become properly segregated into the daughter cells.

No mitotic apparatus for this segregation of daughter DNA molecules exists in bacteria; however, the cytoplasmic membrane, or the central mesosome derived from it, may play an equivalent role (Fig. 7-4).

2   What causes the septum to form at approximately the middle of the cell—why not near one of the poles? (Indeed, some mutants of *Escherichia coli* and *B. subtilis* have been obtained which often do form the septum near a pole, resulting

in a very small daughter cell (termed a **minicell**) which lacks DNA and therefore cannot multiply.)

3 How does the completion of DNA replication initiate septum formation?

It is apparent from these and other questions that although transverse binary fission may be a primitive means of reproduction compared to that which occurs in eucaryotes, it is by no means a simple process; rather, it is the result of a precisely orchestrated series of interdependent events, many of which are not yet completely understood.

GROWTH

The most common means of bacterial reproduction is binary fission; one cell divides, producing two cells. Thus, if we start with a single bacterium, the increase in population is by geometric progression:

$$1 \rightarrow 2 \rightarrow 2^2 \rightarrow 2^3 \rightarrow 2^4 \rightarrow 2^5 \dots 2^n$$

where n = the number of generations. Each succeeding generation, assuming no cell death, doubles the population. The total population N at the end of a given time period would be expressed

$$N = 1 \times 2^n \tag{1}$$

However, under practical conditions, the number of bacteria N_0 inoculated at time zero is not 1 but more likely several thousand, so the formula now becomes

$$N = N_0 \times 2^n \tag{2}$$

Solving Eq. (2) for n, we have

$$\log_{10}N = \log_{10}N_0 + n \log_{10}2$$

$$n = \frac{\log_{10} N - \log_{10}N_0}{\log_{10}2} \tag{3}$$

If we now substitute the value of $\log_{10} 2$, which is 0.301, in the above equation, we can simplify the equation to

$$n = \frac{\log_{10}N - \log_{10}N_0}{0.301}$$

$$n = 3.3 (\log_{10}N - \log_{10}N_0) \tag{4}$$

Thus, by use of Eq. (4), we can calculate the number of generations that have taken place, providing we know the initial population and the population after growth has occurred.

Normal Growth Cycle (Growth Curve) of Bacteria

Assume that a single bacterium has been inoculated into a flask of liquid culture medium which is subsequently incubated. Eventually the bacterium will undergo binary fission and a period of rapid growth will ensue in which the cells multiply at an exponential rate. During this period of rapid growth, if we used the theoretical number of bacteria which should be present at various intervals of time and then plotted the data in two ways (logarithm of number of bacteria and arithmetic number of bacteria versus time), we would obtain the curve shown in Fig. 7-5. Here, the population increases regularly, doubling at regular

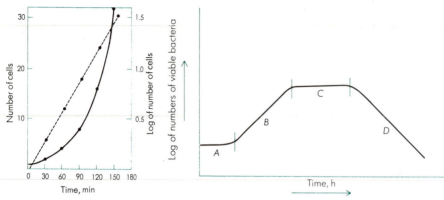

Figure 7-6. Typical bacterial growth curve. A, lag phase; B, log (logarithmic), or exponential, phase; C, stationary phase; D, death or decline phase.

Figure 7-5. Hypothetical bacterial growth curve, assuming that one bacterial cell is inoculated into a medium and divisions occur regularly at 30-min intervals (generation time). _ _ _ = logarithm of number of bacteria versus time; _____ = arithmetic number of bacteria versus time.

time intervals (the generation time) during incubation. However, exponential growth represents only one specific portion of the growth cycle of a population. In reality, when we inoculate a fresh medium with a given number of cells, determine the bacterial population intermittently during an incubation period of 24 h (more or less), and plot the logarithms of the number of cells versus time, we obtain a curve of the type illustrated in Fig. 7-6. From this it can be seen that there is an initial period of what appears to be no growth (the lag phase), followed by rapid growth (the exponential or logarithmic phase), then a leveling off (stationary phase), and finally a decline in the viable population (death or decline phase). Between each of these phases there is a transitional period (curved portion). This represents the time required before all cells enter the new phase. Let us examine what happens to the bacterial cells during each of the phases of the growth curve.

The Lag Phase

The addition of inoculum to a new medium is not followed immediately by a doubling of the population. Instead, the population remains temporarily unchanged, as illustrated in Fig. 7-6. But this does not mean that the cells are quiescent or dormant; on the contrary, during this stage the individual cells increase in size beyond their normal dimensions. Physiologically they are very active and are synthesizing new protoplasm. The bacteria in this new environment may be deficient in enzymes or coenzymes which must first be synthesized in amounts required for optimal operation of the chemical machinery of the cell. Time for adjustments in the physical environment around each cell may be required. The organisms are metabolizing, but there is a lag in cell division.

At the end of the lag phase, each organism divides. However, since not all organisms complete the lag period simultaneously, there is a gradual increase in the population until the end of this period, when all cells are capable of dividing at regular intervals.

The Logarithmic or Exponential Phase

During this period the cells divide steadily at a constant rate, and the log of the number of cells plotted against time results in a straight line (Figs. 7-5 and 7-6). Moreover, the population is most nearly uniform in terms of chemical composition of cells, metabolic activity, and other physiological characteristics.

Table 7-1. Generation Times of Several Species of Bacteria

Bacterium	Medium	Temperature, °C	Generation Time, min
Escherichia coli	Milk	37	12.5
	Broth	37	17
Bacillus thermophilus	Broth	55	18.3
Streptococcus lactis	Milk	37	26
	Lactose broth	37	48
Staphylococcus aureus	Broth	37	27–30
Bacillus mycoides	Broth	37	28
Lactobacillus acidophilus	Milk	37	66–87
Bradyrhizobium japonicum	Mineral salts + yeast extract + mannitol	25	344–461
Mycobacterium tuberculosis	Synthetic	37	792–932
Treponema pallidum	Rabbit testes	37	1,980

SOURCE: W. B. Spector (ed.): *Handbook of Biological Data*, table 75, Saunders, Philadelphia, 1956.

The generation time g (the time required for the population to double) can be determined from the number of generations n that occur in a particular time interval t. Using Eq. (4) for n, the generation time can be calculated by the following formula:

$$g = \frac{t}{n} = \frac{t}{3.3\,(\log_{10}N - \log_{10}N_0)} \tag{5}$$

Not all bacteria have the same generation time; for some, such as *E. coli*, it may be 15 to 20 minutes; for others it may be many hours (see Table 7-1). Similarly, the generation time is not the same for a particular species under all conditions. It is strongly dependent upon the nutrients in the medium and on prevailing physical conditions, such as those outlined in Chap. 6.

During exponential growth, the **growth rate** (i.e., the number of generations per hour), termed R, is the reciprocal of the generation time g. It is also the slope of the straight line obtained when the log number of cells is plotted against time:

$$R = \frac{3.3(\log_{10} N - \log_{10}N_0)}{t} \tag{6}$$

You may ask how this growth rate can remain constant during the logarithmic phase of growth even though the concentration of substrate (i.e., some essential nutrient in the culture medium, usually the carbon and energy source) is continually decreasing through utilization by the organisms. To understand this, one must recognize that the relationship between R and substrate concentration is not a simple linear relationship, as shown in Fig. 7-7. When the substrate concentration is high, a change in the concentration has very little effect on the growth rate. It is only when the substrate concentration becomes quite low that the growth rate begins to decrease significantly. Since bacteria are commonly "overfed" in laboratory culture, (i.e., are supplied with far greater substrate concentrations then they need), they can multiply at a constant exponential rate for many generations before the substrate level becomes low enough to affect this rate.

Figure 7-7. The effect of nutrient (substrate) concentration upon the growth rate of a bacterial culture. The level of substrate commonly provided in a bacterial culture is sufficiently high (right portion of curve) so that, even though the bacteria use up some substrate during the log phase of growth, the growth rate does not decrease appreciably. It is only when substrate levels become very low (left portion of curve) that the growth rate begins to be severely affected.

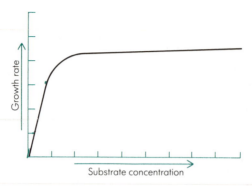

A microbiologist must be able to calculate growth rates and generation times. For example, it is often essential to predict how long it will take a certain population to grow to a given level. An appreciation of the full meaning of the normal growth curve is also necessary; it must be understood that during some phases of growth the cells are young and actively metabolizing while during others they are dying, so that there may be enormous structural and physiological differences between cells harvested at different times. Physical conditions and chemical substances may also affect organisms differently during different phases. Because, in general, cells in the logarithmic phase of growth are the most uniform and are in a more clearly defined condition than in any other phase, log-phase cultures are commonly used for studies of microbial metabolism.

The Stationary Phase

The logarithmic phase of growth begins to taper off after several hours, again in a gradual fashion represented by the transition from a straight line through a curve to another straight line, the stationary phase, as shown in Fig. 7-6. This trend toward cessation of growth can be attributed to a variety of circumstances, particularly the exhaustion of some nutrients, and, less often, the production of toxic products during growth. The population remains constant for a time, perhaps as a result of complete cessation of division or perhaps because the reproduction rate is balanced by an equivalent death rate.

The Phase of Decline or Death

Following the stationary phase the bacteria may die faster than new cells are produced, if indeed some cells are still reproducing. Undoubtedly a variety of conditions contribute to bacterial death, but the most important are the depletion of essential nutrients and the accumulation of inhibitory products, such as acids. During the death phase, the number of viable cells decreases exponentially, essentially the inverse of growth during the log phase. Bacteria die at different rates, just as they grow at different rates. Some species of Gram-negative cocci die very rapidly, so that there may be very few viable cells left in a culture after 72 h or less. Other species die so slowly that viable cells may persist for months or even years.

Transitional Periods Between Growth Phases

Note that a culture proceeds gradually from one phase of growth to the next (Fig. 7-6). This means that not all the cells are in an identical physiological condition toward the end of a given phase of growth. Time is required for some to catch up with others.

Synchronous Growth

There are many aspects of microbiological research where it would be desirable to relate the various aspects of bacterial growth, organization, and differentiation to a particular stage of the cell division cycle. It is not feasible to analyze a single bacterium because of its small size. However, if all the cells in a culture were to be in the same stage of the division cycle, the result from analysis of the cell crop could be interpreted as that for a single cell. There are several laboratory techniques by which we can manipulate the growth of cultures so that all the cells will be in the same stage of their growth cycle, i.e., growing **synchronously.**

The synchrony generally lasts only a few generations, since even the daughters of a single cell soon get out of phase with one another. A population can be synchronized by manipulating the physical environment or the chemical composition of the medium. For example, the cells may be inoculated into a medium at a suboptimal temperature; if they are kept in this condition for some time, they will metabolize slowly but will not divide. When the temperature is subsequently raised, the cells will undergo a synchronized division. The most common method of synchronization takes advantage of the fact that the smallest cells in a log-phase culture are those which have just divided. When these cells are separated out by filtration or by differential centrifugation, they are reasonably well synchronized with each other. Figure 7-8 shows the growth pattern of a population of synchronous cells.

Figure 7-8. Synchronous growth of bacteria. The steplike growth pattern indicates that all the cells of the population divide at about the same time.

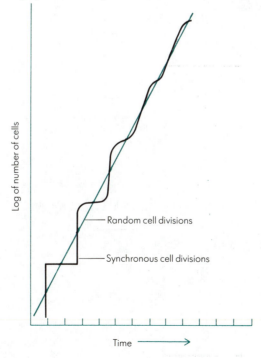

Random cell divisions

Synchronous cell divisions

Log of number of cells

Time ⟶

Continuous Culture

In both experimental research and in industrial processes, it is often desirable to maintain a bacterial population growing at a particular rate in the exponential or log phase. This condition is known as **steady-state** growth. The culture volume and the cell concentration are both kept constant by allowing fresh sterile medium to enter the culture vessel at the same rate that "spent" medium, containing cells, is removed from the growing culture (see Fig. 7-9). Under these conditions, the rate at which new cells are produced in the culture vessel is exactly balanced by the rate at which cells are being lost through the overflow from the culture vessel.

One type of system that is widely employed for continuous cultivation is the **chemostat.** This system depends on the fact that the concentration of an essential nutrient (substrate) within the culture vessel will control the growth rate of the cells. The concentration of substrate within the culture vessel is in turn controlled by the dilution rate, i.e., the rate at which fresh medium is being added to the culture (flow rate) divided by the volume of the culture vessel. Therefore, by adjusting the dilution rate we can control the growth rate. For example, suppose that the dilution rate is very low. The cells reach a high density because they are leaving the culture vessel at a very slow rate; moreover, they have time to use the substrate almost completely. Therefore, the substrate concentration is maintained at a low level within the vessel. This low substrate concentration permits the cells to grow at only a slow rate. On the other hand, if the dilution rate is high, the cell density is low because the cells are leaving the vessel at a

Figure 7-9. Apparatus for continuous cultivation of bacteria. The system can be regulated for continuous addition of fresh sterile medium to and removal of spent medium (and cells) from the culture vessel.

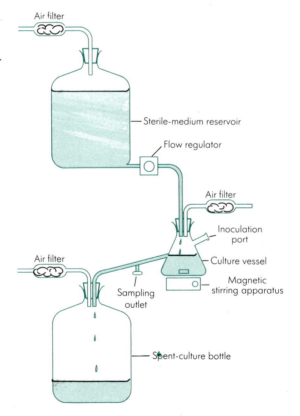

Air filter

Sterile-medium reservoir

Flow regulator

Air filter

Inoculation port

Air filter

Culture vessel

Magnetic stirring apparatus

Sampling outlet

Spent-culture bottle

high rate; moreover, they have little time to utilize the substrate that is entering the vessel, and therefore the substrate concentration is maintained at a high level within the vessel (but still less than that in the sterile-medium reservoir). This high concentration allows the cells to grow at a high rate. In each case the growth rate automatically adjusts to match the dilution rate. However, if the dilution rate is increased to the point where it exceeds the maximum growth rate of the cells, then washout occurs; that is, the cells cannot grow as fast as the rate at which the culture is being diluted by fresh medium, and they are soon eliminated from the culture vessel.

A second type of continuous culture apparatus is the turbidostat. Here a photoelectric device continuously monitors the cell density within the culture vessel and controls the dilution rate to maintain the cell density at a constant value. If the density becomes too high the dilution rate is increased; if the density becomes too low, the dilution rate is decreased.

QUANTITATIVE MEASUREMENT OF BACTERIAL GROWTH

We have seen that the term *growth* as commonly applied in microbiology refers to the magnitude of the total population. Growth in this sense can be determined by numerous techniques based on one or more of the following types of measurement:

1 *Cell count.* Directly by microscopy or by using an electronic particle counter, or indirectly by a colony count
2 *Cell mass.* Directly by weighing or by a measurement of cell nitrogen, or indirectly by turbidity
3 *Cell activity.* Indirectly by relating the degree of biochemical activity to the size of the population

Certain specific procedures will illustrate the application of each type of measurement.

Direct Microscopic Count

Bacteria can be counted easily and accurately with the Petroff-Hausser counting chamber. This is a special slide accurately ruled into squares that are $1/400$ mm^2 in area; a glass cover slip rests $1/50$ mm above the slide, so that the volume over a square is $1/20,000$ mm^3 ($1/20,000,000$ cm^3). A suspension of unstained bacteria can be counted in the chamber, using a phase-contrast microscope. If, for example, an average of five bacteria is present in each ruled square, there are $5 \times 20,000,000$, or 10^8, bacteria per milliliter. Direct microscopic counts can be made rapidly and simply with a minimum of equipment; moreover, the morphology of the bacteria can be observed as they are counted. Very dense suspensions can be counted if they are diluted appropriately; however, suspensions having low numbers of bacteria, e.g., at the beginning of a growth curve, cannot be counted accurately.

Electronic Enumeration of Cell Numbers

In this method, the bacterial suspension is placed inside an electronic particle counter, within which the bacteria are passed through a tiny orifice 10 to 30 μm in diameter. This orifice connects the two compartments of the counter which contain an electrically conductive solution. As each bacterium passes through the orifice, the electrical resistance between the two compartments increases momentarily. This generates an electrical signal which is automatically counted. Although this method is rapid, it requires sophisticated electronic equipment; moreover, the orifice tends to become clogged.

The main disadvantage of direct counting of cell numbers is that there is no way to determine whether the cells being counted are *viable*. To determine the viable count of a culture, we must use a technique that allows viable cells to multiply, such as the plate-count method or the membrane-filter method.

The Plate-Count Method

This method, illustrated in Fig. 7-10, allows determination of the number of cells that will multiply under certain defined conditions. A measured amount of the bacterial suspension is introduced into a Petri dish, after which the agar medium (maintained in liquid form at 45°C) is added and the two thoroughly mixed by rotating the plate. When the medium solidifies, the organisms are trapped in the gel. Each organism grows, reproducing itself until a visible mass of organisms—a colony—develops; i.e., one organism gives rise to one colony. Hence, a colony count performed on the plate reveals the viable microbial population of the inoculum. The original sample is usually diluted so that the number of colonies developing on the plate will fall in the range of 30 to 300. Within this range the count can be accurate, and the possibility of interference of the growth of one organism with that of another is minimized. Colonies are usually counted by illuminating them from below (dark-field illumination) so that they are easily visible, and a large magnifying lens is often used (see Fig. 7-11A). Various electronic techniques have been developed for the counting of colonies (Fig. 7-11B).

Figure 7-10. The plate-count technique, in which the sample is diluted quantitatively and measured amounts of the dilutions are cultured in Petri dishes.

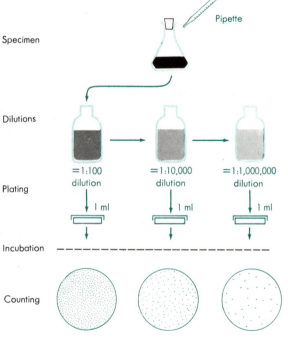

Specimen — Culture of bacteria or any other sample containing bacteria in suspension.

Dilutions — 1 ml transferred to 99 ml dilution blank; 1 ml transferred to 2d 99 ml dilution blank; 1 ml transferred to 3d 99 ml dilution blank.

=1:100 dilution   =1:10,000 dilution   =1:1,000,000 dilution

Plating — After addition of inoculum to plate, 15 to 20 ml of agar medium is poured into each plate. The plate is gently rotated for thorough distribution of inoculum throughout the medium.

Incubation — Plates are placed, inverted, in an incubator for 24 hr or longer.

Counting — A plate is selected which contains from 30 to 300 colonies.

Calculation of count — Number of colonies counted on plate × dilution of sample = number of bacteria per ml.

Figure 7-11. Bacterial colony counters. (A) Quebec colony counter. A Petri dish fits into the recess in the platform. The Petri dish is illuminated from beneath while the lens provides X1.5 magnification. *(Courtesy of American Optical Corporation.)* (B) An electronic colony counter. The Petri dish is placed on the illuminated stage, the count bar is depressed, and the precise number of colonies is instantly displayed on a digital readout. *(Courtesy of New Brunswick Scientific Company, Inc.)*

One limitation of the plate-count technique is that the only bacteria that will be counted are those which can grow on the medium used and under the conditions of incubation provided. This can be an important consideration if a mixture of bacteria is to be counted. Another limitation is that each viable organism that is capable of growing under the culture conditions provided may not necessarily give rise to one colony. The development of one colony from one cell can occur when the bacterial suspension is homogeneous and no aggregates of cells are present; however, if the cells have a tendency to aggregate, e.g., cocci in clusters (staphylococci), chains (streptococci), or pairs (diplococci), the resulting counts will be lower than the number of individual cells. For this reason the "counts" are often reported as **colony-forming units** per milliliter rather than number of bacteria per milliliter.

The plate-count technique is used routinely and with satisfactory results for the estimation of bacterial populations in milk, water, foods, and many other materials. It is easy to perform and can be adapted to the measurement of populations of any magnitude. It has the advantage of sensitivity, since very small numbers of organisms can be counted. Theoretically, if a specimen contains as few as one bacterium per milliliter, one colony should develop upon the plating of 1 ml.

Membrane-Filter Count

A very useful variation on the plate-count technique is based on the use of molecular or membrane filters. These filters have a known uniform porosity of predetermined size sufficiently small to trap microorganisms. This technique is particularly valuable in determining the number of bacteria in a large sample that has a very small number of viable cells; e.g., the bacteria in a large volume of air or water can be collected simply by filtering them through an assembly as illustrated in Fig. 7-12A. The membrane, with its trapped bacteria, is then placed in a special plate containing a pad saturated with the appropriate medium. Special media and dyes can be used to make it easier to detect certain types of organisms than with the conventional plate count. During incubation,

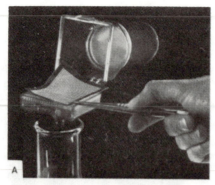
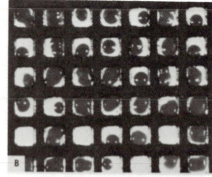

Figure 7-12. (A) Filtration apparatus for use with a membrane filter. After placing the filter on the support, the hinged upper part of the apparatus clamps it in place. A known volume of the bacteria-containing sample is then passed through the filter. (B) The filter is then incubated on a suitable culture medium. In the particular type of filter shown, a grid divides the filter into 1,600 small square compartments, and colony growth is restricted to these compartments. This greatly facilitates the counting of the colonies. *(Courtesy of New Brunswick Scientific Company, Inc.)*

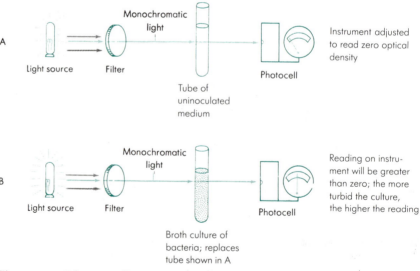

Figure 7-13. Schematic illustration of the use of a photoelectric colorimeter for measuring bacterial populations. The instrument measures optical density (also termed absorbance), a function of light intensity which is almost linearly proportional to cell mass. (A) Adjustment of instrument. A glass tube (cuvette) filled with uninoculated culture medium is used to set the instrument to give a basal optical density reading of 0. (B) The "blank" cuvette is replaced by a similar cuvette containing the broth culture (i.e., medium + cells), and the increase in optical density is recorded.

the organisms grow into colonies which appear on the membrane surface (see Fig. 7-12B).

Turbidimetric Methods

Anyone who has tried to see through a fog realizes that visibility is reduced in proportion to the density of the fog and the distance between the observer and the object that he or she is looking at. This is because each droplet of water in the fog absorbs and scatters the light passing through it, and the more droplets in the light path, the less one can see. Similarly, bacteria in a suspension absorb and scatter the light passing through them, so that a culture of more than 10^7 to 10^8 cells per milliliter appears turbid to the naked eye. A spectrophotometer or colorimeter can be used for turbidimetric measurements of cell mass (see Fig. 7-13). Turbidimetry is a simple, rapid method for following growth; however, the culture to be measured must be dense enough to register some turbidity on the instrument. Moreover, it may not be possible to measure cultures grown in deeply colored media or cultures that contain suspended material other than bacteria. It must also be recognized that dead as well as living cells contribute to turbidity.

Determination of Nitrogen Content

The major constituent of cell material is protein, and since nitrogen is a characteristic part of proteins, one can measure a bacterial population or cell crop in terms of bacterial nitrogen. Bacteria average approximately 14 percent nitrogen on a dry-weight basis, although this figure is subject to some variation introduced by changes in cultural conditions or differences between species. To measure growth by this technique, you must first harvest the cells and wash them free of medium and then perform a quantitative chemical analysis for nitrogen. Bacterial nitrogen determinations are somewhat laborious and can be performed only on specimens free of all other sources of nitrogen. Furthermore, the method is applicable only for concentrated populations. For these and other reasons, this procedure is used primarily in research.

Determination of the Dry Weight of Cells

This is the most direct approach for quantitative measurement of a mass of cells. However, it can be used only with very dense suspensions, and the cells must be washed free of all extraneous matter. Moreover, dry weight may not always be indicative of the amount of living material in cells. For example, the intracellular reserve material poly-β-hydroxybutyrate can accumulate in *Azotobacter beijerinckii* at the end of the log phase of growth and during the stationary phase and finally can comprise up to 74 percent of the dry weight of the cells; thus, the dry weight may continue to increase without corresponding cell growth. Yet, for many organisms the determination of dry weight is an accurate and reliable way to measure growth and is widely used in research.

Measurement of a Specific Chemical Change Produced on a Constituent of the Medium

As an example of this method of estimating cell mass, we may take a species that produces an organic acid from glucose fermentation. The assumption is that the amount of acid produced, under specified conditions and during a fixed period of time, is proportional to the magnitude of the bacterial population. Admittedly, the measurement of acid or any other end product is a very indirect approach to the measurement of growth and is applicable only in special circumstances.

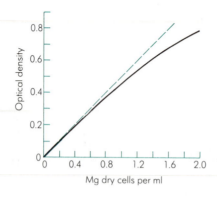

Figure 7-14. Turbidity of a culture serves as a convenient indirect measure of the dry weight of bacterial cells in the culture, as shown in this typical curve (———); however, some deviation from a theoretical linear relationship (_ _ _ _) does occur, particularly at optical densities greater than 0.4.

The Relation of Turbidity Measurements to Direct Expressions of Growth

It is frequently desirable to relate measurements of growth made by an indirect method, e.g., turbidity, to a direct measurement, e.g., dry weight of the cell crop. This can be done conveniently by measuring the bacterial suspension simultaneously by the two methods and establishing a relationship between the values obtained, as in the following example. Samples are removed from a cell suspension and dried under predetermined conditions, and the weight of cells per milliliter is determined. From the same cell suspension, dilutions are prepared and turbidity measurements are made. We can calculate the weight of the bacteria in each dilution, since the weight of cells per milliliter of the original sample was determined. Two sets of data will be obtained, which can then be plotted (cell weight against turbidity), as illustrated in Fig. 7-14, to obtain a standard curve. For practical purposes, and within a certain range of concentrations, a nearly linear relationship exists. When the standard curve has been

Table 7-2. Summary of Methods for Measuring Bacterial Growth

Method	Some Applications	Manner in Which Growth Is Expressed
Microscopic count	Enumeration of bacteria in vaccines and cultures	Number of cells per ml
Electronic enumeration	Same as for microscopic count	Same as for microscopic count
Plate count	Enumeration of bacteria in milk, water, foods, soil, cultures, etc.	Colony-forming units per ml
Membrane filter	Same as plate count	Same as plate count
Turbidimetric measurement	Microbiological assay, estimation of cell crop in broth, cultures, or aqueous suspensions	Optical density (absorbance)
Nitrogen determination	Measurement of cell crop from heavy culture suspensions to be used for research in metabolism	Mg nitrogen per ml
Dry weight determination	Same as for nitrogen determination	Mg dry weight of cells per ml
Measurements of biochemical activity, e.g., acid production by cultures	Microbiological assays	Milliequivalents of acid per ml or per culture

established, we can measure the turbidity of a bacterial suspension and convert this value to bacterial weight. Similarly, we can prepare a standard curve correlating other direct measurements (numbers of bacteria or bacterial nitrogen) with turbidity. Thus it is possible to use the convenient indirect measurement (turbidity) and convert the value to a direct expression of growth.

The Selection of a Procedure to Measure Growth

Table 7-2 summarizes the methods described above for measuring bacterial growth. Each has its particular advantages and limitations, and no one method can be recommended universally. The best procedure for your work can be selected only after these factors are considered in relation to the problem at hand. The colony count is the most widely used procedure for general microbiological work, and complete familiarity with this technique, both in principle and practice, is essential. It should be emphasized that the colony count is theoretically the only technique that reflects the viable population. Furthermore, it is not at all unlikely that discrepancies may occur in results of growth of a bacterial population when measured by two different methods. For example, a microscopic count of a culture in the stationary phase would include all cells, viable and nonviable, whereas the colony count would reveal only the viable population.

Importance of Quantitative Measurement of Growth

Before we can evaluate or interpret growth responses of bacteria in different media or under various conditions, growth must be expressed in quantitative terms. In microbiology the term *growth* is used in several ways. For example, we may judge a certain set of conditions as being good because the bacteria grow rapidly, but the final total cell crop may not be as large as under another set of conditions where the growth proceeds at a slower rate but continues to increase over a longer time period. Such a situation is shown schematically in Fig. 7-15, where the growth of the same bacterial species is compared in two different media. If we measured growth at time *A*, we should conclude that growth is best in medium II; measured at time *B*, growth would be equally good in both media; and at time *C*, growth would be better in medium I. If we were primarily interested in a large cell crop, we should select medium I. In any

Figure 7-15. Quantitative measurement of growth is significant for interpretation of various growth responses. Hypothetical growth response of same bacterium in media of two different compositions. Compare the cell crops, or amount of growth, at times *A*, *B*, and *C*.

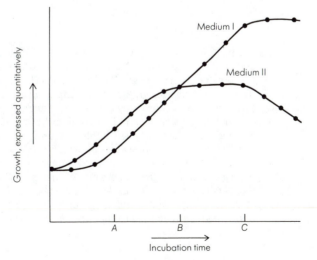

event, we must have knowledge of growth in quantitative terms to make the correct choice.

1 How does the term *growth* as used in microbiology differ from the same term as applied to higher plants and animals?

2 Describe the differences between the various modes of cell division in bacteria.

3 How is it that the septum formation in *Streptococcus faecalis* accounts for all the new cell wall material of newly formed daughter cells, yet in *Bacillus subtilis* it accounts for only 15 percent?

4 In a culture of bacteria that is actively multiplying by transverse binary fission, are there any "old" cells? Explain.

5 During log-phase growth of a bacterial culture, a sample is taken at 8:00 a.m. and found to contain 1,000 cells per milliliter. A second sample is taken at 5:54 p.m. and is found to contain 1,000,000 cells per milliliter. What is the generation time in hours?

6 Would you expect generation time to be a constant characteristic of a bacterial species? Explain.

7 In the lag phase of growth the number of bacteria remains constant. Does this mean the cells are dormant and inert? Explain.

8 Draw a typical bacterial growth curve and label the various phases. Discuss those factors which determine the beginning and end of each phase.

9 When the appropriate data are plotted for the log-phase growth of a bacterial culture, how is it that we obtain a steady increase in the number of cells rather than a series of stepwise increases? Explain.

10 How can synchronous growth of a bacterial culture be obtained? In what way could a synchronously growing culture be useful for the electron microscopist who is trying to determine the cytological changes associated with bacterial growth?

11 What is steady-state growth and what advantages does it offer? Describe how steady-state growth at a certain rate could be obtained by use of a chemostat.

12 Compare the direct and indirect methods for estimating bacterial populations on the basis of (a) practical applications, (b) advantages and (c) limitations of use.

REFERENCES

Gerhardt, P., R. G. E. Murray, R. N. Costilow, E. W. Nester, W. A. Wood, N. R. Krieg, and G. B. Phillips (eds.): *Manual of Methods for General Bacteriology*, American Society for Microbiology, Washington, D.C., 1981. *Section II of this book is devoted to bacterial growth, including measurement methods, continuous culture methods, synchronous cultures, and many other aspects.*

Ingraham, J. L., O. Maaløe, and F. C. Neidhardt: *Growth of the Bacterial Cell*, Sinauer, Sunderland, Mass., 1983. *Chapters 5 and 6 discuss the growth of bacterial populations and the processes by which it is regulated.*

Lamanna, C., M. F. Mallette, and L. N. Zimmerman: *Basic Bacteriology. Its Biological and Chemical Background*, 4th ed., Williams & Wilkins, Baltimore, 1973. *Chapter 7 of this text covers the fundamental aspects of bacterial growth.*

Moat, A. G.: *Microbial Physiology*, John Wiley & Sons, New York, 1979. *Chapter 9 of this book provides a succinct discussion of the principles of bacterial growth.*

Chapter 8   Pure Cultures and Cultural Characteristics

In natural environments a single kind of bacterium, i.e., a bacterial species, usually occurs as only one component of a large and complex population containing many other species. To study the characteristics of one species, that species must be separated from all the other species, i.e., it must be **isolated** in pure culture. However, before attempting isolation, it is often helpful to use a **selective method** first. Such a method can increase the relative proportion of the desired species in the population so that it can be more easily isolated. Once obtained, a pure culture can be maintained or preserved in a **culture collection.** Different species of bacteria growing on the same kind of medium may appear quite different; thus knowledge of the appearance, or the **cultural characteristics,** of a species is useful for the recognition of certain kinds of bacteria and may also serve as an aid to the identification of species. In this chapter we shall describe methods for selection, isolation, and preservation of bacteria, as well as their cultural characteristics on various media.

NATURAL MICROBIAL POPULATIONS (MIXED CULTURES)

The microbial population in our environment is large and complex. Many different microbial species normally inhabit various parts of our bodies, such as the oral cavity, the intestinal tract, and the skin. These microbes may be present in extremely large numbers. For example, a single sneeze may disperse from 10,000 to 100,000 bacteria. One gram of feces may contain 10^{11} bacteria. Our environment—air, soil, water—likewise consists of mixed populations of

133

bacteria plus other microbes. In fertile garden soil, microorganisms may number several billions per gram and include many species of bacteria, fungi, algae, and protozoa. A study of the microorganisms in these habitats requires knowledge of the specific microbes present. This, in turn, requires unraveling the complex mixed population into pure cultures of separate, distinct species.

SELECTIVE METHODS

A particular bacterial species is often present in small numbers compared to the total population of a mixed culture. Moreover, the species may be one that grows less rapidly on ordinary culture media than other species. In order to achieve its isolation into pure culture, it is helpful—and often necessary—to first achieve an increase in the relative number of the species, preferably to the point where the species becomes the numerically dominant component of the population. This can be accomplished by the use of selective methods. These methods favor the growth of the desired species while discouraging, or even killing, the other organisms present in the mixed culture. Chemical, physical, or biological methods are used in order to achieve selection of a particular kind of bacterium.

Chemical Methods of Selection

Use of a Special Carbon or Nitrogen Source

One type of chemical method is to provide in the culture medium a substrate, i.e., a single carbon or nitrogen source, that can be used only by the species being sought (Fig. 8-1). This particular kind of selection is often referred to by a special name, **enrichment.** For example, if we wish to isolate, from soil, bacteria capable of utilizing a very complex organic compound like α-conidendrin, a constituent of wood, we find that when we inoculate a medium such as nutrient agar directly with the soil sample, our chances of finding α-conidendrin–utilizing bacteria will be very limited. There are so many other rapidly growing bacteria present that the more slowly growing kind we wish to obtain will be soon overgrown. Consequently, we prepare a liquid **enrichment medium** in which α-conidendrin is the sole source of carbon. Under these conditions, only organisms capable of utilizing this compound will be able to grow well. However, it is important to recognize that other bacteria may be able to grow to some extent by utilizing organic compounds made by the conidendrin-utilizing

Soil

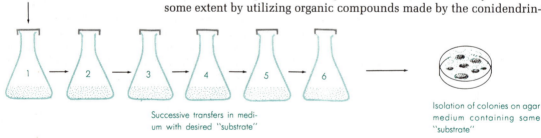

Successive transfers in medium with desired "substrate"

Isolation of colonies on agar medium containing same "substrate"

Isolation of pure cultures and verification of ability of isolates to utilize "substrate"

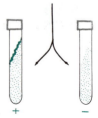

Figure 8-1. Enrichment method for bacteria capable of utilizing a particular substrate.

organisms and that the method is not completely specific. As another example, if we wish to select for nitrogen-fixing bacteria, nitrogen gas (N_2) can be supplied as the sole nitrogen source, since only nitrogen-fixing bacteria will be able to grow well. Other bacteria may grow, but to a lesser degree, by using the nitrogenous products made by the nitrogen-fixers.

Use of Dilute Media

Certain aquatic bacteria, such as *Caulobacter* species, are capable of growing with very low levels of carbon or nitrogen sources. Consequently, one way to select for such bacteria is to inoculate a mixed culture into a very dilute medium, e.g., a broth containing only 0.01 percent peptone. The medium must have low enough levels of nutrients that other kinds of organisms will not be able to grow well in it.

Use of Inhibitory or Toxic Chemicals

The addition of low levels of certain chemicals, such as dyes, bile salts, salts of heavy metals, or antibiotics, to culture media can be useful for the selection of certain kinds of bacteria. The following are examples of this type of selection:

1 Many Gram-negative bacteria can grow in the presence of low concentrations of various dyes that inhibit the growth of Gram-positive bacteria. Similarly, intestinal bacteria can grow in the presence of bile salts such as sodium deoxycholate, whereas nonintestinal bacteria are usually inhibited. Consequently, a medium containing crystal violet dye plus sodium deoxycholate will allow Gram-negative intestinal bacteria to grow but will inhibit most other kinds of bacteria. An example of such a medium is MacConkey agar, which is widely used to select for Gram-negative intestinal pathogens such as *Salmonella* and *Shigella* species.
2 *Campylobacter jejuni* is a frequent cause of diarrhea in humans, yet diarrheic stool samples contain many other kinds of bacteria that interfere with the isolation of this species. By incorporating certain antibiotics or other chemotherapeutic agents, such as vancomycin, polymyxin, and trimethoprim, into the culture medium, most of these contaminants can be inhibited without affecting the growth of *C. jejuni*.

Physical Methods of Selection

Heat Treatment

To select for endospore-forming bacteria, a mixed culture can be heated to 80°C for 10 min before being used to inoculate culture media. Vegetative cells will be killed by this treatment but endospores will survive and subsequently germinate and grow.

Incubation Temperature

To select for psychrophilic or psychrotrophic bacteria, cultures are incubated at low temperatures, e.g., 0 to 5°C. For selection of thermophiles, a high incubation temperature is used, e.g., 55°C.

pH of the Medium

To select for acid-tolerant bacteria, a low-pH medium can be used. For example, to select for the lactobacilli present in cheddar cheese, the pH of the medium is maintained at 5.35 with an acetic acid/acetate buffer; other organisms in the cheese cannot grow well at such a low pH. Similarly, to select for alkali-tolerant organisms, a high-pH medium can be used. For example, to select for the cholera-causing bacterium, *Vibrio cholerae*, from a stool sample, we can use a medium with a pH of 8.5; most other intestinal bacteria are unable to grow at this pH.

Cell Size and Motility

We can sometimes make use of a small cell diameter or of bacterial motility to achieve selection. For instance, *Treponema* species from the human oral cavity can be selected by taking advantage of both of these properties. A membrane filter having a pore size of 0.15 μm is placed on the surface of an agar plate and gingival scrapings are placed on the filter. The unusually small size of treponemes allows them to penetrate the pores of the filter to reach the underlying agar. Moreover, treponemes have the ability to swim through solid agar media; consequently, they migrate away from the filter and grow to form a hazy zone within the agar, from which they can be subcultured. Other bacteria from the oral cavity are either too large to penetrate the membrane filter or, if they can penetrate it, are unable to migrate away through the agar.

Biological Methods of Selection

A disease-producing species occurring in a mixed culture can often be selected by taking advantage of its pathogenic properties. For example, a sputum sample containing *Streptococcus pneumoniae* is ordinarily contaminated by many other bacterial species. However, laboratory mice are extremely susceptible to infection by *S. pneumoniae*, and if the sputum sample is injected into a mouse the pathogen will multiply extensively. Nonpathogenic bacteria present in the sample will either be inhibited or killed by the defense mechanisms of the animal. In a sense, the animal serves as the selective medium.

How can the microbiologist know what selective media or conditions to use for a given species? Many selective methods are given in the references at the end of this chapter. Moreover, you can often devise a satisfactory selective procedure by comparing the characteristics of the species sought with those of the accompanying contaminants. Differences in these characteristics, e.g., in susceptibility to certain antibiotics, can provide the basis for a suitable selective procedure.

Selection in Nature

It is important to realize that the principle of selection is not limited to the laboratory; it also commonly operates in nature. For instance, the occurrence of high salt concentrations in bodies of water such as the Dead Sea selects for extreme halophiles such as those of the genus *Halobacterium*. In lakes, the anaerobic, sulfide-containing zone on or above the sediment mud provides conditions that often favor the mass development of green or purple sulfide-oxidizing, phototrophic bacteria. The nodules that occur on the roots of leguminous plants contain bacteria of the genus *Rhizobium*, which are uniquely suited for nitrogen fixation in association with these plants. In many types of natural infections of humans or animals, a single, uncontaminated pathogenic bacterial species can often be obtained from a blood sample; blood from a healthy animal or human is normally free of bacteria. Numerous other examples of selective conditions in nature exist.

PURE CULTURES

If the bacterial species being sought comprises a suitably high proportion of the mixed population, it can be isolated in pure culture. The descendents of a single isolation in pure culture comprise a **strain.** A strain is usually made up of a succession of cultures and is often derived from a single colony; however, the number of bacteria which gave rise to the original colony is usually unknown. If a strain is derived from a single parent cell, it is termed a **clone.** Each strain is designated by an identifying number and its history is recorded (the source from which the isolation was made, the name of the person who made the isolation, the date of the isolation, and the culture collection in which the strain is maintained and from which it can be obtained for study).

A variety of techniques have been developed whereby isolation into pure culture can be accomplished. Each technique has certain advantages and limitations, and there is no one method that can be used for all bacteria.

Methods of Isolating Pure Cultures

The Streak-Plate Technique

By means of a transfer loop, a portion of the mixed culture is placed on the surface of an agar medium and streaked across the surface. This manipulation "thins out" the bacteria on the agar surface so that some individual bacteria are separated from each other. Figure 8-2 illustrates a nutrient-agar plate culture that has been streaked to provide isolated colonies. When streaking is properly performed, the bacterial cells will be sufficiently far apart in some areas of the plate to ensure that the colony developing from one cell will not merge with that growing from another. Figure 8-3 illustrates a modification known as the **roll-tube technique** that is used for the isolation of stringent anaerobes.

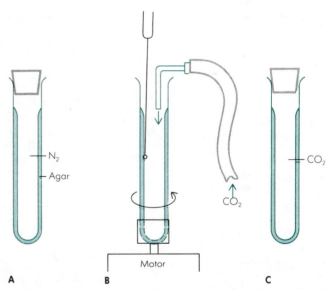

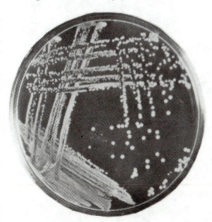

Figure 8-2. Streak-plate culture showing areas of isolated colonial growth. Note that where the colonies are sparse they are larger than when crowded together. *(Courtesy of Naval Biological Laboratory.)*

Figure 8-3. Roll-tube method for isolating stringent anaerobes. (A) Stoppered anaerobic culture tube whose inner walls have been coated with a prereduced agar medium. The tube contains an atmosphere of oxygen-free N_2. (B) When the stopper is removed the tube is kept anaerobic by continuously flushing it with oxygen-free CO_2 from a gas cannula. Inoculation is done with a transfer loop held against the agar surface as the tube is being rotated by a motor. By starting at the bottom and drawing the loop gradually upward, the inoculum becomes "thinned" to the point where well-isolated colonies can develop. (C) After inoculation the tube is restoppered and incubated.

The assumption is often made that a colony is derived from a single cell and, therefore, that the colony is a clone. However, this is not necessarily true. With species in which the cells form a characteristic grouping during cell division (for example, clumps of staphylococci or chains of streptococci), the colony may develop from a group of cells rather than from a single cell. Although not a clone, such a colony is nevertheless a pure culture if it contains only one kind of organism.

One should recognize that subculturing a colony from a single streak plate does not automatically assure purity. The colony may have been derived from two or more different kinds of bacteria. For example, when we attempt to isolate slime- or chain-producing bacteria, contaminants may be found to have adhered to the slime or to have been enmeshed in the network of chains, thereby resulting in impure colonies. The use of selective media can also lead to impure colonies. Although the growth of contaminants is inhibited on selective media, low numbers of viable cells may still be present, and such cells can be subcultured along with a colony. For these reasons, it is advisable to streak a culture several times in succession, preferably on nonselective media, in order to ensure purity.

The Pour-Plate and Spread-Plate Techniques

In both of these methods the mixed culture is first diluted to provide only a few cells per milliliter before being used to inoculate media. Since the number of bacteria in the specimen is not known beforehand, a *series of dilutions* is made so that at least one of the dilutions will contain a suitably sparse concentration of cells.

In the **pour-plate method** the mixed culture is diluted directly in tubes of liquid (cooled) agar medium (see Fig. 8-4). The medium is maintained in a liquid state at a temperature of 45°C to allow thorough distribution of the inoculum. The inoculated medium is dispensed into Petri dishes, allowed to solidify, and then incubated. A series of agar plates showing decreasing numbers of colonies resulting from the dilution procedure in the pour-plate technique is shown in Fig. 8-4. The pour-plate technique has certain disadvantages. For

Figure 8-4. Pour-plate technique is used for isolation of pure cultures of bacteria. Step 1: One loopful of original suspension is transferred to tube A (liquid, cooled agar medium). Tube A is rolled between the hands to effect thorough mixing of inoculum. Similar transfers are made from A to B to C. Step 2: Contents of each tube are poured into separate Petri dishes. Step 3: After incubation, plates are examined for the one which contains well-separated colonies. From this plate, pure cultures of bacteria can be isolated by transferring a portion of a colony to a tube of sterile medium.

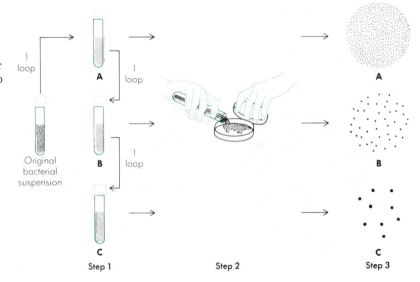

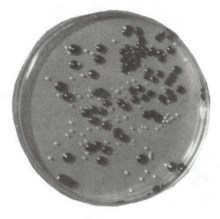

Figure 8-5. Spread plate showing colonies of two different bacterial species. A dilution of the mixed culture was spread over the surface with a glass rod. The large, dark colonies are *Serratia marcescens*, which has a brick-red pigment, and the smaller, light colonies are *Micrococcus luteus*, which has a lemon-yellow pigment. *(Courtesy of Naval Biological Laboratory.)*

instance, some of the organisms are trapped beneath the surface of the medium when it gels, and therefore both *surface* and *subsurface* colonies develop. The subsurface colonies can be transferred to fresh media only by first digging them out of the agar with a sterile instrument. Another disadvantage is that the organisms being isolated must be able to withstand temporary exposure to the 45°C temperature of the liquid agar medium; for instance, the pour-plate method would be unsuitable for isolating psychrophilic bacteria.

In the **spread-plate method** the mixed culture is not diluted in the culture medium; instead it is diluted in a series of tubes containing a sterile liquid, usually water or physiological saline. A sample is removed from each tube, placed onto the surface of an agar plate, and spread evenly over the surface by means of a sterile, bent glass rod. On at least one plate of the series the bacteria will be in numbers sufficiently low as to allow the development of well-separated colonies (see Fig. 8-5). In contrast to the pour-plate technique, only surface colonies develop; moreover, the organisms are not required to withstand the temperature of liquid agar.

Unlike the streak-plate technique, the pour-plate and the spread-plate techniques may be performed in a *quantitative* manner to determine the number of bacteria (of a particular type) present in a specimen (see Chap. 7).

Micromanipulator Techniques

A device called the **micromanipulator** can be used in conjunction with a microscope to pick a single bacterial cell from a mixed culture. The micromanipulator permits the operator to control the movements of a **micropipette** or a **microprobe** (a fine needle) so that a single cell can be isolated (see Fig. 8-6). This technique requires a skilled operator and is reserved for studies in which a clone must be obtained unequivocally.

MAINTENANCE AND PRESERVATION OF PURE CULTURES

Most microbiology laboratories maintain a large collection of strains, frequently referred to as a **stock-culture collection.** These organisms are needed for laboratory classes and research work, as test agents for particular procedures, or as reference strains for taxonomic studies. Most major biological companies maintain large culture collections. The strains are used for screening of new, potentially effective chemotherapeutic agents; as assay tools for vitamins and amino

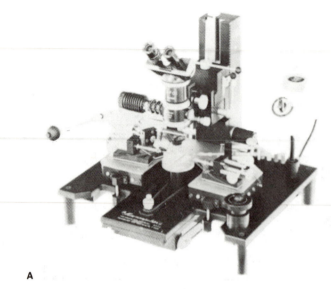

A

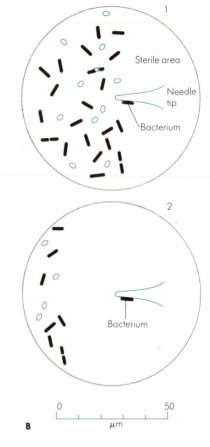

B

Figure 8-6. Isolation of single bacterial cells. (A) Micromanipulator equipment and microscope. The micromanipulator is equipped with probes that can position on objects a few μm in size. The manipulation of the probes is done while viewing the specimen through the microscope. *(Micromanipulator Company.)* (B) Schematic illustration showing isolation of a single bacterium from a mixture of cells on an agar gel. The microscope field (1) shows the point of the microprobe touching the agar gel near a bacterial cell, causing the organism to float in the small amount of water exuded from the gel. The microscope stage is then moved to the left (2) while the microprobe remains fixed; this causes the bacterium to follow the probe and become separated from the other cells. The agar block is subsequently dissected and the small piece containing the isolated bacterium is transferred to a sterile culture medium. *(Courtesy of K. I. Johnstone, Manipulation of Bacteria, Churchill Livingstone, Edinburgh, 1973.)*

acids; as agents for the production of vaccines, antisera, antitumor agents, enzymes, and organic chemicals; and as reference cultures that are cited in company patents. For these and other purposes it is extremely important to have properly identified and cataloged strains of bacteria available. Consequently, a considerable amount of research has been performed to develop methods whereby bacterial strains can be preserved and stored until they are needed. Several different methods have been developed, since not all bacteria respond in a similar manner to a specific method. Moreover, there are various practical considerations such as the amount of labor involved and the amount of storage space required. However, all the methods which we will now describe have the same objective: to maintain strains alive and uncontaminated and to prevent any change in their characteristics.

Methods of Maintenance and Preservation

Strains can be maintained by periodically preparing a fresh stock culture from the previous stock culture. The culture medium, the storage temperature, and the time interval at which the transfers are made vary with the species and must

Periodic Transfer to Fresh Media

be ascertained beforehand. The temperature and the type of medium chosen should support a slow rather than a rapid rate of growth so that the time interval between transfers can be as long as possible. Many of the more common heterotrophs remain viable for several weeks or months on a medium like nutrient agar. The transfer method has the disadvantage of failing to prevent changes in the characteristics of a strain due to the development of variants and mutants.

Preservation by Overlaying Cultures with Mineral Oil

Many bacteria can be successfully preserved by covering the growth on an agar slant with sterile mineral oil. The oil must cover the slant completely; to ensure this, the oil should be about ½ in above the tip of the slanted surface. Maintenance of viability under this treatment varies with the species (1 month to 2 years). This method of maintenance has the unique advantage that you can remove some of the growth under the oil with a transfer needle, inoculate a fresh medium, and still preserve the original culture. The simplicity of the method makes it attractive, but changes in the characteristics of a strain can still occur. Figure 8-7 illustrates a culture collection maintained by this technique.

Preservation by Lyophilization (Freeze-Drying)

Most bacteria die if cultures are allowed to become dry, although spore- and cyst-formers can remain viable for many years. However, freeze-drying can satisfactorily preserve many kinds of bacteria that would be killed by ordinary

Figure 8-7. A culture collection maintained by overlaying cultures with mineral oil. *(Courtesy of U.S. Department of Agriculture.)*

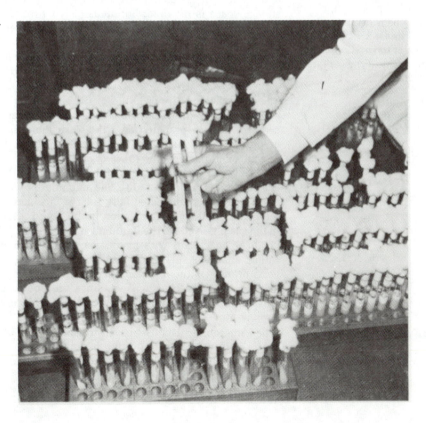

Figure 8-8. Lyophilization process for preservation of cultures. (A) A simple apparatus for lyophilization. Small cotton-plugged vials containing frozen suspensions of bacteria are placed in the glass flask, which is attached to a condenser. The condenser is connected to a high-vacuum pump. The bacteria become desiccated as the ice in the frozen suspension sublimes directly to water vapor. The vapor is trapped on the cold surface of the condenser, thereby preventing it from entering the vacuum line and contaminating the pump oil. (B) After desiccation of the cultures as in (A), the vials are removed and each is placed in a larger tube. After insulating the vial with a plug of glass wool packing, the outer tube is hermetically sealed under a vacuum by means of torch. *(Courtesy of American Type Culture Collection.)*

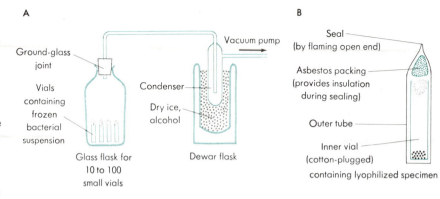

A

Ground-glass joint

Vials containing frozen bacterial suspension

Glass flask for 10 to 100 small vials

Vacuum pump

Condenser

Dry ice, alcohol

Dewar flask

B

Seal (by flaming open end)

Asbestos packing (provides insulation during sealing)

Outer tube

Inner vial (cotton-plugged)

containing lyophilized specimen

Storage at Low Temperatures

drying. In this process a dense cell suspension is placed in small vials and frozen at -60 to $-78°C$. The vials are then connected to a high-vacuum line. The ice present in the frozen suspension sublimes under the vacuum, i.e., evaporates without first going through a liquid water phase. This results in dehydration of the bacteria with a minimum of damage to delicate cell structures. The vials are then sealed off under a vacuum and stored in a refrigerator. One arrangement of equipment employed to lyophilize cultures is shown in Fig. 8-8. Many species of bacteria preserved by this method have remained viable and unchanged in their characteristics for more than 30 years. Only minimal storage space is required; hundreds of lyophilized cultures can be stored in a small area. Furthermore, the small vials can be sent conveniently through the mail to other microbiology laboratories when packaged in special sealed mailing containers. Lyophilized cultures are revived by opening the vials, adding liquid medium, and transferring the rehydrated culture to a suitable growth medium.

The ready availability of liquid nitrogen has provided the microbiologist with another very useful means for long-term preservation of cultures. In this procedure cells are prepared as a dense suspension in a medium containing a cryoprotective agent such as glycerol or dimethyl sulfoxide (DMSO), which prevents cell damage due to ice crystal formation during the subsequent steps. The cell suspension is sealed into small ampoules or vials and then frozen at a controlled rate to $-150°C$. The ampoules or vials are then stored in a liquid nitrogen refrigerator (essentially a large tank having vacuum-insulated walls; see Fig. 8-9) either by immersion in the liquid nitrogen ($-196°C$) or by storage in the gas phase above the liquid nitrogen ($-150°C$). The liquid nitrogen method has been successful with many species that cannot be preserved by lyophilization, and most species can remain viable under these conditions for 10 to 30 years or more without undergoing change in their characteristics. However, the method is relatively expensive, since the liquid nitrogen in the refrigerators must be replenished at regular intervals to replace the loss due to evaporation.

Culture Collections

When microbiologists first began to isolate pure cultures, each microbiologist kept a personal collection of those strains having special interest. Subcultures of some strains were often sent to other microbiologists; other subcultures were received and added to the scientist's own collection. Certain strains had taxonomic importance because they formed the basis for descriptions of species and genera. Others had special properties useful for various purposes. However, many important strains became lost or were inadequately maintained. Thus, it became imperative to establish large **central collections** whose main purpose would be the acquisition, preservation, and distribution of authentic cultures of living microorganisms.

Many countries have at least one central collection. As examples, in France a collection of bacteria is maintained at the Institut Pasteur in Paris; in England the National Collection of Type Cultures is in London; the Federal Republic of Germany maintains the Deutsche Sammlung von Mikroorganismen in Darmstadt; and Japan maintains a large collection at the Institute for Fermentation in Osaka. Many other such collections exist.

In the United States the major collection is the American Type Culture Collection (ATCC), located in Rockville, Maryland. In 1980 the collection included the following numbers of strains: bacteria, 11,500; bacteriophages, 300; fungi and fungal viruses, 13,700; protozoa, 720; algae, 130; animal-cell cultures, 500; animal viruses, rickettsiae, and chlamydiae, 1,135; and plant viruses, 220. More than 1 million ampoules of lyophilized or frozen living strains are inventoried and stored at the ATCC. Other large collections in the United States are more specialized in scope. For example, the Northern Utilization Research and Development Division, USDA, at Peoria, Illinois, maintains a collection of yeasts, molds, and bacteria especially for use in fermentations. The Quartermaster Research and Development Center, U.S. Army, Natick, Massachusetts, maintains a collection of microbial strains that are associated with deterioration processes. A number of smaller collections of a specialized nature also exist, such as the collection of anaerobic bacteria maintained by the Department of Anaerobic Microbiology, Virginia Polytechnic Institute and State University, Blacksburg.

One of the major functions of a large national collection is the preservation of type strains. As discussed in Chap. 3, the type strain of a species has great taxonomic importance because it is the "name-bearer" strain, or permanent example, of the species. Microbiologists who propose a new species are expected to deposit the type strain with one or more national collections so that it can be preserved and so that subcultures can be distributed to other workers for study and comparison with other microorganisms.

Figure 8-9. (A) Liquid-nitrogen refrigerators used for preservation of bacteria. Each container holds many thousands of ampoules. *(Courtesy of American Type Culture Collection.)* (B) Preservation of bacterial cultures in the gas phase (−150°C) of liquid-nitrogen refrigerators. For preservation, a bacterial suspension is placed in a vial which is then sealed. This picture shows six vials, attached to a metal cane, being removed from storage. *(Courtesy of Alma Dietz, The Upjohn Company.)*

CULTURAL CHARACTERISTICS

One of the major features of a bacterial strain is its appearance following growth on various media. Such commonplace characteristics as the abundance of the growth, the size of the colonies, and the color (or chromogenesis) of the colonies provide useful clues for identification.

To determine the growth characteristics of a bacterial strain, it is customary to observe the features of colonies and broth cultures. Inoculation of agar plates to obtain isolated colonies has already been described. Tubes of broth can be inoculated with the transfer needle or the loop; generally the loop is used when the inoculum is a liquid.

After inoculation of the medium and subsequent incubation, the cultural characteristics can be determined. The main features can be summarized as follows.

Colony Characteristics

Size. Colonies range in size from extremely small (pinpoint), measuring only a fraction of a millimeter in diameter, to large colonies measuring 5 to 10 mm in diameter. Although the colonies of a given species have a characteristic diameter, one must be aware of certain factors affecting colony diameter. For instance, only well-separated colonies should be measured, since such colonies tend to have a larger diameter than those which are crowded together (for example, see Fig. 8-2). This is because widely separated colonies are subject to less competition for nutrients and less inhibition by toxic products of metabolism. Moreover, young colonies are smaller than older colonies; therefore the time at which measurements are made must be stated. There is generally an upper limit to the final size of the colonies of a given species; i.e., a point is reached where further incubation no longer results in a corresponding increase in size. However, some bacteria (e.g., certain species of *Proteus* and *Bacillus*) can spread across the entire agar surface, and the colony size is limited only by the dimensions of the Petri dish!

Margin or Edge. The periphery of bacterial colonies may take one of several different patterns, depending on the species. It may be evenly circular like the edge of a droplet or it may show irregularities such as rounded projections, notches, and threadlike or rootlike projections.

Surface Texture. Depending on the species, the colony surface may be smooth (shiny, glistening); rough (dull, granular, or matte); or mucoid (slimy or gummy). Certain species have colonies posessing a highly wrinkled surface.

For a pure culture, all the colonies on the plate should have a similar type of surface; however, you should bear in mind that some pure cultures may exhibit surface variation. One of the commonest variations is known as the S → R variation. This is due to the presence of mutant cells that give rise to some rough (R) colonies in a population that otherwise produces smooth (S) colonies. Some R mutants produce rough colonies because they lack the ability to make capsules, or, if the species is Gram-negative, they may no longer be able to form O antigens.

For several species of pathogenic bacteria, the surface texture of colonies may bear a relation to virulence. For instance, S colonies of S. *pneumoniae* or of *Salmonella* species are usually virulent, whereas R colonies are not. On the other hand, for strains of *Mycobacterium tuberculosis* a rough surface showing serpentine cords is usually a good indicator of virulence.

Elevation. Depending on the species, colonies may be **thin** to **thick,** and the surface may be **flat** or it may exhibit varying degrees of **convexity.**

Consistency. This can be determined by touching a transfer needle to the colony. Some bacterial species form colonies having a **butyrous** or butterlike consistency. Others may form colonies that are **viscous, stringy,** or **rubbery;** in the latter type the whole colony, rather than just a portion of it, may come off the agar surface with the transfer needle. Still other species may form **dry, brittle,** or **powdery** colonies that break up when touched with the needle.

Optical features. Colonies may be **opaque, translucent,** or **opalescent.**

Chromogenesis or Pigmentation. Some bacterial species produce and retain **water-insoluble pigments** intracellularly, thus causing the colonies to become colored **(pigmented).** Some species which form pigmented colonies are:

Flectobacillus major	Pink
Serratia marcescens	Red
Chromobacterium violaceum	Violet
Staphylococcus aureus	Gold
Micrococcus luteus	Yellow
Derxia gummosa	Brown
Bacteroides melaninogenicus	Black

Some colonies produce pigments that are **water-soluble;** these diffuse into the surrounding agar and stain it. For instance, *Pseudomonas aeruginosa* forms a blue water-soluble pigment called pyocyanin. Some pigments are only sparingly water-soluble and may precipitate in the medium. For example, *Pseudomonas chlororaphis* forms a pigment called chlororaphin which accumulates in the form of green crystals around the colonies.

Certain water-soluble pigments are **fluorescent;** i.e., the agar medium around the colonies glows white or blue-green when exposed to ultraviolet light. For example, *P. aeruginosa* produces not only the nonfluorescent pigment pyocyanin but also a fluorescent pigment, pyoverdin.

For a bacterial strain to exhibit its characteristic pigmentation, special media, incubation temperatures, or other conditions may be required. For instance, *Mycobacterium kansasii* forms a characteristic yellow pigment (β-carotene) only when the colonies are exposed to light.

Several types of bacterial colonies are shown in Fig. 8-10.

Characteristics of Broth Cultures

1 *Amount of growth.* Scanty, moderate, or abundant.
2 *Distribution and type of growth.* The growth may be uniformly distributed throughout the medium (evenly turbid). Alternatively, it may be confined to the surface of the broth as a scum or film (pellicle), or it may accumulate as a sediment, which may be granular or viscous.

The scheme for interpreting the appearance of bacterial growth has been described in some detail to emphasize the fact that many differences in cultural characteristics do occur among bacteria. With experience, familiarity with such characteristics becomes very helpful as a guide for the recognition of major groups of bacteria. Too often students pay little attention to these features of

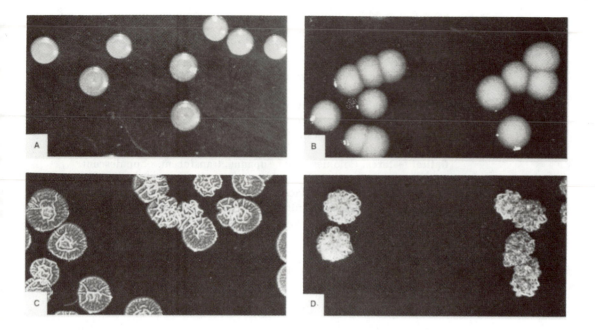

Figure 8-10. Bacterial colonies illustrating differences in characteristics. (A) Circular, raised, smooth surface; (B) circular, raised, finely granular surface; (C) irregular edge, flat, elevated folds in surface; (D) undulate edge, raised, irregularly elevated surface. *(Courtesy of Naval Biological Laboratory.)*

bacterial growth and thus deprive themselves of much useful information in the laboratory study of cultures.

To further emphasize the importance of cultural characteristics, suppose that we have prepared plate and broth cultures of an unidentified strain designated as strain 24. The colonies are irregular and raised and appear dry, with a roughened, granular surface. When we touch a colony with a transfer needle it proves to be brittle, and when a portion of the colony is removed it will not emulsify easily when spread in a drop of water. Growth in broth cultures occurs mainly in the form of a heavy surface pellicle, and the medium below the pellicle is only slightly turbid. Familiarity with the cultural appearance of bacteria would suggest that strain 24 might be an acid-fact bacterium (*Mycobacterium*). Additional tests must be performed to verify this possibility, but the cultural characteristics have provided a clue to the type of organism we are working with.

QUESTIONS

1 Devise an enrichment procedure for an aerobic bacterial species that can use methane gas as a sole carbon and energy source.

2 During an epidemic of meningitis caused by *Neisseria meningitidis*, many people become healthy carriers (i.e., harbor the organism in their nasopharynx but do not have meningitis) and can spread the organism by coughing and sneezing. Suppose you are given the task of determining how many people in the epidemic region are healthy carriers. You soon discover that the human nasopharynx is inhabited by many different kinds of microorganisms. However, you learn from selected references that *N. meningitidis*

is resistant to the chemotherapeutic agents vancomycin, colistin, and nystatin but is susceptible to penicillin. On the basis of this information devise a selective medium that could help you in your task.

3 What selective procedure would you use in the process of isolating from a soil sample (a) an endospore-forming organism, (b) a nonsporeforming gliding organism, (c) a psychrophilic organism?

4 Give three examples of selective mechanisms occurring in nature that lead to the predominance of particular kinds of bacteria.

5 Distinguish between the meanings of the terms mixed culture, pure culture, clone, and strain.

6 Compare the advantages and disadvantages of the various techniques for the isolation of microorganisms in pure culture.

7 What are the advantages and disadvantages of the various methods for preservation of pure cultures?

8 Why have organizations been established to maintain pure cultures? Of what use are such collections?

9 What difficulty might exist in subculturing the colony of a desired organism from a selective agar medium? What additional steps should be taken to help assure culture purity?

10 How could you acquire a subculture of the type strain of *Streptococcus lactis* so that you could compare its characteristics with another strain that you have isolated from milk?

11 What general categories of pigments are produced by bacteria? For each category give an example of an organism that makes such a pigment.

12 Give several reasons why industrial biological companies maintain large stock-culture collections.

REFERENCES

American Type Culture Collection: *Catalog of Strains*, vol. 1, 15th ed., American Type Culture Collection, Rockville, Md., 1982. *Contains a listing of the extensive holdings of the collection for algae, bacteria, bacteriophages, fungi and fungal viruses, plant viruses, protozoa, plasmids, and recombinant DNA vectors. References, conditions for cultivation, media formulations, and other information are provided.*

Gherna, R. L.: "Preservation," in P. Gerhardt, R. G. E. Murray, R. N. Costilow, E. W. Nester, W. A. Wood, N. R. Krieg, and G. B. Phillips (eds.): *Manual of Methods for General Bacteriology*, American Society for Microbiology, Washington, D.C., 1981. *In this chapter the head of the bacteriology department at the American Type Culture Collection gives details of the various procedures for preserving bacterial cultures.*

Gibbons, N. E. (revised by P. H. A. Sneath and S. P. Lapage): "Reference Collections of Bacteria—The Need and Requirements for Type Strains," in N. R. Krieg (ed.): *Bergey's Manual of Systematic Bacteriology*, vol. 1, Williams & Wilkins, Baltimore, 1984. *This chapter provides a brief history of culture collections and emphasizes their function as repositories for type strains.*

Krieg, N.R.: "Enrichment and Isolation," in P. Gerhardt, R. G. E. Murray, R. N. Costilow, E. W. Nester, W. A. Wood, N. R. Krieg, and G. B. Phillips (eds.), *Manual of Methods for General Bacteriology*, American Society for Microbiology, Washington, D.C., 1981. *Numerous specific physical, chemical, and biological selec-*

tive methods are given for various species of bacteria. Also presented are the details of the methods for isolating bacteria into pure culture.

Krieg, N. R. (ed.): *Bergey's Manual of Systematic Bacteriology*, vol. 1, Williams & Wilkins, Baltimore, 1984. *This together with subsequent volumes of this international reference work provides the characteristics of the genera and species of bacteria, including methods for the selection, isolation, and maintenance of each group.*

Starr, M. P., H. Stolp, H. G. Trüper, A. Balows, and H. G. Schlegel (eds.): *The Prokaryotes: A Handbook on Habitats, Isolation, and Identification of Bacteria*, Springer-Verlag, New York, 1981. *This monumental work provides specific information about the isolation and cultivation of nearly every bacterial group.*

PART THREE
MICROBIAL PHYSIOLOGY AND GENETICS

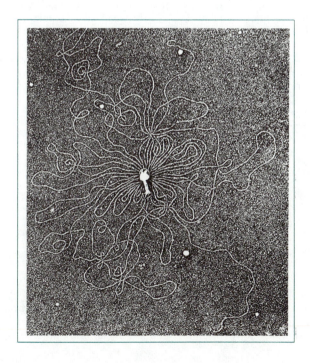

The many parts of a cell

Lewis Thomas, in his fascinating book *The Lives of a Cell*, describes the interesting protozoan *Myxotricha paradoxa* that lives within the digestive tract of Australian termites. Therein it thrives symbiotically, enzymatically degrading the tough woody cellulose down to assimilable carbohydrate for the termites. It is also a very motile creature, able to swim from place to place by means of surface appendages. A cursory examination would deduce these appendages to be cilia or flagella. However, upon careful scrutiny, these locomotor organelles turn out to be fully formed, whole bacterial spirochetes that have attached themselves over the entire surface of the protozoan. It is the sum total of the individual movements of the spirochetal cells that contributes to the locomotion of the protozoan. Furthermore, close to the point of attachment of the spirochetes and embedded in the surface of the protozoan, as well as elsewhere in the protozoan cytosol, are oval organelles. Under the electron microscope, these turn out to be true bacteria existing symbiotically with the spirochetes and the protozoan and probably helping in the degradation of cellulose by the excretion of enzymes. Thus these various creatures thrive together in an ecosystem (the termite digestive tract), each type contributing in its own way to the benefit of the whole.

It now appears that the existence of a creature like *Myxotricha paradoxa* is not so strange after all. There is a theory in biology called the **endosymbiotic theory.** It ascribes the origin of the eucaryotes to a series of endosymbioses, with the endosymbionts becoming the organelles of the host cell. One of the most articulate proponents of this theory is Lynn Margulis of Boston University. According to this theory, the "modern" eucaryotic nucleated cell is the result of the "coming together" of endosymbionts and their host in the evolutionary process. There was presumed, in primordial time, a moderately large, amoeboid, heterotrophic, anaerobic procaryote that customarily fed on smaller procaryotes. Some of the latter were aerobic organisms that were not digested; they became stabilized as endosymbionts and continued to function as aerobes within an anaerobic host. In this capacity they conferred the advantages of aerobic respiration upon the anaerobic host. With time, they became so well integrated with the host cell that they became the mitochondria of the host. In the same way, the photosynthetic cyanobacteria integrated so well with their procaryotic host cell in evolutionary time to become the chloroplasts of the first plant eucaryotic cell. Similarly, flagellated procaryotes, like the ones attached to *M. paradoxa*, became integrated with the host cell to become its locomotor organelles. The centrioles and centromeres of the mitotic process also presumably had similar origins (much of the symbiont parts having been lost except for the basal portions including the DNA).

Much of the credence of the endosymbiotic theory rests on two contemporary findings: demonstrations that many organelles like the mitochondrion have their own DNA, and the widespread occurrence of symbiosis, including endosymbiosis. The metabolic and genetic processes carried out by procaryotic cells—progenitors of the nucleated cell—form the subject matter of this part of the text.

Preceding page. DNA content of a T2 phage appears as a single thread in this electron photomicrograph. Measurements of the DNA thread system indicate a length of $49 \pm 4 \, \mu$. The phage "ghost" (the flask-shaped object at the center) is about $0.1 \, \mu$ long. *(From A. K. Kleinschmidt et al., in Biochim Biophys Acta, 61:857–864, 1962.)*

Chapter 9   Enzymes and Their Regulation

A cell must be capable of performing a multitude of chemical changes in order to stay alive, grow, and reproduce. It may have to alter complex nutrients in the medium before they can enter the cell. It must effect additional changes on the nutrients once they are in the cell. These nutrients are broken down chemically to provide energy for the cell and simple precursors for the synthesis of cell material. The chemical changes involved are exceedingly complex, considering the diversity of materials used as foods on the one hand and the variety of substances synthesized into cell constituents on the other. How does the cell accomplish these changes? The answer lies in the activity of **enzymes,** substances present in the cell in minute amounts and capable of speeding up chemical reactions associated with life processes. Any impairment of enzyme activity is reflected by some change in the cell, or even by death. There can be no life without enzymes.

Within a living cell, enzymes function in sequences of reactions called **pathways.** For a cell to grow normally, it is essential that the flow of chemical substances, or **metabolites,** through these pathways be under a high degree of regulation or control. This regulation ensures that no products are deficient or in excess, and is exerted either on enzyme activity or on enzyme synthesis. Our purpose in this chapter is to describe some of the major characteristics of enzymes and how the enzymes are regulated.

SOME CHARACTERISTICS OF ENZYMES

Certain substances in small amounts have the unique capacity of speeding up chemical reactions without themselves being altered after the reaction; they accelerate the velocity of the reaction without necessarily initiating it. Substances that behave in this manner are called **catalysts** or **catalytic agents.** For example, hydrogen and oxygen do not combine to any appreciable extent under normal atmospheric conditions. If, however, the two gases are allowed to touch colloidal platinum, they react instantaneously to produce water. The platinum greatly increases the speed at which this reaction takes place without being used up in the reaction. Nor do these substances ordinarily have any effect on the equilibrium of a reversible chemical reaction; they merely speed it up until it reaches equilibrium. They also exhibit specificity: a given substance will only affect a certain type of reaction. Enzymes are catalysts. However, unlike platinum, which is inorganic, enzymes are organic compounds produced by living organisms. Thus we may define an enzyme as an **organic catalyst** produced by a living cell. For example, in the cell the oxidation of a fatty acid to carbon dioxide and water takes place smoothly and rapidly within a narrow range of temperature and pH because of enzymes. Without these organic catalysts, as in a test tube, this same process requires extremes of pH, high temperatures, and corrosive chemicals.

Although all enzymes are initially produced in the cell, some are secreted through the cell wall and function in the cell's environment. Thus we recognize two types of enzymes on the basis of site of action: **intracellular enzymes,** or **endoenzymes** (functioning in the cell), and **extracellular enzymes,** or **exoenzymes** (functioning outside the cell). The principal function of the extracellular enzymes is to perform necessary changes on the nutrients in the medium to allow them to enter the cell. Intracellular enzymes synthesize cellular material

---

The Origin of the Word *Enzyme*

The word *enzyme* was coined in 1878 by Kühne from a Greek term meaning "in yeast." Earlier, enzymes were referred to as "ferments" because their actions were similar to yeast fermentation. A major controversy on this subject occurred between Pasteur and the great German chemist Liebig. Liebig maintained that fermentations were caused by chemical substances not associated with living cells, but Pasteur held that the fermentation process was inseparable from living cells. As we know today, neither position was strictly wrong. The Pasteur-Liebig controversy was resolved in 1897 by Büchner, who demonstrated that a cell-free juice, prepared from yeasts by filtration, contained active enzymes.

and also perform catabolic reactions which provide the energy required by the cell.

The general characteristics of enzymes are the same whether the enzymes are produced by the cells of microbes, by people, or by other forms of life. In fact, cells from organisms that are very different may contain some enzymes that are identical or at least have identical functions.

In any one cell there are a thousand or more different enzymes. The enzymes present in a microbial cell are determined by the environmental conditions and by the cell's genetic constitution. This means that at any one moment, the enzyme content of a microbe is a reflection of the manner in which that cell copes with the environment.

CHEMICAL AND PHYSICAL PROPERTIES OF ENZYMES

Enzymes are proteins or proteins combined with other chemical groups. Enzymes therefore possess the properties characteristic of proteins: they are denatured by heat, are precipitated by ethanol or high concentrations of inorganic salts like ammonium sulfate, and do not dialyze (pass through semipermeable membranes).

Many enzymes consist of a protein combined with a low-molecular-weight organic molecule called a coenzyme. The protein portion in this instance is referred to as the apoenzyme. When united, the two form the complete enzyme, identified as the holoenzyme, as shown below:

Apoenzyme + coenzyme	→ holoenzyme	
Inactive	Inactive	Active
Protein	Organic molecule	
High molecular weight	Low molecular weight	
Nondialyzable	Dialyzable	

The integral part of some coenzymes is a vitamin. Several of the B vitamins, as listed in Table 9-1, have been identified as the main components of coenzymes.

A specific example of the relation of a vitamin to a coenzyme (vitamin B_6 in pyridoxal phosphate and pyridoxamine phosphate) is shown in Fig. 9-1.

In some instances the nonprotein portion of an enzyme may be a metal, e.g., iron in the enzyme catalase. The metal may be tightly bound to the protein or loosely bound and easily dissociable, depending on the specific enzyme. Many enzymes require the addition of metal ions (Mg^{2+}, Mn^{2+}, Fe^{2+}, Zn^{2+}, etc.) in order to be "activated." It is assumed that these metal ions function in combination with the enzyme protein, and they are regarded as inorganic coenzymes, or cofactors. Sometimes both a cofactor and a coenzyme (organic) are required before an enzyme becomes active.

A large number of enzymes have been extracted from cells and, by a combi-

Table 9-1. Some vitamins and their coenzyme forms

Vitamin	Coenzyme
Thiamine (B_1)	Cocarboxylase
Riboflavin (B_2)	Riboflavin adenine dinucleotide
Niacin	Nicotinamide adenine dinucleotide
Pyridoxine (B_6)	Pyridoxal phosphate
Folic acid	Tetrahydrofolic acid

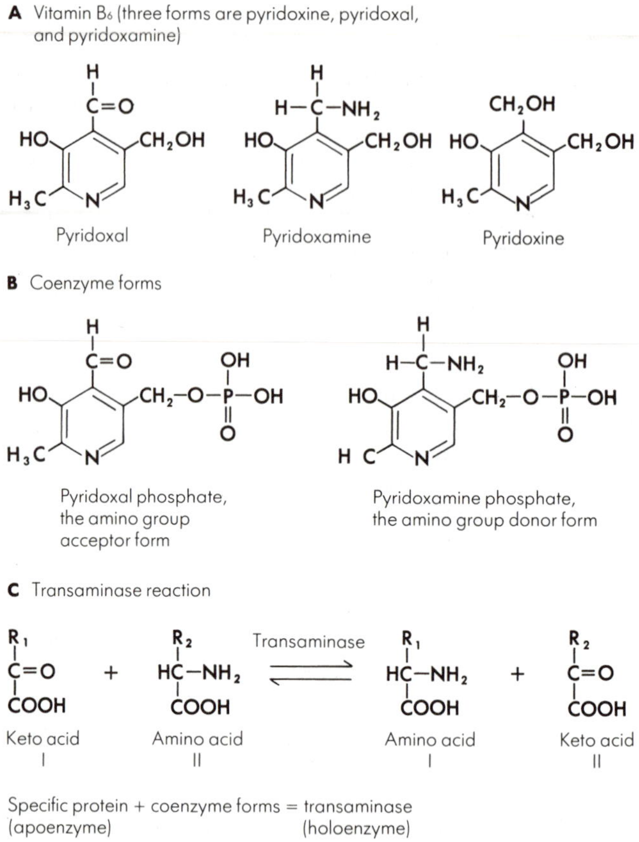

A Vitamin B₆ (three forms are pyridoxine, pyridoxal, and pyridoxamine)

Pyridoxal   Pyridoxamine   Pyridoxine

B Coenzyme forms

Pyridoxal phosphate, the amino group acceptor form

Pyridoxamine phosphate, the amino group donor form

C Transaminase reaction

$$R_1\text{—}C\text{=O}\ \ +\ \ R_2\text{—}HC\text{—}NH_2\ \ \underset{}{\overset{\text{Transaminase}}{\rightleftharpoons}}\ \ R_1\text{—}HC\text{—}NH_2\ \ +\ \ R_2\text{—}C\text{=O}$$

COOH   COOH   COOH   COOH

Keto acid   Amino acid   Amino acid   Keto acid
I   II   I   II

Specific protein + coenzyme forms = transaminase
(apoenzyme)   (holoenzyme)

Figure 9-1. (A) Vitamin B₆ exists in three forms. (B) Coenzyme forms of the vitamin. (C) The transaminase reaction.

nation of physical and chemical techniques, have been obtained in chemically pure form. Urease was the first enzyme isolated in crystalline form; J. B. Sumner of Cornell University received the Nobel Prize in 1947 for this achievement. The protein nature of enzymes was accepted widely only after John Northrop and his colleagues, in the 1930s, crystallized pepsin and trypsin and found them also to be proteins. Enzymes are very large protein molecules; molecular weights from approximately 10,000 to 1 million have been determined for different enzymes. The enzyme catalase, which splits hydrogen peroxide into water and oxygen, has a molecular weight of 250,000. Hydrogen peroxide, on the other hand, has a molecular weight of 34.

Enzyme molecules are exceedingly efficient in accelerating the transformation of **substrate** (substance acted on by enzymes) to end product. A single enzyme molecule can effect the change of as many as 10,000 to 1 million molecules of substrate per minute. This ability, together with the fact that the enzyme is not consumed or altered in the reaction, reveals why very small quantities of enzymes are sufficient for cellular processes.

Enzymes are vulnerable to various environmental factors. Their activity may be significantly diminished or destroyed by a variety of physical or chemical conditions, as will be shown later, but great differences exist among enzymes in this respect. Some may become inactivated by very minor alterations in the environment. The destruction of enzymes by physical or chemical agents results in a loss to the cell of the functions performed by the enzymes—a further revelation of their essential nature.

The two most striking characteristics of enzymes are (1) their high catalytic efficiency and (2) their high degree of specificity for substrates. One ounce of pure crystalline pepsin, for instance, can digest nearly two tons of egg white in only a few hours; whereas it would take 10 to 20 tons of strong acid 24 to 48 h at elevated temperatures without the enzyme. An enzyme molecule may transform 10^2 to 10^6 molecules of substrate per minute. A single enzyme may react with only a single substrate or, in some instances, with a particular chemical grouping on chemically related substrates. Essentially, this means that cells usually produce different enzymes for every compound they metabolize. Furthermore, each enzyme causes a one-step change in the substrate. For example, yeasts ferment glucose to alcohol and carbon dioxide. The initial reactants and the final products of the reaction are shown in the following equation:

$$\text{Glucose} + \text{yeast cells} \rightarrow \text{alcohol} + \text{carbon dioxide}$$

Substrate   Source of enzymes   End products

This transformation is accomplished not by a single enzyme but by a group of enzymes, an **enzyme system.** More than a dozen single enzymes work in sequence, each performing a single specific change in the product formed by the preceding enzyme reaction. The last reaction in the system yields the final products. (Examples of enzyme systems will be found in Chap. 10.)

Today over 1,000 different enzymes are known, and well over 150 have been crystallized. Many more remain to be discovered. Exciting new areas of enzyme involvement have been discovered in recent years, including the self-regulating nature of many enzyme systems, the genetic control of enzyme function and synthesis, and the role of enzymes in development and differentiation. E.g., Thomas Cech and his colleagues at the University of Colorado reported in the journal *Cell* (**31**:147–157, 1982) the discovery of a biochemical reaction mediated by RNA in the absence of protein. The notion of catalytic RNA unfolds new perspectives on the basic components required for life, and also on the evolution of life. This finding has been described as "one of the most exciting discoveries of the decade."

NOMENCLATURE OF ENZYMES

In 1956 the International Union of Biochemistry established an international Commission on Enzymes to work on a systematic arrangement and nomenclature for the large and rapidly increasing number of enzymes. Its recommendations were first published in 1961, and they have been adapted and are universally used (see References). Briefly, these recommendations follow.

How Enzymes Are Named and Classified

Except for a few of the originally studied enzymes such as pepsin, renin, and trypsin, most enzyme names end in *ase*, e.g., succinate dehydrogenase. When naming a complex of several enzymes on the basis of the overall reaction catalyzed by it, the word *system* should be used, e.g., the succinate oxidase system, which catalyzes the oxidation of succinate by O_2 and consists of succinate dehydrogenase, cytochrome oxidase, and several intermediate carriers. For classification, only single enzymes, and not enzyme systems, are considered.

The type of chemical reaction catalyzed is the basis for the classification and naming of enzymes, because it is this specific property that distinguishes one enzyme from another. Two names are recommended for each enzyme: a working or trivial name and a systematic name. The trivial name is shorter and more convenient to use; in many cases, it is the name already in current use. The systematic name is formed in accordance with definite rules, identifies the substrate or substrates, and specifies the type of reaction catalyzed. Furthermore, each enzyme is given an identifying classification number in addition to its trivial and systematic names. For example, E.C.1.1.1.1 is the Enzyme Commission number for alcohol dehydrogenase (trivial name); its systematic name is alcohol:NAD oxidoreductase. According to international classification, enzymes are grouped into six major classes (see Table 9-2). These groupings are based on the type of chemical reaction the enzymes catalyze.

Many enzymes occur in different structural forms but possess identical (or nearly so) catalytic properties. Such enzymes are called isozymes or isoenzymes. The present nomenclatural system makes no provision for structural diversity with similar catalytic function.

THE NATURE AND MECHANISM OF ENZYME ACTION

Most enzyme reactions may be represented by the following overall reaction:

Enzyme E + substrate S ⇆ enzyme-substrate complex ES ⇆

product P + enzyme E

The enzyme E and substrate S combine to give an enzyme-substrate complex ES, which then breaks up to yield the product P. The enzyme is not used up in the reaction but is released for further reaction with another substrate molecule. This process may be repeated many times until all the available substrate molecules are consumed. However, equilibrium, a steady-state condition, is reached when the forward reaction rates equal the backward reaction rates. This is the basic equation upon which most enzymatic studies are based.

Table 9-2. Major classes of enzymes

Class No.	Class	Catalytic Reaction
1	Oxidoreductases	Electron-transfer reactions (transfer of electrons or hydrogen atoms)
2	Transferases	Transfer of functional groups (functional groups include phosphate, amino, methyl, etc.)
3	Hydrolases	Hydrolysis reactions (addition of a water molecule to break a chemical bond)
4	Lyases	Addition to double bonds in a molecule as well as nonhydrolytic removal of chemical groups
5	Isomerases	Isomerization reactions (reactions in which one compound is changed into an isomer, i.e., a compound having the same atoms but differing in molecular structure)
6	Ligases	Formation of bonds with cleavage or breakage of ATP (adenosine triphosphate)

Figure 9-2. (A) Enzyme-substrate reaction, depicted schematically. The substrate is attracted to some site on the surface of the enzyme molecule. The chemical groupings of the substrate are strained by this attraction, and cleavage results. The cleavage products are released from the enzyme, and the enzyme is free to combine with more substrate and continue the process. (B) The drawing shows a schematic representation of an active site. Solid circles represent "contact" amino acid residues whose fit with substrate determines specificity; triangles are catalytic residues acting on substrate bond, indicated by a jagged line; open circles are nonessential residues on the surface; squares are residues whose interaction with each other maintains the three-dimensional structure of the protein. (*Redrawn from D. E. Koshland, Jr., "Correlation of Structure and Function in Enzyme Action," Science, **142**:1533–1541, 1963.*)

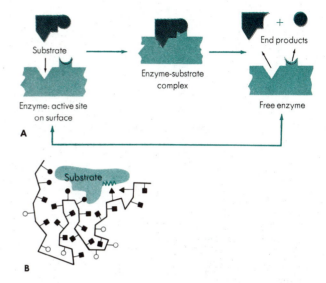

There is a high chemical affinity of the substrate for certain areas of the enzyme surface called the **active sites.** A strain or distortion is produced at some linkage in the substrate molecule, making it **labile** (unstable), and it undergoes a change determined by the particular enzyme. The altered substrate molecule then lacks affinity for the active site and hence is released. The enzyme is then free to combine with more substrate to repeat the action (see Fig. 9-2). Almost all intracellular enzymes have more than one active site per molecule. For example, lactate dehydrogenase has four. In contrast, α-chymotrypsin, like other secreted extracellular enzymes, has only one active site.

The main function of an enzyme is to lower the **activation energy** barrier to a chemical reaction. Activation energy refers to the amount of energy required to bring a substance to the reactive state. The enzyme combines with the substance (substrate) to produce a transition state requiring less activation energy for the chemical reaction to proceed. Figure 9-3 illustrates this concept.

The above discussion applies to substrates undergoing degradation and being utilized for energy. The same type of explanation could be used to describe **synthesis,** or building up of complex compounds from simpler ones. For example, two different molecules might become attached to adjacent sites on the enzyme surface. A unique activation by the enzyme would lead to establishment of a bond between the two molecules, thereby creating a new compound from the original two substrates. The product has little affinity for the active sites and hence is released, thereby freeing the active sites to repeat the process with another two molecules of substrate.

The active site on an enzyme surface is actually a very small area, which means that large regions of the enzyme protein (which has hundreds of amino acids) do not contribute to enzyme specificity or enzyme action (see Fig. 9-2). Thus only a relatively few amino acids are directly involved in the catalytic process—perhaps even less than five! It should also be emphasized that the "fit" between an active site of the enzyme surface and the substrate is not a static

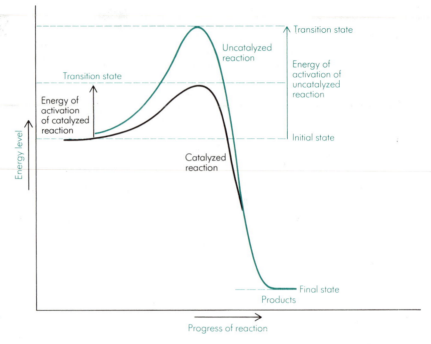

Figure 9-3. The main function of an enzyme is to lower the activation energy, the energy necessary to attain the transition state of a chemical reaction. As can be seen, an enzyme-catalyzed reaction has a lower activation energy and so requires less energy for the reaction to proceed.

one. Rather it is a dynamic interaction in which the substrate induces a structural change in the enzyme molecule, as a hand changes the shape of a glove. This phenomenon is illustrated in Fig. 9-4.

CONDITIONS AFFECTING ENZYME ACTIVITY

Among the conditions affecting the activity of an enzyme are the following:

1 Concentration of enzyme
2 Concentration of substrate
3 pH
4 Temperature

Generally speaking, there is an optimum relation between the concentrations of enzyme and substrate for maximum activity. But in order to study the effect of increasing enzyme concentrations upon the reaction rate, the substrate concentration must be in *excess*. This means that the reaction must be independent of the substrate concentration so that any variation in the amount of product formed is a function only of the enzyme concentration present. This is illustrated in Fig. 9-5.

If the amount of enzyme is kept *constant* and the substrate concentration is then gradually increased, the velocity of reaction will increase until it reaches a maximum. Any further increases in substrate concentration will not increase the reaction velocity. This is shown in Fig. 9-6.

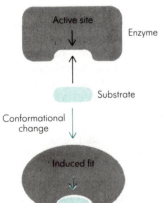

Figure 9-4. Induced fit of the active site of the enzyme molecule to the substrate molecule.

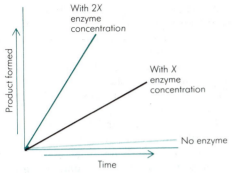

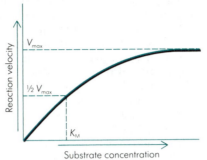

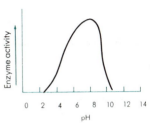

Figure 9-5. Effect of enzyme concentration on rate of enzyme activity (in the presence of excess substrate). These reactions shown are said to be "zero order" because the rates are *independent* of substrate concentration. The formation of product proceeds at a rate which is linear with time (e.g., doubling the assay time results in doubling the amount of products formed).

Figure 9-6. Effect of substrate concentration on enzyme activity. When the maximum velocity has been reached, all of the available enzyme has been converted to ES, the enzyme-substrate complex (the point is designated V_{max} on graph). K_m shown refers to the Michaelis constant and is defined as the substrate concentration at one-half the maximum velocity.

Figure 9-7. Effect of pH on enzyme activity. Maximum activity occurs at a particular pH, and deviations from it result in decreased activity. (Not all enzymes exhibit optimum activity at the same pH.)

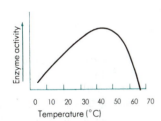

Figure 9-8. Effect of temperature on enzyme activity. Starting at a low temperature, the enzyme activity increases with increasing temperature until the optimum activity is reached. Further increases in temperature result in decreased enzyme activity and eventual destruction of the enzyme. (Not all enzymes exhibit optimum activity at the same temperature.)

Each enzyme functions optimally at a particular pH and temperature (Figs. 9-7 and 9-8). From these illustrations, it is clear that deviations from the optimal conditions result in significant reduction of enzyme activity. This is characteristic of all enzymes. Extreme variations in pH can even destroy the enzyme, as can high temperatures; boiling for a few minutes will **denature** (destroy) most enzymes. Extremely low temperatures, for all practical purposes, stop enzyme activity but do not destroy the enzymes. Many enzymes can be preserved by holding them at temperatures around 0°C or lower.

We know that the growth of microorganisms is influenced by a variety of physical and chemical conditions. Since enzymes are responsible for catalyzing the reactions associated with life processes, it follows that the conditions mentioned above affect the enzymes and thereby the growth response. Just as there is an optimal pH and temperature for growth, there is also an optimal pH and temperature for the activity of each enzyme and for the production of each enzyme by the cell. This does not mean, however, that the values are the same for each enzyme in the same cell; they may not be. The reason for the differences is that during growth, activity or response is measured in terms of the total activities required for growth when all enzymes and enzyme systems are functioning harmoniously in the cell. Optimum conditions must be estimated in terms of what is best for the entire cell. In assessing the activity of an isolated and purified specific enzyme, the situation is entirely different. The enzyme is no longer in its normal environment and thus is not influenced by, or integrated into, the multitude of reactions which occur within the cell. Hence, the optimum conditions for activity of any one enzyme in vitro are not necessarily optimum for the same enzyme in vivo.

INHIBITION OF ENZYME ACTION

The activity of an enzyme can be inhibited (slowed down or stopped) by chemical agents in several different ways. We are concerned here with chemical substances that inhibit in a more subtle fashion than denaturation (destruction) of the protein portion of the enzyme. Specific information about how these chemical agents (naturally occurring or otherwise) exert their detrimental action contributes to our knowledge of how enzymes work, suggests new types of chemicals which may be useful for inhibiting microorganisms, and in other ways contributes to our understanding of life processes.

Enzyme inhibition may be classified as either **nonreversible** or **reversible.** Nonreversible inhibition usually involves the modification or inactivation of one or more functional groups of the enzyme so that it is no longer active.

There are two major types of reversible inhibition, namely, **competitive inhibition** and **noncompetitive inhibition.** Competitive inhibition can be reversed by increasing the substrate concentration, whereas noncompetitive inhibition cannot. We will present a few examples.

The enzyme succinic dehydrogenase accomplishes the transfer of hydrogen atoms from succinic acid to a suitable acceptor compound (in the reaction below, the acceptor, methylene blue, is a compound that does not occur naturally within a cell):

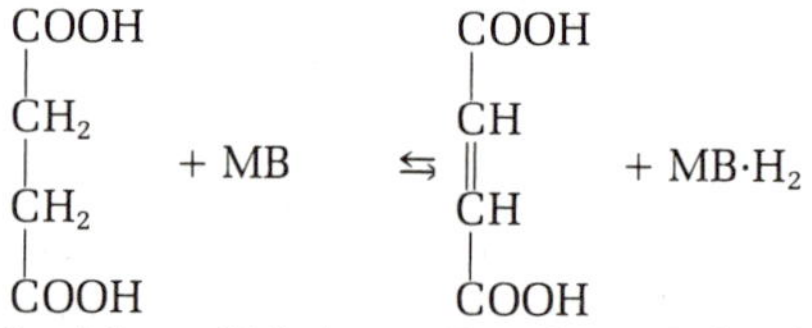

$$
\begin{array}{ccccc}
\text{COOH} & & \text{COOH} & & \\
| & & | & & \\
\text{CH}_2 & + \text{ MB} & \text{CH} & + \text{ MB·H}_2 & \\
| & \rightleftarrows & \| & & \\
\text{CH}_2 & & \text{CH} & & \\
| & & | & & \\
\text{COOH} & & \text{COOH} & & \\
\text{Succinic} & \text{Methylene} & \text{Fumaric} & \text{Reduced} & \\
\text{acid} & \text{blue} & \text{acid} & \text{methylene} & \\
& & & \text{blue (colorless)} &
\end{array}
$$

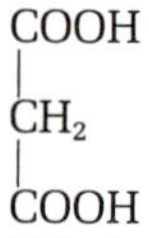

This reaction can be inhibited by chemical compounds which have a structure similar to that of succinic acid. One of these inhibitors is malonic acid, which has the structure shown in the margin. Because of its structural similarity to succinic acid, malonic acid attaches itself to the enzyme active site where succinic acid normally attaches. The succinic acid is "blocked out" by the malonic acid, but since the malonic acid is not activated by the enzyme, there is no reaction. In this example there is *competition* for the same active site by two different molecules (see Fig. 9-9). This type of enzyme inhibition is referred to as competitive inhibition and can be relieved simply by increasing the substrate concentration.

Certain chemical substances, such as cyanide, have a high affinity for metal

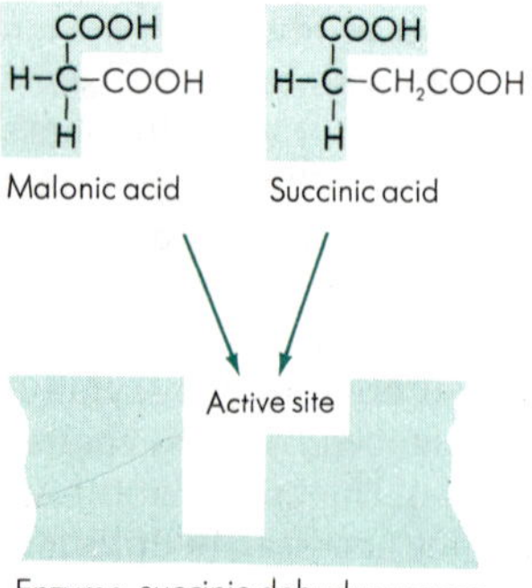

$$
\begin{array}{l}
\text{COOH} \\
| \\
\text{CH}_2 \\
| \\
\text{COOH} \\
\text{Malonic} \\
\text{acid}
\end{array}
$$

Figure 9-9. Competitive inhibition (schematic) between malonic acid and succinic acid. Note that each molecule has a structurally similar fragment. Since this portion of either molecule can fit or combine with the active site on the enzyme surface, there is competition between the two substrates for this site. Because this enzyme is specific for succinic acid, if the malonic acid occupies the site, further activity is blocked, as malonic acid is not changed by this enzyme.

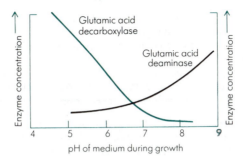

Figure 9-10. Variations in the concentrations of glutamic acid decarboxylase and glutamic acid deaminase present in *Escherichia coli* with variations in the pH of the medium during growth. Note that at a low pH (acidic medium) glutamic acid decarboxylase predominates and at a high pH (basic medium), glutamic acid deaminase predominates.

ions, and form complexes with the metal. As already mentioned, many enzymes require a metal ion for their activity. Cyanide is a strong inhibitor of iron-containing enzymes because it "ties up" the iron, depriving the enzyme of an essential component. Similarly, fluoride inhibits enzymes which require calcium or magnesium by binding these metals. (It may be noted that the relatively small amount of fluoride which we find in tooth enamel as a result of water fluoridation does not exert this inhibitory action in vivo.) But the most important noncompetitive inhibitors are naturally occurring metabolic intermediates that can combine reversibly with specific sites on certain regulatory enzymes, thus changing the activity of their catalytic sites. This type of enzyme inhibition is referred to as noncompetitive inhibition since the inhibitor is not competing with the substrate for an active site on the enzyme surface and therefore is not reversed by increasing substrate concentration.

CONDITIONS AFFECTING ENZYME FORMATION

The enzymatic content of animal-tissue cells is relatively constant, since they are in an environment which in terms of physical and chemical conditions is subjected to little change. But the bacterial cell is exposed to an ever-changing environment. For example, *Escherichia coli* may grow in an acid or alkaline medium (from pH 4.5 to pH 9.5), at room temperature or above body temperature, aerobically or anaerobically. Cells of *E. coli* grown at the extremes of these conditions do not contain the same kinds or amounts of enzymes (see Fig. 9-10). But the fact that the environment influences the formation of enzymes should not be misconstrued to mean that there is no consistent enzyme pattern for a given organism. Organisms manifest changes in reaction to environment only within certain limits. It is important to recognize their capacity for undergoing these changes when examining organisms. When organisms are studied physiologically, these studies must be performed under certain established conditions, which include the composition of the medium in which the cells are grown as well as the physical conditions during incubation.

DETERMINATION OF ENZYME ACTIVITY

The enzymatic activity of bacteria (or of any cells or tissues) can be determined by a variety of techniques. Some procedures require special, elaborate instruments; others require only a test tube and a few reagents. All are based on a few simple principles. In order to carry out a quantitative assay of enzyme activity, it is necessary to know the following:

1 The nature of the reaction catalyzed
2 What cofactors and coenzymes are required
3 The required concentrations of both substrate and cofactor or coenzyme
4 The optimum pH
5 The optimum temperature
6 A simple analytical method for determining the disappearance of substrate or the appearance of products of the reaction

The substrate concentration should be above the saturation level so that the initial reaction rate is proportional to enzyme concentration alone. Coenzymes and cofactors should also be added in excess. Doing this ensures that the true limiting factor is the enzyme concentration (at its optimum pH and temperature). Generally, measurement of reaction product formation is more accurate than measurement of the disappearance of substrate. Such determinations of enzymatic activity have provided us with a wealth of knowledge concerning cells and their chemical reactions.

ENZYME PREPARATIONS

The source of the enzymes used in any microbiological experiment may be one of the following: (1) cells in a growing culture, (2) cells removed from a growing culture and resuspended in a nonnutrient solution (resting cells, or nongrowing but viable cells), and (3) enzymes extracted from cells (cell-free enzyme preparation). Cell-free enzyme preparations may be of varying degrees of purity. They may be very crude extracts, as in the case when the material contains all the cellular contents recovered after fracturing the cell walls. However, this crude extract can be refined to obtain some particular fraction, e.g., ribosomal material, soluble cytoplasmic substance, or material of a certain molecular size range. In the last-mentioned instance, the objective may be to discover which subcellular entities house all the enzymes (enzyme system) responsible for certain cellular functions.

The manner in which each of these is used is as follows:

GROWING-CULTURE TECHNIQUE
1 Inoculation of bacteria into medium containing substrate.
2 Incubation of bacteria (1 or more days of growth).
3 Examination for change or disappearance of substrate and presence of end products.

RESTING-CELL TECHNIQUE
1 Growth of bacteria in a suitable medium.
2 Preparation of a resting-cell suspension. The cells are harvested from the medium by centrifugation and resuspended in a nonnutrient solution. The process is repeated to free the cells of all material from the medium and is known as **washing the cells.** The final suspension is referred to as a **resting-cell suspension.**
3 Addition of resting cells and substrate to special testing unit, e.g., Warburg apparatus or Thunberg tube.
4 Incubation.
5 Examination (or testing) for disappearance of substrate and appearance of end products.

CELL-FREE ENZYME TECHNIQUE

1 Preparation of concentrated resting-cell suspension.

2 Disintegration of cells by special techniques to release enzymes, e.g., grinding, sonic treatment. (Any remaining whole cells are removed by centrifugation or filtration to obtain cell-free enzymes.)

3 Addition of cell-free enzymes and substrate to special testing unit.

4 Incubation.

5 Examination (or testing) for disappearance of substrate and appearance of end products.

The growing-culture technique is used routinely for the characterization of the enzymatic activities of microorganisms. Results of such tests provide information necessary for their identification. The resting-cell technique and cell-free enzyme preparations are principally used in research work where the object is to determine how the organism accomplishes each specific change. They provide favorable tools for the meticulous examination of the events that occur when a substrate undergoes change.

REGULATION OF ENZYMES

A living cell contains upward of a thousand different enzymes, each of which is an effective catalyst for some chemical reaction. But these enzymes act together in a coordinated manner so that all the chemical activities in a living cell are integrated with one another. One consequence of this enzyme coordination is that the living cell synthesizes and degrades materials as required for normal growth and metabolism.

The control of cellular metabolism ultimately is accomplished by the regulation of enzymes. In a microbial cell, such as a bacterium, the existence of cellular regulatory mechanisms is all the more important because of the absence of supracellular controls, such as neural and hormonal controls, which are present in the tissue cells of higher organisms. Microorganisms have evolved a variety of enzyme regulatory mechanisms which accommodate the changing needs of the microbial cell in a changing environment.

MECHANISMS OF REGULATION OF ENZYMES

Within the cell, there are two different regulatory mechanisms: the regulation of enzyme activity and the regulation of enzyme synthesis. Both mechanisms share the following properties:

1 They are mediated or governed by low-molecular-weight compounds (effector molecules). These are either formed in the cell during metabolism or are found in the environment.

2 They involve the participation of a special class of control proteins. Such proteins are mediators of metabolic change as directed by the small effector molecules. There are generally two types of control proteins, namely, allosteric enzymes and regulatory proteins.

The activities of allosteric enzymes are enhanced or inhibited by combination with their effector molecules. (Allosteric enzymes are so called because the site on the enzyme molecule where an effector molecule acts *is different from* the catalytic site. It is called the allosteric site, and it regulates the activity of the enzyme. Allosteric enzymes are generally larger and more complex than simple enzymes.) Regulatory proteins have no catalytic activity, but they modulate the biosynthesis of enzymes by attaching to the bacterial chromosome at specific sites and thus regulate gene expression. This attachment is also affected by the binding of small effector molecules.

REGULATION OF ENZYME ACTIVITY

Metabolic pathways that supply energy, precursor substances (building blocks), and reduced compounds for other biosynthetic processes are usually controlled by modulation of enzyme activity (rather than by control of enzyme synthesis). However, not every enzyme in each pathway needs to be controlled. But in each enzyme system (group of enzymes working together to carry out a given metabolic process) there is at least one enzyme that is controlled and sets the rate of the overall sequence. Different effector molecules modulate an enzyme's activity depending on the need of the particular organism either for energy or for precursors used in biosynthesis.

Energy-Link Control

In this type of regulation the effector molecules involved are adenylates, such as adenosine triphosphate (ATP), or other purine or pyrimidine nucleotides. Some enzymes appear sensitive to the absolute concentrations of ATP, adenosine diphosphate (ADP), or adenosine monophosphate (AMP); others seem to respond to the ratio of a given two of these nucleotides. (Such compounds are discussed in Chap. 10.) In general, enzymes responsible for energy production are inhibited by energy charge (e.g., high concentration of ATP), while certain key biosynthetic enzymes are stimulated. Such regulation is very important in balancing energy production and energy utilization.

Feedback Inhibition

For a biosynthetic pathway with one major end product, such as an amino acid, control is exerted by the final concentration of the product in the cell. This product, the effector molecule, typically inhibits the activity of the enzyme in the first reaction of the biosynthetic pathway. This method of regulation is called feedback or end-product inhibition. By this device, microorganisms and other cells prevent the overproduction of low-molecular-weight intermediates such as amino acids and purine and pyrimidine nucleotides. Glutamine synthetase, for example, is influenced by feedback inhibition from nine different compounds.

Precursor Activation

Sometimes the precursor, or the first metabolite of a pathway, is the effector molecule. It activates, or stimulates the activity of, the last or later enzyme in the sequence of reactions of a pathway. This is called precursor activation.

A scheme showing the above three mechanisms in the regulation of enzyme activity is given in Fig. 9-11. Two techniques that have been especially useful in elucidating these mechanisms are the analysis of the properties of isolated enzymes (enzymes removed from cells) and observation of the behavior of mutant bacteria with specific enzyme defects. Mutant studies have been important in establishing that the regulatory mechanisms postulated from isolated enzyme studies do indeed function with the cell.

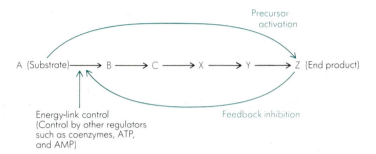

Figure 9-11. Some mechanisms for the regulation of enzyme activity by direct control through a coupling of the catalytic mechanism with other processes. Feedback inhibition, precursor activation, and energy-link control are shown. See text for fuller explanation.

General Processes Regulating Enzyme Activity

The cell can regulate enzyme activity by *less specific* or *more general* processes than those just described. These controls do not require the participation of effector molecules. The following cases illustrate this.

The cell membrane is a barrier to most hydrophilic molecules, but it has systems for the transport of specific compounds. Most of these systems require metabolic energy to function and therefore can be controlled by the availability of ATP.

The enzymatic reaction rate can be controlled by substrate concentration. As the substrate concentration increases, the reaction rate increases until a limiting value is reached when all the enzymes are saturated. And as the product accumulates, the reaction rate decreases. In addition, the concentrations of coenzymes and cofactors can exert controlling influences.

Control can also be effected by compartmentalization within the cell. Enzymes may be bound to various internal structures, especially membranes and macromolecules, so that enzymes and substrates are not in direct contact. The limiting physical access of enzymes to their substrates is more evident in the eucaryotic cell. Substrates can exist in separate **pools** because of their location within various membrane-bound organelles.

In some microbes, highly specific **proteolytic** (protein-degrading) **enzymes,** or **proteases,** break down other enzymes which are no longer required for metabolic reactions.

In a few instances, alteration in enzyme activity is brought about by a phenomenon called **covalent modification** of the regulatory enzyme molecule itself so that it can switch back and forth from an active to an inactive form. This modification is accomplished by the action of other enzymes. For example, the activity of phosphorylase a is increased by the hydrolytic action of phosphorylase phosphatase:

$$\text{Phosphorylase } a + 2H_2O \xrightarrow{\text{phosphorylase phosphatase}} \text{phosphorylase } b + 2P_i$$
$$\text{(More active)} \qquad\qquad\qquad\qquad\qquad\qquad\qquad \text{(Less active)}$$

In turn, phosphorylase b can be changed back into the more active form, phosphorylase a, by another enzyme in the following manner:

$$\text{Phosphorylase } b + 2ATP \xrightarrow{\text{phosphorylase kinase}} \text{phosphorylase } a + 2ADP$$
$$\text{(Less active)} \qquad\qquad\qquad\qquad\qquad\qquad \text{(More active)}$$

Figure 9-12 indicates the general processes involved in the regulation of enzyme activity.

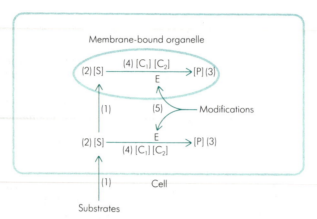

Figure 9-12. General processes for control of enzyme activity in a cell. (1) Membrane barrier controlled by energy (ATP) supply (*active transport*). (2) Substrate concentration [S]. (3) Product concentration [P]. (4) Concentrations of coenzymes [C_1] and cofactors [C_2]. (5) Modifications to enzymes (E) by other enzymes, e.g., proteases.

The regulation of enzyme activity provides a continuous control of metabolite concentration and is more responsive to fluctuations in a cell's environment than is regulation of enzyme synthesis, which we will now discuss.

REGULATION OF ENZYME SYNTHESIS

When the product of an enzymatic pathway is no longer required by a cell, the enzymes that catalyze the reactions of the pathway become unnecessary. Control mechanisms that modulate the enzymatic composition of the cell can come into play. This regulation is effected at the level of gene expression.

Induction and Repression of Enzyme Synthesis

Enzymes may be divided into two groups as follows:

1 *Constitutive enzymes.* These are always produced by the cell. Some of the enzymes of *glycolysis*, or sugar breakdown, are constitutive enzymes. They are found in essentially the same amounts regardless of the concentrations of their substrates in the medium.
2 *Inducible enzymes.* These are produced by the cell only in response to the presence of a particular substrate; they are produced, in a sense, only when needed. The process is referred to as **enzyme induction,** and the substrate (or a compound structurally similar to the substrate) responsible for evoking formation of the enzyme is an **inducer.** An example of an inducible enzyme is β-galactosidase; its inducer is the sugar lactose.

In reality, the distinction presented by these definitions is more operational than literal. Inducible enzymes are believed to exist in noninduced cells (in the absence of an inducer) but in relatively low quantities. Likewise, production of constitutive enzymes can often be enhanced when their specific substrates are present. The technique of evoking new enzyme formation through use of inducers has been extensively exploited in research designed to elucidate the mechanism of enzyme formation.

Induction is the process that occurs when an **inducer** (the effector molecule), which is either the substrate or a compound related to the substrate of the enzyme-catalyzed reaction, is required for enzyme synthesis to occur. **Repression** is the process that takes place when a regulatory protein, the **repressor,**

binds to a specific segment on the DNA called the **operator,** thereby preventing or repressing the synthesis of specific enzymes. Effector molecules, either products or related compounds of the particular reaction, act as **corepressors** in preventing synthesis of the enzyme. Corepressors function by combining with the repressor to form an active complex which combines with the operator gene to prevent messenger ribonucleic acid (mRNA) synthesis by the **structural genes** (see below). The repressor is also capable of combining with an inducer to form an inactive complex incapable of binding to the operator gene, in which case synthesis of mRNA can proceed. (The essential role of mRNA in protein synthesis is discussed in Chap. 11.) It is seen that the operator gene is one of the **regulator genes** on a deoxyribonucleic acid (DNA) chromosome. As discussed above, the operator gene prevents gene expression by **negative control** (see Fig. 9-13). In other cases, it can enhance gene expression by **positive control.** In this case, the repressor binds to the inducer, undergoes a conformational change, and is converted into an activator, which triggers gene expression.

To synthesize a specific protein, an organism must also have structural genes for that protein on the chromosome. A structural gene determines the amino acid sequence of a protein molecule (see Chap. 11). Unlike the regulator genes, structural genes do not control the rate at which enzymes are produced. Genetic control of the rate of enzyme synthesis is directed by the regulator genes.

In many bacteria the structural genes governing the biosynthesis of proteins are positioned in the exact order of the sequence of reactions in the particular metabolic pathway. This means that the ordering of the sequential reactions in the metabolic pathway is directed by the chromosome. A group of such consecutive genes forming an operational unit was named an **operon** by Francois Jacob and Jacques Monod. The operon includes both the structural and associated regulator genes. The regulator genes function primarily at the level of **transcription** (enzymatic process whereby the genetic information in DNA is used to specify a complementary sequence of bases in a mRNA chain) and not at the level of **translation** (process in which the genetic information present in a mRNA molecule directs the sequence of amino acids during protein synthesis). Clus-

Figure 9-13. Regulation of enzyme synthesis by negative control.

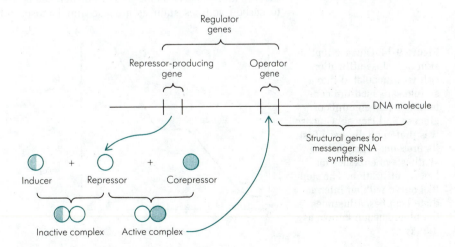

tering of related genes provides a simple way of coordinating the response to a particular environmental change.

End-Product Repression

In biosynthetic pathways, the important metabolite in overall regulation is the end product and not the substrate of the pathway. In many bacteria, the addition of such an end product (e.g., an amino acid) to the culture medium results in inhibition of the *synthesis* of the enzymes of the particular pathway. This process is termed **end-product repression** or **feedback repression.**

Enzymes that are usually subject to end-product repression can be **derepressed** when the intracellular concentration of the end product falls to a low level.

We see then that biosynthetic pathways are subject to two types of feedback control: feedback inhibition of enzyme activity discussed previously, and end-product repression of enzyme synthesis. Both mechanisms are highly complementary, bringing about highly efficient regulation of biosynthetic pathways.

Catabolite Repression

Induction and end-product repression of enzyme synthesis are *specific* responses to a particular metabolite or closely related group of metabolites which we have called effector molecules. Two examples are induction of the lactose utilization system by some β-galactosides (see Chap. 12), and end-product repression of the enzymes of arginine biosynthesis by arginine. There are other important controls that are more *general* in their action and affect many operons. These controls allow cells to use the substrate that supports the most rapid rate of growth (the "preferred" substrate) in the presence of several others. A good example of this is the **glucose effect** (see Fig. 9-14). In a medium containing both glucose and lactose, *E. coli* uses glucose preferentially. Lactose is not metabolized until all the glucose is used up. The enzymes for catabolism of lactose are not synthesized (even though the substrate, or inducer, is present) until the glucose in the medium is exhausted. This type of regulation is called **catabolite repression** of enzyme synthesis and takes place whenever rapidly metabolizable energy sources are available in the presence of energy sources that are more slowly metabolized. Further, catabolite repression is not restricted to carbon sources such as glucose and lactose; enzymes degrading nitrogen-

Figure 9-14. Glucose repression of lactose utilization. *E. coli* was inoculated into a simple salts medium containing equal amounts of glucose and lactose. Glucose was preferentially used first; its presence repressed the synthesis of enzymes for lactose utilization. The steplike curve with an intermediate lag phase illustrates the phenomenon known as *diauxy.*

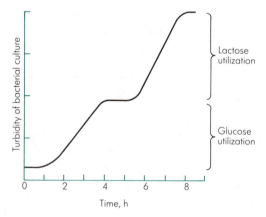

containing metabolites are also subject to catabolite repression if preferred nitrogen sources such as ammonium ion or glutamine are present. It is apparent that catabolite repression allows cells to save energy by not expending it on the synthesis of enzymes used in less efficient pathways.

SOME DIFFERENCES BETWEEN PROCARYOTIC AND EUCARYOTIC ENZYME REGULATION

There are some aspects of enzyme regulation that are not the same between procaryotes and eucaryotes. For example, the difference between constitutive (basal) and fully induced enzyme levels is usually less for eucaryotes than it is for procaryotes. In the eucaryotic yeast *Saccharomyces cerevisiae* arginase is induced about 10-fold over the basal level while in *E. coli* the enzyme can be induced to about 100 times over basal level. In addition, in those eucaryotic organisms examined so far in sufficient genetic detail, there is no significant gene clustering into operons. Instead, the structural genes for enzymes in a specific biosynthetic pathway generally are scattered over many chromosomes and are not linked to each other. However, regulator genes can act at separate regulatory sites to coordinate enzyme synthesis.

QUESTIONS

1 Why are enzymes important to a cell?
2 Define the following terms: apoenzyme, holoenzyme, coenzyme, cofactor, and pathway.
3 Describe two most striking characteristics of enzymes.
4 What is meant by an enzyme system?
5 Distinguish between a trivial name and a systematic name in enzyme nomenclature.
6 What are the six major classes of enzymes and their catalytic reactions?
7 What is the main function of an enzyme and how is it accomplished?
8 What is an active site of an enzyme?
9 Discuss the conditions that affect the activity of an enzyme.
10 Distinguish between the following types of inhibition of enzyme action: nonreversible, reversible, competitive, and noncompetitive inhibition.
11 Differentiate between constitutive and induced enzymes.
12 Describe the technique you would use to identify a microorganism by its enzymatic activities.
13 What advantages might possibly be gained by performing studies of enzyme action by means of resting cells in preference to a growing culture? Cell-free extracts in preference to whole cells?
14 Why is the existence of cellular regulatory mechanisms more important in a microbial cell than in a cell of a higher organism?
15 Describe the two different regulatory mechanisms of enzymes in a cell and the two general properties that they share.
16 What are allosteric enzymes?
17 Describe three specific types of regulation of enzyme activity.
18 Explain what is meant by the negative control of enzyme induction.
19 Give the meaning of the term *operon*.
20 What are the two types of feedback control in biosynthetic pathways?
21 How does the phenomenon of diauxy reflect the meaning of catabolite repression?

REFERENCES

Dawes, I. W., and I. W. Sutherland: *Microbial Physiology*, Blackwell Scientific Publications, Oxford, 1976. *A short paperback book serving as an introduction to microbial physiology. It provides a foundation for further reading in specific aspects of the discipline. Chapter 7, "Regulation of Metabolism," is pertinent to our discussion on enzyme regulation.*

International Union of Biochemistry: *Enzyme Nomenclature 1978*, Academic, New York, 1979. *This book contains the recommendations of the Nomenclature Committee of the International Union of Biochemistry on the nomenclature and classification of enzymes. Chapter 1 is an historical introduction. Chapter 2 contains the discussion on the classification and nomenclature of enzymes, such as the general principles and rules. Chapter 3 lists the enzymes. Chapter 4 has the references to the enzyme list, and Chapter 5 is an index to the list.*

Lehninger, A. L.: *Principles of Biochemistry*, Worth Publishers, New York, 1982. *This is a superb text in biochemistry written specifically for students taking their first course in biochemistry. It describes the biochemical aspects of cell structure and function with a molecular biology approach. Chapter 9 gives a clear, general treatment of the nature and characteristics of enzyme-catalyzed reactions.*

Mandelstam, J., and K. McQuillen (eds.): *Biochemistry of Bacterial Growth*, 2d ed., Blackwell Scientific Publications, Oxford, 1973. *There is a good chapter on regulation entitled "Coordination of Metabolism." Many of the metabolic pathways depicted have arrows indicating modulator control. This innovation adds greatly to the reality that metabolic maps are intended to convey.*

Stanier, R. Y., E. A. Adelberg, and J. Ingraham: *The Microbial World*, 4th ed., Prentice-Hall, Englewood Cliffs, N.J., 1976. *An advanced general text on microbiology with emphasis on the physiology and molecular biology of microorganisms. Chapter 8 is very relevant to enzyme regulation.*

Chapter 10   Microbial Metabolism: Energy Production

The multiplicity of processes performed by all biological systems can be traced, directly or indirectly, to certain chemical reactions. Even the shape of a bacterium depends on such reactions: shape is determined by the geometric structure of the rigid peptidoglycan component of the cell wall, the peptidoglycan in turn being determined by the series of chemical reactions involved in its synthesis. The term **metabolism** denotes all the organized chemical activities performed by a cell, which comprise two general types, **energy production** and **energy utilization.** Energy is the ability to do work, and the work of a bacterial cell is extensive and varied. Energy is utilized for the construction of the physical parts of the cell such as wall or membrane; it is required for synthesis of enzymes, nucleic acids, polysaccharides, and other chemical components; it is required for repair of damage and mere maintenance of the status quo, as well as for growth and multiplication; it is required for accumulating certain nutrients in high concentration in the cell and for keeping certain other substances out of the cell; and it is required for motility. To support such extensive activities, vast amounts of energy must be provided. Under certain optimal conditions, some bacteria have been found capable of metabolizing an amount of nutrient equivalent to their own weight *every few seconds* to provide such energy! A bacterial cell can be pictured as a dynamo of tremendous energy production. In this chapter, some basic principles of energetics will be discussed, followed by descriptions of just a few of the energy-producing mechanisms used by bacteria. Many of these mechanisms are also used by other microorganisms and by higher organisms, including human beings.

SOME PRINCIPLES OF BIOENERGETICS

Most cells obtain energy by carrying out chemical reactions which liberate energy. Some cells are able to use light as a source of their energy, but even here the light energy must be converted into chemical energy to be in a form useful for the work of the cell.

In the course of any chemical reaction, energy available for the performance of *useful* work is either released or absorbed. The amount of energy liberated or taken up during the course of a reaction is referred to as the **free-energy change** (ΔG) of the reaction. Thus free-energy change can be defined as useful energy. ΔG is expressed in terms of calories; however, this is merely a convenience since the free energy is not always in the form of heat but can, instead, be in the form of chemical energy. If the ΔG of a chemical reaction has a negative value (such as -8000 cal), the reaction releases energy (an **exergonic** reaction). If the ΔG of a reaction has a positive value (such as $+3000$ cal), the reaction requires energy (an **endergonic** reaction).

Concentration of reactants affects the value of ΔG for a chemical reaction, and to make valid comparisons between the energetics of various reactions, a basis of reference must be used. For purposes of comparison, it is assumed that the concentration of all reactants is 1.0 M in the steady state; this is referred to as **standard concentration.**

Under conditions of standard concentration, the free-energy change (ΔG) of a reaction is referred to by a special term, $\Delta G°$. In other words, $\Delta G°$ is the amount of free energy released (or absorbed) when one mole of the reactant is converted to one mole of product at 25°C and one atmosphere of pressure, and, under (hypothetical) conditions, where all reactants and products are maintained at 1 M concentration.

The $\Delta G°$ or standard free-energy change is related to the equilibrium constant, K_{eq}, of a chemical reaction by the equation

$$\Delta G° = -RT \ln K_{eq}$$

where R is the gas constant and T is the absolute temperature. If $\Delta G°$ is a negative value, the K_{eq} is greater than 1.0 and the formation of products is favored. If $\Delta G°$ is positive in value, K_{eq} is less than 1.0 and the chemical reaction tends to proceed in the reverse direction.

The value of $\Delta G°$ for a reaction can be calculated from the equilibrium constant of the reaction by using the above equation. For example, let us calculate $\Delta G°$ for the reaction catalyzed by phosphoglucomutase at 25°C:

$$\text{Glucose-1-P} \rightleftharpoons \text{glucose-6-P}$$

$$K_{eq} \text{ of } \frac{\text{glucose-6-P}}{\text{glucose-1-P}} = 17$$

The equation $\Delta G^\circ = -RT \ln K_{eq}$ may be rewritten as:

$$\Delta G^\circ = -2.303RT \log K_{eq}$$

Substituting values for these terms, we get

$$\Delta G^\circ = -2.303RT \log 17$$
$$\Delta G^\circ = -2.303(1.987)(298) \log 17$$
$$\Delta G^\circ = -1,680 \text{ cal/mole}$$

Since the value of ΔG° is negative, the reaction can proceed from left to right under standard conditions.

Let us now calculate ΔG under physiological conditions. (ΔG is the actual free-energy change of a given chemical reaction under the conditions of concentration, pH, and temperature actually prevailing during the reaction, which are not necessarily the standard conditions as defined above.) Suppose that at 38°C the concentrations of glucose-6-P is 1×10^{-4} M and glucose-1-P are 3×10^{-5} M. We substitute these values in the following equation:

$$\Delta G = \Delta G^\circ + 2.303RT \log K_{eq}$$

We get

$$\Delta G = -1,680 + 2.303(1.987)(311) \log \frac{(1 \times 10^{-4})}{(3 \times 10^{-5})}$$
$$\Delta G = -1,680 + 1,423 \log 3.3$$
$$\Delta G = -1,680 + 740$$
$$\Delta G = -940 \text{ cal/mole}$$

Thus, under physiological conditions, the reaction still proceeds from left to right.

It is also possible to obtain energy from a chemical reaction in the form of electric potential. Conversely, it is possible to use electric potential energy to drive a chemical reaction. Electric energy is generated when oxidations occur by the removal of electrons (as we will soon see). When the electrons fall through a potential difference or drop, energy is produced. The relationship between an oxidation-reduction potential difference and the standard free-energy change is

$$\Delta G^\circ = -nFE^\circ$$

where n is the number of moles of electrons transferred in the reaction, F is Faraday's constant (23,061 cals/V per equivalent), and E° is the standard oxidation-reduction potential difference.

For example, let us calculate ΔG° for the reaction in which cytochrome c is oxidized by oxygen from the ferrous to the ferric state with $E^\circ = +0.56$V:

$$2 \text{ Cytochrome } c\text{–}Fe^{2+} \text{ (reduced)} + \tfrac{1}{2}O_2 \rightleftharpoons 2 \text{ cytochrome } c\text{–}Fe^{3+} \text{ (oxidized)} + O^{2-}$$
$$\Delta G^\circ = -2(23,061)(0.56)$$
$$\Delta G^\circ = -25,828 \text{ cal/mole}$$

In order for life to continue, it is essential that the energy released from exergonic reactions be used to drive endergonic reactions, and living organisms have developed characteristic ways of *coupling* exergonic reactions with endergonic reactions. The basic principle involved is that there be a *common reactant*. This can be best understood by the following example.

Consider the two general reactions

$$A \rightarrow B \qquad \Delta G^{\circ\prime} = -10,000 \text{ cal (at pH 7)} \qquad (1)$$

$$C \rightarrow D \qquad \Delta G^{\circ\prime} = +5,000 \text{ cal (at pH 7)} \qquad (2)$$

(ΔG° at pH 7 is designated $\Delta G^{\circ\prime}$.) The energy liberated by the first reaction (exergonic) can be used to drive the second reaction (endergonic) by coupling the two reactions in the following manner.

$$A + X \rightarrow B + Y \qquad \Delta G^{\circ\prime} = -2,000 \text{ cal} \qquad (3)$$

$$C + Y \rightarrow D + X \qquad \Delta G^{\circ\prime} = -3,000 \text{ cal} \qquad (4)$$

Here, Y is a reactant common to both reactions. In the first reaction (3), the overall $\Delta G^{\circ\prime}$ of $-2,000$ cal indicates that 8,000 of the original 10,000 cal was used for the conversion of X to Y. In the second reaction, Y was converted back to X, thereby releasing the previously trapped 8,000 cal to drive the endergonic conversion of C→D. Thus, the overall $\Delta G^{\circ\prime}$ of the second reaction (4) is $+5,000$ $-8,000$, or $-3,000$ cal. The common reactant Y is referred to as an **energy-rich** or **energy-transfer compound.**

The common reactants of greatest use to the cell are those capable of transferring large amounts of free energy, called **high-energy-transfer compounds.** A variety of such compounds exists in cells; and although such compounds may possess no more total energy than other compounds, the energy is distributed in such a way within the molecule that one portion opposes another, resulting in considerable molecular strain. Triggering the breakdown of the molecule (by the catalytic action of an enzyme) causes release of the energy no longer held in check. A high-energy-transfer molecule is analogous to a mousetrap. When set, the trap has great energy, but the energy of the spring is opposed by the action of the catch. Tripping the catch is analogous to catalyzing the breakdown of the transfer molecule, with subsequent release of energy of the spring.

Table 10-1 lists some of the high-energy-transfer compounds found in cells, of which ATP is by far the most important. Just as money constitutes a common medium of exchange in our society, so ATP constitutes the "energy currency" of the cell in the exchange of energy between exergonic and endergonic reactions. It should be noted that all the compounds in Table 10-1 can transfer their energy directly or indirectly to ATP synthesis; for example:

1,3-Diphosphoglyceric acid + ADP → 3-phosphoglyceric acid + ATP

Energy is released from ATP by hydrolysis (Fig. 10-1). The amount of energy so

Table 10-1. Some High-Energy-Transfer Compounds Found in Cells with Their Standard Free-Energy Changes upon Hydrolysis

Compound	$\Delta G^{\circ\prime}$, kcal mol^{-1}
Adenosine triphosphate (ATP)	-7.3
Guanosine triphosphate (GTP)	-7.3
Uridine triphosphate (UTP)	-7.3
Cytidine triphosphate (CTP)	-7.3
Acetyl phosphate	-10.1
1,3-Diphosphoglyceric acid	-11.8
Phosphoenolpyruvic acid (PEP)	-14.8

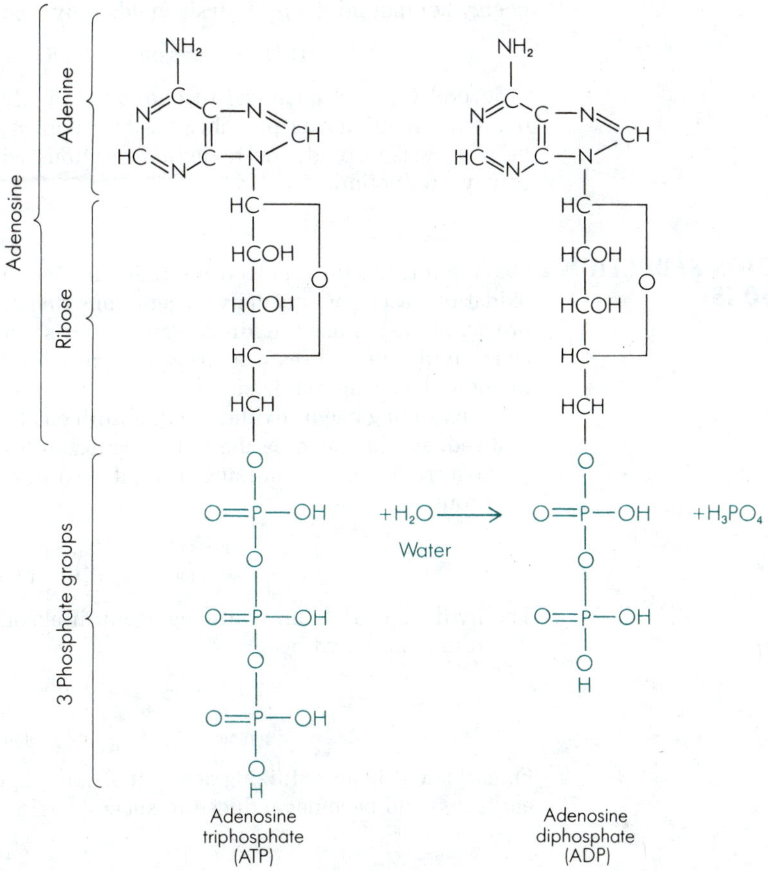

Adenosine

Adenine

Ribose

3 Phosphate groups

NH₂ ... Adenosine triphosphate (ATP)

$+H_2O \longrightarrow$

Water

NH₂ ... Adenosine diphosphate (ADP)

$+H_3PO_4$

Figure 10-1. Hydrolysis of adenosine triphosphate.

Overall reaction:
$$ATP + H_2O \rightarrow ADP + H_3PO_4; \quad \Delta G^{\circ\prime} = -7.3 \text{ kcal mol}^{-1}$$

released is a measure of the transfer energy capacity of ATP. It should be remembered, however, that the $\Delta G^{\circ\prime}$ of -7.3 kcal for ATP is not necessarily that existing in the intact cell. The pH and the concentrations of ATP, ADP, Mg^{2+}, etc., in the cell are not identical to the standard conditions employed for determining $\Delta G^{\circ\prime}$. If appropriate corrections are made, the free energy of hydrolysis within the cell is closer to -12.5 kcal, although this value can vary depending upon intracellular concentrations of various materials. However, for consistency and comparison, thermodynamic calculations of biological energy exchanges must be carried out under arbitrarily defined standard conditions.

The compound ADP (adenosine diphosphate) is also a high-energy-transfer compound, since its hydrolysis also liberates a large quantity of energy:

$$ADP + H_2O \rightarrow AMP + H_3PO_4 \qquad \Delta G^{\circ\prime} = -7.3 \text{ kcal}$$

where AMP stands for adenosine monophosphate. AMP, however, is a low-energy compound; its hydrolysis yields only a small quantity of energy:

$$AMP + H_2O \rightarrow \text{adenosine} + H_3PO_4 \qquad \Delta G^{\circ\prime} = -2 \text{ kcal}$$

Several types of chemical reactions are involved in energy production, but oxidation-reduction is probably the commonest. A discussion of some of the basic aspects of oxidation-reduction reactions will clarify their relationship to energy production.

OXIDATION-REDUCTION REACTIONS

Oxidation is the loss of electrons; reduction is the gain of electrons. Frequently, oxidation reactions are dehydrogenations (reactions involving the loss of hydrogen atoms); since a hydrogen atom consists of a proton plus an electron, a compound which loses a hydrogen atom has essentially lost an electron and therefore has been oxidized.

An oxidizing agent (oxidant) will absorb electrons and will therefore become reduced, as illustrated by the following examples.

The ferric ion is an oxidizing agent; it absorbs electrons and becomes reduced to ferrous ion:

$$Fe^{3+} \quad + \quad e^- \quad \rightarrow \quad Fe^{2+}$$

Ferric ion　　　Electron　　　Ferrous ion

The hydrogen ion is an oxidizing agent; it absorbs electrons and becomes reduced to atomic hydrogen:

$$2H^+ \quad + \quad 2e^- \quad \rightarrow \quad 2H$$

Hydrogen ion:　　Electron　　　Hydrogen atom:
proton　　　　　　　　　　　proton plus electron

Fumaric acid is an oxidizing agent; it absorbs hydrogen atoms (which contain electrons) and becomes reduced to succinic acid:

$$
\begin{array}{ccc}
\text{COOH} & & \text{COOH} \\
| & & | \\
\text{CH} & & \text{CH}_2 \\
\| & + 2e^- + 2H^+ \rightarrow & | \\
\text{HC} & & \text{CH}_2 \\
| & & | \\
\text{HOOC} & & \text{COOH} \\
\text{Fumaric acid} & & \text{Succinic acid}
\end{array}
$$

A reducing agent (reductant) donates electrons, becoming oxidized in the process. The ferrous ion is a reducing agent, it donates electrons and becomes oxidized to ferric ion:

$$Fe^{2+} \rightarrow Fe^{3+} + e^-$$

From this example, one can see that the reverse of each oxidation reaction is a reduction and the reverse of a reduction reaction is an oxidation. Moreover, in each reaction, a *pair* of substances is involved: one is the *reduced form*, the other the *oxidized form*, e.g., ferrous ion and ferric ion, succinic acid and fumaric acid. Each such pair of substances is referred to as an oxidation-reduction (O/R) system.

One O/R system may tend to absorb electrons from another O/R system; i.e., the first system will oxidize the second. On the other hand, the tendency of the first system to absorb electrons may be so low that the second system may oxidize the first. This power (the tendency to absorb electrons) is expressed by the standard oxidation-reduction potential or the electromotive potential (E_0') of an O/R system, which is measured electrically under standardized conditions of comparison (electron donor and its conjugate at 1.0 M concentration, 25°C, and pH 7.0) and expressed in volts. The more positive the E_0', the greater the oxidizing ability of the system. Consequently, any system listed in Table 10-2 can oxidize any other system listed above it, but not below it, under the standard conditions. Such relationships are very important in understanding the orderly sequence in which biological oxidations occur.

As discussed previously, when one O/R system oxidizes another, energy is released. It is important to know the values of E_0' for each system, because the $\Delta G^{\circ \prime}$ of the overall reaction is directly proportional to the difference in E_0' values. If the voltage difference is large, an amount of free energy sufficient to drive the synthesis of ATP may be liberated.

In respiration, an oxidizable substrate is the primary electron donor. In aerobic respiration the terminal electron acceptor is oxygen; in anaerobic respiration the final electron acceptor is a compound like fumarate, NO_3^-, SO_4^{2-}, or CO_3^{2-}. In fermentation, an organic compound is the final electron acceptor; an oxidizable substrate is the electron donor. In photosynthesis carried out by bacteria, bacteriochlorophylls serve as both electron donors and acceptors. In photosynthesis by green plants, algae, and cyanobacteria, water serves as a primary electron donor and $NADP^+$ (nicotinamide adenine dinucleotide phosphate) as a terminal electron acceptor. The paths through which these electrons flow in the various processes are called electron-transport chains.

Electron-transport chains are sequences of oxidation-reduction reactions that occur in cells. These reactions are mediated by a number of electron carriers and electron-carrier enzymes (discussed later). As the electrons flow through the chains, much of their free energy is conserved in the form of ATP; this process is called oxidative phosphorylation.

The multicomponent electron-transport chains are always associated with membranes. In eucaryotes, they are in mitochondrial or chloroplast membranes; in procaryotes, they are in the cytoplasmic membrane.

THE RESPIRATORY CHAIN

A respiratory chain is an electron-transport chain. When a pair of electrons or hydrogen atoms (which contain electrons) from an oxidizable substrate is coupled with the reduction of an ultimate electron acceptor, such as oxygen, there is a large free-energy change (ΔG°). The flow of electrons through the transport chain allows a stepwise release of this energy, some of which is conserved in the form of ATP at several steps in the chain. At these specific steps the difference in E_0' values is great enough to permit sufficient energy to be liberated for oxidative phosphorylation to occur.

The component O/R systems of a common type of respiratory chain are shown in Table 10-2 and are illustrated in Fig. 10-2.

A respiratory chain consists of enzymes having prosthetic groups or coenzymes. These can be regarded as the working parts of the enzymes, and in the case of the respiratory chain each is in fact an O/R system. The oxidized form of each prosthetic group or coenzyme has an absorption spectrum different from

Table 10-2. Component O/R Systems of a Respiratory Chain, with Their Corresponding E_0' Values

O/R System	E_0',V
$NAD^+/NADH + H^+$ *	-0.32
Flavoprotein/flavoprotein-H_2	-0.03
CoQ/CoQ-H_2	$+0.04$
Cyt b–Fe^{3+}/cyt b–Fe^{2+}	$+0.07$
Cyt c_1–Fe^{3+}/cyt c_1–Fe^{2+}	$+0.21$
Cyt c–Fe^{3+}/cyt c–Fe^{2+}	$+0.23$
Cyt a–Fe^{3+}/cyt a–Fe^{2+}	$+0.29$
Cyt a_3–Fe^{3+}/cyt a_3–Fe^{2+}	$+0.53$
Oxygen/water	$+0.82$

* $NADH + H^+$ may also be designated as $NADH_2$ or simply as NADH since the other hydrogen atom appears as a free H^+ ion.

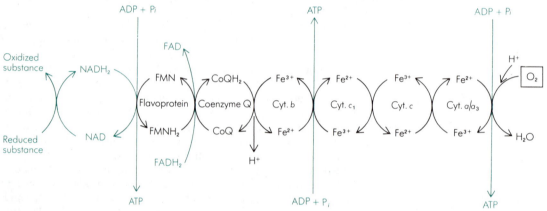

Figure 10-2. A respiratory chain, showing sequential oxidation steps and points where sufficient energy is liberated to permit synthesis of ATP. Electron transfer is accompanied by a flow of protons (H^+) from $NADH_2$ through coenzyme Q but not in later steps involving cytochromes. Note that three ATP are formed per molecule of $NADH_2$ reoxidized but only two ATP per molecule of $FADH_2$ reoxidized. (Note that this is an abbreviated representation. Actually, there are more than 15 chemical substances in the chain.)

that of the reduced form, so that the two states can be distinguished by spectrophotometry. A summary of these coenzymes follows.

Nicotinamide Adenine Dinucleotide (NAD) and Nicotinamide Adenine Dinucleotide Phosphate (NADP)

Certain enzymes which remove electrons and hydrogen ions from reduced substrates (referred to as **dehydrogenases**) have NAD^+ or $NADP^+$ as their coenzyme. NAD^+ can exist in a reduced form, $NADH + H^+$, to form an O/R system:

$$NAD^+ + 2H^+ + 2e^- \rightleftharpoons NADH + H^+$$

In the same way, $NADP^+$ can exist in a reduced state.

The vitamin niacin (nicotinic acid) forms part of the structure of NAD and NADP and is a precursor (building block) in their biosynthesis.

Flavin Adenine Dinucleotide (FAD) and Flavin Mononucleotide (FMN)

Another class of dehydrogenases known as **flavoproteins** exists and contains either FAD or FMN as prosthetic groups. One of the basic parts of their coenzyme structure is the vitamin **riboflavin.** Riboflavin can exist in either an oxidized or reduced form: Riboflavin + 2H $\rightleftharpoons$ riboflavin-H_2

The reduced forms of the coenzymes are $FADH_2$ and $FMNH_2$.

Coenzyme Q

Coenzyme Q is also called **ubiquinone** because it is a quinone and is present in all cells. Coenzyme Q is a fat-soluble coenzyme. It functions as an acceptor of reducing power from the flavin-linked dehydrogenases:

Flavoprotein-H_2 + ubiquinone $\rightleftharpoons$ ubiquinone-H_2 + flavoprotein

NAD^+, $NADP^+$, flavoproteins, and ubiquinones carry $2H^+$ and $2e^-$, but the cytochromes (discussed below) transfer only electrons, the protons being associated with an —NH_2 group or a —COO^- group and eventually transferred to O_2 (see Fig. 10-2).

Cytochromes

Another major class of oxidative enzymes in the respiratory chain is the **cytochromes.** The prosthetic group of a cytochrome is a derivative of **heme** and contains a single iron atom, which is responsible for the oxidative or reductive properties of the enzyme. On the basis of differences in absorption spectra, cytochromes can be divided into three main categories: cytochromes a, cytochromes b, and cytochromes c. Each of these groups has a different function in the respiratory chain and can be further subdivided on the basis of minor differences in absorption spectra, e.g., cytochromes c and c_1 or cytochromes a and a_3. Each cytochrome type can exist in either an oxidized or reduced form, depending on the state of the iron atom contained in their structure:

$$\underset{\text{Ferric}}{\text{Cyt } b\text{–}Fe^{3+}} + e^- \rightleftharpoons \underset{\text{Ferrous}}{\text{cyt } b\text{–}Fe^{2+}}$$

The cytochromes act sequentially to transport electrons from coenzyme Q to O_2. Cytochromes a and a_3 together are called **cytochrome oxidase.** Both of them also contain copper. But only cytochrome a_3 can react directly with oxygen.

Sequence of Oxidation

The arrangement of O/R systems in Table 10-2 according to E_0' values is based on the experimentally determined sequence of oxidation reactions in the respiratory chain illustrated in Fig. 10-2. Sufficient energy for ATP synthesis is liberated at three points along the chain. The incremental release of energy in the respiratory chain results in a more efficient trapping of energy in ATP than would be true of direct oxidation of the reduced substrate by oxygen.

The respiratory chain of bacteria is associated with the *cytoplasmic membrane;* that of eucaryotes is present in mitochondrial membranes. Much of the electron transfer in membranes is accomplished within highly integrated particles or complexes.

ENERGY PRODUCTION BY ANAEROBIC PROCESSES

Heterotrophic bacteria can use a variety of organic compounds as energy sources. These compounds include carbohydrates, organic and fatty acids, and amino acids. For many microorganisms the preferred compounds are carbohydrates, especially the 6-carbon sugar glucose.

Glycolysis

The most common pathway of glucose catabolism is the Embden-Meyerhof pathway of glycolysis ("splitting of sugar"). This process occurs very widely and is found in microorganisms as well as in animals and plants. Glycolysis does not require the presence of oxygen and therefore can occur in both aerobic and anaerobic cells. Aerobic cells degrade glucose by glycolysis, and this process constitutes the preparatory stage for the aerobic phase of glucose oxidation. Thus, under anaerobic conditions this situation prevails:

$$\text{Glucose} \xrightarrow{\text{fermentation}} \text{fermentation products}$$

whereas under aerobic conditions, the following occurs:

$$\text{Glucose} \xrightarrow{\text{fermentation}} \underset{\text{(pyruvate)}}{\text{intermediate}} \xrightarrow[\text{O}_2]{\text{respiration}} CO_2 + H_2O$$

In glycolysis, as shown in Fig. 10-3, fructose-1,6-diphosphate formed from glucose is split into two 3-carbon units (dihydroxyacetone phosphate and glyceraldehyde-3-phosphate), and they are subsequently oxidized to pyruvic acid. At the step where glyceraldehyde-3-phosphate is oxidized, a pair of electrons (two hydrogen atoms) is removed. In the absence of oxygen, this pair of electrons may be used to reduce pyruvic acid to lactic acid or ethanol. In the presence of oxygen, this pair of electrons may enter the respiratory chain.

Many of the reactions of the glycolytic pathway are freely reversible and can be used for the synthesis of glucose as well as for its breakdown. Only three of the reactions are not reversible by common enzymes; but the presence of other enzymes can reverse them for glucose synthesis to occur. Thus phosphoenolpyruvate is synthesized from pyruvate by the action of phosphoenolpyruvate synthase and specific phosphatases hydrolyze fructose-1,6-diphosphate and glucose-6-phosphate in the biosynthetic direction. The enzymes at these steps in the degradative direction are kinases and require ATP. (See Fig. 10-3.) For each molecule of glucose metabolized, two molecules of ATP are used up and four molecules of ATP are formed. Therefore for each molecule of glucose metabolized by glycolysis, there is a net yield of two ATP molecules. This is shown in Fig. 10-3.

The overall reaction of glycolysis can be summarized as follows:

$$\underset{\text{Glucose}}{C_6H_{12}O_6} + 2NAD + 2ADP + \underset{\substack{\text{Inorganic}\\\text{phosphate}}}{2P_i} \rightarrow \underset{\text{Pyruvic acid}}{2CH_3COCOOH} + 2NADH_2 + 2ATP$$

The Pentose Phosphate Pathway

The pentose phosphate pathway, like the glycolytic one, is another catabolic reaction pathway that exists in both procaryotic and eucaryotic cells. Since it involves some reactions of the glycolytic pathway, it has been viewed as a "shunt" of glycolysis; hence it may also be called the hexose monophosphate shunt. Its other synonym is the phosphogluconate pathway.

Glucose can be oxidized by the pentose phosphate pathway with the liberation of electron pairs, which may enter the respiratory chain. However, this cycle is not generally considered a major energy-yielding pathway in most microorganisms. It provides reducing power in the form of NADPH + H⁺, which is required in many biosynthetic reactions of the cell, and it provides pentose phosphates

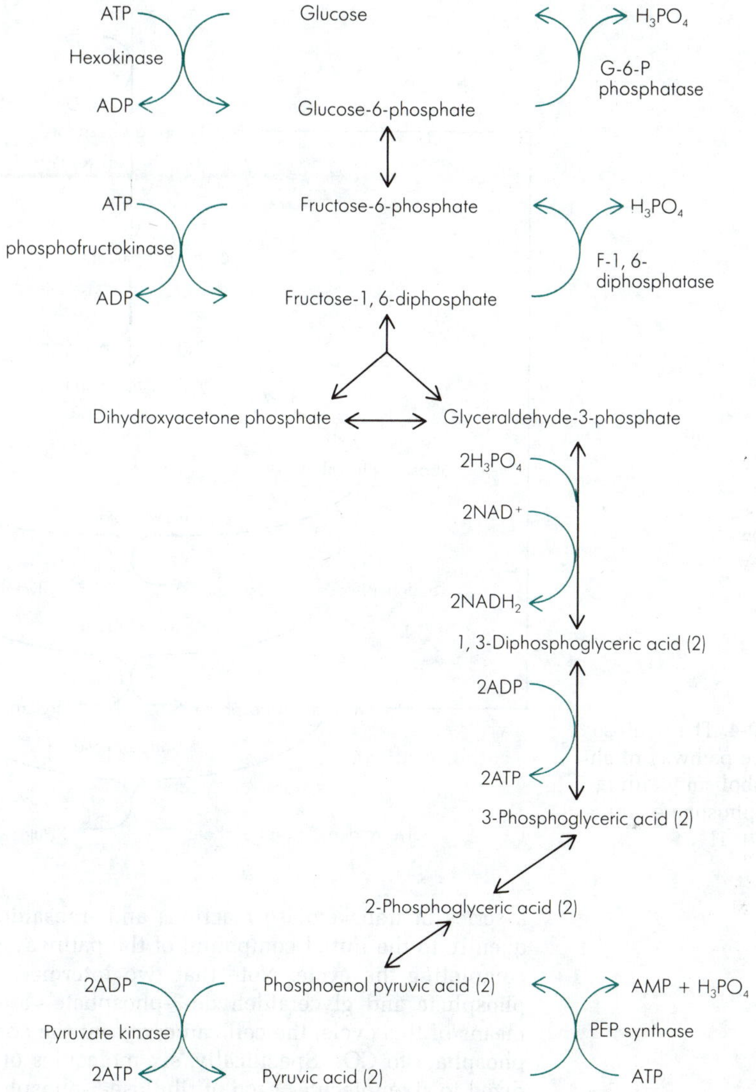

Figure 10-3. The Embden-Meyerhof (glycolytic) pathway of glucose catabolism. Enzymes shown are for those steps which are not freely reversible by a common enzyme.

for use in nucleotide synthesis. Although it can produce energy for the cell as an alternate pathway for the oxidation of glucose, it is also a mechanism for obtaining energy from 5-carbon sugars.

As seen in Fig. 10-4, the pentose phosphate pathway involves the initial phosphorylation of glucose to form glucose-6-phosphate; the latter is oxidized to 6-phosphogluconic acid with the simultaneous production of NADPH. Decarboxylation of 6-phosphogluconic acid, together with a yield of NADPH, produces ribulose-6-phosphate. Epimerization reactions yield xylulose-5-phosphate and ribose-5-phosphate. These two compounds are the starting point for

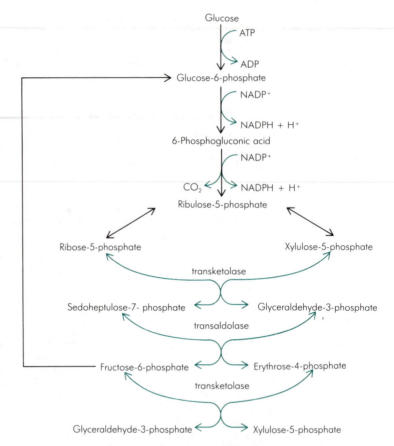

Figure 10-4. The pentose phosphate pathway of glucose catabolism yielding ribose-5-phosphate and $NADPH + H^+$.

a series of transketolase reactions and transaldolase reactions leading subsequently to the initial compound of the pathway, 6-phosphogluconic acid, thus completing the cycle. Note that two intermediates of glycolysis—fructose-6-phosphate and glyceraldehyde-3-phosphate—are generated. Theoretically, by means of this cycle, the cell can carry out the complete oxidation of glucose-6-phosphate to CO_2. Specifically, six molecules of glucose-6-phosphate are oxidized to six molecules each of ribulose-5-phosphate and CO_2; five molecules of glucose-6-phosphate are then regenerated from the six molecules of ribulose-5-phosphate. The overall equation is as follows:

$$6 \text{ Glucose-6-phosphate} + 12NADP^+ \rightarrow$$
$$5 \text{ glucose-6-phosphate} + 6CO_2 + 12NADPH + 12H^+ + P_i$$

The net equation is therefore:

$$\text{Glucose-6-phosphate} + 12NADP^+ \rightarrow 6CO_2 + 12NADPH + 12H^+ + P_i$$

In the real situation, it is more probable that the pentose phosphate pathway feeds into the glycolytic pathway by means of fructose-6-phosphate and glyceraldehyde-3-phosphate.

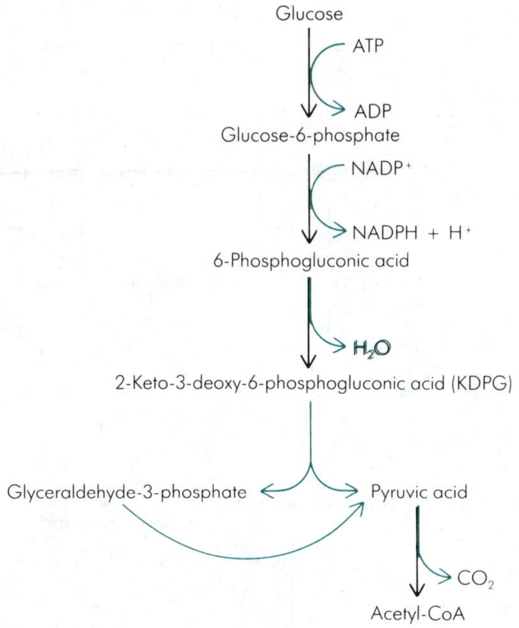

Glucose

ATP

ADP

Glucose-6-phosphate

NADP$^+$

NADPH + H$^+$

6-Phosphogluconic acid

H$_2$O

2-Keto-3-deoxy-6-phosphogluconic acid (KDPG)

Glyceraldehyde-3-phosphate ⟷ Pyruvic acid

CO$_2$

Acetyl-CoA

Figure 10-5. The Entner-Doudoroff pathway of glucose catabolism for some bacteria.

The Entner-Doudoroff Pathway

Another pathway of glucose catabolism is called the Entner-Doudoroff pathway. It is found in both aerobic and anaerobic procaryotes but not in eucaryotes. It is fairly widespread, particularly among Gram-negative bacteria. As shown in Fig. 10-5, glucose is phosphorylated to glucose-6-phosphate. It is then oxidized to 6-phosphogluconic acid. A dehydration step follows to yield 2-keto-3-deoxy-6-phosphogluconic acid (KDPG); the latter is cleaved to pyruvic acid and glyceraldehyde-3-phosphate, which is metabolized via some Embden-Meyerhof pathway enzymes to produce a second molecule of pyruvic acid. In the aerobic pseudomonads the catabolism is completed via acetyl-CoA and the tricarboxylic acid cycle (see later in this chapter).

Fermentation

Anaerobes also produce energy by reactions called **fermentations,** which use organic compounds as electron donors and acceptors. Facultative anaerobic bacteria and obligately anaerobic bacteria employ many different kinds of fermentations to produce energy. The lactic fermentation is a typical example. *Streptococcus lactis*, the bacterium responsible for the normal souring of raw milk, dissimilates glucose to lactic acid, which accumulates in the medium as the sole fermentation product. How does this happen? By glycolysis (Fig. 10-3), one molecule of glucose is converted to two molecules of pyruvic acid with concomitant production of two NADH + H$^+$. The pyruvic acid is converted to lactic acid in the following reaction:

$$2 \begin{array}{c} \text{COOH} \\ | \\ \text{C}{=}\text{O} \\ | \\ \text{CH}_3 \end{array} + 2\text{NADH} + 2\text{H}^+ \rightleftarrows 2 \begin{array}{c} \text{COOH} \\ | \\ \text{H}{-}\text{C}{-}\text{OH} \\ | \\ \text{CH}_3 \end{array} + 2\text{NAD}^+$$

Pyruvic acid   Lactic acid

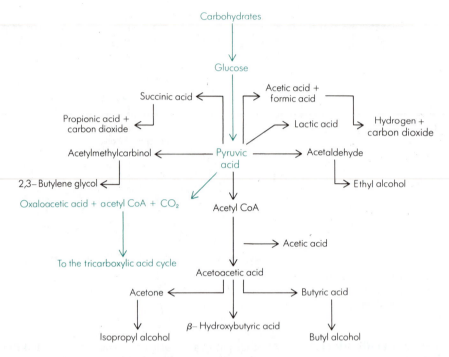

Figure 10-6. Pyruvic acid is regarded as the key compound in the dissimilation of glucose, as shown in this schematic illustration.

Insufficient energy for ATP synthesis results from this reaction; however, NAD^+ is regenerated for further use as an oxidant.

In other carbohydrate fermentations, the initial stages of glucose dissimilation frequently, but not always, follow the scheme of glycolysis. Differences in carbohydrate fermentations usually occur in the ways the resulting pyruvic acid is used. Thus pyruvic acid is the "hub" of carbohydrate fermentations. Figure 10-6 illustrates the variety of products resulting from the metabolism of pyruvic acid.

Most heterotrophic bacteria produce several end products of the types indicated in Fig. 10-6 from glucose dissimilation, but no single species produces all these end products. The types listed represent a summary of what can be expected when one takes an inventory of the end products of glucose dissimilation by all heterotrophs. Actually, it is possible to group microorganisms on the basis of their products of fermentation (the lactic acid group or the propionic acid group of bacteria, for example, as shown in Table 10-3). Such designations are established on the basis of the major end products of carbohydrate fermentation. From this it is evident that not all microorganisms metabolize the same substrate in exactly the same manner. For example, *Streptococcus lactis* and *Escherichia coli* both ferment glucose but by quite different pathways of fermentation, as shown in Fig. 10-7.

However, some anaerobes do not have a functional glycolytic system. They may have carbohydrate fermentation pathways that use the pentose phosphate pathway and the Entner-Doudoroff pathway. Fermentations of noncarbohydrate substrates, such as amino acids, involve highly specific pathways.

Table 10-3. Bacteria Grouped According to Major Products of Glucose Dissimilation

Groups with Examples of Some Genera	Representative Products

Lactic acid bacteria
    Streptococcus
    Lactobacillus
    Leuconostoc
	Lactic acid only or lactic acid plus acetic acid, formic acid, and ethyl alcohol; species producing only lactic acid are homofermentative, and those producing lactic acid plus other compounds are heterofermentative

Propionic acid bacteria
    Propionibacterium
    Veillonella
	Propionic acid plus acetic acid and carbon dioxide

Coli-aerogenes-typhoid bacteria
    Escherichia
    Enterobacter
    Salmonella
	Formic acid, acetic acid, lactic acid, succinic acid, ethyl alcohol, carbon dioxide, hydrogen, 2,3-butylene glycol (produced in various combinations and amounts depending on genus and species)

Acetone, butyl alcohol bacteria
    Clostridium
    Eubacterium
    Bacillus
	Butyric acid, butyl alcohol, acetone, isopropyl alcohol, acetic acid, formic acid, ethyl alcohol, hydrogen, and carbon dioxide (produced in various combinations and amounts depending on species)

Acetic acid bacteria Acetobacter	Acetic acid, gluconic acid, kojic acid

Figure 10-7. Glucose is fermented by many different bacteria and in many different ways. (A) *Escherichia coli* fermentation of glucose results in a mixture of products, whereas (B) *Streptococcus lactis* fermentation of glucose produces lactic acid almost exclusively.

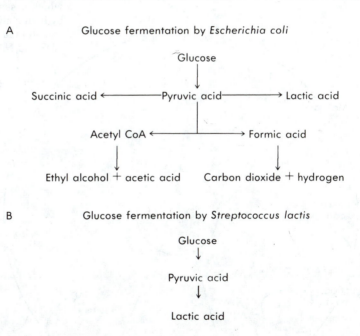

A   Glucose fermentation by *Escherichia coli*

B   Glucose fermentation by *Streptococcus lactis*

ENERGY PRODUCTION BY AEROBIC PROCESSES

The Tricarboxylic Acid Cycle

The tricarboxylic acid (TCA) cycle is a sequence of reactions that generate energy in the form of ATP and reduced coenzyme molecules ($NADH_2$ and $FADH_2$). It also performs other functions. Many intermediates in the cycle are precursors in the biosynthesis of amino acids, purines, pyrimidines, etc. For example, oxaloacetic acid and α-ketoglutaric acid are amino acid precursors as shown in the following:

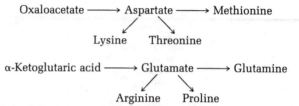

Thus the TCA cycle is an **amphibolic cycle,** which means that it functions not only in catabolic (breakdown) but also in anabolic (synthesis) reactions. The cycle is shown in Fig. 10-8.

The overall reaction of the TCA cycle can be summarized as follows:

$$\text{Acetyl-CoA} + 3H_2O + 3NAD^+ + FAD + ADP + P_i \rightarrow$$
$$2CO_2 + CoA + 3NADH_2 + FADH_2 + ATP$$

Since the breakdown of glucose by glycolysis yields two acetyl-CoA molecules

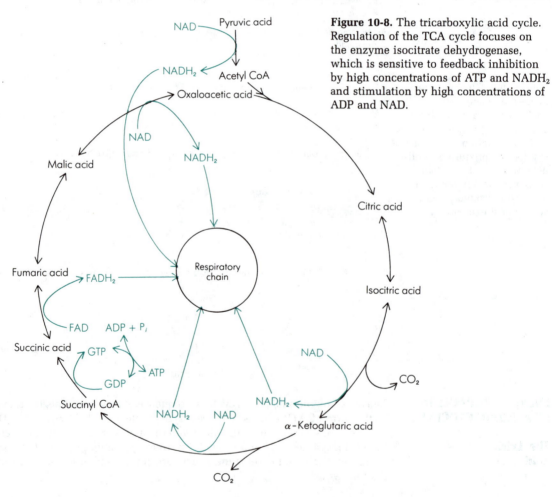

Figure 10-8. The tricarboxylic acid cycle. Regulation of the TCA cycle focuses on the enzyme isocitrate dehydrogenase, which is sensitive to feedback inhibition by high concentrations of ATP and $NADH_2$ and stimulation by high concentrations of ADP and NAD.

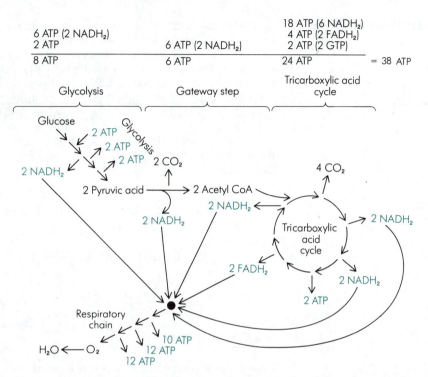

Figure 10-9. ATP yield per glucose molecule broken down in aerobic respiration.

which can enter this cycle, the overall equation for the cycle, per glucose molecule broken down, is twice the above.

Energy Yield in Aerobic Respiration

We may now look at the energy yield from the aerobic breakdown of one molecule of glucose when the electrons stored in the reduced coenzyme molecules are fed into the electron-transport chain. As shown previously, the electrons are transferred stepwise from the coenzyme carriers to molecular oxygen, and this transfer is coupled to the generation of ATP by oxidative phosphorylation.

For each glucose molecule broken down, there are 12 reduced coenzymes to be oxidized: 2 $FADH_2$ (1 from each turn of the TCA cycle) and 10 $NADH_2$ (2 from glycolysis; 2 from the gateway step between glycolysis and the TCA cycle, i.e., pyruvic acid to acetyl-CoA; and 6 from two turns of the TCA cycle). Since 3 ATP are produced from each $NADH_2$ and 2 ATP from each $FADH_2$, there are 34 ATP generated from the reduced coenzymes via oxidative phosphorylation through the respiratory chain. But the total yield of ATP from the aerobic respiration of 1 glucose molecule is 38: 34 from the oxidation of reduced coenzymes, 2 from glycolysis, and 2 from the side reaction of the TCA cycle, that is, from 2 GTP. The total ATP yield per glucose molecule from aerobic respiration is summarized in Fig. 10-9.

The complete oxidation of glucose via glycolysis, the TCA cycle, and the respiratory chain is summarized in this overall reaction:

$$C_6H_{12}O_6 + 6O_2 \rightarrow 6CO_2 + 6H_2O$$
Glucose

Catabolism of Lipids

Glucose is the single most important source of energy for most cells. However, for many microorganisms, other substances, such as lipids and proteins, may be used as alternate sources of energy. There is a general rule that governs their utilization: they are converted as quickly and efficiently as possible into intermediates of the glycolytic and TCA pathways so that a minimum number of additional enzymes is required to effect complete breakdown. This rule highlights the fact that the glycolytic pathway and the TCA cycle serve as a common center around which other catabolic pathways are built.

The breakdown of lipids or fats begins with the cleavage of triglycerides by the addition of water to form glycerol and fatty acids by means of enzymes called **lipases**:

$$
\begin{array}{l}
H_2C-O-\overset{\displaystyle O}{\overset{\|}{C}}-R_1 \\[2mm]
HC-O-\overset{\displaystyle O}{\overset{\|}{C}}-R_2 + 3H_2O \xrightarrow{\text{lipase}} \\[2mm]
H_2C-O-\overset{\displaystyle O}{\overset{\|}{C}}-R_3 \\
\end{array}
\qquad
\begin{array}{ll}
H_2C-OH & HO-\overset{\displaystyle O}{\overset{\|}{C}}-R_1 \\[2mm]
HC-OH & + \; HO-\overset{\displaystyle O}{\overset{\|}{C}}-R_2 \\[2mm]
H_2C-OH & HO-\overset{\displaystyle O}{\overset{\|}{C}}-R_3 \\
\end{array}
$$

Triglyceride
(—R = hydrocarbon chain)
    Glycerol   Fatty acids

Glycerol as a component of fats can be converted into an intermediate of the glycolytic pathway (dihydroxyacetone phosphate) by the following reactions:

$$\text{Glycerol} + \text{ATP} \xrightarrow[\text{Mg}^{2+}]{\text{glycerol kinase}} \text{ADP} + \text{glycerol-3-phosphate}$$

$$\text{Glycerol-3-phosphate} + \text{NAD}^+ \xrightarrow{\substack{\text{glycerol phosphate} \\ \text{dehydrogenase}}} \text{dihydroxyacetone phosphate} + \text{NADH}_2$$

The dihydroxyacetone phosphate formed would be broken down by the mechanisms shown in Fig. 10-3. Fatty acids are oxidized by the successive removal of 2-carbon fragments in the form of acetyl-CoA, a process known as **β-oxidation.** The acetyl-CoA formed can then enter the TCA cycle, and the hydrogen atoms and their electrons enter the respiratory transport chain, leading to oxidative phosphorylation.

There is more energy yield per gram of fat than per gram of carbohydrate. However, relatively few microbial species are effective in breaking down lipids of either simple or complex types, partly because of the limited solubility of lipids.

Catabolism of Proteins

Many heterotrophic microorganisms can degrade exogenous proteins, using the products as carbon and nitrogen energy sources. Since protein molecules are too large to pass into the cell, bacteria secrete exoenzymes called proteases that hydrolyze exogenous proteins to peptides, which are then transported into the cell cytoplasm.

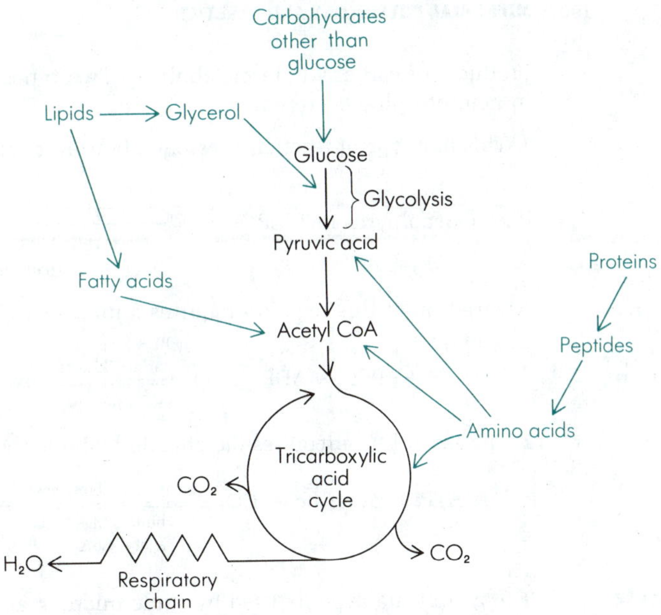

Figure 10-10. Metabolism of carbohydrates, lipids, and amino acids. As can be seen from the diagram, acetyl-CoA is a common intermediate of carbohydrate and lipid metabolism, and the TCA cycle is the common pathway for oxidation of carbohydrates, lipids, and amino acids.

Bacteria produce peptidases that break down peptides to the individual amino acids, which are then broken down according to the specific amino acid and the species or strain of bacteria breaking it down. This process may be shown as follows:

$$\text{Proteins} \xrightarrow{\text{proteases}} \text{peptides} \xrightarrow{\text{peptidases}} \text{amino acids}$$

Where amino acids are broken down, the carbon skeletons of the amino acids undergo oxidation to compounds that may enter the TCA cycle for further oxidation. Entry into the TCA cycle can be via acetyl-CoA, α-ketoglutaric acid, succinic acid, fumaric acid, or oxaloacetic acid.

An overall view of the dissimilation of carbohydrates, lipids, and proteins is shown in Fig. 10-10.

Respiration without Oxygen in Some Bacteria

Some bacteria which are ordinarily aerobic can grow anaerobically if nitrate is present. For example, *Aquaspirillum itersonii*, an aquatic bacterium, is dependent on oxygen unless potassium nitrate is added to the medium. In such cases nitrate essentially substitutes for oxygen as the final electron acceptor in the respiratory chain. This process is termed **anaerobic respiration**. The pathways for the dissimilation of the carbon and energy sources are identical with those in aerobic respiration, and electron transport occurs via a respiratory chain similar to that in aerobic cells. Oxygen is replaced as the terminal electron acceptor by nitrate. However, in some strict anaerobes, other compounds, such as carbon dioxide, or ions, such as sulfate ion, can be the terminal electron acceptors.

Heterotrophic CO_2 Fixation

This phenomenon (unrelated to autotrophic CO_2 fixation) is important because it provides a mechanism for synthesis of compounds of the TCA cycle from the products of carbohydrate metabolism. Two types of CO_2-ixing reactions occur in heterotrophic bacteria.

1 This first type of reaction is essentially irreversible and occurs in many bacteria:

$$\text{Phosphoenolpyruvate (PEP)} + CO_2 \xrightarrow[\text{acetyl-CoA}]{\substack{\text{PEP} \\ \text{carboxylase}}}$$
$$\text{oxaloacetate} + P \text{ (inorganic phosphate)}$$

A variation of this reaction requires a nucleoside diphos; hate:

$$\text{PEP} + \text{ADP} + CO_2 \underset{\text{Mg}^{2+}}{\overset{\substack{\text{PEP} \\ \text{carboxykinase}}}{\rightleftarrows}} \text{oxaloaceta} + \text{ATP}$$

2 The second type requires the vitamin biotin for activity:

$$\text{ATP} + \text{pyruvate} + CO_2 \underset{\substack{\text{biotin, Mg}^{2+}, \\ \text{acetyl-CoA}}}{\overset{\text{pyruvate carboxylase}}{\rightleftarrows}} \text{oxaloacetate} + \text{ADP} + P_i$$

The Glyoxylate Cycle

The glyoxylate cycle is used by some microorganisms when acetate is the sole carbon source or during oxidation of primary substrates (such as higher fatty acids) that are cleaved to acetyl-CoA without the intermediate formation of pyruvic acid. This pathway does not occur in higher organisms because they are never forced to feed on 2-carbon molecules alone.

The specific enzymes of the glyoxylate cycle are isocitrate lyase and malate synthase. Figure 10-11 shows how these two enzymes fit together with other

Figure 10-11. The glyoxylic acid cycle or bypass. Its reactions permit the replenishment of the pool of intermediates of the TCA cycle. The specific enzymes are isocitrate lyase and malate synthase.

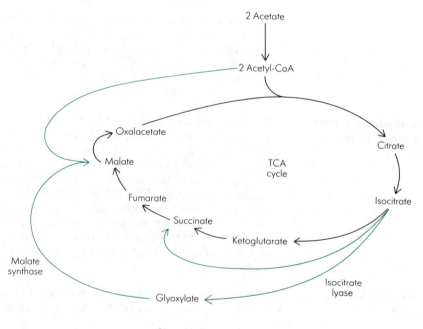

reactions of the tricarboxylic acid cycle to provide a bypass around some of the TCA-cycle reactions. The overall reaction of the glyoxylate cycle is

$$2 \text{ Acetyl-CoA} \rightarrow \text{succinate} + 2\text{H} + 2\text{CoA}$$

As seen in Fig. 10-11, acetyl-CoA enters the cycle at two places. It condenses with oxalacetate to give citrate, which is the entry point for the TCA cycle, and the further reaction leads to the formation of isocitrate. Isocitrate lyase is a splitting enzyme that produces succinate and glyoxylate. The second acetyl-CoA molecule condenses with glyoxylate to give malate by the action of malate synthase. Enzymes which carry out replenishment reactions such as this are known as **anaplerotic** enzymes; their function is to maintain the pool of essential intermediates for biosynthesis.

ENERGY PRODUCTION BY PHOTOSYNTHESIS

Plants, algae, and cyanobacteria are photoautotrophs. They use light as their source of energy and carbon dioxide as their sole source of carbon. In order for carbon dioxide to be useful for metabolism, it must first be reduced to carbohydrate. This process, by which light is used to convert carbon dioxide to carbohydrate, is called **photosynthesis.** The overall reaction can be written as

$$2\text{H}_2\text{O} + \text{CO}_2 \rightarrow (\text{CH}_2\text{O})_x + \text{O}_2 + \text{H}_2\text{O}$$

In the presence of light (radiant energy) and the green pigment chlorophyll   Carbohydrate

Here $(\text{CH}_2\text{O})_x$ is a formula representing any carbohydrate.

Photosynthesis has two important requirements: (1) a large amount of energy in the form of ATP, and (2) a large quantity of a chemical reductant, in this case water.

Several groups of bacteria—the photoautotrophic green and purple bacteria—are also characterized by their ability to perform photosynthesis. But unlike plants, algae, and cyanobacteria, they do not use water as their chemical reductant, nor do they produce oxygen as one of their end products of photosynthesis. The general equation for bacterial photosynthesis is:

$$2\text{H}_2\text{A} + \text{CO}_2 \rightarrow (\text{CH}_2\text{O})_x + 2\text{A} + \text{H}_2\text{O}$$

In the presence of light (radiant energy) and the green pigment bacteriochlorophyll   Carbohydrate

Here H_2A represents the chemical reductant, such as the inorganic compounds H_2, H_2S, or $\text{H}_2\text{S}_2\text{O}_3$, or the organic compounds lactate or succinate. If H_2A in this equation stood for H_2S, then A would stand for S.

Both of the preceding equations represent the overall results of photosynthesis. A great deal has been learned about the specific chemical reactions involved in bacterial and plant photosynthesis. What follows is a look at the light-dependent energy-yielding processes involving bacteriochlorophyll in bacteria and chlorophyll in plants, algae, and cyanobacteria. What is presented is in accord with the latest results of many investigators but may require modification as further evidence is accumulated.

Cyclic and Noncyclic Photophosphorylation

Anoxygenic photosynthetic bacteria possess chlorophylls, called bacteriochlorophylls, that differ from the chlorophylls of plants in structure and in light-absorbing properties. Bacteriochlorophylls absorb light in the infrared region (725 to 1,035 nm). They are not contained in chloroplasts but are found in extensive membrane systems throughout the bacterial cell.

When a molecule of bacteriochlorophyll absorbs a quantum of light, the energy of the light raises the molecule to an excited state. In this excited state an electron is given off by bacteriochlorophyll. Bacteriochlorophyll thus becomes positively charged. It then serves as an electron trap or strong oxidizing agent.

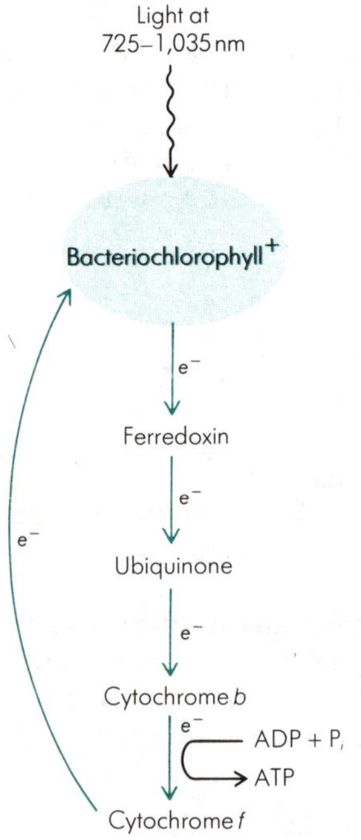

The electron, carrying some of the energy absorbed from light, is transferred to an iron-containing heme protein known as **ferredoxin.** From there it is passed successively to ubiquinone, to cytochrome b, and to cytochrome f, and finally back to the positively charged bacteriochlorophyll. Essentially, the electron has gone around in a cycle, beginning with, and returning to, bacteriochlorophyll. This relatively simple process is illustrated in Fig. 10-12.

The energy released in the step between cytochrome b and cytochrome f is used for **photophosphorylation**—the generation of ATP from ADP and inorganic phosphate.

Note that no $NADP^+$ has been reduced in these reactions. The reduction of $NADP^+$ in photosynthetic bacteria is accomplished not by photosynthesis but by using reducing power from constituents of the environment, such as H_2S and other inorganic and organic compounds. Such reduced compounds usually abound in the anaerobic environment of photosynthetic bacteria.

It may be added that light of higher energy than that absorbed by bacteriochlorophylls can contribute to bacterial photosynthesis since there are carotenoids and other accessory pigments in the bacterial cells which absorb light at shorter wavelengths and transfer the energy to the bacteriochlorophylls.

In plants, algae, and cyanobacteria (oxygenic photosynthetic bacteria), noncyclic photophosphorylation occurs in photosynthesis. In this process, when a molecule in pigment system II (one of two systems of light reactions) absorbs light, this energy raises the molecule to an excited state and the molecule releases an electron. This electron is transferred to plastoquinone, to cytochrome b, to cytochrome f, and finally to pigment system I. Photophosphorylation occurs with generation of ATP from ADP and inorganic phosphate in the step between cytochrome b and cytochrome f. When pigment system I absorbs light, it releases an electron. This electron is transferred from ferredoxin, to flavoprotein, to $NADP^+$. Photophosphorylation occurs again between the release of the electron from pigment system I to ferredoxin. Also note that $NADP^+$ is reduced in this part of the process (see Fig. 10-13). This process differs from cyclic photophosphorylation because the electron lost by pigment system II is not cycled back to it. Instead, electrons are replaced in pigment system II by the light-generated breakdown of water, called **photolysis.** There is some evidence that this scheme of noncyclic photophosphorylation, shown in Fig. 10-13, may have to be modified. It appears that system II pigments alone can carry out the entire process

Light at
725–1,035 nm

Bacteriochlorophyll$^+$

e^-

Ferredoxin

e^-

Ubiquinone

e^-

Cytochrome b

e^- ADP + P_i

ATP

Cytochrome f

e^-

Figure 10-12. Cyclic photophosphorylation as it occurs in anoxygenic photosynthetic bacteria. The electron returns, at a lower energy state, to the bacteriochlorophyll, which had become positively charged after the initial ejection of the electron. No NADP is reduced and no external donor is necessary for this process.

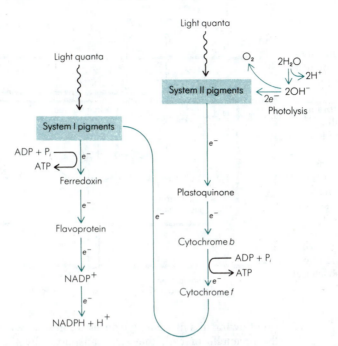

Figure 10-13. Noncyclic photophosphorylation as it occurs in green plants, algae, and cyanobacteria. In this process, electrons raised to a high energy state ultimately reduce NADP$^+$ and are not recycled to the light-pigment systems. The protons necessary for reduction come from the dissociation of water, which results in evolution of oxygen. Electrons are restored to the pigments of system II from the OH$^-$ ion of H$_2$O. The OH$^-$ ion is split to e^-, H$^+$, and ½ O$_2$ by photolysis.

of noncyclic photophosphorylation. Thus the noncyclic reduction of ferredoxin need not involve system I pigments. Further, the most important role of plastoquinone is in the transport of protons originating from water. This modified process has been termed oxygenic photophosphorylation. Further evidence is needed for its confirmation.

THE MECHANISM OF ATP SYNTHESIS

The chemical reactions that lead to the synthesis of ATP are now well understood. But how the transfer of electrons through the respiratory transport chain is coupled to the synthesis of ATP is not very clear. Several alternate hypotheses have been proposed to explain how energy released during electron transport is conserved in the form of ATP. The prevailing theory is the **chemiosmotic hypothesis** advanced in 1961 by Peter Mitchell, a British biochemist. Mitchell was awarded the Nobel prize for his work in this field in 1978. According to this theory, the flow of electrons through the system of carrier molecules releases energy which drives positively charged hydrogen ions (H$^+$), or protons, across the membranes of chloroplasts, mitochondria, and bacterial cells (Fig. 10-14). This movement of hydrogen ions results in the acidification of the surrounding medium and the generation of a **pH gradient** (a difference in pH) across the organelle or cell membrane. In addition, such hydrogen-ion movements lead to the formation of an **electric potential gradient** (a difference in charge) across the membrane (since an electric charge is carried by the proton). In this way, energy released during the transfer of electrons through the respiratory chain is conserved as a "protonmotive force"; the electric potential gradients are produced by pumping hydrogen ions across the membrane.

Figure 10-14. Mechanism of ATP synthesis. Flow of electrons through the respiratory chain drives hydrogen ions across the membrane. This results in a high hydrogen-ion concentration outside the cell and a low concentration inside the cell. This produces a pH and electrochemical gradient. ATP synthesis at the site of the ATPase complex (a knobbed structure on the membrane) is driven by the release of energy when hydrogen reenters the bacterial cell.

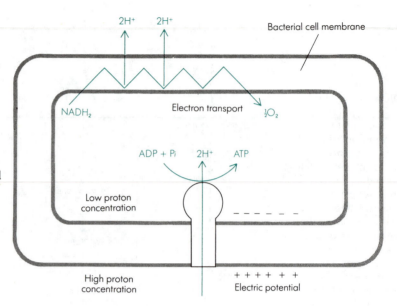

Following this first energy-conservation step, when the hydrogen ions reenter the organelle or cell, they are transported by the membrane-bound enzyme adenosine triphosphatase. The energy released on reentry drives the synthesis of ATP, the second energy-conservation step. This process is shown in Fig. 10-14.

QUESTIONS

1 What is meant by the following?
  (a) Free-energy change
  (b) Exergonic reaction
  (c) Endergonic reaction
  (d) ΔG, $\Delta G°$, $\Delta G°'$
2 Explain the relationship between $\Delta G°$ and the equilibrium constant, and the relationship between $\Delta G°$ and an oxidation-reduction potential difference.
3 What role does ATP play in energy exchanges in cells?
4 In what way is the coupling of exergonic reactions with endergonic reactions important in living organisms?
5 Define the meaning of a high-energy-transfer compound. Name those that occur in the glycolytic pathway.
6 Explain what is meant by an oxidation-reduction system.
7 What is oxidative phosphorylation? Where does it occur in the respiratory chain?
8 Briefly explain how glycolysis fits into the metabolism of glucose in aerobic cells.
9 Compare the disposition of electrons (or hydrogen atoms) obtained from the oxidation of glyceraldehyde-3-phosphate in aerobic and anaerobic cells.
10 Identify the three reactions in the glycolytic pathway that are not freely reversible by the same specific enzymes.

11 Account for the ATP yield per glucose molecule in glycolysis.

12 Describe the various ways in which the pentose phosphate cycle is useful to a cell.

13 Is the Entner-Doudoroff pathway found in (a) both aerobes and anaerobes; (b) both eucaryotes and procaryotes?

14 Explain why fermentation is a less efficient process for obtaining energy than aerobic respiration.

15 Explain how fermentation products can be used for the identification of bacteria. Provide specific examples to support your answer.

16 Why is the TCA cycle called an amphibolic cycle?

17 What general rule governs the utilization of substances other than carbohydrates for the production of energy?

18 Name the specific enzymes of the glyoxylate cycle and describe the reactions they catalyze.

19 What are the essential differences between photosynthesis by bacteria and by algae?

20 Explain why photophosphorylation in procaryotes is termed *cyclic*.

21 Describe the prevailing model for the mechanism of ATP synthesis.

REFERENCES

Note that the references of Chap. 9 are also relevant to this chapter.

Cohen, P.: *Control of Enzyme Activity*, Chapman & Hall, London, 1976. (Distributed in the United States by Wiley, New York.) *A short paperback that deals with direct control of enzyme activity by changing the activity of preexisting enzyme molecules. Detailed examples of each type of control are given.*

Doelle, H. W.: *Bacterial Metabolism*, 2d ed., Academic, New York, 1975. *Biochemical reactions and processes of the particular microorganisms are developed in detail. Studies of microbial systematics and microbial chemistry are included, with emphasis on catabolic events.*

Gottschalk, G.: *Bacterial Metabolism*, Springer-Verlag, New York, 1979. *A text that not only describes metabolic pathways and enzyme reactions but also covers the physiology of the microorganisms which carry out all these metabolic reactions.*

Holloway, M. R.: *The Mechanism of Enzyme Action*, Blackwell Scientific, Oxford, 1976. *A short monograph summarizing the state of knowledge about the mechanism of enzyme action. Well illustrated with structural diagrams.*

Jones, C. W.: *Bacterial Respiration and Photosynthesis*, American Society for Microbiology, Washington, D.C., 1982. *One of a series of short books on current topics of microbiology. This particular one contains highly condensed and current information on respiration and photosynthesis in bacteria.*

Moat, A. G.: *Microbial Physiology*, Wiley, New York, 1979. *A textbook on microbial physiology that covers the nature of microorganisms and their physiological activities.*

Rose, A. H.: *Chemical Microbiology*, 3d ed., Plenum, New York, 1976. *A general text on the subject of microbial physiology. Quite comprehensive and well referenced.*

Chapter 11   Microbial Metabolism: Utilization of Energy and Biosynthesis

In the preceding chapter, we discussed some of the chemical mechanisms by which microorganisms obtain energy. In this chapter, we provide the sequel by considering some of the ways in which energy, once obtained, is utilized by microorganisms. We shall look at some examples of how this is done.

As we have seen in Chap. 10, energy is stored in the form of high-energy-transfer compounds (of which ATP is the most important). But energy is also available in the form of the **protonmotive force** (or electrochemical proton gradient). In these forms energy is used to drive the many endergonic reactions required for the life of the cell.

The principle of coupling exergonic reactions to endergonic reactions requires the utilization of high-energy-transfer compounds like ATP. An electrochemical proton gradient may result in ATP synthesis (see Fig. 10-14), but it can also be used for other biological purposes without the synthesis of ATP. For example, it can be used to generate heat and the rotation of bacterial flagella.

UTILIZATION OF ENERGY IN NONBIO-SYNTHETIC PROCESSES

The ATP formed by the energy-producing reactions of the bacterial cell is expended in various ways. Much of it is used in the biosynthesis of *new* cell components, including energy-storage inclusion granules such as glycogen and poly-β-hydroxybutyrate. Other metabolic processes which require phosphate-bound energy or the energy of the protonmotive force include maintenance of the physical and chemical integrity of the cell, transport of solutes across membranes, and activity of locomotor organelles.

Maintenance of the physical and chemical integrity of the cell is mainly through reactions that lead to biosynthesis of macromolecules, such as nucleic acids and proteins, that are continuously being broken down and need replacement. The extent of this degradation varies with the environmental conditions. Transport of solutes across membranes also requires energy, as does mechanical work such as motility by means of flagella. In general, the rate of utilization of energy in ATP determines the rate at which ATP is regenerated from ADP at the expense of energy from the environment.

Bacterial Motility

Bacterial flagella filaments appear to have no machinery for interconverting chemical and mechanical energy. For example, flagellin, the flagellar protein molecule, has no enzymatic activity, i.e., no detectable ATPase activity (such as is present in cilia and flagella of eucaryotic microorganisms). Bacterial flagella thus differ markedly from the much larger and more complex cilia and flagella that propel eucaryotic cells such as protozoa.

It is therefore not surprising that ATP has been demonstrated *not* to be the immediate source of energy for flagella rotation. Instead the flagellar motor (that rotates the flagellum) is driven by the protonmotive force, i.e., the force derived from the electric potential and the hydrogen gradient across the cytoplasmic membrane.

The rotary motor is believed to be the two elements in the basal body, the M ring and the S ring (see Chap. 5). The rod (which is connected to the filament by the hook) is fixed rigidly to the M ring, which rotates freely in the cytoplasmic membrane. The S ring is mounted rigidly on the cell wall. The inward flux of protons drives the flagellar motor (Fig. 11-1). Exactly what molecular events cause the conversion of protonmotive force into mechanical rotation are still unknown. However, it is clear that in the case of flagellar rotation, proton movements, and not ATP, constitute the energy currency.

Transport of Nutrients by Bacteria

We shall now give an account of the various processes by which ions or molecules cross the cytoplasmic membrane. It is the cytoplasmic membrane that allows the passive passage of certain small molecules and actively concentrates others within the cell.

Passive Diffusion

Except for water and some lipid-soluble molecules, few compounds can pass through the cytoplasmic membrane (a lipid-protein, semipermeable cell mem-

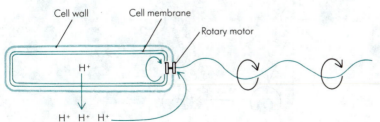

Cell wall   Cell membrane

Rotary motor

H+

H+ H+ H+

Figure 11-1. The flagellum of bacteria is driven by the protonmotive force. Protons flowing back into the cell through the basal body rings of each flagellum cause it to rotate; these rings constitute the rotary motor.

Figure 11-2. Mechanisms of nutrient transport into cells: (A) passive diffusion; (B) facilitated diffusion; (C) group translocation; (D) active transport.

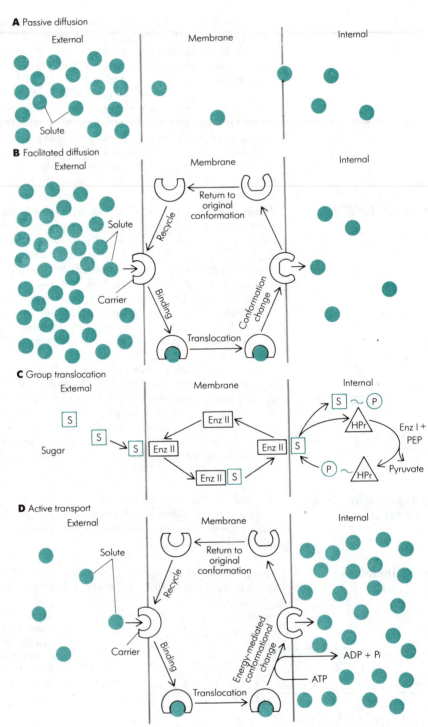

brane) by **simple,** or **passive, diffusion.** In this process solute molecules cross the membrane as a result of a difference in concentration of the molecules across the membrane. The difference in concentration (higher outside the membrane than inside) governs the rate of inward flow of the solute molecule. With time, this concentration gradient diminishes until equilibrium is reached. In passive diffusion no substance in the membrane interacts specifically with the solute molecule as illustrated in Fig. 11-2A.

Facilitated Diffusion

Another mechanism by which substances cross the semipermeable cell membrane is **facilitated diffusion.** This process is similar to passive diffusion in that the solute molecule also flows from a higher to a lower concentration. But it is different from passive diffusion because it involves a specific protein carrier molecule (called a **porter** or **permease**) located in the cytoplasmic membrane. The carrier molecule combines reversibly with the solute molecule, and the carrier-solute complex moves between the outer and inner surfaces of the membrane, releasing one solute molecule on the inner surface and returning to bind a new one on the outer surface. This process is shown in Fig. 11-2B. The entry of glycerol into bacterial cells is by facilitated diffusion. Although this mechanism of transport is common in eucaryotic cells (e.g., sugars enter them in this way), it is relatively rare in procaryotic cells. Neither of the above two mechanisms, passive diffusion or facilitated diffusion, require metabolic energy. Nor do they result in *concentration* or *accumulation* of solute against an electrochemical (with ions) or osmotic (with nonelectrolytes) gradient. Of greater interest to us in the context of this chapter are the two other mechanisms by which solutes cross membranes, both of which require metabolic energy and accumulate substrates against concentration gradients. Solutes can be concentrated within the cell several thousand times greater than outside the cell. These two mechanisms are **group translocation** and **active transport.**

Group Translocation

In group translocation the solute is altered chemically during transport. The best-studied group-translocation system is the **phosphoenolpyruvate–dependent sugar–phosphotransferase system.** It is widely distributed in many bacterial genera and mediates the translocation of many sugars and sugar derivatives. These solutes enter the cell as sugar phosphates and are accumulated in the cell in this form.

Phosphotransferase system (PTS) sugar uptake and phosphorylation require the participation of several soluble and membrane-bound enzymes. These proteins catalyze the transfer of the phosphoryl group of phosphoenolpyruvate to the sugar molecule. The products formed are therefore sugar phosphate and pyruvate; the overall reaction requires Mg^{2+}.

Specifically, a relatively heat-stable carrier protein (HPr) is activated first by transfer of a phosphate group from the high-energy compound phosphoenolpyruvate (PEP) inside the cell, as shown in Fig. 11-2C:

$$PEP + HPr \xrightleftharpoons{\text{enzyme I}} pyruvate + phospho\text{-}HPr$$

Enzyme I and HPr are soluble proteins and are nonspecific components of the process.

At the same time, the sugar combines with enzyme II at the outer membrane surface and is transported to the inner membrane surface (see Fig. 11-2C). Enzyme II is specific for a particular sugar and is an integral component of the cytoplasmic membrane. Here it combines with the phosphate group carried by the activated HPr. The sugar-phosphate is released by enzyme II and enters the cell. This is illustrated by the reaction equation shown below. Some investigators have reported a peripheral membrane enzyme III that mediates between enzyme II and phospho-HPr in translocating the sugar.

$$\text{Sugar} + \text{phospho-HPr} \xrightarrow{\text{enzyme II}} \text{sugar-phosphate} + \text{HPr}$$
(Outside cell) (Inside cell)

Enzyme I has been partially purified from several bacteria including *Escherichia coli* and *Salmonella typhimurium*. HPr has been purified to homogeneity from several bacteria. Mannitol enzyme II has been purified from *E. coli*.

The net chemical reaction of PTS sugar uptake is therefore:

$$\text{PEP} + \text{sugar} \xrightarrow[\substack{\text{enzyme II}}]{\substack{\text{enzyme I} \\ \text{HPr}}} \text{sugar-phosphate} + \text{pyruvate}$$
(Outside cell) (Inside cell)

Other known group-translocation processes include the uptake of adenine and butyrate at the exterior surface of the cell and their conversion at the interior membrane surface to adenosine monophosphate and butyryl-coenzyme A, respectively.

Active Transport

Almost all solutes, including sugars, amino acids, peptides, nucleosides, and ions, are taken up by cells through active transport. The three steps of active transport are:

1 Binding of a solute to a receptor site on a membrane-bound carrier protein.
2 Translocation of the solute-carrier complex across the membrane.
3 Coupling of translocation to an energy-yielding reaction to lower the affinity of the carrier protein for the solute at the inner membrane surface so that the carrier protein will release solute to the cell interior. This process is illustrated in Fig. 11-2D.

Several mechanisms have been proposed to explain the molecular basis of active transport of solutes in microorganisms. The accumulated evidence suggests that active transport may also be explained by Mitchell's chemiosmotic theory (see Chap. 10). In this case, energy released during the flow of electrons through the electron-transport chain or the splitting of a phosphate group from ATP drives protons out of the cell. This generates a difference in pH value and electric potential between the inside and the outside of the cell or across the membrane. This proton gradient gives rise to a protonmotive force which can be used to pump the solutes into the cell. When protons reenter the cell, the energy released on reentry drives the transport mechanism in the cell membrane, probably by inducing a conformational change in the carrier molecule so that its affinity for the solute is decreased and the solute is released into the cell interior. The link between active transport and metabolic energy generation is illustrated in Fig. 11-3.

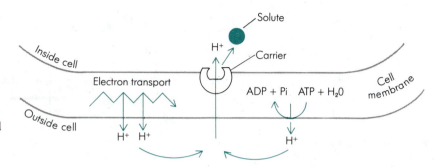

Figure 11-3. How release of metabolic energy is coupled to and drives active transport.

Many active-transport systems of Gram-negative bacteria are associated with binding proteins in the periplasmic space. These binding proteins have very high affinities for specific nutrients, including amino acids, sugars, and inorganic ions. Over one hundred different binding proteins have been isolated and characterized. They are essential for active transport of their specific substrates. However, they are not porters since they are located in the periplasmic space rather than in the cell membrane itself. But binding proteins function in conjunction with porters in the active transport of specific nutrients.

UTILIZATION OF ENERGY IN BIOSYN-THETIC PROCESSES

We have seen how energy is utilized for motility and transport of nutrients into bacterial cells. These are nonbiosynthetic processes. Biosynthetic processes in the cell also require energy; energy from ATP is used to convert one chemical substance into another and to synthesize complex substances from simpler ones.

Synthesis of Small Molecules: The Amino Acids

Amino acids, of which there are about 20 (Table 11-1), are the building blocks of proteins. The sequence and manner in which they are linked (i.e., their three-dimensional structure) determine the type of protein they form.

A microorganism growing in a medium may have all 20 of the amino acids present in the medium; that is, they are available for the microbe, preformed in the medium. If they are not available freely in the medium, the microorganism may have to liberate amino acids from proteins by the action of intracellular or extracellular proteolytic enzymes. In this way, the amino acids become available for use as nutritional building blocks. Sometimes, only a few amino acids are present in a medium, in which case the microbe has to convert other amino acids from the available ones into those that are missing. In yet other instances, the medium contains only inorganic sources of nitrogen, such as ammonium salts. The microorganism then has to synthesize all the required amino acids from these sources of available nitrogen, provided, of course, that it has this

Table 11-1. Amino Acid Building Blocks of Protein, with Standard Abbreviations

Alanine (Ala)	Glycine (Gly)	Proline (Pro)
Arginine (Arg)	Histidine (His)	Serine (Ser)
Asparagine (Asp-NH$_2$, Asn)	Isoleucine (Ile)	Threonine (Thr)
Aspartic acid (Asp)	Leucine (Leu)	Tryptophan (Trp)
Cysteine (Cys)	Lysine (Lys)	Tyrosine (Tyr)
Glutamic acid (Glu)	Methionine (Met)	Valine (Val)
Glutamine (Glu-NH$_2$, Gln)	Phenylalanine (Phe)	

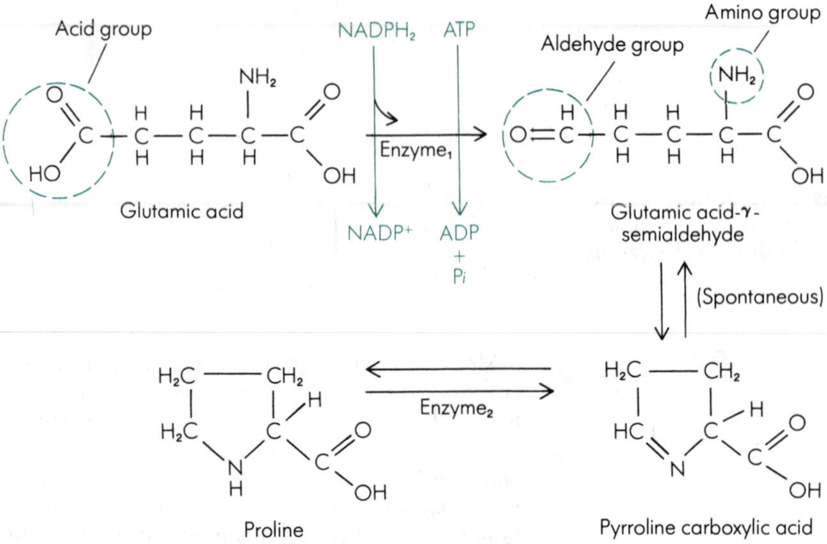

Figure 11-4. The biosynthesis of proline from glutamic acid in *E. coli*. Note the utilization of metabolic energy in the form of ATP in the initial step. (See text for details of synthetic steps.) Sections of the molecule reacting in each step are enclosed in color dashed lines.

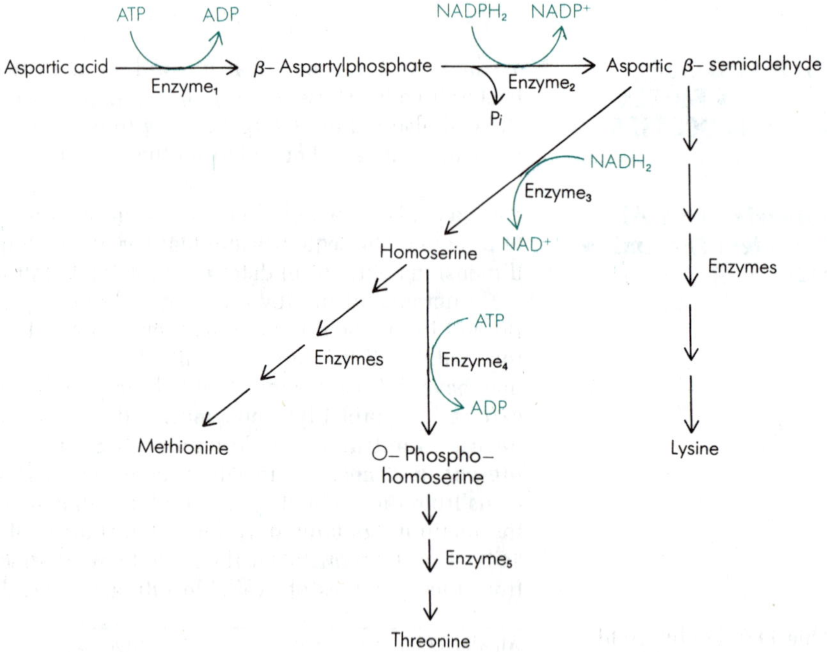

Figure 11-5. The biosynthesis of methionine, threonine, and lysine: another example of metabolic energy utilization in the form of ATP in the interconversion of substances.

ability. All these processes, the interconversion and biosynthesis of chemical substances, require the expenditure of energy.

Consider the specific example of the synthesis of the amino acid proline by the bacterium *E. coli*. Glutamic acid is the initial reactant, itself formed from the reductive amination of α-ketoglutaric acid derived from the citric acid cycle. The steps involved are shown in Fig. 11-4. In the first step an acid group (—COOH) is reduced to an aldehyde group (—CHO). This requires two electrons from $NADPH_2$ and energy from ATP. The aldehyde group then spontaneously reacts with the amino group (—NH$_2$) on the same molecule, forming a ring. This step is followed by ring reduction to form proline.

Another example is the pathway for conversion of aspartic acid to lysine, methionine, and threonine. The conversion utilizes metabolic energy in the form of ATP, as shown in Fig. 11-5. These two examples serve to illustrate the expenditure of energy in the synthesis of amino acids. It may be added that the biosynthetic pathways for each of the 20 amino acids are well understood.

Just as amino acids are used by microorganisms to form proteins (see below for the process of protein synthesis), other low-molecular-weight organic precursors are polymerized to form other macromolecules. Nucleotides form nucleic acids (discussed later) and monosaccharides form polysaccharides. The precursors of lipids, especially the complex ones, include fatty acids, polyalcohols, amines, simple sugars, and even amino acids. It has been estimated that about 150 different small molecules are used to synthesize a new cell.

These small molecules are, in turn, synthesized from intermediates in the catabolic pathways of the microbes (Chap. 10). The most important of these intermediates are pyruvate, acetate, oxalacetate, succinate, α-ketoglutarate, and the sugar-phosphates.

Synthesis of Macromolecules: The Structure and Biosynthesis of a Cell-Wall Peptidoglycan

In all cells the major end products of biosynthesis are proteins and nucleic acids. However, there are other macromolecules peculiar to the procaryotes which require specialized biosynthetic processes. The utilization of energy in one of these processes is illustrated by the biosynthesis of bacterial cell-wall peptidoglycan. This particular biosynthetic process also serves as an example of how polymers are synthesized outside the membrane. Synthesis of cell-wall components is of interest because polymerization takes place outside the cell membrane by enzymes located on the membrane's outer surface.

Structure of Peptidoglycan

As discussed in Chap. 5, the rigid portion of a bacterial cell wall is a polymeric structure known as a murein, peptidoglycan, or mucopeptide. The walls of Gram-positive bacteria contain a large proportion of peptidoglycan; those of Gram-negative bacteria have a much smaller proportion. Peptidoglycans vary in their chemical composition and structure from species to species, but there are basic similarities. Peptidoglycans are very large polymers composed of three kinds of building blocks: (1) *acetylglucosamine* (AGA or GlcNAc), (2) *acetylmuramic acid* (AMA or MurNAc), and (3) a *peptide* consisting of *four* or *five* amino acids of limited variety. Several of the amino acids exist in the D configuration, not usually found elsewhere in nature. A peptidoglycan can best be thought of as consisting of polysaccharide backbone chains composed of alternating units of AGA and AMA linked by β(1→4) bonds, with the short peptide chains projecting from the AMA units. Many of these peptide chains are cross-linked with each other, imparting great rigidity to the total structure. Figure 11-6A illustrates the basic structure of peptidoglycans, and Fig. 11-6B shows a building block of the *E. coli* peptidoglycan. Some peptidoglycans differ in that

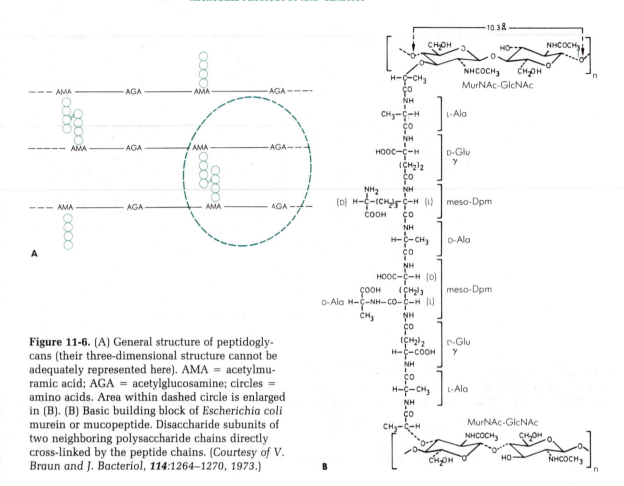

Figure 11-6. (A) General structure of peptidoglycans (their three-dimensional structure cannot be adequately represented here). AMA = acetylmuramic acid; AGA = acetylglucosamine; circles = amino acids. Area within dashed circle is enlarged in (B). (B) Basic building block of *Escherichia coli* murein or mucopeptide. Disaccharide subunits of two neighboring polysaccharide chains directly cross-linked by the peptide chains. (*Courtesy of V. Braun and J. Bacteriol,* **114***:1264–1270, 1973.*)

the peptide chains may not be directly cross-linked to each other, being linked instead by another kind of peptide which forms a bridge between the terminal carboxyl group of one side chain with the free amino group of lysine or diaminopimelic acid (DPM or DAP) on the other side chain; e.g., in *Staphylococcus aureus* a bridge composed of five glycine molecules can link two muramic acid peptides together. This is shown in Fig. 11-7.

Activation of a Peptidoglycan Precursor

Escherichia coli can synthesize cell-wall peptidoglycan when grown in a simple medium of glucose, ammonium sulfate, and mineral salts. One of the early steps in this synthesis is the formation of an activated derivative of AMA. This process, which is shown in Fig. 11-8, requires energy at several points and occurs in the cytoplasm. The activation of sugars, such as acetylglucosamine, by the attachment of a uridine diphosphate (UDP) to form a sugar-UDP precursor is not peculiar to AMA but is a general method involved in the biosynthesis of many kinds of polysaccharides.

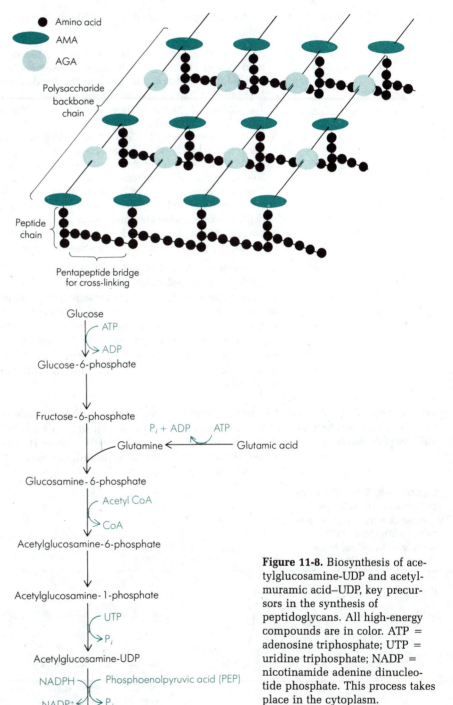

Figure 11-7. General structure of peptidoglycans. Note the pentapeptide bridge for cross-linking. In some peptidoglycans, peptides extending from AMA are linked directly to each other without pentapeptide bridges. AMA = acetylmuramic acid; AGA = acetylglucosamine.

Figure 11-8. Biosynthesis of acetylglucosamine-UDP and acetylmuramic acid–UDP, key precursors in the synthesis of peptidoglycans. All high-energy compounds are in color. ATP = adenosine triphosphate; UTP = uridine triphosphate; NADP = nicotinamide adenine dinucleotide phosphate. This process takes place in the cytoplasm.

Synthesis of Peptidoglycan

After formation of the activated AMA, the synthesis of peptidoglycan proceeds as follows:

1 Amino acids are sequentially linked to the AMA portion of the activated precursor to form a short pentapeptide chain. Ribosomes are not involved, but each amino acid addition requires energy from the breakdown of ATP and the presence of Mg^{2+} or Mn^{2+} and a specific enzyme. These reactions occur in the cytoplasm.

2 The AMA-UDP precursor is coupled to a membrane phospholipid called bactoprenol (undecaprenol phosphate).

3 The AGA couples with AMA of the AMA-UDP precursor. This reaction requires the *activated* form of AGA, that is, the AGA-UDP derivative. In some organisms, the addition of bridging peptides takes place at this step. Reactions of steps 2 and 3 occur in the cell membrane.

4 The precursor, still linked to bactoprenol, is carried out of the cell through the cell membrane and is linked to a growing peptidoglycan chain in the cell wall. Peptide cross-linking may now occur, and the incorporation of the precursor into the growing peptidoglycan is thus completed. These reactions occur in the periplasm. Figure 11-9 illustrates the steps in a typical peptidoglycan biosynthesis.

The synthesis of peptidoglycan illustrates the utilization of energy in joining together smaller molecules into larger ones. Note that all the energy needed for polymerization is used in the cytosolic (cytoplasmic) reactions in synthesizing the activated precursors. Later on in the chapter, we will discuss the biosyntheses of those macromolecules that require a *template* which, acting like a tape, provides information about the *order* in which the smaller pieces are assembled into larger ones. Such processes include DNA synthesis (another piece of DNA is the template) and protein synthesis (a molecule of RNA serves as template).

Synthesis of Organic Cell Material in Chemoautotrophic Bacteria

Chemoautotrophic bacteria utilize carbon dioxide as their sole source of carbon. These bacteria oxidize inorganic nutrients such as hydrogen, ammonia, nitrite, and thiosulfate to produce metabolic energy (in the form of ATP) and reducing power (in the form of $NADPH_2$) in order to reduce CO_2 and convert it to organic cell material.

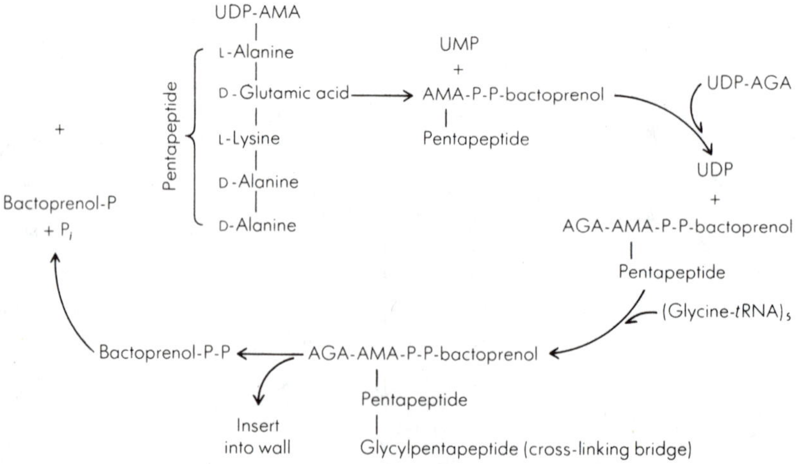

Figure 11-9. Biosynthesis of peptidoglycan in *Staphylococcus aureus*. AMA = acetylmuramic acid; AGA = acetylglucosamine. Note that unlike *E. coli* (Fig. 11-6B) *S. aureus* has a cross-linking pentaglycine bridge.

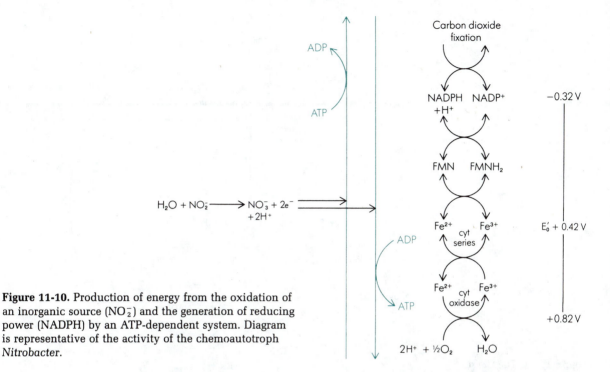

Figure 11-10. Production of energy from the oxidation of an inorganic source (NO_2^-) and the generation of reducing power (NADPH) by an ATP-dependent system. Diagram is representative of the activity of the chemoautotroph *Nitrobacter*.

Compared with other bacteria, such as the heterotrophs, chemoautotrophic bacteria are at a considerable energetic disadvantage for the following reason. Electrons entering the respiratory chain from oxidations of inorganic substrates by chemoautotrophic bacteria usually enter the chain at a higher point (E_0') than those electrons from oxidations of organic substrates by chemoheterotrophs. Consider the typical example of inorganic substrate oxidation by *Nitrobacter*. Since electrons enter the respiratory chain at cytochrome a_1 from the oxidation of nitrite, *Nitrobacter* produces much less ATP than heterotrophs and no reducing power in the form of $NADPH_2$.

Since the E_0' of NO_3^-/NO_2^- is higher than the E_0' of $NADP/NADPH_2$ ($+0.35$ V and -0.32 V, respectively), it is not possible for oxidation of nitrite by *Nitrobacter* to be coupled with the production of reducing power in the form of $NADPH_2$ at the beginning of the electron-transport chain. How then do chemoautotrophs like *Nitrobacter* generate $NADPH_2$ for use with ATP in carbon dioxide fixation—the beginning of the biosynthesis of all organic compounds they require? These chemoautotrophs use a process called **reversed electron flow**, or **ATP-dependent $NADPH_2$ production.** In this process, energy released on breakdown of ATP is used to drive electrons from the oxidation of the inorganic energy source to an E_0' at which they can subsequently reduce NAD^+ or $NADP^+$. In this case of *Nitrobacter*, $NADP^+$ is reduced by ATP-driven electron transport from cytochrome a_1 via cytochromes c, b, and flavoprotein (FMN); this process is shown in Fig. 11-10.

The principal method of carbon dioxide fixation in autotrophic bacteria is the

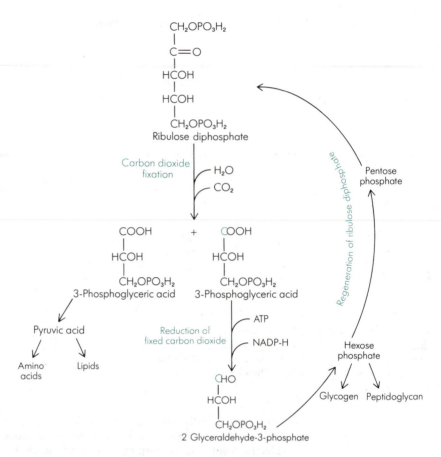

$$CH_2OPO_3H_2$$
$$|$$
$$C=O$$
$$|$$
$$HCOH$$
$$|$$
$$HCOH$$
$$|$$
$$CH_2OPO_3H_2$$

Ribulose diphosphate

Carbon dioxide fixation ⎰ H₂O / CO₂

COOH + COOH
| |
HCOH HCOH
| |
CH₂OPO₃H₂ CH₂OPO₃H₂
3-Phosphoglyceric acid   3-Phosphoglyceric acid

Pyruvic acid

Amino acids   Lipids

Reduction of fixed carbon dioxide ⎰ ATP / NADP-H

CHO
|
HCOH
|
CH₂OPO₃H₂
2 Glyceraldehyde-3-phosphate

Regeneration of ribulose diphosphate

Pentose phosphate

Hexose phosphate

Glycogen   Peptidoglycan

Figure 11-11. The Calvin cycle for carbon dioxide fixation in autotrophic organisms.

Calvin cycle, which is illustrated in Fig. 11-11. In the Calvin cycle, carbon dioxide is fixed in a reaction with the acceptor molecule ribulose diphosphate. The primary product of carbon dioxide fixation is 3-phosphoglyceric acid, from which all other organic molecules of the cell are synthesized. However, carbon dioxide fixation is dependent on a supply of the acceptor molecule, ribulose diphosphate, and so most of the 3-phosphoglyceric acid produced must be used to regenerate ribulose diphosphate. Thus the process of carbon dioxide fixation is cyclic. Each turn of the cycle results in the fixation of one molecule of carbon dioxide. Various intermediates of the cycle are drawn off and enter different biosynthetic pathways.

This cycle of carbon dioxide fixation is complex. It shares certain reactions of the glycolytic and pentose phosphate pathways discussed in Chap. 10. Two reactions are specific to the cycle: the carbon dioxide fixation reaction and the reaction which generates the carbon dioxide acceptor ribulose diphosphate.

The overall reaction for the Calvin cycle is

$$6CO_2 + 12NADPH + 12H^+ + 18ATP + 12H_2O \rightarrow$$
$$C_6H_{12}O_6 + 12NADP + 18ADP + 18P_i$$
(Glucose)

Note the high utilization of reducing power and energy in this cycle.

THE BIOSYNTHESIS OF DEOXYRIBONUCLEIC ACID

We now begin a discussion on the biosynthesis of those macromolecules that require a template, which provides information on the order in which smaller pieces are assembled into larger ones. In the biosynthesis of deoxyribonucleic acid (DNA) another DNA is the template; in the biosynthesis of protein a molecule of ribonucleic acid (RNA) is the template. A discussion of these biosynthetic processes must be preceded by an account of the structures of the specific macromolecules.

The Structure of Deoxyribonucleic Acid

Deoxyribonucleic acid (DNA) from any cell is a long ropelike molecule (Fig. 11-12) composed of two strands, each wound around the other to form a double helix (Fig. 11-13). The model for this structure was first proposed by James Watson and Francis Crick in 1953 when they published an article entitled "A Structure for Deoxyribonucleic Acid." They received the Nobel prize for these studies in 1962. Each strand of the DNA helix is made up of nucleotides linked together to form a chain, a polynucleotide.

Each nucleotide is constructed of three parts:

1 A heterocyclic (with more than one kind of atom) ring of carbon and nitrogen atoms called a nitrogenous base which is either a *purine* or a *pyrimidine*

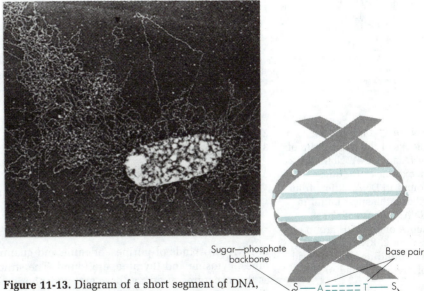

Figure 11-12. A disrupted *E. coli* cell showing the ropelike DNA that has spilled out. Note the plasmid (top center pointer), a circular piece of DNA which is not part of the *E. coli* chromosome and which replicates separately from it. (*By permission from J. D. Griffith, The University of North Carolina.*)

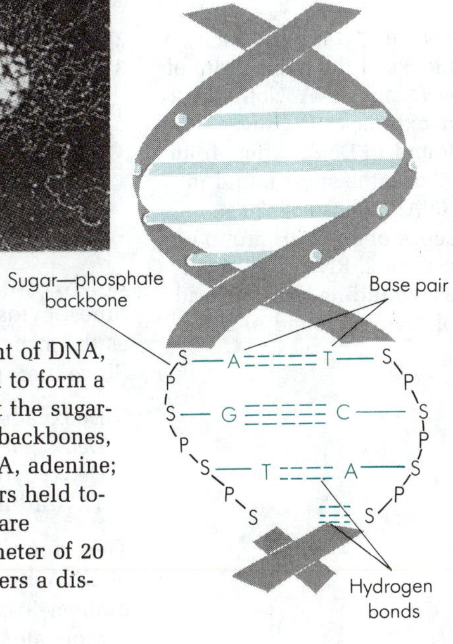

Sugar—phosphate backbone

Base pair

Hydrogen bonds

Figure 11-13. Diagram of a short segment of DNA, showing how its two strands are wound to form a double helix. The two ribbons represent the sugar-phosphate (S = sugar; P = phosphate) backbones, and the horizontal bars represent base (A, adenine; G, guanine; C, cytosine; T, thymine) pairs held together by hydrogen bonds. The strands are wrapped around an axis that has a diameter of 20 Å. Every complete turn of the helix covers a distance of 34 Å.

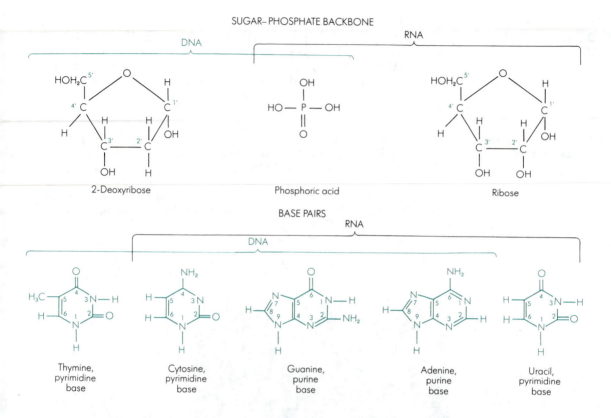

SUGAR– PHOSPHATE BACKBONE

2-Deoxyribose

Phosphoric acid

Ribose

BASE PAIRS

Thymine,
pyrimidine
base

Cytosine,
pyrimidine
base

Guanine,
purine
base

Adenine,
purine
base

Uracil,
pyrimidine
base

Figure 11-14. Building blocks of the nucleotides of DNA and RNA. Note that 2-deoxyribose, the sugar found in DNA, differs from ribose, the sugar found in RNA, because of the absence of an —OH group on carbon 2. RNA uses uracil as a building block instead of thymine found in DNA.

2 A 5-carbon sugar (pentose) called *deoxyribose*
3 A *phosphoric acid*

These parts are linked together as follows:

Heterocyclic base-deoxyribose-phosphate
├────Nucleoside────┤
├────────Nucleotide────────┤

In DNA two kinds of purines, adenine and guanine, and two kinds of pyrimidines, cytosine and thymine, are found. The structures of these bases, as well as the structures of deoxyribose and phosphoric acid, are shown in Fig. 11-14. Since there are four kinds of bases, four kinds of nucleotides are found in DNA:

Deoxyadenosine-5′-monophosphate (adenine + deoxyribose + phosphate)
Deoxyguanosine-5′-monophosphate (guanine + deoxyribose + phosphate)
Deoxycytidine-5′-monophosphate (cytosine + deoxyribose + phosphate)
Thymidine-5′-monophosphate (thymine + deoxyribose + phosphate)

These four kinds of nucleotides are joined together in the polynucleotide strands of DNA by **phosphodiester linkages;** that is, each phosphate group links the number-3 carbon atom of one deoxyribose of a nucleotide to the number-5 carbon atom of the deoxyribose of the next nucleotide, with the phosphate

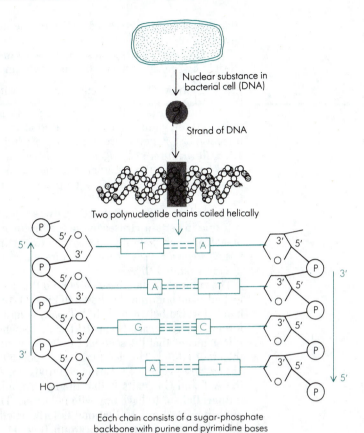

Figure 11-15. DNA location in the cell, molecular configuration, and chemical structure as viewed in progressively greater detail. The double helix maintains a constant width because purines always face pyrimidines in specific G-C and A-T base pairs.

group on the outside of the chain (Fig. 11-15). The result is a chain of alternating phosphate and sugar groups, with the nitrogenous bases projecting from the sugar groups (see Fig. 11-15). Weak bonds, known as **hydrogen bonds,** link the base on one chain and the base on the other chain. Two bases so linked are called a **complementary base pair.** Only two kinds of complementary base pairs are found in double-stranded DNA because of their hydrogen-bonding properties: adenine (A) and thymine (T), and guanine (G) and cytosine (C). As a consequence, the ratio of adenine to thymine or of guanine to cytosine in double-stranded DNA is always 1:1. That is, the amount of purines is equal to the amount of pyrimidines.

The DNA of each species shows a characteristic composition that is not affected by age, growth, conditions, environmental changes, etc. The molar ratio $\frac{[A] + [T]}{[G] + [C]}$ indicates a characteristic composition of DNA of each species. For example, in humans this value is 1.52; in sheep, 1.36; in wheat germ, 1.19. Indeed, these values can even be used at the species level in bacteria for identification or taxonomic grouping: *Escherichia coli*, 0.93; *Staphylococcus aureus*, 1.50; *Clostridium perfringens*, 2.70; *Micrococcus lutea*, 0.35. (However, in bacterial taxonomy, the differences in base composition between species are more commonly expressed as [G + C] percent of total bases. This was explained in Chap. 3.)

The failure to find a 1:1 ratio between adenine and thymine or guanine and cytosine in certain viruses led to the discovery of single-stranded DNA in these organisms. For example, phage φX174 contains a single strand of DNA in a ring form. However, these cases are rare exceptions to the general occurrence of double-stranded DNA.

The complementary base pairs hold the two strands of the DNA helix together by hydrogen bonding. As shown in Figs. 11-13 and 11-15, there are two hydrogen bonds formed between each A-T pair, whereas there are three hydrogen bonds formed between each G-C pair. The complementarity of the purines and pyrimidines means that the sequence of bases on one strand dictates the sequence on the other strand. This is of critical importance in the synthesis, or **replication,** of new strands of DNA during cell division. A consequence of the formation of the A-T and G-C pairs is that the two strands of the DNA helix are said to be **antiparallel,** or to have opposite polarities. This means that each strand runs in opposite directions so that one is terminated by a free 3'-hydroxyl group and the other by a 5'-phosphate group (Fig. 11-15), where 3' and 5' refer to the numbering of the carbon atoms of the deoxyribose molecule. If you examine the nucleotides in Fig. 11-15, you will find that on the left strand the phosphate on the fifth carbon (5') of the sugar points up. On the right strand the phosphate on the fifth carbon points down. Each strand keeps the same polarity as it winds around the molecule.

The relationship of DNA to its low-molecular-weight components is shown

Figure 11-16. Breakdown of deoxyribonucleic acid into lower-molecular-weight components. DNA is made up of nucleotides which, when you remove phosphoric acid, yield nucleosides. Nucleosides can be broken down into bases and sugar (2-deoxyribose).

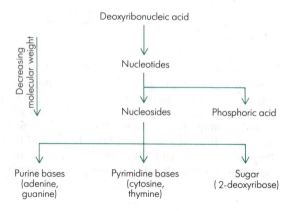

in Fig. 11-16. The removal of the phosphate group from the nucleotide yields a nucleoside consisting of a pentose sugar linked to a heterocyclic base.

The helical structure of the DNA molecule proposed by Watson and Crick is a right-handed, double-stranded one. This double helix is right-handed because the turns run clockwise looking along the helical axis. It represents what is known as the B form of DNA. But recent evidence indicates that DNA may be able to exist in other types of double-helical structures. There are A and C forms of the DNA helix; they differ from the B form in several features, such as the numbers of base pairs per turn, the vertical rise per base pair, or the helical diameter.

However, the Z form exhibits the most contrast with the other forms because it is a left-handed helix (sinistral DNA). It has the most base pairs per turn. Its name is taken from the anticlockwise zigzag path that the sugar-phosphate backbone follows along the helix. The structures of Z-DNA and B-DNA are compared in Fig. 11-17.

Figure 11-17. Left-handed (Z-form) and right-handed (B form) DNA models. The superimposed line traces the sugar-phosphate backbone; in the B form it is a continuous helix, while in the Z form the backbone zig-zags. (*Reproduced from Mosaic, **14**:2, 1983.*)

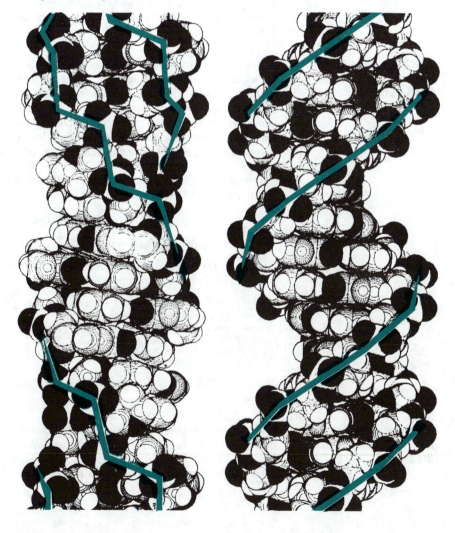

Z-DNA has been found in a variety of living organisms, including rats, rabbits, several species of plants, and protozoa. It appears that wherever there is regulation of gene action there is found Z-DNA. That is, Z-DNA seems to be important for biological development and control.

The Structure of Ribonucleic Acid

The other naturally occurring nucleic acid is ribonucleic acid (RNA). It plays a fundamental role by making it possible for the sequence of chemical groups in DNA to dictate the sequence of amino acids in proteins. It differs from DNA in these respects:

1 The sugar component of the nucleotides which make up RNA is ribose, instead of deoxyribose as in DNA. Ribose is similar to deoxyribose except for the presence of a hydroxyl group at the number-2 (2′) carbon atom (see Fig. 11-14).
2 The pyrimidine nitrogenous base uracil, instead of thymine, is found in the nucleotides that make up RNA (see Fig. 11-14).
3 It is usually single-stranded. The single strands can bend themselves backward and base-pair to form three-dimensional structures.
4 RNA is degraded to its nucleotides much more easily than DNA.

The Biosynthesis of Nucleotides in DNA Synthesis

Before the polynucleotide chains of DNA can be synthesized by bacteria (or any other organisms), an intracellular pool of nucleotides must be available. In some bacteria, these nucleotides must be supplied preformed in the medium, but in bacteria with relatively simple nutritional requirements, this reservoir of nucleotides can be synthesized from glucose, ammonium sulfate, and some minerals. The conversion of simple nutrients into nucleotides for DNA synthesis involves a complex series of enzymatically catalyzed reactions, several of which require energy in the form of ATP. Some of these reactions form *activated* nucleotides as direct precursors for synthesis of the polynucleotide chains of double stranded DNA:

$$\text{Nucleotide} + \text{ATP} \xrightarrow{\text{kinase}} \text{nucleotide-phosphate} + \text{ADP}$$

$$\text{Nucleotide-phosphate} + \text{ATP} \xrightarrow{\text{kinase}} \text{nucleotide-diphosphate} + \text{ADP}$$

As seen in the above equations, energy in the form of ATP is utilized.
The overall chemical reaction is

Nucleotide diphosphates or deoxynucleoside triphosphates $\xrightarrow[\substack{\text{Mg}^{2+} \\ \text{DNA polymerase}}]{\text{preformed DNA}}$ PP_i + new DNA strand
(dATP, deoxyadenosine triphosphate;   (Inorganic
dGTP, deoxyguanosine triphosphate;   pyrophos-
dCTP, deoxycytidine triphosphate;   phate)
dTTP, thymidine triphosphate)

The new DNA strand is complementary to the preformed or template DNA strand.

Semiconservative Replication of DNA Strands

Prior to cell division, the two DNA chains of a DNA molecule separate by unwinding, and each serves as a template for the synthesis of a new complementary chain, thus forming two new helices, each exactly like the original. Each half of the dividing cell then receives one of these helices. This type of duplication in which one polynucleotide chain acts as a template to direct the synthesis of a new chain complementary to itself is termed **semiconservative replication;** it results in two daughter helices, each containing one old template strand and one new complementary strand. In other words, only one of the old strands is conserved in each daughter helix.

Replication of the DNA Molecule

Bacteria are almost always **haploid,** which means that their chromosomes are unpaired. In contrast, most eucaryotic cells (except for gametes) are **diploid;** they have paired homologous chromosomes which may be **heterozygous** (carry different alleles or genes occupying the same relative locus on homologous chromosomes). All bacteria studied to date have their genetic loci in a single linkage group; that is, they have a single chromosome per genome.

The chromosome of a typical bacterium is a circular double-stranded DNA molecule; that is, the double helix for a complete genome has no free ends. It has an approximate molecular weight of 2.5×10^9 daltons (a dalton is equal to the mass of one hydrogen atom) and has about 4×10^6 base pairs. The circular chromosome is further twisted on itself in the bacterial cell to form a supercoil. (This is also the case in many viruses.) If, instead, the chromosome were extended linearly, it would measure approximately 1250 μm (1.25 mm), which is several hundred times longer than the bacterial cell that contains it. A circular form is typical of the DNA molecules of procaryotic microorganisms, of viruses, and of organelles in eucaryotic organisms. However, not all DNA molecules are circular; the chromosomes of eucaryotic organisms and of many viruses consist of linear DNA molecules.

There are three general methods of replication of the DNA molecule.

θ (Theta) Mode

The replication of a circular DNA molecule is initiated at a certain point called the **origin,** which is specific for each bacterial species. Replication proceeds in two directions around the chromosome, leading to the formation of a "bubble," which increases in size as replication proceeds. This mode is called **theta** because intermediate structures resemble the Greek letter θ (see Fig. 11-18). In this process, a circular parental chromosome is replicated to two circular daughter chromosomes, in each of which one strand of the parental DNA molecule is conserved and a complementary strand is newly synthesized. Figure 11-18 illustrates this θ mode of replication of the DNA molecule.

σ (Sigma) or "Rolling Circle" Mode

Replication begins with the cleavage of a phosphodiester bond in one strand of the circular DNA molecule to produce a nick with $3'$-OH and $5'$-PO_4 ends on that strand. The complementary circular strand then serves as a template for the synthesis of a new strand, which is covalently linked to the $3'$-OH end of the nicked parental strand. As this strand grows at the $3'$-OH end, the $5'$-PO_4 end of the same strand is displaced to form a "tail" on the circle. As replication proceeds, a circular parental molecule is converted to two daughter molecules, one circular and the other linear. This mode is called **sigma** because intermediate structures have the Greek letter σ conformation (see Fig. 11-19). The sigma mode of DNA molecule replication is carried out by some bacteriophages, such as λ and φX174 whose progeny viral DNA is linear; by bacteria involved in sexual conjugation; and by certain eucaryotes during oogenesis.

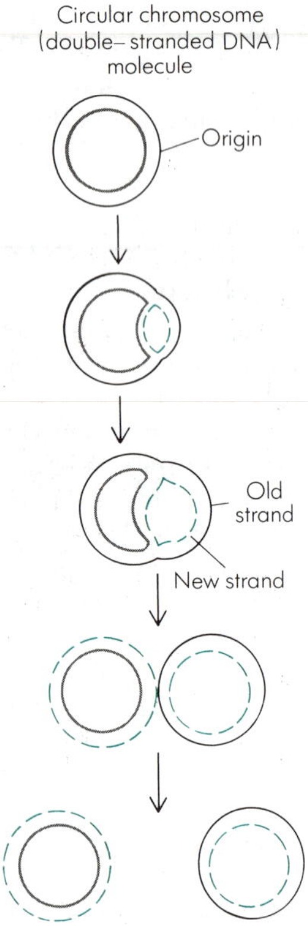

Circular chromosome
(double–stranded DNA)
molecule

Figure 11-18. Replication of bacterial DNA by the theta mode. As shown, there is one origin of replication and two directions of replication from the origin; i.e., there are two growing points. (Note that the DNA strands are not drawn in double-helix form for the sake of simplicity.) In *E. coli* the average rate at which these two growing points, or *replication forks*, move during replication is about 45 kilobases per minute per fork at 37°C. Since there are 10 base pairs per turn of the helix, the rate of unwinding of the parental duplex at each fork must be approximately 4500 turns per minute.

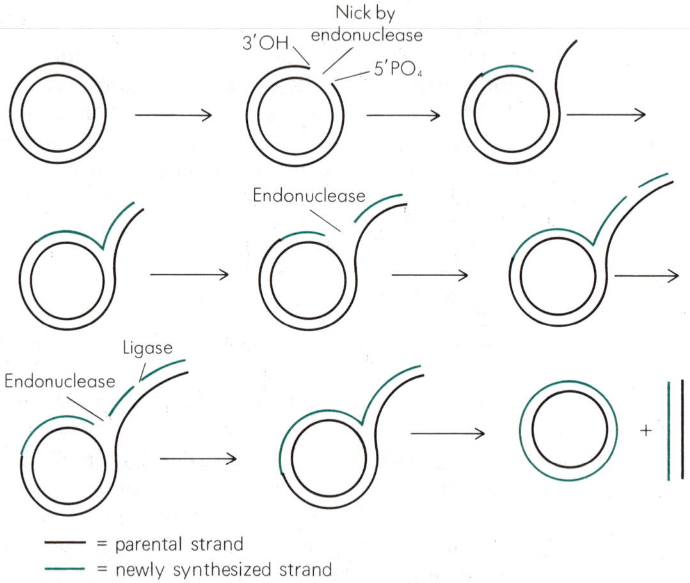

—— = parental strand
—— = newly synthesized strand

Figure 11-19. Sigma mode of DNA replication. The double-stranded circular DNA molecule is nicked at a specific point exposing 3′-hydroxyl and 5′-phosphate terminal groups. DNA replication begins at the 3′-hydroxyl terminus with the addition of nucleotides by DNA polymerase; therefore, synthesis is in the 5′ → 3′ direction. It then switches over to its complementary strand. After a short period an endonuclease nicks the new DNA strand at the growing point with further unwinding of unreplicated portions of parental helix. The process is repeated again and again. The short fragments are joined together by polynucleotide ligase. A specific nuclease then cuts unit-length segments. (For the sake of simplicity, the double-helix form of DNA strands is not shown.)

Linear Mode

All eucaryotic organisms and some viruses have linear DNA molecules. Replication of these chromosomes is initiated at specific sites by the formation of replication bubbles. Small viral linear DNA molecules may have only one point of initiation per molecule. Large DNA molecules of eucaryotes may have hundreds of initiation points per molecule. A replication bubble grows in size as DNA replication proceeds from the point of initiation. Adjacent bubbles fuse to form larger ones as replication proceeds. Upon completion of replication, two linear double-helical daughter molecules are formed from one linear parental molecule. Each daughter molecule, of course, contains a conserved strand from the parental molecule and a newly synthesized strand. The linear mode of replication is illustrated in Fig. 11-20.

It may be noted that procaryotes replicate their DNA from one origin or growing point per molecule while eucaryotes replicate from many origins per molecule. Replication may occur in either a unidirectional or bidirectional manner from each origin.

Events at the Growing Point (Replication Fork)

It is at the growing point, or replication fork (see Fig. 11-20), that both DNA strands are duplicated. Let us examine the events that take place at the growing point.

Procaryotic DNA strands are synthesized at the rate of about 1000 nucleotides per second at a replication fork. Eucaryotic DNA strands are synthesized more slowly at about 100 nucleotides per second. Many enzymes are involved in this synthesis. Initial enzymatic activities are carried out by a helix-unwinding protein (ATP-dependent), a helix-destabilizing protein, and a helix-relaxing protein (DNA gyrase). These enzymes participate in opening the parental DNA helix ahead of the replication fork.

DNA replication is **discontinuous**; that is, the strands are replicated in small fragments, called **Okazaki fragments** after their discoverer, in the $5' \rightarrow 3'$ direction. This process is shown schematically in Fig. 11-21.

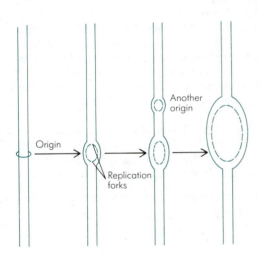

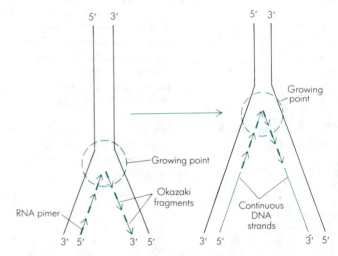

Figure 11-20. Linear mode of DNA replication and bidirectional replication from multiple origins. (Double-helical form of DNA strands is not shown, for the sake of simplicity.)

Figure 11-21. The discontinuous synthesis of DNA. Short segments, Okazaki fragments, of DNA are synthesized in the $5' \rightarrow 3'$ direction. These fragments are represented by short colored arrows. Subsequently the fragments are linked together by ligase to form continuous strands as the growing point moves farther along the DNA. (See text for details.)

Initiation of DNA replication requires a **primer,** a short sequence of RNA that is synthesized by RNA polymerase and is complementary to the DNA which serves as a template. (These RNA primers are used because RNA polymerase, unlike DNA polymerase, requires only template but not primer. "Polymerase" implies a polymerization action.) The RNA primer's 3′-OH end serves for the addition of nucleotide diphosphates by DNA polymerase III. DNA polymerase I then removes the RNA primer with its 5′-nuclease activity; simultaneously it fills in the gap with DNA via its 3′-polymerase activity. This DNA polymerase I uses the 3′-OH end of the preceding DNA fragment as its primer. Thus as it removes the RNA, it extends or elongates the preceding DNA molecule that served as its primer. When the entire RNA is removed, a single-stranded nick remains between the two DNA fragments and is sealed by DNA ligase, a DNA-joining enzyme. (DNA polymerase II is found also in *E. coli,* but, as yet, no specific function has been ascribed to it.)

TRANSCRIPTION AND TRANSLATION OF GENETIC INFORMATION

We have examined how DNA is replicated as a cell grows and before it divides into two cells. One of these DNA copies is then transmitted to the progeny or sister cell. In this way genetic information programmed in the genes is passed on from an organism to its offspring. Thus, DNA, or the genes, is a very important molecule in the cell. It is carefully guarded from damage, and it is repaired when damage is detected. (On the other hand, RNA and proteins are subject to degradation and replacement rather than to repair; that is, they undergo **turn-over.**) We will now see how genetic information is coded in the DNA, and how the genetic code is deciphered and used in the production of proteins. All these processes constitute what is known as the **central dogma** of molecular genetics, which may be described in three steps. The first is DNA **replication** (which we have just studied); the second is **transcription,** the process in which the genetic message on DNA is transcribed to RNA; and the third is **translation,** in which the genetic message coded by RNA is translated by the ribosomes into a protein structure. The outline of the central dogma is shown in Fig. 11-22.

The Building Blocks of Proteins

Just as nucleotides are building blocks of DNA, amino acids are building blocks of proteins. However, DNA consists of only four kinds of nucleotides, whereas proteins consist of about 20 kinds of amino acids. Microorganisms differ widely in their ability to synthesize amino acids. For example, *E. coli* can synthesize

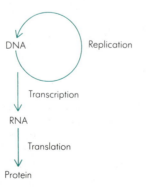

Figure 11-22. Processes in the central dogma of molecular genetics.

all of the amino acids required for protein synthesis, but lactic acid bacteria cannot and therefore must be supplied with preformed amino acids.

There are thousands of different proteins in a bacterial cell. Each type of protein has its own specific sequence of amino acids and three-dimensional structure. The amino acids are joined together by *peptide bonds* to form a long chain. A peptide bond is formed as shown:

$$
\underset{\substack{\text{Amino acid I}}}{R-\overset{\displaystyle H}{\underset{\displaystyle NH_2}{C}}-\overset{\displaystyle O}{C}-OH} + \underset{\substack{\text{Amino acid II}}}{H-N-\overset{\displaystyle H}{\underset{\displaystyle C=O}{C}}-R'} \rightarrow \underset{\substack{\text{Dipeptide}}}{R-\overset{\displaystyle H}{\underset{\displaystyle NH_2}{C}}-\overset{\displaystyle O}{C}-N-\overset{\displaystyle H}{\underset{\displaystyle C=O}{C}}-R'} + \underset{\substack{\text{Water}}}{H_2O}
$$

Peptide bond

Figure 11-23. The amino acid sequence of the enzyme ribonuclease. The shaded areas between cysteines represent disulfide bridges. This illustration is diagrammatic; the polypeptide chain is actually folded to give a complex three-dimensional configuration. (*Courtesy of D. G. Smyth, W. H. Stein, and S. Moore, J. Biol. Chem.,* **238**:227, 1963.)

The chain of amino acids formed when a large number of amino acids are joined together by peptide bonds is called a **polypeptide chain.** Proteins consist of one or more polypeptide chains. Polypeptide chains range from fewer than a hundred amino acid **monomers** or **residues** (e.g., the hormone insulin) to over a thousand residues (e.g., DNA polymerase). The sequence of amino acids in the enzyme ribonuclease (RNAase) is shown in Fig. 11-23.

The sequence of amino acids is characteristic for each protein. It is determined by the sequence of bases in the DNA of a gene. That is, each kind of polypeptide is specified by a particular gene, giving rise to the dictum "one gene, one polypeptide chain."

The sequence of amino acid residues in a polypeptide constitutes its **primary structure.** Polypeptide chains can take on specific shapes by folding; this folding

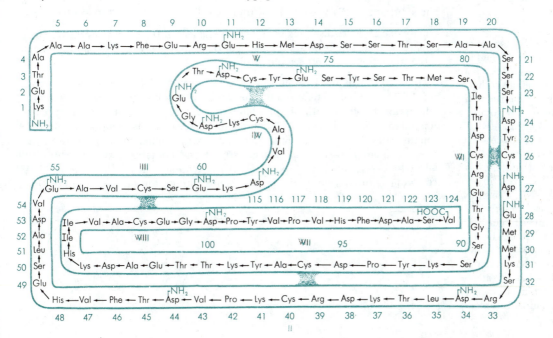

pattern constitutes the **secondary structure** of the protein. The amino acid side chains or groups of a polypeptide (e.g., cysteine residues) contribute further to the folding process by forces of attraction and repulsion; this process gives a protein its **tertiary structure.** In Fig. 11-23, the tertiary structure of RNAase is seen to be contributed by four disulfide bridges. As may be observed, RNAase is a single polypeptide chain of 124 amino acid residues; it is folded, bent, and twisted into a globular shape in its active form. Finally, the overall shape resulting from the interaction of two or more polypeptide chains constitutes the **quaternary structure** of a protein.

Transcription

Protein synthesis takes place on the ribosomes, which are large RNA-protein particles in the cytoplasm of the bacterial cell. (In eucaryotic cells the ribosomes are attached to the endoplasmic reticulum.) **Ribosomal RNA (rRNA)** constitutes about 90 percent of the total cellular RNA. Before protein synthesis can proceed, the coding of DNA must first be transferred to a substance that passes information from the DNA in the nuclear region to the ribosomes in the cytoplasm. This substance is known as **messenger RNA (mRNA).** The process or step in which a single-stranded mRNA is synthesized complementary to one DNA strand is called transcription. The synthesis of the polynucleotide chain of mRNA is catalyzed by the enzyme RNA polymerase. Just as activated deoxyribonucleotides are required in DNA synthesis, activated ribonucleotides are required as substrates for this enzyme:

$$
\begin{array}{l}
\text{ATP (adenosine triphosphates)} \\
\text{CTP (cytidine triphosphates)} \\
\text{GTP (guanosine triphosphates)} \\
\text{UTP (uridine triphosphates)}
\end{array}
\xrightarrow[\text{RNA polymerase}]{\text{DNA template}}
\quad \text{mRNA} + \quad
\begin{array}{l}
\text{PP}_i \\
\text{(Inorganic} \\
\text{pyrophosphate)}
\end{array}
$$

The synthesis of RNA is different from that of DNA in the following features:

1 Only *one* of the two strands of any given segment of DNA serves as the template.
2 Only specific, relatively short lengths of DNA are transcribed; i.e., an RNA chain is a transcript of a short section of DNA.

Transcription is the first step in gene expression. This process, as shown in Fig. 11-24, involves separation of the two DNA strands, one of which serves as a template for the synthesis of a complementary strand of mRNA by DNA-dependent RNA polymerase. When a short RNA chain is completed, the DNA double helix closes again. The strand of DNA selected for transcription in a given segment is called the "sense" strand and contains a specific **initiation site,** which is a regulatory sequence of DNA nucleotides called the **promoter region.** (The other strand may be the "sense" strand for another group of genes

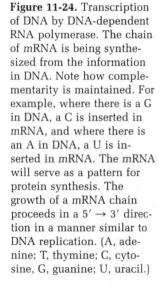

Figure 11-24. Transcription of DNA by DNA-dependent RNA polymerase. The chain of mRNA is being synthesized from the information in DNA. Note how complementarity is maintained. For example, where there is a G in DNA, a C is inserted in mRNA, and where there is an A in DNA, a U is inserted in mRNA. The mRNA will serve as a pattern for protein synthesis. The growth of a mRNA chain proceeds in a 5′ → 3′ direction in a manner similar to DNA replication. (A, adenine; T, thymine; C, cytosine, G, guanine; U, uracil.)

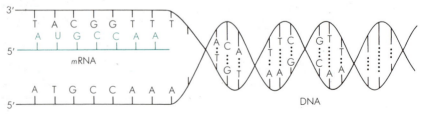

in another segment.) In bacteria, the initiation of RNA polymerase activity at this site is due to an initiation factor called the sigma factor, which is a component of the enzyme. Termination of mRNA synthesis is also at specific regulatory sequences of DNA nucleotides along the DNA molecule which are recognized by the RNA polymerase. Furthermore, a tetrameric protein factor called the rho factor binds to RNA polymerase and promotes its termination. When transcription has been completed, rho dissociates from the RNA polymerase-DNA complex.

Translation

Translation, the next step in gene expression, is the process in which the genetic information now present in the mRNA molecule directs protein synthesis.

When the four different bases of the nucleotides of mRNA are arranged in sequences of three, each base triplet, called a codon, is capable of specifying a particular amino acid. Since there are four different bases, the number of sequences of three of them is 4^3, or 64. These base triplets, each of which specifies a particular amino acid, constitute the genetic code (Table 11-2). The code is probably universal for all species of living organisms.

How is this code translated? Using Table 11-2, suppose the base sequence of mRNA is

CUUAGAAAAUUUAGUGGGACUUCU

The translation of this code into amino acids in a polypeptide chain, at a ribosome, would be

Leu-Arg-Lys-Phe-Ser-Gly-Thr-Ser

Table 11-2. The Genetic Code for the Base Triplets of mRNA and the Amino Acids They Code for*

First Base	Second Base				Third Base
	U	C	A	G	
U	UUU UUC } Phenylalanine UUA UUG } Leucine	UCU UCC UCA UCG } Serine	UAU UAC } Tyrosine UAA "Ochre" UAG "Amber"	UGU UGC } Cysteine UGA "Umber" UGG Tryptophan	U C A G
C	CUU CUC CUA CUG } Leucine	CCU CCC CCA CCG } Proline	CAU CAC } Histidine CAA CAG } Glutamine	CGU CGC CGA CGG } Arginine	U C A G
A	AUU AUC AUA } Isoleucine AUG Methionine	ACU ACC ACA ACG } Threonine	AAU AAC } Asparagine AAA AAG } Lysine	AGU AGC } Serine AGA AGG } Arginine	U C A G
G	GUU GUC GUA GUG } Valine	GCU GCC GCA GCG } Alanine	GAU GAC } Aspartic acid GAA GAG } Glutamic acid	GGU GGC GGA GGG } Glycine	U C A G

* The codons read in the 5′ to 3′ direction (left to right) on the mRNA. Codons UAA (ochre), UAG (amber), and UGA (umber) cause termination of synthesis of a protein chain. AUG and GUG are chain-initiating codons. Note that the same amino acid may be coded for by more than one codon (such a code is called *degenerate*). But no codon codes for more than one amino acid.

AUG and GUG are polypeptide-chain-initiating codons, and UAA, UAG, and UGA are polypeptide-chain-terminating codons. The latter three are called **nonsense codons.**

Another distinctive property of the genetic code is that the same amino acid may be coded for by more than one codon; that is, the code is **degenerate.** Furthermore, no "punctuation," or signal, is necessary to indicate the end of one codon and the beginning of the next. Therefore, the reading frame, or the sequence in which the genetic code is deciphered, must be correctly set at the beginning of the readout of a mRNA molecule. Reading then moves sequentially from one triplet to the next one without pause. If the reading frame is incorrectly set in the beginning, all codons will be out of step and lead to the formation of a **missense protein** with a deranged amino acid sequence.

The events occurring from DNA to RNA to protein are shown in Fig. 11-25. This figure leads us to a more detailed discussion of protein synthesis, which is a very elaborate biosynthetic process.

THE PROCESS OF PROTEIN SYNTHESIS

The first step in protein synthesis is the activation of amino acids. The amino acids are activated by amino acid–activating enzymes called **aminoacyl-tRNA synthetases.** This activation reaction requires energy in the form of ATP:

$$\text{Amino acid} + \text{ATP} + \underset{\substack{\text{Amino acid-}\\\text{activating}\\\text{enzyme}}}{\text{E}} \rightleftarrows \underset{\substack{\text{Aminoacyl-}\\\text{adenylate-E}}}{\text{amino acid-AMP-E}} + \underset{\text{Pyrophosphate}}{\text{PP}_i}$$

There is a specific activating enzyme for each kind of amino acid. The activated amino acid remains tightly bound to the enzyme after activation.

Next the activated amino acid binds to an RNA molecule called **transfer RNA (tRNA).** This reaction is catalyzed by the enzyme that was originally bound to the amino acid:

$$\text{Amino acid-AMP-E} + t\text{RNA} \rightarrow \text{amino acid-}t\text{RNA} + \text{AMP} + \text{E}$$

Figure 11-26. The cloverleaf structure of tRNA.

Figure 11-25. The events from DNA to mRNA to protein.

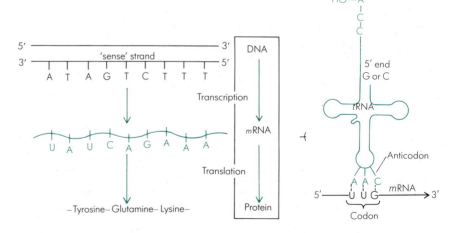

The tRNA functions in protein synthesis to carry amino acids to, and recognize codons in, mRNA.

Transfer RNA is a single chain of about 80 nucleotides that is folded back upon itself and held in a cloverleaf arrangement by means of hydrogen bonding due to complementary base pairing. The general structure of tRNA is shown in Fig. 11-26. Three of the unpaired bases in tRNA form an **anticodon** triplet which specifically recognizes the complementary codon in mRNA for a specific amino acid. The terminal sequence of nucleotides is adenylic-cytidylic-cytidylic (ACC) and is found in all tRNA. The amino acid to be carried is linked to the terminal nucleotide containing adenine.

Like mRNA, tRNA is transcribed from a certain region of the DNA molecule by RNA polymerase. The tRNA molecules function as "adapters" into which specific amino acids are "plugged" so that they can be adapted to the nucleotide triplet language of the genetic code which is transcribed on the mRNA.

The tRNA now carries the amino acid to the mRNA attached to the surface of the ribosome. Here the amino acid is added to a growing polypeptide chain. The surface of the ribosome may be viewed as the assembly point for protein synthesis.

Assembly of the Protein Chain on the Ribosome

As mentioned before, the ribosome is the site of protein synthesis. Its rRNA is transcribed from certain portions of DNA by the same energy-requiring process used for synthesis of mRNA and tRNA. A ribosome is analogous to a videotape-playing machine; just as the latter will produce any kind of image, depending on the videotape played, the ribosome will manufacture any kind of protein, depending on the kind of mRNA supplied.

The ribosomes of E. coli have been studied extensively and have been found to consist of two subunits, each of which is identified by a sedimentation constant (S) determined by ultracentrifugation studies. The larger subunit is a 50S particle, while the smaller unit is a 30S particle. Ribosomal subunits may associate or dissociate with each other. A 30S and a 50S subunit associate to form a 70S ribosome. (S is the Svedberg unit, a measure of how fast a particle sediments during ultracentrifugation. The association of a 30S and a 50S subunit to form a 70S ribosome shows that the sedimentation behavior of the 70S ribosome is not a simple addition of the units of the smaller particles.) Each subunit is made up of ribosomal RNA molecules and numerous proteins.

The synthesis of a protein chain on a ribosome is carried out as follows:

Step 1   A ribosome binds to one end of a mRNA molecule at a specific site. (The specificity here is important because it starts the translation of the mRNA in the correct reading sequence. See Fig. 11-27.)

Step 2   A charged tRNA carrying the first amino acid molecule then attaches to the chain-initiating codon (X) of the mRNA.

Step 3   Another tRNA carrying the second amino acid binds at the next codon (Y).

Step 4   The amino group of the amino acid on the second tRNA reacts with the active terminal carboxyl group on the amino acid of the first tRNA to form a dipeptide; the first tRNA is then released. The mRNA is moved along the ribosome to position the next codon (Z) in readiness for the tRNA carrying the third amino acid.

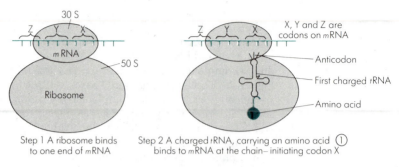

Step 1 A ribosome binds to one end of *m*RNA

Step 2 A charged *t*RNA, carrying an amino acid ① binds to *m*RNA at the chain–initiating codon X

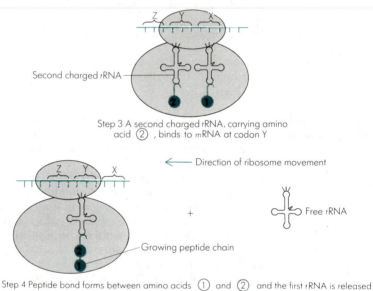

Step 3 A second charged *t*RNA, carrying amino acid ② , binds to *m*RNA at codon Y

Step 4 Peptide bond forms between amino acids ① and ② and the first *t*RNA is released

Figure 11-27. Synthesis of a protein chain on a ribosome.

Figure 11-28. Schematic representation of a polysome during protein synthesis. The *m*RNA moves from right to left. The ribosomes move from the 5′ end to the 3′ end (left to right) of the *m*RNA, all reading simultaneously. In procaryotes, the life span of mRNA is very short, only about 2 min. Solid dots (circles) are amino acids.

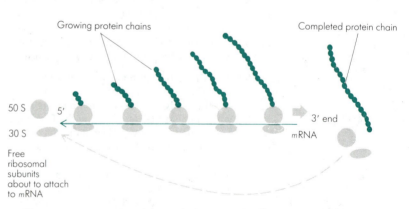

The process is continued until the peptide chain is complete. Termination takes place at one of the nonsense codons UAA, UAG, or UGA, and the chain dissociates from the last *t*RNA molecule.

A single molecule of mRNA is long enough for several ribosomes to read the molecule at the same time. When a number of 70S ribosomes are actively engaged in protein synthesis on a strand of mRNA, this is called a **polysome** (see Fig. 11-28).

Protein synthesis in eucaryotic cells differs in some details from protein synthesis in procaryotic cells, but the main features are similar.

QUESTIONS

1 Describe in general terms the ways in which bacteria utilize energy other than for biosynthesis.
2 What form of energy drives the rotation of the bacterial flagellum? Explain how this is accomplished.
3 Describe the process of passive diffusion.
4 How is passive diffusion differentiated from facilitated diffusion?
5 How is group translocation similar to and different from active transport of solutes?
6 What are binding proteins?
7 Explain how the molecular mechanisms of active transport and ATP synthesis are coupled on the basis of Mitchell's chemiosmotic theory.
8 Give an example to illustrate the expenditure of energy in the biosynthesis of amino acids.
9 Draw a diagram to show the structure of peptidoglycan in the bacterial cell wall.
10 What is meant by an activated precursor? What activated precursors are involved in bacterial peptidoglycan biosynthesis?
11 Is a template involved in bacterial peptidoglycan synthesis? Explain.
12 How is energy utilized in the biosynthesis of peptidoglycan?
13 Why are chemolithotrophs at a considerable energetic disadvantage compared with chemoheterotrophs in the oxidation of substrates?
14 How is energy used in the synthesis of organic cell material by chemoautotrophic bacteria?
15 Explain why reducing power is required by chemoautotrophs in the fixation of carbon dioxide.
16 What is reversed electron flow? Why is this process important for carbon dioxide fixation?
17 What two chemical reactions are specific to the Calvin cycle?
18 What are the three parts of a nucleotide?
19 Explain how nucleotides serve as building blocks of DNA.
20 What are complementary base pairs in DNA and how are they bonded together?
21 Explain how molar base ratios are used in the taxonomic grouping of bacteria. (Refer also to Chap. 3.)
22 Why are the two strands in a DNA molecule said to be antiparallel?
23 Write a general description of the nature of the bacterial chromosome.
24 Compare and contrast the mode of DNA replication in a bacterial cell, the phage lambda (λ), and a mammalian cell.
25 Discuss the role of primers in DNA replication.
26 What is meant by the primary, secondary, tertiary, and quaternary structures of a protein?

27 What is meant by the central dogma of molecular genetics?

28 Why is the genetic code said to be degenerate? How does it determine the initiation and termination of protein synthesis?

29 Describe the different kinds of RNA that participate in protein synthesis.

30 Outline the process of protein synthesis on a ribosome.

REFERENCES

Note that some references of Chaps. 9 and 10 are also relevant to this chapter.

Dills, S. S., A. Apperson, M. R. Schmidt, and M. H. Saier, Jr.: "Carbohydrate Transport in Bacteria," *Microbiol Rev*, **44**:385–418, 1980. *A comprehensive review on the aspects of bacterial transport systems.*

Rogers, H. J., H. R. Perkins, and J. B. Ward: *Microbial Cell Walls and Membranes*, Chapman and Hall, London, 1980. *A textbook for final-year undergraduate and postgraduate students as well as research workers in microbiology. Covers and discusses in detail the structures and functions of the microbial envelope components of microorganisms.*

Smith, E. L., R. L. Hill, I. R. Lehman, R. J. Lefkowitz, P. Handler, and A. White: *Principles of Biochemistry: General Aspects*, 7th ed., McGraw-Hill, New York, 1983. *A classic text (editions spanning the last 30 years) giving the principles of biochemistry. Emphasis of the book is on mammalian biochemistry, particularly that of humans. Also contains information on biological structures and processes. Well illustrated with diagrams.*

Chapter 12   Bacterial Genetics

It was only a little more than 100 years ago that the first serious study of genetics was undertaken by the Austrian monk Gregor Mendel in his pea patch. He crossed strains of peas and studied the results of these crosses—changes in color, shape, size, and other properties of peas. He published his work entitled *Experiments with Plant Hybrids* in 1865. In it he provided the first description of the segregation of parental characters from one generation to the next. From these studies he developed the basic laws of heredity. The laws of heredity, first formulated by Mendel, are common to all forms of life. They apply to humans just as well as to the "lowly" bacterium.

Genetics is the study of the **inheritance** (heredity) and the **variability** of the characteristics of an organism. Inheritance concerns the exact transmission of genetic information from parents to their progeny. Variability of the inherited characteristics can be accounted for by a change either in the genetic makeup of a cell or in environmental conditions.

Another great step forward was the recognition of deoxyribonucleic acid (DNA) as the chemical substance responsible for heredity in all cells. It was identified as a compound bearing genetic information when, in 1944, Avery, MacLeod, and McCarty discovered that a nonvirulent strain of the bacterium *Streptococcus pneumoniae* could be transformed in a heritable manner into a virulent strain by simply adding DNA extracted from a dead virulent strain into the medium. That is, the now virulent bacterium could transmit that virulence indefinitely to its progeny. The DNA derived from dead nonvirulent bacteria had no effect under the same conditions.

THE STUDY OF MICROBIAL GENETICS

Like biochemical principles, genetic principles are universal. The study of microbial genetics has contributed much to what we know about the genetics of all organisms. At first, geneticists were very reluctant to believe that procaryotes, being so small and devoid of complex structures, were also genetic beings like higher organisms. But it was not long before bacteria (and their viruses) became important subjects for genetic research. Indeed, they became the principal experimental tools for unraveling the basic knowledge of genetics at the molecular level.

There are distinct advantages in the use of bacteria for genetic experiments. Bacterial cultures contain millions of individual cells. Therefore, by appropriate selective techniques, rare genetic events can be discovered. Further, a procaryotic cell contains a single chromosome; thus a change in the genetic material of a procaryote results in an immediate, observable change in characteristics. (There is no masking effect due to the presence of an unaffected member of a paired chromosome.) Other obvious advantages in using microbes in genetic studies include the rapid growth rates of microbes, the relative ease of growing bacteria and their viruses in a constant, controlled environment, and the great diversity of metabolic types among microorganisms.

THE INHERITANCE OF CHARACTERISTICS AND VARIABILITY

A characteristic of all forms of life, from the standpoint of genetics, is the general stability or "likeness" in the characteristics of progeny and parent. We readily observe in our own species, for example, that some families regularly have black hair, brown eyes, and a certain shape of nose and chin, whereas other families have blond hair, blue eyes, and a different facial structure. In the same way, and in spite of their small dimension, bacteria and other protists also transmit characteristics to their progeny. The very fact that we can identify species and even strains of bacteria implies that they are capable of transmitting genetic information from generation to generation with great accuracy.

However, in addition to the inheritance of characteristics, which accounts for the constancy exhibited by biological species, there is variability or change expressed in the progeny. These changes are associated with two fundamental properties of the cell or organism, namely, the **genotype** and the **phenotype.** The genotype refers to the genetic constitution of the cell. The phenotype is the expression of the genotype in observable properties characteristic of the cell or organism.

The genotype of a culture of cells remains relatively constant during growth. However, it can change by mutation. This change can result in an alteration in the observable properties, or phenotype, of the cells. So the genotype represents the heritable total *potential* characteristics of a cell, whereas the phenotype represents the characteristics *expressed*.

PHENOTYPIC CHANGES DUE TO ENVIRONMENTAL ALTERATIONS

Bacteria, like the cells of higher organisms, carry more genetic information—their genotype—than is utilized or expressed at any one time. The extent to which this information is expressed depends on the environment. For instance, a facultatively anaerobic bacterium will produce different end products of metabolism, depending on the presence or absence of oxygen during growth. The presence or absence of oxygen determines which enzymes function and which do not. Indeed in a given environment, knowledge of the factors which regulate genetic activity constitutes a very important aspect in the understanding of cell metabolism.

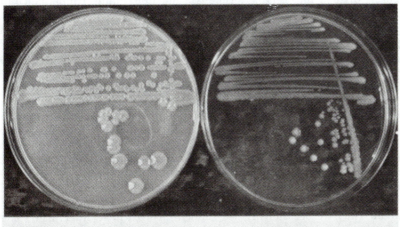

Figure 12-1. *Agrobacterium radiobacter* grown on two different media. Left: Mucoid colonies on sucrose-salts medium; right: nonmucoid colonies on trypticase-soy agar medium.

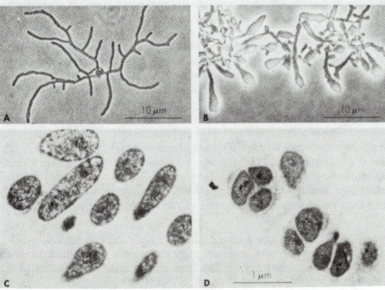

Figure 12-2. Morphological modifications (phenotypic changes) resulting from changes in media composition. (A) and (B) are phase-contrast micrographs of *Nocardia* sp. in (A) tryptone agar culture; (B) brain-heart infusion agar; both cultures 12 h at 30°C. (*Courtesy of B. L. Beaman and D. M. Shankel, and J. Bacteriol,* **99**:*876, 1969.*) (C) and (D) are electron micrographs of thin sections of *Arthrobacter globiformis* 425 grown in (C) nutritionally complete medium and (D) biotin-deficient medium resulting in abnormal forms of the bacterium (several protoplasts embedded in an amorphous matrix). Note that the aberrant cells are devoid of cell walls and no longer exhibit the typical shape of the species as shown in (C). Incubation was at 25°C for 36 h. (*Courtesy of Margaret Gomersall and E. C. S. Chan.*)

The outstanding feature of a phenotypic change due to environment is that it involves most of the cells in the culture. A phenotypic change of this type is not inherited; rather, it occurs when some condition of the environment changes. A return to the original phenotype occurs when the original environmental conditions are restored.

Figures 12-1 and 12-2 show some phenotypic changes due to alterations in environmental conditions. Such phenotypic changes have also been called **phys-iological adaptations** (to the environment) and so are differentiated from phenotypic changes as a consequence of mutation (discussed below).

GENOTYPIC CHANGES

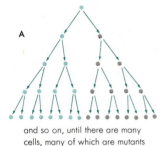

A

and so on, until there are many cells, many of which are mutants

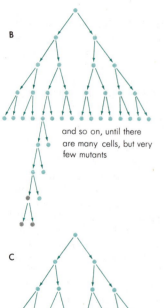

B

and so on, until there are many cells, but very few mutants

C

and so on, until there are many cells, but no mutants

The genotype of a cell is determined by the genetic information contained in its chromosome (or chromosomes, in the case of a eucaryotic cell). The chromosome is divided into genes. A **gene** is a functional unit of inheritance; it specifies the formation of a particular polypeptide as well as various types of RNA. Each gene consists of hundreds of nucleotide pairs. For instance, if a polypeptide chain contains 300 amino acids, then the gene coding for this polypeptide must contain 900 base pairs (three bases for each amino acid). This is the basis for the one gene–one polypeptide relationship mentioned in Chap. 11. It has been estimated that the bacterial chromosome has the capacity to code for approximately 3,500 different proteins.

Any gene is capable of changing or *mutating* to a different form so that it specifies formation of an altered or new protein which may in turn change the characteristics of the cell (sometimes leading to its death). For example, the substitution of even one amino acid among several hundred in a polypeptide chain may cause the protein to be nonfunctional. A **mutation** is a change in the nucleotide sequence of a gene. This gives rise to a new genetic trait, or a changed genotype. A cell or an organism which shows the effects of a mutation is called a **mutant.** Thus we occasionally see sudden changes in familiar plants and animals. Now and then, an albino cat appears in a black litter, or a yellow pea appears among many green peas. The same sort of phenomenon occurs among microorganisms.

In nature, mutations are rare events which occur at random and arise spontaneously with no regard to environmental conditions. Spontaneous bacterial mutations may occur at a rate of only one mutation in 1 million bacterial cells to a rate of only one mutation in 10 billion bacterial cells. Generally, the mutants in a cell population are masked by the greater numbers of unmutated cells. Isolating a mutant cell is like looking for the proverbial needle in a haystack. However, microbiologists have developed techniques which facilitate isolation of the few mutants from a large population of nonmutated **(wild-type)** cells. For example, an antibiotic can be incorporated in a medium to select for antibiotic-resistant mutants.

As mentioned previously, for a long time many geneticists were very reluctant to believe that the observable changes in bacterial cultures were ever anything

Figure 12-3. Formation of phage-resistant bacteria by mutation. In some cultures (A) a mutation occurs very early, and thus many of the cells are resistant. In others (B) a mutation may occur very late, so that there are few resistant cells. In still others (C) there is no mutation and hence no resistant cells.

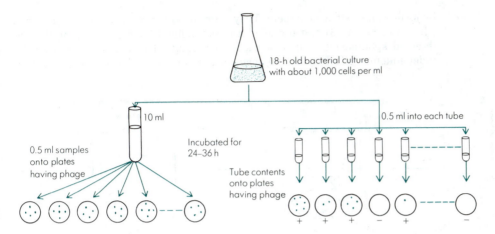

18-h old bacterial culture
with about 1,000 cells per ml

10 ml

0.5 ml into each tube

0.5 ml samples
onto plates
having phage

Incubated for
24–36 h

Tube contents
onto plates
having phage

+ + + − + −

Figure 12-4. The fluctuation test was performed essentially as follows. A series of tubes containing 0.5 ml of cells was incubated without phage until a certain poulation size was reached. The cultures were then exposed to phage by pouring the contents of each tube into an agar plate containing phage. The number of phage-resistant mutants in each tube was thus determined. The colony counts from such a *series of similar cultures* were then compared with the results of a series of samples taken from *one culture* started with a similar density of cells per milliliter and allowed to reach a similar population number per milliliter. The results showed that resistant bacteria arise spontaneously prior to the exposure to phage since a series of similar cultures yielded results different from those obtained with a series of samples from one culture. (See text for further explanation.)

more than physiological adaptations or phenotypic changes. That is, they refused to believe that bacteria have stable hereditary systems and could undergo permanent changes or mutations and assumed that any changes in the characteristics of a culture were simply due to environmental influences. However, Max Delbrück and Salvador Luria believed that bacteria have stable hereditary mechanisms, and in 1943 they performed an elegant experiment that proved the point.

Bacterial viruses called **bacteriophages,** or simply **phages,** are capable of killing bacteria. When susceptible bacteria are exposed to a phage, some of the bacterial cells survive, and they and their descendants are resistant to the phage. Some microbiologists assumed that these cells were modified by their contact with the phage so they became resistant by physiological adaptation. But Luria and Delbrück believed that phage-resistant bacteria were the result of mutations that occurred before the bacteria came into contact with the phages.

Let us suppose that resistant cells are the result of contact with phages; then if we set up a large number of identical bacterial cultures and expose them all to identical batches of phage, approximately the same number of resistant cells should appear in all cultures. However, if the resistant cells are really the result of mutations and since mutations occur entirely at random, then when we grow many identical cultures and expose them all to identical batches of phage, we should find a great fluctuation in the numbers of resistant cells. In some cultures, there may be no mutations and therefore no resistant cells, but in other cultures a mutation might have occurred very early, so that nearly all the cells are resistant (Fig. 12-3).

Thus Luria and Delbrück predicted that physiological adaptation to the phage by the bacteria should give about the same number of resistant bacteria (within sampling error) in each culture. But if phage resistance was due to mutations, preexisting and merely selected by the addition of phage, the number of resistant bacteria should fluctuate widely in each culture. Luria and Delbrück found a much greater fluctuation in numbers than could be accounted for by physiological adaptation, and thus they proved statistically by their **fluctuation test** that phage resistance was really the result of mutation (see Fig. 12-4).

Shortly after, in 1952, more direct proof of preexisting mutants was provided by Joshua and Esther Lederberg. They introduced the replica plating technique (shown in Fig. 12-5) and provided a direct method for demonstrating the undirected spontaneous origin of bacterial mutants; i.e., the mutants occurred independently of any selective agent or environment. The procedure made it

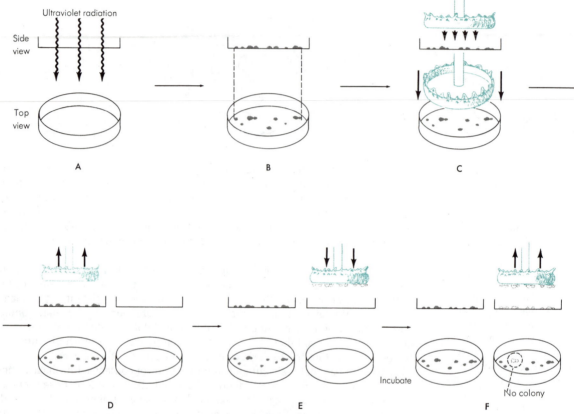

Figure 12-5. Replica plating is used for isolating nutritional mutants of *Escherichia coli*. (A) Bacterial suspension placed in open half of Petri dish and exposed to mutagenic agent, such as ultraviolet radiation. (B) Sample from (A) plated on surface of a "complete" medium such as nutrient agar. The plate is incubated; after incubation, the exact position of colonies on the plate is noted. (C) A sterile replica plating unit is gently pressed to the surface of plate (B), then raised (D), and then pressed to the surface of a plate of "minimal" agar medium (E). The positioning of the replica plating unit on the minimal agar must be precise, so that colony locations will be comparable on each of the two plates. The plates will be replicas of one another. The minimal agar in the plate in (E) consists of inorganic salts and glucose, nutrients which normally permit growth of *E. coli*. After incubation (F), colonies appear on the new plate at most, but not all, of the positions corresponding to locations of colonies on the original plate. It may be assumed that the organisms that failed to develop are nutritional mutants; that is, they are not able to grow on an inorganic salts–glucose medium, a characteristic which they originally possessed.

practical to examine large numbers of **clones** (populations of cells descending from a single cell) for a particular characteristic. By using sufficiently large samples, one could, for example, demonstrate the occurrence of phage-resistant mutants in a culture which was known to be phage-sensitive. The mutant types developed and could be located on an agar-plate culture which had not en-

Figure 12-6. Simplified representation of the experiment of the Lederbergs showing the spontaneous nature of mutation in bacteria. The drawing shows that isolation of a pure colony of phage-resistant bacteria from a medium is possible without prior exposure to the virus.

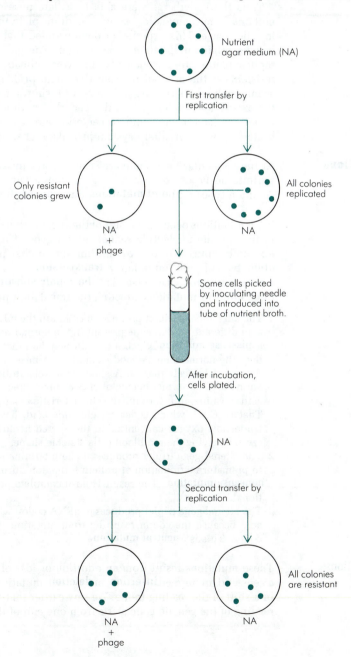

Nutrient agar medium (NA)

First transfer by replication

Only resistant colonies grew

NA + phage

All colonies replicated

NA

Some cells picked by inoculating needle and introduced into tube of nutrient broth.

After incubation, cells plated.

NA

Second transfer by replication

NA + phage

NA

All colonies are resistant

countered phage previously. Similarly, the spontaneous appearance of anti-biotic-resistant strains could be demonstrated without previous exposure of the culture to the antibiotic. As Fig. 12-5 shows, replica plating can also be used for isolating nutritional mutants. In essence the technique provides a practical means for finding the one cell in a million (more or less) which has mutated.

Thus the Lederbergs were able to isolate pure colonies of resistant *E. coli* mutants from many colonies on a medium plate that had *never been exposed* to lytic phage. This experiment demonstrated that mutation against the phage had its origin in *spontaneous* mutation. The growth of resistant colonies on replica plates arose from cells that were already present and were already resistant on the original nonselective plate *prior* to exposure to the selective agent, such as a lytic phage. Figure 12-6 shows a simplified representation of the experiment performed by the Lederbergs. Since then many other types of mutations have been found in bacteria, and it is now firmly established that bacteria have a hereditary system just like higher organisms.

Types of Mutations

At the molecular level there are several ways in which changes in the purine-pyrimidine base sequence of a gene can occur, resulting in mutation. Two common types are **point mutations** and **frameshift mutations.**

Point Mutations

Point mutations occur as a result of the substitution of one nucleotide for another in the specific nucleotide sequence of a gene. The substitution of one purine for another purine or one pyrimidine for another pyrimidine is termed a **transition** type of point mutation. A **transversion** is the replacement of a purine by a pyrimidine, or vice versa. This **base-pair substitution** may result in one of three kinds of mutations affecting the translation process:

1 The altered gene triplet produces a codon in the mRNA which specifies an amino acid different from the one present in the normal protein. This mutation is called a **missense mutation.** Such a protein may be functionally inactive or less active than the normal one. A good example of a missense mutation in humans is the disease sickle cell anemia. A single base substitution in the codon for the sixth amino acid of normal hemoglobin A changes the sixth amino acid from glutamic acid to valine, thus forming the characteristic hemoglobin S of sickle cell anemia. That is, GAG, which codes for glutamic acid, has changed to GUG for valine. Under low oxygen concentration the altered hemoglobin S molecules stack into crystals, giving the red blood cells a sickle shape.

2 The altered gene triplet produces a chain terminating codon in mRNA, resulting in premature termination of protein formation during translation. This is called a **nonsense mutation.** The result is an incomplete polypeptide which is nonfunctional.

3 The altered gene triplet produces a mRNA codon which specifies the same amino acid because the codon resulting from mutation is a synonym for the original codon. This is a **neutral mutation.**

Frameshift Mutations

These mutations result from an *addition* or *loss* of one or more nucleotides in a gene and are termed **insertion** or **deletion** mutations, respectively. This results in a shift of the reading frame. We saw earlier that during protein synthesis the reading of the genetic code starts from one end of the protein template, mRNA,

Figure 12-7. Frameshift mutation, as a result of insertion of a nucleotide in a gene. Insertion of a nucleotide in a gene results in the transcription of an additional nucleotide in mRNA. This results in a frameshift when codons are read during translation, so all codons following the insertion are altered and all amino acids coded for are changed. A framshift mutation as a result of deletion of a nucleotide would have essentially the same effect.

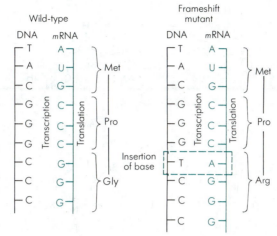

and is read in consecutive blocks of three bases. Frameshift mutations, therefore, generally lead to nonfunctional proteins, because an entirely new sequence of amino acids is synthesized from a frameshift reading of the nucleotide sequences of mRNA (which was transcribed from a mutation in the DNA of the cell). This type of mutation is illustrated in Fig. 12-7.

How Mutations Occur

Mutations most commonly occur during DNA replication. Some mutations occur as the result of damages inflicted by ultraviolet (UV) light or x-rays. Since these agents are an inescapable part of the environment (for example, UV light is a component of sunlight), they probably account for many spontaneous mutations. However, mutation rates can be increased substantially by deliberately exposing a culture to such radiation. Any agent that increases the mutation rate is called a **mutagen.** Mutations obtained by use of a mutagen are said to be **induced,** rather than spontaneous, though they may differ only in frequency, not in kind. For example, UV light causes mutation under both natural and laboratory conditions. The number of mutants obtained by laboratory conditions is much higher, however, because of the high dosage of UV light used.

The major effect of UV light is to cause the formation of **dimers** by cross-linking between adjacent pyrimidine, especially thymine, residues in DNA. These cross-linked residues disrupt the normal process of replication by preventing the various polymerases from functioning. When x-rays interact with DNA, the result is usually a break in the phosphodiester backbone of the nucleic acid.

The most revealing findings about mutation in recent years have come from studies on the mutagenic effects of various chemicals. There are three main types of mutagenic chemicals. The first consists of *compounds that can react chemically with DNA.* Since specificity of DNA replication depends upon purine-pyrimidine bonding, which results from hydrogen bonding between the amino and hydroxyl groups of the purines and pyrimidines, chemical modification of these amino and hydroxyl groups can cause mutation. Nitrous acid, which can remove amino groups from purines and pyrimidines, is such a

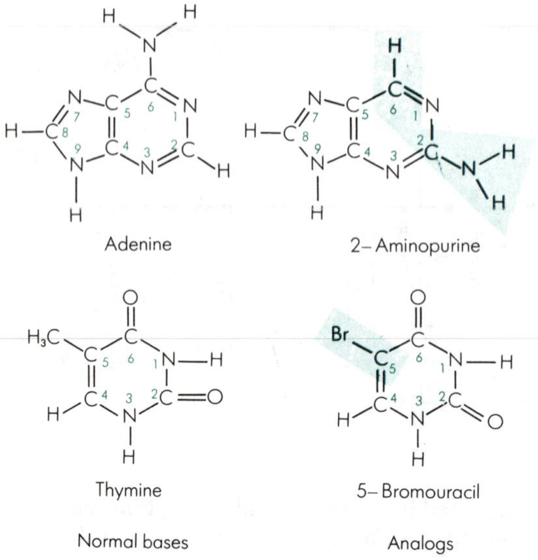

Figure 12-8. Two normal DNA bases and two base analog mutagens. 2-Aminopurine is an analog of adenine and can pair with thymine or cytosine. 5-Bromouracil is an analog of thymine and can pair with adenine or guanine. A color box highlights the part of the analog which differs from the normal base.

mutagen. The second type of mutagenic chemicals consists of *base analogs*. These are chemicals sufficiently similar in structure to normal DNA bases to be substituted for them during DNA replication (Fig. 12-8). Although similar in structure, base analogs do not have the same hydrogen-bonding properties as the normal bases. They can therefore introduce errors in replication which result in mutation. A third type of mutagenic chemicals is *intercalating agents*. These are flat molecules that can intercalate (slip in) between base pairs in the central stack of the DNA helix. By this means they distort the structures and cause subsequent replication errors. Examples of such agents are acridine orange, proflavin, and nitrogen mustards.

Recently, it was shown that mutations can occur because of transposons. **Transposons** are units of DNA which move from one DNA molecule to another, inserting themselves nearly at random. They are also capable of causing DNA rearrangements such as deletions or inversions. For example, one such transposon is from the bacterial virus called Mu, which may be considered a mutagen.

How Mutations Are Repaired

We have said that DNA damage can occur by UV radiation, x-rays, and certain chemicals. Fortunately, cells contain specific enzymes which can repair damaged DNA. In this way, some affected cells can continue to function normally.

Many kinds of bacterial cells and yeasts have been shown to possess an efficient photoreactivating mechanism for repairing damage caused by UV radiation. This **photoreactivation** occurs when cells exposed to lethal doses of UV light are immediately exposed to visible light. A special enzyme designated PRE, induced by visible light, splits or unlinks the dimers formed because of exposure to UV light and restores the DNA to its original state.

Some bacteria have enzymes, called **endonucleases** and **exonucleases,** that excise or cut out a damaged segment of DNA. Then the other enzymes, polymerases and ligases, repair the resulting break by filling in the gap and joining the fragments together. This mechanism is illustrated in Fig. 12-9 and is called **excision repair.**

The process by which E. coli repairs large amounts of DNA damage is called **inducible** or **SOS repair.** This process is not a single discrete mechanism but includes diverse responses such as the ability to repair pyrimidine dimers, to induce various prophages, to shut off respiration, and to delay septum formation during cell division. But all the responses are coordinately regulated. The process is a very efficient one; however, it tends to insert mismatched bases and thus is error-prone and introduces additional mutation.

Mutation Rate

The rate of mutation is the probability that a gene will mutate at any particular cell division. Thus the mutation rate is generally defined as the average number of mutations per cell per division. It is expressed as a negative exponent per cell division. For example, if there is one chance in a million that a gene will mutate when the cell divides, the mutation rate for any single gene equals 10^{-6} per cell division. Generally, the mutation rate for any single gene ranges between 10^{-3} and 10^{-9} per cell division. Thus a mutation rate for E. coli may be given as 5.8×10^{-8} mutations per bacterium per cell division.

The mutation rate has some practical implications. Since genes mutate at random and independently of each other, the chance of two mutations in the same cell is the product of the single mutation rates for each. So, for example, if the mutation rate to penicillin resistance is 10^{-8} per cell division and that to streptomycin resistance is 10^{-6} per cell division, the probability that both mutations will occur in the same cell is $10^{-6} \times 10^{-8}$, or 10^{-14}. This mutation rate is very low. For this reason, it is a common practice to give two antibiotics simultaneously in the treatment of some diseases. For example, a combination of penicillin G and streptomycin has been of proven value in treating streptococcal infections. A cell which has become resistant to one antibiotic is still likely to be inhibited or killed by the other.

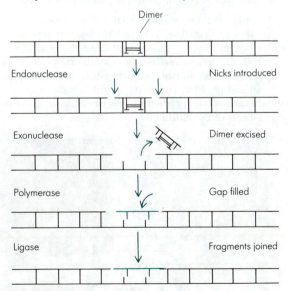

Figure 12-9. Excision repair of UV-light-damaged DNA containing a thymine-thymine dimer generated by covalent links between adjacent bases.

Phenotypes of Bacterial Mutants

Since all properties of living cells are ultimately gene-controlled, any cell characteristic may be changed by mutation. A large variety of bacterial mutants have been isolated and studied intensively. Some of the major phenotypes of mutants are as follows:

1 Mutants that exhibit an increased tolerance to inhibitory agents, particularly antibiotics (antibiotic- or drug-resistant mutants)
2 Mutants that demonstrate an altered fermentation ability or increased or decreased capacity to produce some end product
3 Mutants that are nutritionally deficient, that is, that require a more complex medium for growth than the original culture from which they were derived (auxotrophic mutants)
4 Mutants that exhibit changes in colonial form or ability to produce pigments
5 Mutants that show a change in the surface structure and composition of the microbial cell (antigenic mutants)
6 Mutants that are resistant to the action of bacteriophages
7 Mutants that exhibit some change in morphological features, for example, the loss of ability to produce spores, capsules, or flagella (Fig. 12-10)
8 Mutants that have lost a particular function but retain the intracellular enzymatic activities to catalyze the reactions of the function, for example, loss of a permease (cryptic mutants)
9 Mutants that yield a wild-type phenotype under one set of conditions and a mutant phenotype under another (conditionally expressed mutants)

It is evident from our list of bacterial mutants that all the characteristic features of bacteria are subject to alteration by the process of mutation. It is also apparent that some of the specific changes caused by mutation are similar or the same as those resulting from a change in environmental conditions. It is therefore necessary to ascertain experimentally that a change is really due to a mutation and not to a response to the environment.

There are many practical implications associated with the occurrence of microbial mutants. The following examples illustrate this.

1 Some microorganisms are known to develop resistance to certain antibiotics because of mutation. This fact is of great importance in the treatment of disease, since antibiotics originally effective for the control of a bacterial infection become less effective or ineffective as antibiotic-resistant mutants appear (Fig. 12-11).
2 It is possible to isolate biochemical mutants capable of producing large yields of an end product. This is important in industry. For example, the yield of penicillin in commercial production was dramatically increased through selection of mutant strains of *Penicillium*.

Figure 12-10. Some mutants exhibit morphological changes. Mutants of *Bacillus subtilis* that are grossly deficient in the enzymes needed to separate daughter cells after division grow at normal rates as very long chains of unseparated cells. Under certain growth conditions these mutants also form helical structures. (A) Wild-type, phase-contrast photomicrograph; (B) mutants, phase-contrast photomicrograph; (C) mutants, scanning electron micrograph. Note helical structures in (B) and (C). *(Courtesy of Jared E. Fein, McGill University.)*

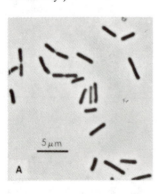

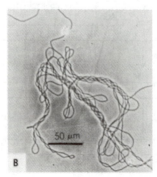

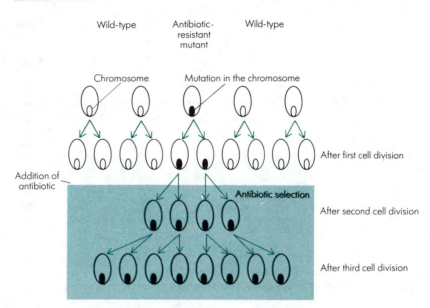

Wild-type　　Antibiotic-resistant mutant　　Wild-type

Chromosome　　Mutation in the chromosome

Addition of antibiotic

Antibiotic selection

After first cell division

After second cell division

After third cell division

Figure 12-11. Antibiotics originally effective for control of a bacterial infection become less effective or ineffective as antibiotic-resistant mutants appear. Note how the use of antibiotics (color screen) actually selects for bacteria that have become antibiotic-resistant due to chromosomal mutation. Bacteria sensitive to the antibiotic are killed or prevented from reproducing. *(Courtesy of SANDORAMA 1978, Sandoz Ltd., Basel, Switzerland, and G. Lebek, University of Berne.)*

3　The maintenance of pure cultures of typical microorganism species requires that occurrence of mutation be prevented; otherwise, the culture will no longer be typical.

4　Microbial mutants have been extensively used in the investigation of various biochemical processes, particularly biosynthetic reactions. For example, mutants with blocks or impairment at different enzymatic steps have been used to unravel metabolic sequences.

Many mutants, perhaps a majority, are able to revert to the wild-type condition by **reverse mutation.** This is a return to the original phenotype by the mutant cells. However, this may not *necessarily* be due to a precise reversal of the original mutation. Sometimes, the effect of the original mutation may be partially or entirely suppressed by a second mutation at a different site on the chromosome.

Designation of Bacterial Mutants

The conventional designations used for bacterial mutants may be described briefly as follows. Each genotype is given a lowercase, italicized, three-letter code. For instance, a mutation which affects proline synthesis is designated *pro*. Since mutation in a number of different genes may exhibit identical phenotypes, discrete genetic loci can be differentiated by means of capital letters, for example, *proA*, *proB*, and *proC*. Numbers may be added sequentially to designate particular mutations; that is, as each new mutation is isolated, it is assigned a number that identifies it in bacterial pedigrees, for example, *proA52* is the 52d isolate of the *Escherichia coli* Genetic Stock Center at Yale University. A further example is given in Fig. 12-12. Table 12-1 shows some frequently encountered genotype abbreviations.

In referring to the phenotype of a bacterium, we use the same three-letter abbreviation, except that it is not italicized and its first letter is capitalized.

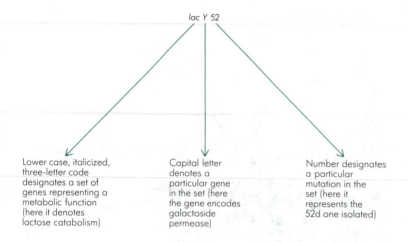

lac Y 52

Lower case, italicized, three-letter code designates a set of genes representing a metabolic function (here it denotes lactose catabolism)

Capital letter denotes a particular gene in the set (here the gene encodes galactoside permease)

Number designates a particular mutation in the set (here it represents the 52d one isolated)

Figure 12-12. Designation of bacterial mutations. The example used here is that of a mutation in β-galactosidase synthesis.

Table 12-1. Some Frequently Encountered Genotypes

Genotype	Mutation
ala	Alanine requirement
azi	Azide resistance
div	Cell division
fla	Flagella biosynthesis
gal	Galactose utilization
lac	Lactose utilization
met	Methionine requirement
pur	Purine biosynthesis
str	Streptomycin resistance
thi	Thiamine requirement
ton	Phage T1 resistance
uvr	Ultraviolet radiation sensitivity

Thus a mutant strain designated *pro* would be phenotypically Pro⁻ (which means inability to synthesize the amino acid proline). The superscript "+" would designate the wild type, for example, Pro⁺. The "−" superscript represents the mutant.

BACTERIAL RECOMBINATION

Genetic recombination is the formation of a new genotype by reassortment of genes following an exchange of genetic material between two different chromosomes which have similar genes at corresponding sites. These are called **homologous chromosomes** and are from different individuals. Progeny from recombination have combinations of genes different from those that are present in the parents. In bacteria, genetic recombination results from three types of gene transfer:

1 *Conjugation.* Transfer of genes between cells that are in physical contact with one another
2 *Transduction.* Transfer of genes from one cell to another by a bacteriophage
3 *Transformation.* Transfer of cell-free or "naked" DNA from one cell to another

These three types of gene transfer are shown in Fig. 12-13.

In bacterial recombination the cells do not fuse, and usually only a portion of the chromosome from the **donor cell** (male) is transferred to the **recipient cell** (female). The recipient cell thus becomes a **merozygote,** a zygote that is a partial diploid. Once merozygote transformation has occurred, recombination can take place.

The general mechanism for bacterial recombination is believed to take place as follows. Inside the recipient cell the donor DNA fragment is positioned alongside the recipient DNA in such a way that homologous genes are adjacent. Enzymes act on the recipient DNA, causing nicks and excision of a fragment.

Figure 12-13. (*Right*) Three types of gene transfer from which genetic recombination results. (A) Conjugation, the transfer of genes between cells in physical contact with each other, perhaps by a sex pilus; (B) transduction, the transfer of genes between cells by a bacteriophage; (C) transformation, the transfer of cell-free or "naked" DNA from one cell to another.

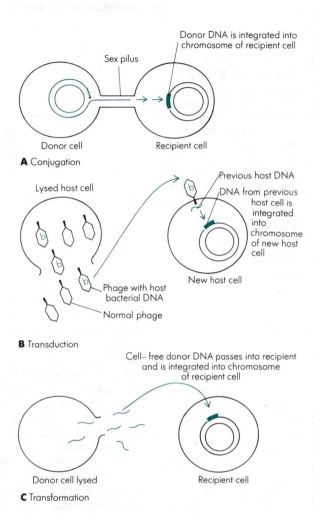

A Conjugation

B Transduction

C Transformation

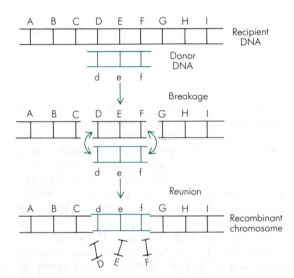

Figure 12-14. (*Above*) The breakage and reunion model of bacterial recombination. The donor DNA becomes integrated into the recipient DNA.

Then the donor DNA is integrated into the recipient chromosome in place of the excised DNA. The recipient cell then becomes the recombinant cell because its chromosome contains DNA of both the donor and the recipient cell. (The excised DNA pieces from the recipient chromosome are probably broken down by specific enzymes.) This general recombination mechanism is seen in Fig. 12-14.

**BACTERIAL
CONJUGATION**

Even though Luria and Delbrück had demonstrated in 1943 that bacteria have a stable hereditary system, it was impossible at that time to explore the system experimentally because of the lack of knowledge of any mating system in bacteria. The genetics of plants and animals depends upon the regular cycle of sexual reproduction in these organisms; once each generation, there is an opportunity for different mutants of a species to mate with each other and produce

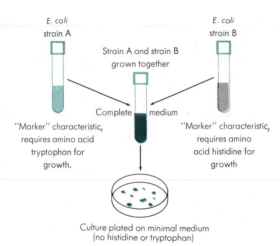

Figure 12-15. Evidence for conjugation in bacteria. The two specific characteristics of the E. coli strains are (A) trp⁻, a tryptophan auxotroph, and (B) his⁻, a histidine auxotroph. The mixture is plated on a minimal medium; growth allows for the selection of prototrophic recombinants (organisms which can synthesize all their amino acid requirements).

new individuals with new combinations of mutations, i.e., to *recombine* with each other, or to produce **recombinants.** For example, a plant that produces smooth, yellow peas can be bred with one that produces wrinkled, green peas. Some of the next generation will be plants that produce the parental types—smooth and yellow or wrinkled and green. But other plants will produce the recombinant types smooth and green or wrinkled and yellow. Only by performing such crosses and observing the progeny can genetic work be done. The first demonstration of recombination in bacteria was achieved by Lederberg and Tatum in 1946 in a brilliant and remarkable experiment that opened the door to a whole new world of microbiology. Lederberg and Tatum knew that conjugation in bacteria must be quite rare, since no one had found it in spite of many attempts, and so they determined to *select* the few possible recombinants out of a large population. They combined two different auxotrophic strains of E. coli and gave them an opportunity to mate. Then they plated the combined cultures on a minimal medium, where only prototrophs could grow; when they found prototrophic colonies growing there, they knew that these must have been the result of a recombination between the auxotrophs.

Figure 12-15 shows the principle of their experiments in simplified form. When Lederberg and Tatum did their experiments, they used **polyauxotrophs** (mutants with more than one nutritional requirement) so that back mutation or spontaneous reversion to the wild type would not occur to confuse their results. For example, the probability of simultaneous reversion of three different mutations is of the order of $(10^{-6})^3$.

It had to be shown also that the prototrophs which arose could not have arisen by the phenomenon of transformation (discussed in the next section) since transformation had been discovered by Griffith in 1928. Cell-to-cell contact as a precondition was shown by using a U tube where the auxotrophic parental cells could be cultured together yet separated physically by a microporous fritted glass disk. This disk was permeable to macromolecules like DNA but not to cells. Culture fluid and soluble material could pass freely between the two parent cultures. Prototrophs or recombinants were never recovered from such cultures. Result like these indicated that bacterial recombination by conjugation is indeed a true sexual process.

It is apparent that mating or conjugation in E. coli is radically different from sexual mating in higher organisms. It is not a reproductive process that occurs regularly at each generation. It does not involve meiosis since bacterial cells are haploid, nor does it involve the fusion of gametes. Instead, it involves the transfer of some DNA from one cell to another followed by separation of the mating pair of cells. While only very small fragments of the bacterial chromosome are transferred in transduction and transformation (discussed later), in conjugation it is possible for large segments of the chromosome, and in special cases the entire chromosome, to be transferred.

Sex Factors

A clearer understanding of conjugation in bacteria came about with the discovery that there is sexual differentiation in E. coli; in other words, different mating types of the bacterium exist. Male cells contain a small circular piece of DNA, which is in the cytoplasm and not part of the chromosome, called the **sex factor** or **F factor** (fertility factor). These cells are referred to as F^+ and are donors in mating. Female cells lack this factor and are labeled F^-. They are recipient cells.

Crosses between two F^- strains do not yield recombinants. However, in F^+ × F^- crosses, the male replicates its sex factor, and one copy of it is almost always transferred to the female recipient. The F^- cell is converted to an F^+ cell and is itself capable of serving as a donor (see Fig. 12-16). Therefore, as long as the cells grow, the conjugation process can continue in an infectious way with repeated transfer of the sex factor. The transfer of the F factor in an F^+ × F^- cross occurs with a frequency that approaches 100 percent. But the formation of recombinants in an F^+ × F^- cross occurs at a low frequency— about one recombinant per 10^4 to 10^5 cells. Thus we see that the transfer of the F factor is independent of the transfer of chromosomal genes.

Since the transfer of the F factor is independent, it follows that the F factor DNA replicates independently of the F^+ donor cell's normal chromosome. The F factor DNA is only sufficient to specify about 40 genes which control sex-factor replication and synthesis of sex pili. One or more sex pili are produced by each F^+ cell (Fig. 12-17). Sex pili seem to act to bind an F^- cell to an F^+ cell and then to retract into the F^+ cell, pulling the F^- cell into close contact. There is also some evidence that sex pili are tubules through which DNA passes

Figure 12-16. During mating of an F^+ and F^- cell, the F^+ cell replicates the sex or F factor and the copy is almost always transferred to the F^- cell. Thus an F^- cell usually becomes an F^+ cell during mating.

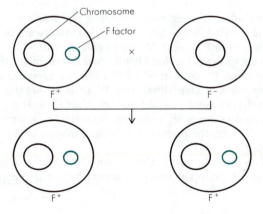

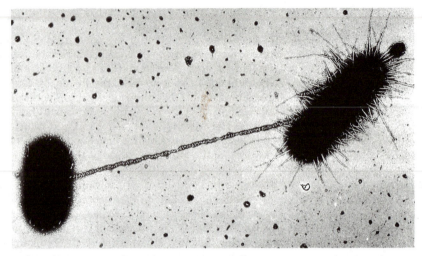

Figure 12-17. Sex pilus holding together a mating pair of *E. coli*. The male cell (on the right) also has another type of pili besides the sex pilus. Small RNA bacteriophages adsorbed to the sex pilus may be seen as dots. (X25,000.) *(Courtesy of C. Brinton, Jr., University of Pittsburgh.)*

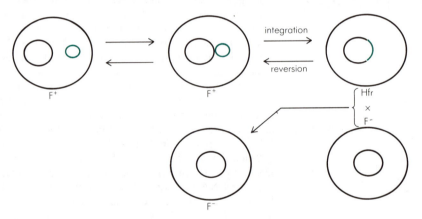

Figure 12-18. An Hfr cell arises from an F$^+$ cell in which the F factor becomes integrated into the bacterial chromosome. During mating of an Hfr and F$^-$ cell, the F$^-$ cell almost always remains F$^-$. This results because Hfr rarely transfers an entire F factor to the F$^-$ cell. But the recombination frequency is high.

from an F$^+$ to an F$^-$ cell during conjugation, although the DNA may be passed from one cell to another at sites of contact between them.

High-Frequency Recombination Strains

The study of conjugation in bacteria was made easier when new strains of cells were isolated from F$^+$ cultures which underwent sexual recombination with F$^-$ cells at a rate at least 10^3 times greater than F$^+$ × F$^-$ cells. These new donor strains were thus called **high-frequency recombination, or Hfr, strains.** Hfr cells arise from F$^+$ cells in which the F factor becomes integrated into the bacterial chromosome. They differ from F$^+$ cells in that the F factor of the Hfr is rarely transferred during recombination. Thus in an Hfr × F$^-$ cross, the frequency of recombination is high and the transfer of F factor is low (Fig. 12-18); in an F$^+$ × F$^-$ cross, the frequency of recombination is low and the transfer of F factor is high.

The order in which chromosomal material is transferred from an Hfr donor to an F$^-$ recipient was determined by the *interrupted mating* experiments of

Elie Wollman and François Jacob. An Hfr strain was mixed with an F⁻ strain, and at various times the conjugation was interrupted by breaking the cells apart in a high-speed blender. The cells were then plated on various types of selective agar media in order to select for recombinant cells which had received donor genes before mating was interrupted.

Interrupted mating experiments reveal the order of genes on a chromosome by the time of entry and the frequency of recombination of each **marker,** which is a detectable mutation serving to identify the gene at the **locus** or site where it occurs. Each gene enters the F⁻ cell at a characteristic time, and a **linkage map** of the Hfr chromosome can be constructed using time of entry as a measure. This is the principal method of learning where the genes are on a bacterial chromosome (Fig. 12-19). This is all possible because the Hfr chromosome is transferred to the F⁻ cell in a *linear* fashion (Fig. 12-20) even though it is a circular chromosome. During transfer the Hfr chromosome begins replicating at the point of insertion of the F factor. Since the F factor can integrate in different positions of the Hfr bacterial chromosome, the first genes to enter an F⁻ cell will vary with different Hfr strains (Fig. 12-20). This means that the integrated F factor serves as the point of chromosomal opening, and *part* of it serves as the origin of transfer. The 5′ end of the single DNA strand enters the F⁻ cell first. Data from an interrupted mating experiment is shown in Fig. 12-21.

It takes about 100 min to inject a copy of the whole Hfr *E. coli* genome (i.e., the chromosome and the integrated F factor). Since conjugation is usually in-

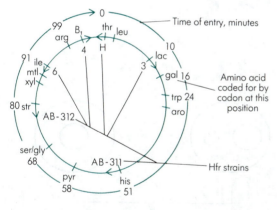

Figure 12-19. Simplified linkage map (color) of the circular chromosome of *E. coli* constructed from interrupted mating experiments using different Hfr strains. The arrows on the linkage map indicate the leading end and direction of entry of the chromosomes injected by each of the Hfr strains, the designations of which are shown inside the circle. This is determined by the position of the F factor in each of the Hfr strains. The numbers around the outside of the map show distances as a function of time, in minutes, based on time of entry of each codon in experiments. (Note that the map distances in minutes are drawn relative to the Hfr strain H so that the *thr* gene has been arbitrarily chosen as the origin.)

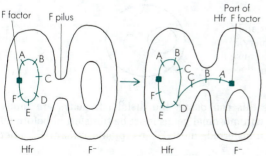

Figure 12-20. Mechanism for DNA transfer between Hfr and F⁻ cells. The Hfr chromosome begins replicating at the point of insertion of the F factor. Since the F factor can integrate in different positions of the Hfr bacterial chromosome, the first genes to enter an F⁻ cell will vary with different Hfr strains. As shown, the order of genes transferred is ABCDEF. In another Hfr strain, the F factor might be integrated between B and C. In this case the order of genes transferred would be CDEFAB.

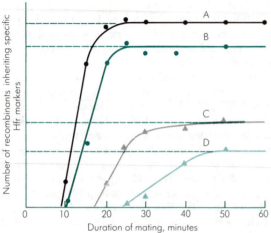

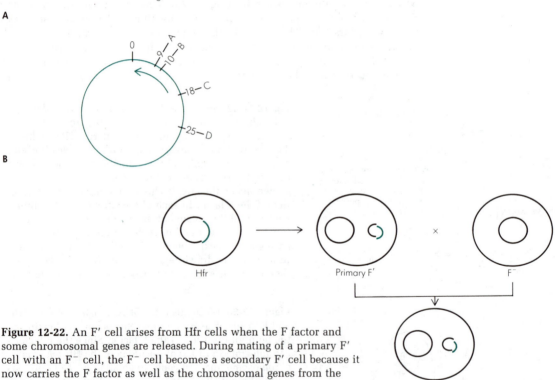

Figure 12-21. An interrupted mating experiment. (A) At intervals during the mating between Hfr and F⁻ cells, samples were removed from the culture and the conjugation interrupted. The cells were then plated on selective media that permitted growth of those F⁻ cells that inherited specific genetic markers received from the Hfr cells. In the graph, the number of recombinant colonies is plotted as a function of the interval of mating allowed before interruption. Extrapolation of each curve to zero recombinants gives the time of entry of the Hfr marker. As may be seen in the graph, the order of genes on the DNA is ABCD, and that A is very close to B. (B) Analysis of the data in (A) allows the plotting of a map as shown. Numbers denote minutes.

Figure 12-22. An F′ cell arises from Hfr cells when the F factor and some chromosomal genes are released. During mating of a primary F′ cell with an F⁻ cell, the F⁻ cell becomes a secondary F′ cell because it now carries the F factor as well as the chromosomal genes from the primary F′ cell.

terrupted by accident before this can occur, the distal Hfr genes are rarely transferred. Since *all* the Hfr chromosomal genes must be transferred before all the F factor genes are also transferred, most F⁻ recipients remain F⁻ after conjugation with Hfr cells.

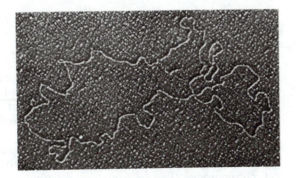

Figure 12-23. Bacterial plasmid shown as a molecule of looped DNA. The drug-resistant plasmid shown is called R28K, carries ampicillin resistance, and has a length of 21 μm. (*Courtesy of Michiko Egel-Mitani.*)

Hfr cells can revert to the F$^+$ state. When this occurs, the sex factor is released from the chromosome and resumes its autonomous replication. Sometimes this detachment is not cleanly accomplished, so the F factor carries along with it some chromosomal genes. In this state it is termed an F′ factor, and the cell in which this has occurred is called an F′ cell (see Fig. 12-22). When such primary F′ cells are mated with F$^-$ recipients, the sex factor is transferred very efficiently *together* with the added bacterial genes. The recipient cell then becomes a secondary F′ cell; it is a **partial diploid** for those genes it receives from the primary F′ cell. This process whereby bacterial genes are transmitted from donor to recipient as part of the sex factor has been termed **sexduction** by Jacob and Wollman (Fig. 12-22).

Extrachromosomal Genetic Elements (Plasmids)

In addition to the normal DNA chromosome, extrachromosomal genetic elements are often found in bacteria. These elements are called **plasmids** and are capable of autonomous replication in the cytoplasm of the bacterial cell (Figs. 11-12 and 12-23). Plasmids are circular pieces of DNA that are extra genes. Some plasmids are capable of either replicating autonomously or integrating into the bacterial DNA chromosome and are called *episomes*. Thus the F factor of *E. coli* was called an episome because it can alternately exist in the F$^+$ or Hfr state.

Some bacteria have plasmids that are **bacteriocinogenic factors.** They determine the formation of **bacteriocins,** which are proteins that kill the same or other closely related species of bacteria. The bacteriocins of *E. coli* are called **colicins;** those of *Pseudomonas aeruginosa* are called **pyocins,** and so on. Bacteriocins have proved useful for distinguishing between certain strains of the same species of bacteria in medical bacteriological diagnosis. Bacteria possess other kinds of plasmids called **R plasmids** which confer resistance to a number of antibiotics. Some of the R plasmids can be transferred to other cells by conjugation, hence the term **infectious resistance.** Each form of resistance is due to a gene whose product is an enzyme that destroys a specific antibiotic.

TRANSDUCTION

Most bacteriophages, the virulent phages, undergo a rapid *lytic* growth cycle in their host cells. They inject their nucleic acid, usually DNA, into the bacterium, where it replicates rapidly and also directs the synthesis of new phage proteins.

Within 10 to 20 min, depending on the phage, the new DNA combines with the new proteins to make whole phage particles, which are released by destruction of the cell wall and lysis of the cell. However, some bacterial viruses, the **temperate** phages, which ordinarily do not lyse the cell, carry DNA that can behave as a kind of episome in bacteria; like other episomes, such as the F factor, these viral genomes can become integrated into the bacterial genome; they are then known as **prophages.** Bacteria that carry prophages (**lysogenic** bacteria) can be *induced* with ultraviolet light and other agents to make the prophages start to replicate rapidly and go through a lytic growth cycle, resulting in lysis of the cell with release of new phage particles. (Bacteriophages are discussed in greater detail in Chap. 20.) Phage particles may become filled with cell chromosomal DNA or a mixture of chromosomal and phage DNA (rather than completely with phage DNA, as is normally the case). Such aberrant phages can attach to other bacteria and introduce bacterial, rather than just phage, DNA into them. By this means they transfer bacterial DNA from one cell to another. Thus we can define **bacterial transduction** as the transfer by a bacteriophage, serving as a vector, of a portion of DNA from one bacterium (a donor) to another (a recipient).

This phenomenon was discovered by Zinder and Lederberg in 1952 when they searched for sexual conjugation among *Salmonella* species. They mixed auxotrophic mutants together and isolated prototrophic recombinant colonies from selective nutritional media. When the U-tube experiment was carried out with a parental auxotrophic strain in each arm and separated by a microporous fritted glass filter, prototrophs appeared in one arm of the tube. Since the filter prevented cell-to-cell contact but allowed free passage of fluid between the cultures, it could be concluded that some phenomenon other than conjugation was involved. Furthermore, the phenomenon could not be prevented by DNAase activity, thus eliminating transformation as the process for changing the recipient auxotrophs to prototrophy. Further experiments implicated a bacteriophage as the vector or transducing agent in the following manner. The bacteriophage was released from a lysogenic (recipient) culture. The phage passed through the filter and infected the other strain (donor), lysing it. During replication in the donor strain, the phage adventitiously included parts of the bacterial chromosome with it. It then passed through the filter again, carrying part of the donor's genetic information and imparting it to the recipient strain.

Generalized Transduction

If all fragments of bacterial DNA (i.e., from any region of the bacterial chromosome) have a chance to enter a transducing phage, the process is called **generalized transduction.** In this process, as the phage begins the lytic cycle, viral enzymes hydrolyze the bacterial chromosome into many small pieces of DNA. Any part of the bacterial chromosome may be incorporated into the phage head during phage assembly and is usually not associated with any viral DNA. For example, coliphage P1 can transduce a variety of genes in the bacterial chromosome. (This means that in a large population of phages there will be transducing phages carrying different fragments of the bacterial genome.) After infection a small proportion of the phages carry only bacterial DNA (see Fig. 12-24). The frequency of such defective phage particles is about 10^{-5} to 10^{-7} of the progeny phage produced. Since this DNA matches the DNA of the new

Figure 12-24. Generalized transduction. The phage P1 chromosome, after injection into the host cell, causes degradation of host chromosome into small fragments. During maturation of the virus particles, a few phage heads may envelop fragments of bacterial DNA instead of phage DNA. When this bacterial DNA is introduced into a new host cell, it can become integrated into the bacterial chromosome, thereby transferring several genes from one host cell to another.

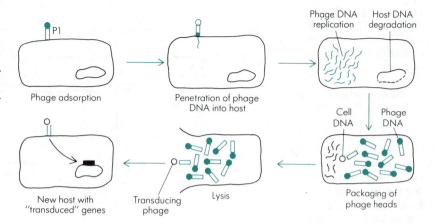

bacterium infected, the recipient bacterium will not become lysogenic for P1 phage. Instead, the injected DNA will be integrated into the chromosome of the recipient cell. Defective P1 phages bearing E. coli DNA can be detected by the genetic markers present in that DNA. For instance, if a thr^- cell is infected by a phage carrying a fragment of E. coli DNA with a thr^+ gene, this thr^+ gene may be integrated (recombined) into the bacterial chromosome to result in a prototrophic recombinant detectable by growth in a medium devoid of threonine.

Generalized transduction, like bacterial conjugation and transformation, also provides a means for mapping bacterial genes, since the fragments transferred by a bacteriophage are often large enough to contain hundreds of genes. The mapping technique involves providing to the phage-infected bacteria a growth medium that *selects for* those recombinants that have inherited a given genetic marker from bacterial DNA carried by a transducing phage. Growth on other media can then be used to test how many of these recombinants have also inherited other donor markers. The closer two markers are together on the bacterial chromosome, the more likely they are to be inherited together by means of a single transducing phage. For example, when coliphage P1 is grown in a $thr^+ leu^+ azi^R$ host and then used to infect a $thr^- leu^- azi^-$ recipient bacterial cell, only 3 percent of the selected Thr^+ recombinants are also Leu^+, and none are Az^R, or azide-resistant. But, if Leu^+ recombinants are selected, about 50 percent of these are also Az^R. This means that leu^+ is more closely linked to azi^R than it is to thr^+ and the suggested order is therefore $thr^+ leu^+ azi^R$. Thus the degree of linkage of genes can be measured by the frequencies of cotransduction of markers. The fact that only 3 percent of thr^+ transducing phages also carry leu^+ shows that these two genes are so far apart that they are rarely included at the same time in a DNA fragment that goes into the P1 head. (The P1 head carries a DNA molecule of slightly less than 10^5 nucleotide pairs.)

Specialized Transduction

Bacterial genes can also be transduced by bacteriophage in another process called **specialized transduction** in which certain temperate phage strains can transfer only a few restricted genes of the bacterial chromosome. More specifi-

Figure 12-25. Specialized transduction. When phage λ infects a cell, its DNA is inserted into the bacterial genome next to the genes for galactose metabolism (*gal* genes). Usually when such a cell is induced, the λ DNA comes out, replicates, and makes normal phage. However, occasionally the λ DNA is excised imperfectly, taking *gal* genes with it and leaving some of itself behind, leading to λdg (defective, galactose transducing phage).

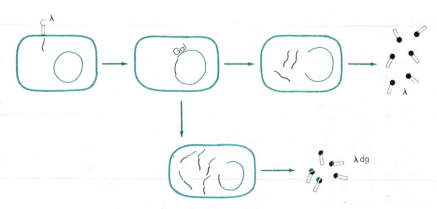

cally, the phages transduce only those bacterial genes adjacent to the prophage in the bacterial chromosome. Thus the process is also called **restricted transduction.** It occurs when a bacteriophage genome, after becoming integrated as prophage in the DNA of the host bacterium, again becomes free upon induction and takes with it into the phage head a small adjacent piece of the bacterial chromosome. In this way, when such a phage infects a cell, it carries with it the group of bacterial genes that has become part of it. Such genes can then recombine with the homologous DNA of the infected cell.

The best-studied specialized transducing phage is the phage lambda (λ) of *E. coli.* The location of the λ prophage in the bacterial chromosome is almost always between the bacterial genes *gal* and *bio.* Whenever the phage genome comes out of, or is *excised* from, the bacterial chromosome, it sometimes takes with it *gal* or *bio* genes. When phages carrying *gal* or *bio* genes infect a new host, recombination with the *gal* or *bio* genes of the host can occur. This process is illustrated in Fig. 12-25. It should be noted that almost all phages that carry some bacterial genes because of "incorrect" excision are defective in certain viral functions because they are missing a piece of phage genetic information taken up by the bacterial genes. They cannot proceed through their entire replicative cycle, but the cell will yield phages if it is also infected with a complete phage that can code for the missing functions of the defective phages.

BACTERIAL TRANSFORMATION

In 1928, an English health officer named Griffith injected mice with a mixture consisting of a few rough (noncapsulated and nonpathogenic) pneumococci and a large number of *heat-killed* smooth (capsulated and pathogenic) cells. (Living smooth pneumococci cause pneumonia in humans and other animals. "Rough" and "smooth" refer to the surface texture of the colonies of the respective cells.) The mice subsequently died of pneumonia, and *live* smooth cells were isolated from their blood. Apparently, some factor responsible for the pathogenicity of the smooth bacteria (even though they were dead) had been transferred to the living rough bacteria and had *transformed* them into pathogenic smooth ones. Griffith also showed that the transforming factor could be passed from the transformed cells to their progeny and thus had the characteristics of a gene. This experiment of Griffith is illustrated in Fig. 12-26.

This "transforming principle" was identified as DNA by Avery, MacLeod, and McCarty in 1944. They defined DNA as the chemical substance responsible for heredity.

Thus transformation is the process whereby *cell-free*, or "naked," DNA containing a limited amount of genetic information is transferred from one bacterial cell to another. The DNA is obtained from the donor cell by natural cell lysis or by chemical extraction. Once the DNA is taken up by the recipient cell,

Figure 12-26. The Griffith experiment. *(From the Office of Technology Assessment.)*

There are two types of pneumococcus, each of which can exist in two forms:

where R represents the rough, nonencapsulated, benign form; and

S represents the smooth, encapsulated, virulent form.

The experiment consists of four steps:

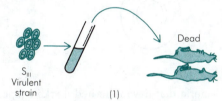

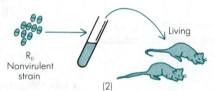

Mice injected with the virulent S_{III} die. (1)

Mice injected with nonvirulent R_{II} do no become infected. (2)

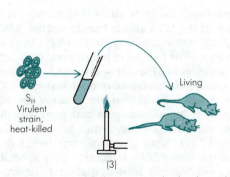

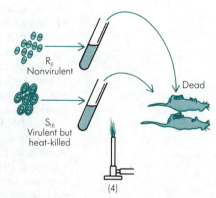

The virulent S_{III} is heat-killed. Mice injected with it do not die. (3)

When mice are injected with the nonvirulent R_{II} and the heat-killed S_{III}, they die. Type II bacteria wrapped in type III capsules are recovered from these mice. (4)

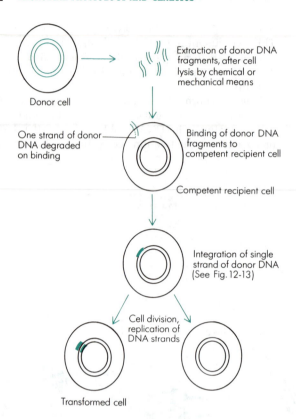

Extraction of donor DNA fragments, after cell lysis by chemical or mechanical means

Donor cell

One strand of donor DNA degraded on binding

Binding of donor DNA fragments to competent recipient cell

Competent recipient cell

Integration of single strand of donor DNA (See Fig. 12-13)

Cell division, replication of DNA strands

Transformed cell

Figure 12-27. Principal steps in bacterial transformation.

recombination occurs. Bacteria that have inherited markers (specific characteristics) from the donor cells are said to be transformed. Thus certain bacteria, when grown in the presence of dead cells, culture filtrates, or cell extracts of a closely related strain, will acquire and subsequently transmit a characteristic(s) of the related strain.

The DNA is taken in through the cell wall and cell membrane of the recipient cell. The molecular size of the DNA affects transformation. Molecular weights of DNA in the range of 300,000 to 8 million daltons have been shown to result in successful transformation. The number of transformed cells increased linearly with increasing concentration of DNA. However, each transformation results from the transfer of a single DNA molecule of double-stranded DNA.

After DNA entry into a cell, one strand is immediately degraded by deoxyribonucleases, while the other strand undergoes base pairing with a homologous portion of the recipient cell chromosome; it then becomes integrated into the recipient DNA (see Fig. 12-27). Since complementary base pairing takes place between one strand of the donor DNA fragment and a specific region of the recipient chromosome, only closely related strains of bacteria can be transformed.

The principal steps of transformation are shown in Fig. 12-27. Bacterial species that have been transformed include, besides *Streptococcus pneumoniae* (pneumococcus), those in the genera *Bacillus, Haemophilus, Neisseria,* and *Rhizobium.*

Properties of Recipient Cells

Conditions suitable for uptake of donor DNA into recipient cells occur only during the late logarithmic phase of growth. During this period, the transformable bacteria are said to be **competent** to take up and incorporate donor DNA. Competent cultures probably produce an extracellular protein factor that apparently acts by binding or trapping donor DNA fragments at specific sites on the bacterial surface. The uptake process has been found to be an energy-requiring mechanism because it can be inhibited by agents that interfere with energy metabolism.

The significance of transformation as a natural mechanism of genetic change is questionable. It probably occurs following the lysis of a microbe and the release of its DNA into the environment. It is conceivable that transformation between bacterial strains of low **virulence** (disease-producing power of a microorganism) can give rise to transformed cells of high virulence. In any case, the phenomenon of transformation has proved to be extremely useful in genetic studies of bacteria in the laboratory, particularly in mapping the bacterial chromosome. This is because when DNA enters a recipient cell during transformation, the entering fragments of DNA are not unlike the DNA fragments transferred in a mating between Hfr and F⁻ cells of *E. coli.* They will undergo **crossing-over** (exchange of portions of homologous chromosomes) with the homologous DNA segments of the recipient cells, and recombinants will be formed. As in conjugation, the frequency of transformation of two genes at the same time is an indication of the distance between these genes on the chromosome.

THE REGULATION AND EXPRESSION OF GENE ACTIVITY

The regulation of gene activity is best controlled at the level of gene transcription. Many examples of such regulation have been discovered in *E. coli* and other bacteria.

Recall that in the bacterial chromosome the genes controlling the enzymes of a metabolic pathway are adjacent to each other. Several adjacent genes code for a single, long mRNA molecule that directs the synthesis of several enzymes of a specific metabolic pathway. The consequence of such an arrangement is that the amount of synthesis of gene products is **coordinately** regulated. Therefore, if a cell is stimulated to synthesize a large amount of one of the enzymes of a group, it will also make large amounts of the other enzymes of the same group. This kind of regulation involves the *induction* and *repression* of enzyme synthesis at the gene level and was discussed in Chap. 9. Maintenance of induction requires the continued synthesis of mRNA to balance its degradation. Thus this mRNA instability, coupled with transcriptional control, assumes that only necessary proteins are synthesized by the cell.

We can better understand the regulation of gene expression in procaryotes by discussing the lactose (*lac*) operon of *E. coli.* This operon is now by far the best understood part of any cellular genome. There are other bacterial operons, but they are less well understood and are different in detail from the *E. coli lac* operon.

Figure 12-28. The Jacob-Monod model of gene control for the *lac* operon. (A) Repression of mRNA synthesis from *lac* operon. In the absence of inducer, the repressor (the product of the *i* gene) binds to the *o* gene to prevent transcription of the *z*, *y*, and *a* genes. (B) Induction of mRNA synthesis from *lac* operon. In the presence of inducer, the repressor binds to the inducer and can no longer combine with the *o* gene. The *lac* operon is no longer repressed, and transcription of the *z*, *y*, and *a* genes takes place. (C) Positive control of enzyme synthesis. Presence of cylic AMP (cAMP) activates the catabolite gene activator protein (CAP), which in turn activates transcription of the *lac* operon.

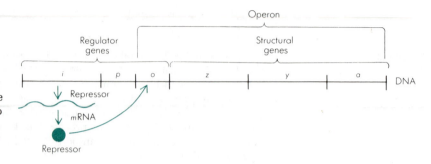

A Repression of mRNA synthesis

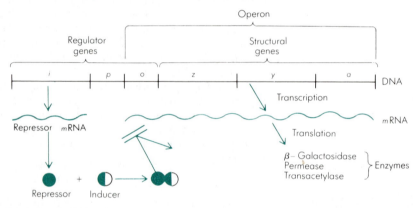

B Induction of mRNA synthesis

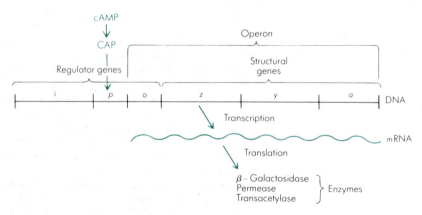

C Positive control of enzyme synthesis

The *lac* Operon

When inducers such as lactose or other β-galactosides are added to a culture of *E. coli*, there is a 1,000-fold increase in the rate of synthesis of the enzymes β-galactosidase (which hydrolyzes lactose to glucose and galactose), β-galactoside permease (which transports lactose into the cell), and thiogalactoside transacetylase (which plays no role in lactose utilization but may play a role in detoxifying certain thiogalactosides). The genes for these proteins are linked together on the *E. coli* chromosome. These are shown in Fig. 12-28 as z, y, and a, coding for β-galactosidase, galactoside permease, and transacetylase, respectively. In the absence of control, the rate of enzyme production would be constant and depend only on the structural genes (such as z, y, and a), amino acid levels, activating enzymes, and other substances. However, the control of the rate of enzyme synthesis is directed by the regulator genes designated i, p, and o, shown in Fig. 12-28, where i is the repressor gene, p the promoter gene, and o the operator gene. The i gene codes for a repressor protein which binds to the DNA of the operator o gene, thus preventing transcription, that is, the synthesis of mRNA (see Fig. 12-28). The promoter gene p is considered to be the site on the DNA where the RNA polymerase enzyme, catalyzing the synthesis of mRNA, binds, and is thus the site where the specific *lac* mRNA (responsible for the biosynthesis of the specific enzymes of the operon) synthesis begins. Let us discuss the functioning of the Jacob-Monod model of gene control for the *lac* operon as it is now understood.

1 Genes function as templates or blueprints for the transcription of mRNA. Using the protein-synthesizing machinery of the cell (ribosomes), the mRNA directs the synthesis of polypeptides (long chains of amino acids) in a process called translation.

2 The genes z, y, and a operate as a single unit of transcription, which is initiated at p.

3 Transcription of the operon is both negatively and positively controlled.

Negative control is mediated by the *lac* repressor which binds to the o gene and blocks transcription. Inducers, such as lactose, stimulate *lac* mRNA synthesis by binding to the repressor and reducing its affinity for the operator (see Fig. 12-28). Both repression and induction of enzyme synthesis are negative control systems because, in either case, the synthesis of enzyme can proceed only when the repressor is removed from its blocking site on the o gene.

Positive control of enzyme synthesis is said to occur when an association between a protein and a part of the regulatory region of an operon is essential for expression of related structural genes in the operon. Expression of the *lac* operon is inhibited when a more efficient source of energy, such as glucose, is present in the medium. The presence of glucose results in a decreased concentration of intracellular cyclic AMP (adenosine-3′,5′-monophosphate). Cyclic AMP is necessary for efficient expression of the *lac* operon since it activates the catabolite gene activator protein (CAP), which in turn activates transcription of *lac* mRNA by RNA polymerase at the promoter site (see Fig. 12-28).

Thus both cyclic AMP and a specific inducer acting in concert are necessary for the synthesis of many inducible enzymes in *E. coli*. Little enzyme is made if either is absent.

The Jacob-Monod model of the *lac* operon has given biologists an insight into the molecular events of gene regulation. It shows up impressively the precision by which regulatory proteins modulate gene function: the repressors must recognize the specific nucleotide sequences of the operator gene on the one hand, and on the other hand they must recognize specific inducer molecules like lactose. An understanding of such regulatory mechanisms has been extended into the study of bacterial viruses.

Upon entering a bacterial host cell, the DNA genome of phage λ may either proceed to a developmental cycle leading to host-cell lysis or integrate into the chromosome of the host bacterium, making it lysogenic or temperate (see Fig. 12-25). Several phage genes are involved in deciding how fast a critical level of a specific repressor can be produced. When sufficient repressor is available, it blocks transcription of all the other phage genes by combining with two separate operators that control two important operons. In such a circumstance, no λ phage proteins are made, the host cell does not lyse, and the circular λ DNA is capable of being integrated into the host chromosome. If the repressors are destroyed or inactivated at any time during lysogeny of the bacterial cell, then the phage operons become derepressed and start functioning, the λ genome replicates, and the cell lyses. Thus it is seen that the repressor gene controls the fate of both the bacterial cell and the bacteriophage.

The mechanisms of the operon model of gene regulation are applicable only to procaryotic organisms and viruses. The same mechanisms have not been found to occur in eucaryotic cells in which the situation is more complex. For example, animal and human genes are full of "gibberish": segments of DNA that serve as coded genetic instructions are interrupted by other segments that have no function whatsoever. These extraneous pieces of DNA are called **introns** and often make up a larger portion of a gene than the actual code-bearing sequences (called **exons**). Thus the introns must all be spliced out of the genetic message before the cell can use it. It has been suggested that in eucaryotic cells regulation must involve controlling the functioning of the mRNA rather than its synthesis; that is, translation is controlled rather than transcription.

Furthermore, since cells with the same genome function differently in different organs in multicellular eucaryotic organisms, there must be some means for switching on and off whole sets of genes in particular cells. Since the chromosomes of eucaryotic cells do not exhibit any clustering of functionally related genes, any mechanism for controlling transcription must act, directly or indirectly, on many genes distant from each other. That is, there is coordinate control of many genes in different chromosomes.

In addition, it has been found that the mRNA that codes for a polypeptide does not always have the full sequence of nucleotides of the corresponding gene. Such mRNAs have been modified, after transcription, by *splicing* with specific enzymes.

Our knowledge of the mechanism of gene control in eucaryotic organisms is still fragmentary. But progress is slowly being made.

GENETIC ENGINEERING

Genetic engineering refers to the development of organisms with genetic structure altered by biochemical manipulation. This kind of biochemical procedure is termed **recombinant DNA technology** and involves the use of plasmids as well as certain bacteriophages. In a nutshell, this technology, developed in the early 1970s, consists of isolating, purifying, and identifying genetic material from one source; tailoring it for insertion into a new host; and isolating a colony of cells with the desired new genes. Since much of the results of genetic engineering have been utilized in biotechnology, this subject will be discussed in greater detail in Chap. 29.

QUESTIONS

1 Explain how variability can arise in a bacterial culture.
2 Describe the advantages in using microorganisms for the study of genetics.
3 How do phenotypic changes differ from genotypic changes?
4 Compare the contribution of Max Delbrück and Salvador Luria with that of Esther and Joshua Lederberg in establishing that bacteria have a hereditary system.
5 What are point mutations, and what are their consequences?
6 What are frameshift mutations, and what causes them?
7 Explain the mechanism of photoreactivation.
8 What are transposons?
9 Why would it be desirable to give two antibiotics simultaneously in the treatment of some diseases?
10 Explain the notation used to describe bacterial mutants with respect to phenotypes and genotypes.
11 Describe some practical implications associated with the occurrence of microbial mutants.
12 What are the distinguishing characteristics of each of the three types of gene transfer?
13 How is sexual mating in bacteria like *Escherichia coli* different from that in higher organisms?
14 In what way did the discovery of sexual differentiation in *E. coli* contribute to an understanding of conjugation in bacteria?
15 Explain how interrupted mating experiments are used for determining the location of genes on a bacterial chromosome.
16 Describe the relationship between plasmids and episomes.
17 What are some conditions that facilitate the occurrence of transformation?
18 Differentiate between generalized transduction and specialized transduction.
19 Describe the events that take place during conjugation between an Hfr cell and an F^- cell of *E. coli*.
20 In what manner is the formation of a λgal transducing phage similar to the formation of an F' factor?
21 Briefly explain how gene activity is controlled at the level of gene transcription.
22 Describe the molecular mechanism which governs the fate of the host cell upon infection with λ phage.

REFERENCES

Alberts, B., D. Bray, J. Lewis, M. Raff, K. Roberts, and J. D. Watson: *Molecular Biology of the Cell*, Garland Publishing, Inc., New York, 1983. *This text is chiefly concerned with the molecular biology of eucaryotic cells and is intended for students* taking a first course in cell biology. A good reference for microbiology students who wish to do further readings into eucaryotic genetic systems.

Ayala, F. J., and J. A. Kiger, Jr.: *Modern Genetics*, Benjamin/Cummings, Menlo Park, Calif., 1980. *A basic genetics text that has the molecular basis of heredity as a dominant theme. Chapter 6 covers the viral genome; chap. 7 is on the bacterial genome.*

Birge, E. A.: *Bacteriophage and Bacteriophage Genetics*, Springer-Verlag, New York, 1981. *A superb introduction to the subject. A text intended for the student taking a first course in the genetics of bacteria and their viruses. Subject presented as a logical development of concepts.*

Grobstein, C.: *A Double Image of the Double Helix*, Freeman, San Francisco, Calif., 1979. *A good introductory account of the nature and origins of the recombinant DNA controversy.*

Hayes, W.: *The Genetics of Bacteria and Their Viruses*, 2d ed., Wiley, New York, 1968. *An older but classic advanced text on the principles of microbial genetics.*

Ingraham, J. L., D. Maaløe, and F. C. Neidhardt: *Growth of the Bacterial Cell*, Sinauer Associates, Sunderland, Mass., 1983. *This text is about the process of growth in bacteria with the concept that growth is the core of bacterial physiology. Material is highly condensed and advanced. Chapter 4 is on the bacterial genome, and chap. 7 is on the molecular genetics of selected operons.*

Lewin, B.: *Genes*, Wiley, New York, 1983. *This text admirably answers the questions as to what is a gene, how it is reproduced, how it is expressed, and what controls this expression. An excellent comprehensive introduction to the molecular biology of the gene. Well written for easy comprehension of a difficult subject.*

Luria, S. E., S. J. Gould, and S. Singer: *A View of Life*, Benjamin/Cummings, Menlo Park, Calif., 1981. *An excellent modern textbook of biology with the subject matter organized around a unifying theme—that all the characteristics of life are prescribed by a single program inscribed in DNA. Not only the classic content of biology is covered, but also those aspects that have made biology the most rapidly developing science in the last half century.*

Watson, J. D.: *Molecular Biology of the Gene*, 3d ed., W. A. Benjamin, Inc., New York, 1976. *Classical text on the structure and function of DNA. It is extremely well written with superb illustrations.*

PART FOUR
THE WORLD OF BACTERIA

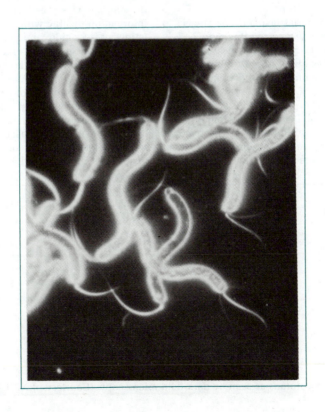

A mirror for the world of bacteria

Prior to 1923, many identification schemes for bacteria had been devised, but these schemes were usually fragmentary. There was need for a single, practical scheme that could cover all the described bacteria. From 1916 to 1918, Robert E. Buchanan was the first to prepare such a comprehensive scheme in a series of papers. In 1917, the Society for American Bacteriologists (now called the American Society for Microbiology) appointed a committee to coordinate all this information, and the final report from this committee in 1920, based largely on Buchanan's work, provided the beginning of a new outline for bacterial classification.

During this period, David H. Bergey began preparing a more complete review of the enormous literature of bacterial taxonomy. To aid the publication of this work, the Society of American Bacteriologists appointed an Editorial Board, chaired by Bergey. This resulted in the publication of the first edition of *Bergey's Manual of Determinative Bacteriology* in 1923.

Every taxonomic scheme for bacteria reflects the knowledge that exists at the time, and since knowledge constantly increases, bacterial taxonomy is subject to continual change. No one recognized this more clearly than Bergey and his colleagues. To cope with these changes, a second edition of the *Manual* was published in 1925, and a third edition in 1930. Five subsequent editions have appeared.

In 1934, the Society of American Bacteriologists transferred to Dr. Bergey all its rights, title, and interests in the *Manual* in order to allow Bergey to create an independent, nonprofit trust—The Bergey's Manual Trust. Throughout the years, this trust continues to prepare and publish successive editions of the *Manual* and promotes research in the field of bacterial taxonomy.

Until 1974, the *Manual* was perceived largely as an "American classification" for bacteria (although it was becoming used in other countries as well), and it was prepared by a relatively small group of microbiologists. It owed its popularity to its breadth of coverage: it was the only book available that attempted to describe all the genera and species of bacteria. But knowledge about the properties of bacteria continued to accumulate at a nearly exponential rate. No longer were just a few people able to cope with the enormous number of bacterial taxa. Thus in 1974 the *Manual* began to become a truly international cooperative effort. Authorities from all over the world were invited to prepare the descriptions of the various genera and species, and the eighth edition of the *Manual* contained contributions from 135 authors.

In 1984, another major change occurred: the scope of the *Manual* was greatly broadened to bring together information dealing with the ecology, enrichment, isolation, preservation, and characteristics of bacteria, all of which concerned bacterial classification and identification. The new breadth of coverage was reflected by a new name—*Bergey's Manual of Systematic Bacteriology*. The new edition of this work is presently being prepared as four volumes, with contributions from hundreds of microbiologists.

As knowledge about bacteria continues to increase, so will *Bergey's Manual* continue to change with it and to act as a "mirror" for the world of bacteria.

Preceding page. *Spirillum volutans* as seen by dark-field microscopy. (*Courtesy of N. R. Krieg.*)

Chapter 13

The World of Bacteria I: "Ordinary" Gram-Negative Bacteria

The most widely used reference for bacterial classification is *Bergey's Manual of Systematic Bacteriology,* now published in four volumes. Volume 1 includes mainly the familiar or "ordinary" Gram-negative chemoheterotrophic eubacteria, many of which have clinical, industrial, or agricultural importance. Most of the organisms have a relatively simple morphology and cellular arrangement. They do not form complex structures such as prosthecae (semirigid extensions of the cell wall and cytoplasmic membrane), sheaths (hollow tubes that enclose chains or trichomes), or, with rare exceptions, endospores (heat-resistant, thick-walled refractive forms) and cysts (thick-walled forms that are desiccation-resistant but not heat-resistant). The cells are not arranged in trichomes. They reproduce mainly by transverse binary fission rather than by budding, fragmentation, or spore formation. Motility, if present, is of the free-swimming type rather than the gliding type. The organisms are mainly heterotrophic, but some can grow autotrophically in the presence of H_2. Some are **saprophytes** (live on dead organic matter), others are **parasites** (living in association with a host). Some are **highly pathogenic** (often cause disease), others may be **opportunistic pathogens** (cause disease only in a patient whose defense mechanisms against infection have been weakened or compromised), and others are not known to cause disease. In this chapter we shall describe briefly some of the organisms included in the first volume of the *Manual*.

BERGEY'S MANUAL OF SYSTEMATIC BACTERI-OLOGY, VOLUME 1

Each volume of *Bergey's Manual* is divided into a number of major sections, each bearing a descriptive common name rather than a formal taxonomic name. The major sections of Volume 1 are listed in Table 13-1. Within each section the bacteria are divided into formal taxa at various levels, most attention being given to families, genera, and species. In the remaining pages of this chapter we present a thumbnail sketch of the organisms included in the major sections of Volume 1. The purpose is to acquaint the student with the diversity of organisms that belong to the bacterial world.

THE SPIROCHETES

The distinguishing features of spirochetes are:

1 A helical shape
2 An ability to twist or contort their shape (i.e., flexibility)
3 The occurrence of a special kind of flagella termed *periplasmic flagella* (also called *axial fibrils* or *endoflagella*)

The major difference in structure from other Gram-negative eubacteria is in the location of the periplasmic flagella. As indicated in Fig. 13-1, they are located between the outer membrane (often termed the *outer sheath* in spirochetes) and the *protoplasmic cylinder* (i.e., the protoplast plus the overlying peptidoglycan layer); thus they are are located in the periplasmic space of the cell. The periplasmic flagella have an ultrastructure similar to that of ordinary flagella, including a basal body with disks, and they are responsible for the swimming motility of spirochetes; however, since periplasmic flagella do not extend outward from the cells as do ordinary flagella, the exact mechanism by which they accomplish this motility is not yet clear. Spirochetes swim best in media of high viscosity, whereas bacteria with ordinary flagella swim best in media of low viscosity; spirochetes can also exhibit a creeping or crawling motility when in contact with solid surfaces. Most spirochetes are so thin that they cannot be easily seen by light microscopy, even when Gram-stained; however, dark-field microscopy does provide sufficient contrast and is the method of choice for visualizing these organisms.

Table 13-1. Gram-Negative Bacteria Included in *Bergey's Manual*, Volume 1

Section	Other Major Characteristics
THE SPIROCHETES	Flexible; helical; have periplasmic flagella; saprophytes or parasites
AEROBIC/MICROAEROPHILIC, MOTILE, HELICAL/VIBRIOID, GRAM-NEGATIVE BACTERIA	Rigid; motile by polar flagella; oxidative type of metabolism; saprophytes or parasites
NONMOTILE (OR RARELY MOTILE), GRAM-NEGATIVE CURVED BACTERIA	Rigid; curved, ring-shaped, or helical cells lacking flagella; saprophytes
AEROBIC GRAM-NEGATIVE RODS AND COCCI	Rigid; straight or slightly curved (but not helical) rods, and cocci; oxidative type of metabolism; saprophytes and parasites
FACULTATIVELY ANAEROBIC GRAM-NEGATIVE RODS	Rigid; straight or curved rods; have both an oxidative and a fermentative type of metabolism; saprophytes and parasites
ANAEROBIC GRAM-NEGATIVE STRAIGHT, CURVED AND HELICAL RODS	Rigid; obtain energy by fermentation or by an anaerobic respiration that does not use sulfur compounds as electron acceptors; parasites
DISSIMILATORY SULFATE- OR SULFUR-REDUCING BACTERIA	Rigid; anaerobic; use sulfur compounds as electron acceptors; saprophytes and parasites
ANAEROBIC GRAM-NEGATIVE COCCI	Rigid; nonmotile; fermentative; parasites
THE RICKETTSIAS AND CHLAMYDIAS	Rigid; tiny cells; intracellular parasites of humans, other animals, and arthropods; can be isolated and cultivated in host cells and sometimes on culture media
THE MYCOPLASMAS	Soft and plastic; nonmotile; lack cell walls; parasites and saprophytes
ENDOSYMBIONTS	Bacteria-like forms that are obligate parasites of protozoa, arthropods, or other hosts; often beneficial to their hosts; have not been isolated or cultivated

Figure 13-1. Basic anatomical components of spirochetes as interpreted from electron micrographs; surface views. *(Courtesy of S. C. Holt, Microbiol Rev* **42:**114, 1978.)

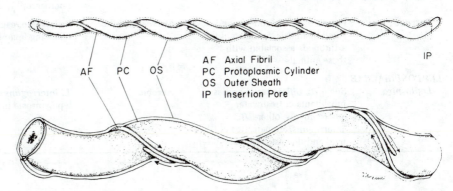

AF Axial Fibril
PC Protoplasmic Cylinder
OS Outer Sheath
IP Insertion Pore

The spirochetes comprise the order *SPIROCHAETALES.* This order is divided into two families, which are distinguished as follows:

THE FAMILY *SPIROCHAETACEAE*

1 They are stringent anaerobes, facultative anaerobes, or microaerophiles.
2 Carbohydrates or amino acids are used as carbon and energy sources.

THE FAMILY *LEPTOSPIRACEAE*

1 They are aerobes.
2 Long-chain fatty acids are used as the carbon and energy source.

Characteristics of the genera of these two families are indicated in Table 13-2.

AEROBIC/MICROAERO-PHILIC, MOTILE, HELI-CAL/VIBRIOID, GRAM-NEGATIVE BACTERIA

These bacteria have the following characteristics:

1 The cells are rigid (unlike spirochetes) and range from vibrioid (having less than one turn or twist) to helical (having one to many turns or twists).
2 They swim by means of polar flagella.

Table 13-2. Characteristics of the Genera of Spirochetes

Family and Genus	Habitat	Oxygen Relationships	Major Characteristics
SPIROCHAETACEAE			
Spirochaeta	Harmless inhabitants of water, mud, and sediments of marine and freshwater environments	Anaerobic and facultatively anaerobic	Use carbohydrates but not amino acids as carbon and energy sources
Cristispira	Harmless parasites of freshwater and marine molluscs	Unknown	Have never been isolated; have unusually large number of periplasmic flagella (>100)
Treponema	Mouth, intestinal tract, and genital areas of humans and animals; some are pathogenic	Anaerobic and microaerophilic	Use carbohydrate and amino acids; some have been cultivated in vitro (on nonliving media) and are stringent anaerobes; these are mainly harmless parasites, but one species, *T. hyodysenteriae,* causes hog dysentery. Some species have not been cultivated in vitro, e.g., *T. pallidum* subsp. *pallidum,* which causes syphilis in humans and is microaerophilic
Borrelia	Parasites of wild rodents and small mammals, and also of the arthropods associated with these animals	Microaerophilic	Pathogenic, causing louseborne or tickborne relapsing fever in humans
LEPTOSPIRACEAE			
Leptospira	Some *(L. biflexa)* are harmless inhabitants of freshwater environments; others *(L. interrogans)* are parasites of wild and domestic animals	Aerobic	*L. interrogans* is pathogenic and causes leptospirosis in animals and humans

3 They are aerobic or microaerophilic.

4 They attack few or no carbohydrates.

5 The organisms usually give a positive reaction by the oxidase test (a laboratory test based on the presence of cytochrome c).

Most of the organisms are harmless saprophytes and occur in freshwater or marine environments, but a few are parasitic and can be pathogenic for humans and animals or for other bacteria. Some examples of genera in this section of *Bergey's Manual* are as follows:

Aquaspirillum

Aquaspirilla are helical or vibrioid organisms that typically possess bipolar tufts of flagella (see Fig. 13-2). These harmless saprophytes are aerobic to microaerophilic and occur in stagnant stream or pond water. No growth occurs in the presence of 3% NaCl or sea water.

Azospirillum

The cells are plump and vibrioid (see Fig. 13-3) with a single polar flagellum and, if grown on solid media, with numerous lateral flagella as well. Azospirilla occur within the roots of grasses, wheat, corn, and many other plants or as free-living soil organisms. They fix N_2 within plant roots or in laboratory cultures. Under N_2-fixing conditions they are microaerophilic, but they are aerobic if supplied with a source of fixed nitrogen such as an ammonium salt. One species, *A. lipoferum,* can grow autotrophically with hydrogen gas as the energy source.

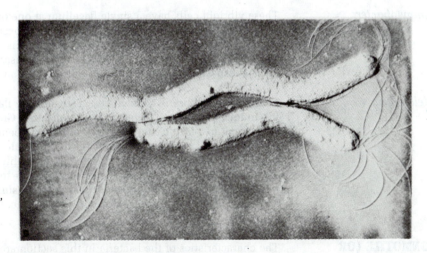

Figure 13-2. *Aquaspirillum bengal* (X9,800). *(Courtesy of R. Kumar, A. K. Banerjee, J. H. Bowdre, L. J. McElroy, and N. R. Krieg, Int J Syst Bacteriol,* **42:***453, 1974.)*

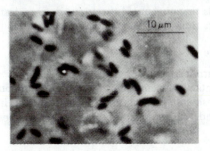

Figure 13-3. Plump vibrioid and straight cells of *Azospirillum brasilense* from a 48-h-old culture grown under nitrogen-fixing conditions. The cells are 1 μm in width. *(Courtesy of N. R. Krieg, Bacteriol Rev,* **40:***55, 1976.)*

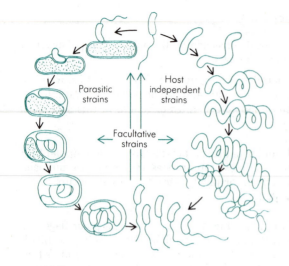

Figure 13-4. Schematic representation of the life cycles of *Bdellovibrio*. All strains isolated from nature require host bacteria for growth (i.e., they are obligately parasitic). In the cycle depicted at the left, a bdellovibrio attaches to a host cell, penetrates the wall, grows within the periplasmic space into a long, coiled form that eventually fragments into new bdellovibrio progeny. Certain mutants can grow only on nonliving culture media (host-independent cycle, shown on the right). Rare strains are apparently facultative parasites and will grow in culture media or in host cells. *(Courtesy of J. C. Burnham, T. Hashimoto, and S. F. Conti, J Bacteriol **101**:997, 1970.)*

Oceanospirillum

The cells are helical, usually with bipolar tufts of flagella. Oceanospirilla are aerobic and are harmless saprophytes, occurring in coastal marine waters. Sea water is required for their growth.

Campylobacter

These vibrioid cells have a single flagellum at one or both poles (see Fig. 36-4). Campylobacters are microaerophilic parasites, occurring in the reproductive organs, intestinal tract, and oral cavity of humans and other mammals. Some species are pathogenic, e.g., *C. jejuni,* which causes diarrhea in humans, or *fetus* subspecies *venerealis,* which causes abortion in cattle.

Bdellovibrio

These aerobic, vibrioid cells possess a single polar flagellum. Bdellovibrios have the unique property of being parasitic on other Gram-negative bacteria. After attachment to a host bacterium, the bdellovibrio penetrates the outer membrane of the cell wall and grows within the periplasmic space. Eventually the host bacterium becomes an empty "ghost" cell. This life cycle is illustrated in Fig. 13-4. Bdellovibrios occur in soil, sewage, and in freshwater and marine environments. The genus *Vampirovibrio* has certain similarities to the genus *Bdellovibrio*, but the organisms attack eucaryotic algae, not bacteria.

NONMOTILE (OR RARELY MOTILE), GRAM-NEGATIVE, CURVED BACTERIA

The characteristics of the bacteria in this section are as follows:

1 Rigid cells that are curved to various degrees, forming coils, helical spirals, and sometimes rings (i.e., cells that are curved around so that the ends overlap (Fig. 13-5)
2 Nonmotile

These harmless saprophytes occur mainly in soil, freshwater, and marine environments. One example is the family **Spirosomaceae,** which contains three genera whose cells, which are aerobic, form no intracellular gas vacuoles, are catalase- and oxidase-positive, and form colonies that are yellow (the genus

Spirosoma) or pink (the genera *Runella* and *Flectobacillus*). Another common genus, not included in the family *Spirosomaceae*, is *Microcyclus*, which forms prominent intracellular gas vacuoles and whose colonies have no pigment.

AEROBIC, GRAM-NEGATIVE RODS AND COCCI

This section forms one of the largest and most diverse groups of bacteria. Two general features are as follows:

1 The cells are mainly straight or slightly curved (but not helical) rods, but some are cocci.
2 They have a strictly respiratory type of metabolism.

Several families and some additional genera that are not assigned to any family are represented.

The Family PSEUDOMONADACEAE

The features of this family include the following:

1 Straight or slightly curved rods
2 Motile by polar flagella (Fig. 13-6)
3 Catalase-positive and usually oxidase-positive

Some of the genera included in this family are described here.

Pseudomonas

These bacteria are widely distributed in soil and water. The genus contains five genetically distinct groups (each of which might be considered as a separate genus) as well as a large number of additional, less well studied organisms. All pseudomonads can grow aerobically, but some can also grow anaerobically by using nitrate as an electron acceptor. Several species are pathogenic for humans or animals; others are important plant pathogens. Some cause spoilage of meats and other foods. Species identification is based on many physiological and nutritional characteristics, such as the ability to use certain compounds as carbon sources for growth. Some species can use any of over 100 compounds as a carbon source. Sugar-containing media are acidified only weakly, and acid

Figure 13-5. (*Below*) Phase-contrast photomicrograph of *Spirosoma* cells showing rings and coils. The bar indicates 10 μm. (*Courtesy of J. M. Larkin, P. M. Williams, and R. Taylor, In J Syst Bacteriol 27:147, 1977.*)

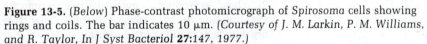

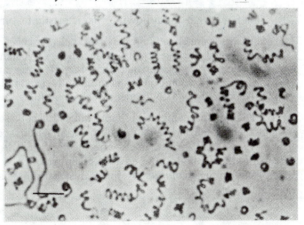

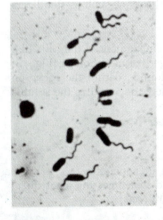

Figure 13-6. (*Above*) Flagella stain (X2,000) of cells of a *Pseudomonas* strain showing the characteristic polar flagella. (*Courtesy of General Biological Supply House, Inc.*)

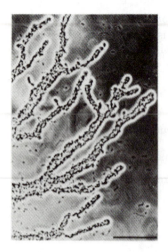

Figure 13-7. Cells of *Zoogloea ramigera* embedded in a matrix of exopolymer. The fingerlike projections of the slimy mass are characteristic of the genus. The bar indicates 50 μm. *(Courtesy of R. Unz, from N. Palleroni in The Prokaryotes: A Handbook on Habitats, Isolation, and Identification of Bacteria, Springer-Verlag, 1981.)*

production from sugars is generally not useful in differentiating species (in contrast, for example, to the family *Enterobacteriaceae*). A few examples of *Pseudomonas* species are listed below:

P. aeruginosa produces a water-soluble blue pigment, pyocyanin, and a water-soluble fluorescent pigment, pyoverdin. The organism is mainly a soil and water saprophyte, but it is also frequently an opportunistic pathogen and can often be isolated from wound, burn, and urinary tract infections.

P. maltophilia is a nonfluorescent species that is also frequently isolated from clinical specimens.

P. fluorescens is a common saprophytic soil and water organism that makes only a fluorescent pigment.

P. syringae and several other species are important plant pathogens, causing diseases such as leaf spot, leaf stripe, wilt, and necrosis.

P. mallei is the causative agent of glanders and farcy, diseases of horses and donkeys that are transmissible to humans. *P. pseudomallei* causes melioidosis in animals and humans.

Xanthomonas

This genus forms characteristic yellow pigments called xanthomonadins. All species are pathogenic for plants, causing diseases such as spots, streaks, cankers, wilts, and rots. Xanthomonads make viscous exocellular polysaccharides (xanthan gums) which are useful for industrial applications such as stabilizers in foods and antidrip agents in paints.

Zoogloea

The outstanding characteristic of this genus is the embedment of the cells in a gelatinous matrix to form slimy masses with a fingerlike morphology (Fig. 13-7). These saprophytic organisms are commonly found coating the rocks on trickling-filter beds in sewage treatment plants, where they oxidize the soluble organic components of the sewage.

The Family AZOTOBACTERACEAE

This family has the following characteristic features:

1 The cells are large blunt rods, oval cells, or cocci.
2 Their motility and flagellar arrangement vary; some are nonmotile.
3 They are saprophytes that occur in soil, water, and sometimes the plant rhizosphere (soil region subjected to the influence of plant roots).
4 The organisms fix N_2 under aerobic conditions.

The unusually high respiratory rate of the cells serves to use up oxygen rapidly at the cell surface and maintain an anaerobic cell interior; this protects the oxygen-sensitive nitrogenase enzyme complex (responsible for N_2 fixation) from being inactivated under an air atmosphere.

One genus, *Azotobacter,* forms desiccation-resistant cysts (see Figs. 5-32 and 5-33).

The Family RHIZOBIACEAE

This family contains rod-shaped cells that incite hypertrophies on plants (root nodules, leaf nodules, or tumors). Three genera of this family are described below.

Rhizobium and Bradyrhizobium

These bacteria fix N_2 by means of a complex, highly evolved symbiosis with the roots of leguminous plants. The bacteria attach to the root hairs, penetrate the root, and induce proliferation of the root cells (see Figs. 25-11 and 25-12). Within the resulting root nodules the bacteria exist as highly pleomorphic N_2-fixing forms called *bacteroids*. *Leghemoglobin* occurs within the root nodules and serves to protect the nitrogenase enzyme complex from being destroyed by excess oxygen. The species and strains of rhizobia and bradyrhizobia exhibit a range of specificities for various legumes.

Agrobacterium

Agrobacteria do not fix N_2. The organisms are plant pathogens that incite tumors when they invade the crown, roots, and stems of a great variety of dicotyledonous and some gymnospermous plants (see Fig. 36-20). Tumor induction is associated with the presence in the bacteria of a particular plasmid (see Chap. 36 for further details).

The Family METHYLOCOCCACEAE

This family consists of a diverse group of rods, vibrios, and cocci having in common the ability to use methane gas as a sole carbon and energy source under aerobic or microaerophilic conditions. These harmless organisms occur in soil, mud, or water adjacent to or overlying the anaerobic environments where methane is formed. Some members of the family fix nitrogen under microaerophilic conditions. Some form *Azotobacter*-like cysts. The genera presently included—*Methylococcus* and *Methylomonas*—are all *obligate* methane-oxidizers (i.e., carbon sources such as glucose cannot be used for growth); however, the definition of the family permits inclusion of facultative methane-oxidizers as well.

The Family ACETOBACTERACEAE

This family contains ellipsoidal to rod-shaped cells that oxidize ethanol to acetic acid in neutral or acidic (pH 4.5) media. Two genera are included, *Acetobacter* and *Gluconobacter,* which are differentiated by certain biochemical characteristics and by the occurrence of peritrichous flagella (*Acetobacter*) or polar flagella (*Gluconobacter*).

Members of these two genera are saprophytes that occur in sugar- or alcohol-enriched, acidic environments such as flowers, fruits, beer, wine, cider, vinegar, souring fruit juices, bees, and honey. Some have industrial importance: acetobacters are used to make vinegar, and gluconobacters are involved in the manufacture of chemicals such as dihydroxyacetone, sorbose, and 5-ketogluconic acid (see Table 29-1). Some strains of *Acetobacter* have the highly unusual ability (for bacteria) to make exocellular cellulose fibrils that accumulate around the cells.

The Family LEGIONELLACEAE

These rod-shaped bacteria require L-cysteine, iron salts, and activated powdered charcoal for growth (the charcoal destroys toxic hydrogen peroxide in the medium). All belong to a single genus, *Legionella.* The organisms are motile by polar or lateral flagella. They occur in surface water, thermally polluted lakes and streams, water from air-conditioning cooling towers and evaporative condensers, and in moist soil adjacent to a body of water. All species are opportunistic pathogens of humans, causing legionellosis.

The Family
NEISSERIACEAE

This family contains nonmotile rods and cocci that are catalase-positive and/or oxidase-positive. Examples are the genera *Neisseria* and *Acinetobacter.*

Neisseria

This genus has traditionally consisted of oxidase- and catalase-positive cocci that occur most often in pairs with the adjacent sides flattened (Fig. 13-8). However, one rod-shaped species *(N. elongata)* is now included because of its genetic relatedness. The neisseriae are parasites that inhabit the mucous membranes of humans and animals. Two species are highly pathogenic for humans, e.g., *N. gonorrhoeae,* the causative agent of gonorrhea, and *N. meningitidis,* the causative agent of epidemic cerebrospinal meningitis.

Acinetobacter

These diplobacilli are catalase-positive but oxidase-negative. They are saprophytes that occur in soil, water, and sewage, but they are also opportunistic human pathogens that can cause a variety of infections, particularly in hospitalized patients.

Other Genera of Aerobic Gram-Negative Rods and Cocci Not Assigned to Any Family

Several genera are not assigned to any family, yet are included in this general section of *Bergey's Manual.* Some of these are described in Table 13-3.

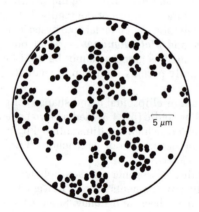

Figure 13-8. Drawing of *Neisseria gonorrhoeae,* showing the characteristic diplococcus arrangement of the cells. *(Erwin F. Lessel, illustrator.)*

FACULTATIVELY ANAEROBIC GRAM-NEGATIVE RODS

The organisms in this section form a very diverse group of straight or curved rods that can grow aerobically by respiring with oxygen and also under anaerobic conditions by fermenting various carbohydrates. Most genera are associated with animals or plants, but some occur in soil and water. The following are examples of some of the organisms included in this section.

The Family
ENTEROBACTERIACEAE

Some distinctive features of this family are:

1 Cell diameter is 0.3 to 1.5 μm.
2 Cell shape is that of a straight rod.

Table 13-3. Some Genera of Aerobic Gram-Negative Rods and Cocci Not Affiliated with Any Family

Genus	Shape	Flagella	Major Characteristics
Beijerinckia and *Derxia*	Rods	Peritrichous and polar, respectively	Tropical soil bacteria that can fix N_2 aerobically; *Derxia* can grow autotrophically with H_2 as the energy source
Xanthobacter	Rods	Usually none	Soil bacteria that fix N_2 under microaerophilic conditions; can grow autotrophically with H_2 as the energy source; form yellow colonies; cells stain Gram-positive or Gram-variable
Thermus and *Thermo-microbium*	Rods	None	Occur in hot springs; obligate thermophiles; optimum temperature is 70–75°C
Alteromonas	Straight or curved rods	Polar	Harmless marine organisms; require sea water for growth; oxidase-positive
Flavobacterium	Rods	None	Mainly saprophytes, widely distributed in nature; may often occur in hospital environments; form yellow to orange colonies; oxidase-positive; one species, *F. meningosepticum,* can cause a severe meningitis in newborn infants
Alcaligenes	Very short rods	Peritrichous	Occur in soil, freshwater, and marine environments but can be opportunistic human pathogens. Form non-pigmented colonies. Oxidase-positive. Some can grow autotrophically with H_2 as the energy source.
Brucella	Very short rods	None	Parasites and pathogens of animals. Three species are highly pathogenic for animals and humans, causing brucellosis.
Bordetella	Very short rods	None	Parasitic and pathogenic for various mammalian hosts. *B. pertussis* occurs only in humans and causes whooping cough.
Francisella	Very short rods	None	The major species, *F. tularensis,* is a parasite of wild animals but can also cause tularemia in humans.
Lampropedia	Cocci, nearly cubical	None	Harmless saprophytes occurring in aquatic environments. Occur in distinctive, flat, square tablets of 16–64 cells (see Fig. 13-9).

3 Motility, if present, is by means of lateral flagella.

4 They are oxidase-negative.

5 Na^+ is not required or stimulatory for growth.

6 Cells contain a characteristic antigen, called the enterobacterial common antigen.

7 The organisms have simple nutritional requirements.

The family contains a large number of genera that are biochemically and genetically related to one another. Many of the more traditional or familiar bacteria are to be found in this family. Differentiation of the various genera is based on characteristic patterns obtained from a large number of biochemical tests; a few properties of some of the genera are listed in Table 13-4. Because

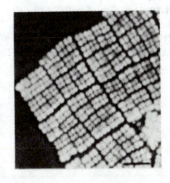

Figure 13-9. Negatively stained preparation of *Lampropedia hyalina* showing a sheet of actively growing tablets of cells. The bar indicates 5 μm. (*Courtesy of R. G. E. Murray, from Bergey's Manual of Determinative Bacteriology, 8th ed., Williams & Wilkins, Baltimore, 1974.*)

Table 13-4. Typical Characteristics of Some *Enterobacteriaceae*

Property	Escherichia	Shigella	Salmonella	Enterobacter	Serratia	Proteus	Yersinia
Motility	d	−	+	+	+	+	−
Voges-Proskauer test	−	−	−	+	d	d	−
Indole from tryptophan	d	d	−	−	−	d	d
Hydrogen sulfide produced	−	−	+	−	−	d	−
Lysine decarboxylase	+	−	+	d	d	−	−
Gas from glucose	+	−	+	+	d	+	−
Acid from lactose	d	−	−	+	d	−	d
Urease	−	−	−	d	−	+	d
Phenylalanine deaminase	−	−	−	−	−	+	−
Deoxyribonuclease	−	−	−	−	+	−	−

+ = most or all species positive; − = most or all species negative; d = different reactions occur among species.

sugar-containing media are strongly acidified, acidic reactions from various sugars are used extensively to differentiate the genera and species (in contrast, for example, to the genus *Pseudomonas*). Some selected genera are listed below.

Escherichia

(See Fig. 13-10.) The major species, *E. coli,* occurs in the lower portion of the intestine of humans and warm-blooded animals, where it is part of the normal flora. Some strains can cause gastroenteritis; others can cause urinary tract infections.

Shigella

This genus is very closely related to *Escherichia* but differs in a few characteristics (Table 13-4). Moreover, all strains are pathogenic, causing bacillary dysentery in humans.

Salmonella

(See Fig. 13-10.) This is a group of organisms that are closely related to one another and probably should be considered as a single species. All strains are pathogenic for humans, causing enteric fevers (such as typhoid and paratyphoid fevers), gastroenteritis, and septicemia; many strains also infect a variety of animals. Over 2,000 antigenic types of salmonellae occur.

Enterobacter

Unlike most other *Enterobacteriaceae*, *Enterobacter* species grow best at 30°C rather than at 37°C. They occur mainly in water, sewage, soil, meat, plants, and

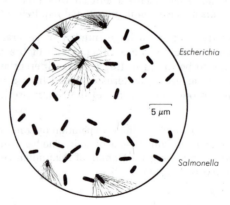

Figure 13-10. Drawing of cells of the genera *Escherichia* and *Salmonella*. The peritrichous flagella are not visible by ordinary staining. (*Erwin F. Lessel, illustrator.*)

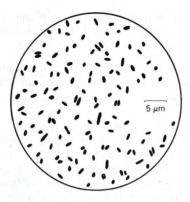

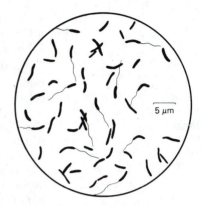

Figure 13-11. Drawing of cells of *Yersinia pestis*, the causative agent of plague. (*Erwin F. Lessel, illustrator.*)

Figure 13-12. Drawing of cells of the genus *Vibrio*. The polar flagella are not visible by ordinary staining. (*Erwin F. Lessel, illustrator.*)

vegetables. Some species also occur in human and animal feces, and some can be opportunistic human pathogens.

Erwinia

These bacteria differ from most other *Enterobacteriaceae* by being mainly associated with plants. They are often plant pathogens, causing diseases such as blights, cankers, die back, leaf spot, wilts, discoloration of plant tissues, and soft rots. They are seldom isolated from animals or humans.

Serratia

The serratiae are widely distributed in soil, in water, and on plant surfaces. Many strains produce pink or red colonies. These organisms were once thought to be harmless; however, it is now clear that they can be opportunistic human pathogens and are particularly prone to infect hospitalized patients.

Proteus

These organisms can *swarm* on agar media; that is, they spread over the plates in a thin film resulting from periodic cycles of migration. Such swarming often makes it difficult to obtain pure cultures of other bacteria from streak plates. *Proteus* strains occur in the intestine of humans and a wide variety of animals, in polluted waters, and in soil, and they can be opportunistic human pathogens. Like *E. coli*, *P. mirabilis* is one of the leading causes of urinary tract infections in humans.

Yersinia

Yersiniae are parasites of animals but can also cause infections in humans. For example, *Y. pestis* (see Fig. 13-11) is the causative agent of plague, and *Y. enterocolitica* is a frequent cause of gastroenteritis in children.

The Family VIBRIONACEAE

Some distinctive features of this family are:

1 Cell diameter is 0.3 to 1.3 μm.
2 Cell shape is that of curved or straight rods (Fig. 13-12).
3 The cells are motile by means of polar flagella.
4 They are usually oxidase-positive.

5 Na$^+$ is required or stimulatory for growth of some genera.

6 Cells do not contain the enterobacterial common antigen.

7 The organisms have simple nutritional requirements.

The *Vibrionaceae* occur in marine and freshwater environments or in association with animals living in those environments. Two genera are described below:

Vibrio

These are distinguished from other members of the family by having flagella that are covered by a membrane (sheathed flagella). The organisms occur in aquatic habitats with a wide range of salinities. Some species can emit light of a blue-green color *(bioluminescence)*, an oxygen-dependent reaction catalyzed by the enzyme luciferase. One such species, *V. fischeri,* occurs in the specialized luminous organs of certain deep-sea fishes. Most *Vibrio* species are harmless saprophytes, but some species are pathogenic; examples are *V. cholerae,* the causative agent of cholera in humans, *V. parahaemolyticus,* which causes gastroenteritis in humans, and *V. anguillarum,* which is a pathogen of marine fish and eels.

Aeromonas

Cells are straight rods that have nonsheathed flagella. The organisms occur in fresh water sources and sewage. Some species are pathogenic for frogs and fish; e.g., *A. salmonicida* is the causative agent of furunculosis in salmon and trout.

The Family PASTEURELLACEAE

Distinctive features of this family are:

1 The cell diameter is small (0.2 to 0.4 μm).

2 Cell shape is that of a straight rod.

3 The cells are nonmotile.

4 They are usually oxidase-positive.

5 Na$^+$ is not required or stimulatory for growth.

6 Cells do not contain the enterobacterial common antigen.

7 The organisms often have complex nutritional requirements.

8 The family occurs as parasites of vertebrates.

Some genera included in the family are described below:

Pasteurella

These organisms are parasitic on the mucous membranes of the upper respiratory tract of mammals (rarely humans) and birds. The major pathogen is *P. multocida,* which causes hemorrhagic septicemia in cattle and fowl cholera in domestic and wild birds.

Haemophilus

These bacteria are distinguished by unusual nutritional requirements: the X *factor* (heme, occurring in blood) and/or the V *factor* (the coenzyme nicotinamide adenine dinucleotide). *Haemophilus* species occur as parasites of the mucous membranes of humans and animals. Some are pathogenic for humans: for example, *H. influenzae* (see Fig. 13-13) is a leading cause of meningitis in children.

Actinobacillus

These bacteria are also parasitic on mammals and birds. The organisms are only occasionally pathogenic for humans, but several species are pathogenic for

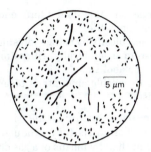

Figure 13-13. Drawing of cells of *Haemophilus influenzae. (Erwin F. Lessel, illustrator.)*

animals, e.g., *A. lignieresii,* which causes granulomatous lesions in cattle and sheep, and *A. suis,* which causes septicemia, pneumonia, and arthritis in pigs.

Other Genera of Facultatively Anaerobic Gram-Negative Rods Not Assigned to Any Family

Several genera of facultatively anaerobic rods are not assigned to any family but belong to this section of *Bergey's Manual.* A few examples are listed below.

Zymomonas

These are saprophytic rods that occur in honey and as spoilage organisms in beer and cider. They are unusual because they form large amounts of ethanol from glucose.

Chromobacterium

These motile, rod-shaped organisms have the unusual property of forming violet colonies, due to a pigment called violacein. The species *C. violaceum* occurs as a saprophyte in soil and water but can occasionally cause infections of humans and other mammals.

Gardnerella

These nonmotile pleomorphic rods stain Gram-negative to Gram-variable. Whether they should be classified with Gram-negative bacteria based on studies of cell-wall ultrastructure and chemical composition is still inconclusive. The only species included in the genus, *G. vaginalis,* occurs in the human genito-urinary tract and is a major cause of bacterial "nonspecific" vaginitis.

Streptobacillus

During cultivation of these pleomorphic rods, *L-phase variants* may occur spontaneously; these have a defective cell wall, are more or less spherical in shape, and form tiny "fried-egg" colonies similar to those formed by mycoplasmas (discussed later in this chapter). The single species of the genus, *S. moniliformis,* is a parasite of rats and causes one form of rat-bite fever in humans.

ANAEROBIC, GRAM-NEGATIVE STRAIGHT, CURVED, AND HELICAL RODS

The organisms in this section are placed within a single family, which is described as follows.

The Family *BACTEROIDACEAE*

This family is a diverse assemblage of bacteria that exhibit the following features:

1 They are anaerobic organisms that do not form endospores.
2 They may be motile or nonmotile.
3 They do not respire anaerobically by using inorganic sulfur compounds as electron acceptors; some do respire anaerobically with nitrate or fumarate; most have only a fermentative type of metabolism.
4 Most species produce detectable amounts of organic acids as the result of their fermentative metabolism.

The genera are differentiated mainly on the basis of their morphology and the kinds of organic acid end products they produce, as indicated in Table 13-5. The organic acids are identified by means of a gas chromatograph. The organisms occur in the oral cavity and intestinal tract of humans and animals, in the rumen of cattle, sheep, and other ruminant animals, or in other anaerobic environments. Some genera contain species that are pathogenic for humans: e.g., *Bacteroides fragilis* (Fig. 13-14) is the most common anaerobic bacterium isolated from human soft tissue infections and anaerobic blood infections.

Table 13-5. Characteristics of Some Genera of Anaerobic, Gram-Negative Straight, Curved, and Helical Rods

Genus	Morphology	Major Organic Acid End Products of Fermentation
Bacteroides	Straight rods; nonmotile or motile by peritrichous flagella	Mixtures including succinate, acetate, formate, lactate, and propionate; butyrate is either not formed or, if produced, is accompanied by isobutyrate and isovalerate
Fusobacterium	Straight rods; nonmotile	Butyrate
Succinimonas	Short rods or coccobacilli; motile by a single polar flagellum	Succinate and acetate
Wolinella	Helical, curved, or straight rods; motile by a single polar flagellum	Do not have a fermentative type of metabolism; respire anaerobically with H_2 or formate as electron donors and fumarate or nitrate as electron acceptors; the formate is oxidized to CO_2, and the fumarate is reduced to succinate
Selenomonas	Crescent-shaped cells; motile by a tuft of flagella located at the middle of the concave side	Acetate, propionate, and sometimes lactate
Anaerovibrio	Slightly curved rods; motile by a single polar flagellum	Propionate and acetate

Figure 13-14. Drawing of cells of *Bacteroides fragilis*. (*Erwin F. Lessel, illustrator.*)

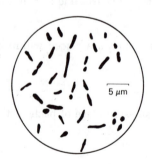

5 μm

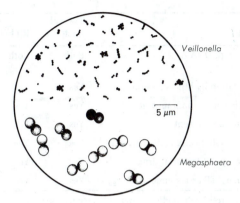

Figure 13-15. Drawing of the cells of two genera of anaerobic cocci, *Veillonella* and *Megasphaera*. (*Erwin F. Lessel, illustrator.*)

DISSIMILATORY SULFATE- OR SULFUR-REDUCING BACTERIA

The characteristic feature of this diverse group of anaerobic bacteria is that they all respire anaerobically by using inorganic sulfur compounds as electron acceptors with the consequent formation of large amounts of H_2S.

The various genera occur in mud from freshwater, marine, or brackish environments, and in the intestinal tract of humans and animals. One genus (*Desulfuromonas*) can respire with elemental sulfur as the electron acceptor; the other genera cannot use sulfur but can use sulfate, thiosulfate, or other oxidized sulfur compounds. The latter genera are differentiated mainly on the basis of morphology; e.g., *Desulfovibrio*—vibrioid or helical cells, *Desulfococcus* spherical cells, etc.

ANAEROBIC GRAM-NEGATIVE COCCI

The bacteria in this section constitute a single family, *VEILLONELLACEAE*, which consists of nonmotile cocci, typically occurring in pairs, often with the adjacent sides flattened. They are placed into three genera, *Veillonella, Acidaminococcus,* and *Megasphaera,* which differ with regard to their size (see Fig. 13-15) and their energy sources and fermentation products. They are inhabitants of the oral cavity, respiratory tract, or intestinal tract of humans, ruminants, rodents, and pigs.

THE RICKETTSIAS AND CHLAMYDIAS

Most of these tiny, nonmotile Gram-negative bacteria are obligate parasites, able to grow only within host cells. Although their size in some instances approaches that of some large viruses (which are also obligate parasites) they are classified as bacteria on the basis of characteristics indicated in Table 13-6.

The Order *RICKETTSIALES*

This order contains the rickettsias, which differ from the chlamydias (order *CHLAMYDIALES*) by (1) having a more complex metabolism that allows them to synthesize ATP, and (2) lacking a complex developmental cycle. Rickettsias are associated with various arthropods which may serve as major hosts or which can act as vectors for transmission of the rickettsias to vertebrates. In some instances there may be a mutualistic relationship in which the rickettsias provide factors essential for the growth and reproduction of their arthropod host.

Table 13-6. Characteristics of Typical Bacteria, Rickettsias, Chlamydias, and Viruses

Property	Typical Bacteria	Rickettsias	Chlamydias	Viruses
Nucleic acid, DNA or RNA	Both	Both	Both	One or the other
Multiplication by binary fission	Yes	Yes	Yes*	No
Cell walls contain muramic acid	Yes	Yes	No†	No
Ribosomes are present	Yes	Yes	Yes	No
Contain metabolically active enzymes	Yes	Yes	Yes	No
Inhibited by antibacterial drugs	Yes	Yes	Yes	No
Synthesize ATP as a source of energy	Yes	Yes	No	No

* Only the reticulate bodies undergo binary fission.
† However, inhibitors of bacterial cell-wall synthesis, such as penicillin and D-cycloserine, do prevent reorganization of reticulate bodies into elementary bodies, suggesting that some form of peptidoglycan may be present.

Those rickettsias which are capable of multiplying within vertebrates grow within the reticuloendothelial cells, vascular endothelial cells, or erythrocytes of these animals.

The Family RICKETTSIACEAE

This family consists of small rod-shaped or coccoid rickettsias which are often pleomorphic. The family is subdivided into three tribes on the following basis:

The tribe *RICKETTSIEAE*   Pathogenic for humans
The tribe *EHRLICHIEAE*   Pathogenic for vertebrate hosts other than humans (usually domestic animals)
The tribe *WOLBACHIEAE*   Not pathogenic for vertebrates; infect arthropods only

The tribe *RICKETTSIEAE* contains three genera, which are described as follows.

Rickettsia

The features of this genus are:

1 Transmission to humans occurs via an arthropod vector (lice, fleas, ticks, or mites, depending on the species).
2 The organisms multiply within the cytoplasm, and sometimes within the nucleus, of host cells.
3 In the laboratory they are cultured in (a) host animals, such as guinea pigs or mice (see Fig. 13-16A and B), (b) embryonated chicken eggs, particularly within the cells of the yolk sac membrane (Fig. 13-16C and D); (c) tissue cell cultures, usually cells from 10-day-old chicken embryos.

Diseases caused by *Rickettsia* species, and the arthropod vectors which transmit them, include: Rocky Mountain spotted fever (ticks), classical typhus fever (lice), murine typhus fever (fleas), rickettsialpox (mites), and scrub typhus (mites). (See also Table 35-5.)

Rochalimaea

This genus is similar to *Rickettsia* except for the following features:

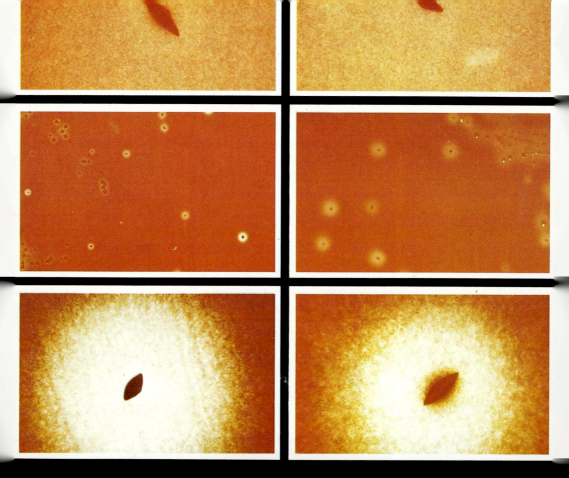

Hemolytic reactions of streptococci:

Alpha hemolysis (α): an indistinct zone of partial destruction of red blood cells around the colony. This illustration shows a typical colony of an alpha hemolytic streptococcus observed microscopically within a blood agar pour plate.

2. a, b Beta hemolysis (β): a clear, colorless zone around the

cells lying adjacent to the bacterial colony with a zone of complete hemolysis extending further out into the medium. When examined macroscopically α′ hemolysis can be confused with β hemolysis. (The similarity of hemolysis in slides 2 and 4a is striking and accounts for confusion of α′ and β hemolysis. However, the hemolysis revealed by macroscopic observation of the surface and subsurface colonies in slide 4a is not beta

Cyanobacteria:

a *Merismopedia glauca.*
b *Oscillatoria* sp.
c *Microcystis aeruginosa.*
d *Anabaena planktonica.*
e *Lyngbya aestuarii.*
f *Gloeotrichia echinulata.*
g *Arthrospira jennir.*
h *Gloeocapsa repestis.*

(Courtesy of G. J. Schumacher, Harpur College, SUNY-Binghamton, New York.)

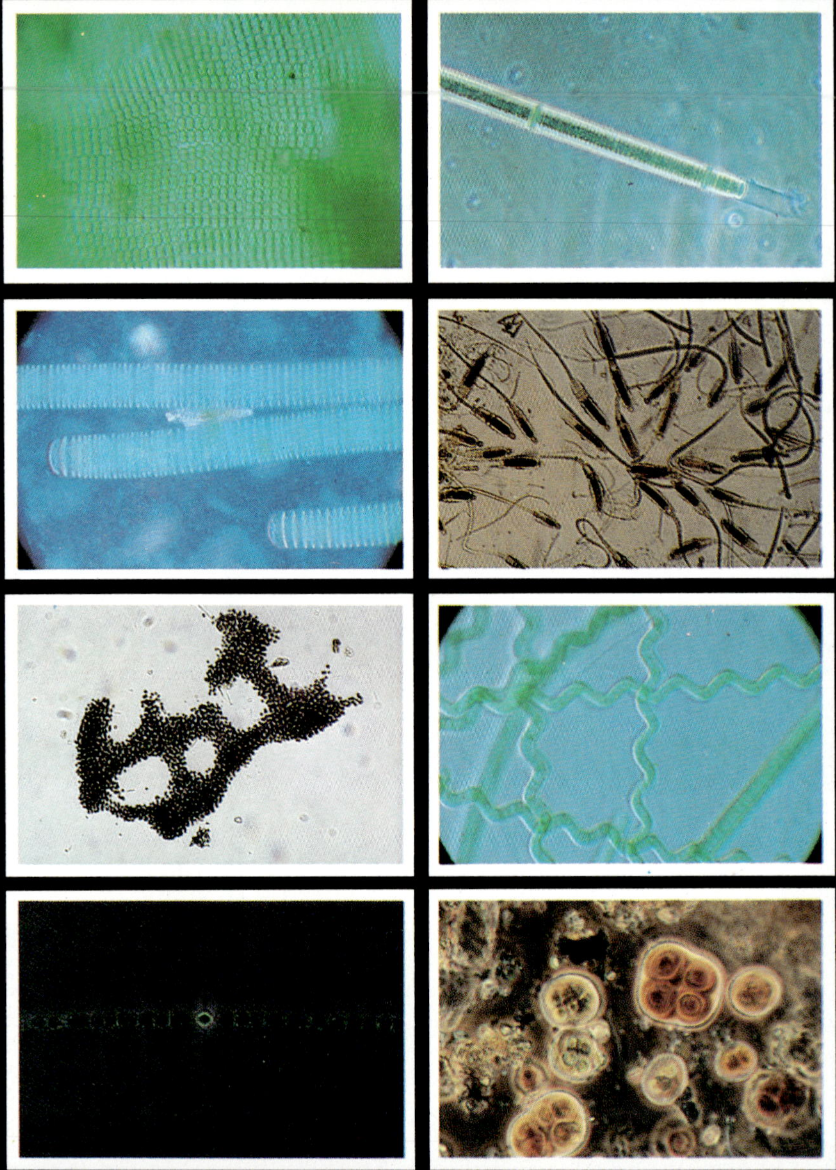

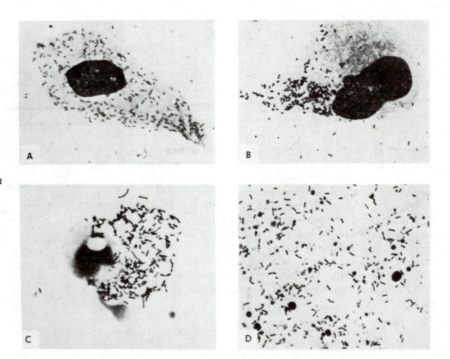

Figure 13-16. Photomicrographs of rickettsias grown under various conditions in the laboratory. (A) *Rickettsia akari* in smear of peritoneal scraping of infected laboratory mouse (X940). (B) *R. tsutsugamushi* in cytoplasm of infected cell (X940). (C) *R. prowazekii* in yolk-sac culture (X1,500). (D) *R. typhi* in yolk-sac culture (X1,000). *(Courtesy of N. J. Kramis and the Rocky Mountain Laboratory, U.S. Public Health Service.)*

1 Although the organisms are mainly parasites of humans and human body lice, they can be cultivated in vitro on laboratory media (e.g., a blood-based agar).

2 They grow *epicellularly* (i.e., on the surface of host cells) rather than in the cytoplasm or nucleus.

The single species of the genus, *R. quintana,* causes a louseborne disease, trench fever, in humans.

Coxiella

This genus is distinguished by several unusual properties:

1 Growth occurs preferentially within membrane-bounded vacuoles of host cells rather than free in the cytoplasm or nucleus.

2 The organisms have an unusually high resistance to heat [may survive a temperature of 62°C (143°F) for 30 min], probably due to the occurrence of endospore-like structures in the cells.

3 Although transmission to vertebrates can occur via an arthropod vector, it occurs mainly by inhalation of airborne infectious dust. The organisms can also be acquired by drinking contaminated unpasteurized milk.

The single species of the genus, *C. burnetii,* is the causative agent of Q fever, a type of pneumonia.

The Family BARTONELLACEAE

This family consists of parasites of the red blood cells of humans and other vertebrates. The organisms can be cultivated on nonliving laboratory media. The genus *Bartonella* causes Oroya fever in humans and is transmitted by biting

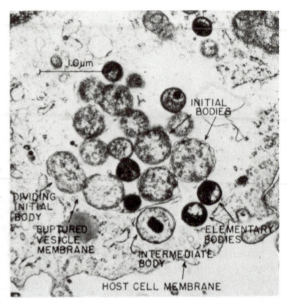

Figure 13-17. Electron micrograph of ultrathin section through a microcolony of *Chlamydia psittaci* in cytoplasm of a McCoy cell after 48-h incubation. The various developmental forms are labeled. The membrane of the vacuole containing the chlamydias has been ruptured, and the chlamydias are being released into the cytoplasm. The multi-laminated nature of the wall of the elementary bodies and the double-unit membrane surrounding the reticulate (initial) bodies and the intermediate bodies (intermediate between the elementary and reticulate forms) are visible (X28,700). *(Courtesy of Randall C. Cutlip, Inf and Immun, 1:500, 1970.)*

flies that occur along the western slopes of the Andes mountains in South America.

The Family
ANAPLASMATACEAE

These organisms grow within or on erythrocytes or occur free in the plasma of various wild and domestic animals. None have been cultivated on nonliving laboratory media.

The Order
CHLAMYDIALES

This order contains the chlamydias, intracellular parasites that are distinguished from rickettsias by (1) an inability to make ATP (they have an absolute reliance on host cells for this compound and are sometimes termed "energy parasites"), and (2) occurrence of a characteristic developmental cycle. In the laboratory, chlamydias are cultivated in the yolk sac membrane of embryonated chicken eggs or in tissue cultures of mammalian cells, such as McCoy and HeLa cells.

Reproduction of chlamydias usually proceeds according to the following sequence.

1 An infectious small particle, or *elementary body* (see Fig. 13-17), having an electron-dense nucleoid, is taken into the host cell by phagocytosis.

2 The elementary body is enclosed within a membrane-bounded vacuole in the cytoplasm of the host cell.

3 Within the vacuole the elementary body is reorganized into a *reticulate body* (also termed *initial body*), which is two or three times the size of the elementary body and contains a less dense arrangement of nucleoid material (see Fig. 13-17). The reticulate body is not infectious (i.e., when cells are disrupted at this stage of chlamydial development the reticulate bodies that are liberated cannot infect other host cells).

4 The reticulate body undergoes *binary fission* until a number of reticulate bodies are formed, which then undergo reorganization into elementary bodies. This

aggregate of reticulate bodies and elementary bodies within the vacuole of the host cell forms a *host cell inclusion* which can be seen by light microscopy.

5 The progeny elementary bodies are then liberated from the host cell and proceed to infect other cells.

All chlamydias belong to a single family, *CHLAMYDIACEAE*, which contains a single genus, *Chlamydia*. Some strains of *C. trachomatis* cause a type of keratoconjunctivitis, trachoma, that often results in blindness. Other strains cause nongonococcal urethritis, which is the most prevalent sexually transmitted disease in the United States today. Still other strains cause the sexually transmitted disease lymphogranuloma venereum. The species *C. psittaci* is mainly a pathogen of birds and domestic and wild mammals but can also cause psittacosis in humans.

THE MYCOPLASMAS

The mycoplasmas are distinguished by their lack of a cell wall, the outer boundary of the cells being the cytoplasmic membrane. As a result, the cells have plasticity and can assume many different shapes ranging from spheres to branched filaments (Fig. 13-18A). The plasticity allows many of the cells to pass through bacteriological filters even though the smallest cells are about 0.3 μm in diameter. They are susceptible to lysis by osmotic shock caused by sudden dilution of the medium with water. Because of the lack of a cell wall, mycoplasmas are not inhibited by even high levels of penicillin; however, they can be inhibited by antibiotics that affect protein synthesis, such as tetracyclines or chloramphenicol. Mycoplasmas can be cultivated in vitro on nonliving media (of rich composition) as facultative anaerobes or obligate anaerobes. They have

Figure 13-18. Mycoplasma cells and colonies. (A) Scanning electron micrograph of *Mycoplasma pneumoniae* from a 6-day culture showing irregular forms, crossing filaments, and piling up of spherical organisms probably representing an early stage of colony formation (X27,600). *(Courtesy of G. Biberfeld and P. Biberfeld, J Bacteriol, 102:855, 1970.)* (B) Colonies of *M. molare* showing the typical "fried-egg" appearance (X35). *(Courtesy of S. Rosendal and Int J Syst Bacteriol 24:125, 1974.)*

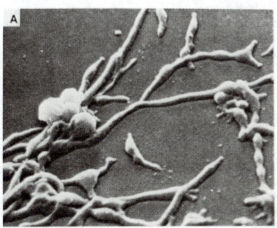

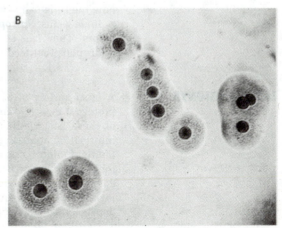

genomes that are about one-fifth to one-half the size of those of other bacteria capable of growth on nonliving media, which explains why these organisms have complex nutritional requirements and limited biosynthetic abilities. Colonies on agar plates are usually tiny and require observation by means of a low-power microscope. The colonies are embedded in the agar surface and usually have a characteristic fried-egg appearance (Fig. 13-18B).

Mycoplasmas differ from the "L-phase variants" that can develop from other bacteria. Such variants are osmotically fragile, cell-wall-defective forms that may occur spontaneously (as in the genus *Streptobacillus*) or as the result of continuous exposure to sublethal levels of penicillin. They form fried-egg colonies resembling those of mycoplasmas. However, L-phase variants are derived from walled bacteria and can usually revert to the normal walled bacterial form (e.g., when penicillin is removed), whereas mycoplasmas do not give rise to walled forms. Moreover, penicillin-binding proteins and peptidoglycan precursors can be demonstrated in the membranes of L-phase variants but not in the membranes of mycoplasmas.

At present, mycoplasmas are placed in the taxonomic class *MOLLICUTES*, which contains the single order *MYCOPLASMATALES*. Three families are included in this order.

The Family MYCOPLASMATACEAE

These mycoplasmas are parasites of the mucous membranes and joints of humans or animals and require cholesterol for growth. Many species of the genus *Mycoplasma* are pathogenic for animals; the species *M. pneumoniae* has the most significance for humans and is the causative agent of primary atypical pneumonia. Members of the genus *Ureaplasma* require urea for growth and cause urethritis in humans, pneumonia in cattle, and urogenital disease in cattle and other animal species.

The Family ACHOLEPLASMATACEAE

These mycoplasmas do not require cholesterol for growth. They are widely distributed in vertebrates, in sewage and soil, and possibly on plants. Their pathogenicity is unknown. Only a single genus, *Acholeplasma,* occurs.

The Family SPIROPLASMATACEAE

These organisms are unusual in that they are helical and exhibit a swimming motility. How a helical shape is maintained in the absence of a cell wall and how the cells can swim while lacking flagella is unknown. The family contains a single genus, *Spiroplasma*. The organisms are pathogenic for citrus and other plants. They can be isolated from plant fluids and plant surfaces and from arthropods that feed on plants.

ENDOSYMBIONTS

A great variety of bacteria-like forms have been observed within the cells of protozoa, insects, fungi, sponges, coelenterates, helminths, and annelids. Most of these bacteria-like forms have not been cultivated in the laboratory, and information about them has generally been restricted to observations made with the light microscope or electron microscope. Most endosymbionts do not harm their hosts; indeed, many appear to be beneficial or even necessary for the growth and development of their hosts. This is suggested, for example, by the inability of certain insects to develop without vitamin supplements after being deprived of their endosymbionts by treatment with antibiotics or other means.

Figure 13-19. (A) Osmium-lacto-orcein preparation of the protozoan *Paramecium tetraurelia* bearing ensymbiont *Lyticum flagellatum*. The numerous black rods throughout the cytoplasm are the endosymbionts. Bar indicates 20 μm. (B) *L. flagellatum* separated from its protozoan host, showing the peritrichous flagella; negatively stained with phosphotungstic acid. Bar indicates 1 μm. *(Courtesy of J. R. Preer, Jr., L. B. Preer, and A. Jurand, Bacteriol Rev 38:113, 1974.)*

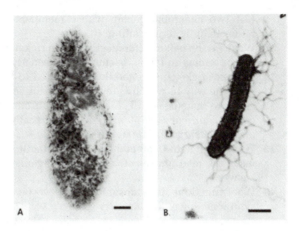

The greatest amount of information has been obtained for endosymbionts of protozoa, even to the point of classifying many of them by separating their DNA from that of their host cells and performing base composition and homology experiments with the DNA. Indeed, many of the protozoan endosymbionts now bear formal genus and species names. For example, **Lyticum flagellatum** is an endosymbiont carried by certain strains of the protozoan *Paramecium tetraurelia* (see Fig. 13-19). One function of *L. flagellatum* is to synthesize the vitamin folic acid for its host; symbiont-free lines of the same strains of the protozoan need to be supplied with this vitamin. Another function of *L. flagellatum* is to produce a toxin that is liberated into the culture medium: when an endosymbiont-bearing strain of *P. tetraurelia* (called a *killer* strain) is mixed with certain strains lacking it (called *sensitive* strains), the latter protozoa are rapidly killed and lysed. The endosymbiont-bearing strains are resistant to this toxin.

QUESTIONS

1 In what ways do spirochetes differ from other bacteria? What combination of characteristics sets them apart?

2 Which genera of Gram-negative bacteria are associated with plants as nitrogen fixers? As plant pathogens?

3 What general kinds of Gram-negative bacteria (i.e., aerobes, facultative anaerobes, or anaerobes) are associated with the ability to grow autotrophically with H_2 as the energy source? With the ability to use methane gas as a carbon source?

4 List four genera of Gram-negative bacteria that produce distinctive pigments.

5 How are various sugars and other carbon sources used in the laboratory differentiation of *Pseudomonas* species? How does this differ from the way sugars are used to differentiate the genera of *Enterobacteriaceae*?

6 On what bases are the genera of the family *Bacteroidaceae* differentiated?

7 What is an opportunistic pathogen? List four Gram-negative bacteria that are opportunistic pathogens.

8 In what type of environment is one most likely to find (a) *Aquaspirillum*, (b) *Escherichia coli*, (c) *Thermus*, (d) *Erwinia*, (e) *Vibrio fischeri*, (f) *Zymomonas*?

9 List two anaerobic genera whose members obtain energy by respiration rather than by fermentation.

10 What are the major differences between rickettsias and chlamydias? Between mycoplasmas and other bacteria? Between *Mycoplasma* and *Acholeplasma*?

11 How are rickettsias generally transmitted to humans? How does this differ from the way in which Q fever is transmitted?

12 How are rickettsias and chlamydias cultured in the laboratory? In what way can *Rochalimaea quintana* be cultured that differs from the methods used for *Rickettsia* and *Chlamydia*?

13 What agent could you add to a growth medium for mycoplasmas that would make the medium selective for these organisms? Explain the basis for your answer.

14 What functions do endosymbionts serve for their hosts? How might endosymbionts have originated?

REFERENCES

Barile, M. F., S. Razin, J. G. Tully and R. F. Whitcomb (eds.): *The Mycoplasmas* (in three volumes), Academic, New York, 1979. *A comprehensive treatment of the biology and taxonomy of mycoplasmas.*

Krieg, N. R. (ed.): *Bergey's Manual of Systematic Bacteriology*, vol. 1, Williams & Wilkins, Baltimore, 1984. *This international reference work provides detailed descriptions of the genera and species of bacteria discussed in Chap. 13, as well as descriptions of endosymbionts and unclassified bacteria-like organisms.*

Preer, J. R., Jr., L. B. Preer, and A. Jurand: "Kappa and Other Endosymbionts in Paramecium aurelia," *Bacteriol Rev*, **38**:113–163, 1974. *This review contains detailed information about the biological properties of the endosymbionts of these protozoa and is profusely illustrated.*

Starr, M. P., H. Stolp, H. G. Trüper, A. Balows, and H. G. Schlegel (eds.): *The Prokaryotes: A Handbook of Habitats, Isolation, and Identification of Bacteria.* Springer-Verlag, New York, 1981. *Although not primarily concerned with classification, this monumental reference work provides a wealth of descriptive information and illustrations concerning the various genera of bacteria.*

Weiss, E.: "The Biology of Rickettsiae," *Annu Rev Microbiol*, **36**:345–370, 1982. *A comprehensive review of the morphology and structure, growth and metabolism, and pathogenicity and immunology of the rickettsias.*

Chapter 14

The World of Bacteria II: "Ordinary" Gram-Positive Bacteria

Like Volume 1 of *Bergey's Manual,* Volume 2 includes many of the familiar or "ordinary" chemoheterotrophic eubacteria; however, the organisms are Gram-positive rather than Gram-negative. Most have a simple morphology and none form prosthecae, sheaths, or cysts. Some genera form heat-resistant endospores. Although most of the organisms have a regular, uniform shape, some are pleomorphic. Most of the organisms occur singly, in pairs, or in chains, but some are arranged in trichomes and others form branching hyphae. Motility, if present, is by means of flagella. In general, reproduction occurs mainly by transverse binary fission; however, multiplication by fragmentation or by conidiospore production can occur in several genera. Saprophytes and parasites are included; some of the parasitic organisms can be highly pathogenic for humans, animals, or plants. In this chapter we shall describe briefly some of the organisms included in the second volume of *Bergey's Manual.*

BERGEY'S MANUAL OF SYSTEMATIC BACTERIOLOGY, VOLUME 2

As in Chap. 13, the arrangement of the bacteria continues to be based primarily on practical considerations rather than on phylogenetic relationships. The organisms are divided into major sections, each bearing a descriptive common name. The major sections of Volume 2 are listed in Table 14-1. Formal taxonomic names are emphasized mainly at the family, genus, and species level of classification. In this chapter we shall indicate the great diversity of the organisms included in Volume 2 of *Bergey's Manual* by highlighting their important characteristics.

GRAM-POSITIVE COCCI

All the cocci in this group have the following features:

Aerobic/Facultatively Anaerobic Cocci

1 They possess cytochromes.
2 They are able to respire with oxygen, i.e., have an oxidative type of metabolism.
3 Some can also obtain energy under anaerobic conditions by fermentation (have a fermentative type of metabolism).

The organisms have been placed in two families. Members of the family *DEINOCOCCACEAE* exhibit the following characteristics:

1 The cocci occur mainly in tetrads or cubical packets.

Table 14-1. Gram-Positive Bacteria Included in *Bergey's Manual*, Volume 2

Section	Other Major Characteristics
GRAM-POSITIVE COCCI	May have a strictly respiratory type of metabolism, a respiratory plus a fermentative metabolism, or a strictly fermentative metabolism; in the latter category they may be able to grow in air (aerotolerant), or they may be anaerobic
ENDOSPORE-FORMING GRAM-POSITIVE BACTERIA	Mainly rod-shaped, but some are cocci; range from aerobic to facultatively anaerobic to anaerobic; most of the anaerobes live by fermentation, but some respire anaerobically with sulfate
NONSPOREFORMING GRAM-POSITIVE RODS OF REGULAR SHAPE	The cells have a uniform appearance without swellings, branching, or other types of variation; some occur in characteristic trichomes; aerobes, facultative anaerobes, or aerotolerant anaerobes are included
NONSPOREFORMING GRAM-POSITIVE RODS OF IRREGULAR SHAPE	The cells may exhibit swellings, Y or V shapes, rod/coccus cycles, or other deviations from a uniform morphology; some are filamentous during at least some stage of their growth; aerobic, facultatively anaerobic, and anaerobic genera are included
MYCOBACTERIA	Aerobic, slightly curved or straight rods which sometimes show branching; stain acid-fast
NOCARDIOFORMS	Aerobic organisms that tend to form a substrate mycelium and sometimes an aerial mycelium; the hyphae fragment into rod-shaped or coccoid elements; conidiospores may develop from the aerial hyphae

2 The organisms have an unusually high resistance to gamma and ultraviolet radiation.

The family contains a single genus, **Deinococcus,** which forms red colonies. The radiation resistance of the genus is reflected by the name of one of the species, **D. radiodurans.** The organisms can often be isolated as spoilage agents from foods preserved by treatment with ionizing radiation.

The family **MICROCOCCACEAE** exhibits the following features:

1 The cocci occur mainly in clusters, tetrads, or cubical packets of eight cells.
2 The cells do not exhibit any unusual resistance to gamma and ultraviolet radiation.

Three of the genera included in the family are described below.

Micrococcus

These nonmotile cocci are aerobic, oxidative, and are catalase-positive. Their colonies may be red, orange, yellow, or nonpigmented. Micrococci are harmless saprophytes occurring in soil and freshwater, but they can also be found on the skin of humans and animals.

Planococcus

These organisms are also aerobic, oxidative, catalase-positive cocci; however, the cells are motile and possess one to three flagella. The colonies are yellow-brown. Planococci are harmless saprophytes that occur in marine environments.

Staphylococcus

See Fig. 14-1. Staphylococci are nonmotile cocci that are catalase-positive and facultatively anaerobic, having both an oxidative and a fermentative type of metabolism. They are parasites, occurring on the skin and mucous membranes of humans and warm-blooded animals. The major pathogenic species is **S. aureus,** which can cause boils, abscesses, wound infections, postoperative infections, toxic shock syndrome, and food poisoning in humans, and infections in animals, such as mastitis in cattle. In the laboratory, S. aureus produces white to golden-colored colonies and is positive for the *coagulase test* (a test for the ability of bacteria to cause blood plasma to clot). **S. epidermidis** and **S. saprophyticus,** which are coagulase-negative, can cause wound infections, endocarditis, and urinary tract infections.

Aerotolerant Fermentative Cocci

These cocci have the following characteristics:

1 They do not possess cytochromes.
2 They have only a fermentative type of metabolism and do not respire; yet they can grow anaerobically or aerobically.
3 The cells are arranged in pairs, chains, or tetrads.

Some representative genera are described below.

Streptococcus

This genus has the following features:

1 The cells are arranged in pairs or chains (see Fig. 14-1).
2 They are catalase-negative.
3 The organisms are *homofermentative*, i.e., the predominant end product of sugar fermentation is lactic acid. (In the case of Streptococcus, it is the L(+) optical isomer of lactic acid.)

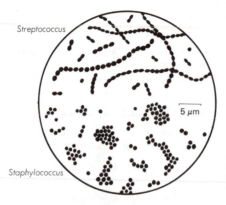

Streptococcus

5 µm

Staphylococcus

Figure 14-1. Drawing of the cells of *Staphylococcus* and *Streptococcus*. *(Erwin F. Lessel, illustrator.)*

Figure 14-2. (A) α-hemolysis. Enzymes produced by some streptococci, such as S. pneumoniae, only partially hemolyze red blood cells of certain species of animals. Colonies on blood agar plates are surrounded by a greenish-colored zone which is due to the reduction of hemoglobin in the red blood cells to methemoglobin. (B) β-hemolysis. Enzymes produced by some streptococci, such as S. pyogenes, completely hemolyze red blood cells of certain species of animals. Colonies on blood agar plates are surrounded by a clear, colorless zone. *(Courtesy of Liliane Therrien and E. C. S. Chan, McGill University.)*

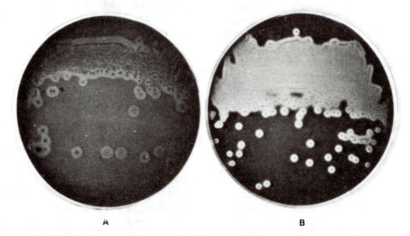

A

B

Although the genus is usually considered aerotolerant, some strains can tolerate only low levels of oxygen and some are anaerobic. Nutritional requirements are complex, including several amino acids and vitamins. The streptococci are divided into categories known as the *Lancefield groups* based on differences in their cell-wall polysaccharides.

Some streptococci are β-*hemolytic* on blood agar: the colonies are surrounded by a clear, colorless zone that indicates complete lysis of the erythrocytes (Fig. 14-2). Other streptococci are α-*hemolytic* (colonies surrounded by a cloudy, colorless or greenish zone of partially lysed erythrocytes; see Fig. 14-2) or are *nonhemolytic*.

Most streptococci are parasites of humans and animals, and several species are pathogenic. There are many species of streptococci; a few examples follow.

S. pyogenes (β-hemolytic; Lancefield group A) is the most clinically important species. It causes streptococcal sore throat, scarlet fever, erysipelas, acute glomerulonephritis, rheumatic fever, and other human infections.

S. mutans (nonhemolytic; not placed in any Lancefield group) inhabits the human oral cavity and is the major causative agent of dental caries.

S. faecalis (α-, β-, or nonhemolytic; Lancefield group D) occurs normally in the intestinal tracts of humans and animals and is therefore called an "entero-coccus"; it can be an opportunistic pathogen, causing urinary tract infections and endocarditis.

S. lactis and *S. cremoris* (Lancefield group N) are harmless contaminants of milk and dairy products; they cause rapid souring and curdling of milk, and because of this are widely used as "starter cultures" in the manufacture of buttermilk and cheeses.

S. pneumoniae (α-hemolytic; not placed in any Lancefield group) is collo-quially called the "pneumococcus" and has great clinical significance, causing nearly 70 percent of all cases of lobar pneumonia in humans.

Leuconostoc

This genus has the following characteristics:

1 The cocci are arranged in pairs and chains.
2 They are catalase-negative.
3 The organisms are *heterofermentative:* they form CO_2 and ethanol or acetic acid in addition to lactic acid; moreover, the lactic acid is of the D($-$) type.

Leuconostocs are harmless saprophytes and are isolated from diverse sources such as grass, silage, grape leaves, sauerkraut, and spoiled food. They are often used in "starter cultures" for the manufacture of butter, buttermilk, and cheese because of their formation of the flavor compound diacetyl (2,3-butanedione) from citrate.

Pediococcus

This genus has the following features:

1 Cocci occur in pairs and tetrads (see Fig. 14-3).
2 They are catalase-negative.
3 They exhibit a homolactic type of fermentation, forming optically inactive lactic acid, i.e., a mixture of the L($+$) and D($-$) types.

Pediococci are saprophytes and are particularly noted for their ability to form capsular material that causes beer to become ropy and viscous.

Anaerobic Gram-Positive Cocci

These cocci have a fermentative type of metabolism. Some genera must be supplied with a fermentable carbohydrate in order to grow; others can ferment amino acids and do not require carbohydrates. Lactic acid, if formed, is not a major fermentation product (unlike the genus *Streptococcus*). Most genera form CO_2, H_2, short-chain fatty acids, and in some cases ethanol or succinic acid.

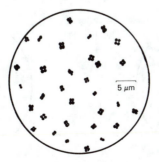

Figure 14-3. Drawing of the cells of *Pediococcus cerevisiae*. (*Erwin F. Lessel, illustrator.*)

5 μm

Table 14-2. Characteristics of Anaerobic Gram-Positive Cocci

Genus	Arrangement of Cells	Main Sources of Carbon and Energy	Occurrence
Peptococcus	Pairs, clusters, tetrads, and short or long chains	Peptone or amino acids	Human intestine and respiratory tract; clinical specimens; tidal bay mud
Peptostreptococcus	Pairs and short or long chains	Peptone or amino acids	Human clinical specimens
Ruminococcus	Pairs and short or long chains	Carbohydrates	Bovine and ovine rumen; animal ceca
Coprococcus	Pairs and short or long chains	Carbohydrates	Human feces
Sarcina	Cubical packets of eight cells	Carbohydrates	Soil; mud; cereal grains; diseased human stomachs

Table 14-2 summarizes some of the characteristics for the various genera included in the group, and the morphological features of two of the genera are depicted in Fig. 14-4.

ENDOSPORE-FORMING GRAM-POSITIVE BACTERIA

Most of the endospore-forming bacteria are rod-shaped, but some are cocci. The majority stain Gram-positive but some species stain Gram-negative. Motility, if present, is by means of peritrichous flagella. Some of the genera included in the group are described briefly here.

Aerobic/Facultatively Anaerobic Sporeforming Rods and Cocci

This genus contains rod-shaped bacteria. Most species are harmless saprophytes occurring in soil, freshwater, or sea water. Many form exocellular enzymes that hydrolyze proteins or complex polysaccharides, activities that are often important causes of food spoilage. Moreover, because of the heat resistance of the endospores, *Bacillus* species may survive milk pasteurization or inadequate heat treatment during canning of foods. A few examples of the many species are listed below.

Bacillus

B. subtilis and **B. cereus.** (See Fig. 5-29.) These common mesophilic sapro-

Figure 14-4. Anaerobic Gram-positive cocci. (A) Drawing of the cells of *Peptostreptococcus*. (B) Drawing of the cells of *Sarcina*. (Erwin F. Lessel, illustrator.)

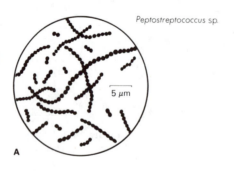

Peptostreptococcus sp.

5 μm

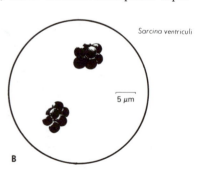

Sarcina ventriculi

5 μm

A

B

phytes are widely distributed in nature. Both species produce exoenzymes that hydrolyze starch and casein. *B. cereus* can cause a type of food poisoning.

B. stearothermophilus. This is a thermophilic species having a minimum growth temperature of 30 to 45°C and a maximum of 65 to 75°C. The endospores are highly resistant to heat and, therefore, this species is one of those associated with spoilage of canned goods.

B. polymyxa. Unlike many other *Bacillus* species, *B. polymyxa* has the ability to form gas during sugar fermentation. Another unusual characteristic is its ability to fix N_2 under anaerobic conditions.

B. thuringiensis. This species is noted for its pathogenicity for insects. Ingestion of the sporulated cultures of *B. thuringiensis* by larvae of *Lepidoptera* results in a paralytic disease. Other *Bacillus* species that are pathogenic for insects include *B. popilliae* ("milky disease" of Japanese beetle grubs) and certain strains of *B. sphaericus* (lethal for mosquito larvae).

B. anthracis. This is the only *Bacillus* species that is highly pathogenic for animals and humans; it is the causative agent of anthrax.

Sporosarcina

This genus contains cocci that are arranged in tetrads or cubical packets of eight cells (see Fig. 14-5). Sporosarcinae are widely distributed in fertile soil, where they play an active role in the decomposition of urea.

Anaerobic Spore-forming Rods

Clostridium

The members of this genus have a fermentative type of metabolism. They are widely distributed in soil, in marine and freshwater anaerobic sediments, and in the intestinal tract of humans and animals. The species are differentiated on the basis of their proteolytic activity, various enzyme activities, acid production from carbohydrates, and the kinds and amounts of organic acid end products of fermentation (the latter being determined by gas chromatography). A few of the many species in the genus are listed below:

C. botulinum causes a severe and often fatal type of food poisoning known as botulism.

C. tetani is the causative agent of tetanus. The characteristic terminal spores formed by this species are illustrated in Fig. 5-29.

C. perfringens is the major causative agent of the wound infection known as gas gangrene. Some strains of *C. perfringens* (enterotoxigenic strains) can cause a type of food poisoning.

C. difficile causes pseudomembranous colitis, a severe disease of the bowel.

C. thermosaccharolyticum is thermophilic, growing optimally at 55°C (minimum temperature 45°C; maximum, 67°C). The spores are extremely heat-resistant, and this species is often able to survive inadequate heat treatment during canning of foods and subsequently can cause spoilage of canned goods.

C. pasteurianum is a mesophilic soil clostridium that is particularly noted for its ability to fix N_2.

Desulfotomaculum

Unlike clostridia, members of this genus obtain energy by anaerobic respiration, with sulfate serving as the terminal electron acceptor and organic substrates such as lactic or pyruvic acid serving as the electron donors. Large amounts of H_2S are formed during growth. The organisms occur in soil, freshwater, intestines of insects, and the rumen of animals. See Fig. 14-5.

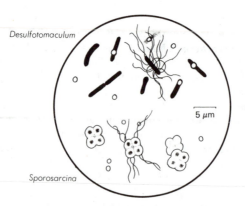

Figure 14-5. Drawing of the cells of *Sporosarcina* and *Desulfotomaculum*, showing the endospores and flagella. The flagella are not visible by ordinary staining. *(Erwin F. Lessel, illustrator.)*

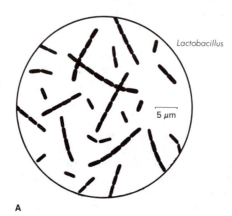

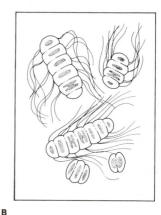

A   **B**

Figure 14-6. (A) Drawing of the cells of *Lactobacillus*. *(Erwin F. Lessel, illustrator.)* (B) Sketch of *Caryophanon* showing trichomes 3 μm in diameter composed of disk-shaped cells, together with numerous peritrichous flagella. *(Redrawn from K. A. Bisset, Bacteria, E. and S. Livingstone, Ltd,, Edinburgh, 1952.)*

NONSPOREFORMING GRAM-POSITIVE RODS OF REGULAR SHAPE

This heterogenous group is composed of harmless saprophytes as well as parasitic and pathogenic organisms. The cells range from long rods to very short rods as, for example, in the genus *Lactobacillus* (Fig. 14-6A). One genus, *Caryophanon,* is unusual in that it is composed of large, disk-shaped cells arranged in trichomes (Fig. 14-6B). Some characteristics of the various genera included in the group are indicated in Table 14-3.

NONSPOREFORMING GRAM-POSITIVE RODS OF IRREGULAR SHAPE

Aerobic/Facultatively Anaerobic Nonfilamentous Rods

This group contains a heterogenous variety of bacteria, the few common features being:

1 Straight to slightly curved rods that exhibit swellings, club shapes, or other deviations from a uniform rod shape
2 An aerobic or facultatively anaerobic nature, being capable of a respiratory type of metabolism and in some instances also of a fermentative type of metabolism

Some examples of the genera included in this group follow.

Corynebacterium

This genus contains rod-shaped cells which are pleomorphic and frequently

Table 14-3. Genera of Nonsporeforming Gram-Positive Rods of Regular Shape

Genus	Morphology and Motility	Oxygen Relationships	Catalase Test	Other Characteristics
Lactobacillus	Long to very short rods, often in chains; usually nonmotile	Strictly fermentative organisms, but can usually tolerate air; some strains are anaerobic	−	Large amounts of lactic acid formed; homo- or heterofermentative; occur as saprophytes in fermenting animal or plant products or as parasites in the mouth, vagina, and intestinal tract of humans and warm-blooded animals
Listeria	Very short rods, often in chains; motile by peritrichous flagella when grown at 25°C; few flagella are formed at 37°C	Aerobic to microaerophilic	+	The species *L. monocytogenes* is a parasite and pathogen of a wide variety of animals; in humans it causes meningitis in adults and prenatal or postnatal disease in infants
Erysipelothrix	Filament-forming rods; nonmotile	Aerobic	−	Parasitic on mammals, birds, and fish; causes erysipelas in swine and erysipeloid in humans
Brocothrix	Rods, often occurring in long, kinked filaments; nonmotile	Facultatively anaerobic	+	Best growth occurs at 20 to 22°C; none at 37°C; saprophytes, found in meat and meat products
Renibacterium	Short rods; nonmotile	Aerobic	+	Best growth occurs at 15 to 18°C; parasites of salmonid fishes, causing a kidney disease
Kurthia	Rods in chains; motile by peritrichous flagella	Aerobic	+	Harmless saprophytes occurring in meat and meat products and in animal dung
Caryophanon	Large disk-shaped cells arranged in trichomes; motile by peritrichous flagella	Aerobic	Not reported	The morphology is unusual and distinctive; saprophytic, occurring in ruminant dung

exhibit club-shaped swellings and a palisade arrangement (see Fig. 14-7). The cells accumulate intracellular volutin granules (metachromatic granules) which

Figure 14-7. Drawing of the cells of *Corynebacterium* and *Mycobacterium*. (*Erwin F. Lessel, illustrator.*)

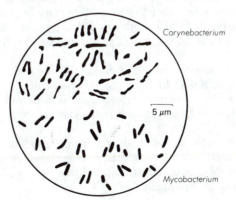

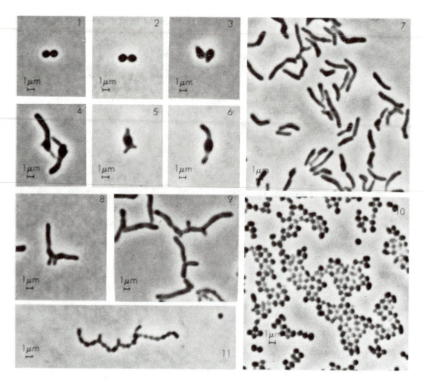

Figure 14-8. Morphology of *Arthrobacter globiformis* at different times of growth and on different media. Insets 1, 2, 3, and 4, slide cultures grown on yeast extract–soil extract medium for 1.5, 4.25, 6, and 9.75 h, respectively; insets 7 and 8, after growth for 24 h. Insets 5 and 6, cultures grown on yeast extract–mineral salts medium for 5 h; inset 11, grown for 3 days. Insets 9 and 10, grown on yeast extract–peptone–soil extract medium for 11.5 h and 3 days. *(Courtesy R. M. Keddie and Bergey's Manual of Determinative Bacteriology, 8th ed., Williams & Wilkins, Baltimore, 1974.)*

stain reddish purple with dilute methylene blue. The cell walls contain mycolic acids (corynemycolic acids) containing 32 to 36 carbon atoms.

The corynebacteria are divided into three large groups: saprophytes occurring in soil and water, the animal or human parasites and pathogens, and the plant pathogens. Of the animal or human pathogens, the major species is *C. diphtheriae,* the causative agent of diphtheria in humans.

Arthrobacter

This genus of saprophytic soil organisms is characterized by an unusual "rod-coccus" cycle. Cells in the log phase of growth are irregularly shaped rods that may show a tendency toward rudimentary branching. In contrast, cells in the stationary phase of growth are distinctly coccoid; when these are inoculated into fresh media, they give rise to rod-shaped cells (see Fig. 14-8).

Brevibacterium

Like arthrobacters, brevibacteria exhibit a rod-coccus cycle. The only recognized species, *B. linens,* forms orange colonies and is salt-tolerant; its usual habitat is on the surface of certain cheeses such as brick and Limburger, where it produces proteolytic enzymes that aid in the cheese-ripening process.

Microbacterium

These bacteria are small, slender, irregularly shaped rods that do not exhibit a rod-coccus cycle (although the rods become shorter in the stationary phase of growth). Microbacteria are saprophytes that occur in milk, in dairy products, and on dairy equipment.

Cellulomonas

This genus contains irregularly shaped rods that may be slightly filamentous and show rudimentary branching. No *Arthrobacter*-like rod-coccus cycle occurs, although a few cells in old cultures may be coccoid. The outstanding characteristic of the genus is the ability to degrade cellulose and to use it as a major carbon and energy source.

Aerobic/Facultatively Anaerobic Branched Filamentous Rods

The bacteria of this group form colonies which at first are microscopic in size (microcolonies) and contain branched filamentous cells. As the colonies develop to macroscopic size, many of the cells become diphtheroid (i.e., resemble corynebacteria) or coccoid in shape. One genus, *Agromyces,* is microaerophilic to aerobic, catalase-negative, and is a saprophyte that occurs in soil. The genus *Arachnia* is facultatively anaerobic, catalase-negative, and parasitic and pathogenic for humans and animals, being one of the causative agents of actinomycosis. The genus *Rothia* is aerobic, catalase-positive, and a normal inhabitant of the human mouth.

Anaerobic Nonfilamentous or Filamentous Rods

The organisms of this group are either anaerobes or, if facultatively anaerobic, are preferentially anaerobic. They are differentiated by their morphology and by their fermentation end products as determined by gas chromatography. Table 14-4 indicates the characteristics of several genera included in the group, and the morphological features of two genera, *Propionibacterium* and *Actinomyces,* are depicted in Figs. 14-9 and 14-10.

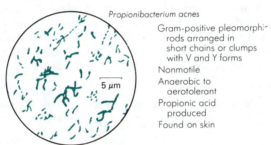

Propionibacterium acnes

Gram-positive pleomorphic rods arranged in short chains or clumps with V and Y forms
Nonmotile
Anaerobic to aerotolerant
Propionic acid produced
Found on skin

5 μm

Figure 14-9. (*Right*) Drawing of the cells of *Propionibacterium acnes.* (*Erwin F. Lessel, illustrator.*)

Figure 14-10. (*Below*) *Actinomyces israelii.* (A) Dark-field preparation showing V and Y forms. (B) Gram-stain preparation showing elongated filaments, branching, and irregular staining. (C) Gram-stain preparation showing mass of intertwined filaments. Some filaments are irregularly stained, and some have bulbous ends. (Approximately X800.) *From J. M. Slack, S. Landfried, and M. A. Gerencser, J Bacteriol,* **97:***873, 1969. By permission.*)

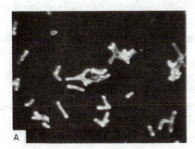

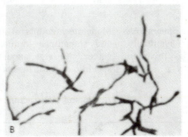

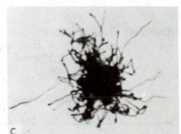

MYCOBACTERIA

This group of aerobic bacteria contains a single genus, *Mycobacterium,* which consists of slightly curved or straight rods (see Fig. 14-7) that may show branching. Mycolic acids having about 90 carbon atoms occur in the cell walls. A major characteristic of mycobacteria is that they are acid-fast; that is, once stained with aniline dyes, they are difficult to decolorize, even when treated with a mixture of acid and alcohol. Some species such as *M. phlei* and *M. smegmatis* are harmless saprophytes. Many species are pathogenic, for example, *M. tuberculosis,* the causative agent of tuberculosis in humans; *M. kansasii* and *M. intracellulare,* which cause noncontagious tuberculosis-like infections; *M. scrofulaceum,* which causes lymphadenitis in children; and *M. leprae,* the causative agent of leprosy.

NOCARDIOFORMS

This group contains aerobic bacteria that produce a substrate mycelium, i.e., a mat of branching hyphae formed under the surface of the agar medium. This mycelium may range from rudimentary to extensively developed. The substrate mycelium usually *fragments* into rod-shaped or coccoid cells. Some genera also form an aerial mycelium that may give rise to conidiospores. Because the various genera of the group resemble the genus *Nocardia* with respect to their morphology and cell-wall composition, they are referred to as the nocardioforms. Several genera possess mycolic acids (nocardomycolic acids) in the cell walls; however, only the genus *Nocardia* contains acid-fast species.

The taxonomic placement of mycelium-forming bacteria, including the nocardioforms, is confused and controversial. Among the methods that have been used to establish various groupings are studies of the chemical composition of the cell walls. The walls of the nocardioform bacteria have the following characteristics, which define the type IV chemotype:

Table 14-4. Some Anaerobic, Irregularly Shaped, Nonsporeforming Gram-Positive Bacteria

Genus	Morphology	Organic Acids from Fermentation	Occurrence
Propionibacterium	Pleomorphic, nonmotile	Mainly propionic + acetic acids	Some species occur in dairy products; others are normal flora of human skin and of the intestines of humans and animals; *P. acnes* may be related to the skin disease acne vulgaris
Eubacterium	Pleomorphic, motile or nonmotile	Either butyric + other acids; acetic + formic; or no major acids	Human oral cavity, intestinal tract of humans and animals, infected tissues, soil, water, spoiled food; usually not pathogenic
Actinomyces	Initially, cells are filamentous with branching; eventually, diphtheroid cells predominate	Moderate amounts of acetic and sometimes formic, together with large amounts of succinic or lactic, or both	Oral cavity of humans and animals and human female genital tract; *A. israelii* and other species can cause human actinomycosis; *A. bovis* causes actinomycosis (lumpy jaw) in cattle
Bifidobacterium	Pleomorphic, nonmotile	Acetic and lactic acids	Intestinal tract of humans and animals; not known to be pathogenic

1 The peptidoglycan contains *meso*-diaminopimelic acid.
2 No glycine interpeptide bridges occur between the peptidoglycan chains.
3 The walls contain the sugars arabinose and galactose.

Other groups of mycelium-forming bacteria differ in these respects; for example, the walls of members of the genus *Streptomyces* (see Chap. 16) contain LL-diaminopimelic acid, glycine interpeptide bridges, and no distinctive sugars (type I chemotype).

A few examples of the nocardioform group are described below.

Nocardia

The morphological features of this genus are illustrated in Figs. 14-11 and 14-12A. Some nocardias form only a limited mycelium because the center of the colony undergoes *early fragmentation* into rod-shaped or coccoid cells. Other

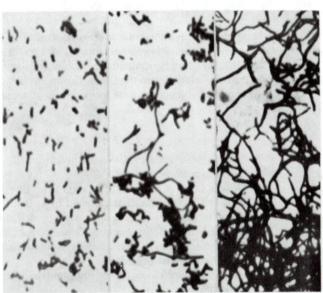

Figure 14-11. *Nocardia asteroides.* Three different strains of the same species, showing variations in morphology. (X700) *(Courtesy of Ruth E. Gordon.)*

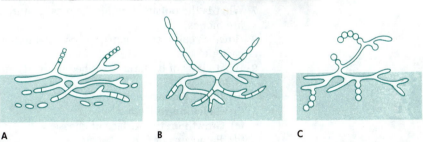

A   B   C

Figure 14-12. Schematic representation of the morphology of some nocardioform bacteria. Both substrate and aerial mycelia are illustrated. (A) *Nocardia*, showing fragmentation and aerial chains of spores. (B) *Pseudonocardia*, showing budding of the hyphae and aerial chains of cylindrical spores. (C) *Micropolyspora*, showing chains of round spores on both aerial and substrate mycelium.

nocardias have *delayed fragmentation* and are thus able to establish an extensive substrate mycelium, and sometimes an aerial mycelium that gives the surface of the colonies a dull, fuzzy appearance. The aerial hyphae of some species may bear chains of spores. Nocardias are saprophytes that are widely distributed in soil and water, but they can be opportunistic pathogens, causing nocardiosis and actinomycetoma in humans and animals.

Pseudonocardia

This genus does not contain nocardomycolic acids, and, in contrast to the hyphae of nocardias, the hyphae of pseudonocardias grow by a distinctive *budding process.* A constriction occurs behind the hyphal tip; the tip elongates to form a new hyphal segment; this segment develops a constriction behind the tip, and in turn undergoes budding, etc. The aerial mycelium bears long chains of cylindrical conidiospores, which arise terminally or laterally from the hyphae (see Fig. 14-12B). Pseudonocardias occur in soil and in fresh or rotten manure.

This genus forms an extensive aerial mycelium. The hyphae of both the substrate and aerial mycelia bear short chains of 1 to 20 round conidiospores (see Fig. 14-12C). The spores occur in moldy hay or silage or in the air of farm buildings and can be inhaled into the lungs; however, they are apparently not pathogenic. Nocardomycolic acids occur in the walls of some *Micropolyspora* species.

QUESTIONS

1 How is *Micrococcus* distinguished from *Staphylococcus*? From *Deinococcus*? From *Planococcus*?

2 How is *Leuconostoc* distinguished from *Streptococcus*? From *Pediococcus*?

3 How is *Clostridium* distinguished from *Bacillus*? From *Desulfotomaculum*?

4 What is the outstanding morphological feature of each of the following? *Caryophanon. Arthrobacter. Nocardia. Sarcina. Corynebacterium.*

5 Name three genera that contain mycolic acids in their cell walls. Which contain acid-fast organisms?

6 How can *Staphylococcus* be distinguished from *Streptococcus*? Give at least two characteristics.

7 What is the outstanding biochemical feature of *Propionibacterium*? Of *Cellulomonas*?

8 How has cell-wall chemistry been used in the classification of nocardioform bacteria?

9 What kinds of bacteria described in this chapter are noted for the following characteristics?
   (a) Insect pathogenicity
   (b) Radiation resistance
   (c) Growth on the surface of cheeses
   (d) Pathogens of fish
   (e) Occurrence of metachromatic granules
   (f) Cellulose degradation

10 Most strains of *Streptococcus* can grow in air, but some strains are anaerobic. The genus *Peptostreptococcus* cannot grow in air, and all strains are anaer-

obic. How could you differentiate an anaerobic strain of *Streptococcus* from a strain of *Peptostreptococcus*?

11 Some strains of *Actinomyces* produce large amounts of lactic acid. All strains of *Lactobacillus* produce large amounts of lactic acid. How could you differentiate a strain of *Lactobacillus* from *Actinomyces*?

REFERENCES

Holdeman, L. V., E. P. Cato, and W. E. C. Moore (eds.): *Anaerobe Laboratory Manual*, 4th ed., Anaerobe Laboratory, Virginia Polytechnic Institute and State University, Blacksburg, 1977. *This authoritative work provides the identifying features of many of the anaerobic bacteria discussed in Chap. 14.*

Joklik, W. K., H. P. Willett, and D. B. Amos: *Zinsser Microbiology*, 18th ed., Appleton-Century-Crofts, New York, 1984. *This textbook of medical microbiology gives the characteristics of many of the clinically important bacteria mentioned in Chap. 14.*

Sneath, P. H. A. (ed.): *Bergey's Manual of Systematic Bacteriology*, vol. 2, Williams & Wilkins, Baltimore (in press). *This international reference work provides detailed descriptions of the genera and species of bacteria discussed in Chap. 14.*

Stanier, R. Y., E. A. Adelberg, and J. L. Ingraham: *The Microbial World*, Prentice-Hall, Englewood Cliffs, N.J., 1976. *Chapters 22 and 23 of this work provide a survey of the biological properties of many of the organisms discussed in volume 2 of Bergey's Manual.*

Starr, M. P., H. Stolp, H. G. Trüper, A. Balows, and H. G. Schlegel (eds.): *The Prokaryotes: a Handbook of Habitats, Isolation, and Identification of Bacteria*, Springer-Verlag, New York, 1981. *Although not primarily concerned with classification, this monumental reference work provides a wealth of descriptive information and illustrations concerning the various genera of bacteria.*

Chapter 15

The World of Bacteria III: Bacteria with Unusual Properties

The organisms in Volume 3 of *Bergey's Manual* have properties that are quite different from those of the bacteria described in Volumes 1 and 2. Some are distinguished by their unusual type of metabolism. For instance, some of the organisms are **phototrophic,** able to use light as an energy source. Others are **chemolithotrophic,** able to obtain energy by oxidizing inorganic compounds such as ammonia, nitrite, hydrogen sulfide, or ferrous iron. Still others are not distinguished by an unusual metabolism but rather by other features such as the occurrence of **gliding motility** rather than motility by the action of flagella, reproduction by **budding** rather than by binary fission, or special morphological structures such as **sheaths, prosthecae,** and **stalks.** Most of the organisms are Gram-negative **eubacteria,** but some, such as those that form methane gas, belong to the major bacterial group known as the **archaeobacteria,** which may stain Gram-negative or Gram-positive. In the present chapter we will describe the properties of several groups of unusual eubacteria, and also the major features of various archaeobacteria.

BERGEY'S MANUAL OF SYSTEMATIC BACTERIOLOGY, VOLUME 3

The bacteria included in Volume 3 of *Bergey's Manual* are extremely diverse in their properties. As in Chaps. 14 and 15, they are grouped largely on the basis of practical considerations. The major sections of Volume 3 are listed in Table 15-1. The archaeobacteria are considered together as a group despite the diversity of their characteristics because they are all phylogenetically distinct from eubacteria.

ANOXYGENIC PHOTOTROPHIC BACTERIA

The bacteria of this group belong to the order *RHODOSPIRILLALES.* They are Gram-negative and are all capable of carrying out a photolithotrophic and/or photoorganotrophic type of metabolism, and they contain bacteriochlorophyll (which differs in structure from chlorophyll such as occurs in cyanobacteria and eucaryotic algae). Also present are various water-insoluble carotenoid pigments, which can also trap or absorb light energy and transmit it to the bacteriochlorophyll. The anoxygenic bacteria grow phototrophically only under anaerobic conditions and are incapable of forming O_2 (i.e., are *anoxygenic*) because they possess only photosystem I (see Chap. 10). They are believed to be more

Table 15-1. Bacteria Included in *Bergey's Manual*, Volume 3

Section	Some Major Characteristics
ANOXYGENIC PHOTOTROPHIC BACTERIA	Gram-negative bacteria that contain bacteriochlorophyll and can use light as an energy source; the organisms are anaerobic and do not evolve O_2 during photosynthesis
OXYGENIC PHOTOTROPHIC BACTERIA	Bacteria that contain chlorophyll, can use light as an energy source, and evolve O_2 in a manner similar to that of green plants; the group includes the cyanobacteria ("blue-green algae")
GLIDING, FRUITING BACTERIA	Gram-negative nonphototrophic bacteria that lack flagella, yet can glide across solid surfaces; they have a complex life cycle in which the cells swarm together in masses and form fruiting bodies
GLIDING, NONFRUITING BACTERIA	Gram-negative nonphototrophic rods, filaments, or multicellular trichomes that glide across solid surfaces; fruiting bodies are not produced
THE SHEATHED BACTERIA	Gram-negative nonphototrophic bacteria that form an external sheath that covers the chains or trichomes
BUDDING AND/OR APPENDAGED BACTERIA	Gram-negative nonphototrophic bacteria that reproduce asymmetrically by budding and/or form prosthecae or stalks
CHEMOLITHOTROPHIC BACTERIA	Gram-negative nonphototrophic bacteria that obtain energy for carbon dioxide fixation from the oxidation of ammonia, nitrite, reduced sulfur compounds, or ferrous iron
ARCHAEOBACTERIA	Gram-positive or Gram-negative bacteria that are phylogenetically distinct from eubacteria; some produce methane gas; some require unusually high levels of NaCl for growth; others are distinguished by their ability to grow at a low pH and a high temperature

primitive than oxygenic (O$_2$-evolving) phototrophic organisms: geological studies have provided evidence that the atmosphere of our planet was anaerobic at the time life began to develop and that oxygen did not appear in appreciable quantities in the atmosphere until oxygenic bacteria evolved (i.e., bacteria having both photosystem I and photosystem II).

Anoxygenic phototrophic bacteria occur in anaerobic freshwater or marine environments. They may occur beneath the surface of shallow aquatic environments rich in organic matter, such as stagnant ponds, ditches, and salt marsh pools, or, in some instances, they may have a much deeper habitat, as at the bottom of a lake. The bacteriochlorophyll absorbs light most strongly when the light is of long wavelength—about 725 to 745 nm (far red light, at the extreme end of the visible spectrum) to 1035 nm (infrared light, invisible to the human eye). This light is of longer wavelength than that absorbed by the chlorophyll of oxygenic bacteria or oxygenic eucaryotic algae (about 680 nm). Although oxygenic organisms may grow at the surface of a shallow, stagnant pond, they do not absorb far red or infrared light and thus do not prevent it from reaching the anoxygenic phototrophs below. The bacteriochlorophyll and the carotenoid pigments of anoxygenic bacteria can also absorb some light in the blue to blue-green range (between 400 and 550 nm). This becomes important when anoxygenic bacteria occur in the depths of a lake, because blue light can penetrate water to greater distances than red light can.

The color of anoxygenic phototrophic bacteria is determined mainly by the carotenoid pigments rather than by the bacteriochlorophyll, and the anoxygenic phototrophs can be divided into two major groups on the basis of their pigmentation: *purple bacteria* and *green bacteria*. Motility, if present in these two groups, is by means of polar flagella, except for the family *Chloroflexaceae* which exhibits a gliding type of motility. *Nitrogen can be fixed by purple or green bacteria, but usually only under anaerobic conditions and illumination.*

Purple Phototrophic Bacteria

These organisms contain bacteriochlorophyll types *a* or *b*. The pigments that harvest the energy of light (i.e., bacteriochlorophyll and auxiliary carotenoid pigments) are located in the cytoplasmic membrane, which may be greatly

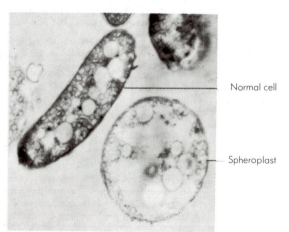

Figure 15-1. Electron micrograph of a purple nonsulfur bacterium, *Rhodospirillum rubrum*, showing the small intracellular vesicles that contain the photosynthetic apparatus. The large clear areas are poly-β-hydroxybutyrate granules (X14,000). [*From E. S. Boatman and H. C. Douglas, Electron Microscopy, vol. 2, Fifth International Congress of Electron Microscopy (Philadelphia), Academic, New York, 1962.*]

Normal cell

Spheroplast

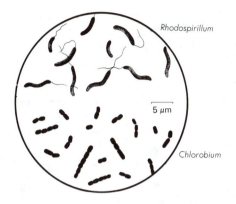

Figure 15-2. Drawing of the cells of *Rhodospirillum* (family *Rhodospirillaceae*) and *Chlorobium* (family *Chlorobiaceae*). The flagella of *Rhodospirillum* cannot be seen by ordinary staining. *(Erwin F. Lessel, illustrator.)*

Figure 15-3. Some species of the family *Rhodospirillaceae*. (A) *Rhodomicrobium vannielii*, a prosthecate budding species. The bud forms at the tip of a prostheca and eventually reaches the size of the mother cell. (B) *Rhodopseudomonas acidophila*, a nonprosthecate budding species. The bud is sessile at the pole of the mother cell and separates by constriction when the bud reaches the size of the mother cell. Some bundles of polar flagella can be seen in the field. (C) *Rhodopseudomonas palustris*, a nonprosthecate budding species. The cells are narrower than those in (B). Phase-contrast (X1,464). *(From N. Pfennig, J Bacteriol,* **99:**597, 1969.)

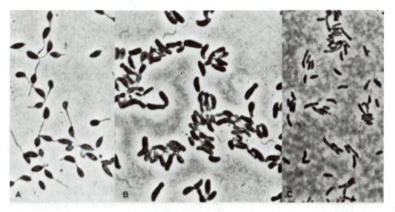

invaginated to form vesicles (see Fig. 15-1), folded layers (lamellae), or tubules. Two families are recognized, as follows.

The family **RHODOSPIRILLACEAE** contains the **purple nonsulfur bacteria.** Cultures appear orange-brown to purple-red under aerobic conditions. Some may be similarly pigmented under anaerobic conditions, but others may be greenish-yellow. The purple nonsulfur bacteria exhibit a diversity of shapes: helical (e.g., **Rhodospirillum;** see Fig. 15-2), nonprosthecate rod-shaped, ovoid, or spherical cells that multiply by binary fission or budding (e.g., **Rhodopseudomonas;** see Figs. 15-3B and C), or ovoid cells that multiply by the formation of buds at the end of prosthecae (e.g., **Rhodomicrobium**; see Figs. 15-3A and 15-4).

The purple nonsulfur bacteria are *photoorganotrophs:* organic substances serve both as carbon sources and as electron donors for the reduction of carbon dioxide. Some species can grow autotrophically by using H_2S as the electron donor, but only if very low concentrations are provided; none can use elemental sulfur as an electron donor. Photosynthesis occurs only under anaerobic conditions in the light. Some species can also grow under aerobic or microaerophilic conditions in the dark by respiration with organic compounds.

The family **CHROMATIACEAE** contains the **purple sulfur bacteria.** Cultures appear orange-brown to purple-violet. Purple sulfur bacteria may be ovoid to

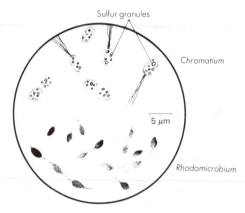

Sulfur granules

Chromatium

5 μm

Rhodomicrobium

Figure 15-4. Drawing of the cells of Chromatium (family Chromatiaceae) and Rhodomicrobium (family Rhodospirillaceae). The intracellular sulfur granules of Chromatium are indicated. (Erwin F. Lessel, illustrator.)

rod-shaped (e.g., *Chromatium;* see Fig. 15-4), coccoid (e.g., *Thiocystis*), or helical (e.g., *Thiospirillum*). Coccal species may be arranged as diplococci (e.g., *Lamprocystis*), in cubical packets (e.g., *Thiosarcina*), or in flat sheets (e.g., *Thiopedia*). Some species contain gas vacuoles.

All genera are capable of *photolithotrophic* growth, using H_2S or elemental sulfur as the electron donor for CO_2 fixation. When H_2S is the electron donor, globules of elemental sulfur are formed, usually within the cells. The sulfur is eventually oxidized to sulfate. Some strains can also grow photoorganotrophically. Most species are anaerobic and cannot grow in the dark even under microaerophilic conditions.

Green Phototrophic Bacteria

In contrast to the purple bacteria, these organisms contain bacteriochlorophyll types *c* or *d* and minor amounts of *a*. Moreover, cultures are green or brown. The pigments involved in photosynthesis are located in membrane-bound vesicles within the cell; some of these may be attached to the cytoplasmic membrane. Two families are recognized, as follows.

The family *CHLOROBIACEAE* contains the **green sulfur bacteria.** The cells are ovoid, bean-shaped, or rod-shaped (e.g., see the genus *Chlorobium,* Fig. 15-2) and multiply only by binary fission. One genus, *Prosthecochloris,* consists of star-shaped cells, this shape being caused by the production of about 20 prosthecae per cell. Gas vacuoles may occur in some genera.

Green sulfur bacteria live as *photolithotrophs,* using H_2S as the electron donor for CO_2 fixation. Granules of elemental sulfur are deposited *outside* the cells, never within the cells; the sulfur can eventually be oxidized to $SO_4{}^{2-}$ (soluble sulfur). The organisms are anaerobic, being incapable of growing in the dark even under microaerophilic conditions.

The family *Chloroflexaceae* contains the **green nonsulfur bacteria.** The main genus, *Chloroflexus,* is thermophilic (optimum temperature 52 to 60°C) and occurs in hot springs where it forms green or orange mats. *Chloroflexus* cells occur as filaments or trichomes and exhibit gliding motility. The organisms are mainly *photoorganotrophic,* as the purple nonsulfur bacteria, but they can also grow as photolithotrophs with H_2S as the electron donor. In the dark they can grow aerobically as chemoheterotrophs.

OXYGENIC PHOTOTROPHIC BACTERIA

Cyanobacteria (Blue-Green Algae)

These organisms exhibit an enormous diversity of shapes and arrangements, from unicellular cocci (Fig. 15-5) or rods to long trichomes (Fig. 15-6). Gas vacuoles may be formed by many species (Fig. 15-6B). Some cyanobacteria are surrounded by a sheath that surrounds the aggregates (Fig. 15-5B) or trichomes. Unicellular cyanobacteria are usually nonmotile, but trichome-formers generally possess gliding motility. Flagella are absent. Cyanobacteria are widespread in soil, freshwater, and marine habitats. Some are thermophilic, growing in hot springs. Cyanobacteria can grow as mats on the surface of bare soil as primary colonizers. They are important in adding organic matter to the soil and in preventing incipient erosion. Some cyanobacteria grow in symbiosis with other organisms. For example, they may occur as algal symbionts of lichens (see Chap. 18). Some live within the plant bodies of certain liverworts, water ferns, cycads (a class of naked-seed plants), and angiosperms (plants whose seeds are borne within a fruit) where they fix nitrogen. Cyanobacteria have also been associated with certain protozoa, where they are called *cyanellae*.

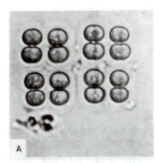

Figure 15-5. Examples of coccoid cyanobacteria. (A) *Merismopedia glauca.* The cells are 3 to 5 μm in diameter, and the colony is a flat plate. This species is a common member of the plankton of soft water lakes. (B) *Gloeocapsa rupestris.* The cells are 6 to 9 μm in diameter and are surrounded by a sheath. This species occurs on moist rocks, in soil, or on submerged objects and forms gelatinous masses that are often colored yellow, red, or brown. *(Courtesy of George J. Schumacher, State University of New York at Binghamton.)*

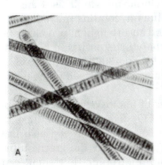

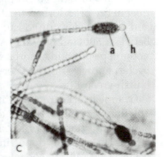
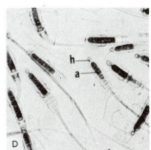

Figure 15-6. Examples of trichome-forming cyanobacteria. (A) *Oscillatoria limosa.* The trichomes consist only of vegetative cells, which are 12 to 18 μm wide. (B) *Anabaena planktonica.* The vegetative cells are 10 to 15 μm wide and contain gas vacuoles (bright areas). A heterocyst (h) is also shown. The heterocysts of this species are unique in that they possess lateral mucilaginous winglike structures, as shown in this photograph. (C) *Cylindro-spermum majus.* The vegetative cells are 3 to 5 μm wide and the heterocysts (h), which are always terminal in location, are slightly larger. The akinetes (a) are much larger and are 25 to 30 μm in length. (D) *Gloeotrichia echinulata.* The vegetative cells are 8 to 10 μm wide but decrease in width along the long, tapering trichome. The terminal heterocysts (h) are 8 to 10 μm in diameter, and the akinetes (a) are 10 to 20 μm wide by 45 to 50 μm long. *(Courtesy of George J. Schumacher, State University of New York at Binghamton.)*

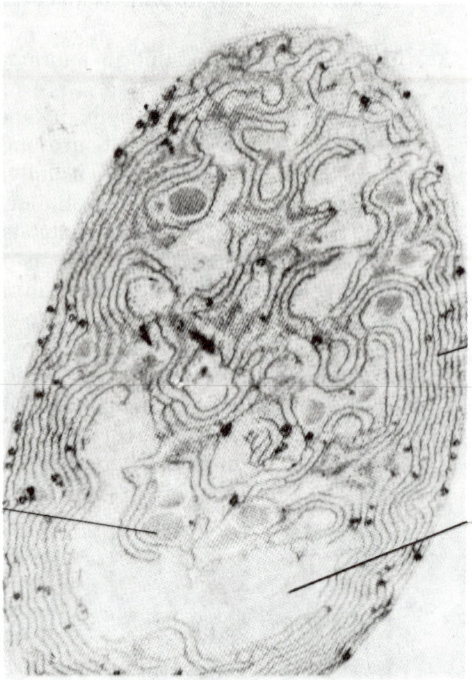

Figure 15-7. Section of a vegetative cell of the cyanobacterium *Anabaena azollae*. Most of the photosynthetic lamellae (thylakoids) are peripheral, but some extend into the midportions of the cell. *(Courtesy of Norma J. Lang and J Phycol,* **1**:127–134, 1965.)

Photosynthetic lamellae (thylakoids)

Polyhedral bodies

Nucleoplasmic regions

Cyanobacteria contain chlorophyll *a* rather than bacteriochlorophyll; because of this chlorophyll the cells absorb red light of 680 to 683 nm. Other pigments include water-insoluble carotenoids and also water-soluble *phycobilins*, which are the major light-absorbing pigments in cyanobacteria and which can transmit the energy of absorbed light to the chlorophyll. Blue phycobilins (phycocyanin and allophycocyanin) occur in all cyanobacteria and absorb light at wavelengths between 500 and 650 nm. A red phycobilin, phycoerythrin, occurs in some but not all species and absorbs shorter wavelengths between 470 to 600 nm. Cyanobacteria possessing phycoerythrin have a red or brown color instead of the usual bluish-green hue.

Cyanobacteria are *photolithotrophs,* and because of photosystem II (see Chap. 10) they can use H_2O as an electron donor for CO_2 fixation, in contrast to anoxygenic phototrophic bacteria. However, some cyanobacteria can also use H_2S as an electron donor in a manner similar to that used by the green sulfur bacteria. Many cyanobacteria are obligately photolithotrophic, but some can also grow as chemoorganotrophs at a slow rate in the dark.

The photosynthetic apparatus (i.e., chlorophyll *a*, carotenoid pigments, photochemical reaction centers, and the photosynthetic electron transport chain) is contained in the *thylakoids*—flattened membranous sacs located within the cell (see Fig. 15-7). The surface of the thylakoids is studded with granules called *phycobilisomes*, which contain the phycobilin pigments.

Many trichome-forming cyanobacteria can fix N_2. It seems strange that oxygen-evolving organisms can do this, since nitrogenase is highly oxygen-sensitive; however, in most instances it is not the vegetative cyanobacterial cells that carry out nitrogen fixation but rather specialized cells called *heterocysts* (see Fig. 15-6), which occur periodically along or at the ends of the trichome. Nitrogen fixation is possible in heterocysts because they lack photosystem II and therefore do not evolve oxygen; moreover, their walls are much thicker than those of the vegetative cells, which may prevent rapid diffusion of oxygen into the heterocyst. Some cyanobacteria that form heterocysts also form large, thick-walled cystlike cells called *akinetes* (Fig. 15-6), which are resistant to desiccation.

The Prochlorophytes

Unlike cyanobacteria, these unicellular spherical organisms contain chlorophyll *b* in addition to chlorophyll *a*. Moreover, they lack phycobilin pigments; consequently, the cells appear grass-green rather than blue-green. Prochlorophytes live in association with certain marine invertebrates (ascidians) and have not yet been cultured; information about them has been obtained only by studying them in their natural symbiotic state or by analysis of cells mechanically removed from their hosts. Only a single genus is recognized, *Prochloron.*

GLIDING, FRUITING BACTERIA

The organisms of this group are Gram-negative, nonphototrophic, and nonflagellated. They exhibit a creeping or gliding motility on solid surfaces. The mechanism of gliding is not known. The organisms are not filamentous; instead, the vegetative cells are unicellular short rods resembling typical bacteria except the walls are flexible. This flexibility may be due to the fact that the peptidoglycan layer is thin and occurs in patches that are connected by nonpeptidoglycan material. When the organisms are placed on a solid culture medium, growth, which is frequently of a slimy consistency, spreads rapidly over the surface. As the cells glide across the surface of the medium, they leave a layer of slime behind them. Indeed, the prefix *myxo*, which occurs in the name of the order to which these bacteria belong, *Myxobacterales*, reflects this property, being derived from the Greek noun *myxa*, meaning "mucus, slime."

Although there are some exceptions, a remarkable and striking feature of the myxobacters is that the vegetative cells at some stage of growth *swarm* together in masses and form *fruiting bodies*. [This behavior bears some resemblance to that of slime molds; however, the latter are eucaryotic organisms (see Chap. 17).] The fruiting bodies contain *myxospores*, which are shorter and thicker than the vegetative cells and are resistant to desiccation and ultraviolet radiation but not to heat. The fruiting bodies range from simple to complex. The simplest fruiting bodies are merely heaps of myxospores embedded in a mass of slime. The most elaborate fruiting bodies may have a *stalk* composed of slime, and walled containers (*sporangioles*, sometimes termed *cysts*) which enclose the myxospores. The life cycle of one species that forms complex fruiting bodies, *Stigmatella aurantiaca,* is shown schematically in Fig. 15-8; electron micrographs of various stages in the growth cycle of another species, *Chondromyces crocatus,* are shown in Fig. 15-9.

Myxobacters are strictly aerobic organisms found in surface layers of soil, compost, manure, rotting wood, and animal dung. Many myxobacters develop very colorful pigments in their normal environment. Some species produce exocellular enzymes that degrade complex substrates such as cellulose, agar, chitin, and even the cell walls of other bacteria.

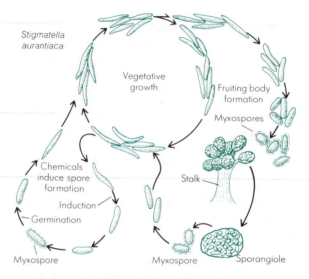

Figure 15-8. Life cycle of *Stigmatella aurantiaca* showing vegetative cells, myxospores, and fruiting body. [After H. Reichenbach, from Martin Dworkin, "The Myxobacterales," in A. I. Laskin and H. A. Lechevalier (eds.), Handbook of Microbiology, CRC Press, Inc., Boca Raton, Fla., 1974.]

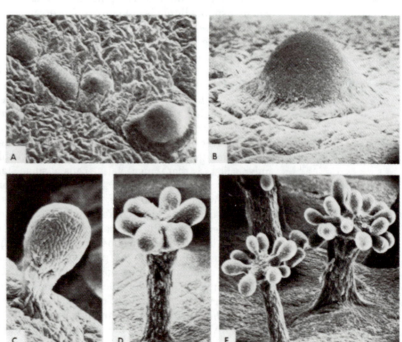

Figure 15-9. Stages in the fruiting body formation of the myxobacter *Chondromyces crocatus*. Early stages: (A) Initial stages of vegetative cell aggregation; (B) "fried-egg" stage showing orientation of peripheral cells; (C) bulb formation and development of stalk. Late stages: (D) Initial stages of sporangiole formation; (E) sporangiole formation after elongation of stalk to maximum length. Structures range in size from approximately 10 to 60 μm. (From P. L. Grilione and J. Pangborn and J Bacteriol, **124**:1558, 1975.)

GLIDING, NONFRUITING BACTERIA

These bacteria resemble myxobacters in their gliding motility; however, fruiting bodies are not formed, and only one genus, ***Sporocytophaga***, forms myxospores. Most of the genera in the order are aerobic or microaerophilic organisms that live in soil or water. Many species can degrade natural polymers such as cellulose, chitin, pectin, keratin, or even agar. One genus, *Capnocytophaga,* is unusual because of its occurrence in the oral cavity of humans—in the gingival crevice (the space between the surface of the enamel of a tooth and the gum)—and may be involved in periodontal disease. Some aquatic genera such as *Beggiatoa* occur mainly in microaerophilic environments containing H_2S, which is oxidized by the cells to elemental sulfur; the sulfur accumulates as granules within the cells. Thus, *Beggiatoa* may possibly be an autotrophic organism.

Gliding, nonfruiting bacteria may appear as individual rods or filaments (e.g., *Cytophaga, Flexibacter,* or *Vitreoscilla*), or as multicellular trichomes (e.g., *Beggiatoa, Simonsiella, Saprospira,* or *Thiothrix*). Some of the cells are very long; for example, *Flexibacter* cells may reach 50 μm in length (see Fig. 15-10). Some genera such as *Herpetosiphon* and *Flexithrix* produce a sheath which

Figure 15-10. Species of gliding, nonfruiting bacteria. (A) *Flexibacter polymorphus.* Cells collected on the surface of a Nucleopore membrane filter (X730). (B) Colony of *F. polymorphus* growing on surface of Nucleopore membrane filter layered over a nutrient agar surface (X100). The holes in the filter are 5.0 μm in diameter. *(Courtesy of H. F. Ridgeway, Jr., Scripps Institution of Oceanography.)* (C) Filaments of the gliding bacterium *Herpetosiphon giganteus* on agar showing "bulbs" (bright spherical enlarged regions) (X500). (D) Same as (C) but at lower magnification (X330). *(Courtesy Hans Reichenbach.)* (E, F) *Simonsiella* sp. showing cells arranged in apposition to form trichomes with free faces of terminal cells rounded. (E) Scanning electron micrograph (X2,200); (F) transmission electron micrograph of thin section (X20,000). *(Courtesy J. Pangborn and Daisy Kuhn.)*

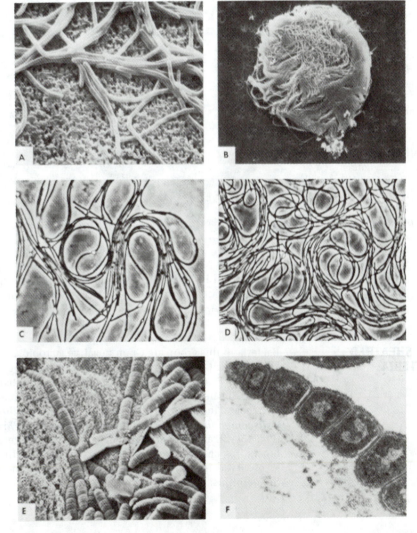

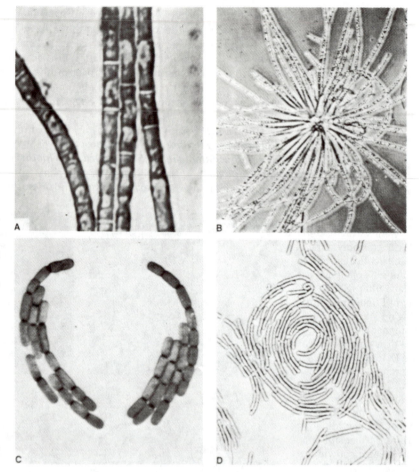

Figure 15-11. Gliding, non-fruiting bacteria. (A) Trichomes of *Beggiatoa* stained to demonstrate presence of a cell membrane (X2,250). *(From H. L. Scotten and J. L. Stokes, Arch Mikrobiol,* **42***:353, 1962.)* (B) Trichomes attached to a common object are illustrated in this photomicrograph of *Thiothrix* sp.. (X420). *(Courtesy of F. E. Palmer and E. J. Ordal.)* (C) *Vitreoscilla* cell morphology. *(Courtesy of G. J. Hageage, Jr.)* (D) Pattern of cell arrangement of *Vitreoscilla. (Courtesy of V. B. D. Skerman.)*

encloses the individual cells. Examples of the morphological features characterisic of gliding, nonfruiting bacteria are shown in Figs. 15-10 and 15-11 (see also Fig. 5-4).

THE SHEATHED BACTERIA

Bacteria in this group are Gram-negative, aerobic, and nonphototrophic, and are characterized by the formation of a sheath surrounding a chain of cells or a trichome. The sheaths of some genera are encrusted with ferric and manganic oxides. Sheathed bacteria inhabit freshwater and marine environments. Among the genera included in the group are: *Sphaerotilus, Leptothrix* (Fig. 15-12), *Haliscomenobacter, Streptothrix, Lieskeella, Phragmidiothrix, Crenothrix* (Fig. 15-12), and *Clonothrix.* Only the first three genera have been isolated; the others are characterized solely on the basis of their distinctive morphology as observed in samples from natural sources. One of the cultivatable genera, *Sphaerotilus,* is discussed here.

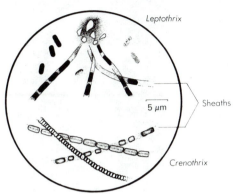

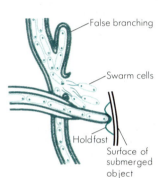

Figure 15-12. Drawing of sheathed bacteria of the genus *Leptothrix* and *Crenothrix*. *(Erwin F. Lessel, illustrator.)*

Figure 15-13. Sheathed bacteria. Drawing of *Sphaerotilus* showing sheath, false branching, and motile swarmers. When a swarmer encounters a solid object, it can become attached by a holdfast, formed by secreting a sticky substance that hardens. Sheath formation begins at the holdfast. *(Redrawn from K. A. Bisset, Bacteria, E. and S. Livingstone, Ltd., Edinburgh, 1952.)*

Sphaerotilus

The cells typically occur as chains of rods enclosed within a sheath, as shown in Fig. 5-17. The sheath may branch, giving the impression that the cells are branched; however, this is recognized as "false branching" (Fig. 15-13). The cells that emerge from an open end or a break in the sheath are called *swarm cells*. The hollow sheaths accumulate. The swarm cells are rod-shaped and possess polar or subpolar flagella; they thus resemble pseudomonads in appearance. ***S. natans*** is a common species that normally occurs in polluted waters, and its sheaths, of organic composition, are thin and colorless. In unpolluted water containing iron, iron hydroxide may be deposited in or on the sheaths, which turn yellow-brown and may become encrusted with ferric iron. Hence, these organisms are sometimes referred to as "iron bacteria."

BUDDING AND/OR APPENDAGED BACTERIA

This group of nonphototrophic Gram-negative bacteria is characterized by the formation of *prosthecae* (extensions of the cell wall and cytoplasmic membrane) or *stalks* (nonliving ribbonlike or tubular appendages that are excreted by the cell) and/or by the asymmetric mode of reproduction called budding. The organisms range from aerobic to microaerophilic to facultatively anaerobic. Although nonphototrophic, some genera have morphologically similar counterparts in the phototrophic group of bacteria. A few examples of budding and/or appendaged bacteria are described below.

Prosthecate Budding Bacteria

Hyphomicrobium

Individual cells are initially coccoid and flagellated but mature into oval or bean-shaped cells. Prosthecae are produced at either one or both ends of a cell. Buds develop at the tips of the prosthecae (see Fig. 15-14A and also Fig. 7-1E), and as they mature they separate from the prosthecae. Hyphomicrobia are aerobic and chemoorganotrophic; however, their morphology is similar in many ways to that of the phototrophic genus *Rhodomicrobium* (see Fig. 15-3). Hy-

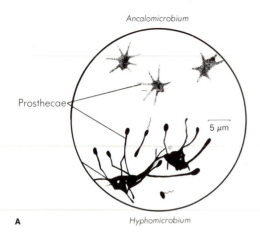
Ancalomicrobium

Prosthecae

5 µm

Hyphomicrobium

A

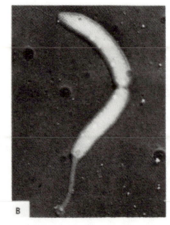
B

C

Figure 15-14. Budding and/or appendaged bacteria. (A) Drawing of cells of *Hyphomicrobium* and *Ancalomicrobium*. *(Erwin F. Lessel, illustrator.)* (B) *Caulobacter*, undergoing binary fission. The upper cell possesses a polar flagellum; the lower cell has a prostheca with terminal holdfast (X13,000). *(Courtesy of A. L. Houwink and W. van Iterson, Biochem Biophys Acta, **5**:10, 1950.)* (C) *Caulobacter* cells attached to a common holdfast and exhibiting a rosette pattern. *(Courtesy of V. B. D. Skerman.)*

phomicrobia have been found widely in soils of all continents, as well as in numerous aquatic environments.

Ancalomicrobium

These facultatively anaerobic aquatic bacteria have three to eight, long, tapering prosthecae per cell (see Fig. 15-14A and also Fig. 5-18). Buds are formed directly from one position on the mother cell, never from the prosthecae. A genus of phototrophic green sulfur bacteria, **Ancalochloris,** bears a morphological resemblance to *Ancalomicrobium*.

Prosthecate Nonbudding Bacteria

Caulobacter

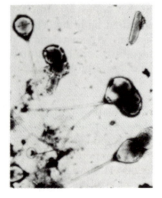

In this genus of aerobic organisms, an individual cell is either a straight or curved rod with a single prostheca. The daughter cell arises by binary fission; it possesses a single polar flagellum (Fig. 15-14B) and is termed a *swarm cell.* The free motile swarm cell secretes an adhesive material (*holdfast*) at the end of the cell where the flagellum is located; eventually, a prostheca is produced at the same pole and the holdfast becomes located at the end of the prostheca. The flagellum is usually lost during formation of the prostheca. By means of the holdfast, cells may become attached to each other to form rosettes (Fig. 15-14C) or they may become attached to some other substance. Caulobacters are normally found in salt water and freshwater and have the ability to grow in environments with very low concentrations of nutrients.

Figure 15-15. Electron micrograph of stalked cells of a member of the *Blastocaulis-Planctomyces* group, morphotype II. A bud developing from a mother cell can be seen at the lower right (X4,400). *(Courtesy of Jean M. Schmidt, Arizona State University.)*

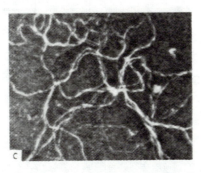

Figure 15-16. *Gallionella*, showing the kidney-shaped cells (A) bearing twisted stalks (B). (C) *Gallionella* sp. from salt water showing long entangled stalks (dark-phase microscopic preparation; X147). *(From J. M. Sharpley, Appl Petrol Microbiol, 9:380, 1961.)*

Nonprosthecate Budding Bacteria

Bacteria of the **Blastocaulis-Planctomyces** group occur in freshwater, brackish, or marine environments. The cells are spherical or ovoid and possess a stalk with a holdfast at the distal end. Budding occurs directly from a mother cell, as shown in Fig. 15-15.

Nonprosthecate, Nonbudding Bacteria

Gallionella

A twisted ribbonlike stalk extends from the middle of the curved or kidney-shaped cells (Fig. 15-16). The cells reproduce by binary fission. The organisms grow under microaerophilic conditions where both O_2 and a supply of ferrous iron are being continuously supplied. Since ferrous iron spontaneously oxidizes in the presence of O_2, these environmental conditions are unusual but can be met in aerated, iron-removal water-treatment plants, drainage from certain coal mines, or in various thermal springs. It is likely that *Gallionella* is autotrophic, obtaining energy by oxidizing ferrous iron to the ferric form. Because of its ability to form insoluble oxidized iron compounds, *Gallionella* may cause problems, such as clogging, in pipelines of water systems.

CHEMOLITHOTROPHIC BACTERIA

Three distinct metabolic types constitute this category of Gram-negative autotrophic bacteria, namely:

1 Obtain energy by oxidizing ammonia or nitrite (family *Nitrobacteraceae*)
2 Obtain energy by oxidizing sulfur or sulfur compounds; not assigned to any family
3 Deposit iron and/or manganese oxides (family *Siderocapsaceae*)

The Family NITROBACTERACEAE

The "nitrifying bacteria," as these organisms are called, include species of diverse morphological types—rods, cocci, and helical cells. They are nonmotile or motile by subpolar or peritrichous flagella. They are aerobic autotrophs, incapable of chemoheterotrophic growth with the exception of one species, **Nitrobacter winogradskyi.** The nitrifying bacteria comprise two distinct metabolic groups in terms of reactions that provide energy: (1) those which oxidize

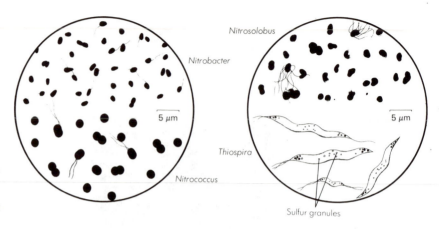

Figure 15-17. Drawing of chemolithoautotrophic bacteria. Nitrifying bacteria of the genera *Nitrobacter*, *Nitrococcus*, and *Nitrosolobus*, and sulfide-oxidizing bacteria of the genus *Thiospira*. *(Erwin F. Lessel, illustrator.)*

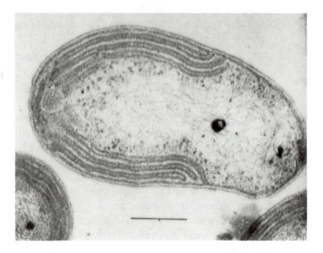

Figure 15-18. Electron micrograph of a longitudinal section of *Nitrobacter winogradskyi* showing intracellular membranes derived from invagination of the cytoplasmic membrane. This species reproduces by budding. The bar indicates 0.25 μm. *(From S. W. Watson, Int J Syst Bacteriol, **21**:254–270, 1971.)*

nitrite to nitrate, the generic names beginning with the prefix *nitro*, e.g., ***Nitrobacter*** (Fig. 15-17), ***Nitrococcus*** (Fig. 15-17), and ***Nitrospina;*** and (2) those which oxidize ammonia to nitrite, the generic names beginning with the prefix *nitroso*, e.g., ***Nitrosolobus*** (Fig. 15-17), ***Nitrosomonas, Nitrosovibrio, Nitrosococcus,*** and ***Nitrosospira.*** Some species have extensive invaginations of the cytoplasmic membrane [e.g., see N. *winogradskyi* (Fig. 15-18) and ***Nitrosococcus oceanus*** (Fig. 5-27)]. Nitrifying bacteria are commonly found in soils, where they play an important role in the nitrogen cycle and in maintaining the fertility of soil. They are discussed in more detail in Chap. 25.

Sulfur- and Sulfur-Compound-Metabolizing Bacteria

The bacteria of this group can be divided into those genera which have been isolated into pure culture and those which have not yet been isolated (noncultivatable). The cultivatable genera contain Gram-negative short rods (***Thiobacillus***) or helical cells (***Thiomicrospira***); most species are motile by means of polar flagella. Both genera are widely distributed in soil, freshwaters, coal-mine drainage waters, and marine environments. They derive energy from the oxidation of reduced sulfur compounds, including sulfides, elemental sulfur, thiosulfate, polythionates, and sulfite. The final oxidation product is sulfate. One species, *Thiobacillus ferrooxidans,* can also derive energy by oxidizing ferrous iron to the ferric form. Some species are obligate autotrophs (e.g., *Thiobacillus thioparus, Thiobacillus thiooxidans,* and *Thiomicrospira pelophila*), and others are facultative autotrophs (e.g., *Thiobacillus novellus*). Most species are strictly aerobic, but a few can also grow anaerobically with nitrate as the terminal electron acceptor (e.g., *Thiobacillus denitrificans* and *Thiomicrospira denitrificans*). Some species are *acidophilic.* For instance, *T. thiooxidans* grows best at pH values of 2 to 5; moreover, it produces so much sulfuric acid that the pH of the medium may decrease to 0 or lower.

The noncultivatable genera include *Thiobacterium, Macromonas, Thiovulum, Achromatium,* and *Thiospira.* These genera are recognized by their distinctive morphological features as observed in samples from natural sources. Because these organisms have not been isolated, their metabolic nature cannot yet be described with certainty; however, they are probably chemolithotrophic because (1) they occur only in environments having a continuous supply of O_2 and H_2S, and (2) they accumulate sulfur granules intracellularly (e.g., see *Thiospira,* Fig. 15-17). H_2S is unstable (oxidized nonbiologically) in the presence of O_2; consequently, the environments that support growth are those located between the aerobic and anaerobic zones of freshwater or marine environments or in H_2S-bearing springs.

The Family *SIDEROCAPSACEAE*

This group includes unicellular, aerobic to microaerophilic organisms which do not form prosthecae or stalks and which deposit iron or manganese oxides on or in capsules or exocellular slime. The group includes such genera as *Siderocapsa, Siderococcus, Siderocystis,* and *Naumanniella.* (The prefix *sidero* which occurs in some of these names is derived from the Greek noun *siderus* meaning "iron.") There is some question as to whether they are indeed chemolithotrophs, since they generally occur in environments containing not only iron and manganese but also organic matter (stagnant waters, swamp ditches, mud, etc.). Only three species, *Siderocapsa eusphaera, Siderocapsa geminata,* and *Naumanniella polymorpha* have apparently been isolated (although their morphology differs from that seen in nature), but only *N. polymorpha* seems to be capable of autotrophic growth. The other species have not been isolated and are recognized solely by their morphological features as observed in samples from natural sources.

ARCHAEOBACTERIA

Evidence based on studies of ribosomal RNA indicates that archaeobacteria and eubacteria diverged at a very early stage in the evolution of life on earth (see Chap. 3). The phylogenetic gap that exists between the two groups is reflected by certain phenotypic differences, some of which are summarized in Table 15-2. One should recognize, however, that archaeobacteria do not comprise a homogeneous group. Just as great heterogeneity occurs among the eubacteria, so do the various kinds of archaeobacteria differ from each other in terms of morphology, chemical composition, metabolism, and habitat. At present, three

Table 15-2. Some Differences between Archaeobacteria and Eubacteria

Characteristic	Archaeobacteria	Eubacteria
Cell Walls		
Peptidoglycan containing muramic acid and D-amino acids is present	−	+
Lipids of Cytoplasmic Membrane		
Long-chain fatty acids bound to glycerol by ester linkages	−	+
Long-chain branched alcohols (phytanols) bound to glycerol by ether linkages	+	−
Properties Related to Protein Synthesis		
First amino acid to initiate a new polypeptide chain is		
Methionine	+	−
N-Formylmethionine	−	+
Translation process sensitive to action of		
Diphtheria toxin*	+	−
Chloramphenicol†	−	+

* For the action of diphtheria toxin, see Chap. 31.
† For the action of the antibiotic chloramphenicol, see Chap. 24.

main categories of archaeobacteria are recognized: the methane-producers (methanogens), the red extreme halophiles, and the thermoacidophiles.

Methanogenic Bacteria

These archaeobacteria are stringent anaerobes that share an ability to obtain energy for growth by oxidizing compounds such as H_2 or formate, and utilizing the electrons thus generated to reduce CO_2 with the formation of methane gas (CH_4). Some genera can grow as autotrophs, using H_2 and CO_2 as sole sources of carbon and energy; others require additional substances such as vitamins, acetate, amino acids, or organic sulfur compounds. Most species grow better in complex media (e.g., containing yeast extract) than in inorganic media.

At least two unusual coenzymes occur in almost all methanogens that have not been found in other bacteria: *Coenzyme M,* involved in methyl transfer reactions, and *Coenzyme F_{420},* a flavin-like compound involved in the anaerobic electron transport system of these bacteria. The latter coenzyme fluoresces under ultraviolet light. Its presence can be detected by observing the organisms with a fluorescence microscope; this provides a convenient means to identify methanogens.

The genera of methane-producing bacteria are differentiated on the basis of morphology and Gram reaction (see Table 15-3 and Fig. 15-19). Differences in cell wall composition have been found to correlate with these genera (Table 15-3). The cell walls of two genera consist of *pseudomurein,* which differs from eubacterial peptidoglycan by (a) substitution of N-acetyltalosaminuronic acid for N-acetylmuramic acid, and (b) by a tetrapeptide composed entirely of L-amino acids, with glutamic acid at the C-terminal end.

Methanogens occur in various anaerobic habitats rich in organic matter which nonmethanogenic bacteria ferment to produce H_2 and CO_2. Such habitats include marshes, swamps, pond and lake mud, marine sediments, the intestinal tract of humans and animals, the rumen of cattle, and anaerobic sludge digesters in sewage-treatment plants.

Table 15-3. Methanogenic
Bacteria

Genus	Morphology	Motility	Wall Composition
Methanobacterium	Gram-positive to Gram-variable long rods	−	Pseudomurein
Methanobrevibacter	Gram-positive lancet-shaped cocci or short rods	−	Pseudomurein
Methanomicrobium	Gram-negative short rods	+, single polar flagellum	Protein
Methanogenium	Gram-negative pleomorphic cocci	+, peritrichous flagella	Protein
Methanospirillum	Gram-negative curved rods or long wavy filaments	+, polar flagella	Protein; an external sheath is present
Methanosarcina	Gram-positive cocci in clusters	−	Heteropolysaccharide
Methanococcus	Gram-negative pleomorphic cocci	+, one flagellar tuft	Protein with trace of glucosamine

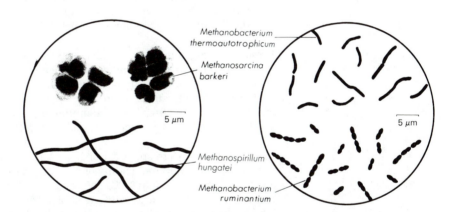

Figure 15-19. Drawing of the cells of various methane-producing bacteria. *(Erwin F. Lessel, illustrator.)*

Extreme Halophiles

These chemoorganotrophic, aerobic bacteria require approximately 17 to 23 percent NaCl for good growth. They stain Gram-negative and range from rod- or disk-shaped cells (the genus *Halobacterium*) to cocci (the genus *Halococcus*). They occur in salt lakes (e.g., the Dead Sea and the Great Salt Lake), industrial plants that produce salt by solar evaporation of sea water, and salted proteinaceous materials such as salted fish (in which they may cause spoilage). The colonies are a red to orange color due to carotenoids which seem to protect the cells against the damaging effect of sunlight.

At high NaCl concentrations the cells resist dehydration by maintaining a high intracellular osmotic concentration of KCl. The cytoplasmic membrane and ribosomes are stable only at high concentrations of KCl, and the enzymes are active only at high levels of either KCl or NaCl. *Halobacterium* cell walls are composed of protein subunits that are held together only in the presence of salt; thus if the level of NaCl falls below about 10 percent, the cells lyse. On the other hand, the walls of *Halococcus* cells are composed of a complex hetero-polysaccharide that is stable even at low salt concentrations.

ATP Synthesis

Halobacteria are mainly aerobic. As in other aerobic organisms, an electron transport chain generates a protonmotive force which in turn drives ATP synthesis (see Chap. 10). Halobacteria can also generate ATP by fermenting the amino acid arginine; this allows them to grow anaerobically. However, a third method of ATP generation is unique to halobacteria. At low oxygen levels, patches of a purple pigment called *bacteriorhodopsin* are formed in the cell membrane. (This pigment is so named because of its similarity to the photosensitive pigment rhodopsin that occurs in the retinal rods of higher vertebrates.) When cells containing the bacteriorhodopsin are exposed to light, the pigment bleaches. During this bleaching, protons (hydrogen ions, H^+) are extruded to the outside of the membrane, thus creating a protonmotive force which in turn drives ATP synthesis. Thus, like the phototrophic bacteria considered earlier in this chapter, halobacteria possess a mechanism for light-driven synthesis of ATP; however, *they possess no bacteriochlorophyll.*

Thermoacidophiles

These aerobic Gram-negative archaeobacteria are characterized by a remarkable ability to grow under *highly acidic conditions at high temperatures.* Two genera included in this group are described below.

Thermoplasma

These chemoorganotrophic organisms resemble mycoplasmas (see Chap. 13) by *lacking a cell wall and forming tiny "fried-egg" colonies.* Like mycoplasmas, the cells are pleomorphic, ranging from spherical to filamentous. The optimum temperature for growth is 55 to 59°C (maximum, 62°C; minimum, 40°C), and the optimum pH is 2 (maximum, 4; minimum, 1). Cells undergo lysis at a neutral pH. Thermoplasmas have been isolated from piles of burning coal refuse.

Sulfolobus

Cells of this genus are spherical or lobe-shaped. Unlike thermoplasmas, a cell wall is present (composed mainly of protein). Various species have temperature optima ranging from 70 to 87°C. The optimum pH is 2 (maximum, 4; minimum, 1). *Sulfolobus* species are facultatively autotrophic. They can grow as chemolithotrophs when supplied with elemental sulfur as an electron donor. Alternatively, they can grow as chemoorganotrophs in media containing organic substrates. In nature, the organisms are predominant in acidic hot springs.

QUESTIONS

1 List the major differences between the families of anoxygenic phototrophic bacteria.
2 In what ways do cyanobacteria differ from other phototrophic bacteria?
3 If cyanobacteria evolve oxygen, and if nitrogenase is oxygen-labile, how can cyanobacteria fix nitrogen?
4 How do the members of the order *Myxobacterales* differ from other gliding bacteria?
5 Besides the organisms listed under Gliding, Nonfruiting Bacteria, what other kinds of bacteria may exhibit gliding motility?

6 Prosthecae occur in bacteria other than those described in the section entitled Budding and/or Appendaged Bacteria. Give two examples.

7 Give an example of a genus of budding bacteria in which (a) the bud forms on a prostheca, (b) the bud forms directly on the mother cell.

8 How does "false branching" differ from "true branching"?

9 In what environments would one expect to find the following bacteria?

(a) Halobacterium   (d) Sulfolobus
(b) Chromatium   (e) Gallionella
(c) Capnocytophaga   (f) Sphaerotilus

10 What is the most significant difference between the following?

(a) Methylococcus (see Chap. 13) and Methanococcus

(b) Rhodomicrobium and Hyphomicrobium

(c) Nitrococcus and Nitrosococcus

(d) Aquaspirillum (see Chap. 13), Rhodospirillum, Thiospirillum, and Methanospirillum

11 List the features of archaeobacteria that distinguish them from eubacteria.

12 What property of Halobacterium might allow one to consider this genus as facultatively phototrophic? What differences exist between halobacteria and the phototrophic bacteria described in the first section of this chapter?

13 Define the following terms:

Bacteriorhodopsin	Akinete	Fruiting body
Phycobilins	Coenzyme F_{420}	Swarm Cells
Thylakoids	Carotenoids	Thermoacidophile
Sporangiole	Heterocyst	Pseudomurein

REFERENCES

Balch, W. E., G. E. Fox, L. J. Magrum, C. R. Woese, and R. S. Wolfe: "Methanogens: Reevaluation of a Unique Biological Group," Microbiol Rev, 43:260–296, 1979. This review provides detailed information about the morphology, chemical composition, and taxonomy of the methane-producing bacteria.

Rippka, R., J. Deruelles, J. B. Waterbury, M. Herdman, and R. Y. Stanier: "Generic Assignments, Strain Histories and Properties of Pure Cultures of Cyanobacteria," J Gen Microbiol, 111:1–161, 1979. This article indicates the differential characteristics of many genera of cyanobacteria and is accompanied by illustrations of cyanobacterial morphology.

Staley, J. T. (ed.): Bergey's Manual of Systematic Bacteriology, vol. 3, Williams & Wilkins, Baltimore (in preparation). Volume 3 of this international work is presently being prepared and will provide detailed descriptions of the genera and species of bacteria discussed in Chap. 15.

Stanier, R. Y., E. A. Adelberg, and J. L. Ingraham: The Microbial World, Prentice-Hall, Englewood Cliffs, N.J., 1976. Chapters 17, 18, and 21 of this work provide a survey of the biological properties of many of the organisms discussed in Volume 3 of Bergey's Manual.

Starr, M. P., H. Stolp, H. G. Trüper, A. Balows, and H. G. Schlegel (eds.): The Prokaryotes: A Handbook of Habitats, Isolation, and Identification of Bacteria, Springer-Verlag, New York, 1981. Although not primarily concerned with classification, this monumental reference work provides a wealth of descriptive information and illustrations concerning the various genera of bacteria.

Chapter 16

The World of Bacteria IV: Gram-Positive Filamentous Bacteria of Complex Morphology

Volume 4 of *Bergey's Manual* is devoted entirely to certain aerobic, Gram-positive bacteria which form structures similar to those that are characteristically found in the microscopic eucaryotic fungi. These structures include (1) a **mycelium** composed of branched filamentous hyphae, these generally being of much smaller diameter (approximately 1 μm) than the hyphae formed by microscopic fungi (5 to 10 μm in diameter), and (2) asexual spores, which are termed **conidiospores** (or simply **conidia**) if they are naked, or **sporangiospores** if they are enclosed within a specialized sac **(sporangium).** The spores represent a major means of reproduction, since they are produced in large numbers, each capable of giving rise to a new organism. Although they are not heat-resistant, the spores are resistant to desiccation and aid survival of the species during periods of drought. The Gram-positive filamentous bacteria are mainly harmless soil organisms, although a few are pathogenic for humans, other animals, or plants. In the soil they are saprophytic and chemoorganotrophic, and they have the important function of degrading plant or animal residues; however, some are best known for their ability to produce a wide range of antibiotics useful in treating human disease. One genus fixes N_2 symbiotically in woody plants. In this chapter we shall consider some of the genera included in the fourth volume of *Bergey's Manual.*

BERGEY'S MANUAL OF SYSTEMATIC BACTERIOLOGY, VOLUME 4

Table 16-1 lists the major groups of bacteria included in Volume 4 of the *Manual.* The reader should also note that some bacteria, the nocardioforms, are included in Volume 2 of the *Manual* rather than Volume 4 and have been described previously in this textbook (see Chap. 14). The dividing line between the no-

Table 16-1. Gram-Positive Filamentous Bacteria Included in *Bergey's Manual*, Volume 4

Section	Some Major Characteristics
FILAMENTOUS BACTERIA THAT DIVIDE IN MORE THAN ONE PLANE	The hyphae divide not only transversely but also longitudinally to produce clusters or packets of cells or spores; cell-wall type III*; soil organisms, animal pathogens, and symbiotic nitrogen-fixers are represented
FILAMENTOUS BACTERIA THAT FORM TRUE SPORANGIA	Harmless soil and water organisms whose hyphae divide in a single plane; the spores are formed within special sacs; cell-wall types II or III*
STREPTOMYCES AND SIMILAR GENERA	The hyphae divide in a single plane; long chains of conidiospores are formed at the tips of sporogenic hyphae; the organisms are mainly harmless soil organisms that are noted for production of antibiotics; a few are human or plant pathogens; cell-wall type I*
ADDITIONAL FILAMENTOUS BACTERIA HAVING UNCERTAIN TAXONOMIC PLACEMENT	A heterogeneous collection of organisms whose relationships to the major groups of Gram-positive filamentous bacteria is not yet agreed upon; some have remarkable morphological or physiological properties; a few organisms are pathogenic for humans; the cell-wall types vary

* For cell-wall types, see Table 16-2.

cardioforms and the bacteria described in the present chapter is indistinct and controversial, and is based upon such factors as (1) whether or not the hyphae can reproduce by fragmentation and (2) the chemical composition of the cells, particularly the walls. In general, the organisms described in Volume 4 of the *Manual* do not undergo fragmentation of the mycelium. With regard to chemical composition, Table 16-2 indicates the major patterns that occur among the Gram-positive filamentous bacteria with regard to cell-wall amino acid composition

Table 16-2. Amino Acid and Sugar Patterns of Aerobic, Gram-Positive, Filamentous Bacteria

Composition of Peptidoglycan	Amino Acid Pattern Designations for Cell-Wall Types			
	I	II	III	IV*
Optical isomer of diaminopimelic acid present	LL	*meso*	*meso*	*meso*
Presence of glycine interpeptide bridges linking the tetrapeptides	+	+	−	−
Sugars Present in Hydrolysates of Whole Cells	Sugar Pattern Designations for Cell Types			
	A	B	C	D
Arabinose	+	−	−	+
Galactose	+	−	−	−
Xylose	−	−	−	+
Madurose	−	+	−	−

*Type IV is distinguished from Type III by the occurrence of the sugars arabinose and galactose in the cell walls of Type IV.

and to the sugars present in whole cell hydrolysates; these differences, as well as differences in DNA base composition (mol % G + C values; see Chap. 3) and in the phospholipid composition of the cell membranes, have been used extensively to classify these organisms, in addition to the morphological features. For example, the genus *Streptomyces* has a type I cell-wall amino acid pattern and a type C sugar pattern (i.e., chemotype I, C), whereas *Dermatophilus* has a type III amino acid pattern and a type B sugar pattern (chemotype III, B). The bacteria included in Volume 4 of *Bergey's Manual* are mainly of the following types: I, C; II, D; III, B; or III, C, whereas the nocardioforms included in Volume 2 of the *Manual* are mainly type 4, A. Moreover, the organisms included in Volume 4 do not possess mycolic acids in their cell walls, whereas some nocardioforms do have these lipids.

FILAMENTOUS BACTERIA THAT DIVIDE IN MORE THAN ONE PLANE

The organisms in this group have hyphae that divide not only transversely but also longitudinally to produce clusters or packets of coccoid or cuboid cells or spores. One of the three genera, *Geodermatophilus* (chemotype III, C) is a soil organism; the other two genera are quite different, as indicated below.

Dermatophilus

The members of this genus belong to *chemotype III, B*. The narrow tapering hyphal filaments develop septa that are formed in transverse and in horizontal and vertical planes; this results in the formation of mulberrylike clusters of cocci (see Fig. 16-1). Each coccus is released as a motile spore bearing a tuft of flagella. The single species of the genus, *D. congolensis,* is not a soil organism; rather, it is a parasite and pathogen of wild and domestic mammals, causing a skin infection (streptothrichosis).

Frankia

The members of this genus belong to *Chemotype III, D*. Like *Rhizobium* species

Figure 16-1. Phase-contrast photomicrographs of *Dermatophilus congolensis* suspended in a 28% albumin solution to reveal cellular detail (X1,100, approx.). (A) Branching hyphae in early stages of division that is mainly transverse, although a few longitudinal septa are already evident (arrow). (B) Mulberrylike clusters of cocci enveloped in mucoid capsular material that appears light against the denser, more refractive, albumin solution. [*Courtesy of D. S. Roberts, in M. P. Starr, H. Stolp, H. G. Trüper, A. Balows, and H. G. Schlegel (eds.), The Prokaryotes: A Handbook on Habitats, Isolation, and Identification of Bacteria, Springer-Verlag, New York, 1981, p. 2011.*]

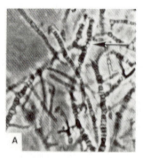 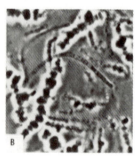

Figure 16-2. (A) Scanning electron micrograph of a large sporogenic body of *Frankia* strain AvcII from broth culture; thickened spore-bearing hypha (arrow) attaches the spore-bearing body to a vegetative hypha. Bar indicates 10 μm. (*Courtesy of D. Baker and J. G. Torrey, Can J Microbiol* **26**:*1066, 1980.*) (B) Scanning electron micrograph illustrating numerous *Frankia* vesicles and hyphae within the infected cell of a root nodule of *Elaeagnus umbellata.* Bar indicates 10 μm. (*Courtesy of D. Baker, W. Newcomb, and J. G. Torrey, Can J Microbiol,* **26**:*1072, 1980.*)

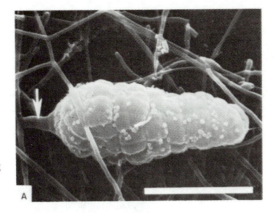
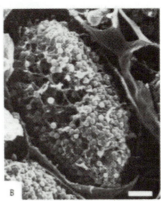

(see Chap. 13), these bacteria are highly efficient microaerophilic N_2-fixers that occur within the root nodules of plants. Unlike *Rhizobium* species, however, they infect nonleguminous woody plants, such as alders. Isolation of the symbiont defied all efforts for years, but in 1978–1980 cultures were isolated by means of elaborate procedures such as (1) microdissection of nodules to remove the bacterial clusters, or (2) sucrose gradient centrifugation of nodule suspensions, which allowed separation of the organisms from extraneous material on the basis of differences in density. Both within root nodules and in laboratory culture, *Frankia* strains form branching hyphae from the ends of which are produced *vesicles* (globular structures on short stalks; see Fig. 16-2); circumstantial evidence suggests that these may be the sites of N_2 fixation. Also produced are *sporogenic bodies* (often called sporangia, but whether they are true sporangia is uncertain) which contain masses of spores that are cemented together (Fig. 16-2). The spores arise by hyphal divisions that occur in several planes.

FILAMENTOUS BACTERIA THAT FORM TRUE SPORANGIA

The members of this group form an extensive substrate mycelium; some genera also form an aerial mycelium. In either instance spores are formed within a sac called a *sporangium*, which is formed aerially at the tip of a sporogenic hypha (*sporangiophore*).

The sporangium may be either of two major kinds, depending on the genus: (1) round, rod-shaped, or irregular, containing *masses of spores*, or (2) fingerlike, club-shaped, or pear-shaped, containing *one to four spores arranged linearly* within the sporangium. The wall of the sporangium is derived from an expansion of the outer layer of the bilayered wall of the hypha, forming a sac. The inner hypha grows into this sac and subsequently undergoes septation to form spores. Some examples of genera forming multispored sporangia are illustrated in Figs. 16-3 and 16-4; in some instances the spores are motile by means of flagella and can swim away from ruptured sporangia (see Fig. 16-5). *Chemotypes of sporangia-forming bacteria are either III, B or II, D.* Sporangia-forming bacteria are widespread in nature, occurring in humus-containing soils, dead plant parts such as pollen and leaves, or on shed animal material such as hair.

Figure 16-3. Semidiagram-matic drawings of three genera of bacteria that form multispored sporangia. The bacteria are growing on the pollen of *Liquidambar styraciflua*. The pollen floats at the surface of the water, the hyphae and sporangiophores emerging into the air through the small circular pits in the wall of the pollen grains. (A) *Actinoplanes*. (B) *Ampullariella*. (C) *Spirillospora* (X350). (From McGraw-Hill Encyclopedia of Science and Technology, vol. 1, 1968.)

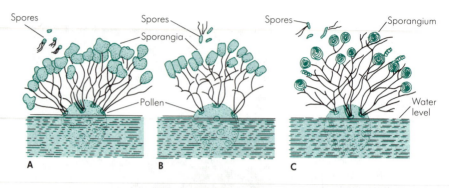

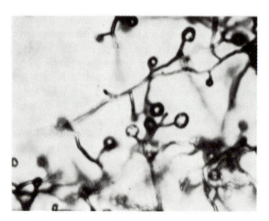

Figure 16-4. *Streptosporangium roseum*, showing sporangia, containing nonmotile spores, on aerial hyphae. The sporangia range from 7 to 19 μm in diameter, usually 8 to 9 μm. (Courtesy of Mary P. Lechevalier.)

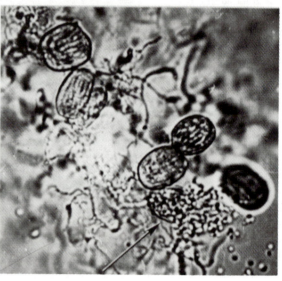

Figure 16-5. Sporangia of *Actinoplanes rectilineatus* in a water mount showing spores in longitudinal rows within the sporangia and one ruptured sporangium (arrow) with spores swimming away (X970). (Courtesy of M. P. Lechevalier and H. A. Lechevalier, Int J Syst Bacteriol, **25**:371, 1975.)

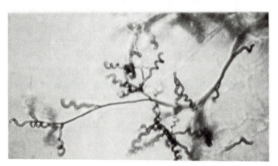

Figure 16-6. *Streptomyces viridochromogenes* showing spiral chains of spores on aerial mycelium. Specimen from culture grown on yeast extract–malt extract medium for 7 days at 25°C (X450). (Courtesy of Tom Cross.)

Species of the genus *Actinoplanes* can polymerize amino acids to form polypeptide and depsipeptide antibiotics and can also synthesize polyene-type macrolides, glycolipid, and aromatic polycyclic antibiotics.

STREPTOMYCES AND SIMILAR GENERA

The members of this group belong to *chemotype I, C*. They usually form an abundant aerial mycelium bearing long chains of conidiospores (5 to 50 or more per chain; see Fig. 16-6). Unlike the hyphae of the substrate mycelium (i.e., submerged below the surface of the medium), the aerial hyphae possess an extra cell-wall layer (sheath). The hyphal tip undergoes septation within this sheath to form a chain of conidia (see Fig. 16-7).

Several genera occur within the group, including *Streptoverticillium, Actinopycnidium, Actinosporangium, Chainia, Elytrosporangium, Kitasatoa,* and *Microellobosporia,* but the most familiar genus is *Streptomyces.* Indeed, the other genera—except for *Streptoverticillium*—may deserve to be merged with *Streptomyces* to form a single genus. Species and subspecies are differentiated by the following characteristics:

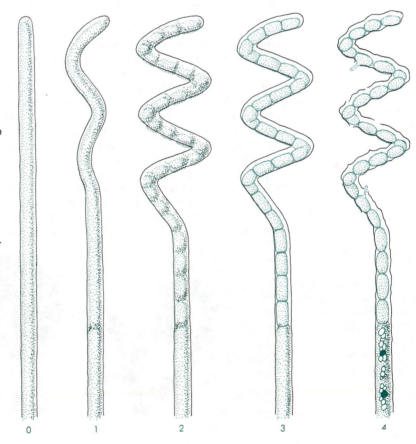

Figure 16-7. Diagram of sporulation stages in *Streptomyces coelicolor.* After a phase of growth (0) the sporulating hyphae are divided into long cells by ordinary cross walls, and the tips begin to coil (1). The apex is then partitioned into spore-sized compartments by sporulation septa (2). The cell walls thicken and constrictions appear between the young spores (3). As spores mature, they round off and separate (4). Some spores begin to germinate immediately after maturation. *(Redrawn from H. Wildermuth and D. A. Hopwood, J Gen Microbiol,* **60:***51–59, 1970; by permission.)*

0   1   2   3   4

1 *Morphology of spore chain.* The chains may be straight, flexuous (wavy), or coiled to various degrees. They also vary in the number of spores per chain.
2 *Spore surface.* By electron microscopy, the surface of the conidia may be smooth, warty, spiny, or hairy, the texture depending on the kind of adherent sheath material (for example, see Fig. 16-8).
3 *Color of aerial mycelium.* The pigmentation of the aerial mycelium falls within any of seven color groups: white, yellow, violet, red, blue, green, or gray. These terms cannot precisely describe the many hues that can occur; therefore, a system of published color standards is used for this purpose.
4 *Color of substrate mycelium.* This may differ from that of the aerial mycelium and is determined by observing the reverse side of the growth after removal of most of the agar medium with a razor blade.
5 *Color of the medium.* Many streptomycetes may form pigments that are excreted into the medium. These may be water-soluble, or they may precipitate in the medium close to the cells.
6 *Physiological characteristics.* This refers to characteristics such as utilization of various carbohydrates and organic acids, nitrate reduction, urea and esculin hydrolysis.

No other genus in bacteriology has been divided into as many species as has *Streptomyces.* The names of nearly 340 species and 39 subspecies are officially recognized, but thousands of additional names have been unofficially assigned (mostly in literature dealing with patents; see below). It is likely that DNA homology and other techniques of molecular biology will eventually indicate that a much smaller number of species exists than is represented by these names. Why then have so many names been used? The main reason rests with the ability of streptomycetes to make a *great number and variety of antibiotics* (see Table 16-3 for a few examples). As a result of the screening of soil samples by pharmaceutical companies to obtain new antibiotics, thousands of new antibiotics have been discovered, and the great majority have been made by streptomycetes. The discoverer of a new antibiotic had to have a name for the organism producing it in order to meet scientific publication or patent requirements. The organism might have been identified as belonging to an established species; however, many isolates failed to agree in all details with the description of any established species. Although some variation in characteristics does occur among the strains of any bacterial species, an easier (but much less satisfactory) solution to such a taxonomic problem was simply to give the isolate a new

Figure 16-8. Hairy conidia of *Streptomyces acrimycini* as seen by (A) transmission electron microscopy and (B) scanning electron microscopy. The bar indicates 1 μm. *(Courtesy of A. Dietz and J. Mathews, Int J Syst Bacteriol,* **27**:282, 1977.)

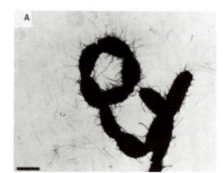
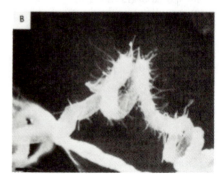

Table 16-3. Some Anti-biotics Made by *Strepto-myces* Species

Antibiotic*	Species
Amphotericin B	*S. nodosus* and others
Chloramphenicol	*S. venezuelae* and others
Chlortetracycline	*S. aureofaciens* and others
Cycloserine	*S. garyphalus* and others
Erythromycin	*S. erythraeus*
Kanamycin	*S. kanamyceticus*
Neomycin	*S. fradiae* and others
Nystatin	*S. noursei* and others
Oxytetracycline	*S. rimosus* and others
Streptomycin	*S. griseus* and others
Tetracycline	*S. viridifaciens* and others

* The modes of action of these antibiotics are described in Chap. 24.

Table 16-4. Filamentous Bacteria Having Uncertain Taxonomic Placement

Genus	Chemotype	Spores	Major Characteristics
Actinomadura	III, B	Short chains of conidia formed on aerial hyphae	Mainly soil saprophytes, but can cause actinomycetoma in humans
Nocardiopsis	III, C	Long chains of conidia formed on aerial hyphae	As in *Nocardia* (Chap. 14), the substrate mycelia undergo fragmentation; unlike *Nocardia*, mycolic acids are absent and the cell wall chemotype is different; soil saprophytes, but occasionally isolated from human or animal infection
Actinopolyspora	IV, A	Long chains of conidia formed on aerial hyphae.	Extreme halophiles: growth occurs only in the presence of 10–30% NaCl; isolated from dairy salt
Actinosynnema	III; galactose and mannose are present	Chains of rodshaped conidia are formed on aerial hyphae	Form compacted groups of erect hyphae called *synnemata*; the hyphae are often fused to form a stemlike structure which bears chains of conidia; isolated from grass
Thermomonospora	III, B or C	Single conidia formed at the tips of very short sporogenic hyphae	Some grow best at 40–45°C, others at 35–40°C; they occur in high temperature habitats such as composts and can decompose cellulose
Thermoactinomyces	III, B or C	Single spores formed either at the tips of short sporogenic hyphae or are sessile (directly attached to vegetative hyphae)	The spores are endospores and survive 95°C for 10 min, are highly refractile, and contain dipicolinic acid; the organisms grow best at about 50°C; they occur in high-temperature habitats such as damp hay, composts and piles of moist grain
Sporichthya	I	Forms upright aerial chains of conidia	Strikingly different from all other filamentous bacteria in that aerial hyphae are formed *but no substrate hyphae*; a single cell attaches to the agar surface and divides to form an upright chain of conidia, each with a basal, collarlike structure

species name. Thus many species came to be distinguished from one another on the basis of only minor differences.

Most streptomycetes are harmless saprophytes occurring mainly in soils of neutral pH, although some species prefer acidic or alkaline soil. They can grow in soils having less water content than that needed for most other bacteria, and because of their spores they can survive well in dry soil. Many streptomycetes can degrade polymeric organic substances in soil that are refractory to being

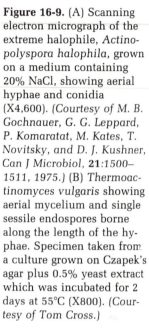
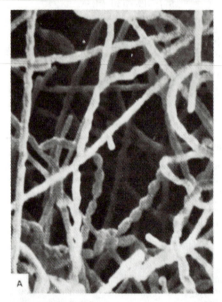
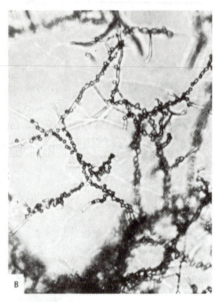

Figure 16-9. (A) Scanning electron micrograph of the extreme halophile, *Actinopolyspora halophila*, grown on a medium containing 20% NaCl, showing aerial hyphae and conidia (X4,600). *(Courtesy of M. B. Gochnauer, G. G. Leppard, P. Komaratat, M. Kates, T. Novitsky, and D. J. Kushner, Can J Microbiol, 21:1500–1511, 1975.)* (B) Thermoactinomyces vulgaris showing aerial mycelium and single sessile endospores borne along the length of the hyphae. Specimen taken from a culture grown on Czapek's agar plus 0.5% yeast extract which was incubated for 2 days at 55°C (X800). *(Courtesy of Tom Cross.)*

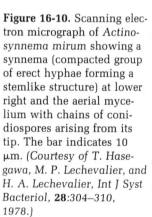
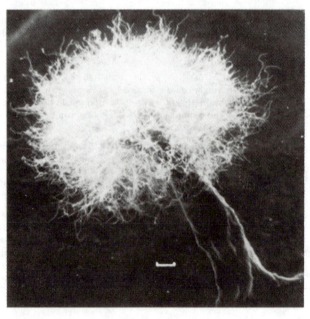

Figure 16-10. Scanning electron micrograph of *Actinosynnema mirum* showing a synnema (compacted group of erect hyphae forming a stemlike structure) at lower right and the aerial mycelium with chains of conidiospores arising from its tip. The bar indicates 10 μm. *(Courtesy of T. Hasegawa, M. P. Lechevalier, and H. A. Lechevalier, Int J Syst Bacteriol, 28:304–310, 1978.)*

Figure 16-11. (A) *Thermo-monospora mesophila,* showing monospores on the aerial hyphae. The spores are about 1 μm in diameter. *(Courtesy of Mary P. Lechevalier.)* (B) Scanning electron micrograph of *Sporichthya polymorpha* showing a general view of developing upright cells. This genus grows only at the surface of the medium and forms no substrate mycelium (X3,900). *(Courtesy of S. T. Williams, J Gen Microbiol,* **62**:*67–73, 1970.)*

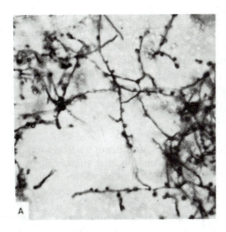 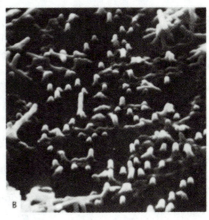

decomposed by many other microorganisms, e.g., starch, pectin, and chitin. A few species are plant pathogens, the most important of these being *Streptomyces scabies,* the causative agent of "common scab" in potatoes and sugar beets. One species, *Streptomyces somaliensis,* is pathogenic for humans, causing actinomycetoma. Nonstreptomycetes can also cause actinomycetoma: see the genera *Nocardia* (Chap. 14) and *Actinomadura* (Table 16-4).

ADDITIONAL FILAMENTOUS BACTERIA HAVING UNCERTAIN TAXONOMIC PLACEMENT

The taxonomic placement of the bacteria of this heterogeneous group is not yet agreed upon. Many of the organisms have unusual and striking morphological or physiological characteristics, such as the extreme halophilism exhibited by *Actinopolyspora* (Fig. 16-9A), formation of heat-resistant endospores by *Thermoactinomyces* (Fig. 16-9B), formation of compacted groups of erect hyphae which have a stemlike appearance (*synnemata;* singular, synnema) by *Actinosynnema* (Fig. 16-10), the thermophilic and cellulolytic properties of *Thermomonospora* (Fig. 16-11A), and development of upright cells by *Sporichthya* (Fig. 16-11B). The major characteristics of these and other genera are summarized in Table 16-4.

QUESTIONS

1 Besides *Frankia,* what other genera of symbiotic, nodule-inducing nitrogen-fixing bacteria can you list? Are these genera also Gram-positive? (Refer also to Chap. 13.)

2 Distinguish between *Nocardia, Pseudonocardia,* and *Nocardiopsis.* (The first two are described in Chap. 14.)

3 Give the outstanding features of each of the following genera: (a) *Thermoactinomyces,* (b) *Dermatophilus,* (c) *Actinoplanes,* (d) *Actinosynnema,* (e) *Sporichthya.*

4 Why is it that the genus *Streptomyces* contains so many named species?

5 What is responsible for the surface texture (i.e., smooth, warty, hairy, spiny) of the conidia of streptomycetes?

6 The cell walls of many soil fungi contain chitin. Therefore, in addition to the

production of certain antifungal antibiotics, how might streptomycetes be able to compete successfully with fungi when growing in soil?

7 Despite its extreme halophilic nature, *Actinopolyspora* belongs to the eubacteria and not to the archaeobacteria (in contrast, for example, to *Halobacterium*). What sort of evidence might have led to this conclusion?

8 Define the following terms: synnemata, conidia, sporangiophore, sessile spore, sporangium, chemotype.

REFERENCES

Okami, Y.: "Antibiotics produced by Actinomycetes," in A. I. Laskin and H. A. Lechevalier (eds.), *Handbook of Microbiology*, Vol. III, CRC Press, Cleveland, pp. 717–972, 1973. *A compilation of the antibiotics produced by filamentous bacteria, the species producing them, and the activity of the antibiotics against various microorganisms.*

Starr, M. P., H. Stolp, H. G. Trüper, A. Balows, and H. G. Schlegel (eds.): *The Prokaryotes: A Handbook of Habitats, Isolation, and Identification of Bacteria*, Springer-Verlag, New York, 1981. *Volume II of this monumental reference work provides a wealth of descriptive information and illustrations concerning the various genera of Gram-positive filamentous bacteria.*

Williams, S. T. (ed.): *Bergey's Manual of Systematic Bacteriology*, vol. 4, Williams & Wilkins, Baltimore (in preparation). *Volume 4 of this international work is presently being prepared and will provide detailed descriptions of the genera and species of the bacteria discussed in Chap. 16.*

PART FIVE

MICROORGANSIMS—FUNGI, ALGAE, PROTOZOA, AND VIRUSES

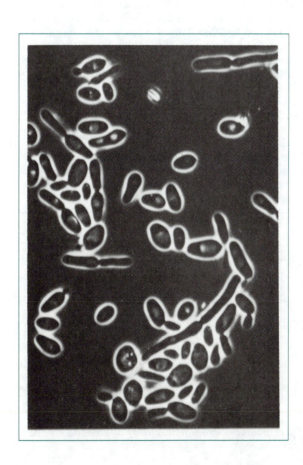

The many kinds of microbes

Professor John O. Corliss of the University of Maryland is recognized for his scholarly knowledge of the protozoa. His recent contribution entitled "A Puddle of Protists" (in *The Sciences*, New York Acadamy of Science, May–June, 1983) is recommended reading to introduce this part of the text. And so in the discourse that follows, we borrow freely from his article the historical development of human knowledge on the protists and other "wee beasties."

It all began in the nineteenth century when the German evolutionist Ernst Haeckel proposed a third biological kingdom, the Protista, to include all the many microbes, large and small, especially those which could not be neatly classified with either the *plants* or the *animals*. While this was a convenient taxonomic device, it had its own problems because the microbes assembled together included members which were fundamentally dissimilar. For instance, the bacteria, the protozoa, the algae, and the fungi were all lumped together. This is not to say that there was no common thread running through them all; indeed, to be counted as a protist, an organism must be unicellular in at least one stage of its life history and must at no stage develop organized tissues. Nonetheless, Haeckel's revolutionary proposal essentially fell flat and has laid dormant for all these years.

After Haeckel, the next breakthrough in the classification of the largely unicellular forms came a century later. Roger Y. Stanier and his colleagues at the University of California at Berkeley proposed two assemblages of organisms based mainly on the type of nucleus present within their cells. These two kingdoms of organisms are called the *procaryotes* (prenucleated cells) and the *eucaryotes* (cells with a true nucleus). The number of their differential characteristics (some of which we have propounded in the beginning of this text) has grown as biologists probe into the molecular properties of the cells of these two groups. Such findings have strengthened the major dichotomy between them.

Very recently, Carl Woese, of the University of Illinois, and other molecular geneticists working primarily with bacteria, have proposed that the procaryotes themselves should be split into two groups. This split results in three kingdoms of organisms: the **Eucaryota,** the **Archaeobacteria** of Woese, and the rest of the **Procaryota.** The archaeobacteria are distinguished from the other procaryotes by the sequence of bases in their ribosomal RNA, the chemical composition of their lipids, and their cell wall structure. They constitute a group of procaryotes that are found in specialized or extreme habitats—the hot acidic niches of the thermoacidophilic bacteria, the nearly saturated salt solutions of the extreme halophilic bacteria, and the highly anaerobic environment required for the growth of the methane-producing bacteria.

In this part of the text we shall look at the world of the eucaryotic protists—the fungi (molds and yeasts), the algae, the protozoa—as well as that group of taxonomically elusive infectious entities which we call the *viruses* and which we are hard pressed to know where to place in the evolved classification schemes for all the life forms.

Preceding page. The yeast *Saccharomyces uvarum. (Courtesy of I. Benda and Avi Publishing Co., Inc.)*

Chapter 17   Fungi—Molds and Yeasts

The **fungi** (singular, **fungus**) are a group of eucaryotic organisms that are of great practical and scientific interest to microbiologists. One good reason for this is obviously that many fungi are of microscopic cellular dimensions. The gross appearance of many multicellular fungi is familiar to each and every one of us. We have seen the velvety blue-and-green growth on rotting oranges and lemons as well as on stale cheeses, the whitish-gray furry outgrowths on bread and jam, and the mushrooms in the fields. These are the bodies of various fungi. Thus fungi have a diversity of morphological appearances, depending on the species. Fungi comprise the molds and yeasts. Whereas molds are filamentous and multicellular, yeasts are usually unicellular. Fungi are eucaryotic spore-bearing protists that lack chlorophyll. They generally reproduce both sexually and asexually.

THE IMPORTANCE OF FUNGI

The fungi are heterotrophic organisms—they require organic compounds for nutrition. When they feed on dead organic matter, they are known as **saprophytes.** Saprophytes decompose complex plant and animal remains, breaking them down into simpler chemical substances that are returned to the soil, thereby increasing its fertility. Thus they can be quite beneficial to humans. But they can also be undesirable when they decompose timber, textiles, food, and other materials.

Saprophytic fungi are also important in industrial fermentations: for example, the brewing of beer, the making of wine, and the production of antibiotics such as penicillin. The leavening of dough and the ripening of some cheeses also depend on fungal activity.

As **parasites** (i.e., when living in or on another organism), fungi cause diseases in plants, humans, and other animals. Although fungal diseases are less commonly encountered than bacterial or virus diseases in humans and other animals, they are of great importance in causing diseases of plants.

Quite apart from the applied aspects of the study of fungi, these microorganisms are to be studied in their own right as biological entities. The science or study of fungi is called **mycology.** Fungi have also become tools for the physiologist, biophysicist, geneticist, and biochemist, who find them highly suitable subjects for the study of some biological processes.

DISTINGUISHING CHARACTERISTICS OF FUNGI

Fungi are eucaryotic chemoorganotrophic organisms that have no chlorophyll. The **thallus** (plural, **thalli**), or body of a fungus may consist of a single cell as in the yeasts; more typically the thallus consists of filaments, 5 to 10 μm across, which are commonly branched. The yeast cell or mold filament is surrounded by a true cell wall (the exception being the slime molds, which have a thallus

Figure 17-1. Actively budding yeasts produce mature daughter cells in about 30 min. These photographs were taken at 10-min intervals. The typical unicellular nature of yeast cells is shown. (*Courtesy of Carl C. Lindegren.*)

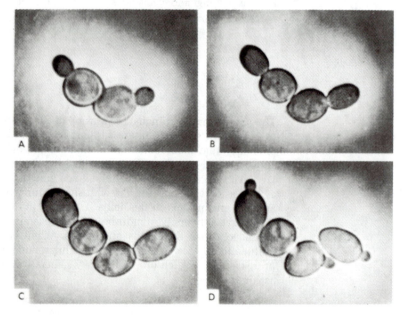

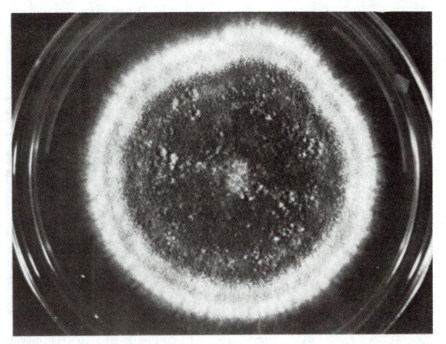

Figure 17-2. A mold colony growing in a Petri dish. Note the filamentous growth of the organism. The powdery appearance is due to the presence of thousands of asexual spores, or conidia. The species shown belongs to the genus *Penicillium,* the same genus of mold that produces the antibiotic penicillin.

consisting of a naked amoeboid mass of protoplasm). Fig. 17-1 shows the typical morphology of a yeast cell.

Some fungi are **dimorphic;** that is, they exist in two forms. Some pathogenic fungi of humans and other animals have a unicellular and yeastlike form in their host, but when growing saprophytically in soil or on a laboratory medium they have a filamentous mold form. The laboratory identification of such fungal pathogens is often dependent on the demonstration of dimorphism. The opposite dimorphic phenomenon occurs in some plant pathogens. In *Taphrina* (which causes peach leaf curl) or in smuts (which cause diseases of cereal crops), the mycelial form occurs in the host and the unicellular yeastlike form occurs in laboratory culture.

Thus a fungal colony may be a mass of yeast cells (not unlike a bacterial colony except for surface texture), or it may be a filamentous mat of mold as shown in Fig. 17-2.

MORPHOLOGY

In general, yeast cells are larger than most bacteria. Yeasts vary considerably in size, ranging from 1 to 5 μm in width and from 5 to 30 μm or more in length. They are commonly egg-shaped, but some are elongated and some spherical. Each species has a characteristic shape, but even in pure culture there is considerable variation in size and shape of individual cells, depending on age and environment. Yeasts have no flagella or other organelles of locomotion.

The thallus of a mold consists essentially of two parts: the **mycelium** (plural, **mycelia**) and the **spores** (resistant, resting, or dormant cells). The diversity of

Figure 17-3. Scanning electron micrographs of *Rhizopus stolonifer* spores at sequential stages of germination, with corresponding phase-contrast micrographs (X1,200) inserted. (A) Ungerminated spore; (B) swollen spore; (C) elongated spore; (D) germ-tube emergence; and (E) germ-tube elongation. *(Courtesy of James L. van Etten, Lee A. Bulla, Jr., and Grant St. Julian, "Physiological and Morphological Correlation of Rhizopus stolonifer Spore Germination," J Bacteriol,* **117***:882–887, 1974.)*

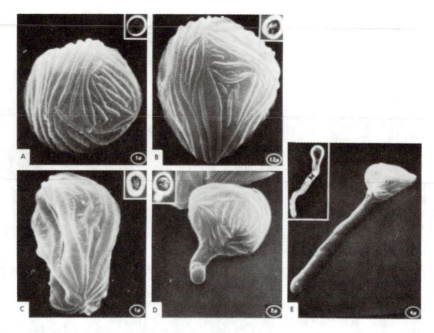

these spores will be discussed later. The mycelium is a complex of several filaments called **hyphae** (singular, **hypha**). New hyphae generally arise from a spore which on germination puts out a germ tube or tubes (Fig. 17-3). These germ tubes elongate and branch to form hyphae.

Each hypha is about 5 to 10 μm wide, as compared with a bacterial cell which is usually 1 μm in diameter. Hyphae are composed of an outer tubelike wall surrounding a cavity, the **lumen,** which is filled or lined by protoplasm. Between the protoplasm and the wall is the **plasmalemma,** a double-layered membrane which surrounds the protoplasm. The hyphal wall consists of microfibrils composed for the most part of hemicelluloses or chitin; true cellulose occurs only in the walls of lower fungi. Wall matrix material in which the microfibrils are embedded consists of proteins, lipids, and other substances.

Growth of a hypha is distal, near the tip. The major region of elongation takes place in the region just behind the tip. The young hypha may become divided into cells by crosswalls which are formed by **centripetal invagination** (inward growth) from the existing cell wall. These crosswalls constrict the plasmalemma and grow inward to form generally an incomplete **septum** (plural, **septa**) that has a **central pore** which allows for protoplasmic streaming. Even nuclei may migrate from cell to cell in the hypha.

Hyphae occur in three forms (Fig. 17-4):

1 Nonseptate, or **coenocytic.** Such hyphae have no septa.
2 Septate with uninucleate cells.
3 Septate with multinucleate cells. Each cell has more than one nucleus in each compartment.

Mycelia can be either vegetative or reproductive. Some hyphae of the vege-

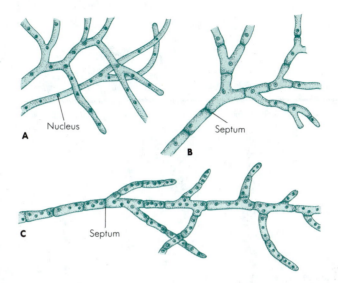

Figure 17-4. Three types of hyphae. (A) Nonseptate (coenocytic), (B) septate with uninucleate cells, (C) septate with multinucleate cells.

tative mycelium penetrate into the medium in order to obtain nutrients; soluble nutrients are absorbed through the walls. (Insoluble nutrients are first digested externally by secreted enzymes.) Reproductive mycelia are responsible for spore production and usually extend from the medium into the air. The mycelium of a mold may be a loosely woven network or it may be an organized, compact structure, as in mushrooms.

REPRODUCTION

Asexual Reproduction

Fungi reproduce naturally by a variety of means. Asexual reproduction (also called **somatic** or **vegetative reproduction**) does not involve the union of nuclei, sex cells, or sex organs. It may be accomplished by (1) fission of somatic cells yielding two similar daughter cells; (2) budding of somatic cells or spores, each bud a small outgrowth of the parent cell developing into a new individual; (3) fragmentation or disjointing of the hyphal cells, each fragment becoming a new organism; or (4) spore formation.

Asexual spores, whose function is to disseminate the species, are produced in large numbers. There are many kinds of asexual spores (Fig. 17-5):

1 *Sporangiospores.* These single-celled spores are formed within sacs called **sporangia** (singular, **sporangium**) at the end of special hyphae (**sporangiophores**). **Aplanospores** are nonmotile sporangiospores. **Zoospores** are motile sporangiospores, their motility being due to the presence of flagella.
2 *Conidiospores* or **conidia** (singular, **conidium**). Small, single-celled conidia are called **microconidia.** Large, multicelled conidia are called **macroconidia.** Conidia are formed at the tip or side of a hypha (see Fig. 17-6).
3 *Oidia* (singular, **oidium**) or **arthrospores.** These single-celled spores are formed by disjointing of hyphal cells. (See Fig. 17-7.)
4 *Chlamydospores.* These thick-walled, single-celled spores are highly resistant to adverse conditions. They are formed from cells of the vegetative hypha.
5 *Blastospores.* These are spores formed by budding.

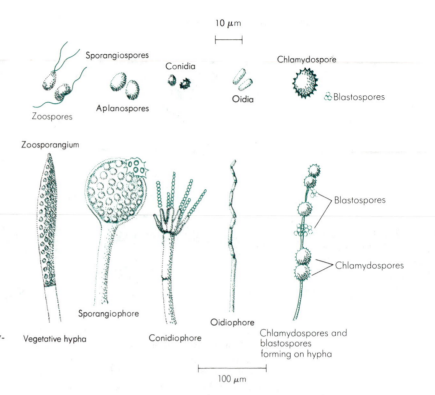

10 μm

Sporangiospores

Conidia

Chlamydospore

Zoospores

Aplanospores

Oidia

Blastospores

Zoosporangium

Blastospores

Chlamydospores

Figure 17-5. Asexual spore types in fungi. *(Redrawn from the McGraw-Hill Encyclopedia of Science and Technology, 1977, vol. 5, p. 117.)*

Vegetative hypha

Sporangiophore

Conidiophore

Oidiophore

Chlamydospores and blastospores forming on hypha

100 μm

Figure 17-6. Conidia are produced in large numbers as exemplified here by a species of *Penicillium* (X1,000). *(Courtesy of Douglas F. Lawson.)*

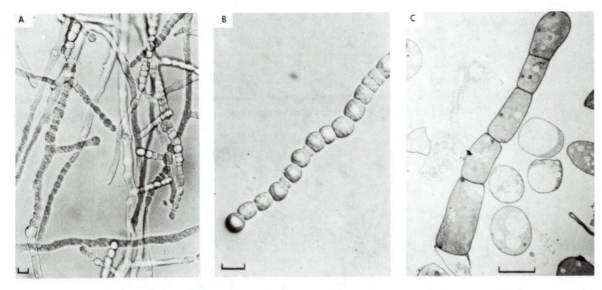

Figure 17-7. Gross appearance of *Mucor rouxii* arthrospores. (A) Light microscopy showing extensive arthrospore formation. (B) High-magnification light microscopy showing variation in arthrospore size and shape. (C) Transmission electron micrograph of arthrospore chain. All bars represent 10 μm. *(Courtesy of C. R. Barrera, New Mexico State University, and the American Society for Microbiology.)*

Sexual Reproduction

Sexual reproduction is carried out by fusion of the compatible nuclei of two parent cells. The process of sexual reproduction begins with the joining of two cells and fusion of their protoplasts (**plasmogamy**), thus enabling the two haploid nuclei of two mating types to fuse together (**karyogamy**) to form a diploid nucleus. This is followed by **meiosis,** which again reduces the number of chromosomes to the haploid number.

The sex organelles of fungi, if they are present, are called **gametangia** (singular, **gametangium**). They may form differentiated *sex cells* (**gametes**) or may contain instead one or more gamete nuclei. If the male and female gametangia are morphologically different, the male gametangium is called the **antheridium** (plural, **antheridia**) and the female gametangium is called the **oogonium** (plural, **oogonia**).

The various methods of sexual reproduction (by which compatible nuclei are brought together in plasmogamy) may be summarized as follows (Fig. 17-8):

1 Gametic copulation. Fusion of naked gametes, one or both of which are motile.
2 Gamete-gametangial copulation. Two gametangia come into contact but do not fuse; the male nucleus migrates through a pore or fertilization tube into the female gametangium.
3 Gametangial copulation. Two gametangia or their protoplasts fuse and give rise to a zygote that develops into a resting spore.
4 Somatic copulation. Fusion of somatic or vegetative cells.

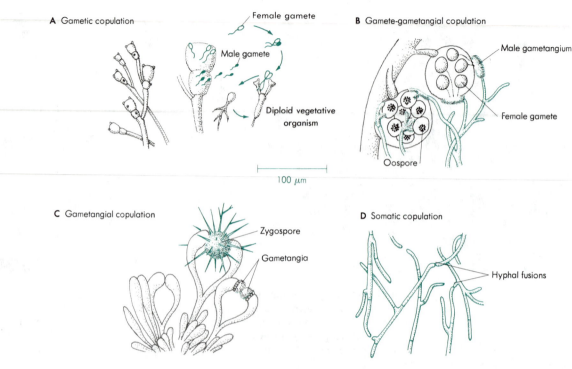

Figure 17-8. Some sexual mechanisms in fungi. (A) Gametic copulation: the fusion in pairs of sexual cells or gametes, formed in specialized sporangialike gametangia; (B) gamete-gametangial copulation: the fusion of a differentiated gamete of one sex with a gametangium of the other sex; (C) gametangial copulation: the direct fusion of gametangia without differentiation of gametes; (D) somatic copulation: the sexual fusion of undifferentiated vegetative cells; (E) spermatization: spermatia uniting with receptive hyphae of the opposite (compatible) strain. *(A–D: From the McGraw-Hill Encyclopedia of Science and Technology, 1971, vol. 5, p. 118.)*

5 Spermatization. Union of a special male structure called a **spermatium** (plural, **spermatia** with a female receptive structure. The spermatium empties its contents into the latter during plasmogamy.

Sexual spores, which are produced by the fusion of two nuclei, occur less frequently, later, and in smaller numbers than do asexual spores. There are several types of sexual spores:

1 *Ascospores.* These single-celled spores are produced in a sac called an **ascus** (plural, **asci**). There are usually eight ascospores in each ascus (Fig. 17-9).

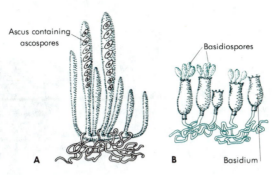

Figure 17-9. Some sexual spore types and the structure of the corresponding reproductive mycelia in fungi. (A) Ascospores. Nuclear fusion takes place in the ascus. The diploid zygote nucleus divides by meiosis almost immediately after fusion, and produces four haploid nuclei. These haploid nuclei divide once more by mitosis, forming the eight ascospores which are typically produced in each ascus. (B) Basidiospores. Nuclear fusion and meiosis take place in the basidium. Basidiospores are then formed exogenously at the tips of the special outgrowths called **sterigmata** (singular, **sterigma**). Usually four spores are formed, one at the tip of each sterigma, but if meiosis is followed by a mitotic division then eight basidial nuclei are formed (although rarely do all develop into basidiospores).

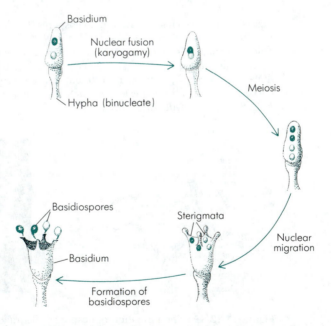

Figure 17-10. The sexual process of basidiospore formation. A basidium begins with one nucleus from each parent. The basidium assumes the shape characteristic of that species and generally produces four tapering projections, the sterigmata. The four nuclei, produced after nuclear fission from meiosis, move toward the sterigmata and form the basidiospores.

2 *Basidiospores.* These single-celled spores are borne on a club-shaped structure called a **basidium** (plural, **basidia**) (Fig. 17-9). Their formation is illustrated in Fig. 17-10).

3 *Zygospores.* Zygospores are large, thick-walled spores formed when the tips of

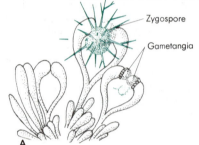

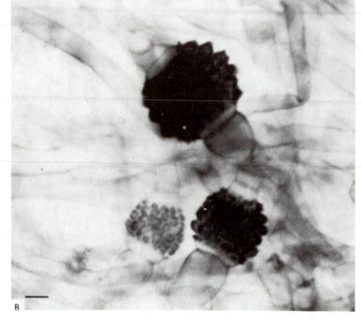

Figure 17-11. (A) The formation of zygospores. *(Redrawn from the McGraw-Hill Encyclopedia of Science and Technology, 1977, vol. 5, p. 118.)* (B) Zygospores in *Mucor hiemalis*. Sexual reproduction in *Mucor hiemalis* occurs when two sexually compatible mating types, + and −, come into contact with each other and produce zygospores. Zygospores of different ages are shown, the oldest one being darkest, largest, and roughest. Bar equals 0.1 mm. *(Courtesy of L. Kapica and E. C. S. Chan, McGill University.)*

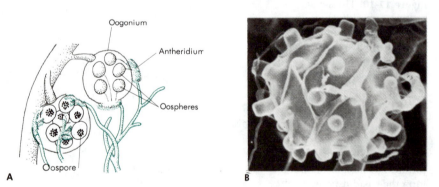

Figure 17-12. (A) The formation of oospores. *(Redrawn from the McGraw-Hill Encyclopedia of Science and Technology, 1977, vol. 5, p. 118.)* (B) Scanning electron micrograph of an oogonium of *Achlya recurva* with appressed antheridia (X780). *(Courtesy of H. E. Huizar, J. T. Ellzey, and W. L. Steffens, The University of Texas at El Paso.)*

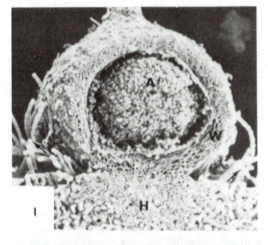

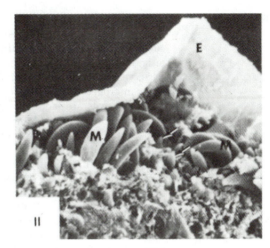

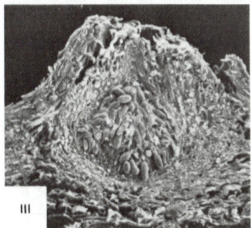

Figure 17-13. Fungal fruiting bodies as seen by scanning electron microscopy. (I) Longitudinal section of a perithecium (ascocarp) of *Ceratocystis fimbriata*. The perithecial wall (W) and hyphae (H) are clearly seen. Other structures seen are ascospores (A) within the perithecial cavity and conidiophores (arrows), which are specialized aerial conidia-bearing hyphae (X250). (II) Cross section of a subepidermal acervulus of *Marsonina juglandis* in black walnut leaf. Mature (M) and immature (arrows) conidia are exhibited. The host epidermis (E) is clearly seen (X1,500). (III) Section of a pycnidium of *Dothiorella ribis* in apple bark tissue, showing conidia compacted in the mucilaginous matrix (X200). *(Courtesy of M. F. Brown and H. G. Brotzman, University of Missouri.)*

two sexually compatible hyphae, or gametangia, of certain fungi fuse together (Fig. 17-11).

4 *Oospores.* These are formed within a special female structure, the **oogonium.** Fertilization of the eggs, or **oospheres,** by male gametes formed in an **antheridium** gives rise to oospores. There are one or more oospheres in each oogonium. This process is illustrated in Fig. 17-12.

Asexual and sexual spores may be surrounded by highly organized protective structures called **fruiting bodies.** Asexual fruiting bodies have names such as **acervulus** and **pycnidium.** Sexual fruiting bodies have names such as **perithecium** and **apothecium.** Some of these structures are shown in Fig. 17-13.

Although a single fungus may produce asexual and sexual spores by several methods at different times and under different conditions, the spores are sufficiently constant in their structures and in the method by which they are produced to be used in identification and classification.

PHYSIOLOGY

Fungi are better able to withstand certain extreme environmental conditions than most other microorganisms. For example, yeasts and molds can grow in a substrate or medium containing concentrations of sugars that inhibit most bacteria; this is why jams and jellies may be spoiled by molds but not by bacteria. Also, yeasts and molds generally can tolerate more acidic conditions than most other microbes.

Some yeasts are facultative; that is, they can grow under both aerobic and anaerobic conditions. Molds and many yeasts are usually aerobic microorganisms. Fungi grow over a wide range of temperature, with the optimum for most saprophytic species from 22 to 30°C; pathogenic species have a higher temperature optimum, generally 30 to 37°C. Some fungi will grow at or near 0°C and thus can cause spoilage of meat and vegetables in cold storage.

Fungi are capable of using a wide variety of materials for nutrition. However, they are heterotrophic. Unlike some bacteria, they cannot use inorganic carbon compounds, such as carbon dioxide, as their sole carbon source. Carbon must come from an organic source, such as glucose. Some species can use inorganic compounds of nitrogen, such as ammonium salts. But all fungi can use organic nitrogen; this is why culture media for fungi usually contain peptone, a hydrolyzed protein product. A summary of the physiological characteristics of fungi in comparison with those of bacteria is found in Table 17-1.

CULTIVATION OF FUNGI

Molds and yeasts can be studied by the same general cultural methods used for bacteria. Nearly all of them grow aerobically on the usual bacteriological culture media at temperatures ranging from 20 to 30°C. Most of them grow more slowly than bacteria, so that media which support bacteria as well as fungi may be overgrown by bacterial contaminants in a mixed inoculum. Where fungi are to be isolated, it is good practice to use a medium that favors their growth but is

Table 17-1. Comparative Physiology of Fungi and Bacteria

Characteristic	Fungi	Bacteria
Cell type	Eucaryotic	Procaryotic
Optimum pH	3.8–5.6	6.5–7.5
Optimum temperature	22–30°C (saprophytes) 30–37°C (parasites)	20–37°C (mesophiles)
Oxygen requirement	Strictly aerobic (molds) Facultative (some yeasts)	Aerobic to anaerobic
Light requirement	None	Some photosynthetic groups occur
Sugar concentration in laboratory media	4–5%	0.5–1%
Carbon requirement	Organic	Inorganic and/or organic
Cell-wall structural components	Chitin, cellulose, or hemicellulose	Peptidoglycan
Antibiotic susceptibility	Resistant to penicillins, tetracyclines, chloramphenicol; sensitive to griseofulvin	Resistant to griseofulvin; sensitive to penicillins, tetracyclines, chloramphenicol

not optimal for the growth of bacteria. Acidic (pH 5.6) media that incorporate a relatively high concentration of sugar are tolerated by molds but are inhibitory to many bacteria.

One of the best-known and oldest media for the growth of fungi was devised by Sabouraud and contains maltose and peptone as its principal ingredients. The most common modification used in America now contains glucose and any one of several specified peptones. This medium is widely used for the isolation of molds and certain yeasts and is especially useful for growing pathogenic fungi from infected body fluids and exudates. Its partial selective action is due to the high sugar concentration and low pH.

Most parts of a mold are potentially capable of growth. Inoculation of a small fragment of mycelium on a medium is sufficient to start a new mold colony. This is done by planting the inoculum on a fresh medium with the aid of a transfer needle, a method similar to that used for bacteria. One difference is that the needle used for molds is stiffer and has a flattened tip for cutting the mycelium. However, an ordinary inoculating loop used for bacteria is suitable for the inoculation of yeasts.

CLASSIFICATION OF FUNGI

The classification of fungi, unlike that of bacteria, is based primarily on the characteristics of the sexual spores and fruiting bodies present during the sexual stages of their life cycles. However, many fungi produce sexual spores and fruiting bodies only under certain environmental conditions, if they are known to produce them at all. Thus the complete or **perfect** life cycles are as yet unknown for many fungi. Consequently, these **imperfectly** described fungi must be classified on bases other than the characteristics of their sexual stages. The morphology of their asexual spores and thalli then becomes important—until such time that their sexual traits are observed, which requires reevaluation of their taxonomic status. Thus imperfect higher fungi are provisionally placed in a special class called form-class *Deuteromycetes*. Most of the sexual stages found subsequently among members of this class have been of the ascomycete type. When sexual stages are found, the organisms are reclassified and placed in with the ascomycetes or basidiomycetes (discussed below).

You should be aware that all is not settled in mycological classification and that differences of opinion on classification are very numerous. Divergent views arise because of our incomplete knowledge on the structure, development, and physiology of fungi. Therefore, as our knowledge of fungi increases, our classification schemes are bound to change.

Taxonomy of the fungi follows the recommendations of the Committee on International Rules of Botanical Nomenclature. Accordingly, the various taxa have endings as follows:

Divisions:	-*mycota*
Subdivisions:	-*mycotina*
Classes:	-*mycetes*
Subclasses:	-*mycetidae*
Orders:	-*ales*
Families:	-*aceae*

Genera and species have no standard endings.

In this chapter, the classification scheme proposed by Alexopoulos, the eminent American mycologist, is used. Its outline is shown in Fig. 17-14. A summary of the major organisms in the kingdom *Fungi* is given in Table 17-2.

Table 17-2. Summary of the Major Organisms of the Kingdom *Fungi* (Modified after Alexopoulos, 1979)

Division/Class	Major Distinguishing Characteristics	Representative Organisms
Division *Gymnomycota* (slime molds)	Organisms which ingest particulate nutrients and which lack cell walls in the vegetative stage	
Class *Acrasiomycetes* (cellular slime molds)	Vegetative stage: free-living amoebae which aggregate to form a stalked sorocarp bearing spores in a mucilaginous matrix (Figs. 17-15 and 17-16)	*Dictyostelium discoideum* *Polysphondelium violaceum*
Class *Myxomycetes* (acellular slime molds)	Vegetative stage: a multicellular, wall-less plasmodium, which transforms into highly organized sporangia bearing sporangiospores (Fig 17-17)	*Physarum polycephalum* *Didimium iridis*
Division *Mastigomycota* (flagellated lower fungi)	Aquatic fungi producing motile, flagellated cells	
Class *Chytridiomycetes*	Motile cells bearing a single, posteriorly positioned, whiplash type flagellum	*Allomyces macrogynus*
Class *Hyphochrytridiomycetes*	Motile cells bearing a single, anteriorly positioned, tinsel-type flagellum	*Rhizidiomyces arbuscula* *Hyphochrytrium catenoides*
Class *Plasmodiophoromycetes*	Obligate parasites on higher plants; vegetative stage a plasmodium; motile cells with two unequal anterior whiplash flagella	*Plasmodiophora brassica*
Class *Oomycetes*	Motile cells with two laterally inserted flagella, one tinsel and anteriorly directed, the other whiplash and posteriorly directed	*Saprolegnia ferax*
Division *Amastigomycota* (terrestrial fungi)	Flagella absent	
Class *Zygomycetes*	Sexual reproduction by gametangial fusion; zygote transformed into a thick-walled resting spore, the *zygospore*; vegetative reproduction by means of mitospores produced within a sporangium (Fig. 17-18)	*Rhizopus stolonifer* *Phycomyces blakesleanus* *Mucor rouxii*
Class *Ascomycetes*	Sexual spores produced endogenously in a saclike ascus typically produced in a well-differentiated ascocarp; vegetative reproduction by conidiospores (Figs. 17-19 and 17-20)	The yeasts; e.g., *Saccharomyces cerevisiae* Morels and truffles *Neurospora crassa*
Class *Basidiomycetes*	Sexual spores produced exogenously on clublike cells called basidia; basidia formed on well-differentiated basidiocarps (Fig. 17-21)	Mushrooms and puffballs Rusts, smuts, and jelly fungi Bracket fungi
Form-Class *Deuteromycetes*	Sexual reproduction absent; vegetative reproduction by means of conidiospores arising from well-defined conidiogenous cells	Molds and mildews Causal agents of ringworm and athlete's foot and many plant diseases *Trichophyton rubra* *Candida albicans* *Alternaria tenuis*

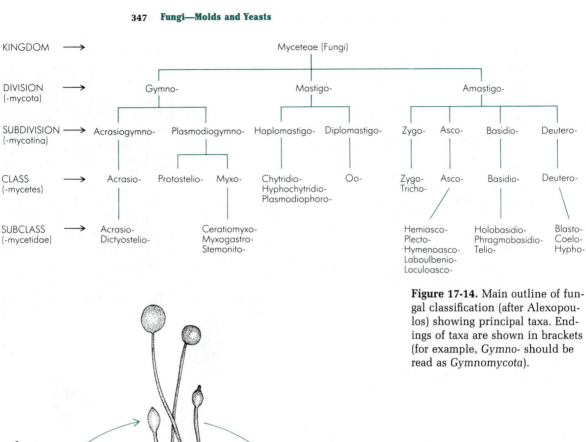

KINGDOM ⟶   Myceteae (Fungi)

DIVISION ⟶   Gymno-   Mastigo-   Amastigo-
(-mycota)

SUBDIVISION ⟶ Acrasiogymno-   Plasmodiogymno-   Haplomastigo-   Diplomastigo-   Zygo-   Asco-   Basidio-   Deutero-
(-mycotina)

CLASS ⟶ Acrasio-   Protostelio-   Myxo-   Chytridio-   Oo-   Zygo-   Asco-   Basidio-   Deutero-
(-mycetes)   Hyphochytridio-   Tricho-
                        Plasmodiophoro-

SUBCLASS ⟶ Acrasio-   Ceratiomyxo-   Hemiasco-   Holobasidio-   Blasto-
(-mycetidae) Dictyostelio-   Myxogastro-   Plecto-   Phragmobasidio-   Coelo-
               Stemonito-   Hymenoasco-   Telio-   Hypho-
                                 Laboulbenio-
                                 Loculoasco-

Figure 17-14. Main outline of fungal classification (after Alexopoulos) showing principal taxa. Endings of taxa are shown in brackets (for example, *Gymno-* should be read as *Gymnomycota*).

Spore

Pseudoplasmodium

Germination

Amoebas

Aggregation

Figure 17-15. The life cycle of the cellular slime mold *Dictyostelium discoideum*. Amoebas migrate into aggregation centers, becoming associated end to end in chains (see Fig. 17-16). They go through several multicellular stages (see Fig. 17-16) to form a fruiting body. The spores disperse to find a suitable environment before they germinate to form amoebas which begin the life cycle anew.

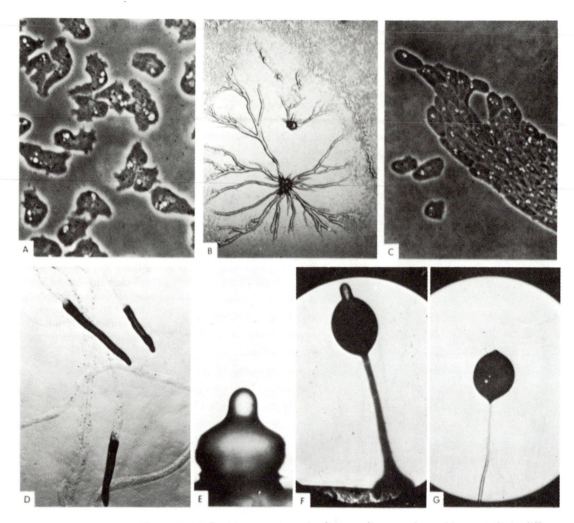

Figure 17-16. Fruiting structures in dictyostelium are formed by an orderly differentiation of myxamoebas, starting with (A) cells which have no regular shape or
orientation (phase-contrast microscopy, X442). In (B) (X11) we see a pseudoplasmodium in a well-advanced stage composed of many myxamoebas which are now
oriented into the stream of flow (C) (phase-contrast microscopy, X300). As the
pseudoplasmodia migrate, they leave trails of slime on the surface of the agar (D)
(X12). Fruiting bodies of the slime molds are called sori (masses of spores). An
early stage in the formation of a fruiting body in dictyostelium is seen in (E),
where the organisms have assumed a vertical orientation (X53). In this figure the
stalk is just beginning to form, but (F) shows the entire sorocarp, including the
basal disk bearing an immature sorus (X42). A mature sorus (X42) is shown in (G).
[*Courtesy of Kenneth B. Raper, Proc Am Phil Soc,* **104**(6):579–604, December
1960.]

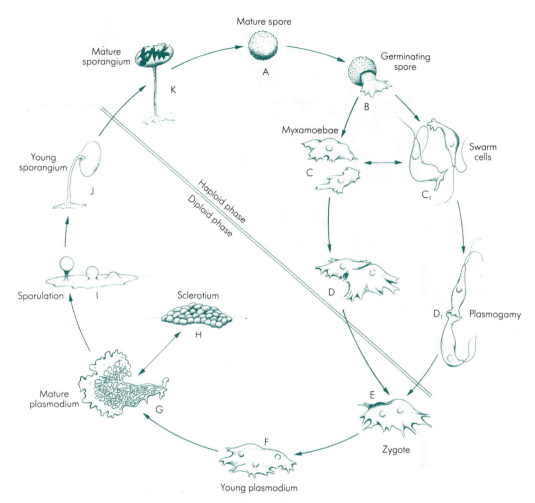

Figure 17-17. Life cycle of a typical myxomycete. (A) Mature haploid spore, (B) germinating spore, (C) myxamoebae, (C₁) swarm cells, (D) fusing myxamoebae, (D₁) fusing swarm cells, (E) young zygote, (F) young plasmodium, (G) mature plasmodium, (H) sclerotium, (I) sporulation—sporangial initials, (J) young premeiotic sporangium with spores, and (K) mature postmeiotic sporangium. *(Erwin F. Lessel, illustrator.)*

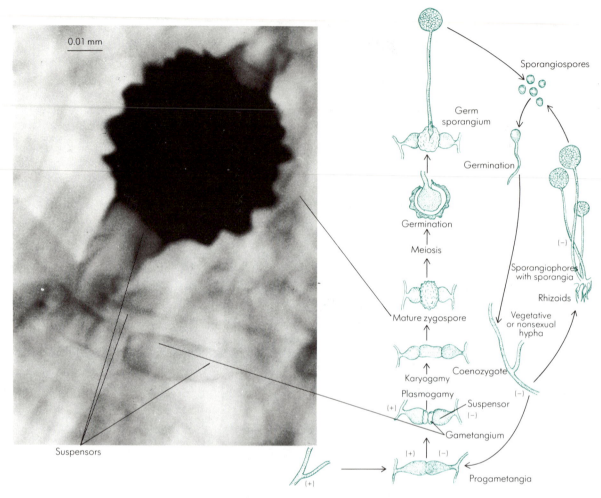

Figure 17-18. Life cycle of *Rhizopus stolonifer*. When the wall of the sporangium disintegrates, sporangiospores are released. A sporangiospore germinates to develop into an organism with many vegetative hyphae. Rhizoids, which penetrate into the medium, are also formed. Directly above the rhizoids, one or more sporangiophores are produced. The top of each sporangiophore develops into a sporangium containing sporangiospores. This completes the asexual portion of the life cycle. Sexual reproduction requires two sexually compatible types (+ and −). When these types come into contact with one another, copulating branches called progametangia are formed. A septum then forms near the tip of each progametangium, separating it into two cells, a terminal gametangium and a suspensor cell. The walls of the two contacting gametangia dissolve at the point of contact, the two protoplasts mix (plasmogamy), and the + and − nuclei fuse (karyogamy) to form many zygote nuclei. The structure which contains them is called the coenozygote. The wall around the coenozygote thickens and its surface becomes black and warty, forming the mature zygospore, which lies dormant for 1 to 3 or more months. The zygospore germinates to form a new organism; meiosis takes place during the germination process. Bar equals 0.01 mm. (*Photomicrograph courtesy of L. Kapica and E. C. S. Chan, McGill University.*)

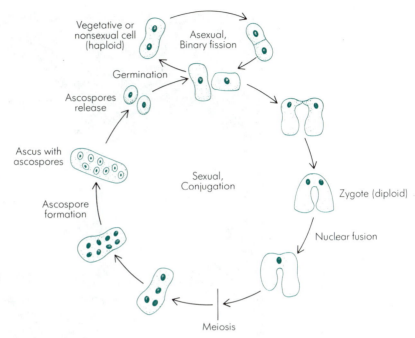

Figure 17-19. Life cycle of the common yeast *Schizosaccharomyces.* Asexual reproduction is by binary fission. Sexual reproduction is by conjugation of compatible cells with the subsequent formation of ascospores. (Haploid: half the number of chromosomes characteristic of a species; diploid: the number of chromosomes characteristic of a species; zygote: a diploid cell resulting from the fusion of two haploid cells.)

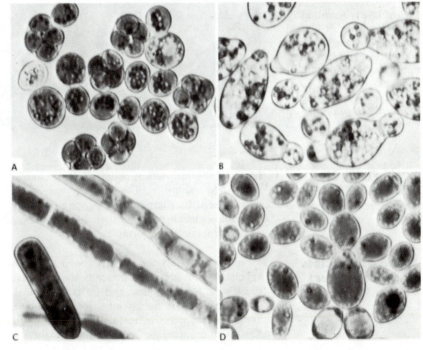

Figure 17-20. The morphology of yeasts varies widely. (A) *Saccharomyces cerevisiae* with cells appearing as vegetative forms, budding cells, and spores (X73). *(Courtesy of George Svihla and with permission of The Microscope and Crystal Front.)* (B) *Saccharomyces ludwigii* (about X95.) *(From George Svihla, Argonne Natl Lab Annu Rep, 1965.)* (C) *Geotrichum candidum* (about X110). (D) *Pichia membranaefaciens* (X88). *[(C) and (D), courtesy of George Svihla.]*

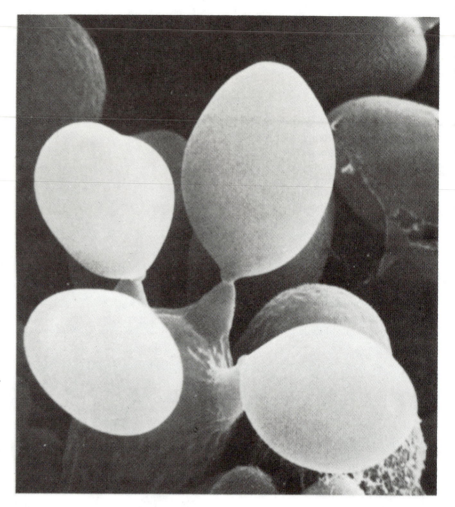

Figure 17-21. Scanning electron micrograph of a basidium bearing four basidiospores (magnification approximately X8,000). *(Courtesy of Stanley F. Flegler, Pesticide Research Center, Michigan State University.)*

SOME FUNGI OF SPECIAL INTEREST

Many of the properties, activities, and characteristics of fungi can best be described by examples. The following are some of the genera most frequently encountered in microbiology. Among them are some of the most interesting to microbiologists because of their unique biological properties or particular economic or medical importance.

Synchytrium (Class Chytridiomycetes)

More than 100 species of this genus are parasitic on flowering plants. Many species are not very destructive to the host plant, but they induce the formation of galls on leaves, stems, and fruits. The most serious parasite is *S. endobioticum*, the cause of black wart disease of potato. The dark warts on the potatoes are galls in which the host cells have been stimulated to divide by the fungus. The resting sporangia are released by the decay of the warts and they may remain alive for many years in the soil. The zoospores released from a sporangium are capable of swimming for about two hours in soil water until a new host tuber is found to infect. The zoospores may also fuse to form zygotes which retain their flagella and swim actively for a time. The zygote then encysts on the surface of the host epidermis and penetrates the host cell to infect it. The results of zygote infection differ from infection by zoospores. When infected by a zoospore, the host cell reacts by undergoing **hypertrophy,** i.e., increase in cell volume, and adjacent cells also enlarge to form the characteristic rosette. When infected by a zygote, the host cell undergoes **hyperplasia,** i.e., repeated cell division resulting in the characteristic wart. Control of wart disease is based largely on the breeding of resistant varieties of potatoes. Figure 17-22 shows the life cycle of *Synchytrium*.

Saprolegnia (Class *Oomycetes*)

Species of *Saprolegnia* are common in soil and fresh water; hence they are commonly called water molds. They are saprophytic on plant and animal remains, but a number of species such as *S. ferax* and *S. parasitica* have been

Figure 17-22. Life cycle of *Synchytrium endobioticum.* (Erwin F. Lessel, illustrator.)

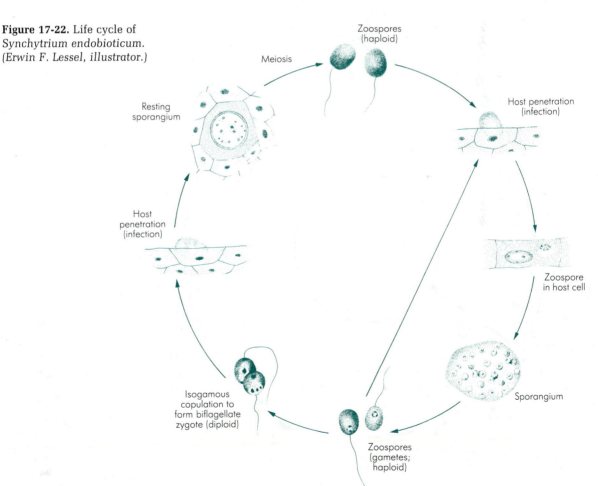

Zoospores (haploid)

Meiosis

Host penetration (infection)

Resting sporangium

Host penetration (infection)

Zoospore in host cell

Sporangium

Isogamous copulation to form biflagellate zygote (diploid)

Zoospores (gametes; haploid)

implicated in diseases of fish and their eggs. *S. parasitica* causes severe epidemics of disease among fish in the natural environment.

The somatic portion of the organism consists of two types of hyphae: the rhizoidal hyphae which enter the substratum and serve to anchor the organism and to absorb nutrients, and the branched hyphae on which the reproductive organs are formed. Elongated, tapering sporangia are formed at the tips of somatic hyphae. The nuclei present differentiate into zoospores. An opening develops at the tip of the sporangium and the pear-shaped primary zoospores escape into the surrounding aqueous environment. They swim for some time (from a minute to over an hour), then withdraw their flagella and encyst. This cyst, after a period of rest (2 to 3 h, depending on the species), germinates to release a further bean-shaped secondary zoospore. The secondary zoospore may

Figure 17-23. Life cycle of *Saprolegnia. (Erwin F. Lessel, illustrator.)*

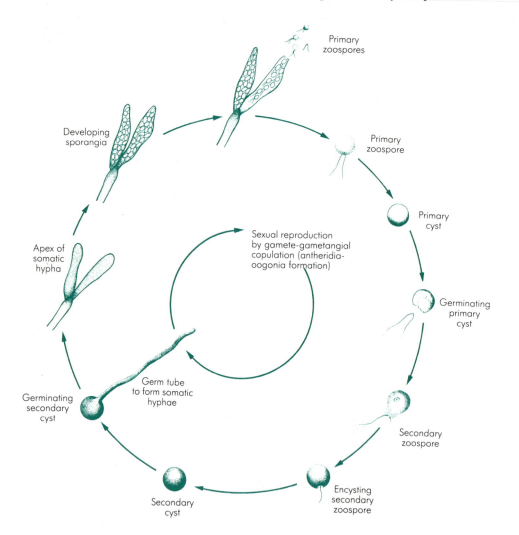

Primary zoospores

Developing sporangia

Primary zoospore

Apex of somatic hypha

Sexual reproduction by gamete-gametangial copulation (antheridia-oogonia formation)

Primary cyst

Germinating primary cyst

Germ tube to form somatic hyphae

Germinating secondary cyst

Secondary zoospore

Secondary cyst

Encysting secondary zoospore

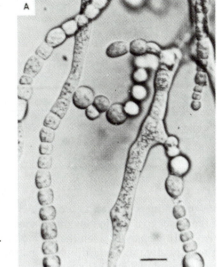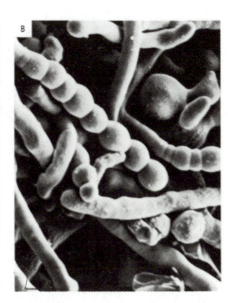

Figure 17-24. Formation of arthrospores (yeastlike forms) by *Mucor rouxii.* (A) As seen by light microscopy. (B) Scanning electron micrograph. Bars equal 10 μm. *(Courtesy of C. R. Barrera, New Mexico State University, and the American Society for Microbiology.)*

swim vigorously for several hours before encysting again. This encysted spore now germinates by sending a germ tube that develops into a hypha, forming a new colony (see Fig. 17-23). When conditions are favorable for sexual reproduction, the somatic hyphae give rise to oogonia and antheridia.

Mucor (Class *Zygomycetes*)

Members of this genus occur abundantly in soil and manure and on fruits, vegetables, and starchy foods. Some are responsible for food spoilage, but others are used in the manufacture of some cheeses and other foods. Common species are *M. racemosus* and *M. rouxii.* Morphologically, their mycelia are usually white or gray and are nonseptate. Sporangiophores may be branched. The **columellae** (sterile structures in sporangium) are round, cylindrical, or oval. Spores are black or brown and are smooth in appearance. Zygospores are produced when plus and minus strains (so called because there is no morphological differentiation between the male and female strains even though there is physiological heterothallism) of the organism are both present. Zygospores are also formed in growth on artificial media in the laboratory. No stolons or rhizoids are produced. It is interesting that in the absence of a fermentable carbon source in a medium consisting of yeast extract and peptone supplemented with potassium acetate, *Mucor* may grow in a yeastlike instead of a filamentous form (Fig. 17-24).

Rhizopus (Class *Zygomycetes*)

These are common bread molds (Fig. 17-25), which cause much food spoilage. They grow on bread, vegetables, fruits, and other food products. The most commonly encountered species is *R. stolonifer.* Morphologically, they have nonseptate, cottony mycelia with sporangiophores arising at the nodes where the rhizoids form. Their sporangia are usually quite large and black; their columellae are hemispherical. The base of the sporangium, or **apophysis,** is cup-shaped. These molds produce clusters of rootlike holdfasts called **rhizoids,** as

Figure 17-25. *Rhizopus stolonifer,* the common bread mold. This fungus forms rootlike hyphae (rhizoids), vegetative hyphae which penetrate the substrate, and fertile hyphae, which produce sporangia at the tips of sporangiophores (spore-producing hyphae). Stolons are rootlike filaments which connect individual organisms.

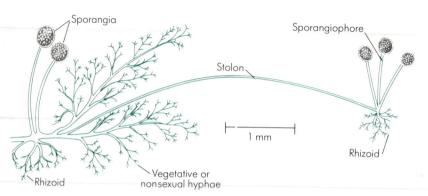

well as stolons or "runners" (like those of strawberry plants) capable of taking "root" where they may give rise to new organisms. The life cycle of *R. stolonifer* is given in Fig. 17-18. It may be seen that *R. stolonifer* is heterothallic. Sexual reproduction requires two thalli of different mating types. Since the thalli are indistinguishable morphologically, they are designated plus and minus instead of male and female. (Generally, with gametes that differ morphologically, the larger gamete of a pair is regarded as female, the smaller as male. Also, a gamete that leaves the structure where it was formed and later fuses with a relatively stationary gamete is regarded as male.) When both mating thalli are present, the hyphal tips differentiate into **progametangia** which come into contact and develop into **gametangia** by the formation of septa. The walls between two gametangia then dissolve and their protoplasts coalesce. Nuclei of both mating types fuse in pairs, producing many zygote nuclei. The structure that contains them is then called a **coenozygote.** The wall of the coenozygote soon thickens, turns black, and becomes rough; it develops into the **zygospore.** The resting stage of the zygospore lasts from 1 to 3 months, and sometimes even longer. Upon germination, meiosis takes place, the zygospore breaks open, and a single sporangiophore bearing a **germ sporangium** at its tip emerges. The germ sporangium is similar to an asexually produced sporangium. Some germ sporangia contain spores all of one mating type (either plus or minus), but others contain spores of both mating types in about equal numbers.

Schizosaccharomyces
(Class *Ascomycetes*)

All fission yeasts belong to this genus. They reproduce by transverse division as well as by ascospores (see Fig. 17-19 for the life cycle of *Schizosaccharomyces*). The best known species is S. *octosporus,* which has been isolated from currants and honey. Its cells are globose to cylindrical, uninucleate and haploid. S. *pombe* is the fermenting yeast of some kinds of beer (such as African millet beer), and it has been isolated from sugar molasses and from grape juice. S. *versatilis,* isolated from grape juice, grows like a yeast, but it can also form a true mycelium. Fig. 17-26 illustrates *Schizosaccharomyces.*

Saccharomyces
(Class *Ascomycetes*)

There are about thirty species of *Saccharomyces.* The best known is S. *cerevisiae,* strains of which are used in the fermentation of beer and wine and in baking. It is found in nature on ripe fruit. Grape wines are often made by spontaneous fermentation by yeasts growing on the surface of the fruit. Thus *S. cerevisiae* is a yeast of great economic importance. Its cells are elliptical, measuring about 6 to 8 by 5 μm. (See Fig. 17-20 for the morphology of these cells.) They multiply asexually by budding. Where a bud has formed on a cell, a raised scar remains (Fig. 17-27). As many as 23 bud scars have been seen on a single cell. During budding, the nucleus divides by constriction and a portion of it enters the bud along with other organelles. The cytoplasmic connection is closed by the laying down of cell-wall material. Under appropriate conditions, *S. cerevisiae* forms asci. The cytoplasm of the cell differentiates into four thick-walled spherical spores, although the number of spores can be fewer (Fig.

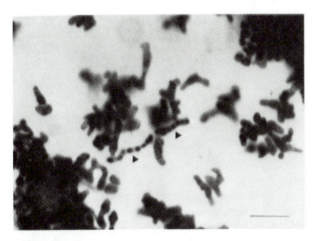

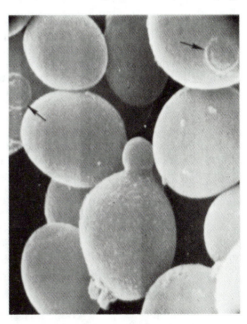

Figure 17-26. The fission yeast *Schizosaccharomyces octosporus*. Vegetative cells as well as ascospores in asci (▶) are clearly seen. Bar equals 10 μm. *(Courtesy of L. Kapica and E. C. S. Chan, McGill University.)*

Figure 17-27. *Saccharomyces cerevisiae* showing bud and bud scars (arrows) (X1,600). *(Courtesy of R. G. Kessel and C. Y. Shih, Scanning Electron Microscopy in Biology, Springer-Verlag, Berlin, 1974.)*

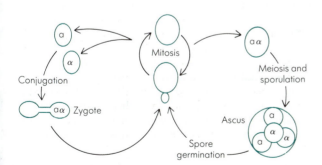

Figure 17-28. The life cycle of *Saccharomyces cerevisiae*; *a* and *α* refer to mating-type alleles; unlabeled cells may be *a*, *α*, or *aα*.

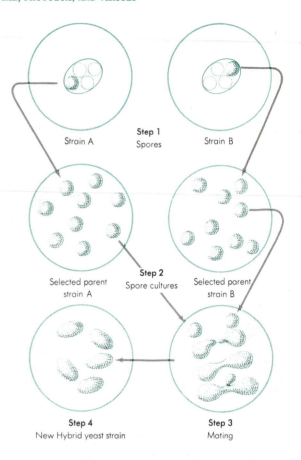

Figure 17-29. The four steps in the process of hybridization. Step 1: Yeast cells from two strains A and B are shown after they have been induced to form spores. The spores are removed by microdissection by means of a micromanipulator. Step 2: Each individual spore is planted in a nutrient medium and allowed to multiply to produce separate spore cultures. After careful testing for the desirable characteristics, two spore cultures of different sexes are selected for mating or crossbreeding. Step 3: When these two spore cultures are brought together, instead of budding they fuse in pairs and produce a completely new yet stable combination of inherited qualities. Step 4: The new hybrid strain shown is a combination of the best qualities of strains A and B. The new hybrid grows and reproduces by budding. As it grows, the new inherited qualities are equally transmitted to all cells reproduced. *(From A. J. Salle, Fundamental Principles of Bacteriology, 7th ed., McGraw-Hill, New York, 1973.)*

17-20A). The cells from which asci develop are diploid and the nuclear divisions which precede spore formation are meiotic. Figure 17-28 shows the life cycle of *S. cerevisiae*. It may be noted that many strains of the yeast are heterothallic, and the ascospores are of two mating types. Mating type is specifically controlled by a single gene which exists in two allelic states a and α, and segregation at reduction-division preceding ascospore formation gives rise to two a and two α ascospores (Fig. 17-28). Fusion normally occurs only between cells of differing mating types, a process termed **legitimate copulation.** Such fusions result in diploid cells which form asci containing viable ascospores. Many studies have been made of yeast genetics. For instance, by means of **hybridization** (crossing of different yeasts) it is possible to develop strains of yeasts (hybrids) with desirable characteristics from two genetically different strains. This technique is illustrated in Fig. 17-29.

Neurospora (Class *Ascomycetes*)

This genus of *Ascomycetes* is of particular interest to biologists because of its wide use in the study of genetics and metabolic pathways. Two well-known species are *N. crassa* and *N. sitophila*. Some species are responsible for food

Figure 17-30. *Agaricus* sp. *(Courtesy of B. Kendrick, University of Waterloo.)*

spoilage, and some are used in industrial fermentations. Certain species produce ascospores, and some species are heterothallic. Morphologically, *Neurospora* produces a loose network of long strands of septate aerial mycelia. Conidia, usually oval, form branched chains at the tips of the aerial hyphae. Because of the characteristic pink or red color of the conidia and the common occurrence of *Neurospora* species on farinaceous foods, they are often called pink bread molds. The ascospores of neurospora are produced in perithecia.

Agaricus (Class Basidiomycetes)

The best known species are *A. campestris*, the field mushroom, and *A. bisporus*, the cultivated mushroom. The pink coloration of the young gills is due to cytoplasmic pigment in the spores. The gills later turn a purplish brown because of the deposition of dark pigments in the spore wall. Most of the larger species of *Agaricus* are edible. (See Fig. 17-30.)

Filobasidiella neoformans (Class Basidiomycetes)

Of the approximately 12,000 species of *Basidiomycetes*, none was implicated in human disease until recent times. The perfect stage of *Cryptococcus neoformans* was discovered in 1975; it is now called *Filobasidiella neoformans* (Fig. 17-31). It is an important basidiomycetous pathogen of humans, causing cryptococcosis, a generalized (systemic) mycotic infection involving the bloodstream as well as the lungs, central nervous system, and other organs.

Figure 17-31. *Filobasidiella neoformans* cell. A budding cell is seen emerging from the parental cell. B = bud, C = capsule, CW = cell wall, M = mitochondria, N = nucleus, V = vacuole. The bar indicates 1 μm. *(Courtesy of Phyllis C. Braun, Georgetown University.)*

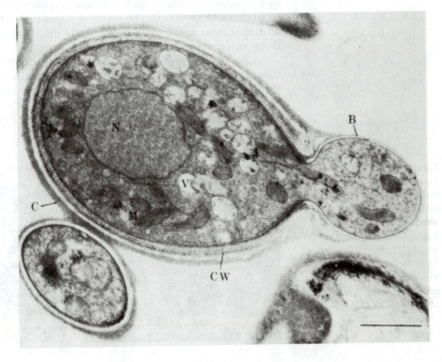

Aspergillus (Class Deuteromycetes)

The aspergilli are widespread in nature, being found on fruits, vegetables, and other substrates which may provide nutriment. Some species are involved in food spoilage. They are important economically because they are used in a number of industrial fermentations, including the production of citric acid and gluconic acid, both produced in abundance by *A. niger*. Morphologically, the aspergilli produce septate, branching mycelia with the vegetative portions submerged in nutrient. The conidiophores, or fertile hyphae, arise from foot cells which may also be submerged. Conidiophores may be septate or nonseptate. At the apex, the conidiophore inflates to form a vesicle. This in turn gives rise to sterigmata, which may be single-layered (primary) or double-layered (primary and secondary). Conidia arise from the sterigmata and are borne in chains. The vesicles vary in size and shape, depending on the species. Conidia are produced within the tubular sterigmata and are extruded to form spore chains (Fig. 17-32). Conidia are of various colors and are quite characteristic of the species; the most common colors are black, brown, and green. Aspergilli grow in high concentrations of sugar and salt, indicating that they can extract water required for their growth from relatively dry substances.

Penicillium (Class Deuteromycetes)

Members of this group (Fig. 17-6), also *Deuteromycetes*, occur widely in nature. Some species cause rot or other spoilage of fruits, vegetables, preserves, grains, and grasses. Others are used in the ripening of cheese, e.g., Roquefort, blue, and Camembert. Some are used in industrial fermentations, and one of the best-known antibiotics (penicillin) is produced by *P. notatum* and *P. chrysogenum*. These molds are closely related to the aspergilli, and some reproduce sexually by ascospore formation. Ascus formation is observed in the penicillia less commonly than in aspergilli. The conidial stages of *Penicillium* and *Aspergillus* are so distinct and well known that they are discussed on that basis rather than on their perfect stages; see Fig. 17-33.

Penicillia have septate vegetative mycelia which penetrate the substrate and then produce aerial hyphae on which conidiophores develop. Conidiophores may be branched and have brushlike heads bearing spores. Clusters of sterigmata are usually in one place, and from each is formed a chain of conidia. The color of the mature plants is useful in helping to identify species. They grow best at temperatures ranging from 15 to 30°C.

Figure 17-32. *Aspergillus flavus* as seen by scanning electron microscopy. Left: Conidiophores bearing large heads of conidia (X150). Right: Conidial head of the mold (X1,000). *(Courtesy of M. F. Brown, University of Missouri. From E. J. King and M. F. Brown, Can J Microbiol,* **29***:653–658, 1983.)*

Figure 17-33. Conidial heads of *Aspergillus* and *Penicillium*. Note the different arrangements of conidia, which are useful in identification.

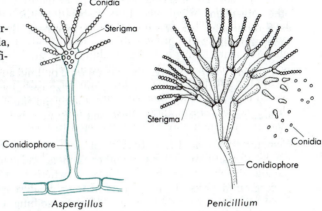

Candida (Class Deuteromycetes)

Candida albicans is often isolated from warm-blooded animals, including humans, where it exists as part of the normal microbiota of mucous membranes. Sometimes this fungus may beome pathogenic, causing candidiasis, a disease of the mucous membranes of the mouth (thrush), vagina, and alimentary tract. More serious infections can involve the heart (endocarditis), the blood (septicemia), and the brain (meningitis). It seems that predisposition factors such as other diseases, physiological disorders, obesity, alcoholism, and prolonged use of broad-spectrum antibiotics and steroids can create conditions in which *C. albicans* can cause disease. This makes the fungus an opportunistic pathogen. Fig. 17-34 shows the morphology of *C. albicans*.

Figure 17-34. *Candida albicans*, a yeast pathogenic for humans. (A) Note pseudomycelia and blastospores (▶) in a urine sample from an infected patient. (B) The yeast also forms chlamydospores (▷) as well as pseudomycelia and blastospores (▶), when grown on a special medium in the laboratory. *(Courtesy of L. Kapica and E. C. S. Chan, McGill University.)*

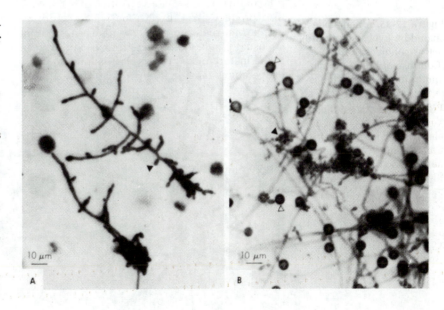

MOLDS AND THEIR ASSOCIATIONS WITH OTHER ORGANISMS

There are some interesting partnerships in nature involving a mold and some other organism. In some of these associations the partners are dependent on each other and cannot survive alone. In others, the individuals can survive independently.

Lichens

Lichens are composite organisms composed of fungi and algae, each contributing to the benefit of both. The algae synthesize carbohydrates by photosynthesis and obtain other nutrients from the fungi; the fungi depend on the algae for organic carbon. Lichens are discussed in greater detail in Chap. 18.

Fungi and Nematodes

Nematodes are small roundworms predominantly pathogenic for plants; some may be parasitic in humans and other animals, and some are saprophytes. Plant-pathogenic nematodes live in the soil, where they destroy roots and other underground parts of plants. One of their greatest natural enemies is certain fungi that thrive on a diet of nematodes. These fungi use interesting techniques to capture and overcome the much larger nematodes. One of the most unusual techniques employs a loop of fungal hyphae. Upon contact by the nematode, the loop constricts around the body of the worm and holds it firmly while penetrating the prey with **haustoria** (special hyphal branches that penetrate a host cell) which will digest the nematode's body (Fig. 17-35). Other trapping devices used by fungi to capture nematodes, rotifers, and protozoa in the soil involve an adhesive on the mycelium. Still another fungal snare is a complex network of mycelia which grasps the worm as it tries to penetrate the fungal barrier.

Fungi as Parasites of Insects

Not all associations of fungi with higher organisms are beneficial to the host. Common among the detrimental associations are fungal infections of insects, of which more than 50 are known. At times, some appear as epidemics, when they

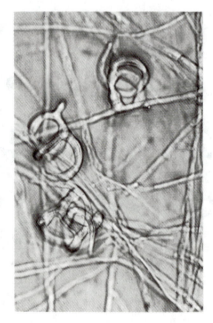

Figure 17-35. A nematode snared by hyphae of *Arthrobotrys conoides* (X400). *(Courtesy of David Pramer and Shimpei Kuyama, Bacteriol Rev, 27:282–292, 1963.)*

destroy large numbers of insect hosts. The fungus *Entomophthora muscae* has been the cause of mild epidemics among common houseflies, crickets, and grasshoppers. Such epidemics are self-limiting and can only rarely be used effectively to destroy insect pests; e.g., pathogenic fungi which attack aphid infestations of citrus orchards have been used in attempts to develop biological control methods for insect pests.

Mycorrhizas

A **mycorrhiza** is an infected root system arising from the rootlets of a seed plant. The word *mycorrhiza* is derived from the Greek, meaning "fungus root." These associations are usually beneficial to the host plant as well as the symbiont, and sometimes the host cannot thrive without the benefits derived from the fungus. Mycorrhizas enhance mineral absorption by the green plants. Host plants usually have a built-in check on the fungus to prevent it from causing injury to the rootlets. Truffles are subterranean fruiting bodies of certain *Ascomycetes* which grow in association with some trees, oak and beech being common symbionts in this mycorrhizal partnership. The fungus provides certain nutrients to the tree, which in turn provides essential growth substances to the fungus. Such relationships are quite common. Truffles consist of a mass of ascospores and mycelia, covered with a thick knobby rind of mycelium. They have a pleasant odor, flavor, and texture which make them highly prized by gourmets.

In plants that do not have root hairs, e.g., orchids, the fungi absorb nutrients and water for their hosts. Under natural conditions, orchids rarely grow without fungal partners. Experimentally, orchids will grow alone if appropriate nutrients are provided.

Another type of mycorrhiza is found in *Monotropa uniflora* (Indian pipe), which has no chlorophyll but grows well in the heavy shade of the forest floor as long as the fungi are on its roots to gather food for both host and parasite.

QUESTIONS

1 Yeasts, like molds, are fungi. How are they morphologically different from molds? How do they resemble bacteria?

2 In what ways are fungi important to humans?

3 Describe the phenomenon of dimorphism as it exists in certain pathogenic fungi.

4 Draw a diagram of a hypha and label it with the following: wall, lumen, plasmalemma, matrix, nucleus, septum, microfibrils, and central pore.

5 Describe the different ways by which fungi reproduce asexually.

6 Describe sexual reproduction as it occurs in fungi.

7 Explain the difference between sexual spores and asexual spores, with particular reference to their formation.

8 Compare the physiology of fungi with that of bacteria. Which physiological features are used to advantage in the preparation (formulation) of selective media for fungi?

9 What is unique about the class *Deuteromycetes*, making it different from other classes of fungi?

10 Compare the criteria used to identify molds with those used to identify bacteria.

11 What are the unique features of the slime molds that distinguish them from the true fungi?

12 Describe the life cycle of the cellular slime mold *Dictyostelium discoideum*.

13 Describe how a typical myxomycete reproduces sexually.

14 What main features distinguish the members of the division *Mastigomycota* from the members of the division *Gymnomycota*?

15 Compare the sexual and asexual reproduction of *Saccharomyces* and *Schizosaccharomyces*.

16 Define these terms: basidiocarp, sterigma, and basidium.

17 Describe how *Synchytrium endobioticum* infects a new host.

18 Describe the process of sexual reproduction in the mold *Saprolegnia*.

19 Describe the following:

  (a) Dimorphism in *Mucor*

  (b) Plus and minus strains in *Rhizopus stolonifer*

  (c) Asexual reproduction of *Schizosaccharomyces*

  (d) Budding in *Saccharomyces*

  (e) Yeast hybridization

20 Why are the following fungi important to our economy or health: (a) *Aspergillus niger*, (b) *Penicillium notatum*, (c) *Neurospora sitophila*, (d) *Agaricus bisporus*, (e) *Candida albicans*, and (f) *Filobasidiella neoformans*?

21 In what manner are the following associated with fungi: (a) lichens, (b) nematodes, and (c) orchids?

REFERENCES

Alexopoulos, C. J., and C. W. Mims: *Introductory Mycology*, 3d ed., Wiley, New York, 1979. *A widely used and classic textbook in introductory mycology courses. It is based on the morphology/taxonomy approach to the study of fungi. A very useful reference.*

Emmons, C. W., C. H. Binford, J. P. Utz, and K. J. Kwon-Chun: *Medical Mycology*, 3d ed., Lea & Febiger, Philadelphia, 1977. *A classic text on medical mycology. The biology of the fungi is quite well covered also.*

Moore-Landecker, E.: *The Fundamentals of Fungi*, 2d ed., Prentice-Hall, Englewood Cliffs, New Jersey, 1982. *A good introductory text for a first course in mycology.*

Phaff, H. J., M. W. Miller, and E. M. Mrak: *The Life of Yeasts*, 2d ed., Harvard, Cambridge, Mass., 1978. A book written to serve nonspecialists who occasionally encounter yeasts in their work. It is a clearly written "encyclopedia" of all aspects of yeast biology as well as the applied aspects of yeasts.

Ross, I. K.: *Biology of the Fungi*, McGraw-Hill, New York, 1979. *An undergraduate text on the biology of fungi, with emphasis on their developmental and regulatory aspects.*

Chapter 18 Algae

This chapter introduces the **algae** (singular, **alga**), many of which are unicellular microorganisms. These organisms are ubiquitous; many live in aquatic environments but many also thrive as terrestrial and subterranean algae. Algae contain chlorophyll and are photosynthetic. They differ from other green plants in having simple reproductive structures for sexual reproduction. For example, the unicellular algae themselves may function as gametes. In their asexual reproduction, many algae produce flagellated spores and/or nonmotile spores in sporangia.

Algae are of great general interest to all biologists because single algal cells are complete organisms capable of photosynthesis and the synthesis of a multitude of other compounds which constitute the cell. Geneticists and evolutionists find algae interesting and useful for study because of their conspicuous pattern of evolutionary specialization. The convenient techniques used for the study of bacteria and molds are applicable to microscopic algae discussed in this chapter.

In size, habitat, and reproductive processes algae are a heterogeneous group. They range from microscopic unicellular forms comparable in size to bacteria, to seaweeds that may grow to many feet in length. The study of this unique group of organisms is called **phycology.**

365

OCCURRENCE OF ALGAE

Many thousands of algal species occur in nature. There are few places on earth where algae of some kind cannot be found. They occur in great abundance in the oceans, seas, salt lakes, freshwater lakes, ponds, and streams. Many are found in damp soil, on rocks, stones, and tree bark, and on other plants and animals.

Small aquatic forms make up a large part of the free-floating microscopic life in water called **plankton,** which is the principal food for aquatic animals, including such large ones as whales. (Plankton is generally considered to be composed of both algae and microscopic animal forms. **Phytoplankton** is made up of plants, i.e., algal forms, and **zooplankton** is composed of animal organisms.) Algae are found where there are sufficient light, moisture, and simple nutrients to sustain them.

Some species of algae grow on the snow and ice of polar regions and mountain peaks, sometimes occurring in such abundance that the landscape becomes colored by the red pigments in their cells. At the other extreme, some algae grow in hot springs at temperatures as high as 55°C. Some freshwater algae have adapted their metabolism to the high salt concentrations found in the brine lakes of the arid southwestern United States. The salinity of marine environments varies from place to place, but marine algae adapt to the variations in salt concentration. Others adjust to air and to drying when they are exposed at low tide. Marine algae are not normally found in northern waters of the ocean at depths greater than 150 to 180 ft. In the clearer, warmer tropical waters, however, where the sunlight is more direct and has a longer daily period, they may be found at depths as great as 600 ft. These and other factors are responsible for the **zonation** phenomenon, which is the stratification of certain kinds of algae at certain depths and locations in the ocean.

Some algae are adapted to moist soil, the bark of trees, and the surface of rocks, which the algae degrade. The decomposition products are made available for soil building and enrichment.

Algae are often a problem in water supplies because they produce undesirable tastes and odors. An especially notorious offender is a flagellated brown alga belonging to the genus *Synura.* Heavy algal growths may form blankets or mats which interfere with the use of some natural waters for recreational purposes. These algal mats may act as barriers to the penetration of oxygen into the water; they prevent photosynthesis by excluding light from deeper water and thus may cause fish and other marine animals to suffocate. Algae may even increase the corrosive quality of water and cause the disintegration of concrete.

On the other hand, when dispersed in natural waters algae increase the oxygen concentration through photosynthesis. Heavy growth of some algae reduces hardness of water and removes salts which are the cause of brackishness. An installation using this principle on a large-scale experimental basis has been built in South Dakota in an area where the natural water contains a high concentration of organic salts.

Some algae are **endophytic;** that is, they are not free-living but live in other organisms. Such algae are widespread in protozoa, molluscs, sponges, and corals.

THE BIOLOGICAL AND ECONOMIC IMPORTANCE OF ALGAE

Algae as Primary Producers

Most algae are aquatic organisms. Since 70 percent of the earth's surface is covered with water, it is probable that as much carbon is fixed (captured as carbon dioxide and changed to organic carbon compounds such as sugars) through photosynthesis by algae as is fixed by all land flora. As mentioned before, tiny floating algae constitute the phytoplankton of the sea and serve as an important food source for other organisms. These algae form the base or beginning of most aquatic food chains because of their photosynthetic activities and are therefore called primary producers of organic matter.

Algae are also present in soil even if their presence is not so obvious. They are probably important in stabilizing and improving the physical properties of soil by aggregating particles and adding organic matter. In addition, in many countries where the large red and brown seaweeds are plentiful, they are used as fertilizer.

Commercial Products from Algae

Many products of economic value are derived from algal cell walls. Three of these, agar, alginic acid, and carrageenan, are extracted from the walls of algae. Another, diatomaceous earth, is composed of millions upon millions of diatom glass walls deposited over time in either freshwater or the ocean.

Agar and carrageenan are polymers of galactose, or galactose-containing compounds, with sulfate groups. Some of the sulfate groups are involved in the bonds between individual sugar residues. Agar and carrageenan are both called sulfated galactans, with carrageenan the more sulfated compound. Alginic acid consists of uronic acid residues. All three compounds are used either to make gels or to make solutions viscous.

Carrageenan is extracted from the walls of several red algae. Species of *Chondrus*, *Gigartina*, and *Eucheuma* are most frequently used. Carrageenan has been used as a stabilizer or emulsifier in foods such as ice cream and other milk products. It is also used as a binder in toothpaste or in pharmaceutical products, as well as an agent in ulcer therapy. Carrageenan is also useful as a finishing compound in the textile and paper industries, as a thickening agent in shaving creams and lotions, and in the soap industry.

Agar is well known as a solidifying agent in the preparation of microbiological media. It is obtained from red algae. Species of *Gelidium* and *Gracilaria* are used extensively. It is also important in the food industry where it is valuable in the manufacture of processed cheese, mayonnaise, puddings, jellies, baking products, and canned goods. In the pharmaceutical industry agar can be used as a carrier for a drug. Lotions and ointments can contain some agar.

Alginic acid and its salts are obtained from the walls of brown algae, where they may represent as much as 25 percent of the dry weight. Species of brown algae producing this compound include *Macrocystis*, *Agarum*, *Laminaria*, *Fucus*, and *Ascophyllum*. About 50 percent of the ice cream in the U.S. contains alginates, which provide a smooth consistency and eliminate ice crystal formation. Alginates are also incorporated into cheeses and bakery products, especially frostings. Other industrial applications include paper manufacturing, the printing of fabrics, and paint thickening.

An alginate material is used by dentists for making impressions of the teeth for crowns, etc. The stipe (stemlike part) of the brown alga *Laminaria japonica* may be used by physicians for cervical dilatation and/or softening of the cervix (for example, in performing an abortion or placing a radium implant). The stipe is cut into sections, dried, and sterilized by ethylene oxide gas; such prepared stipes are available commercially. The section is put into place and swells gradually, rendering symmetrical dilatation of the cervical canal and softening of cervical tissue. Japanese physicians have used this method for more than a century.

Diatomaceous earth is used primarily for filters or filter aids. It is especially suitable because it is not chemically reactive, is not readily compacted or compressed during use, and is available in many grades. The material is so finely divided that one gram of diatomaceous earth has 120 square meters of surface area, and yet in use up to 90 percent of the volume of the filter cake is open space. Diatomaceous earth is also used for polishing delicate surfaces because the diatom walls are so lightly silicified that they collapse under pressure and do not damage the surfaces.

Algae as Food

Many species of algae (mostly red and brown algae) are used as food in the Far East. Of the red algae one of the most important is *Porphyra*; it is used as a food in Japan, where it is called "nori" and is usually processed into dried sheets. The algae are collected and washed in fresh water to remove debris, then chopped and spread on frames to dry into thin sheets. Nori is commonly toasted over a flame and sprinkled in soup or rice, or it is rolled around flavored rice with fish or vegetables to make a popular luncheon snack called sushi.

Although several species are still collected wild, *Porphyra* has been cultivated since 1570. Today Japan's seaweed cultivation is the most advanced in the world. The nori industry is based mainly on *P. tenera* and *P. yezoensis* (Fig. 18-1A), although up to seven other species have been cultivated in Tokyo Bay in the past.

Other red algae, such as *Chondrus, Acanthopeltis, Nemalion,* and *Eucheuma,*

Figure 18-1. Some algae used as food. (A) *Porphyra yezoensis* Ueda. (B) *Chondrus crispus* Stackhouse. (C) *Palmaria palmata* (L.) O. Küntze. Bar = 1 cm. *(Erwin F. Lessel, illustrator)*

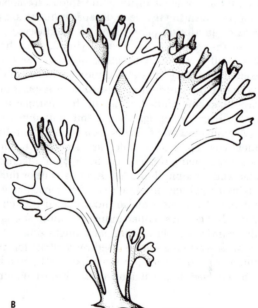
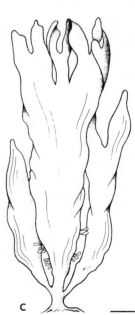

A   B   C

are locally collected and prepared. Most of them are eaten as vegetables or in soups or prepared as sweetened jellies.

Red algae continue to be a significant food in China. Such algae have been a food staple and delicacy for the Chinese for thousands of years. For example, *Gracilaria* is mentioned in Chinese *materia medica* dating back to 600 B.C. In the past two decades, China has developed an impressive seaweed cultivation program. The species farmed are the edible brown alga *L. japonica* as well as *Porphyra* and *Gracilaria*. *Porphyra* and other red seaweeds are expected to be increasingly important in the Chinese diet.

In contrast to the diversity of species eaten by Asian and Polynesian peoples, red algal food usage in Europe and North America has centered around three genera: *Porphyra* or laver, *Chondrus* (Fig. 18-1B) or Irish moss, and *Palmaria* or dulse (Fig. 18-1C).

Laver is used extensively in the British Isles. The miners of southern Wales are the major laver consumers. About 40 to 50 tons of dried dulse are produced each year in Canada. The alga is collected in the Bay of Fundy and on the shores of Nova Scotia. Dulse is commonly eaten as a snack in taverns.

There is increasing interest in the use of the smaller forms of algae, especially *Chlorella*, as food for humans and domestic animals. When these organisms are grown under suitable conditions, they provide a rich source of protein comprising all the amino acids essential for animal growth. They are also a good source of carbohydrates and fats. The nutritive value of the microscopic algae has been demonstrated in tests with rats and chicks. Methods for mass cultivation of these plants, using waste products and sewage for their nutrition, have been developed. After algae have been grown on waste products, the residues can be disposed of in streams and lakes without causing pollution that would destroy aquatic animals. Although much can be said in favor of using algae in place of higher plants for human food, general acceptance of their use in this country is not to be expected until food from other sources is in short supply. In the meantime algae will doubtless find wide application as animal feeds or feed supplements.

Algae and Diseases

Although few algae are pathogenic, one, *Prototheca*, has been found to be a probable pathogen of humans. It has been found in systemic and subcutaneous infections, as well as in bursitis (an inflammation of the joints). *Prototheca* is a colorless *Chlorella*-like organism which superficially resembles yeasts (Fig. 18-2).

Several species are parasitic on higher plants; e.g., the green alga *Cephaleuros* attacks leaves of tea, coffee, pepper, and other tropical plants, causing considerable damage. Some algae live in the roots and fleshy parts of higher plants; liverworts, duckweeds, and other hosts to such algae do not seem to be harmed by their presence.

Some of the extracellular inhibitors produced by algae have been shown by chemical analysis to be simple chemical substances; e.g., acrylic acid is produced by a unicellular alga in plankton. It is quite possible that as we learn more about algae and their extracellular secretions, their usefulness will become more apparent.

Some planktonic algae produce toxins which are lethal to fish and other

Figure 18-2. Prototheca cells are colorless and do not carry out photosynthesis. However, the presence of starch-containing inclusions consitutes significant evidence for their algal nature. They resemble yeast cells in their gross growth characteristics on media, which can lead to errors in laboratory diagnosis. (A) Colonies of *P. wickerhamii* on Sabouraud agar. (B) Streaked cultures of *P. wickerhamii* (Proto.), *Candida albicans* (Candida), and *Filobasidiella* (*Cryptococcus*) *neoformans* (Crypto.). The latter two species are yeasts. (*Courtesy of L. Kapica, McGill University.*)

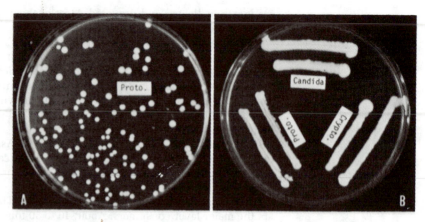

animals. These toxins are extracellular, or they are liberated from the alga by bacterial decomposition of water blooms. It is known, however, that certain marine dinoflagellates belonging to the genera *Gymnodinium* and *Gonyaulax* cause death of aquatic animals by producing a high molecular weight neurotoxin. Those few dinoflagellate toxins which have been identified are among the strongest toxins known—50 times stronger than curare, a poison with which certain primitive peoples tipped their darts.

Shellfish poisoning occurs along the northeastern coast of North America as well as in the North Pacific. The organisms are *Gonyaulax catenella* on the west coast and *Gonyaulax excavata* on the east coast. Yearly outbreaks occur around Nova Scotia. The bloom of these dinoflagellates usually lasts just a few weeks, and often it is safe to eat shellfish about two weeks after the end of the bloom. The poisoning to humans comes from eating filter feeders, i.e., clams, scallops, or mussels, which filter the plankton from seawater as a source of food and accumulate the poison (toxin).

After ingesting sufficient toxin, the victim first experiences a numbing of the lips, tongue, and fingertips, usually within 30 min of eating shellfish. The diaphragm is soon affected and in serious cases respiratory failure can result.

CHARACTERISTICS OF ALGAE

Morphology

Algae have a wide range of sizes and shapes (Fig. 18-3). Many species occur as single cells that may be spherical, rod-shaped, club-shaped, or spindle-shaped. Others are multicellular and appear in every conceivable form, shape, and degree of complexity, including membranous colonies, filaments grouped singly or in clusters with individual strands that may be branched or unbranched, and tubes (which may or may not be divided by cell walls). Some colonies are simply aggregations of single, "identical" cells that cling together after division; others are composed of different kinds of cells specializing in particular functions. These colonies become quite complex and superficially resemble higher plants in structure.

Algal cells are eucaryotic. In most species the cell wall is thin and rigid. Cell walls of diatoms are impregnated with silica, making them thick and very rigid; they are often delicately sculptured with intricate designs characteristic of the

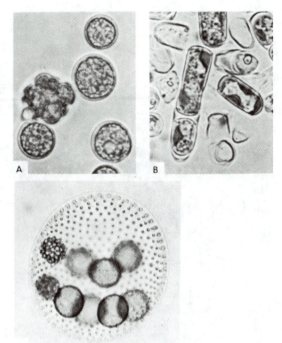

Figure 18-3. Algae occur in a wide variety of sizes, shapes, and arrangements. (A) The green alga *Chlorococcum scabellum.* This photomicrograph shows vegetative cells and a cluster of aplanospores (X760). (B) Another alga, *Pseudobumilleriopsis* sp. (X912). [(A) and (B) courtesy of Harold C. Bold.] (C) *Volvox spermatosphaera,* illustrating parental colony with developing daughter colonies. (*Courtesy of Richard C. Starr.*)

species or variety. The motile algae such as *Euglena* have flexible cell membranes called **periplasts.** The cell walls of many algae are surrounded by a flexible, gelatinous **outer matrix** secreted through the cell wall, reminiscent of bacterial capsules. As the cells age, the outer matrix often becomes pigmented and stratified.

Algae contain a discrete nucleus. Other inclusions are starch grains, oil droplets, and vacuoles. Chlorophyll and other pigments are found in membrane-bound organelles known as **chloroplasts.** These chloroplasts may be massive structures situated near the wall (parietal) or embedded in the midst of the cytoplasm (asteroidal). They may occur as one, two, or many per cell; they may be ribbonlike, barlike, netlike, or in the form of discrete disks, as in green plants. Within the plastid (chloroplast) matrix or stroma are found flattened membranous vesicles called **thylakoids.** The fine structure of a eucaryotic algal cell is shown in Fig. 18-4.

Algal Pigments

The chloroplasts of different divisions of algae containing similar pigments appear to have similar thylakoid arrangements. Chloroplast ultrastructure and pigment chemistry have been used as markers for algal phylogeny. Table 18-1 shows the **divisions** (primary taxa in algal classification) of algae, with some of their photosynthetic pigments.

It should be pointed out that several divisions of algae include colorless members, e.g., certain species and genera in the *Euglenophycophyta,* the *Pyrrophycophyta,* and the *Chlorophycophyta.* These are sometimes considered

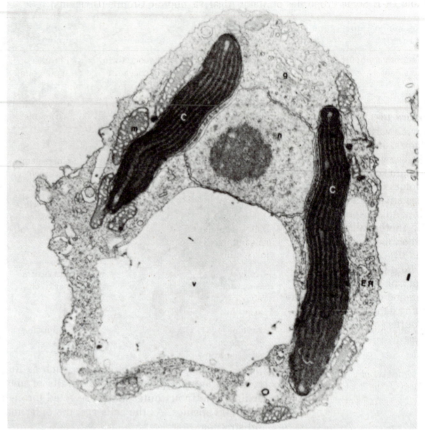

Figure 18-4. The fine structure of an algal cell is revealed by this near-median longitudinal section of a cell of *Ochromonas danica*. The single Golgi body (g) lies anterior to the nucleus (n). The endoplasmic reticulum (ER) is visible on the right side of the section. A single chloroplast (C) is shown, but it appears to be two chloroplasts because of the plane of sectioning. A large starch-containing vacuole (v) occupies almost half of the cell. Numerous mitochondria (m) are present in the peripheral cytoplasm of the cell (X12,120). (*Courtesy of Sarah P. Gibbs, McGill University.*)

protozoa. Nevertheless, some colorless flagellates have been shown to possess chloroplasts. There are three kinds of photosynthetic pigments in algae: chlorophylls, carotenoids, and biloproteins (also called phycobilins). Their distribution is shown in Table 18-1.

Chlorophylls

There are five chlorophylls: *a*, *b*, *c*, *d*, and *e*. Chlorophyll *a* is present in all algae, as it is in all photosynthetic organisms other than anoxygenic photosynthetic bacteria. Chlorophyll *b* is found in the *Euglenophycophyta* and *Chlorophycophyta* and in no other algal division. Chlorophyll *c* is more widespread and is present in members of *Xanthophycophyta*, *Bacillariophycophyta*, *Chrysophycophyta*, *Pyrrophycophyta*, *Cryptophycophyta*, and *Phaeophycophyta*. Chlorophyll *d* appears to be present only in the *Rhodophycophyta*. Chlorophyll *e* is rare and has been identified in only two genera of *Xanthophycophyta*, namely, *Triboneara* and the zoospores of *Vaucheria*.

Carotenoids

There are two kinds of carotenoids: carotenes and xanthophylls. Carotenes are linear, unsaturated hydrocarbons, and xanthophylls are oxygenated derivatives

Table 18-1. Some Properties of Major Algal Taxonomic Groups (Divisions)

Taxonomic Group (Division)	Chloro-phyll	Carotenoids*	Biloproteins	Storage‡ Products	Flagellation and Details of Cell Structure
Rhodophycophyta† (red algae)	*a*, rarely *d*	β-carotene, zeaxanthine ±α-carotene	Phycoerythrin, phycocyanin	Floridean starch, oils	Flagella absent
Xanthophycophyta† (yellow-green algae)	*a, c*, rarely *e*	β-carotene, diadinoxanthin, heteroxanthin, vaucheriaxanthin ester		Chrysolami-narin, oils	2 unequal, apical flagella
Chrysophycophyta (golden algae)	*a, c*	β-carotene, fucoxanthin		Chrysolami-narin, oils	1 or 2 equal or unequal, apical flagella; in some, cell surface covered by characteristic scales
Phaeophycophyta (brown algae)	*a, c*	β-carotene ±α-carotene, rarely ε-carotene, fucoxanthin, violaxanthin		Laminarin, soluble carbo-hydrates, oils	2 lateral flagella
Bacillariophycophyta (diatoms)	*a, c*	β-carotene ±α- carotene, rarely ε-carotene, fucoxanthin		Chrysolami-narin, oils	1 apical flagellum in male gametes; cell in two halves; the walls silicified with elaborate markings
Euglenophycophyta (euglenoids)	*a, b*	β-carotene ±γ-carotene diadinoxanthin		Paramylon, oils	1, 2, or 3 equal, slightly apical flagella; gullet present
Chlorophycophyta (green algae)	*a, b*	β-carotene ±α-carotene, rarely γ-carotene and lycopene, lutein		Starch, oils	1, 2, 4 to many, equal, apical or subapical flagella
Cryptophycophyta (cryptomonads)	*a, c*	α-carotene ±β-carotene, rarely ε-carotene, alloxanthin	Phycoerythrin, phycocyanin	Starch, oils	2 lateral flagella; gullet present in some species
Pyrrophycophyta (dinoflagellates)	*a, c*	β-carotene, peridinin, neoperidinin, dinoxanthin, neodinoxanthin		Starch, oils	2 lateral, 1 trailing, 1 girdling flagellum; in most, there is a longitudinal and transverse furrow and angular plates

* Only predominant xanthophylls are included.
† Some workers have recently separated a new division, *Eustimatophycophyta*, from these.
‡ These may be polymers of glucose molecules with variations in chemical linkages, oils, or cyclic alcohols.

of these. The predominant carotenes and xanthophylls present in the various algal divisions are shown in Table 18-1.

Biloproteins (Phycobilins)

These are water-soluble pigments, whereas chlorophylls and carotenoids are lipid-soluble. Phycobilins are pigment-protein complexes and are present in only two algal divisions: the *Rhodophycophyta* and *Cryptophycophyta*. There are two kinds of phycobilins: phycocyanin and phycoerythrin. The proportion of one kind of pigment to another can vary considerably with changes in environmental conditions. Pigment quantitation is therefore not too reliable for use as a taxonomic feature.

Motility

The motile algae, also called the **swimming algae,** have flagella occurring singly, in pairs, or in clusters at the anterior or posterior ends of the cell. Since the advent of electron microscopy, considerable variation of taxonomic significance has been found in algal flagella. It will suffice for this discussion to mention three types: whiplash (cylindrical and smooth); tinsel (cylindrical and with hairlike appendages); and ribbon, or straplike. Some algae have no means of locomotion and are carried about by tides, waves, and currents. Others move about by means other than flagella. In some forms only the zoospores, the asexual reproductive cells, are motile. Some attach themselves to the substrate in the body of water where they are living and are occasionally broken loose by currents, which move them to new locations.

A small red or orange body, the **eyespot,** is often present near the anterior end of motile algae. Other structures occurring in certain algae include spines or knobs on their exteriors and gelatinous stalks by which they may be anchored to some object.

Reproduction

Algae may reproduce either asexually or sexually. Some species are limited to one of these processes, but many have complicated life cycles employing both means of propagation.

Asexual reproductive processes in algae include the purely vegetative type of cell division by which bacteria reproduce. A new algal colony or filament may even start from a fragment of an old multicellular type from which it has broken. However, most asexual reproduction in algae is more complex than this and involves the production of unicellular spores, many of which, especially in the aquatic forms, have flagella and are motile; these are called **zoospores.** The nonmotile spores, or **aplanospores,** are more likely to be formed by the terrestrial types of algae. However, some aplanospores can develop into zoospores.

All forms of sexual reproduction are found among the algae. In these processes

Figure 18-5. *Polysiphonia lanosa,* a species of red alga, found in the littoral and sublittoral zones in eastern Canada. It is an erect plant, growing attached to other vegetation and other algae (as shown here growing attached to the brown alga *Ascophyllum nodosum*) as well as shells, stones, and woodwork. The bushy filamentous thallus is P. *lanosa* (▶). (*Courtesy of E. C. S. Chan, McGill University.*)

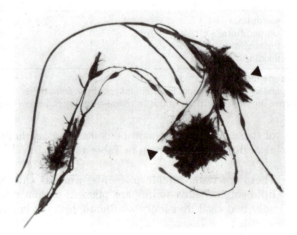

there is a fusion (conjugation) of sex cells, called gametes, to form a union in which "blending" of nuclear material occurs before new generations are formed. The union of gametes forms a zygote. If the gametes are "identical," i.e., if there is no visible sex differentiation, the fusion process is isogamous. If the two gametes are unlike, differing in size (male and female), the process is heterogamous. As we proceed to the higher, though not necessarily larger, forms of algae, the sexual cells become more characteristically male and female. The ovum (female egg cell) is large and nonmotile, and the male gamete (sperm cell) is small and actively motile. This type of sexual process is termed oogamy. Exclusively male or exclusively female thalli also exist. Although these thalli may look alike, they are of opposite sex types, since one produces male gametes and the other ova. Such plants are called unisexual or dioecious. Plants in which gametes from the same individual can unite are said to be bisexual or monoecious.

CLASSIFICATION

Although specialists do not agree on the details of algal classification, algae are generally classified on the basis of the following characteristics:

1 Nature and properties of pigments
2 Chemistry of reserve food products or assimilatory products of photosynthesis
3 Type and number, insertion (point of attachment), and morphology of flagella
4 Chemistry and physical features of cell walls
5 Morphological characteristics of cells and thalli
6 Life history, reproductive structures, and methods of reproduction

The major divisions of algae (Table 18-1) will be discussed briefly. Only a few of the outstanding characteristics of each group will be mentioned.

Rhodophycophyta— The Red Algae

The Rhodophycophyta, or red algae, are marine forms found in the warmer seas and oceans, but some of them grow in colder water as well as in freshwater. Most red algae grow in the subtidal (submerged) zone, only a few being able to survive desiccation or exposure. Rhodophycophyta are smaller than most Phaeophycophyta, rarely becoming more than 2 or 4 ft long. Their reproductive process is highly specialized. Asexual reproduction is accomplished by nonmotile spores. Sexually, however, they reproduce heterogamously by the union of well-differentiated nonmotile male and female germ cells, the spermatia and carpogonia (female sex organs), respectively. Some species deposit upon their surfaces lime from seawater; ultimately this results in deposition of lime in the ocean and plays a part in the formation of algal reefs. Several red algae are of economic importance, particularly Gelidium, from which agar is made. Irish moss (Chondrus crispus), the source of carrageenan, is another species of Rhodophycophyta. A genus (Polysiphonia) that is commonly found in marine waters is illustrated in Fig. 18-5.

Xanthophycophyta— The Yellow-Green Algae

These yellow-green algae were once classified with the green algae. However, their pale green or yellow-green coloration indicates that they have a unique group of pigments. They are found more frequently in temperate regions in freshwater and marine habitats, as well as on and in soil. But they are not dominant organisms in any environment.

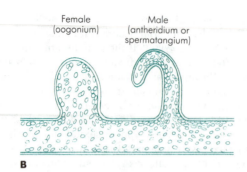

Zoospore

Female (oogonium)

Male (antheridium or spermatangium)

A

B

Figure 18-6. Asexual and sexual reproduction in *Vaucheria*. (A) Release of a single zoospore from a zoosporangium. The zoospore is multinucleated and has many pairs of slightly unequal flagella covering its surface. (B) Female (left) and male (right) reproductive structures are shown. These structures are macroscopic but the reproductive cells (sperms and eggs) are microscopic. Numerous disclike plastids are also shown. (*Erwin F. Lessel, illustrator*)

Xanthophytes exist as single cells, colonies, and both branched and unbranched filaments. Motile genera are not common, although some reproduce asexually by motile reproductive cells (zoospores). Flagella are of unequal length; the longer flagellum usually has hairs in two rows. Some have nonmotile spores. Asexual reproduction can also occur by cell division and fragmentation of filaments and colonies. Sexual reproduction is rarely observed in certain xanthophytes and is not known for others.

The xanthophyte walls are typically of cellulose and pectin. The cellular storage product is an oil or (a branched glucan) **chrysolaminarin**.

Vaucheria, the water felt, is a well-known member of this division and is widely distributed on moist soil and in both quiet and rapidly flowing water. Both freshwater and marine species are known. *Vaucheria* has a very large macroscopic filamentous form. The coenocytic filaments form a felty mass. Septa are laid down only when gametangia and zoosporangia are formed. In addition to the many nuclei in the filaments, there are thousands of ellipsoidal chromatophores or plastids.

Zoospores are formed singly in terminal sporangia in asexual reproduction. Their flagella are inserted in pairs all over the zoospore surface. Sex organs are large and develop as side branches (Fig. 18-6). Sperms are formed in elongate, curved spermatangia. They have a pair of flagella which are inserted laterally. The egg, formed singly within an erect oogonium, is uninucleate and sperms get within the wall through a pore.

Chrysophycophyta— The Golden Algae

Species of *Chrysophycophyta* are predominantly flagellates; some are amoeboid, with pseudopodial extensions of the protoplasm. The naked amoeboid forms can ingest particulate food with the pseudopodia. Nonmotile coccoid and filamentous forms are also included in the division.

The *Chrysophycophyta* differ from the green algae in the nature of their pigments, in storing reserve food as oil or chrysolaminarin rather than as starch, and in their frequent incorporation of silica. Most forms are unicellular, but some form colonies. Their characteristic color is due to the masking of their chlorophyll by brown pigments. Reproduction is commonly asexual (binary fission) but occasionally isogamous.

Ochromonas is an interesting unicellular genus with unequal flagellation. One species is remarkably versatile in nutrition in that it may grow photoautotrophically, heterotrophically, or phagotrophically. Although many chryso-

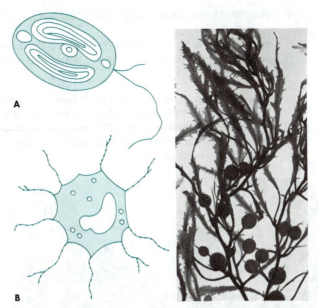

Figure 18-7. Algae of *Chrysophycophyta*. (A) *Ochromonas* sp. A commonly occurring unicellular flagellate with flagella markedly unequal in length. There are two distinctive plastids in the photosynthetic forms. (B) *Chrysamoeba* sp. It is an amoeboid cell with a number of long rhizopodia radiating from the central mass of protoplasm.

A

B

Figure 18-8. Sargassum is a type of brown alga which gets its name from the Sargasso Sea, where it is found in great abundance. The ball-like structures are air bladders. X1. (*Courtesy of General Biological Supply House.*)

phytes are to some degree amoeboid, *Chrysamoeba* is a strongly amoeboid genus with flagella (see Fig. 18-7).

Phaeophycophyta—The Brown Algae

These algae are multicellular and contain a brown pigment which gives them their characteristic color and common name of **brown algae,** or brown seaweeds. Nearly all are marine and most frequently found in the cool ocean waters. They are structurally quite complex, and some—the kelps—are large, the individual plants reaching a length of several hundred feet. Many have holdfasts; and some have air bladders, which give them buoyancy. The brown algae reproduce asexually, by zoospores, and sexually, both isogamously and heterogamously. This group includes algae used in commerce, such as the many varieties of kelp. They are used as food for humans, other animals, and fish; in medicinal preparations; in fertilizers; and as a source of iodine and mineral salts. Others, such as *Sargassum natans*, occur as a great floating population in that part of the Atlantic Ocean known as the Sargasso Sea (Fig. 18-8).

Bacillariophycophyta—The Diatoms

Members of this group are the diatoms, found in both fresh and salt water and in moist soil. Abundant in cold waters, diatoms are the most plentiful form of plankton in the Arctic. The thousands of species of diatoms provide an ever-present and abundant food supply for aquatic animals. Diatoms are either unicellular, colonial, or filamentous and occur in a wide variety of shapes (Fig. 18-9). Each cell has a single prominent nucleus and massive ribbonlike, or smaller lenslike, plastids. They produce shells (cell walls) containing silica, some of which are very beautiful. Shells of diatoms are called **frustules.** Deposits of these shells resulting from centuries of growth are called **diatomite** or **diatomaceous earth.** Although diatomite from prehistoric times is found in Oregon, Nevada, Washington, Florida, and New Jersey, the world's largest and most productive commercial source is at Lompoc, California. It is used in insulating materials; as a filter for clarifying fruit juices, cane sugar, wine, beverages, and swimming pool water; in cosmetic bases; and as polishing material.

Figure 18-9. Diatoms are unicellular algae found abundantly in fresh and salt water. Their hard silica-containing walls consist of two halves which fit together. like a Petri dish and its cover. They occur in myriads of shapes, many with beautiful surface designs. Magnifications range from X400 to X800. (*Courtesy of Johns-Manville Research Center.*)

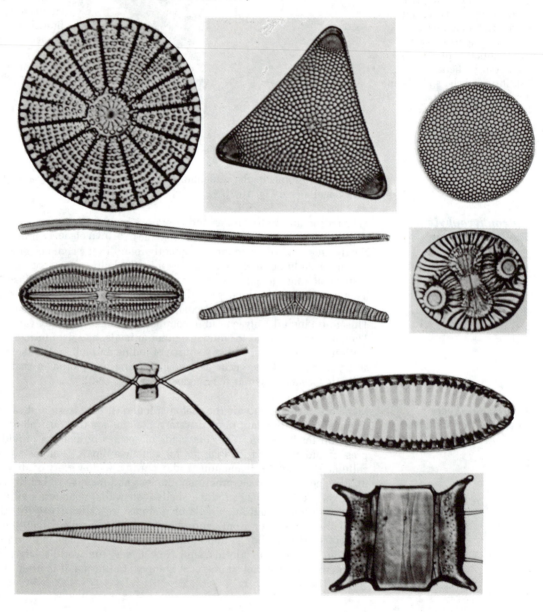

Euglenophycophyta— The Euglenoids

These unicellular organisms are actively motile by means of flagella, and they reproduce by cell division. Of particular interest is the genus *Euglena*, which is representative of a group designated as animals by some zoologists but as plants by many botanists. There are sound arguments to support either side. H. C. Bold brings them together by saying that *Euglena* may be regarded "as an organism descended from animallike ancestors and developing in the direction of the photosynthetic self-sufficiency of plants." (See Fig. 18-10.) *Euglena* is widely distributed and occurs in soil as well as in water, where it often forms a velvety film or bloom.

The *Euglena* cell is not rigid; it is pliable. There is no cell wall containing cellulose. The outer membrane is an organized periplast. An anterior "gullet" (Fig. 18-10) is present even though no food is ingested through it. Certain species develop a prominent stigma or red eyespot. Contractile vacuoles and fibrils are also present in the cell. All these are animal attributes. On the other hand, the organism carries out photosynthesis in chloroplasts and is facultatively autotrophic; these are plant attributes. The majority can assimilate organic substances during photosynthesis. A few types can even ingest particulate food through transient openings adjacent to the gullet.

Reproduction is by longitudinal binary fission. Dormant cysts are formed by all types. (Cysts are dormant stages whose walls are physically and sometimes chemically distinct from the wall of the vegetative cells from which they are derived.)

Chlorophycophyta— The Green Algae

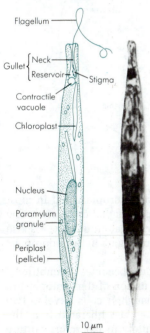

Flagellum

Gullet { Neck
         Reservoir

Stigma

Contractile vacuole

Chloroplast

Nucleus

Paramylum granule

Periplast (pellicle)

10 μm

Members of this large and diverse group of organisms, called green algae, are principally freshwater species. They are also found in seawater, and many of them are terrestrial. The cells of the *Chlorophycophyta* have a well-defined nucleus and, usually, a cell wall, and the chlorophyll and other pigments are in chloroplasts, as in higher plants. The majority of green algae contain one chloroplast per cell. It may be laminate, cup-shaped, or reticulate. The chloroplasts also often contain dense regions called pyrenoids, on which surface starch granules are formed. Food reserves are stored as starch, a product of photosynthesis.

There are many single-celled forms and many colonial types of green algae. Many unicellular green algae are motile by flagella action. Colonial types occur as spheres, filaments, or plates. Some species have special structures called holdfasts, which anchor them to submerged objects or aquatic plants.

Chlorophycophyta reproduce by zoospores, fission, and other asexual methods or by isogamous and heterogamous sexual means. From the evolutionary standpoint, they represent not only the primitive plants but intermediate forms progressing to those which reproduce by advanced heterogamy involving the union of differentiated sex cells, eggs, and sperm.

Chlamydomonas is considered a typical green alga (Fig. 18-11). It is a typical unicellular, motile, green alga and is widely distributed in soils and freshwater. Cell organization is shown in Fig. 18-12. It varies from 3 to 29 μm in common forms and is motile except during cell division. Motility is by means of two

Figure 18-10. Left: Schematic representation of a euglenoid; right: *Euglena acus.* (*Courtesy of Carolina Biological Supply Company.*)

Figure 18-11. *Chlamydomonas* in vegetative and palmelloid state. Usually the cells in the palmelloid state are nonflagellated and are embedded in a gelatinous matrix. Flagella reappear and the cells swim away when favorable conditions return (X1,500). (*Courtesy of R. G. Kessel and C. Y. Shih, Scanning Electron Microscopy in Biology, Springer-Verlag, Berlin, 1974.*)

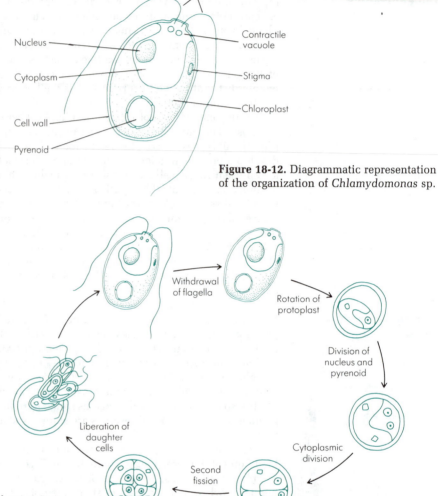

Figure 18-12. Diagrammatic representation of the organization of *Chlamydomonas* sp.

Figure 18-13. Asexual reproduction in a species of *Chlamydomonas*.

flagella. Each cell has one nucleus and a single large chloroplast that in most species is cup-shaped, although in some it may be star-shaped or layered. The cell wall contains cellulose; in many species an external gelatinous layer is also present. There is some evidence that the red eyespot or stigma in the chloroplast is the site of light perception.

In asexual reproduction the free-swimming individual becomes nonmotile by withdrawing its flagella and undergoes longitudinal fission of the protoplast to form two, four, or eight daughter protoplasts. The daughter cells develop two flagella each and construct new cell walls. They are then liberated from the parent cell wall. This cycle may be repeated indefinitely in laboratory culture or in nature (see Fig. 18-13).

In some cases, the daughter cells do not develop flagella and escape. Instead, they keep on multiplying within a more or less gelatinized matrix. The masses of cells so formed are called **palmelloid** stages (Fig. 18-11). The formation of such stages is determined by environmental conditions which are generally favorable to growth but not to motility. Any individual cell, however, can develop flagella and escape from the mass. Palmelloid stages occur in many eucaryotic algae as predominant or as occasional phases of development.

Under certain conditions, sexual reproduction occurs in many species of *Chlamydomonas*, by isogamy, heterogamy, or oogamy. The first is by far the most common. In each type the diploid zygote remains dormant for some time. Meiosis occurs as the zygote germinates. Typically, four or eight motile haploid cells are released. One of the conditions for sexual reproduction is the compatibility of mating types. When cells of compatible mating types are present, they aggregate. From these aggregates, pairs of gametes emerge and unite to form zygotes.

In addition to motile unicellular algae like *Chlamydomonas*, other, nonmotile unicellular green algae are widely distributed. One of the most important of these is *Chlorella*, which has served as a useful tool in many investigations on photosynthesis and supplemental food supply.

Volvox is a colonial green alga which may form water blooms. Its colonies are visible to the naked eye. Each colony contains from 500 to thousands of cells arranged at the surface of a watery colloidal matrix. The individual cells are biflagellate and are morphologically similar to that of *Chlamydomonas* (see Fig. 18-3C).

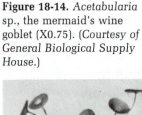

Figure 18-14. *Acetabularia* sp., the mermaid's wine goblet (X0.75). (*Courtesy of General Biological Supply House.*)

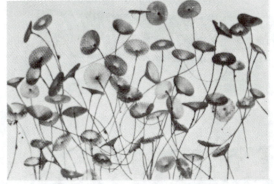

Figure 18-15. Scanning electron micrograph of a desmid, *Cosmarium* sp. The cells are uninucleate and composed of two halves with a constriction in the middle of cells. A single nucleus lies embedded in the middle of the cytoplasm that separates the two chloroplasts into semicells (SC) (X800). (*Courtesy of R. G. Kessel and C. Y. Shih, Scanning Electron Microscopy in Biology, Springer-Verlag, Berlin, 1974.*)

Acetabularia is one of the most interesting radially symmetrical forms. Figure 18-14 shows why it is called the mermaid's wine goblet.

Desmids (Fig. 18-15) are interesting green algae found in a wide variety of attractive shapes and designs. Each cell is made up of two symmetrical halves (semicells) containing one or more chloroplasts. In some, the semicells are joined by an isthmus where the nucleus lies.

Ulothrix is a filamentous form (Fig. 18-16) found in flowing streams, attached to twigs or stones by holdfasts at the bases of the filaments. It reproduces asexually by means of flagellated zoospores, which are produced in pairs or in multiples of 2 up to 16 per cell. Each zoospore may become attached to a solid object in the water and develop into a filament. Sexual reproduction may also occur, in which 32 or 64 biflagellate (with two flagella) isogametes are produced.

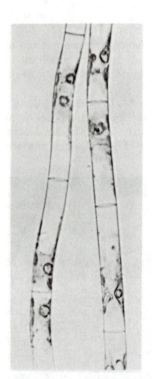

Figure 18-16. *Ulothrix.* Segments of filaments are shown. (*Courtesy of Carolina Biological Supply Company.*)

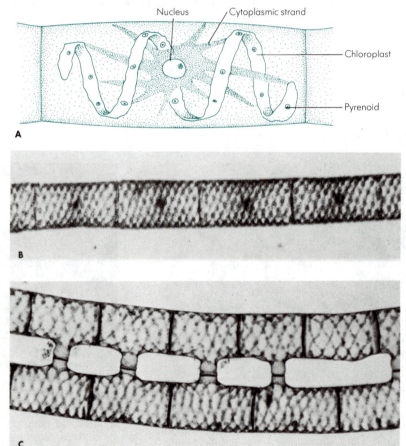

Figure 18-17. Spirogyra (A) Diagram of a vegetative cell. (B) A vegetative filament. (C) Conjugation. Formation of fertilization tubes between conjugating filaments. (*Courtesy of Carolina Biological Supply Company.*)

By fusion of a pair of gametes a zygote is formed, which germinates to form four zoospores; these then attach to a stone or another object in the water and develop into filaments.

A very common green alga is Spirogyra (Fig. 18-17), a filamentous form seen in the scums that cover ponds and slow-moving water. It is of interest because of its common occurrence and its possession of unusual chloroplasts, which are arranged spirally.

Cryptophycophyta—The Cryptomonads

The cryptomonads are a small group of biflagellate organisms. They have two unequal flagella, which arise from the base of a groove; both are of the tinsel type, with stiff hairs. The cells are slipper-shaped, are flattened into a dorsal-ventral plane, and occur singly. Some forms have a cellulose wall while others are naked, being surrounded only by a plasmalemma with a thin granular material on the outside. There are one or two plastids, with or without pyrenoids, per cell. Food reserve is stored as a true starch as well as oil. Reproduction is either by means of longitudinal cell division or the formation of zoospores or cysts. Sexual reproduction has been confirmed in the genus *Cryptomonas*. A

Figure 18-18. *Cryptomonas* cell. It has a flattened oval shape with two flagella coming from the anterior reservoir. *(Erwin F. Lessel, illustrator)*

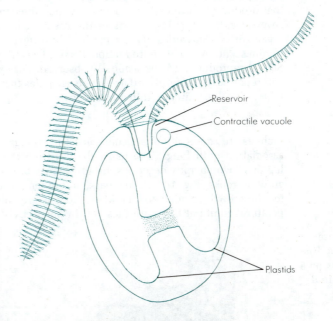

Reservoir

Contractile vacuole

Plastids

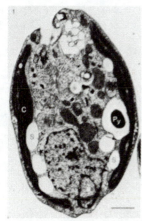

Figure 18-19. Longitudinal section of a cell of *Cryptomonas*. The periplastidal compartment which surrounds the chloroplast (C) contains starch (S), ribosomes, dense globules (arrows), and a nucleomorph (Nm) or nucleus. The chloroplast endoplasmic reticulum (CER) is continuous at places with the nuclear envelope (stars). Other structures seen are the Golgi (Go), pyrenoid (Py), and basal bodies (Bb). Bar represents 1 um. *(Courtesy of Sarah P. Gibbs, McGill University. From L. McKerracher and Sarah P. Gibbs, "Cell and Nucleomorph Division in the Alga Cryptomonas," Can J Botany,* **60***:2440–2452, 1982.)*

Cryptomonas cell is shown in Fig. 18-18. The fine structure of a cryptomonad cell is shown in Fig. 18-19.

Pyrrophycophyta— The Dinoflagellates

This division includes the dinoflagellates, a diverse group of biflagellated unicellular organisms. The dinoflagellates are so named because of their twirling motion rather than their morphology. These organisms constitute an important component of marine, brackish, and fresh bodies of water. This is another group of organisms that has both plantlike and animallike characteristics. The cells are typically flattened and have a transverse constriction, the girdle, usually around the cell equator. Distinctive features of dinoflagellates are that the flagella are inserted in the girdle and that the flagella are arranged with one encircling the cell and one trailing. Hairs project from the flagellar surface. Dinoflagellates can move a distance 100 times their own length each second! Many dinoflagellates are covered only by a plasmalemma (e.g., *Gymnodinium* is a naked organism). In some forms there is a wall made of cellulose. Still others have a series of cellulose plates within the plasmalemma. These are termed thecal plates and dinoflagellates with them are said to be armored. Dinoflagellates are an important constituent of plankton. They are best known as the organisms that produce "red tides," or blooms in which the concentration of cells may be so great as to color large areas of the ocean red, brown, or yellow. Such an organism is *Gonyaulax*. Other marine dinoflagellates such as *Noctiluca* are luminescent. Asexual reproduction takes place through division of the cell. Sexual reproduction is occasionally observed. For example, a single *Noctiluca* cell reportedly can form uniflagellate isogametes that can fuse to form a zygote, which develops directly into a *Noctiluca*.

LICHENS

Lichens are composite organisms consisting of algae or cyanobacteria living in association with fungi. The name derives from the Greek meaning "scaly or leprous," which aptly describes the appearance of many species in nature. They grow on rocks (Fig. 18-20), tree bark, and other substrates generally unsuitable for the growth of other plants. Many lichens are able to grow at the low temperatures found at high altitudes and in polar environments. Reindeer mosses

Figure 18-20. A crustose (flat, appressed) lichen, *Lasillia papulosa*, growing on a rock. (*Courtesy of Alvin R. Grove.*)

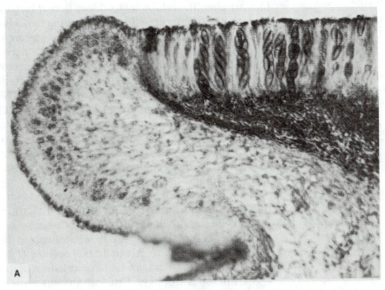

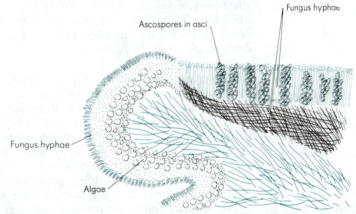

Figure 18-21. The structure of lichens is suggested by the two parts of this illustration. (A) A portion of a vertical section through a foliose lichen thallus, *Physica* sp. (X150). (*From General Biological Supply House.*) (B) Detail of algal and fungal relationships in a foliose lichen.

are lichens which furnish forage and fodder for herbivorous animals in the Arctic regions; they may cover many kilometers of land with ankle-deep growth. The colors of lichens range from white to black through shades of red, orange, yellow, and green.

Morphology

Morphologically, a simple lichen is made up of a top layer of tightly woven fungal mycelia, below this a layer of photosynthetic cells, and below that another layer of fungus (Fig. 18-21). The bottom layer may attach to the substrate directly or by means of short twisted strands of hyphae, called **rhizines,** which serve as anchors. Some lichens are more complex than this. For example, in some a midlayer of fungus directly below the alga appears to be a reservoir for stored food.

Not all species of algae (or cyanobacteria) or all species of fungi can enter into a lichenlike relationship. Most lichen fungi are *Ascomycetes*, although a few are *Basidiomycetes*. The algae belong to division *Chlorophycophyta* (green algae). Lichen fungi are unique in their preference for growth in association with algae or cyanobacteria. Typically, each lichen thallus consists of a single species of fungus associated with a single species of alga or cyanobacterium. About 30 different genera of algae and cyanobacteria have been found in lichens. The commonest is the unicellular green alga *Trebouxia*. The other two most common genera are *Trentepohlia* (a filamentous alga) and *Nostoc* (a cyanobacterium). These three genera account for the photosynthetic partner in over 90 percent of all lichen species. However, each lichen species does not necessarily have a specific photosynthetic partner. For example, one *Calothrix* species is found in *Lichina confinis* in Norway or Spain, but another photosynthetic species is found in *L. confinis* in Sweden.

The assignment of a single Latin binomial to an "organism" consisting of two different microorganisms raises obvious difficulties. Consequently, virtually all the morphological characteristics now used in lichen classification are fungal; and it is acceptable to speak of a "lichenized fungus," with the Latin name referring to the fungus. About 18,000 species of lichens have been described.

There are three kinds of lichens: **crustose** (flat, appressed); **foliose** (leaflike); and **fruticose** (shrublike). Crustose lichens grow closely appressed to the substrate (rock, wood, etc.), or even within its surface. Foliose lichens are flattened like leaves but may not be connected to the substrate at all points. Fruticose lichens have an erect shrublike or filamentous morphology and can be about 10 cm high. (See Figs. 18-20 and 18-22.)

Reproduction

Figure 18-22. Examples of two types of lichens. (A) foliose lichen; (B) fruticose lichen. (*Courtesy of Carolina Biological Supply Company.*)

Lichens reproduce predominantly by vegetative processes. Propagation by fragmentation occurs when bits of the thallus are broken off from the parent plant and fall on a suitable substrate. Lichens may produce asexual "reproductive bodies" called **soredia** which are knots of hyphae containing a few algal cells. During sexual reproduction, the components of lichens reproduce independently of each other. The fungal components produce ascospores, and when they germinate they must come in contact with algal cells or they will not be able to survive. The algal partner reproduces by cell division or occasionally by spores. Lichens grow slowly because of their low metabolic rate. Many species grow less than a centimeter per year. Some lichens have therefore reached great age in nature, for example, 4,500 years for some species in the Arctic regions.

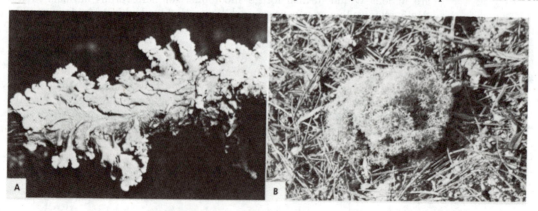

Such longevity suggests a very well-balanced association between the symbionts. They are also very resistant to heat and desiccation.

Symbiotic Nature

It is generally agreed that lichens are the product of a mutualistic symbiosis, in which each partner of the association derives something essential or useful for its survival. Apparently the alga or cyanobacterium provides the fungus with food, particularly carbohydrates produced by photosynthesis, and possibly with vitamins. The fungus probably absorbs, stores, and supplies water and minerals required by the alga or cyanobacterium, as well as providing carbon dioxide (a needed growth factor) and a moist, sheltered supporting framework for the photosynthetic partner. The fungus is able to obtain nourishment from its partner by means of tiny rootlike projections called **haustoria** (singular, **haustorium**) which penetrate the photosynthetic cell.

Chemical Interaction

The chemical interaction of lichens is interesting because many of their organic products ("lichen substances") are unique. Unusual fats and phenolic compounds make up from 2 to 20 percent of the dry weight of lichen bodies. Litmus, the well-known pigment indicator, is obtained from lichens. Essential oils from some species are used in perfumes. Lichen pigments are used in England to color many of the woolen fabrics of Harris tweeds. Interest has grown in lichens as a source of new chemical substances that show promise for a variety of applications.

QUESTIONS

1 How are algae similar to and different from higher green plants?
2 How would you distinguish unicellular algae from photosynthetic bacteria?
3 Where are algae found?
4 Why are algae important in the food chain of aquatic environments?
5 Discuss the various uses of algae that make them commercially important.
6 Describe some of the algae used as food for human consumption.
7 Explain the etiology of shellfish poisoning.
8 Discuss the characteristics of algae that are used as a basis for algal classification.
9 What is diatomaceous earth? Why is it used commercially?
10 What are the attributes *Euglena* shares with plants and with animals?
11 Describe briefly the life history of *Chlamydomonas*.
12 Why are dinoflagellates of medical concern?
13 Describe the microbial association that exists in a lichen.

REFERENCES

Bold, H. D., and M. J. Wynne: *Introduction to the Algae. Structure and Reproduction,* Prentice-Hall, Englewood Cliffs, N.J., 1978. *A comprehensive, well-written, and very well referenced scholarly text for the teaching of general phycology. Very well illustrated with line drawings and both photomicrographs and electron micrographs.*

Pickett-Heaps, J. D.: *Green Algae. Structure, Reproduction and Evolution in Selected Genera,* Sinauer Associates, Sunderland, Mass., 1975. *This book describes the green algae as seen by light- and electron-microscopy in reasonably complete detail as they pass through various stages of their life cycles. Beautifully illustrated and clearly written.*

Smith, D. C.: *The Lichen Symbiosis*, Oxford, London, 1973. *A short monograph on the biology of lichens.*

Trainor, F. R.: *Introductory Phycology*, Wiley, New York, 1978. *A well-written and well-organized textbook on the algae, written for undergraduate students at the university level. Good line drawings as well as good photographs.*

Chapter 19   Protozoa

Protozoa (singular, **protozoan**), from the Greek *protos* and *zoon*, meaning "first animal," are eucaryotic protists. They occur generally as single cells and may be distinguished from other eucaryotic protists by their ability to move at some stage of their life cycle and by their lack of cell walls. Protozoa are predominantly microscopic in size. The majority are between 5 and 250 μm in diameter. Colonies of protozoa also occur. In a protozoan colony, the individual cells are joined by cytoplasmic threads or are embedded in a common matrix. Thus colonies of protozoa are essentially aggregates of independent cells. The study of these eucaryotic protists is called **protozoology.**

By a conservative estimate, there are more than 65,000 described species of protozoa distributed among seven named phyla. (In 1964, there were about 48,000 members of what was then known as a single phylum *Protozoa*.) Slightly more than 50 percent of the species are fossil forms. Of the remaining 50 percent, some 22,000 are free-living species while 10,000 species are parasitic. Of the latter, only a few species cause disease in humans, but these few inflict much misery and death on millions of people. Even though the present species count appears staggering, there are hundreds of thousands of species yet to be described (even in 1983, an average of two new protozoan species were discovered daily).

OCCURRENCE OF PROTOZOA

Protozoa are found in all moist habitats. They are common in the sea, in soil, and in freshwater. Free-living protozoa have even been found in the polar regions and at very high altitudes. Parasitic protozoa may be found in association with most animal groups. Many protozoa survive dry conditions by the formation of a resistant cyst, or dormant stage. For example, the soil amoeba *Naegleria* is a resistant cyst in dry weather, is a naked amoeba in moist soil, and becomes flagellated when flooded with water. Parasitic protozoa can modify their morphology and physiology to cope with a change in host. For example, the malarial parasite *Plasmodium* produces male gametes in response to a drop in temperature on transfer from a warm-blooded mammalian host to a mosquito.

The distribution of trophic (vegetative) forms of protozoa in the sea and freshwater and of cyst forms in the atmosphere has resulted in the spread of free-living species throughout the world.

ECOLOGY OF PROTOZOA

From the ecological standpoint, protozoa may be divided into free-living forms and those living on or in other organisms. The latter group is referred to as the symbiotic protozoa. Some of the symbiotic ones are parasitic and may cause disease. Others such as those found in the gut of the termite are beneficial to the host (live in a mutualistic association).

Free-Living Protozoa

Free-living protozoa are found in a variety of habitats. The factors which influence the distribution and number of free-living protozoa in a habitat are moisture, temperature, light, available nutrients, and other physical and chemical conditions. The vegetative, or trophic stages of free-living protozoa occur in every type of salt water, freshwater, sand, soil, and decaying organic matter.

Light

Obviously, for those protozoa which bear chromatophores (these protozoa are considered algae by phycologists) and carry out photosynthesis, sunlight is essential. It follows also that those protozoa which feed on photosynthetic microorganisms also require sunlight, albeit indirectly. On the other hand, some protozoa avoid light and thrive in a environment where it is absent.

Hydrogen-Ion Concentration

Some protozoa can tolerate a wide range of pH, for example, pH 3.2 to 8.7. However, for the majority a pH range of 6.0 to 8.0 is optimal for maximum metabolic activity.

Nutrients

The protozoan population in an aquatic environment is influenced by the chemical constituents of the water. Some protozoa thrive in water rich in oxygen but low in organic matter (mountain springs, brooks, or ponds); others require water rich in minerals. Some grow in water where there is active oxidation and degradation of organic matter (the majority of freshwater protozoa, such as the ciliates). Still others prefer water with little oxygen but many decomposition products (e.g., black bottom slime and sewage). Some species have been found to live in both salt water and freshwater.

The nutrient supply in a habitat is a major determining factor in the distribution and number of protozoa within it. Species of *Paramecium* and other holozoic protozoa (protozoa that eat other organisms) must have a supply of

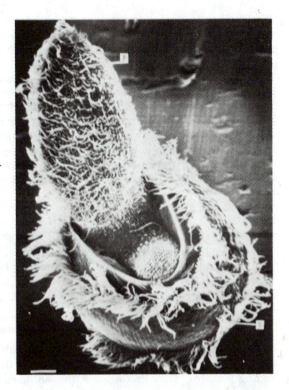

Figure 19-1. A scanning electron micrograph of a paramecium (P) being ingested by a *Didinium* (D). Note that the cilia on the paramecium are being lysed as the paramecium enters the cytopharynx of the predator. The paramecium is about twice the size of the *Didinium*. But it is gulped down as shown, a situation comparable to a human grabbing a cow and swallowing it whole! The bar represents 10 μm. (*Courtesy of Eugene B. Small and Donald S. Marszalek, Science,* **63:**1064–1065, 1969. *Copyright 1969 by the American Association for the Advancement of Science.*)

bacteria or other protozoa (Fig. 19-1). As a general rule, holozoic protozoa which feed on a variety of organisms are widely distributed; those that are more selective and feed only a few species are limited in their distribution.

Temperature

Most protozoa have an optimum temperature of between 16 and 25°C; the maximum is between 36 and 40°C. The minimum temperature is less detrimental. The temperature tolerance varies with different environmental conditions. Even warm waters (30 to 56°C) of hot springs have been known to contain protozoa. The so-called red snow of high altitudes is due to the presence of several hematochrome-bearing flagellates (considered algae by some biologists). In the **encysted** stage (a thick-walled structure in an inactive stage), protozoa can withstand a far greater temperature variation than in the trophic stage.

Symbiotic Protozoa

The association between these protozoa and their hosts or other organisms can differ in various ways. The term *symbiotic* describes any type of coexistence between different organisms.

In *commensalism* the host is neither injured nor benefited, but the commensal is benefited. **Ectocommensalism** is often represented by protozoa which attach themselves to a host's body. **Endocommensalism** is the association when the protozoan is inside the host's body, e.g., the protozoa which live in the lumen of the alimentary tract.

Mutualism occurs between some protozoa and their hosts. For example, certain flagellates are present in the gut of termites and digest the woody material eaten by the termite to a glycogenous substance which can be used by the host cells. If deprived of these flagellates, the termite dies; if the flagellates are removed from the termite gut, they too perish.

In parasitism, one organism—the parasite—lives at the expense of the other. The parasite feeds on the host cells or cell fragments by pseudopodia or cytostome (an opening for ingestion of food; see Fig. 19-4), or enters the host tissues and cells, living upon the cytoplasm and even the nuclei. As a result, the host may develop pathological conditions. The sporozoa are strictly parasitic and are among the most important of the disease-producing protozoa.

Some parasitic protozoa parasitize other protozoan or metazoan (animals whose bodies consist of many cells) parasites. Such an association is termed hyperparasitism.

THE IMPORTANCE OF PROTOZOA

Protozoa serve as an important link in the food chain of communities in aquatic environments (see Chap. 26). For example, in marine waters, zooplankton (animallike organisms) are protozoa that feed on the photosynthetic phytoplankton (plantlike organisms). They in turn become food for larger marine organisms. This food chain can be represented as follows:

Light energy → Phytoplankton → Zooplankton → Carnivores
(Primary producers)   (Primary consumers)   (Secondary consumers)

Also of particular importance in the ecological balance of many communities, in wetlands as well as aquatic environments, are the saprophytic and bacteria-feeding protozoa. They make use of the substances produced and organisms involved in the final decomposition stage of organic matter. This can be represented by the following sequence:

Dead bodies of producers → Decomposition by → Ingestion of
and consumers, and their   fungi and bacteria   bacteria by
excretion products,   protozoa
including feces

It follows that microorganisms are important in the degradation of sewage. Although bacteria participate in a prominent way in the process, the role of protozoa is becoming more completely understood and appreciated. Biological sewage treatment involves both anaerobic digestion and/or aeration (see Chap. 27). Anaerobic protozoa such as species of *Metopus, Saprodinium,* and *Epalxis* are active in the anaerobic steps, while those treatment steps requiring aeration and flocculation include the aerobic protozoa such as *Bodo, Paramecium, Aspidisca,* and *Vorticella.*

In the treatment of industrial wastes, where there is an accumulation of nitrates and phosphates, the settling tanks are illuminated to promote the growth of algae and protozoa. These protists remove the inorganic material from the water for their own synthesis. Water quality is improved, and the autotrophs are skimmed from the water surface, dried, and used as fertilizer.

Some protozoa cause disease in animals, including humans. They have caused untold misery. Such parasitic protozoa multiply within the host much as bacteria do. Some live only as obligate parasites and may produce chronic or acute diseases in humans. Some well-known protozoan diseases in humans are intestinal amoebiasis, African sleeping sickness, and malaria (see Chap. 38).

Protozoa have also become important research organisms for biologists and biochemists for the following reasons. Many protozoa are easily cultured and maintained in the laboratory. Their capacity to reproduce asexually enables clones to be established with the same genetic makeup.

Studies of mating types and killer particles in *Paramecium* have shown a relationship between genotype and the maintenance of cytoplasmic inclusions and endosymbionts. *Tetrahymena*, *Euplotes*, and *Paramecium* species have been used to study cell cycles and nucleic acid biosynthesis during cell division.

MORPHOLOGY OF PROTOZOA

The size and shape of these organisms show considerable variation. For example, *Leishmania donovani*, the cause of the human disease kala azar, measures 1 to 4 μm in length. *Amoeba proteus* measures 600 μm or more (Fig. 19-2). Certain common ciliates reach 2,000 μm, or 2 mm, and the tests (a kind of protective envelope) of some extinct (fossilized) members of *Foraminiferida*, the nummulites, measure up to 15 cm in diameter.

Intracellular Structures

Like all eucaryotic cells, the protozoan cell also consists of cytoplasm, separated

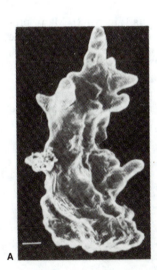

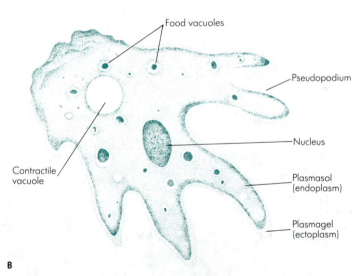

A

B

Figure 19.2. (A) *Amoeba proteus* photographed with a scanning electron microscope. By this technique the pseudopodia and other structures of the cell are remarkably clear. The bar represents 10 μm. (*Courtesy of Eugene B. Small and Donald S. Marszalek, Science, 63:1064–1065, 1969. Copyright 1969 by the American Association for the Advancement of Science. By permission.*) (B) Anatomy of *Amoeba proteus*. (*Erwin F. Lessel, illustrator.*)

from the surrounding medium by a special cell envelope, and the nucleus or nuclei.

Cytoplasm

The cytoplasm is a more or less homogeneous substance consisting of globular protein molecules loosely linked together to form a three-dimensional molecular framework. Embedded within it are the various structures that give protozoan cells their characteristic appearance.

Submicroscopic protein fibrils **(fibrillar bundles, myonemes,** and **microtubules)** are groups of parallel fibrils in the cytoplasm. Protozoan contractility is probably due to these fibrils.

In several forms of protozoa, pigments are diffused throughout the cytoplasm. The hues are numerous; e.g., they can be green, brown, blue, purple, or rose.

In the majority of protozoa, the cytoplasm is differentiated into the **ectoplasm** and the **endoplasm.** The ectoplasm is more gel-like and the endoplasm is more voluminous and fluid, but the change from one layer to another is gradual. Structures are predominantly found in the endoplasm (see Fig. 19-2).

Like other eucaryotic cells, protozoa have membrane systems in the cytoplasm. They form a more or less continuous network of canals and lacunae giving rise to the **endoplasmic reticulum** of the cell. Other structures in the cytoplasm include ribosomes, Golgi complexes or **dictyosomes** (piles of membranous sacs), mitochrondria, **kinetosomes** or **blepharoplasts** (intracytoplasmic basal bodies of cilia or flagella), food vacuoles, contractile vacuoles, and nuclei (Fig. 19-2).

Nucleus

The protozoan cell has at least one eucaryotic nucleus (Fig. 19-3). Many protozoa, however, have multiple nuclei (e.g., almost all ciliates) throughout the greater part of the life cycle. The protozoan nuclei are of various forms, sizes, and structures. In several species, each individual organism has two similar nuclei. In the ciliates, two dissimilar nuclei, one large (macronucleus) and one small (micronucleus), are present (Fig. 19-4). The macronucleus controls the metabolic activities and regeneration processes; the micronucleus is concerned with reproductive activity.

Figure 19-3. Section of the amoeba *Entamoeba histolytica.* (*Courtesy of Z. Ali-Khan and Margaret Gomersall, McGill University.*)

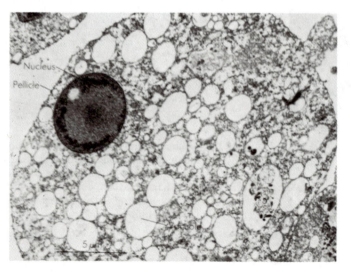

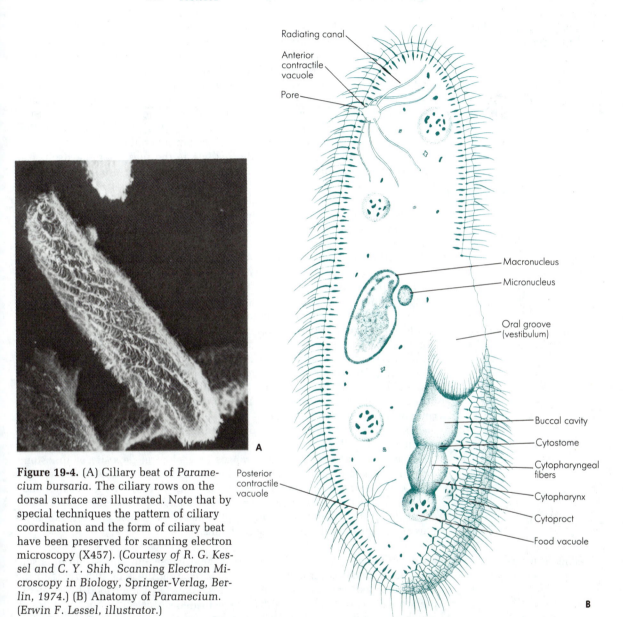

Figure 19-4. (A) Ciliary beat of *Paramecium bursaria*. The ciliary rows on the dorsal surface are illustrated. Note that by special techniques the pattern of ciliary coordination and the form of ciliary beat have been preserved for scanning electron microscopy (X457). (*Courtesy of R. G. Kessel and C. Y. Shih, Scanning Electron Microscopy in Biology, Springer-Verlag, Berlin, 1974.*) (B) Anatomy of *Paramecium*. (*Erwin F. Lessel, illustrator.*)

The essential structural elements of the nucleus are the chromosomes, the nucleolar substance, the nuclear membrane, and the karyoplasm (nucleoplasm). It has been shown that the number of chromosomes is constant for a particular species of protozoan. For example, *Spirotrichonympha polygyra* has 2 haploid chromosomes; *Spirotrichosoma magnum* has 60. Some protozoa divide only asexually (by mitosis). Others may divide either asexually or sexually (by meiosis).

Plasmalemma and Other Cell Coverings

The cytoplasm with its various structures is separated from the external environment by a cell unit membrane **(plasmalemma)**. The plasmalemma not only provides protection but also controls exchange of substances (semipermeable); it is the site of perception of chemical and mechanical stimuli as well as the establishment of contact with other cells (cell sensitivity to external factors). Although all protozoa possess a cell membrane, many protozoa have compound coverings of membranes modified for protection, support, and movement. Such combinations of membranes are referred to as a **pellicle.** In *Euglena* the pellicle is organized to ensure flexibility; in *Paramecium* it is quite rigid.

Actually, in its simplest form, the pellicle is the plasmalemma itself, e.g., amoebas are surrounded by a plasmalemma only. However, even in these, some species (e.g., *A. proteus*) have a diffuse layer of mucopolysaccharides over the plasmalemma. This layer is thought to play an important role in **pinocytosis** (uptake of fluids and soluble nutrients through small invaginations in the cell membrane that subsequently form intracellular vesicles) or in adhesion of the cell to the substratum.

The pellicle of a ciliate is thick and often variously ridged and sculptured. There may even be rows of elevated platelets and nodular thickenings. For example, the pellicle of *Paramecium* consists of three membranes, the outer one sculptured in a series of evenly distributed polygons. The contact between adjacent polygons results in a series of ridges, giving a latticed pattern.

Additional protective coverings external to the pellicle have also evolved for some protozoa. This results in the great diversity of forms exhibited by protozoa. These coverings are known variously as **thecae, shells, tests,** or **loricae** and occur in almost all major groups of protozoa. The theca is a secreted layer directly apposed to the cell surface. Tests, shells, and lorica are coverings that are loose-fitting. Special openings provide the connection with the environment. The coverings consist of very different materials. In general, they have an organic matrix reinforced by incrustation of inorganic substances such as calcium carbonate or silica.

We are familiar with mountain ranges or geological deposits of limestone, fusuline chalk, and green sandstone. These were formed by continuous sinking of calcareous shells and silicon skeletons of planktonic amoebas and other protozoa to the bottom of ancient oceans. For example, the white cliffs of Dover are made up of billions of scales of the phytoflagellates called coccolithopharids plus the shells of millions of foraminiferans.

Feeding Structures

Food-gathering structures in the protozoa are diverse and range from the pseudopodia of amoebas through the tentacular feeding tubes of suctorians to the well-developed "mouths" of many ciliates. Amoebas gather food by means of pseudopodial engulfment. In ciliates the **cytostome** is the actual opening through which food is ingested (Fig. 19-5). It ranges from a simple round opening to a slitlike structure surrounded by feeding membranelles. It is usually found anteriorly and remains open all the time in some groups; in other species it can be opened and closed.

An **oral groove** (Fig. 19-4B) is an indentation in the pellicle of certain ciliates. It guides food toward the cytostome and acts as a concentrating device. The addition of membranelles to the oral groove makes it a **peristome** (Fig. 19-5).

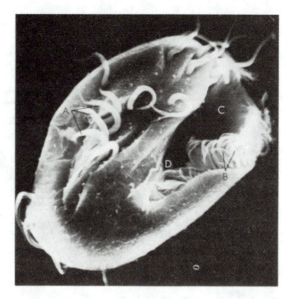

Figure 19-5. Ventral surface of a *Euplotes aediculatus* cell. (A) Cirri. Each cirrus is composed of 80 to 100 individual cilia that are not fused but beat as a functional unit in locomotion. (B) Cilia. Two to three long rows of cilia that function in locomotion as well as in food collection. (C) Peristome (buccal cavity). (D) Cytostome (cell "mouth") (X512). (*Courtesy of John A. Kloetzel, University of Maryland, Baltimore County.*)

On its edges are located cilia that function to facilitate feeding. The **cytopharynx** is a region through which nutrients must pass to be enclosed in a food vacuole.

Cysts

Many protozoa form resistant cysts at certain times of their life cycle. As indicated before, these cysts are able to survive adverse environmental conditions such as desiccation, low nutrient supply, and even anaerobiosis. In parasitic protozoa, the developmental stages are often transmitted from host to host within a cyst. Other kinds of cysts (e.g., reproductive) are also known.

Intestinal protozoa comprise about 50 percent of the parasitic species. In most instances they enter the alimentary tract as resting cysts, hatch in a suitable region, and then leave the host again as dormant cysts. By this means they can survive for long periods outside the host. Asexual reproduction in some ciliates and flagellates is associated with cyst formation. Sexual reproduction of sporozoa invariably results in a cyst.

The cyst wall is secreted as a closely fitting extracellular coat or structure. The cytoplasm is commonly attached to the cyst wall at one or several points; it is reduced in size and dormant.

Other Protective Structures

Some protozoa protect themselves with structures other than external coverings. These include various structures or materials formed within membrane-bound vesicles. For example, certain ciliates secrete a mucilage from subpellicular vesicles called **mucocysts**. Several protozoa probably defend themselves by the expulsion of harpoonlike **trichocysts**, although their function has not been actually proved. Other protozoa, such as *Didinium* sp. (Fig. 19-1), have **toxicysts**. These have a threadlike tubular structure with an occlusion at the distal end which may contain toxin. When the toxicyst is discharged the toxin is distributed along the surface of the thread. Toxicysts are used to paralyze and capture prey; the toxin causes paralysis and cytolysis when it contacts protozoan prey.

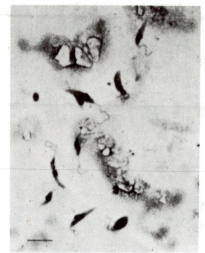
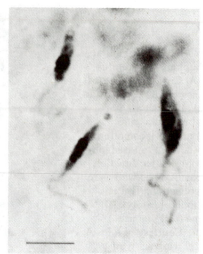

Figure 19-6. *Crithidia* sp., a flagellate, at two magnifications. Bar represents 10 μm. (*Courtesy of Z. Ali-Khan and E. C. S. Chan, McGill University.*)

Figure 19-7. *Trypanosoma brucei* exhibiting a flagellum with undulating membrane on each cell. Giemsa stain (X1,560). (*Courtesy of P. Gardiner, International Laboratory for Research on Animal Diseases, Nairobi, Kenya, and the Society of Protozoologists. From P. R. Gardiner, L. C. Lamont, T. W. Jones, and I. Cunningham, J Protozool,* **27:**182–185, 1980.)

In a similar manner, **haptocysts** occur in the tentacles of suctorian protozoa and are used to contact and immobilize prey.

Locomotor Organelles

Protozoa may move by three types of specialized organelles: pseudopodia, flagella, and cilia. In addition, a few protozoa without such organelles can carry out a gliding movement by body flexion.

Pseudopodia

A pseudopodium is a temporary projection of part of the cytoplasm of those protozoa which do not have a rigid pellicle (Fig. 19-2). Pseudopodia are therefore characteristic of the amoebas (*Sarcodina*). These organelles are also used for capturing food substances.

Flagella and Cilia

The flagellum is an extremely fine filamentous extension of the cell (Fig. 19-6). As a rule, the number of flagella present in an individual protozoan varies from one to eight; one or two is the most frequent number. A flagellum is composed of two parts: an elastic filament called an **axoneme** and the contractile cytoplasmic sheath that surrounds the axoneme.

In certain parasitic *Mastigophora*, such as *Trypanosoma* (Fig. 19-7), there is

Figure 19-8. Transverse section of an infective form of *Trypanosoma brucei* showing the uniform, electron-dense surface coat covering the cell body and the flagellum (with the typical "9 + 2" structure). Surface coat or pellicle = SC; F = flagellum; MT = microtubules. (X92,500). (*Courtesy of P. Gardiner, International Laboratory for Research on Animal Diseases, Nairobi, Kenya, and the Society of Protozoologists. From P. R. Gardiner, L. C. Lamont, T. W. Jones, and I. Cunningham, J Protozool, 27:182–185, 1980.*)

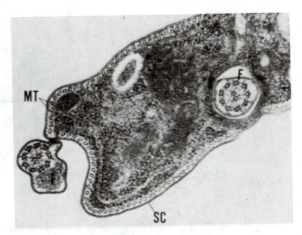

a very delicate membrane that extends out from the side of the body with a flagellum bordering its outer margin. When the membrane vibrates, it shows a characteristic undulating movement; thus it is called the **undulating membrane.**

Cilia, in addition to their locomotor function, also aid in the ingestion of food and serve often as a tactile organelle. They are fine and short threadlike extensions from the cell (Figs. 19-4 and 19-5). They may be uniform in length or may be of different lengths depending on their location. Generally, cilia are arranged in longitudinal, oblique, or spiral rows, inserted either on the ridges or in the furrows.

Electron microscopy has shown that the fine structure of the flagella and cilia of all eucaryotes follows the same basic design. Sections show two central and nine double peripheral microtubules ("9 + 2" structure) along most of the length of the shaft, which is enveloped by a membrane continuous with the pellicle (Fig. 19-8).

REPRODUCTION OF PROTOZOA

As a general rule, protozoa multiply by asexual reproduction; the majority of higher animals reproduce by sexual means. This is not to say that sexual processes are absent in the protozoa. Indeed, many protozoa are able to carry out both asexual and sexual processes. Some parasitic forms may have an asexual phase in one host and a sexual phase in another host.

Asexual Reproduction

Asexual reproduction occurs by simple cell division, which can be **equal** or **unequal**—the daughter cells are of equal or unequal sizes, respectively. If two daughter cells are formed, then the process is called **binary fission;** if many daughter cells are formed, it is **multiple fission. Budding** is a variation of unequal cell division.

Binary Fission

The simplest form of binary fission is found in the amoebas (Fig. 19-9). The pseudopodia are withdrawn before the nucleus divides. After the nucleus divides, the organism elongates and constricts in the center in order to form two daughter cells.

Amoebas with special protective coverings are more complex in their manner

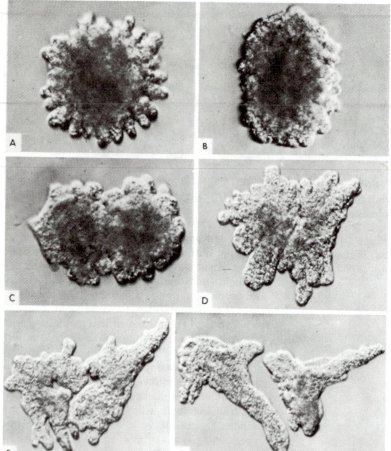

Figure 19-9. Amoebas reproduce by fission as shown in this series of photomicrographs of the dividing amoeba. Total elapsed time from (A) through (F) is 21 min. Intermediate stages photographed at (B) 6 min, (C) 8 min, (D) 15 min, (E) 18 min, and (F) 21 min. Reproductive material is concentrated in a band at the equator of the nucleus; the chromosomes divide, half of each going to one of the two new nuclei. As the cell divides one of the newly formed nuclei goes to each daughter cell; reproduction is complete when the cells are completely separated. (*Courtesy of Carolina Biological Supply Company.*)

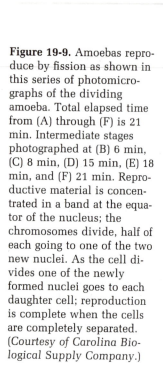

of binary fission, which is directly related to the type of covering they possess. In those with soft coverings, the division plane is longitudinal along the body axis and the covering constricts into two halves. In those with more rigid coverings, part of the cytoplasm protrudes from the aperture (opening in covering) to secrete a new covering over its surface. Only after the formation of the new covering does nuclear division proceed, and binary fission is completed by cytoplasmic division.

In flagellates, with the exception of the dinoflagellates, fission is longitudinal along the major body axis. Since the flagella themselves are incapable of division, they must be regenerated from basal bodies (the blepharoplasts which arise in the vicinity of the old basal bodies. Thus multiplication of basal bodies usually precedes cell division. In dinoflagellates, division is at right angles to the cell axis because the flagella which determine the plane of division are not at the front end but at the side of the cell.

Transverse fission is characteristic for ciliates. Fission occurs at a right angle

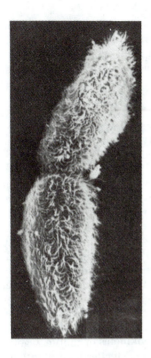

Figure 19-10. Binary fission in *Paramecium multimicronucleatum*. Note the transverse constriction furrow which extends inward from the equator so that one organism is eventually divided into two daughter organisms, each of which eventually achieves the volume of the parent (X550). (*Courtesy of R. G. Kessel and C. Y. Shih, Scanning Electron Microscopy in Biology, Springer-Verlag, Berlin, 1974.*)

to the long axis of the cell (Fig. 19-10). In the *simplest* case of transverse fission, an equatorial furrow appears first which separates the surface cell layer into an anterior and a posterior half. A constriction follows, leading to the separation of two daughter cells. Their form and structure usually indicate from which half of the mother cell they developed.

The presence of cilia and other complex organelles has a profound influence on cell division in the ciliates. For example, ciliates have two types of nuclei—the macronucleus, which determines vegetative processes, and the micronucleus, which is involved in sexual processes. During asexual binary fission the diploid micronucleus divides normally by mitotic division. The macronucleus undergoes DNA synthesis and divides into two portions without the regular reduplication of the chromosomes (amitosis). However, each daughter macronucleus contains the full complement of genes (in fact, in multiple sets—polyploidy). In some primitive ciliates found in marine sands of the intertidal zone, the macronucleus is incapable of division; after fission of the organism a new one is formed from the micronucleus. In general, with regard to the other organelles, those originally present in each half are retained, and those which are lost during division are regenerated. Depending on the extent of differentiation of the species, divisions may involve extensive reorganizations, including transformation of preexisting structures and formation of new ones.

Multiple Fission

In multiple fission, a single mother (parental) cell divides to form many daughter (filial) cells. Division is usually preceded by formation of multiple nuclei within the mother cell, which then cleaves rapidly to form a corresponding number of daughter cells. Multiple fission is not as widespread as binary fission but it often takes place in addition to the latter process. In ciliates and flagellates, this type of fission is found in relatively few species. Multiple fission occurs commonly in the foraminifera, the radiolaria, and the heliozoa. Perhaps the best-known examples of multiple fission are found in the sporozoa, e.g., in the malarial parasite *Plasmodium* where it is known as schizogony and serves to spread the parasite quickly in the host.

Budding

The term *budding* is not used in the same sense as mycologists use it in describing the asexual reproductive process in the yeast *Saccharomyces*. Instead, in protozoology it is often used to describe the varied processes by which sessile protozoa produce motile offspring. That is, the mother cell remains sessile and releases one or more swarming daughter cells. The swarmer differs from the parent cell not only in a lower degree of differentiation but also in the possession of special locomotor organelles. Some form of budding is found in all sessile ciliates and is used to disseminate the species while the mother cell remains in situ.

Budding can be **exogenous** or **endogenous**. The former involves formation and separation of the bud toward the outside. Definite exogenous budding is seen in suctorians when a portion or portions of an adult sessile suctorian bud from the parent, develop cilia, and swim away. In endogenous budding, which also occurs in many suctorian species, the swarmer is formed inside the mother cell (see Fig. 19-17F).

Sexual Reproduction

Various types of sexual reproduction have been observed among protozoa. Sexual fusion of two gametes (**syngamy** or **gametogamy**) occurs in various groups of protozoa. Conjugation, which is generally a temporary union of two individuals for the purpose of exchanging nuclear material, is a sexual process found exclusively in the ciliates (see Fig. 19-11). After exchange of nuclei, the conjugants separate and each of them gives rise to its respective progeny by fission or budding. However, some ciliates show "total conjugation," with complete fusion of the two organisms.

When the gametes (which develop from trophozoites) are morphologically alike, they are called **isogametes.** When they are unlike in morphology (as well as physiology), they are **anisogametes** and can be either **microgametes** or **macrogametes.** That is, they are like the spermatozoa and the ova of metazoa, respectively. Thus microgametes are motile, relatively small, and usually numerous in comparison to macrogametes. Anisogametes are common among the sporozoa. For example, in *Plasmodium vivax* (a sporozoan that causes a type of malaria), anisogamy results in the formation of **ookinetes** or motile zygotes which give rise to a large number of **sporozoites** (long, slender bodies with an oval nucleus and firm cuticle, capable of producing new infection).

Regeneration

The capacity to regenerate lost parts is characteristic of all protozoa, from simple forms to those with highly complex structures. When a protozoan is cut in two,

Figure 19-11. Pair of *Euplotes aediculatus* cells during conjugation. They are united by their apposed ventral surfaces (X710). (*Courtesy of John A. Kloetzel, University of Maryland, Baltimore County.*)

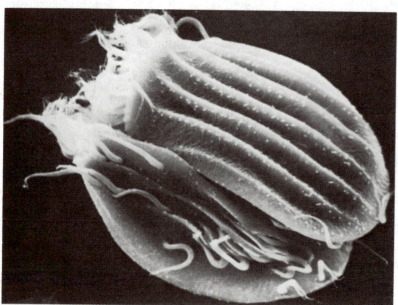

the nucleated portion regenerates but the anucleated portion degenerates. In general, the nucleus is necessary for regeneration. In ciliates, the macronucleus alone (or even just a portion of it) is sufficient for this process.

CLASSIFICATION OF PROTOZOA

The myriad forms of protozoa have been grouped together, not because they are all related in an evolutionary way, but simply as a matter of convenience. The classification scheme of the protozoa now in common use was developed by the Committee on Systematics and Evolution of the Society of Protozoologists and published in 1980 (see reference at the end of the chapter). Electron microscopy has played a significant role in the revision of this classification scheme. The old classification schemes of protozoa were based primarily on organelles of locomotion.

The protozoa might be considered a subkingdom of the kingdom *Protista*. The major groups are called **phyla** (singular, **phylum**). An abbreviated classification scheme of these organisms is shown in Table 19-1.

CHARACTERISTICS OF THE MAJOR GROUPS OF PROTOZOA

The Flagellates (Subphylum *Mastigophora*)

These protozoa are conventionally divided into two groups: the plantlike forms (class *Phytomastigophorea*, the phytoflagellates) and the animallike forms (class *Zoomastigophorea*, the zooflagellates). Plantlike protozoa usually contain green or yellow chloroplasts as well as flagella and are photosynthetic. Many are colonial in nature. These organisms (considered as algae by some biologists) are discussed in Chap. 18. The zooflagellates have no chlorophyll and must obtain nutrition heterotrophically. All members of this group have one or more flagella; some members are capable of forming pseudopodia.

Asexual reproduction in the zooflagellates occurs by longitudinal binary fission. A form of multiple fission takes place in some organisms. Encystment is common. Sexual reproduction is not common.

Table 19-1. Abbreviated Classification of the Subkingdom *Protozoa*, Describing the Major Taxa and Their Characteristics

Taxonomic Group		Characteristics
Phylum I.	*Sarcomastigophora*	Single type of nucleus; sexuality, when present, essentially syngamy (union of gametes); flagella, pseudopodia, or both types of locomotor organelles
	Subphylum *Mastigophora*	One or more flagella typically present in trophozoites; division by longitudinal binary fission; sexual reproduction known in some groups
	Class *Phytomastigophorea*	Plantlike flagellates typically with chromatophores; amoeboid forms in some groups; sexuality well studied in some groups; mainly free-living; e.g., *Euglena*, *Chlamydomonas*
	Class *Zoomastigophorea*	Animallike flagellates; chromatophores absent; one to many flagella; amoeboid forms, with or without flagella, in some groups; mainly parasitic; e.g., *Leishmania*, *Trypanosoma*, *Giardia*, *Trichomonas*
	Subphylum *Opalinata*	Binary fission occurs between rows of flagella which cover the entire body surface; known life cycles involve syngamy with anisogamous flagellated gametes; all parasitic; e.g., *Opalina*

Table 19-1. (*continued*)

Taxonomic Group		Characteristics
	Subphylum *Sarcodina*	Pseudopodia, or locomotive protoplasmic flow without discrete pseudopodia; flagella restricted to developmental stages when present; asexual reproduction by fission; sexuality, if present, associated with flagellate or, more rarely, amoeboid gametes; mostly free-living species
	Superclass *Rhizopoda*	Locomotion by pseudopodia; includes the naked (*Amoeba*) and testate amoebae (*Arcella*) and the foraminifera (*Allogromia*)
	Superclass *Actinopoda*	Often spherical; usually planktonic; pseudopodia delicate, with axopodia (i.e., with a supportive axial filament); some naked, others with tests of chitin, silica, or strontium sulfate; reproduction asexual and/or sexual; trophic cells rarely flagellated; includes the classes *Acantharea* and *Heliozoea*
Phylum II.	*Labyrinthomorpha*	Spindle-shaped or spherical vegetative cells capable of producing a network of mucus tracks; parasitic on marine plants; e.g., *Labyrinthula*
Phylum III.	*Apicomplexa*	The apicomplexans may be regarded as the sporozoa (organisms with a spore-forming stage in their life history); organisms are grouped together by the possession of an apical complex—the total combination of anteriorly located organelles in the sporozoan; these include the polar ring (within which is a conoid—an electron-dense conical structure composed of coiled filaments); rhoptries (paired tubular organelles enlarged at the posterior end); and micronemes (elongate, electron-dense organelles); (all these organelles are shown in Fig. 19-12); micropores generally present at some stage; cilia absent; sexuality by syngamy; all species parasitic
	Class *Sporozoea*	Reproduction generally both sexual and asexual; oocysts generally containing infective sporozoites which result from sporogony; locomotion of mature organisms by body flexion, gliding, or undulation of longitudinal ridges; flagella present only in microgametes of some groups
	Subclass *Gregarinia*	Trophozoites large and extracellular; parasites of digestive tracts and body cavities of invertebrates; some glide by body flexion
	Subclass *Coccidia*	Trophozoites small and typically intracellular; e.g., *Toxoplasma*, *Plasmodium*
	Subclass *Piroplasmia*	Small, piriform, round, rod-shaped, or amoeboid parasites of vertebrate blood cells, with ticks as vectors; locomotion by body flexion or gliding; e.g., *Babesia*
Phylum IV.	*Microspora*	Intracellular parasites of invertebrates, particularly arthropods; invaded host cells are hypertrophied; spores are small, up to 6 μm, and some have a coiled filament; single sporoplasm with simple or complex extrusion apparatus; spores have polar tube and cap but none have polar capsules (see Fig. 19-13); e.g., *Nosema*
Phylum V.	*Acetospora*	Spores with one or more sporoplasms; without polar capsules or polar filaments; all parasitic; e.g., *Haplosporidium*
Phylum VI.	*Myxozoa*	Spores of multicellular origin, with one or more polar capsules and sporoplasms (Fig. 19-13); all species parasitic; cysts develop in infected internal organs of vertebrates, particularly fish; e.g., *Ceratomyxa*
Phylum VII.	*Ciliophora*	Organisms possess cilia or compound ciliary organelles in at least one stage of the life cycle; two types of nuclei are present—typically, a large macronucleus and a smaller micronucleus; reproduction is by asexual fission (binary transverse); the sexual process of conjugation is also exhibited; most are free-living, though many are commensalistic and a few parasitic

Table 19-1. (*continued*)

Taxonomic Group	Characteristics
Class *Kinetofragminophorea*	Oral infraciliature (linear orientation of subpellicular basal granules and associated subpellicular tubular fibrils) only slightly distinct from somatic infraciliature; cytostome often apical (or subapical) or midventral, on surface of body; cytopharyngeal apparatus commonly prominent; compound ciliature, oral or somatic, typically absent; e.g., *Acineta, Didinium, Balantidinium*
Class *Oligohymenophorea*	Oral apparatus generally well defined; oral ciliature clearly distinct from somatic ciliature; cytostome usually ventral and/or near anterior end; cysts common; colony formation in some groups; e.g., *Tetrahymena, Paramecium, Vorticella*
Class *Polymenophorea*	Well-developed adoral zone of membranelles (these are oriented to the left of the oral cavity and beat in a circular and clockwise manner, driving food toward the cytostome); one or several lines of "paroral" ciliature; somatic ciliature complete or reduced or appearing as cirri (see Fig. 19-5); cytostome at bottom of buccal cavity; cysts very common in some groups; often large and free-living forms in a great variety of habitats; e.g., *Stentor, Euplotes*

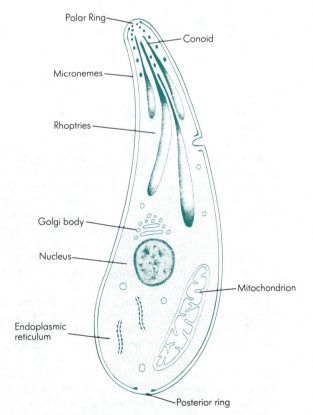

Figure 19-12. Diagram of a merozoite or sporozoite of an apicomplexan showing the apical complex and other organelles. (*Erwin F. Lessel, illustrator*)

Polar Ring

Conoid

Micronemes

Rhoptries

Golgi body

Nucleus

Mitochondrion

Endoplasmic reticulum

Posterior ring

The Zooflagellates (Class *Zoomastigophorea*)

The choanoflagellates (order *Choanoflagellida*) are distinctive in that they are either stalked or embedded in jelly, and each cell has a thin transparent collar

Figure 19-13. (A) Diagram of a microsporidian spore in which there are no polar capsules. (Scale = 1 μm.) (B) Diagram of a typical myxosporidian spore showing two polar capsules: one contains a coiled polar filament; the other, an extruded filament. A longitudinal suture is present, as is a binucleate sporoplasm. (Scale = 10 μm.) *(Erwin F. Lessel, illustrator)*

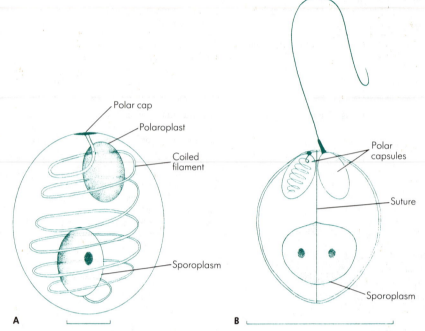

that encircles a single flagellum (Fig. 19-14). The collar functions as a food-catching device.

Organisms in the order *Kinetoplastida* are grouped together because of the presence of a **kinetoplast** (an extranuclear region of DNA associated with the mitochondrion). The single mitochondrion itself is extensive, traversing the length of the body as single tube, hoop, or network of branching tubes. One or two flagella may be present; if there are two, one is either trailing free or attached to the body, with undulating membranes occurring in some cases. Reproduction is by longitudinal binary fission. Both free-living and parasitic forms are included in the group. Parasitic forms are found in plants and in invertebrate and vertebrate hosts. Some of these are of great economic importance because of their disease potential in domestic animals and humans. Species in this group include *Leishmania* and *Trypanosoma*.

Other groups of zooflagellates are characterized by variations in morphology. For example, members of order *Retortamonadida* have 2 to 4 flagella; those of order *Diplomonadida* have bilateral symmetry and 6 to 8 flagella; and those of order *Trichomonadida* have 2 to 8 flagella and an undulating membrane.

Importance of Some Zooflagellates

Trichonympha campanula (Fig. 19-14G) is interesting because of its habitat, its complex structure, and its unusual ability to digest cellulose. These tiny creatures live in the intestines of termites, which by themselves are unable to digest cellulose, the principal constituent of wood. *T. campanula*, however, ingests wood particles and enzymatically converts the cellulose into soluble carbohydrates easily digested by the insect.

Figure 19-14. Flagellated protozoa. (A) *Euglena gracilis* are solitary, free-living flagellates, with chlorophyll. (B) *Giardia intestinalis*, a parasite found in the human intestine, where it may cause dysentery. (C) *Trichomonas hominis*, also found in the human intestine. Its role in the causation of disease has not been established. (D) *Trypanosoma rhodesiense*, which causes African sleeping sickness. (E) *Trypanosoma cruzi*, the causative agent of Chagas' disease, or South American trypanosomiasis. (F) *Codosiga*, a colonial flagellate with a transparent protoplasmic collar into which food particles are whipped by the action of the flagellum. (G) *Trichonympha*, a complex protozoan that inhabits the intestines of termites, where it converts wood cellulose into soluble carbohydrates that can be utilized by the female. (*Redrawn after Ralph Buchsbaum, Animals Without Backbones, rev. ed., The University of Chicago Press, Chicago, 1948.*)

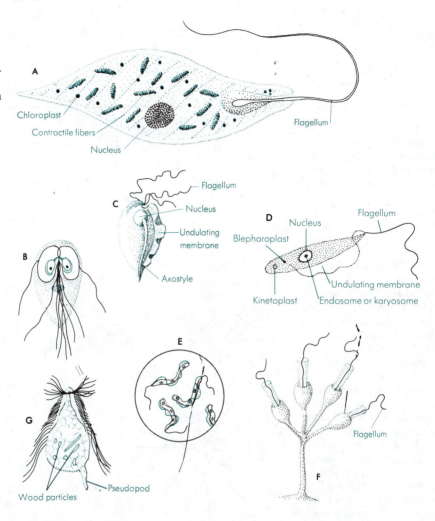

Several of the flagellated protozoa are responsible for disease in humans. *Giardia intestinalis* (formerly called *G. lamblia*) (Fig. 19-14B) is associated with diarrhea in children and, infrequently, in adults. Its feeding (trophozoite) form has eight flagella and a ventral sucker with which it attaches to the intestinal mucosa. Trichomonads are found in the mouth, where they may contribute to gingivitis; in the intestine, where they may be associated with diarrheal conditions; and in the urethra and vagina, where they cause an inflammation and purulent discharge. Although these organisms are morphologically similar, they are designated as three distinct species on the basis of habitat. Those found in the mouth are called *Trichomonas buccalis*, those in the intestine are *T. hominis* (Fig. 19-14C), and those in the urogenital tract are *T. vaginalis*. Several species of trypanosomes are responsible for disease in humans. They are characterized by a narrow, elongated body with an undulating membrane and a flagellum that

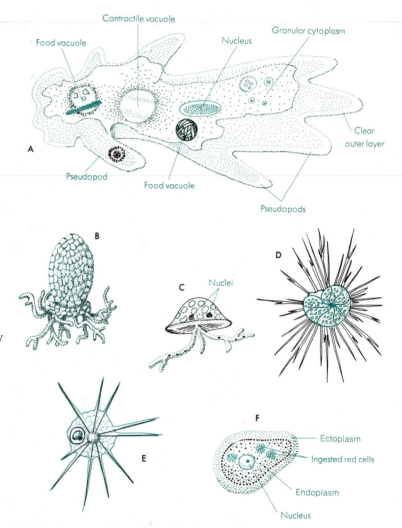

Figure 19-15. Amoeboid protozoa. (A) Diagrammatic sketch showing principal structures of an amoeba. (B) *Difflugia* lives in freshwater and builds a shell of sand-grains and a cement secreted by the cell. (C) *Arcella* is similar to *Difflugia* except that the shell is made of a chitinlike material secreted by the amoeba. It has two nuclei. (D) Foraminifera are saltwater types that build chalky shells with several chambers. They resemble snail shells. Pseudopodia are extended through pores in the shell wall. (E) *Heliozoans* may produce skeletons of silica covering the whole cell, or the body may be covered with a gelatinlike material with rigid pseudopodia made of silica. (F) *Entamoeba histolytica*, a pathogenic amoeba which causes dysentery in humans.

extends the length of the cell and beyond (Fig. 19-14D). They have a single nucleus and reproduce asexually. Some species pass through a complex life cycle, part of which is spent in a bloodsucking insect that transmits the parasite to humans and vertebrate animals. Important species of this group include *Trypanosoma gambiense* and *T. rhodesiense* (Fig. 19-14D), which are transmitted by the tsetse fly and cause African sleeping sickness; and *T. cruzi* (Fig. 19-14E), which is carried to humans by biting insects (e.g., *Triatoma*, or kissing bug) and causes a condition known as Chagas' disease, endemic in South America. Other species are pathogenic for a limited host range. For example *T. equiperdum* causes infections in equines only; transmission is by sexual means.

The genus *Leishmania* includes species which have both nonmotile and motile stages in their life cycles. Bloodsucking insects transmit the motile forms to humans; nonmotile forms lacking flagella are produced inside the cells of the spleen and other organs of the body and sometimes inside white blood corpuscles. *L. donovani* causes a tropical disease known as kala azar, and *L. tropica* causes a skin disease known as oriental sore. *L. brasiliensis*, common in South America, causes a disease characterized by ulceration of the mucous membranes of the mouth and nose.

The Amoebas (Subphylum *Sarcodina*)

Amoebas get their name from the Greek word *amoibe*, meaning "change," because their shapes are constantly changing. A typical example is *A. proteus*.

Morphology

Amoebas are composed of protoplasm differentiated into a cell membrane, cytoplasm, and a nucleus (Figs. 19-2 and 19-15). The cytoplasm shows granules as well as vacuoles containing food, wastes, water, and possibly gases. The outer membrane is selective and permits the passage of certain soluble nutrients into the cell and waste material out of the cell. Solid food is ingested with the help of pseudopods (see below). The nucleus functions in reproduction, metabolism, and the transmission of hereditary characteristics. Amoebas react to various physical and chemical stimuli in their surroundings. This is an **irritability response** which is at least superficially analogous to responses of higher organisms to their environment.

Amoebas are almost constantly in motion. They move by sending out portions of their bodies in one direction so that the whole cell moves into the location of the *pseudopodium*, or false foot, as the projection is called. Several pseudopodia may be sent out at one time from a single cell.

Nutrition and Excretion

Amoebas also use pseudopodia to capture food. The projections surround the food particles, which become enclosed in food vacuoles inside the cytoplasm

Figure 19-16. Amoeba ingesting food. As the amoeba encounters a food particle (A), a flagellate, pseudopodia form (B) and close in on the particle (C). An opening in the cell membrane allows the food particle to pass into the cell (D and E), where it is digested in a food vacuole or, if not accepted, expelled. (*Redrawn after Ralph Buchsbaum, Animals Without Backbones, rev. ed., The University of Chicago Press, Chicago, 1948.*)

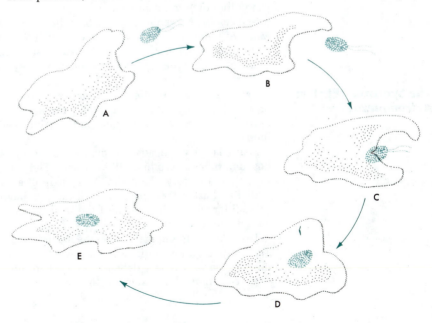

(Fig. 19-16). As the enzymes (and possibly hydrochloric acid) from the cell enter the vacuole, the food is digested and assimilated. Wastes are disposed of through the cell membrane into the surrounding medium. It is possible that amoebas carry on a special kind of respiration in the form of an exchange of gases between the cell and the surrounding fluid, for at rather regular intervals vacuoles apparently containing some carbon dioxide in solution are expelled from the cell through an opening that develops in the cell membrane. Oxygen in the medium is absorbed through the surface of the cell.

Reproduction

Reproduction in amoebas takes places by the simple asexual process of binary fission (Fig. 19-9). For protection during periods that are unfavorable for normal growth, some amoebas have the capacity of encysting. In this condition their metabolic activities are minimized, and the cyst form enables them to survive until conditions again become favorable for growth.

Other Amoeboid Protozoa

Some interesting free-living amoeboid protozoa (Fig. 19-15) are the foraminiferans, many of which produce a chalky shell with numerous chambers. These organisms obtain their food by the action of pseudopodia that extend through pores in the tiny shells. As mentioned before, the famous white (chalk) cliffs of Dover, England and the chalk beds in Mississippi and Georgia contain deposits of these shells. Fossils of extinct foraminiferans in rocks are of value to geologists (because they indicate oil-bearing deposits) and paleontologists. Radiolarians, like foraminiferans, are marine forms, but most of them construct shells of silica. Deposits of their skeletons are incorporated in rocks formed in areas where they have occurred in abundance: Indiana limestone, for example, contains many of them. Freshwater counterparts of the radiolarians are the heliozoida, which are sometimes covered by a gummy substance or needles of silica rather than a skeleton. The genera *Arcella* and *Difflugia* contain species that make shells they can withdraw into for protection.

Species of the genus *Entamoeba* inhabit the intestinal tract of vertebrates. Many of them, such as *E. gingivalis*, which lives in the human mouth, and *E. coli*, which inhabits the human intestine, are harmless. However, one species, *E. histolytica*, is the cause of amoebic dysentery in human beings.

The Sporozoa (Phylum *Apicomplexa*)

All sporozoa are parasitic for one or more animal species. Adult forms have no organs of motility but all are probably motile by gliding at one stage of their life cycle. They cannot engulf solid particles but feed on the host's cells or body fluids.

Many have complicated life cycles, certain stages of which may occur in one host and other stages in a different host. They all produce spores at some time in their life history. Their life cycles exhibit an alternation of generations of sexual and asexual forms, such that the intermediate host usually harbors the asexual forms and the final host the sexual forms. Sometimes humans serve as hosts to both forms.

Toxoplasmosis and malaria are the major human diseases caused by sporozoa. *Toxoplasma gondii* is the etiologic agent of toxoplasmosis. The symptoms of this disease vary greatly depending on the location of the parasites in the body. They can mimic, for example, the symptoms of meningitis and hepatitis. *T.*

gondii is the most widespread of the parasites that infect vertebrates. More than 50 percent of adults in the United States have been infected at some time, but the disease in humans is usually mild and asymptomatic. Spontaneous recovery usually follows. However, transplacental infection, that is, infection of the human embryo, may occur with serious consequences. The result may be a stillborn child or a child with mental retardation and other disorders. It is of interest that the parasite will undergo its sexual reproductive cycle only in the intestinal cells of members of the cat family, including the domestic cat.

The most important sporozoa are those that cause malaria (see also Chap. 38). Malaria is a mosquito-borne disease of humans caused by sporozoa belonging to the genus *Plasmodium* which infect the liver and red blood cells. The final host for the parasite is the female anopheline mosquito; sexual reproduction of the parasite occurs in this host. Malaria has been one of the greatest killers of humans through the ages. At the present time, it has been conservatively estimated that 300 million people in the world have the disease and that about 3 million of them will die of it.

The Ciliates (Phylum Ciliophora)

Common examples of the ciliated protozoa are included in the genus *Paramecium*, found in freshwater ponds and lakes where adequate food supplies exist.

Morphology and Structure

Like most amoebas, paramecia are microscopic; some, however, are just barely visible to the unaided eye. The outer layer of the cell is less flexible than the outer membrane of the amoeba, and the interior is composed of semifluid, granular protoplasm containing nuclei and vacuoles of several kinds. Paramecia are easily distinguished by their characteristic shape, which has been likened to that of a slipper (Figs. 19-4 and 19-17). The anterior (front) end of the cell is rounded, and the posterior (rear) end is slightly pointed. The entire cell is covered with hundreds of short hairlike projections called cilia, which are the organs of locomotion and also serve to direct food into the cytostome.

Motility

Paramecia move very rapidly by a rhythmic beating of the cilia. Since cilia can beat in either direction, reverse as well as forward motion is possible and the organism can turn in any direction. The motion of cilia has been compared to that of the arms of a swimmer doing the crawl stroke.

Nutrition and Excretion

Paramecia take in food through a fixed cytostome at the base of the gullet. The food particles are directed by cilia through the oral groove into the **gullet** and are collected in a food vacuole inside the cell at the end of the cytopharynx, where they are digested by enzymes, as in the amoeba. Undigested particles are eliminated from the cell through the **cytoproct**.

As in amoebas and other single-celled microorganisms, oxygen enters the cell through the cell membrane, and carbon dioxide diffuses out. Waste fluids are collected in the **contractile vacuoles,** which have fixed positions in paramecia instead of appearing anywhere in the cell, as in amoebas.

There are two types of nuclei in each paramecium, a large macronucleus and one or more small micronuclei. The former regulates the ordinary activities of the cell's metabolism, and the latter is/are associated with reproduction.

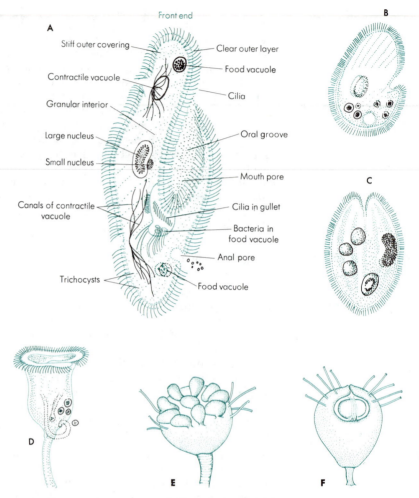

Front end

A

Stiff outer covering

Contractile vacuole

Granular interior

Large nucleus

Small nucleus

Canals of contractile
vacuole

Trichocysts

Clear outer layer

Food vacuole

Cilia

Oral groove

Mouth pore

Cilia in gullet

Bacteria in
food vacuole

Anal pore

Food vacuole

B

C

D

E

F

Figure 19-17. Ciliated protozoa. (A) Diagrammatic sketch showing principal structures of a paramecium. (B) *Colpoda*, a common freshwater protozoan. (C) *Balantidium coli*, the only disease-producing parasitic ciliate found in humans. (D) *Vorticella* has a bell-shaped body attached to a contractile stem. Cilia on the outer edge of the bell sweep bacteria and other food particles into the gullet by setting a miniature whirlpool in motion. (E) A protozoan of the subclass *Suctoria*, *Ephelota*, showing external or exogenous budding. (F) Another member of subclass *Suctoria*, *Tokophyra*, showing endogenous budding with larval form free within brood pouch. [(E) and (F) redrawn after those of John O. Corliss, University of Maryland at College Park, The Ciliated Protozoa: Characterization, Classification, and Guide to the Literature, 2d ed., Pergamon, New York, 1979.]

Reproduction

Paramecia reproduce asexually by binary fission, in which each of the nuclei elongates and divides into halves (Fig. 19-10). The cell itself constricts across the midline, producing two daughter cells, each of which receives a pair of nuclei. In this type of division, the rear daughter cell receives the gullet of the parent, and a new gullet is formed for the other new cell. Other parts are formed as required. Under favorable conditions, one or more divisions may take place within a 24-h period.

Conjugation may occur when unlike mating strains of paramecia happen to come into contact with one another. Although these cells are not differentiated as male and female, under certain circumstances two individuals come together and unite along their oral grooves. While they are joined in this fashion, their nuclei undergo divisions; then the cells exchange haploid nuclei derived from their micronuclei. In each conjugant, the two haploid nuclei fuse to form a diploid nucleus. The ciliates then separate, and nuclear divisions and fissions result in the asexual form.

Other Ciliated Protozoa

The ciliated protozoa (Fig. 19-17) are represented by many forms other than the paramecia. *Colpoda* is a common freshwater genus. The genus *Didinium* lives on a diet of paramecia, which are captured by a special structure and swallowed whole (Fig. 19-1). The genus *Stentor* comprises large cone-shaped protozoa that move about freely but attach to some object by a tapered lower end while feeding. The rapid movement of the cilia at *Stentor's* upper end sweeps smaller protozoa and bacteria into its gullet. *Vorticella* is bell-shaped. Cilia ring the outer edge of the bell and wash food, principally bacteria, into its gullet. Related colonial ciliates occur in groups attached to a long fibrous stalk with the ability to recoil when disturbed. Clusters of the organisms are comprised of many daughter cells. The only human ciliate parasite is *Balantidium coli*. It causes a type of dysentery.

QUESTIONS

1 What is the derivation of the name protozoa? How are protozoa distinguished from other eucaryotic protists?
2 What is meant by the statement that protozoa have attributes of higher organisms?
3 Give two specific examples of how protozoa adapt or respond to changing environmental conditions.
4 Define the following terms: parasitism, hyperparasitism, ectocommensalism, and endocommensalism.
5 Describe the roles played by protozoa in the natural environment.
6 Describe the roles played by protozoa in the treatment of industrial and domestic wastes.
7 Briefly describe the different organelles found in the cytoplasm of a protozoan such as *Amoeba proteus* or *Paramecium bursaria*.
8 What is a pellicle? Is it the same as the plasmalemma? Explain.
9 What protective coverings, other than the pellicle, do some protozoa possess?
10 Describe the food-gathering structures found in the ciliated protozoa.

11 What are the collective and individual functions of mucocysts, trichocysts, toxicysts, and haptocysts?

12 Describe the various ways by which protozoa move.

13 Compare and contrast asexual reproduction in the protozoa and in the bacteria.

14 Describe the process of nuclear division in the ciliates during binary fission.

15 What is multiple fission? Cite an example of its occurrence in protozoa.

16 How does budding in the protozoa differ from budding in yeasts?

17 Define the following terms: syngamy, isogametes, anisogametes, microgametes, macrogametes.

18 How many phyla are there in the new scheme of protozoan classification (1980)? Do they reflect any evolutionary relationship?

19 In what taxa are the following protozoa found: (a) flagellates, (b) ciliates, (c) amoebas, and (d) sporozoa?

20 What is a zooflagellate? Name several zooflagellates responsible for disease in humans.

21 How do the digestive and reproductive processes of a paramecium compare with those of an amoeba?

22 Why is *Trichonympha campanula* an interesting proozoan?

23 Explain with specific examples why the sporozoa are of importance to humans.

24 Write a short description of a common ciliated protozoan.

REFERENCES

Curds, C. R., and C. G. Ogden: "Form and Function—IV. Protozoa," in *Essays in Microbiology*, J. R. Norris and M. H. Richmond (eds.): Wiley, New York, 1978. *A good general essay on the biology of the protozoa.*

Dales, R. P. (ed.): *Practical Invertebrate Zoology*, University of Washington Press, Seattle, 1970. *A laboratory manual for the study of the major groups of invertebrates.*

Grell, Karl G.: *Protozoology*, Springer-Verlag, Heidelberg, 1973. *An advanced text on protozoology. It is superbly illustrated and referenced and contains comprehensive information on the biology of the protozoa.*

Levine, N. D., et al.: "A Newly Revised Classification of the Protozoa," *J Protozool*, **27**:37–58, 1980. *A commonly used classification scheme of the protozoa. Based on electron-microscope features rather than primarily on modes of locomotion.*

Sleigh, M. A.: *The Biology of the Protozoa*, American Elsevier, New York, 1973. *A general text with emphasis on the physiology and ecology of protozoa. Well illustrated.*

Chapter 20 Viruses of Bacteria

Viruses are infectious agents so small that they can only be seen at magnifications provided by the electron microscope. They are 10 to 100 times smaller than most bacteria, with an approximate size range of 20 to 300 nm. Thus they pass through the pores of filters which do not permit the passage of most bacteria.

Viruses are incapable of independent growth in artificial media. They can grow only in animal or plant cells or in microorganisms. They reproduce in these cells by **replication** (a process in which many copies or replicas are made of each viral component and are then assembled to produce progeny virus). Thus viruses are referred to as **obligate intracellular parasites**. (If the least requirement for life is that an organism duplicate itself, then viruses may be viewed as microorganisms.)

Viruses largely lack metabolic machinery of their own to generate energy or to synthesize proteins. They depend on the host cells to carry out these vital functions. However, like the host cells, viruses have the genetic information for replication; and viruses have information in their genes for usurping the host cell's energy-generating and protein-synthesizing systems.

Actually, viruses in transit from one host cell to another are small packets of genes. The viral genetic material is either DNA or RNA, but the virus does not have both. (Host cells have both DNA and RNA.) The nucleic acid is enclosed in a highly specialized protein coat of varying design. The coat protects the genetic material when the virus is outside of any host cell and serves as a vehicle for entry into another specific host cell. The structurally complete mature and infectious virus is called the **virion**.

During reproduction in the host cells viruses may cause disease. In fact, viruses incite the most common acute infectious diseases of humans (like the "cold" or "flu"), and there is growing evidence that they may cause many chronic diseases as well. Significantly, all viruses are generally insensitive to the broad range of available antibiotics such as penicillin, streptomycin, and others.

From the above discussion of what does or does not constitute a virus, we may now attempt a definition for this group of infectious agents. We can define viruses as *noncellular infectious entities whose genomes are a nucleic acid, either DNA or RNA; which reproduce only in living cells; and which use the cells' biosynthetic machinery to direct the synthesis of specialized particles (virions), which contain the viral genomes and transfer them efficiently to other cells.*

Bacterial viruses, or **bacteriophages** (or simply **phages**) have provided the microbiologist with a model for **virology** (the study of viruses) and **molecular biology** (a discipline which examines the structure, function, and organization of the macromolecules in which biological specificity is encoded). We shall discuss the phages in this chapter and follow with a discussion of the viruses that infect animals and plants in Chapter 21.

BACTERIOPHAGES: DISCOVERY AND SIGNIFICANCE

Bacteriophages, viruses that infect bacteria, were discovered independently by Frederick W. Twort in England in 1915 and by Felix d'Herelle at the Pasteur Institute in Paris in 1917. Twort observed that bacterial colonies sometimes underwent lysis (dissolved and disappeared) and that this lytic effect could be transmitted from colony to colony. Even high dilutions of material from a lysed colony that had been passed through a bacterial filter could transmit the lytic effect. However, heating the filtrate destroyed its lytic property. From these observations Twort cautiously suggested that the lytic agent might be a virus. D'Herelle rediscovered this phenomenon in 1917 (hence the term *Twort-d'Herelle phenomenon*) and coined the word bacteriophage, which means "bacteria eater." He considered the filtrable agent to be an invisible microbe—for example, a virus—that was parasitic for bacteria.

Since the bacterial hosts of phages are easily cultivated under controlled conditions, demanding relatively little in terms of time, labor, and space compared with the maintenance of plant and animal hosts, bacteriophages have received considerable attention in viral research. Furthermore, since bacteriophages are the smallest and simplest biological entities known which are capable of self-replication (making copies of themselves), they have been used widely in genetic research. Of importance too have been studies on the bacterium-bacteriophage interaction. Much has been learned about host-parasite relationships from these studies, which have provided a better understanding of plant and animal infections with viral pathogens. Thus the bacterium-bacteriophage interaction has become the model system for the study of viral pathogenicity.

GENERAL CHARACTERISTICS

Bacterial viruses are widely distributed in nature. Phages exist for most, if not all, bacteria. With the proper techniques, these phages can be isolated quite easily in the laboratory.

Bacteriophages, like all viruses, are composed of a nucleic acid core surrounded by a protein coat. Bacterial viruses occur in different shapes, although many have a tail through which they inoculate the host cell with viral nucleic acid (Figs. 20-1 and 20-2).

There are two main types of bacterial viruses: lytic, or virulent, and temperate or avirulent. When lytic phages infect cells, the cells respond by producing large numbers of new viruses. That is, at the end of the incubation period the host cell bursts or lyses, releasing new phages to infect other host cells. This is called a lytic cycle. In the temperate type of infection, the result is not so readily apparent. The viral nucleic acid is carried and replicated in the host bacterial

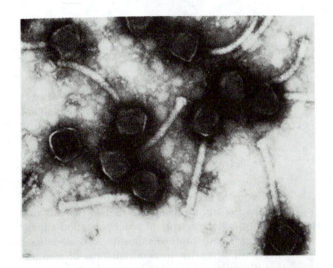

Figure 20-1. A bacteriophage of *Bacillus thuringiensis* (X297,000). (*Courtesy of H.-W. Ackermann, Laval University.*)

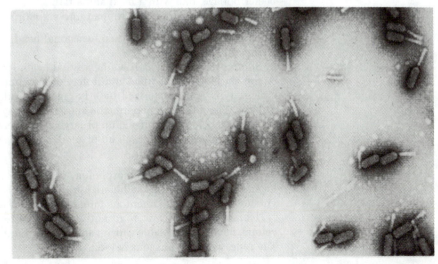

Figure 20-2. The unique morphology of many bacteriophages is exemplified by this phage of *Salmonella newport*. Each phage has a tail through which it inoculates the host cell with viral nucleic acid. Note that some tails have broken off from the phage heads in this preparation (X59,400). (*Courtesy of H.-W. Ackermann, Laval University.*)

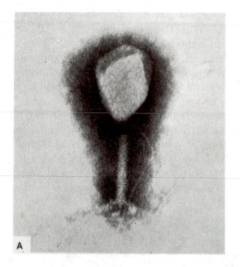

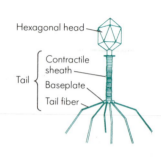

Figure 20-3. (A) Electron micrograph and (B) diagrammatic representation of the coliphage T2 virion. Magnification in micrograph X189,000. (*Courtesy of H.-W. Ackermann, Laval University.*)

cells from one generation to another without any cell lysis. However, temperate phages may spontaneously become virulent at some subsequent generation and lyse the host cells. In addition, there are some filamentous phages which simply "leak" out of cells without killing them.

MORPHOLOGY AND STRUCTURE

Morphological Groups of Phages

The electron microscope has made it possible to determine the structural characteristics of bacterial viruses. All phages have a nucleic acid core covered by a protein coat, or **capsid.** The capsid is made up of morphological subunits (as seen under the electron microscope) called **capsomeres.** The capsomeres consist of a number of protein subunits or molecules called **protomers.** Figure 20-3 shows the fine structure and anatomy of a common morphological form of the bacteriophage, one with a head and a tail.

Bacterial viruses may be grouped into six morphological types (Fig. 20-4).

A This most complex type has a hexagonal head, a rigid tail with a contractile sheath, and tail fibers.

B Similar to A, this type has a hexagonal head. However, it lacks a contractile sheath, its tail is flexible, and it may or may not have tail fibers.

C This type is characterized by a hexagonal head and a tail shorter than the head. The tail has no contractile sheath and may or may not have tail fibers.

D This type has a head made up of large capsomeres, but has no tail.

E This type has a head made up of small capsomeres, but has no tail.

F This type is filamentous.

Types A, B, and C show a morphology unique to bacteriophages. The morphological types in groups D and E are found in plant and animal (including insect) viruses as well. The filamentous form of group F is found in some plant viruses.

Pleomorphic viruses were recently discovered to have a lipid-containing envelope, have no detectable capsid, and possess double-stranded DNA (ds-DNA). The representative phage is MV-L2.

Phage Structure

Most phages occur in one of two structural forms, having either cubic or helical symmetry. In overall appearance, cubic phages are regular solids or, more specifically, polyhedra (singular, polyhedron); helical phages are rod-shaped.

Polyhedral phages are icosahedral in shape. (The icosahedron is a regular polyhedron with 20 triangular facets and 12 vertices.) This means that the capsid has 20 facets, each of which is an equilateral triangle; these facets come together to form the 12 corners. In the simplest capsid, there is a capsomere at each of the 12 vertices; this capsomere, which is surrounded by five other capsomeres, is termed a penton. (See Fig. 20-5A.) For example, the phage φX174 exhibits the simplest capsid. In larger and more complex capsids, the triangular facets are subdivided into a progressively larger number of equilateral triangles. Thus a capsid may be composed of hundreds of capsomeres but it is still based on the simple icosahedron model.

The elongated heads of some tailed phages are derivatives of the icosahedron. For example the head of the T2 and T4 phages is an icosahedron elongated by one or two extra bands of hexons.

Rod-shaped viruses have their capsomeres arranged helically and not in stacked rings (Fig. 20-5B). An example is the bacteriophage M13.

Some bacteriophages, such as the T-even coliphages (T2, T4, and T6), have very complex structures, including a head and a tail. They are said to have binal symmetry because each virion has both an icosahedral head and a hollow helical tail (Fig. 20-6).

Phage Nucleic Acids

Different morphological types of phages are also characterized by having different nucleic acid types (Fig. 20-4). All tailed phages contain double-stranded

Figure 20-4. Schematic representation of morphological types of bacteriophages. [As described by D. E. Bradley, "Ultrastructures of Bacteriophages and Bacteriocins," Bacteriol Rev, **31**:230–314 (1967).] A further group, G, may be added; these phages have a lipid envelope and are pleomorphic.

Type	A	B	C	D	E	F	G
Morphology	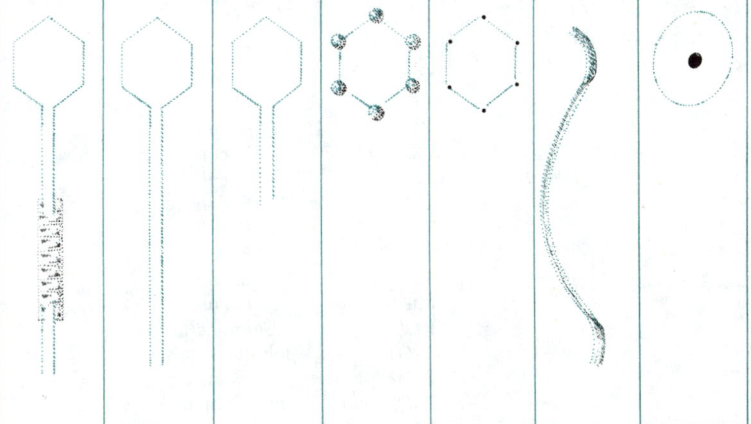						
Nucleic acid type and number of strands	DNA, 2	DNA, 2	DNA, 2	DNA, 1	RNA, 1	DNA, 1	DNA, 2
Examples of phages	Coli-phages T2, T4, T6	Coli-phages T1, T5	Coli-phages T3, T7	Coli-phage φ×174, S13	Coli-phages f2,MS2	Coli-phages fd, f1	MV-L2

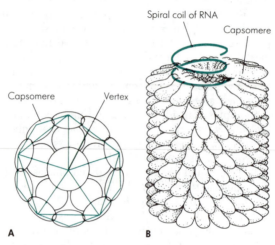

Figure 20-5. (A) Diagram of the simplest icosahedral capsid. The colored triangular outlines delineate the icosahedral symmetry. The circles represent capsomeres. (B) Diagram of a rod-shaped virus with helical symmetry. The capsomeres are arranged helically around a hollow core containing a spiral coil of RNA.

Figure 20-6. The morphology of phage lambda as seen by electron microscopy (X297,000). (*Courtesy of H.-W. Ackermann, Laval University.*)

Table 20-1. Molecular Weights and Type of Nucleic Acid of Commonly Studied Phages

Phage	Host	DNA Type	RNA Type	MW*
φX174	*Escherichia coli*	single-stranded (ss), circular (circ.)		1.7
fd	*E. coli*	ss, circ.		1.7–2.0
M13	*E. coli*	ss, circ.		2.4
T1	*E. coli*	double-stranded (ds)		30
T3	*E. coli*	ds		24
T5	*E. coli*	ds		75
T7	*E. coli*	ds		38
λ	*E. coli*	ds		30
T2	*E. coli*	ds		120
T4	*E. coli*	ds		110
T6	*E. coli*	ds		120
Mu	*E. coli*	ds		25
P22	*Salmonella*	ds		26
SPO1	*Bacillus subtilis*	ds		100
SPO2	*B. subtilis*	ds		25
PM2	*Pseudomonas*	ds, circ.		6
N1	Cyanobacteria	ds		43
f2	*E. coli*		ss	1.2
R17	*E. coli*		ss	1.2
QB	*E. coli*		ss	1.2

* MW = molecular weight, expressed in 10^6 daltons.

Figure 20-7. A DNA molecule isolated from *Bacillus subtilis* bacteriophage SP 50 appears in this electron micrograph as a tangled thread (X90,000). (*Courtesy of William S. Reznikoff and C. A. Thomas, Jr.*)

DNA. The phages with large capsomeres (group D) and the filamentous ones (group F) have single-stranded DNA. Group E phages have single-stranded RNA. The DNAs of some phages are circular under certain conditions. (Circular simply means a *closed loop*; the molecule is in the form of a loose folded coil packed inside the capsid.) The DNA of phage φX174 is circular both in the virion and in the host cell. The DNA of phage lambda (λ) is linear in the virion, but on entering the host cell the cohesive ends join to form a circle. The molecular weights and nucleic acid types of commonly studied bacteriophages are shown in Table 20-1. Figure 20-7 shows a DNA molecule isolated from a phage.

THE CLASSIFICATION AND NOMENCLATURE OF BACTERIOPHAGES

It may be apparent by now that the common names of bacteriophages do not follow particular guidelines. They are simply designations or code symbols assigned by investigators. Although serving the practical needs of the laboratories, this is a haphazard way of naming a group of microorganisms.

Consequently, the International Committee on Taxonomy of Viruses (ICTV) has a Bacterial Virus Subcommittee working on the classification and nomenclature of bacteriophages. However, taxonomic development within the bacterial viruses remains slow and difficult for two main reasons. First, some 1,900 descriptions of bacterial virus isolates of known morphology have been published. About 150 new descriptions are published each year. Many of these descriptions give a characterization of the virus that is quite inadequate for establishing its relationships with other phages. (This difficulty may be surmounted in the near future since guidelines have now been established for the characterization of a new phage isolate.) Second, a substantial proportion of the scientists who work with bacterial viruses are molecular biologists rather than virologists. They work with a relatively small selection of extremely well characterized viruses and are quite satisfied with simple code designations for these

Figure 20-8. Schematic representation of the families of bacterial viruses. Note that all the diagrams have been drawn to the same scale and provide a good indication of the shapes and relative sizes of the virions. To aid recognition, a well-known member of the family is given in parentheses, but the dimensions and shape used for the drawing may not be exactly those of the virus example given. (*Drawings by H.-W. Ackermann, Laval University. Reproduced from R. E. F. Matthews, "Classification and Nomenclature of Viruses," Intervirology,* **17:**1–199, 1982. By permission from S. Karger AG, Basel, Switzerland.)

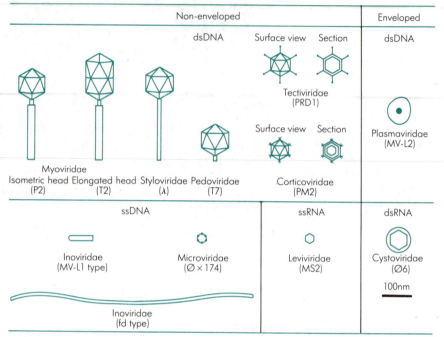

THE FAMILIES OF BACTERIAL VIRUSES

viruses because these biologists have little need for viral classification, nomenclature, or natural relationships.

The Bacterial Virus Subcommittee has now recommended names for families of phages (names ending in -*viridae*). These families are illustrated in Fig. 20-8.

SOME BACTERIOPHAGES OF *Escherichia coli*

The most extensively studied group of bacteriophages are the **coliphages,** so called since they infect the nonmotile strain B of *Escherichia coli*. They are designated T1 to T7. All of these phages are composed almost exclusively of DNA and protein in approximately equal amounts. Except for T3 and T7, all have tadpole shapes, with polyhedral heads and long tails. The tails of T3 and T7 are very short (see Fig. 20-4). The T phages range from about 65 to 200 nm in length and 50 to 80 nm in width. The continuous, or circular, molecule of double-stranded DNA (about 50 μm long, or about 1,000 times as long as the phage itself) is tightly packed in the protein head.

There are other bacteriophages for *E. coli* whose morphology and chemical composition are very different from those of the T phages. The f2 phage, for example, is much smaller than the T phages and has a single-stranded linear molecule of RNA, rather than DNA. It has no visible tail.

There are also coliphages that possess single-stranded DNA. Morphologically they can be either icosahedral (cubic symmetry) or filamentous (helical symmetry). An icosahedral phage with circular single-stranded DNA is φX174.

Figure 20-9. Electron micrograph of If1 phage that infects *Escherichia coli*. The phage absorbs specifically to the I pilus, a type of surface appendage of cells (X24,750). (*Courtesy of R. L. Wiseman, The Public Health Research Institute of the City of New York, Inc.*)

Filamentous coliphages (Fig. 20-9) were discovered long after the tadpole-shaped phages were known. Filamentous phages are continuously produced by viable and reproducing bacteria (and without the host-cell lysis characteristic of the virulent icosahedral phages). Such filamentous phages for *E. coli* include the fd and f1 bacteriophages (Fig. 20-4). They all have circular single-stranded DNA.

REPLICATION OF BACTERIAL VIRUSES

Much of what is known about bacteriophage replication has come from studies of the virulent even-numbered T phages (T2, T4, T6) of *E. coli*. However, it is apparent that the basic sequence of events during phage replication is similar for most phages. Variation occurs with respect to the number of phage proteins made and the degree of usurping of host functions. We shall use the T-even phages as a model for discussing phage replication.

Adsorption

The first step in infection of a host bacterial cell by a phage is adsorption (Fig. 20-10). The tip of the virus tail becomes attached to the cell via specific receptor sites on the cell surface. Attachment is specific in that certain viruses and susceptible bacteria have complementary molecular configurations at their opposing receptor sites. In some cases, the specific receptor of the bacterium is part of the bacterial lipopolysaccharide, although any surface structure can function as a specific phage receptor, including flagella, pili, and carbohydrates and proteins in the membrane or cell wall. Fig. 20-11 shows the adsorption of phage T4 to its host cell by means of the interaction of the tip of the phage **core** (tail tube within the sheath) and the cytoplasmic membrane of the host spheroplast.

It should be noted that infection of a host bacterial cell cannot occur without adsorption. Some bacterial mutants have lost the ability to synthesize specific receptors; they also become resistant to infection by the specific phage.

Initial adsorption of the phage to the receptor is reversible (i.e., the phage can be washed away) when only the tips of the tail fibers attach first to the cell surface. But this soon becomes irreversible when the **tail pins** attach; this is shown in Fig. 20-12. (Six tail pins protrude from the base of the tail.)

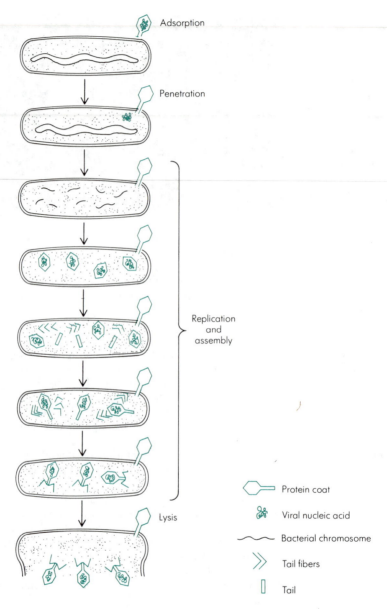

Figure 20-10. Viruses infect bacteria by injecting the contents of the virus head (viral nucleic acid) through a puncture hole in the cell wall. The viral nucleic acid then takes control of the cell metabolism and "directs" the bacterium in the synthesis of more viral nucleic acid and other materials needed for making complete virus particles. In a short time the newly formed virus particles are released by a sudden rupture of the cell wall (lysis) and are free to infect other susceptible bacteria.

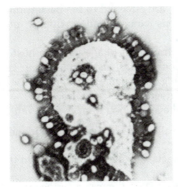
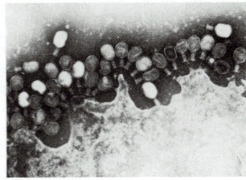

Figure 20-11. Phage T4 adsorption and DNA injection into *Escherichia coli* sphero-plasts. The phage particles were incubated with spheroplasts at 37°C for 10 min, directly negatively stained, and examined under an electron microscope. Left: X63,500; right: X150,000. (*Courtesy of S. Mizushima, Nagoya University. From H. Furukawa, T. Kuroiwa, and S. Mizushima, "DNA Injection During Bacterio-phage T4 Infection of Escherichia coli," J Bacteriol,* **154:***938–945, 1983.*)

Figure 20-12. Summary of process of phage T4 infection. OM, outer membrane; IM, inner membrane; PG, peptidoglycan layer. (*Reproduced from H. Furukawa, T. Kuroiwa, and S. Mizushima, "DNA Injection During Bacteriophage T4 Infection of Escherichia coli," J Bacteriol,* **154:***938–945, 1983. Courtesy of S. Mizushima, Nagoya University.*)

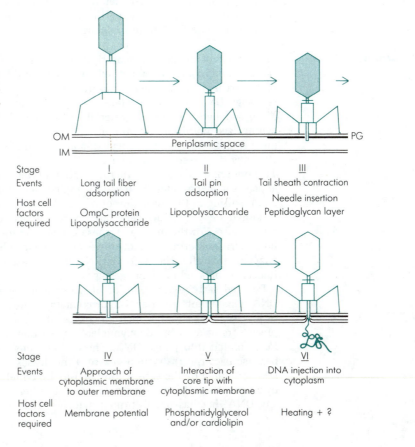

Penetration

If too many phages are attached to the bacterium and penetrate it, there may be premature lysis (lysis from without), which is not accompanied by the production of new virus.

The actual penetration of phage into the host cell is mechanical. But it may be facilitated by localized digestion of certain cell surface structures either by phage enzymes (e.g., lysozyme) carried on the tail of the phage or by viral activation of host degradative enzymes.

In the T-even phages, penetration is achieved when:

1 The tail fibers of the virus attach to the cell and hold the tail firmly against the cell wall;
2 The sheath contracts, driving the tail core into the cell through the cell wall and membrane; and
3 The virus injects its DNA the way a syringe injects a vaccine.

The protein coat, which forms the phage head, and the tail structure of the virus remain outside the cell. The processes of adsorption and penetration are shown in Fig. 20-12.

Phages, such as T1 and T5, that do not have a contractile sheath also inject their nucleic acid through the cell envelope, possibly at adhesion sites between the inner and outer membranes. That is, sheath contraction is not a prerequisite for phage infection. The filamentous, rod-shaped DNA phages (such as fd and M13), like animal viruses, enter the bacterial cell as discrete virions prior to the liberation of the DNA from the phage capsid.

Transcription

In the case of phage T4, transcription occurs in several stages leading to the formation of immediate early, delayed early, and late gene products, so named on the basis of their time of appearance. (The sequence of transcription events of phage DNA in other bacteriophages may vary from that of the T4 model discussed here.) In brief, bacterial mRNA and bacterial proteins stop being synthesized within a few minutes after entry of phage DNA. Bacterial DNA is quickly degraded to small fragments and the nucleoid region of the bacterium becomes dispersed. Some phage mRNA is made immediately after infection. The amount of phage DNA increases after a brief delay. Virion-specific proteins appear somewhat later, followed by appearance of organized capsid precursors and resulting in the formation of mature infectious capsids.

Immediate early phage genes are transcribed using the existing bacterial RNA polymerase. For the most part, these genes code for nucleases that break down host DNA (rendering its nucleotides available for phage DNA synthesis) and for enzymes that alter the bacterial RNA polymerase so that it will preferentially transcribe delayed early phage genes.

Delayed early genes code for phage enzymes which produce unique phage DNA constituents such as 5-hydroxymethylcytosines (which replace cytosine in the bacterial DNA); which glucosylate these nucleotides; or which destroy precursors of cytosine deoxynucleotides so that no bacterial cytosine will be incorporated into phage DNA. These alterations enable the phage to survive because bacterial restriction enzymes (nucleases) are unable to degrade phage DNA modified by the substitution of 5-hydroxymethylcytosine for cytosine and by glucosylation of this substituted base. Further, a phage nuclease will destroy any DNA that has unsubstituted cytosine. Delayed early genes also code for polymerases and ligases that play specific roles in phage DNA replication and recombination and for a second altered RNA polymerase that will transcribe the late genes.

Late gene products include the structural components of new phage particles (e.g., heads, tails, and fibers). They also include a phage lysozyme (an endolysin) which will lyse the bacterial cell, releasing the mature virions.

Assembly and Release

Only after the synthesis of both structural proteins and nucleic acid is well under way do the phage components begin to assemble into mature phages. About 25 min after initial infection, some 200 new bacteriophages have been assembled (see Fig. 20-13), and the bacterial cell bursts, releasing the new phages

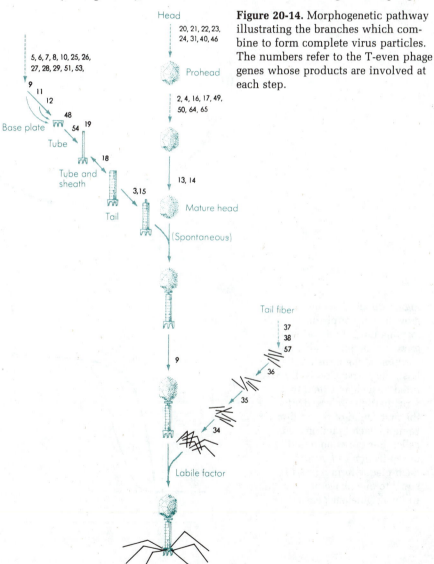

Figure 20-14. Morphogenetic pathway illustrating the branches which combine to form complete virus particles. The numbers refer to the T-even phage genes whose products are involved at each step.

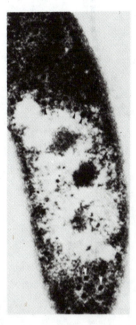

Figure 20-13. The intracellular location of a phage is shown. This cell of *Pseudomonas cepacia* 383 was sampled 5 min prior to culture lysis. Electron-dense particles (arrows) occupy the ribosome-rich regions of the cell. Note the hexagonal shape of some particles. (*Courtesy of T. G. Lessie, University of Massachusetts.*)

Figure 20-15. A simplified genetic map of the T4 virus. The squares show the morphological element whose production is governed by a particular gene. (*Based on data from W. B. Wood and H. R. Revel, Bacteriol Rev,* **40**:*847, 1976.*)

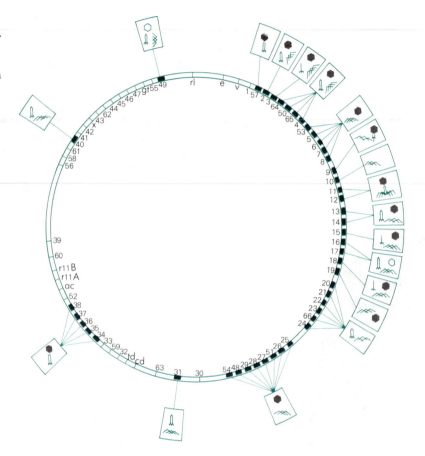

Figure 20-16. One-step growth curve of plaque-forming units. In a one-step growth experiment, after adsorption of the virus on the host, the suspension is diluted to such an extent that virus particles released after the first round of replication cannot attach to uninfected cells; thus only one round of replication can occur. Each plaque-forming unit is equal to one phage particle in the original suspension.

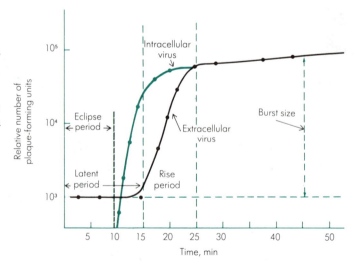

to infect other bacteria and begin the cycle over again. The overall process of phage infection and replication is shown in Fig. 20-10. The assembly of phage T4 is illustrated in Fig. 20-14. As may be seen from this figure, three different pathways are involved in the formation of a head, tail, and tail fibers, each being genetically controlled. Fig. 20-15 is a simplified genetic map of the T4 virus showing the 50 or so "architectural" genes which control the synthesis and assembly of new virus.

THE VIRAL MULTIPLI-CATION (REPLICATION) CYCLE

The sequence of events initiated by the injection of the phage nucleic acid and culminating in the release of newly synthesized virions is termed the viral multiplication or replication cycle. It can be plotted as a one-step multiplication cycle as shown in Fig. 20-16, which describes the production of progeny virions by cells as a function of time after infection under one-step conditions (i.e., the cells are infected simultaneously and secondary infection by progeny virus is eliminated by dilution).

Specifically, during the first 10 minutes or so after injection of phage DNA, no phage can be recovered by disrupting the infected bacterium. This is termed the eclipse period. At the end of this period, mature phages begin to accumulate intracellularly (Fig. 20-13) until they are released by cell lysis. No newly released extracellular phages can be seen until lysis begins; the time from infection until lysis is the latent period (see Fig. 20-16). The extracellular phage number increases until it reaches a constant titer (number assayed) at the end of the multiplication cycle; this time interval is termed the rise period (Fig. 20-16). The yield of phage per bacterium is called the burst size and is shown also in Fig. 20-16. This procedure affords observation of a single cycle of bacteriophage growth.

It may be mentioned that there are other phages which can replicate without drastic interruption of host cell physiology. For example, the filamentous single-stranded DNA phage fd can replicate within a host cell and is released from it

---

Medical Use of Virulent Phages

Since virulent or lytic phages can destroy their host bacterial cells, it was logical to think that inoculation of such phages into a bacteria-infected individual would result in elimination of the pathogens. However, after numerous studies, there is no evidence to show that phages can be used therapeutically to destroy bacterial pathogens in the human body, principally because phages do not persist in the body. Consequently, the primary uses of bacteriophages are in the identification of bacterial strains and as genetic models in molecular biology.

Thus lytic phages have been used in the detection and identification of pathogenic bacteria. Strains of bacteria (of a single species) may be characterized by their resistance or susceptibility to lysis by specific virulent phages. The resulting pattern of lysis (from visible plaques on a lawn of bacterial growth) of a bacterial strain by different phage types gives an indication of the identity of the bacterium. This laboratory procedure is termed bacteriophage-typing and is used routinely for the identification of certain strains of bacterial pathogens such as the staphylococci and the typhoid bacilli. In this way phages serve as a tool for medical diagnosis and for tracing of the source of a disease spreading in a community.

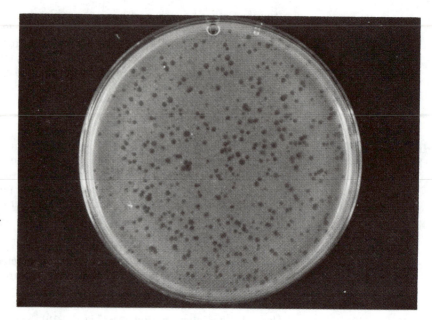

Figure 20-17. Plaques (clear zones) are formed when a bacterial growth in a Petri dish is lysed by a bacteriophage. (*Courtesy of C. Alfieri and E. C. S. Chan, McGill University.*)

Isolation and Cultivation of Bacterial Viruses

Bacterial viruses are easily isolated and cultivated in young, actively growing cultures of bacteria in broth or on agar plates. In liquid cultures, lysing of the bacteria may cause a cloudy culture to become clear, whereas in agar-plate cultures, clear zones, or **plaques,** become visible to the unaided eye (Fig. 20-17).

The principal requirement for isolation and cultivation of phages is that optimal conditions for growth of the host organisms be provided. The best and most usual source of bacteriophages is the host habitat. For example, coliphages or other phages pathogenic for other bacteria found in the intestinal tract can best be isolated from sewage or manure. This is done by centrifugation or filtration of the source material and addition of chloroform to kill the bacterial cells. A small amount (such as 0.1 ml) of this preparation is mixed with the host organism and spread on an agar medium. Growth of phage is indicated by the appearance of plaques in the otherwise opaque growth of the host bacterium as shown in Fig. 20-17.

---

as mature virions without accompanying cell lysis or death. That is, the infected cells continue to reproduce themselves as well as the virus, and the mature virions are extruded fom the cell surface continuously over a long period of time. This type of release mechanism is called a **productive infection.**

LYSOGENY

Not all infections of bacterial cells by phages proceed as described above to produce more viral particles and terminate in lysis. An entirely different relationship, known as **lysogeny,** may develop between the virus and its bacterial host (Fig. 20-18). In lysogeny the viral DNA of the temperate phage, instead of taking over the functions of the cell's genes, is incorporated into the host DNA and becomes a **prophage** in the bacterial chromosome, acting as a gene. In this situation the bacterium metabolizes and reproduces normally, the viral DNA being transmitted to each daughter cell through all successive generations. (Thus prophages behave as plasmids and are in fact considered as such by most molecular biologists.) Sometimes, however, for reasons unknown, the viral DNA is removed from the host's chromosome and the lytic cycle occurs. This process is called **spontaneous induction.** Infection of a bacterium with a temperate phage can be detected by the observations that the bacterium is resistant to infection by the same or related phages and that it can be induced to produce phage particles. A change from lysogeny to lysis can sometimes be induced by irradiation with ultraviolet light or by exposure to some chemical.

Mechanism of Lysogeny

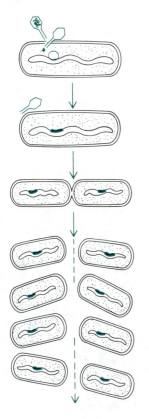

A good part of our knowledge on lysogeny comes from studies on the coliphage **lambda** (λ). It is generally considered to be typical of the temperate phages.

When a sensitive bacterium is infected by a temperate phage, two things may happen. In some of the infected cells, multiplication of the phage occurs and a lytic cycle takes place. In the other infected cells (ranging from a few to 100 percent, depending on both the host and the phage), the multiplication of the phage is repressed (because late genes required for phage multiplication and host lysis are switched off) and lysogenization occurs. Specifically, the temperate phage possesses a gene that codes for a repressor protein which makes the cell resistant to lysis initiated either by the prophage or by lytic infection by other viruses. (Radiation or chemicals may induce release of the prophage from the host genome so that a lytic cycle can ensue.)

The repressor protein (also called **immunity repressor** since the cell is resistant to lysis from externally infecting phage) from λ phage has been isolated and purified. It is an acidic protein with a molecular weight of 26,000. It reacts with two different operator sites on the λ phage genome to prevent the expression of phage lytic functions and the formation of mature phage particles. Thus the repression of phage genes is very much like the repression of bacterial operons (discussed in Chap. 12).

More specifically, the lysogenic state is governed by the activity of the regulatory region of the λ phage genome, which both bestows immunity to externally infecting phages and causes integration of the phage genome into cellular DNA. This region is termed the **immunity operon.** Upon infection by λ phage, the phage *cro* gene is transcribed, resulting in the synthesis of a protein repressor that *inhibits* the synthesis of the immunity repressor. Thus the basic mechanism in the production and maintenance of the lysogenic state is the antagonism of two repressors—the immunity repressor and the *cro* repressor, which prevents immunity.

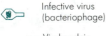

Infective virus (bacteriophage)

Viral nucleic acid

Bacterial chromosome

Figure 20-18. Lysogeny is a process in which the viral nucleic acid does not usurp the functions of the host bacterium's synthetic processes but becomes an integral part of the bacterial chromosome. As the bacterium reproduces, viral nucleic acid is transmitted to the daughter cells at each cell division. In the lysogenic state the virus is simply one of the bacterial genes. Under certain natural conditions or artificial stimuli (such as exposure to ultraviolet light), the synthesis of virus may take over, and lysis occurs.

As previously mentioned, the lytic cycle of bacteriophage λ can be induced by radiation, e.g., ultraviolet light. At the molecular level, this induces the synthesis of a host cell protein encoded in the *recA* gene of *E. coli*. This protein has proteolytic activity; once induced to accumulate, it cleaves the immunity repressor, preventing the latter from binding to the λ prophage. It is suggested that spontaneous induction of lysis may involve the same mechanism.

No RNA phages have yet been shown to be temperate. It is possible that temperate RNA phages exist; the phage could form a DNA copy of the RNA genome, which can then be integrated into the bacterial chromosome.

Figure 20-19. Conversion of phage λ linear DNA molecule into covalently closed ring and its insertion into the *Escherichia coli* chromosome.

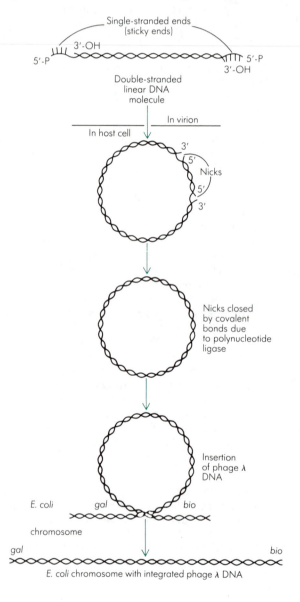

Medical Aspects of Lysogeny

Diphtheria is caused by the bacterial pathogen *Corynebacterium diphtheriae*. Its capacity to cause disease is directly related to its ability to produce a toxin. It can only produce toxin when it carries a temperate phage. In the same way, only those streptococci which carry a temperate phage can produce the erythrogenic (rash-producing) toxin of scarlet fever. In another known instance, some types of botulism toxin are produced by *Clostridium botulinum* as a result of lysogeny. This phenomenon, in which a prophage is able to make changes in the properties of a host bacterium in lysogeny, is termed **lysogenic conversion**. Its occurrence is based on the fact that not all phage genes in lysogeny are blocked by the immunity repressor. Part of the phage DNA can be transcribed to form new proteins.

Bacterial lysogeny is a good conceptual model for the study of oncogenic, or cancer-producing viruses since these viruses also have the capacity of perpetuating their genomes in infected cells.

Prophage DNA Replication

A mechanism exists to allow the prophage DNA to replicate in step with the host cell DNA. When phage λ DNA is injected from the capsid (in which it exists as a linear double-stranded form), it rapidly forms into a closed circular molecule (Fig. 20-19). A λ phage enzyme, coded by the *int* gene, directs the insertion of the phage DNA into the bacterial chromosome. This enzyme recognizes a specific site on both the phage DNA and the bacterial chromosome and catalyzes a single site-specific recombination event, resulting in the insertion of the phage DNA into the cell genome (between the *gal* and *bio* genes) as shown in Fig. 20-19. It should be noted that different temperate phages have their own specific sites of integration on the bacterial chromosome. However, some temperate phages, such as phage Mu, have no site specificity for insertion and may even be able to insert multiple copies of their DNA into a single bacterial chromosome. Wherever insertion occurs, inactivation of the specific bacterial gene gives rise to a *mutant*; hence the name *phage Mu* was derived.

QUESTIONS

1　Describe the nature of a bacterial virus with respect to the following: (*a*) size, (*b*) growth, (*c*) metabolism, and (*d*) genome.
2　Define a virus and explain why it is defined that way.
3　Who coined the word *bacteriophage* and what does it mean?
4　Sketch seven morphological groups of phages and give an example of a phage for each group.
5　Why are bacteriophages suitable subjects for research in genetics?
6　What features distinguish a bacteriophage from a bacterium?
7　Define the following terms: lysogeny, lysis, temperate, and virulent.
8　Distinguish between the two structural forms of phages.
9　Explain why the icosahedron is considered the basic structure for all cubic viruses.
10　How would you describe the structure of the heads of the T2 and T4 phages? Why do we say that these phages have binal symmetry?

11 Give the names of five commonly studied phages and the nature of their nucleic acids.

12 Give some reasons for the slow and difficult development of phage taxonomy.

13 List the family names for the following viruses: T2, λ, T7, fd, and φX174.

14 Describe the differences between the filamentous coliphages and the tadpole-shaped ones.

15 Describe the technique by which phages are isolated.

16 What is the T series of bacteriophages?

17 Explain the process and application of phage-typing.

18 What are some unique features of phage adsorption on a host cell surface?

19 Describe the sequence of events that occurs when T-even phages penetrate an *E. coli* cell.

20 Describe the process of transcription in phage T4 infection with respect to immediate early, delayed early, and late gene products.

21 Explain the meaning of the following terms with reference to the viral multiplication cycle: eclipse period, latent period, rise period, and burst size.

22 In what way does the immunity operon of phage λ regulate lysogeny?

23 Explain the mechanism that allows prophage DNA to replicate in step with host DNA.

24 What is the relationship of temperate phages to the disease-producing ability of some bacteria? Why does this occur?

REFERENCES

Fraenkel-Conrat, H., and P. C. Kimball: *Virology*, Prentice-Hall, Englewood Cliffs, N.J., 1982. *This textbook attempts to cover the entire field of virology, focusing mainly on the biophysical and biochemical aspects of viruses. Virus families are discussed in a systematic manner advancing from the smallest and simplest to the largest and most complex. A general knowledge of biochemistry is a prerequisite for the use of this text.*

Hughes, S. S.: *The Virus: A History of the Concept*, Science History Publications, New York, 1977. *A concise and highly readable history of virology in relatively nontechnical language.*

Knight, A. C.: *Molecular Virology*, McGraw-Hill, New York, 1974. *A condensed and integrated presentation of the basic principles of molecular virology for the serious student of microbiology.*

Luria, S. E., J. E. Darnell, Jr., D. Baltimore, and A. Campbell: *General Virology*, 3d ed., Wiley, New York, 1978. *A well-organized text on the biology of viruses and their activities.*

Watson, D. H.: "Form and Function—III. Viruses," in *Essays in Microbiology*, J. R. Norris and M. H. Richmond (eds.): Wiley, New York, 1978. *An article that surveys the structural principles of icosahedral and helical viruses; a concise essay on the morphological aspects of viruses.*

Chapter 21   Viruses of Animals and Plants

In the last chapter we discussed the viruses of bacteria, the bacteriophages. Viruses also exist that infect other living cells. There are viruses that infect fungi, algae, and insects. And of course there are viruses that infect animal and plant cells.

Animal and plant viruses vary greatly in size and shape (Figs. 21-1 and 21-2), but they do not have the tadpole morphology characteristic of some bacteriophages discussed in the previous chapter. Size and shape are characteristic properties of each type of virus. Virions range in size from 20 to 350 nm and represent the smallest and simplest infectious agents. Since most viruses measure less than 150 nm, they are beyond the limit of resolution of the light microscope and are only visible by means of the electron microscope.

Much of the basic biology of bacteriophages discussed in Chap. 20 is also applicable to animal and plant viruses. Accordingly, only the special properties of animal and plant viruses are discussed in this chapter.

Figure 21-1. Electron micrographs of some animal viruses. (A) A cluster of polioviruses, the cause of poliomyelitis. (B) Rotavirus particles, the cause of acute infectious diarrhea, a major cause of death in very young infants. (C) An enveloped herpesvirus. It is a persistent virus in humans and occasionally manifests its presence in "fever blisters" or "cold sores" of mucous membranes. (A, B, and C are the same magnification, X200,000.) (*Courtesy of Margaret Gomersall, McGill University.*) (D) A negatively stained parainfluenza virus from the nasopharyngeal secretions of a patient with an acute respiratory disease. Within the fringed outer envelope is contained the tightly packed helical nucleocapsid (X125,000). (*Courtesy of Frances Doane, University of Toronto.*)

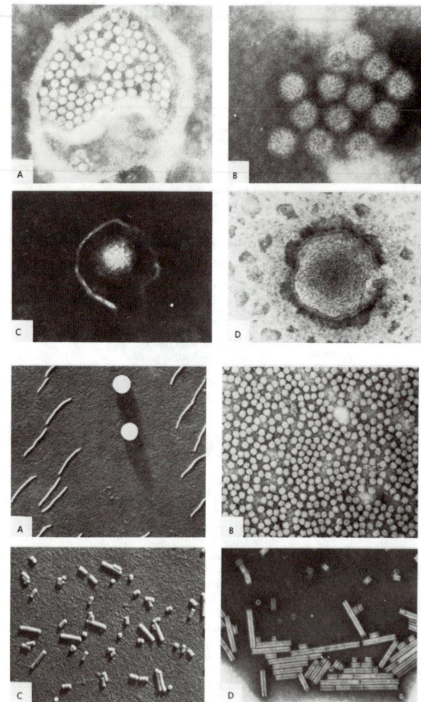

Figure 21-2. Electron micrographs of plant viruses. (A) Potato virus X particles appear as flexous rods 513 nm long (X17,000). Also shown are two latex spheres used in electron microscopy to show relative sizes. (B) Tomato ringspot virus is icosahedral in structure (X150,000). (C) and (D) Tobacco rattle virus particles appear as both long and short rods. Both lengths are necessary to establish infection. (C: X23,600. D: X25,400.) (*Courtesy of M. K. Corbett, University of Maryland.*)

HISTORY

Viruses and Vaccination

Some virus diseases have been known clinically for centuries. Indeed, the first infectious disease for which a practical and effective method of prevention was developed was a virus disease. In 1796 Jenner first vaccinated an 8-year-old boy named James Phipps with material removed from a cowpox lesion on the hand of a milkmaid. Proof that the inoculation gave protection against smallpox was obtained 6 weeks later, when the boy was inoculated with pus from a smallpox victim and did not develop the disease.

Jenner found that persons successfully inoculated intracutaneously with virus isolated from cowpox lesions developed a small scab at the site of application which dropped off after about 2 weeks, leaving a single small scar. Before making his report in 1798, he had successfully vaccinated 23 persons. The material he used came from cows (Latin, *vacca*); hence the term *vaccination*, in contrast to *variolation*, which identified an earlier procedure in which variola (smallpox) virus was artificially introduced into a subject to provide protection against natural smallpox infection. Today, smallpox vaccine is made from virus grown on the skin of healthy calves or sheep or in embryonated eggs, thus eliminating the possibility of transmitting other human diseases in the process. Jenner never saw the causative agent of smallpox; indeed, his discovery and its application came more than half a century before the establishment of the germ theory of disease.

The first American known to have been vaccinated was Daniel Oliver Waterhouse, whose father, Dr. Benjamin Waterhouse, a Boston physician, obtained the cowpox virus from Europe. The 5-year-old Daniel and two servants were vaccinated July 8, 1800; they were later inoculated with smallpox virus and found to be fully protected. Such daring and dramatic demonstrations proved the value of vaccination and brought it to the attention of officials and the public. Recognizing its usefulness, Thomas Jefferson wrote to Jenner in 1806, "Future nations will know by history only that the loathsome smallpox has existed and by you has been extirpated."

Half a century passed before Louis Pasteur became interested in infectious diseases and the role of microorganisms in causing them. His successes in controlling undesirable organisms in fermentation reactions and in diseases of silkworms led him into some problems of human health. One health problem that opportunity dropped on his doorstep was rabies, a disease transmitted to humans by bites of rabid dogs, foxes, wolves, cats, bats, and other animals. Pasteur did not think of this agent as being particularly different from the microorganisms (bacteria, yeasts, and protozoa) he had previously worked with, and he applied his previous experience to the problem at hand. Through laboratory manipulations he was able to attenuate, i.e., make less virulent, the virus from rabid animals. When inoculated into an animal, the attenuated virus produced an active immunity against the disease. It was not until several years later that the viral nature of the disease was established. The method Pasteur used in developing a vaccine against rabies provided the basis for the production of vaccines for other virus diseases.

One of Pasteur's associates, Charles Chamberland, had meanwhile learned that porcelain filters would prevent passage of bacteria but allow passage of the causative agent of rabies. Because in those days the word *virus* was loosely used to describe any toxic substance that caused disease, it was only natural to describe those unseen filter-passing agents of disease as **filterable viruses.**

Tobacco Mosaic Virus and Filterability

In 1892, Dmitrii Ivanowski discovered that the causative agent of tobacco mosaic disease was filterable. By placing some bacteria-free filtrate from ground-up infected plants on healthy tobacco leaves, Ivanowski produced the disease in healthy plants (Fig. 21-3). Beijerinck (1898) confirmed this work, and in the same year Loeffler and Frosch filtered fluid from vesicles in the mouths of cattle with foot-and-mouth disease and transmitted the virus of that disease to healthy animals. These experiments marked the beginning of a new phase of microbiology, the study of infectious agents invisible even through the most powerful microscopes then available. Although they cannot be grown on nonliving culture media, viruses are capable of causing disease in plants and animals.

One of the most important scientific contributions to the field of virology was the discovery in 1935 that the tobacco mosaic virus (TMV) can be crystallized. For this fundamental research, Wendell Stanley, then at the Rockefeller Institute for Medical Research, shared the 1946 Nobel prize in chemistry. Even before Stanley's work, many people had questioned whether viruses are truly living organisms. When it was shown that the "inanimate" crystals of TMV could produce disease in healthy plants, the controversy was renewed with more vigor. One thing these virus crystals (Fig. 21-4) do that other pure toxic chemical compounds cannot do is replicate themselves. In that sense, viruses seem to be alive, because only living things can reproduce and multiply. On the other hand, they seem to be nonliving because they have no intrinsic metabolism and depend on their host for survival and multiplication. The question has never been settled to everyone's satisfaction, and little is to be gained by arguing about it. What is more important is that thanks to the intensive study of microbiologists, chemists, and physicists, these particles we call viruses have given and will continue to give us important information about life processes.

Yellow Fever Virus and Vaccines

Another milestone in virology occurred when Max Theiler, 1951 Nobel laureate in physiology or medicine, found in 1937 that virulent yellow fever virus can be attenuated by serial passage on cultures of chick embryo tissue. Later investigators modified this technique to produce vaccines against other virus diseases. For example, Enders, Robbins, and Weller laid the foundation for the development of effective poliomyelitis vaccines by culturing the virus of poliomyelitis on monkey kidney cells in 1949. This is not attenuation but a method for mass growth of the virus. Indeed, tissue-culture techniques made it possible to cultivate many mammalian viruses in the laboratory (Fig. 21-5) and will undoubtedly lead to the development of methods for controlling the diseases they cause.

Among the virus diseases for which vaccines have recently been developed through the use of tissue cultures is measles (rubeola). The first live attenuated strain of measles virus was isolated in 1962 by Enders after passage of the virus through human kidney cells, human amnion cells, and finally, chick embryo tissue culture. The attenuated live measles vaccine licensed for use in the United States is prepared from chick embryo cell cultures infected with attenuated measles virus. Mumps vaccine is prepared from cultures of chick fibroblasts infected with an attenuated live mumps virus; it has been available since 1968. After extensive clinical trials a vaccine to protect against German measles (rubella) was approved in 1969 by the U.S. Public Health Service for clinical use. Rubella vaccines contain viruses either isolated in African green monkey cells

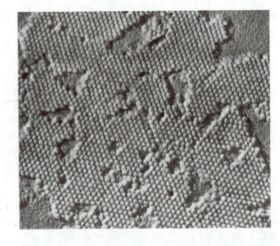

Figure 21-4. Surface of a crystal of tobacco necrosis virus (X55,200). (*Courtesy of M. K. Corbett, University of Maryland.*)

Figure 21-3. Transmission of a plant virus. By grinding or macerating leaves from naturally infected plants and passing the fluid through a porcelain filter, Ivanowski demonstrated that a virus causes tobacco mosaic disease. The filtrate containing the virus causes infection when placed on the leaves of healthy plants.

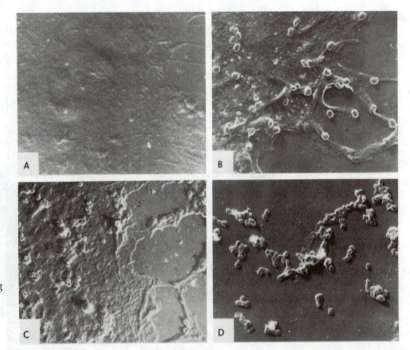

Figure 21-5. Tissue culture of rabbit kidney cells destroyed by vaccinia virus. (A) Uninfected cells appear as smooth flat sheets; 6 h after inoculation, some virus particles are released and lie on the surface of the cells (B). After 24 h of incubation the number of virus particles increases, and the tissue cells are beginning to disintegrate (C). After 48 h the virus appears as clumps dissociated from the cellular material (D). Magnifications are (A) X5,180, (B) X6,280, (C) X3,810, and (D) X4,940. (*Courtesy of John Mathews and the Upjohn Company.*)

and attenuated by further cell passage (in primary duck embryo cells) or isolated and passed in diploid human embryo cells.

STRUCTURE AND COMPOSITION

Like bacteriophages, animal and plant viruses are composed of a central core of nucleic acid surrounded by a capsid, which is made up of capsomeres. Most plant and animal viruses exhibit a characteristic symmetry: (1) icosahedral in the case of spherical viruses, (2) helical in the case of the rod-shaped viruses, and (3) complex in the case of a miscellaneous group. This symmetry is a basic criterion of viral classification, which will be discussed later. But in some animal viruses the **nucleocapsid** (nucleic acid and capsid) is covered by an outer membranelike structure called the **envelope,** which is made of lipoproteins and conceals this symmetry. Virions that have envelopes are sensitive to lipid solvents such as ether and chloroform. Their capacity to infect cells is inactivated by these solvents. Nonenveloped viruses are referred to as **naked** virions. Such viruses are not affected by lipid solvents.

Morphology

Icosahedral Symmetry

In searching for a simple, stable architectural structure, R. Buckminster Fuller, the American architect, engineer, and inventor, discovered that an icosahedral shell (discussed in Chap. 20) is easy to assemble and provides an enclosure possessing minimum stress. This is the idea behind Fuller's geodesic domes, the design of which he patented in 1947. These domes usually look spherical and cover more space with less material than any other buildings ever designed. These domes are actually subtriangulated icosahedra constructed of almost

Figure 21-6. Examples of subtriangulated icosahedra constructed from pentamers and hexamers. (a) Geodesic dome in a New Brunswick, Canada, campground. Note the clusters of triangular facets in 5's and 6's. (*Courtesy of Frances Doane, University of Toronto.*) (B) Geodesic dome of the U.S. pavilion in the World Exposition of 1967 in Montreal, Canada. (*Courtesy of E. C. S. Chan, McGill University.*) Thousands of geodesic domes are used as theaters, auditoriums, defense facilities (DEW Line stations in the Arctic), and dwelling places. One of the most recent such structures is the 180-foot-high polished aluminum sphere of Spaceship Earth in Epcot Center at Disney World in Florida.

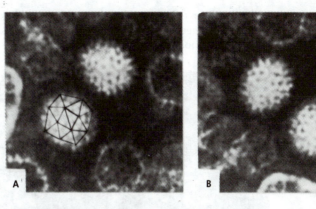

Figure 21-7. Bluetongue virus particles (A) and (B). Part of the icosahedral surface lattice has been drawn on one of the particles (X312,000). (*From H. J. Ehis and D. W. Verwoerd, Virology,* **38**:213–219, 1969; by permission.)

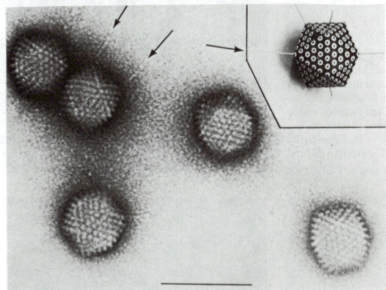

Figure 21-8. A simian adenovirus, SV 15, in an electron micrograph and with a model of an icosahedral particle. The bar equals 100 nm. (*Courtesy of Kendall O. Smith and Melvin D. Trousdale.*)

identical triangular units in clusters of fives and sixes (see triangular outline in Fig. 20-5A and illustrations of geodesic domes in Fig. 21-6). Viruses, too, long ago "discovered" this kind of structure, and so spherical viruses are in reality icosahedral in symmetry (Fig. 21-7). Examples of icosahedral viruses are polioviruses and adenoviruses (Fig. 21-8), which cause poliomyelitis and respiratory infections, respectively.

Helical Symmetry

Plant viruses with helical symmetry are typically rod-shaped (Fig. 21-9). One of the first viruses studied by electron microscopy was the tobacco mosaic virus (Fig. 21-10). Its nucleic acid core is covered by a capsid consisting of closely packed capsomeres arranged in a regular helix (see Fig. 20-5B). Animal viruses with capsids displaying helical symmetry include measles, mumps, influenza, and rabies. In these viruses, the nucleocapsid is a flexible structure packed within a fringed lipoprotein envelope (Fig. 21-11). The fringes are actually spiked projections made of glycoproteins.

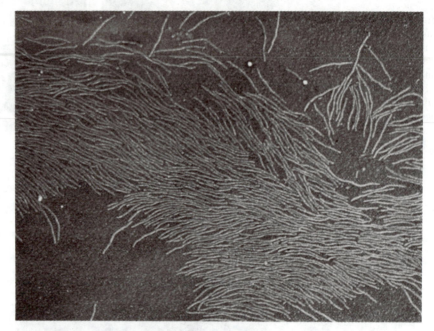

Figure 21-9. Electron micrograph of a chromium-shadowed preparation of cymbidium mosaic virus. Virus particles are flexuous rods 480 nm long (X33,000). *(Courtesy of M. Kenneth Corbett, University of Maryland.)*

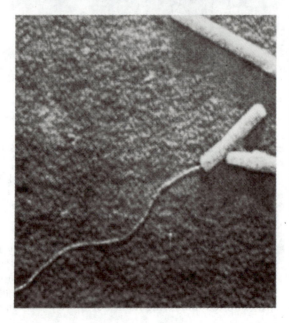

Figure 21-10. Tobacco mosaic virus particles partially degraded by phenol, showing protein coat and RNA strand (X120,000). *(From M. K. Corbett, Virology, 22:539–543, 1964; by permission.)*

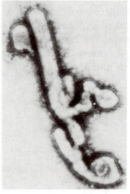

Figure 21-11. Influenza virus. Note the fringes of spikes on the surfaces of the virions (X55,300). *(Courtesy of Margaret Gomersall, McGill University.)*

Figure 21-12. Vaccinia virus, a poxvirus with complex morphology. (A) Whole (coated) virus showing surface tubules (X200,000). (*Courtesy of Margaret Gommersall, McGill University.*) (B) Immature virion obtained from an infected cell showing a bounding membrane with subunit projections (X160,000). (*Courtesy of K. B. Easterbrook, Dalhousie University.*)

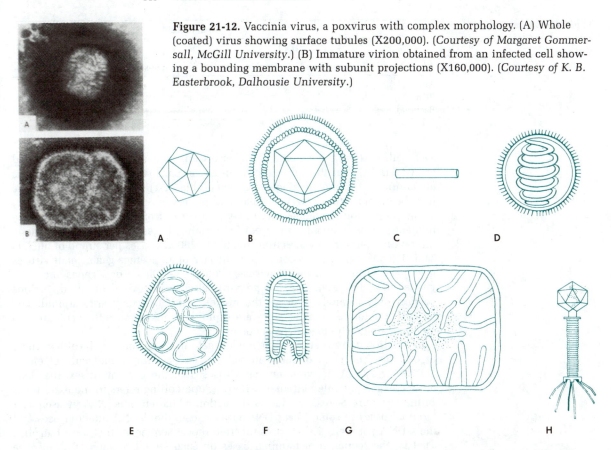

Figure 21-13. Morphology of some well-known viruses. Icosahedral symmetry: (A) polio, wart, adeno, rota; (B) herpes. Helical symmetry: (C) tobacco mosaic; (D) influenza; (E) measles, mumps, parainfluenza; (F) rabies. Complex or uncertain symmetry: (G) poxviruses; (H) T-even bacteriophages. (*Redrawn by Erwin F. Lessel after a drawing by Frances Doane, University of Toronto.*)

Complex-Structured Viruses

There are viruses with complex or uncertain symmetries. For example, the poxviruses, such as smallpox and molluscum contagiosum, have the most complex virion structure known (Fig. 21-12). They consist of many different proteins and lipoproteins. (The tailed phages discussed in Chap. 20 may be considered to be complex-structured viruses.)

A schematic representation of the morphology of viruses is given in Fig. 21-13.

Nucleic Acids

The viral genome, containing all the genetic information, is composed of nucleic acid. Like bacteriophages, animal and plant viruses contain either DNA or RNA, but never both in the same virion. This, of course, is in contrast to all *cellular* forms of life, which without exception contain *both* types of nucleic acid in

Table 21-1. Occurrence of Types of Nucleic Acids in Viruses

	Nucleic Acid			
	DNA		RNA	
Virus	Single-stranded	Double-stranded	Single-stranded	Double-stranded
Animal	+	+	+	+
Plant	+	+	+	+
Bacterial	+	+	+	+

each cell. Further, the genome of higher organisms consists of double-stranded DNA (dsDNA). But the genome of a virus can consist of DNA or RNA that is either double-stranded or single-stranded (Table 21-1). All four types of genome have been found in bacterial, animal, and plant viruses.

The proportion of nucleic acid in a virion varies from about 1 percent for the influenza virus to about 50 percent for certain phages. The content of genetic information per virion varies from about 3 to 300 kilobases per strand of nucleic acid. Thus if 1 kilobase is considered the size of an average gene, small viruses (e.g., parvoviruses and picornaviruses) contain perhaps 3 or 4 genes and large viruses (e.g., herpesviruses and poxviruses) contain several hundred. Virions contain only a single copy of the nucleic acid; i.e., they are **haploid.** The exception is the retroviruses; they are **diploid** virions because they contain two identical single-stranded RNA genomes.

In addition, the structure of the nucleic acid in the virion may be either linear or circular. The DNA of most animal viruses is a linear molecule of either dsDNA or ssDNA. In some animal viruses, like the papovaviruses, the DNA occurs as a supercoiled circular dsDNA. (Supercoiling refers to the extra turns in the structure of dsDNA due to the action of the enzyme DNA gyrase.) The terminal bases of some linear DNA viruses, e.g., the dsDNA adenoviruses and the ssDNA parvoviruses, exist as inverted repeat sequences that form hairpins. That is, the sequence of terminal bases on each strand of such DNA may be represented as ABCD . . . D'C'B'A', where ABCD are complementary to A'B'C' D'. Such sequences are important in replication or in the cyclization of linear progeny strands.

Some plant viruses appear to have a genome of circular dsRNA. But the RNA in animal viruses exists only as linear double-stranded or single-stranded molecules. Unlike the DNA genome, the RNA genome within a virion may exist as a **segmented genome** (divided into several units). Thus reovirus contains 10 different segments of dsRNA and influenza virus has 8 separate segments of ssRNA.

Since the replication of nucleic acids requires that one nucleic acid strand be used as a template for the synthesis of the new nucleic acid strand (see Chap. 12), it follows that both of the original strands must be duplicated so that a virus can replicate its nucleic acid. Except for the RNA tumor viruses, viruses containing single-stranded nucleic acid package only one of the two strands within the capsid. Single-stranded viral RNA molecules which function directly as mRNA in the host cells have been designated as positive, or plus (+) strands. Viruses with negative, or minus-strand (−) RNA molecules must first replicate their RNA (using RNA transcriptase carried within the virion) to form a complementary strand which then acts as the mRNA. RNA tumor viruses have two equal positive-strand RNA molecules.

Other Chemical Components

Protein. Besides nucleic acid, protein is the other major chemical component of the virion. The capsid is made up of protein. Virions also contain internal proteins. Some are basic proteins bound to the nucleic acids. In papovaviruses these basic proteins are regular cellular histones; in adenoviruses they are histonelike but are specified by the viruses. Small peptides and polyamines are found in phages. These cationic compounds are presumed to aid in the folding of the nucleic acids by linking together different loops.

In addition, many viruses have now been found to contain one or more enzymes that function in the replication of their nucleic acid components. The most common viral enzyme is an RNA polymerase. Except for the RNA viruses carrying the single positive strand (mRNA), all RNA viruses must possess their own RNA polymerase. The positive-strand RNA viruses code for their RNA polymerase, which is synthesized by host cell translation of the viral mRNA. This RNA polymerase transcribes the viral RNA, allowing viral replication to proceed. The RNA tumor viruses contain an enzyme (RNA-dependent DNA polymerase, or reverse transcriptase) that synthesizes a DNA strand, using the viral RNA genome as a template.

Lipid. A wide variety of lipid (fatty) compounds have been found in viruses. These include phospholipids, glycolipids, neutral fats, fatty acids, fatty aldehydes, and cholesterol. Phospholipid is the predominant lipid substance and is found in the viral envelope.

Carbohydrate. All viruses contain carbohydrate since the nucleic acid itself contains ribose or deoxyribose. Some enveloped animal viruses, such as the influenza virus and other myxoviruses, have spikes made of glycoprotein on the envelope.

VIRUS REPLICATION

Virus particles outside a host cell have no independent metabolic activity and are incapable of reproduction by processes characteristic of other microorganisms (Table 21-2). Multiplication takes place by **replication,** in which the viral protein and nucleic acid components are reproduced within susceptible host cells.

The entire process of infection can be generalized as follows. The virion attaches to a susceptible host cell at more or less specific sites. Either whole virus or viral nucleic acid penetrates to the inside of the cell. If whole virus has penetrated the cell, uncoating of the virus must take place to release the nucleic acid. Reproduction of the virus takes place in the cytoplasm, the nucleus, or both. The viral protein and nucleic acid components are assembled into virus particles and released from the host cell. The steps of virus infection and replication are therefore: (1) adsorption, (2) penetration and uncoating, (3) component replication and biosynthesis, (4) assembly, and (5) release.

Adsorption

The adsorption process occurs in two steps. The first step involves preliminary

Table 21-2. Comparison of Viruses with Some Bacteria

Microorganism	Multiplication	Diameter, nm	Chemical Composition	Inhibition by Antibiotics
Typical bacteria	In vitro* in fluid and solid media; on cell surfaces; or intracellularly; by binary fission	1000–3000	Complex proteins, carbohydrates, fats, etc.; DNA and RNA; peptidoglycan in cell wall	Yes
Mycoplasmas	Like typical bacteria but by budding rather than fission	150–1000	Like typical bacteria but without cell walls	Yes
Rickettsias	In living cells only and by binary fission	250–400	Like typical bacteria	Yes
Viruses	In living cells only and by synthesis from pools of constituent chemicals	10–300	Either RNA or DNA, plus protein; some may have lipid and/or carbohydrate components	No

* *In vitro* means "in glass," i.e., in laboratory vessels. (This is in contrast to *in vivo*, which means "within a living organism.")

attachment by ionic bonds or charges and is easily reversed by a shift in pH or salt concentration. The second step appears to involve firmer, more specific attachment and to be irreversible. Molecular entities on the surface of cells act as receptors for viruses, interacting with specific proteins on the capsids of naked virions or on the envelopes of enveloped virions. In contrast to the marked specificities of adsorption of animal and bacterial viruses, plant viruses do not seem to require specific receptor sites.

Penetration and Uncoating

The penetration of animal viruses into attached cells occurs by one of two mechanisms. One mechanism consists of engulfment of whole virions by the cells in a phagocytic process called **viropexis,** followed by uncoating or removal of the capsid. This takes place in the phagocytic vacuoles and is due to the action of enzymes called **lysosomal proteases.** The other mechanism occurs in the enveloped viruses; the viral lipoprotein envelope fuses with the host cell's surface membrane. This fusion results in the release of the viral nucleocapsid material into the cytoplasm of the host cell. Uncoating again occurs within the host cell.

Plant viruses penetrate host cells through transient pores (called **ectodesmata**) which protrude through the cell wall at intervals and communicate to the exterior of the cells. These pores function for the purpose of water and nutrient uptake as well as for secretion of substances such as waxes. Whole virus particles are apparently engulfed at these points. Also, insects can inoculate plant viruses into cells during feeding. Sometimes this is a purely mechanical process; at other times the virus is found in the insect tissue and may even multiply there. Insect feeding is probably the most important means of plant virus transfer in nature. Once the virus is inside the plant cell, uncoating occurs.

Replication and Biosynthesis of Virus-Specific Molecules

Shortly after penetration, there follows an interval of time called the **latent period** (discussed in Chap. 20). It is during the latent period that uncoating of the virion takes place; the viral nucleic acid is freed from the capsid and is accessible to enzymes required to translate, transcribe, or replicate it. The uncoating process varies from one virus to another. Some viruses are uncoated in the cytoplasm, while others are uncoated in the nucleus.

Transcription of the viral nucleic acid into mRNA is the usual next step for all viruses except those RNA viruses (e.g., picornaviruses) whose viral RNA acts directly as mRNA. (Recall that RNA viruses that carry minus-strand RNA, such as orthomyxoviruses, paramyxoviruses, and rhabdoviruses, must first transcribe their RNA to form the plus strand that can function as mRNA. This transcription is catalyzed by a viral RNA polymerase released during uncoating.)

The biosynthetic processes of virus-specific molecules can be divided into early and late events. Thus in most viruses, only part of the nucleic acid of the infecting virion is initially transcribed into mRNA. Generally, early mRNA codes for early enzymes required for nucleic acid replication and for proteins that inhibit synthesis of cellular macromolecules and break down cellular polyribosomes, making them available for the transcription of viral genes. The exact mechanism for the replication of new copies of the virus genome varies with different types of viruses, i.e., whether their nucleic acid is RNA or DNA, single-stranded or double-stranded, and of positive or negative polarity. This replication process may take place in the cell nucleus or cytoplasm, depending on the specific virus. Generally, the dsDNA viruses include those capable of using host cell polymerases. Viruses with RNA genomes require viral-coded polymerases.

Nucleic acid replication (which follows early protein synthesis) serves as the demarcation line between early and late events in the viral replication process. Since late mRNA is not synthesized until after viral DNA replication has begun, it is transcribed from progeny genomes. Thus late protein synthesis may be defined as taking place after nucleic acid replication. Late mRNA usually specifies the structural proteins (such as those for protomer formation) of the virion. Translation of mRNA into proteins takes place in the host cell cytoplasm and uses ribosomes, transfer RNAs, and enzymes in the cytoplasm. (If the mRNA is synthesized in the host nucleus, it first enters the cytoplasm before translation.)

Assembly

When a critical number of the various viral components has been synthesized, the components (virus-specific molecules) are assembled into mature virus particles in the nucleus and/or cytoplasm of the infected cell (depending on the type of virus). The DNA viruses, with the exception of the poxviruses, are assembled in the nucleus. The RNA viruses are generally assembled in the cytoplasm, as are the poxviruses.

Release

Release of the completed virions from the host cell is the final step in virus multiplication. The mechanism of release varies with the type of virus. In some animal virus infections, the host cells lyse, releasing the virions. Naked virions

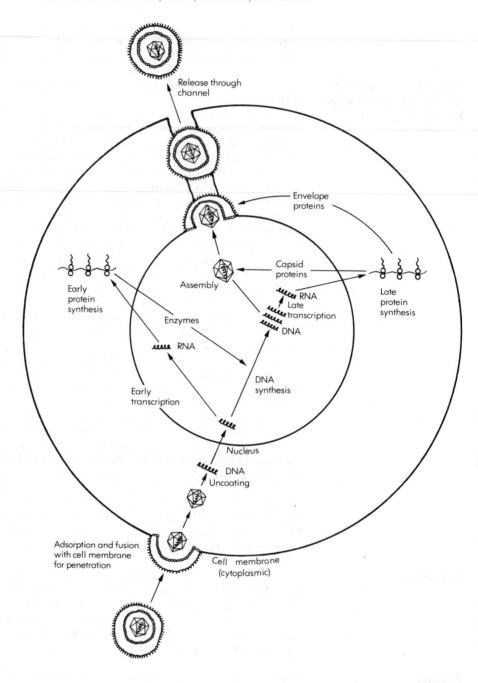

Figure 21-14. Replication of herpes simplex virus. Specific glycoproteins in the viral envelope are essential for optimal adsorption to host cell receptors in the cytoplasmic membrane. Viral envelope and membrane fuse and the nucleocapsid of the virion is released into the cytoplasm. The virion is uncoated and the freed viral DNA is transported to the nucleus. Early transcription and mRNA processing are apparently executed by host cell enzymes. The resulting enzymes are used in viral DNA replication. Nuclear RNA transcripts synthesized after DNA replication are responsible for the synthesis of structural proteins that go to form the capsid and envelope as well as the glycoproteins in both the cytoplasmic and nuclear membranes. The structural proteins enter the nucleus to participate in the assembly of the virion. The nucleocapsids are enveloped by budding through the nuclear membrane. The virus is released from the cell through cytoplasmic channels. (*Erwin F. Lessel, illustrator.*)

are generally released in a burstlike fashion from the cells as they lyse. Alternatively, they may be extruded by a process that is essentially the reverse of phagocytosis. Enveloped animal viruses (and presumably enveloped plant viruses also) are released by budding through special areas of the host-cell membrane coded for by the virus, and in doing so the virions acquire a portion of the host membrane. In a few animal and plant viruses, the host cells are not destroyed. The virions leave the cells by special channels (tubules) over an extended period of time.

The yield of virus particles per cell varies with the virus, the type of cell, and the growth conditions. The average yield of plant and animal virions ranges from several thousand to about a million per cell, compared with a yield of a few hundred bacterial T phages.

As an example of the viral replicative process in the eucaryotic cell, Fig. 21-14 shows the replication of the herpes simplex virus, which is the cause of "fever blisters" or "cold sores." As can be seen in the figure, the events related to biochemical replication occur in both the nucleus and the cytoplasm, with assembly of the virion initiated in the nucleus. The nucleocapsids of these viruses then migrate to the cytoplasmic membrane (after envelopment by budding through the nuclear membrane), where the mature enveloped viruses appear to reach the surface of the cell through cytoplasmic channels.

CLASSIFICATION OF VIRUSES

Many attempts to classify viruses have been made. One of the earliest systems, which still has limited use, established classification according to the kind of host the viruses normally infect (e.g., hog cholera virus, swine influenza virus, fowl plague virus, cucumber mosaic virus, tobacco mosaic virus, and others). Another early means of classification was based on virus tissue affinities, for example, neurotropic (nerve-tissue) viruses and dermatropic (skin-tissue) viruses. Apparently this method was useful for physicians, epidemiologists, and some allied health investigators. However, as methods of measuring physical, chemical, and biological characteristics of viruses have been developed, information has been accumulated that allows formulation of a classification scheme for all viruses on the basis of these properties. Such properties are summarized in Table 21-3.

Table 21-3. Properties Used for Classification of Viruses

Primary Characteristics	Secondary Characteristics
Chemical nature of nucleic acid: RNA or DNA; single- or double-stranded; single or segmented genome; (+) or (−) strand; molecular weight	Host range: Host species; specific host tissues or cell types
Structure of virion: Helical, icosahedral, or complex; naked or enveloped; complexity; number of capsomeres for icosahedral virions; diameter of nucleocapsids for helical viruses	Mode of transmission: e.g., feces
	Specific surface structures: e.g., antigenic properties
Site of replication: Nucleus or cytoplasm	

Table 21-4. Classification of Viruses on the Basis of Differences in Their Transcription Processes

Class	Type of Virus	Characteristics
I	Viruses containing dsDNA (most DNA viruses)	Viral DNA (except for poxvirus) enters nucleus of host cell, where mRNA is synthesized with no intermediate form; the mRNA then passes into the cytoplasm (poxvirus DNA does not enter the host cell nucleus but remains in the cytoplasm); protein synthesis and assembly of virus take place in the cytoplasm
II	Viruses containing ssDNA (some bacterial viruses and the parvovirus group of animal viruses)	The single-stranded DNA molecule serves as a template for the synthesis of a complementary strand of DNA, resulting in a dsDNA molecule called a replicative form; not known whether mRNA is a transcription product of the replicative form or of the parental ssDNA
III	Viruses containing dsRNA (+ and − strands) (reoviruses; the reovirus genome consists of 10 dsRNA segments)	During the uncoating process not all of the nucleoprotein core is degraded—the viral genome remains enclosed; ten mRNA molecules are transcribed from one strand of the dsRNA and pass into the host cell cytoplasm; this mRNA may be used to produce double-stranded progeny RNA or structural proteins
IV	ssRNA viruses containing the (+) strand (poliovirus)	Following uncoating in the cytoplasm the parental (+) strand RNA acts as mRNA and directs the synthesis of viral proteins; one protein is an RNA polymerase that catalyzes the formation of a (−)RNA strand called a **replicative intermediate** (RI); the RI is a partial dsRNA since there is a complete (−)ssRNA molecule to which are attached partially synthesized (+)RNA strands; from the RI a complete dsRNA molecule (replicative form) will be produced, consisting of one (+) and one (−) strand; the other (+)RNA strands that were part of the RI may become genomes for new virus or they act as mRNA
V	ssRNA viruses containing the (−) strand (rhabdoviruses, paramyxoviruses, orthomyxoviruses)	Following uncoating in the cytoplasm the (−)RNA strand remains associated with an RNA transcriptase, which catalyzes the synthesis of mRNA from the (−)RNA parental strand; the mRNAs code for structural proteins as well as polymerase and transcriptase; the transcriptase also catalyzes the synthesis of a (+)RNA strand; from the (+) strand more (−)RNA molecules will be transcribed which can act as mRNA or as genomes for future virus progeny

Table 21-4. (*continued*)

Class	Type of Virus	Characteristics
VI	RNA tumor viruses requiring a DNA intermediate for replication	Following uncoating, viral reverse transcriptase (RNA-dependent DNA polymerase) uses the RNA genome as a template to transcribe a complementary DNA strand, resulting in the formation of a DNA-RNA hybrid; from the DNA strand of the hybrid a complementary DNA strand is synthesized using a conventional DNA polymerase enzyme, and a dsDNA molecule called a provirus is formed; the provirus is integrated into the genome of the host—the mechanism for viral transformation of animal cells

Table 21-5. Typical Classification of Animal and Plant Viruses on the Basis of Their Inherent Properties

Nucleic Acid	Capsid Symmetry	Envelope Presence	Virion Size (nm)	Family or Group	Genus or Typical Member
RNA	Icosahedral	−	28	*Tymovirus*	*Turnip yellow mosaic virus*
			70	*Reoviridae*	*Reovirus*
			28	*Picornaviridae*	*Enterovirus*
			30	*Tombusvirus*	*Tomato bushy stunt virus*
		+	35–80	*Togaviridae*	*Alphavirus*
			100–120	*Retroviridae*	*Rous sarcoma virus*
	Helical	−	17.5 by 300	*Tobamovirus*	*Tobacco mosaic virus*
			10 by 1250	*Closterovirus*	*Beet yellows virus*
		+	80–120	*Orthomyxoviridae*	*Influenzavirus*
			150–300	*Paramyxoviridae*	*Mobillivirus*
			60–180	*Rhabdoviridae*	*Lyssavirus; lettuce necrotic yellows virus*
			90–100	*Bunyaviridae*	*Bunyamwera*
			100	*Coronaviridae*	*Avian infectious bronchitis virus*
			50–300	*Arenaviridae*	*Lymphocytic choriomeningitis virus*
DNA	Icosahedral	−	50	*Caulimovirus*	*Cauliflower mosaic virus*
			18–24	*Parvoviridae*	*Parvovirus*
			40–50	*Papovaviridae*	*Papillomavirus*
			70–80	*Adenoviridae*	*Mastadenovirus*
		+	180–200	*Herpesviridae*	*Herpesvirus*
	Complex	Coat	230–300	*Poxviridae*	*Orthopoxvirus*

As can be seen from Table 21-3, the chemical nature of the nucleic acid plays an important role in virus classification. Thus in 1971, David Baltimore (a Nobel prize winner for his work on tumor viruses) proposed a classification of viruses based on how the viral genome is replicated and expressed. All viruses are divided into six classes on the basis of their method of mRNA synthesis or transcription (see Table 21-4). The division is based primarily on (1) whether the genome is a single- or double-stranded nucleic acid and (2) whether the genome is or is not converted to an intermediate form before the mRNA (plus strand) is produced. Although this scheme groups together viruses with similar replicative steps, it also groups very different virions together in the same class, e.g., bacteriophages and animal viruses. Such a scheme finds favor with molecular biologists who study viruses. On the other hand, biology-minded virologists prefer a more general approach modeled after Linnaeus's classification scheme with nomenclature for families, genera, and species.

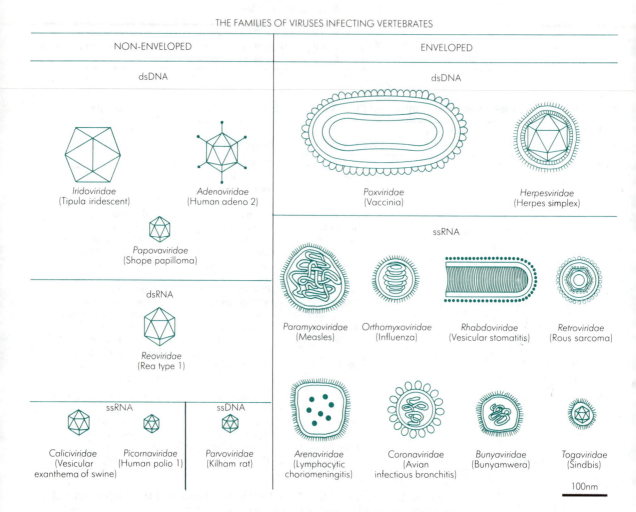

THE FAMILIES OF VIRUSES INFECTING VERTEBRATES

NON-ENVELOPED	ENVELOPED

dsDNA / dsDNA

Iridoviridae (Tipula iridescent)

Adenoviridae (Human adeno 2)

Poxviridae (Vaccinia)

Herpesviridae (Herpes simplex)

Papovaviridae (Shope papilloma)

ssRNA

dsRNA

Reoviridae (Rea type 1)

Paramyxoviridae (Measles)

Orthomyxoviridae (Influenza)

Rhabdoviridae (Vesicular stomatitis)

Retroviridae (Rous sarcoma)

ssRNA / ssDNA

Caliciviridae (Vesicular exanthema of swine)

Picornaviridae (Human polio 1)

Parvoviridae (Kilham rat)

Arenaviridae (Lymphocytic choriomeningitis)

Coronaviridae (Avian infectious bronchitis)

Bunyaviridae (Bunyamwera)

Togaviridae (Sindbis)

100nm

Figure 21-15. Line drawings of the families of viruses infecting vertebrates. All diagrams have been drawn to the same scale. For each drawing the family name is given together with a well-known member of it (but the dimensions and shape used for the drawing may not be exactly those of the virus named). (*Reproduced from drawings by Mrs. J. Keeling in R. E. F. Matthews, "Classification and Nomenclature of Viruses," Intervirology, 17:1–99, 1982. By permission from S. Karger AG, Basel, Switzerland.*)

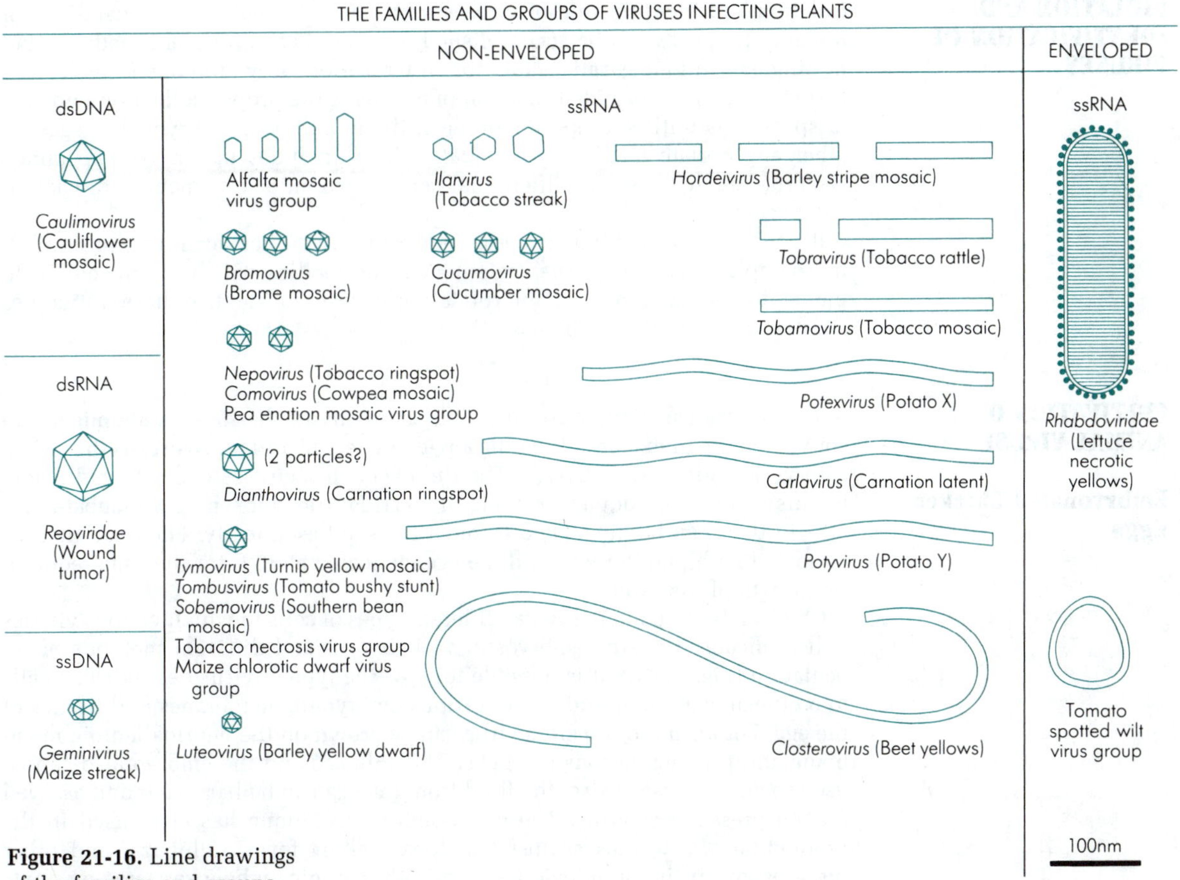

THE FAMILIES AND GROUPS OF VIRUSES INFECTING PLANTS

Figure 21-16. Line drawings of the families and groups of viruses infecting plants. All diagrams have been drawn to the same scale. For each drawing the group name is given together with a well-known member of it (but the dimensions and shape used for the drawing may not be exactly those of the virus named). (*Reproduced from drawings by Mrs. J. Keeling in R. E. F. Matthews, "Classifications and Nomenclature of Viruses," Intervirology,* **17**:1–99, 1982. By permission from S. Karger AG, Basel, Switzerland.)

In order to prevent chaos in the naming of new viruses, an International Committee on Nomenclature of Viruses (ICNV) of the International Association of Microbiological Societies was formed at the Ninth International Congress for Microbiology in 1966. Since that time, this committee, now called the International Committee on Taxonomy of Viruses, has worked constructively towards a meaningful and practical scheme of viral classification. So far, the animal and bacterial virologists have subscribed to a nomenclature for various taxa. Family names agreed upon end in -*viridae*, subfamily names in -*virinae*, and genera, like species, in -*virus*. However, the plant virologists have not classified all their viruses in terms of families and genera. They use **groups** of viruses that share similar properties. Names for these groups are usually derived from the name of the prototype of the group. For example, the name of the group of viruses related to *tobacco mosaic virus* is the tobamo group or tobamovirus. Table 21-5 shows the manner in which the criteria of Table 21-3 are used for the classification of viruses. Figs. 21-15 and 21-16 are line drawings of the families of animal viruses and of the groups of plant viruses, respectively.

ISOLATION AND IDENTIFICATION OF VIRUSES

Isolation and identification of viruses from clinical specimens or materials for research purposes can be accomplished by a number of different methods, but no single technique is satisfactory for all viruses or every kind of specimen. The first step in laboratory identification of a virus is the proper collection and care of specimens until susceptible animals, tissue cultures, embryonated eggs, or other appropriate media are inoculated. This includes making the specimen bacteria-free by filtration, differential centrifugation, or treatment with bactericidal agents.

If a virus is present, characteristic antibodies, i.e., hemagglutination-inhibiting, complement-fixing, or neutralizing viral antibodies, may be produced. Techniques for the conduct and interpretation of tests to identify these antibodies are described briefly in Chap. 34.

CULTIVATION OF ANIMAL VIRUSES

Embryonated Chicken Eggs

Since viruses can grow only in living cells, one of the most economical and convenient methods for cultivating a wide variety of animal viruses is the chick-embryo technique (Fig. 21-17). The discovery that viruses could be cultivated by this simple technique was made in 1931. Fertile chicken eggs incubated for 5 to 12 days can be inoculated through the shell aseptically. The opening may be sealed with paraffin wax and the egg incubated at 36°C for the time required for growth of the virus.

Chick embryos contain several different types of cells in which various viruses will replicate. By using embryos of various ages and different methods of inoculation (Fig. 21-18), it is possible to grow the type of virus desired. The cells to be inoculated are found in the various embryonic membranes and tissues of the egg. For example, vaccinia virus can be grown on the chorioallantoic membrane and produce lesions or pocks. The yolk sac and the embryo also can be used to grow viruses. Also, the fluid from the egg can be harvested and assayed for the presence of virus. The chick-embryo technique has been used in the production of vaccines against smallpox, yellow fever, influenza, and other diseases and in immunologic tests and other studies whenever large amounts of virus are required.

Tissue Cultures

Cell cultures are today the method of choice for the propagation of viruses for many reasons. Among them are convenience, relative economy of maintenance compared to animals, observable cytopathic effects, and choice of cells for their susceptibility to particular viruses. On the basis of their origin and character-

Figure 21-17. Embryonated hen's egg is used for the cultivation of many mammalian viruses.

Figure 21-18. Diagrammatic representation in sagittal section of the embryonated hen's egg 10 to 12 days old. The hypodermic needles show the routes of inoculation of the yolk sac, allantoic cavity, and embryo (head). The chorioallantoic membrane is inoculated after it has been dropped by removing the air from the air sac.

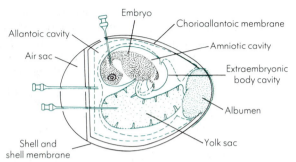

istics, cell cultures are of three types: primary cell cultures, diploid cell strains, and continuous cell lines. Primary cell cultures are derived from normal tissue of an animal (such as mouse, hamster, chicken or monkey tissue) or a human (e.g., gingival tissue). When cells from these tissues are processed and cultured, the first monolayer is referred to as a **primary culture.** (A **monolayer** is a confluent layer of cells covering the surface of a culture vessel.) The cells from subcultures are called **secondary cultures.** (See box for the technique for processing such primary cell cultures.) Cell cultures prepared from fresh tissue resemble more closely the cells in the whole animal than do the cells in continuous cell lines. Unfortunately, cells derived in this manner can be subcultured only a limited number of times before dying. For some types of cells only a few divisions are possible. For others, 50 to 100 divisions occur. Cell cultures derived from embryonic tissue are generally capable of a greater number of divisions in vitro than those derived from adult tissue.

Diploid cell strains are derived from primary cell cultures established from a particular type of tissue, such as lung or kidney, which is of embryonic origin. They are of a single cell type and can undergo 50 to 100 divisions before dying. They possess the normal diploid **karyotype** (appearance of the set of chromosomes). Such diploid cell strains are the host of choice for many viral studies, especially in the production of human vaccine virus. Vaccines prepared from tissue cultures have an advantage over those prepared from embryonated chicken eggs in minimizing the possibility of a patient developing hypersensitivity or allergy to egg albumen. The Salk poliomyelitis vaccine, which is produced in tissue culture, was developed after basic research had shown that the poliovirus would grow satisfactorily on monkey kidney cell cultures.

Processing of Primary Cell Cultures

Primary cell cultures are prepared from fresh tissue, which is usually minced with a sharp sterile razor and dissociated with the aid of proteolytic enzymes (such as trypsin) into a cell suspension. The cells are washed with a physiological buffer (to remove the proteolytic enzymes used) and then suspended in a special growth medium containing a balanced salt solution, a buffer, necessary nutrients (vitamins, coenzymes, amino acids, glucose), and serum. Antibiotics may be added to inhibit bacterial growth. The cell suspension in the growth medium is placed in a tissue-culture vessel and incubated. The cells settle on the surface of the vessel and grow into a monolayer.

Continuous cell lines appear to be capable of an infinite number of doublings. Such cell lines may arise with the mutation of a cell strain, or more commonly from the establishment of cell cultures from malignant tissue. The karyotype of these cells is aneuploid (a variable multiple of the haploid chromosome number) and not diploid. These cells are also different morphologically from the cells of origin. They are usually less fastidious in their nutritional requirements. They do not attach as strongly as other cell cultures to the surface of the culture vessel, so under certain circumstances they can grow in suspension. They also have a tendency to grow on top of each other in multilayers on culture-vessel surfaces.

Even though cells from continuous cell lines are very different from normal cells in both genotype and phenotype, they are very useful in studies where large numbers of cells are required. Furthermore, they are easy to propagate serially. But because of their derivation from malignant tissue or their possession of malignant characteristics, such cells obviously are not used in virus production for human vaccines. Nevertheless, continuous cell lines have been extremely useful in cultivating many viruses previously difficult or impossible to grow.

Growth of viruses in tissue culture is shown in Fig. 21-19. It can be seen that the tissue structure deteriorates as the virus multiplies. This deterioration is called the cytopathic effect (CPE).

Animals

Some viruses cannot be cultivated in cell culture or in embryonated chicken eggs and must be propagated in living animals. Mice, guinea pigs, rabbits, and primates are used for this purpose. Animal inoculation is also a good diagnostic tool because the animal can show typical disease symptoms and histological (tissue) sections of infected tissue can be examined microscopically.

The Origin of HeLa Cells

It was in the winter of 1951 when Henrietta Lacks, a young black woman of 31, went to the medical clinic of Johns Hopkins University in Baltimore to seek medical treatment. The examining physician found a malignant tumor within her cervix. Some of this cancerous tissue was taken to a laboratory for cultivation.

In spite of intensive radiation treatment, the tumor continued to grow. Eight months after her first visit to the clinic, the cancer had spread throughout her whole body and she died. But the tumor cells taken from Henrietta Lacks thrived; they divided and doubled their number every 24 hours. Cells taken previously from the tumors of dozens of other patients had not grown at all, or grew only poorly and then died off.

The cancer cells of Henrietta Lacks continued to flourish in culture in Petri dishes. These cells, now code-named HeLa cells, became one of the best-known continuous tissue-culture cell lines. HeLa cells are widely used in research because they are so readily available, so versatile, and so easy to propagate serially. They have been dubbed the "cells that would not die." Thus Henrietta Lacks left behind her the first widely available model of human tissue in vitro for scientific investigation. Perhaps her legacy will help to conquer the disease that vanquished her in 1952.

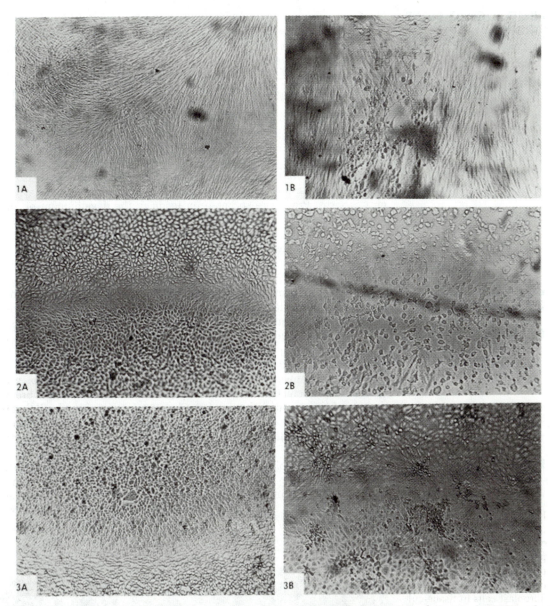

Figure 21-19. Light-microscope view of tissue cultures used for the cultivation of viruses. (1A) Normal human lung fibroblast cell culture. (1B) Human lung fibroblast cell culture infected with varicella (chicken pox) virus. Characteristic cytopathic effect (CPE) seen: rounded and enlarged cells with splitting of cell sheet (layer) and stranding. (2A) Normal monkey kidney cell, line GL V3A. (2B) Monkey kidney cell line GL V3A, infected with poliovirus. CPE seen: cell shrinkage with retracted margins exhibiting angular shapes. (3A) Normal rabbit cell line RK 13. (3B) Rabbit cell line RK 13, infected with rubella (German measles) virus. CPE seen: discrete foci or centers of aggregated cells. Magnification approx (X850). (*Courtesy of A. F. Doss, McGill University.*)

CULTIVATION OF PLANT VIRUSES

Plant viruses can be cultivated by direct mechanical inoculation of virus suspensions by rubbing on leaves of living plants. Rubbing is accomplished with the aid of an abrasive such as carborundum. This can lead to formation of local lesions as well as general infection. Some plant viruses can replicate to large numbers in infected plants. For example, a single hair cell of an infected tobacco plant may contain over 10^7 tobacco mosaic virions. Indeed, as much as 10 percent of the dry weight of infected leaves may be tobacco mosaic virus. Transfer of infection from cell to cell occurs mainly by direct transfer of virions via **plasmodesmata** (intercellular bridges). Also, in most plant diseases, infected cells can continue to manufacture virus without either lysing or dying.

In the last decade some progress has been made in the preparation of plant cell protoplasts for the cultivation of plant viruses. For instance, isolated protoplasts from mesophyll cells of tobacco can be infected directly with tobacco mosaic virus. Progress also has been made in the development of monolayer cultures of susceptible cells from the insects that are vectors of some plant viral diseases. For example, rhabdovirus has been cultivated in leafhopper cell culture giving a yield of over 10,000 virions per cell.

EFFECTS OF VIRUS INFECTION ON CELLS

Cell Death

Disease symptoms in the host due to virus infection vary from none to massive destruction of infected cells leading to cell death. In cell tissue culture, groups of killed cells **(plaques)** have been used in the enumeration of viruses because the number of plaques is proportional to the number of infectious virus particles present. Each virion gives rise to a single plaque, just as a bacterium gives rise to a single colony (Fig. 21-20).

Other effects of cell infection by viruses include formation of giant cells **(polykaryotes)**, creation of genetic changes such as chromosomal breakage, induction of **interferon** production (see Chap. 32) by the infected cell that prevents infection of healthy cells, appearance of **pocks** or necrotic lesions on the chorioallantoic membrane of embryonated eggs, and formation of **inclusion bodies.**

Inclusion Bodies

Before it was possible to study the morphology of viruses at the high magnifications provided by the electron microscope, investigators using light microscopy had observed intracellular structures, or inclusion bodies, associated with virus diseases (Fig. 21-21). In 1887 J. B. Buist noted small particles in the

Figure 21-20. Virus particles in a suspension may be enumerated by means of a plaque assay. Shown are plaques of reovirus on a monolayer of L-929 mouse fibroblast cells in tissue-culture dishes. Note that increasing dilution of the virus suspension results in decreasing plaque numbers in the dishes. (*Courtesy of Collette Oblin, McGill University.*)

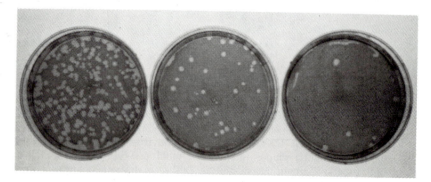

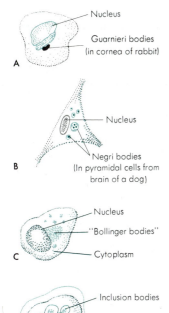

Figure 21-21. Inclusion bodies produced by viruses in certain host tissues. (A) Guarnieri bodies of variola (smallpox) virus in the cytoplasm of rabbit corneal cells. (B) Negri bodies in the cytoplasm of Purkinje cells (nerve cells of the brain) infected with rabies virus. (C) Bollinger bodies in the cytoplasm of cells infected with fowl pox virus. (D) Intranuclear inclusions in epithelial cells of rabbit cornea inoculated with herpesvirus.

cytoplasm of cells surrounding the lesions of smallpox. These he called **elementary bodies.** E. Paschen made the same observation independently in 1906. It is now known that these **Paschen bodies** are aggregates, or colonies, of virions growing in the cytoplasm of the host cell. In 1892, G. Guarnieri reported having seen small round particles in the cytoplasm of similar cells. These **Guarnieri bodies** are also thought to consist of aggregates of unassembled virus subunits and intact virions.

Characteristic inclusion bodies are found in the cytoplasm of the Purkinje cells of the cerebellum and in certain other nerve cells in cases of rabies infection. Finding these typical inclusions (called **Negri bodies** after their discoverer) is diagnostic of the disease.

Inclusion bodies have been found in connection with many other virus diseases. They occur in the cytoplasm in most pox diseases (smallpox, sheep pox, fowl pox), rabies, molluscum contagiosum, and others. Intranuclear inclusions are found in chicken pox, herpes, and the polyhedral diseases of insects. Intranuclear and intracytoplasmic inclusions may be found in the same cell in instances of multiple infection. Some inclusions are useful in establishing diagnosis while the significance of others is not yet known.

Inclusion bodies are for the most part characteristic of the virus causing the infection and even suggest definite pathological changes in the cell. It is generally true, however, that inclusion bodies are aggregates of unassembled virus subunits and intact virions in infected cells. It is experimentally possible to remove inclusion bodies from the cell and use them as inoculum to infect other cells.

PROGRESSIVE AND FATAL DISEASES ASSOCIATED WITH VIRUSES IN ANIMALS

Some progressive or gradually extending diseases which usually terminate in death are poorly understood and require much research. Some of these are or may be caused by viruses, such as classic slow virus diseases and cancer. (The conventional viral diseases are discussed in Chap. 37.)

Classic Slow Virus Diseases

These diseases have a *slow* progressive course, usually with a fatal outcome. (Incubation periods are measured in years!) They are caused by transmissible agents whose properties and behavior (for example, unusual resistance to ultraviolet radiation and heat) suggest an **unconventional** or **atypical** virus. Indeed, these agents have been called **prions** (proteinaceous infectious particles) because they appear to have no nucleic acids at all, protein being their only detectable component. If prions in fact do not contain nucleic acids, their ability to replicate would seem to pose a challenge to the central dogma of molecular biology.

There are four classic slow virus diseases. Each may be described as a neurological disease. They are kuru and Creutzfeldt-Jakob disease of humans and scrapie and transmissible mink encephalopathy (TME) of animals. (The term *encephalopathy* describes these diseases well because each involves widespread destructive cerebral changes due to degeneration without inflammation.)

Scrapie is so named because the diseased animal tends to scrape against fixed objects. It is a chronic infection of the central nervous system of sheep. TME, a disease found in mink farms in the United States, may have arisen from mink fed on the meat of sheep contaminated with scrapie virus. It is also possible that skunks and raccoons harbor this virus naturally.

Kuru and Creutzfeldt-Jakob disease are similar degenerative diseases of the human central nervous system. Symptoms include tremors, progressive uncoordinated movements, and mental deterioration. Fortunately, both are rare, and kuru is restricted to the Foré people of New Guinea. Kuru was spread among these people by ingestion of infectious human brain tissue during ritual cannibalism of the dead as a mourning ceremony. Now that ritual cannibalism has been eliminated, the disease has been declining in prevalence and should soon disappear. (Dr. D. Carleton Gajdusek of the National Institute of Neurological Diseases and Strokes studied this disease extensively and was awarded the 1976 Nobel prize in medicine for his pioneering work.) Unlike kuru, which is restricted geographically, Creutzfeldt-Jakob disease has a worldwide distribution. The onset generally occurs between the ages of 35 and 65.

Cancer and Viruses

More than 100 clinically distinct types of cancer are recognized, each having a unique set of symptoms and requiring a specific course of therapy. However, most of them can be grouped into four major categories:

1 *Leukemias.* Abnormal numbers of white cells (leukocytes) are produced by the bone marrow.
2 *Lymphomas.* Abnormal numbers of lymphocytes (a type of leukocyte) are produced by the spleen and lymph nodes.
3 *Sarcomas.* Solid tumors grow from derivatives of embryonal mesoderm, such as connective tissues, cartilage, bone, muscle, and fat.
4 *Carcinomas.* Solid tumors grow from epithelial tissues, the most common form of cancer; epithelial tissues are the internal and external body surface coverings and their derivatives and thus include skin, glands, nerves, breasts, and the linings of the respiratory, gastrointestinal, urinary, and genital systems.

Cancer has three major characteristics: hyperplasia, anaplasia, and metastasis. Hyperplasia is the uncontrolled proliferation of cells. Anaplasia is structural abnormality of cells (these cells also have a loss of or reduction in their functions). Metastasis is the ability of a malignant cell to detach itself from a tumor and establish a new tumor at another site within the host.

For a long time microbiologists entertained the idea that cancer might be caused by viruses. This was because some early experiments associated viruses with some cancers. In the early 1900s, both leukemias and sarcomas in chickens were shown to be caused by viruses. But for many years these discoveries were not considered relevant to the cause of human cancer. Furthermore, human cancer did not appear to be infectious, and there were no confirmed isolations of an etiologic (causative) virus; the idea that viruses are the cause of human cancer became less attractive.

However, in recent years significant evidence has accumulated to show that some viruses do in fact cause cancer in animals. These findings revived the idea that human cancers might also be caused by viruses.

Both RNA and DNA viruses have been found capable of inducing cancer in animals; in these animals the affected cells are transformed, resulting in the formation of tumors. (A transformed cell acquires phenotypic, biochemical, and other properties distinctly different from uninfected cells or infected cells in which tumors are not produced. Such changes arise from modification of a cell's genome induced by its incorporation of DNA from other cells or from viruses.) Such tumor-inducing viruses are called **oncogenic** viruses.

Among the viruses whose virions contain RNA, only some members of the family *Retroviridae* (more commonly known as the RNA tumor viruses) cause cancer in animals. These members (of the subfamily *Oncovirinae*) replicate through a DNA intermediate. None of the other RNA viruses, all of which replicate through an RNA intermediate, are known to cause cancer. In contrast, among the DNA viruses, members of at least three families (*Herpesviridae, Adenoviridae,* and *Papovaviridae*) can cause cancer in animals.

All the viruses that can cause cancer in animals induce synthesis of DNA that is present in the nucleus and that codes for proteins which are not virion structural proteins. A common characteristic of all oncogenic viruses is that the viral genome in some way becomes either integrated or intimately associated with the host DNA. The host cell does not lyse—a situation similar to the model of lysogeny in bacteria infected with temperate phages. As mentioned earlier in this chapter, if the viral genome is RNA, it serves as a template for the synthesis of a DNA molecule complementary to it; the enzyme **reverse transcriptase** is responsible for this synthesis giving rise to a DNA-RNA hybrid. A conventional DNA polymerase enzyme is then used to synthesize a strand of DNA complementary to the first strand. This results in a double-stranded DNA molecule synthesized from the viral RNA and called a **provirus,** which now can be integrated into the host DNA. In this way transformation and tumors are induced in host cells.

There are differences in the ability of oncogenic viruses to cause cancer. Some RNA tumor viruses, some herpesviruses, and one papovavirus cause cancer in nonlaboratory situations. In contrast, some other RNA tumor viruses, adenoviruses, and most papovaviruses are found to cause cancer only in laboratory animals. The DNA tumor viruses primarily induce cancers only in foreign hosts, that is, hosts the viruses do not infect outside the laboratory.

Furthermore, the transforming viruses differ from each other in the probability that only one virus particle can cause transformation and in the fraction of cells that are transformed in a genetically susceptible population. This means that, in the laboratory, transforming viruses differ in the efficiency with which they cause cancer. They also differ as to how dependent the transformation is on a specific type of differentiated cell.

Oncogenic DNA Viruses

DNA tumor viruses, such as adenovirus 12, polyoma virus, and simian virus 40 (SV 40), can transform susceptible cells rapidly. But usually only a small percentage of cells can be transformed.

Polyoma virus is endemic in wild and laboratory mouse populations. SV 40 cannot induce tumors in the monkey (natural host) but can do so in rodents in the laboratory. The Epstein-Barr virus (EBV), a herpesvirus, has been consistently associated with certain human neoplasias, or tumors. It was discovered by Epstein and Barr in 1964 in cultured Burkitt's lymphoma cells. Burkitt's lymphoma is a cancer of the lymphoid system. EBV has also been implicated in infectious mononucleosis (regarded by some authorities as a self-limiting leukemia), in Hodgkin's disease (a form of lymphoma), and in nasopharyngeal carcinoma. Other herpesviruses, such as herpes virus (HSV) types 1 and 2, have also been implicated in certain human cancers: cancers of the lip of mouth have been associated with HSV1, and cervical cancer has been associated with HSV2.

Much of the mechanism of transformation has been learned from the SV40 virus infection of mouse cells. Infection of certain cells leads to integration of the viral DNA with cell DNA under conditions that do not allow full virus expression, which would normally lead to virus production and cell death. But sometimes this nonproductive infection (no progeny virus and consequently no cell death) allows formation of an early gene product essential for virus replication. Transformation is the accidental result of the continued presence of this early viral product.

Oncogenic RNA Viruses

The RNA sarcoma viruses are the most strongly transforming viruses. They transform many different types of cells, e.g., fibroblasts, myoblasts, iris epithelial cells, etc.; they can also transform all of the cells in a culture.

The genetic basis for transformation has been elucidated with Rous sarcoma virus infection of chicken fibroblast cells. The RNA viral genome contains a gene that is responsible for transformation and is not essential for virus replication. Upon infection of susceptible cells, the transformation gene, along with all the other viral genes, is synthesized as DNA, using the viral genome as template, and is integrated at a unique site in the cellular DNA. The viral gene for transformation, together with all other viral genes, is transcribed, giving rise to a product that causes cell transformation. In addition, progeny virus is produced (a situation different from that of the DNA oncogenic viruses). The existence of a specific gene for transformation and its integration to become a stable component of the cell genome make Rous sarcoma virus a strongly transforming virus capable of transforming many types of cells. This explains why Rous sarcoma virus has a greater probability of transforming susceptible cells than do oncogenic DNA viruses.

Until recently RNA viruses were not isolated from human cancer cells, although some circumstantial evidence implicated them in this disease. For example, molecular components were found in human leukemic cells which are related to similar components in RNA tumor viruses. However, recently Robert Gallo and coworkers of the National Cancer Institute in Bethesda, Maryland, and Yorio Hinuma and colleagues at Kyoto University separately isolated the first identified human cancer virus associated with a rare form of human leukemia. As of 1982, the virus had been discovered in four cutaneous T-cell leukemia or lymphoma patients in the United States. The virus is a new RNA tumor virus (retrovirus).

VIROIDS

Viroids are nucleic acid entities of relatively low molecular weight (1.1 to 1.3 $\times 10^5$) and unique structure that cause several important diseases of cultivated

Oncogenes, Retroviruses, and Cancer

Oncogenes (from *oncos*, the Greek word for "mass" or "tumor") were originally defined as the genetic elements (genomic sequences) of retroviruses responsible for the malignant transformation of host cells. These oncogenes have also been found in spontaneous tumors. Furthermore, DNA sequences homologous to the transforming genes of certain retroviruses have been found in normal, untransformed cells. Thus these oncogenes are not themselves of viral origin but are cellular genes that the viruses picked up by recombination events during the course of infection. This means that most of the oncogenes found in viruses and cancer cells have counterparts, either identical or closely similar, among the normal genes of the human and animal body. In fact, oncogenes are found in all the cells of the mammalian body except the red blood cells, which do not possess nuclei; but the majority of cells never become cancerous. Oncogenes have the potential of causing cancerous transformation of cells when appropriately activated. Some 20 oncogenes have been identified, most of which were originally discovered in viruses that cause cancers in laboratory animals.

The structures of the cellular counterparts of the viral oncogenes (termed cellular oncogenes) have been conserved throughout evolutionary history. This suggests that they have essential roles to play in cellular physiology, probably in cell differentiation or regulation of cell division. At the present time, we still do not know what makes these genes go awry and induce uncontrolled cell division and abnormal cell differentiation patterns. It is speculated that an oncogene can be induced to act in the cancer process in many different ways: (1) by undergoing mutation, (2) by being abnormally activated when thrown into association with some other genes, (3) by being removed from natural repressors that regulate its activities, (4) by producing oncogene products that may be growth factors that stimulate cell division and contribute to the uncontrollable growth of cancer cells, and (5) by overexpression due to the accidental integration of a retrovirus next to it. The evidence implicating cellular oncogenes in the development of human cancers is still largely circumstantial. Nevertheless, these oncogenes are currently the subject of many exciting and promising studies.

plants. Such diseases include potato spindle tuber, citrus exocortis, chrysanthemum stunt, and cucumber pale fruit. Viroids are the smallest known agents of infectious disease. Unlike viruses, viroids do not possess a protein coat and exist only as short, infectious molecules of RNA. In spite of their small size, viroids replicate autonomously in cells of susceptible plant species. Some known viroids are single-stranded, covalently closed circular RNA molecules, and others are single-stranded linear RNA molecules.

It is not known how viroids cause disease. It is thought that their location in the nucleus (since most infectivity of viroids is associated with cell chromatin material) as well as their apparent inability to act as mRNA suggest that they cause host symptoms by interference with gene regulation in the infected host cells. Present evidence also suggests that viroidlike nucleic acids may exist in organisms other than higher plants. They may be responsible for certain diseases of humans and animals (although how they escape the destructive action of nucleases in animal cells is unexplained at the present time).

QUESTIONS

1 What association did the following people have with virology?

Jenner	Loeffler and Frosch
Waterhouse	Stanley
Pasteur	Theiler
Chamberland	Enders
Ivanowski	Baltimore
Beijerinck	Lacks

2 Cite three types of viral vaccines prepared from tissue cultures.

3 Discuss the characteristic symmetry of spherical and of rod-shaped viruses.

4 What is the difference between a naked and an enveloped virion?

5 Draw a simple virion, identifying the following structures: capsid, capsomere, nucleic acid core, and envelope.

6 How is the genome of a virus different from that of a cell of a higher organism?

7 What components of a virion are made of protein?

8 Explain why some single-stranded RNA viruses have plus strands while others have minus strands.

9 Compare the enzymatic activity of RNA polymerase with that of RNA-dependent DNA polymerase.

10 Compare the mode of multiplication of viruses with that of some representative bacteria.

11 Explain how the mechanism of adsorption of virus to animal cells is different from that of adsorption to plant cells.

12 Describe the mechanisms of penetration of animal viruses into host cells.

13 Compare the functions of early mRNA and late mRNA in the process of replication of viral components.

14 Describe the process by which virions are released from host cells.

15 How does the yield of virus particles per cell compare between a phage-infected bacterium and a virus-infected animal or plant cell?

16 Describe the temporal events in the replication of herpesvirus.

17 Discuss the primary characteristics used in the classification of viruses.

18 How do the plant virologists differ from the animal and bacterial virologists in their approach to viral classification?

19 Prepare a separate stylized drawing of a virion to represent each of the following: (a) Reoviridae, (b) Poxviridae, (c) Orthomyxoviridae, (d) tobamovirus, (e) closterovirus, and (f) Rhabdoviridae.

20 Describe three techniques for cultivating viruses in the laboratory. How do these methods differ from methods for culturing bacteria?

21 What are the advantages and disadvantages of primary cell cultures and continuous cell lines for the cultivation of viruses?

22 What type of cell culture is generally used for viral vaccine production?

23 In what ways have plant viruses been cultivated in vitro?

24 What are inclusion bodies? Where are they found? What is their significance?

25 What new concepts have we learned about virology from an understanding of slow virus diseases?

26 Which are the RNA-containing viruses that can cause cancer in animals?

27 Describe some unique characteristics of the viruses that cause cancer in animals.

28 What are some of the differences among oncogenic viruses in their ability to cause cancer?

29 Compare the known mechanism of transformation by an oncogenic DNA virus with that by an oncogenic RNA virus. Which virus is more efficient in transforming susceptible cells?

30 Has a human cancer virus been isolated and identified? Explain.

31 What is a viroid? Why is it unique?

32 What are oncogenes and where are they found?

33 Describe some ways by which oncogenes are activated to cause cancer.

REFERENCES   The references cited for Chap. 20 are also applicable to this chapter.

PART SIX
CONTROL OF MICROORGANISMS

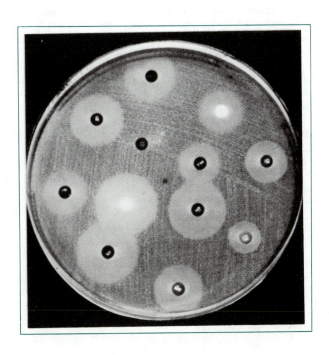

Penicillin—Its Discovery and Development

Penicillin was discovered in 1927 by Alexander Fleming, a microbiologist working at St. Mary's Hospital in London.

Fleming was conducting experiments in search of new antibacterial agents, particularly ones that would be effective against wound infections. In the course of these experiments, he observed a plate culture of *Staphylococcus aureus* that had been contaminated by a mold. The area around the edges of the mold colony was clear—no bacterial colonies. Apparently, the staphlyococcus cells near the mold growth were inhibited or killed by the mold. Further studies of this phenomenon revealed that it was a mold of the genus *Penicillium* that produced a substance which was very potent against staphylococci and hence very attractive as a potential chemotherapeutic agent. Fleming named the substance *penicillin*.

Penicillin was effective against bacteria in laboratory cultures, but was it effective in the human body? This question needed to be answered. Unfortunately, the amount of penicillin produced in the mold cultures was extremely small. In addition, serious problems were encountered in attempts to isolate and purify penicillin.

The first clinical trial with a crude penicillin preparation was conducted on February 12, 1941. The patient, an Oxford policeman, was dying from a staphylococcus infection. The administration of penicillin resulted in an initial dramatic improvement, but five days later, when the supply of penicillin was exhausted, the staphylococci reemerged, the infection spread, and the patient died. This was a tragic end to a trial that did not succeed only because there was not enough penicillin available to treat the patient. The major problem continued to be the failure of large-scale production and recovery of penicillin.

Britain, at this point (1940–41), was engaged in a grim war. There was little likelihood that a major share of her national resources could be diverted to an intensive program for development of penicillin.

Fortunately, the British reports of penicillin attracted the attention of Americans. As a result, the Rockefeller Foundation invited Harold W. Florey, a professor of pathology at Oxford University, who had investigated the development of penicillin as a chemotherapeutic agent, and N. G. Heatley, his colleague, to the United States to explore means of large-scale production of penicillin. They arrived in the United States on July 2, 1941. Meetings were arranged with members of the National Research Council, Charles Thom, a world class mycologist with the U.S. Department of Agriculture, and others. Work on penicillin production began immediately at the U.S. Department of Agriculture's Northern Regional Research Laboratory in Peoria, Illinois, which had a record of achievements in microbial fermentations.

The U.S. Office of Scientific Research and Development, aware of the tremendous potential of penicillin for treatment of casualties of war, gave this project top priority. Major pharmaceutical companies and universities were called in to cooperate in the research and development of penicillin production. The results proved dramatic. Fleming's original mold cultures produced 2 units/ml; within a matter of months improvements in technology increased the yield to 900 units/ml; today the yield is approximately 50,000 units/ml!

In the autumn of 1941 there was little penicillin available in the United States for treatment of patients. One year later, as a result of the collaborative efforts of governments, universities, and industry, appreciable quantities were available.

Few other discoveries have contributed as much as that of penicillin (and the antibiotics that followed) to the health and welfare of people worldwide. The story of penicillin provides an elegant example of the benefits gained from collaboration of scientists from government agencies, industry, and universities.

Preceding page. Inhibition of bacteria by chemotherapeutic chemicals. The clear zone around the disks is evidence that the chemical agent in the disk, which diffuses into the inoculated agar medium, inhibits growth of the bacteria. (*Courtesy of BBL Microbiology Systems.*)

Chapter 22

The Control of Microorganisms by Physical Agents

The term **control** as used here refers to the reduction in numbers and/or activity of the total microbial flora. The principal reasons for controlling microorganisms are: to prevent transmission of disease and infection, to prevent contamination by or growth of undesirable microorganisms, and to prevent deterioration and spoilage of materials by microorganisms.

Microorganisms can be removed, inhibited, or killed by various **physical agents, physical processes,** or **chemical agents.** A variety of techniques and agents are available; they act in many different ways, and each has its own limits of application.

In the first part of this chapter we will consider the fundamentals of control which are applicable when either physical or chemical methods are used. In the second part, we will describe the physical agents and physical methods for reducing or eliminating the viable microbial population.

Before considering the physical means or chemical agents by which microorganisms are controlled, it is important to understand some of the characteristics of a microbial population when exposed to a destructive agent. We shall use bacteria as an example.

FUNDAMENTALS OF CONTROL

The term death, as used in microbiology, is defined as the irreversible loss of the ability to reproduce. Viable microorganisms are capable of multiplying; dead microorganisms do not multiply (grow). The determination of death requires laboratory techniques that indicate whether growth occurs when the sample is inoculated into a suitable medium. The failure of a microorganism to grow when inoculated into an appropriate medium indicates that the organism is no longer able to reproduce, and the failure to reproduce is the criterion of death. A complicating factor in this definition is that the response of the organism may not be the same in all media. For example, a suspension of *Escherichia coli* exposed to a heat treatment may yield a greater number of survivors if a plating medium of trypticase soy agar is used rather than a medium containing bile salts such as deoxycholate agar.

The Rate of Death of Bacteria

When one drops a suspension of bacteria into a bottle of hot acid or an incinerator, the bacteria may all be killed so fast that it is not possible to measure the death rate. However, less drastic treatment may result in the cells being killed over a longer period of time, at a constant *exponential* rate that is essentially the inverse of their exponential growth pattern.

Exponential death can be understood easily in terms of a simple model. Imagine that each cell is a *target* and that a large number of *bullets* (i.e., units of a physical or chemical agent) are being sprayed at them in a random manner, as with a machine gun; that is, no one is aiming the gun directly at a target. Common sense dictates some rules about the way bacteria die under these conditions. To begin, we assume that a single hit kills a bacterium.

The probability of hitting a target is proportional to the number of targets, i.e., the number of bacteria, *present.* Intuition tells us that if we shoot randomly at many targets, we have a good chance of hitting one, but as time goes on, the number of targets not yet hit decreases steadily and it becomes harder and harder to hit the remaining ones. (Hitting a target again and again does not count; a bacterium can be killed only once.) Let us take a simple numerical example. Assume that we have an initial population of 1 million targets. We shower them with bullets for 1 min and manage to hit 90 percent, so that there are now 100,000 survivors left. Then we shower them with bullets 1 min more, but since we only have one-tenth as many targets as in the first round, we hit only one-tenth as many. In other words, this time we hit 90,000 of the targets and have 10,000 survivors. We shower these with bullets another minute, and since we have only one-tenth as many as in the last minute, we again hit only one-tenth as many, or 9,000. This pattern repeats itself until there are no targets left, as shown in Table 22-1. But notice that it is just as hard to kill the last nine bacteria as it was to kill the first 900,000. In fact, we can never be sure that we have killed the last one; all that we can do is give the targets enough overkill for there to be a good chance that the last has been hit.

The pattern of death among *Bacillus anthracis* spores exposed to 5% phenol is shown in Fig. 22-1. The number of survivors is plotted against time; however, curve A is an arithmetic plot and curve B is a logarithmic plot. Both curves show that some portion of the population dies during any given unit of time,

Table 22-1. A Theoretical Case of the Order of Death of Bacteria When Exposed to a Lethal Agent

Time	Survivors	Deaths per Unit Time	Total Deaths
0	1,000,000	0	0
1	100,000	900,000 = 90%	900,000
2	10,000	90,000 = 90%	990,000
3	1,000	9,000 = 90%	999,000
4	100	900 = 90%	999,900
5	10	90 = 90%	999,990
6	1	9 = 90%	999,999

SOURCE: O. Rahn, *Physiology of Bacteria*, McGraw-Hill, New York, 1932.

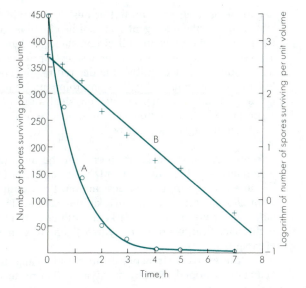

Figure 22-1. The death curve of *Bacillus anthracis* spores exposed to 5% phenol. Curve A: Number of survivors expressed arithmetically per unit volume plotted against time. Curve B: Logarithm of number of surviving bacteria plotted against time.

but what the logarithmic plot also reveals is that *the death rate is constant.* The points fall on a straight line, and the slope of the curve is the measure of the death rate.

Results such as those shown in Fig. 22-1 are obtainable only when all conditions are kept strictly uniform, including the age and the physiological condition of all the microorganisms in the population. If the cells in the microbial population vary in age or physiological stage of growth, they will exhibit differences in susceptibility to the agent. As a consequence the logarithmic plot of survivors will not fall in a straight line. Both the slope of the death curve and its form are affected by the species of microorganism and the homogeneity of the cells in the population.

The probability of hitting a target is also *proportional to the number of bullets shot,* i.e., concentration of the chemical or intensity of the physical agent. Intuition again tells us that the more bullets we shoot in a given time, the faster the targets will be hit. If the targets are bacteria and the bullets are x-rays or ultraviolet light, it stands to reason that the cells will be killed faster as the *intensity* of the radiation increases. If the bullets are molecules of some chemical agent, the cells will be killed more rapidly as the *concentration* of the agent increases (up to a certain limit, of course).

The longer we shoot, the more targets we hit, but the more targets we have, the longer it takes to hit them all. This is an obvious restatement of the exponential death pattern. It simply means that it takes time to kill the population, and if we have many cells, we must treat them for a longer time to be reasonably sure that all of the bacteria are dead.

Conditions Influencing Antimicrobial Action

Microorganisms are not simple physical targets. Many biological characteristics influence the rate at which microorganisms are killed or inactivated by various agents. Many factors must be considered in the application of any physical or chemical agent used to inhibit or destroy microbial populations. It is not possible to prescribe one agent that will be effective for the control of microorganisms for all materials and all circumstances. Hence it is necessary to evaluate each situation separately in order to select a process that research and experience have shown will accomplish the desired result. Some of the biological characteristics of the cells, as well as environmental conditions which influence the efficacy of antimicrobial agents (physical and chemical), are outlined below. More specific information is presented where the application of a particular antimicrobial agent is described.

Environment

The physical or chemical properties of the medium or substance carrying the organisms, i.e., the **environment,** has a profound influence on the rate as well as the efficacy of microbial destruction. For example, the effectiveness of heat is much greater in acid than in alkaline material. The consistency of the material (aqueous or viscous) will markedly influence the penetration of the agent, and high concentrations of carbohydrates generally increase the thermal resistance of organisms.

The presence of extraneous organic matter can significantly reduce the efficacy of an antimicrobial agent by inactivating it or protecting the microorganism from it.

An increase in temperature, when used with another agent such as a chemical, hastens the destruction of microorganisms. This phenomenon is illustrated in Fig. 22-2.

Kinds of Microorganisms

Species of microorganisms differ in their susceptibility to physical and chemical agents. In sporeforming species, the growing vegetative cells are much more susceptible than the spore forms; bacterial spores are extremely resistant. In fact, bacterial spores are the most resistant of all living organisms in their capacity to survive under adverse physical and chemical conditions. The relative resistance of bacterial spores in comparison with other microorganisms is shown in Table 22-2.

Figure 22-2. Increasing temperature also decreases bacterial survival when the concentration of the disinfectant remains constant. In this experiment *Escherichia coli* was exposed to phenol at a concentration of 4.62 g/liter at temperatures between 30 and 42°C. The number of survivors, expressed logarithmically, is plotted against time. (*From R. C. Jordan and S. E. Jacobs, J Hyg,* **44:**210, 1945. *Courtesy of Cambridge University Press.*)

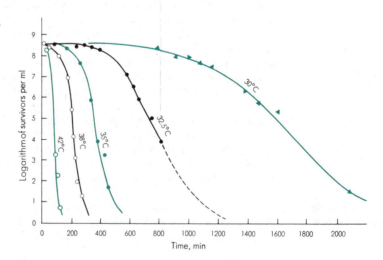

Table 22-2. Resistances of Bacterial and Mold Spores and of Viruses, Relative to the Resistance of *Escherichia coli* as Unity

Sterilizing Agent	*Escherichia coli*	Bacterial Spores	Mold Spores	Viruses and Bacteriophages
Phenol	1	100,000,000	1–2	30
Formaldehyde	1	250		2
Dry heat	1	1,000	2–10	±1
Moist heat	1	3,000,000	2–10	1–5
Ultraviolet	1	2–5	5–100	5–10

SOURCE: O. Rahn, *Bacteriol Rev*, **9**:1, 1945.

Figure 22-3. Comparative susceptibility of young and old cells to a lethal agent. The young cells are all killed within 25 min, but a considerable part of the more resistant, older cells still survives. (*Courtesy of Martin Frobisher et al., Fundamentals of Microbiology, Saunders, Philadelphia, 1974.*)

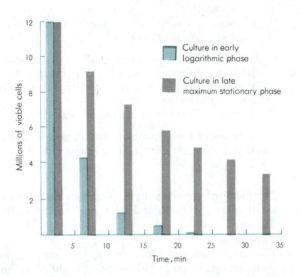

Physiological State of Cells

The physiological state of cells may influence susceptibility to an antimicrobial agent. Young, actively metabolizing cells are apt to be more easily destroyed than old, dormant cells in the case of an agent that causes damage through the interference with metabolism; nongrowing cells would not be affected. A comparison of the susceptibility of young and old cells to a lethal agent is shown in Fig. 22-3.

Mode of Action of Antimicrobial Agents

The many processes and substances used as antimicrobial agents manifest their activity in one of several ways. For both academic and practical reasons, it is important to know how microorganisms are inhibited or killed. Knowledge of the mode of action of a particular agent may make it possible to predict the conditions under which it will function most effectively as well as the kinds of microorganisms it will be most effective against. A great deal of research has been performed to determine the specific site of action of various agents. Such investigations are complicated by the fact that when a cell is exposed to a lethal agent, many changes can be observed. A domino effect occurs once the initial inhibitory or lethal process is inflicted. The real problem is to establish the *primary site of damage* responsible for inhibition or death.

In a general way, one may view the possible sites of action of an antimicrobial agent by recalling certain features of the microbial cell. A normal living cell contains the multitude of enzymes responsible for metabolic processes. A semipermeable membrane (cytoplasmic membrane) maintains the integrity of the cellular contents; the membrane selectively controls the passage of substances between the cell and its external environment and is also the site of some enzyme reactions. The cell wall proper provides a protective covering to the cell in addition to participating in certain physiological processes. Damage at any of these areas may initiate a number of subsequent changes leading to the death of the cell. It is necessary to bear in mind that there are many sites of damage to the cell and that the damage may be caused by one or more of a variety of agents.

The manner in which antimicrobial agents inhibit or kill can be attributed to the following kinds of actions:

Damage to the cell wall or inhibition of cell-wall synthesis
Alteration of the permeability of the cytoplasmic membrane
Alteration of the physical or chemical state of proteins and nucleic acids
Inhibition of enzyme action
Inhibition of protein or nucleic acid synthesis

Examples of these modes of action will be described as we discuss specific physical agents in this chapter and chemical agents in Chaps. 23 and 24.

PHYSICAL AGENTS

The major physical agents or processes used for the control of microorganisms are temperature (high and low), desiccation, osmotic pressure, radiation, and filtration.

High Temperatures

Microorganisms can grow over a wide range of temperatures, from very low temperatures characteristic of psychrophiles to the very high growth temperatures characteristic of thermophiles. Every type has an optimum, minimum, and maximum growth temperature. Temperatures above the maximum generally kill, while those below the minimum usually produce stasis (inhibition of metabolism) and may even be considered preservative. The amount of water present in the environment at any temperature has a significant effect upon microorganisms in terms of their survival.

High temperatures combined with high moisture is one of the most effective methods of killing microorganisms. It is important to distinguish between dry heat and moist heat in any procedure for microbial control. Moist heat kills microorganisms by coagulating their proteins and is much more rapid and effective than dry heat, which destroys microorganisms by oxidizing their chemical constituents. Two examples will illustrate the difference. Spores of *Clostridium botulinum* are killed in 4 to 20 min by moist heat at 120°C, whereas a 2-h exposure to dry heat at the same temperature is required. Spores of *B. anthracis* are destroyed in 2 to 15 min by moist heat at 100°C, but with dry heat 1 to 2 h at 150°C is required to achieve the same result (Tables 22-3 and 22-4).

Vegetative cells are much more sensitive to heat than are spores; the higher level of "water activity" in the vegetative cells accounts for this. Cells of most bacteria are killed in 5 to 10 min at 60 to 70°C (moist heat). Vegetative cells of yeasts and other fungi are usually killed in 5 to 10 min by moist heat at 50 to

Table 22-3. Some Quoted Destruction Times of Bacterial Spores by Moist Heat

Organism	Destruction times, min							
	At 100°C	At 105°C	At 110°C	At 115°C	At 120°C	At 125°C	At 130°C	At 134°C
Bacillus anthracis	2–15	5–10						
B. subtilis	Many hours			40				
A putrefactive anaerobe	780	170	41	15	5.6			
Clostridium tetani	5–90	5–25						
Cl. welchii	5–45	5–27	10–15	4	1			
Cl. botulinum	300–530	40–120	32–90	10–40	4–20			
Soil bacteria	Many hours	420	120	15	6–30	4		1.5–10
Thermophilic bacteria		400	100–300	40–110	11–35	3.9–8.0	3.5	1
Cl. sporogenes	150	45	12					

SOURCE: G. Sykes, *Distinfection and Sterilization*, 2d ed., Lippincott, Philadelphia, 1965.

Table 22-4. Some Quoted Killing Times of Bacterial Spores by Dry Heat

Organism	Destruction times, min						
	At 120°C	At 130°C	At 140°C	At 150°C	At 160°C	At 170°C	At 180°C
Bacillus anthracis			Up to 180	60–120	9–90		3
Clostridium botulinum	120	60	15–60	25	20–25	10–15	5–10
Cl. welchii	50	15–35	5				
Cl. tetani		20–40	5–15	30	12	5	1
Soil spores				180	30–90	15–60	15

SOURCE: G. Sykes, *Disinfection and Sterilization*, 2d ed., Lippincott, Philadelphia, 1965.

60°C; their spores are killed in the same time but at temperatures of 70 to 80°C. Bacterial spores are much more resistant to high temperatures as shown in Tables 22-3 and 22-4. The susceptibility of viruses to heat is generally similar to that of mesophilic vegetative bacterial cells.

Thermal Death Time and Decimal Reduction Time

Thermal death time refers to the shortest period of time to kill a suspension of bacteria (or spores) at a prescribed temperature and under specific conditions. Another unit of measurement of the destruction of microorganisms by heat is the **decimal reduction time.** This is the time in minutes to reduce the population by 90 percent, or stated differently, it is the time in minutes for the thermal-death-time curve to pass through one log cycle (see Fig. 22-4). Figure 22-5 illustrates a thermal-death-time curve for spores of bacterial species responsible for a type of canned-food spoilage, e.g., flat sour spoilage.

From the definition of these terms, it is clear that they express a time-temperature relationship to killing. In thermal death time, the temperature is selected as the fixed point and the time varied. Decimal reduction time is a modification of thermal death time which measures a 90 percent rather than 100 percent kill rate. In the experimental determination of these values, it is an absolute requirement that the conditions be rigidly controlled. Attention must

Figure 22-4. Graph illustrating the concept of decimal reduction time (D value), the time in minutes to reduce the microbial population by 90 percent. Stated differently, it is the time in minutes for the thermal-death-time curve to pass through one log cycle. The D value is independent of time when the response is logarithmic, that is, when the same length of time is required to accomplish any given log decrease in number of survivors. For example, the D value in this illustration is approximately 20 min., the time required to reduce the survivors from 10^8 to 10^7, or from 10^7 to 10^6, and so on.

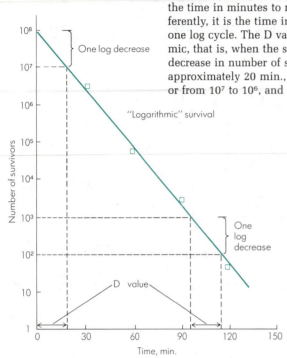

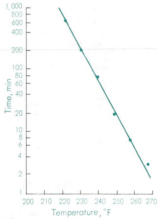

Figure 22-5. Thermal-death-time curve for spores of a type of bacterium encountered in food spoilage. (*Courtesy of W. C. Frazier, Food Microbiology, Mc-Graw-Hill, New York, 1958.*)

be given to the nature of the medium, pH, and the number of organisms, since all these factors have a bearing on the susceptibility of the microorganisms to heat.

Thermal-death-time data and decimal-reduction-time data are extremely important in many applications of microbiology. The canning industry, for example, carries out extensive studies on this subject to establish satisfactory processing temperatures for the preservation of canned foods.

Application of High Temperatures for Destruction of Microorganisms

The killing action of heat is, as we have seen, a time-temperature relationship affected by numerous conditions that must be taken into consideration in selecting the time and temperature required to reduce the microbial population to the desired level. Practical procedures by which heat is employed are conveniently divided into two categories: **moist heat** and **dry heat.**

Moist Heat. The application of moist heat for inhibiting or destroying microorganisms is discussed by the method used to obtain the desired result.

Steam Under Pressure. Heat in the form of saturated steam under pressure is the most practical and dependable agent for sterilization. Steam under pressure provides temperatures above those obtainable by boiling, as shown in Table 22-5. In addition, it has the advantages of rapid heating, penetration, and moisture in abundance, which facilitates the coagulation of proteins.

The laboratory apparatus designed to use steam under regulated pressure is

Table 22-5. Temperature of Steam under Pressure

Steam Pressure, lb/in²	Temperature, °C
0	100.0
5	109.0
10	115.0
15	121.5
20	126.5

SOURCE: J. J. Perkins, *Principles and Methods of Sterilization*, Charles C. Thomas, Springfield, Ill., 1956.

called an **autoclave** (see Fig. 22-6). It is essentially a double-jacketed steam chamber equipped with devices which permit the chamber to be filled with saturated steam and maintained at a designated temperature and pressure for any period of time. In the operation of an autoclave it is absolutely essential that the air in the chamber be completely replaced by saturated steam. If air is present, it will reduce the temperature obtained within the chamber substantially below that which would be realized if pure saturated steam were under the same pressure. It is not the *pressure* that kills the organisms but the *temperature* of the steam.

The autoclave is an essential unit of equipment in every microbiology laboratory. Many media, solutions, discarded cultures, and contaminated materials are routinely sterilized with this apparatus. Generally, but not always, the autoclave is operated at a pressure of approximately 15 lb/in² (at 121°C). The time of operation to achieve sterility depends on the nature of the material being sterilized, the type of the container, and the volume. For example, 1000 test tubes containing 10 ml each of a liquid medium can be sterilized in 10 to 15 min at 121°C; 10 liters of the same medium contained in a single container would require 1 h or more at the same temperature to ensure sterilization.

Fractional Sterilization. Some microbiological media, solutions of chemicals, and biological materials cannot be heated above 100°C without being damaged. If, however, they can withstand the temperature of free-flowing steam (100°C), it is possible to sterilize them by **fractional sterilization (tyndallization)**. This method involves heating the material at 100°C on three successive days with incubation periods in between. Resistant spores germinate during the incubation

Figure 22-6. Pressure steam sterilizer (autoclave), cross-sectional view illustrating operational parts and path of steam flow. *(Courtesy of Wilmont Castle Company.)*

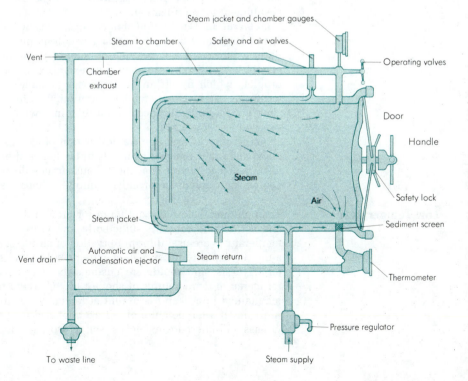

periods; on subsequent exposure to heat, the vegetative cells will be destroyed. If spores are present and do not germinate during the incubation periods, the material will not be sterilized. An apparatus known as the **Steam Arnold** is used for this technique; however, it is also possible to operate an autoclave with free-flowing steam for this purpose.

Boiling Water. Contaminated materials or objects exposed to boiling water cannot be sterilized with certainty. It is true that all vegetative cells will be destroyed within minutes by exposure to boiling water, but some bacterial spores can withstand this condition for many hours. The practice of exposing instruments for short periods of time in boiling water is more likely to bring about disinfection (destruction of vegetative cells of disease-producing microorganisms) rather than sterilization. Boiling water cannot be (and is not) used in the laboratory as a method of sterilization.

Pasteurization. Milk, cream, and certain alcoholic beverages (beer and wine) are subjected to a controlled heat treatment (called **pasteurization**) which kills microorganisms of certain types but does not destroy all organisms. Pasteurized milk is not sterile milk. The pasteurization of milk is discussed in Chap. 28.

Dry-Heat: *Hot-Air Sterilization.* Dry-heat, or hot-air, sterilization is recommended where it is either undesirable or unlikely that steam under pressure will make direct and complete contact with the materials to be sterilized. This is true of certain items of laboratory glassware, such as Petri dishes and pipettes, as well as oils, powders, and similar substances. The apparatus employed for this type of sterilization may be a special electric or gas oven or even the kitchen stove oven. For laboratory glassware, a 2-h exposure to a temperature of 160°C is sufficient for sterilization.

Incineration. Destruction of microorganisms by burning is practiced routinely in the laboratory when the transfer needle is introduced into the flame of the Bunsen burner. A note of caution should be added here. When the transfer needle is sterilized, care should be exercised to prevent *spattering*, since the droplets which fly off are likely to carry viable organisms. The danger from spattering can be greatly reduced or eliminated by using a Bunsen burner or an electric heat coil equipped with a tube into which the transfer needle can be inserted (see Fig. 22-7).

Incineration is used for the destruction of carcasses, infected laboratory animals, and other infected materials to be disposed of. Special precautions need to be taken to ensure that the exhaust fumes do not carry particulate matter containing viable microorganisms into the atmosphere.

Low Temperatures

Temperatures below the optimum for growth depress the rate of metabolism, and if the temperature is sufficiently low, growth and metabolism cease. Low temperatures are useful for preservation of cultures, since microorganisms have a unique capacity for surviving extreme cold. Agar-slant cultures of some bacteria, yeasts, and molds are customarily stored for long periods of time at refrigeration temperatures of about 4 to 7°C. Many bacteria and viruses can be maintained in a deep-freeze unit at temperatures from −20 to −70°C. Liquid nitrogen, at a temperature of −196°C, is used for preserving cultures of many viruses and microorganisms, as well as stocks of mammalian tissue cells used

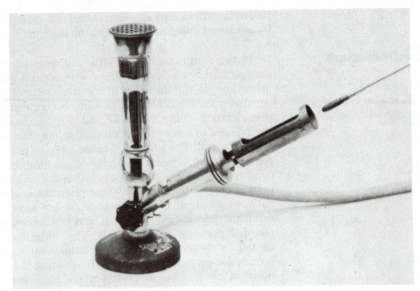

Figure 22-7. When a transfer needle is placed in a flame, spattering may occur with resultant spread of living organisms. To prevent this, one can use a Bunsen burner which is modified so that the transfer needle is exposed to a flame within a tubular space, as shown here.

in animal virology and many other types of research. In all these procedures, the initial freezing kills a fraction of the population, but the survivors may remain viable for long periods. (See Chap. 8.)

From these facts it is immediately apparent that low temperatures, however extreme, cannot be depended upon for disinfection or sterilization. Microorganisms maintained at freezing or subfreezing temperatures may be considered dormant; they perform no detectable metabolic activity. This static condition is the basis of successful application of low temperatures for the preservation of foods. Thus from a practical standpoint, high temperatures may be considered as microbicidal and low temperatures (freezing or lower) as microbistatic.

Desiccation

Desiccation of the microbial cell causes a cessation of metabolic activity, followed by a decline in the total viable population. In general, the time of survival of microorganisms after desiccation varies, depending on the following factors:

1 The kind of microorganism
2 The material in or on which the organisms are dried
3 The completeness of the drying process
4 The physical conditions to which the dried organisms are exposed, e.g., light, temperature, and humidity

Species of Gram-negative cocci such as gonococci and meningococci are very sensitive to desiccation; they die in a matter of hours. Streptococci are much more resistant; some survive weeks after being dried. The tubercle bacillus (*Mycobacterium tuberculosis*) dried in sputum remains viable for even longer periods of time. Dried spores of microorganisms are known to remain viable indefinitely.

In the process of **lyophilization,** organisms are subjected to extreme dehydration in the frozen state and then sealed in a vacuum. In this condition, desiccated (lyophilized) cultures of microorganisms remain viable for many years.

Osmotic Pressure

When two solutions with differing concentrations of solute are separated by a semipermeable membrane, there will occur a passage of water, through the membrane, in the direction of the higher concentration. The trend is toward equalizing the concentration of solute on both sides of a membrane. The solute concentration within microbial cells is approximately 0.95 percent. Thus if cells are exposed to solutions with higher solute concentration, water will be drawn out of the cell. The process is called plasmolysis. The reverse process, that is, passage of water from a low solute concentration into the cell, is termed plasmoptysis. The pressure built up within the cell as a result of this water intake is termed osmotic pressure. These phenomena can be observed more conveniently with animal cells since they do not have rigid cell walls. Plasmolysis results in dehydration of the cell, and as a consequence metabolic processes are retarded partially or completely. The antimicrobial effect is similar to that caused by desiccation. Because of the great rigidity of microbial cell walls (except for protozoa), the cell-wall structure does not exhibit distortions as a result of plasmolysis or plasmoptysis. However, changes in the cytoplasmic membrane, and particularly shrinkage of the protoplast from the cell wall, can be observed during plasmolysis.

Radiation

Energy transmitted through space in a variety of forms is generally called radiation. For our purposes, the most significant type of radiation is probably electromagnetic radiation, of which light and x-rays are examples. Electromagnetic radiation has the dual properties of a *continuous* wave phenomenon and a *discontinuous* particle phenomenon; the particles are packets, or quanta, of energy, sometimes called photons, which vibrate at different frequencies. Radiation of a given frequency can also be described by its wavelength, λ; it is measured in angstroms, where $10,000$ Å $= 1$ μm, and the energy of the radiation in electron volts (ev) is given by $12,350/\lambda$. The various parts of the electromagnetic spectrum, distinguished by their wavelengths, are shown in Fig. 22-8.

Electromagnetic radiation can interact with matter in one of two general ways.

Figure 22-8. Spectrum of radiant energy. (1 Å = 0.1 nm.)

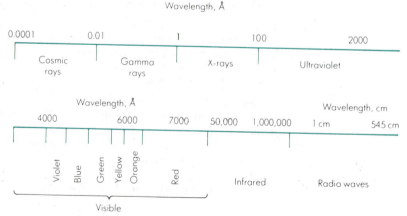

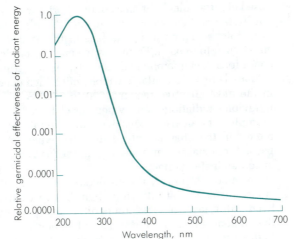

Figure 22-9. Relative germicidal effectiveness of radiant energy between 2000 and 7000 Å. *(Courtesy of General Electric Company, Lamp Division, Publication LD-11.)*

Gamma rays and x-rays, which have energies of more than about 10 eV, are called **ionizing radiations** because they have enough energy to knock electrons away from molecules and ionize them. When such radiations pass through cells, they create free hydrogen radicals, hydroxyl radicals, and some peroxides, which in turn can cause different kinds of intracellular damage. Moreover, since this damage is produced in a variety of materials, ionizing radiations are rather nonspecific in their effects. Less energetic radiation, particularly ultraviolet light, does not ionize; it is **absorbed** quite specifically by different compounds because it excites electrons and raises them to higher energy levels, thus creating different chemical species that can engage in a variety of chemical reactions not possible for unexcited molecules.

In addition to electromagnetic radiation, organisms may be subjected to acoustic radiation (sound waves) and to subatomic particles, such as those released in radioactive decay. The atomic era has alerted us to the damaging potential of radiation. Consequently, a tremendous expenditure of research effort is being directed toward determining the minimum dosage which affects cells, how radiations damage cells, and how the damage can be prevented. Microorganisms have been used for the major part of this research for the same reasons they are used in so many other areas of basic biological research; they are easy to grow and lend themselves to rapid, efficient experimentation.

Besides the fundamental research in radiation microbiology, there have been many developments in the application of ionizing radiation to sterilize biological materials. This method is called **cold sterilization** because ionizing radiations produce relatively little heat in the material being irradiated. Thus it is possible to sterilize heat-sensitive substances, and such techniques are being developed in the food and pharmaceutical industries.

Ultraviolet Light

The ultraviolet portion of the spectrum (Fig. 22-8) includes all radiations from 150 to 3900 Å. Wavelengths around 2650 Å have the highest bactericidal efficiency (see Fig. 22-9). Although the radiant energy of sunlight is partly composed of ultraviolet light, most of the shorter wavelengths of this type are filtered out by the earth's atmosphere (ozone, clouds, and smoke). Consequently, the ultraviolet radiation at the surface of the earth is restricted to the span from about 2670 to 3900 Å. From this we may conclude that sunlight, under certain conditions, has microbicidal capacity, but to a limited degree.

Many lamps are available which emit a high concentration of ultraviolet light in the most effective region, 2600 to 2700 Å. Germicidal lamps, which emit ultraviolet radiations, are widely used to reduce microbial populations. For example, they are used extensively in hospital operating rooms, in aseptic filling rooms, in the pharmaceutical industry, where sterile products are being dispensed into vials or ampules, and in the food and dairy industries for treatment of contaminated surfaces.

An important practical consideration in using this means of destroying microorganisms is that ultraviolet light has very little ability to penetrate matter. Even a thin layer of glass filters off a large percentage of the light. Thus, only the microorganisms on the surface of an object where they are exposed directly to the ultraviolet light are susceptible to destruction.

Mode of Action. Ultraviolet light is absorbed by many cellular materials but most significantly by the nucleic acids, where it does the most damage. The absorption and subsequent reactions are predominantly in the pyrimidines of the nucleic acid. One important alteration is the formation of a *pyrimidine dimer* in which two adjacent pyrimidines become bonded. Unless dimers are removed by specific intracellular enzymes, DNA replication can be inhibited and mutations can result. (See Chap. 12.)

X-Rays (Roentgen Rays)

X-rays are lethal to microorganisms and higher forms of life (see Table 22-6). Unlike ultraviolet radiations, they have considerable energy and penetration ability. However, they are impractical for purposes of controlling microbial populations because (1) they are very expensive to produce in quantity and (2) they are difficult to utilize efficiently, since radiations are given off in all directions from their point of origin. However, x-rays have been widely employed experimentally to produce microbial mutants, as mentioned in Chap. 12.

Gamma Rays

Gamma radiations are high-energy radiations emitted from certain radioactive isotopes such as ^{60}Co. As a result of the major research programs with atomic energy, large quantities of radioisotopes have become available as by-products of atomic fission. These isotopes are potential sources of gamma radiations. Gamma rays are similar to x-rays but are of shorter wavelength and higher energy. They are capable of great penetration into matter, and they are lethal to all life, including microorganisms.

Because of their great penetrating power and their microbicidal effect, gamma rays are attractive for use in commercial sterilization of materials of considerable thickness or volume, e.g., packaged foods and medical devices. However, certain technical problems must be resolved for practical applications, e.g., development of radiation sources for large-scale use and the design of equipment to eliminate any possible hazards to the operators.

Results of quantitative studies on the effect of ionizing radiations on the cells

A rad (radiation absorbed dose)

Table 22-6. Median Lethal Dose of X-Rays for Various Species of Organisms

Organism		Median Lethal Dose, rd*
Viruses:	Tobacco mosaic	200,000
	Rabbit papilloma	100,000
Bacteria:	*Escherichia coli*	5,000
	Bacillus mesentericus	130,000
Algae:	*Mesotenium*	8,500
	Pandorina	4,000
Protozoa:	*Colpidium*	330,000
	Paramecium	300,000
Vertebrates:	Goldfish	750
	Mouse	450
	Rabbit	800
	Rat	600
	Monkey	450
	Humans (?)	400

* A rad (radiation absorbed dose), abbreviated rd, is the dose which delivers 100 ergs/g of irradiated material; it is equal to 6×10^{13} eV.
SOURCE: Modified from E. Paterson, in R. Paterson (ed.), *The Treatment of Malignant Disease by Radium and X Rays*, E. Arnold, London, 1948; *McGraw-Hill Encyclopedia of Science and Technology*, vol. 11, p. 244, McGraw-Hill, New York, 1971.

have resulted in the establishment of the "target" theory of action. This implies that the radiant-energy particle makes a "direct hit" on some essential substance such as DNA within the bacterial cell, causing ionization which results in the death of the cell.

Cathode Rays (Electron-Beam Radiation)

When a high-voltage potential is established between a cathode and an anode in an evacuated tube, the cathode emits beams of electrons, called **cathode rays** or **electron beams.** Special types of equipment have been designed which produce electrons of very high intensities (millions of volts), and these electrons are accelerated to extremely high velocities. These intense beams of accelerated electrons are microbicidal as well as having other effects on biological and nonbiological materials.

The electron accelerator, a type of equipment which produces the high-voltage electron beam, is used today for the sterilization of surgical supplies, drugs, and other materials. One of the unique features of the process is that the material can be sterilized after it has been packaged (the radiations penetrate the wrappings) and at room temperature. Electron-beam radiation has limited power of penetration; but within its limits of penetration, sterilization is accomplished on very brief exposure.

The susceptibility of microorganisms to doses of different radiations is shown in Table 22-7.

Surface Tension and Interfacial Tension

The interface, or boundary, between a liquid and a gas is characterized by unbalanced forces of attraction between the molecules in the surface of the liquid and in the interior. A molecule at the surface of the liquid-air interface is pulled strongly toward the interior of the liquid beneath it, whereas

Table 22-7. Lethal Doses of Different Radiations

Type of Organism	Lethal Doses, mrd			
	Cathode Rays from van de Graaff Accelerators*	From Capacitron Pulsed Beam†	Gamma Rays from ^{60}Co	X-rays from 3-MeV Source
Vegetative:				
Nonpathogenic	0.1–0.25			
Pathogenic	0.45–0.55	0.1–0.25	0.15–0.25	0.03–0.5
Bacterial spores	0.5–2.1	0.2–0.4	1.5‡	0.5–2.0
Molds	0.25–1.15	0.35–0.4	0.2–0.3	0.25–1.0
Yeasts	0.5–1.0		0.3	0.25–1.5

* Various authors.
† Huber and colleagues, quoted by Hannan, 1955.
‡ Approximate.
SOURCE: G. Sykes, "Methods and Equipment for Sterilization of Laboratory Apparatus and Media," in J. R. Norris and D. W. Ribbons (eds.), *Methods in Microbiology*, vol. 1, Academic, New York, 1969.

molecules in the interior of the liquid are attracted uniformly in all directions. This behavior of molecular forces at the liquid-air interface imparts a distinctive characteristic to the surface of a liquid, known as surface tension. Surface forces also exist between two immiscible liquids and at the interface between a solid and a liquid. Here they are referred to as interfacial tension. Changes in surface tension may alter the permeability charcteristics of the cytoplasmic membrane, causing leakage of cellular substances, which results in damage to the cell.

Filtration

Bacteriological Filters

For many years a variety of filters have been available to the microbiologist which can remove microorganisms from liquids or gases. These filters are made of different materials—an asbestos pad in the Seitz filter, diatomaceous earth in the Berkefeld filter, porcelain in the Chamberland-Pasteur filter, and sintered glass disks in other filters.

The mean pore diameter in these bacteriological filters ranges from approximately one to several micrometers; most filters are available in several grades, based on the average size of the pores. However, it should be understood that these filters do not act as mere mechanical sieves; porosity alone is not the only factor preventing the passage of organisms. Other factors, such as the electric charge of the filter, the electric charge carried by the organisms, and the nature of the fluid being filtered, can influence the efficiency of filtration.

In recent years a new type of filter termed the membrane or molecular filter has been developed whose pores are of a uniform and specific predetermined size. Membrane or molecular filters are composed of biologically inert cellulose esters. They are prepared as circular membranes of about 150-μm thickness and contain millions of microscopic pores of very uniform diameter (see Fig. 22-10). Filters of this type can be produced with known porosities ranging from approximately 0.01 to 10 μm. Membrane filters are used extensively in the laboratory and in industry to sterilize fluid materials. They have been adapted to microbiological procedures for the identification and enumeration of microorganisms from water samples and other materials (see Chap. 26.)

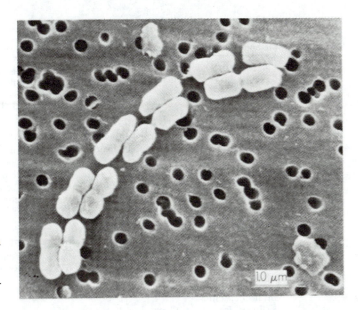

Figure 22-10. Bacteria from a marine water sample are retained by this membrane filter. (*Courtesy of Pall Corporation.*)

It is customary to force the fluid through the filter by applying a negative pressure to the filter flask by use of a vacuum or water pump or to impose a positive pressure above the fluid in the filter chamber, thus forcing it through. Upon completion of filtration, precautions must be taken to prevent contamination of the filtered material when it is transferred to other containers.

The development of high-efficiency particulate air (HEPA) filters has made it possible to deliver clean air to an enclosure such as a cubicle or a room. This type of air filtration together with a system of laminar airflow is now used extensively to produce dust- and bacteria-free air. (See. Fig. 22-11.)

A summary of the application of physical agents for the control of microorganisms is provided in Table 22-8.

Table 22-8. Application of Physical Agents for Controlling Microorganisms

Method	Recommended Uses	Limitations
Moist heat Autoclave	Sterilizing instruments, linens, utensils, and treatment trays, media and other liquids	Ineffective against organisms in materials impervious to steam; cannot be used for heat-sensitive articles
Free-flowing steam or boiling water	Destruction of nonsporeforming pathogens; sanitizes bedding, clothing, and dishes	Cannot be guaranteed to produce sterilization on one exposure
Dry heat Hot-air oven	Sterilizing materials impermeable to or damaged by moisture, e.g., oils, glass, sharp instruments, metals	Destructive to materials which cannot withstand high temperatures for long periods

Table 22.8. *(continued)*

Method	Recommended Uses	Limitations
Incineration	Disposal of contaminated objects that cannot be reused.	Size of incinerator must be adequate to burn largest load promptly and completely; potential of air pollution
Radiation		
Ultraviolet light	Control of airborne infection; disinfection of surfaces	Must be absorbed to be effective (does not pass through transparent glass or opaque objects); irritating to eyes and skin; low penetration
X-ray, gamma, and cathode radiation	Sterilization of heat-sensitive surgical materials and other medical devices	Expensive and requires special facilities for use
Filtration		
Membrane filters	Sterilization of heat-sensitive biological fluids	Fluid must be relatively free of suspended particulate matter
Fiberglass filters (HEPA)	Air disinfection	Expensive
Physical cleaning		
Ultrasonics	Effective in decontaminating delicate cleaning instruments	Not effective alone, but as adjunct procedure enhances effectiveness of other methods
Washing	Hands, skin, objects	Sanitizes; reduces microbial flora

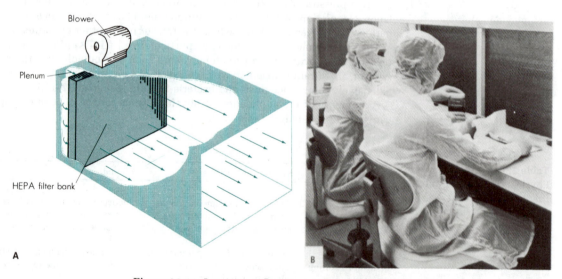

Figure 22-11. Laminar-airflow system. (A) Schematic drawing of horizontal laminar-flow tunnel. Arrows in tunnel denote parallel flow of air through a room. (*Redrawn from M. S. Favero, "Industrial Applications of Laminar Air-flow," in Developments in Industrial Microbiology, American Institute of Biological Sciences, Washington, 1970, vol. 11.*) (B) Laboratory personnel performing sterility test in laminar-airflow unit. (*Courtesy of B. Phillips, Becton, Dickinson & Company.*)

QUESTIONS

1 Describe the death-rate pattern of bacteria when exposed to a lethal agent.

2 Enumerate the conditions which affect the death of microorganisms when they are exposed to an antimicrobial agent.

3 What are the terms by which one can express, quantitatively, the resistance of microorganisms to high temperatures? Distinguish between the meaning of each of these terms.

4 Compare vegetative cells of bacteria with bacterial spores in terms of resistance to heat. What is thought to account for the difference?

5 Describe the process of fractional sterilization, or tyndallization.

6 List several physical agents (or processes) that produce a microbistatic condition.

7 How are microorganisms affected by subzero temperatures?

8 The mechanism of antimicrobial action caused by desiccation is similar to that caused by plasmolysis. Explain why.

9 List several different kinds of radiations that are destructive to microorganisms. Comment on the practical application of each.

10 What is a membrane or molecular filter? How does it differ from older types of bacteriological filters in terms of how it removes microorganisms?

11 What method of sterilization would be appropriate for each of the following?

(a) Petri dishes for laboratory use.

(b) Water

(c) Usual laboratory media, e.g., nutrient agar

(d) A dry powder product

(e) A heat-labile solution of vitamins

(f) A heat-labile antibiotic solution

(g) Contaminated hospital linens

REFERENCES

See Chap. 23.

Chapter 23 — Control by Chemical Agents

A large number of chemical compounds have the ability to inhibit the growth and metabolism of microorganisms or to kill them. Commercial products which incorporate these compounds are available for controlling microbial populations in many different circumstances. For example, solutions of some chemical compounds are used to reduce the microbial flora of the oral cavity; other chemical compounds are recommended for reducing the microbial population in the dust of hospital floors. No single chemical agent is best for any and all purposes. This is not surprising in view of the variety of conditions under which they are used, their differences in mode of action, and the differences of resistance among microbial species. Experience and research have shown that certain kinds of chemicals are more appropriate and effective for certain situations. As a result, several classes of chemical substances have been identified and new compounds developed within these classes which have destructive properties in terms of their suitability for practical application.

In this chapter we will identify and characterize the major classes of chemical compounds used for the practical control of microorganisms. It is important to bear in mind that the efficacy of an antimicrobial agent is influenced by many factors, as described in Chap. 22.

CHARACTERISTICS OF AN IDEAL ANTIMICROBIAL CHEMICAL AGENT

As already stated, there is not a single chemical agent which is best for the control of microorganisms for any and all purposes. If there were an ideal general-purpose chemical antimicrobial agent, it would have to possess an extremely elaborate array of characteristics. It is unlikely that such a compound will be found. Nevertheless, the specifications for such an ideal compound, as they are described below, can be aimed for in the preparation of new compounds. They should also be taken into consideration in the evaluation of new chemical agents proposed for practical use.

1 *Antimicrobial activity.* The capacity of the substance to kill or inhibit microorganisms is the first requirement. The chemical, at a low concentration, should have a broad spectrum of antimicrobial activity.

2 *Solubility.* The substance must be soluble in water or other solvents to the extent necessary for effective use.

3 *Stability.* Changes in the substance upon standing should be minimal and should not result in significant loss of germicidal action.

4 *Nontoxicity to humans and other animals.* Ideally, the compound should be lethal to microorganisms and noninjurious to humans and other animals.

5 *Homogeneity.* The preparation must be uniform in composition so that active ingredients are present in each application. Pure chemicals are uniform, but mixtures of materials may lack homogeneity.

6 *Noncombination with extraneous organic material.* Many disinfectants have an affinity for proteins or other organic material. When such disinfectants are used in situations where there is considerable organic material besides that of the microbial cells, little, if any, of the disinfectant will be available for action against the microorganisms.

7 *Toxicity to microorganisms at room or body temperatures.* In using the compound, it should not be necessary to raise the temperature beyond that normally found in the environment where it is to be used.

8 *Capacity to penetrate.* Unless the substance can penetrate through surfaces, its germicidal action is limited solely to the site of application. Sometimes, of course, surface action is all that is required.

9 *Noncorroding and nonstaining.* It should not rust or otherwise disfigure metals nor stain or damage fabrics.

10 *Deodorizing ability.* Deodorizing while disinfecting is a desirable attribute. Ideally the disinfectant itself should either be odorless or have a pleasant smell.

11 *Detergent capacities.* A disinfectant which is also a detergent (cleaning agent) accomplishes two objectives, and the cleansing action improves the effectiveness of the disinfectant.

12 *Availability.* The compound must be available in large quantities at a reasonable price.

DEFINITION OF TERMS

The following terms are used to describe the processes and chemical agents employed in controlling microorganisms.

Sterilization. The process of destroying all forms of microbial life. A *sterile* object, in the microbiological sense, is free of living microorganisms. The terms *sterile, sterilize,* and *sterilization* therefore refer to the complete absence or destruction of all microorganisms and should not be used in a relative sense. An object or substance is sterile or nonsterile; it can never be semisterile or almost sterile.

Disinfectant. An agent, usually a chemical, that kills the growing forms but not necessarily the resistant spore forms of disease-producing microorganisms. The term is commonly applied to substances used on inanimate objects. **Disinfection** is the process of destroying infectious agents.

Antiseptic. A substance that opposes sepsis, i.e., prevents the growth or action of microorganisms either by destroying microorganisms or by inhibiting their growth and metabolism. Usually associated with substances applied to the body.

Sanitizer. An agent that reduces the microbial population to safe levels as judged by public health requirements. Usually it is a chemical agent that kills 99.9 percent of the growing bacteria. Sanitizers are commonly applied to inanimate objects and are generally employed in the daily care of equipment and utensils in dairies and food plants and for glasses, dishes, and utensils in restaurants. The process of disinfection would produce sanitization; however, in the strict sense, sanitization implies a sanitary condition which disinfection does not necessarily imply.

Germicide (Microbicide). An agent that kills the growing forms but not necessarily the resistant spore forms of germs; in practice a germicide is almost the same thing as a disinfectant, but germicides are commonly used for all kinds of germs (microbes) for any application.

Bactericide. An agent that kills bacteria (adjective, **bactericidal**). Similarly, the terms **fungicide, virucide,** and **sporicide** refer to agents that kill fungi, viruses, and spores, respectively.

Bacteriostasis. A condition in which the growth of bacteria is prevented (adjective, **bacteriostatic**). Similarly, **fungistatic** describes an agent that stops the growth of fungi. Agents that have in common the ability to inhibit growth of microorganisms are collectively designated **microbistatic** agents.

Antimicrobial Agent. One that interferes with the growth and metabolism of microbes. In common usage the term denotes inhibition of growth, and with reference to specific groups of organisms such terms as **antibacterial** or **antifungal** are frequently employed. Some antimicrobial agents are used to treat infections, and they are called **chemotherapeutic** agents.

SELECTION OF A CHEMICAL AGENT FOR PRACTICAL APPLICATIONS

The major factors that need to be assessed in the process of selecting the most appropriate chemical agent for a specific practical application are:

1 *Nature of the material to be treated.* To cite an extreme example, a chemical agent used to disinfect contaminated utensils might be quite unsatisfactory for application to the skin; i.e., it might do serious injury to the tissue cells. Consequently, the substance selected must be compatible with the material to which it is applied.

2 *Types of microorganisms.* Chemical agents are not all equally effective against

bacteria, fungi, viruses, and other microorganisms. Spores are more resistant than vegetative cells. Differences exist between Gram-positive and Gram-negative bacteria; *Escherichia coli* is much more resistant to cationic disinfectants than *Staphylococcus aureus*. Differences in action also exist between strains of the same species. Therefore, the agent selected must be known to be effective against the type of organism to be destroyed.

3 *Environmental conditions.* The factors discussed in Chap. 22, e.g., temperature, pH, time, concentration, and presence of extraneous organic material, may all have a bearing on the rate and efficiency of antimicrobial action. The successful use of an antimicrobial agent requires an understanding of the influence of these conditions on the particular agent, so it can be employed under the most favorable circumstances.

MAJOR GROUPS OF CHEMICAL ANTIMICRO-BIAL AGENTS

The major antimicrobial agents can be grouped as shown below. We will discuss representative examples from each of these groups.

1 Phenol and phenolic compounds
2 Alcohols
3 Halogens
4 Heavy metals and their compounds
5 Dyes
6 Detergents
7 Quaternary ammonium compounds
8 Aldehydes
9 Gaseous agents

Phenol and Phenolic Compounds

Phenol has the distinction of being used successfully in the 1880s by Joseph Lister, a surgeon, to reduce infection of surgical incisions and surgical wounds. Lister became aware of Pasteur's studies which incriminated germs as the cause of infection. Accordingly, he instituted the practice of applying a solution of phenol (carbolic acid) to surgical incisions and wounds. The reduction in infections was striking. Later he developed the practice of spraying phenol into the operating room area to control infection.

Phenol has the additional distinction of being the standard against which other disinfectants of a similar chemical structure are compared to determine their antimicrobial activity. The procedure used is called the **phenol-coefficient technique.** This technique is described later in this chapter.

Phenol and phenolic compounds are very effective disinfectants. A 5% aqueous solution of phenol rapidly kills the vegetative cells of microorganisms; spores are much more resistant. Many derivatives of phenol have been prepared and evaluated for their antimicrobial activity. The chemical structures of phenol and a few phenol derivatives are shown in Fig. 23-1. Antimicrobial activity is enhanced by the addition of chemical substitutions in the phenol ring structure as shown in Table 23-1.

Hexylresorcinol, a derivative of phenol, is marketed in a solution of glycerin and water. It is a strong surface-tension reductant, which may account in part for its high bactericidal activity. A commercial product containing hexylresorcinol, S.T. 37, is so named because of its surface-tension value. Hexylresorcinol preparations are employed as general antiseptics.

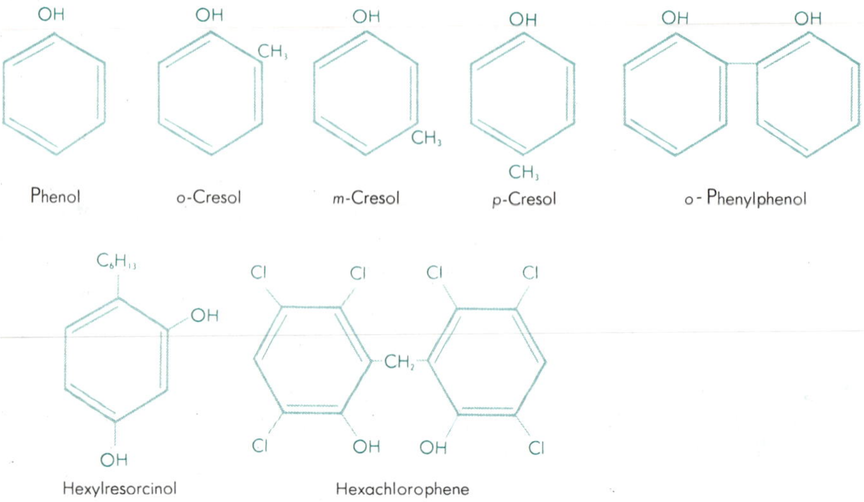

Figure 23-1. Phenol and some phenolic compounds which are used as disinfectants.

Phenol   o-Cresol   m-Cresol   p-Cresol   o-Phenylphenol

Hexylresorcinol   Hexachlorophene

Table 23-1. Microbicidal Action of Phenol Derivatives (Phenol Coefficients at 37°C)*

Name	Salmonella typhi	Staphylococcus aureus	Mycobacterium tuberculosis	Candida albicans
Phenol	1.0	1.0	1.0	1.0
o-Cresol	2.3	2.3	2.0	2.0
m-Cresol	2.3	2.3	2.0	2.0
p-Cresol	2.3	2.3	2.0	2.0
4-Ethylphenol	6.3	6.3	6.7	7.8
2,4-Dimethylphenol	5.0	4.4	4.0	5.0
2,5-Dimethylphenol	5.0	4.4	4.0	4.0
3,4-Dimethylphenol	5.0	3.8	4.0	4.0
2,6-Dimethylphenol	3.8	4.4	4.0	3.5
4-n-Propylphenol	18.3	16.3	17.8	17.8
4-n-Butylphenol	46.7	43.7	44.4	44.4
4-n-Amylphenol	53.3	125.0	133.0	156.0
4-tert-Amylphenol	30.0	93.8	111.1	100.0
4-n-Hexylphenol	33.3	313.0	389.0	333.0
4-n-Heptylphenol	16.7†	625.0	667.0	556.0

* The higher the value the greater the microbicidal activity.
† Approximate.
SOURCE: From R. F. Prindle and E. S. Wright, "Phenolic Compounds," in C. A. Lawrence and S. S. Black, *Disinfection, Sterilization, and Preservation*, Lea & Febiger, Philadelphia, 1968.

Practical Applications. Phenolic substances may be either bactericidal or bacteriostatic, depending upon the concentration used. Bacterial spores and viruses are more resistant than are vegetative cells. Some phenolics are highly fungicidal. The antimicrobial activity of phenolics is reduced at an alkaline pH and by organic material. Low temperatures and the presence of soap also reduce antimicrobial activity.

Pure crystalline phenol is colorless. Aqueous solutions of from 2 to 5% can be employed to disinfect such materials as sputum, urine, feces, and contaminated instruments or utensils. Solutions of pure phenol have limited application. However, derivatives of phenol diluted in detergents or some other carrier find use in many commercial antiseptic and disinfectant preparations. One of the widely used phenolic derivatives is o-phenylphenol. Combination of compounds of this class with detergents results in products with good disinfectant as well as detergent properties.

Mode of Action. Exposure of microbial cells to phenolic compounds produces a variety of effects. Depending upon the concentration of the phenolic compound to which microbial cells were exposed, researchers have described results such as disruption of cells, precipitation of cell protein, inactivation of enzymes, and leakage of amino acids from the cells. Although the specific mode of action is not clear, there is a consensus that the lethal effect is associated with physical damage to the membrane structures in the cell surface, which initiates further deterioration.

Alcohols

Ethyl alcohol, CH_3CH_2OH, in concentrations between 50 and 90%, is effective against vegetative or nonsporeforming cells. For practical application a 70% concentration of alcohol is generally used.

Ethyl alcohol cannot be relied upon to produce a sterile condition. Concentrations which are effective against vegetative cells are practically inert against bacterial spores. In his book *Disinfection and Sterilization*, Sykes notes that "There is one record of survival of anthrax spores in alcohol for 20 years and another one of the *Bacillus subtilis* for 9 years."

Methyl alcohol is less bactericidal than ethyl alcohol; furthermore, it is highly poisonous. Even the fumes of this compound may produce permanent injury to the eyes, and is not generally employed for the destruction of microorganisms. The higher alcohols—propyl, butyl, amyl, and others—are more germicidal than ethyl alcohol. In fact, there is a progressive increase in germicidal power as the molecular weight of alcohols increases (as shown in Table 23-2). Since alcohols of molecular weight higher than that of propyl alcohol are not miscible in all proportions with water, they are not commonly used in disinfectants. Propyl and isopropyl alcohols in concentrations ranging from 40 to 80% are bactericidal for vegetative cells.

Table 23-2. Phenol Coefficients of Alcohols

Alcohol	Phenol Coefficient	
	Against *Salmonella typhi*	Against *Staphylococcus aureus*
Methyl, CH_3OH	0.026	0.03
Ethyl, CH_3CH_2OH	0.04	0.039
n-Propyl, $CH_3CH_2CH_2OH$	0.102	0.082
Isopropyl, $(CH_3)_2CHOH$	0.064	0.054
n-Butyl, $CH_3(CH_2)_2CH_2OH$	0.273	0.22
n-Amyl, $CH_3(CH_2)_3CH_2OH$	0.78	0.63
n-Hexyl, $CH_3(CH_2)_4CH_2OH$	2.3	
n-Heptyl, $CH_3(CH_2)_5CH_2OH$	6.8	
n-Octyl, $CH_3(CH_2)_6CH_2OH$	21.0	0.63

SOURCE: G. Sykes, *Disinfection and Sterilization*, 2d ed., Lippincott, Philadelphia, 1965.

Figure 23-2. Comparative effectiveness of washing with various antiseptic solutions. This chart summarizes a large number of tests. In each test, the calculated bacterial flora immediately before the antiseptic was applied was considered as 100 percent; the residual flora immediately after use of the antiseptic is shown as a proportion of the original one. The steeper the curve, the greater the effect. (Note: 1:1000 means 1 part in 1000.) [*Courtesy of P. B. Price, "Skin Antisepsis," in J. H. Brewer (ed.), Lectures on Sterilization, Duke, Durham, N.C., 1957.*]

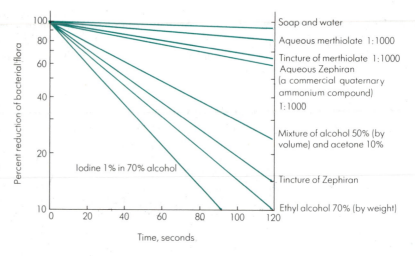

Practical Applications. Alcohol is effective in reducing the microbial flora of skin and for the disinfection of clinical oral thermometers. The comparative effectiveness of alcohol and other disinfectants applied to skin is shown in Fig. 23-2. Alcohol concentrations above 60% are effective against viruses; however, the effectiveness is influenced considerably by the amount of extraneous protein material in the mixture. The extraneous protein reacts with the alcohol and thus protects the virus.

Mode of Action. Alcohols are protein denaturants, and this property may, to a large extent, account for their antimicrobial activity. Alcohols are also solvents for lipids, and hence they may damage lipid complexes in the cell membrane. They are also dehydrating agents. This may account for the relative ineffectiveness of absolute alcohol on "dry" cells; it is possible that very high concentrations remove so much water from the cell that the alcohol is unable to penetrate. The severe dehydration occurring under these conditions would result in a bacteriostatic condition. Some of the effectiveness of alcohol for surface disinfection can be attributed to its cleansing or detergent action which results in mechanical removal of microorganisms.

Halogens

Iodine

Iodine is one of the oldest and most effective germicidal agents. It has been in use for more than a century, having been recognized by the U.S. Pharmacopoeia in 1830. Pure iodine is a bluish-black crystalline element having a metallic luster. It is only slightly soluble in water but readily soluble in alcohol and aqueous solutions of potassium or sodium iodide. The element is traditionally used as a germicidal agent in a form referred to as **tincture of iodine.** There are several preparations available, such as 2% iodine plus 2% sodium iodide diluted in alcohol, 7% iodine plus 5% potassium iodide in 83% alcohol, and 5% iodine and 10% potassium iodide in aqueous solution. Iodine is also used in the form of substances known as **iodophors.** Iodophors are mixtures of iodine with surface-active agents which act as carriers and solubilizers for the iodine. One of these agents is polyvinylpyrrolidone (PVP); the complex can be expressed as

PVP-I. Iodine is released slowly from this complex. Iodophors possess the germicidal characteristics of iodine and have the additional advantages of non-staining and low irritant properties.

Practical Applications. Iodine is a highly effective bactericidal agent and is unique in that it is effective against all kinds of bacteria. Iodine also possesses sporicidal activity; however, the rate at which the spores are killed is markedly influenced by the conditions under which they are exposed, e.g., amount of organic material and extent of dehydration. In addition, it is highly fungicidal and is to some extent virucidal.

Iodine solutions are chiefly used for the disinfection of skin, and for this purpose they rank among the best disinfectants. The effectiveness of iodine preparations for the reduction of the microbial flora of skin is well recognized (see Fig. 23-2). Iodine preparations are effective for other purposes, such as disinfection of water, disinfection of air (iodine vapors), and sanitization of food utensils.

Mode of Action. The mechanism by which iodine exerts its antimicrobial activity is not specifically understood. Iodine is an oxidizing agent, and this fact may account for its antimicrobial action. Oxidizing agents can irreversibly oxidize and thus inactivate essential metabolic compounds such as proteins with sulfhydryl groups. It has also been suggested that the action may involve the halogenation of tyrosine units of enzymes and other cellular proteins requiring tyrosine for activity.

Tyrosine   Diiodotyrosine

Chlorine and Chlorine Compounds

Chlorine, either in the form of gas or in certain chemical combinations, represents one of the most widely used disinfectants. The compressed gas in liquid form is almost universally employed for the purification of municipal water supplies. Chlorine gas is difficult to handle unless special equipment is available to dispense it. Hence, its usefulness in the gaseous state is limited to large-scale operations such as water-purification plants, where it is feasible for installing suitable equipment for safe handling.

There are available many compounds of chlorine which can be handled more conveniently than free chlorine and which, under proper conditions of use, are equally effective as disinfectants. One class of compounds in this category is the **hypochlorites.** Calcium hypochlorite, $Ca(OCl)_2$ (also known as chlorinated lime), and sodium hypochlorite, $NaOCl$, are popular compounds.

The **chloramines** represent another category of chlorine compounds used as disinfectants, sanitizing agents, or antiseptics. Chemically they are characterized by the fact that one or more of the hydrogen atoms in an amino group of a compound are replaced with chlorine. The simplest of these is monochloramine, NH_2Cl. Chloramine-T and azochloramide, two of several germicidal compounds in this category, have more complex chemical structures.

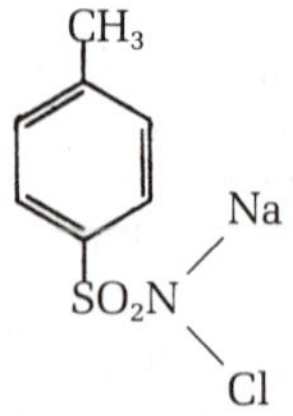

Chloramine-T 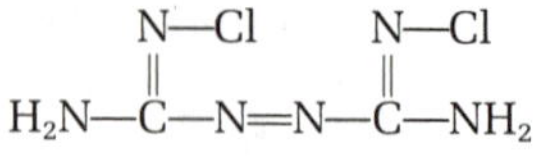   Azochloramide

One of the advantages of the chloramines is stability; they are more stable than the hypochlorites in terms of prolonged release of chlorine.

Practical Applications. Semmelweis is credited with having used hypochlorites in 1846 to 1848 in an attempt to reduce the incidence of childbed fever. Medical students were required to wash their hands and soak them in a hypochlorite solution before examining patients.

Chlorine compounds are very widely used to control microorganisms. Major applications are in water treatment, in the food industry, for domestic uses, and in medicine.

Products containing calcium hypochlorite are used for sanitizing dairy equipment and eating utensils in restaurants. Solutions of sodium hypochlorite of a 1% concentration are used for personal hygiene and as a household disinfectant; higher concentrations of 5 to 12% are also employed as household bleaches and disinfectants and for use as sanitizing agents in dairy and food-processing establishments. The amount of hypochlorite added should provide a residual concentration of approximately 1 mg per liter of free chlorine.

Chlorine compounds have been used to disinfect open wounds, to treat athlete's foot, to treat other infections, and as a general disinfectant.

Mode of Action. The antimicrobial action of chlorine and its compounds comes through the hypochlorous acid formed when free chlorine is added to water:

$$Cl_2 + H_2O \rightarrow HCl + HClO \text{ (hypochlorous acid)}$$

Similarly, hypochlorites and chloramines undergo hydrolysis, with the formation of hypochlorous acid. The hypochlorous acid formed in each instance is further decomposed:

$$HClO \rightarrow HCl + O$$

Formed from
chlorine,
hypochlorites,
chloramines

The oxygen released in this reaction (nascent oxygen) is a strong oxidizing agent, and through its action on cellular constituents, microorganisms are destroyed. The killing of microorganisms by chlorine and its compounds is also

Table 23-3. Some Compounds of Heavy Metals That Have Antimicrobial Activity

Heavy Metal	Examples of Compounds	Applications
Mercury	Inorganic compounds: Mercuric chloride (bichloride of mercury) Mercurous chloride Mercuric oxide Ammoniated mercury	Bactericidal in dilutions of 1:1,000; limited use because of corrosive action, high toxicity to animals, and reduction of effectiveness in presence of organic material; insoluble compounds, used in ointments as antiseptics
	Organic compounds: Mercurochrome Metaphen Merthiolate Mercresin	Less irritating and less toxic than the inorganic mercury compounds; employed as antiseptics on cutaneous and mucosal surfaces; may be bactericidal or bacteriostatic
Silver	Colloidal silver compounds: Silver nitrate Silver lactate Silver picrate	Consist of protein in combination with metallic silver or silver oxide (colloidal solution); bacteriostatic or bactericidal effect is a function of the free silver ions released from the combination; used as antiseptics, silver nitrate is the most widely used of these compounds, all of which are germicidal and employed as antiseptics in specific conditions; silver nitrate is bactericidal for most organisms at a dilution of 1:1,000; many states require that the eyes of newborns be treated with a few drops of 1% silver nitrate solution to prevent ophthalmia neonatorum, a gonococcal infection of eyes
Copper	Copper sulfate	Much more effective against algae and molds than bacteria; 2 ppm in water sufficient to prevent algal growth; used in swimming pools and open water reservoirs; used in the form of Bordeaux mixture as a fungicide for prevention of certain plant diseases

due in part to the direct combination of chlorine with proteins of the cell membranes and enzymes.

Heavy Metals and Their Compounds

Most of the heavy metals, either alone or in certain compounds, exert a detrimental effect upon microorganisms. The most effective are mercury, silver, and copper. Examples of these are summarized in Table 23-3.

Mode of Action. Heavy metals and their compounds act antimicrobially by combining with cellular proteins and inactivating them. For example, in the case of mercuric chloride the inhibition is directed at enzymes which contain the *sulfhydryl* grouping. Note below that the effect is upon the sulfhydryl group.

$$\text{Enzyme} \begin{array}{c} \diagup \text{SH} \\ \diagdown \text{SH} \end{array} + \text{HgCl}_2 \rightarrow \text{enzyme} \begin{array}{c} \diagup \text{S} \\ | \\ \text{Hg} \\ | \\ \diagdown \text{S} \end{array} + 2\text{HCl}$$

Active enzyme   Mercuric chloride   Inactive enzyme

High concentrations of salts of heavy metals like mercury, copper, and silver coagulate cytoplasmic proteins, resulting in damage or death to the cell. Salts of heavy metals are also precipitants, and in high concentrations such salts could cause the death of a cell.

Dyes

Two classes of dye compounds which have antimicrobial properties are of special interest to microbiologists. These are triphenylmethane and acridine dyes.

Triphenylmethane Dyes

Included in this category are malachite green, brilliant green, and crystal violet. As a rule Gram-positive organisms are more susceptible to lower concentrations of these compounds than are Gram-negative ones. Crystal violet will inhibit Gram-positive cocci at a dilution of 1:200,000 to 1:300,000; 10 times this concentration is required to inhibit *Escherichia coli*. *Staphylococcus aureus* is inhibited by malachite green at a concentration of 1:1,000,000; a concentration of about 1:30,000 is required to inhibit *E. coli*. This general relationship between Gram reaction and susceptibility to triphenylmethane dyes has a number of practical applications.

Practical Applications. Certain media can be made selective by the incorporation of low concentrations (about 1:100,000) of the dyes crystal violet, brilliant green, or malachite green. Gram-positive bacteria will be inhibited. Media of this kind are used extensively in public health microbiology, where detection of *E. coli* is important. Susceptibility to various dyes can also be used for identification of bacteria. Three species of *Brucella* can be distinguished by their patterns of resistance to several dyes. Crystal violet has also been used as a fungicide. A concentration of 1:10,000 is lethal for *Monilia* and *Torula*, and a concentration of 1:1,000,000 is inhibitory.

Mode of Action. The mode of action of triphenylmethane dyes is uncertain, but there is speculation that they exert their inhibitory effect by interfering with cellular oxidation processes.

Acridine Dyes

Two examples of dyes derived from acridine are acriflavine and tryptoflavine. These compounds exhibit selective inhibition against bacteria, particularly staphylococci and gonococci. Gonococci are inhibited by tryptoflavine in dilutions of 1:10,000,000 to 1:50,000,000. They possess little, if any, antifungal activity. Presently, they have less application than before the advent of antibiotics and other chemotherapeutic agents. They are used to some extent for the treatment of burns and wounds and for ophthalmic application and bladder irrigation.

Synthetic Detergents

Surface-tension depressants, or wetting agents, employed primarily for cleansing surfaces are called **detergents.** Soap is an example. However, soap is a poor detergent in hard water. For this reason many new more efficient cleaning agents have been developed, called *surfactants* or **synthetic detergents,** many of which are superior to soap. They do not form precipitates in alkaline or acid water, nor do they produce deposits with minerals found in hard water. They are extensively used in laundry and dishwashing powders, shampoos, and other washing preparations. Some are also highly bactericidal.

Chemically, detergents are classified as follows:

1 Those which ionize with the detergent property resident in the *anion* are referred to as **anionic detergents.** For example,

$$[C_9H_{19}COO]^- \ Na^+ \qquad\qquad [C_{12}H_{25}OSO_3]^- \ Na^+$$

A soap   Sodium lauryl sulfate

2 Those which ionize with the detergent property resident in the *cation* are referred to as **cationic detergents.** For example,

Cetylpyridinium chloride (Ceepryn)

3 A third category of detergents is **nonionic;** i.e., they do not ionize. However, these substances do not possess significant antimicrobial activity.

Insofar as reduction of the microbial flora from surfaces such as skin and clothing is concerned, the real value of ordinary soaps lies in the mechanical removal of microorganisms. Soaps reduce surface tension and thereby increase the wetting power of the water in which they are dissolved. Soapy water has the ability to emulsify and disperse oils and dirt. The microorganisms become enmeshed in the soap lather and are removed by the rinse water. Various chemicals have been incorporated into soaps to enhance their germicidal activity.

Cationic detergents are regarded as more germicidal than anionic compounds and will be discussed separately here as quaternary ammonium compounds.

Quaternary Ammonium Compounds

Most compounds of the germicidal cationic-detergent class are **quaternary ammonium salts.** Their characteristic structure with reference to a common inorganic ammonium salt such as ammonium chloride is shown in Fig. 23-3. The R_1, R_2, R_3, and R_4 groups are carbon groups linked to the nitrogen atom as shown in Fig. 23-3C. The R groups may be any one of a large number of different alkyl groups. Accordingly, a very large number of different quaternary ammonium compounds have been synthesized and evaluated for their antimicrobial activity. Several are available commercially as effective antimicrobial agents for a variety of uses (see Fig. 23-4).

The bactericidal power of the quaternaries is exceptionally high against Gram-positive bacteria, and they are also quite active against Gram-negative organisms. Bactericidal concentrations range from dilutions of one part in a few thousand to one part in several hundred thousand, as shown in Table 23-4. Another of their characteristics is the ability to manifest bacteriostatic action far beyond their bactericidal concentration. For example, the limit of bactericidal action for a given compound may be at a dilution of 1:30,000; yet it may be bacteriostatic in dilutions as high as 1:200,000. The action of these compounds demonstrates the need to distinguish between static and lethal activity in test procedures for the evaluation of disinfectants.

Table 23-4. Some Bactericidal Concentrations* of Three Quaternary Ammonium Compounds

Organism	Lethal Concentrations†		
	Cetrimide	Ceepryn	Zephirol
Staphylococcus	20,000‡	83,000	18,000
	35,000	218,000	20,000
	218,000		38,000
			50,000
			200,000
Streptococcus pyogenes	20,000	42,000	40,000
		127,000	
Escherichia coli	3,000	66,000	12,000
	27,500	67,000	27,000
	30,000		
Salmonella typhi	13,000	15,000	10,000
		48,000	20,000
		62,000	
Pseudomonas aeruginosa	3,500		2,500
	5,000		
Proteus vulgaris	7,500	34,000	1,300

* These figures have been collected from various published sources; they were therefore obtained with different testing techniques.
† Expressed as 1 part quaternary ammonium compound in stated volume of diluent; e.g., see footnote ‡.
‡ 1 part cetrimide in 20,000 parts of diluent.
SOURCE: G. Sykes, *Disinfection and Sterilization*, 2d ed., Lippincott, Philadelphia, 1965.

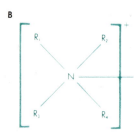

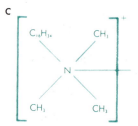

Figure 23-3. Chemical structure of quaternary ammonium compounds shown in relation to the structure of ammonium chloride. (A) Ammonium chloride. (B) The general structure of a quaternary ammonium compound. R_1, R_2, R_3, and R_4 are carbon-containing groups, and the X^- is a negatively charged ion such as Br^- or Cl^-. (C) The quaternary ammonium compound CTAB, or cetrimide.

Zephiran, benzalkonium chloride

Ceepryn chloride

Figure 23-4. Some examples of quaternary disinfectants.

Diaparene chloride

Quaternaries have been shown to be fungicidal as well as destructive to certain of the pathogenic protozoa. Viruses appear to be more resistant than bacteria and fungi.

Practical Applications. The combined properties of germicidal activity and detergent action, plus such other features as low toxicity, high solubility in water, stability in solution, and noncorrosiveness, have resulted in many applications of quaternaries as disinfectants and sanitizing agents. They are used as skin disinfectants, as a preservative in ophthalmic solutions, and in cosmetic preparations. Quaternaries are widely used for control of microorganisms on floors, walls, and other surfaces in hospitals, nursing homes, and other public places. They are used to sanitize food and beverage utensils in restaurants as well as surfaces and certain equipment in food-processing plants. Other applications are to be found in the dairy, egg, and fishing industries to control microbial growth on surfaces of equipment and the environment in general.

Mode of Action. A variety of damaging effects of quaternaries upon microorganisms have been observed. These include denaturation of proteins, interference with glycolysis, and membrane damage. Experimental evidence suggests that the most likely site of the damage to the cell is the cytoplasmic membrane; the quaternaries alter the vital permeability features of this cell structure.

Aldehydes

Among the class of chemicals with the general formula RCHO (aldehydes), several of the low-molecular-weight compounds are antimicrobial. Two of the most effective are formaldehyde and glutaraldehyde. Both are highly microbicidal, and both have the ability to kill spores (sporicidal).

Formaldehyde

Formaldehyde (HCHO) is the simplest compound in the aldehyde series. It is a gas that is stable only in high concentrations and at elevated temperatures. At room temperature it polymerizes, forming a solid substance. The important polymer is paraformaldehyde, a colorless substance which rapidly yields formaldehyde upon heating. Formaldehyde is also marketed in aqueous solution as formalin, which contains 37 to 40% formaldehyde.

The fumes of formaldehyde are noxious; they are irritating to tissues and eyes.

Practical Applications. Formaldehyde in solution is useful for sterilization of certain instruments. Formaldehyde in gaseous form can be used for disinfection and sterilization of enclosed areas. Formalin and paraformaldehyde are two principal sources of formaldehyde when it is used for gaseous disinfection. Vaporization of formaldehyde from either of these sources into an enclosed area for an adequate time will cause sterilization, vegetative cells being killed more quickly than spores. Humidity and temperature have a pronounced effect on the microbicidal action of formaldehyde; in order to sterilize an enclosure the temperature must be about room temperature (22°C) and the relative humidity between 60 to 80 percent. One of the disadvantages of this process is the limited ability of the formaldehyde vapors to penetrate covered surfaces.

Mode of Action. Formaldehyde is an extremely reactive chemical. It combines readily with vital organic nitrogen compounds such as proteins and nucleic acids. It is likely that interaction of formaldehyde with these cellular substances accounts for its antimicrobial action.

Glutaraldehyde

Glutaraldehyde is a saturated dialdehyde with the formula

$$O=\overset{\overset{\textstyle H}{|}}{C}-CH_2-CH_2-CH_2-\overset{\overset{\textstyle H}{|}}{C}=O$$

A 2% solution of this chemical agent exhibits a wide spectrum of antimicrobial activity. It is effective against vegetative bacteria, fungi, bacterial and fungal spores, and viruses. It is used in the medical field for sterilizing urological instruments, lensed instruments, respiratory therapy equipment, and other special equipment.

Gaseous Agents

Certain kinds of medical devices that need to be available in a sterile condition are made of materials that are damaged by heat. Examples are plastic syringes, blood transfusion apparatus, and catheterization equipment. The same is true for routinely used laboratory ware, such as plastic pipettes, Petri dishes, and other equipment, that is packaged and sterilized ready for use. On occasion there is need to disinfect or sterilize an enclosed area. Sterilization by means of gaseous agents is effective and practical for such situations. The main agents currently used for gaseous sterilization are ethylene oxide, β-propiolactone, and formaldehyde. We have discussed formaldehyde in the preceding section under aldehydes.

Ethylene Oxide

Ethylene oxide is a relatively simple organic compound having the formula

$$H_2C—CH_2$$
$$\diagdown \diagup$$
$$O$$

It is a liquid at temperatures below 10.8°C (51.4°F). Above this temperature it vaporizes rapidly. Vapors of this compound in air are highly flammable even in low concentrations. In this respect it is very much like diethyl ether. This objectionable feature was overcome by preparing mixtures of ethylene oxide in carbon dioxide or Freon, which are now available commercially. The carbon dioxide–ethylene oxide or Freon–ethylene oxide mixtures are nonflammable, and there is no alteration of the microbicidal activity of the ethylene oxide. The carbon dioxide and the Freon merely serve as inert diluents which prevent flammability.

Ethylene oxide is a unique and powerful sterilizing agent. Its use for sterilizing heat- or moisture-sensitive materials in hospitals, industry, and laboratories has become universal. Bacterial spores, which are many times more resistant than vegetative cells as measured by other antimicrobial agents, show little resistance to destruction by this agent. Figure 23-5 illustrates the sporicidal action of this gas. An outstanding and desirable feature of ethylene oxide is its power to

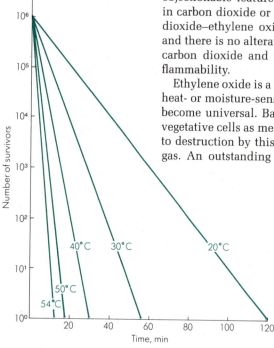

Figure 23-5. Decrease in numbers of *Bacillus subtilis* spores on paper strips surviving at various temperatures in gaseous ethylene oxide at 1200 mg/liter and 40 percent relative humidity. (*Courtesy of R. R. Ernst, "Ethylene Oxide Gaseous Sterilization for Industrial Applications," in G. B. Phillips and W. S. Miller (eds.) Industrial Sterilization, Duke, Durham, N.C., 1973.*)

penetrate. It will pass through and sterilize large packages of materials, bundles of cloth, and even certain plastics. It must be used with caution, although devices are available for its safe, routine laboratory use. The commercially available apparatus for this purpose is essentially an autoclave modified to allow the chamber to be filled with the gas under controlled conditions. The concentration of ethylene oxide, as well as the temperature and humidity, are critical factors which together determine the time required to achieve sterilization. Modern autoclaves are equipped with controls to maintain the desired concentration of ethylene oxide and the proper temperature and humidity.

An evaluation of the antimicrobial action of ethylene oxide and a comparison of its efficacy with other agents is shown in Table 23-5.

Practical Application. Ethylene oxide has been established as an effective sterilizing agent for heat- and moisture-sensitive materials. Effective usage requires careful control of three parameters: ethylene oxide concentration, temperature, and moisture. The varieties of materials on which it is used include spices, biological preparations, soil, plastics, certain medical preparations, and contaminated laboratory equipment. It has been used in the space program by both the Americans and the Russians for decontaminating spacecraft components. Among the advantages already mentioned for this agent as a sterilant is its remarkable penetration and its broad spectrum of activity against microorganisms, including spores. In addition, it is effective at relatively low temperatures,

Table 23-5. Evaluation of Selected Germicides

Class	Use Concentration	Activity Level*
Ethylene oxide (gas) (in autoclave-type equipment at 55 to 60°C)	450–800 mg/l	High
Glutaraldehyde, aq.	2%	High
Formaldehyde + alcohol	8% + 60–70%	High
Formaldehyde, aq.	3–8%	High to intermediate
Iodine + alcohol	0.5–70%	Intermediate
Alcohols	70–95%	Intermediate
Chlorine compounds	4–5%	Intermediate
Phenolic compounds	0.5–3%	Intermediate to low
Iodophors	75–150 ppm	Intermediate to low
Quaternary ammonia compounds	1:750	Low
Mercurial compounds	1:500–1:1000	Low

Interpretation of Activity Level

	Test Organism					
	Bacteria			Fungi‡	Viruses	
	Vegetative†	Tubercle Bacillus	Spores		Lipid and Medium-Sized	Nonlipid and Small
High	+	+	+	+	+	+
Intermediate	+	+	−	+	+	+
Low	+	−	−	+	+	−

* + means that cidal effect can be expected.
† Common forms of bacterial cells, e.g., *Staphylococcus*.
‡ Includes usual asexual spores but not necessarily dried chlamydospores and sexual spores.
SOURCE: Courtesy of E. H. Spaulding, Temple University.

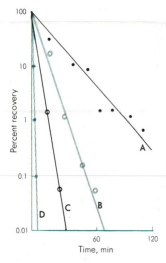

Percent recovery

Time, min

β-Propiolactone

Figure 23-6. Effect of β-propiolactone concentration on death rate of spores of *Bacillus subtilis* var. *niger*. Relative humidity, 80 ± 5 percent; temperature, 27 ± 2°C. Concentrations: (A) 0.1, (B) 0.2, (C) 0.4, (D) 1.6 mg/liter. (*Courtesy of R. K. Hoffman and B. Warshowsky, Appl Microbiol, 6:358, 1958.*)

and it does not damage materials exposed to it. One of its disadvantages is its comparatively slow action upon microorganisms.

Mode of Action. The mode of action of ethylene oxide is believed to be alkylation reactions with organic compounds such as enzymes and other proteins. Alkylation consists in the replacement of an active hydrogen atom in an organic compound, e.g., the hydrogen atom in a free carboxyl, amino, or sulfhydryl group, with an alkyl group. In this reaction the ring in the ethylene oxide molecule splits and attaches itself where the hydrogen was originally. This reaction would inactivate an enzyme with a sulfhydryl group:

$$H_2C-CH_2 + enzyme-SH \rightarrow enzyme-S-CH_2CH_2OH$$
$$\underset{O}{\diagdown\diagup}$$
$$\text{Inactive}$$

This compound is a colorless liquid at room temperature with a high boiling point (155°C) and has the formula

$$\begin{array}{c} CH_2-CH_2 \\ | \quad\quad | \\ O\quad\quad C \\ \text{β-Propiolactone} \end{array}$$

It is not flammable like ethylene oxide but is a vesicant and lachrymator and consequently must be handled with care. It lacks the penetrating power of ethylene oxide but is considerably more active against microorganisms; it is sporicidal, fungicidal, and virucidal. The sporicidal activity of β-propiolactone is shown in Fig. 23-6. Whereas the usual concentration of ethylene oxide for sterilization purposes is 400 to 800 mg/liter, only 2 to 5 mg of β-propiolactone is required. β-propiolactone is very effective in destroying microorganisms on surfaces. However, the fact that it has a low power of penetration coupled with its alleged carcinogenic properties has restricted its use as a practical sterilizing agent.

EVALUATION OF ANTIMICROBIAL CHEMICAL AGENTS

Laboratory techniques for the evaluation of antimicrobial chemical agents are conducted by one of three general procedures. In each the chemical agent is tested against a specified microorganism referred to as the **test organism.**

Tube-Dilution and Agar-Plate Techniques

1 Liquid water-soluble substances appropriately diluted are dispensed into sterile test tubes, to which are added a measured amount of the test organism. At specified intervals, a transfer is made from this tube into tubes of sterile media that are then incubated and observed for the appearance of growth. It is necessary in this type of procedure to ascertain whether the inhibitory action is bactericidal and not bacteriostatic. This approach can also be used to determine the number of organisms killed per unit time by performing a plate count on samples taken at appropriate intervals.

2 The chemical agent is incorporated into an agar medium or broth, inoculated with the test organism, incubated, and then observed for (a) decrease in the amount of growth or (b) complete absence of growth.

3 A plate of agar medium is inoculated with the test organism, and the chemical agent is placed on this medium. Following incubation, the plate is observed for a zone of inhibition (no growth) around the chemical agent. This is particularly suitable for semisolid preparations. It may also be used for liquid solutions, in which case the solution is first impregnated in absorbent paper or confined by a hollow cylinder placed on the agar surface.

For evaluation of gaseous substances, paper strips impregnated with known numbers of bacterial spores are exposed to the gas under prescribed conditions, after which they are cultured for survivors. These general procedures are illustrated in Fig. 23-7.

Phenol-Coefficient Method

A specific official test method based on the principle outlined in the first procedure above is the AOAC phenol-coefficient method, sometimes called the FDA method. (These abbreviations refer to the Association of Official Agricultural Chemists and the Food and Drug Administration, respectively.) This procedure is suitable for testing disinfectants miscible with water and exerting their antimicrobial action in a manner similar to that of phenol. The test organism employed in this procedure is a specific strain of either *Salmonella typhi* or *Staphylococcus aureus*. The temperature at which the test is performed, the manner of making subcultures, the composition of the subculture medium, the size of the test tubes, and other details of the test are spelled out in the official procedure. Briefly, the test is performed as follows.

To a series of dilutions of the disinfectant being tested (5 ml per tube), 0.5 ml of 24-h broth culture of the test organism is added. At the same time, similar additions, in the same amounts, are made to a series of dilutions of phenol. All tubes (disinfectant + organisms and phenol + organisms) are placed in a 20°C water bath. At intervals of 5, 10, and 15 min, subcultures are made with a loop transfer needle into sterile tubes of medium. The inoculated subculture tubes are incubated and subsequently examined for growth. The greatest dilution of the disinfectant killing the test organism in 10 min but not in 5 min is divided by the greatest dilution of phenol showing the same result. The number obtained by this division is the phenol coefficient of the substance tested. An example of the type of result obtained in this test and the method of calculation of the phenol coefficient are shown in Table 23-6.

It should be emphasized that no single microbiological test method is suitable for the evaluation of all germicidal chemicals for all applications recommended. Therefore, one must exercise care in selecting a test method for a specific chemical agent, so the results obtained will be meaningful and reproducible and lend themselves to some degree of practical interpretation. The ultimate criterion for the effectiveness of a germicidal agent is its performance under practical conditions. However, the laboratory test should provide a reliable index of its practical value.

GENERAL OBSERVATIONS

From the foregoing description of antimicrobial chemical agents, it is apparent that a large variety of substances is available for a diversity of applications. This

Figure 23-7. Laboratory evaluation of chemical antimicrobial agents: (A) No growth or growth in broth; (B) increased growth in broth as concentration of chemical agent is decreased; (C) increased growth in nutrient agar plates as concentration of chemical agent is decreased; (D) inhibition of growth by chemical agent applied to center of inoculated medium in Petri dish; zone of inhibition develops if compound is active. (*Courtesy of Procter and Gamble Company.*)

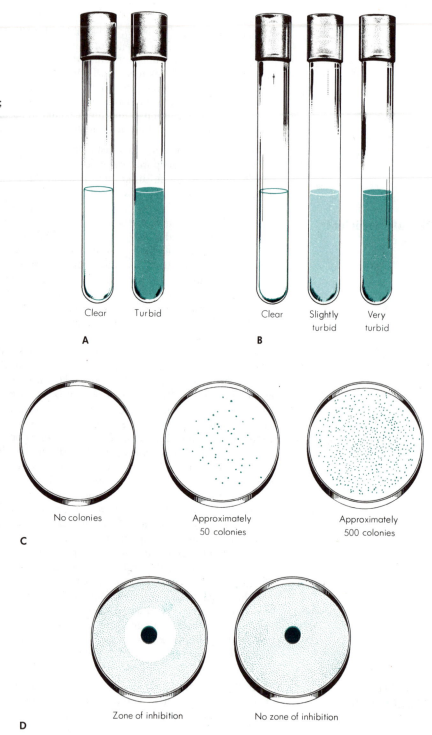

fact is compounded by the many proprietary products available to the public. Table 23-7 provides a summary of the various antimicrobial agents (or classes of agents) with a general statement of their applicability for practical usage. A summary of their modes of attack upon microbial cells is shown in Table 23-8.

Table 23-6. An Example of the Type of Result Obtained in the Phenol-Coefficient Method for Testing Disinfectants; Test Organism = *Salmonella typhi*

		Subculture Tubes*		
	Dilution	5 min	10 min	15 min
Disinfectant (X)	1:100	0	0	0
	1:125	+	0	0
	1:150	+	0	0
	1:175	+	+	0
	1:200	+	+	+
Phenol	1:90	+	0	0
	1:100	+	+	+

Phenol coefficient of (X) = $^{150}\!/_{90}$ = 1.6

NOTE: The phenol resistance of the test cultures must adhere to the following pattern:

	Phenol Dilution	5 min	10 min	15 min
Salmonella	1:90	+ or 0	+ or 0	0
typhi	1:100	+	+	+
Staphylococcus	1:60	+	0	0
aureus	1:70	+	+	

* 0 = no growth; + = growth.

Table 23-7. Application of Chemical Agents for Controlling Microorganisms

Chemical Agent	Recommended Use	Limitations
Phenol and phenolic compounds	General disinfectant	Microbial effectiveness limited; irritating and corrosive
Alcohols: ethyl and isopropyl	Skin and thermometer antiseptic	Antiseptic
Iodines	Disinfect skin	Irritating to mucous membranes
Chlorine	Water disinfection	Inactivated by organic material; pH dependent for effectiveness; objectionable taste and odor unless strictly controlled
Silver nitrate	Treating burns	Possible irritation
Mercurials	Skin disinfection	Slow-acting; toxic
Quaternaries	Skin disinfection	Not sporicidal
Formaldehyde	Sterilizing instruments; fumigation	Permeation poor; corrosive
Glutaraldehyde	Sterilizing instruments; fumigation	Stability limited
Ethylene oxide	Sterilizing heat-sensitive materials, instruments, and large equipment	Flammable; potentially explosive in pure form
β-propiolactone	Sterilizing instruments and heat-sensitive materials	Lacks penetrating power

Table 23-8. Sites of Action of Antimicrobial Chemical Agents Other Than Antibiotics

Sites of Action*	Acridine Dyes	Alcohols	Chlorine and Chlorine Compounds	Ethylene Oxide	Formaldehyde	Glutaraldehyde	Heavy-metal Salts	Iodine	Phenols	β-Propiolactone	Quaternary Compounds
Cell wall									+		
Cytoplasmic membrane		+							+		+
Proteins (denaturation)		+					+		+		+
Nucleic acids	+					+					
Enzymes with sulfhydryl (SH) groups			+	+		+	+	+		+	
Amino acids			+	+	+	+				+	

* In some instances the site of action is dependent upon the concentration of the chemical.

QUESTIONS

1 What is the distinction between the following terms?
   (a) Bactericidal and bacteriostatic
   (b) Sterile and disinfected
   (c) Virucidal and fungicidal
   (d) Germicidal and bactericidal
   (e) Sporicidal and bactericidal
2 List the major conditions influencing the effectiveness of antimicrobial chemical agents.
3 Compare the bactericidal property of phenol with that of other disinfectants of the phenolic type.
4 What relationship exists between various concentrations of ethyl alcohol and the higher alcohols?
5 List several halogens and compounds of halogens that are used to control microbial populations. Describe several practical applications for these agents. What is their mode of action upon microorganisms?
6 Give examples of the selective inhibition of dyes on microorganisms. Describe the use of such dyes in bacteriological media.
7 Explain the term cationic detergents. Describe their chemical structure; their practical use; and their mode of antimicrobial action.
8 What are the attractive features of sterilization by ethylene oxide? What kinds of materials are sterilized with ethylene oxide? Why?
9 A disinfectant is found to have a phenol coefficient of 3.0. What does this mean?
10 How would you demonstrate the antimicrobial capacity of an antiseptic ointment?

11 As a sterilizing agent, how does β-propiolactone compare with ethylene oxide?

12 Describe a laboratory experiment by which you could prove that the antimicrobial activity of a chemical in high dilutions was bacteriostatic.

13 Describe five different modes of antimicrobial activity exhibited by chemical antimicrobial agents. Identify the chemical agent involved in each example.

14 Why is it unlikely that a single "ideal" disinfectant agent will become available?

REFERENCES

Block, S. S. (ed.): *Disinfection, Sterilization, and Preservation*, 3d ed., Lea & Febiger, Philadelphia, 1983. *Fundamental principles and practical aspects of the control of microorganisms by chemical and physical methods to achieve disinfection, sterilization, and preservation. The mode of action of antimicrobial agents, bacterial resistance, and methods of testing are discussed. Each major category of chemical substances and physical agents is examined in terms of use and efficacy.*

Collins, C. H., M. C. Allwood, S. F. Bloomfield, and A. Fox (eds.): *Disinfectants: Their Use and Evaluation of Effectiveness*, Academic, New York, 1981. *This volume is one of a technical series (No. 16) published by the British Society for Applied Bacteriology. It provides brief descriptions of methods of testing disinfectants and describes the manner of using disinfectants in numerous environments such as the home, hospital, swimming pools, farm buildings, and milking areas.*

Favero, M. S.: *Sterilization, Disinfection, and Antisepsis in the Hospital*, in Lennette, E. H., et al. (eds.): *Manual of Clinical Microbiology*, 4th ed., Washington, American Society for Microbiology, 1985. *Practical information for control of microorganisms in hospitals.*

Hugo, W. B. (ed.): *Inhibition and Destruction of the Microbial Cell*, Academic, New York, 1971. *Covers inhibition and destruction of various groups of microorganisms (bacteria, fungi, viruses) by physical and chemical agents.*

Russell, A. D.: *The Destruction of Bacterial Spores*, Academic, New York, 1982. *A condensed, extensively referenced volume on the physical processes and chemical agents available for destruction of bacterial spores.*

Russell, A. D., W. B. Hugo, and G. A. J. Ayliffe (eds.): *Principles and Practice of Disinfection*, Blackwell Scientific Publications, Oxford, 1982. *A good general reference for methods used in disinfection.*

Sykes, G.: *Disinfection and Sterilization*, 2d ed., Lippincott, Philadelphia, 1965. *Contains practical information on the methods available for accomplishing sterilization and the major categories of chemical substances used for disinfectants.*

Chapter 24

Antibiotics and Other Chemotherapeutic Agents

The treatment of a disease with a chemical substance is known as **chemotherapy;** the chemical substance is called a **chemotherapeutic agent.** Chemotherapy has been practiced for centuries, but it was only early in the present century (the mid-1930s) that this kind of therapy revolutionized the field of medicine. This turn in events is attributed to two discoveries. The first was the finding that sulfonamide compounds (sulfa drugs) could be used successfully for the treatment of certain bacterial diseases. The second was the discovery of a new and potent class of antibacterially active chemotherapeutic agents, namely, antibiotics. In this chapter we present a brief history of chemotherapy and then discuss

510

the modern chemotherapeutic substances. Emphasis is given to antibiotics and particularly the manner in which they inhibit or kill microbial cells.

CHEMOTHERAPEUTIC AGENTS AND CHEMO-THERAPY

Chemotherapeutic agents are chemical substances used for the treatment of infectious diseases or diseases caused by the proliferation of malignant cells. These substances are prepared in the chemical laboratory or obtained from microorganisms and some plants and animals. In general, naturally occurring substances are distinguished from synthetic compounds by the name antibiotics. Some antibiotics are prepared synthetically, but most of them are prepared commercially by microbial biosynthesis. Antitoxins and other substances produced by the bodies of infected animals are not considered to be chemotherapeutic agents; the compounds discussed in Chap. 23 used for killing or inhibiting microbial growth in vitro are not classified as chemotherapeutic agents but usually as disinfectants, antiseptics, or germicides.

To be useful as a chemotherapeutic agent a substance *must* have **selective toxicity** for the parasite, which means a low toxicity for host cells and high toxicity for the parasite. In other words, the substance must damage the parasite and cause little or no damage to the cells of the host. For this and other reasons antiseptics and germicides such as phenol, coal-tar derivatives, and many mercurial compounds are unsatisfactory as chemotherapeutic agents. Germicides are not selective in their action on cells, and they interfere with such natural defense mechanisms as **phagocytosis;** since they do not penetrate cells and tissues well, they do not come into contact with the parasites; because they are inactivated by protein, their effectiveness is destroyed by body fluids rich in protein; and, finally, the tissues killed by the germicide or antiseptic provide an excellent medium for microorganisms to grow.

Since the reasons enumerated above make some chemical compounds useless in treating microbial infections, it follows that a satisfactory chemotherapeutic agent must:

1 Destroy or prevent the activity of a parasite without injuring the cells of the host or with only minor injury to its cells
2 Be able to come in contact with the parasite by penetrating the cells and tissues of the host in effective concentrations
3 Leave unaltered the host's natural defense mechanisms, such as phagocytosis and the production of antibodies

HISTORICAL HIGH-LIGHTS OF CHEMO-THERAPY

Quinine and Malaria

Europeans used natural quinine from the bark of the cinchona tree to treat malaria as early as 1630. It was used even earlier by South American Indians, who relieved symptoms of malarial fever by chewing the bark of the cinchona tree.

Ehrlich, Salvarsan, and Syphilis

Syphilis is the first known disease for which a chemotherapeutic agent was used. Mercury was used to treat syphilis as early as 1495, but it was not until about 1910, when an arsenical compound known as Salvarsan was synthesized by Paul Ehrlich (Fig. 24-1), that a specific drug capable of curing disease without great danger to the patient was developed. Ehrlich's contributions were especially important because his was the first systematic and deliberate search for a compound that had potent microbicidal properties, low toxicity for humans and other animals, and good chemical stability. For this important discovery he was awarded a share, with Elie Metchnikoff, of the 1908 Nobel prize in physiology and medicine. Ehrlich's compound has now been replaced in syphilis therapy by arsphenamine, neoarsphenamine, and other arsenical compounds and antibiotics.

Domagk, Sulfonamides, and Bacterial Infections

Because of the effectiveness of Ehrlich's drug, it is rather surprising that no further significant development in the synthesis of chemotherapeutic agents occurred until 1935, when Domagk showed the therapeutic value of a group of compounds known as the **sulfonamides.** These substances are not specific for a special group of organisms, as arsphenamine is for *Treponema*, but are effective against a large variety of pathogenic organisms. Sulfanilamide, the first compound in this group to be synthesized, was made by Gelmo in 1908, and in 1913 Eisenberg studied the bactericidal properties of azo dyes with a sulfonamide grouping. Possibly progress in this field was delayed because of the hesitancy of physicians to accept a type of therapy that Ehrlich referred to as "the chemical knife," since it really amounted to "cutting out" microbes from tissues. Ehrlich was justified in this designation, because the use of such drugs was attended by danger to the patient comparable to surgical risk. Fortunately, the new chemotherapeutic agents are not dangerous drugs since they are carefully studied in the laboratory followed by extensive clinical trials. A more realistic explanation for nonacceptance of Ehrlich's concept of chemotherapy is that it was then generally believed that the struggle between host and parasite was too complex to permit such a direct attack. It would be better, his opponents reasoned, to stimulate the host's defenses.

After Domagk's reports in 1935 in Germany and confirmatory work by investigators in other countries—notably England, France, and later the United States—interest in chemotherapy reached an all-time high. The compound on which Domagk reported was known as Prontosil. French chemists at the Pasteur Institute who studied its action on bacteria and attempted to improve it discovered that its antibacterial activity is due to the sulfanilamide moiety, previously synthesized and reported by Gelmo in 1908. This observation lighted the fuse for an explosive search for related compounds having therapeutic value. By 1945 it was estimated that several thousand derivatives of sulfanilamide had been made.

The important result of the search for new varieties of sulfonamides has been the development of drugs with increased antibacterial activities and fewer unfavorable reactions in the host animal. Some have been especially useful in certain types of infections, but sulfadiazine and sulfamerazine are extensively used because of their antibacterial effectiveness in a wide range of bacterial infections and because they are least likely to produce toxic reactions in the patient.

The sulfonamides are especially useful in the treatment of infections caused by meningococci and *Shigella*, respiratory infections caused by streptococci and staphylococci, and urinary infections due to Gram-negative organisms. They are

Figure 24-1. Paul Ehrlich is generally regarded as having established chemotherapy as a science. His research in the early 1900s resulted in the synthesis of an arsenical compound (Salvarsan) for the treatment of syphilis. His research represented a major contribution to the systematic search for new drugs.

useful in the prevention of rheumatic fever, bacterial endocarditis, wound infections, and urinary-tract infections following surgery or catheterization.

Antibiotics, Fleming, and Penicillin

Antibiotics are a special kind of chemotherapeutic agent usually obtained from living organisms. The word **antibiotic** has come to refer to a metabolic product of one microorganism that in very small amounts is detrimental or inhibitory to other microorganisms. It has been known for many years that antagonisms can exist between microorganisms growing in a common environment. The term **antibiosis** was first defined by Vuillemin in 1889 as a condition in which "one creature destroys the life of another in order to sustain his own, the first being entirely active and the second entirely passive; one is in unrestricted opposition to the life of the other." However, it can be seen that this definition is not entirely compatible with the present-day use of the term *antibiotics* proposed by Waksman in 1945 as applying to *those chemical substances of microbial origin which in small amounts exert antimicrobial activity*.

Antibiotics were known by their activities long before they were given the name by which we know them. Many years ago the Chinese used moldy soybean curd for the treatment of boils and controlled foot infections by wearing sandals furry with mold. In 1881, Tyndall reported that culture media cloudy with bacterial growth became clear when mold grew on the surface. Pasteur and Joubert found that pure cultures of anthrax bacilli grew well in urine but that when certain other organisms were present, the anthrax bacilli disappeared. This observation was related to that of Emmerich and Low, who demonstrated in 1901 that when liquid cultures of *Pseudomonas aeruginosa* were injected into rabbits, the animals were protected against anthrax. They called this material *pyocyanase* because they thought its activity was due to enzymes from *Bacillus pyocyaneus*, as *Ps. aeruginosa* was then called.

An early clinical application of bacterial antagonism was the use of lactobacilli in the treatment of dysentery, as recommended by Metchnikoff in 1899. This was an example of replacement therapy; i.e., a harmless microbe was able to eliminate and replace one that could cause disease. Modern antibiosis is based not on replacement but on utilization of an active inhibitory principle obtained from the antibiotic-producing microbes.

The first systematic search for, and study of, antibiotics, made by Gratia and Dath about 1924, resulted in the discovery of *actinomycetin* in strains of actinomycetes, soil organisms that are representative of the group that has given us a number of antibiotics since 1940. Actinomycetin was never used for the treatment of patients but was used to lyse cultures of bacteria for the production of vaccines.

In 1929 Alexander Fleming (Fig. 24-2A) noticed that an agar plate inoculated with *Staphylococcus aureus* had become contaminated with a mold and that the mold colony was surrounded by a clear zone, indicating inhibition of bacterial growth, or lysis of the bacteria (Fig. 24-2B). He was inspired to isolate and identify the mold and study its activities, but not until there was an urgent need for a better means of preventing death from infection of war wounds was the importance of Fleming's observation realized. With the aid of many investigators in England and the United States, and at a great deal of expense, the inhibitory substance from Fleming's "contaminant mold" became a "miracle drug." Be-

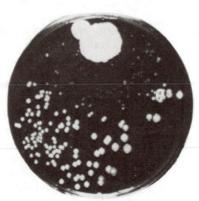

A   B

Figure 24-2. (A) Sir Alexander Fleming discovered the bacterial inhibitory properties of a metabolic product of *Penicillium notatum*. He called the substance penicillin. This discovery, in 1929, opened the era of antibiotics. For his contributions Fleming was knighted and shared the Nobel prize in physiology and medicine for 1945 with Ernst B. Chain, a chemist, and Sir Howard W. Florey, a physician. (B) Fleming's original plate demonstrated the inhibition of *Staphylococcus* (colonies at bottom) by a colony of *Penicillium notatum* (large white circle at top). This led to the discovery of penicillin. (*Courtesy of Robert Cruickshank.*)

cause the mold was identified as a *Penicillium* sp., Fleming called the antibiotic *penicillin*.

In 1939, René Dubos (Fig. 24-3) isolated from New Jersey soil a culture of *Bacillus brevis* which produced a substance that killed many Gram-positive bacteria. The cell-free extract produced from *B. brevis* by Dubos was found to contain two active principles now known as *gramicidin* and *tyrocidine*. These successes were followed closely by the discovery of *streptomycin* by Selman Waksman and associates.

Several thousand antibiotic substances have been isolated and identified since 1940. Many of them are of no practical importance as yet, but a few have changed the entire concept of chemotherapy. The popularity of antibiotics is due to their ability to destroy many kinds of pathogens and to their relatively nontoxic

Figure 24-3. In 1939, René Dubos isolated two antibiotics, gramicidin and tyrocidine, from a soil bacterium, *Bacillus brevis*. (*Courtesy of National Library of Medicine.*)

properties to the host when given systemically. Few developments in the field of medicine have had as dramatic an effect as have antibiotics in the treatment of microbial infections.

CHARACTERISTICS OF ANTIBIOTICS THAT QUALIFY THEM AS CHEMOTHERAPEUTIC AGENTS

To be useful as chemotherapeutic agents antibiotics must have the following qualities:

1 They should have the ability to destroy or inhibit many different species of pathogenic microorganisms. This is what is meant by a "broad-spectrum" antibiotic.
2 They should prevent the ready development of resistant forms of the parasites.
3 They should not produce undesirable side effects in the host, such as sensitivity or allergic reactions, nerve damage, or irritation of the kidneys and gastrointestinal tract.
4 They should not eliminate the normal microbial flora of the host, because doing so may upset the "balance of nature" and permit normally nonpathogenic microbes, or particularly pathogenic forms normally restrained by the usual flora, to establish a new infection. The broad-spectrum antibiotics, for example, may eliminate the normal bacterial flora but not *Monilia* from the intestinal tract. Under these conditions the *Monilia* may establish an infection that is not controlled by antibiotic therapy.

ANTIBIOTICS AND THEIR MODE OF ACTION

Antibiotics can be classified in several ways. For example, some are bactericidal and others are bacteriostatic. They may be grouped on the basis of chemical structure. A third way of classifying antibiotics is on the basis of their **mode of action,** that is the manner in which they manifest their damage upon microbial cells. Our discussion in this section will be organized on the basis of this latter manner of grouping antibiotics, i.e., their mode of action.

The major points of attack of antibiotics on microorganisms include:

Inhibition of cell-wall synthesis
Damage to the cytoplasmic membrane
Inhibition of nucleic acid and protein synthesis
Inhibition of specific enzyme systems

In the discussion which follows we will characterize only a few of the many antibiotics in each of these categories.

Inhibition of Cell-Wall Synthesis

Among the antibiotics whose antimicrobial activity is expressed by inhibition of the biosynthesis of the peptidoglycan cell-wall structure are the penicillins, cephalosporins, cycloserine, vancomycin, and bacitracin.

As you may recall from Chap. 11, the substance that gives rigidity to the cell wall is the peptidoglycan (see Figs. 11-6 and 11-7). The structure of this compound is essentially that of a series of strands (polymers with repeating units of N-acetylglucosamine and N-acetylmuramic acid) that are cross-linked with small peptides, with a frequency and in a manner that imparts considerable rigidity to the cell wall. It is a protective covering for the bacterial cell.

As you will note from the description of the biosynthesis of peptidoglycan in

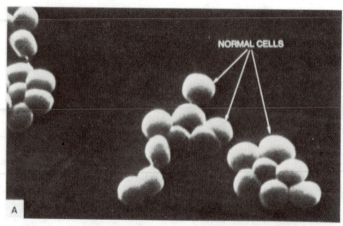

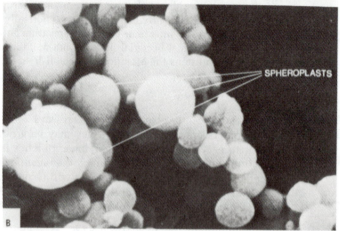

Figure 24-4. The morphologic changes, which occur in *Escherichia coli* as a result of exposure to penicillin, shown in this illustration, provide visual evidence that this antibiotic acts upon the cell wall. (A) Normal cells of *E. coli*. (B) *E. coli* cells after exposure to penicillin. Enlarged (bulged) defect indicates alteration of cell wall. (*Courtesy of Lilly Research Laboratories, Division of Eli Lilly and Company.*)

Chap. 11, the process involves numerous steps. Interference with any step in the sequence may inhibit cell-wall synthesis and result in the inability of the bacterium to survive because of the absence of a protective covering (cell wall).

Early experimental evidence suggested that some antibiotics exert their antimicrobial effect by inhibiting biosynthesis of the peptidoglycan polymer, resulting in the inhibition of cell-wall formation. Subsequent research identified the sequence of reactions in the biosynthetic pathway of the peptidoglycan and demonstrated that antibiotics like penicillin inhibited its formation. Some experimental observations that led to this conclusion can be summarized as follows:

1 Bacterial cells susceptible to penicillin can be protected from destruction if the medium in which they are exposed is of high osmotic pressure. The high osmotic pressure prevents the cells from bursting. Rod-shaped cells become spherical because they lack the cell structure which imparts shape. These cells without cell walls are called **spheroplasts.** (See Fig. 24-4).

2 Some species of bacteria such as the mycoplasmas lack the peptidoglycan structure and are not inhibited by penicillin.

3 Concentrations of penicillin below that which kill susceptible bacteria results in the accumulation of compounds that are precursors to peptidoglycan formation.

The Penicillins

The first of the modern antibiotics, and still one of the most useful, penicillin is produced by *Penicillium notatum* (Fig. 24-5), *Penicillium chrysogenum*, and by other species of molds. As previously noted, the first of these was isolated by Fleming in 1929, when he found it as a contaminant on a culture plate. Florey and his associates at Oxford University isolated the active ingredient and used the crude material clinically in 1940. Penicillin is selective for Gram-positive bacteria, some spirochetes, and the Gram-negative diplococci (*Neisseria*). Although it is rarely toxic in human patients, it may give rise to sensitivity reactions which vary from a mild skin reaction to severe anaphylaxis.

Penicillins are a class of β-lactam antibiotics of related structure with slightly different properties and activities. All penicillins have a common basic nucleus, a fused β-lactam-thiazolidine ring with different side chains which give each its unique properties (Fig. 24-6). Several chemically different penicillins are produced by biosynthesis in a single fermentation.

Natural penicillins can be prepared as salts of sodium, potassium, procaine, and other bases. The crystalline sodium or potassium salts are freely soluble in water, ethyl alcohol, ether, esters, and dioxane but only slightly soluble in chloroform and benzene. In pure crystalline form penicillins are colorless. The natural penicillins are inactivated by heat, cysteine, sodium hydroxide, penicillinase, and hydrochloric acid. They are not affected by the action of saliva or bile. Penicillin V exhibits greater stability than others in acids. Some of

Figure 24-5. *Penicillium notatum.* (A) A colony on agar-plate culture; magnification X3. (B) A microscopic view showing spores and mycelia; magnification X400.

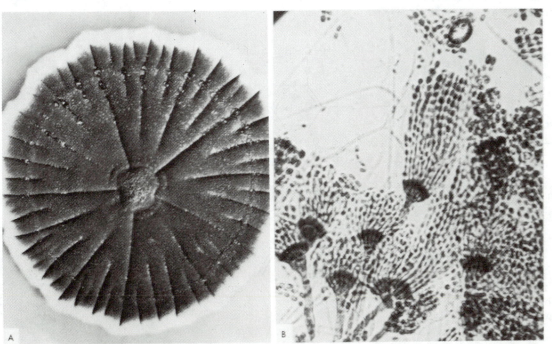

A   B

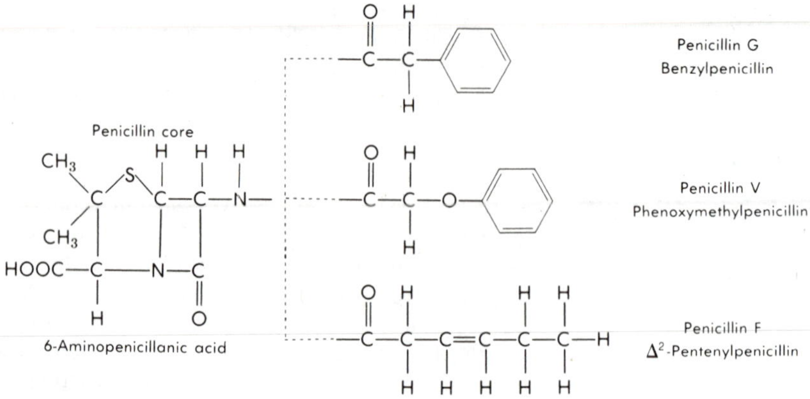

Penicillin G
Benzylpenicillin

Penicillin core

Penicillin V
Phenoxymethylpenicillin

Figure 24-6. Some "natural" penicillins showing the basic core of 6-aminopenicillanic acid with side chains which differ, thereby conferring special properties.

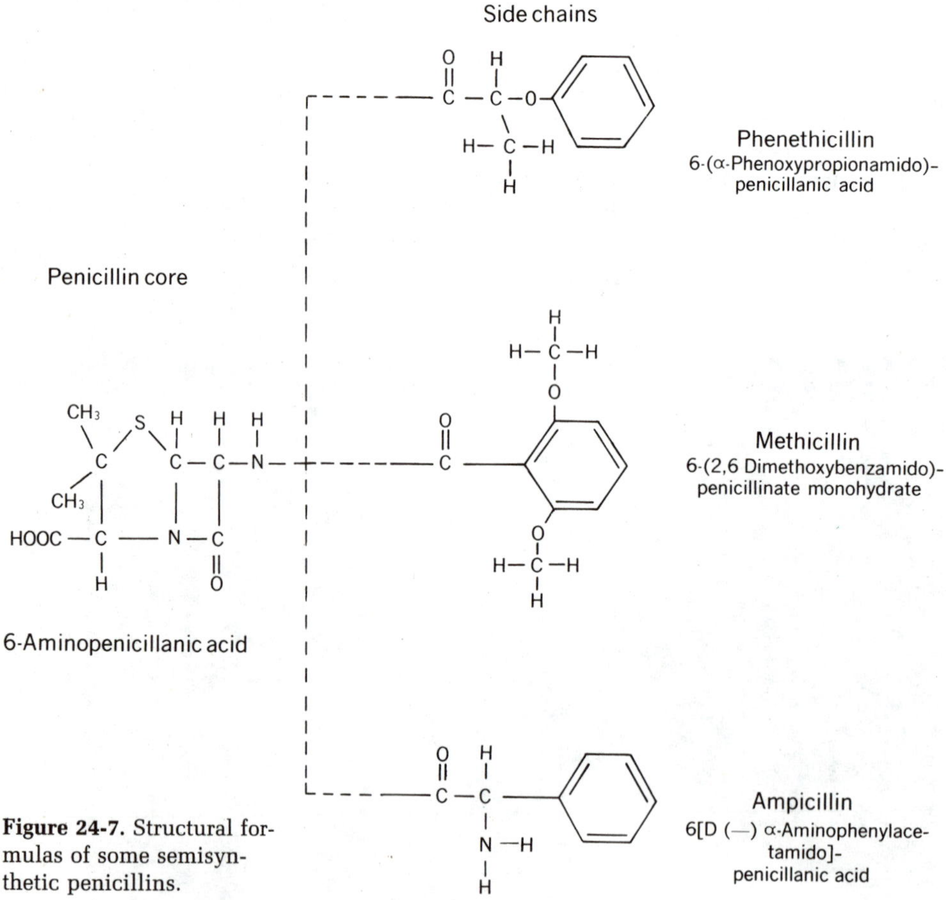

6-Aminopenicillanic acid

Penicillin F
Δ^2-Pentenylpenicillin

Side chains

Phenethicillin
6-(α-Phenoxypropionamido)-penicillanic acid

Penicillin core

Methicillin
6-(2,6 Dimethoxybenzamido)-penicillinate monohydrate

6-Aminopenicillanic acid

Figure 24-7. Structural formulas of some semisynthetic penicillins.

Ampicillin
6[D (—) α-Aminophenylacetamido]-penicillanic acid

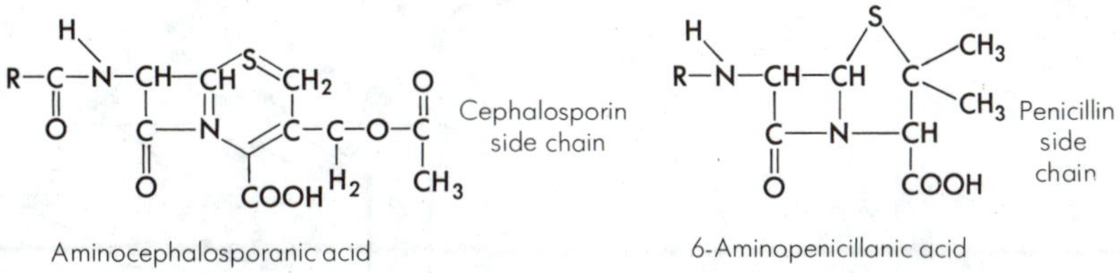

Figure 24-8. A comparison of the nucleus of cephalosporin (aminocephalosporanic acid) with the nucleus of penicillin (6-aminopenicillanic acid).

Aminocephalosporanic acid

6-Aminopenicillanic acid

the new semisynthetic penicillins may be much more stable than those produced by biosynthesis, the "natural penicillins."

The first break in the production of the new semisynthetic penicillins was the discovery that the basic nucleus of the molecule, common to all penicillins, is 6-aminopenicillanic acid. The next step was to obtain 6-aminopenicillanic acid in quantity so that suitable side chains could be attached to it. This was a very difficult task, but it was then discovered that under suitable conditions *P. chrysogenum* would produce the basic nucleus in abundance by "interrupted biosynthesis," and that the side chains could be removed from penicillin G, produced by biosynthesis, by amidase enzymes, leaving the 6-aminopenicillanic acid free for attaching new side chains as desired.

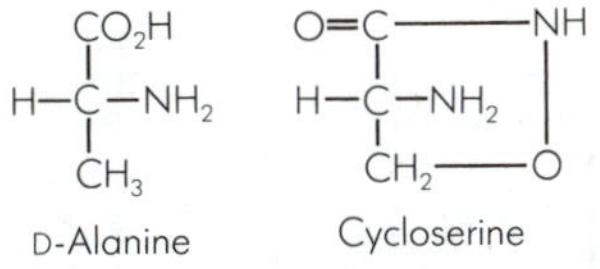

Figure 24-9. The chemical structures of D-alanine and the antibiotic cycloserine.

One of the first semisynthetic penicillins to be produced for clinical use was phenethicillin. It is more readily absorbed than penicillin V and just as effective as penicillin G. Another of the semisynthetic penicillins, methicillin, is more resistant to penicillinase and therefore is less likely to be inactivated.

Ampicillin

Ampicillin, another semisynthetic penicillin, acts against a broad spectrum of bacteria. It is strongly bactericidal and lacks toxicity, but it is not resistant to penicillinases. It is relatively stable to gastric acid and hence can be administered orally. The chemical structures of these three penicillins are shown in Fig. 24-7. Several additional semisynthetic penicillins have been developed for chemotherapeutic use.

Penicillins interfere with the final stages of peptidoglycan biosynthesis. The penicillins inhibit the transpeptidase reaction, namely, the cross-linking of the two linear polymers. The penicillins are bactericidal to growing cells.

Cephalosporins

Cephalosporins are a group of antibiotcs produced by a species of marine fungus, *Cephalosporium acremonium*, which bears considerable resemblance to *Penicillium* spp. They are effective against Gram-positive and Gram-negative bacteria. The cephalosporins have antibacterial properties similar to those of the semisynthetic penicillins. They are effective therapeutically and have a low toxisity. The nucleus of the cephalosporins (Fig. 24-8) resembles that of penicillin. As with penicillin, several semisynthetic cephalosporins have been manufactured commercially for therapeutic use.

As would be anticipated from the similarity in chemical structure of penicillin and cephalosporin, the mode of action of the cephalosporins is that of inhibition of the cross-linking transpeptidase. They are bactericidal to growing cells.

Cycloserine

Cycloserine, a relative simple compound, is related in structure to alanine (see Fig. 24-9). It was originally discovered as an antibiotic produced by streptomyces and is now manufactured through chemical synthesis. The main use of

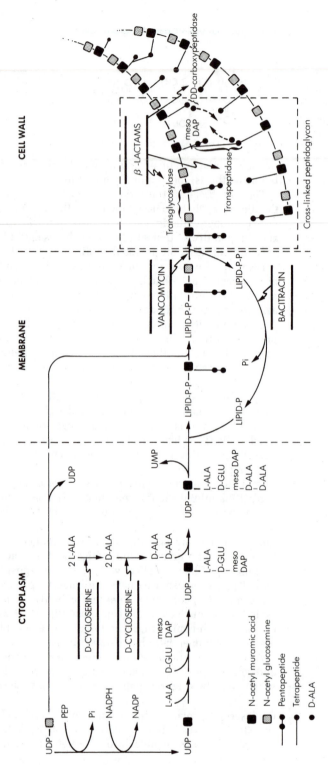

Figure 24-10. Schematic illustration of sites of attack of antibiotics on cell-wall synthesis (formation). (*Erwin F. Lessel, illustrator.*)

this antibiotic is in tuberculosis therapy. However, because of potential undesirable side effects, its utilization is limited.

Cycloserine manifests its inhibitory effect on peptidoglycan synthesis by interference with synthesis of the peptide moiety of the peptidoglycan. Specifically, it inhibits both alanine racemase and D-alanyl-D-alanine synthetase, the enzymes involved in the synthesis of the pentapeptide side chains.

Bacitracin

Bacitracin is a product of *Bacillus subtilis* and chemically is a polypeptide. Because of its toxicity to animal and human cells it cannot be used for systemic chemotherapy. It does have application for topical treatment of infections caused by Gram-positive bacteria.

Bacitracin interferes with regeneration of the monophosphate form of bactoprenol from the pyrophosphate form (lipid-P-P in Fig. 24-10).

Vancomycin

Vancomycin is an antibiotic produced by *Streptomyces orientalis*. It is a complex chemical entity consisting of amino acids and sugars.

Vancomycin inhibits peptidoglycan synthesis by binding the D-alanyl-D-alanine group on the peptide side chain of one of the membrane-bound intermediates.

A schematic summary of the modes of action of some antibiotics that exert their antibacterial effect through interference with cell-wall synthesis is shown in Fig. 24-10.

Damage to Cytoplasmic Membrane

Several polypeptide antibiotics produced by *Bacillus* spp. have the ability to damage cell-membrane structure. They adversely affect the normal permeability characteristics of the cell membrane. Included in this category are the **polymyxins, gramicidins,** and **tyrocidines** (see Fig. 24-11).

The polymyxins are particularly effective against Gram-negative organisms, while the tyrocidines and gramicidins are more effective against Gram-positive organisms.

These agents are bactericidal; they cause a leakage from the cytoplasmic content of the cell. Because of their toxicity to tissue they have limited application in chemotherapy.

Another category referred to as **polyene** antibiotics are large ring structures with many double bonds. Examples are *nystatin*, produced by *Streptomyces noursei*, and *amphotericin*, produced by *Streptomyces nodosus*. Polyene anti-

Figure 24-11. The structural formulas of tyrocidine A and gramicidin S, polypeptide antibiotics which exert their antibacterial action through binding with the cytoplasmic membrane. (Amino acid configuration is L except for those marked D.)

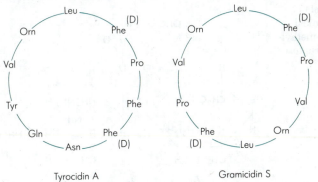

Tyrocidin A

Gramicidin S

biotics act upon cells which have sterols in their cytoplasmic membrane. They act upon fungi (including yeasts) and animal cells but do not affect bacteria. Their antimicrobial action is attributed to their ability to increase cell permeability.

Inhibition of Nucleic Acid and Protein Synthesis

The process by which the cell synthesizes nucleic acids and proteins was described in Chap. 11. As you will recall, synthesis of these substances involves a number of intricate biochemical reactions. It is recommended that these reactions be reviewed to better comprehend the mode of action of those antibiotics that interfere with these metabolic processes.

Examples of the major categories of antibiotics affecting nucleic acid and protein synthesis are described below.

Streptomycin

Streptomycin is produced by *Streptomyces griseus* (Fig. 24-12), a soil organism isolated by Schatz, Bugie, and Waksman, who reported on its antibiotic activities in 1944. It is particularly important because it inhibits many organisms resistant to sulfonamides and penicillin. Its antibacterial spectrum includes many Gram-negative bacteria, including *Francisella tularensis* and some organisms in the

Figure 24-12. *Streptomyces griseus*, the organism that produces streptomycin. This antibiotic inhibits the growth of certain Gram-negative pathogens and *Mycobacterium tuberculosis*. (*Courtesy of the Institute of Microbiology, Rutgers University.*)

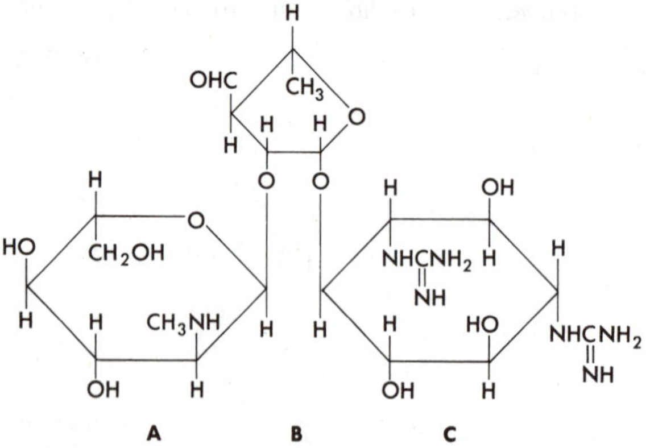

Figure 24-13. Streptomycin. This antibiotic consists of three components, linked glycosidically: (A) N-methyl-L-glycosamine, (B) streptose, and (C) streptidine.

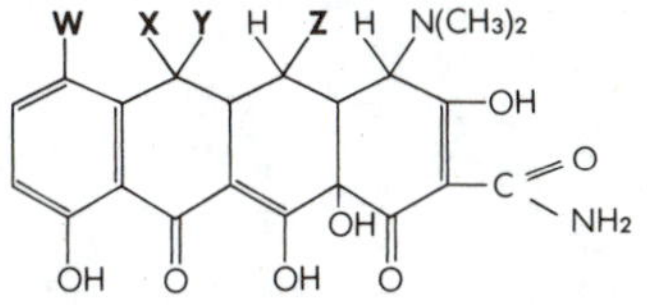

Figure 24-14. Tetracyclines, broad-spectrum antibiotics produced from *Strepto-myces*, differ slightly in chemical structure as shown in the positions labeled "w," "x," "y," and "z" in the above molecule.

Antibiotic	Position on molecule			
	W	X	Y	Z
Tetracycline	−H	−CH₃	−OH	−H
Oxytetracycline	−H	−CH₃	−OH	−OH
Chlortetracycline	−Cl	−CH₃	−OH	−H
Minocycline	−N(CH₃)₂	−H	−H	−H
Doxycycline	−H	−CH₃	−H	−OH

salmonella group. It is inhibitory for several species of *Mycobacterium*, including *Mycobacterium tuberculosis*. Highly purified streptomycin is nontoxic to humans and other animals when given in small doses, but it appears to have a cumulative detrimental effect on a specific region of the nervous system when given as a medication over long periods of time.

Streptomycin is characterized chemically as an **aminoglycoside** antibiotic; its structure is shown in Fig. 24-13. Other aminoglycoside antibiotics are kanamycin, produced by *Streptomyces kanamyceticus*, and neomycin, produced by *Streptomyces fradiae* and other species of streptomyces.

Streptomycin and other aminoglycoside antibiotics inhibit protein synthesis by combining irreversibly with the 30S subunit mRNA. Thus the normal synthetic sequence is disrupted.

Tetracyclines

Chlortetracycline, oxytetracycline, tetracycline, doxycycline, and *minocycline* are generic names for five antibiotics having similar biological and chemical properties. As a group they are commonly called **tetracyclines.** Their structural formulas are shown in Fig. 24-14. Note that the antibiotic produced by *Streptomyces aureofaciens* is chlortetracycline, while *Streptomyces rimosus* produces oxytetracycline. They are broad-spectrum antibiotics with similar antimicrobial spectra, and cross resistance of bacteria to them is common.

Hydrochlorides and bases of the tetracyclines are extremely stable as dry powders. In solutions, tetracycline retains its activity for 3 weeks or more, whereas chlortetracycline and oxytetracycline are less stable. As shown in Fig. 24-14, tetracycline, oxytetracycline, chlortetracycline, minocycline, and doxycycline are chemically very similar. It is not surprising, therefore, that there are no great differences in their activity. The antimicrobial spectra are similar, and all are bacteriostatic in their action. Organisms that are resistant to one are likewise resistant to the others. Tetracycline has a low order of toxicity in laboratory animals. It is readily absorbed from the intestinal tract; hence it is effective when given orally.

The tetracyclines inhibit protein synthesis through interference with the binding of aminoacyl-*t*RNA to the 30S subunit ribosome.

Chloramphenicol

Chloramphenicol is a broad-spectrum antibiotic active against many Gram-positive and Gram-negative bacteria. Its antimicrobial spectrum is similar to

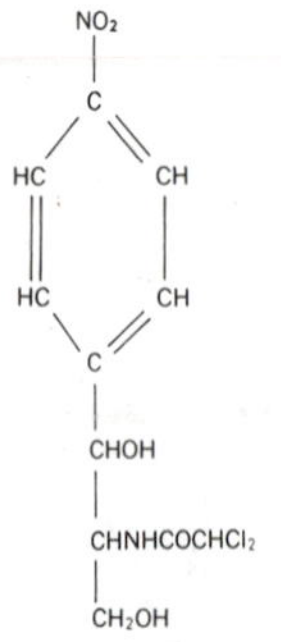

Figure 24-15. Structure of chloramphenicol, a broad-spectrum antibiotic from *Streptomyces venezuelae.*

that of tetracycline. It is also bacteriostatic. Chemically, it is a nitrobenzene ring with nonionic chlorine (Fig. 24-15). The possibility of serious side affects such as blood dyscrasias have limited the use of this antibiotic as a general antibacterial agent.

Chloramphenicol inhibits protein synthesis by combining with the 50S subunit ribosome. The transpeptidation and translocation functions associated with this site are blocked.

Erythromycin

Erythromycin is produced by a strain of *Streptomyces erythraeus* isolated from soil collected in the Philippines. Erythromycin is active against the Gram-positive bacteria, some Gram-negative bacteria, and pathogenic spirochetes. With regard to antimicrobial spectrum and clinical usefulness, it resembles penicillin, but it is also active against organisms that become resistant to penicillin and streptomycin. It is, therefore, often prescribed to those patients with allergies when penicillin is indicated.

Erythromycin belongs to the chemical class of antibiotics known as **macrolides.** Structurally it contains a large lactone ring linked with amino sugars through glycosidic bonds. Erythromycin inhibits protein synthesis as a result of binding on the 50S subunit ribosome; the steps of transpeptidation and translocation in protein synthesis are blocked.

The specific steps in protein synthesis which are interrupted by various antibiotics are summarized in Fig. 24-16.

Inhibition of Specific Enzyme Systems

The sulfonamides, which were discussed earlier in this chapter for their role in the development of chemotherapy, represent a category of compounds whose antibacterial attack is directed toward a specific essential enzyme. There are numerous sulfonamides as shown in Table 24-1. All of them have the same basic structure. This structure is related to the compound p-aminobenzoic acid. Many bacteria require p-aminobenzoic acid (PABA) as a precursor to their synthesis of the essential coenzyme tetrahydrofolic acid (THFA). PABA is a structural part of the THFA acid molecule. The selective action of sulfonamides is explained by the fact that the PABA molecule and a sulfonamide molecule are so very similar that the sulfonamide may enter the reaction in place of the PABA and block the synthesis of an essential cellular constituent, which in this case is THFA, as shown in Fig. 24-17. The cellular functions of the THFA coenzyme include amino acid synthesis, thymidine synthesis, etc. Lack of this

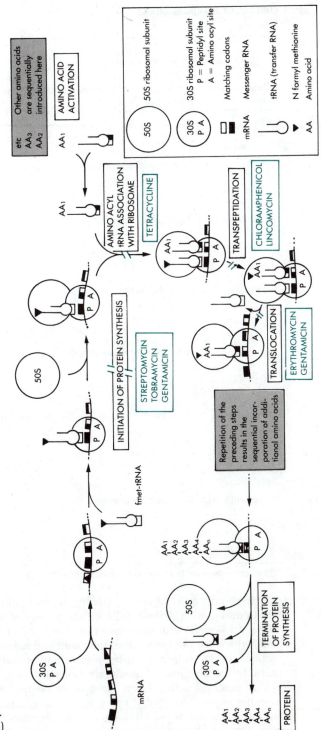

Figure 24-16. Schematic illustration of sites of action of antibiotics on the sequence of protein synthesis. (*Erwin F. Lessel, illustrator.*)

Table 24-1. Some Examples of Sulfonamides

The basic structure for the sulfonamides is H_2N—⟨ring⟩—SO_2—$N\begin{smallmatrix}H\\R'\end{smallmatrix}$ which is a p-aminobenzenesulfonamide. The sulfonamides differ primarily by virtue of the different substituents in the R' position, as indicated.

Name*	R'
Sulfanilamide	H
Sulfapyridine (N'-2-Pyridylsulfanilamide)	Pyridine
Sulfathiazole (N'-2-Thiazolysulfanilamide)	Thiazole
Sulfadiazine (N'-2-Sulfanilamidopyrimidine)	Pyrimidine
Sulfamerazine [N'-(4-Methyl-2-pyrimidyl)-sulfanilamide]	4-Methylpyrimidine
Sulfamethazine [N'-(4,6-Dimethyl-2-pyrimidyl)-sulfanilamide]	4,6-Dimethylpyrimidine
Sulfaguanidine (N'-Guanylsulfanilamide)	Guanidyl

* The common name is followed by the systematic name.

coenzyme will quite obviously disrupt normal cellular activity. Sulfonamides will inhibit growth of those cells which synthesize their THFA from PABA and will not interfere with the growth of those cells (including mammalian host cells) which require the vitamin folic acid and reduce it directly to THFA. This accounts for the selective antibacterial action of sulfonamides and makes them useful in the treatment of many infectious diseases.

This mode of action is an example of competitive inhibition between an essential metabolite (PABA) and a metabolic analog (a sulfonamide). After the antimicrobial activity of sulfonamides was discovered, D. D. Woods, an English bacteriologist observed that its effect could be reversed by PABA. Although, in

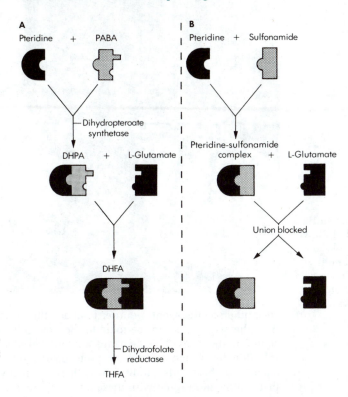

Figure 24-17. The mode of action of sulfonamides in inhibition of tetrahydrofolic acid synthesis. (*Erwin F. Lessel, illustrator.*)

1940, PABA was unknown as a bacterial metabolite, Woods predicted the mode of action described above.

ANTIFUNGAL ANTIBIOTICS

Nystatin is an antifungal agent useful in the therapy of nonsystemic fungal infections. It is produced during fermentation by a strain of *Streptomyces noursei*. This antibiotic was discovered in 1950 by Elizabeth Hazen and Rachel Brown (see Fig. 24-18).

Nystatin

The antimicrobial activity of nystatin is restricted to yeasts and other fungi, e.g., *Candida, Aspergillus, Penicillium,* and *Botrytis*; it is fungicidal in action. Chemically, nystatin is a polyene with an empirical formula of $C_{46}H_{75}NO_{18}$.

Griseofulvin

Griseofulvin is obtained from *Penicillium griseofulvin*. It is used in the treatment of many superficial fungus infections of the skin and body surfaces and is also effective in the treatment of some systemic (deep-seated) mycoses. The drug is administered orally.

ANTIVIRAL CHEMO-THERAPEUTIC AGENTS

Antibiotics such as those that we have discussed are generally not effective against viruses. You will recall that viruses are intracellular, and hence the

Figure 24-18. Elizabeth Hazen (left) and Rachel Brown examine early samples of the first antifungal antibiotic, nystatin. (*Courtesy of Research Corporation.*)

chemotherapeutic agent, in order to attack the virus, must enter the host cells. Also the agent must not be toxic to the host cell while exerting an inhibiting action on the virus. This demands a high level of selective toxicity. In cases of infection by bacteria, fungi, or protozoa the infectious agent is acted upon outside the host cells. Additionally, there are many more metabolic processes that can be interrupted with these microorganisms.

Among the more promising of the chemotherapeutic agents for treating viral diseases is **interferon.** Interferons are small glycoprotein substances of which two types are leukocytic interferon and fibroblast interferon. Cells exposed to interferon develop antiviral properties. The antiviral action of interferon is attributed to interference of protein synthesis.

Natural interferons are in very short supply and are expensive. Recent advances in recombinant DNA techniques (genetically engineered bacteria like *Escherichia coli* to produce interferon on a large scale commercially) have increased the availability of interferon for both chemotherapeutic and experimental use.

Acycloguanosine is a nucleoside analog which is active against the herpes virus in animals. Its mode of action appears to be that of inhibition of nucleotide utilization. A synthetic nucleotide analog, 5'-iododeoxyuridine (see Fig. 24-19),

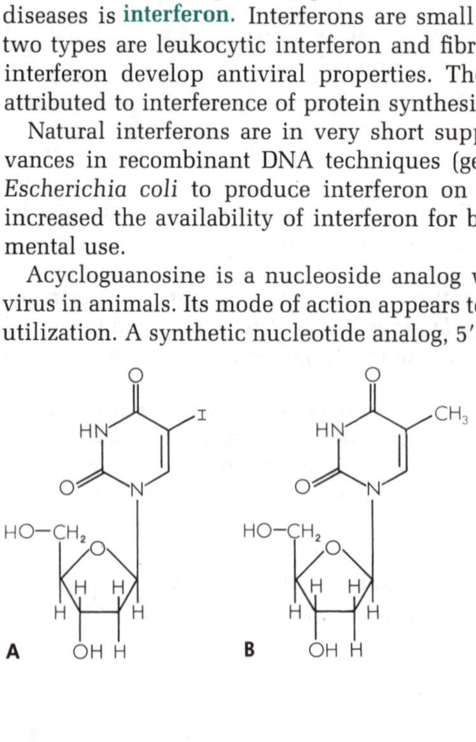

Figure 24-19. (A) The pyrimidine analogue 5'-iododeoxyuridine. (B) Thymidine, the pyrimidine which the analog resembles. This pyrimidine analog exhibits antiviral activity.

has been shown to have antiviral activity and promise as an antiviral chemotherapeutic agent. Its mode of action is most likely that of inhibition of nucleic acid synthesis—preventing the incorporation of thymidine into DNA.

Amantadine is a low-molecular-weight compound which is very effective against influenza A virus; it is not effective against influenza B. The incidence of influenza A infections is greatly reduced by use of this drug. The mode of action of amantadine is that of interfering with the uncoating of virus particles and the subsequent release of their nucleic acids.

ANTITUMOR ANTIBIOTICS

Some antibiotics have been found to possess antitumor activity. The **anthramycin** group (anthramycin, sibromycin, tomaymycin, and neothramycin) is an example of potent antitumor agents. See Fig. 24-20. One of the complicating factors associated with the potential use of these anticancer agents is that they are also cardiotoxic, a fact that illustrates the need for a high level of specificity in a chemotherapeutic agent. The antitumor action of these antibiotics is directed toward DNA structure and function. One of the problems is that of determining whether, through the manipulation of the structure of an antibiotic, e.g. anthramycin, one can cut out the cardiotoxic property without destroying the antitumor property.

SYNTHETIC CHEMOTHERAPEUTIC AGENTS

Nitrofurans

The **nitrofurans** are antimicrobial drugs which differ from the antibiotics in that they do not occur naturally. The prototype of the nitrofuran derivatives is furfural, which can be prepared from corncobs and cornstalks, oat hulls, beet pulp, peanut hulls, and other vegetable by-products. Furfural, an aldehyde derivative of furan chemically known as 2-furaldehyde, was identified in 1832 as an accidental finding during sugar-distillation studies.

However, it was not until 1944 that the American investigators Dodd and Stillman reported on the discovery of the antimicrobial properties of nitrated

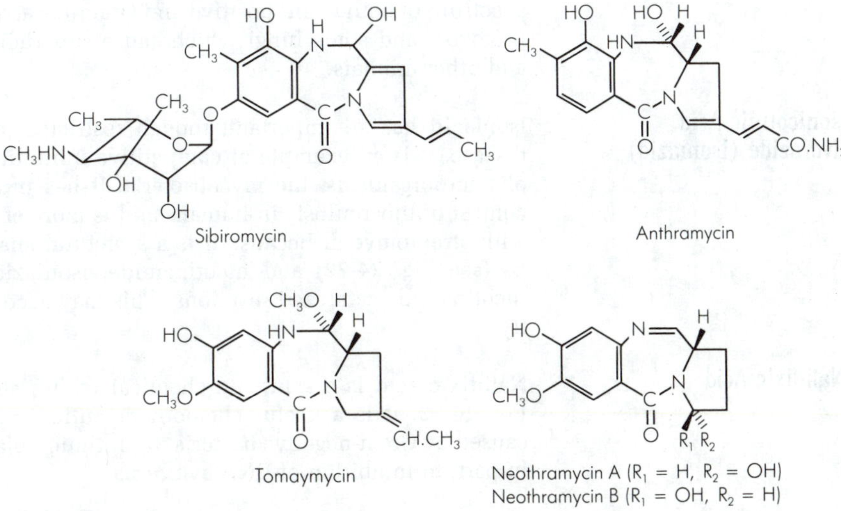

Figure 24-20. Structure of the anthramycin group of antitumor antibiotics.

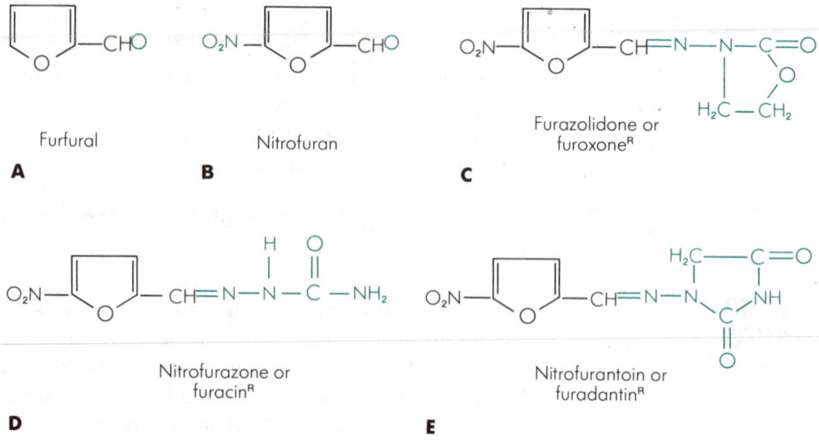

Furfural

A

Nitrofuran

B

Furazolidone or
furoxone^R

C

Figure 24-21. Furfural (A) is the prototype of nitrofuran compounds, and (B), (C), (D), and (E) are chemotherapeutic derivatives of furfural.

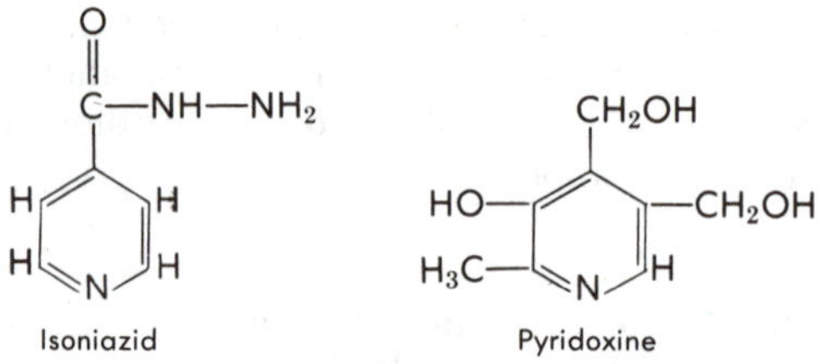

Nitrofurazone or
furacin^R

D

Nitrofurantoin or
furadantin^R

E

Figure 24-22. Isoniazid, a structural analog of pyridoxine (vitamin B$_6$), may prevent the growth of microorganisms by blocking pyridoxine-catalyzed reactions in the microbial cell.

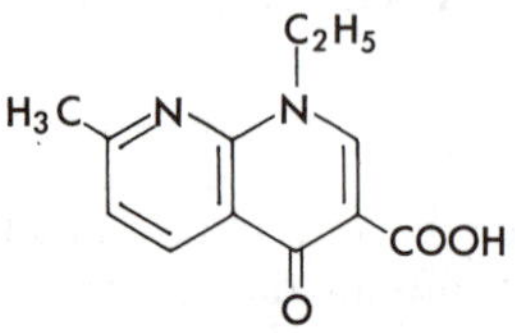

Isoniazid

Pyridoxine

Figure 24-23. Nalidixic acid is a synthetic antibacterial drug with a selective action against bacterial DNA synthesis.

furan derivatives. They found that a surprisingly high antibacterial effect was conferred upon furans by the addition of a nitro group in the 5-position of the furan ring. Variations in the side chain in the 2-position of 5-nitrofuran provide a wide spectrum of compounds in this group. Over 1,000 compounds have been synthesized and studied.

The chemical structures of some chemotherapeutic nitrofurans are shown in Fig. 24-21. As a class, the nitrofurans generally are effective against a broad spectrum of both Gram-positive and Gram-negative bacteria, several pathogenic protozoa, and some fungi which cause superficial infections in both humans and other animals.

Isonicotinic Acid Hydrazide (Isoniazid)

Isoniazid has an important, though restricted, application in the therapy of disease. It is an example of competitive inhibition affecting a restricted group of microorganisms, the mycobacteria. It has proved to be very useful in the control of tuberculosis in humans and is more effective when given alternately with streptomycin. Because it is a structural analog of pyridoxine, or vitamin B$_6$ (see Fig. 24-22) and nicotinamide, isoniazid can block pyridoxine- and nicotinamide-catalyzed reactions. This may account for its antimicrobial activity.

Nalidixic Acid

Nalidixic acid is a synthetic chemical with a structural formula as shown in Fig. 24-23. It is a useful chemotherapeutic agent for urinary-tract infections caused by Gram-negative bacteria. Its antimicrobial activity is attributed, at least in part, to inhibition of DNA synthesis.

DEVELOPMENT OF RESISTANCE TO ANTIBIOTICS

Drug resistance is one of nature's never-ending processes whereby organisms develop a tolerance for new environmental conditions. Drug resistance may be due to a preexisting factor in the microorganism, or it may be due to some acquired factor(s). Penicillin resistance, for example, may result from the production of penicillinase by resistant organisms, which converts penicillin into inactive penicilloic acid. On the other hand, some normally susceptible strains of bacteria may acquire resistance to penicillin. Acquired resistance is also due to penicillinase production in genetically adapted varieties of microorganisms. In cultures of penicillin-sensitive bacteria, perhaps one organism in a hundred million may be a penicillin-resistant mutant. Normally the ratio of sensitive to resistant organisms is maintained, and no problem develops. When penicillin is present, the sensitive strains do not reproduce whereas the resistant mutants do and eventually dominate the population. This has important clinical implications and is one of the practical reasons why research effort has been made to develop synthetic penicillins which are resistant to the action of penicillinase.

Many organisms which do not produce penicillinase are also resistant to penicillin. This suggests an alternative metabolic pathway or enzyme reaction not susceptible to inhibition by penicillin.

Other mechanisms of drug or antibiotic resistance may be due to (1) competitive inhibition between an essential metabolite and a metabolic analog (drug), (2) development of an alternate metabolic pathway which bypasses some reaction that would normally be inhibited by the drug, (3) production of an enzyme altered in such a way that it functions on behalf of the cell but is not affected by the drug, (4) synthesis of excess enzyme over the amount that can be inactivated by the antibiotic or drug, (5) inability of the drug to penetrate the cell due to some alteration of the cell membrane, (6) alteration of ribosomal protein structure.

An illustration of the pattern of resistance of bacteria to antibiotics over a three-year period is shown in Table 24-2.

Table 24-2. Occurrence of Single and Multiple Resistance in S. aureus in Hospitalized Patients during the Period 1977–1980

Resistance Pattern*	Resistant, %	Resistance Pattern*	Resistant, %
XX	91	XX CT	3.6
CT	4	XX CX	4.6
TC	44	XX EM FU LM	0.012
EM	10	XX EM FU LM GM	0.00083
SU	94	XX LM	0.91
CX	5	CX LM	0.05
FU	13	CX GM	0.35
LM	1	CT EM	0.40
GM	7	TC EM	4.4
XX TC EM	4	TC EM SU	4.1
XX TC CX	1.9	TC GM	3.1
XX TC EM CX	0.20	TC FU	5.7
XX TC EM CX FU	0.026	EM CX	0.5
XX EM	9.1	EM LM	0.10
XX TC	40	FU LM	0.13

* XX = β-lactamase production; CT = cephalothin; TC = oxytetracycline; EM = erythromycin; SU = sulfadimidine; CX = isoxazolylpenicillins (and methicillin); FU = fusidic acid; LM = lincomycin; GM = gentamicin; CL = chloramphenicol.

SOURCE: T. Bergan and J. Lernestedt, "Antibiotic Resistance in Staphylococci from a Hospital Environment," *Chemotherapy,* **29:**28–36, 1983.

Transmission of Drug Resistance

When chemotherapeutic agents such as the sulfonamides and antibiotics were first used, development of bacterial resistance was rather infrequent. Resistance became much more of a problem as the widespread use of antibiotics led to the elimination of sensitive organisms from the population with the accompanying increase in the numbers of resistant organisms.

The initial appearance of resistant organisms was thought to be the result of a change in a single bacterial gene that conferred resistance to the bacterium. The evidence that this takes place during sulfonamide therapy is convincing. Another, more recent, explanation for the development of resistance, at least in some Gram-negative bacteria, is that resistant organisms have an additional gene whose function is to protect the bacterium from the bactericidal effect of the drug or antibiotic. For example, such a gene is responsible for penicillinase production by penicillin-resistant staphylococci. In some instances bacteria carry the resistant gene at the time of infection, and their propagation is encouraged while sensitive strains are inhibited or killed. At other times the resistant gene is transmitted by conjugation from other bacteria during treatment.

Gene transfer between cells has been explained in Chap. 12 as being accomplished by transformation, transduction, or conjugation. Here we are interested in the transfer of antibiotic resistance by conjugation. This phenomenon was first reported independently in 1958 by two Japanese scientists, Akiba and Ochiai. They isolated both antibiotic-sensitive and antibiotic-resistant organisms of the same serotype from patients with enteric infections being treated with sulfonamides, tetracyclines, streptomycin, or chloramphenicol. They went on to demonstrate that this was due to resistant genes in a reservoir of *Escherichia coli* in the intestinal tract being transferred to *Shigella dysenteriae* that caused the infection. Since then transfer of antibiotic resistance by bacterial conjugation has been observed in other organisms and in other parts of the world.

We now know that this resistance factor, or R factor, is present in plasmids, which are small extrachromosomal self-replicating extranuclear DNA units (Chap. 12). The transmission of resistance factors in enteric infections is especially important in places where such infections are common. Organisms that are good recipients of the R factor from an *E. coli* donor include species of *Enterobacter, Klebsiella, Salmonella*, and *Shigella*. Weak recipients are species of *Pasteurella, Proteus*, and *Serratia*.

Antibiotic resistance represents a serious problem for clinicians, and great effort is being made to understand the mechanisms involved and to prevent its occurrence. The development of resistance can be minimized by (1) avoiding the indiscriminate use of antibiotics where they are of no real clinical value, (2) refraining from the use of antibiotics commonly employed for generalized infections for topical applications, (3) using correct dosages of the proper antibiotic to overcome an infection quickly, (4) using combinations of antibiotics of proved effectiveness, and (5) using a different antibiotic when an organism gives evidence of becoming resistant to the one used initially.

MICROBIOLOGICAL ASSAY OF ANTIBIOTICS

The potency of antibiotic content in samples can be determined by chemical, physical, and biological means. An **assay** is made to determine the ability of an antibiotic to kill or inhibit the growth of living microorganisms. Biological tests offer the most convenient means of making an assay.

Chemical Assay

Where antibiotics exist in pure chemical form, their concentration can be expressed in micrograms of the pure chemical per milligram of the specimen. To be of value, such tests must give results that correlate well with those obtained in biological assays. Chemical-assay methods are generally more accurate and require less time than biological methods, but they are less sensitive, and caution must be used lest biologically inactive degradation products give misleading results.

Biological Assay

Biological potency is expressed in terms of either micrograms or units determined by comparing the amount of killing, or bacteriostasis, of a test organism caused by the substance under test with that caused by a standard preparation under rigidly controlled conditions (see Fig. 24-24).

Although the unit of measurement for some antibiotics is arbitrary, it is established by international agreement for some antibiotics and by the FDA regulation for others. For example, the international unit of penicillin is the

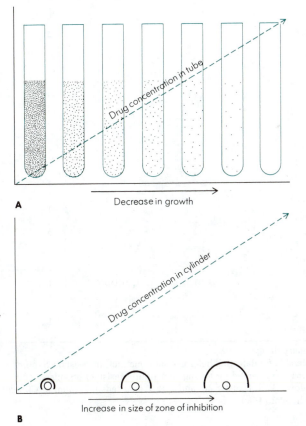

Figure 24-24. Microbiological assay of antibiotics and some other chemotherapeutic agents is accomplished by either the tube-dilution or the cylinder-plate method (a variation of the paper-disk-plate technique). (A) In the tube-dilution technique, the inhibition of growth (decrease in turbidity) produced by the unknown sample is compared with that produced by the known or standard sample. The amount of antibiotic present in the unknown sample can then be calculated. (B) The cylinder-plate technique follows much the same procedure for determining antibiotic potency, except that inhibition of growth is measured in terms of the size of the zones of inhibition.

amount of activity produced under defined conditions by 0.5988 μg of the International Standard, which is a sample of pure benzyl-penicillin (1 mg = 1,667 units). Some modifications of the method referred to above measure the interference of the antibiotic with the production by the test organism of a characteristic metabolic product such as acid, hemolysin, or the enzyme reductase.

The assay of antibiotics in blood serum, urine, tissues, and other similar substances presents some special problems because (1) the amounts present are generally very small as compared with other substances, (2) the antibiotic may be bound to proteinaceous materials in the specimen, (3) normal inhibitory substances may be present in the blood or other body fluids. Therefore, the techniques described above are modified to make them more sensitive for these specimens.

Table 24-3. Basic Sets of Antimicrobial Agents to be Tested Routinely Against Rapidly Growing Aerobic and Facultatively Anaerobic Bacteria*

Antimicrobics	Staphylococci and Streptococci	Enterococci	Enteric Gram-Negative Bacilli Other than *P. aeruginosa*		*P. aeruginosa*
			Urinary	Other	
Penicillins					
Ampicillin		1	1	1	
Carbenicillin			1	1	1
Nafcillin, oxacillin, or methicillin†	1				
Penicillin G	1	1			
Cephalosporins					
Cefamandole	2		1‡	1‡	
Cefoxitin	2		1‡	1‡	
Cephalothin	1		1	1	
Chloramphenicol	2	2		2	
Clindamycin	1				
Erythromycin	1	1			
Aminoglycosides					
Amikacin			1	1	1
Gentamicin	2		1	1	1
Kanamycin	2		1	1	
Tobramycin			1	1	1
Polymyxin B or E			2	2	1
Tetracycline	2	2	1	1	
Vancomycin	2				
Urinary tract agents					
Nalidixic acid			1		
Nitrofurantoin			1		
Sulfonamides			1		
Sulfamethoxazole/tri-methoprim			1		

* 1 = primary set; 2 = secondary drugs.

† Oxacillin or nafcillin is preferable for detecting heteroresistant methicillin-resistant *S. aureus*.

‡ Cefamandole and cefoxitin may be reserved for testing cephalothin-resistant organisms only.

SOURCE: E. H. Lennette, A. Balows, W. J. Hausler, Jr., and J. P. Truant (eds.): *Manual of Clinical Microbiology*, 3d ed., American Society for Microbiology, Washington, 1980.

MICROBIAL SUSCEPTI- BILITY TO CHEMO- THERAPEUTIC AGENTS

Species and strains of species of microorganisms have varying degrees of susceptibility to different antibiotics. Furthermore, the susceptibility of an organism to a given antibiotic may change, especially during treatment. It is therefore important for the clinician to know the identity of the microbe and the specific antibiotic which may be expected to give the most satisfactory results in treatment (Table 24-3). For this information the microbiologist will be called upon to make an accurate microbiological diagnosis and to determine the susceptibility of the organism to various antibiotics. From time to time during the course of therapy the microbiologist may be required to make estimates of any change in the susceptibility of the pathogen to the drug, and possibly even to assay the antibiotic concentration in the body fluids.

Tube-Dilution Technique

The susceptibility of a microorganism to antibiotics and other chemotherapeutic agents can be determined by either the **tube-dilution** or the **paper-disk-plate** technique. By the tube-dilution technique, one can determine the smallest amount of chemotherapeutic agent required to inhibit the growth of the organism in vitro (see Fig. 24-25). This amount is referred to as the **MIC (minimal inhibitory concentration).**

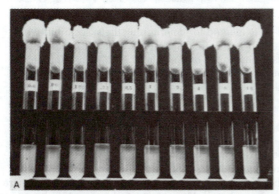

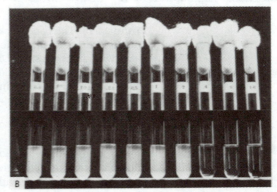

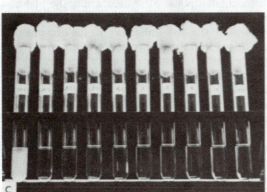

Figure 24-25. Susceptibility of microorganisms to antibiotics can be determined by the tube-dilution technique. Increasing amounts of the antibiotics under examination are placed in a series of culture tubes containing a suitable broth medium inoculated with the test organism. The dosage of drugs used in the test illustrated is indicated by the labels on the tubes. The tube at the extreme left is the control tube and contains no antibiotics. After incubation, the concentration of drugs required to inhibit the growth of the organism used is determined by observing the absence of growth. In the upper rack of tubes (A), no inhibition is observed at any drug concentration used. In (B) growth was inhibited by 4 μg of drug. (C) The organism was inhibited by all concentrations of the drug. (*Courtesy of Abbot Laboratories.*)

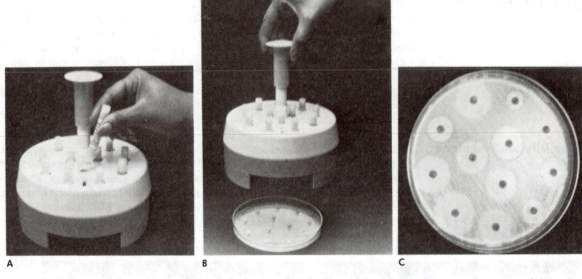

A   B   C

Figure 24-26. The paper-disk-plate method for determining the susceptibility of microorganisms to antibiotics. (A) Automatic dispenser of paper disks impregnated with antibiotics. (B) Disks positioned on inoculated Petri dish before incubation. (C) Zones of inhibition develop after incubation around each disk that contains an antibiotic that inhibits growth of microorganisms. (*Courtesy of Becton-Dickinson, BBL Microbiology Systems.*)

Disk-Plate Technique

The paper-disk-plate method is the most commonly used technique for determining susceptibility of microorganisms to chemotherapeutic agents. Small paper disks impregnated with known amounts of chemotherapeutic agents are placed upon the surface of an inoculated plate. After incubation, the plates are observed for any zones of inhibition surrounding the disks (see Fig. 24-26). A zone of inhibition (a clear area) around the disk indicates that the organism was inhibited by the drug, which diffused into the agar from the disk.

The single-disk method for susceptibility testing currently recommended by the FDA is a slight modification of the procedure developed by Bauer, Kirby, Sherris, and Turck in 1966. This is a highly standardized technique; the amount of antimicrobial agent contained in the disk is specified as well as the test medium, size of the inoculum, conditions of incubation, and other details. When the susceptibility test is performed in conformity with the FDA procedure, one can correlate the sizes of the zones of inhibition with the MIC of the drug for the microorganism in question; it is possible to determine whether the microorganism is resistant or susceptible to the antimicrobial agent. The relationship of MIC's to zone-of-inhibition diameters for the antibiotic cephalothin against several bacteria is shown in Fig. 24-27.

Figure 24-27. The relationship between the dilution and diffusion methods of testing the ability of an antibiotic to inhibit bacterial growth is demonstrated here for cephalothin. The size of the inhibition zone produced by an antibiotic disk goes up as the MIC goes down. All test conditions must be held constant. ▲ Enterococci, ● *Staphylococcus aureus*, • *Escherichia coli*, △ *Enterobacter-Klebsiella*, ■ *Haemophilus*. (Courtesy of K. J. Ryan, F. D. Schoenknecht, and W. M. M. Kirby, Hosp Pract, p. 99, 1970.)

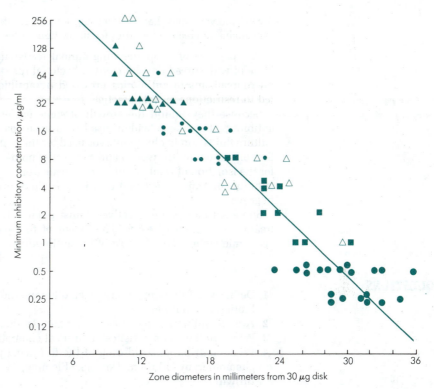

NONMEDICAL USES OF ANTIBIOTICS

Antibiotics are used as growth stimulants in poultry and livestock feeds. After the discovery that many domestic food-producing animals require vitamin B_{12} for optimum growth when fed a diet consisting of plant protein, it developed that by adding wastes from fermentation by-products to feeds, growth was stimulated more than could be accounted for by B_{12} alone. Even when adequate amounts of B_{12} were present in the diet, more rapid growth of young animals was noted when they were fed mash from the antibiotic fermenters. Use of pure antibiotics has given similar results. Commercially, the addition of aureomycin, terramycin, or penicillin to swine or poultry feeds at the rate of 5 to 20 g per ton of feed increases the rate of growth of young animals by at least 10 percent and sometimes by as much as 50 percent.

The stimulating effect of antibiotics on growth of domestic animals may be explained in several ways:

1 The antibiotics may destroy bacteria and other intestinal parasites that cause subclinical disease and retard growth and development. For example, it has been suggested that pigs respond dramatically to the addition of oxytetracycline to their diet because the antibiotic inhibits the growth of *Clostridium perfringens* in their intestines and prevents or reduces a chronic but subclinical toxemia.

2 Removal of the saprophytic bacteria from the intestinal tract may have a beneficial effect on the nutrition of the animals.

3 Streptomycin may have a "sparing effect" on the B_{12} in the diet, making it available in greater quantities for utilization by the animals.

The practice of supplementing animal feed with antibiotics has raised the issue of widespread development of bacterial resistance. The broad exposure of microorganisms to antibiotics provided by antibiotic supplemental feeds has led to restrictions on this practice.

Because they inhibit the growth of some bacteria and do not affect others, antibiotics have been widely used to prevent bacterial growth in tissue and culture fluids and in chick embryos used for the cultivation of viruses. Fleming's first use of penicillin was to add the crude mold-culture filtrate to media used for the isolation of *Haemophilus influenzae* from nose-and-throat washings. The antibiotic inhibited the Gram-positive cocci present but permitted *H. influenzae* to grow.

Some antibiotics are effective against plant pathogens and are attractive for treatment of plant diseases. The extent of this practice is limited mainly by economic factors, i.e., the cost of the antibiotic.

QUESTIONS

1 Define the following terms: chemotherapeutic agent, chemotherapy, antibiotics, and antibiosis.

2 Are all antibiotics useful as chemotherapeutic agents? Explain.

3 What are the characteristics of an ideal chemotherapeutic agent?

4 What contributions did each of the following persons make to the field of chemotherapy: Ehrlich, Domagk, Fleming, Waksman, Dubos, Hazen, and Brown?

5 Antibiotics are generally effective against bacterial infections and ineffective against viral infections. What are some of the reasons for this?

6 What are the major modes of antibacterial action of chemotherapeutic agents?

7 Describe the mode of antimicrobial action of the following chemotherapeutic agents: penicillins, cephalosporins, streptomycin, chlortetracycline, bacitracin, and sulfonamides.

8 How can the potency (or units) of an unknown sample of penicillin be determined?

9 Why is it important to determine the susceptibility of an infectious microorganism to chemotherapeutic agents? How can this be done?

10 Why are some antibiotics used to supplement animal foods?

11 In terms of their mode of action, explain why some chemotherapeutic agents are bacteriostatic and others bactericidal.

12 Which genera of microorganisms produce the most antibiotics?

13 What is the objection to using antibiotics as a food preservative?

14 Outline an experiment by which you could proceed to search for new antibiotics.

15 Assuming that you discovered one or more new antibiotics, what more would you need to know to ascertain whether any of them were good for chemotherapy?

REFERENCES   Abraham, E. P.: "Some Aspects of the Development of the Penicillins and Cephalosporins," in E. Don Murray and A. W. Bourquin (eds.), *Developments in Industrial Microbiology*, Society for Industrial Microbiology, Washington, D.C., 1974. *An interesting short story of the early development of antibiotics.*

Franklin, D. J., and G. A. Snow: *Biochemistry of Antimicrobial Action*, 3d ed., Chapman and Hall, Ltd., London, 1981. *A short book (215 pages) which presents in a very direct manner the characteristics of medically important antimicrobial compounds with a description of their mode of action.*

Gale, E. F., et al.: *The Molecular Basis of Antibiotic Action*, 2d ed., Wiley, New York, 1981. *Highly technical discussion of the mode of action of antibiotics in terms of modern molecular biology.*

Garrod, L. P., H. P. Lambert, and F. O'Grady (eds.): *Antibiotics and Chemotherapy*, 5th ed., Churchill Livingstone, Edinburgh, 1981. *Mainly about antibiotics, but it also deals with sulfonamides and other synthetic drugs used in the treatment of infectious diseases. The discussion of modes of action of antibiotics and chemotherapeutic drugs is very instructive.*

Korzybski, Tadeusz, et al.: *Antibiotics: Origin, Nature, and Properties*, American Society for Microbiology, Washington, D.C., 1978. *A systematic presentation of a large amount of information on the sources, varieties, and properties of antibiotics in a three-volume publication.*

Lancini, G., and F. Parenti, with M. Starr (ed.): *Antibiotics: An Integrated View* (translated from the Italian by Betty Rubin), Springer-Verlag, New York, 1982. *A small, concise book (253 pages), suitable for students, in which the authors attempt to present the basic facts and concepts on all aspects of the science of antibiotics.*

Lennette, E. H., A. Balows, W., J. Hausler, Jr., J. P. Truant, and H. J. Shadomy (eds.): *Manual of Clinical Microbiology*, 4th ed., American Society for Microbiology, Washington, D.C., 1985. *Section eleven of this comprehensive manual covers in considerable detail the methodology of determining the susceptibility of microorganisms to antimicrobial agents.*

PART SEVEN
ENVIRONMENTAL AND INDUSTRIAL MICROBIOLOGY

Life under extreme conditions

New discoveries are continually being made that expand our ideas about the ability of life to exist under harsh environmental conditions. For instance, although the surface waters of the ocean contain many microorganisms, it has long been thought that relatively few bacteria exist on the deep ocean floor because of the low temperature of the water, the scarcity of nutrients, and absence of light to support the growth of phototrophic organisms, and the enormous hydrostatic pressure due to the great depth of the overlying water column. This is indeed true for most regions of the ocean floor, but a startling exception was found in the late 1970s during exploration of deep-sea hydrothermal "vents" (hot submarine springs) located along submarine tectonic rifts and ridges of the ocean floor at a depth of 2,500 to 2,600 meters. In the areas surrounding these vents, living organisms were discovered in amazing and unexpected abundance, ranging from various bacteria to invertebrate animals such as giant clams, mussels, and tube worms.

The occurrence of this proliferation of organisms was partially accounted for by the increased temperature of the water in the regions surrounding the vents (10 to 20°C above the normal seawater temperature of 2.1°C), but what was the primary source of carbon and energy on which all of this life depended? The answer came when further studies indicated that the vents discharge heated water rich in geothermally produced H_2S and other reduced inorganic compounds into the surrounding waters. Moreover, geological and geochemical evidence indicated that oxygenated seawater percolates through nearby porous lava and mixes with the heated water issuing from the vents. Thus the supply of a reduced compound (H_2S) as well as the additional occurrence of oxygen might support the growth of certain chemolithotrophic bacteria that obtain energy by oxidizing H_2S and which use CO_2 as their carbon source.

Further support for this idea came when samples were obtained from the seawater and from the invertebrate animals near vent discharges by means of a submersible, deep-sea research vessel. From these samples, microbiologists from the Woods Hole Oceanographic Institution isolated cultures of chemolithotropic, sulfide-oxidizing bacteria such as *Thiomicrospira* and *Thiobacillus*-like organisms. It is likely that these and possibly other autotrophic bacteria represent the predominant or sole primary food source for the various animal populations living near the vents.

The discovery of an unexpected abundance of life occurring in these remote regions of the ocean floor has opened an entirely new and exciting area for biological research and is currently the subject of intense investigation.

Preceding page. Cultures of microorganisms being incubated on shaker platforms prior to testing for pharmaceutical products. (*Courtesy Cetus Corporation.*)

Chapter 25　Microbiology of Soil

However dead the earth may look and be considered in our thoughtless moments, the experience of man far back beyond his written records has led him to associate trouble capable of multiplying itself as coming from dirt. Bacillus tetanus, amoebic dysentery, thermophilic spoilage, actinomycosis, and botulism are new terms, but the need of freedom of earth in wounds, in food, and in clothing is no recent discovery. The demonstration that soil, instead of being all dead, harbors millions of organisms releases that flight of imagination which pictures the soil as a sort of Lilliputian Zoo in which some magic hand has eliminated all barriers and set free every grade of minute but rapacious monster to go roaring after the next lesser grade as its lawful prey. Thus the soil is pictured to us in terms that lead us to ask what manner of thing it is.

This paragraph is from the introduction of an address entitled "A Microbiologist Digs in the Soil" given by the late Charles Thom, one of America's great mycologists and soil microbiologists.

Directly or indirectly the wastes of humans and other animals, their bodies,

543

and the tissues of plants are dumped onto or buried in the soil. Somehow they all disappear, transformed into the substances that make up the soil. It is the microbes that make these changes—the conversion of organic matter into simple inorganic substances that provide the nutrient material for the plant world. Thus microorganisms play a key role in maintaining life on earth as we know it.

PHYSICAL CHARACTERISTICS OF SOIL

Soil has been defined as *that region on the earth's crust where geology and biology meet.* From a functional viewpoint, the soil may be considered as the land surface of the earth which provides the substratum for plant and animal life. The characteristics of the soil environment vary with locale and climate. Soils differ in depth, physical properties, chemical composition, and origin. A profile of soil is shown in Fig. 25-1.

Mineral Particles

The dominant mineral particles in most soils are compounds of silicon, aluminum, and iron, and lesser amounts of other minerals, including calcium, magnesium, potassium, titanium, manganese, sodium, nitrogen, phosphorus, and sulfur. The mineral constituents of soil range in size from small clay particles (0.002 mm or less) to large pebbles and gravel. The physical structure, aeration, water-holding capacity, and availability of nutrients are determined by the proportion of these particles, which are formed by the weathering of rock and the degradative metabolic activities of microorganisms.

Soils can be classified as **mineral soils,** which have solid matter that is largely inorganic, and **organic soils,** which have very little inorganic material. The latter are typically found in bogs and marshes. Most of the information in this chapter is concerned with mineral soils.

Figure 25-1. A schematic illustration showing the profiles of soil.

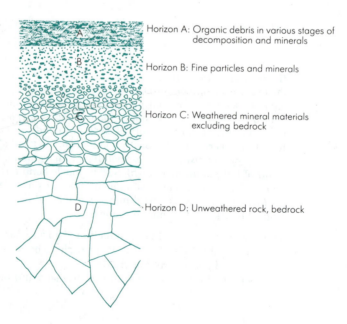

Horizon A: Organic debris in various stages of decomposition and minerals

Horizon B: Fine particles and minerals

Horizon C: Weathered mineral materials excluding bedrock

Horizon D: Unweathered rock, bedrock

Organic Residue

The plant and animal remains deposited on or in the soil contribute organic substances. Their decomposition will be described later in this chapter. In the last stages of decomposition, such material is referred to as **humus**, a dark-colored, amorphous substance composed of residual organic matter not readily decomposed by microorganisms. Indeed, the microbial population, both dead cells and living cells, is of such a large magnitude that it contributes significantly to the organic matter of soil. Certain agriculturally important properties are contributed to the soil by humus, which improves the texture and structure of the soil, contributes to its buffering capacity, and increases its water-holding capacity.

Water

The amount of water in soil depends on the amount of precipitation and other climatic conditions, drainage, soil composition, and the living population of the soil. Water is retained as free H_2O in the spaces between soil particles and adsorbed to the surfaces of particles. Various organic and inorganic components of soil are dissolved in soil water and thus are made available as nutrients for soil inhabitants.

Gases

The soil atmosphere is derived from air but differs in composition from it because of the biological processes occurring in soil. The gaseous phase of soil consists mainly of carbon dioxide, oxygen, and nitrogen. These gases exist primarily in the spaces between soil particles which are not filled with water, although a small amount of gas, especially carbon dioxide, is dissolved in water. Obviously, then, the amount of gases in the soil is related to the amount of moisture.

MICROBIAL FLORA OF SOIL

Fertile soil is inhabited by the root systems of higher plants, by many animal forms (e.g., rodents, insects, and worms), and by tremendous numbers of microorganisms.

The vast differences in the composition of soils, together with differences in their physical characteristics and the agricultural practices by which they are cultivated, result in corresponding large differences in the microbial population both in total numbers and in kinds.

The conditions described earlier as influencing the growth of organisms under laboratory cultivation are equally applicable to the soil. With specific reference to soil, these conditions can be summarized as follows: (1) amount and type of nutrients, (2) available moisture, (3) degree of aeration, (4) temperature, (5) pH, (6) practices and occurrences which contribute large numbers of organisms to the soil, e.g., floods or addition of manure. The existence of roots and the extensiveness of the root system in soil also influence the numbers and kinds of microorganisms present.

Variations of climatic conditions may selectively favor certain physiological types. Interactions between and among microbial species no doubt has an important effect on the members of the population. This is an extremely complex situation. Predatory protozoa and antibiotic-producing actinomycetes may eliminate certain groups of microorganisms. Cellulolytic and proteolytic organisms, on the other hand, may provide nutrients for less versatile biochemical species.

Few environments on earth have as great a variety of microorganisms as fertile soil. Bacteria, fungi, algae, protozoa, and viruses make up this microscopic menagerie, which may reach a total of billions of organisms per gram (Table 25-1). The great diversity of the microbial flora makes it extremely difficult to determine accurately the total number of microorganisms present. Cultural methods will reveal only those physiological and nutritional types compatible with the cultural environment. Direct microscopic counts theoretically should permit enumeration of all except the viruses, but this technique also has its limitations, especially in distinguishing living from dead microorganisms. Very often the microbiological analysis of soil is concerned with the isolation and identification of specific physiological types of microorganisms. For this purpose enrichment-culture techniques are appropriate.

Bacteria

Table 25-1. Soil Population in a Fertile Agricultural Soil

Type	Number per Gram
Bacteria:	
Direct count	2,500,000,000
Dilution plate	15,000,000
Actinomycetes	700,000
Fungi	400,000
Algae	50,000
Protozoa	30,000

SOURCE: A. Burges, *Microorganisms in the Soil*, Hutchinson, London, 1958.

The bacterial population of the soil exceeds the population of all other groups of microorganisms in both number and variety (see Table 25-2). Direct microscopic counts as high as several billions per gram have been reported; plate counts from the same samples yield only a fraction of this number (millions). The reason for this discrepancy is that there is such a great variety of nutritional and physiological types of bacteria in soil that no single laboratory environment (i.e., composition of medium and conditions of incubation) supports the growth of every viable cell in the inoculum. The following are all likely to be found in soil: autotrophs and heterotrophs; mesophiles, thermophiles, and psychrophiles; aerobes and anaerobes; cellulose digesters and sulfur oxidizers; nitrogen fixers and protein digesters; and other kinds of bacteria. It is generally agreed that there are many species of bacteria in soil yet to be discovered.

Large numbers of actinomycetes, as many as millions per gram, are present in dry warm soils. The most predominant genera of this group are *Nocardia*, *Streptomyces*, and *Micromonospora*. These organisms are responsible for the characteristic musty or earthy odor of a freshly plowed field. They are capable of degrading many complex organic substances and consequently play an important role in building soil fertility. The actinomycetes are also noted for their ability to synthesize and excrete antibiotics. The presence of antibiotic substances in soil can rarely be detected, but this does not exclude the possibility that they may be present and active in the microenvironment.

Fungi

Hundreds of different species of fungi inhabit the soil. They are most abundant near the surface, where an aerobic condition is likely to prevail. They exist in both the mycelial and spore stage. Since growth can take place from either a spore or a fragment of a mycelium, it is difficult to estimate their numbers; however, counts ranging from thousands to hundreds of thousands per gram of soil have been reported. Fungi are active in decomposing the major constituents of plant tissues, namely, cellulose, lignin, and pectin. The physical structure of soil is improved by the accumulation of mold mycelium within it. One of the characteristics of soil of considerable agricultural importance is its **crumble structure,** the binding together of fine soil particles to form water-stable aggregates. This is accomplished by the penetration of mycelium through the soil, forming a network which entangles the small particles.

Table 25-2. Physiological Groups of Bacteria in Various Types of Soil (Numbers of Bacteria per Gram of Soil)

Soil Type	Garden	Field	Meadow	Coniferous Forest	Marshland
Moisture content in percent of moist soil	17.9	18.1	17.0	21.2	37.2
Percent calcium carbonate	4.7	5.0	11.4	0	7.6
Bacteria developing on nutrient-gelatin plates	8,400,000	8,100,000	8,100,000	1,500,000	1,500,000
Bacteria developing on nutrient-agar plates	2,800,000	3,500,000	3,000,000	900,000	1,700,000
Bacteria growing in deep cultures of glucose agar (anaerobes)	280,000	137,000	620,000	345,000	2,180,000
Urea-decomposing bacteria	37,000	8,500	5,200	8,800	2,500
Denitrifying bacteria	830	400	850	380	370
Pectin-decomposing bacteria	535,000	70,000	235,000	810,000	3,700
Anaerobic butyric acid bacteria	368,000	50,300	83,500	203,000	235,000
Anaerobic protein-decomposing bacteria	35,000	22,000	36,800	17,000	2,000
Anaerobic cellulose-decomposing bacteria	367	350	367	17.7	1.1
Aerobic nitrogen-fixing bacteria	2,350	1,885	18	0	17
Anaerobic nitrogen-fixing bacteria	5,500	700	370,000	2,020	67
Nitrifying bacteria	880	1,701	37	0	34

SOURCE: M. Düggeli in S. A. Waksman, *Principles of Soil Microbiology*, Williams & Wilkins, Baltimore, 1932.

Yeasts are likely to be more prevalent in soils of vineyards, orchards, and apiaries, where special conditions, particularly the presence of sugars, favor their growth.

Algae

The population of algae in soil is generally smaller than that of either bacteria or fungi. The major types present are the green algae and diatoms. Their photosynthetic nature accounts for their predominance on the surface or just below the surface layer of soil. In a rich fertile soil, the biochemical activities of algae are dwarfed by those of bacteria and fungi. In some situations, however, algae perform prominent and beneficial changes. For example, on barren and eroded lands they may initiate the accumulation of organic matter because of their ability to carry out photosynthesis and other metabolic activities. This has been observed in some desert soils.

Cyanobacteria, the oxygenic photosynthetic bacteria, are known to grow on the surfaces of freshly exposed rocks where the accumulation of their cells results in simultaneous deposition of organic matter. This establishes a nutrient base that will support growth of other bacterial species. The growth and activities of the initial algae and bacteria pave the way for the growth of other bacteria and fungi. The mineral nutrients of the rock are slowly dissolved by acids resulting from microbial metabolism. This process continues with a gradual accumulation of organic matter and dissolved minerals until a condition results that supports growth of lichens, then mosses, then higher plants. The cyanobacteria play a key role in the transformation of rock to soil, a first step in rock-plant succession.

Protozoa

Most soil protozoa are flagellates or amoebas; the number per gram of soil ranges from a few hundred to several hundred thousand in moist soils rich in organic matter. From a microbiological standpoint they are of significance since their dominant mode of nutrition involves ingestion of bacteria. Of academic interest

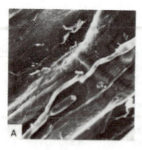  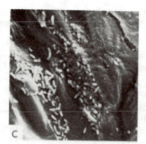 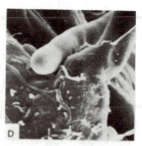

Figure 25-2. The microbial flora in the rhizosphere as seen in scanning electron micrographs of epidermal root cells. In this study plants were removed from a forest preserve, and special precautions were observed to recover root systems with as little mechanical damage as possible to the tissues. The root systems were dissected from the plants and representative specimens selected for electron microscopy. (A and B) Fungal hyphae and bacteria on the epidermal cells of *Ammophilia arenaria*. [(A) × 900; (B) × 800.] (C) Bacteria on the epidermis of a barley root, × 1,000. (D) Root hair, epidermal cell, and bacteria, × 850. (*Courtesy of K. M. Old and New Phytol*, **74**:51, 1975.)

is the fact that they demonstrate a preference for certain microbial species. Since not all bacteria are suitable as food for protozoa, the protozoa may be a factor in maintaining some equilibrium of microorganisms in soil.

Viruses

Bacterial viruses (bacteriophages), as well as plant and animal viruses, periodically find their way into soils through additions of plant and animal wastes. Also, soil microorganisms themselves may harbor viruses.

The Rhizosphere

The region where the soil and roots make contact is designated the **rhizosphere.** The microbial population on and around roots is considerably higher than that of root-free soil; the differences are both quantitative and qualitative. Bacteria predominate, and their growth is enhanced by nutritional substances released from the plant tissue, e.g., amino acids, vitamins, and other nutrients; the growth of the plant is influenced by the products of microbial metabolism that are released into the soil. It has been reported that amino acid–requiring bacteria exist in the rhizosphere in larger numbers than in the root-free soil. It has been demonstrated that the microbiota of the rhizosphere is more active physiologically than that of nonrhizosphere soil. The rhizosphere represents a tremendously complex biological system, and there is a great deal yet to be learned about the interactions which occur between the plant and the microorganisms intimately associated with its root system.

Electron-microscope techniques have been developed to observe microorganisms directly on the root surfaces (see Fig. 25-2).

INTERACTIONS AMONG SOIL MICROORGANISMS

The microbial **ecosystem** of soil includes the total microbial flora together with the physical composition and physical characteristics of the soil. It is the sum of the **biotic** and the **abiotic** components of soil.

The microorganisms that inhabit the soil exhibit many different types of associations or interactions. Some of the associations are indifferent or neutral; some are beneficial or positive; others are detrimental or negative. As each different type of association or interaction is discovered, it has been given a specific descriptive label. As you might presume, many of these associations do not fall neatly into discrete categories. Furthermore, and likewise not unexpected, there is the existence of some confusion and contradiction in the use of terms. The term symbiosis, for example, as first proposed, referred to the "living together of dissimilarly named organisms"; it was used as a general term. Later it took on a more specific meaning, namely, an association between bacteria and plants referred to as symbiotic nitrogen fixation which is described later in this chapter. Currently the trend is to use the term *symbiosis* as originally intended, that is, merely as a condition in which the individuals of a species live in close association with individuals of another species. We shall describe the following types of microbial associations:

Neutral: neutralism
Positive or beneficial: mutualism, commensalism
Negative or detrimental: antagonism, competition, parasitism, predation

Neutral Associations

Neutralism

It is conceivable that two different species of microorganisms occupy the same environment without affecting each other. For example, each could utilize different nutrients without producing metabolic end products that are inhibitory. Such a condition might be transitory; as conditions change in the environment, particularly availability of nutrients, the relationship might change.

Positive Associations

Mutualism

Mutualism is an example of a symbiotic relationship in which each organism benefits from the association. The manner in which benefit is derived varies. One type of mutualistic association is that involving the exchange of nutrients between two species, a phenomenon called syntrophism. Many microorganisms synthesize vitamins and amino acids in excess of their nutritional requirements. Others have a requirement for one or more of these nutrients. Still others synthesize a particular essential nutrient in suboptimal amounts. Hence, certain combinations of species will grow together but not apart when nutrient levels are very low.

Another mutalistic association is characterized by different metabolic products from the association as compared with the sum of the products of the separate species. Figure 25-3A illustrates a mutualism between *Thiobacillus ferrooxidans* and *Beijerinckia lacticogenes* in a medium which lacks carbon and nitrogen sources. The growth of the two species in association and the resulting effect on the rate and extent of leaching copper from an ore is shown. Leaching is one of the processes for recovering metals from ore; microorganisms play the important role of oxidizing insoluble metal sulfides to soluble sulfates. The interactions that occur in this mutualistic association are shown in Fig. 25-3B.

Commensalism

The phenomenon of commensalism refers to a relationship between organisms in which one species of a pair benefits; the other is not affected. This occurs commonly in soil with respect to degradation of complex molecules like cellulose and lignin. For example, many fungi are able to dissimilate cellulose to

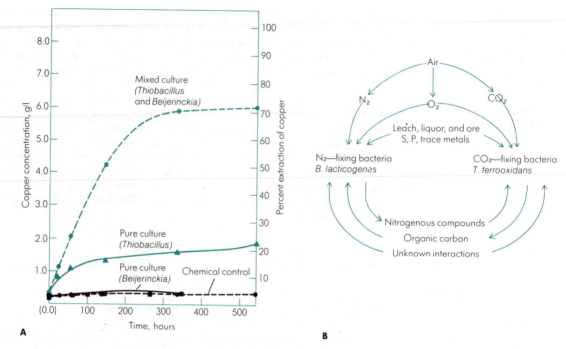

Figure 25-3. An example of microbial mutualism in ore leaching. *Thiobacillus ferrooxidans* and *Beijerinckia lacticogenes* were inoculated into a medium free of added carbon and nitrogen sources; sterile ore concentrate was added to the medium. (A) Results of leaching of copper with pure and mixed culture. (B) Proposed mutualistic interactions between the two species in a leaching environment devoid of fixed carbon or fixed nitrogen. (*Courtesy of N. C. Trivedi and H. M. Tsuchiya, Int J Miner Process,* **2**:1, *1975.*)

glucose and beyond (they are cellulolytic). Many bacteria are unable to utilize cellulose, but they can and do utilize the fungal breakdown products of cellulose, e.g. glucose and organic acids. The existence of many nutritionally fastidious bacteria in soil suggests that their growth and survival is dependent upon the synthesis and excretion of vitamins and amino acids by less fastidious species.

Another example of commensalism is that of a change in the substrate produced by a combination of species and not by individual species. For example, lignin, a major constituent of wood, is generally resistant to degradation by pure cultures of microorganisms under laboratory conditions. However, the lignin in forest soil is degraded by the soil microbial flora, particularly fungi.

Negative Associations

Antagonism

When one species adversely affects the environment for another species, it is said to be **antagonistic.** Such organisms may be of great practical importance, since they often produce antibiotics or other inhibitory substances which affect the normal growth processes or survival of other organisms. Antagonistic relationships are quite common in nature. For instance, both *Staphylococcus aureus* and *Pseudomonas aeruginosa* are antagonistic toward *Aspergillus terreus*. Certain *Pseudomonas* pigments inhibit germination of *Aspergillus* spores. *S. aureus* produces a diffusable antifungal material that causes distortions and hyphal swellings in *A. terreous* (Fig. 25-4). Although microorganisms from a variety of natural habitats produce antibiotics, soil microorganisms are the most common producers. It is not unusual for one organism to produce five or six different antimicrobial agents. There is some question about the role of antibiotics in nature; production in the laboratory is under conditions quite different from those in nature. Production of antibiotics in soil may enable the antibiotic-producing organism to thrive successfully in a competitive environment. For example, large populations of actinomycetes have been found in the chitinaceous shells of dead crustaceans in the sea. Their existence, in the environment free of other microorganisms, is likely due to the antibiotics they produce.

Organisms that elaborate antibiotics represent the classic example of this phenomenon; however, antibiosis may result from a variety of other conditions

Figure 25-4. *Aspergillus-Staphylococcus* interaction. (A) *Aspergillus terreus* 20-h culture in nutrient broth with *Staphyloccus aureus* added after 8 h (X500). (B) *A. terreus* and *S. aureus* as in (A) in glucose-peptone broth (X600). (C) *A. fumigatus* and *S. aureus* in glucose-peptone broth (X600). (D) *A. terreus* 20-h culture in nutrient broth (X800). *(Courtesy of A. Mangan, J Gen Microbiol, **58**:261, 1969.)*

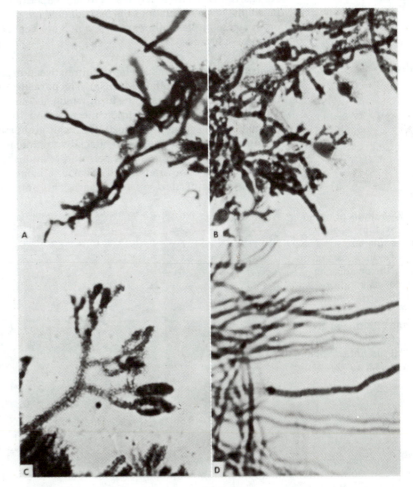

operative in mixed populations. Cyanide is produced by certain fungi in concentrations toxic to other microorganisms, and the algae elaborate fatty acids which exhibit a marked antibacterial activity. Other metabolic products that may result from microbial activity in the soil which are likely to be inhibitory to other species are methane, sulfides, and other volatile sulfur compounds.

Many soil microorganisms—important examples are the myxobacteria (slime bacteria) and streptomycetes—are antagonistic because they secrete potent lytic enzymes which destroy other cells by digesting their cell wall or other protective surface layers, as shown in Fig. 25-5. Presumably the degraded cellular material, as well as the released protoplasmic material, serves as nutrients. Although it might be assumed that organisms producing lytic substances have a selective advantage over sensitive microbes, microbial interactions of this type are difficult to interpret. It appears that in the natural environment producers of lytic substances are often found in close proximity with sensitive organisms and do not predominate over them.

Competition

A negative association may result from competition among species for essential nutrients. In such situations the best adapted microbial species will predominate or, in fact, eliminate other species which are dependent upon the same limited nutrient substance.

Parasitism

Parasitism is defined as a relationship between organisms in which one organism lives in or on another organism. The **parasite** feeds on the cells, tissues, or fluids of another organism, the **host,** which is commonly harmed in the process. The parasite is dependent upon the host and lives in intimate physical and metabolic contact with the host. All major groups of plants, animals, and microorganisms are susceptible to attack by microbial parasites.

An interesting example of a parasitic relationship between microbes is the bacterial parasite of Gram-negative bacteria named *Bdellovibrio bacteriovorus,* which is widespread in soil and sewage. This unusual motile bacterium attaches

Figure 25-5. Lysis of cyanobacteria by a myxobacter. Shown in this series is a sequence of lysis of *Nostoc* filament by the myxobacter. The myxobacter culture used in this experiment was isolated from fish ponds and is capable of lysing many species of bacteria. (*Courtesy of Mirian Shilo, J Bacteriol,* **104**:453, 1970.)

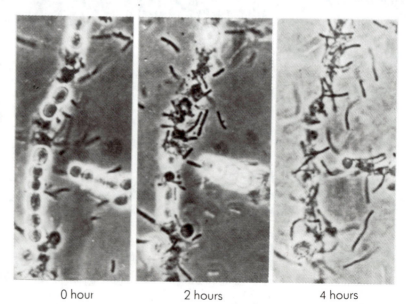

0 hour   2 hours   4 hours

Figure 25-6. Bacteriolysis produced by *Bdellovibrio bacteriovorus.* (A) Plate culture of *B. bacteriovorus* on a lawn of *E. coli* showing whitish-gray colony surrounded by circular plaquelike clearing zone. Central colony consists of bdellovibrios and the clear zone contains a few intact *E. coli* cells and spheroplasts of the host cells (*E. coli*). (B) Electron micrograph thin section showing *B. bacteriovorus* penetration into *E. coli* cell (X48,000). (*Courtesy of J. C. Burnham, T. Hashimoto, and S. F. Conti, J Bacteriol,* **96**:*1366, 1968.*)

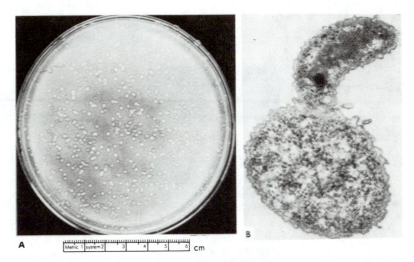

to a host cell at a special region and eventually causes the lysis of that cell (see Chap. 13). As a consequence, plaquelike areas of lysis (Fig. 25-6) appear when these parasites are plated along with their host bacteria. There are also many strains of fungi which are parasitic on algae and other fungi by penetration into the host.

Viruses which attack bacteria, fungi, and algae are strict intracellular parasites since they cannot be cultivated as free-living forms. The phenomenon of lysogeny is quite important because of the possibility for genetic recombination in natural populations and the subsequent expression of new characteristics.

BIOGEOCHEMICAL ROLE OF SOIL MICROORGANISMS

Soil microorganisms serve as biogeochemical agents for the conversion of complex organic compounds into simple inorganic compounds or into their constituent elements. The overall process is called **mineralization.** This conversion of complex organic compounds into inorganic compounds or elements provides for the continuity of elements (or their compounds) as nutrients for plants and animals including people.

It is possible to construct a sequence of reactions to illustrate that microorganisms perform an essential role in maintaining a cyclic process for the reutilization of elements under natural conditions. In this respect we can view the planet earth as a closed system dependent upon the process of recycling for maintenance of life as we know it.

In the following paragraphs we shall discuss the role of soil microorganisms with respect to the transformations they bring about on nitrogen, carbon, sulfur, phosphorus, and their compounds.

BIOCHEMICAL TRANSFORMATIONS OF NITROGEN AND NITROGEN COMPOUNDS: THE NITROGEN CYCLE

Because of the importance of nitrogen for plant nutrition, the biochemical events that make up the nitrogen cycle have been studied in considerable detail.

The sequence of changes from free atmospheric nitrogen to fixed inorganic nitrogen, to simple organic compounds, to complex organic compounds in the tissues of plants, animals, and microorganisms, and the eventual release of this

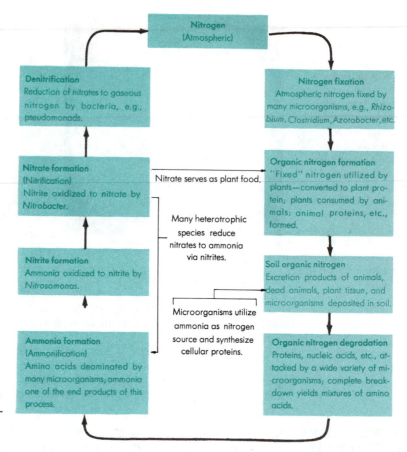

Figure 25-7. Nitrogen cycle in nature (schematic), showing the role of microorganisms.

nitrogen back to atmospheric nitrogen is summarized in Fig. 25-7, the **nitrogen cycle.**

Proteins, nucleic acids, purine and pyrimidine bases, and amino sugars (glucosamine and galactosamine) represent the complex organic nitrogenous substances which are deposited in soil in the form of animal and plant wastes or their tissues. Synthetic processes of microorganisms also contribute some amount of complex organic nitrogen compounds.

The simplest form of nitrogen involved in biological transformations is gaseous elementary nitrogen. The overall transformations in which microorganisms are involved range from nitrogen gas to protein. A great many intermediate products and a corresponding large number of intricate enzymatic reactions are involved in bringing about these changes.

Some of the biochemical events in the nitrogen cycle are summarized below.

Proteolysis

The nitrogen in proteins (as well as in nucleic acids) may be regarded as the end of the line as far as synthesis of nitrogenous compounds is concerned. The nitrogen in proteins is "locked" and is not available as a nutrient to plants. In order to set this organically bound nitrogen free for reuse, the first process that must take place is the enzymatic hydrolysis of proteins (proteolysis). This is accomplished by microorganisms capable of elaborating extracellular protein-ases that convert the protein to smaller units (peptides). The peptides are then attacked by peptidases, resulting ultimately in the release of individual amino acids. The overall reactions may be summarized:

$$\text{Proteins} \xrightarrow{\text{proteinases}} \text{peptides} \xrightarrow{\text{peptidases}} \text{amino acids}$$

Some bacterial species elaborate large amounts of proteolytic enzymes. Among the most active in this respect are some of the clostridia, e.g., *Clostridium histolyticum* and *C. sporogenes*; a lesser degree of activity is found in species of the genera *Proteus*, *Pseudomonas*, and *Bacillus*. Many fungi and soil acti-nomycetes are extremely proteolytic. Peptidases, however, occur widely in mi-croorganisms as demonstrated by the fact that peptones (partially hydrolyzed proteins) are a common constituent of bacteriological media and provide a readily available source of nitrogen.

Amino Acid Degradation: Ammonification

The end products of proteolysis are amino acids. Their fate in the soil may be utilization as nutrients by microorganisms or degradation by microbial attack. Amino acids are subject to a variety of pathways for microbial decomposition. We are concerned here with the liberation of nitrogen from these compounds, which is accomplished by deamination, i.e., removal of the amino group. Al-though several variations of deamination reactions are exhibited by microorgan-isms, one of the end products is always ammonia, NH_3. An example of a specific deamination reaction is

$$\underset{\text{Alanine}}{CH_3CHNH_2COOH} + \tfrac{1}{2}O_2 \xrightarrow{\underset{\text{deaminase}}{\text{alanine}}} \underset{\text{Pyruvic acid}}{CH_3COCOOH} + \underset{\text{Ammonia}}{NH_3}$$

This reaction is classified as an oxidative deamination. Many microorganisms can deaminate amino acids. The production of ammonia is referred to as **am-monification**. The fate of the ammonia thus produced varies, depending upon conditions in the soil. Ammonia is volatile and, as such, leaves the soil; however, if solubilized, NH_4^+ is formed. Some of the subsequent possibilities include accumulation and utilization by plants and microorganisms and, under favora-ble conditions, oxidation to nitrates.

Nitrification

Microorganisms convert ammonia to nitrate, and the process is called **nitrifi-cation**. The process occurs in two steps, each step performed by a different group of bacteria.

1 Oxidation of ammonia to nitrite by ammonia-oxidizing bacteria

$$2NH_3 + 3O_2 \rightarrow 2HNO_2 + 2H_2O$$

2 Oxidation of nitrite to nitrate by nitrite-oxidizing bacteria

$$HNO_2 + \tfrac{1}{2}O_2 \rightarrow HNO_3$$

Bacteria of both physiological groups, ammonia oxidizers and nitrite oxidiz-

Table 25-3. Composition of Medium for Isolation of Nitrifying Bacteria Using Enrichment Culture Technique

Ingredients	g/L
$(NH_4)_2SO_4$	2.0
K_2HPO_4	1.0
$MgSO_4$	0.5
$FeSO_4$	0.4
NaCl	0.4
$CaCO_3$	1.0
$MgCO_3$	1.0

ers, are Gram-negative chemolithotrophs. Their main source of carbon is obtained through carbon dioxide fixation; energy is derived by the oxidation of NH_3 or NO_2^- depending upon the group. Nitrifying bacteria occur widely in nature in a variety of habitats, including soil, sewage, and aquatic environments.

Nitrifying bacteria cannot be isolated directly by the usual techniques employed to isolate heterotrophic bacteria. Some of the reasons for this are: they are slow-growing compared with heterotrophs; and they may be present in very small numbers compared with other physiological types. Accordingly, enrichment cultures are used for their isolation. An example of a medium for this purpose is shown in Table 25-3. A relatively large inoculum is used, and incubation is in the dark at 25 to 30°C for a period of 1 to 4 months.

Species of ammonia-oxidizing bacteria vary in morphology (rod, spherical, spiral, or lobular) and usually have an extensive membrane system within their cytoplasm. They frequently form cysts and zooglea. See Fig. 25-8. The following species have been recognized as ammonia oxidizers:

Nitrosomonas europaea
Nitrosovibrio tenuis
Nitrosococcus nitrosus
Nitrosococcus oceanus

Species of nitrite-oxidizing bacteria exhibit some of the same morphological characteristics as the ammonia oxidizers. Only a few species have been isolated and described. These include *Nitrobacter winogradskyi* and *Nitrospina gracilis*.

An interesting historical event involving nitrification and production of gunpowder may be cited. During the Napoleonic wars, France was unable to import nitrate, which was needed for the manufacture of gunpowder. To solve this

Figure 25-8. Ultrastructure of *Nitrobacter winogradskyi.* (A) Thin section from cell grown chemoautotrophically and harvested during exponential phase of growth, showing lamellar membrane system (L) at the swollen end of the cell and electron-dense polyhedral bodies (B). (B) Thin section from cell grown on nitrite mineral-salts medium supplemented with 5 mmol sodium acetate and harvested during exponential phase of growth. Section shows lamellas (L), polyhedral bodies (B), and electron-transparent bodies believed to be poly-β-hydroxybutyrate (PHB) reserve material. (*Courtesy of L. M. Pope, D. S. Hoare, and A. J. Smith, J Bacteriol,* **97**:936, 1969.)

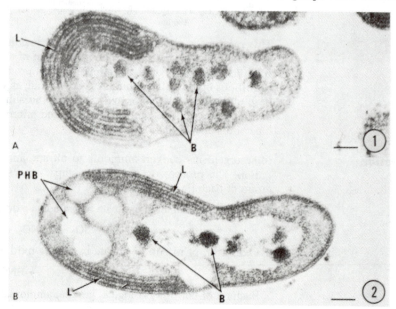

dilemma, artificial niter beds were made, consisting of soil mixed with animal waste and vegetable materials, ashes, etc. Aeration was performed by turning the heap over from time to time. After a long period of incubation, crude saltpeter (mineral nitrates) was extracted with hot water. This occurred, of course, long before the specific activities of microorganisms were known. Nitrification was discovered to be a biological process by Schloesing and Muntz in 1877; Winogradsky isolated the bacteria responsible for the process in 1890.

Reduction of Nitrate to Ammonia

Several heterotrophic bacteria are capable of converting nitrates into nitrites or ammonia. This normally occurs under anaerobic conditions, e.g., in waterlogged soil. The oxygen of the nitrate serves as an acceptor for electrons and hydrogen. The process involves several reactions, and the overall result is

$$HNO_3 + 4H_2 \rightarrow NH_3 + 3H_2O$$

This reaction is not of major significance in well-cultivated agricultural soil.

Denitrification

The transformation of nitrates to gaseous nitrogen is accomplished by microorganisms in a series of biochemical reactions. The process is known as **denitrification.** From the standpoint of agriculture, this is an undesirable process in that it results in loss of nitrogen from the soil and hence a decline in nutrients for plant growth.

Species of several genera of bacteria are capable of transforming NO_3^- to N_2, e.g., *Achromobacter, Agrobacterium, Alcaligenes, Bacillus, Chromobacterium, Flavobacterium, Hyphomicrobium, Pseudomonas, Thiobacillus,* and *Vibrio.*

The overall biochemical reaction which expresses the process of denitrification is

$$2NO_3^- \rightarrow 2NO_2^- \rightarrow 2NO \rightarrow N_2O \rightarrow N_2$$
$$\text{Nitrate} \quad \text{Nitrite} \quad \text{Nitric oxide} \quad \text{Nitrous oxide} \quad \text{Nitrogen}$$

Experimental results which illustrate the order in which products rise and fall during denitrification are shown in Fig. 25-9.

Environmental conditions in a soil have a significant effect on the level of denitrification. For example, the process is enhanced in soils (1) by an abundance of organic matter, (2) by elevated temperatures (25 to 60°C), and (3) by neutral or alkaline pH. Availability of oxygen has a dual effect. Denitrification proceeds only when the oxygen supply is limited. However, oxygen is necessary for nitrite and nitrate formation.

Nitrogen Fixation

A number of microorganisms are able to use molecular nitrogen in the atmosphere as their source of nitrogen. The conversion of molecular nitrogen into ammonia is known as **nitrogen fixation.** Two groups of microorganisms are involved in this process: (1) **nonsymbiotic** microorganisms, those living freely and independently in the soil; and (2) **symbiotic** microorganisms, those living in roots of plants. Several types of experiments are used to detect nitrogen fixation by microorganisms. One approach is to demonstrate growth in a nitrogen-free medium. More specific evidence of fixation can be obtained by cultivating the microorganism in the presence of nitrogen labeled with isotopic nitrogen. $^{15}N_2$ can be measured by using a mass spectrometer. In essence, after

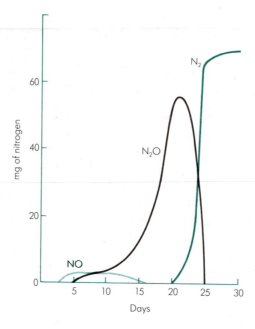

Figure 25-9. Sequence of products during denitrification in Norfolk sandy loam. (*Courtesy of F. B. Cady and W. V. Bartholomew, Soil Sci Soc Am Proc,* **24**:*477, 1960.*)

Table 25-4. Some Examples of Nitrogen-Fixing Bacteria

Cyanobacteria
  Anabaena spp.
  Nostoc spp.
  Gloeotrichia spp.
  Synechococcus spp.
  Plectonema spp.
  Oscillatoria spp.

Phototrophic bacteria
  Rhodospirillum rubrum
  Rhodopseudomonas palustris
  Rhodomicrobium vannielii
  Chromatium vinosum
  Chlorobium thiosulfatophilum

Chemotrophic bacteria
  Azospirillum lipoferum
  Azotobacter chroococcum
  Beijerinckia indica
  Rhizobium leguminosarum
  Methylomonas methanitrificans
  Escherichia coli
  Enterobacter aerogenes
  Bacillus macerans
  Clostridium butyricum
  Xanthobacter autotrophicus

the organism is grown in the mixture of atmospheric nitrogen and $^{15}N_2$, the culture is examined for evidence of $^{15}N_2$ incorporated in any compounds. Its presence is positive proof that nitrogen has been fixed. Under suitable conditions an increase of as little as 0.001 μg of nitrogen can be detected by this technique.

The capability of the nitrogen-fixing enzyme to act upon acetylene, discovered in the mid-1960s, has led to the development of a simple, rapid, relatively inexpensive technique now widely used to measure nitrogen fixation. The test is based on the observation that the nitrogen-fixing enzyme (*nitrogenase*) interacts with triple-bonded compounds, e.g., acetylene, to form ethylene as follows:

$$\underset{\text{Acetylene}}{HC\equiv CH} \xrightarrow[\substack{\text{enzyme}\\\text{nitrogenase}}]{2H} \underset{\text{Ethylene}}{H_2C=CH_2}$$

The comparable reaction with nitrogen is

$$\underset{\text{Nitrogen}}{N\equiv N} \xrightarrow[\substack{\text{enzyme}\\\text{nitrogenase}}]{6H} \underset{\text{Ammonia}}{2NH_3}$$

The technique involves exposing the specimen being assayed for nitrogenase activity to acetylene in a suitable vessel and, after a period of incubation, analyzing the gas phase for ethylene by gas-liquid chromatography. The amount of ethylene produced is a measure of nitrogenase activity.

The essential reactants in the bacterial nitrogen fixation process are:

1 The **nitrogenase enzyme complex.** This has been characterized as two components, and neither is active without the other. Component I is nitrogenase and component II is nitrogenase reductase. Component I is known as the **MoFe protein** (Mo for molybdenum, Fe for iron). Component II, which is a smaller molecule is designated the **Fe protein.** Both molecules contain sulfur.

2 A strong reducing agent such as ferredoxin or flavodoxin

3 ATP

4 A regulating system for NH_3 production and utilization

5 A system that protects the nitrogen-fixing system from inhibition by molecular oxygen

The overall biochemical reaction for nitrogen fixation can be expressed as:

$$N_2 + 6e^- + 12ATP + 12H_2O \xrightarrow[\text{complex}]{\text{nitrogenase}} 2NH_4^+ + 12ADP + 12P_i + 4H^+$$

Nonsymbiotic Nitrogen Fixation

Nonsymbiotic nitrogen fixation has been studied extensively with *Clostridium pasteurianum* and species of *Azotobacter*. For many years, these bacteria were the only ones known to be capable of this activity. The former is an anaerobic bacillus, and the latter are aerobic oval to spherical cells; both are widely distributed in soils. The nitrogen-fixing capacity of the *Azotobacter* species is greater than that of *Cl. pasteurianum*. In recent years many other microorganisms have been found to fix nitrogen (see Table 25-4).

It has been estimated that the amount of nitrogen fixed by the nonsymbiotic process ranges between 20 and 50 lb/acre annually. This estimate is no doubt subject to much variation depending upon the conditions peculiar to a particular soil.

Symbiotic Nitrogen Fixation

Symbiotic nitrogen fixation is accomplished by bacteria of the genus *Rhizobium* in association with legumes (plants that bear seeds in pods, e.g., soybeans, clover, and peas). Before these bacteria can fix nitrogen, they must establish themselves in the cells of root tissue of the host plant. Infection of the root system by the rhizobia bacteria is closely associated with the formation of an "infection thread" that develops in certain root hairs (see Fig. 25-10). The nitrogen-fixing bacteria invade the host plant cells via this infection thread. Some of the cells of the plant are thus infected, causing cell enlargement and an increased rate of cell division, leading to the formation of abnormal growths (nodules) on the root system. Several types of nodulation are illustrated in Fig. 25-11.

The legume, the bacteria, and the nodule constitute the system for this type of nitrogen fixation. It is a process where both the bacteria and the plant benefit by the association. The bacteria convert atmospheric nitrogen to fixed nitrogen which is available to the plant, and in turn, the bacteria derive nutrients from the tissues of the plant.

Not all species of *Rhizobium* produce nodulation and nitrogen fixation with any legume. There is a degree of specificity between the bacteria and legumes. For purposes of inoculation with commercial preparations of these bacteria, legumes are divided into seven major categories as follows: alfalfa, clover, peas and vetch, cowpeas, beans, lupines, and soybeans. *Rhizobium* species or strains effective for one group are less effective or ineffective for other groups. Even within a species, certain strains are more effective than others with a given host plant. Evidence of this specificity is demonstrated in Fig. 25-12.

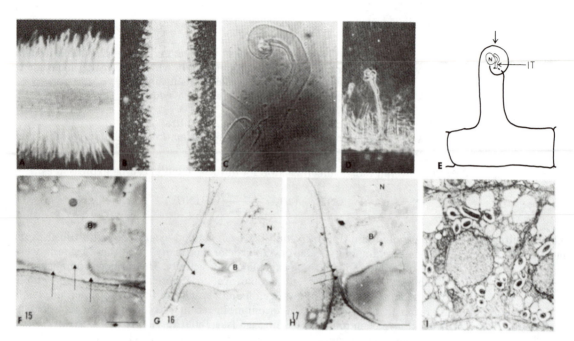

Figure 25-10. Nodule formation by *Rhizobium* on legume plants. The first stage in the establishment of the *Rhizobium*-legume N₂-fixing symbiosis is the infection of the host legume by the appropriate *Rhizobium* species. Root hairs are the site of infection. The first microscopically visible indication of the bacteria-plant interaction is deformation and curling of the normally straight root hairs. Aseptically cultured clover seedling with undeformed root hairs is shown (A), × 40. (B) A clover seedling (× 40) inoculated with *R. trifolii*. The bacteria are in clumps (flocs) in the rhizosphere. Note the change in appearance of the root hairs. A characteristic deformation is curling at the root hair tip to produce a "shepherd's crook" (C). The bacteria enter the root hair and are enclosed in a tubular structure, the infection thread (C and D), which is the first microscopically visible sign of a successful infection. The bacteria appear to enter the root hair by a process of invagination. Root hair cell-wall growth is redirected at a localized point resulting in the wall growing back into the root hair to form the tubular infection thread. There is no direct penetration through the root hair cell wall, and the bacteria remain extracellular within the infection thread. (E, F, G, H) A serial section sequence through a root hair which had a shepherd's crook at the origin of the infection thread. (E) A diagrammatic illustration of a serial sectioned root hair showing the infection thread (IT), nucleus (N), and the initiation of sectioning (top arrow). (F) A section before the invagination showing the infection thread (IT) which contained bacteria (B). The arrows indicate the region of the root hair cell wall where the invagination process has begun (× 4,500). (G) A section through the middle of the invagination showing the infection thread wall (arrows) of the pore, bacteria (B) within the infection thread, and the root hair nucleus (× 4,500). (H) A section past the pore; the arrows point out where the wall of the pore is grazed by the knife (× 4,500). Bacteroids within a nodule (I) are surrounded by membrane which is believed to be derived from the plant. The bacteroids contain electron-dense, unidentified inclusions. (*Courtesy of C. A. Napoli and H. Hubbell, Appl Microbiol,* **30**:1003, 1975.)

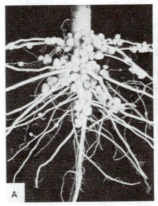

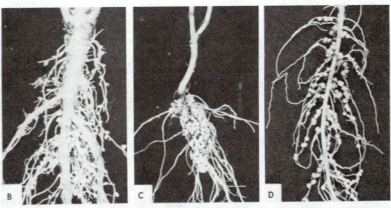

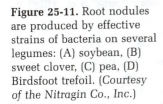

Figure 25-11. Root nodules are produced by effective strains of bacteria on several legumes: (A) soybean, (B) sweet clover, (C) pea, (D) Birdsfoot trefoil. (*Courtesy of the Nitragin Co., Inc.*)

Figure 25-12. Different strains of rhizobia have different effects on the growth of clover. Tests are carried out on Crimson clover in the following manner: Seeds are planted in sterile sand contained in a jar. The sand is then inoculated with the bacteria. Each jar contains a solution of nutrients—except nitrogen—which diffuse through the sand. Thus the extent of growth is indicative of the amount of nitrogen being supplied by the bacteria. (A) was not inoculated; (B), (C), and (D) were inoculated with different strains of rhizobia. Note the difference in growth response. (*Courtesy of L. W. Erdman, USDA.*)

Inoculation of seeds before planting is a desirable practice, since not all agricultural soils contain the right kinds of bacteria for optimum symbiotic nitrogen fixation with legume crops. Most of the commercial preparations consist of selected strains of bacteria dispensed in moist humus. This material is mixed with water and sprinkled over the seeds prior to planting.

Recombinant DNA and Nitrogen Fixation (Genetic Engineering)

The vast amount of knowledge that has accumulated in the last decade about microbial genetics, including the development of highly sophisticated techniques for gene splicing and cloning, has led to some dramatic developments in the field commonly referred to as **genetic engineering.** Applied aspects of this development are discussed later in Chap. 29, Industrial Microbiology. Nevertheless, we wish to mention here that many laboratories and research scientists are concentrating their efforts on the possibility of developing new systems for nitrogen fixation using recombinant DNA technology.

One area of research is directed toward introducing the "package" of nitrogen-fixing genes from bacteria into plant cells. If this were achieved, plants might be capable of directly fixing nitrogen from the atmosphere. This would be a tremendous advance not only for agriculture but for the world at large in terms of producing food more economically and abundantly. Obviously, considerably more research is necessary before this kind of genetic engineering can be attempted at a practical level. For instance, nitrogenase is easily destroyed by oxygen, and some means of protection of this enzyme complex from oxygen would have to be provided in order for a plant cell to be able to fix nitrogen.

Alternatively, it may be possible to modify certain bacteria in a manner so that they would develop a relationship with the root system of other plants, as the *Rhizobium* species grow with legumes. For example, a symbiotic bacterial nitrogen-fixing system with cereal grains would have a tremendous effect on grain production both in yield and cost.

BIOCHEMICAL TRANSFORMATIONS OF CARBON AND CARBON COMPOUNDS: THE CARBON CYCLE

Carbon Dioxide Fixation

The ultimate source of organic carbon compounds in nature is the carbon dioxide present in the atmosphere (or dissolved in water). The process, carbon dioxide fixation, was discussed in Chap. 11. Although green plants and algae are the most important agents of carbon dioxide fixation, bacteria are also capable of synthesizing organic matter from inorganic carbon. The occurrence of photosynthesis among microorganisms has already been described. Other examples of carbon dioxide transformation or incorporation into organic compounds by bacteria are:

1 Utilization of carbon dioxide by autotrophic bacteria; the carbon dioxide represents the sole source of carbon for these organisms and is transformed by a reduction reaction to carbohydrates. The general reaction is

$$CO_2 + 4H \rightarrow (CH_2O)_x + H_2O$$

2 Carbon dioxide fixation by heterotrophic microorganisms is common among bacteria. A specific example of this type of reaction is

$$\underset{\text{Pyruvic acid}}{CH_3COCOOH} + CO_2 \rightarrow \underset{\text{Oxalacetic acid}}{HOOCCH_2COCOOH}$$

Organic Carbon Compound Degradation

The organic carbon compounds that eventually are deposited in the soil are degraded by microbial activity. The end product, carbon dioxide, is released into the air and soil. Fresh air contains approximately 0.03 percent carbon dioxide by volume. Bacteria and fungi are the principal microorganisms that degrade organic carbon compounds.

Under most natural systems of vegetation, e.g., forests, the amount of organic material in the soil remains approximately the same from year to year. This results from a balance established between the annual litter fall and death of the plants and the capacity of microorganisms to degrade these tissues.

The most abundant organic material in plants is cellulose. It is readily attacked by many species of bacteria and fungi. The initial enzymatic attack is by cellulase which splits this long-chain polymer of glucose to cellobiose, which contains two glucose units. In turn, the cellobiose is split to glucose by the enzyme β-glucosidase; glucose is metabolized readily by many microorganisms. Complete oxidation yields CO_2 and H_2O. The process can be summarized as follows:

1 Cellulose $\xrightarrow[\text{cellulase}]{\text{enzyme}}$ cellobiose

2 Cellobiose $\xrightarrow[\text{β-glucosidase}]{\text{enzyme}}$ glucose

3 Glucose $\xrightarrow[\substack{\text{of many micro-}\\\text{organisms}}]{\text{enzyme systems}}$ carbon dioxide, water, and/or other end products

An example of the breakdown rate of glucose (and microbial growth) by soil microorganisms is shown in Fig. 25-13. Similar degradation pathways occur for the other major plant tissue substances such as hemicellulose, lignin, and pectin. Carbon dioxide may also originate from the decarboxylation of amino acids, as well as from the dissimilation of fatty acids. All of these transformations may occur in the soil.

A general summary of the carbon cycle is shown in Fig. 25-14.

Figure 25-13. Plate counts of bacteria and fungi and cumulative CO_2 evolution during the incubation of soil treated with glucose. (*Courtesy of B. Behera and G. H. Wagner, Soil Sci Soc Am Proc,* **38**:*591, 1974.*)

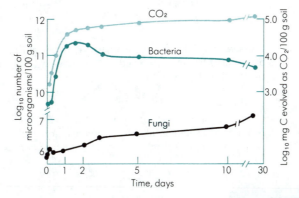

Figure 25-14. Carbon cycle (schematic), showing the role of microorganisms.

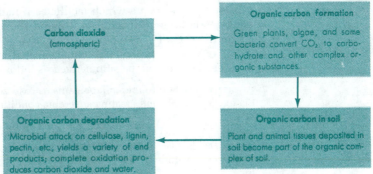

BIOCHEMICAL TRANSFORMATIONS OF SULFUR AND SULFUR COMPOUNDS: THE SULFUR CYCLE

Sulfur, like nitrogen and carbon, passes through a cycle of transformations mediated by microorganisms (see Fig. 25-15). Some species oxidize and others reduce various sulfur compounds. The microbial transformations of sulfur have counterparts in the microbial transformation of nitrogen. For example, sulfide and ammonia are reduction products of the dissimilation of some organic compounds; both may be oxidized by various bacterial species. Some of the biochemical changes by microorganisms involved in this cycle may be summarized as follows:

1 Sulfur in its elemental form cannot be utilized by plants or animals. Certain bacteria, however, are capable of oxidizing sulfur to sulfates. The classical example is *Thiobacillus thiooxidans*, an autotroph; the reaction involved is

$$2S + 2H_2O + 3O_2 \rightarrow 2H_2SO_4$$

2 Sulfate is assimilated by plants and is incorporated into sulfur-containing amino acids and then into proteins. Degradation of proteins (proteolysis) liberates amino acids, some of which contain sulfur. This sulfur is released from the amino acids by enzymatic activity of many heterotrophic bacteria. The following reaction is an example:

$$
\underset{\text{Cysteine}}{
\begin{matrix}
CH_2SH \\
| \\
CHNH_2 \\
| \\
COOH
\end{matrix}}
+ H_2O
\xrightarrow[\text{desulfurase}]{\text{cysteine}}
\underset{\substack{\text{Pyruvic} \\ \text{acid}}}{
\begin{matrix}
CH_3 \\
| \\
C=O \\
| \\
COOH
\end{matrix}}
+ \underset{\substack{\text{Hydrogen} \\ \text{sulfide}}}{H_2S}
+ NH_3
$$

3 Sulfates may also be reduced to hydrogen sulfide by soil microorganisms. An example of bacteria involved in this process is the genus *Desulfotomaculum*, and the reaction suggested is

$$4H_2 + CaSO_4 \rightarrow H_2S + Ca(OH)_2 + 2H_2O$$

4 Hydrogen sulfide resulting from sulfate reduction and amino acid decomposition is oxidized to elemental sulfur. This reaction is characteristic of certain pigmented (photosynthetic) sulfur bacteria and is expressed as

$$CO_2 + 2H_2S \xrightarrow{\text{light}} \underset{\text{Carbohydrate}}{(CH_2O)_x} + H_2O + 2S$$

A laboratory technique which facilitates isolation of various sulfur-metabolizing bacteria is the Winogradsky column shown in Fig. 25-16. The column contains mud, $CaSO_4$, plant tissue (a source of carbohydrate-cellulose), and water. It is exposed to daylight and incubated at room temperature. The microbiological events can be summarized as follows:

1 A variety of heterotrophic microorganisms oxidizes various substrates, depleting the oxygen supply and creating anaerobic conditions:

$$\text{Organic matter} + O_2 \rightarrow \text{organic acids} + CO_2$$

2 Organic acids serve as the electron donors for the reduction of sulfates and sulfites to hydrogen sulfide by anaerobic sulfate-reducing bacteria, e.g., *Desulfotomaculum*:

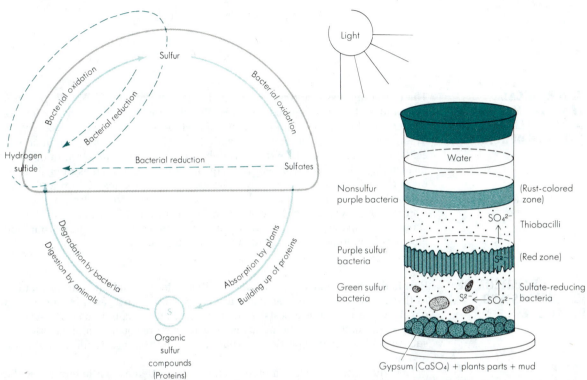

Figure 25-15. Sulfur cycle (schematic), showing the role of microorganisms. (*After Bunker.*)

Figure 25-16. A Winogradsky column showing areas of localization of sulfur-metabolizing bacteria. See text for explanation of sequential developments. (*Courtesy of T. Hattori, Microbial Life in the Soil, Marcel Dekker, Inc., New York, 1973.*)

$$\text{Organic acids} + SO_4^{2-} \rightarrow H_2S + CO_2$$

3 Photosynthetic microorganisms such as the purple and green sulfur bacteria (*Chromatium* and *Chlorobium*) use hydrogen sulfide as the electron donor to reduce CO_2:

$$CO_2 + H_2S \xrightarrow{\text{sunlight}} (CH_2O)_x + S$$

4 The aerobic sulfur-metabolizing bacteria, *Thiobacillus* spp., develop in the upper portion of the column and oxidize reduced sulfur compounds (sulfides, elemental sulfur, sulfite). Final oxidation product is sulfate; sulfur accumulates:

$$\text{Reduced sulfur compounds} \rightarrow SO_4^{2-} + \text{accumulation of S}$$

5 The nonsulfur purple bacteria (*Rhodospirillum*, *Rhodopseudomonas*, and *Rhodomicrobium*) are facultative phototrophs; they grow aerobically in the dark and anaerobically in the light and can utilize sulfide at low levels. They are capable of utilizing hydrogen gas as an electron donor in photosynthesis:

$$CO_2 + H_2S \xrightarrow{light} (CH_2O)_x + S$$

$$CO_2 + 2H_2 \xrightarrow{light} (CH_2O)_x + H_2O$$

BIOCHEMICAL TRANSFORMATIONS OF OTHER ELEMENTS AND THEIR COMPOUNDS

The preceding discussion was concerned with transformations of nitrogen, carbon, and sulfur and their compounds. But this represents only a part of the elements and their compounds that are subject to assimilation and dissimilation by microorganisms. The metabolic activity of microorganisms (production of acids) solubilizes phosphate from insoluble calcium, iron, and aluminum phosphates. Phosphates are released from organic compounds such as nucleic acids by microbial degradation. Bacteria change insoluble oxides of iron and manganese to soluble manganous and ferrous salts. The reverse is also possible.

From these examples of biogeochemical changes that take place in the soil, it should be apparent that microorganisms do, indeed, perform numerous and essential functions that contribute to the productivity of soil.

BIODEGRADATION OF HERBICIDES AND PESTICIDES

Herbicides are chemical substances that kill plants, especially weeds; **pesticides,** as the term denotes, are chemical substances that destroy pests. In the context of soil, we think of those pests which adversely affect economic crops—weeds, insects, and pathogenic microorganisms. Thus a more specific nomenclature for substances classified as pesticides would be herbicides, insecticides, fungicides, and nematocides.

The wide-scale application of herbicides and pesticides, while improving the crop yield, raises questions as to the short- and long-range effects as they are deposited in the soil. Are they degraded by soil microorganisms, and if so how rapidly? Do they have a temporary (or permanent) effect upon the soil microbiota? Do they constitute a form of runoff pollution to streams and rivers and as such affect aquatic plant life? These are some of the questions that concern the soil microbiologist as well as other soil scientists, biologists, and environmentalists. Naturally, a major research effort is directed toward answering the questions asked above, as well as others. An ideal pesticide compound would be one that destroys the pest quickly, and, in turn, the pesticide compound would be degraded to more elementary nontoxic substances. The soil is the

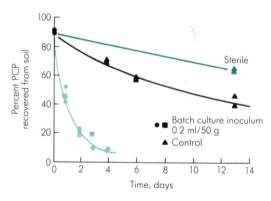

Figure 25-17. Degradation of PCP (pentachlorophenol) in soil by indigenous and inoculated bacteria under laboratory conditions at 30°C. (*Courtesy of R. U. Edgehill and R. K. Finn, Appl Environ Microbiol, 45:1122, 1983.*)

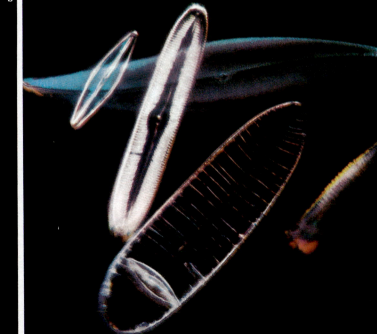

Marine plankton:

a Diatoms, copepods, crustacean larvae, protozoa, animal eggs, and other organisms.
b Diatoms.

(Figure a courtesy of D. P. Wilson; Fig. b courtesy of Dr. Boris Gueft.)

a Exposure of agar medium in petri dish to room air.

b Collection of microorganisms by special air sampling device.

(Figures a and b courtesy of Environmental Services Branch, National Institutes of Health, Public Health Service.)

"sink" which receives the pesticide, and it is the soil microbiota that we depend upon to degrade the compound. As an example to illustrate the results of research on this topic, Fig. 25-17 shows the rate of disappearance (degradation) of a pesticide deposited in the soil. This aspect of soil microbiology, namely, the impact of, and the fate of, pesticides deposited in the soil, is a subject of growing concern.

QUESTIONS

1 Describe how the physical composition of soil influences the magnitude and diversity of the microbial flora.
2 Describe one contribution made by Winogradsky and one by Beijerinck to our knowledge of soil microbiology.
3 Assume that you made a microscopic count on a soil sample and a standard nutrient agar plate count from the same sample. What generalizations are likely with respect to the comparability of the counts?
4 How could one proceed to enumerate, by cultural techniques, the various physiological groups of microorganisms present in soil?
5 Compare the microbial flora of soil in the region of the rhizosphere to that in an area at a distance from the rhizosphere.
6 What is meant by the term *mineralization*? Give an example.
7 Assume that some protein material is buried in the soil. Trace the changes it may undergo as a result of microbial attack. Identify bacteria capable of bringing about each of the changes.
8 Distinguish between symbiotic and nonsymbiotic nitrogen fixation. Name several genera of bacteria that are nonsymbiotic nitrogen fixers.
9 What are the components of the bacterial nitrogen-fixing system?
10 How may the process of nitrogen fixation be determined experimentally?
11 Describe the process by which *Rhizobium* spp. invade the root system of a leguminous plant.
12 How is recombinant DNA technology being explored to develop new means of nitrogen fixation?
13 Tremendous amounts of plant material, largely cellulose, are deposited annually on the earth's surface. Insofar as microbiological events are concerned, what happens to this cellulose?
14 Illustrate, with reactions, the manner in which organically bound sulfur is released by microbial dissimilation.
15 Describe the mineralization process as related to the dissimilation of organic phosphorous compounds.

REFERENCES

Alexander, M.: *Introduction to Soil Microbiology*, Wiley, New York, 1977. *A general book useful for more detailed information on the topics presented in this chapter as well as other subjects related to soil microbiology.*

Gray, T. R. G., and D. Parkinson (eds.): *The Ecology of Soil Bacteria: An International Symposium*, University of Toronto Press, Toronto, 1968. *The papers include identification of soil bacteria, factors influencing their development, effects on soil and metabolic interrelations between bacteria and between bacteria and other microorganisms, as well as the microbial flora in the plant-root environment.*

Griffen, D. M.: *Ecology of Soil Fungi,* Syracuse University Press, Syracuse, New York, 1972. *A view of general problems of soil fungal ecology is presented in the first section. The second part provides a critical review on the physical ecology of soil fungi.*

Hattori, Tsutomu: *Microbial Life in the Soil, an Introduction,* Marcel Dekker, New York, 1973. *The subject matter is presented by interweaving relevant material from other disciplines. The coverage of microbial distribution, physiology, and interactions is good, and interrelationships between colloidal systems of the soil and microbes are examined in depth.*

Payne, W. J.: *Denitrification,* Wiley, New York, 1981. *A small book (214 pages) which provides an in-depth discussion of microbiological denitrification.*

Richards, B. N.: *Introduction to the Soil Ecosystem,* Longman, Inc., 1974. *A good overview of the characteristics and activities of organisms found in soil is given. The ecosystem concept and soil as an ecosystem are presented in detail.*

Starr, M. P., et al. (eds.): *The Prokaryotes: A Handbook on Habitats, Isolation and Identification of Bacteria,* vols. 1 and 2, Springer-Verlag, New York, 1981. *These two volumes provide extremely comprehensive coverage of all groups of procaryotes. One can obtain good coverage of groups of soil bacteria in these volumes.*

Stewart, W. D. P. (ed.): *Nitrogen Fixation of Free-Living Microorganisms,* Cambridge, New York, 1976. *A series of papers discussing the distribution of the nitrogenase enzyme in certain bacteria and blue-green algae and their role in the soil and aquatic ecosystems and methods of measuring nitrogen fixation.*

Chapter 26　Aquatic Microbiology

Aquatic microbiology is the study of microorganisms and their activities in fresh, estuarine, and marine waters, including springs, lakes, rivers, bays, and seas. It is the study of the microorganisms—viruses, bacteria, algae, protozoa, and microscopic fungi—which inhabit these natural waters. Some of these microorganisms are indigenous to natural bodies of water; others are transient, entering the water from air or soil or from industrial or domestic wastes. For example, wastewater can be pumped into rivers and coastal waters, or it can be disposed of in deep ocean dump sites. Wastewater usually contains microorganisms which will influence the activities of microorganisms already present in the receiving waters. This important aspect of aquatic microbiology, wastewater, will be treated in greater detail in the following chapter. The present chapter will deal primarily with aquatic microorganisms and their habitat.

Aquatic microorganisms and their activities are of great importance in many ways. They may affect the health of humans and other animal life; they occupy a key position in the food chain by providing rich nourishment for the next higher level of aquatic life; they are instrumental in the chain of biochemical reactions which accomplish recycling of elements, e.g., in mineralization. Aquatic microbiology, which in previous decades was studied by a relatively few microbiologists, has emerged as one of the more important areas of applied microbiology. Urbanization and consequently the growing demand for water by communities, the importance of natural water as a major food source, the offshore exploration for oil and minerals, and other developments have resulted in the establishment of federal agencies which exercise jurisdiction over many aspects of natural bodies of water. The Environmental Protection Agency (EPA) and the National Oceanic and Atmospheric Administration (NOAA) are two of these agencies.

NATURAL WATERS

The earth's moisture is in continuous circulation, a process known as the **water cycle** or **hydrologic cycle** (see Fig. 26-1). It has been estimated that about 80,000 cubic miles of water from oceans and 15,000 cubic miles from lakes and land surfaces evaporate annually. The total evaporation is equaled by the total precipitation, of which about 24,000 cubic miles fall on land surfaces. Microorganisms of various kinds are present at different stages of this cyclic process—in **atmospheric water, surface water,** and **groundwater.** Because the kinds of aquatic environments are so different, it is not surprising that different species of microbes are considered to be indigenous to specific habitats.

Atmospheric Water

The moisture contained in clouds and precipitated as snow, sleet, hail, and rain constitutes atmospheric water. The microbial flora of this water is contributed by the air. In effect, the air is "washed" by atmospheric water, which carries with it the particles of dust to which microorganisms are attached. Most of the microorganisms are thus removed from air during the early stages of precipitation.

Surface Water

Bodies of water such as lakes, streams, rivers, and oceans represent surface water. To a greater or lesser degree, these waters are susceptible to contamination

Figure 26-1. Diagram of hydrologic cycle. Water returns chiefly to dry continental air through transpiration (A) and evaporation from soil (B), lakes and ponds (C), and streams (D). Continental air moves over ocean to become more moist (E) with conversion to maritime air with precipitation over the oceans. (*Adapted from B. Holzman.*) (*Courtesy of McGraw-Hill Encyclopedia of Science and Technology, vol. 6, McGraw-Hill, New York, 1982, p. 251.*)

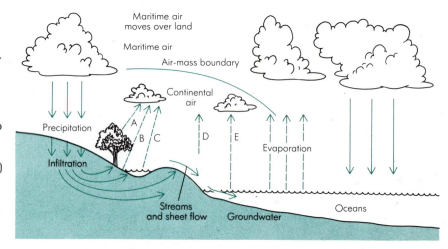

with microorganisms from atmospheric water (precipitation), the surface runoff from soil, and any wastes deliberately dumped into them. Microbial populations vary in both number and kind with the source of water, with composition of the water in terms of microbial nutrients, and with geographical, biological, and climatic conditions.

Groundwater

Groundwater is subterranean water that occurs where all pores in the soil or rock-containing materials are saturated. Bacteria as well as suspended particles are removed by filtration, in varying degrees, depending on the permeability characteristics of the soil and the depth to which the water penetrates. Springs consist of groundwater that reaches the surface through a rock fissure or exposed porous soil. Wells are made by sinking a shaft into the ground to penetrate the groundwater level. Wells less than 100 ft deep are considered to be shallow. Bacteriologically speaking, wells and springs that are properly located produce water of very good quality. If precautions are taken to avoid contamination, the microbial content is negligible.

THE AQUATIC ENVIRONMENT

The microbial population in a body of natural water is, to a large extent, determined by the physical and chemical conditions which prevail in that habitat. This generalization applies to all habitats. It is readily apparent that these conditions vary over wide extremes when one compares streams, estuaries, and the open sea. Some of these conditions are described below.

Temperature

The temperature of surface waters varies from near 0°C in polar regions to 30 to 40°C in equatorial regions. More than 90 percent of the marine environment (by far the major aquatic habitat) is below 5°C, a condition favorable for the growth of psychrophilic microoganisms. Microorganisms do occur in natural hot springs where temperatures as high as 75 to 80°C prevail (*Thermus aquaticus*, a common bacterial inhabitant of hot springs, has an optimum growth temperature of 70 to 72°C). Recently, microbiologists have reported extreme thermophilic microorganisms associated with geothermal vents in the Pacific Ocean floor. These unusual microbes are said to be capable of growing at 250°C and 265 atm of pressure. If these reports can be confirmed, the existence of such organisms will cause all biologists to reevaluate their concept of the maximum temperature that can be tolerated by life on earth! Aside from this extreme, the temperature in lakes, streams, and estuaries is influenced by the seasons, and there are corresponding shifts in the microbial flora.

Hydrostatic Pressure

There are striking differences in the hydrostatic pressure of surface waters and of water in oceanic depths. Hydrostatic pressure affects chemical equilibrium, which, in turn, results in lowering the pH of seawater, resulting in a change in the solubility of nutrients such as bicarbonate, HCO_3^-. Hydrostatic pressure also increases the boiling point of water, thereby maintaining water in its liquid state at high temperatures and pressures. By definition hydrostatic pressure increases with depth at the rate of 1 atm per 10 m. **Barophilic** microorganisms, organisms which cannot grow at normal atmospheric pressures, have been isolated from Pacific trenches (depth 1000 to 10,000 m), where enormous hydro-

Table 26-1. Composition of Some Typical Natural Waters, g/liter

Water	Na$^+$	K$^+$	Ca^{2+}	Mg^{2+}	Cl$^-$	SO$_4^{2-}$	CO$_3^{2-}$
Seawater	10.7	0.39	0.42	1.34	19.3	2.69	0.073
Freshwater:							
Hard	0.021	0.016	0.065	0.014	0.041	0.025	0.119
Soft	0.016		0.010	0.00053	0.019	0.007	0.012

SOURCE: Data from E. Baldwin, *An Introduction to Comparative Biochemistry*, Cambridge, London, 1948.

Figure 26-2. Growth curves in salt solutions for (A) a halotolerant freshwater bacterium, (B) an obligatory halophilic bacterium of brackish water, and (C) an obligatory halophilic marine bacterium, all isolated from the Baltic Sea. (*Courtesy of G. Rheinheimer, Aquatic Microbiology, Wiley, New York, 1974.*)

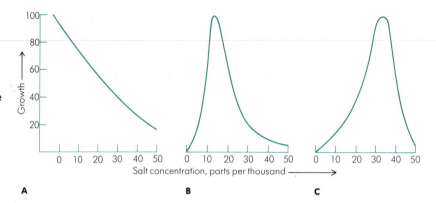

static pressures exist (>100 atm). Hydrostatic pressure of the deep sea is an important factor in the occurrence and growth of marine microorganisms in this environment. Deep sea bacteria have now been isolated from sediment, water, and deep sea animals by using special pressure-retaining sampling devices. In general, barophilic bacteria grow best at pressures slightly less than the pressure of the site from which they were isolated, and almost all must be grown under psychrophilic conditions (about 2°C).

Light

Most forms of aquatic life depend, directly or indirectly, upon the metabolic products of photosynthetic organisms. In most aquatic habitats these **primary producers** are algae, and their growth is restricted to the upper layers of waters through which light can penetrate. The depth of the **photic zone** varies depending on such local conditions as latitude, season, and particularly the turbidity of the water. Generally, the photosynthetic activity is confined to the upper 50 to 125 m. Carbon dioxide is available largely from HCO$_3^-$, although some gaseous CO$_2$ is available.

Salinity

The degree of salinity in natural waters ranges from near zero in freshwater to saturation in salt lakes. A distinctive characteristic of sea water is its high salt content, which is remarkably constant. The concentration of dissolved salts varies between 33 and 37 g/kg water. The major mineral constituents of sea water are listed in Table 26-1. The principal salts are the chlorides, sulfates, and carbonates of sodium, potassium, calcium, and magnesium. The concentration of salts is usually less in shallow offshore regions and near river mouths. Most marine microorganisms are halophilic; they grow best at salt concentrations of 2.5 to 4.0 percent, whereas those from lakes and rivers are salt sensitive and do not grow at a salt concentration of more than 1 percent. The growth response of three bacterial types to different levels of salinity is shown in Fig. 26-2.

Turbidity

There is marked variation in the clarity of surface waters. The Adriatic Sea is sparkling clear at great depths, whereas some near-shore rivers are often turbid. The suspended material responsible for the turbidity includes (1) particles of mineral material which originate from land; (2) detritus, predominantly particulate organic material, such as cellulose, hemicellulose, and chitin fragments; and (3) suspended microorganisms. As previously mentioned, turbidity of the water influences the penetration of light, which in turn affects the photosynthetic zone. Particulate matter also serves as a substrate to which microorganisms adhere or as substrates that are metabolized. Many species of marine bacteria characteristically grow while attached to a solid surface (see Fig. 26-3) and are called epibacteria or periphytes.

Figure 26-3. Marine bacteria attached to particulate materials are referred to as epibacteria. Shown here are marine bacteria attached to (A) agar particles, (B) chitin particles, and (C) cellulose fragments. (*Courtesy of W. A. Corpe, L. Matsuuchi, and B. Armbruster and Proceedings of Third International Biodegradation Symposium, 1976.*)

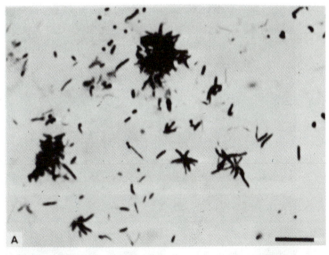
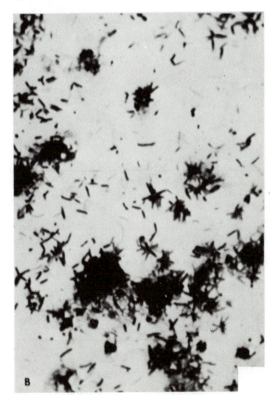

Protozoa
Left: The dinoflagellate protozoan *Noctiluca* that causes nontoxic red tides in Puget Sound. Note the flagellum. Actual diameter approximately 100 micrometers.

Right: the lorica of the ciliate protozoan *Tintinnopsis*. Actual length approximately 100 micrometers. (Photos courtesy Alexander J. Chester)

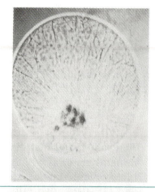 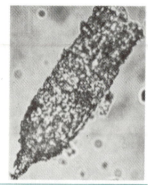

Crustaceans
Left: The herbivorous copepod *Calanus* showing the antennae and feeding (upper) and swimming (lower) appendages. Actual length approximately 3 millimeters.

Right: The carnivorous copepod *Euchaeta*, with coarse feeding appendages adapted for grasping, in contrast to the filtering apparatus of *Calanus*. Actual length approximately 1 centimeter. (Photos courtesy Charles H. Greene)

  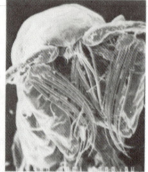

The euphausiid "krill" *Euphausia*, which is mostly herbivorous in Puget Sound. Actual length approximately 2 centimeters. (Courtesy Mark D. Ohman)

Rotifers
The brackish-water rotifer *Brachionus*. The blur is caused by beating cilia. Actual length approximately 200 micrometers. (Courtesy Richard Kaiser)

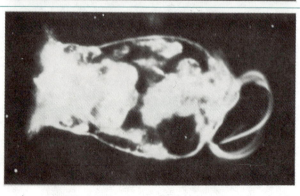

Figure 26-4. (A) Zooplankton comprises an extremely diverse animal population ranging from microscopic unicellular organisms (protozoans) to multicellular metazoans as shown in this figure. Several species found on Puget Sound on the Pacific Coast are shown here.

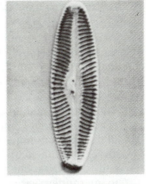

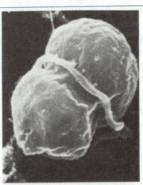

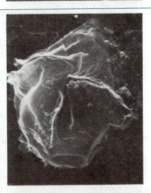

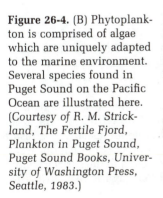

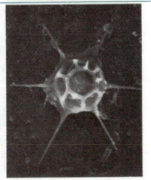

Diatoms

Top left: Separated pieces of the frustule of the solitary centric diatom *Coscinodiscus*. Actual diameter approximately 100 micrometers. (Courtesy National Marine Fisheries Service (NMFS), NOAA, micrograph by Michael Eng)

Top right: The solitary pennate diatom *Navicula*. Actual length approximately 100 micrometers. (Courtesy School of Oceanography, University of Washington)

Bottom left: A portion of a chain of the centric diatom *Skeletonema*. Actual cell diameter approximately 25 micrometers. (Courtesy B. Dumbauld)

Bottom right: A segment of the centric diatom *Chaetoceros*. Actual cell diameter approximately 35 micrometers. (Courtesy Beatrice C. Booth)

Dinoflagellates

Left: The unarmored dinoflagellate *Gymnodinium*. Note the flagellum in the transverse groove. Actual cell length approximately 6 micrometers.

Right: The armored dinoflagellate *Gonyaulax*, cause of paralytic shellfish poisoning. Actual size approximately 30 micrometers. (Photos courtesy Susan B. Stanton)

Phytoflagellates

Left: A green flagellate (probably *Pyramimonas*). Note flagella and surface scales. Actual size approximately 8 micrometers.

Right: The skeleton of the golden-brown silicoflagellate *Dichtyocha*. Actual size including spines approximately 50 micrometers. (Photos courtesy Beatrice C. Booth)

Figure 26-4. (B) Phytoplankton is comprised of algae which are uniquely adapted to the marine environment. Several species found in Puget Sound on the Pacific Ocean are illustrated here. (*Courtesy of R. M. Strickland, The Fertile Fjord, Plankton in Puget Sound, Puget Sound Books, University of Washington Press, Seattle, 1983.*)

Hydrogen-Ion Concentration (pH)

Aquatic microorganisms, in general, can be grown at pH 6.5 to 8.5. The pH of the sea is 7.5 to 8.5. Optimum growth of most marine species is obtained on media adjusted to pH 7.2 to 7.6. Lakes and rivers may show a wider range in pH depending upon local conditions.

Inorganic and Organic Constituents

The quantity and type of inorganic and organic materials present in the aquatic environment are important in determining the microbial flora. Nitrates and phosphates are important inorganic constituents, particularly for the growth of algae. Organic compounds are required for the growth of saprophytic bacteria and fungi. Near-shore waters, which receive domestic wastewater, are subject to intermittent variations in their nutrient load, whereas the nutrient load of the open sea is very low and stable. Industrial wastes may contribute antimicrobial substances to estuaries and coastal waters. Mercury and other heavy metals in small concentrations may inhibit growth of some microorganisms while simultaneously permitting the growth of resistant forms. Resistance is usually coded for by genes associated with R (resistance) plasmids. For example, many pseudomonads and staphylococci are capable of volatilizing mercury, thereby removing its toxic effects from their immediate environment.

DISTRIBUTION OF MICROORGANISMS IN THE AQUATIC ENVIRONMENT

Microorganisms in the aquatic environment may occur at all depths ranging from the surface region to the very bottom of ocean trenches. The top "layers," especially the surface film, and the bottom sediments harbor the higher concentrations of microorganisms, particularly in deep waters.

Plankton (Phytoplankton, Zooplankton)

The aggregation of floating and drifting microbial life in the surface region of the aquatic ecosystem is called plankton. Plankton may be composed primarily of algae (phytoplankton), or it may be predominantly protozoa and other minute animal life (zooplankton). Phototrophic microorganisms are regarded as the most important plankton since they are the primary producers of organic matter via photosynthesis. Most phytoplanktonic organisms are motile, possess some structural feature, or contain oil droplets which give them buoyancy; all these features aid the organisms in maintaining their location in the photosynthetic zone (see Figs. 26-4A and B). The multitude of physical conditions which influence the plankton population quantitatively and qualitatively are shown in Fig. 26-5.

Benthic Microorganisms

Microbial inhabitants of the bottom region of a body of water are referred to as the benthic organisms. The richest region of an aquatic system in terms of numbers and kinds of organisms is the benthic region. Many aquatic microorganisms inhabit the gut of marine animals, an even richer habitat.

Mixing of Waters (Upwelling)

The movement of water by wind, tide, or currents accomplishes some redistribution of the microbial flora. A phenomenon called upwelling occurs in an ocean when water rises from a deeper to a shallower depth, usually as a result of divergence of offshore currents or winds. In this process the bottom water carries with it a rich supply of nutrients that are delivered to the surface region. Upwelling occurs off the coasts of California and Peru and is responsible for

Figure 26-5. Physical forces produce both positive and negative effects on phytoplankton growth, and blooms occur when the forces are balanced. Phytoplankton standing stock is increased by photosynthesis and nutrient uptake, which are regulated by sunshine, stratification, and mixing. Standing stock is decreased by animal consumption, sinking, and flushing by winds, tides, and runoff. (Courtesy of R. M. Strickland, The Fertile Fjord, Plankton in Puget Sound, Puget Sound Books, University of Washington Press, Seattle, 1983.)

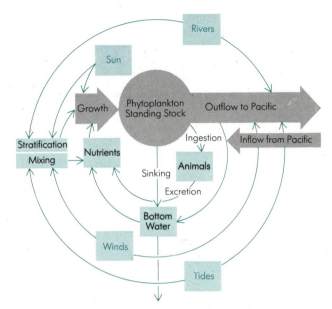

the high productivity of these regions. Geothermal vents also contribute to the total nutrient budget of the ocean. It has been calculated that vents such as the one near the Galapagos Islands account for most of the nutrients dissolved in the oceans of the world. Prior to discovery of these vents, oceanographers were unable to account for nutrients on the basis of precipitation, input from rivers and streams, and other obvious sources. Another interesting feature of the oceans is the gyre, large spiraling surface currents in the ocean that tend to aggregate and retain nutrients, wastes, and microorganisms. Gyres have only been appreciated in recent years, through the use of satellite imagery (Fig. 26-6).

TECHNIQUES FOR THE STUDY OF AQUATIC MICROORGANISMS

Numerous problems are associated with attempts to characterize the microbial flora of aquatic evironments. This is particularly true of samples from the open sea. Among these problems are the following:

1 Many aquatic microorganisms will not grow on the usual laboratory media such as nutrient agar or nutrient broth and consequently cannot be isolated. It is generally acknowledged that the estuaries and oceans contain a large number of microbial species that await discovery.

2 A high percentage of aquatic bacteria have a natural affinity to grow attached to solid surfaces, either on particulate material or on large organisms. See Fig. 26-3.

3 During the time which elapses between sample collection and transport back to the home-based laboratory, there is a loss of viability of many organisms. Accordingly, a laboratory-equipped ship is desirable for on-location culturing of specimens. This is a very costly facility.

4 The collection of samples from the depth of the estuary or ocean requires specialized sampling equipment (Fig. 26-7).

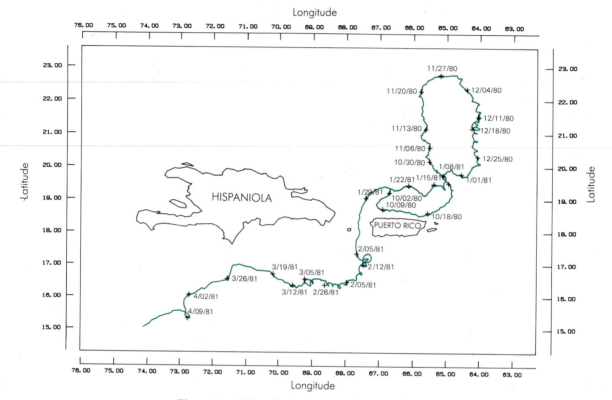

Figure 26-6. Meandering course taken by a buoy released at an ocean waste disposal site, 64 km north of Arecibo, Puerto Rico, on 2 October 1980. The buoy was carried by large surface gyres north toward Bermuda, then back through the original point of release and, finally, into the Caribbean Sea. Gyres can concentrate nutrients, wastes, and microorganisms in the aquatic environment. (*Courtesy of D. J. Grimes, University of Maryland, and W. G. Williams, Clearwater Consultants Inc.*)

5 Routine dependable techniques are not available for isolation of aquatic viruses.

A variety of procedures is used for the microbiological examination of aquatic specimens. The choice of method is determined by the purpose of the examination, e.g.:

1 Microsopic examination for identification and enumeration of algae, bacteria, protozoa, and many fungi.
  (*a*) Direct viable counting of physiologically responsive bacteria (see Fig. 26-8).
  (*b*) Detection of specific bacteria by means of epifluorescent microscopy and fluorescent antibody techniques (see Fig. 26-9).

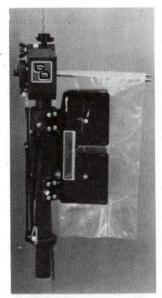

Figure 26-7. Special equipment like the "chopstick" apparatus shown here is used to collect water samples aseptically from the marine environment. Shown here is a sterile plastic bag sealed and ready to be lowered to a desired depth, at which point it is tripped (opened) with a messenger. The sample of water is collected, after which the bag is again sealed and the sample is raised to the surface. (*Courtesy of General Oceanics.*)

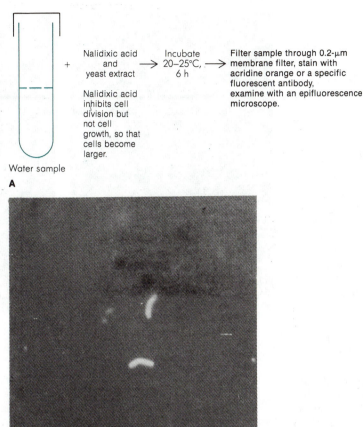

Nalidixic acid and yeast extract $+$ $\longrightarrow$ Incubate 20–25°C, 6 h $\longrightarrow$ Filter sample through 0.2-μm membrane filter, stain with acridine orange or a specific fluorescent antibody, examine with an epifluorescence microscope.

Nalidixic acid inhibits cell division but not cell growth, so that cells become larger.

Water sample

A

B

Figure 26-8. The isolation of physiologically responsive bacteria from marine samples. (A) Outline of technique. (B) Photomicrograph of physiologically responsive *Vibrio cholerae* stained with specific fluorescent antiserum (X1,000). Physiologically responsive bacteria are those which have an increased size; compare these cells to the small *V. cholerae* cells in Fig. 26-9 that have been stained with fluorescent antibody without a preincubation with nalidixic acid and yeast extract. (*Courtesy of P. Brayton and D. J. Grimes, University of Maryland.*)

(c) Detection of epibacteria by the submerged-slide technique (see Fig. 26-10).

2 Isolation and/or enumeration of certain groups of bacteria, e.g., *Escherichia*, *Pseudomonas*, *Flavobacterium*, *Proteus*, and *Vibrio*. Many species of bacteria can be cultured by the plate technique on the usual bacteriological media. The membrane-filter technique (see Chap. 27) is applicable for the examination and culti-

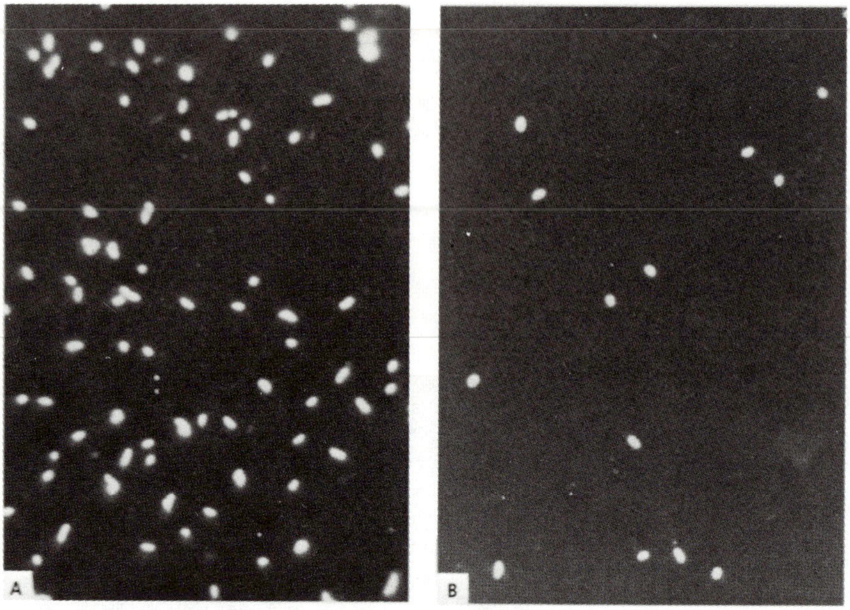

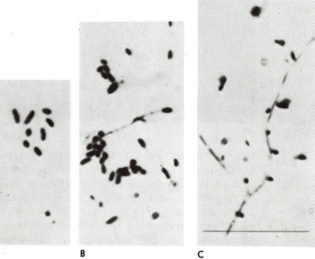

Figure 26-9. Demonstration of marine bacteria *(Vibrio cholerae)* by epifluorescence microscopy (X1,000). (A) Cells stained with acridine orange. (B) Cells stained with fluorescent antibody. *(Courtesy of P. Brayton and D. J. Grimes, University of Maryland.)*

Figure 26-10. The attachment of bacteria to glass slides. Slides were submerged in a regulated marine aquarium, fixed with 2% acetic acid, and stained with Hucker's cyrstal violet solution. (A), (B), and (C) are from slides submerged for 24 h, 48 h, and 72 h, respectively. Note the variety of morophological forms. The scale shown is 5 μm. *(Courtesy of W. A. Corpe, "Attachment of Marine Bacteria to Solid Surfaces," in Adhesion in Biological Systems, Academic, New York, 1970.)*

vation of many bacteria from the aquatic environments. It can also be used to separate different-size fractions of the aquatic microbial community.

3 Enrichment-culture technique for isolation of specific physiological or metabolic types of microorganisms as described in Chap. 6.

4 Measurement of total mass or biochemical activity.

    (a) ***Biomass determination.*** Dry weight determination of cell mass.

    (b) ***Carbon-14 uptake.*** Supplementation of the CO_2 in water with radioactive $^{14}CO_2$ supplied as radioactive sodium bicarbonate ($NaH^{14}CO_3$) and measurement of ^{14}C assimilated by cells.

    (c) ***ATP synthesis.*** Measurement of amount of ATP as a function of the rate of microbial activity or total biomass.

    (d) ***Chlorophyll determination.*** In the case of algae one can make a measure of chlorophyll.

AQUATIC MICROORGANISMS

As previously stated, aquatic microbiology is the study of microbial life in fresh, estuarine, and marine waters. It includes the microbiology of lakes, ponds, streams, estuaries, and the sea. it is an all-inclusive term for the study of microorganisms in natural waters. The microbiology of freshwaters constitutes a part of the science of limnology, which is the study of the flora and conditions for life in lakes, ponds, and streams.

Lakes and Ponds

Lakes and ponds have a characteristic zonation and stratification (see Fig. 26-11). There is usually a fairly large littoral zone along the shore, which has considerable rooted vegetation and includes regions where light penetrates to the bottom. In open areas, the limnetic zone is determined by the light-compensation level (depth of effective light penetration). Photosynthetic activity decreases progressively in the deeper regions of the open water (profundal

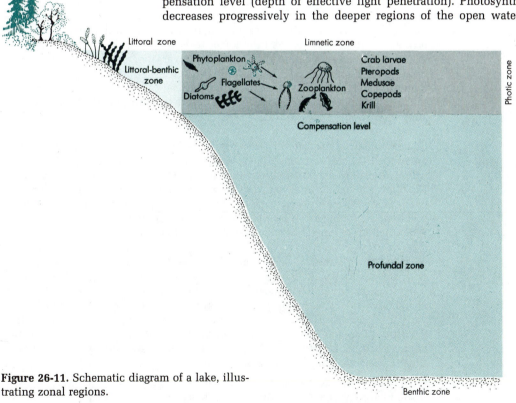

Figure 26-11. Schematic diagram of a lake, illustrating zonal regions.

Figure 26-12. Occurrence of "blooms" or phytoplankton "pulses" in northern temperate lakes in spring and autumn. The combination of conditions—nutrient concentration, light, and temperature—accounts for this phenomenon. *(Courtesy of E. P. Odum, Fundamentals of Ecology, 3d ed., Saunders, Philadelphia, 1971.)*

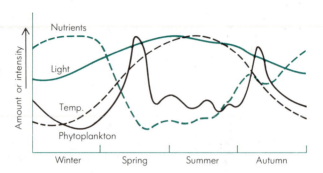

zone). The **benthic zone** is composed of soft mud or ooze at the bottom. Together with the profundal zone, the benthic region is largely populated by heterotrophic organisms. When the benthic zone is composed primarily of organic material, the majority of organisms will be anaerobic decomposers. The greatest variety of physiological types is found in the limnetic and littoral zones, and in addition they constitute the most productive regions. Productivity, of course, is affected by the chemical nature of the basin and the nature of imported materials from streams and rivers. Lakes and ponds of the temperate region exhibit interesting seasonal changes in their microbial populations due to stratification of the water as a result of temperature differences. Such stratification acts as a barrier to nutrient and oxygen exchange, especially in still water. In the summer, the top layers tend to be warmer than the lower regions, but in the winter, ice, which is less dense, collects on the top; therefore, a reversal of temperature and mixing occurs in the spring and in the fall, often resulting in massive growth of algae (**bloom**) (see Fig. 26-12). Lakes or ponds enriched with nutrients, particularly nitrogen and phosphorus, a process referred to as **eutrophication,** are likely to support excessive algal growth.

Streams

Streams obtain a majority of nutrients from the flow of inorganic and organic materials from the surrounding terrestrial system or lakes or ponds. To a major extent, the microbial flora reflects the immediate terrestrial conditions, including the effects of agricultural and industrial practices. The drastic environmental changes in streams and rivers created by rapidly expanding urbanization on the one hand and changes in farming practices on the other make it impossible to generalize upon typical or characteristic microbial flora.

Estuaries

An **estuary** is a semienclosed coastal body of water which has a free connection with the open sea. Stated differently, it is the coastal adjunct of the marine ecosystem. Compared with ocean waters the estuary lacks constancy in many characteristics. Estuaries are complex systems which receive inputs from a variety of sources. Temperature, salinity, turbidity, nutrient load, and other conditions fluctuate over a wide gradient in space and time. Some characteristics of the Chesapeake Bay, one of the world's major estuarine systems, serve to illustrate the variations in an estuary.

1 The Bay serves as the receiving basin for nine major rivers, draining much of southern New York State, Pennsylvania, Maryland, and Virginia.

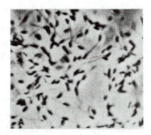

Figure 26-13. (A) Colonization of *Hyphomicrobium* sp. from a marine environment on a glass slide. *Hyphomicrobium* spp. and *Caulobacter* spp. are important film-forming marine bacteria. Phase microscopy, X600. *(Courtesy of W. A. Corpe.)* (B) Scanning electron micrograph of *Leucothrix mucor* growing on the carapace of a 3-day-old lobster (X1,860). *(Courtesy of J. M. Sieburth, Microbial Seascapes, University Park Press, Baltimore, 1975.)*

2 It has a shoreline of 4,600 miles, which includes highly industrialized areas, residential areas, farmlands, and uninhabited marshlands.

3 The salinity varies from less than 1 percent in the tributaries to 3.5 percent (normal seawater) at the mouth of the Bay. It is estimated that an estuary system is filled approximately half by seawater and half by river water. Unlike rivers, estuaries are subject to tidal changes.

4 The Bay, directly or indirectly, is subject to the activities of a multimillion population of humans who reside within the region. Practices associated with agriculture, commerce, industry, and recreation influence the conditions of the Bay.

From these observations it is apparent that the microbial flora of the estuary is subject to considerable fluctuation. Some species are indigenous to specific ecological niches of the estuary; others are transient, having been added from domestic, industrial, agricultural, or atmospheric sources. In areas receiving domestic pollution rich with organic nutrients, predominant bacteria include coliforms, fecal streptococci, and species of *Bacillus, Proteus, Clostridium, Sphaerotilus, Beggiatoa, Thiothrix, Thiobacillus,* and many others. Viruses of the enteric group are also likely to be found. In regions of the estuary that are nutritionally poor, one is likely to find the budding and/or the appendaged bacteria, e.g., *Hyphomicrobium, Caulobacter,* and *Gallionella,* in addition to pseudomonads (see Fig. 26-13). Soil bacteria, e.g., *Azotobacter, Nitrosomonas,* and *Nitrobacter,* are also likely to be present. Numerous fungi (*Ascomycetes, Phycomycetes,* and *Fungi Imperfecti*) occur in various regions of the estuary.

The Sea

Microorganisms are found at all depths and at all latitudes in seawater. They occur in plankton and in the sediment of the ocean floor. The great volume of the open sea provides an environment with less variations in conditions than the other aquatic waters discussed. By the same token, the process of obtaining samples for study from the sea, including bottom sediments, presents significant technical problems.

Marine Plankton

The phytoplankton population comprises numerous species of diatoms, cyanobacteria, dinoflagellates, coccolithophores, silicoflagellates, chrysomonads, cryptomonads, and chlamydomonads. This group of microorganisms is chiefly responsible for the conversion of radiant energy to chemical energy—energy stored in chemical substances that accumulate in the sea. The magnitude of this accomplishment is revealed by a calculation that suggests a requirement of 50 billion metric tons of phytoplankton to support the growth of the potential world fish catch, estimated at 50 million metric tons.

Planktonic algae, under certain environmental conditions, may grow into enormous populations with resultant discoloration of the water—a condition referred to as **bloom** (see Chap. 18). The characteristic color of the Red Sea is associated with heavy blooms of a cyanobacterium, *Oscillatoria erythraea*, which contains the pigments phycoerythrin and phycocyanin. "Red tides" are likewise due to the explosive growth of certain planktonic species. Brown, amber, or greenish-yellow discoloration of extensive areas of water occur as a result of blooms by other microorganisms.

Figure 26-14. Bioluminescence, the chemical emission of light by organisms, is characteristic of many marine forms of life. (A) A saltwater squid is visible in the dark as a result of the development of luminous bacteria on its surface. (B) Colonies of luminous bacteria as they appear in the dark. *(From W. D. McElroy and H. H. Seliger, "Biological Luminescence," Sci Am,* **207**:*76–89, December 1962. Reprinted with permission. Copyright © 1962 by Scientific American, Inc. All rights reserved.)* (C) A photograph taken by the light of the luminous bacterium *Photobacterium phosphoreum.* This bacterium is found in seawater, on the surface and in the alimentary tract of some marine fishes, and in the luminous organs of some fish and cephalopods. *(Courtesy of F. H. Johnson, Princeton University.)*

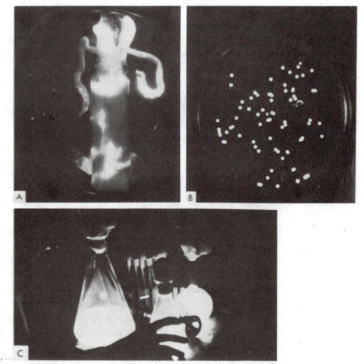

The bacterial population throughout the photosynthetic zone is closely related to the distribution of phytoplankton. The beneficial effect of the plankton may be attributed to both the organic substances they elaborate and the solid surfaces provided for bacterial aggregation. Bacterial populations differ widely with prevailing conditions. The temperature of the marine environment and the degree of salinity would be suitable for the growth of psychrophilic and halophilic physiological types. Most marine bacteria, however, are halotolerant or slightly to moderately halophilic. Among the psychrophilic forms are certain luminous bacteria which can produce light in the presence of oxygen. These bacteria may exist in symbiotic association with certain species of marine animals (see Fig. 26-14). In general, aquatic bacteria tend to be Gram-negative. It is thought that the Gram-negative envelope provides a structure better suited to support life in nutritionally dilute aquatic environments than the Gram-positive cell wall. Important hydrolytic enzymes are retained in the periplasmic space, rather than being excreted and lost to the aquatic environment, as would be the case for Gram-positive bacteria. In addition, lipopolysaccharide (LPS) of the outer membrane affords Gram-negative bacteria protection from certain toxic molecules, e.g., fatty acids and antibiotics, and it may serve to sequester important nutrients from the water. Species of the following genera are commonly found in freshwater: *Pseudomonas, Flavobacterium, Aeromonas,* and *Alcaligenes.* Common to marine and estuarine waters are *Vibrio, Acinetobacter, Pseudomonas, Flavobacterium, Alteromonas,* and *Staphylococcus.* Bacteria in the surface region of the marine environment are often pigmented, a characteristic which may afford protection from the lethal portion of solar radiation. Mold spores and mycelium fragments are present in seawater throughout the photosynthetic zone. Species of *Deuteromycetes, Phycomycetes,* and *Myxomycetes* have been isolated from marine environments. Although the yeasts have not been extensively studied, they probably have a role similar to bacteria.

Protozoa (species of *Foraminifera* and *Radiolaria,* as well as many flagellated and ciliated species) are present in large numbers in the region inhabited by the phytoplankton. These zooplankton animals feed ("graze") upon phytoplankton organisms, bacteria, or detritus. Observations indicate that many zooplankton avoid light, exhibiting diurnal migrations. At night the animals graze on phytoplankton at the surface, and during the day they sink below the photic zone.

The microbial population is sparse near the surface of the sea because the intensity of illumination is inhibitory. Beneath the region of photic activity there exists another region inhabited during the day by vertically migrating zooplankton. The bacterial population is distributed more or less uniformly throughout and below these layers, feeding on descending organic material and other nutrients.

The area between these upper strata and the area just above the sea floor is relatively barren, a vast microbiological oceanic desert region. The bottom of the sea is populated by a variety of microorganisms described below.

The Benthic Population

Offshore sediments are inhabited by bacteria and protozoa. Large numbers are present at the mud-water interface; the bacterial population may range from a few hundred to millions per milliliter. The counts in sediments are as high as

Table 26-2. Bacteria Isolated from the Pacific Ocean, Grouped into Physiological Types

Bacteria per Gram of Sediment (Wet Basis)	Sample 8160 at 32°51.2′ N, 117°28.3′ W and depth of 780 m	Sample 8330 at 33°25.9′ N, 118°06.5′ W and depth of 505 m	Sample 9309 at 33°44.2′ N, 118°46.1′ W and depth of 1,322 m
Total aerobes, plant count	930,000	31,000,000	8,800,000
Total anaerobes, oval-tube count	190,000	2,600,000	1,070,000
Ammonification:			
Peptone → NH$_4$	100,000	1,000,000	1,000,000
Nutrose → NH$_4$	10,000	1,000,000	100,000
Urea fermentation, urea → NH$_4$	100	+	1,000
Proteolysis:			
Gelatin liquefaction	100,000	10,000,000	1,000,000
Peptone → H$_2$S	10,000	1,000,000	100,000
Denitrification, NO$_3$ → N$_2$	100	10,000	10,000
Nitrate reduction, NO$_3$ → NO$_2$	100,000	10,000,000	10,000
Nitrogen fixation	0	0	0
Nitrification, NH$_4$ → NO$_2$	0	0	0
Sulfate reduction, SO$_4$ → H$_2$S	1,000	1,000	10,000
Dextrose fermentation	10,000	100,000	1,000
Xylose fermentation	10,000	+	10,000
Starch hydrolysis	10,000	100,000	10,000
Cellulose decomposition	1,000	+	1,000
Fat hydrolysis (lipoclastic)	1,000	+	+
Chitin digestion	100	+	+

SOURCE: C. E. Zobell, *J Sediment Petrol* **8**:10, 1938.

10^8 bacteria per gram. Bacteria from marine bottom deposits represent a variety of physiological types, as indicated by the data in Table 26-2. Many are facultative or strict anaerobes.

THE ROLE AND IMPORTANCE OF AQUATIC MICROBIAL ECOSYSTEMS

Aquatic life exhibits a vast complex of interactions among microorganisms and between microorganisms and macroorganisms—both plant and animal. Microorganisms, particularly algae and protozoa, occupy a key role in the food chain of the aquatic environment. Numerous bacterial species perform a variety of biochemical changes in various substrates that allow the recycling of elements and nutrients.

PRODUCTIVITY OF AQUATIC ECOSYSTEMS

The **primary producers** in the marine system are cyanobacteria and eucaryotic algae which predominate in the phytoplankton. Through photosynthesis they are capable of transforming radiant energy into chemical energy (organic compounds). The biological activity of an aquatic ecosystem is indeed dependent upon the rate of primary production performed by the photosynthetic organisms. In shallow estuaries the role of photosynthetic organisms as primary producers is considerably reduced. Plant growth from the shoreline contributes leaves, stems, and roots of vegetation and other organic detritus.

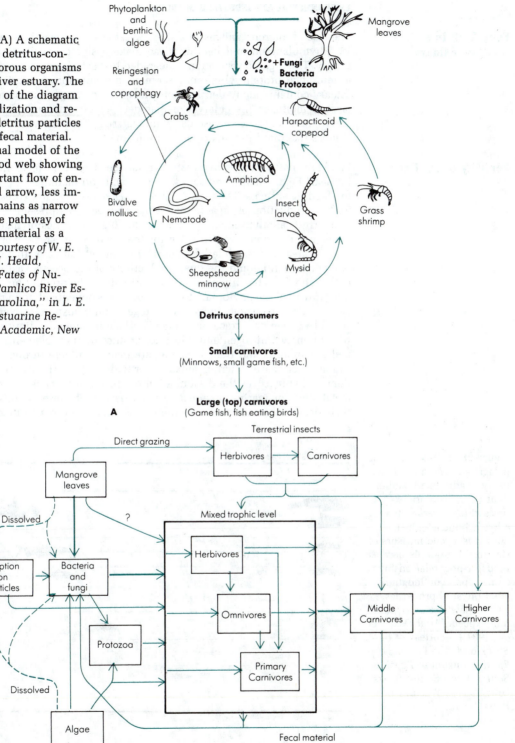

Figure 26-15. (A) A schematic diagram of the detritus-consuming omnivorous organisms of the North River estuary. The cyclical nature of the diagram depicts the utilization and reutilization of detritus particles in the form of fecal material. (B) A conceptual model of the North River food web showing the most important flow of energy as a broad arrow, less important food chains as narrow arrows, and the pathway of dissolved-leaf material as a dashed line. [*Courtesy of W. E. Odum and E. J. Heald, "Sources and Fates of Nutrients of the Pamlico River Estuary, North Carolina," in L. E. Cronin (ed.), Estuarine Research, vol. 1, Academic, New York, 1975.*]

Food Web in a Shallow Estuary

The role of microorganisms in a shallow-estuary food web is shown in Fig. 26-15. Phytoplankton and benthic algae make a small contribution to the food supply. The organic vegetation is degraded by bacteria and fungi and converted to microbial protein which may serve as nutrients for protozoa. However, the estuary contains many detritus consumers (herbivorous and omnivorous crustaceans, mollusks, insect larvae, nematodes, polychaetas, and a few fishes). They derive their energy from vascular plant material. The food web in a deep estuary is more like that in the marine environment.

Fertility of the Ocean

Plankton, particularly phytoplankton, has been referred to as the "pasture of the sea." Fish, whales, and squids feed directly on plankton or on larger plankton-feeding animals. The term **fertility of oceans** is used to express the capacity for the production of organic matter by the organisms present in these waters. The terrestrial environment produces 1 to 10 g of dry organic matter per square meter per day compared to 0.5 g for the deep ocean areas. Nevertheless, the oceanic area is so much larger than the productive land area that this difference is inconsequential and the total productivity of the oceans vastly exceeds that of the land. This fertility depends primarily upon the production of phytoplankton. Growth of the phytoplanktonic organisms is dependent upon radiant energy, carbon dioxide, water, inorganic nitrogen and phosphorus compounds, and several elements in trace amounts. The factors generally limiting growth are radiant energy, nitrogen and phosphorus, and the trace elements. The nitrogen and phosphorus and trace elements are made available through the mineralization reactions of microorganisms, particularly bacteria. As previously mentioned, this involves the dissimilation of organic substrates contributed by the metabolism of marine organisms or the tissues of these organisms following their death. Thus we note that there is a cycle of events in meeting the food

Figure 26-16. Size ranges of pelagic animals from immaturity to adulthood, and of their prey. Sizes are expressed in powers of ten on a logarithmic length scale. Length is a good measure of size in this example because predator-prey relationships often depend on length-related physical properties, such as jaw size or swimming speed. *(After Dexter et al., 1981.) (Courtesy of R. M. Strickland, The Fertile Fjord, Plankton in Puget Sound, Puget Sound Books, University of Washington Press, Seattle, 1983.)*

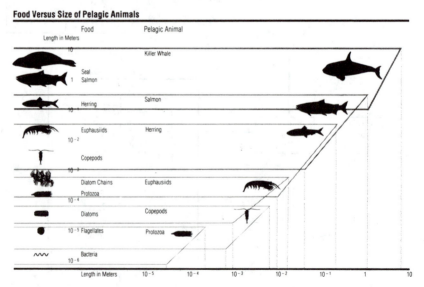

Food Versus Size of Pelagic Animals

requirements of various levels of life in the sea and that microorganisms are indispensable to the process.

The Antarctic Ocean is described as being richer in life than any other major oceanic area. The richness of nutrients in this region is attributable to the mixing of the waters of the Atlantic, Pacific, and Indian Oceans. This mixing action is brought about by the movement of the bottom current of cold water that runs outward from the continental shelf of the Antarctica.

An abundance of nutrients in the upper layers results in a large phytoplankton crop. The red shrimplike crustacean *Euphausia superba* (commonly called krill) feeds upon the phytoplankton; in turn, this organism serves as food for fish, penguins, sea birds, seals, and whales. In this environment the krill represents the key organism in the food chain; it is the link between photosynthetic planktonic life and higher forms of life characteristic of the region. The sequence of organisms which serve as food for prey in the food chain is shown in Fig. 26-16, where the food organism is identified and the pelagic animal that feeds upon it. Although coastal and estuarine regions provide a less stable physical environment, their fertility exceeds that of the open sea because of the large amounts of nutrients available.

BIOGEOCHEMICAL TRANSFORMATIONS

A major biochemical activity of bacterial flora is the dissimilation and mineralization of organic matter, i.e., the dissimilation of organic compounds to carbon dioxide, water, and various inorganic salts. All classes of organic substrates are available; they originate from the metabolic processes of the plant and animal life and from the cells of the dead organisms. Under aerobic conditions the principal products resulting from dissimilation of organic compounds are ammonium, carbon dioxide, sulfate, and phosphate. These can be recognized as nutrients for plant growth, including the phytoplankton. Anaerobic degradation yields more reduced products such as methane, hydrogen, and hydrogen sulfide in addition to ammonia, carbon dioxide, and phosphate.

Biochemical Cycles

The transformation of elements to organic compounds and the subsequent release of elements for conversion to inorganic compounds and reutilization occurs in the aquatic environment. Among the better understood cyclic processes are:

    The nitrogen cycle
    The carbon cycle
    The sulfur cycle
    The phosphorus cycle

The principal microbial reactions that occur in these cyclic processes were described in Chap. 25. In each instance, microorganisms are responsible for releasing the element (e.g., nitrogen) from an organic compound through a series of reactions mediated by various species of microorganisms, particularly bacteria. The element, in the form of an inorganic compound, becomes available again as a plant nutrient.

In the biogeochemical cycles referred to above, many complex materials are dissimilated by microorganisms. For example, cellulosic material is abundant

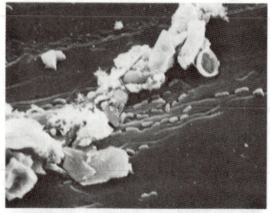

Figure 26-17. Marine bacteria attacking cellulose. The bacteria shown here by scanning electron microscopy are degrading a cellulose (dialysis) membrane (X 1,700). *(Courtesy of J. M. Sieburth, Microbial Seascapes, University Park Press, Baltimore, 1975.)*

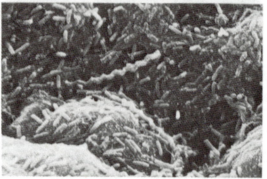

Figure 26-18. Marine bacteria (bacilli and spirochetes) growing on a salmon tail and degrading the tissue. *(Courtesy of J. M. Sieburth, Microbial Seascapes, University Park Press, Baltimore, 1975.)*

in waters and is also important in the sea. Cellulolytic aerobic bacteria degrade the cellulose to the sugars cellobiose or even glucose. These sugars are readily metabolized by many other bacteria. An illustration of aquatic microorganisms attacking cellulose is shown in Fig. 26-17. Similarly, proteolytic microorganisms degrade proteins to peptides and amino acids; further dissimilation results in liberation of nitrogen as ammonia. Figure 26-18 illustrates the microbial growth on the flesh of a salmon tail.

Marine Sediments

Large amounts of diatomaceous materials settle to the bottom of marine environments. Diatoms are characterized by a cell wall of silica. The thickness of the siliceous wall may vary greatly from very thin to very thick depending on the benthic species. Radiolaria and silicoflagellates also have siliceous skeletons. Foraminifera and coccolithophores produce calcareous skeletons. These organisms are important members of the plankton. When they die, they gradually sink to the bottom, where the undissolved calcareous and siliceous skeletons accumulate. Vast beds of these materials, sometimes thousands of feet thick, occur in various regions of the sea. The chalk beds of England and France are chiefly the remains of foraminifera, and a large part of the Monterey series at Lompoc, California, consists of the remains of diatoms. Bacteria, particularly cyanobacteria, are involved in the precipitation of calcium carbonate, which also leads to limestone formation.

The deposition and transformation of iron and manganese in sediments and sulfur deposits, like those found in the Gulf Coast areas of Texas and Louisiana, are closely linked to microbial activity. Petroleum deposits are formed from accumulated and buried organic materials. Microorganisms are believed to play the major role in these processes.

QUESTIONS

1 What is plankton? Phytoplankton? Zooplankton? Name some organisms that belong to each category.

2 What is the Gram reaction of most aquatic bacteria? What explanation is suggested to account for this?

3 Define an estuary.

4 Discuss some of the physical and chemical properties of the aquatic environment in terms of their effects upon microbial growth and survival.

5 What is meant by the term *upwelling*? What effect does upwelling have on productivity of an ocean area? What is a gyre?

6 Describe several techniques that are particularly useful for enumeration and identification of microorganisms from aquatic sources.

7 What occurs when a lake becomes eutrophic?

8 What comparisons can be made between the open sea and offshore oceanic areas with respect to microbial flora?

9 Describe several types of biochemical changes brought about by microorganisms in marine environments. What occurs during the process of mineralization?

10 What contributes to the fertility of the ocean? What does the term *pasture of the sea* refer to?

11 What is referred to by the term *primary producers*?

12 Compare and contrast the fertility of the oceans and inland lakes and rivers.

13 Blooms in lakes usually occur in the spring and fall. Why?

14 Name several geologically important transformations caused by marine microorganisms.

REFERENCES

Cairns, J., Jr. (ed.): *Fresh-Water Microbial Communities*, Research Division Monograph 3, Virginia Polytechnic Institute and State University, Blacksburg, Va., 1969. *Covers the subject of microcosms and microsystems with emphasis on the ecological response of the freshwater environment to the activities of microorganisms.*

Colwell, R. R., and R. Y. Morita (eds.): *Effect of the Ocean Environment on Microbial Activities*, University Park Press, Baltimore, 1974. *Provides information on microorganisms in the deep ocean environment and the effects of physical conditions and ecological interactions of microorganisms in the ocean.*

Cronin, L. Eugene (ed.): *Estuarine Research: Chemistry, Biology and the Estuarine System*, vol. 1, Academic, New York, 1975. *Volume 1 of a two-volume series which provides information about the chemistry and biology of estuarine systems.*

Droop, M. R., and H. W. Jannasch (eds.): *Advances in Aquatic Microbiology*, Academic, New York. *A series that covers all aspects of aquatic microbiology. Volume 1 appeared in 1977.*

Fogg, G. E.: *Algal Cultures and Phytoplankton Ecology*, 3d ed., University of Wisconsin Press, Madison, 1975. *The characteristics of algae, i.e., growth and metabolism, are covered in the first chapters. The second part (four chapters) is devoted to their ecology in lakes and seas.*

McErlean, A. J., C. Kerby, and M. L. Wass (eds.): "Biota of the Chesapeake Bay," *Chesapeake Sci Suppl*, **13**, 1972. *A series of reports by investigators which summarizes information on the biota of the Chesapeake Bay.*

Mitchell, R.: *Water Pollution Microbiology*, Wiley, New York, vol. 1, 1972 and vol. 2, 1978. *These two volumes provide comprehensive information on the role of pollutants in contributing microorganisms to water.*

Rheinheimer, G.: *Aquatic Microbiology*, 2d ed. Wiley, New York, 1980. *A good introductory book to aquatic microbiology, a survey of aquatic bacteria and fungi and their role in the life of lakes, rivers, and seas.*

Sieburth, J. M.: *Microbial Seascapes*, University Park Press, Baltimore, 1975. *This "pictorial essay" on marine microorganisms and their environments is a study of undersea life seen using the scanning electron microscope.*

Sieburth, J. M.: *Sea Microbes*, Oxford University Press, New York, 1979. *A lavishly illustrated text that focuses on marine microorganisms, their ecology, taxonomy, and physiology.*

Strickland, R. M.: *The Fertile Fjord; Plankton in Puget Sound*, Puget Sound Books, University of Washington Press, Seattle, 1983. *An excellent introductory discussion of plankton which provides descriptions of planktonic organisms, their form and function. Very well illustrated.*

Chapter 27 Microbiology of Domestic Water and Wastewater

The drinking water of most communities and municipalities is obtained from surface sources—rivers, streams, and lakes. Such natural water supplies, particularly streams and rivers, are likely to be polluted with domestic and industrial wastes, i.e., the used water of a community (wastewater). Many city dwellers (whose water comes from the rivers) are not aware that a considerable portion of their drinking water may have been used earlier for domestic or industrial purposes. Municipal water-purification systems have been very effective in protecting the inhabitants against polluted water. At the same time, as population centers grow, pollution problems become more serious. A greater quantity of water is required, and the used water must be disposed of, generally by returning it to a natural body of water in the vicinity, which in turn may be the water supply source of another community or municipality.

As a potential carrier of pathogenic microorganisms, water can endanger health and life. The pathogens most frequently transmitted through water are those which cause infections of the intestinal tract; namely, typhoid and paratyphoid bacteria, dysentery (bacillary and amoebic) and cholera bacteria, and enteric viruses. The causative organisms of these diseases are present in the feces or urine of an infected person, and when discharged may gain entrance into a body of water that ultimately serves as a source of drinking water. The incidence of waterborne diseases is discussed in Chap. 35.

It is therefore necessary to employ (1) treatment facilities that purify wastewater prior to its disposal and (2) water-purification methods that provide safe drinking water.

In the first part of this chapter we shall discuss water purification, i.e., ways of providing safe drinking water, the bacteriological examination of water, and methods of determining whether drinking water is free of infectious microorganisms. The second part deals with wastewater disposal systems.

Wastewater treatment procedures, both natural and artificial, are largely dependent upon microbial activity to eliminate or greatly reduce the development of hazardous or objectionable situations.

WATER PURIFICATION

Water that is free of disease-producing microorganisms and chemical substances deleterious to health is called **potable** water. Water contaminated with either domestic or industrial wastes is called **nonpotable** or **polluted** water. The objectives of primary concern in providing potable water are freedom from harmful microorganisms and freedom from undesirable or harmful chemicals. These standards apply both to wells or springs serving single families and to water systems serving hundreds of thousands of persons. However, the processes of purifying water are quite different in these two extreme situations.

Individual Water Supplies

Underground sources—wells and springs—provide most of the water for individual homes in rural areas. Rainwater caught and stored in cisterns is also used to a limited extent; surface water, however, should not be used for drinking purposes unless it is subjected to purification, since there is constant danger of contamination and consequent transmission of disease. Since it is generally impractical to treat surface waters so as to ensure a continuous safe supply for rural domestic purposes, our discussion is limited to water obtained from wells and springs.

As it penetrates through the layers of soil, water from wells and springs undergoes filtration, which removes suspended particles, including microorganisms. It is of prime importance that the supply of groundwater selected be located a safe distance from possible sources of contamination, e.g., pit privies, cesspools, septic tanks, and barnyards.

Municipal Water Supplies

The following information about American public water systems, taken from an article by H. E. Jordan which appeared in the 1955 Yearbook of Agriculture, "Water," illustrates the progress that has been made in this area:

At Bethlehem, Pa., in 1754, Hans Christiansen, who was an immigrant millwright, began work on the first pumped public water works, which was put into regular operation in 1755. The supply provided at Winchester, Va., in 1799 appears to

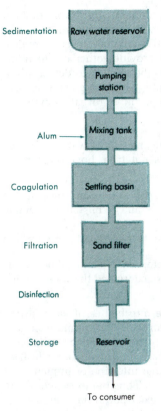

Sedimentation — Raw water reservoir

Pumping station

Alum → Mixing tank

Coagulation — Settling basin

Filtration — Sand filter

Disinfection

Storage — Reservoir

To consumer

Figure 27-1. Flow diagram of usual procedures in municipal water-purification plant.

have been the first complete works developed as a public enterprise. Seventeen cities were served by public water works in 1800 and 83 at the end of 1850.

In operation in 1900 were 3,300 water works. In 1950 approximately 17,000 communities were served by public systems. . . .

Two-thirds of the nation's population is now served by public water works systems. . . . That service, which involves well-supervised production and distribution of safe water, must be given some of the credit for all but wiping out the scourge of typhoid fever of 50 years ago and for virtually eliminating epidemics of water-borne dysentery.

The principal operations employed in a municipal water-purification plant to produce water of a quality safe for human consumption are sedimentation, filtration, and disinfection (see Fig. 27-1). Sedimentation occurs in large reservoirs, where the water remains for a holding period; large particulate matter settles to the bottom. Sedimentation is enhanced by the addition of alum (aluminum sulfate) at the treatment plant, which produces a sticky flocculent precipitate. Many microorganisms, as well as finely suspended matter, are removed as this precipitate descends through the water in the settling basins. The water is next passed through sand filter beds, a process which removes 99 percent of the bacteria. Subsequently, the water is disinfected to ensure its potability. The majority of municipal water-treatment facilities employ chlorination for disinfection. The chlorine dosage must be sufficient to leave a residual of 0.2 to 2.0 mg/liter free chlorine. Other disinfection methods include ozonation and irradiation with ultraviolet light.

The purification process may include procedures for removing minerals that cause the water to be hard, adjusting the pH if the water is too acid or alkaline, removing undesirable colors or tastes, and adding fluoride for the prevention of dental caries.

DETERMINING SANITARY QUALITY

Water can be perfectly clear in appearance, free from peculiarities of odor and taste, and yet be contaminated. Obviously special procedures are necessary to determine its sanitary quality.

Sanitary Surveys

Inspection of a water-producing system by a qualified sanitarian or engineer is called a **sanitary survey.** It includes inspection of (1) the source of the raw water and the conditions that may influence its quality, (2) the operation of the water-purification plant or the construction of the well, and (3) the mechanism for distributing the water to the consumers. Conditions in a community or municipality that may influence the quality of water are not static. There may be changes in population, types of industry, and the quantity of sewage and the manner in which it is disposed; consequently, periodic and comprehensive sanitary surveys are necessary. Data obtained from these surveys are of considerable value, both from the standpoint of indicating any changes in operation that may be necessary to prevent complications from developing and of finding the sources of difficulties that may appear unexpectedly.

Sanitary surveys reveal whether water is being produced under conditions in which potable water would ordinarily be produced. However, potability can be determined only by chemical and bacteriological laboratory tests. Chemical analysis indicates whether water is polluted and provides other useful information as well; however, it is not sensitive or specific enough to detect minor degrees of sewage contamination. On the other hand, bacteriological tests have been designed which are extremely sensitive and specific in revealing evidence of pollution.

Bacteriological Evidence of Pollution

One might assume that the objective in the routine analysis of water would be to search for pathogenic microorganisms. This is not true, however, for the following reasons:

1 Pathogens are likely to gain entrance into water sporadically, and they do not survive for long periods of time; consequently, they could be missed in a sample submitted to the laboratory.
2 If they are present in very small numbers, pathogens are likely to escape detection by laboratory procedures.
3 It takes 24 h or longer to obtain results from a laboratory examination. If pathogens were found to be present, many people would have consumed the water by this time and would be exposed to infection.

It is known that the pathogens that gain entrance into bodies of water arrive there via intestinal discharges of humans or other animals. Furthermore, certain bacterial species, particularly *Escherichia coli* and related organisms designated as **coliforms,** fecal streptococci (e.g., *Streptococcus faecalis*), and *Clostridium perfringens*, are normal inhabitants of the large intestine of humans and other animals and are consequently present in feces. Thus the presence of any of these bacterial species in water is evidence of fecal pollution of human or animal origin. If these organisms are present in water, the way is also open for intestinal pathogens to gain entrance, since they, too, occur in feces. Since the laboratory examination of water for pathogens is beset with the disadvantages already enumerated, techniques have been developed for the demonstration of bacterial species of known excretal origin, particularly organisms of the coliform group. This approach, which has proved satisfactory in practice, has the following advantages:

1 Coliform organisms, particularly *E. coli*, are constantly present in the human intestine in large numbers. It is estimated that billions of these organisms are excreted by an average person in one day.
2 These organisms generally live longer in water than intestinal pathogens.
3 A healthy person would not normally excrete pathogenic organisms, but should an intestinal-tract infection develop, the pathogen is likely to appear in the feces. Thus, the presence of coliforms in water is regarded as a warning signal: the water is subject to potentially dangerous pollution.

The Coliform Group

The coliform group of bacteria includes all the aerobic and facultatively anaerobic, Gram-negative, nonsporulating bacilli that produce acid and gas from the fermentation of lactose. The classical species of this group are *Escherichia coli* and *Enterobacter aerogenes*. The relationship of these organisms to others of

Table 27-1. Differentiation of Typical Strains of *Escherichia coli* from *Enterobacter aerogenes* on the Basis of IMViC Reactions

Organism	Test			
	Indole	Methyl Red	Voges-Proskauer	Citrate
Escherichia coli	+	+	−	−
Enterobacter aerogenes	−	−	+	+

the enteric group—*Salmonella, Shigella, Klebsiella, Proteus, Serratia,* and other genera—all of which are Gram-negative, nonsporulating bacilli, is discussed in Chap. 13. *E. coli,* as we have already pointed out, is a normal inhabitant of the intestinal tract of humans and other animals. *Ent. aerogenes* is most frequently found on grains and plants but may occur in human and animal feces. These species bear a very close resemblance to each other in their morphological and cultural characteristics. Consequently, it is necessary to resort to biochemical tests for differentiation. Tests with the following four characteristics are especially important for this purpose:

1 Ability to produce indole from tryptophan. *E. coli* does, and *Ent. aerogenes* does not.
2 Amount of acidity produced in a special glucose-broth medium and detected by the pH indicator methyl red. Both organisms produce acid from glucose. However, *E. coli* produces a lower pH, which turns the indicator red, whereas *Ent. aerogenes* cultures do not produce as large an amount of acid and thus do not produce the color change.
3 Ability to produce the compound acetylmethylcarbinol in a glucose-peptone medium. This chemical is detected by the Voges-Proskauer test procedure. *E. coli* does not produce acetylmethylcarbinol, but *Ent. aerogenes* does.
4 Utilization of sodium citrate. *Ent. aerogenes* is capable of utilizing sodium citrate as its sole source of carbon; i.e., it will grow in a chemically defined medium in which sodium citrate is the only carbon compound. *E. coli* does not grow under the same circumstances.

For convenience, these tests collectively are designated as the **IMViC reactions** (I = indole, M = methyl red, Vi = Voges-Proskauer reaction, and C = citrate). The reactions for a typical strain of each species are shown in Table 27-1. The reactions for all coli-aerogenes organisms are unfortunately not as clear-cut as those described here. Some cultures give other combinations of reactions to this scheme of testing and are usually referred to as intermediate types. Furthermore, there are additional coliform species in the genera *Klebsiella* and *Citrobacter* for which more detailed biochemical, genetic, and immunologic data are needed for identification.

Bacteriological Techniques

Methods of examining water bacteriologically are set forth in the book *Standard Methods for the Examination of Water and Wastewater,* prepared and published jointly by the American Public Health Association, the American Water Works Association, and the Water Pollution Control Federation. The Environmental Protection Agency (EPA) has also developed a manual for this purpose under the title *Microbiological Methods for Monitoring the Environment—Water and Wastewater.* It is essential that strict attention be given to the following details when water samples are submitted for bacteriological analysis:

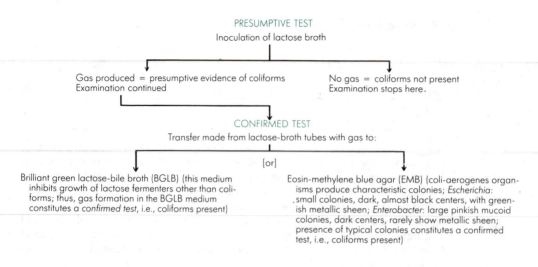

PRESUMPTIVE TEST
Inoculation of lactose broth

Gas produced = presumptive evidence of coliforms
Examination continued

No gas = coliforms not present
Examination stops here.

CONFIRMED TEST
Transfer made from lactose-broth tubes with gas to:

[or]

Brilliant green lactose-bile broth (BGLB) (this medium inhibits growth of lactose fermenters other than coliforms; thus, gas formation in the BGLB medium constitutes a *confirmed test*, i.e., coliforms present)

Eosin-methylene blue agar (EMB) (coli-aerogenes organisms produce characteristic colonies; *Escherichia*: small colonies, dark, almost black centers, with greenish metallic sheen; *Enterobacter*: large pinkish mucoid colonies, dark centers, rarely show metallic sheen; presence of typical colonies constitutes a confirmed test, i.e., coliforms present)

COMPLETED TEST
The most typical colonies are selected from EMB plate (if BGLB used in confirmed test, it is first streaked to EMB) and inoculated into:

Lactose broth (coliforms produce gas)

Agar slant (Gram stain prepared from growth; coliforms are Gram-negative, nonsporulating bacilli)

Fermentation of lactose broth and demonstration of Gram-negative, nonsporulating bacilli constitute a positive completed test demonstrating the presence of some member of the coliform group in the volume of sample examined

Figure 27-2. General scheme of laboratory testing for detection of coliform group in water.

1 The sample must be collected in a sterile bottle.
2 The sample must be representative of the supply from which it is taken.
3 Contamination of the sample must be avoided during and after sampling.
4 The sample should be tested as promptly as possible after collection.
5 If there is a delay in examination of the sample, it should be stored at a temperature between 0 and 10°C.

The routine bacteriological procedures consist of (1) a plate count to determine the number of bacteria present and (2) tests to reveal the presence of coliform bacteria.

Standard Plate Count

Colony counts are performed after plating aliquots of the water sample. The interpretation of the results of the **standard plate count** must take into account the fact that the presence of a few pathogenic microorganisms is more significant than water containing many saprophytic bacteria. However, water of good quality is expected to give a low count, less than 100 per milliliter. Plate counts are useful in determining the efficiency of operations for removing or destroying organisms, e.g., sedimentation, filtration, and chlorination. A count can be made before and after specific treatment. The results indicate the extent to which the microbial population has been reduced.

Tests for the Presence of Coliform Bacteria

Several selective and differential media greatly expedite the process of examining water for coliform organisms. The **standard microbiological** technique involves three successive steps: (1) the **presumptive** test, (2) the **confirmed** test, and (3) the **completed** test (see Fig. 27-2).

Membrane-Filter Technique   The membrane-filter technique for the bacteriological examination of water is illustrated in Fig. 27-3 and consists of the following steps:

1 A sterile disk is placed in a filtration unit.
2 A volume of the water to be tested is drawn through this filter disk; the bacteria are retained on the surface of the membrane.
3 The filter disk is removed and placed upon an absorbent pad that has previously been saturated with the appropriate medium. Alternatively, the disk can be placed

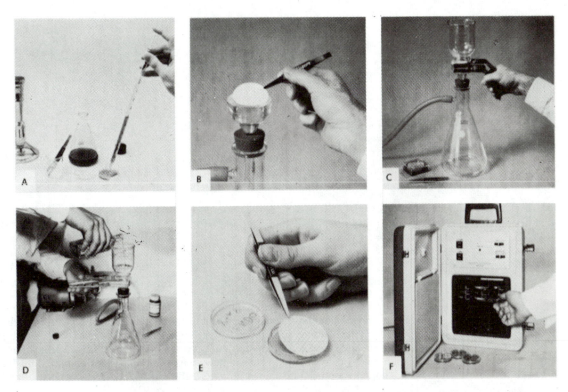

Figure 27-3. Analysis of water with the Millipore filter membrane. (A) Approximately 2 ml of Endo medium is added to the pad contained in the dish. The dish is then covered until the water sample has been filtered through the membrane. (B, C, D) The filter is placed on a filter holder and clamped in position below the funnel, and the water sample (100 ml for testing potability) is poured into the funnel and passed through the Millipore filter by the aid of a vacuum pump. (E) The funnel is removed, and the filter disk, handled with sterile forceps, is placed on the pad previously impregnated with medium (A). (F) The plates are incubated at 35°C for 20 h, at which time the number of coliform colonies can be determined. (*Courtesy of Millipore Filter Corporation.*)

Figure 27-4. (A) Coliform colonies from a water sample grown on a membrane filter with MF-Endo medium acquire an easily distinguishable green metallic "sheen" and are pink to rose red in color. *(Courtesy of Millipore Filter Corporation.)* (B) Verification of total coliform colonies on the membrane filter. *(From Microbiological Methods for Monitoring the Environment: Water and Waste, Environmental Protection Agency, Washington, D.C., 1978.)*

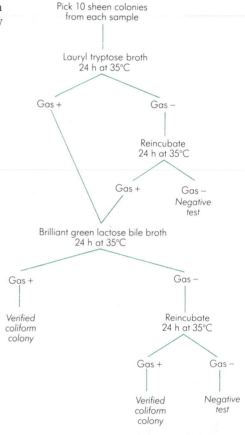

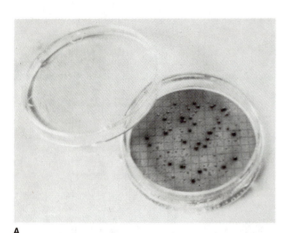

A

B

on the surface of an agar medium in a Petri dish. Special Petri dishes of a size to accommodate the absorbent pad and filtration disk are employed for incubation.

4 Upon incubation, colonies will develop upon the filter disk wherever bacteria were entrapped during the filtration process. (See Fig. 27-4A). Verification of the "sheen" colonies (distinctive appearance of colonies on a differential medium) is done by a procedure shown in Fig. 27-4B.

The membrane filter technique has several desirable features:

1 A large volume of water sample can be examined. Theoretically almost any volume of nonturbid water could be filtered through the disk, the organisms from any given volume being deposited on the disk.

2 The membrane can be transferred from one medium to another for purposes of selection or differentiation of organisms.

3 Results can be obtained more rapidly than by the conventional MPN standard methods.

4 Quantitative estimations of certain bacterial types, e.g., coliforms, can be accomplished when appropriate media are used.

This technique, with modifications, has been adopted for many microbiological procedures other than the examination of water.

Microorganisms Other Than Coliform Bacteria

Some microorganisms besides coliform bacteria are of intestinal origin and could also be used as indicators of fecal contamination of water, such as the fecal streptococci. Other intestinal microorganisms, such as intestinal viruses, are frank pathogens and can cause serious diseases. Still other microorganisms are regarded mainly as nuisance organisms, because they create problems of odor, color, and taste, or cause obstruction of water flow.

Fecal Streptococci

Fecal streptococci are enteric bacteria found in the intestines of warm-blooded animals, including humans. *Streptococcus faecalis* is representative of this group; other species are *S. faecium*, *S. bovis*, and *S. equinus*. Because fecal streptococci, particularly, *S. faecalis*, are abundantly present in the large intestines of humans, their occurrence in water is indicative of fecal pollution.

Slime-Forming Bacteria

Many bacteria are capable of elaborating gummy or mucilaginous materials, either as capsular structures or as extracellular excretion products. The organic and inorganic constituents of the water, which provide nutrients for the bacteria, help to determine whether slime is produced and what organisms are responsible for its production.

Iron Bacteria

The iron bacteria are one of the most important types of nuisance organisms in water. They transform soluble compounds of iron to insoluble compounds of iron (ferric hydroxide) which may be deposited in a sheath around the organism (*Sphaerotilus*) (Fig. 27-5) or secreted so as to form stalks or ribbons attached to the cell (*Gallionella*). This deposition and accumulation of insoluble material in the piping system may eventually have a significant effect on the rate of water flow. Iron bacteria can also produce slime, discolor water, and cause undesirable odors and tastes.

Sulfur Bacteria

Some of the sulfur bacteria are capable of producing and tolerating extreme acidity. Organisms of the genus *Thiobacillus* oxidize elemental sulfur to sulfuric acid and can produce an acidity in the range of pH 1; thus they may be

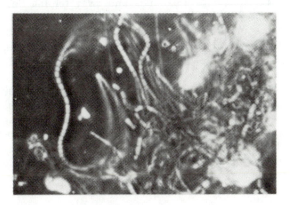

Figure 27-5. Long trichomes of *Sphaerotilus* appear in this photomicrograph of a slime mass. This organism causes many difficulties in industrial water systems because it forms large slime masses and iron deposits. Dark phase contrast (X640). *(Courtesy of J. M. Sharpley, Applied Petroleum Microbiology, Buckman Laboratories, Inc., Memphis, Tenn., 1961.)*

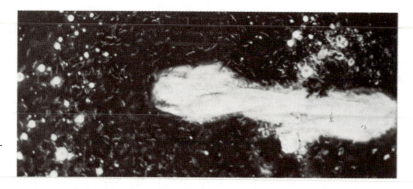

Figure 27-6. Sulfate-reducing bacteria *(Desulfovibrio)* are visible as white curved lines in this photomicrograph from a mine-water specimen. Dark phase contrast (X1,700). *(Courtesy of J. M. Sharpley, Applied Petroleum Microbiology, Buckman Laboratories, Inc., Memphis, Tenn., 1961.)*

responsible for the corrosion of pipes. *Desulfovibrio desulfuricans* (Fig. 27-6) reduces sulfates and other sulfur compounds to hydrogen sulfide.

Algae

When water is exposed to sunlight, algal growth often results; the occurrence of algae in water is much like the growth of weeds in a garden. Algae are present in all natural aquatic environments. Their nuisance characteristics involve production of turbidity, discoloration, odor, and taste in water. Algae are frequently the primary cause for the clogging of filters during water purification. The diatoms are the most important in this respect, although green and yellow algae are also involved. Aside from these nuisance characteristics, some algae are capable of producing substances toxic to humans and animals.

Viruses

Many viruses are known to be excreted from humans through the intestinal tract, and these may find their way via sewage into sources of drinking water. The enteroviruses are the ones most commonly found in sewage; they include the polio, coxsackie, and echo viruses. The virus that causes infectious hepatitis has been isolated from polluted water and shellfish; occurrence of this disease has been traced to these sources. Rotaviruses are also of major importance. The possibility that virus diseases, particularly the enteric virus diseases, may be waterborne indicates that methods for evaluating the potability of a water supply from a virological standpoint should be developed.

Considerable research is underway for the development of a routine test method for the detection of viruses in water and wastewater. At the same time more attention is being given to the assessment of the effectiveness of water-treatment processes for the removal and/or inactivation of viruses.

SWIMMING POOLS

Water in swimming places, and particularly in public swimming pools, may be a health hazard. Swimming pools and surrounding areas may be involved in the transmission of infections of the eye, nose, and throat, infections of the intestinal tract, and the spread of athlete's foot, impetigo, and other dermatoses. These facts, together with the extensive increase in the use of public pools, make it imperative that constant attention be given to the sanitary quality of the water.

Water in swimming pools is disinfected with chlorine, the only disinfectant that has been approved by all state health departments for use in treating bathing waters. Bromine, which belongs to the same chemical family, the halogens, is also an effective disinfectant for this purpose and has been approved by some states. In using these chemicals for treatment of bathing water, dosage must be carefully regulated. There must be enough *residual* chemical in the water to destroy microorganisms effectively, and in this concentration the agent must be nontoxic and nonirritating to swimmers' eyes, skin, and mucous membranes. Various types of filtration units are also used for the treatment of swimming-pool waters.

WATER POLLUTION

The 1972 Federal Water Pollution Control Act and Amendments represents a major national program directed to restoring the quality of the United States streams and lakes as well as the coastal waters. Administered by the Environmental Protection Agency (EPA), the law established a stringent regulatory system stating precise and detailed pollution abatement requirements with heavy penalties for violations. A few of the provisions of the law are as follows:

1 The EPA is required to establish effluent limitations and national performance standards for sources of water pollution, including factories, power plants, wastewater-treatment plants, and animal feedlots.
2 The law requires publicly owned waste-treatment plants to meet treatment effluent limits by July 1988, wherever possible.

WASTEWATER

Wastewater is the used water supply of a community and consists of:

1 Domestic waterborne wastes, wastewater or sewage, including human excrement and wash waters—everything that goes down the drains of a home and into a sewerage system
2 Industrial waterborne wastes such as acids, oil, greases, and animal and vegetable matter discharged by factories
3 Ground, surface, and atmospheric waters that enter the sewerage system

The wastewater of a city is collected through a sewerage system, which carries the used water to its ultimate point of treatment and disposal. There are three kinds of sewerage systems: (1) **sanitary sewers,** which carry domestic and industrial wastewater; (2) **storm sewers,** designed to carry off surface and storm (atmospheric) water; and (3) **combined sewers,** which carry all the wastewater (sewage) through a single system of sewers.

Chemical Characteristics

Domestic wastewater or sewage consists of approximately 99.9 percent water, 0.02 to 0.03 percent suspended solids, and other soluble organic and inorganic substances. On a percentage basis, the amount of solids appears small; however, the tremendous volume of material handled daily by a major municipal plant (e.g., several hundred millions of gallons) contains as much as 100 tons of solids. The chemical constituents, present in low concentrations, nevertheless are extremely important and are subject to variations, between communities as well as within a community, even from hour to hour. Inorganic chemicals initially present in the water supply will likewise be present in the sewage;

organic compounds are contributed through human excrement and other domestic wastes, and both organic and inorganic compounds are added by industrial wastes. For example, slaughterhouses, sugar factories, paper mills, and creameries add organic substances; mines and metal industries contribute acids and salts of metals and other inorganic wastes. The organic compounds in sewage are classified as nitrogenous or nonnitrogenous. The principal nitrogenous compounds are urea, proteins, amines, and amino acids; the nonnitrogenous substances include carbohydrates, fats, and soaps.

Modern technology may produce significant changes in sewage characteristics. The increased use of household garbage-disposal units has increased the total organic load. Some synthetic detergents displacing soaps are resistant to microbial degradation.

Biochemical Oxygen Demand (BOD)

The **biochemical oxygen demand (BOD)** is a measure of the amount of oxygen used in the respiratory processes of microorganisms in oxidizing the organic matter in the sewage and for the further metabolism (oxidation) of cellular components synthesized from the wastes. One of the primary reasons for treating wastewater prior to its being returned to the water resource (e.g., stream or lake) is to reduce the drain on dissolved oxygen supply of the receiving body of water. The magnitude of the BOD is related to the amount of organic material in the wastewater—i.e., the more oxidizable organic material, the higher the BOD. The "strength" of wastewater is expressed in terms of BOD level.

Microbiological Characteristics

Since the composition of wastewater varies, it is to be expected that the types and numbers of organisms will fluctuate. Fungi, protozoa, algae, bacteria, and viruses are present. Raw sewage may contain millions of bacteria per milliliter, including the coliforms, streptococci, anaerobic spore-forming bacilli, the *Proteus* group, and other types originating in the intestinal tract of humans. Sewage is also a potential source of pathogenic protozoa, bacteria, and viruses. The causative agents of dysentery, cholera, and typhoid fever may occur in sewage. The poliomyelitis virus, the virus of infectious hepatitis, and the coxsackie viruses are excreted in the feces of infected hosts and thus may appear in sewage. Certain bacterial viruses are readily isolated from sewage.

Predominant physiological types of bacteria may shift during the course of sewage digestion. In an anaerobic digester, facultative types (*Enterobacter*, *Alcaligenes*, *Escherichia*, *Pseudomonas*, etc.) predominate during initial stages. This is followed by methane producers, which are strict anaerobes, e.g., *Methanobacterium*, *Methanosarcina*, and *Methanococcus*. Organic acids produced by the facultative bacteria are metabolized by the methane formers; the end products are methane and carbon dioxide. Large amounts of these gases are produced in anaerobic digesters.

The various processes associated with treatment of wastewater bring about pronounced changes in the predominant types of organisms. These changes and their significance will be discussed later.

WASTEWATER TREAT-MENT AND DISPOSAL

Wastewater treatment is necessary before wastewater can be disposed of without producing significant undesirable or even harmful effects. However, some communities and municipalities still dispose of inadequately treated wastewater into natural bodies of water, either because they are indifferent to the consequences or because it is assumed that the body of water is sufficiently large and so located that dilution prevents hazards. Communities and municipalities can no longer rely on disposal of wastewater by dilution. There is an ever-increasing demand for domestic and industrial water, necessitating more reuse of waters that receive wastewaters. Disposal of inadequately treated wastewater leads to:

1 Greater possibility for dissemination of pathogenic microorganisms
2 Increased danger in using natural bodies of water for drinking supplies
3 Contamination of oysters and other shellfish by the pollution, making them unsafe for human consumption
4 Large losses in the waterfowl population, chargeable to pollution of their winter feeding grounds
5 Increased danger of swimming in the water and diminished value of the water for other recreational purposes
6 Depletion of oxygen supply of the water by unstable organic matter in sewage, killing aquatic life
7 Creation of miscellaneous objectionable conditions such as offensive odors and accumulation of debris, which decrease property values
8 Accumulation and dissemination of toxic chemicals that endanger ecosystems and threaten public health

WASTEWATER-TREATMENT PROCESSES

Wastewater-treatment processes are many and varied. We will discuss the treatment processes as they are applicable to two separate situations: (1) a single dwelling or unit structure, and (2) a community or municipality.

Single Dwelling Units

Treatment and disposal of wastewater and sewage from individual dwellings or other unit structures (e.g., some motels or shopping centers) can be accomplished by anaerobic digestion and/or by aerobic metabolic processes. One of the more common installations used to accomplish this is the septic tank, an anaerobic digesting system.

Septic Tank

A septic tank (see Fig. 27-7) is a sewage-settling tank designed to retain the solids of the sewage entering the tank long enough to permit adequate decomposition of the sludge. Thus the unit accomplishes two processes: sedimentation and biological degradation of the sludge. As sewage enters this type of tank, sedimentation occurs from the upper portion, permitting a liquid with fewer suspended solids to be discharged from the tank. The sedimented solids are subject to degradation by anaerobic bacteria; hence the end products are still very unstable, i.e., high in BOD and odorous. The effluent from the septic tank is distributed under the soil surface through a disposal field, as shown in Fig. 27-7.

Septic tanks are the most satisfactory method for disposing of sewage from small installations, especially individual dwellings and isolated rural buildings where public sewers are not available. They cannot, however, be relied upon to eliminate pathogenic microorganisms carried in the sewage. Consequently, it is imperative that the drainage from the tank be prevented from contaminating the drinking-water supply.

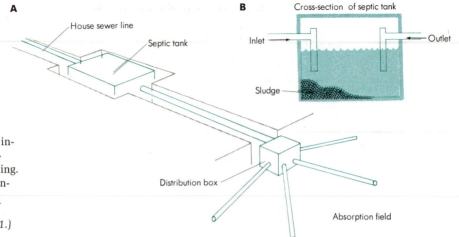

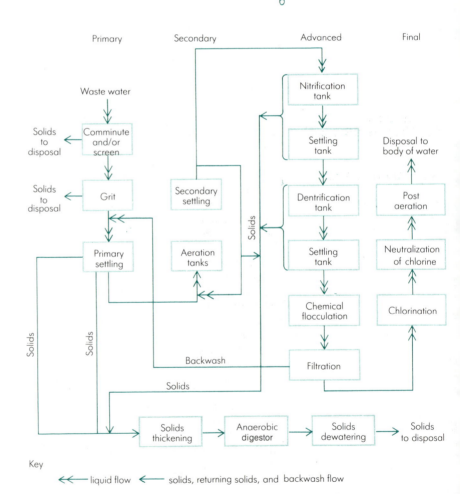

Figure 27-7. Septic-tank installation for sewage disposal from private dwelling. (A) Overall installation including absorption field. *(Redrawn from Public Health Rep, Reprint 2361.)* (B) Cross-sectional view.

Figure 27-8. Modern advanced wastewater treatment facility.

Municipal Treatment Processes

Municipal wastewater–treatment plants carry out a series of treatment processes (see Fig. 27-8) which may be summarized as follows:

1 *Primary treatment.* To remove coarse solids and to accomplish removal of "settleable" solids
2 *Secondary (biological) treatment.* To adsorb and ultimately oxidize organic constituents of the wastewater, i.e., to reduce the BOD
3 *Advanced treatment.* To remove additional objectionable substances to further reduce BOD; includes removal of nutrients such as phosphorus and nitrogen
4 *Final treatment.* To disinfect and dispose of liquid effluent
5 *Solids processing.* To stabilize solids removed from liquid processes, to dewater solids, and ultimately to dispose of solids (land application, landfill, incineration)

Each of these processes is briefly described below:

Primary Treatment

Wastewater as it arrives at a wastewater-disposal plant is first treated to remove coarse solid materials by a variety of mechanical techniques including screening, grinding, and grit chambers. Subsequent to this it is treated to remove settleable solids.

Sedimentation. Sedimentation units (tanks, basins, or mechanical devices) provide the means for concentrating and collecting the particulate material referred to as **sludge.** Following sedimentation, the sludge and the liquid effluent are processed separately.

Secondary (Biological) Treatment

Secondary treatment processes accomplish oxidation of the organic material in the liquid wastewater by microbial activity. The oxidation methods employed are:

1 Filtration by intermittent sand filters, contact filters, and trickling filters
2 Aeration by the activated-sludge process or by contact aerators
3 Oxidation ponds

Only the more commonly used secondary treatment processes will be described.

The Trickling Filter. The trickling filter consists of a bed of crushed stone, gravel, slag, or synthetic material with drains at the bottom of the tank. Trickling filters have been described as "a pile of rocks over which sewage or organic wastes slowly trickle." The liquid sewage is sprayed over the surface of the bed either by a rotating arm or through nozzles. The spraying saturates the liquid with oxygen. Intermittent application of the sewage permits maintenance of aerobic conditions in the bed. The filtering medium of the tank becomes coated with a microbial flora, the **zoogloeal film,** which consists of bacteria, fungi, protozoa, and algae. As the sewage seeps over these surfaces, the microorganisms adsorb and metabolize the organic constituents to more stable end products. This operation may be regarded as a stationary microbial culture (on the stones) fed by a continuous supply of nutrients (organic constituents of the sewage). A newly constructed bed must acquire the zoogloeal film before it can function efficiently. This requires operation over a period of a few weeks. Figures 27-9 and 27-10 illustrate the design and operation of a trickling filter. Examples of zoogloeas from trickling filters are shown in Fig. 27-11.

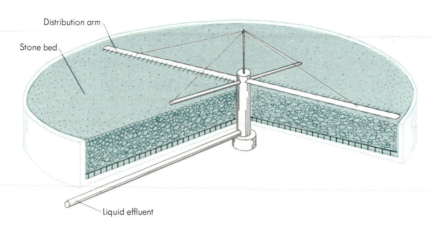

Figure 27-9. Commercial distributor (trickling filter) is shown in this cut-away drawing. Note the distributor arm which applies the liquid sewage, the filter bed through which the liquid travels, and the trough for collection of the effluent. *(Courtesy of Door-Oliver, Inc.)*

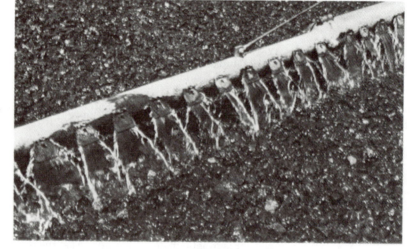

Figure 27-10. Distribution of wastewater onto filter stones from rotary distributor. Filter stones are coated with microbial growth. *(Courtesy of R. E. McKinny, Microbiology for Sanitary Engineers, McGraw-Hill, New York. Copyright 1962. Used by permission.)*

The upper region of the trickling filter is favorable for the growth of algae, and at times their growth may become so extensive that it impairs the operation of the filter. Many species of protozoa and fungi occur throughout the filter; their numbers are influenced in part by the availability of oxygen and nutrients. It is apparent that the microbial activities and interactions are extremely complex in such a heterogeneous environment.

The Activated-Sludge Process. Vigorous aeration of sewage (Fig. 27-12) results in the formation of a **floc**; the finely suspended and colloidal matter of sewage forms aggregates designated as **floccules**. If this floc is permitted to settle and then added to fresh sewage that is again vigorously aerated, flocculation occurs in a shorter time than before. By repetition of this process, i.e., addition of sedimented floc to fresh sewage, aeration, sedimentation, addition of sediment

Figure 27-11. Branched and amorphous zoogloeas collected from the surface of trickling filters receiving primarily domestic wastewaters. (A) Natural, branched, trickling-filter zoogloea. (B) Portion of specimen shown in (A) illustrating morphological similarity of bacterial cells and their concentration at anterior points of zoogloeal branches. (C) Natural, amorphous, trickling-filter zoogloea. (D) Portion of specimen shown in (C) illustrating presence of morphologically different bacteria. (All specimens treated with 10% skim milk to accentuate zoogloeal matrix. Phase-contrast microscopy.) *(Courtesy of R. F. Unz and N. C. Dondero, Water Res, 4:575, 1970.)*

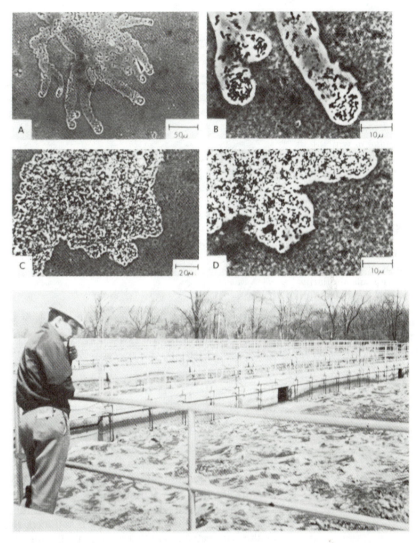

Figure 27-12. An activated-sludge basin with a full load of sewage in the process of aerobic microbiological treatment. *(Courtesy of Washington Suburban Sanitary Commission.)*

to fresh sewage, aeration, etc., a stage is reached where complete flocculation of the fresh sewage occurs very quickly, e.g., a few hours. These particles of floc, i.e., "activated sludge," contain large numbers of very actively metabolizing bacteria, together with yeasts, molds, and protozoa. This combination of microbial growth is very effective in the oxidation of organic compounds. A poor settlement of activated-sludge flocs adversely affects performance of a sewage-treatment plant. The sludge becomes more voluminous and is difficult to con-

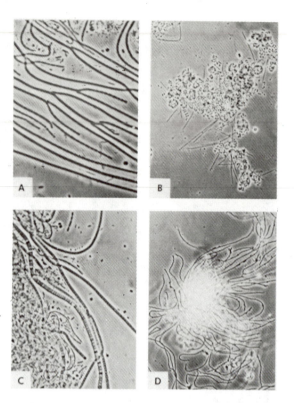

Figure 27-13. Filamentous organisms observed in activated sludge in which bulking has occurred. (A) *Sphaerotilus natans*. The cells occur within sheaths (X825). (B) *Streptothrix hyalina*, thin filaments protruding from the flocs (X406). (C) Unidentified long, unbranched, nonmotile multicellular filaments. Cells have some similarity to *Leucothrix* and *Thiothrix* (X825). (D) *Microthrix parvicella*. Filamentous cells, strongly curled (X310). (*Courtesy of D. H. Eikelboom and Water Res, 9:365, 1975.*)

trol. The principal reason for poor settling (bulking) of activated sludge is the growth of filamentous microorganisms. Many different species of microorganisms have been isolated from sludge in this condition. A few types are shown in Fig. 27-13.

The use of activated sludge is of great importance in wastewater treatment. This process usually employs an aeration period of 4 to 8 h, after which the mixture is piped to a sedimentation tank. The effluent from these tanks represents wastewater treated to secondary levels; there is a considerable reduction of suspended solids, and the BOD is reduced. Some of the advantages of this method of treatment are that relatively little land space is needed and the quality of final effluent is such that it does not require high dilution for disposal.

Oxidation Ponds (Lagoons). Oxidation ponds (also called lagoons or stabilization ponds) are shallow ponds (2 to 4 ft in depth) designed to allow algal growth on the wastewater effluent. Use of oxidation ponds should be preceded by primary treatment. Oxygen for biochemical oxidation of nutrients is supplied from the air, but the release of O_2 during photosynthesis by the alga *Chlorella pyrenoidosa* provides an additional important source of oxygen.

Advanced Treatment

Advanced wastewater treatment is required when removal of substances beyond the limits normally achieved by conventional primary and secondary processes are necessary. Unit processes have been developed to remove nutrients, simple organic substances, and complex synthetic organic compounds. Processes include biological treatment; however, physical-chemical methods predominate. Common unit processes include biological nitrification-denitrification, filtration, reverse osmosis, carbon adsorption, chemical addition, and ion exchange. The major disadvantage of advanced treatment processes is high cost.

Final Treatment

The liquid effluent, upon completion of other treatments, is disinfected and usually discharged into a body of water. Disinfection of wastewaters is necessary to protect public health when the receiving waters are used for purposes such as downstream water supply, recreation, irrigation, or shellfish harvesting. Most facilities use chlorination for disinfection. Current research has proved the serious impact chlorinated waters have on the aquatic life of the receiving water. This has led to the development of several disinfection alternatives. The use of ozone and ultraviolet light is becoming more prevalent. Many facilities that continue to employ chlorine for disinfection now include dechlorination prior to discharge to a body of water.

Dissolved oxygen may also be added to the treated wastewater prior to final discharge. This process is termed **post aeration** and is accomplished by mechanical means or a cascading slow technique. Post aeration minimizes the decrease in receiving water-dissolved oxygen that normally occurs where treated wastewater effluent is discharged.

Although disposal into surface waters is widely practiced, land application has been and continues to be a feasible alternative to surface-water discharge. Land application of wastewaters was given substantial recognition in the Federal Water Pollution Control Amendment of 1972 to implement the "national goal that discharge of pollutants into navigable waters be eliminated by 1985." It is a fact that municipal and industrial wastewaters have been used nationwide for crop irrigation. During the next decade we are likely to give more consideration to land application of the treated liquid waste from wastewater-treatment plants.

Solids Processing

A major cost at modern wastewater-treatment facilities is associated with solids processing—the thickening, stabilization, dewatering, and disposal of sludge. Solids are removed from the primary, secondary, and the advanced stages of the treatment process. Thickening is generally employed to further concentrate the solids or sludge prior to stabilization. Thickening may be accomplished by gravity thickening, similar to sedimentation, or by dissolved air flotation.

Many stabilization processes are employed, including aerobic and anaerobic digestion, composting, chemical addition, and heat treatment. The most common process for modern municipal treatment facilities is the anaerobic digestion system.

Dewatering is accomplished by physical methods and is often enhanced by the addition of polymer or other chemical coagulant aids. Equipment used for dewatering includes vacuum filters (Fig. 27-14), belt filter presses, plate and frame presses, and centrifuges. Small treatment facilities and older plants continue to use sand filter beds for dewatering.

Anaerobic Sludge Digestion. The solids which accumulate during sedimentation

Figure 27-14. The dewatering (drying) of sludge can be accomplished in several ways. This illustration shows sludge being dried by a vacuum-filter process; the liquid in the sludge is drawn through steel coils of a rotating drum by suction. The sludge "cake" remains on the outside of the drum and when dry falls off onto a conveyor belt. *(Courtesy of Washington Suburban Sanitary Commission.)*

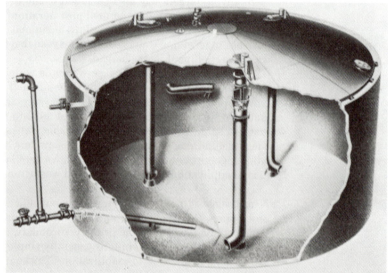

Figure 27-15. Sludge digestor is a special tank designed to process sludge under controlled conditions. *(Courtesy of Door-Oliver, Inc.)*

are pumped into a separate tank designed especially for the digestion of sludge under controlled conditions (Fig. 27-15 and Fig. 27-16). Solids recovered from the aerobic treatment processes may also be returned to the sludge digestor.

The microbial action on the constituents of sludge is termed **sludge digestion.** An anaerobic condition prevails in these tanks; and the anaerobic and facultative types of bacteria are active. These microorganisms bring about a decrease in organic solids by degrading them to soluble substances and gaseous products. Large amounts of gases consisting mainly of methane (60 to 70 percent) and carbon dioxide (20 to 30 percent), with smaller amounts of hydrogen and nitrogen, are produced during sludge digestion. This gas mixture can be used as a fuel for heating purposes and for operating power.

Figure 27-16. Anaerobic sludge digestors. Gases formed in the process of microbiological metabolism can be cycled through the treatment plant to provide fuel for heating. *(Courtesy of Washington Suburban Sanitary Commission.)*

Conditions affecting microbial growth and metabolism will be reflected in the degree of digestion of the sludge, e.g., inoculum, pH, and temperature. Fresh sludge entering the tanks is inoculated, or **seeded,** with a portion of **ripe sludge** (sludge which has already undergone digestion). The ripe sludge may be regarded as an actively growing culture of the types of microorganisms required for rapid digestion of sludge. The proportion of inoculum used and the thoroughness of mixing with the fresh sludge are important. Most rapid digestion of sludge occurs at temperatures which favor growth of thermophilic bacteria (50 to 60°C). A neutral pH (7.0) is optimum for sludge digestion. These conditions can be controlled only in specially designed tanks as shown in Fig. 27-15. The temperature is maintained around 30°C, which is below the optimum stated above for most rapid digestion, because it has not been found practical to operate at thermophilic temperatures. Complete digestion requires 2 to 3 weeks or even longer.

In a modern wastewater treatment plant the sludge remaining in the digestor is removed to a mechanical drying unit (see Fig. 27-14). The sludge dries into "cakes" which are carried by conveyor belts to incinerators. The incinerator reduces the sludge cake to ash for final disposal in a landfill. An alternative is to spread stabilized dewatered sludge on land. Several municipalities are engaged in land-utilization techniques for disposal of treated sludge. However, such practice requires inexpensive and available land. Furthermore, the volume of sludge generated by a large plant can be staggering.

Composting. Composting is a process where dewatered sludge undergoes decomposition, usually within the thermophilic temperature range. Dewatered sludge is mixed with a bulking agent such as wood chips. The bulking material is added to enhance circulation of air throughout the sludge to improve the stabilization process. The mixture of sludge and bulking material is placed in aerated piles. Oxygen is furnished by forced aeration. The mixture is allowed

Table 27-2. Generalized Scheme of Microbial Degradation of the Organic Constituents in Sewage

Substrates	+	Enzymes of Microorganisms	→	Anaerobic Conditions	Aerobic Conditions
					Representative End Products
Proteins and other organic nitrogen compounds				Amino acids Ammonia Nitrogen Hydrogen sulfide Methane Carbon dioxide Hydrogen Alcohols Organic acids Indole	Amino acids Ammonia → nitrites → nitrates Hydrogen sulfide → sulfuric acid Alcohols $\Big\}$→ CO_2 + H_2O Organic acids
Carbohydrates				Carbon dioxide Hydrogen Alcohols Fatty acids Neutral compounds	Alcohols $\Big\}$→ CO_2 + H_2O Fatty acids
Fats and related substances				Fatty acids + glycerol Carbon dioxide Hydrogen Alcohols Lower fatty acids	Alcohols $\Big\}$→ CO_2 + H_2O Lower fatty acids

to "cure" or biologically decompose for a period of time. For effective stabilization this period of time is normally considered to be 21 days.

After 21 days, the bulking agent is separated from the sludge and the sludge is allowed to cure further for several weeks. Upon final curing, the sludge has been transformed to a humus-type material and is suitable for use as a soil conditioner.

MICROORGANISMS AND WASTEWATER-TREATMENT PROCEDURES

The effectiveness of secondary wastewater-treatment processes is almost entirely dependent upon microbial growth and metabolism. The chemical activities of the microorganisms are responsible, to a major degree, for stabilization of the final effluent. The same is true for liquefaction, gasification, and mineralization of the constituents of sludge. From the standpoint of ecology, wastewater represents one of the most complex microbiological environments. Total populations, as well as the distribution of physiological types, are subject to wide fluctuations. The variety of substrates are susceptible to dissimilation by this mixed microbial flora. Interactions may occur among species, producing results that are not characteristic of pure cultures. However, the types of change accomplished by these microorganisms may be summarized as shown in Table 27-2.

From the reactions shown in Table 27-2, it can be seen that anaerobic conditions result in a variety of incompletely oxidized products. Aerobic conditions yield products that are more highly oxidized. The efficiency of aeration, the metabolic capacity of the microorganisms, and the time allowed for treatment will determine the extent of oxidation; e.g., carbohydrates under suitable conditions can be degraded and oxidized to carbon dioxide and water.

EFFICIENCY OF WASTE-WATER-TREATMENT PROCEDURES

The efficiency of wastewater treatment obtained by the several procedures described is shown in Tables 27-3 and 27-4. It should be noted that the efficiency reported for any single process shows considerable variation.

This variation may be attributed to the design of the unit, the method of operation, the nature of the wastewater, or other differences. Most importantly, however, the data in these tables demonstrate that methods are available for accomplishing a high degree of wastewater purification.

THE POLLUTION PROBLEM

Despite the fact that modern technology has provided effective methods for wastewater treatment, many communities and municipalities do not employ adequate treatment procedures; but, even worse, some perform no treatment: they dump raw wastewater into the waterways. Fortunately significant progress is being made in upgrading wastewater-treatment processes.

Table 27-3. Efficiency of Various Sewage-Treatment Methods

Method	Percentage of Removal of Suspended Solids	Gallons of Sludge per Million Gallons of Sewage	Percentage of Removal of Bacteria	BOD
Plain sedimentation	40–95	1,000–5,000	40–75	30–75
Chemical precipitation	75–95	5,000–10,000	80–90	60–80
Septic tank	40–75	500–1,500	40–75	25–65
Imhoff tank	35–80	250–750	40–75	25–65
Intermittent sand filter	95–98		98–99 +	70–96
Contact bed	55–90		50–75	60–80
Trickling filter	0–80	250–750	70–85	60–90
Activated sludge	70–97	10,000–30,000	95–99 +	70–96

SOURCE: From P. L. Gainey and P. H. Lord, *Microbiology of Water and Sewage*, Prentice-Hall, Englewood Cliffs, N.J., 1952.

Table 27-4. The Transformation of Pollutants in Sewage after Processing in a Modern Sewage-Treatment Plant

	Influent →	Primary and Secondary →	Nitrification →	Denitrification →	Filtration and Disinfection →	Effluent
		Pollutant Concentrations, mg/liter				
BOD	206	35	10	6	4	
Phosphorus, total	8.4	2.0	1.0	0.5	0.2	
Nitrogen:						
Organic	8.6	3.0	1.0	1.0	0.5	
NH_3—N	13.7	14.8	1.5	1.0	1.0	
NO_2 and NO_3—N	0	0.2	11.1	1.0	0.5	
Total N	22.3	18.0	13.6	3.0	2.0	

SOURCE: *Environ Sci Technol,* **8**(10), October 1974.

The demand for larger quantities of potable water continues to grow world-wide. Accordingly, there is more need for conservation together with more efficient wastewater treatment processes so that the cycle between "used" water and its reuse can be shortened. The gravity of the water pollution problem is reflected in the following quotation taken from the article written by Nandor Porges, which is included in the references at the end of this chapter.

Waste treatment is a means of maintaining or recovering man's most precious and most abused natural resource, fresh water. Fresh water supplies were all-important in the establishment and growth of civilizations. Much of man's bitterest fighting has been incited by altercations over water rights, and the course of history may well be written around the theme of primitive and modern man's need for water.

QUESTIONS

1 Define the following terms as they refer to the sanitary quality of water: potable, polluted, contaminated, sanitary survey, wastewater.
2 Why is *Escherichia coli* considered an indicator of pollution? What are coliforms?
3 Name several species of pathogenic organisms which may be present in polluted water.
4 Why isn't the routine bacteriological examination of water directed toward isolation and identification of specific pathogens?
5 Outline the process by which a municipality produces potable water. How is the microbial population of the raw water affected at each step?
6 What is the significance of finding *Streptococcus faecalis* in water? What are fecal streptococci?
7 Is fermentation of lactose with production of acid and gas positive evidence for the presence of *E. coli*? Of the coliforms? Explain.
8 Describe how selective and differential media facilitate the bacteriological analysis of water samples.
9 How can one distinguish among members of the coliform group?
10 What advantages does the membrane filter technique offer for microbiological analysis of water?
11 What are the morphological and cultural characteristics of *Escherichia coli, Enterobacter aerogenes, Proteus vulgaris, Salmonella typhi,* and *Shigella dysenteriae?*
12 Where are septic tanks used? Describe the microbiological activities that take place in a septic tank.
13 Outline the process of wastewater treatment which is followed in most large cities. Which steps in the process depend upon microbial activity for successful performance? Explain.
14 Define the following terms: effluent, sludge, activated sludge, BOD, lagoon.
15 In terms of sewage treatment, what problem has accompanied the wide usage of detergents? The increase in use of home garbage-disposal units?
16 What is activated sludge? Compare the microbial activity in the activated-sludge process with that which occurs in a septic tank.

REFERENCES

Dugan, P. R.: *Biochemical Ecology of Water Pollution,* Plenum, New York, 1972. *This small volume (152 pages) provides broad coverage of water pollution in-*

cluding sources of pollution, biochemical events in purification processes, and ecological phenomena associated with pollution.

Environmental Protection Agency: *Microbiological Methods for Monitoring the Environment, Water and Wastes.* U. S. Environmental Protection Agency, Washington, D.C. 1978. *A manual of laboratory procedures.*

Metcalf and Eddy, Inc. (revised by G. Tchobanoglous): *Wastewater Engineering: Treatment, Disposal, and Reuse,* McGraw-Hill, New York, 1979. *Principles and practices of wastewater (sewage) collection, treatment, and disposal especially for persons in the environmental engineering profession.*

Mitchell, R. (ed.): *Water Pollution Microbiology,* Wiley-Interscience, New York, vol. 1, 1972 and vol. 2, 1978. *A series of essays on modern microbiological concepts and their application to the solution of problems in water-pollution control.*

National Research Council: *Drinking Water and Health,* National Academy Press, Washington, D.C., vol. 1, 1977; vol. 2, 1980; vol. 3, 1980; vol. 4, 1982; and vol. 5, 1983. *This series of reports by the Safe Drinking Water Committee of the National Research Council provides a current assessment of the following aspects of potable water: biological and chemical contaminants, the processes and chemicals available for disinfecting water, and the epidemiological, toxicological, and nutritional aspects of drinking water and public health.*

National Research Council: *Quality Criteria for Water Reuse,* National Academy Press, Washington, D.C., 1982. *This is a report of the Panel on Quality Criteria of Water Reuse of the Commission on Life Sciences of the National Research Council. It addresses issues concerned with quality criteria, chemical and microbiological, that should be applied if wastewater is to be reused for potable purposes.*

National Research Council: *Groundwater Contamination,* National Academy Press, Washington, D.C., 1984. *A comprehensive analysis of problems associated with the disposal of large quantities of wastes in terms of contamination of groundwater, which is the major source of freshwater in the United States.*

Palmer, C. M.: "Algae in Water Supplies," *Public Health Serv Publ 657,* 1959. *An illustrated manual on the identification, significance, and control of algae in water supplies.*

Porges, N.: "Newer Aspects in Wastewater Treatment," in W. W. Umbreit (ed.), *Advances in Applied Microbiology,* vol. 2, Academic, New York, 1960.

Standard Methods for the Examination of Water and Wastewater, 15th ed., American Public Health Association, New York, 1980. *An extensive compendium of physical, chemical, and biological "standard methods" for the examination of water.*

Water Pollution Control Federation: Literature Review Issue, Journal-Water Pollution Control Federation, vol. 55, no. 6, Washington, D.C., June, 1984. *Each year the entire June issue of the Journal-Water Pollution Control Federation is devoted to the review of the current literature on topics such as wastewater treatment, industrial wastes, and water pollution.*

Chapter 28   Microbiology of Foods

Microorganisms are associated, in a variety of ways, with all of the food we eat. They may influence the quality, availability, and quantity of our food. Naturally occurring foods such as fruits and vegetables normally contain some microorganisms and may be contaminated with additional organisms during handling. Food can serve as a medium for the growth of microorganisms, and this growth may cause the food to undergo decomposition and spoilage. Foods may also carry pathogenic microorganisms and as a result transmit disease. Other microorganisms, if allowed to grow in certain food products, produce toxic substances that result in food poisoning when the food is ingested. Still other microorganisms are used in the preparation and preservation of food products, such as yogurt and sauerkraut.

From these general remarks it is clearly evident that the major concern in food microbiology is the control of microorganisms. In this chapter we will discuss the microbiology of foods in terms of the normal flora of foods, their significance, and the manner in which foods may be protected from microbial contamination and microbial spoilage. We will also describe food products manufactured by microbial fermentation. The role of foods in transmission of disease and in food poisoning will be discussed in Chap. 35.

MICROBIAL FLORA OF FRESH FOODS

The inner tissues of healthy plants and animals are free of microorganisms (sterile). However, the surfaces of raw vegetables and meats are contaminated with a variety of microorganisms. The magnitude of this microbial contamination reflects one or more of the following: the microbial population of the environment from which the food was taken, the condition of the raw product, the method of handling, the time and conditions of storage. It is desirable to maintain a very low microbial level of contamination on raw foods; the presence of extremely large numbers of microorganisms suggests that some undesirable events have occurred and that the food is indeed susceptible to further deterioration.

Meats

The carcass of a healthy animal slaughtered for meat and held in a refrigerated room is likely to have only nominal surface contamination while the inner tissues are sterile. Fresh meat cut from the chilled carcass has its surface contaminated with microorganisms characteristic of the environment and the implements (saws or knives) used to cut the meat. Each new surface of meat, resulting from a new cut, adds more microorganisms to the exposed tissue. The ultimate in providing new surfaces and potential contamination of meat surfaces occurs in the process of making hamburger.

To improve the microbiological quality of meats, particularly ground beef (in addition to cold cuts and frankfurters), most states have adopted standards, or are considering establishing regulations, to require microbiological standards for these products at the time of purchase.

Among the more common species of bacteria occurring on fresh meats are pseudomonads, staphylococci, micrococci, enterococci, and coliforms. The low temperature at which fresh meats are held favors the growth of psychrophilic microorganisms.

Poultry

Freshly dressed eviscerated poultry have a bacterial flora on their surface (skin) that originates from the bacteria normally present on the live birds and from the manipulations during killing, defeathering, and evisceration. Under good sanitary conditions the bacterial count has been reported to be from 100 to 1,000 bacteria per square centimeter of skin surface, whereas under less sanitary conditions the count may increase 100-fold or more. Pseudomonads constitute the major contaminants on the skin of freshly dressed poultry.

Eggs

The interior of a freshly laid egg is usually free of microorganisms; its subsequent microbial content is determined by the sanitary conditions under which it is held, as well as the conditions of storage, i.e., temperature and humidity. Microorganisms, particularly bacteria and molds, may enter the egg through cracks in the shells or penetrate the shells when the "bloom" (thin protein coat) covering the shell deteriorates. The types of microorganisms involved reflect those present in the environment.

Fruits and Vegetables

Fruits and vegetables are normally susceptible to infection by bacteria, fungi, and viruses. Microbial invasion of plant tissue can occur during various stages of fruit and vegetable development, and, hence, to the extent that the tissues are invaded, the likelihood of spoilage is increased. A second factor contributing to the microbial contamination of fruits and vegetables pertains to their post-harvest handling. Mechanical handling is likely to produce breaks in the tissue which facilitates invasion by microorganisms. The pH of fruits is relatively acid, ranging from 2.3 for lemons to 5.0 for bananas. This restricts bacterial growth but does not retard fungal growth. The pH range for vegetables is slightly higher, pH 5.0 to 7.0, and hence they are more susceptible than fruits to attack by bacteria.

Shellfish and Finfish

The microbial flora of freshly caught oysters, clams, fish, and other aquatic specimens is very largely a reflection of the microbial quality of the waters where they are harvested. Of particular significance is whether the water is sewage-polluted, in which case the seafood is potentially capable of transmitting various pathogenic microorganisms. The marine bacterium *Vibrio parahaemolyticus* has been responsible for a number of gastroenteritis epidemics in the United States due to consumption of raw or inadequately cooked seafood. This organism occurs widely in the Atlantic, Pacific, and Gulf Coast waters and has been isolated from seafood samples including fish, shellfish, and crustaceans. Shellfish that grow in contaminated water can concentrate viruses and may be sources of hepatitis infection. For example, raw oysters and clams from polluted waters have caused numerous epidemics in various parts of the world.

Milk

At the time it is drawn from the udder of a healthy animal, milk contains organisms that have entered the teat canal through the teat opening. They are mechanically flushed out during milking. The number present at the time of milking has been reported to range between several hundred and several thousand per milliliter. The counts vary among cows and among the quarters of the same cow and are highest during the initial stages of milking. From the time the milk leaves the udder until it is dispensed into containers, everything with which it comes into contact is a potential source of more microorganisms. This includes the air in the environment, the milking equipment, and the personnel. Disregard of sanitary practices will result in heavily contaminated milk that spoils rapidly. However, milking performed under hygienic conditions with strict attention to sanitary practices will result in a product with low bacterial content and good keeping quality.

We shall discuss the microorganisms found in milk on the basis of their major characteristics, namely:

1 Biochemical types
2 Temperature response
3 Ability to cause infection and disease

Biochemical Types of Bacteria in Milk

If maintained under conditions that permit bacterial growth, raw milk of a good sanitary quality will develop a clean, sour flavor. This change is brought about mainly by *Streptococcus lactis* and *S. cremoris* (Fig. 28-1A, B) and certain lactobacilli (Fig. 28-1C). The principal change is lactose fermentation to lactic acid; evidence of proteolysis or lipolysis is not detectable by taste or smell. This type of change is sometimes referred to as the **normal fermentation** of milk. However, other organisms may produce changes beyond mere production of acid as shown in Table 28-1.

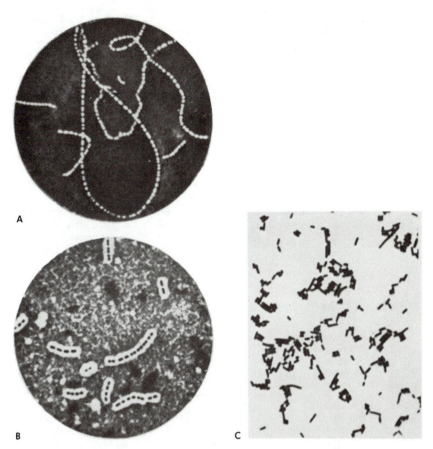

Figure 28-1. *Streptococcus lactis* (A) and *S. cremoris* (B), two important fermentative bacterial species in milk and milk products. These species, along with *Lactobacillus fermenti* (C), cause the so-called normal fermentation of milk; they are not pathogens. *(Courtesy of S. Orla-Jensen, The Lactic Acid Bacteria, Ejnar Munksgarrd, Copenhagen, 1919.)* (C) *Lactobacillus fermenti*, one of the heterofermentative lactobacilli. It produces a mixture of acids and is involved in the normal fermentation of milk; it is not pathogenic. Its cells vary in length and are Gram-positive, nonmotile, and nonsporeforming. *(Courtesy of A. P. Harrison.)*

Temperature Characteristics of Bacteria in Milk

Bacteria that gain entrance into milk may be classified according to their optimum temperature for growth and heat resistance. Temperature is a very practical consideration, since low temperatures are used to prevent changes due to microorganisms and high temperatures (pasteurization) are used to reduce the microbial population, destroy pathogens, and in general improve the keeping quality of the milk. Collectively, the bacteria encountered in milk are of the following four temperature types: psychrophilic, mesophilic, thermophilic, and thermoduric.

Since certain psychrophiles grow at temperatures just above freezing and some thermophiles grow at temperatures in excess of 65°C, it follows that the temperature at which milk is held will determine which species grow and predominate. Pasteurized milk stored in a refrigerator may be satisfactorily preserved for a week or even longer. But eventually microbial deterioration, manifested by "off" flavor or odor, will become evident because of the accumulation of metabolic products of psychrophilic bacteria. Thermophiles present a problem at the other extreme of the temperature scale. The holding method of pasteuri-

Table 28-1. Biochemical Types of Microorganisms in Milk

Biochemical Types	Representative Microorganisms	Source of Microorganisms	Substrate Acted Upon and End Products	Additional Remarks
Acid producers	Streptococci, e.g., *Streptococcus lactis* *S. cremoris*	Dairy utensils, silage, plants	Lactose fermented to lactic acid or lactic acid and other products such as acetic acid, ethyl alcohol, and carbon dioxide	Acid producers that produce only lactic acid are referred to as *homofermentative* types; those which produce a variety of products are called *heterofermentative* types
	Lactobacilli, e.g., *Lactobacillus casei* *L. plantarum* *L. brevis* *L. fermentum*	Feeds, silage, manure	Lactose is fermented to lactic acid and other products. Some species of lactobacilli are homofermentative; others are heterofermentative	
	Microbacteria, e.g., *Microbacterium lacticum*	Manure, dairy utensils, and dairy products	Lactose fermented to lactic acid and other end products; do not produce as much acid as the streptococci or lactobacilli	Some of these bacteria can survive exposure to very high temperatures, e.g., 80–85°C for 10 min
	Coliforms, e.g., *Escherichia coli* *Enterobacter aerogenes*	Manure, polluted water, soil, and plants	Lactose fermented to a mixture of end products, e.g., acids, gases, and neutral products	The number of coliform bacteria present in milk is an indicator of its sanitary quality
	Micrococci, e.g., *Micrococcus luteus* *M. varians* *M. freudenreichii*	Ducts of cow's mammary glands, dairy utensils	Small amounts of acid produced from lactose (weakly fermentative); micrococci are also weakly proteolytic	Moderately heat resistant; some strains capable of surviving 63°C for 30 min
Gas producers	Coliforms *Clostridium butyricum* *Torula cremoris*	Soil, manure, water, feed	Lactose fermented with accumulation of gas; the gas may be a mixture of carbon dioxide and hydrogen, or only carbon dioxide in the case of yeast fermentation	Bulk containers of milk may have their lids lifted by gas pressure in instances where contamination with gas producers is unusually high.
Ropy or stringy fermentation	*Alcaligenes viscolactis* *Enterobacter aerogenes* *Streptococcus cremoris*	Soil, water, plants, feed	Organisms synthesize a viscous polysaccharide material that forms a slime layer or capsule on the cells	Milk favors the formation of capsular material; sterile skim milk is frequently used as the culture medium when capsule formation is sought
Proteolytic	*Bacillus* spp., e.g., *B. subtilis* *B. cereus* *Pseudomonas* spp. *Proteus* spp. *Streptococcus liquefaciens*	Soil, water, utensils	Proteolytic organisms degrade the casein to peptides which may be further dissimilated to amino acids; proteolysis may be preceded by coagulation of the casein by the enzyme renin	End products of proteolysis may impart abnormal flavor or odor to the milk; *Pseudomonas* spp. may produce coloration of milk.
Lipolytic	*Pseudomonas fluorescens* *Achromobacter lipolyticum* *Candida lipolytica* *Penicillium* spp.	Soil, water, utensils	Lipolytic microorganisms hydrolyze milk fat to glycerol and fatty acids	Some fatty acids impart rancid odor and taste to milk

zation exposes milk to a temperature of 62.8°C for 30 min; the thermophile *Bacillus stearothermophilus*, however, grows at 65°C. Other generalizations relating to bacterial growth and the types of bacteria that predominate in milk held at various temperatures are shown in Table 28-2.

In the dairy industry, thermoduric bacteria are regarded as those which survive pasteurization in considerable numbers but do not grow at pasteurization temperatures. Microorganisms of this category are extremely troublesome from the standpoint of producing raw milk with a low bacterial count. Because they are not killed by pasteurization, the microorganisms may contaminate equipment and accumulate as a result of faulty cleaning. Subsequent batches of milk processed through the same equipment will become heavily contaminated.

Thermoduric bacteria are not restricted to a single species or genus but are found in species of several genera, e.g., *Microbacterium lacticum*, *Micrococcus luteus*, *Streptococcus thermophilus*, and *Bacillus subtilis*.

Pathogenic Types of Bacteria in Milk

In recent years milk has been involved in fewer and fewer outbreaks of illness, to the point that the public and regulatory agencies no longer consider milk a primary source of foodborne illness. Milk and dairy products can now be considered model foods from the standpoint of regulations and surveillance of production, processing, and distribution. Furthermore, there are companion standardized methods for analysis of dairy products. No other food can claim the degree of standardized surveillance and analysis that is practiced for milk and milk products.

A variety of diseases are potentially transmissible through milk. The source of a pathogenic agent occurring in milk may be either a cow or a human, and

Table 28-2. Effect of Holding Temperature of Raw Milk on Numbers and Types of Bacteria

Holding Temperature, °C	Changes in Numbers	Predominant Organisms
1–4	Slow decline first few days followed by gradual increase after 7 to 10 days	True psychrophiles, e.g., species of *Flavobacterium*, *Pseudomonas*, and *Alcaligenes*
4–10	Little change in number during first few days followed by rapid increase in numbers; large populations present after 7 to 10 days or more	As above; changes produced on holding are of the following types: ropiness, sweet curdling, proteolysis, etc.
10–20	Very rapid increase in numbers; excessive populations reached within few days or less	Mainly acid-producing types such as lactic streptococci
20–30	High populations develop within hours	Lactic streptococci, coliforms, and other mesophilic types; in addition to acid there may be gas, off flavors, etc.
30–37	High populations develop within hours	Coliform group favored
Above 37	High populations develop within hours	Some mesophiles, thermophiles, e.g., *Bacillus coagulans* and *B. stearothermophilus*

it may be transmitted to either. The following modes of transmission are possibilities.

1 Pathogen from infected cow → milk → human or cow, e.g., tuberculosis, brucellosis, mastitis

2 Pathogen from human (infected or carrier) → milk → human, e.g., typhoid fever, diphtheria, dysentery, scarlet fever

It is also possible for humans to infect cows. For example, mastitis may be caused by a variety of organisms, including *Staphylococcus aureus*. The infecting organism, in some cases, has been traced to humans.

More specific aspects of disease transmission by milk and other foods are discussed in Part Eight.

MICROBIAL SPOILAGE OF FOODS

Considering the variety of natural food substances and the methods by which each is handled during processing, it is apparent that practically all kinds of microorganisms are potential contaminants. The type of food substance and the method by which it is processed and preserved may favor contamination by certain groups of microorganisms. Most foodstuffs serve as good media for the growth of many different microorganisms. Given a chance to grow, the organisms will produce changes in appearance, flavor, odor, and other qualities of foods. These degradation processes may be described as follows:

Putrefaction:
Protein foods + proteolytic microorganisms → amino acids + amines + ammonia + hydrogen sulfide

Fermentation:
Carbohydrate foods + carbohydrate-fermenting microorganisms → acids + alcohols + gases

Rancidity:
Fatty foods + lipolytic microorganisms → fatty acids + glycerol

The changes that microbes cause in foods are not limited to the results of degradation; they may also be caused by products of microbial synthesis. Some microorganisms discolor foods as a result of pigment production. Slimes may be developed in or on foods by microorganisms capable of synthesizing certain polysaccharides.

Fresh Foods

Examples of types of food spoilage (other than canned-food spoilage), together with some of the microorganisms involved, are shown in Table 28-3.

Fresh Milk

Milk is an excellent bacteriological medium. In fact, sterile skimmed milk is used routinely as a culture medium. Fresh whole milk contains protein (casein), carbohydrate (lactose), and fat. All of these substrates can be degraded enzymatically by microorganisms. If the degradation of these substrates is extensive, the accumulation of end products will impart undesirable characteristics to the milk. Some microorganisms can synthesize compounds like pigments and slimes which also give undesirable characteristics to the milk. A summary of

Table 28-3. Types of Food Spoilage (Other than Canned Foods) with Some Examples of Causative Organisms

Food	Type of Spoilage	Some Microorganisms Involved
Bread	Moldy	*Rhizopus nigricans* *Penicillium* *Aspergillus niger*
	Ropy	*Bacillus subtilis*
Maple sap and syrup	Ropy	*Enterobacter aerogenes*
	Yeasty	*Saccharomyces* *Zygosaccharomyces*
	Pink	*Micrococcus roseus*
	Moldy	*Aspergillus* *Penicillium*
Fresh fruits and vegetables	Soft rot	*Rhizopus* *Erwinia*
	Gray mold rot	*Botrytis*
	Black mold rot	*A. niger*
Pickles, sauerkraut	Film yeasts, pink yeasts	*Rhodotorula*
Fresh meat	Putrefaction	*Alcaligenes* *Clostridium* *Proteus vulgaris* *Pseudomonas fluorescens*
Cured meat	Moldy	*Aspergillus* *Rhizopus* *Penicillium*
	Souring	*Pseudomonas* *Micrococcus*
	Greening, slime	*Lactobacillus* *Leuconostoc*
Fish	Discoloration	*Pseudomonas*
	Putrefaction	*Alcaligenes* *Flavobacterium*
Eggs	Green rot	*P. fluorescens*
	Colorless rots	*Pseudomonas* *Alcaligenes*
	Black rots	*Proteus*
Concentrated orange juice	"Off" flavor	*Lactobacillus* *Leuconostoc* *Acetobacter*
Poultry	Slime, odor	*Pseudomonas* *Alcaligenes*

microbial biochemical types that may occur in milk, their source, and the changes they produce are shown in Table 28-1.

Table 28-4. Microbiology of Canned-Food Spoilage

Types of Product	Type of Spoilage Organisms, with Examples	Signs of Spoilage	
		Can	Contents of Can
Low and medium acid products, pH above 4.6, e.g., corn, peas, spinach, asparagus	Flat sour (*Bacillus stearothermophilus*)	Possible loss of vacuum on storage	Appearance not usually altered; pH markedly lowered; sour; may have slightly abnormal odor; sometimes cloudy liquor
	Thermophilic anaerobe (*Clostridium thermosaccharolyticum*)	Can swells, may burst	Fermented, sour, cheesy, or butyric odor
	Sulfide spoilage (*Clostridium nigrificans*)	Can flat, hydrogen sulfide gas absorbed by product	Usually blackened, "rotten egg" odor
	Putrefactive anaerobe (*Clostridium sporogenes*)	Can swells, may burst	May be partially digested; pH slightly above normal; typical putrid odor; may be toxic
	Aerobic sporeformers (odd types) (*Bacillus* spp.)	Usually no swelling, except in cured meats when nitrate and sugar are present	Coagulated evaporated milk, black beets
Acid products, pH below 4.6, e.g., tomato juice, fruits, and fruit juices	Flat sour (*Bacillus thermoacidurans*)	Can flat, little change in vacuum	Slight pH change; off odor and flavor
	Butyric anaerobes (*Clostridium butyricum*)	Can swells, may burst	Fermented, butyric odor
	Nonsporeformers (mostly lactic acid types of bacteria)	Can swells, usually bursts, but swelling may be arrested	Acid odor
	Yeasts	Can swells, may burst	Fermented; yeasty odor
	Molds	Can flat	Surface growth; musty odor

SOURCE: Data from the National Food Processors Association.

Canned Foods

Because of their heat resistance, sporeformers (species of *Clostridium* and *Bacillus*) constitute the most important group of microorganisms in the canning industry. The three most important types of microbiological spoilage of commercially canned foods are (1) flat sour spoilage, (2) thermophilic anaerobe (TA) spoilage, and (3) putrefaction. Table 28-4 presents a summary of organisms involved in spoilage of canned food, together with the changes they produce.

MICROBIOLOGICAL EXAMINATION OF FOODS

Microbiological examination of foods may provide information concerning the quality of the raw food and the sanitary conditions under which the food was processed as well as the effectiveness of the method of preservation. In the case of spoiled foods, it is possible to identify the agent responsible for the spoilage; having discovered the agent, it may be possible to trace the source of contamination and the conditions which permitted spoilage to occur. Corrective measures can then be instituted to prevent further spoilage.

Microbiological food examination takes advantage of special microscopic techniques and cultural procedures. Extensive use is made of selective and differential media to facilitate the enumeration and isolation of certain types of microorganisms. The particular procedure used is determined by the type of food product being examined as well as by the specific purpose of the examination. For example, a food sample being investigated for possible contamination by *Clostridium botulinum* would be subject to different laboratory tests than one being examined for coliform organisms. The increasing significance of salmonellas in foodborne disease has made it mandatory to develop more rapid, reliable, and reproducible methods for the detection of salmonellas in foods.

A schematic summary of the various approaches to consider in the microbiological examination of a food sample is shown in Fig. 28-2. The procedures selected for examination of a particular sample are, of course, determined by the facts relating to that sample and the purpose of the examination.

Microscopic Techniques

Standard microscopic techniques are available for the examination of some food products. For example, a procedure known as the Breed smear is used to make a direct microscopic count of microorganisms in milk. The essential procedures of this technique are: (1) spreading a measured amount of milk over a known area on a glass slide, (2) staining the film of milk with methylene blue, (3) making a microscopic count of organisms or clumps of organisms in several microscopic fields, and (4) calculating the total number of bacteria per unit volume.

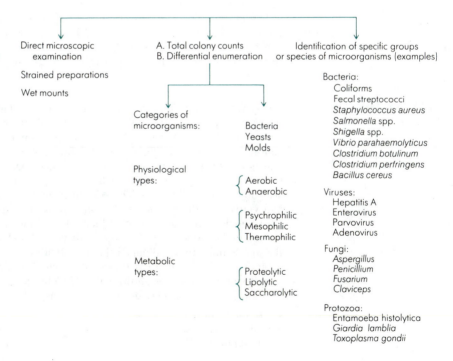

Figure 28-2. Generalized scheme for microbiological examination of foods.

A slide designed with a chamber, known as the Howard mold slide, is used, as its name suggests, to enumerate mold filaments in food products such as fruits, juices, and vegetables. When the mold counts obtained by this procedure exceed certain limits, it indicates raw material of poor quality or unsatisfactory sanitary processing.

Protozoa can be identified and enumerated by direct microscopic examination. Since the protozoa may be present in small numbers, it is frequently necessary to use a procedure which will concentrate these organisms in the food sample prior to microscopic examination.

Culture Techniques

The numerous techniques for cultivating microorganisms described in earlier chapters of this book find application, sometimes with modifications, for the examination of foods. For example, plate culture techniques are available for the enumeration of the "total" microbial population or some particular group of microorganisms, as illustrated in Fig. 28-2. The word *total*, of course, needs qualification; the microorganisms enumerated by a cultural technique are only that portion of the total population which will grow into colonies under the conditions provided, namely, the composition of the medium and the physical conditions of incubation. For example, the standard procedure for counting microorganisms in milk is designed to enumerate bacteria by the standard plate count (SPC). The conditions for the procedure are very specifically articulated in a volume entitled *Standard Methods for the Examination of Dairy Products*. It is mandatory that the procedure be carried out precisely as specified in the publication. Other culture procedures are available for particular physiological or biochemical types of microorganisms.

The cultivation of viruses from food specimens requires the use of tissue-culture techniques as described in Chap. 21. Prior concentration of the food specimen suspected to be contaminated with viruses may be necessary. Additional provisions are necessary to inhibit bacterial growth in the tissue culture.

PRESERVATION OF FOODS

Today we associate food preservation with the refrigerator, the deep freeze, and the canning process, all developments of the nineteenth and twentieth centuries. However, humans have grappled with the problem of food preservation for many centuries. The ancient Egyptians and Romans were aware of the preservative effects of salting, drying, and smoking. It has been suggested that the first salt preservation was accomplished by burying the food along the shore, where seawater effected the cure. The American Indians placed strips of fresh bison and venison at the top of a teepee or over a campfire, where preservation was accomplished through drying and smoking. Dried salt cod was a common food for colonial Americans. Perishable foods were stored in caves and springs, where the low temperature prolonged the preservation.

Modern methods of food preservation employ elaborate refinements of the primitive processes plus additional new techniques. The various practices used for food preservation may be summarized as follows:

1 Aseptic handling
2 High temperatures
   (a) Boiling

(b) Steam under pressure
(c) Pasteurization
(d) Sterilization (of milk)
(e) Aseptic processing
3 Low temperatures
(a) Refrigeration
(b) Freezing
4 Dehydration
5 Osmotic pressure
(a) In concentrated sugar
(b) With brine
6 Chemicals
(a) Organic acids
(b) Substances developing during processing (smoking)
(c) Substances contributed by microbial fermentation (acids)
7 Radiation
(a) Ultraviolet
(b) Ionizing radiations

All methods of food preservation are based upon one or more of the following principles: (1) prevention or removal of contamination, (2) inhibition of microbial growth and metabolism (microbistatic action), and (3) killing of microorganisms (microbicidal action).

Aseptic Handling

Food items undergo considerable handling prior to being processed by some specific method of preservation such as canning, freezing, or dehydration. Each step in the preparation of a food for its final treatment is a potential source of contamination. For example, the shell of an egg provides a protective covering which normally excludes microorganisms. However, when the eggs are cracked open in the process of preparing dehydrated egg powder it is likely that the interior of the egg may become contaminated. The extent of the contamination will depend upon the cleanliness of the eggs and the level of aseptic precautions observed in the process.

One can recognize more vividly the importance of asepic technique in the processing of more perishable foods like oysters and crabmeat, each of which requires considerable handling by people.

High Temperatures

High temperature is one of the safest and most reliable methods of food preservation. Heat is widely used to destroy organisms in food products in cans, jars, or other types of containers that restrict the entrance of microorganisms after processing.

Steam under pressure, such as in a pressure cooker, is the most effective method of high-temperature food preservation since it can kill all vegetative cells and spores. Food preservation by heat requires knowledge of the heat resistance of microorganisms, particularly spores. In addition, one must consider the rate at which heat penetrates through foods of different consistencies as well as the size of the containers in which they are packed. Killing microorganisms by heat involves a time-temperature relationship, as discussed in Chap. 22, and considerable experimentation has been performed to determine the thermal death times of bacteria likely to cause spoilage. From such information it is possible to establish satisfactory heat-processing conditions. Much research has been done on this subject, and this accounts for the highly successful results achieved in food preservation by canning. Special laboratory equipment has been designed to determine with precision the heat resistance of various bacterial species, particularly the sporeformers.

Canning

Canning has been the basic method of food preservation for approximately 175 years. In 1810 Nicholas Appert, a Frenchman, published *L'Art de Conserver*, which described his successful researches in food preservation; and in the same year Peter Durand was granted an English patent describing the use of tin containers for food preservation.

The temperatures used for canning foods ranges from 100°C for high-acid foods to 121°C for low-acid foods. The canning process does not guarantee a sterile product. For example, spores of some bacterial species may survive these temperatures.

The most important organism to be eliminated in canned foods is the spore-forming anaerobe *Cl. botulinum*, which is capable of producing a very potent lethal toxin.

Pasteurization of Milk

The "Milk Ordinance and Code" of the U.S. Public Health Service comments on the word *pasteurization* as follows:

The terms *pasteurization, pasteurized,* and similar terms shall mean the process of heating every particle of milk or milk product to at least 145°F., and holding it continuously at or above this temperature for at least 30 minutes, or to at least 161°F., and holding it continuously at or above this temperature for at least 15 seconds, in equipment which is properly operated and approved by the health authority . . . .

The original time-temperature relationships for pasteurization were worked out with *Mycobacterium tuberculosis* since this was regarded as the most heat-resistant pathogen likely to occur in milk (see Fig. 28-3). This organism is

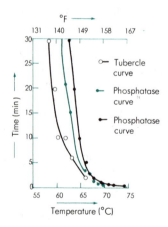

Figure 28-3. Time-temperature curve for the killing of *Mycobacterium tuberculosis* compared with the time and temperature required for the inactivation of the enzyme phosphatase. The two phosphatase curves are plotted from different experimental data. (*Courtesy of McGraw-Hill Encyclopedia of Science and Technology, p. 502, vol. 8. Copyright 1971. McGraw-Hill Book Company.*)

destroyed when exposed to a temperature of 140°F for 10 min. The pasteurization temperature was set at 143°F for 30 min. Later it was discovered that *Coxiella burnetii*, the causative agent of Q fever which can be transmitted by milk, can survive in milk heated to 143°F for 30 min. This observation resulted in the establishment of the present time and temperature for pasteurization.

Pasteurization Processes. Methods of pasteurization of milk used commercially include a low-temperature holding (LTH) method and a high-temperature short-time (HTST) method. The holding method, or vat pasteurization, exposes milk to 145°F (62.8°C) for 30 min in appropriately designed equipment. The HTST process employs equipment capable of exposing milk to a temperature of 161°F (71.7°C) for 15 s (seconds). In either method of pasteurization it is essential that the equipment be designed and operated so that every particle of milk is heated to the required temperature and held for the specified time. Precautions must be taken to prevent recontamination after pasteurization. The finished product should be stored at a low temperature to retard growth of microorganisms which survived pasteurization.

In addition to milk numerous other food products and some fermented beverages like beers and wines are commercially pasteurized.

The Phosphatase Test. Phosphatase is an enzyme, present in raw milk and in many tissues, which is destroyed by adequate pasteurization (see Fig. 28-3). Thus one can determine whether milk has been properly pasteurized by testing for the absence of this enzyme. The principle of the test is illustrated by the following reaction. Milk, which in its raw condition contains the enzyme phosphatase, is added to a substrate upon which the enzyme will react:

$$\underset{\text{Substrate}}{\text{Disodiumphenyl phosphate}} + \underset{\substack{\text{Enzyme from} \\ \text{raw milk}}}{\text{phosphatase}} \rightarrow \underset{\text{Products}}{\text{phenol} + \text{sodium phosphate}}$$

The amount of phenol liberated can be conveniently estimated by the addition of a reagent which turns blue in the presence of phenol. Color standards are used to interpret the results of this test. This is a very simple testing procedure, yet it provides valuable information about the heat treatment milk has received.

Sterilization

Commercial milk-sterilization techniques have been developed which expose milk to ultrahigh temperatures for very short periods of time, for example, 300°F (148.9°C) for 1 to 2 s. In addition, the sterilization process includes steps that eliminate any traces of cooked flavor. The final product is comparable in flavor and nutritional quality to pasteurized milk. The sterile milk product has several attractive features: it does not require refrigeration and it has an indefinite shelf life.

Aseptic Processing

A relatively new commercial development in the food industry is known as **aseptic processing.** The food item is commercially sterilized and packaged into previously sterilized containers under aseptic conditions.

This process has the advantage that it uses containers other than cans. This provides significant economic and user advantages.

Low Temperatures

Temperatures approaching 0°C and lower retard the growth and metabolic activities of microorganisms. Modern refrigeration and freezing equipment has made it possible to transport and store perishable foods for long periods of time. Refrigerated trucks and railway cars, ships' storage vaults, and the home refrigerator and freezer have improved the quality of the human diet and increased the variety of foods available. Frozen-food production in the United States almost doubled from 11 billion pounds in 1965 to 20 billion in 1975 and is expected to more than double to 48 billion pounds by 1985. Much of this increase will be in prepared frozen foods, whose quantity tripled over the last 10 years and is expected to approach 50 percent of all frozen foods by 1985. The growth and importance of this segment of the food industry places greater emphasis on the study of microorganisms at low temperatures, e.g., their survival, growth, and metabolic activity.

Before freezing, the fresh produce is steamed (blanched) to inactivate enzymes that would alter the product even at low temperatures. Quick-freeze methods, using temperatures of −32°C or lower, are considered most satisfactory; smaller crystals of ice are formed, and cell structures in the food are not disrupted. It should be emphasized that freezing foods, no matter how low the temperature, cannot be relied upon to kill all microorganisms. The number and types of viable and nonviable microorganisms present in frozen foods reflect the degree of contamination of the raw product, the sanitation in the processing plant, and the speed and care with which the product was processed. The microbial count of most frozen foods decreases during storage; but many organisms, including pathogens, e.g., species of *Salmonella*, survive for long periods of time at −9 and −17°C. The temperature ranges at which food-poisoning bacteria and psychrophilic microorganisms are capable of growing are shown in Fig. 28-4.

The increased use of precooked ready-to-serve foods and the prevalence of automatic vending machines for dispensing perishable foods have made it necessary to obtain more data on microbial growth and survival at low temperatures. Figure 28-5 illustrates the growth of salmonellas and staphylococci in prepared foods at various temperatures and times of incubation. Note that the type of

Figure 28-4. Food-poisoning organisms grow in a somewhat higher temperature range than psychrophilic microorganisms. (*Courtesy of R. P Elliot and H. D. Michener, Review of the Microbiology of Frozen Foods, in "Conference on Frozen Food Quality," ARS-74-21, USDA, 1960.*)

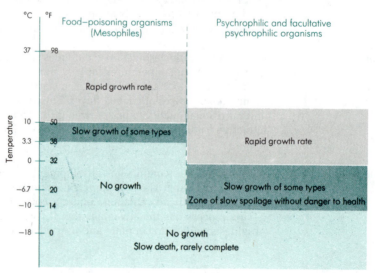

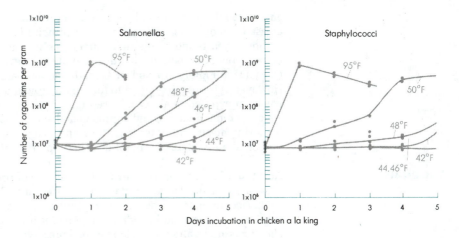

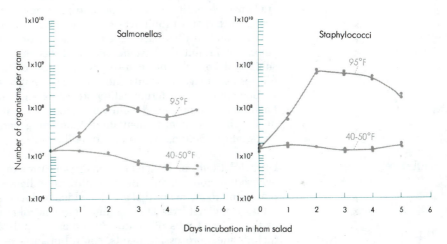

Figure 28-5. Salmonellas and staphylococci multiply rapidly in chicken a la king and ham salad incubated at room temperature. Curves also show growth at other temperatures. (*Courtesy of R. Angelotti, M. J. Foter, and K. H. Lewis, "Time-Temperature Effects on Salmonellae and Staphylococci in Foods," Am J Public Health,* **51:**76–88, 1961.)

food product has considerable influence on the rate of bacterial growth at the lower temperatures.

Dehydration

Dried foods have been used for centuries, and they are more common throughout the world than frozen foods. The removal of water by drying in the sun and air or with applied heat causes **dehydration.** The preservative effect of dehydration is due mainly to microbistasis; the microorganisms are not necessarily killed. Growth of all microorganisms can be prevented by reducing the moisture content of their environment below a critical level. The critical level is determined by the characteristics of the particular organism and the capacity of the food item to bind water so that it is not available as free moisture. It will be recalled that lyophilized cultures of microorganisms survive for years.

Osmotic Pressure

Water is withdrawn from microorganisms placed in solutions containing large amounts of dissolved substances such as sugar or salt. The cells are plasmolyzed, and metabolism is arrested. Thus the antimicrobial condition imposed by increased osmotic pressure is similar in principle to inhibition by dehydration. Although yeasts and molds are relatively resistant to osmotic changes, processes of food preservation based on this principle are, nevertheless, very useful. Jellies and jams are rarely affected by bacterial action because of high sugar content. However, it is not uncommon to find mold growth on the surface of jelly which has been exposed to air. Condensed milk is preserved in part by the increased concentration of lactose and supplemental sucrose. Similar results are obtained by curing meats and other foods in brines. High osmotic pressure may inhibit microbial growth, but it cannot be relied upon to kill all organisms.

Chemicals

Addition of chemicals to foods for the purpose of preservation is subject to the provisions of the United States Food, Drug, and Cosmetic Act as revised in 1972. According to this act, a food is adulterated if any poisonous or deleterious substance has been added which may render it injurious to health. Only a few chemicals are legally acceptable for food preservation. Among the most effective are benzoic, sorbic, acetic, lactic, and propionic acids, all of which are organic acids. Sorbic and propionic acids are used to inhibit mold growth in bread. Nitrates and nitrites used in curing meats, primarily for the preservation of color, are inhibitory to some anaerobic bacteria. This practice has been the subject of considerable controversy because of the potential of nitrates and nitrites as mutagenic agents and the subsequent relationship to carcinogenesis.

Foods prepared by fermentation processes, e.g., sauerkraut, pickles, and silage for animals, are preserved mainly by acetic, lactic, and propionic acids produced during the microbial fermentation. Smoking generates cresols and other antibacterial compounds which penetrate the meat.

Radiation

Ultraviolet light of sufficient intensity and time of exposure is microbicidal to exposed microorganisms. Because ultraviolet light has very limited penetration power, microorganisms that are embedded or covered are unlikely to be affected. Thus, ultraviolet irradiation is limited to control of microorganisms on surfaces or thin, clear layers of liquid. Examples of applications in the food industry include meat-processing plants, control of surface growth on bakery products, sanitation of equipment, and treatment of water used for the depuration (cleansing) of shellfish.

Ionizing radiations are lethal to microorganisms. The fact that they are microbicidal at room temperature and have the ability to penetrate are characteristics that make them attractive candidates for control of microorganisms in foods. Gamma rays and electron beams (beta and cathode rays) have been experimented with extensively for use in the food industry.

Canned and packaged foods can be sterilized by an appropriate radiation dosage. This "cold sterilization" produces a rise in temperature of the product of only a few degrees. **Radiation pasteurization** is a term describing the killing of over 98—but not 100—percent of the organisms by intermediate doses of ionizing radiation.

The ionizing radiation resistance of microorganisms does not correspond to their thermal resistance. *Clostridium botulinum* appears to be the most radioresistant organism of importance to the food technologist. Figure 28-6 illus-

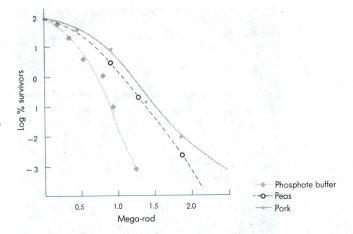

Figure 28-6. Gamma radiation kills spores of *Clostridium botulinum* in frozen foods. Curves show the effect on spores in pork, peas, and phosphate buffer. (*Courtesy of C. B. Denny, C. W. Bohrer, W. E. Perkins, and C. T. Townsend, "Destruction of Clostridium botulinum by Ionizing Radiation," Food Res, 24:44–50, 1959.*)

trates the lethal effect of gamma radiation on spores of *C. botulinum.* Note that the survival of spores is influenced by the material in which they are suspended and that time is not a factor. In the case of radiation, unlike temperature, the radiation death dose rather than radiation death time is determined.

Ionizing radiation sterilization provides the possibility of an entirely new approach to food preservation; it could bring about a radical change in industrial methods of food processing. However, despite the extensive research and documentation on the effectiveness of ionization radiation for the preservation of foods, this method of preservation has not been approved in the United States. This is due in part to economic factors as well as to some lingering uncertainties about the effect of the radiation on the food material. In addition, the United States already has well-developed systems for food preservation. This is not the case for all countries. The World Health Organization approved (1976) radiation of poultry at a specified level, as has Canada for controlling salmonellas. In July 1983, the U.S. Food and Drug Administration approved the use of ionizing radiation for sterilization of specific spices and vegetable seasonings.

FERMENTED FOODS

Thus far we have stressed the undesirable characteristics of microorganisms in food. However, there are many useful applications of microorganisms in the food industry. A variety of important products in our diet are produced with the aid of microbial activity.

Fermented Dairy Products

In the dairy industry, fermented milks are produced by inoculating pasteurized milk with a known culture of microorganisms, sometimes referred to as a **starter culture,** which can be relied on to produce the desired fermentation, thus assuring a uniformly good product. (See Fig. 28-7, which shows *Streptococcus thermophilus* and *Lactobacillus bulgaricus,* organisms used as starter cultures in the preparation of yogurt.)

Several hundred varieties of cheese are manufactured, and with few exceptions, most of them can be made from the same batch of milk. Microorganisms—

Figure 28-7. Photomicrograph of yogurt, illustrating microbial flora, *Streptococcus thermophilus* and *Lactobacillus bulgaricus* (X800). (*Courtesy of K. J. Demeter, Bakteriologische Untersuchungsmethoden der Milchwirtschaft, Eugen Ulmer, Stuttgart, 1967.*)

bacteria or molds—convert the curd of the milk into the desired cheese. For the manufacture of some cheeses, such as blue cheese or Roquefort (blue cheese made in Roquefort, France), it is necessary to inoculate the curd with the microorganism which brings about the changes (in this case, *Penicillium roqueforti*). Some of the steps in the process of making Roquefort cheese are shown in Fig. 28-8.

Other Fermented Foods

Important food items produced in whole or in part by microbial fermentations include pickles, sauerkraut, olives, and certain types of sausage. Lactic acid bacteria are chiefly responsible for the desirable type of fermentation required for the production of each of these products. The microorganisms that produce the changes may be the natural flora on the material to be fermented or may be something added as a starter culture. Most commercial sour, sweet, mustard, and mixed pickles are made from fermented salt-stock pickles. The other major type of pickled cucumber is the fermented dill pickle. An illustration of a commercial fermentation process for the production of dill pickles is shown in Fig. 28-9.

The list of food products produced by microbial fermentation is very long. A few examples are shown in Tables 28-5 and 28-6.

MICROORGANISMS AS FOOD—SINGLE-CELL PROTEIN

Bacteria, yeasts, and algae, produced in massive quantities, are attractive sources of food for animals as well as humans. These microorganisms can be cultivated on industrial wastes or by-products as nutrients and yield a large cell crop that is rich in protein (single-cell protein). Bacterial cells grown on hydrocarbon wastes from the petroleum industry are a source of protein in France, Japan, Taiwan, and India. Yeast-cell crops harvested from the vats used to produce

Figure 28-8. Roquefort and blue cheese. (A) Cubes of sterile whole wheat bread are inoculated with *Penicillium roqueforti*. After extensive growth of the mold on the bread cubes, the cubes are removed, dried, and powdered and used as inoculum for making cheese. (*Courtesy of the Borden Company.*) (B) The addition of a lactic culture and rennet curdles the milk. The curd is cut when it becomes firm. (C) The curd particles are removed and placed in metal hoops. The addition of the spores of *P. roqueforti* may take place in either of these steps. (D) The hoops are placed on a draining board to facilitate whey drainage and matting of the curd, after which the curd is removed, salted periodically, and (E) eventually placed in an area of high humidity (95 to 98 percent) and low temperature (9 to 12°C), where the ripening process occurs over a period of several months. The hoops of cheese shown ripening here are wrapped in foil. [(B to E) Courtesy of Roquefort Association, Inc.]

alcoholic beverages have been used as a food supplement for generations. The attractiveness of single-cell protein as a food substitute or supplement is apparent from the following characteristics of the process.

1 Microorganisms grow very rapidly and produce a high yield. It has been calculated that one can obtain a gain of 1 lb of protein in 1 day's growth from a 1000-lb steer; 1000 lb of yeast would produce several tons of protein in one day! Algae grown in ponds can produce 20 tons (dry weight) of protein per acre per year.

Figure 28-9. (A) Typical tank yard for fermentation and storage of brined cucumbers. The wooden tanks shown are 600- to 1000-bushel capacity. Some pickle companies now use fiberglass tanks. Note the tall, white tank in the background, which is for storage of liquid nitrogen. Nitrogen gas is piped to each brine tank for use in purging of dissolved CO_2 from fermenting brines to prevent bloater damage in the cucumbers. (*Courtesy of H. P. Fleming, USDA.*) (B) Surface of a cucumber brine tank being nitrogen-purged with a sidearm purger. Nitrogen gas purges dissolved CO_2 from the brine and also serves to circulate the brine. The white frothing on the surface is caused by the purging action. (*Courtesy of H. P. Fleming, USDA.*)

Table 28-5. Some Characteristics of Fermented Milks

Fermented Product	Principal Microorganisms Responsible for Fermentation	General Remarks
Cultured buttermilk	A mixture of lactic streptococci (*Streptococcus lactis* or *S. cremoris*) with aroma-producing bacteria (*Leuconostoc citrovorum* or *L. dextranicum*)	The function of the lactic acid streptococci is to produce lactic acid that gives the sour taste and to curdle the milk; the function of the leuconostocs is to produce volatile and neutral products that impart a characteristic desirable odor; the starter culture must contain vigorously growing bacteria; incubation is performed at 21°C.
Cultured sour cream	Same as used for cultured buttermilk, i.e., streptococci and leuconostocs	Not strictly a fermented milk but manufacture resembles that of cultured buttermilk; cream is inoculated and incubated until the desired acidity develops; flavor and aroma compounds are also contributed by the starter culture
Bulgarian milk	*Lactobacillus bulgaricus*	Incubation of inoculated milk at 37°C, but otherwise similar to cultured buttermilk; product differs from commercial buttermilk in having higher acidity and lacking aroma
Acidophilus milk	*L. acidophilus*	Milk for propagation of *L. acidophilus* and the bulk milk to be fermented is sterilized, since this organism is eaily overgrown by contaminating bacteria; incubation is at 37°C; acidity allowed to develop to 0.6 to 0.7%

Table 28-5. (continued)

Fermented Product	Principal Microorganisms Responsible for Fermentation	General Remarks
Yogurt	Streptococcus thermophilus L. bulgaricus	Made from milk in which solids are concentrated by evaporation of some water and addition of skim milk solids; product has consistency resembling custard; now common in Europe and North America; similar products with different names are produced elsewhere (see Fig. 28-7)
Kefir	S. lactis L. bulgaricus Lactose-fermenting yeasts	A mixed lactic acid and alcoholic fermentation; bacteria produce acid (0.6 to 1.0% lactic acid), and yeasts produce alcohol (0.5 to 1.0% ethanol); the organisms conglomerate to form small granules called kefir grains; the granules are used as the starter culture; in the Balkans, the fermentation is carried out in leather bags made of goatskin; the fermentation process may be continuous by adding fresh milk as the fermented product is removed; Kefir is made from cow, goat, or sheep milk
Kumiss	Similar to those found in kefir grains	A mixed acid-alcoholic fermentation product made from mares' milk in some parts of Russia

Table 28-6. Some Examples of Fermented Food Products

Fermented Food	Starting Product	Microorganisms Involved

Sauerkraut	Shredded cabbage	Early stage:
		    Enterobacter cloacae
		    Erwinia herbicola
		
		Intermediate stage:
		    Leuconostoc mesenteroides
		Final stage:
		    Lactobacillus plantarum

Pickles	Cucumbers	Early fermentation:
		    L. mesenteroides
		    Streptococcus faecalis
		    Pediococcus cerevisiae
		Later fermentation:
		    Lactobacillus brevis
		    L. plantarum

Green olives	Olives	Early stage:
		    L. mesenteroides
		Intermediate stage:
		    L. plantarum
		    L. brevis
		Final stage:
		    L. plantarum

Sausage	Beef and pork	Pediococcus cerevisiae Micrococcus spp.

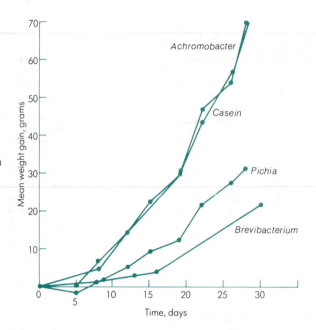

Figure 28-10. The nutritional quality of microbial proteins. This experiment shows the mean weight gain (per group) of rats fed various proteins: casein, bacterial proteins *(Achromobacter, Brevibacterium)*, or yeast protein *(Pichia)*. Note that the growth response to casein and *Achromobacter* protein was very similar. *(Courtesy of V. F. Coty and R. I. Leavitt, Dev Ind Microbiol, 12, 1971.)*

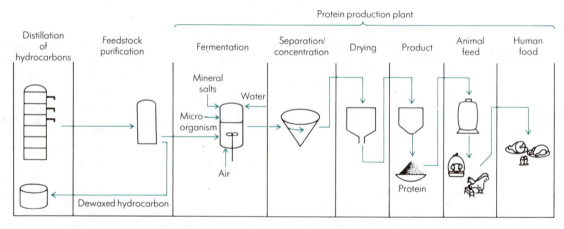

Figure 28-11. The process used by British Petroleum to produce single-cell protein from hydrocarbons. *n*-Alkanes are distilled for use in the fermenter. Minerals are added. Following fermentation, the cells are separated and dried for use as animal feed. *(Courtesy of British Petroleum Co. Ltd.)*

This yield is 10 to 15 times higher than soybeans and 25 to 50 times higher than corn.

2 The protein content of the microbial cells is very high. Dried cells of *Pseudomonas* spp. grown on petroleum products have 69 percent protein; yeast cells have a protein content in a 40 to 50 percent range; for algae, the range is from 20 to 40 percent.

3 The proteins of selected microorganisms contain all the essential amino acids. An example of the nutritional quality of microbial proteins is shown in Fig. 28-10.

4 Some microorganisms, particularly yeasts, have a high vitamin content.

5 The medium (nutrients) for growth of microorganisms may contain industrial wastes or by-products, e.g., liquid paraffins (hydrocarbons) from oil refineries, spent sulfite liquors from the pulp and paper industry, beet molasses, and wood hydrolysates.

A fermentation system using yeast cells for single-cell protein production is shown in Fig. 28-11. The growth medium consists of hydrocarbons (n-alkanes) supplemented with mineral salts. The cell crop is harvested by centrifugation, dried, and used as animal feed.

Despite the very attractive features of single-cell protein as a nutrient for humans there are problems which deter its adoption on a global basis. For example, individual tastes and customs make microorganisms unattractive as a food substance to many persons. More specifically, the high nucleic acid content of microbial cells can produce intestinal disturbances. There is also the need to ascertain if the amino acid composition and content of the microbial protein meet the dietary requirements of the consumer.

QUESTIONS

1 List and describe the principles upon which methods of food preservation are based.

2 Compare the antimicrobial action of the following methods of food preservation: canning, refrigeration, dehydration, and increased osmotic pressure.

3 What is the lowest temperature range at which food-poisoning bacteria will grow?

4 What physiological types of bacteria are most likely to be present when canned food spoils?

5 Compare the types of microorganism that might be involved in the spoilage of refrigerated foods with those incriminated in the spoilage of canned foods.

6 List several types of microbial food spoilage, and name the organisms responsible in each instance.

7 Why is milk an excellent bacteriological culture medium?

8 Is milk sterile as it is drawn from the cow? Explain.

9 List the major sources of bacterial contamination of milk.

10 Describe the various types of biochemical changes brought about in milk by microorganisms. Identify the predominant types of bacteria responsible for each of these changes.

11 Of what particular significance are psychrophilic, thermoduric, and thermophilic bacteria in milk and milk products?

12 Is pasteurized milk sterile milk? Explain.

13 Compare the heat resistance of *Mycobacterium tuberculosis* and *Coxiella burnetii*. What bearing does this have on requirements for adequate pasteurization time and temperature?

14 What information does the phosphatase test reveal about milk?

15 What are the attractive features of food preservation through use of radiation?

16 Outline a procedure suitable for enumeration, isolation, and identification

of the following groups of microorganisms from a sample of food: thermophilic sporeformers, coliforms, and viruses.

17 Compare the microscopic and cultural techniques for microbiological analysis of foods. What are the advantages and limitations of each of these procedures?

18 Name several foods that are prepared by microbial fermentations. Describe the role of microorganisms in each example.

REFERENCES

Ayres, J. C., J. O. Mundt, and W. E. Sandine: *Microbiology of Foods*, W. H. Freeman and Company, San Francisco, 1980.

Banwart, G. J.: *Basic Food Microbiology*, AVI Publishing Company, Westport, Conn., 1980.

BioScience: *Food from Microbes* (special issue), American Institute of Biological Sciences, vol. 30, no. 6, American Institute of Biological Sciences, Arlington, Va., 1980. *A series of six essays which present an excellent assessment of present-day uses of microorganisms for the production of foods.*

Frazier, W. C. and D. C. Westhoff: *Food Microbiology*, 3d ed., McGraw-Hill, New York, 1978.

The three books and the special issue of BioScience cited above provide comprehensive general coverage of the major topics in food microbiology. They cover the subject material of a university-level course in food microbiology.

Helferich, W. and D. C. Westhoff: *All About Yogurt*, Prentice-Hall, Englewood Cliffs, N.J., 1980. *A popular treatment of a food item that has become a national favorite in the last two decades. This is a short (145-page), nontechnical essay which answers many questions that the consumer might have about yogurt.*

International Commission on Microbiological Specifications for Foods: *Microbial Ecology*, vol. 1: *Factors Affecting Life and Death of Microorganisms*; vol. 2: *Food Commodities*, Academic, New York, 1980. *Volume 1 discusses the effects of physical and chemical agents on microorganisms. Volume 2 discusses the microbial flora of various food products, both natural food substances and processed foods.*

Marth, E. H. (ed.): *Standard Methods for Examination of Dairy Products*, 14th ed., American Public Health Association, Washington, D.C., 1978. *This publication describes laboratory test procedures for milk and milk products in meticulous detail. It is regarded as the basic reference in the field of dairy products.*

Mossel, D. A. A.: *Microbiology of Foods*, The University of Utrecht, Faculty of Veterinary Medicine, University of Utrecht, The Netherlands, 1982. *A brief (188-page) paperback that covers general principles of controlling microorganisms, transmission of diseases by food, microbial deterioration of foods, microbiological sampling, and other aspects of food microbiology.*

Rose, A. A. (ed.): *Fermented Foods*, vol. 7: *Economic Microbiology*, Academic, New York, 1982. *Comprehensive coverage of the subject of foods that are prepared through microbial fermentation processes.*

Chapter 29   Industrial Microbiology

Microorganisms were exploited for useful purposes long before anything was known about their existence or their characteristics. As early as 6000 B.C., the Babylonians and Sumerians used yeast to make alcohol. History reveals many other applications of microbial processes that resulted in the production of desirable materials, particularly foods and beverages. However, as we have mentioned earlier, it was not until the studies of Louis Pasteur in the second half of the nineteenth century that the role of microorganisms in these processes was understood.

From the standpoint of industrial microbiology, microorganisms can be considered chemical factories in miniature. They have the capacity to convert a raw material (nutrient or substrate) into end products. If the end products have value for human use, then it becomes attractive to exploit the microbiological process, that is, to produce the end products on a commercial scale.

Industrial microbiology has experienced two dramatic explosions during the last few decades. In the 1940s the discovery of antibiotics, led by penicillin, initiated a major new industry built upon the products of microorganisms. More recently, as a result of the great advances in our knowledge of microbial genetics, it is possible to manipulate microorganisms genetically to produce new products. The process is called **recombinant DNA technology.** This development, namely, the "engineering" of microorganisms to produce needed valuable chemical substances, is likely to revolutionize the field of industrial microbiology.

MICROORGANISMS AND INDUSTRY

Microorganisms, under natural conditions, produce an extremely large number, as well as a very large variety, of chemical substances. Some of these substances are very useful for the treatment of diseases and disorders of people and other animals and hence are attractive to the pharmaceutical industry; others are valuable as raw materials for the chemical industry for precursors for other products, solvents, and for other uses. We have already discussed the role of microorganisms in the production of many foods. Other applications of large-scale microbial activity can be found in mining, where microorganisms are used to leach metals from low-grade ores; in dealing with environmental pollution where, for example, microorganisms are used to degrade obnoxious pollutants; and in agriculture for enhancement of plant growth, control of insect pests, and other purposes.

The overall reaction characterizing the industrial application of microorganisms can be summarized as follows:

$$\text{Substrate (raw material)} + \text{microorganisms} \rightarrow \text{new products}$$

Prerequisites to Practical Industrial Microbiological Processes

If a microorganism converts cheap raw materials into a useful product, it may be feasible to perform this reaction on a large industrial scale. Some of the prerequisites to an economically practicable industrial microbiological process are the following, in terms of the organism, the medium, and the product:

1 **The organism.** The organism to be used must be able to produce appreciable amounts of the product. It should have relatively stable characteristics and the ability to grow rapidly and vigorously, and it should be nonpathogenic.
2 **The medium.** The medium, including the substrate from which the organism produces the new product, must be cheap and readily available in large quantities. In several instances it has been found practicable to utilize nutrient-containing wastes from the dairy industry (whey), the paper industry (waste liquors resulting from the cooking of wood, waste sulfite liquors), and other commercial operations.
3 **The product.** Industrial fermentations are performed in large tanks; capacities of 50,000 gal are not unusual. The product formed by the metabolism of the microorganism is present in a heterogenous mixture that includes a tremendous crop of microbial cells and unused constituents of the medium, as well as products other than those being sought. Thus, an efficient and economical mass-scale method of recovery and purification of the desired end product must be developed.

Major Classes of Products and Processes

The major commercial products of microorganisms can be classified as follows: (1) the microbial cells; (2) large molecules like enzymes that are synthesized by the microorganisms; (3) primary metabolic products, i.e., compounds essential for cell growth; (4) secondary metabolic products, i.e., compounds not required for cell growth. Substances in groups 3 and 4 are generally much smaller in molecular size than those in group 2.

One may also group industries on the basis of the type of microbial products they market, as listed below:

1 *Pharmaceutical chemicals.* Most prominent in this category are the antibiotics and steroid drugs, but other substances, such as insulin and interferon, are now being produced by genetically engineered bacteria. Many other new products are likely through genetic biotechnology.

2 *Commercially valuable chemicals.* Solvents, enzymes, and intermediate compounds for the synthesis of other substances are representative of the kinds of substances produced commercially by microorganisms. Specific examples are provided later in this chapter.

3 *Food supplements.* Mass production of yeasts, bacteria, and algae from media containing inorganic nitrogen salts and other readily available and cheap nutrients provides a good source of protein and other organic nutrients useful as food supplements. Large-scale microbial production of amino acids is an attractive industrial process being employed in many parts of the world.

4 *Alcoholic beverages.* Brewing, wine making, and production of other alcoholic beverages constitute some of the oldest and largest microbiological industries.

5 *Vaccines (immunizing antigens).* Some microorganisms are grown in very large quantities for use as vaccines. The whole cell or some part or product of the cell is used for the preparation of vaccines.

6 *Deterioration of materials by microorganisms.* All kinds of material such as leather, textiles, wood, metals, and even optical equipment are subject to deterioration by contamination with and growth of microorganisms. The magnitude of potential destruction with resulting financial losses demands that methods for prevention of this destruction be developed. Industry is responsive to this need and produces many chemicals and treatment processes for this purpose.

7 *Analytical microbiology.* Microbiological techniques have been developed for assaying a variety of products like antibiotics, amino acids, and vitamins. Other microbiological procedures are available for evaluating wood and paint preservatives and for testing the efficacy of sterilization procedures.

Microorganisms Used in Industrial Processes

Industrial microbiological processes have been developed using specific strains of algae, fungi (yeasts and molds), bacteria, protozoa, and viruses. Microbial species which have potential for industrial application are continually being sought. The attractiveness of a microorganism may reside in its ability to produce a new product, e.g. an antibiotic. Or the industrial application might involve the use of a microorganism in a process such as cleaning up oil spills; the microorganism degrades the oil to nonobjectionable compounds.

Once a species has been found to have industrial application, a research program is undertaken to increase the capacity of the microorganism to produce the desirable change, that is, to give a higher yield of the end product or a greater rate of change in the substrate being decomposed. The customary approach to achieve these ends has been through improvements in culture media and cultural conditions, selection of new strains, and development of mutants.

However, research in molecular biology and more specifically research in bacterial genetics, as described in Chap. 12, has provided the knowledge and the technology to deliberately change the genetic makeup of a microorganism. This process, known as **genetic recombination,** has dramatically altered industrial microbiology.

BIOENGINEERING OF MICROORGANISMS FOR INDUSTRIAL PURPOSES

What is commonly referred to today as bioengineering of microorganisms is, in fact, an application of recombinant DNA technology—the in vitro incorporation of segments of genetic material from one cell into another cell. This technology was made possible from the knowledge accumulated over the last few decades in biological research at the molecular level which elucidated the structure and synthesis of DNA. The fundamental aspects of this subject were presented in Chaps. 11 and 12.

Genetic Engineering of Microorganisms

The essential steps in the technology of producing a genetically engineered bacterium are shown in Fig. 29-1A. They can be summarized as follows:

1 *Source of donor genetic material.* DNA containing the genetic code for the property to be transferred into a bacterium is isolated from cells, or it may be synthesized. The DNA is *tailored* to form the gene which contains the genetic information to code for a desired characteristic such as production of human insulin.

2 *Production of hybrid DNA molecule.* The donor genetic material (DNA segment) is incorporated into the DNA molecule of a bacteriophage or a bacterial plasmid. This is accomplished by the use of two enzymes: **restriction endonucleases ligases.** Restriction endonucleases cut double-stranded DNA molecules at particular nucleotide sequences and thus produce a well-defined DNA fragment for a given enzyme and a given DNA. In this process both the donor DNA and the agent **(vehicle)** into which the fragment of the donor DNA is to be incorporated are treated with the same restriction endonuclease. As shown in Fig. 29-1B, the endonuclease *Eco* R1 cuts the plasmid DNA and the donor DNA in a manner such that the ends of each are identical and self-complementary. The fragments can be connected by the addition of an enzyme called DNA ligase.

Hybrid DNA can also be produced by other more elaborate experimental techniques.

3 *Incorporation of hybrid DNA into host cell.* **Transformation** in genetic engineering is the process by which plasmid hybrid DNA molecules are introduced into a competent host bacterial cell. **Transfection** involves the introduction of phage hybrid DNA into the host cell. The most common technique for transformation depends on treating the recipient bacteria with calcium chloride to make the membrane permeable to the DNA. The recipient bacteria are capable of receiving recombinant DNA molecules on the basis of only one molecule per bacterium.

When bacteria are transformed or transfected, a mixture of bacteria of various genotypes is usually produced. But each bacterial cell is capable of binary fission, yielding a colony of identical cells possessing equivalent genetic, and therefore physiological, traits. Once a colony with the proper phenotype is identified, the bacteria in it can be grown in limitless quantity to amplify the gene.

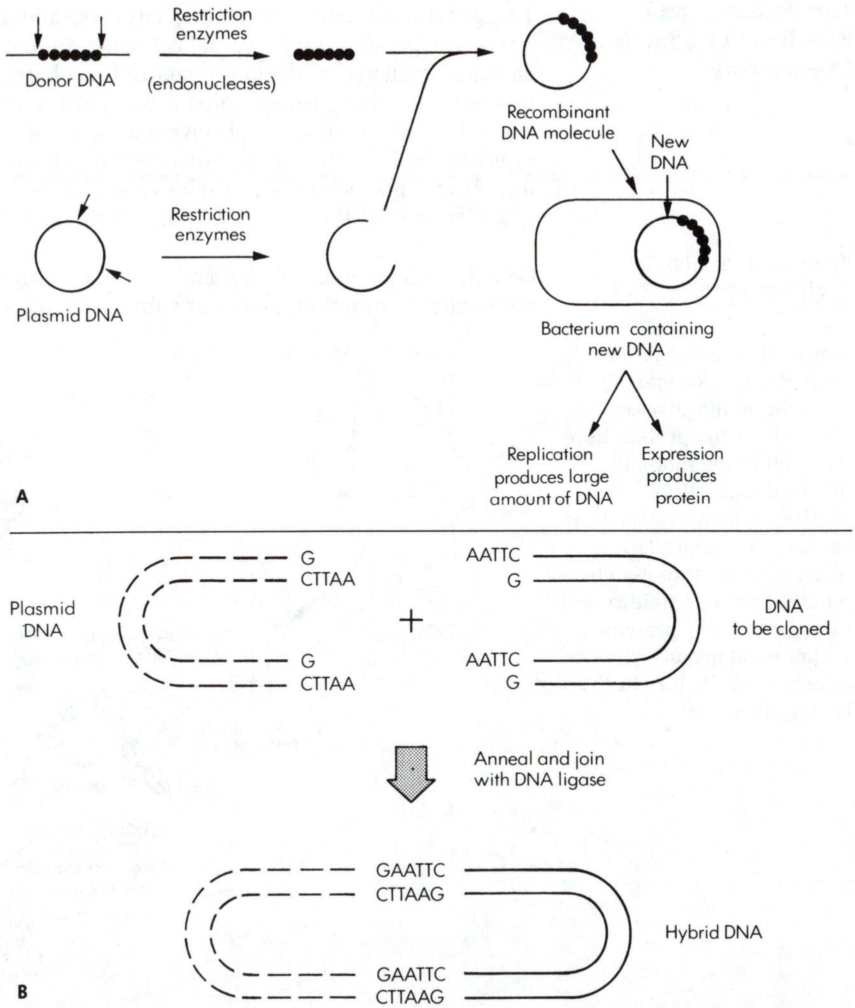

Figure 29-1. (A) The major steps in producing a "genetically engineered" bacterium. (B) Fragments of donor DNA and plasmid DNA excised by endonucleases and formation of hybrid DNA by joining these fragments using DNA ligase. *(Erwin F. Lessel, illustrator.)*

Thus it is seen that the difficult problem of the chemical purification of a gene has been surmounted by the screening of bacterial colonies. **Cloning,** the isolation and proliferation of individual, genetically unique cells, thus provides one type of a high-resolution separation method for DNA molecules which would be almost impossible to fractionate by any other means. The progeny of the selected bacterium constitutes a **clone** and the gene is said to have been **cloned.** Since there is no difficulty in physically separating the plasmid DNA from the rest of the bacterial DNA, it is possible to obtain the DNA of the cloned gene in pure state and in unlimited amounts. Furthermore, one plasmid inserted into an *Escherichia coli* bacterium may generate a hundred or more copies of itself within the cell. Cloned genes have been obtained in great numbers from a wide variety of species, ranging from bacteria through brewing-yeasts, fruit flies, sea urchins, toads, and mice.

The Potential and Problems of Genetic Engineering

The genetic alteration of plants, animals, and microorganisms has been an important practice in many of our major industries, such as agriculture, the beverage industry, and more recently the pharmaceutical industry (antibiotic production). These genetic alterations have been achieved through mutation and selection. As a result of new genetic technologies our capabilities to manipulate the inherited characteristics of all species of life has been enormously increased. This technology provides almost limitless possibilities for the benefit of society and at the same time poses serious problems.

Benefits from Genetic Engineering

Genetic technologies, present and future, can contribute to the improvement of our health, our environment, our supply of food, and many other aspects of our

Figure 29-2. Recombinant DNA strategy for making foot-and-mouth disease vaccine. VP$_3$ is the protein from the shell of the foot-and-mouth disease virus (FMDV), which can act as a vaccine for immunizing livestock against foot-and-mouth disease. The idea is to make this VP$_3$ protein without making any virus or infectious RNA. (*Erwin F. Lessel, illustrator.*)

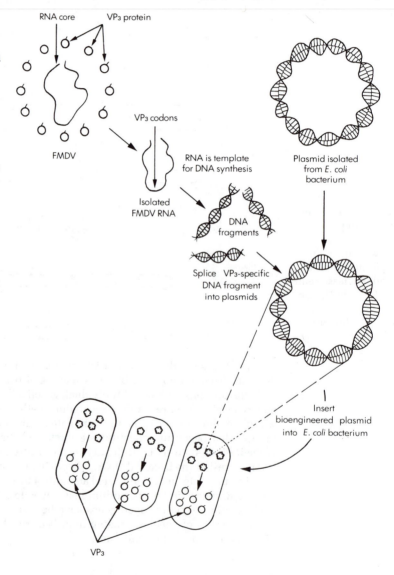

lives. The pharmaceutical industry has already produced several products for human therapy, such as human insulin, interferon, urokinase (for the treatment of blood clots), and somatostatin (a brain hormone), and new techniques for vaccine development have emerged (see Fig. 29-2). A major research effort is underway to produce genetically engineered microorganisms that can fix nitrogen in cereal crops and thus greatly improve soil fertility.

Microorganisms have been genetically engineered to decompose oil in oil spills, and other commercial applications are likely in pollution-control industries, mining, and oil recovery.

Potential Problems of Genetic Engineering

The practical application of molecular genetic technology allows the movement of genes across species lines, such as from animals and fruit flies to bacteria! This results in the creation of new, redesigned organisms. This has raised questions of risks that might be involved.

There is concern that production of recombinant DNA molecules that are functional in vivo could prove biologically hazardous. If they are carried in a microbe like *E. coli*, which is a commensal bacterium in the human gut and can exchange genetic information with other types of bacteria, they might possibly become widely disseminated among human, bacterial, plant, or animal populations, with unpredictable results.

Of special concern is construction of new autonomously replicating bacterial plasmids that could, if not very carefully controlled, introduce genetic determinants for antibiotic resistance or bacterial toxin formation into bacterial strains that do not presently carry such determinants. Experiments to link all, or segments of, DNA from oncogenic or other animal viruses to autonomously replicating DNA elements, such as bacterial plasmids or other viral DNAs, also pose threats.

Because of the concerns associated with genetic engineering, The National Institutes of Health have established guidelines for research involving recombinant DNA molecules. Under these guidelines, The National Institutes of Health serve an overseeing role by sponsoring risk-assessment programs, certifying new host-vector systems, serving as an information clearing house, and coordinating federal and local activities.

INDUSTRIAL USES OF BACTERIA

Some products of bacterial origin together with their uses are shown in Table 29-1. The production processes for lactic acid, vinegar, amino acids (lysine and glutamic acid), and insulin will be described as examples of the many others in operation today.

Lactic Acid Production

Several carbohydrate substances such as corn and potato starch, molasses, and whey can be used for the production of lactic acid. Starch must first be hydrolyzed to glucose by acid or enzymatic treatment. The choice of carbohydrate material depends upon its availability, treatment required prior to fermentation, and cost. We shall describe the production of lactic acid from whey.

Large quantities of whey constitute a waste product in the manufacture of certain dairy products such as cheese. From the standpoint of pollution problems created by the disposal of untreated whey, as well as for economic reasons,

Table 29-1. Some Industrial Products (Other than Antibiotics) Produced by Bacteria

Product	Microorganism	Uses
Acetone-butanol	*Clostridium acetobutylicum* and others	Solvents; chemical manufacturing
2,3-Butanediol	*Bacillus polymyxa* *Enterobacter aerogenes*	Solvent; humectant; chemical intermediate
Dihydroxyacetone	*Gluconobacter suboxydans*	Fine chemical
2-Ketogluconic acid	*Pseudomonas* spp.	Intermediate for D-araboascorbic acid
5-Ketogluconic acid	*G. suboxydans*	Intermediate for tartaric acid
Lactic acid	*Lactobacillus delbrueckii* *L. bulgaricus*	Food products; textile and laundry; chemical manufacturing; deliming hides
Bacterial amylase	*Bacillus subtilis*	Modified starches; sizing paper; desizing textiles
Bacterial protease	*B. subtilis*	Bating hides; desizing fibers; spot remover; tenderizing meat
Dextran	*Leuconostoc mesenteroides*	Stabilizer in food products; blood-plasma substitute
Sorbose	*G. suboxydans*	Manufacture of ascorbic acid
Cobalamin (vitamin B_{12})	*Streptomyces olivaceus* *Propionibacterium freudenreichii*	Treatment of pernicious anemia; food and feed supplementation
Glutamic acid	*Brevibacterium* spp.	Food additive
Lysine	*Micrococcus glutamicus*	Animal-feed additive
Streptokinase-streptodornase	*Streptococcus equisimilis*	Medical use (dissolving blood clots)
Bioinsecticides	*Bacillus thuringiensis* *Bacillus popilliae*	Control of insects
Insulin, interferon, somatostatin (human growth hormone)	Recombinant DNA Varieties of *E. coli*	Human therapy
Microbial protein (SCP)	Methane-oxidizing bacteria	Food supplement

it is desirable to use it to make some useful product. Whey represents a satisfactory medium for the growth of certain bacteria, since it contains carbohydrate (lactose), nitrogenous substances including vitamins, and salts. The first requirement for the development of a method of producing lactic acid is an organism capable of growing in whey and fermenting most if not all the lactose to lactic acid. Lactobacilli are suitable for this purpose, particularly *Lactobacillus bulgaricus*. This organism grows rapidly and is homofermentative and thus is capable of converting the lactose to the single end product—lactic acid. Stock cultures of the organism used are maintained in a skim-milk medium. To prepare a sufficient amount of inoculum for addition to the main fermentation tank, the culture is successively transferred and incubated in increasing amounts of sterile skim milk, pasteurized skim milk, and finally whey. Milk is used in "building up" the inoculum, since it is a superior medium. Inoculum from the whey-incubation tank is added to the fermentation tank in an amount equivalent to 5 to 10 percent of the volume to be fermented. An incubation temperature of 43°C is used and has the desirable effect of inhibiting growth of many extraneous microorganisms. During the fermentation, a slurry of lime, $Ca(OH)_2$, is added intermittently to neutralize the acid, and calcium lactate is formed; otherwise the accumulation of acid would retard fermentation. Upon completion of fermentation (approximately 2 days) the material in the tank is boiled to coagulate the protein, which is then filtered and processed for use as an animal-feed supplement. The filtrate containing the calcium lactate is then concentrated by removal of water under a vacuum, followed by additional treatments to purify the compound. The process is shown schematically in Fig. 29-3.

The biochemical reactions performed by the microorganism in producing the lactic acid can be summarized:

$$\underset{\text{Lactose}}{C_{12}H_{22}O_{11}} + H_2O \xrightarrow{\text{lactase}} \underset{\substack{\text{Glucose} \\ + \text{ galactose}}}{2C_6H_{12}O_6} \xrightarrow[\text{of enzymes}]{\text{system}} \underset{\text{Lactic acid}}{2CH_3CHOHCOOH}$$

Derivatives of lactic acid are used in the treatment of calcium deficiency (calcium lactate) and of anemia (iron lactate), as a solvent in lacquers (n-butyl lactate), and as a plasticizer and humectant (sodium lactate).

Vinegar Production

The word *vinegar* is derived from the French term *vinaigre*, meaning "sour wine." It is prepared by allowing a "wine" to go sour under controlled conditions.

The production of vinegar involves two types of biochemical changes: (1) an alcoholic fermentation of a carbohydrate and (2) oxidation of the alcohol to acetic acid. There are several kinds of vinegars, and the differences among them are primarily associated with the kind of material used in the alcoholic fermentation, e.g., fruit juices, sugar-containing syrups, and hydrolyzed starchy materials. The definition and standards for one type as given by the U.S. Food and Drug Administration are as follows:

Vinegar, cider vinegar, apple vinegar. The product made by the alcoholic and subsequent acetous fermentations of the juice of apples. It contains, in 100 cubic centimeters (20°C), not less than 4 grams of acetic acid.

Figure 29-3. Lactic acid production from whey by *Lactobacillus bulgaricus*.

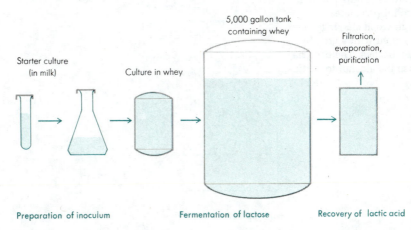

Starter culture (in milk)

Culture in whey

5,000 gallon tank containing whey

Filtration, evaporation, purification

Preparation of inoculum

Fermentation of lactose

Recovery of lactic acid

A yeast fermentation is used for production of the alcohol. The alcohol concentration is adjusted to between 10 and 13 percent and then exposed to the action of acetic acid bacteria. Many types of equipment have been designed for industrial production of vinegar. All depend upon providing a suitable environment for the bacterial oxidation of alcohol to acetic acid. The essential features of one of the industrial processes for vinegar production, the Frings method, is shown in Fig. 29-4 and may be summarized as follows. A mix is prepared which consists of an adjusted solution of alcohol acidified with acetic acid and special nutrients for the growth of acetic acid bacteria. Acetic acid bacteria, species of the genus *Acetobacter*, are inoculated onto the beechwood shavings. The mix is applied in a trough at the top of the chamber and allowed to trickle down over the shavings. As the alcohol solution passes over the shavings, the acetobacters oxidize some of the alcohol to acetic acid. The mix is collected at the bottom of the unit and may be recirculated over the shavings, resulting in more oxidation of alcohol until vinegar of the desired strength is produced.

Since this is an aerobic process, oxygen is required as shown in the following reaction accounting for the formation of acetic acid:

$$2CH_3CH_2OH + 2O_2 \rightarrow 2CH_3COOH + 2H_2O$$
Ethyl alcohol　　　　　　　　Acetic acid

An abundant supply of air must be available throughout the chamber. It is also necessary to keep the temperature between 15 and 34°C, the optimum for growth and metabolism of the acetobacters. The Frings vinegar generator is equipped with various accessories which permit control of these factors. Deviation in temperature below or above this range not only has an adverse effect on the acetobacters but permits growth of other microorganisms with different metabolic characteristics.

Amino Acid Production

Many microorganisms can synthesize amino acids from inorganic nitrogen compounds. The rate and amount of synthesis of some amino acids may exceed the cells' need for protein synthesis, whereupon the amino acids are excreted into the medium. Some microorganisms are capable of producing amounts of certain

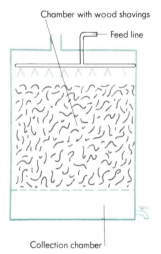

Figure 29-4. Frings vinegar generator. A dilute solution of alcohol percolates through wood shavings that are covered with a growth of acetobacters. The bacteria oxidize the alcohol to acetic acid.

Chamber with wood shavings

Feed line

Collection chamber

amino acids (lysine, glutamic acid, and tryptophan) sufficient to justify their commercial production. Among the advantages of the microbial fermentation processes is that the biologically active forms of the amino acids (L optical isomers) are produced.

L-Lysine Production

One of the commercial methods for production of lysine consists of a two-stage process using two species of bacteria: (1) the formation of diaminopimelic acid (DAP) by *E. coli* and (2) the decarboxylation of the diaminopimelic acid by an enzyme (DAP decarboxylase) obtained from *Enterobacter aerogenes*:

$$
\begin{array}{c}
\text{COOH} \\
| \\
\text{HCNH}_2 \\
| \\
\text{CH}_2 \\
| \\
\text{CH}_2 \\
| \\
\text{CH}_2 \\
| \\
\text{HCNH}_2 \\
| \\
\text{COOH}
\end{array}
\quad
\xrightarrow[\substack{\text{(from Enterobacter aerogenes} \\ \text{ATCC 12409)}}]{\substack{\text{DAP} \\ \text{decarboxylase}}}
\quad
\begin{array}{c}
\text{H} \\
| \\
\text{HCNH}_2 \\
| \\
\text{CH}_2 \\
| \\
\text{CH}_2 \\
| \\
\text{CH}_2 \\
| \\
\text{HCNH}_2 \\
| \\
\text{COOH}
\end{array}
\quad + \text{CO}_2
$$

Diaminopimelic acid   Lysine

E. coli is grown in a medium consisting of glycerol, corn-steep liquor, and $(NH_4)_2HPO_4$ under controlled conditions of aeration, temperature, and pH for optimum production of DAP. After approximately 3 days' incubation, DAP decarboxylase is added to convert the DAP to lysine, as shown in the reaction above.

Lysine is an essential amino acid for the nutrition of humans and is of particular interest since cereal proteins are often deficient in this amino acid. It is used as a supplement for bread and other foodstuffs.

L-Glutamic Acid Production

Many species of microorganisms, especially bacteria and fungi, are capable of producing large amounts of glutamic acid. Species of *Micrococcus*, *Arthrobacter*, and *Brevibacterium* are used for its industrial production. The medium generally consists of a carbohydrate, peptone, inorganic salts, and biotin; the concentration of biotin has a significant influence on the yield of glutamic acid. α-Ketoglutaric acid produced via the tricarboxylic acid cycle (Krebs cycle) is the precursor of glutamic acid.

The conversion of α-ketoglutaric acid to glutamic acid is accomplished by glutamic dehydrogenase.

$$
\begin{array}{c}
\text{COOH} \\
| \\
\text{C}{=}\text{O} \\
| \\
\text{CH}_2 \\
| \\
\text{CH}_2 \\
| \\
\text{COOH}
\end{array}
\quad
\xrightarrow[\text{dehydrogenase}]{\substack{+ \text{NH}_3 + \text{NADH}_2 \\ \text{glutamic acid}}}
\quad
\begin{array}{c}
\text{COOH} \\
| \\
\text{CHNH}_2 \\
| \\
\text{CH}_2 \\
| \\
\text{CH}_2 \\
| \\
\text{COOH}
\end{array}
\quad + \text{NAD}^+ + \text{H}_2\text{O}
$$

α-Ketoglutaric acid   Glutamic acid

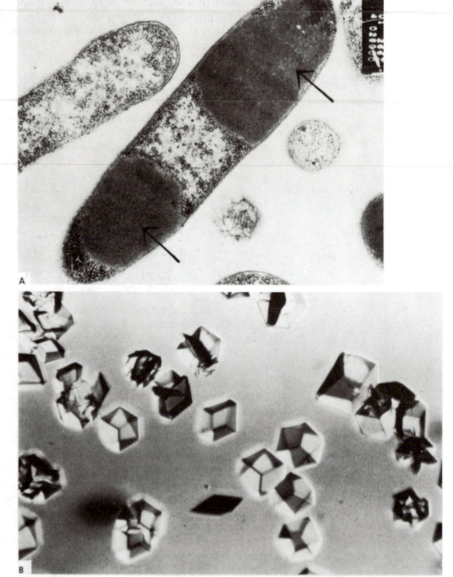

Figure 29-5. (A) Transmission electron micrograph of *E. coli* (X35,000) containing insulin A chain chimeric protein. Arrows indicate concentrations of this chimeric protein in the cells. (B) The first crystals ever obtained of human insulin made by recombinant DNA technology. *(Courtesy of Eli Lilly Co.)*

Glutamic acid is in demand as a condiment and flavor-enhancing agent in the form of monosodium glutamate. Millions of pounds are produced annually.

Insulin is one of the important pharmaceutical products produced commercially by a genetically engineered bacterium. Prior to this development, commercial insulin for the therapy of diabetes was isolated from animal pancreatic tissue. Earlier research on purified insulin isolated from pancreatic tissue led to the

Table 29-2. Some Commercial Products of Yeast

Product	Microorganism	Uses
Bakers' yeast, beer, wine, ale, bread	*Saccharomyces cerevisiae*	Baking industry; brewing industry
Soy sauce	*Saccharomyces rouxii*	Food condiment
Sour French bread	*Candida milleri*	Baking
Commercial alcohol (ethanol)	*S. cerevisiae* *Kluyveromyces fragilis*	Fuel; solvent
Riboflavin	*Eremothecium ashbyi*	Vitamin supplement
Microbial protein	*Candida utilis*	Animal food supplement (single-cell protein) from paper-pulp waste
	Saccharomycopsis lipolytica	Microbial protein from petroleum products

establishment of the amino acid sequence of this protein hormone molecule. From this information it was possible to establish the DNA code for the synthesis of insulin. This was followed by the isolation of the gene from human tissue which controls insulin production. By using recombinant DNA technology, the human insulin gene was introduced into a bacterium (*E. coli*). This genetically engineered bacterium is grown in large quantities, as is characteristic of industrial microbiological processes, to produce human insulin. Following maximum production of insulin in the commercial culture, the insulin is extracted, purified, and evaluated for biological response (see Fig. 29-5A and B).

Human insulin produced by genetically altered bacteria was made available to diabetics in September of 1982.

Two of the major advantages of insulin production by microorganisms is that the resulting insulin is chemically identical to human insulin and it can be made available in unlimited quantities.

INDUSTRIAL USES OF YEASTS

The best known and one of the most important uses of yeasts is in the production of ethyl alcohol from carbohydrate materials. This fermentation process is used by brewers of malt beverages, distillers, bakers, wine makers, chemical manufacturers, homemakers, and many others. A list of some of the commercial products of yeasts is shown in Table 29-2.

Alcohol Fermentations

Next to water, alcohol is the most common solvent and raw material used in the laboratory and chemical industry. The microbiological aspects of the process of ethyl alcohol production can be summarized as follows.

The Substrate

Ethyl alcohol can be produced from any fermentable carbohydrate by yeasts. When starches, such as corn, and other complex carbohydrates are used as the raw material, it is first necessary to hydrolyze them to simple fermentable sugars. The hydrolysis can be accomplished with enzymes from barley malt or molds or by heat treatment of acidified material. Corn, molasses, sugar beets, potatoes, and grapes are some of the common raw materials employed throughout the world.

The Organism

Selected strains of *Saccharomyces cerevisiae* are commonly employed for the fermentation. It is imperative that the culture be one that grows vigorously and has a high tolerance for alcohol as well as a capacity for producing a large yield of alcohol. Much attention has been directed toward the selection and development of strains of yeasts which excel in these particular characteristics.

The Reaction

The biochemical change accomplished by the yeast is as follows:

$$C_6H_{12}O_6 \xrightarrow[\text{enzymes}]{\text{yeast}} 2C_2H_5OH + 2CO_2$$

$$\underset{\substack{\text{Glucose} \\ \text{(fermentable} \\ \text{carbohydrate)}}}{} \qquad \underset{\substack{\text{Ethyl} \\ \text{alcohol}}}{} \underset{\substack{\text{Carbon} \\ \text{dioxide}}}{}$$

Bakers' Yeast

The use of yeast as a leavening agent in baking dates back to the very early histories of the Jews, Egyptians, Greeks, and Romans. In those days leavened bread was made by mixing some leftover dough from the previous batch of bread with fresh dough. Another practice, since the Middle Ages, has been to use excess yeasts from brewing and winemaking operations. The variable quality of such products made this practice unsatisfactory. In modern baking practice, pure cultures of selected strains of *S. cerevisiae* are mixed with the bread dough to bring about desired changes in texture and flavor. Desirable characteristics of *S. cerevisiae* strains selected for commercial production of bakers' yeast include the ability to ferment the sugar in the dough vigorously and to grow rapidly; these as well as other characteristics for which the strain was selected should be relatively stable. The carbon dioxide produced during the fermentation is responsible for the leavening, or rising, of the dough. The quality of the product depends on the proper selection of yeasts and the incubation conditions as well as on the choice of raw materials.

In the manufacture of bakers' yeast the "stock" strain is inoculated into a medium which frequently contains molasses and corn-steep liquor. The medium is adjusted to an acid pH (pH 4 to 5), which helps retard bacterial growth. The inoculated medium is aerated during the incubation period. At the end of incubation the yeast cells are harvested by centrifugation and washed by suspending the cells in water and then centrifuging the cells out. The cells are finally recovered on a filter press, small amounts of vegetable oil are added as a plasticizer, and then this mass of cells is molded into blocks. Some steps in this process are illustrated in Fig. 29-6.

Food Yeasts

Mass cultivation of yeasts, as well as of algae and bacteria, offers a possible source of food supplement or substitute for human and animal consumption. This subject is presented in Chap. 28, where the production of single-cell protein (yeast) from petroleum constituents is discussed. It appears, at the present time, that the major technical problems associated with producing a new type of protein for animal foods have been solved. Thus, massive production of microbial cells may provide the way of bridging the "protein gap" in a protein-hungry world.

INDUSTRIAL USES OF MOLDS

Many substances are produced commercially by molds. Perhaps the most significant is the antibiotic penicillin. Molds are used for the fermentation of rice

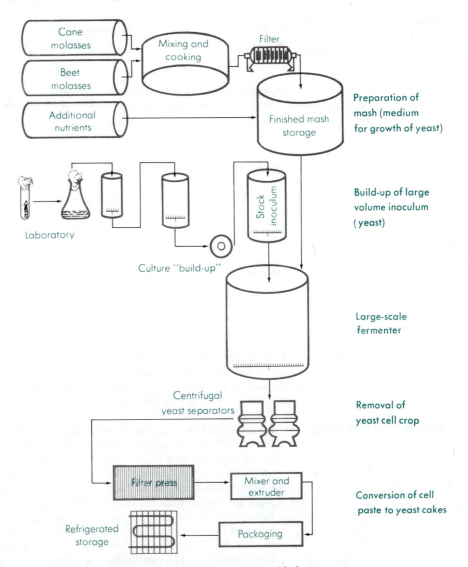

Figure 29-6. Steps in the commercial production of bakers' yeast.

Table 29-3. Some Industrial Products (Other than Antibiotics) Derived from Molds

Product	Microorganism	Uses
Citric acid	Aspergillus niger or Aspergillus wentii	Food products, medicinal citrates; in blood for transfusion
Fumaric acid	Rhizopus nigricans	Manufacture of alkyd resins, wetting agents
Gluconic acid	A. niger	Pharmaceutical products, textiles, leather, photography
Itaconic acid	Aspergillus terreus	Manufacture of alkyd resins, wetting agents
Pectinases, proteases	A. wentii or Aspergillus aureus	Clarifying agents in fruit juice industries
11-γ-Hydroxyprogesterone	Rhizopus arrhizus, R. nigricans, others	Intermediate for 17-γ-hydroxycorticosterone
Gibberellic acid	Fusarium moniliforme	Setting of fruit, seed production
Lactic acid	Rhizopus oryzae	Foods and pharmaceuticals

to produce a variety of oriental foods and food additives. They also produce several enzymes—proteases, amylases, and pectinases—that are manufactured for use in industry. A list of some commercially important products, other than penicillin, is shown in Table 29-3.

Penicillin Production

The commercial production of penicillin and other antibiotics represents one of the most dramatic case histories in the development of industrial microbiology. The antibiotic industry did not exist in 1941, but 10 years later net sales of these products had reached $344 million per year. Data reported in 1983 by the U.S. International Trade Commission revealed that 32.518 million pounds of bulk antibiotics were manufactured in 1982.

Penicillin was the first antibiotic to be produced industrially. Much of what was learned in transforming Fleming's laboratory observations into an economically feasible large-scale operation paved the way for successful production of other chemotherapeutic antibiotics as they were discovered.

The mold isolated by Fleming (*Penicillium notatum*), and as grown in his laboratory, yielded only a few units of penicillin per milliliter, an exceedingly small amount when one considers that a patient may require treatment with millions of units. The remarkable chemotherapeutic effectiveness of penicillin was demonstrated by Florey and Chain during 1939 and 1941. Because of the pressures of war, the British scientists brought the mold to the United States in hope of developing production of the antibiotic on a large scale. An extensive research program having one of the highest wartime priorities was initiated. In a relatively short time the yield of penicillin was increased about a thousand times. The developments contributing to this enormous increase in yield were as follows:

1 Improvements in composition of the medium.
2 Isolation of a better penicillin-producing mold species, *Penicillium chrysogenum.*
3 Development of the submerged-culture technique: cultivation of the mold in large volumes of liquid medium through which sterile air is forced.
4 The production of mutant strains of *P. chrysogenum* which were capable of producing large amounts of penicillin. A series of mutants, produced by x-ray and ultraviolet radiation, resulted in strains with a remarkable capacity for synthesis of penicillin.
5 The addition of chemicals to the medium which served as precursors for synthesis of penicillin.
6 Refinements in methods of recovering penicillin from the fermentation mixture.

The major steps in the commercial production of penicillin are:

1 Preparation of inoculum
2 Preparation and sterilization of medium
3 Inoculation of the medium in the fermenter
4 Forced aeration with sterile air during incubation
5 Removal of the mold mycelium after fermentation
6 Extraction and purification of the penicillin

This process is shown schematically in Fig. 29-7; a commercial production facility is shown in Fig. 29-8. The changes which occur during the fermentation process (growth, synthesis of penicillin, etc.) are shown in Fig. 29-9.

The production of most other antibiotics follows a similar plan. The major differences relate to the organism, composition of medium, and method of extraction. Some manufacturers employ the same fermentation equipment for the production of several different antibiotics.

Figure 29-7. Manufacture of penicillin shown schematically. (A) A medium of corn-steep liquor, lactose, salts, and other ingredients is mixed, sterilized, cooled, and pumped into the fermenter. (B) The mold *Penicillum chrysogenum* is transferred from slant cultures to bran, and spore suspensions from bran are transferred to a sterile vessel with medium, which in turn is used to inoculate the seed tank. (C) The fermenter is inoculated from the seed tank; sterile air is forced through the fermenter during incubation. (D) After the maximum yield of penicillin is produced, the mold mycelium is removed by filtration and the penicillin is recovered in pure form by a series of manipulations which include precipitation, redissolving, and filtration.

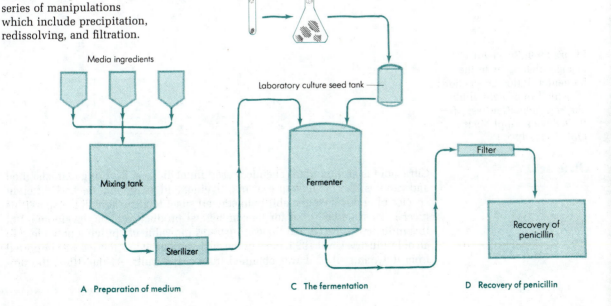

B Preparation of inoculum

Media ingredients

Laboratory culture seed tank

Mixing tank

Fermenter

Filter

Recovery of penicillin

Sterilizer

A Preparation of medium

C The fermentation

D Recovery of penicillin

Figure 29-8. Tops of large fermentation tanks of the type used to produce antibiotics. *(Courtesy of Merck and Co., Inc.)*

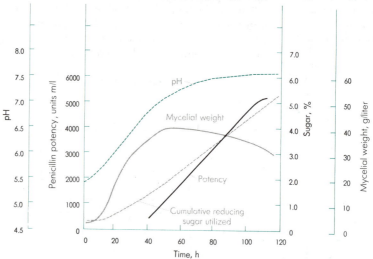

Figure 29-9. Biochemical changes that occur in the fermenter during production of penicillin by *Penicillium chrysogenum*. *(Courtesy of R. Donovick, Appl Microbiol,* **8:***117, 1960.)*

Citric Acid

Citric acid is an important chemical used in medicines, flavoring extracts, food and candies, the manufacture of ink, dyeing, and engraving. Several different species of molds have the ability to convert sugar to citric acid, but *Aspergillus niger* is most widely used for its commercial production. The development of this industry in the United States illustrates the value of applying new ideas in an old industry. Until 1923, most of the citric acid used in America was imported from Italy, and all of it was obtained from citrus fruits. At that time, the production of citric acid by mold fermentation was undertaken, and the industry has grown until today the annual production exceeds 20 million pounds. For several years after production by this method became practical, the United States not only did not import citric acid but exported large quantities. Since other countries now employ a similar method of production, exports have decreased.

Many sugars may serve as the substrate for the production of citric acid; however, molasses is generally used. The carbohydrate is incorporated into a medium containing an inorganic nitrogen compound as well as inorganic salts. The sterile medium is dispensed into shallow pans and inoculated with the mold spores. This is an aerobic process; consequently a large surface area provides an adequate supply of oxygen. An alternative to this method of production is the submerged-culture technique, in which the inoculated medium is contained in large tanks through which a supply of sterile air is forced. The strain of mold employed, the composition of the medium, the degree of aeration, and the temperature of incubation all have an effect on the yield of citric acid.

Enzyme Production

Many molds synthesize and excrete large quantities of enzymes into the surrounding medium. It is industrially feasible to concentrate and purify enzymes from cultures of molds such as *Aspergillus*, *Penicillium*, *Mucor*, and *Rhizopus*. Mold enzymes, e.g., amylases, invertase, proteases, and pectinase, are useful in the processing or refining of a variety of materials. Amylases hydrolyze starch to dextrin and sugars and are used in preparing sizes and adhesives, desizing textiles, clarifying fruit juices, manufacturing pharmaceuticals, and for other purposes. Invertase hydrolyzes sucrose to form glucose and fructose (invert sugar). It is widely used in candy making and production of noncrystallizable syrups from sucrose, which is partially hydrolyzed by this enzyme. The term *protease* refers to a mixture of proteolytic enzymes. Proteases are used for bating (treatment of hides to provide a finer texture and grain) in leather processing, manufacture of liquid glue, degumming of silks, and clarification of beer protein haze, and as an adjunct to soap for cleaning in laundries. For centuries—long before the role of enzymes in the bating of hides was understood—this treatment was accomplished by soaking the hides in suspensions of dog or fowl manure. Today, standard enzyme solutions have replaced the concoctions of dung. Pectinase is used in the clarification of fruit juices and to hydrolyze pectins in the retting of flax for the manufacture of linen.

Immobilized Enzyme Technology

The commercial uses of microbial enzymes have greatly expanded following the development of **immobilized enzyme technology.** The refinements and advances in this technology result from collaboration between the fields of enzymology, engineering, and microbiology. In principle, the enzyme is bound (immobilized) on a material through which the substance to be changed by the enzyme is passed. The process is analogous to passing a solution through a filter pad, the enzyme being present (immobilized) in the filter pad. A variety of substances including paper, wood chips, ceramic and glass beads, and ion-exchange resins have been used to immobilize enzymes. Among the advantages of this technology are (1) reuse of the enzyme and (2) more convenient recovery and purification of the end product of the enzyme reaction since it does not contain the enzyme.

Table 29-4. Net Distribution of Selected Biologics (United States)

Product Description	Net Doses Distributed, January–June, 1983
Influenza vaccine	45,630
Trivalent	
Bivalent	
Diphtheria toxoid and tetanus toxoid (pediatric)	447,213
Diphtheria and tetanus toxoids with pertussis vaccine	8,897,380
Tetanus and diphtheria toxoid (adult)	4,733,327
Tetanus toxoid	3,818,283
Poliomyelitis vaccine, inactivated	17,320
Poliomyelitis vaccine, live, oral, trivalent	9,362,070
Measles virus vaccine, live, attenuated*	2,993,000
Rubella virus vaccine, live*	3,015,030
Mumps virus vaccine, live*	2,605,329
Smallpox vaccine	2,246,507
Immune serum globulin, human (reported in cc)	1,010,602
Tetanus immune globulin, human (reported in cc)	353,380

* All products containing this antigen.

SOURCE: From Rept. No. 86, January–June 1983. Centers for Disease Control, Biologics Surveillance, U.S. Dept. of Health and Human Services, Public Health Service.

HYBRIDOMAS AND MONOCLONAL ANTIBODIES

Genetic research with microorganisms, particularly at the molecular level, has provided techniques which have been applied to studies with mammalian cells. One of the important outcomes of this research has been the fusion of myeloma cells (cancer cells) with antibody-producing white blood cells (B lymphocytes). The resulting hybrid cell is called a **hybridoma** (see Chap. 33). The hybridoma cells can be grown in vitro. Furthermore, as explained in Chap. 33, hybridoma cells can be selected and grown to produce a single, specific antibody. Such an antibody is called a **monoclonal antibody.**

Monoclonal antibodies are now produced on a commercial scale. They have great potential for therapeutic use in combating malignant cells, in immuno-suppression in organ transplantation, and for passive immunization in a variety of infectious diseases. They also serve as powerful analytical reagents for diagnosis of cancer and infectious diseases and for determination of hormone levels. Many investigations of cellular biology involving proteins, antigenic structure, and other phenomena employ monoclonal antibodies as analytical reagents primarily because of their high level of specificity and sensitivity.

BIOLOGICS FOR IMMUNIZATION

Control of infectious diseases through immunization requires the manufacture, on a commercial scale, of a variety of microbiological antigens. The wide practice of disease control through active immunization is discussed in Part 8. Development of effective immunizing antigens together with the stringent test requirements to ensure their safe use constitute major programs in many of the major pharmaceutical houses. The total doses of a selected number of biological products distributed in the United States during January–June 1983 are shown in Table 29-4.

PETROLEUM MICROBIOLOGY

Microorganisms are associated with petroleum in its formation, its recovery by drilling, its decomposition, and its utilization. Only in recent decades has a significant amount of attention been directed to research in this field. Studies in petroleum microbiology require interdisciplinary participation. The microbiologist needs to work closely with chemists, engineers, physicists, and perhaps representatives from other fields of study. Some aspects of microbial involvement in this area are summarized as follows.

Petroleum Formation

Much of the sedimentary material of the marine environment consists of dead microbial cells. Furthermore, biochemical changes in the sedimentary deposit are accomplished by a variety of microorganisms. It is speculated that these changes are associated with the formation of petroleum.

Petroleum Exploration

Soil in the region of a petroleum reservoir may contain vapors of hydrocarbon compounds such as methane or ethane. These may be detected by exposing microbial cultures in a test system which contains all nutrients for growth with the exception of a carbon source. Alternatively, the isolation of a large number of hydrocarbon-oxidizing microorganisms from soil may suggest that their presence is due to continued release of hydrocarbons from a petroleum deposit.

Petroleum Recovery

When an oil well is drilled, the initial recovery is made possible by the pressure within the rock formation. Later, as the original pressure decreases and the oil flow lessens, additional wells are drilled and water or steam is injected to force oil to the surface. Microbial activity has been suggested as a potential means of enhancing the yield of trapped oil. For example, bacteria injected into the oil might produce acid to dissolve rock formations, thereby releasing oil. Through other metabolic activities, microorganisms may decrease the viscosity of the oil.

Figure 29-10. (A) Corroded cast iron pipes from a tidal marsh. Corrosion is due primarily to activities of sulfate-reducing bacteria. (B) *Desulfovibrio* sp. growing on an iron salts-agar medium. The colonies appear black because of iron sulfide formation. *Desulfovibrio* spp. occur widely in fresh, polluted, marine, and brackish waters. *(Courtesy of W. P. Iverson and the National Bureau of Standards, U.S. Department of Commerce.)*

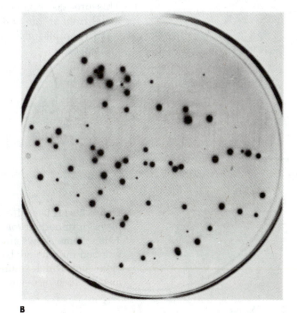

A

B

Corrosion of iron pipe by *Desulfovibrio* spp. is a major problem in the oil industry (see Fig. 29-10). Contamination of drilling fluids by various bacterial species is likewise a serious and costly problem.

Oil Spills

The international traffic of oil in supertankers, with the occasional accidents that result in huge oil spills, has created a major threat to the environment. How do we clean up the oil? One approach is to inoculate the spill area with a microorganism that has the ability to degrade petroleum oil. This concept has been enhanced by genetically engineering a species of *Pseudomonas putida* so that it has the capacity to metabolize the four major hydrocarbons of petroleum: camphor, octane, xylene, and naphthalene. A bacterium with this metabolic capability made legal history by being the first genetically engineered microorganism ever patented.

MICROBIOLOGY AND MINING

Microorganisms play a role in the recovery of minerals from ores. Their importance as agents in the process of extracting metals from ores is likely to increase for the following reasons:

1 The richer mineral deposits are being depleted. Lower-quality ores are being processed, and they require development of techniques which yield more nearly complete extraction of metals.
2 The traditional method of processing ores, namely smelting, is a major cause of air pollution and is under attack from environmental groups.

Microorganisms are capable of improving both these situations. For example, some autotrophic, aerobic bacteria (*Thiobacillus thiooxidans* and *Thiobacillus ferrooxidans*) when grown in the presence of copper ores produce acid and effect oxidation of the ore with subsequent precipitation (removal) of the metal. This process is known as **leaching.** This technique improves the recovery of metal from an ore and is nonpolluting to the atmosphere.

An example of a low-grade ore undergoing bacterial leaching is shown in Fig. 29-11.

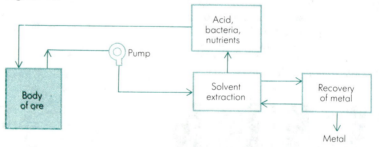

Figure 29-11. Leaching of low-grade ores using bacteria. *Thiobacillus ferrooxidans* plays an important role in the extraction (leaching) of metal from low-grade ores. This scheme shows an arrangement whereby the bacteria, nutrients, and acid are pumped into the ore bed. Continued growth of *Thiobacillus ferrooxidans* produces more acid, which solubilizes the metal content, promoting its extraction. The metal is then recovered from this acid solution.

DETERIORATION OF MATERIALS

The term *materials*, in the sense in which it is used here, refers to all products other than foodstuffs—paper, textiles, wood, rubber, and metals. It has been estimated that deterioration of such materials from all causes represents a loss running into several billions of dollars annually. Microorganisms are responsible for a significant amount of this destruction. Virtually nothing is immune or totally resistant to attack by microorganisms. Metals used in marine environments or the walls of a fuel storage tank on dry land are susceptible to microbiological corrosion. The glass components of optical equipment have been etched by microbial growth on their surfaces. We shall discuss a few examples of the role of microorganisms in deterioration.

Paper

The manufacture of paper involves two major operations. The first consists of the physical or chemical treatment of cellulosic material (e.g., wood, cotton, and linen rags) for the purpose of separating and purifying the cellulose fibers. The second consists of the fabrication of the resulting fibrous pulp, after further refinement, for redeposition of the fibers in the form of a sheet. Microbial deterioration in the form of paper-pulp slime may be encountered in the pulp, and other defects may appear on the finished product.

The development of slime depends largely upon the nature of the pulping operations. Slimes appear in the paper sheet in the form of undesirable slime spots. Bacteria, yeasts, molds, algae, and protozoa have been isolated from pulp slimes. Bacteria, particularly capsulated bacilli, are the most important single group of slime formers.

Finished paper is also subject to microbiological attack. Cellulose, the principal constituent of paper, is susceptible to degradation by a great many species of fungi and some bacteria. Other components of paper, such as glue or casein, may also serve as substrates for microbes. Under conditions permitting growth of microorganisms, the paper may be stained or discolored by the products of microbial metabolism. Growth of cellulolytic microorganisms will produce weakening of fibers, perforations, and even complete destruction of the paper.

Textiles and Cordage

Textiles made from natural fibers—cotton, wool, linen, and silk—are susceptible to deterioration by microorganisms. The same is true of cordage. Estimates on the annual losses due to microbiological attack on fabrics and ropes in the United States extend into millions of dollars. Enormous losses of cellulosic fabrics were experienced during World War II in tropical climates. Molds are the principal microorganisms responsible for this damage; many cellulolytic species inhabit the soil and are readily available as contaminants. *Myrothecium verrucaria* is notorious for its ability to degrade cellulose, and laboratory studies on this subject generally make use of this organism. Mold growth is favored by high humidity, moderate temperature, and diminished light. When this combination of conditions prevails, deterioration is greatly enhanced. For example, a lightweight canvas, when exposed to fungi under ideal conditions for mold growth, can be altered to the extent that it has no measurable strength after a few weeks.

Painted Surfaces

Painted surfaces are not always resistant to microbial disfiguration. Unless the paint film contains an effective fungicidal ingredient, it may under certain

Figure 29-12. Agar-plate test demonstrates the effectiveness of an antifungal agent incorporated into paint. In the plate on the left is paint containing antifungal agent; in the plate on the right is untreated paint overgrown with test fungus. *(Courtesy of Nuodex Products Company, Inc.)*

environmental conditions exhibit evidence of mold spotting, or discoloration. This deterioration is due to products of microbial metabolism of organic constituents of the paint. Many species of molds have been isolated from mildewed or "moldy" painted surfaces. Included among these are species of *Aspergillus, Penicillium, Cladosporium, Pullularia,* and *Alternaria. Pullularia* spp. appear to be the most common cause of mildewed paint. The effectiveness of incorporating an antifungal agent in paint is shown in Fig. 29-12.

Prevention of Microbial Deterioration

To minimize microbiological deterioration of materials, more efficient methods of preservation are continually being developed. Prevention of deterioration is accomplished through application of the principles for controlling microbial populations discussed in Part 6: incorporation of microbicidal agents into the material, packaging that protects material from contamination, and storage under conditions that inhibit microbial growth (e.g., dehumidification).

ANALYTICAL MICROBIOLOGY

Many techniques have been developed whereby a specific microorganism is used to assay quantitatively substances such as vitamins, amino acids, and antibiotics. Microbiological methods are routinely employed to determine the potency of all antibiotic preparations at various stages of development, from their crude forms to the finished product (see Fig. 29-13). This type of assay involves measurement of inhibition of growth caused by the antibiotic. Within established limits of antibiotic concentration there is proportionality between the degree of inhibition and the amount of drug.

Another type of microbiological assay is based on measurement of increase in growth or metabolic activity. The principle of this technique is that a *single* nutrient, e.g., a vitamin or amino acid, may be the limiting factor for growth or metabolic activity of a specific organism in an otherwise complete assay medium. Within limits, the magnitude of the growth or metabolic response is proportional to the amount of the specific nutrient added to the assay medium. The following will serve as an example of this type of assay.

Lactobacillus arabinosus requires the vitamin niacin for growth. When this

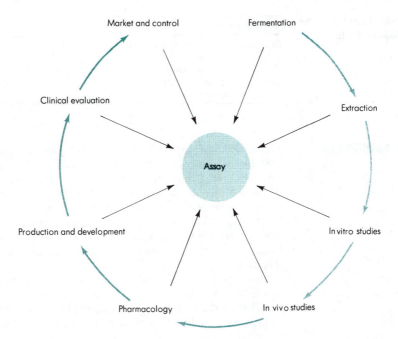

Figure 29-13. Development of an antibiotic involves a number of quantitative assays. (*Prog Ind Microbiol* **1,** 1959.)

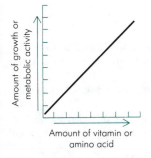

Figure 29-14. Principle of microbiologic assays as used for the measurement of vitamins and amino acids. Within limits of concentration of the substance being assayed, the amount of growth of the organism is proportionate to the amount of substance present.

organism is inoculated into a medium containing all the necessary nutrients (e.g., amino acids and other nitrogen compounds and vitamins other than niacin, glucose, and salts), growth will not occur. If niacin is added to this medium, the organisms will grow and the total growth obtained, within limits, will increase as the niacin is increased. It is therefore possible to prepare a standard curve relating growth to the amount of the vitamin, as shown in Fig. 29-14. If a substance of unknown niacin content is added to the medium and the test is carried out in the usual manner, the amount of growth measured can be referred to the standard curve, and from this the amount of niacin in the unknown sample can be extrapolated.

The measurement of the response of the test organism in these assays varies with the particular tests. It may be growth in terms of turbidity readings, dry weight, or cellular nitrogen. Other assay procedures rely upon measurement of metabolic activity such as production of acid or gas. Many procedures have been developed using bacteria (particularly lactobacilli), yeasts, fungi, algae, and protozoa. Some examples are shown in Table 29-5.

Microbiological techniques are extensively employed for the assay of vitamins and amino acids in pharmaceutical preparations and foods. Theoretically it is possible to assay any chemical to which the organism displays a measurable response. In practice, a wide range of substances, from simple mineral elements to complex organic compounds, are assayed.

Microbiological assays are highly specific and unusually sensitive. For example, as little as 0.1 nanogram (0.0000000001 g) of biotin per milliliter can be detected by using *Lactobacillus casei*.

Table 29-5. Microorganisms Used in Assay Techniques

Microorganisn	Substance Assayed
Streptococcus faecalis (bacterium)	Several amino acids
Tetrahymena geleii (protozoan)	Folic acid
Saccharomyces carlsbergensis (yeast)	Pantothenic acid
Neurospora crassa (mold)	Biotin
Ochromonas malhamensis (alga)	Vitamin B_{12}

FUTURE PROSPECTS

The basic knowledge of molecular biology and genetics accumulated during the past few decades is rapidly being translated into practical objectives and is revolutionizing industrial microbiology. More than 200 companies built upon the new biotechnology have come into being during the last several years, and a large sector of the industry is feverishly pursuing numerous leads suggested by studies in the genetic engineering of microorganisms and other cells. Most state governments, aware of what the new technology promises, have placed a high priority on establishing centers where universities and industry can collaborate.

The impact of the new biotechnology, or applied genetics, has already been significant in many areas and, in time, will affect all industries that are involved with biological systems or their products. Indeed, modern technology is likely to present attractive new options for industries that presently do not use biological systems.

The impact of advanced biotechnology and applied genetics will affect society throughout the world. Agriculture and food production, waste management and environmental quality, raw materials for the chemical industry, new pharmaceutical products, and disease control are some of the areas wherein new accomplishments will occur.

QUESTIONS

1 What is the basis for using microorganisms in a manufacturing industry?
2 What are some of the prerequisites for using microorganisms to manufacture a product?
3 What is meant by a "genetically engineered" bacterium? How is this accomplished?
4 What is a hybridoma? What is the practical significance of hybridomas?
5 Explain how it is possible to produce human insulin from bacteria. What are the advantages of this method of production?
6 Identify the microorganisms and describe the general biochemical processes involved in the production of lactic acid, vinegar, and glutamic acid.
7 What is meant by the term *food yeast? Bakers' yeast?*
8 Outline the procedure for industrial production of penicillin.
9 What developments contributed to the increase in yield of penicillin over that originally obtainable?
10 The technique of immobilized enzymes may increase the use of enzymes in industry for product modification. Why is this likely?
11 What are the desirable features of the use of enzyme preparations for refinement of a product?

12 What is the significance of paper-pulp slime in the manufacture of paper? What is its source?

13 Why is the deterioration of textiles and fabrics of particular concern in tropical climates?

14 What general measures are applicable for the overall control of microbial deterioration?

15 List and describe the various ways in which microorganisms are involved with the petroleum industry.

16 What is the principle upon which microbiological assay techniques are based? What types of microbial response may be measured by these procedures?

17 How has industrial microbiology been affected by the new biotechnology (applied genetics)?

18 What are the prospects for the future of biotechnology?

REFERENCES

Aharonowitz, Y., and G. Cohen: "The Microbiological Production of Pharmaceuticals"; Brill, W. J.: "Agricultural Microbiology"; Demain, A. L., and N. A. Solomon: "Industial Microbiology"; Evleigh, D. E.: "The Microbiological Production of Industrial Chemicals"; Gaden, E. L., Jr.: "Production Methods in Industrial Microbiology"; Hopwood, D. A.: "The Industrial Programming of Industrial Microorganisms"; Phaff, H. J.: "Industrial Microorganisms"; Rose, A. H.: "The Microbiological Production of Food and Drink," *Sci Am*, **245**(3), September 1981, New York. *This series of articles appears in a special issue of the Scientific American entitled "Industrial Microbiology." An excellent coverage of subjects as indicated by the titles of the individual papers.*

Crueger, W., and A. Crueger: *Biotechnology: A Textbook of Industrial Microbiology,* Sinauer Associates, Inc., Sunderland, Mass., 1982. *This is an English translation of a German text. The principles and major processes and products of industrial microbiology are clearly and concisely presented, and the text is generously illustrated.*

Hugo, W. B., and A. D. Russell (eds.): *Pharmaceutical Microbiology,* Blackwell Scientific, Oxford, 1977. *This attractively written and illustrated book covers the manufacture and quality control of pharmaceutical products and the mode of action of antibiotics with an introductory discussion of the biology of microorganisms. It is written for undergraduate pharmacy students as well as for microbiologists going into the pharmaceutical industry.*

Miller, B. M., and W. Litsky: *Industrial Microbiology,* McGraw-Hill, New York, 1976. *A comprehensive coverage of the fundamentals of industrial microbiology and a detailed treatment of the more important modern processes.*

Office of Technology Assessment, Congress of the U.S.: *Impacts of Applied Genetics,* 1981. U.S. Government Printing Office, Washington, D.C. *A comprehensive report of the major findings of the Office of Technology Assessment on the application of modern molecular biology genetic technology to microorganisms, plants, and animals.*

Reed, G. (ed.): *Prescott and Dunn's Industrial Microbiology,* AVI Publishing Co., Inc. Westport, Conn. 1982. *This is the 4th edition of what is considered a standard in the field. The purpose of this book is to present "in a concise but comprehensive manner, the fundamentals of industrial microbiology and to present descriptions of the more important processes within the field."*

Underkofler, L., and C. Nash III (eds.): *Developments in Industrial Microbiology*, vol. 25. Society for Industrial Microbiology, Arlington, Va. 1984. This volume is the latest in a series which reports the proceedings of the annual meetings of the Society for Industrial Microbiology. An excellent reference for subjects in the broad field of industrial microbiology.

PART EIGHT
MICROORGANISMS AND DISEASE

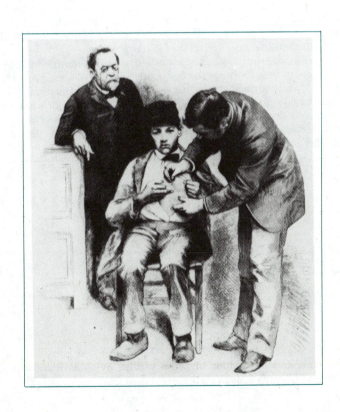

An old "new" agent of human infection

The latter half of the nineteenth century was the "golden age" of microbiology, when the infectious agents of many human diseases were being discovered. However, new causative agents of human disease are still being discovered today. In most instances it is not that a new microorganism suddenly appears on the scene; rather it is that our own knowledge of the occurrence and activities of an already existing organism is new. Such a situation is exemplified by *Campylobacter*, a genus of tiny, Gram-negative, vibrio-like bacteria.

Veterinarians had been aware since 1913 that campylobacters were important pathogens of domestic animals, causing abortion in cattle and sheep. However, before 1972, few clinical microbiologists were aware that some campylobacters were important and widespread human pathogens. In 1957, pioneering studies by Elizabeth King at the Communicable Disease Center in Atlanta had indicated that tiny, vibrio-like bacteria could cause human blood infections. King recognized that the human isolates resembled, but were not identical to, the vibrio-like organisms that veterinary microbiologists frequently isolated, and thus she called them "related vibrios." King suspected that the organisms might be able to cause gastrointestinal infections. However, this could not be confirmed until 1972, when P. DeKeyser and his colleagues in Belgium were the first to isolate "related vibrios" from diarrheic stools. They used a selective method based on (1) the removal of larger bacteria from diarrheic stools by filtration, and (2) the use of antibiotic-containing media that suppressed the growth of other intestinal bacteria.

We know today that the species *Campylobacter jejuni* is a major cause of diarrhea in humans, especially in children and young adults. In fact, it causes as many cases of diarrhea in humans as the more familiar *Salmonella* and *Shigella bacteria*, and it has a worldwide distribution. Why then did it take so long to recognize its clinical significance?

The answer lies not only in the lack of suitable selective methods, but also in the general cultivation methods used by hospitals and public health laboratories. Cultures from diarrheic stools were incubated aerobically on media designed to grow salmonellae and shigellae. However, *C. jejuni* grows neither aerobically nor anaerobically. Veterinary microbiologists had long known that the campylobacters they isolated from animals were *microaerophiles*; i.e., although the organisms required oxygen they could not tolerate the level of oxygen that is present in air (21 percent). The same is true of campylobacters isolated from humans; they grow only when the oxygen level is low—usually about 6 percent. Oxygen is both a blessing and a curse for campylobacters: they need it as a terminal electron acceptor for respiration, but they are poisoned by too much oxygen.

Today it is routine in clinical laboratories to culture diarrheic stool specimens for *C. jejuni*. We now know that successful isolation of this important pathogen depends on the use of suitable selective media and also on incubation of the cultures in jars which have a gaseous atmosphere containing a low level of oxygen.

Preceding page. Pasteur supervises inoculation of rabies vaccine by an assistant into a patient bitten by a rabid animal. *(Courtesy of National Library of Medicine.)*

Chapter 30 Microbial Flora of the Healthy Human Host

We are constantly in contact with a myriad of microorganisms in the environment. However, we are in even more intimate contact with an enormous number of microorganisms that inhabit our bodies. It is estimated that the adult human body is composed of approximately 10^{13} eucaryotic cells; what may be less apparent is that the human body also serves as the natural habitat for 10 times that number of microbial cells! These microorganisms, most of which are bacteria, comprise the **normal flora**, also termed the **normal microbiota.** They inhabit mainly the skin and the inner surfaces of the body such as the mucous membranes of the oral cavity, upper respiratory tract, intestinal tract, and genitourinary tract. Most are highly adapted to survival and growth in these areas despite physical and chemical conditions that discourage many other kinds of microorganisms.

It is useful to know the normal flora of the healthy human body for the following reasons:

1 The term *normal flora* implies that these microbes are harmless, and for the most part they do not cause disease and are even beneficial. Most are **commensals:** they benefit from the association with the host, but the host is not affected. Others have a **mutualistic association** with the host: they benefit the host in some fashion while thriving in the host's body. It is of interest to learn what these beneficial effects are and how they may be lost due to changes in the normal flora caused by the use of antibiotics or other means.

2 Some normal flora organisms can be *opportunistic pathogens;* i.e., they may cause infections if tissue injury occurs at specific sites or if the resistance of the body to infection is decreased. This is especially important because in recent years there has been a rising incidence of infections from these microorganisms.

ORIGIN OF THE NORMAL FLORA

Before birth a healthy human fetus is free of bacteria. Under natural circumstances, the fetus first acquires microorganisms while passing down the birth canal. It acquires them by surface contact, swallowing, or inhaling. These microbes are soon joined by other microbes from many sources in the newborn's immediate surroundings. Microorganisms which find suitable environments, either on the outer or inner body surfaces, quickly multiply and establish themselves. The initial flora may change considerably in composition during the first few days or weeks after birth until a stable flora becomes established and forms the normal flora. Each part of the human body, with its special environmental conditions, has its own particular mixture of microorganisms. For example, the oral cavity acquires a different natural microbial population than the intestines. In a short time, depending on factors such as the frequency of washing, diet, hygienic practices, and living conditions, the child will have the same kind of normal flora as an adult person in the same environment.

Even though an individual has a "normal" flora, it often happens that during his or her life there are fluctuations in the composition of this flora due to general health conditions, diet, hormonal activity, age, and many other factors.

NORMAL FLORA AND THE HUMAN HOST

What effect does the establishment of a normal flora (colonization) have on the body? Three approaches have been used to answer this question:

1 *Use of germfree animals.* If the colonization of experimental animals by microorganisms can be prevented, one can compare the properties of such germfree animals with those of normal animals. The results can be helpful in understanding the functions of the normal flora of human beings.
2 *Use of antimicrobial agents.* If the balance that occurs between the normal flora and the human host is altered by the use of such agents, various effects may occur that are useful for indicating the role of the normal flora for the human body.
3 *Knowledge of certain characteristics of normal flora organisms.* The nature of these characteristics suggests that normal flora organisms may help to discourage the growth of microorganisms that are not part of the indigenous flora.

Germfree and Gnotobiotic Life

Pasteur did not believe that animals could live in the absence of microorganisms, and in 1897, following his suggestion, an abortive attempt was made to rear germfree chickens. Between 1899 and 1908, Schottelius, a German, was successful in raising chickens that were bacteria-free. However, because his birds did not develop normally and died in about a month, he concluded that intestinal bacteria are essential in the nutrition of vertebrates. In 1912, Cohendy at the Pasteur Institute raised 17 germfree chickens for 40 days and concluded that vertebrate life is possible in the absence of microorganisms. We now know that when an adequate diet is provided, germfree chickens live long, healthy lives and reproduce as well as normal birds.

Rearing Germfree Animals

In 1928 James A. Reyniers at the University of Notre Dame started work on germfree chickens. He and his associates developed equipment and techniques for rearing chickens, rats, mice, and other animals in the absence of detectable living microorganisms for several generations. They emphasized the anatomical and physiological description of these animals and made comparisons with conventional nongermfree animals of the same species. As a result of these studies, germfree animals no longer belong to the realm of biological curiosities but have become practical models for solving problems of importance in biology

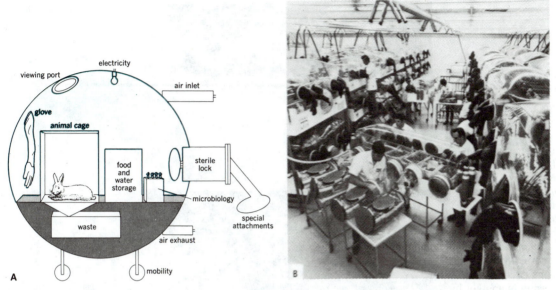

Figure 30-1. Germfree equipment. (A) Schematic diagram of a germfree isolator unit. The interior can be sterilized prior to an experiment and maintained in that condition. *(From McGraw-Hill Encyclopedia of Science and Technology, 3d ed., vol. 6, McGraw-Hill, New York, 1971.)* (B) Germfree colonies of mice reared in plastic isolator units. *(Courtesy of Charles River Breeding Laboratories, Inc.)*

and medicine. Germfree laboratory units are shown in Fig. 30-1. Animals that are either germfree or that live in association with one or more known organisms are said to be **gnotobiotic.**

The first germfree animals reared by Reyniers were chickens obtained by sterilizing the shells of 20-day-old embryonated eggs with an effective germicide and placing them in sterile containers such as glass churn jars or steel tanklike cages into which filtered (sterile) air is passed and from which waste gases are removed. Sterile food and water are placed in the cages prior to introducing the ready-to-hatch chicks. Periodic bacteriological examination of the exhaust air, feathers, excreta, and body orifices are made to confirm the absence of microorganisms in the cages or on the birds. Germfree rats, mice, guinea pigs, and other mammals can be obtained by cesarean section of gravid mothers under sterile operative conditions in a special chamber that allows the young animals to be introduced directly into a rearing cage. These babies must be hand-fed hourly for 2 or 3 weeks with specially devised nipples attached to medicine droppers with a formula containing, as nearly as can be determined, all of the components of the natural mothers' milk. Once established, a colony of germfree animals can be maintained by natural reproduction under germfree conditions.

Germfree Animals versus Normal Animals

Compared with normal animals, germfree animals exhibit an underdeveloped immune system, making them unusually susceptible to infection if subsequently exposed to pathogenic bacteria. They also require higher levels of B vitamins than do normal animals, and they require vitamin K, which normal animals do not require. These findings indicate that the normal flora may make a significant contribution to the vitamin requirements of the host.

Other Uses for Germfree Animals

Gnotobiotic techniques have been used to assess the effect of *particular species* of microorganisms on a host. Here, the germfree animal is reared in the presence of one or more known microbial species to determine the effect of those species

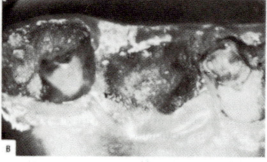

Figure 30-2. Gnotobiotic techniques have been widely used for studies of the role of bacteria in dental caries. (A) Normal noncarious teeth from a noninfected hamster. (B) Extensive plaque formation and caries following inoculation of a hamster with cariogenic streptococci of the *Streptococcus mutans–Streptococcus sanguis* group from a carious lesion in another infected hamster. *(Courtesy of Morrison Rogosa and the National Institute of Dental Research.)*

on growth and development of the animal or on various physiological processes. Similarly, one can inoculate a germfree animal with a known species of microorganism, or a mixture of known species, to determine their ability to produce disease or cause pathological or immunologic changes in the animal. For example, gnotobiotic techniques have helped to elucidate the role of bacteria in causation of dental caries (Fig. 30-2).

Effect of Antimicrobial Agents

Suppression of the normal flora by use of antibiotics or other antimicrobial agents indicates that the normal flora may serve as a defense against colonization by potential pathogens. For instance, treatment of the skin of humans with antibacterial agents such as hexachlorophene results in suppression of the normal Gram-positive flora and promotes colonization and clinical infection by Gram-negative bacilli and other organisms that cannot normally establish themselves on skin. In another example, hospital patients receiving antibiotics may undergo suppression of the normal flora of the large bowel, leading to pseudomembranous colitis, which is a severe disease caused by excessive growth of toxin-producing strains of *Clostridium difficile*. In humans not receiving antibiotic therapy, such strains are ordinarily held in check by the normal flora and do not grow to high numbers. Normal flora organisms can prevent the establishment of pathogens by various means such as successful competition for available nutrients or formation of inhibitory metabolic products and antibiotics.

Characteristics of Normal Flora Organisms

Many species of normal flora organisms have the ability to *adhere* to the surface of host epithelial cells. Thus they have a selective advantage over other micro-

Adherence to Host Cells

organisms in colonizing the host. Adherence occurs as the result of a molecular interaction between the microbial cell surface and a chemical receptor on the body cell. Proteins or polysaccharides on the surface of the microbial cells, as well as the fibrillar structures known as pili which extend out from the microbial cell, have been implicated in adherence; the particular means of adherence varies with the species. Microorganisms may often adhere specifically to one body site. For example, *Streptococcus salivarius* adheres mainly to the surface of the tongue, whereas *Streptococcus mutans* selectively binds to the smooth surface of the teeth.

A phenomenon that bears on microbial adherence is desquamation, the detachment of host epithelial cells from body surfaces and replacement of the lost cells by new cells. In some body sites, e.g., the intestinal tract, the rate of desquamation may be very high. One result of desquamation is the elimination of microorganisms that are not part of the normal flora and that are only feebly attached to the epithelial cells. Normal flora microorganisms, however, have the ability to reattach firmly to the fresh epithelial layer and thus persist at these body sites.

Production of Antimicrobial Substances

In some instances normal flora microorganisms have been shown to be capable of producing metabolic products or other agents that can inhibit other microbes. For example, in the large bowel certain anaerobic bacteria produce various organic acids, e.g., acetic, lactic, or butyric acid, as metabolic waste products which can inhibit the growth of other bacteria. Some strains of skin staphylococci have been shown to produce antibiotics that inhibit a wide variety of other bacteria. In still another example, many strains of *Escherichia coli* in the intestines produce colicins (see Chap. 12) which may help to protect the intestinal tract from closely related pathogenic bacteria.

DISTRIBUTION AND OCCURRENCE OF THE NORMAL FLORA

Bacteria make up most of the normal flora of the human body, and, therefore, this chapter deals mainly with the distribution and numbers of various bacterial genera and species. Although various fungi (mainly yeasts) and protozoa may also inhabit the body, their numbers are usually very low compared to the bacterial flora. As for viruses, it is not clear whether any can be considered as true normal flora even though some may be harbored for long periods in the absence of disease symptoms. For instance, certain human intestinal viruses were discovered only by noting their cytopathogenic effects in tissue cultures; thus they were termed "orphan" viruses (enteric cytopathogenic human orphans, or echoviruses). Similarly, coxsackieviruses, which occur only in human hosts, were initially discovered only because of their pathogenicity for suckling mice; they apparently were not associated with human disease. Many echoviruses and coxsackieviruses have since been found to cause a number of human diseases such as nonspecific febrile illnesses, acute respiratory disease, exanthematous disease, aseptic meningitis, enteritis, and paralytic and encephalitic disease. However, it must be recognized that not all of these viruses have been shown to cause disease in humans. In another example, chronic adenovirus infections are known to occur without disease: the presence of certain adenoviruses in the lymphoid tissue of normal individuals may become evident only after the tissue cells have been cultured in vitro in the laboratory. Consequently,

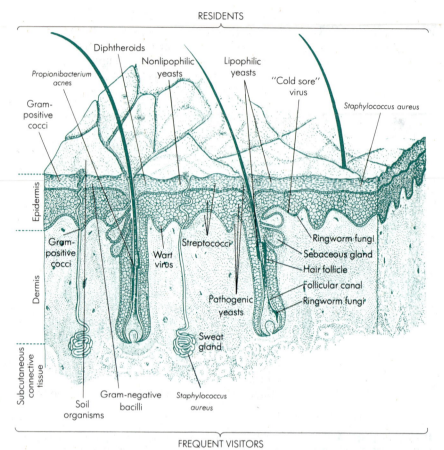

Figure 30-3. The major microbial symbionts found on or in the skin of humans. *(Courtesy of B. C. Block and J. Ducas, Man, Microbes, and Matter, McGraw-Hill, New York, 1975.)*

there do seem to be viruses that cause chronic asymptomatic infections, but whether they are so consistently present as to justify considering them as normal flora is not yet known.

The following sections describe the distribution and occurrence of the bacteria, fungi, and protozoa that comprise the normal flora of different regions of the human body, beginning with the external surfaces.

Skin

The skin is composed of the **epidermis** at the outer surface and of the connective tissue layer, **dermis,** underneath it (Fig. 30-3). The outermost layer of the epidermis consists of a layer of dead, anucleated, horny cells and is constantly in contact with bacteria from the surrounding environment. It is normally impermeable to bacteria; however, cuts, abrasions, or burns can allow bacteria to penetrate. The skin has a wide variation in structure and function in various sites of the body. These differences serve as selective ecological features, determining the types and numbers of microorganisms that occur on each skin site.

The skin surface is hostile to survival and growth of many kinds of bacteria.

For instance, the pathogen *Streptococcus pyogenes* does not survive for more than a few hours when applied to the skin, whereas it may survive for weeks in room dust. Several factors are responsible for discouraging skin colonization:

1 ***Dryness.*** The relatively dry surface of the skin is inhibitory to microbial growth. When allowed to dry, many bacteria remain in a dormant condition; some species die in a matter of hours. Some regions of the skin are more moist than others, e.g., the **axillary region, toe webs** (skin between the toes), and the **perineum** (skin at the lower end of the trunk between the thighs). These regions have higher numbers of normal flora organisms [about 10^6 colony-forming units (cfu)/cm²] than do the drier areas of skin (about 10^2 to 10^4 cfu/cm²).

2 ***Low pH.*** Skin has a normal pH between 3 and 5 (higher in moist regions), which is due in part to lactic or other organic acids produced by normal skin microorganisms such as staphylococci. This low pH can discourage the growth of many kinds of microorganisms.

3 ***Inhibitory substances.*** Several bactericidal or bacteriostatic compounds occur on the skin. For example, **sweat glands** (Fig. 30-3) secrete lysozyme, an enzyme that destroys bacterial cell walls. **Sebaceous glands** (Fig. 30-3) secrete complex lipids, which may be partially degraded by some bacteria such as *Propionibacterium acnes*; the resulting long-chain unsaturated fatty acids, e.g., oleic acid, are highly toxic to other bacteria.

Despite these formidable antimicrobial factors, some bacteria not only survive on the skin but even grow, forming the normal flora. The secretions of the sweat glands and sebaceous glands provide water, amino acids, urea, salts, and fatty acids, which can serve as nutrients for these microorganisms. Most of these bacteria are species of *Staphylococcus* (mainly *Staphylococcus epidermidis*) and aerobic corynebacteria, or diphtheroids. In the deep sebaceous glands are found lipophilic anaerobic bacteria such as *P. acnes*. The latter organism is a normal skin inhabitant and is usually harmless; however, it has been associated with the skin disease known as acne vulgaris. The numbers of these propionibacteria are little affected by washing because of their deep location. The location of various skin bacteria in or on the skin is shown in Fig. 30-3.

Propionibacterium acnes and Acne Vulgaris

Acne vulgaris is a disease of the sebaceous glands of the skin. In the first stage of the disease **comedones** (singular, *comedo*) are formed, i.e., distensions of the glands caused by an accumulation of sebum (fluid secreted by the gland), hair, and bacteria. Comedones may progress no further, or they may become closed (no longer able to eliminate their contents to the skin surface); such comedones can develop into disfiguring inflammatory lesions (papules, pustules, and nodules). *P. acnes* is the predominant organism in comedones; however, since it is also abundant in normal sebaceous glands, it is not yet clear that the organism is actually the causative agent of acne vulgaris. The ability of antibiotics to achieve clinical improvement in acne and at the same time reduce the number of *P. acnes* suggests that the organism may play an important role.

Eye

Lining the eyelids and covering the eyeball is a delicate membrane called the **conjunctiva.** This membrane is continually being washed by a flow of **lachrymal fluid** (tears), which tends to remove microorganisms. Moreover, lysozyme is secreted in tears. Consequently, the conjunctival flora is sparse. The main organisms found are S. epidermidis, Staphylococcus aureus, Corynebacterium species, Streptococcus pneumoniae, Neisseria species, Moraxella species, and Haemophilus parainfluenzae; other organisms may be isolated occasionally.

Respiratory Tract

Upper Respiratory Tract

Although more moist than skin, the mucous membranes of the upper respiratory tract (that portion of the respiratory tract above the larynx) nevertheless represent an environment which is difficult for many kinds of bacteria to colonize. As inspired air containing microorganisms passes along the tortuous nasal passages (sometimes referred to as the nasal baffle) and the nasopharynx (the region of the respiratory tract above the soft palate; Fig. 30-4), it is likely that the microorganisms will impinge on and stick to the thin moist layer of highly viscous mucus that overlies the surfaces. Because of the rhythmic beating of cilia (hairlike appendages) on the surface of the epithelial cells lining the nasopharynx, the mucus layer continuously flows downward toward the oropharynx (Fig. 30-4); thus, trapped bacteria are eventually swallowed and are destroyed by the acid of the stomach. In addition to this mechanical removal of bacteria, a bactericidal effect is exerted by lysozyme present in nasal mucus. Despite these factors, the nose and nasopharynx are inhabited by numerous normal flora microorganisms. Much of their success in colonization is due to their ability to adhere to the epithelial cell layer of the mucous membranes, thereby avoiding being carried away by the flow of mucus. The bacteria most frequently and most consistently found in the nose are S. epidermidis and S. aureus; in the nasopharynx avirulent strains of S. pneumoniae and other α-hemolytic streptococci predominate, but species of the genera Staphylococcus, Corynebacterium, Neisseria, Branhamella, Haemophilus, and Micrococcus are also common.

Lower Respiratory Tract

The mucous membrane surfaces of the trachea (windpipe) and its branches (bronchi) do not have a normal flora, as a result of the mechanical removal of organisms by an *upward*, cilia-driven flow of mucus. Bacteria that manage to traverse the air passages all the way to the alveoli (air sacs) of the lungs are

Figure 30-4. Distribution of normal flora of the human body.

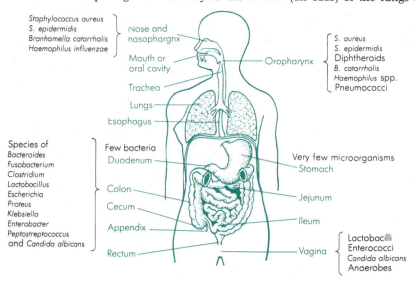

usually engulfed and destroyed by phagocytic body cells known as *alveolar macrophages*.

Mouth

The abundant moisture and constant presence of dissolved food as well as small food particles would seem to make the mouth an ideal environment for bacterial growth. However, the continuous flow of saliva through the mouth exerts a mechanical flushing action that removes many microorganisms, causing them to be swallowed and destroyed by the acid of the stomach. Desquamation of epithelial cells is a second mechanical factor that removes microorganisms from the oral cavity. Consequently, it is not surprising that many of the microbes that constitute the normal flora of the mouth resist such mechanical removal by being able to adhere firmly to various surfaces of the oral cavity.

Acquisition of the Oral Flora

At birth the oral cavity is essentially a sterile, warm, and moist incubator containing a variety of nutritional substances. The saliva is composed of water, amino acids, proteins, lipids, carbohydrates, and inorganic compounds. It is thus a rich and complex medium that can be used as a source of nutrients by microbes at various sites in the mouth. (Saliva itself generally contains transient microbes from other sites of the oral cavity, particularly from the upper surface of the tongue, and generally has a microbial population of about 10^8 bacteria per milliliter.)

The normal flora of a newborn is established within a few days after birth. The predominant bacterial species belong to the genera *Streptococcus*, *Neisseria*, *Veillonella*, *Actinomyces*, and *Lactobacillus*; yeasts are also present. The number and kinds of microbial species found are related to the infant's diet and to association with the mother, attendants, and objects such as towels and feeding bottles. The only species consistently recovered from the oral cavity, even as early as the second day after birth, is S. *salivarius*. This species has an affinity for epithelial tissues and appears in large numbers on the dorsal surface of the tongue.

Normal Flora of the Teeth

Until eruption of the teeth, most microorganisms in the mouth are aerobes and facultative anaerobes. As the first teeth appear, obligate anaerobes, such as species of the genera *Bacteroides* and *Fusobacterium*, become more evident because the tissue surrounding the teeth provides an anaerobic environment.

The teeth themselves can become areas for microbial adherence. *Streptococcus mutans* is associated with the tooth surface and appears to be the major etiological (causative) agent of **dental caries,** or tooth decay. S. *mutans* produces a highly branched, extracellular glucan (polymer of glucose) that acts like a cement which binds the bacterial cells together as well as causing them to adhere to the tooth surface. This glucan is formed only in the presence of the disaccharide sucrose (glucose-1-fructose, the type of sugar found in confections) by means of an enzyme called glycosyl transferase located on the surface of the cocci. The reaction catalyzed by this enzyme is as follows:

$$n \text{ Sucrose} \xrightarrow[\text{transferase}]{\text{Glycosyl}} \underset{\text{Glucan}}{(\text{Glucose})_n} + n \text{ Fructose}$$

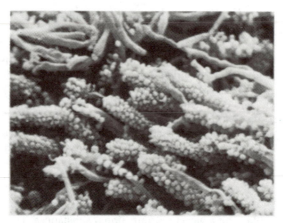

Figure 30-5. Scanning electron micrograph of a human dental plaque, fixed and freeze-dried in situ on an extracted tooth. Seen are cocci-coated filaments in a "corncob" arrangement. Field width is 38 μm. *(Courtesy of Dr. Sheila J. Jones, University College, London.)*

The fructose liberated from the sucrose by this reaction, as well as sucrose itself and other sugars that may be present, can be fermented by the streptococci, giving rise to lactic acid, which can etch the surface of the teeth. Although S. *mutans* initiates dental caries, other bacteria such as *Lactobacillus* and *Actinomyces* species can contribute to caries as secondary invaders. The aggregation of bacteria and organic matter on the surface of teeth is termed **plaque** (Fig. 30-5). Dental plaque contains a very high number of bacteria, about 10^8 cells per milligram.

Once teeth are present, the normal flora in infants appears to be generally similar to that in adults. Then, for reasons which are not well understood at present, but probably as a result of hormonal changes, anaerobic bacteria, especially oral spirochetes (*Treponema* spp.) and *Bacteroides melaninogenicus*, colonize the oxygen-deficient gingival crevices at puberty.

In addition to bacteria, certain commensalistic protozoa may inhabit the oral cavity. For instance, the flagellated protozoan *Trichomonas tenax* may occur in the gingival margins and in the tartar and cavities of teeth. Its presence is usually associated with poor oral hygiene.

Intestinal Tract

Stomach

Although the stomach constantly receives numerous bacteria from the oral cavity, the fluid contents of the healthy stomach generally contain less than 10 bacteria per milliliter because of the bactericidal effect of the hydrochloric acid in the gastric secretion. The few organisms found are mainly lactobacilli and yeasts such as *Candida* spp. Following the ingestion of food the number of bacteria increases (10^3 to 10^6 organisms per gram of contents), but it soon falls as gastric juice is secreted and the pH of the stomach's fluid drops.

Small Intestine

The upper portion (or **duodenum;** Fig. 30-4) of the small intestine has few bacteria (usually $<10^3$ per milliliter of fluid). Of those present, the majority are Gram-positive cocci and bacilli. In the **jejunum** (second part of the small intestine, between the duodenum and ileum; Fig. 30-4) there are occasionally found species of enterococci, lactobacilli, and diphtheroids. The yeast *Candida albicans* may also be found in this part of the small intestine.

In the distal portion (**ileum;** Fig. 30-4) of the small intestine, the flora begins to resemble that of the large intestine. Anaerobic bacteria and members of the family *Enterobacteriaceae* grow extensively here and appear in large numbers.

Large Intestine

In the human body, the **colon**, or large intestine (Fig. 30-4), has the largest microbial population. It has been estimated that the number of microorganisms in stool specimens is about 10^{11} organisms per gram wet weight. (Fifty or sixty percent, dry weight, of fecal material may consist of bacteria and other microorganisms.) Over 300 different bacterial species have been isolated from human feces. It has also been calculated that an adult excretes in the feces 3×10^{13} bacteria daily.

Various factors tend to remove microorganisms from the large intestine. The continual movement of intestinal contents through the channel of the intestine by peristalsis is one factor. Desquamation of surface epithelial cells to which bacteria are attached is another. Mucus is a third factor; just as this substance is important in the mechanical removal of microorganisms from the respiratory tract, it plays a similar role in the intestine. However, the mucus in the intestine forms a discontinuous, meshlike layer rather than a continuous layer. Movement of intestinal contents causes contact and adherence of microorganisms to the mucus, which subsequently rolls up into small masses that, with their attached microorganisms, are eliminated from the body in the feces.

There are about 300 times as many anaerobic bacteria as facultatively anaerobic bacteria (such as *E. coli*) in the large intestine. The anaerobic Gram-negative bacilli present include species of *Bacteroides* (*B. fragilis*, *B. melaninogenicus*, *B. oralis*) and *Fusobacterium*. The Gram-positive bacilli are represented mainly by species of *Bifidobacterium*, *Eubacterium*, and *Lactobacillus*.

The facultatively anaerobic species found in the intestine belong to the genera *Escherichia*, *Proteus*, *Klebsiella*, and *Enterobacter*. Peptostreptococci (anaerobic streptococci) are common. The yeast *C. albicans* is also present.

Some protozoa may also occur as harmless commensals in the intestine, where they grow anaerobically by ingesting the bacteria that are present. For instance, a flagellated protozoan, *Trichomonas hominis* (also known as *Pentatrichomonas hominis*), inhabits the intestinal tract in the area of the cecum. In another example, a number of amoebae belonging to the genera *Entamoeba*, *Endolimax*, and *Iodamoeba* are commensals of the colon. One species, *Entamoeba histolytica*, can live as a commensal but can also be pathogenic, causing amoebic dysentery; it is capable of penetrating the intestinal mucosa and invading various organs of the body.

It is interesting that the intestinal flora of the young breast-fed infant consists mainly of bifidobacteria, which are Gram-positive; in bottle-fed infants, lactobacilli, which are also Gram-positive, predominate. With eventual substitution of solid food and an adult-type diet, a Gram-negative flora consisting mainly of *Bacteroides* spp. predominates.

Factors Influencing the Normal Flora

The composition of the normal flora of the intestine can be influenced by various factors, such as strong emotional stress, changes in air pressure due to altitude, and starvation. It should be noted that in diarrhea, as a result of rapid movements of the intestinal contents, the intestinal flora undergoes considerable change.

Alteration of the flora also occurs in persons receiving antibiotic treatment; organisms susceptible to the antibiotic being administered may be replaced by antibiotic-resistant strains. Other factors that may possibly affect or regulate the normal flora are diet, the bile acids secreted into the duodenum from the gallbladder, and the presence of antibodies secreted into the intestine; however, the importance of these factors is not yet clear, and present evidence suggests that their influence may be negligible.

Implantation of Lactobacilli

As indicated above, prolonged therapy with certain antibiotics may eliminate many normal intestinal microorganisms, permitting antibiotic-resistant species to thrive. This, in turn, may cause gastrointestinal disturbances such as constipation or diarrhea. The oral administration of *Lactobacillus acidophilus* has been found to alleviate the intestinal disorders in some instances. The principle is that ingestion of large numbers of the lactobacilli may result in replacement of undesirable intestinal organisms by harmless or beneficial organisms, a concept first proposed by the Russian bacteriologist Metchnikoff in the early days of bacteriology. The implantation of the lactobacilli seems to depend on ingestion of a large number of the organisms and on supplying a suitable carbohydrate such as lactose that is not readily absorbed by the body but can be easily used by the organisms. A milk product known as **sweet acidophilus milk** has been devised for therapeutic use. This product is made by adding a concentrated suspension of a human strain of *L. acidophilus* to cold pasteurized milk, with subsequent storage at a temperature below 4°C.

Normal Flora and Colon Cancer

An interesting finding that is related to the normal flora of the large intestine is the discovery in 1977 of the presence of a mutagenic compound in the feces of normal humans. Because compounds that are mutagenic may also be carcinogenic, the relationship of the fecal mutagenic compound to the occurrence of colon cancer in humans is being actively investigated. Diet is highly correlated with the occurrence of colon cancer, and populations at high risk for this disease—such as those in the United States, where nearly 125,000 cases of colon cancer occur per year—have been found to have significantly higher levels of this mutagen in their feces than do populations at low risk. Studies in 1982 have demonstrated the ability of several normal flora organisms (species of the genus *Bacteroides*) to produce this mutagen from a precursor compound present in the feces of persons who excrete the mutagen. Whether this precursor arises from the diet of such human excreters, from some product of host metabolism, or from the microbial flora of the intestine, is not known. It must also be emphasized that, as suggestive as these findings are, at the time of this writing it is not yet certain that the mutagenic substance is carcinogenic or is actually responsible for colon cancer.

Genitourinary Tract

In a healthy person, the **kidney, urinary bladder,** and **ureters** (the tubes that convey the urine from the kidney to the bladder) are free of microorganisms. However, bacteria are commonly found in the lower portion of the **urethra** (canal that conveys the urine to the outside of the body) of both males and females. *S. epidermidis, Streptococcus faecalis,* and corynebacteria are found frequently; neisseriae and members of the family *Enterobacteriaceae* are occasionally present. The upper portion of the urethra, near the bladder, has few microorganisms, apparently because of some antibacterial effect exerted by the urethral mucosa and because of the mechanical removal of microorganisms by the frequent flushing of the urethral epithelium by urine. Urine itself is an excellent growth medium for many bacteria; however, as indicated above, urine in the bladder of a healthy person is sterile. It acquires microorganisms as it passes from the bladder to the outside of the body, the source of these organisms being the surface of the lower portion of the urethra. During voiding, if the periurethral surfaces (tip of penis, labial folds, vulva) are first thoroughly cleansed, the first portion of the voiding discarded, and the subsequent urine collected in a sterile container, the urine sample will normally contain less than 10^3 organisms per milliliter. (A count of 10^5 organisms or higher is indicative of a urinary tract infection.)

The adult female genital tract has a very complex normal flora. The character of this population changes with the variation of the menstrual cycle. The main inhabitants of the adult **vagina** are the acid-tolerant lactobacilli; these break down glycogen produced by the vaginal epithelium, forming lactic acid in the process. As a result, the pH in the vagina is maintained at about 4.4 to 4.6. Microorganisms capable of multiplying at this low pH are found in the vagina and include enterococci, diphtheroids, the yeast *C. albicans*, and large numbers of anaerobic bacteria (Fig. 30-4). The accumulation of glycogen in the vaginal wall is due to ovarian activity. Thus it is not present before puberty or after menopause, and in its absence the vaginal secretions are mildly alkaline and contain normal skin microorganisms.

QUESTIONS

1 Define the following terms:

Commensal	Glucan
Lysozyme	Plaque
Desquamation	Gnotobiotic
Nasal baffle	Echovirus
Comedo	Glycosyl transferase

2 Give the major reason or reasons for the occurrence of the following species of bacteria in their natural habitat:
   (a) *Propionibacterium acnes* in skin.
   (b) *Lactobacillus* species in the vagina.
   (c) *Bacteroides melaninogenicus* in the gingival crevice of teeth.
   (d) *Streptococcus mutans* on teeth surfaces.

3 How are germfree animals obtained and reared?

4 What benefits might a human host derive from the normal flora?

5 What factors make the following environments difficult to colonize: (a) skin, (b) upper portion of the urethra, (c) trachea, (d) stomach, (e) conjunctiva?

6 Why is it useful to know about the normal flora of the healthy human body?

7 Provide an example to illustrate that the laws of natural selection also govern the ecology of the normal flora of the human body.

8 What role does microbial adherence play in establishment of the normal flora? Give an example.

9 In what part of the body do most of the normal flora microorganisms occur? What would a pathogenic microorganism need to do to establish itself, if only temporarily, in this part of the body?

10 How might colon cancer possibly be related to the normal flora of the body?

11 What role does mucus play in preventing microorganisms of inhaled air from reaching the alveoli of the lungs?

REFERENCES

Baron, S. (ed.): *Medical Microbiology*, Addison-Wesley, Menlo Park, Calif., 1982. *Chapter 18 of this text discusses the normal flora associated with various areas of the body.*

Gibbons, R. J., and J. van Houte: "Bacterial Adherence in Oral Microbial Ecology," *Annu Rev Microbiol*, **29**:19–44, 1975. *A comprehensive review of the oral environment and factors affecting bacterial growth in the oral cavity, with particular reference to microbial adherence and the formation of dental plaque.*

Hylemon, P. B.: "Current hypotheses of colon cancer etiology," *Cancer News Quarterly* (Medical College of Virginia, Richmond) **3**:1–5, 1983. *A concise review of the possible causes of colon cancer, including a section on detection and identification of fecal mutagens.*

Linton, A. H.: *Microbes, Man and Animals*, John Wiley, New York, 1982. *Chapter 2 discusses the interactions of normal flora microorganisms with healthy hosts.*

Maibach, H. I., and R. Aly (eds.): *Skin Microbiology: Relevance to Clinical Infection*, Springer-Verlag, New York, 1981. *The contributors to this volume provide discussions of such topics as the factors controlling skin bacterial flora, methods of quantifying skin bacteria, microbial interactions and antibiosis, and the role of Propionibacterium acnes in causing acne vulgaris.*

Savage, D. C.: "Microbial Ecology of the Gastrointestinal Tract," *Annu Rev Microbiol*, **31**:107–133, 1975. *A comprehensive review of the microbial flora of the intestinal tract of humans and animals.*

Slack, J. M., and I. S. Snyder: *Bacteria and Human Disease*, Year Book, Chicago, 1978. *Chapter 3 of this text provides a succinct discussion of the indigenous flora of the human body.*

Chapter 31

Host-Microbe Interactions: The Process of Infection

Infectious diseases occur as the result of interactions between **pathogenic** (disease-producing) microorganisms and the host. All infectious diseases begin at some *surface* of the host, whether it be the external surfaces such as the skin and conjunctiva or internal surfaces such as the mucous membranes of the respiratory tract, intestine, or urogenital tract. Many pathogens can selectively attach to particular host surfaces. In most infectious diseases the pathogenic microorganism penetrates the body surface and gains access to the internal tissues. In some kinds of infections the pathogen may remain **localized**, growing near its point of entry into the body. In other instances it may be transported to some other body site. Some pathogens may cause **generalized** infections, in which the microorganism becomes widely distributed and grows throughout the body. Some pathogens may be capable of growth within the cells of the host, causing severe disruption of normal physiological processes. In other infections the pathogen may grow extracellularly; here, damage to body cells usually occurs as the result of elaboration of poisonous substances **(toxins)** by the microorganism.

If a host is to recover from an infection, it must eradicate the pathogenic microorganisms. However, as a group, pathogenic microorganisms exhibit a vast array of weapons, termed **virulence factors,** that can combat the various defense mechanisms of the host; each species of pathogen possesses only one or a few of these factors. Thus, an infection represents a battle between the defenses mounted by the host and the particular armamentarium of virulence factors produced by the pathogen. Often the infection proves lethal to the host; however, it is to the microbe's advantage if the battle is somewhat indecisive, i.e., if the disease the pathogen causes is not so severe as to kill the host. Killing the host would diminish the pathogen's chance of survival; consequently, host-microbe interactions that result in **chronic,** long-lasting infections are regarded as being more highly evolved than interactions that are **acute,** i.e., have a short and relatively severe course.

PATHOGENICITY, VIRULENCE, AND INFECTION

Pathogenicity is the capability of a microbial species to cause disease. However, various strains of a pathogenic species may differ with regard to their *degree of* pathogenicity, i.e., with regard to their **virulence.** For instance, some strains are highly virulent: only a few bacterial cells from a highly virulent strain are needed to cause disease in a host. Other strains may be less virulent, and larger numbers of cells of such strains are needed to cause the disease. Some strains may be avirulent, incapable of causing the disease even when large numbers of cells are inoculated into the host. Virulent strains of many pathogens, when repeatedly cultured on laboratory media or grown in vivo in hosts other than their normal hosts, may lose their virulence: such avirulent strains are called **attenuated** strains and are widely used as vaccines to elicit immunity to various diseases.

The virulence of a pathogen is usually measured by determining its LD_{50} dose for a particular type of laboratory animal. The LD_{50} dose is defined as that number of organisms which, when administered to a number of laboratory animals, will kill 50 percent of them. For example, an LD_{50} dose of 10 cells of strain X compared with 100,000 cells of strain Y would indicate that X is 10,000 times more virulent than Y. The LD_{50} dose can be determined more precisely than other endpoints such as the dose that kills 100 percent of the animals (LD_{100} dose, sometimes also termed **minimum lethal dose** or **MLD**) because the rate of change in mortality versus change in dose is greatest around the point of 50 percent mortality.

Infection represents the most intimate way in which a microorganism may cause disease: the host is invaded by the microorganisms which subsequently multiply in close association with the host's tissues. Most, but not all, microbially caused diseases are infections. An example of one that is not is a type of food poisoning called botulism, in which there is no invasion of the body by the causative microorganism; rather, the disease is contracted by ingesting the poison (toxin) in a food in which the bacterium *Clostridium botulinum* has previously grown.

In order to cause infectious disease a pathogen must accomplish the following:

1 It must enter the host.
2 It must metabolize and multiply on or in the host tissue.
3 It must resist host defenses (see Chaps. 32 and 33).
4 It must damage the host.

Each process is complex, and all four processes must be fulfilled to produce infectious disease. Some infections may result in only a very minor amount of damage to the host, so minor that there are no detectable clinical symptoms of

Table 31-1. Some Types of Infections

Term	Definition	Example
Acute	Has a short and relatively severe course	Streptococcal pharyngitis (sore throat caused by *Streptococcus pyogenes*)
Chronic	Has a long duration	Tuberculosis
Fulminating	Occurs suddenly and with severe intensity	Cerebrospinal meningitis caused by *Neisseria meningitidis*
Localized	Restricted to a limited area of the body	Urinary tract infection caused by *Escherichia coli*
Generalized	Affects many or all parts of the body	Blood infections, such as typhoid fever
Mixed, or polymicrobial	More than one kind of microorganism contributes to the infection	Gaseous gangrene, in which a combination of *Clostridium* species may occur
Primary	An initial localized infection that decreases resistance and thus paves the way for further invasion by the same microorganism or other microorganisms	Viral influenza
Secondary	Infection that is established after a primary infection has caused a decreased resistance	Pneumococcal pneumonia following viral influenza

the infection; such infections are called **subclinical infections.** Other infections vary in regard to severity, location, and the number of microbial species involved (see Table 31-1).

MICROBIAL ADHERENCE

Unless a pathogen is introduced directly into the tissues (as by a wound, injection by an arthropod, or other similar means), the first step in initiation of infection is usually **adherence** or **attachment** of the pathogen to some surface of the host. As indicated in Chap. 30, such surfaces represent hostile environments and the microorganism must compete with normal flora organisms for surface attachment. Moreover, the attachment is selective: various pathogens attach only to certain tissues. For most pathogens, the precise means of attachment are not yet understood, particularly for pathogenic fungi and protozoa.

Examples of Adherence of Pathogenic Bacteria

Neisseria gonorrhoeae, the causative agent of gonorrhea, adheres specifically to the epithelial cell layer of the human cervix, urethra, and conjunctiva by means of pili and thus avoids being washed away by the flow of mucus or tears. *Escherichia coli* strains that cause "scours," a diarrheal disease of newborn pigs, also possess pili that allow the bacteria to attach firmly to the mucosal lining of the small intestine. *Vibrio cholerae* adheres to the epithelial layer of the small intestine of humans (see Fig. 31-1); although the bacterial surface component responsible for the attachment is not yet certain, it may be a **hemagglutinin** (so named because it also permits attachment to erythrocytes in laboratory experiments). In another example, certain proteins located on the outer surface of the bacterial cell wall have been shown to be essential for the initiation of infection. For instance, *Streptococcus pyogenes*, the causative agent of streptococcal sore

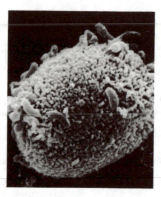

Figure 31-1. Scanning electron micrograph of *Vibrio cholerae* attached to the microvilli of a brush border membrane isolated from the small intestine of a rabbit (X10,000). *(Courtesy of G. W. Jones, G. D. Abrams, and R. Freter, Infect Immun 14:232-239, 1976.)*

throat, attaches specifically to the epithelial cells of the throat by means of cell-wall proteins called **M proteins.**

Examples of Adherence of Viruses

The surface of influenza virus particles is studded with hemagglutinin spikes that can cause attachment of the virus to specific mucoprotein receptors on the surface of host cells. Neuraminidase spikes on the virion surface also may possibly aid attachment by degrading the protective mucus layers of mucous membranes and allowing viral attachment to the underlying epithelial cells. Another example is a protein on the surface of poliovirus, which seems to be critical for attachment of the virus to lipid- and glycoprotein-containing receptors on host cells; the attachment is specific for cells of the intestinal tract and the central nervous system, and subsequent infection of the latter can lead to paralysis. In this regard, it is interesting that the attenuated strains of poliovirus used for vaccination against poliomyelitis (see Chap. 37) can attach to the gastrointestinal tract as the wild-type poliovirus does; however, because of mutation in the genes for the viral surface proteins, these attenuated strains have lost the ability to attach to cells of the central nervous system and thus do not cause the paralysis that is characteristic of poliomyelitis.

PENETRATION OF EPITHELIAL CELL LAYERS

Although penetration of the epithelial layer follows adherence in most infections, this is not always a prerequisite to infection. The microorganism may merely multiply on the epithelial surface and cause damage without penetration into the body. For example, *V. cholerae*, the causative agent of the severe diarrheal disease known as cholera, multiples on the epithelial layer of the small intestine where it produces a toxin that causes the loss of fluid from the epithelial cells and kills the cells.

Passive Penetration into the Body

It should be emphasized that penetration of body surfaces may be achieved not only **actively** (i.e., by the adherence and penetration mechanisms of the pathogen itself) but also **passively,** by mechanisms having nothing to do with the properties of the microorganism. Any mechanically caused breach in the body surfaces can introduce pathogens directly into the underlying tissues. Wounds or burns represent one passive mechanism. For example, soldiers wounded on the battlefield may develop gas gangrene if the wound becomes contaminated by *Clostridium perfringens* present in soil and fecal matter. Burns often become infected by *Pseudomonas aeruginosa* or other aerobic or facultatively anaerobic bacteria from the surrounding environment. Another mode of passive penetration is by arthropods. For example, *Borrelia* species cause relapsing fever in humans when the spirochetes are introduced through the bite of a tick or a body louse.

Active Penetration into the Body

Some pathogenic microorganisms are capable of penetrating the epithelial layer to which they have become attached. For example, in bacillary dysentery, *Shigella* bacteria penetrate into and kill the epithelial cells of the colon, then spread to adjacent cells, which are in turn killed. The result is the formation of lesions (areas of damage) known as **ulcers,** i.e., areas on the intestinal wall which have disintegrating or necrotic (dead) tissue. In another example, the influenza virus penetrates the epithelial cells lining the nasopharynx, trachea, and bronchi. The virus then undergoes replication, and new viral progeny are subsequently liberated from the infected cell. The severity of influenza depends mainly on the degree of host cell destruction during viral multiplication.

After penetration through or between the epithelial cells, some pathogens may penetrate into the deeper tissues of the body and may even become widely disseminated throughout the body, particularly if the organism obtains access to the lymphatic system or the blood vascular system.

EVENTS IN INFECTION FOLLOWING PENETRATION

Whether the means of penetration of a body surface by a microorganism is passive or active, the microorganism multiplies, resists the defense mechanisms of the host, and begins to cause tissue damage.

Growth in Underlying Tissues

In some infections the microorganism may simply grow in the tissue in which it finds itself, causing a localized infection. An example is the type of infection caused by *Staphylococcus aureus*, where the characteristic lesion is an **abscess,** i.e., a walled-off cavity in the tissues containing the staphylococci, numerous white blood cells (that collectively form a pasty mass called pus), and dead, disintegrating tissue cells that have been killed by the toxins elaborated from the staphylococci.

In other infections the organism may not remain localized but may spread through the tissues. An example is the anaerobic bacterium *C. perfringens*, which causes the wound infection gas gangrene. Initiation of gas gangrene depends on the occurrence of anaerobic conditions in the wound, as occurs in crushed tissue or clotted blood. As *C. perfringens* begins to grow, the bacteria elaborate toxins that kill some of the surrounding healthy tissue. This dead tissue becomes anaerobic and can support the growth of more clostridia, which in turn elaborate more toxins that kill more tissue and allow the organisms to spread further. Another factor that contributes to the rapidity of tissue invasion is the production of large amounts of hydrogen gas by the bacteria; the pressure of this gas separates connective tissue sheaths from muscle tissue, forming a space that can rapidly be filled in by clostridia-containing fluid. By this means the clostridia can quickly invade the entire length of a muscle. Amputation of an affected limb is often the only way to stop the spread of *C. perfringens* to the rest of the body.

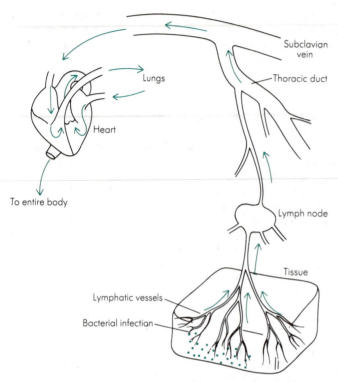

Figure 31-2. Schematic diagram illustrating how infection of the lymphatic system can lead to a bacteremia and generalized infection. Microorganisms infecting a tissue may gain access to thin-walled, finely branched lymphatic vessels which collect excess tissue fluid. The organisms may then pass via the lymphatic vessels to a regional lymph node. If the defenses of the lymph node are overwhelmed, the microorganisms may multiply and then pass to the thoracic duct, which empties lymph into the venous blood circulation via the subclavian vein. Thus the organisms reach the heart and, ultimately, are eventually distributed throughout the body.

Infection of the Lymphatic System

Many pathogenic microorganisms can spread from the initial site of infection to the lymphatic system. Body tissues are permeated by a network of blood capillaries which supply oxygen and nutrients to tissue cells. Capillaries have very thin walls, and some of the fluid portion of the blood leaks out into the tissues. Such fluid would cause tissues to become swollen unless it is returned to the blood vascular system. The function of the lymphatic vascular system is to collect this fluid by means of thin-walled lymphatic vessels (see Fig. 31-2) and return it to the blood vascular system. The fluid within the lymphatic vessels, called lymph, passes to lymph nodes (Fig. 31-2). These are ovoid structures ranging in size from one to several millimeters and are widely distributed throughout the body. Lymph enters a lymph node by any of several lymphatic vessels, passes through the lymph node along tortuous, winding channels, and emerges via a single efferent lymph vessel. From here it passes into larger and larger vessels (lymphatic trunks) and eventually reaches the main lymphatic ducts, which drain into the great veins of the blood vascular system (Fig. 31-2).

Microorganisms present in tissues may also be collected by lymphatic vessels and can infect them. For instance, in erysipelas, an inflammatory disease of the skin caused by S. pyogenes, the painful lesions are spread by invasion of the lymphatic system of the subepidermal tissue.

Pathogenic microorganisms may also be carried by lymphatic vessels to lymph nodes. The winding channel within a node is lined with cells called **macro-**

phages, which can engulf and destroy the bacteria that enter the lymph node (see Phagocytosis, Chap. 32). Thus the lymph node removes nearly all of the bacteria that enter it, and it may become inflamed, enlarged, and sore when infected; in this condition it is sometimes called a bubo. In bubonic plague, such swollen lymph nodes are particularly prominent (hence the name of the disease). The exudates (fluids) draining from such buboes are filled with plague bacilli.

Infection of the Blood

From the initial site of entry into the body by passive or active means, a pathogenic microorganism may be able to enter a blood capillary or venule and thereby gain direct access to the blood vascular system, causing a bacteremia (presence of bacteria in the blood). Once in the bloodstream the organism can be carried to various parts of the body and cause localized infections. For instance, *Neisseria meningitidis* present in the nasopharynx can reach the meninges (membranes that cover the brain and spinal cord) by means of a transient bacteremia, and can thereby cause a severe meningitis. By means of a bacteremia, spirochetes of the genus *Leptospira* can reach the kidneys, where they may eventually cause acute renal failure.

In other instances a pathogenic microorganism may gain access to the bloodstream directly by *first infecting the lymphatic system*. If the defenses of a lymph node are overwhelmed, the organisms may appear in the efferent lymph and ultimately reach the bloodstream. For example, in typhoid fever, *Salmonella typhi* organisms first attach to the epithelium of the small intestine; they penetrate this layer by passing through the epithelial cells, infect the lamina propria mucosae (the underlying connective tissue layer), enter the lymphatic system, and reach the mesenteric lymph nodes. Here, the organisms are not easily destroyed by the macrophages of the lymph node; they multiply and eventually begin to enter the blood circulation, causing a bacteremia. The bacteria eventually localize in various parts of the body, particularly in the macrophages of the liver and spleen, and they also reinfect various lymph nodes, since these are nourished by the blood.

In some infections, bacteria may actively multiply in the bloodstream and produce toxic products—a condition known as septicemia. Septicemic infections range from chronic to acute. One of the most severe is anthrax, a disease of animals and sometimes humans, in which the number of *Bacillus anthracis* organisms may often exceed the number of erythrocytes in the blood! Septicemic infections often begin as localized infections that later become generalized; for example, streptococcal pharyngitis, a staphylococcal abscess, and bubonic plague may all subsequently give rise to a septicemia.

MICROBIAL VIRULENCE FACTORS

Antiphagocytic Factors

The virulence of many pathogenic bacteria is influenced by the presence or absence of a nontoxic polysaccharide material composing the capsules surrounding the cells. Such capsules can prevent the engulfment and destruction of the bacteria by the phagocytic defense mechanisms of the body (see Chap. 32). The importance of capsules can be demonstrated with *Streptococcus pneumoniae*: capsulated cells are virulent, but mutant cells that can no longer make capsules are avirulent.

Other bacteria that produce capsular material with a direct bearing on virulence are *Haemophilus influenzae*, *Klebsiella pneumoniae*, *N. meningitidis*, and *B. anthracis*. The capsule of *B. anthracis* is unusual in that it is composed of a polypeptide rather than a polysaccharide.

Capsular polysaccharides can be isolated in pure form. For instance, when the capsular polysaccharides of pneumococci are injected into humans, they stimulate the production of antibodies that protect against the type of pneumococcus from which they were obtained. This is the basis of the vaccine that protects against pneumococcal pneumonia.

Some antiphagocytic factors made by pathogenic bacteria are not capsules, yet like capsules they are located on the outer surface of the bacterial cell wall. One example is the M protein of *S. pyogenes* (this protein is also responsible for adherence of the organisms to epithelial cells and has been mentioned previously in this chapter).

Exotoxins

Some microorganisms produce poisonous substances known as **toxins.** Their potency can be expressed in various ways: the most precise is in terms of the LD_{50} dose (the dose that kills 50 percent of test animals that are injected with the toxin), but MLD (minimum lethal dose or LD_{100} dose, the dose that kills 100 percent of test animals) is also frequently used. Toxins can be divided into two main categories, **exotoxins** and **endotoxins.** These are differentiated by the properties listed in Table 31-2.

Potency of Exotoxins

Exotoxins are toxic proteins that are secreted by living microorganisms. Some exotoxins have extraordinarily high potency, with only minute amounts being needed to kill animals. For example, *Clostridium botulinum* type A produces the most potent toxin known: 1 MLD for a mouse is 2.5×10^{-5} µg of the purified toxin, about 1 million times more toxic than strychnine. The toxin of *Clostridium tetani* is also highly potent, the MLD for a mouse being 4×10^{-5} µg. The high potency of these two toxins is attributable to their action on the mammalian nervous system. Most toxins affect other kinds of tissues and are less potent; for example, the MLD of diphtheria toxin for a guinea pig is 6×10^{-2} µg, and the MLD of the α toxin of *S. aureus* is 5 µg for a rabbit.

Table 31-2. Some Characteristics of Exotoxins and Endotoxins

Feature	Exotoxins	Endotoxins
Bacterial source	Secreted by living microorganisms	Released from cell walls of lysed Gram-negative bacteria
Chemical nature	Protein	Lipopolysaccharide
Heat tolerance	Inactivated easily by boiling; heat-labile	Will withstand autoclaving; heat-stable
Immunology	Can be converted to toxoids and readily neutralized by antitoxin	Cannot form toxoids; neutralization with antitoxin not possible or only with difficulty
Pharmacology	Each has a highly characteristic mechanism of action	All act similarly to cause their effects; action characterized by pyrogenicity, blood changes, and shock
Lethal dose	Small	Much larger

Fungi:

a *Histoplasma capsulatum*, the causative agent of histoplasmosis, showing macroconidia.

b *Sporotrichum schenkii*, the causative agent of sporotrichosis, as seen in a lacto–phenol blue stained preparation.

c *Candida albicans*, causative agent of thrush, monilia vaginitis, and other infections, showing chlamydospores.

d *C. albicans*, A Gram stain preparation of specimen from patient.

e *Coccidioides immitis*, causative agent of coccidioidomycosis, showing arthrospores.

f Spherules of *C. immitis* in a tissue section from a patient with coccidioidomycosis.

g *Blastomyces dermatitidis*, causative agent of blastomycosis, showing yeast phase.

h *B. dermatitidis*, mycelial phase with conidia.

(Courtesy of National Medical Audiovisual Center, Centers for Disease Control, Atlanta, Georgia.)

Virus infections of plants:

a Aster "Blue Boy" [*Callistephus chinensis* (L.) Nees.] systemically infected with tobacco rattle virus (aster ringspot strain).

b Leaf of *Gomphrena globosa* L. six days after inoculation with tobacco necrosis virus.

c Veinbandung symptom of grapevine fanleaf virus in *Vitis vinifera* var. Gewürztraminer.

d Tobacco leaf *Nicotiana tabacum* L. 19 days after infection with the apricot strain of tomato ringspot virus.

e Necrotic ringspots on a *Grammatophyllum scriptum* orchid leaf infected with a bacilliform virus. (Rhabdovirus)

f Flower of Cattleya orchid exhibiting necrosis resulting from Cymbidium mosaic virus infection.

g Chlorotic ringspots on leaves of *Euonymus fortunei* (Turcz.) Hand.-Mazz. var. *vegetus* inoculated with the euonymus strain of tomato ringspot virus.

h Fruit symptoms (green) on summer crook-neck squash (*Cucurbita pepo* L.) infected with watermelon mosaic virus.

i Leaf of *Nicotiana glutinosa* L. 4 days after inoculation with tobacco mosaic virus.

j Flower of tulip (*Tulipa gesneriana* L.) var. Darwin exhibiting color break resulting from infection by color-breaking virus.

(Courtesy of Kenneth M. Corbett, University of Maryland.)

Not all exotoxins are lethal; some merely cause unpleasant effects. For example, many strains of S. aureus produce a toxin which, when ingested by a human in quantities of as little as 1 μg, gives rise to severe nausea and vomiting (staphylococcal food poisoning). Although this type of food poisoning is most unpleasant, few if any victims have died from it.

Classification of Exotoxins

Exotoxins are often divided into various categories based on the site of the damage that they cause or the kind of cells that are affected. For instance, botulinum and tetanus toxins affect nerve tissue and are termed **neurotoxins.** The toxin made by V. cholerae affects the intestinal tract and is termed an **enterotoxin** (from the Greek word enteron, meaning "intestine"). Diphtheria toxin kills several different kinds of cells and is thus termed a **cytotoxin.** Some cytotoxins may kill leukocytes and hence are known as **leukocidins;** some may cause the lysis of red blood cells and therefore are termed **hemolysins.**

Toxoids

Exotoxins are proteins. They lose their toxicity when treated with formaldehyde, although their antigenic properties are retained; in this form they are called **toxoids** and have the ability to stimulate the production of **antitoxins** (antibodies that react with toxins and neutralize them) in the body of the host animal. This is important in the protection of susceptible hosts from diseases caused by bacteria that produce exotoxins. For instance, toxoids are widely used as vaccines for immunization against tetanus and diphtheria.

Role of Bacteriophages and Plasmids

The ability of a bacterium to make a particular exotoxin **(toxigenicity)** may be due to a chromosomal gene. However, in some instances, such as diphtheria toxin or certain types of botulism toxin, toxigenicity can be conferred on a bacterium as the result of acquiring a temperate bacteriophage that carries the gene for the toxin. This process, called **lysogenic conversion,** has been described in Chap. 20. Bacteria can also become toxigenic by acquiring a plasmid that carries the gene for a toxin. For example, some strains of E. coli are able to cause diarrhea because they contain a plasmid whose DNA codes for an enterotoxin. Another plasmid-mediated toxin is the enterotoxin made by food-poisoning strains of S. aureus.

Mechanisms of Action of Exotoxins

It is becoming increasingly apparent that, just as adherence of microorganisms to tissue cells is an important first step in the process of infection, an initial adherence of exotoxins to tissue cells is an important first step in their toxic activity. The function of attachment is usually assumed by a particular region or subunit of the toxin. Other regions or subunits of the toxin may have enzymatic activity that causes damage to the tissue cell. The mechanisms of action of some exotoxins are well understood; in other instances they are completely unknown.

Diphtheria Toxin

This toxin consists of a single polypeptide chain which consists of two regions, A and B. By means of the B region the toxin becomes bound to the membrane of a tissue cell. The A region is then cleaved from the B region and enters the cell (Fig. 31-3). The A fragment has enzymatic activity and catalyzes a reaction that inactivates **elongation factor 2 (EF-2),** a factor essential for the elongation

Figure 31-3. Hypothetical model showing certain features of diphtheria toxin. The toxin consists of a single polypeptide composed of two regions, A and B. A and B are linked not only by a peptide linkage but also by a disulfide (—S—S—) bridge. The B portion binds to the membrane of a tissue cell. The B portion then allows the A portion to penetrate the membrane. During this transport process the A fragment is liberated by cleavage of the peptide linkage and by reduction of the disulfide bridge. The A fragment then proceeds to inhibit protein synthesis within the host cell.

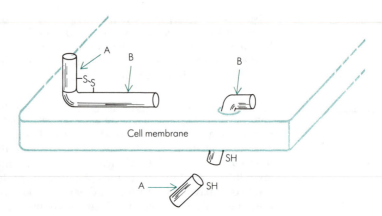

of growing polypeptide chains during protein synthesis by ribosomes. As a result, the tissue cell can no longer make proteins, and the cell eventually dies. Diphtheria antitoxin, i.e., antibodies formed against the toxin, can neutralize the toxicity of diphtheria toxin. More specifically, the antitoxin reacts with the B portion of the toxin, preventing the toxin from attaching to tissue cells. If attachment of the toxin to a host cell has already occurred, however, antitoxin cannot neutralize the toxin. Only toxin that has not yet attached can be neutralized by antitoxin. This is why prompt administration of antitoxin is required in cases of diphtheria. This principle applies to many other exotoxins as well.

Botulinum Toxin

In the normal operation of a voluntary muscle, a nerve impulse sent along the axon of a motor neuron reaches the **neuromuscular junction,** i.e. the junction between the axon and the muscle; secretion of acetylcholine by the end feet of the terminal twigs of the axon then initiates contraction of the muscle. In botulism, the toxin binds to the axon near the neuromuscular junction and prevents the secretion of acetylcholine; thus the muscle cannot contract. If this paralysis extends to the muscles of the chest and diaphragm, death by respiratory failure can result.

Tetanus Toxin

It is common to find that muscular movement involves cooperative action between a set of two opposing muscles. For example, in order to raise one's arm, the biceps muscle must contract and at the same time the opposing muscle, the triceps, must be allowed to stretch. As the triceps begins to stretch, however, stretch receptors in the muscle automatically send signals to the central nervous system that stimulate the firing of the motor neurons controlling the triceps. Thus it is necessary to render those neurons temporarily insensitive to stimulation. This is normally accomplished in the central nervous system by the action of a chemical called **inhibitory transmitter,** viz., the amino acid glycine. In tetanus, the toxin acts on the central nervous system to prevent the release of the glycine, and the result is that opposing sets of muscles often contract simultaneously and in uncontrollable fashion, giving rise to characteristic tetanic spasms that may be powerful enough to break bones and tear tissue. Death usually results from respiratory failure, i.e., an inability to control the muscles of the chest and diaphragm that are involved in breathing.

Cholera Toxin

The toxin consists of *one A subunit* and *five B subunits*. The B subunits are responsible for attachment of the toxin to the surface of the epithelial cells of the small intestine. The A subunit then penetrates the cell membrane and is cleaved to yield a fragment, A_1. This fragment causes an alteration of the regulatory protein that governs the activity of the enzyme **adenylate cyclase** in the cell. The modified regulatory protein becomes "turned on" permanently, causing adenylate cyclase to convert ATP to high levels of **cyclic AMP (cAMP)**:

$$ATP \rightarrow 3',5'\text{-cAMP} + \text{pyrophosphate}$$

The high cAMP level in turn causes a loss of water and electrolytes from the intestinal cells into the lumen of the intestine, i.e. diarrhea. The result of this sequence of events is dramatic: *as much as 10 to 12 liters of fluid may be excreted per day by a patient.* Ultimately the blood is the source of this fluid. Since bicarbonate ions are lost from the blood, the pH of the blood falls, which may lead to death from acidosis. Moreover, as the diarrhea proceeds, proteins and red cells become increasingly concentrated in the blood **(hemoconcentration),** leading eventually to hypovolemic shock and circulatory collapse. The treatment of cholera is to replace the fluids and electrolytes intravenously in amounts equivalent to those being lost (Fig. 31-4).

Streptolysin O (SLO) and Streptolysin S (SLS)

SLO is produced by *Streptococcus pyogenes* and is inactivated by oxygen (hence the "O" in its name). It has multiple activities based on its ability to damage cell membranes. For example, it is a hemolysin, causing β hemolysis around the colonies on blood-agar plates incubated anaerobically. In infections, however, this hemolytic action is much less important than its action as a leukocidin. The streptococci do not need to be ingested by a leukocyte in order to kill it: the soluble toxin penetrates the leukocyte from without and damages the membranes of cytoplasmic granules **(lysosomes)** within the cell. The powerful hydrolytic enzymes contained within the lysosomes are liberated into the cyto-

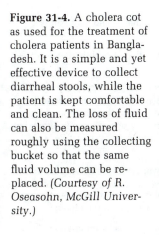

Figure 31-4. A cholera cot as used for the treatment of cholera patients in Bangladesh. It is a simple and yet effective device to collect diarrheal stools, while the patient is kept comfortable and clean. The loss of fluid can also be measured roughly using the collecting bucket so that the same fluid volume can be replaced. *(Courtesy of R. Oseasohn, McGill University.)*

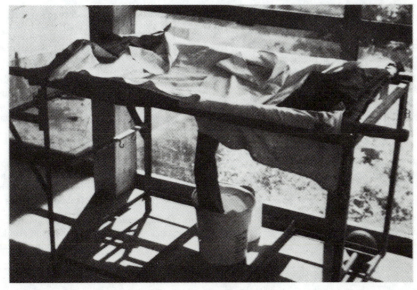

plasm and proceed to destroy the phagocyte. When injected intravenously into rabbits, SLO also acts as a **cardiotoxin,** causing complete disruption of the cardiac pumping cycle within seconds; whether this cardiotoxic activity contributes to human infections is not known.

SLS is also produced by S. *pyogenes* and kills leukocytes in the same manner as SLO. However, SLS is not soluble but rather is bound to the bacterial cells; thus it kills leukocytes only after the bacteria have been engulfed. SLS is oxygen-stable (hence the "S" in the name). Thus, unlike SLO, SLS can cause β hemolysis on *aerobically incubated* blood-agar plates.

Other Exotoxins

Examples of several other exotoxins and their modes of action are indicated in Table 31-3.

Endotoxins

As discussed in Chap. 5, endotoxins are lipopolysaccharides located in the

Table 31-3. Additional Examples of Exotoxins

Toxin	Organism	Action	Mechanism
Clostridial α toxin	*Clostridium perfringens*	Hemolysin, cytotoxin	Phospholipase; hydrolyzes lecithin, thereby damaging the cytoplasmic membranes of blood and tissue cells
Staphylococcal α toxin	*Staphylococcus aureus*	Hemolysin, leukocidin, cytotoxin	Mechanism uncertain; may be due to disruption of hydrophobic regions of cytoplasmic membranes or conversion to an active protease
Staphylococcal enterotoxin	*S. aureus*	Enterotoxin, causing nausea and vomiting (staphylococcal food poisoning)	Mechanism unknown
Panton-Valentine factor	*S. aureus*	Leukocidin	Causes degranulation of leukocytes
LT toxin	*Escherichia coli*	Enterotoxin, causing diarrhea	Mechanism similar to that of cholera toxin (see text)
Exotoxin A	*Pseudomonas aeruginosa*	Cytotoxin	Mechanism similar to that of diphtheria toxin (see text)
Anthrax toxin	*Bacillus anthracis*	Possibly a neurotoxin	Mechanism unknown; the toxin probably acts on the central nervous system, causing respiratory failure; the toxin consists of three distinct proteins: EF, PA, and LF; no single one of these is lethal, but a combination of all three, or even just PA + LF, is lethal
Pertussis toxin (pertussigen)	*Bordetella pertussis*	Cytotoxin	Mechanism unknown; causes increased sensitivity to histamine, causes an increase in the number of leukocytes in the blood, and causes hypoglycemia (subnormal levels of glucose in the blood)
Erythrogenic toxin	*Streptococcus pyogenes*	Cytotoxin, causing red rash in scarlet fever	Damages small blood vessels by an unknown mechanism; also kills macrophages, is pyrogenic (fever-inducing), and enhances susceptibility to endotoxic shock

outer membrane of the cell wall of many Gram-negative bacteria. The presence of an endotoxin in the medium in which Gram-negative bacteria are grown is due to lysis of some of the cells, which may occur during late growth stages of the culture. In Table 31-2 the general properties of endotoxins are differentiated from those of exotoxins. It should be noted that much larger quantities of endotoxins are required to kill experimental animals. All endotoxins exhibit similar pharmacologic effects, which may be described as follows:

1 *Pyrogenicity.* This is the ability to cause a change in body temperature. In humans, endotoxins cause an *increase* in temperature (i.e., a fever response). The pyrogenic effect is indirect: the active chemical agent that causes the temperature change is an endogenous pyrogen that is released from the blood leukocytes under the influence of endotoxins. This pyrogen affects the hypothalamus of the brain, which regulates body temperature.

2 *Blood changes.* When administered to experimental animals, endotoxins first cause a temporary decrease, and later a marked increase, in the number of leukocytes in the blood. Endotoxins also damage blood platelets (thrombocytes), which release factors that may cause blood to clot within blood vessels (intravascular clotting). Moreover, endotoxins cause an increase in the permeability of blood capillaries, causing them to leak the fluid portion of the blood, and sometimes even whole blood (hemorrhage); these effects can cause serious changes in circulation and blood pressure.

3 *Shock.* When Gram-negative bacteria are present in the blood in large numbers, or when endotoxin is injected intravenously, severe shock may occur as evidenced by a decreased blood pressure, feeble rapid pulse, decreased respiration, and sometimes unconsciousness. A high dose can result in circulatory collapse and death.

A sensitive test has been devised for the presence of minute amounts of endotoxin in body fluids, as an aid to the early diagnosis and treatment of Gram-negative bacterial infections. The test is based on the ability of endotoxin to cause gelling of extracts of the amoebocytes (blood cells) of the horseshoe crab, *Limulus polyphemus.*

Other Virulence Factors

A number of factors produced by pathogenic bacteria or occurring as part of their cellular structure may contribute to virulence, although they are not toxins. They may enhance virulence by aiding the spread of the pathogens through tissues, by enhancing abscess formation, by causing tissue damage, or by allowing pathogens to compete more effectively with a host for an essential nutrient. Some of these factors are described below.

Hyaluronidase

This enzyme is produced by the gas gangrene–causing organism *C. perfringens* and has been thought to enhance penetration of the host tissues by hydrolyzing hyaluronic acid, an essential intercellular "tissue cement." However, the failure of antihyaluronidase serum to inhibit the spread of *C. perfringens* through tissues suggests that the enzyme may play only a minor role.

Streptokinase

This substance is produced by the A, C, and G groups of the β-hemolytic streptococci. It converts blood plasminogen to plasmin, a protease that dissolves the fibrin of blood clots. Streptokinase has long been thought to enhance the ability of streptococci to spread through tissue by dissolving the fibrin that tends to wall off areas of tissue damaged by infection; however, its role may be minor in view of the failure of antistreptokinase serum to inhibit streptococcal invasiveness.

Deoxyribonuclease (DNase)

This enzyme is produced by S. pyogenes, S. aureus, C. perfringens, and certain other pathogens. Its ability to destroy DNA would seem to indicate that it is a formidable cytotoxin; however, it cannot penetrate living tissue cells to gain access to the intracellular DNA. It may possibly contribute to the ability of some pathogens to spread through tissue by destroying the viscosity of DNA liberated from cells that have been damaged or killed by other means.

Coagulase

Staphylococcus aureus produces an enzyme commonly called coagulase which reacts with an activator in plasma to cause **clotting,** i.e., conversion of soluble **fibrinogen** to insoluble **fibrin.** The fibrin coats the cell walls of the cocci and has been thought to protect them against phagocytosis. Coagulase is also involved in the walling-off process of staphylococcal abscesses. However, mutants of S. aureus that cannot make coagulase have been found to retain their virulence, indicating that coagulase is probably not a major virulence factor.

Protein A

This protein is located on the cell wall of S. aureus and has the ability to bind antibodies, regardless of their specificity. The antibody molecules are distorted so that their binding site for complement (C) is exposed. (See Chap. 32 for a discussion of complement.) A subsequent series of reactions results in formation of a substance known as **C5a.** This substance, also called **anaphylatoxin,** causes release of the toxic compound, **histamine,** from certain body cells. The histamine causes a variety of damaging effects in tissues.

Hydrogen Peroxide and Ammonia

Bacteria of the genera Mycoplasma and Ureaplasma adhere firmly to epithelial tissue of the respiratory or urogenital tract, where they excrete toxic by-products of their metabolism, viz., H_2O_2 and NH_3. These compounds accumulate locally to high concentrations, thereby damaging the epithelial cells.

Microbial Iron Chelators

The ability of aerobic microbial pathogens to compete with a host for available iron has considerable bearing on microbial virulence. Aerobic or aerotolerant organisms are continually faced with the difficulty of obtaining enough iron for growth, i.e., for biosynthesis of iron-containing enzymes such as cytochromes and catalase. Most of the iron that is available for aerobic or aerotolerant organisms is present in the oxidized ferric form, which is extremely insoluble. (Organisms that can grow under anaerobic conditions have less difficulty in obtaining iron, since in the reduced environments in which anaerobic organisms occur iron is in its reduced, or ferrous form, which is very soluble.) Aerobic organisms have had to evolve ferric iron–binding compounds in order to solubilize and take up ferric iron. The iron-binding compounds formed by microorganisms are termed **siderophores.** These generally belong to two major classes, the phenolates and the hydroxamates. One well-known member of the phenolate group is **enterochelin.** For instance, cells of E. coli secrete enterochelin, which solubilizes polymeric ferric iron and forms complexes with the ferric ions; the ferric-enterochelin complex is then transported into the bacterial cells, where the complex is degraded and the iron reduced to the ferrous form.

The role of microbial siderophores in virulence is to compete with the host for available iron. The host, being aerobic, also possesses iron-binding compounds, proteins known as lactoferrin and transferrin, which can limit the amount of iron available to an invading pathogen (see Chap. 32). Much of the success of an infection appears to be due to the ability of the microbe to "steal" some of this iron, by means of its siderophores. In experimental infections of laboratory animals, strains of bacteria that do not secrete siderophores have markedly reduced virulence.

QUESTIONS

1 Define the following terms:

Pathogenic	Virulent
Avirulent	Attenuated
LD_{50}	MLD
Leukocidin	Primary infection
Secondary infection	Siderophore

2 List two components of microorganisms that are responsible for adherence to host tissue.

3 By what means may pathogenic microorganisms passively penetrate the body to initiate infection? How does passive penetration differ from active penetration?

4 Contrast the type of infection caused by *Staphylococcus aureus* with that caused by *Clostridium perfringens* in terms of the ability of the organisms to spread through tissues.

5 List the various means by which pathogenic microorganisms can become distributed within the human body.

6 Give two examples of bacteria that produce antiphagocytic factors.

7 What role do bacteriophages play in the toxigenicity of *Corynebacterium diphtheriae*? What role do plasmids play in the toxigenicity of diarrhea-causing strains of *Escherichia coli*?

8 Differentiate between exotoxins and endotoxins. Differentiate between neurotoxins and enterotoxins.

9 What accounts for the ability of the α toxin of *C. perfringens* not only to act as a hemolysin but also to destroy tissue cells other than erythrocytes?

10 In what exotoxins does a B region or B subunit play an important role in adherence of the toxin to a host cell?

11 Both the botulinum toxin and the tetanus toxin are neurotoxins; however, they act in entirely different ways. What differences occur between their mechanisms of action?

12 Explain how diphtheria antitoxin neutralizes the toxicity of diphtheria toxin.

13 Why might shock occur as a feature of bacteremias or septicemias caused by Gram-negative bacteria?

14 What sort of evidence suggests that hyaluronidase, coagulase, and streptokinase may not be as important with regard to virulence as once thought?

15 Why might a *chronic* infection represent a more highly evolved host-microbe relationship than an *acute* infection?

REFERENCES

Baron, Samuel (ed.): *Medical Microbiology*, Addison-Wesley, Menlo Park, Calif. 1982. *The chapters of this text were individually prepared by authorities in the various fields of medical microbiology. A useful section on the microbiology of organ systems is included.*

Freeman, Bob A.: *Burrows Textbook of Microbiology*, 21st ed., Saunders, Philadelphia, 1979. *A classical medical microbiology text which deals with the biology of microorganisms and discusses the pathogens in detail.*

Joklik, Wolfgang K., Hilda P. Willett, and D. Bernard Amos: *Zinsser Microbiology*, 18th ed., Appleton Century Crofts, Norwalk, Conn. 1984. *Also a comprehensive reference on medical microbiology with detailed discussions of the pathogens and the factors contributing to their virulence.*

Linton, A. H.: *Microbes, Man and Animals*, John Wiley, New York, 1982. *A treatise on the natural history of microbe-host interactions, with chapters that deal specifically with exotoxins, endotoxins, and mechanisms of microbial pathogenicity.*

Slack, John M., and Irvin S. Snyder: *Bacteria and Human Disease*, Year Book, Chicago, 1978. *A shorter text oriented toward a comprehensive treatment of the properties and activities of bacterial pathogens.*

Chapter 32

Natural Resistance and Nonspecific Defense Mechanisms

While a pathogen uses all the means at its disposal to establish infection, the host's body has a number of defense mechanisms to prevent infection. The intricacies of the host-pathogen relationship are many and varied, and some aspects of the process of infection were discussed in Chap. 31.

This chapter is concerned with **natural** resistance (which provides defense against infection by a number of inherent, mechanical, and chemical barriers) and **nonspecific** defense mechanisms (such as phagocytosis, the complement system, and interferon). Immunity constituting the specific aspects of the body's defense against infection is presented in Chap. 33.

NATURAL RESISTANCE

The metabolic and physiological requirements of a pathogen are important in determining the range of potentially susceptible hosts. Naturally resistant hosts either fail to provide some of the essential environmental factors required by the microorganisms for growth or have defense mechanisms to resist infection. In addition, the host's general health and state of nutrition, social and economic conditions, and other intangible factors (stress, depression, etc.) all play a part. However, these factors are so intertwined that it is difficult to evaluate their individual importance (see Fig. 32-1).

Species Resistance

Resistance to infection varies with the species of animal or plant, but the following discussion will be confined to humans and other animals. In general, species resistance has become physiologically more complex as the species of the animal kingdom have evolved. However, basic physiological characteristics of a species, such as its normal body temperature, can determine whether or

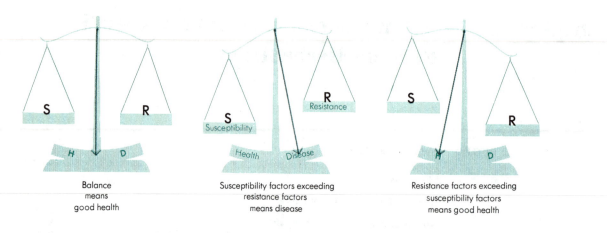

Balance means good health	Susceptibility factors exceeding resistance factors means disease	Resistance factors exceeding susceptibility factors means good health

1 Microbial factors:
Virulence
Invasiveness
Portal of entry
Dosage
Microenvironment
Indigenous microbiota

2 Physiological factors:
Injury, constitutional
   disease
Physiological stress
Psychological stress
Misdirected tissue
   response, inflammation,
   histamine release

3 Cellular factors:
Phagocytosis by macro-
   phages and polymor-
   phonuclear leuko-
   cytes
Lymphocytes

4 Humoral factors:
Antibodies
Complement
Interferon

5 Constitutional factors:
Nutrition
Age
Tissue tone
Hormonal balance
Vascular condition
Lysozyme
Race

6 Socioeconomic factors:
Housing
Hygiene
Occupational hazards
Environment

Figure 32-1. Host-parasite relationship in health and disease: the balance between susceptibility and resistence.

not a microorganism can be pathogenic. For this reason, many diseases of mammals do not affect fish or reptiles, and vice versa. Frogs, being cold-blooded, are resistant to anthrax unless they are warmed to 37°C. Conversely, if the body temperature of chickens is lowered from 39°C to 37°C, they become susceptible to anthrax. Most mammals are resistant to *Mycobacterium avium*, and the type of mycobacterium found in cold-blooded reptiles does not produce characteristic infections in either mammals or fowls. Although generally host-specific, the human and bovine species of the tubercle bacillus will cross-infect between humans and cattle as well as other animals having about the same body temperature.

Metabolic, physiological, and anatomical differences between species affect the ability of a pathogen to cause infection. Herbivorous animals are often resistant to diseases of carnivores and vice versa for a variety of reasons. Intestinal anatomy differs greatly between carnivores and ruminants, which have multiple stomachs and different intestinal microbial flora and digestive juices. Diseases of the skin, to which humans are quite susceptible, are often resisted by animals because they have more hair and thicker hides.

We can conclude that the characteristics of species resistance are (1) the inability of an organism to cause disease in the resistant species under natural conditions, (2) the production in resistant species of a localized or transient infection by experimental inoculation as opposed to a generalized or progressive disease in naturally susceptible species and, (3) the induction of experimental disease in resistant species only by massive doses of the microorganisms, often by an unnatural route or under unnatural conditions.

Racial Resistance

Various animal as well as plant breeds show marked differences in their resistance to certain infectious diseases. A well-known example is that Brahman cattle are resistant to the protozoan parasite responsible for tick fever in other breeds of cattle. That this type of resistance is an inherited racial characteristic is indicated by the fact that crossbreeding the highly susceptible Texas longhorn with Brahman cattle produces a resistant variety. In humans, evidence of racial resistance is explained on the basis of selectivity and survival in most cases.

The presence of a pathogen in isolated races results in a gradual selection for resistant members as the susceptible members die of progressive infection. The introduction of new pathogens, such as the tubercle bacillus by resistant Europeans to an isolated population such as the American Indians (who had not previously developed a resistance to the organism), resulted in epidemics which decimated the population. Similarly, blacks show a relatively high resistance to tropical diseases such as malaria and yellow fever, and Orientals exhibit a reduced susceptibility to syphilis. In addition to physiological factors, racial customs play an important role in racial resistance. Hygienic practices, food taboos, and domestic and migratory habits all affect the opportunity for exposure to infectious disease-causing agents.

Individual Resistance

Some people appear to experience fewer or less severe infections than others, even though they apparently have the same racial background and opportunity for exposure. Such individual resistance is due to a combination of both natural and adaptive resistance factors. Age is important, as young individuals are susceptible to "children's diseases" such as measles and chicken pox prior to the acquisition of immunity, which follows both overt and inapparent infections. Conversely, the aged are susceptible to diseases such as pneumonia, presumably as a result of a decline of immune functions with age. Certain individuals have genetic defects which result in selective or general **immunodeficiencies** (i.e., an inability to develop immunity to pathogens), which greatly increase the susceptibility to disease. Other factors include nutrition, personal hygiene, and the individual's sex. The nature of the workplace and its hazards, the opportunity for contacts with infected individuals, and the individual's hormonal and endocrine balance all affect the selectivity and frequency of certain diseases.

External Defense Mechanisms

External defense mechanisms are another factor in natural resistance. These are largely mechanical, but chemical barriers are also involved. Mechanical barriers, together with host secretions, constitute the body's first line of defense against invading microorganisms.

Mechanical barriers include the unbroken skin and mucous membranes, which are generally impervious to infectious agents. It is possible for microorganisms to enter through hair follicles, openings of sweat glands, or abrasions, but skin and mucous membranes form a generally effective barrier. However, certain fungi will readily produce skin infections when the skin becomes moist and soft, e.g., fungi that cause athlete's foot. Most bacteria are inhibited by lactic acid and fatty acids in sweat and sebaceous glands and by the low pH which these chemicals bring about. **Mucous secretions** of the respiratory tract, the digestive tract, the urogenital tract, and other tissues form a protective covering of mucous membranes and collect and hold many microorganisms until they can be disposed of or lose their infectivity. Peristalsis (the progressive and rhythmic contraction of the intestines) traps microorganisms in mucus and other material present in the intestines and expels them from the body. Small hairlike appendages, or cilia, of epithelial cells lining many of the body cavities and orifices sweep bacteria away from susceptible surfaces. Coughing, sneezing, shedding tears, perspiring, and salivating provide mechanical flushing or clearing that removes microorganisms.

In addition to the mechanical action of mucus, saliva, and tears in removing bacteria, some of these secretions contain chemical substances that inhibit or even destroy microorganisms. An example is lysozyme, an enzyme found in many body fluids and secretions, which has an effective antimicrobial action due to its ability to lyse certain Gram-positive bacteria by hydrolyzing peptidoglycan. Other enzymes and hormones may produce chemical, physiological, or mechanical effects that reduce susceptibility to infection. The acidity or alkalinity of some body fluids has a deleterious effect on many microorganisms and helps to prevent potential pathogens from entering the deeper tissues of the body.

Another protein with known antimicrobial activity is lactoferrin. It is a red, iron-containing protein found in milk (both bovine and human) as well as in most of the secretions that bathe human mucosal surfaces, including bronchial mucus, saliva, nasal discharges, tears, hepatic bile, pancreatic juice, seminal fluid, and urine. It is also an important component of the specific granules of polymorphonuclear leukocytes. Its serum counterpart is transferrin. Both have similar molecular weights (about 78,000 daltons) and metal binding sites. These proteins chelate or tie up available iron in the environment, thus limiting the availability of this essential metal nutrient to invading microorganisms.

INTERNAL DEFENSE MECHANISMS

The second line of defense consists of the body's internal mechanisms, which can be mobilized against invading microorganisms. These mechanisms can be nonspecific in their action (e.g., phagocytosis) or specifically directed against the pathogen (e.g., antibodies and sensitized cells). These two types are called nonspecific defense mechanisms and specific acquired immunity, respectively. Specific acquired immunity is a result of an infection or artificial immunization and is directed to the specific causative organism; it is discussed in Chap. 33. During infection both mechanisms act together to rid the body of the invading microorganisms. This interrelationship, and the interrelationships between the defense mechanisms discussed above, are illustrated in Fig. 32-2.

In the remainder of this chapter the nonspecific defense mechanisms, or cellular defense factors, will be discussed. At this point it is sufficient to define an antibody as a globular serum protein which can specifically combine with an organism, such as a bacterium, parasite, or virus.

NONSPECIFIC DEFENSE MECHANISMS

The body can mobilize many factors which act nonspecifically to the invasion of foreign organisms. These include complement, phagocytic cells, naturally cytotoxic lymphocytes, and interferon, among others.

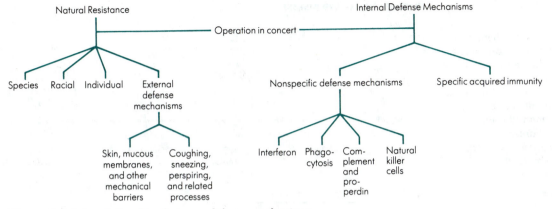

Figure 32-2. Interrelationships between defense mechanisms.

Table 32-1. Functional Activities of the Host Complement System in Host Defense against Infection

Activity	Complement Components or Fragments
Lysis of viruses, virus-infected cells, tumor cells, mycoplasma, protozoa, and bacteria	C1–C9
Endotoxin inactivation	C1–C5
Virus neutralization	C1, C4, C2, C3
Anaphylatoxin release (capillary dilatation)	C3a, C4a, C5a
Opsonization; enhancement of cell-mediated cytotoxicity; stimulation of production of B-cell lymphokines	C3b
Enhanced induction of antibody formation	C3b, C3d
Chemotaxis of neutrophils, monocytes, eosinophils	C5a
Stimulation of macrophage adherence and spreading	Bb

Complement System

The serum of higher animals contains a group of eleven proteins referred to collectively as **complement** because its action complements that of certain antibody-mediated reactions. Complement plays an essential role in resistance against infection and is the principal mediator of the inflammatory response. Upon activation of the first protein of the group of eleven, there occurs a sequential cascade in which active molecules are generated from inactive precursors. Certain of the different proteins activated along the cascade may function as mediators of a particular response as well as activators of the next step. Some of the functional activities of complement are shown in Table 32-1. In general, complement is capable of attacking and killing invading cells only after antibody binds to the cell membrane, thus initiating the complement attachment or fixation; this is illustrated in Fig. 32-3.

There are several characteristics of the complement system. It has a **recognition unit** to respond to the antibody molecules that have identified an invading cell. It has **receptor sites** to combine with the surface of the foreign cell when it is activated. Its activity must be limited in time to minimize damage to the

Figure 32-3. Mechanism of complement action: the bacterial-killing effects of antibody and complement. (*Courtesy of A. J. Vander, J. H. Sherman, and D. S. Luciano, Human Physiology, The Mechanisms of Body Function, McGraw-Hill, New York, 1970.*)

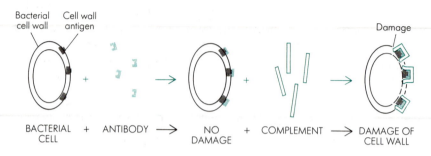

Figure 32-4. Model of complement assembly on the membrane of a bacterial cell in the sequence of events in complement recognition, activation, and cell attack. (*After a diagram suggested by Dr. Russell Siboo, McGill University.*)

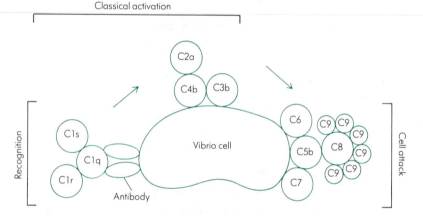

host's own cells. This limitation is brought about partly by the spontaneous decay of activated complement and partly by interference from destructive enzymes and inhibitors.

As stated before, complement is composed of eleven components. These components are named according to the following rules. Each component has been assigned a number in the order of its discovery, and that number is preceded by the capital letter C. Unfortunately, the first four components do not interact in the order of their discovery, but rather in the sequence C1, C4, C2, and C3. But the remaining components do react in the appropriate numerical order: C5, C6, C7, C8, and C9. C1 has three subcomponents, C1q, C1r, and C1s. Fragments of components resulting from cleavage by other components acting as enzymes are assigned lowercase letters a, b, c, d, or e, for example, C3a and C3b. The mechanisms of complement action in bacterial lysis shown in Fig. 32-3 are represented in greater detail in Fig. 32-4 as a cascade of events in complement *recognition* and *activation*, culminating in *cell attack*. This is the *classical* or antibody-dependent pathway requiring activation by antibody, C1, C4, C2, and C3.

The activation of complement may also occur by an alternate pathway sometimes called the **properdin** pathway. Properdin refers to one of the factors in blood serum that, acting together with complement, is a mechanism for the activation of a natural defense when sufficient quantities of specific antibody are unavailable for immune activation by the classical pathway. Components of the alternate pathway have been assigned the letters B, D, and P (properdin). They activate the C3 component of complement directly. That is, the alternate pathway does not use antibody, C1, C4, or C2, which are the early reactants of the classical pathway. Since C3 is a component of both the classical and the properdin pathway, it is apparent that these two pathways of complement activation have much in common, differing only in the initial activation events. The term *complement system* refers generally to both pathways, which interact and are dependent upon each other for their full activity. Among the substances that can activate the alternate pathway are lipopolysaccharides of Gram-negative bacteria, bacterial capsules, teichoic acids of Gram-positive cell walls, inulin, dextran, fungal cell walls, and aggregated globulins high in carbohydrate content.

Since the complement system consists of so many distinct proteins (as well as the many active fragments split during their enzymatic activation), the overall system is very complex, and no further attempt will be made in this chapter to identify the specific roles of individual complement proteins (except as noted in Table 32-1).

Phagocytosis

The importance of phagocytosis in protecting the body from infection was first recognized by the Russian zoologist Elie Metchnikoff in 1883. He called the amoeboid particulate-eating cells *phagocytes* (from *phagein*, "to eat," and *kytos*,

Table 32-2. Cell Types Associated with Host Resistance to Infection

Cell Type	Location	Derivation	Description and Function
1 Leukocytes	Peripheral blood	Bone marrow stem cells	Classified according to structure and affinity for dyes
(a) Polymorphonuclear granulocytes			Lobar nuclei and abundant cytoplasmic granules
(1) Neutrophils			No dye preference; phagocytic
(2) Eosinophils			Stain with eosin, an acid dye; phagocytic
(3) Basophils			Stain with basic stains; bind IgE and produce histamine
(b) Lymphocytes		Lymphoid organs, bone marrow stem cells	Smaller than monocytes; large nucleus; scanty cytoplasm; form T and B cells (see Chap. 33); nonphagocytic
(c) Monocytes		Bone marrow stem cells	Larger than granulocytes; single horseshoe-shaped or oval nucleus; few cytoplasmic granules; phagocytic
2 Plasma cells	Lymphoid organs (lymph nodes, spleen, thymus)	From B lymphocytes	Produce antibodies; eccentric nucleus; nonphagocytic
3 Macrophages	Other tissues	Transformed from monocytes	Numerous cytoplasmic granules; phagocytic
(a) Wandering			
(1) Alveolar	Lung		
(2) Peritoneal	Abdomen		
(b) Fixed			
(1) Histiocytes	Connective tissue		
(2) Kupffer cells	Liver		

"hollow vessel"). He believed that cellular activity, in destroying bacteria and other microorganisms, was the primary defense mechanism of the body. Even though this concept was well entrenched in the minds of people since that time, phagocytes were neglected by research workers from the 1920s on. Little was added to our understanding of phagocytosis after the initial pioneer studies. In recent years, however, there has been a resurgence of interest in phagocytosis as a central resistance process.

It is now known that cellular activity responsible for phagocytosis and other aspects of nonspecific and specific resistance is carried out by a diverse group of blood cells (Table 32-2). In higher organisms, there are two principal classes of phagocytic cells. These are the *polymorphonuclear granulocytes*, or *polymorphs* for short (mainly *neutrophils*), so called because their nuclei come in many shapes, and the *macrophages* (mononuclear phagocytes), large phagocytic cells derived from monocytes. (See Fig. 32-5.)

The polymorphs constitute the front line of internal defense for the host. They are produced in the bone marrow and are discharged into the blood in vast numbers. The 50 to 100 billion polymorphs that are present in normal human blood carry out their functions after leaving the circulation and entering sites of inflammation in tissues. They live for a few days only, and about 10^{11} polymorphs disappear from the blood daily; however, they are replaced by new ones from the bone marrow. Polymorphs contain numerous enzymes and antimicrobial substances for the killing and degradation of bacteria. These substances are contained in membrane-bound organelles called **lysosomes.**

Macrophages are formed from circulating precursor *monocytes* (blood cells with horseshoe-shaped nuclei; see Fig. 32-6) which also arise from the bone marrow. As soon as the monocytes leave the circulation and begin to carry out phagocytosis, they are called macrophages. Unlike the polymorphs, macrophages are long-lived and can persist in tissues for weeks or months. They are

Figure 32-5. (A) Scanning electron micrograph of a peritoneal macrophage from a mouse fixed during locomotion (X2,700). (B, C). Thin sections of cultured mouse peritoneal macrophage showing phagocytosed cells of *Listeria monocytogenes* (a bacterium) indicated by arrows. (B) Horizontal section (X4,000). (C) Vertical section (X3,800). (*Courtesy of P. Gill, McGill University.*)

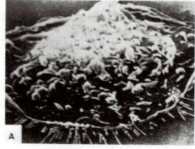

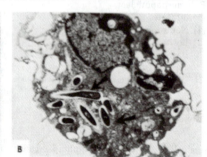

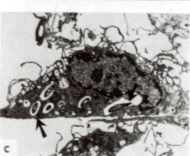

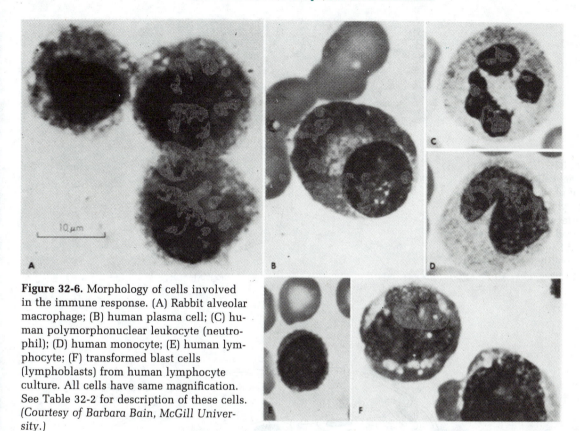

Figure 32-6. Morphology of cells involved in the immune response. (A) Rabbit alveolar macrophage; (B) human plasma cell; (C) human polymorphonuclear leukocyte (neutrophil); (D) human monocyte; (E) human lymphocyte; (F) transformed blast cells (lymphoblasts) from human lymphocyte culture. All cells have same magnification. See Table 32-2 for description of these cells. *(Courtesy of Barbara Bain, McGill University.)*

widely distributed throughout the body, but they are not as numerous as polymorphs (about one-twentieth the concentration). Under certain conditions, macrophages can synthesize DNA and multiply. When they differentiate in connective tissue, they are called *histiocytes*, in the liver, *Kupffer cells*, and in the lung, *alveolar macrophages*.

Two types of mature macrophages are recognized: (1) those wandering in tissues and body spaces (e.g., alveolar and peritoneal macrophages) and (2) those fixed to vascular endothelium (e.g., Kupffer cells and fixed macrophages of the spleen and lymph nodes).

Macrophages, then, are strategically placed throughout the body to combat invading microorganisms. Macrophages also have lysosomes with bactericidal substances. (See Fig. 32-6 for the morphology of the polymorphs and macrophages.)

Mechanism of Phagocytosis

The process of phagocytosis requires a preliminary attachment of the microbe to the phagocytic cell surface. Electrostatic forces are necessary for initial attachment, since divalent cations such as Ca^{2+} and Mg^{2+} are required. The firm attachment is facilitated by serum substances called **opsonins.** Opsonins are antibodies that allow microbes to be more easily ingested by the phagocyte.

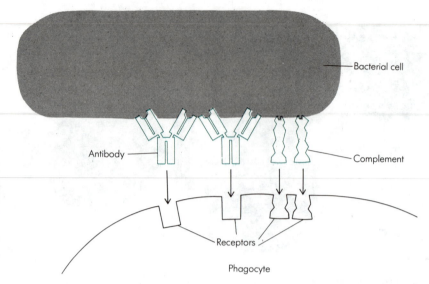

Figure 32-7. Attachment of bacterial cell antigen to surface of phagocyte. The phagocyte has receptors on its surface both for the antibody and for the complement. Phagocytosis occurs more readily with this kind of binding.

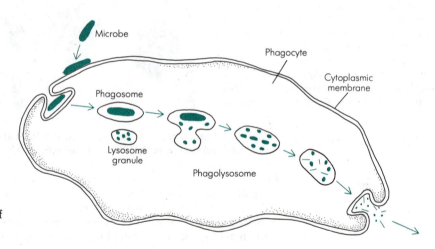

Figure 32-8. Phagocytosis of a microbial cell.

Phagocytic cells have a special receptor affinity for antibody attached to microorganisms. If complement is also bound to the microbial surface, there are complement receptors on the phagocytic cell which provide additional attachment forces. (See Fig. 32-7.)

In phagocytosis, the phagocyte extends small pseudopods around the microbe after adherence. These pseudopods fuse and form a vacuole by means of invagination of the phagocyte plasma membrane engulfing or surrounding the bacterium. The vacuole is now called a **phagosome**. Subsequent events depend on the activity of the lysosomal granules. These move toward the phagosome, fuse with its membrane to form a **phagolysosome**, and discharge their contents (hydrolytic enzymes) into the vacuole, thus initiating the intracellular killing and digestion of the microbe. Within the phagolysosome, most microbes are killed in minutes, although the complete degradation may take a few hours. The process of phagocytosis is shown schematically in Fig. 32-8.

The major mechanism of bacterial killing within phagocytic cells is by the so-called "respiratory burst." This activity is initiated when a particle such as a bacterium binds to a polymorph surface or when a macrophage ingests a particle: the phagocytic cell increases its oxygen consumption—thus the term respiratory burst. This is brought about by an enhanced activity of the hexose monophosphate shunt pathway of metabolism (see Chap. 10). This pathway generates large quantities of $NADPH_2$ (reduced nicotinamide adenine dinucleotide phosphate). When the $NADPH_2$ is oxidized back to NADP, highly toxic oxygen metabolites are generated. These include singlet oxygen, superoxide, hydrogen peroxide, hydroxyl radicals, chloramines, and aldehydes. All of these are effective in killing invading bacteria. The major enzymes involved in the generation of these oxygen metabolites are superoxide dismutase and myeloperoxidase (in neutrophils) or catalase (in macrophages). Their specific activity is shown in Fig. 32-9.

In addition, the lysosome granules of phagocytic cells contain many potent hydrolytic enzymes. These granules migrate through the cytoplasm and fuse with the phagosome containing the invading microbe (Fig. 32-8). The enzymes, now present in the phagolysosome, proceed to destroy the ingested bacteria. Over 60 different enzymes have been found within lysosomes. These include lysozyme, hyaluronidase, and other enzymes that act on carbohydrates; lipases; ribonuclease and deoxyribonuclease; collagenases, elastases, and other enzymes that act on proteins and peptides; and the enzymes of the respiratory burst, namely, myeloperoxidase, superoxide dismutase, and catalase.

The combined actions of the lysosomal enzymes and of the respiratory burst are usually sufficient to destroy all invading microorganisms. However, microbes vary in their response to phagocytic activity. Gram-positive bacteria are rapidly

Figure 32-9. The respiratory burst that produces bactericidal products.

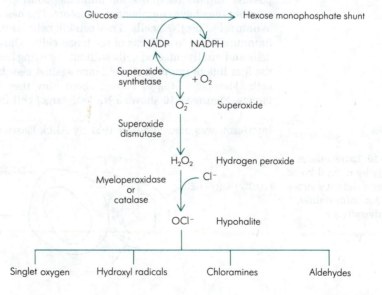

Macrophages Have Many Functions

In addition to maintaining antimicrobial defenses, macrophages play a very important role in the body's day-to-day functioning. Macrophages are involved in the breakdown of coagulated blood, in wound healing, in several pathways of immediate and delayed allergic reactions, and, very importantly, in the digestion of dead cells and tissues. For example, each day, 300 billion senescent red blood cells are removed from the circulation by macrophages in the liver and spleen. In the course of a year, a person's macrophages ingest and digest about 5 lb of hemoglobin.

Macrophages can also be activated to kill cells without ingesting them. Activation is by protein factors called *lymphokines* produced by antigen-stimulated thymus-dependent lymphocytes (T cells; see Chap. 33). Activated (also called angry) macrophages have the capacity to destroy tumor cells, possibly by the secretion of large amounts of hydrogen peroxide and other toxic forms of oxygen.

Macrophages also secrete many molecules into their surrounding environment, affecting the activity of other cells. For instance, one of the macrophage products acts strongly to attract circulating polymorphs and monocytes to the lungs in response to infections and tumors.

---

destroyed. Gram-negative bacteria are somewhat more persistent because their cell wall is relatively resistant to digestion. But some bacteria like *Mycobacterium tuberculosis* are so resistant to phagocytic action that they may literally multiply within the phagocytes.

Natural Killer Cells

Recently, a subpopulation of lymphocytes has been discovered which are referred to as **null cells** because they do not possess certain surface antigens. Morphologically, null cells are about the size of large lymphocytes (12 to 15 μm in diameter) but have a larger cytoplasm-to-nucleus ratio. Null cells generally possess surface receptors for immunoglobulin; reports are conflicting as to whether they have complement receptors. The best-defined null cells are known as **natural killer (NK) cells.** They can kill cellular targets without prior exposure (immunization) to antigens of the target cells. Thus they can directly kill tumor cells and virally infected cells without ingesting them. NK activity is potentially the first line of immune surveillance against neoplastic cells and virus-infected cells. However, little is known about how they recognize and destroy their targets. Figure 32-10 shows a NK cell–target cell interaction.

Interferon

Interferon was discovered in 1957 by Alick Isaacs and Jean Lindenmann of the

Figure 32-10. Extracellular cytotoxicity by natural killer cells. *(After a drawing suggested by Malcolm Baines, McGill University.)*

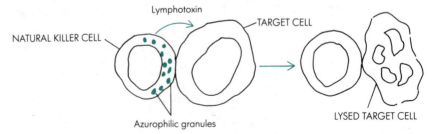

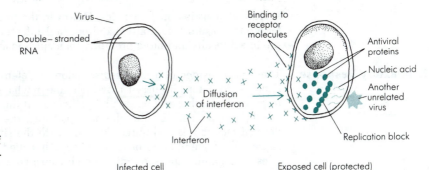

Figure 32-11. Mechanism of interferon induction and action.

National Institute for Medical Research in London while working on the mechanisms of **viral interference** (resistance of an animal or cell infected with one virus to superinfection with a second unrelated virus). Interferon is still virtually the only natural substance with the proven ability to inhibit intracellular viral replication, although results from clinical trials of its therapeutic use are ambiguous. Interferons are small proteins produced by normal eucaryotic cells in response to viral infection or double-stranded RNAs (viral or synthetic). Interferons lack virus specificity since they do not react directly with the virion but instead induce the host cell to synthesize a nonspecific antiviral protein (Fig. 32-11). Conversely, interferons were found to be species-specific with respect to the cells that produced them and so would induce little or no resistance in cells from other species. Thus human interferon was most effective in protecting human cells and poorly protective for mouse or chicken cells. However, recent work with purified interferon shows that its antiviral activity is not strictly species-specific. Though purified interferons are stable at low pH and fairly heat-resistant, they are relatively unstable at physiological pH in tissue fluids. Interferon is secreted by cells only in minute amounts and is extremely difficult to purify. For these reasons interferon has not been particularly useful clinically, although recent advances in molecular genetics have permitted the production and synthesis of interferons in large quantities for experimental use and clinical trials. Specifically, recombinant DNA carrying a gene for human α interferon has been inserted into the bacterium *Escherichia coli* (see Chap. 29 on genetic engineering). Such recombinant *E. coli* cells are grown in large quantities (e.g., in 400-liter fermentation tanks), centrifuged out, and broken open to release intracellular material, including interferon. The cellular debris is separated by centrifugation. Nucleic acids and other viscous materials are removed, and the proteins in the extract are passed through *monoclonal-antibody* columns to yield purified interferon.

Our interest in interferon, in the context of the present chapter, is in its antiviral effect in preventing cell infection. But this antiviral effect is merely one facet of interferon activity. There are other aspects which should interest the microbiologist, e.g., interferon's effect on cell membrane–related events, its inhibition of cell proliferation, and its involvement as a primary regulator of the immune system. Interferon is also capable of promoting the natural cytotoxic activity of natural killer cells, which implies that it has a role in immunologic surveillance against malignancies. However, the primary protective role of interferons is against naturally acquired viral infections because interferons are produced locally and more promptly than specific antibodies.

Inflammatory Response

The final consideration in our discussion of defense mechanisms is how the nonspecific factors discussed above combine in what is termed the *inflammatory response* to combat an invasion by pathogens. If the microbe is able to activate and fix complement by the alternate pathway, the chemotactic complement-derived factors released attract leukocytes to the site, and anaphylatoxin also causes the degranulation of tissue basophils called mast cells. These in turn release histamine and serotonin, which cause constriction of smooth muscles (e.g., in bronchioles and blood vessels) and increased capillary permeability, which promotes the passage of plasma and leukocytes into the affected tissue. The leukocytes pass through the junction between the capillary endothelial cells in response to the chemotactic influence of the complement cleavage fragments. Migration continues until the phagocytes encounter complement fixed to the microbial surface, which causes adherence and facilitates engulfment.

If antibodies specific for the microbe are present, they opsonize it and also increase complement fixation. This greatly enhances leukocyte adherence and promotes phagocytosis. The plasma also contains other microbicidal substances which may inhibit replication and growth of pathogens and modulate the subsequent immune response.

The symptoms of the inflammatory response are local swelling, erythema (reddening), and local and systemic heat. The local swelling is due in part to the accumulation of large numbers of phagocytic cells at the site of infection. The increase in temperature and the erythema are due to the increased blood flow to the local site, enzymatic activity, and the release of bacterial endotoxin. The inflammatory response is usually acute and resolves spontaneously. During long-term infections the inflammatory response may become chronic. The inflammatory response is the most important factor in resisting infection by pyogenic (pus-forming) bacteria.

QUESTIONS

1 What is meant by a naturally resistant host?
2 Discuss species resistance and give several specific examples.
3 Discuss some specific factors governing individual resistance to disease.
4 Describe two chemical factors that form part of the external defense mechanisms.
5 Discuss two functional activities of the host complement system in host defense against infection.
6 Why is complement considered a resistance factor?
7 Explain the terminology for the components of complement.
8 Compare the initial complement activation events of the classical pathway with those of the alternate pathway.
9 Compare and contrast the two principal classes of phagocytic leukocytes in higher organisms.
10 Describe the mechanism of phagocytosis as well as the known mechanism for killing phagocytosed bacteria.

11 What are natural killer cells? What role do they play in the defense of the body?

12 Why is interferon effective as a resistance factor against virus infections?

13 Why is interferon sometimes used in cancer therapy?

14 Explain how nonspecific factors of immunity elicit the inflammatory response.

15 What factors cause the symptoms of the inflammatory response?

REFERENCES

Barrett, J. T.: *Textbook of Immunology*, 4th ed., C. V. Mosby, St. Louis, 1983. *An introduction to immunochemistry and immunobiology written in a style that a student with a limited background can comprehend and appreciate.*

Bier, O. G., W. D. DaSilva, G. Götze, and I. Mota: *Fundamentals of Immunology*, Springer-Verlag, New York, 1981. *A textbook of basic and clinical immunology written primarily for medical and biology students.*

McConnell, I., A. Munro, and H. Waldmann: *The Immune System*, 2d ed., Blackwell Scientific Publications, Oxford, 1981. *An advanced text on the molecular and cellular basis of immunity. The text is divided into three sections: immunochemistry, cellular immunology, and immunopathology.*

Roitt, I.: *Essential Immunology*, 4th ed., Blackwell Scientific Publications, Oxford, 1980. *A very concise paperback that gives an overview of the principles of immunology.*

Tizard, I. R.: *Immunology, an Introduction*, Saunders College Publishing, New York, 1984. *A very well written text that explains immunology simply. Relevance to infectious disease is emphasized.*

Chapter 33

Basic and Theoretical Aspects of the Immune Response

Immunology is a rapidly developing field of study concerned with the discovery of the mechanisms of the immune system, the development of vaccines, and the regulatory procedures for manipulating the immune response. The type of immunity most commonly encountered is naturally acquired active immunity to pathogens (see Fig. 33-1). However, diverse pathological effects have also been shown to arise from immune responses to nontoxic and noninfectious antigens (macromolecules that will induce the formation of immunoglobulins or sensitized cells that react specifically with the antigens). Thus the immune response is directly responsible for allergies, autoimmune diseases (e.g., rheumatoid arthritis), and graft rejection. Also, the failure of various aspects of the immune system can result in the development of malignancies and death due to overwhelming infections (e.g., acquired immunodeficiency syndrome, better known as AIDS, which will be discussed later in this chapter).

The immune system is composed of a single integrated cellular system producing effector products of two types: serum antibodies that constitute part of the humoral immunity and sensitized cells called lymphocytes that constitute cell-mediated immunity. While antibodies are effective in opsonizing bacteria and neutralizing toxins and viruses, lymphocytes are important in eliminating intracellular parasites and viruses and rejecting tumors and transplants. Thus the immune response, in eliciting reactive antibodies and cells in response to antigens, not only forms the principal means of defense in vertebrates against

718

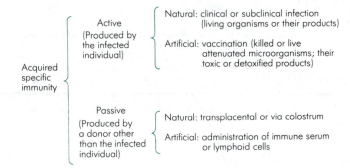

Figure 33-1. The various types of acquired specific immunity.

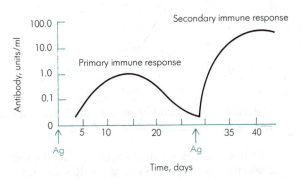

Figure 33-2. Antibody production due to administration of antigen (Ag). Note that the secondary immune response is faster and stronger than the primary immune response because of immunologic memory.

infection by pathogenic microorganisms and larger parasites, but also acts as a surveillance mechanism against the transformation of host cells into cancer cells.

THE IMMUNE RESPONSE

As mentioned above, immune responses are processes in which animals form *specifically reactive* proteins (antibodies) and cells in response to a great variety of *foreign* organic macromolecules and molecules. The generalized immune response has four primary characteristics: **discrimination, specificity, anamnesis,** and **transferability by living cells.** The first primary characteristic refers to the ability of the immune system to *discriminate* between "self" and "nonself," and therefore it responds only to materials which are foreign to the host. Second, the response is highly *specific* for the inducing material or antigen to which the immune antibodies or cells will react in greatest strength. The third characteristic refers to the ability to elicit a larger specific response more quickly when induced by a second exposure to the same foreign antigen. This is called immunologic memory or the *anamnestic* response (Fig. 33-2). Finally, *active* immunity is only transferable from one inbred animal to another by the immune cells or lymphocytes and not by serum. Whereas immune serum is capable of temporarily transferring *passive* immunity, transfer of active immunity requires the long-term regenerative ability of living cells.

Some nonspecific materials, such as mineral oil and alum, have the ability to prolong and intensify the immune response to a specific antigen when they are

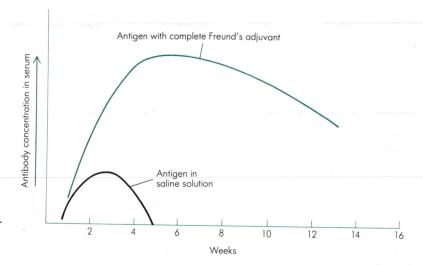

Figure 33-3. Effect of an adjuvant on the antibody response.

injected together with the antigen. Such materials are called **adjuvants** because they help the immune response (Fig. 33-3).

Antigens

An **antigen** is any substance which, when introduced into the vertebrate host, stimulates the production of antibodies and reacts with preformed antibodies, if they are already present. This demonstrates the two properties of antigens: **specificity,** or their ability to react specifically with immune antibodies or cells, and **immunogenicity,** or the capacity to stimulate an immune response. The terms *antigenicity* and *immunogenicity* differ in that the former refers to the "foreignness" of a substance or its specific reactivity while the latter refers to the ability of a substance to stimulate an immune response.

The parts of an antigen responsible for its particular properties are separable into the low-molecular-weight **antigenic determinants** or **haptens** normally linked to the larger supporting portion of the molecule, which may be referred to as the **carrier.** The hapten is the specificity-determining portion which is capable of reacting with a specific antibody but is incapable of inciting an immune response by itself. For this reason haptens are often referred to as **incomplete antigens.** High molecular-weight antigens which possess both immunogenicity and specificity are called **complete antigens.**

Antigens are generally proteins or polysaccharides with molecular weights greater than 10,000 daltons. They may be nucleoproteins, lipoproteins, glycoproteins from any biological source, or synthetic polypeptides or polysaccharides. Antigenic determinants or haptens generally have a molecular weight of less than 1000 daltons. Many naturally occurring substances may act as antigens: bacteria, viruses, and other microorganisms; foreign proteins such as pollens, egg white, and metabolic products of microorganisms; and living cells from different animal species.

Antigens stimulate the production of antibodies that may have prophylactic or therapeutic properties directed against the specific organism that expresses the antigen. The antigenicity of a substance or organism is not, however, related to its ability to produce disease or damage tissue. Antigenicity can be measured only in terms of the antibody response it elicits on the basis of its foreignness to the host's immune system.

Sources of Antigens

When an immune response is elicited in an animal to the multitude of antigens in the tissues of a different species, a variety of antibody specificities are produced. Some antibodies recognize all tissues of the donor species and therefore can identify *species-specific* antigens. Other antibodies recognize the *organ-specific* antigens of the immunizing tissue. Such antibodies are useful in forensic medicine and for product identification in the food industry.

A type of tissue-specific antigen which is present in one individual of a species but not in another is called an **alloantigen** or **isoantigen.** Human sera contain *isoantibodies* against the red blood cell isoantigens of some other persons (Table 33-1). For example, individuals of blood group O contain both anti-A and anti-B isoantibodies in their sera. Individuals of blood group A have only anti-B antibodies, while group B individuals possess anti-A antibodies. It follows that blood group AB individuals have no anti-A and anti-B antibodies in their sera. The designations A, B, AB, and O describe the antigenic phenotype of a person's red blood cells. These antigens are inherited according to a simple Mendelian system involving three allelic genes called A, B, and O. Gene A controls the formation of A antigen, gene B controls the formation of B antigen, while gene O is not expressed (see Table 33-1). Thus a person with AB genes is heterozygous, having inherited an A gene from one parent and a B gene from the other. A person with O gene must be homozygous for O, while persons in A and B groups must be either homozygous (AA or BB) or heterozygous (AO or BO).

There are other isoantigens on human red blood cells. Individuals of blood group O actually produce an antigen called H antigen, which is a precursor of the A and B antigens. The H gene is situated at a different locus from the ABO locus on the chromosomes. The H gene controls the production of the H substance that, in the presence of A and B genes, is converted to A, B, or AB antigens. In addition to the A, B, and H antigens, two other blood group specificities also belong to this group but are inherited at a different locus. These are Lewis a (Le^a) and Lewis b (Le^b) antigens. The three loci, ABO, H, and Le, all code for glycosyl transferases that determine the structure of the blood group substances.

In 1940, Landsteiner and Weiner demonstrated that antibodies produced in rabbits against rhesus monkey erythrocytes agglutinated the red blood cells of 85 percent of a human population. The antibodies were specific for a new

Table 33-1. Human Major Blood Types: The ABO Blood Group System

Blood Group (Isoantigen Present)	Isoantigen on Red Blood Cells	Genotype	Isoantibodies Present
A	A	AO, AA	Anti-B
B	B	BO, BB	Anti-A
AB	AB	AB	No anti-A or anti-B
O	None	OO	Anti-A and anti-B

antigen which came to be called the rhesus (Rh) antigen. Individuals possessing this antigen were called Rh positive. The remaining 15 percent who did not carry this antigen were called Rh negative. Natural antibodies against the Rh antigens are not found. Since then the Rh system has been shown to be highly complex, and approximately 30 different antigenic types have been identified. (Rh incompatibility resulting in hemolytic disease of the newborn is discussed in the next chapter.)

Humans also possess a large number of other blood group systems, e.g., the Kidd group, the Duffy group, and the MNS system. Thus the antigenic mosaic on the red blood cells of humans is extremely complex.

Microbial Antigens

The antigens with which we are primarily concerned are those of microbial origin. The exotoxins produced by *Corynebacterium diphtheriae*, *Clostridium tetani*, and certain other microorganisms are potent antigens that stimulate the production of antitoxins (toxin-specific antibodies) in the body of the host. Most pathogenic bacteria, viruses, and rickettsias are good antigens which stimulate the production of antibodies capable of protecting against infection and which are useful in diagnostic tests. Fungi, however, are not strongly antigenic. Heat or chemical treatment destroys the viability of most microorganisms without necessarily decreasing or altering their antigenicity.

Bacterial cells may contain one antigenic component in the cell wall, another in the body of the cell, and others in the flagella and capsules. The antigen in the capsule of the pneumococcus gives the organism its type specificity. Antigens at the surface of cells of *Salmonella* are somatic antigens called the O antigens. They are used by bacteriologists in classifying related species into serological groups, whereas the H antigens from the flagella of *Salmonella* detect serological differences of species in a group (Fig. 33-4).

Vaccines

Vaccines, which are suspensions of killed, living, or attenuated (having weakened virulence) cultures of microorganisms, are used as antigens to produce immunity against infection due to the particular microorganism. Typhoid fever vaccine consists of killed cells of *Salmonella typhi*; the Salk poliomyelitis vaccine is composed of killed poliomyelitis virus, but the oral type, such as the Sabin vaccine, contains attenuated live virus. Another example of a vaccine made from an attenuated culture is that used for rabies. It is made by treating the rabies virus chemically so that it cannot produce infection but still can stimulate the body to make antibodies. The yellow fever virus is attenuated by growing it in embryonated eggs, thus somehow reducing the virulence of the virus for humans. The vaccine used for immunization against smallpox is different from either of these, since a living culture of a virus, vaccinia, closely

Figure 33-4. External antigens of Gram-negative bacilli such as the salmonellas.

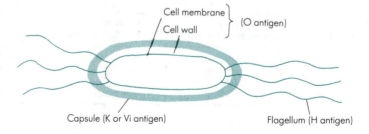

Cell membrane
Cell wall
} (O antigen)

Capsule (K or Vi antigen)

Flagellum (H antigen)

related to variola virus, which is not highly infectious for humans, produces antibodies that protect people from infection with the human pathogen. Hepatitis B vaccine consists of highly purified, formalin-inactivated antigens obtained from the plasma of persistently infected carriers (since the virus has not yet been grown in cell cultures).

Toxoids

Toxoids are made from extracellular toxins (exotoxins) by destroying the poisonous portion with heat, ultraviolet light, or chemicals without altering their antigenic specificity and often enhancing their immunogenicity. Toxoids made from toxin-producing microorganisms such as *C. diphtheriae* and *Cl. tetani* are used to immunize against the harmful effects of diphtheria and tetanus, respectively.

Heterophile Antigens

Substances that stimulate the production of antibodies capable of reacting with tissues of a wide variety of unrelated animals or plants are called heterophile antigens. Heterophile antigen was first demonstrated in 1911 by Forssman, who showed that when emulsions of tissues, including brain, liver, kidney, and adrenals (but not blood or serum), from guinea pigs are injected into rabbits, they stimulate the rabbits to produce hemolysin capable (in the presence of complement) of lysing the red blood cells of sheep. This particular reaction is known as the Forssman reaction. Many other heterophile systems have since been demonstrated. Some pathogenic bacteria contain heterophile antigen, but it is rarely found in saprophytes. Heterophile antigen is found in the tissues of many animals, including horses, cats, dogs, fish, and fowl. A Forssman type of heterophile antigen is of diagnostic importance in mononucleosis, which results in the production of an antibody which agglutinates animal red blood cells as detected in the so-called Paul Bunnell test. Some malignant (cancerous) tissues express Forssman antigens as a consequence of neoplastic transformation.

Antibodies

Specific acquired immunity against infection is primarily a property of a group of serum glycoproteins (proteins with attached carbohydrate) called antibodies. These antibodies have been produced by a subpopulation of white blood cells in the immune system called lymphocytes. These are small round cells, 6 to 7 μm in diameter, with a high nuclear-to-cytoplasmic ratio in their resting stage, and they are capable of expanding greatly in volume and activity in response to an antigen. This process is called lymphocyte activation.

Antibodies are termed immunoglobulins (abbreviated Ig) since they are globular proteins with immune function, the bulk of which generally separate in the gamma region on electrophoresis (Fig. 33-5). All antibody molecules have two basic functions: (1) antigen binding and (2) participation in effector functions depending upon the physical properties of the antibody. These properties can be separated by cleavage of the antibody molecule with a proteolytic enzyme, papain. This produces two antigen-binding fragments (named Fab, or antigen-binding fragment) and a single crystallizable fragment (termed Fc, or crystallizable fragment) which possesses the physical properties of the molecule (Fig. 33-6). Only the Fc fragment has been shown to fix complement and bind to Fc receptors, and the Fab does not interact.

Subsequent biochemical analysis demonstrated that every monomeric anti-

Figure 33-6. (Right) Monomer of the immuno-globulin molecule with two heavy and two light polypeptide chains held together by interchain disulfide bonds. Shaded areas indicate the variable regions or domains.

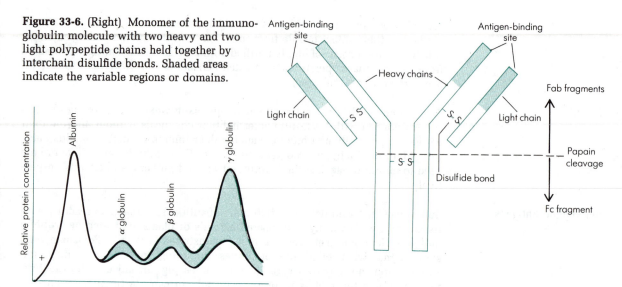

Figure 33-5. Electrophoretic profile of rabbit serum containing antibodies to egg albumin before (shaded zone included) and after absorption with the same antigen (egg albumin). Note the decrease in the globulin peaks, especially that of gamma globulin, demonstrating that the anti-egg albumin antibodies were globulin proteins.

body molecule consisted of two pairs of polypeptide chains linked to each other by disulfide bonds. The larger pair is composed of two heavy chains, each chain having a molecular weight of 50,000 daltons. The smaller pair is composed of two light chains, each with a molecular weight of 25,000 daltons. (See Fig. 33-6.)

Amino acid analysis of the antibody molecule revealed that the antigen-binding site is flanked by a variable amino acid sequence portion of the chains (Fig. 33-6) while the remainder has a relatively constant amino acid sequence. Thus the amino acid sequence of the constant regions determines the common biological and physical properties of the immunoglobulin, and the variable region determines its individual specificity.

In actual fact, there are five separate classes of immunoglobulins: immunoglobulin G (IgG), immunoglobulin M (IgM), immunoglobulin A (IgA), immunoglobulin D (IgD), and immunoglobulin E (IgE). (See Table 33-2.) Their re-

Table 33-2. Some Characteristics of the Different Classes of Human Immunoglobulins

Immunoglobulin	Heavy-Chain Designation	Molecular Weight	Physical State	J Chain	Subclasses
IgG	γ (gamma)	150,000	Monomer	−	4
IgM	μ (mu)	900,000	Pentamer	+	1
IgA	α (alpha)	160,000 (serum)	Monomers	−	2
		(160,000)n (serum)	Polymer	+	
		390,000 (secretions)	Dimer	+	
IgD	δ (delta)	180,000	Monomer	−	1
IgE	ε (epsilon)	185,000	Monomer	−	1

Figure 33-7. Structures of the different classes of immunoglobulins. IgG, IgD, and IgE each consists of a monomer of two light and two heavy polypeptide chains. IgM is a large molecule having five monomers in a star formation joined with a J (joining) polypeptide chain. IgA has three forms. When it appears in the serum, it may consist of one, two, or three monomers (not shown); when it is found in such body fluids as saliva, tears, and nasal secretions, it contains two monomers joined by a special component known as the "secretory piece" (this dimer also has a J chain). The exact location of the J chain in the IgM and IgA molecules is uncertain.

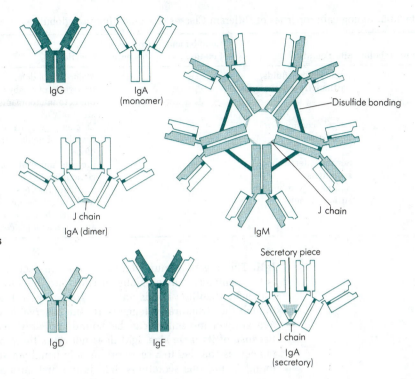

respective heavy chains are designated by the Greek letters gamma (γ), mu (μ), alpha (α), delta (δ), and epsilon (ε). Some of these immunoglobulins are polymers of the basic unit (Fig. 33-6), and some have subclasses. Each immunoglobulin is linked with one of two types of light chains called kappa (κ) or lambda (λ). A primary rule for antibodies is that each basic unit consists of two identical heavy chains and two identical light chains. For example, an IgG antibody could have the formula of either $\gamma_2\lambda_2$ or $\gamma_2\kappa_2$, where γ refers to the type of heavy chain and λ and κ refer to light-chain types. Each class of immunoglobulin has its own characteristic structure and special function (Tables 33-2 and 33-3). Figure 33-7 shows the structures of the different classes of immunoglobulins.

IgG. The most common antibody in serum is IgG, which accounts for 80 percent of the total 10 to 20 mg of immunoglobulin present in 1 ml of serum. IgG can be subdivided on the basis of antigenic differences into four subclasses designated as IgG_1, IgG_2, IgG_3, and IgG_4; they constitute 70, 19, 8, and 3 percent, respectively, of the IgG class. As seen in Table 33-3, IgG with a molecular weight of 150,000 daltons is the only immunoglobulin capable of passing from the mother to the fetus via the placenta, and so it is the major source of passive immunoglobulin protection for the fetus and newborn of many higher vertebrate species, including humans. Monomeric IgG antibody is said to have a valence of 2 since it has two antigen-combining sites per molecule.

Table 33-3. Biological Properties of Different Classes of Human Immunoglobulins

Immunoglobulin	Site Found	Complement Fixation	Crosses Placenta	Functions
IgG	Internal body fluids, particularly extravascular	+	+	Major line of defense against infection during the first few weeks of a baby's life; neutralizes bacterial toxins; binds to microorganisms to enhance their phagocytosis and lysis
IgM	Largely confined to bloodstream	+ + +	−	Efficient agglutinating and cytolytic agent; effective first line of defense in cases of bacteremia (bacteria in blood)
IgA	Serum, external body secretions	−	−	Protects mucosal surfaces from invasion by pathogenic microbes
IgD	Serum, on lymphocyte surface of newborn	−	−	Regulator for the synthesis of other immunoglobulins; fetal antigen receptor
IgE	Serum	−	−	Responsible for severe acute and occasionally fatal allergic reactions; combats parasitic infections

IgM. The largest immunoglobulin with a molecular weight of 900,000 daltons is the pentameric macroglobulin IgM. The five monomers of IgM are held together by disulfide bonds and a single J protein (mol wt 15,000 daltons), and though the structural valence is 10, the observable valence is only 5, i.e., only five antigen molecules can be bound, possibly because of steric hindrence. Because of its large size, IgM does not cross the placenta, but monomeric IgM molecules coupled to a secretory protein usually associated with IgA have been found in mucous secretions. IgM is the first antibody produced in a primary immune response and constitutes 6 percent of the total immunoglobulin pool.

IgA. The protection of respiratory and gastrointestinal mucous membranes and external body secretions such as tears, saliva, seminal fluid, urine, and colostrum is accomplished by IgA, which constitutes 13 percent of the total serum immunoglobulin. Though IgA can exist as monomers, dimers, and polymers in serum, it exists as a dimer in secretions containing a single J chain and a secretory protein with a molecular weight of 60,000 daltons. Whereas the light, heavy, and J chains are produced by an antibody-producing lymphocyte, the secretory piece is added by epithelial cells during secretion. The function of the secretory piece is unknown, but it may be involved in the transport of IgA to an external environment, and it may protect the secreted IgA from proteolytic digestion by extracellular proteases present in gastrointestinal and other secretions. Secretory IgA in colostrum provides passive immune protection to the gut of the newborn against microorganisms which might cause gastroenteritis. In respiratory mucus, IgA may act as a first line of defense against bacteria and viruses and may also neutralize allergens.

IgD. The rare classes of immunoglobulins IgD and IgE were identified as a result of their excess production in patients with a cancer of the antibody-forming cells, called a myeloma. IgD has a molecular weight of 180,000 daltons and constitutes 1 percent of the total serum immunoglobulins. It appears to act as a primary receptor for specific antigen on the surface of fetal lymphocytes destined to become antibody-forming cells in adult life. As such it also may determine to some extent what the immune system will accept as being "self" or respond to as being "foreign" by allowing the removal of specifically autoreactive cells prior to their maturation.

IgE. Even though they make up only 0.002 percent of total serum immunoglobulins, IgE proteins are responsible for some of the most severe immunologic reactions called **hypersensitivity** or **allergic** reactions. IgE is called a **homocytotropic** antibody since it can bind to IgE-specific Fc receptors on **autologous** (one's own) tissue basophils called **mast cells.** When antigen, such as a pollen allergen or parasite, combines with the IgE on a mast cell, the cytoplasmic basophil granules are activated, releasing chemical mediators (e.g., histamine and serotonin). In the respiratory tract these mediators cause the symptoms of allergy: sneezing, wheezing, and excess mucous secretion; in the skin, these antigens cause a typical wheal-and-flare reaction in sensitized individuals. In particularly severe sensitivities, the allergic reaction may produce fatal systemic **anaphylaxis** caused by excessive smooth muscle contraction in, for example, the bronchioles of the lungs.

Even though IgE is noted for its harmful properties, it may have protective functions. The release of histamine may result in the destruction of parasites. This hypothesis arose because it was observed that individuals in tropical areas (where parasites are more prevalent than in northern regions) may have 20 times more IgE antibodies than those who live in colder climates. This, of course, is of little comfort to northerners who suffer the miseries of summer allergies.

Functional Names of Antibodies

Antibodies react against specific microorganisms, their toxic products, and other compounds. They can be used in the treatment of infection caused by the specific micoorganisms, and, more importantly, they *prevent* infection and disease caused by these agents. Antibodies may be designated by names that describe their reaction in vitro or in vivo when they are allowed to act on certain types of antigens: (1) antitoxins, (2) agglutinins, (3) precipitins, (4) lysins, (5) complement-fixing antibodies, and (6) opsonins. Any antibody may be multifunctional and therefore can be called by more than one of the above terms.

These antibodies are all produced as a result of antigenic stimuli. They are humoral antibodies and can be differentiated as follows:

1 Antitoxins neutralize toxins.
2 Agglutinins cause clumping of the bacterial cells for which they are specific.
3 Precipitins cause precipitation or flocculation of extracts of bacterial cells or other soluble antigens.
4 Lysins cause lysis, or breakdown, of bacterial or other cells that are specifically sensitive to their action.
5 Complement-fixing antibodies participate in the complement-fixation reactions described in Chap. 34.
6 Opsonins render microorganisms more susceptible to ingestion by phagocytes.

Monoclonal Antibodies

As mentioned previously, myeloma is a cancer of the antibody-forming' cells (lymphocytes). Because it is derived from a single neoplasic (aberrantly growing) cell, it consists of a huge clone of identical lymphocytes. The antibody produced

Figure 33-8. Electrophoretic distribution of antibodies in serum. (A) Heterogeneous electrophoretic distribution of antibodies typical of a polyclonal response. (Serum from an animal or human producing antibodies to a protein or other macromolecule contains a complex mixture of antibodies.) The heterogeneity has been shown to be due to the many clones of plasma cells, each of which produces a different antibody molecule. (B) Electrophoretic pattern shows the production of homogeneous, or monoclonal, antibody, such as from a myeloma patient. Myelomas (tumors of the antibody-forming cells) produce a single antibody.

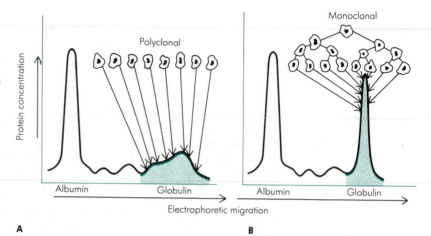

by these lymphocytes is homogeneous because it is all derived from the single homogeneous clone of cells and is therefore called a **monoclonal antibody** or M protein, which forms a single peak on electrophoretic separation of the serum (Fig. 33-8). Unfortunately, the antibody produced by a myeloma tumor is generally to an unknown antigen.

A recent development in cell biology has allowed the fusion of different cells to form a single "hybrid" cell which expresses properties of both "parental" cell lines. The fusing agent can be a defective virus (Sendai virus that characteristically causes cell fusion) or various chemicals (such as polyethylene glycol). If one uses as the fusion partners an immune lymphocyte producing a specific antibody in limited quantity and a myeloma cell, one is able to select fused cells called **hybridomas** (see Fig. 33-9). In these cells the myeloma provides properties of unrestricted proliferation and production of vast quantities of monoclonal antibody and the immune lymphocyte provides the information for the specificity of the antibody, that is, an antibody with known antigen-binding characteristics. Such hybridoma-derived monoclonal antibodies are becoming increasingly important in diagnostic and therapeutic medicine. For example, in cancer therapy, monoclonal antibodies can be used directly to attack and destroy tumor cells. They can be labeled with radioactive isotopes to locate tumors and to deliver specifically lethal doses of radiation to inaccessible tumors. They can also be used to deliver anticancer drugs to tumor cells in a similar manner.

Microbiologists already have generated monoclonal antibodies that allow new differentiations between strains of rabies viruses as well as between strains of influenza viruses. There is no doubt that the diagnosis and epidemiology of many viral, rickettsial, parasitic, and bacterial diseases will be greatly facilitated by the availability of monoclonal antibodies. Such antibodies have been produced against the bacterial pathogens Streptococcus pneumoniae, Mycobacterium leprae, Treponema pallidum, Neisseria gonorrhoeae, N. meningitidis, Pseudomonas aeruginosa, and Haemophilus influenzae.

The most successful use of monoclonal antibodies so far is in diagnostic kits

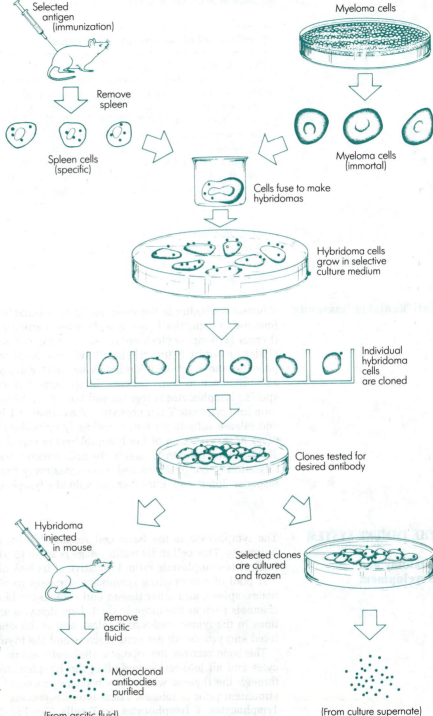

Figure 33-9. How monoclonal antibodies are made. The selected antigen is injected into a mouse. A few days later, the spleen is removed from the mouse. The antibody-synthesizing cells from the spleen are then mixed with fast-growing mouse cancer cells, called myelomas, and a chemical "glue," polyethylene glycol. The result is a hybridoma, a fusion of an antibody-making cell and a cancer cell. The hybridomas are separated, cloned, and tested to select those that make the desired monoclonal antibodies. The chosen hybridomas are injected into the peritoneal cavity of mice, where they induce the accumulation of 2 to 10 ml of ascitic fluid containing large amounts of antibody. Or the hybridoma cells may be cultured in the laboratory. (They may be stored frozen until they are needed to generate additional supplies of monclonal antibody.) The final step is to separate the pure monoclonal antibodies from the hybridomas by differential centrifugation. (*Erwin F. Lessel, illustrator.*)

Selected antigen (immunization)

Myeloma cells

Remove spleen

Spleen cells (specific)

Myeloma cells (immortal)

Cells fuse to make hybridomas

Hybridoma cells grow in selective culture medium

Individual hybridoma cells are cloned

Clones tested for desired antibody

Hybridoma injected in mouse

Selected clones are cultured and frozen

Remove ascitic fluid

Monoclonal antibodies purified

(From ascitic fluid)

(From culture supernate)

Development of the Hybridoma Technique

The hybridoma technique was developed by Georges Köhler and Cesar Milstein in 1975 at the Medical Research Council laboratories in Cambridge, England. Although the technique was never patented, its commercial applications were recognized immediately. In 1982 alone, more than three dozen independent ventures in monoclonal antibodies were initiated, and at least 20 large chemical and pharmaceutical companies are developing monoclonal antibodies for a wide variety of uses. One projection puts the world market value of hybridoma antibodies at a half billion dollars annually by the late 1980s.

For the discovery of the principle that led to production of monoclonal antibodies, Köhler and Milstein were awarded a Nobel prize in medicine in 1984.

---

used to test specimens of blood, urine, or tissues for blood type, viral disease, and other factors. For example, in 1981, the Food and Drug Administration approved four such kits for use in diagnosing allergies, prostate cancer, pregnancy, and anemia.

Cell-Mediated Immunity

Whereas antibodies in the serum and extracellular fluid form part of the humoral immune system, the living lymphocytes, particularly those derived from the thymus (T lymphocytes), are responsible for **cell-mediated immunity.** This includes immunity to tubercle bacilli, viruses, fungi, and many other intracellular parasites, the rejection of foreign tissue grafts, a delayed type of hypersensitivity, and resistance to some cancers. In all these cell-mediated immune reactions, specific lymphocytes recognize and bind the relevant antigen or cell and transform into cytotoxic T lymphocytes. The activated T lymphocytes also proliferate and release soluble mediators called **lymphokines** which recruit and activate other host cells. One of the lymphokines is called **migration inhibition factor (MIF),** which is instrumental in the inflammatory response since it prevents the migration of lymphocytes and phagocytes away from the focus of the immune response. Interferon is another example of a lymphocyte-produced lymphokine.

THE IMMUNE SYSTEM

Anatomy and Development

The lymphocyte is the basic cell responsible for both humoral and cellular immunity. This cell in its resting stage is small (6 μm in diameter), with a high nuclear-to-cytoplasmic ratio, indicative of its lack of activity.

A pool of recirculating lymphocytes passes from the blood into the lymph nodes, spleen, and other tissues and back to the blood by the major lymphatic channels such as the thoracic duct. Lymphocytes are found in high concentrations in the lymph nodes and spleen and at the sites where they are manufactured and processed: the bone marrow and the thymus (Fig. 33-10).

The bone marrow hemopoietic stem cells are the ultimate origin of erythrocytes and all leukocytes including the lymphocytes. Many lymphocytes pass through the thymus where they become processed by the hormonal microenvironment prior to release. These lymphocytes are now called **thymus-derived lymphocytes, T lymphocytes,** or **T cells.** (See Fig. 33-11.) The majority of the

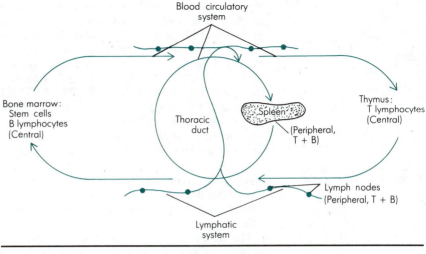

Figure 33-10. Basic anatomy of the immune system. Immune responses occur in the peripheral lymphoid organs, such as the spleen and lymph nodes. The central lymphoid organs (thymus and bone marrow) are the sources of the lymphocyte subpopulations (T and B cells) which seed to the peripheral organs. *(After an original drawing by Malcolm Baines, McGill University.)*

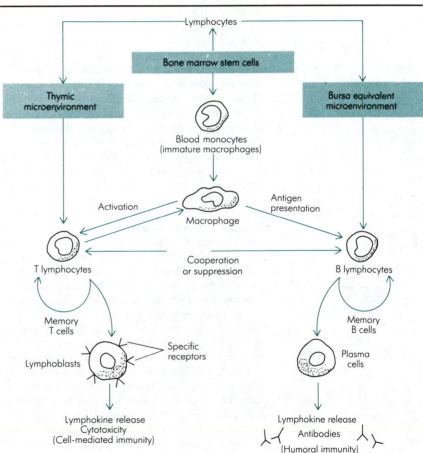

Figure 33-11. The generation of humoral and cell-mediated immune responses. *(After an original drawing by Malcolm Baines, McGill University.)*

bone marrow–derived lymphocytes which do not enter or become processed by the thymus are called B cells. Studies of the chicken have revealed the existence of a central lymphoid organ called the bursa of Fabricius responsible for inducing mature B-lymphocyte function just as the thymus induces T-cell function. It is thought that the site in which human B cells mature (i.e., the bursal equivalent) is the gut-associated lymphoid tissues (GALT), such as the Peyer's patches in the intestines, and the spleen and bone marrow. Though both populations of lymphocytes are morphologically identical, the B cells are destined to synthesize specific humoral antibody and the T cells become primarily responsible for cellular immunity. Lymphocytes which do not express either B- or T-cell markers or functions are called, appropriately, null lymphocytes.

T and B Lymphocytes

The interrelationships between the T and B cells are very complex in that each can augment or regulate the activity of the other (Fig. 33-11). In addition, specific antigen must be presented to the lymphocytes in the proper format by macrophages.

The B cells have on their surface membrane the immunoglobulin receptors representative of the specific antibody which they are capable of producing. One B cell is destined to synthesize only one specificity of antibody, although it may be able to synthesize more than one class of antibody. Thus during an immune response the first or early antibody produced is IgM, and after a few days the predominant antibody synthesized is IgG, even though it may express the same specificity. The B cell must receive a "signal" from an interacting cell (T cell and/or macrophage) in order to be stimulated to proliferate and, eventually, differentiate into the plasma cell form (see Figs. 32-6 and 33-11). Plasma cells are the ultimate antibody-secreting cells; they are also end-stage cells which die after they have served their function. A mature plasma cell can produce immunoglobulins of a unique class and subclass, as well as a unique type of heavy and light chain. Synthesis proceeds at a rate of some 300 molecules per second for its life span of a few days to weeks.

After the immune response wanes, the surviving lymphocytes return to their resting stage and the serum antibodies, mostly IgG, gradually disappear, with a half-life of about 25 days. However, the anamnestic response (Fig. 33-2) suggests the presence of an additional type of differentiated cell (a specific B lymphocyte) that arises following initial exposure to an antigen. It is extremely long-lived, exists in rather low numbers, and is capable of greatly accelerated proliferation into effector cells upon restimulation (boosting) with antigen. This cell is called a memory cell since it is the basis of immunologic memory (Fig. 33-11).

Macrophages have also been shown to produce several lymphokines that serve many of the same functions as those produced by T cells (see below). This release of lymphokines probably serves as a regulatory mechanism to activate or inactivate either T cells, B cells, or null cells.

T cells perform a wide variety of functions. These functions can generally be classified into three types: (1) helper function, (2) suppressor function, and (3) killer function. The helper function is to recognize an antigen and then facilitate a B-cell response to it through the release of lymphokines and interleukins. The suppressor function of T cells is the recognition of antigen followed by the release of factors which result in suppression of B-cell or other T-cell responses.

Table 33-4. Some Known Immunodeficiency Diseases

Immunodeficiency	Cellular Defect	Result
Agammaglobulinemia (Bruton's disease)	B lymphocytes low or absent	Low or absent antibodies; increase in pyogenic infections
Thymic aplasia (DiGeorge's syndrome)	T lymphocytes absent	Low or absent cell-mediated immunity; increase in viral infections; inability to reject grafts; increase in cancers
Phagocytic dysfunction (Chediak-Higashi syndrome)	Phagocytic leukocyte activity	Decreased intracellular killing of microbes; increase in pyogenic infections; decrease in natural killer cell functions; increase in cancers
Complement defects	C3 production	Increase in pyogenic infections

The killer function is cytotoxicity: the ability to cause the death of foreign cells, such as cells from a genetically different donor, parasite cells, single-celled pathogens, and tumor cells. Each of these three types of T-cell functions can be executed either by direct cell-to-cell contact with the interacting cell or by the effect of lymphokines, which were discussed previously in relation to cell-mediated immunity.

Immunodeficiency Diseases

Defects in one or more components of the immune system can lead to immunodeficiency diseases caused by impaired or absent immunity of one type or another. Table 33-4 summarizes a few of the genetic immunodeficiencies currently known. A new immunodeficiency disease of unknown cause and high virulence has come to our attention since late 1978. It is called acquired immunodeficiency syndrome or AIDS. As of April, 1984, physicians and health departments in the United States and Puerto Rico had reported a total of 4,087 cases of the disease. Of these patients, 1,758 (43 percent) are known to have died. About a third of the patients contracted a rare form of cancer called Kaposi's sarcoma. The other major disease manifestation was infection by any of several opportunistic pathogens. But by far the most common life-threatening opportunistic infection in AIDS patients was due to the protozoan *Pneumocystis carinii*, which causes a severe pneumonia.

The underlying problem is a defective immune system, which leaves AIDS patients unable to resist infections and malignancies. Most investigators think that AIDS is caused by an infectious agent, possibly a new virus or a new varient of an existing virus. Recently, virologists at the National Institutes of Health in Bethesda, Maryland, have obtained evidence that the etiologic agent is a member of the retrovirus family, HTLV-III (see Chap. 37). The causative virus is possibly carried by sperm or blood.

Most of the patients are homosexual men and bisexual men (71 percent of cases), intravenous drug users (some 17 percent), and persons born in Haiti and now living in the United States (5 percent of cases). Only 6 percent of the known cases are women.

A defective cellular immunity has been implicated as the cause of AIDS. The patients have low lymphocyte counts, often half or less than half of the normal lower limit of about 1500 lymphocytes per milliliter of blood. The T cells are both low in number and abnormal in composition. Apparently the B lymphocytes (and therefore humoral immunity) are not much affected. However, some recent reports indicate a defect in B-cell function as well; that is, the B cells are not able to respond to a variety of stimuli, presumably because of a lack of helper T cells. In addition to T-cell abnormality, AIDS patients may have a reduced population of natural killer cells, which are important in cancer cell surveillance.

The study of AIDS is providing immunologists with information about the normal workings of the immune system and how its breakdown can result in disease, including cancer.

THEORETICAL CONCEPTS

There are several basic questions concerning the functioning of the immune response which are still only partly understood. They deal with the mechanism of the generation of antibody diversity and how the immune response actually operates at the cellular level.

Generation of Antibody Diversity

It has been estimated that a single individual is capable of producing in excess of a million different specificities of antibody in a normal lifetime. If all this information were stored in every cell, this would occupy an excessive amount of its DNA "memory." Recent studies of the genetics of immunoglobulin inheritance and the biochemistry of antibody specificity have produced a more realistic explanation for the generation of antibody diversity.

This explanation is based on the fact that in the genomes of higher organisms the genes occur in pieces, rather than as continuous stretches of DNA. It is the assembly of the individual genes by combining separate segments of DNA that accounts for much of the great diversity of antibodies. Let us examine this situation in a little more detail.

Each light chain of an antibody molecule consists of about 220 amino acid residues. It consists of two regions of approximately equal length: the variable region, which differs from one chain to the next, and the constant region, which is the same for all light chains of the same class. In a similar manner, heavy chains of antibodies (each containing 440 amino acid residues) have a variable region consisting of 110 amino acids; the rest of the molecule constitutes the constant region. It will be recalled that the variable regions of both the heavy and light chains form the antigen-binding site (see Fig. 33-6).

Two gene segments are needed to produce the variable region of the light chain. One segment (designated V) codes for the first 95 or so amino acids of the variable region. A second segment (designated J because its product *joins* the constant and variable regions) codes for the remaining 15 amino acids. A third segment (designated C) codes for the constant region.

The existence of a separate J-region gene segment increases the amount of diversity that can be programmed into the variable region. For example, the mouse genome is estimated to carry about 300 V gene segments and 4 J gene segments. They can combine to produce variable regions for light chains of the κ class. The κ chains constitute about 95 percent of all the light chains of mice and about 60 percent of those of humans; the remainder belong to the λ class. Simple V-J joining can thus generate about 1200 different κ variable regions. This light-chain gene assembly is shown in Fig. 33-12.

Knowledge about the rearrangement of the genes coding for antibody heavy

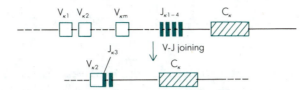

Figure 33-12. Light chain gene assembly. The upper diagram shows the arrangement of κ chain gene segments in the DNA of germ-line cells. During V-J joining, any of the variable gene segments ($V_{\kappa 1}$ through $V_{\kappa m}$) can be connected to any of the joining gene segments ($J_{\kappa 1-4}$), as shown in the lower diagram for the $V_{\kappa 2}$–$J_{\kappa 3}$ combination. The variable region gene thus assembled remains separate from the constant region gene (C_{κ}) until this whole stretch of DNA is copied into RNA transcripts. Then the left end of the variable-region gene is spliced to the right end of the constant-region gene, and the resulting messenger RNA is translated to yield a κ light chain. *(From J. L. Marx, "Antibodies: Getting Their Genes Together," Science, 212:1015–1017, 1981. By permission of the American Association for the Advancement of Science.)*

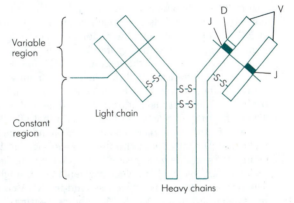

Figure 33-13. Regional structure of the antibody molecule in gene assembly. V, D, and J refer to the variable, diversity, and joining regions of the appropriate chains.

chains has been slower in coming, probably because these proteins are larger than the light chains. However, present evidence shows that heavy-chain genes are also assembled together from separate DNA pieces. Three pieces must be joined together for a complete heavy-chain variable region (unlike the two pieces for the light-chain variable region). A third piece of DNA is inserted between the V and J segments and is designated D for diversity (because it codes for amino acids in one of the most variable sections of the heavy-chain variable region; see Fig. 33-13). Like the V-J joining of light chains, V-D-J joining occurs after an undifferentiated immune cell begins to develop into an antibody-synthesizing cell. The heavy-chain gene assembly is shown in Fig. 33-14. As may be seen, one assembled variable region can be attached to any of five different constant regions. The type of constant region determines the class to which an antibody belongs. For the five classes of immunoglobulin (Ig) M, D, G, A, and E, the corresponding constant-region genes are designated Cμ, Cδ, Cγ, Cα, and Cε, respectively.

The sequence in which the different antibody classes are expressed follows a

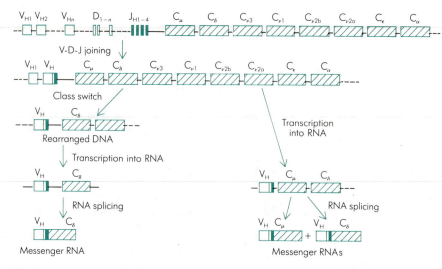

Figure 33-14. Heavy-chain gene assembly. The upper diagram shows the arrangement in germ-line cells of the gene segments for antibody heavy chains. The assembly of a complete variable region gene (V_H) reguires the joining of three gene segments: one of those designated V_{H1} through V_{Hn}, one D, and one J. After V-D-J joining, transcription of the rearranged DNA into RNA, followed by differential splicing of the transcript, allows the simultaneous production of messenger RNAs for two different classes of heavy chain (shown in the right-hand branch of the diagram). The messengers direct the synthesis of either an IgM or IgD heavy chain. The shift to the final class of heavy chain that will be produced by the cell involves a further gene rearrangement (left-hand branch). In this case, the cell will make an IgD heavy chain and the $C\mu$ gene segment has been deleted to bring the completed variable gene nearer to the $C\delta$ gene segment. If the cell were going to make an IgA heavy chain, gene segments $C\mu$ through $C\epsilon$ would have to be deleted. The rearranged DNA is transcribed and the RNA copy spliced to form the final messenger. (From J. L. Marx, "Antibodies: Getting Their Genes Together," Science, **212**:1015–1017, 1981. By permission of the American Association for the Advancement of Science.)

set pattern: the cell first makes IgM, then it may make both an IgM and an IgD at the same time, and finally it becomes committed to producing a single class of heavy chain, which may be any of the five. This necessitates that the completed variable-region gene be switched from one type of constant-region gene to another at the appropriate time during development. That is, as B cells develop, they can replace the $C\mu$ gene with any of the other C genes. This kind of change is termed **class switching** and is effected either at the level of RNA or by gene rearrangement.

As seen in Fig. 33-14, the assorted gene segments that must be rearranged to produce a complete heavy chain are arranged on the chromosome in order of their appearance, with the V, D, and J gene segments followed by the $C\mu$, $C\delta$, $C\gamma$, $C\epsilon$, and $C\alpha$ genes. During the early stages of development, before the antibody-producing cell becomes committed to producing a single type of heavy chain, processing of RNA transcripts is the most important factor in determining the type of chain to be made. For example, after V-D-J joining, a single RNA transcript, which extends from before the beginning of the completed variable-region gene to beyond the end of the $C\delta$ gene, is made. When the noncoding portions of the transcript are excised, the variable-region copy can be attached to either a $C\mu$ or a $C\delta$ gene copy, giving the cell the ability to make an IgM and an IgD at the same time.

In the later stages of development, when the cell becomes committed to production of a particular class of heavy chain to produce, the constant-region genes between the assembled V-D-J region and the desired C gene are apparently removed. This suggests that gene rearrangement is one means of generating the almost limitless diversity of antibodies.

Theories on Antibody Diversity

Historically, three major competing theories were formulated to explain this antibody diversity:

1 The germ-line theory postulated that there was a separate variable gene for every possible antibody chain.
2 The somatic mutation theory held that there need be only one or a few such genes; diversity was brought about by mutations of these genes.
3 The minigene theory held that the genes were assembled from the shuffling of small gene segments in varying combinations.

Present evidence indicates that each of these theories contained some elements of truth. There are at least a few hundred different V gene segments for both light and heavy chains (germ-line theory). Their diversity can be further increased by V-J or V-D-J joining (minigene theory). Mutations also contribute to antibody diversity because there is more variation in the protein structures of the variable regions than can be explained by the genes (somatic mutation theory).

Cooperation and Regulation

As previously stated, many B cells require T-cell help to produce an antibody response to the majority of antigens (viruses, parasites, bacteria, allergens, and transplantation antigens). These antigens are termed thymus-dependent antigens and serve to illustrate how T cells cooperate with B cells by providing a second stimulus to trigger antibody synthesis by B cells. The first stimulus is of course specific antigen. A few antigens are directly *mitogenic* (stimulate cell division) for B cells and do not require helper T cells (T_H), and they are therefore called thymus-independent antigens (e.g., cell-wall lipopolysaccharides of bacteria).

Once an immune response has been initiated, the primary regulator is also specific antigen. Once it has been masked or covered by the specific antibody and removed and degraded by macrophages, the primary stimulus has been eliminated and the immune response stops. In addition, a group of regulatory T cells, called suppressor T cells (T_S), actively turns off the immune response by directly inhibiting the T_H cells. These types of controls are necessary to prevent the generation of uncontrolled immune responses to foreign and autologous (self) substances. The occurrence of *autoimmune diseases*, such as rheumatoid arthritis and hemolytic anemia, serves as evidence that the normal immunologic *tolerance* of self can break down.

HYPERSENSITIVITY

The result of an immune response is that the individual has an increased state of reactivity to the specific antigen or pathogen. Situations in which host immune responses contribute to tissue injury are collectively referred to as **hypersensitivity states.** Hypersensitivity reactions can be conveniently divided into four major types as advocated by two British immunologists, Philip Gel and Robin Coombs.

Type I: Immediate Hypersensitivity Reactions

Type I hypersensitivity is mediated by IgE antibodies which have bound to the Fc receptors of circulating basophils and tissue mast cells, thereby sensitizing them. When antigen contacts the sensitized cells, an *immediate* (within 1 to 10 min) skin reaction occurs leading to intense local inflammation typified by diffuse infiltrations by polymorphonuclear leukocytes, which results in a soft, swollen, red skin reaction. For this reason it is thought to be important in effecting resistance to pathogens.

The chemical basis for this reaction is the release by the sensitized basophils and mast cells of a number of preformed substances (histamine, serotonin, heparin, and chemotactic factors) and unstored (synthesized de novo) mediators (e.g., slow reacting substance of anaphylaxis, leukotactic activity substance, and platelet-activating factors). Besides an immediate skin reaction, the biological effects of these substances or mediators account for a more generalized clinical picture because certain target organs (smooth muscle, vessel walls, etc.) are affected. So-called **anaphylactic reactions** are due to this type of hypersensitivity reaction and are characterized by severe symptoms, including irritation, rash, swelling, wheezing, shock, and occasionally death. Allergic reactions to insect bites (by mosquitoes or wasps) are also due to this type of hypersensitivity reaction.

Type II: Antibody-Dependent Cytotoxic Hypersensitivity Reactions

Type II hypersensitivity is mediated by IgM and IgG antibodies reacting with cellular or particulate antigens. This **antibody-dependent cytotoxic** hypersensitivity results in complement-mediated cytolysis, antibody-dependent cell-mediated cytotoxicity (ADCC), or the opsonization and increased phagocytosis of particles sensitized by antibody or fixed complement. The latter process is particularly effective against bacteria and free viruses in a systemic infection and provides protection against infectious agents. Unfortunately, type II hypersensitivity can also damage the host because of the circulating antibodies. Thus it also is responsible for clinical transfusion reactions (mismatching), erythroblastosis fetalis (Rh isoimmune reactions), and some autoimmune reactions (occurring secondarily to certain infections and diseases). Extrinsic antigens, e.g., penicillin, may bind to red cell surfaces, where they act like haptens and cause the formation of antibodies. These antibodies are capable of fixing complement and lysing the red blood cells (drug-induced hemolytic anemias).

Type II hypersensitivity reactions differ in three ways from type I reactions:

1 Antibody combines with antigenic determinants on the cell surface through its Fab portion before any Fc interactions occur.
2 Interaction of antibody with the target cell directly results in cell death.
3 IgE is not involved.

Type III: Immune-Complex-Mediated Hypersensitivity Reactions

Type III hypersensitivity reactions are also mediated by IgG and IgM antibodies which react with soluble antigens. This causes the formation of circulating immune complexes (IC) which can cause widespread inflammatory responses called **serum sickness** (IC formed intravascularly and deposited at distant sites) or intense local inflammatory responses in the peripheral tissues (IC formed in situ) called an **Arthus reaction** after its original investigator. Bacterial toxins and free viral proteins are presumably the intended targets, but this type of hypersensitivity can cause severe organ dysfunction such as kidney failure, rheumatoid arthritis, and types of toxic shock syndrome.

Type IV: Cell-Mediated Hypersensitivity Reactions

Type IV hypersensitivity is mediated by sensitized T lymphocytes which can cause direct target cell–mediated lysis and the release of soluble lymphokines. Since these cells arrive at a skin site containing antigen 24 to 72 h after infection, producing a typical hard swelling, this type of reaction is called a **delayed reaction.** This contrasts with the 4- to 24-h time course of the Arthus reaction. (Types I, II, and III hypersensitivity reactions have historically been described as immediate hypersensitivity reactions. In addition, these reactions are dependent on the activity of antibody for their pathological effects.) Upon histological examination the delayed reaction site is densely infiltrated by mononuclear cells (lymphocytes, monocytes). The cytotoxic T lymphocytes are important in killing cells infected with intracellular parasites such as viruses, bacteria, and trypanosomes as well as fungi and perhaps cancer cells. On the other hand, T-cell-mediated hypersensitivity is responsible for accelerated graft reactions and allergic skin reactions caused by contact with chemical or metal allergens (e.g., nickel) termed **contact sensitivity.** The cutaneous delayed-type tuberculin reaction (sensitivity test for tuberculosis using partially purified protein derivative of *Mycobacterium tuberculosis*) is the prototype for cell-mediated immunity. Thus nonsensitized individuals (with no prior exposure to the organism) give no reaction.

QUESTIONS

1 What are the two types of effector products of the immune system? Describe the overall division of labor between the two types of effectors with respect to pathogens and transformed (cancer) cells.

2 Explain the four primary characteristics of a generalized immune response.

3 Explain by means of diagrams the significance of (a) anamnestic response, (b) use of adjuvants.

4 Identify the following terms with respect to antigens: (a) specificity, (b) immunogenicity, (c) antigenicity, (d) haptens, (e) carrier, (f) complete antigen.

5 Distinguish between the following pairs of items related to antigens:
   (a) Toxins and toxoids
   (b) H and O antigens of *Salmonella*
   (c) Salk vaccine and Sabin vaccine for poliomyelitis

6 Why are antibodies termed *immunoglobulins*? How do they behave in an electrophoretic field?

7 Provide an account of the general structure of an immunoglobulin molecule giving reference to: variable region, constant region, Fab, and Fc.

8 Compare and contrast the structure and biological functions of IgM and IgE.

9 Write a brief account of the general properties of IgG.

10 Identify the special functional names of antibodies when they are allowed to act on certain types of antigens.

11 What is a monoclonal antibody, and how is it produced?

12 What are the useful properties of hybridomas? How are they developed?

13 Compare and contrast cell-mediated immunity and humoral immunity with respect to their activity.

14 Compare the derivation of the B cells and T cells.

15 Briefly describe the B cells and their production of antibodies.

16 What is the basis of immunologic memory?

17 Describe the three types of functions ascribed to T lymphocytes.

18 What kind of disease is AIDS? Do we know its etiology?

19 Explain the means by which an individual generates the great diversity of antibodies produced in a normal lifetime.

20 What were the theories formulated over the years to explain the immense repertoire of antibodies produced during an individual's lifetime? Are they still sound? Explain.

21 Describe briefly cooperation and regulation between T and B cells.

22 List the four major types of hypersensitivity reactions.

23 Which type of hypersensitivity reaction causes (a) anaphylactic shock, (b) opsonization, (c) transfusion reactions, (d) allergy to penicillin, (e) toxic shock syndrome, (f) tuberculin reaction?

REFERENCES

The references cited in Chap. 32 are also applicable to this chapter.

Eisen, H. N.: "Immunology," in B. D. Davis, R. Dulbecco, H. N. Eisen, and H. S. Ginsberg, *Microbiology*, 3d ed., Harper and Row, Hagerstown, Md., 1980. *A comprehensive coverage of immunology at the advanced level.*

Golub, E. S.: *The Cellular Basis of the Immune Response*, 2d ed., Sinauer Assoc., Sunderland, Mass., 1981. *An advanced book on cellular cooperation and cell-mediated immune responses.*

Chapter 34

Assays and Applications of the Immune Response

As discussed in the previous chapter, the induction of a specific immune response usually results in the production of specific effector cells (cell-mediated immunity) and antibodies (humoral immunity). The detection and quantitation of a specific immune response to a pathogen usually indicates that the patient has had an infection at some time within the past year, and is now immune and no longer at risk. If two blood specimens are taken at 1- to 2-week intervals during a current or recent infection, a rising quantity of antibody in the serum from the first to the second bleeding (called a rising antibody titer), specific for the pathogen, indicates that the pathogen is, in all probability, the causative or etiologic agent. The total lack of any immune antibodies or of a positive reaction to a test dose of the specific antigen indicates a susceptible individual who should be immunized. The above shows the usefulness of immunoassays, and further applications of them will be discussed in this chapter.

Generally, unknown pathogens are identified using antibodies of known specificity, and serodiagnosis is made by reacting patient sera with antigens from known pathogens. When antibodies combine specifically with antigens, usually an immune complex is formed which can be seen or detected using an appropriate technique.

This chapter gives an overall view of assays and applications encountered in the diagnosis of infectious disease and is not intended to be a step-by-step methods manual. For such information, the student should refer to the *Manual of Clinical Immunology* (cited in References).

MEASUREMENT OF HUMORAL ANTIBODIES

Since antibodies alone cannot be seen by the naked eye, we must detect them by what they do in the presence of the specific antigen or pathogen. Many of the in vitro tests depend upon the formation of a visible reaction endpoint by the cross-linking of antigen and antibody to form a large complex.

The Lattice Hypothesis

The most favored model to explain antigen-antibody interactions to form large visible complexes is the lattice hypothesis (Fig. 34-1). In chemical terms, antibody (Ab) usually has a very high binding affinity for the specific antigen (Ag) and can be expressed by the equation shown at the top of the next page.

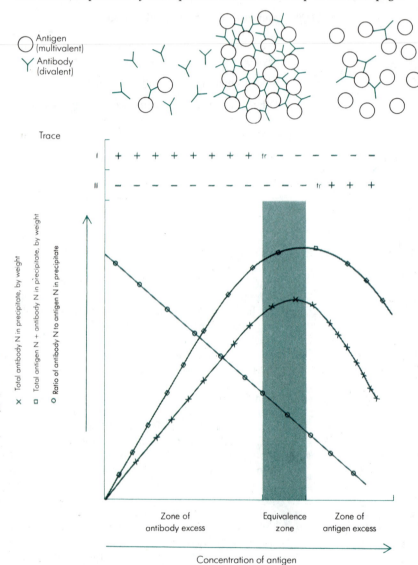

Figure 34-1. Graph and schematic illustration of the events of the precipitin reaction. In this reaction, increasing concentrations of antigen are added to a series of serological tubes containing a constant concentration of antibody (antiserum). The concentration of antigen and antibody present in any tube is determined by assaying for nitrogen (N), a component of both the antigen and antibody. I and II refer to tests for excess antibody and excess antigen, respectively, in the fluid remaining after removal of the precipitate ($+$ = positive; tr = trace; $-$ = negative).

$$_{v1}(Ag) + {}_{v2}(Ab) \rightleftharpoons Ag_{v1}:Ab_{v2}$$

where v_1 = valence of antibody and v_2 = valence of antigen.

The equilibrium constant K is therefore

$$K = \frac{[Ag:Ab]}{[Ag][Ab]}$$

where the brackets indicate the actual concentrations of the reactants and product in moles per liter. As can be seen, the amount of complex formed increases with the K value. In order for the combination of antigen and antibody to form a visible complex their respective valences must be greater than 1. We already know from Chap. 33 that antibodies are bivalent even in their monomeric form (e.g., IgG). However, if the antigen is a monovalent hapten, then two moles of hapten would combine with one mole of antibody forming soluble complexes with the formula $Ag_2:Ab_1$, which would be too small to see. If the antigen were polyvalent (has many antigenic determinants), the antibody now has the opportunity of cross-linking molecules of antigen to form lattices. If these become large enough, then they precipitate out of solution and flocculate or agglutinate particles to form a visible reaction (Fig. 34-1).

To form the greatest amount of precipitate or agglutinate in an immune reaction it is necessary that antigen and antibody be present in optimal or equivalence amounts such that there is just enough antigen to saturate the antibody-combining sites. This is called the **equivalence zone** and can be used to quantitate antibody or antigen. An excess of antigen or antibody will result in the formation of smaller, and probably invisible, soluble immune complexes due to a lack of extensive lattice formation (Fig. 34-1). Some serological tests based on antigen-antibody reactions are shown in Fig. 34-2.

Precipitin Tests

In the **precipitin test** a reaction takes place between a soluble antigen and a solution of its homologous antibody. The reaction is manifested by the formation of a visible precipitate at the interface of the reactants but may be inhibited by an excess of either antigen or antibody. The most useful precipitin tests therefore provide for diffusion of the reactants until optimum concentration is reached (see Fig. 34-1). This equivalence zone represents the concentrations of antigen and antibody where complete precipitation occurs. Culture filtrates or animal sera contain a number of different antigens. If they are used for immunization, they give rise to a correspondingly large number of different antibodies. Unless special precaution is taken to remove all but one precipitin from the solution, several layers of precipitate are formed—one for each precipitin system present. The formation of several precipitate layers is not necessarily a disadvantage since these are used to identify and distinguish between different antigens and antibodies in a solution. Factors which influence precipitin tests are electrolytes, pH, temperature, and time.

Precipitin tests are not limited to the laboratory diagnosis of bacterial infections of humans and other animals by means of specific antigens but are also useful in many other ways, such as (1) serologic screening tests for syphilis where a nonspecific antigen not related to the spirochete is usually used; (2) serological identification of various pathogens; (3) identification of blood or

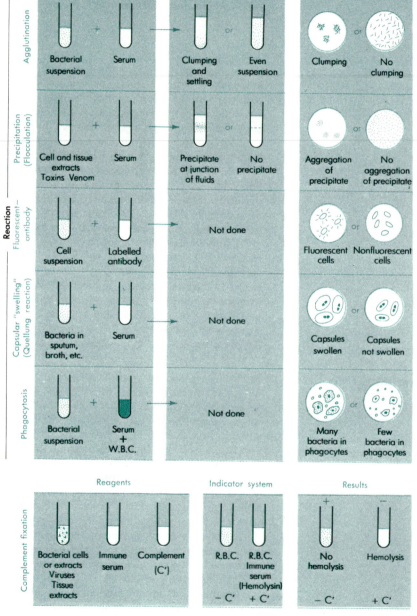

Figure 34-2. Some serological reactions.

seminal fluid in stains on clothing, weapons, or other exhibits for medicolegal purposes; (4) postmortem diagnosis of anthrax from tissue of a dead or decomposed animal by the Ascoli test; (5) determination of the kind of animal a mosquito has recently fed on (this information helps entomologists and epidemiologists to prevent the spread of arthropod-borne diseases); and (6) detection of the adulteration of foods.

Antigen for precipitin tests is prepared by making an extract from bacterial cells, tissues, or other suitable material. For example, in the Ascoli test for anthrax, a small amount of infected spleen is boiled in physiological salt solution. The test consists in layering a milliliter or so of the extract over a similar amount of antiserum prepared by injecting a rabbit with killed anthrax bacilli.

Precipitating antigens for streptococcus typing and diagnosis of certain other bacterial infections are made from 24- to 48-h cultures that are extracted to yield a clear solution of antigenic precipitable material.

One- and Two-Dimensional Immunodiffusion

A tube precipitin test can be performed in a fluid phase. This is called a **ring test** because the precipitate forms a ring or white disk at the interface between the clear solution of antigen and antibody (Fig. 34-3A). For greater accuracy,

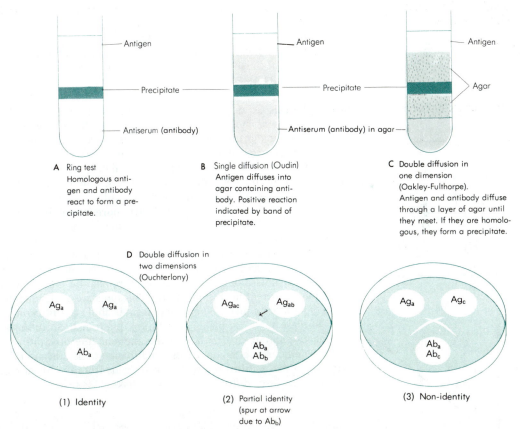

Figure 34-3. Qualitative precipitin tests (Ag = antigen; Ab = antibody).

Antigen and antibody are placed in separate wells cut into agar in a Petri dish. Homologous reactants produce a line of precipitate where they meet in the agar. Like antigens produce bands which meet exactly as in (1). Unlike antigens produce bands which cross as in (2) and (3).

ease of handling, and preservation of the tests the antibody can be contained in agar gel and the antigen allowed to diffuse into it. This is called a **single-diffusion** method. The method of Oudin directly layers antigen over antibody in agar (Fig. 34-3B), and the Oakley-Fulthorpe method separates these two reagents with simple agar (Fig. 34-3C). This results in the diffusion of both reagents into the agar and the formation of precipitin ring(s) with greater accuracy. This method is described as a **double-diffusion** precipitin test.

The simple method for double diffusion in two dimensions developed by Ouchterlony allows the comparison of various antigens and antisera for specific identities. The antigen and antibody diffuse from wells cut in a layer of agar to form a precipitate at the equivalence zone point. By cutting the wells in the agar in two-dimensional patterns, several antigens can be compared. A fusion of immune precipitates of an antibody preparation with antigens from two separate wells indicates identity of the two antigens, whereas the crossing of precipitates shows a lack of identity (Fig. 34-3D).

Single Radial Immuno-diffusion in Agar

Whereas the former precipitin tests described are qualitative, antigens can also be quantitated. The Mancini method allows antigen to diffuse from a well into a layer of agar which contains specific antibodies. Thus, a ring of precipitate forms, the diameter of which is proportional to the antigen concentration. This is a simple but sensitive method for quantitating antigens to concentrations of 1 µg/ml (Fig. 34-4).

Immunoelectrophoresis

Electrophoresis is an electrochemical process in which colloidal (suspended) particles or macromolecules with a net electric charge migrate in a solution or agar gel under the influence of an electric current. A characteristic of living cells in suspension and biological compounds (such as protein antigens) in solution or in a gel is that in an electric field they travel to the positive or negative electrode, depending on the charge on the substances. Positively charged substances travel to the cathode, while negatively charged ones go to the anode; this movement is called **electrophoretic mobility.** When electrophoresis is applied to the study of antigen-antibody reactions, it is called **immunoelectrophoresis.**

When a fluid containing protein antigens is placed in a well in a thin layer of buffered agar, and an electric current is applied, antigens will be distributed in separate spots along a line passing through the well and parallel to the direction of current flow (Fig. 34-5A). The spots are analogous to antigen wells in an Ouchterlony plate. When the current is shut off, diffusion will begin from each of these spots. By placing antiserum in a trench cut in the agar parallel to the electrophoretic distribution of the antigens, the precipitin reaction can be used to demonstrate the nature of the diffusing molecules. In this case, a broad band of antibody diffuses toward the antigens from the linear antibody trench, while the antigens diffuse as expanding disks. This results in a complex pattern

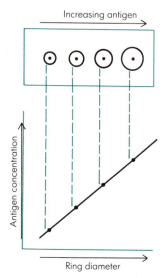

Figure 34-4. Single radial immunodiffusion. The rings, initially faint, become heavier as diffusion nears completion. At equilibrium, the diameter of each ring is directly proportional to the concentration of monospecific antibodies.

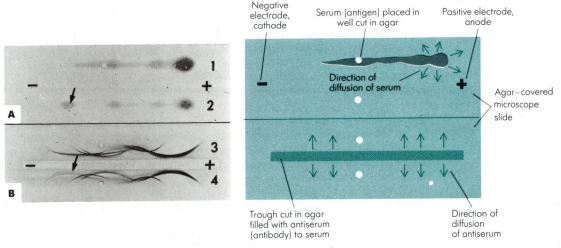

Figure 34-5. Left: Agar-gel electrophoresis (A) and immunoelectrophoresis (B) of sera from normal mice and from mice bearing tumors. 1. Normal mouse serum. 2. Tumor mouse serum. 3. Normal mouse serum. 4. Tumor mouse serum. The precipitin bands were developed with antimouse serum prepared from rabbits inoculated with mouse serum. Those bands corresponding to the tumor protein are indicated with arrows. *(Courtesy of R. A. Murgita, McGill University.)* Right: Explanatory sketches for photographs (A) and (B).

of arc-shaped zones of precipitate (Fig. 34-5B). When the antigen molecules are from human serum, there may be 20 or more such zones, each of which represents at least one distinct precipitin system. Precipitin specificity provides a far more critical indicator of the identity of protein molecules than the physical characteristics which control the movement of the molecules in an electrophoretic field.

Rocket Immuno-electrophoresis

A combination of the immunoelectrophoresis assay and the Mancini assay involves the electrophoretic migration of antigen from wells into an agar gel which contains specific antiserum. This results in the formation of a rocket-shaped precipitate. The height of each "rocket" is proportional to the concentration of antigen in the well. The sensitivity of this assay allows the measurement of antigens in concentrations as low as 0.5 µg/ml. (See Fig. 34-6).

Precipitin Test for Streptococcus Grouping and Typing

Differentiation of the streptococci into serological groups and according to types within some of the groups is accomplished best by a precipitin test. It is an example of how a serological reaction is used for the identification of microorganisms. The serological groups A, B, C, D, E, F, G, H, and K to O of streptococci have been identified on the basis of the carbohydrate antigen known as C substance. Types are determined by the presence of M substance, which is also on the cell wall of the streptococci.

Streptococcus-precipitating antiserum is obtained by immunizing rabbits with heat-killed suspensions of the bacteria or extracts containing both group-specific C and type-specific M substances.

The test is performed by layering a small amount of type-specific antiserum and antigen in a capillary tube (ring test). If the test is positive, a cloudy ring or disk appears at the junction of serum and antigen in a few minutes. In practice, the group is determined, and if it is group A, the antigen is tested against type-specific antisera for type determination.

Agglutination Tests

The agglutination reaction is relatively easy to perform, is simple to interpret, and is the serological method of choice when a suitable cellular antigen can be prepared. Among the human diseases for which the agglutination test is of diagnostic value are typhoid fever, salmonellosis, brucellosis, tularemia, typhus fever, and Rocky Mountain spotted fever. Agglutination tests are useful in the diagnosis of many animal diseases such as brucellosis in cattle and goats, glanders in horses, swine erysipelas, and pullorum disease of chickens.

Not only can the diagnosis of certain infectious diseases be confirmed in the laboratory by agglutination of known antigens by the patient's serum, but unknown cultures of bacteria or other microorganisms can be identified by the capacity of a known antiserum to agglutinate the suspension of unknown cells.

The **Widal test** was devised specifically to aid in diagnosis of typhoid fever by agglutinating typhoid bacilli with patients' serum, but the term is sometimes loosely applied to other agglutination tests using heat-killed cultures of organisms other than *Salmonella typhi*.

Another agglutination test used in laboratory diagnosis is the **Weil-Felix reaction.** This test is based on the fact that several of the rickettsias have a common antigen with strains of *Proteus* spp.; therefore serum from patients with rickettsial infections agglutinates suspensions of the *Proteus* organisms. The strains of *Proteus* used most commonly are *Proteus* OX-19, OX-2, and OX-K. The Weil-Felix reaction is differential for certain rickettsial diseases because of the selective agglutination of these strains.

Agglutination tests are classified as **macroscopic** when the test is carried out in small test tubes referred to as agglutination tubes and **microscopic** when antigen and antiserum are mixed on a slide and examined under a microscope (see Fig. 34-2). Measuring the agglutination of a particulate antigen by its specific antibody (direct agglutination) is the simplest way to estimate the quantity of that antibody in serum. Very small amounts of antibody can be detected by this method, since only a small number of antibody molecules is necessary for the formation of the antigen-antibody lattice. The agglutination reaction has been extended to include a wide variety of antigens by attaching soluble antigens to the surface of inert particles, such as bentonite, latex, or red blood cells. The role of these particles, once coated, is a passive one; i.e., they react as if they themselves possessed the antigenic specificity of the coating antigen.

Red blood cells (erythrocytes) have been found to be extremely convenient carriers of antigen. When specific antibody is added to antigen-coated red blood cells, antibody bridges are formed between neighboring erythrocytes, and large aggregates of erythrocytes are produced which are visible to the naked eye. This agglutination is designated **passive** or **indirect hemagglutination.**

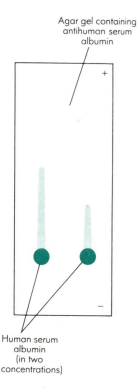

Agar gel containing antihuman serum albumin

Human serum albumin (in two concentrations)

Figure 34-6. Rocket immunoelectrophoresis of human serum albumin into agar containing anti-human serum albumin. At equilibrium, the height of the rocket is proportional to the antigen concentration.

Figure 34-7. Agglutination occurs when bacterial cell antigens are exposed to homologous antibodies. (Note that antibody size is not depicted to scale.)

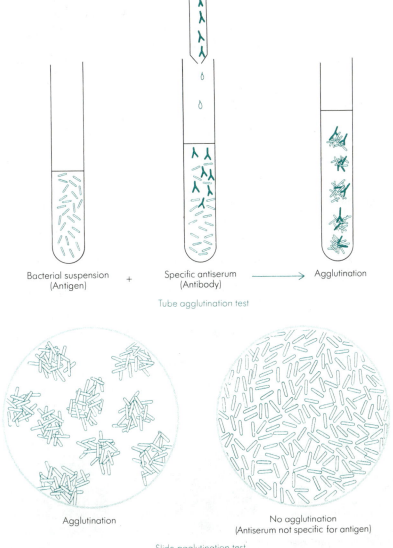

Bacterial suspension   +   Specific antiserum   ⟶   Agglutination
(Antigen)   (Antibody)

Tube agglutination test

Agglutination

No agglutination
(Antiserum not specific for antigen)

Slide agglutination test

Tube Agglutination

The most widely used agglutination test is the macroscopic tube agglutination test (Fig. 34-7). By this method not only can the presence of specific agglutinins in serum be detected, but their approximate concentration can be determined because the antiserum can be diluted serially in a set of test tubes and the antigen (suspension of bacterial cells) mixed with it.

Antigens for agglutination tests are made by preparing a homogeneous suspension of the organisms taken from a young culture in physiological saline

(0.85% NaCl solution). The salt used for dilutions provides an electrolyte without which agglutination will not occur even though the antibodies are specific for the antigen used. A concentrated suspension of the organisms can be stored and diluted for use as required. A standardized suspension of antigen for use in the macroscopic tube agglutination test is made by diluting the cell suspension with physiological salt solution.

The serum is diluted serially with physiological salt solution in several small test tubes. To each tube is added a constant amount of bacterial cell suspension. A control tube contains no serum.

After incubation, bacterial cells (antigen) in the control tubes should not be clumped. If they are, the results of agglutination in other tubes are not reliable. Agglutination in tubes with small amounts of serum indicates a higher titer, i.e., high concentration of antibody in the serum.

Macroscopic Slide Agglutination Test

This is a quick and convenient method for determining the presence of agglutinating antibodies. A drop of a dense suspension of the organisms (antigen) in saline is placed on a clean glass slide. One to three drops of the patient's serum are mixed with the antigen by gentle rotation. Positive agglutination, observable with the unaided eye, occurs in 3 to 5 min.

The test can be made semiquantitative by using exact amounts of antigen and antibody. Graduated amounts of antibody are placed in separate spots on a slide. If a constant amount of antigen is added to each, a range of dilutions results. Positive and negative controls must be set up on separate slides. Since this is a screening test, doubtful or positive reactions should be confirmed.

Microscopic Slide Agglutination Test

For the microscopic agglutination test, serial dilutions of serum in physiological saline are prepared, one loopful of each dilution is placed on a cover glass, and to each a loopful of antigen, which may be a young, living broth culture of bacteria, is added. For the control, one loopful of salt solution (containing no

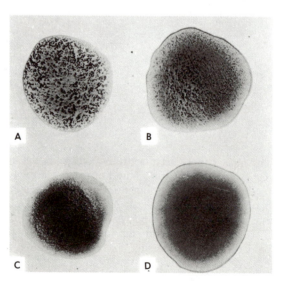

Figure 34-8. Agglutination tests performed with whole blood are particularly useful in testing for certain animal diseases in the field. Agglutination of *Salmonella pullorum* antigen with a drop of whole blood taken from chickens gives prompt reactions and makes it possible to remove infected birds from the flock without delay. (A) Strongly positive, (B) positive, (C) weakly positive, and (D) negative reactions. *(Courtesy of U.S. Department of Agriculture.)*

serum) and a loopful of antigen are used. Each cover glass is placed over a concave slide ringed with petrolatum to prevent drying, incubated for 1 h, and then examined with the high-dry objective of the microscope for clumping (Fig. 34-7). This method can also be used for rapid identification of unknown cultures of bacteria using known antisera.

Whole-Blood Agglutination Tests

Agglutination tests with whole blood are used in pullorum-disease eradication programs and for the diagnosis of swine erysipelas. The test is similar to the rapid serum test except that whole blood is used rather than serum.

The test for pullorum disease is made by mixing a loopful of freshly drawn blood with a drop of the antigen on a glass plate with a white background. Positive serum from infected birds causes clumping within 2 min (Fig. 34-8). It can be carried out quickly and simply in the field.

A positive test with *Erysipelothrix rhusiopathiae* antigen may indicate either infection or contact with the causative organisms of swine erysipelas.

Agglutinin-Adsorption Tests

Related organisms such as those of the genus *Rickettsia* and the organisms in the genus *Salmonella* or the genus *Shigella* may contain common antigens that cause several species or strains to agglutinate with antiserum prepared from a specific species or variety of the group. To illustrate, suppose that antibodies are produced against a given organism X, and by analysis we know that X antigen is composed of fractions A, B, C, and D. The antiserum would then contain corresponding antibodies a, b, c, and d and would agglutinate X antigen at high titer. If, however, organism Y contains antigen fractions A, F, G, and H, because of the common antigen A, it, too, would be agglutinated by antiserum specific for organism X but usually at a lower titer than X, since only fraction A is common. This cross agglutination can be eliminated by adsorbing out a antibodies using Y antigen, i.e., by incubating Y antigen (organism) with the X antiserum. After removal of the Y cells the antiserum would no longer agglutinate organism Y since a antibodies had been adsorbed out, but it would still agglutinate X cells. Unless the phenomenon of cross agglutination is understood, this reaction leads to misinterpretation of results of routine agglutination tests. Proof of cross agglutination can be obtained by agglutinin-adsorption tests, and the results of these tests are valuable aids in the identification of bacteria. By the agglutinin-adsorption test, the antigenic pattern of closely related organisms such as the salmonellas can be determined.

Hemagglutination Test

Certain viruses have the ability to agglutinate red blood cells from certain species of animals, notably chicken, guinea pig, or human type O red blood cells. This is due to an interaction in which the virus is adsorbed on the erythrocytes and then the blood cell–virus aggregates agglutinate (the viral attachment protein cross-links erythrocytes). Because this aggregation dissociates spontaneously, the test must be read in 1 h or less. The amount or intensity of the agglutination depends on the concentration and infective titer of the virus. The procedure consists in making serial dilutions of the virus antigen; then the red blood cell suspension is added to each tube, and the tubes are incubated for 1 h. Agglutination is indicated by an even coating of red blood cells in the bottom of a tube.

In a negative reaction (no agglutination) the red cells merely settle to form a compact button in the bottom of the tubes.

The exact technique for conducting a hemagglutination test varies with the virus, i.e., antigen. This test is useful for serological diagnosis of many viruses including the influenza virus, mumps, and the ECHO viruses. Vaccinia and variola viruses, those of the coxsackie and encephalitides groups, and several viral pathogens of domestic animals also agglutinate erythrocytes.

Hemagglutination Inhibition Test

During the course of many infectious diseases the etiologic agent stimulates the body to produce antibodies. This forms the basis for the use of serological tests used for diagnosis. For example, if two or more serum specimens are taken from the patient at intervals of 10 to 14 days during his or her illness and a serological test reveals an increasing antibody titer, this is an indication that the infective agent is serologically related to—or is the same as—the antigen used in the test. In the case of the influenza or other viruses which cause hemagglutination, an increase in the antibody which inhibits this agglutination (neutralizes the viral hemagglutin) indicates the presence of the virus of the type of antigen employed.

The hemagglutination inhibition test can also be used to identify virus strains isolated from a patient. This requires rather complex techniques for the preparation of specimens and reagents and for excluding false reactions.

Complement-Fixation Tests

Complement-fixation tests are based on the presence of complement-fixing antibodies in serum (see Fig. 34-2). These antibodies are produced in an animal body stimulated by bacterial antigens. In the presence of the antibodies, complement causes lysis of the specific bacterial cells. The purpose of the complement-fixation test is to determine whether specific bacterial antibodies are present in serum. In the actual test, two systems are involved. One is the bacteriolytic or complement-fixing system in which serum, bacterial suspension (or other antigen), and complement are mixed. If the antigen and antibody in the serum are capable of union, the complement is said to be fixed. The second system is simply an indicator system in the test. Rabbit antibodies against sheep red blood cells are added together with sheep red blood cells. If complement is available, the red blood cells are lysed. If, however, the complement is fixed by being used in the reaction between the bacteriolytic antibody and antigen, no hemolysis will occur. Therefore, a hemolytic reaction indicates a *negative* test. Obviously, all reactants in the complement-fixation test must be accurately adjusted.

The Wassermann Test for Syphilis

The complement-fixation test is widely used in the laboratory diagnosis of many infectious diseases, including those of bacterial, viral, protozoan, and fungal etiology. It is also used for the identification of many microorganisms. One of the best-known applications of the complement-fixation test is the Wassermann test for syphilis, which differs from many other serological reactions in that the antigen is not prepared from the causative organism of the disease but is made by extracting beef-heart powder with ether and alcohol. This extract contains a complex phospholipid called cardiolipin. Cholesterol and lecithin are added to the extract to increase the sensitivity and specificity of the antigen. Since the

Positive complement fixation test

Step 1   Step 2

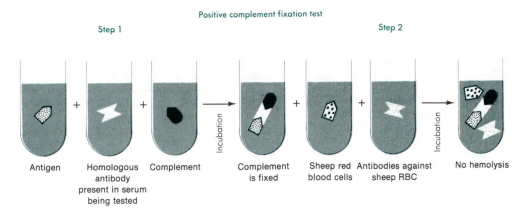

| Antigen | Homologous antibody present in serum being tested | Complement | | Complement is fixed | Sheep red blood cells | Antibodies against sheep RBC | | No hemolysis |

Negative complement fixation test

Step 1   Step 2

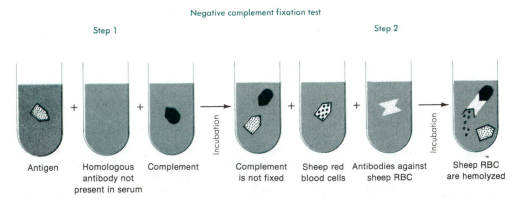

| Antigen | Homologous antibody not present in serum | Complement | | Complement is not fixed | Sheep red blood cells | Antibodies against sheep RBC | | Sheep RBC are hemolyzed |

Figure 34-9. The complement-fixation test is based on the fact that if complement, which is found in blood, is "fixed" by reacting with antigen and its homologous antibody, lysis of sheep red blood cells does not occur even though antibodies against the sheep red blood cells are present.

material is not capable of inducing antibody formation in an animal, it is not a true antigen; but because it combines with syphilitic "antibodies" in the complement-fixation test, it is referred to as a nonspecific antigen.

The principle for all complement-fixation tests is shown in Fig. 34-9.

The Wassermann and other tests for the serodiagnosis of syphilis which employ nonspecific cardiolipin antigens were in use for many years before it was possible to prepare an antigen from the specific organism. Since techniques for cultivating virulent *Treponema pallidum* in vivo have been developed, several new tests have been introduced. One is the ***Treponema pallidum* complement-fixation (TPCF) test,** in which the antigen is an extract of the virulent treponemes. This antigen can be standardized and stored for several months. The principle is the same as that of other complement-fixation tests, but the actual techniques differ somewhat.

Radioimmunoassays (RIAs)

The sensitivity with which antigens can be detected has been increased by the use of radiolabeled (radioactive-labeled) antibodies or pure antigens. The method is also rapid and precise and has an enormous diversity of applications. The direct method employs labeled antibody which can be precipitated with the antigen in immune complexes. In the Farr test, by labeling the antigen, the amount of radioactivity in an antigen-antibody precipitate measures the antigen-binding capacity of the immune sera. A further development often used in quantitating hormone levels is the inhibition assay. In this type of radioimmunoassay the clinical specimen containing an unknown quantity of hormone is reacted with a known quantity of antibody. Labeled pure hormone is then added to the reaction, and its binding is inhibited in direct proportion to the antibody sites occupied by the unknown hormone. In these tests the isotope used to label the proteins is usually ^{125}I, and the final measurement of the radioactive label is made using a solid (gamma) scintillation counter. RIAs are used to detect human chorionic gonadotropin in the serum of women as a test for pregnancy.

Enzyme-Linked Immunosorbent Assays (ELISA)

It is now possible to couple enzymes to antibodies or antigens for use in immunoassays. With the increasing concern over environmental contamination with radioisotopes, all the above radioimmunoassays can be performed using enzyme-labeled reagents. In addition to being as sensitive as radioimmunoassays (antigens and antibodies detectable at levels of about 10^{-10} g/ml or 1 part in 10 billion), enzyme immunoassays are less expensive, safer, and just as reliable and accurate as radioimmunoassays.

The principle of ELISA is based on these two observations:

1 Antibodies and some antigens can attach to polystyrene plastic plates (or other solid-phase supports) and still maintain their full immunologic capabilities.
2 Antigens and antibodies can be bonded to enzymes, and the resulting complexes are still fully functional both immunologically and enzymatically.

It is the enzyme activity which is the measure of the quantity of antigen or antibody present in the test sample. Enzymes used in ELISA include β-galactosidase, glucose oxidase, peroxidase, and alkaline phosphatase.

There are two methods of enzyme immunoassay that have significant clinical value: the double-antibody-sandwich procedure for the detection and measurement of antigen and the indirect-microplate-ELISA procedure for the detection and measurement of antibody.

Double-Antibody-Sandwich Procedure

In this technique (Fig. 34-10A), the wells or depressions in a polystyrene plate receive antiserum. The antibodies in the antiserum adhere to the surface of each well. The test antigen is added, and if the antigen is homologous, it attaches to the antibody immobilized on the well surface. Enzyme-labeled specific antibody is then added; it will bind to the antigen already fixed by the first antibody. This results in an antibody (with enzyme)-antigen-antibody "sandwich." Finally, the enzyme substrate is introduced for reaction with the enzyme. The rate of enzyme action is directly proportional to the quantity of enzyme-labeled antibody present, and that, in turn, is proportional to the amount of test antigen. Enzyme activity may be followed by a color change (brought about by substrate hydrolysis) which can be inspected visually or measured by means of a colorimeter (an instrument used to analyze color changes in a solution). This method has been used to assay hepatitis B antigen with great success.

Indirect ELISA Procedure

The initial step involves the coating of polystyrene wells with *antigen* by passive adsorption. (See Fig. 34-10B.) Test antiserum is added and allowed to incubate.

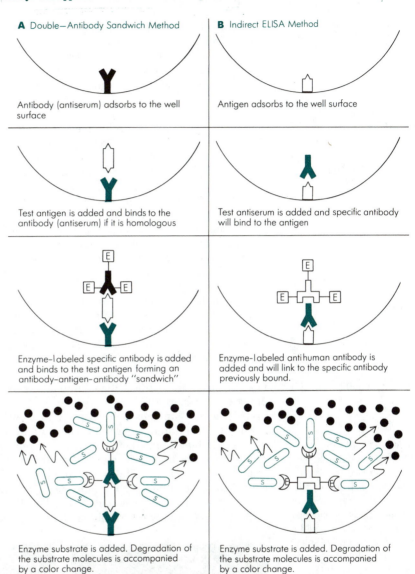

A Double—Antibody Sandwich Method

Antibody (antiserum) adsorbs to the well surface

Test antigen is added and binds to the antibody (antiserum) if it is homologous

Enzyme-labeled specific antibody is added and binds to the test antigen forming an antibody–antigen–antibody "sandwich"

Enzyme substrate is added. Degradation of the substrate molecules is accompanied by a color change.

B Indirect ELISA Method

Antigen adsorbs to the well surface

Test antiserum is added and specific antibody will bind to the antigen

Enzyme-labeled antihuman antibody is added and will link to the specific antibody previously bound.

Enzyme substrate is added. Degradation of the substrate molecules is accompanied by a color change.

Figure 34-10. Enzyme-immunoassay techniques. (A) Double-antibody-sandwich method for the detection and the measurement of antigen. (B) Indirect ELISA method for the detection and measurement of antibody. (Note that the necessary washing after each step is not indicated.)

If the antibodies in the antiserum are homologous, they will bind to the immobilized antigen. Enzyme-labeled antihuman antibodies (antiglobulins or antiantibodies) are added to the system which will link to the antibody-antigen complexes formed in the previous step. Finally, as in the double-antibody-sandwich method above, the enzyme substrate is added and the rate of its degradation (hydrolysis) is associated with a color change proportional to the concentration of antibody present in the test sample. This color change can be monitored visually or by a colorimeter.

This method has been used for the immunodiagnosis of many infectious pathogens such as viruses, parasites, and fungi. For example, one antiserum sample from a pregnant woman can be screened simultaneously for infections of rubella virus (the agent for German measles, which causes congenital malformations and fetal death), type 2 herpesvirus (causes severe congenital nervous system malformations and small heads), and other infections that can affect the fetus.

Special Serological Tests

Fluorescent-Antibody Technique

The fluorescent-antibody technique is a rapid procedure for the identification of an unknown infectious agent. The technique is based on the behavior of certain dyes which fluoresce (glow) when exposed to certain wavelengths of light. Examples of such dyes are fluorescein isothiocyanate and rhodamine isothiocyanate. Antibodies can be conjugated, or tagged, with these dyes and are then termed **labeled** or **fluorescent antibodies.**

If a mixed culture or specimen is placed on a slide, combined with serum containing fluorescent antibodies, and examined by fluorescence microscopy, only those organisms (antigens) that reacted with the specific labeled antibodies will be visible (Fig. 34-11). Thus only a few organisms need to be present to be observed. This is the direct method, in which the fluorescent dye is conjugated with the antibody specific for the antigen.

In the indirect method the initially applied antibody is not labeled. Instead, a second labeled antibody against the globulin of the animal species used for the preparation of the initial specific antibody is applied. This binds the fluorescent label to the specific antibody that has already reacted with antigen in the smear (see Fig. 34-12).

Neufeld Quellung Reaction

Pneumococci can be differentiated into at least 80 serological types on the basis of the polysaccharides in their capsular material. The great majority of clinical cases in adults are caused by types 1 to 8, whereas types 14 and 19 occur frequently in children, along with types 1, 4, 5, and 7. When specific antiserum was used in the treatment of pneumococcus pneumonia (before antibiotics were available), it was necessary to determine the *type of Streptococcus pneumoniae* causing the infection. Agglutination and precipitin reactions could be used, but a more convenient method of typing is based on the swelling of the capsules in the presence of type-specific antiserum (see Fig. 34-2). This reaction was described by Neufeld, who gave it the name *quellung*, from the German meaning

Figure 34-11. Fluorescent-antibody technique is used to detect microorganisms. A vaginal smear from a patient infected with *Neisseria gonorrhoeae*, the etiologic agent of gonorrhea, was treated with gonococcus antibody tagged with fluorescein dye. (A) Slide preparation viewed with dark-field microscopy. Note the presence of several kinds of microorganisms. (B) The same field viewed by fluorescence microscopy. Note the presence only of gonococci, which are paired, oval cells. *(Courtesy of W. E. Deacon, Bull WHO,* **24:***349–355, 1961.)*

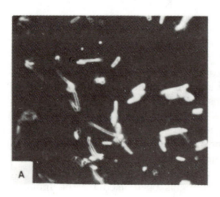
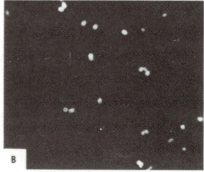

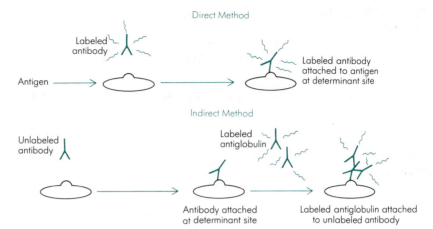

Figure 34-12. Fluorescent-antibody technique—direct and indirect methods. The indirect method is more sensitive as two or more labeled antiglobulin molecules can be attached to each antibody bound to its antigen.

"to swell." This is now considered to be a misnomer because actual swelling takes place only in the presence of complement, whereas in this test the appearance of swelling is due to a combination of the specific antibody with the capsular polysaccharide, which increases the refractive index of the capsular material.

The test can be made on pneumococci in sputum, from cultures, from peritoneal exudate of mice inoculated with sputum, or from other material containing the organisms. Loopfuls of material containing pneumococci are placed on microscope slides and mixed with specific antipneumococcal serums (made by inoculating rabbits with specific types of the organisms) containing a small amount of methylene blue. The dye stains the body of the cell and the surrounding fluid, leaving the capsules clear (Fig. 34-13). A positive test is indicated by a very prominent capsule.

Figure 34-13. Quellung reaction causes the capsules of the pneumococcus to appear enlarged when mixed with type-specific antiserum. This photomicrograph illustrates the appearance of *Streptococcus pneumoniae*, type 3, in the presence of its specific antibody. *(Courtesy of Robert C. Austrian, J Exp Med, **99:**21–34, 1953.)*

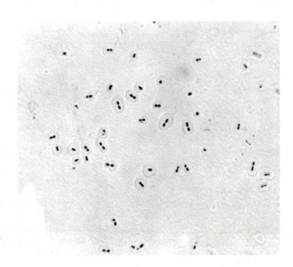

Detection of Heterophile Antibodies

When emulsions of organs from guinea pigs or certain other animals are injected into rabbits, antibodies against the red blood cells of sheep are produced in the rabbit. The antigenic material in this case is called a **heterophile antigen,** or **Forssman antigen,** for the discoverer of the phenomenon. The antibodies produced by injection of sheep cells differ from the heterophile antibody produced as a result of heterophile antigen stimulation.

The test used for the detection of heterophile antibodies in human infectious mononucleosis is based on the agglutination of sheep red blood cells by the serum of the patient.

Virus-Neutralizing Antibodies

Neutralizing antibodies indicate the presence of certain viruses. This test is especially useful in virus infections, but it can also be used to determine antibody titers to other organisms.

Virus obtained from the patient is grown in tissue culture (the brain cells of mice or in some other appropriate tissue), extracted, and exposed to patient's serum and to normal rabbit serum for 2 h. A small amount of each of these mixtures is inoculated into mice. Subsequently, the effectiveness of the patient's serum in neutralizing the virus is determined by the degree of protection afforded the test mice as compared with the lack of protection afforded the mice given normal serum. Neutralizing antibodies can also be detected by tissue-culture or egg-culture techniques.

MEASUREMENT OF CELL-MEDIATED IMMUNITY

The tuberculin, brucellergin, and some other skin tests (Table 34-1) are based on the detection of a delayed type IV hypersensitivity response typical of T-cell-mediated immunity. In contact with specific antigen, the T cells release various substances (see Chap. 33) that may cause tissue necrosis. Delayed-type reactions are apparent only after 18 h and may persist for days. Similar tests can be performed in vitro using these microbial antigens to stimulate the patients' lymphocytes to divide. A positive stimulation indicates the presence of immune T lymphocytes. Similarly, the T lymphocytes of patients who have

Table 34-1. Some Intradermal Tests Based on Cell-Mediated Immunity

Test	Etiologic Agent	Disease
Tuberculin	Bacterium: *Mycobacterium tuberculosis*	Tuberculosis—pulmonary disease with necrotic tubercles in lungs; may spread to other parts of body
Lepromin	Bacterium: *Mycobacterium leprae*	Leprosy—chronic disease that affects peripheral nerves, as well as face, hands, and feet, to cause disfiguring
Brucellergin	Bacterium: *Brucella* spp.	Brucellosis—undulant (fluctuating) fever in humans
Blastomycin	Fungus: *Blastomyces dermatitidis*	Blastomycosis—lesions develop in lungs, skin, and bones
Histoplasmin	Fungus: *Histoplasma capsulatum*	Histoplasmosis—lesions in lung and lymph nodes, but may spread to other organs, including the liver and spleen
Leishmanin	Protozoan: *Leishmania* spp.	Leishmaniasis—any disease caused by these parasitic flagellates, e.g., kala azar, a visceral disease with enlarged liver and spleen

rejected an organ graft can be stimulated to divide by the foreign lymphocytes of the organ donor in a mixed lymphocyte culture (MLC). The cell-mediated immune response in these cultures can be detected either by counting the dividing cells or by allowing them to incorporate radioactive thymidine into their DNA and measuring newly synthesized DNA in a liquid scintillation counter (for counting radioactivity).

The effector cells of the cell-mediated immune system have the ability to kill foreign cells, virally infected cells, and tumor cells within the human body. Cytolytic T lymphocytes (CTL) can be induced in a mixed lymphocyte culture and can lyse radioisotope-labeled cells in vitro; the amount of label released from the target cells is counted in a solid scintillation counter. Thus cytolytic activity is directly proportional to the number of target cells lysed and the amount of label released. This type of cytotoxicity assay is used to quantitate specific CTL-mediated immunity as well as antibody-dependent cellular cytotoxicity (ADCC) and the nonspecific activity of natural killer lymphocytes.

OTHER DIAGNOSTIC APPLICATIONS OF IMMUNOASSAYS

At this point in our discussion, examples have been given of the use of immunoassays to classify microorganisms and identify specific pathogens responsible for disease. But it should be remembered that there are many other uses for immunoassays in biology and medicine.

Classification of Cellular Antigens

One of the most common uses of immunoassays is tissue cross matching. With respect to the ABO blood typing system (see Chapter 33), specific antisera are used in cross-match hemagglutination tests to type the donor and recipient cells for compatibility. Similarly, tissue type (histocompatibility) antigens coded for by the major histocompatibility complex (MHC), a group of genes determining cellular membrane antigens, can be typed using specific antisera as well as the mixed lymphocyte culture reaction. Such tissue typing of the MHC antigens is essential to the successful transplantation of organs and tissue grafts.

Tests to Evaluate Virulence

Antifibrinolysin Test

Most group A, some group C and G, and a few strains of groups B and F hemolytic streptococci produce an extracellular heat-stable protein that has the ability to dissolve fibrin clots. This substance is known as **fibrinolysin** or **streptokinase.** This enzyme has some specificity for fibrin from certain animal species. For example, the fibrinolysin produced by *Streptococcus pyogenes* will dissolve fibrin from humans and bovines but not from rabbits. The ability of streptococcus cultures to produce fibrinolysin (streptokinase) can be determined by mixing oxalated human plasma coagulated with calcium chloride with a broth culture of the streptococci. The time required for the clot to be liquefied is determined.

Antifibrinolysin, or **antistreptokinase,** is the antibody produced by the host stimulated antigenically with fibrinolysin, or streptokinase. Presence of this antibody can be determined by comparing the time required for different sera to dissolve a plasma clot prepared as described above.

When normal blood is used, the plasma clot will be completely dissolved within 1 h. Longer lysing time indicates the presence of the antibody, and the time required for complete lysis of the clot is used as an index of antifibrinolysin (antistreptokinase) concentration.

Table 34-2. Intracutaneous Tests

Name of Test	Material	Administration
Schick test (for determining susceptibility to diphtheria): indicates lack of immunity if positive	Diluted diphtheria toxin; control is heated toxin	0.1 ml intradermally; same amount of control on opposite arm
Dick test (for determining susceptibility to scarlet fever): indicates lack of immunity if positive	Diluted erythrogenic toxin; no control needed	0.1 ml intradermally
Schultz-Charlton test (for diagnosis of scarlet fever)	Antiscarlatinal serum or convalescent serum	0.1 to 0.2 ml injected intradermally into an erythematous area
Frei test (for determining sensitivity to the causative agent of lymphogranuloma venereum)	Chick-embryo culture of *Chlamydia trachomatis*	0.1 ml intradermally; same amount of control on opposite arm
Tuberculin test (for determining allergy to the tubercle bacillus); indicates hypersensitivity	Either PPD (purified protein derivative) or old tuberculin (OT)	0.1 ml intradermally. If PPD is desired, use first dilution (0.002 mg/ml); if OT, use 1:10,000
Ducrey test (for determining sensitivity to *Haemophilus ducreyi*)	A suspension of the specific organisms	0.1 ml intradermally; no control necessary
Brucellergin test (for demonstrating allergy to *Brucella*): indicates hypersensitivity	An extract of brucellas	0.1 ml intradermally; no control necessary
Trichinella test (for determining sensitivity to trichinella protein)	An extract of the worms	0.01 ml intradermally with saline control
Echinococcus test (for determining sensitivity to echinococcus protein)	An extract prepared from hydatid fluid	0.01 ml intradermally with saline control

SOURCE: *Laboratory Guide for Bacteriology*, U.S. Naval Medical School, Bethesda, Md., 1946.

Test for Antistreptolysin O

Many strains of group A streptococci produce a hemolytic, oxygen-labile factor called **streptolysin O.** The specific antibody produced by the body upon stimulation by this antigen is called **antistreptolysin O.** The titer of this antibody is of diagnostic value in streptococcal infections which may lead to rheumatic fever, glomerulonephritis, and other serious complications. Early diagnosis is important, so that treatment to prevent permanent injury to the heart or kidneys can be started. However, a single high antistreptolysin O titer is not significant since it may represent only a "normal" base line for that individual. Antistreptolysin O determinations at 4- to 6-week intervals are highly significant if there

Recording	Remarks
Read at end of 48 h and record as positive or negative; a positive reaction shows edema and usually scaling for 7 days.	The control rules out sensitivity to bacterial protein; in general, a false reaction comes on earlier, fades faster, and leaves less pigmentation than a positive Schick; antibody (IgG) neutralizes toxin
Read between 18 and 24 h; positive test is an erythema over 10 mm in diameter; record as positive or negative	The Dick test reflects only presence or absence of antitoxin to the erythrogenic toxin; it has no significance as a measure of immunity to other streptococcal infections; antibody (IgG) neutralizes toxin
Positive reaction (for scarlet fever) is indicated by a definite and permanent blanching of the surrounding scarlatinal rash in 5 to 6 h	This reaction is due to neutralization of the toxin produced by streptococci that cause scarlet fever; it is useful in making diagnosis in doubtful cases; antibody (IgG) neutralizes toxin
In 48–72 h for first reading, with subsequent check on fourth and seventh days; record presence of a papule greater than 6 mm and without significant reaction of control as positive	This test is supposedly reliable in 90% of cases; a positive reaction may reflect past disease rather than present infection; about 3 weeks must elapse between development of infection and appearance of a positive test; detects cell-mediated immunity
Read in 36–48 h; record amount of redness and of swelling by diameter in millimeters	When results are not clear-cut, it is sometimes advisable to use a dilution of greater strength and repeat the test; detects cell-mediated immunity
A positive test appears in 48 h and is manifested as an area of induration in excess of 7 mm; record as positive or negative	Too little is known of this test for didactic appraisal; an attempt to isolate the organism from local lesion or lymph node should be made; detects cell-mediated immunity
In 24–48 h, inspect for erythema or induration; an area of edema of over 20 mm is recorded as positive	About 15% of people show positive skin test, so that the significance of the test in relation to symptoms must be carefully weighed; detects cell-mediated immunity
A posititve reaction usually appears as a wheal within 20 min	There is evidence that some individuals may display a delayed type of reaction, and tests immediately negative should be checked after 24 h; this test is believed to have about 95% specificity; immediate sensitivity (IgE) to trichinella protein
A positive reaction is indicated by a wheal within 20 min	Some individuals may show a delayed reaction, and all negative tests should be checked at the end of 24 h; immediate sensitivity to echinococcus protein

is a consistent drop or increase in titer, because they indicate recovery or increased inflammation, respectively. The test is based on the ability of the antibody to prevent hemolysis in tubes containing patient's serum, standard buffer, streptolysin O, a suspension of red blood corpuscles, and saline solution.

Intracutaneous Diagnostic Tests

The intradermal injection of a small quantity of a test material is often used to diagnose an infectious disease or determine the susceptibility of a patient (Table 34-2). These tests can be used (1) to detect the presence or absence of an antibody in the patient which reacts with antigenic material from the pathogen (e.g.,

Schick test); (2) to detect microbial antigens in the diseased patient (e.g., Schultz-Charlton test); and (3) to detect cell-mediated immunity to microbial antigens (see Table 34-1).

In Vivo Tests

Animal tests sometimes provide conclusive data on the mechanism of microbial pathogenicity and may aid in the diagnosis of disease. If a human pathogen produces detectable disease in a laboratory animal, it is possible to measure the protective efficiency of antibodies in the patient's serum. The patient's serum is injected into the test animals to provide *passive protection*. The animals are then challenged with the pathogen, and the increase in the numbers of the microorganism required to establish infection is determined. If no increase is seen, then the patient lacks specific antibody. The pathogens can also be injected into previously vaccinated animals to quantitate the immunogenicity of the vaccine and the protective value of the vaccine.

IMMUNOTHERAPY

In the preceding chapters vaccines have been mentioned many times, and this emphasis underlies their importance in safeguarding our health. Vaccines are either killed or attenuated pathogens which, when injected, elicit an immune response which is protective against the original pathogen. In addition, toxins can be inactivated and still retain their antigenic specificity as an immunogenic toxoid. New vaccines are being developed even today.

Once used more often than it is now, passive immunization still has its uses. Now largely confined to the use of hyperimmune human antisera to avoid problems with serum sickness, passive immunization is also used in combatting some infections (e.g., rabies and hepatitis) and in the prevention of erythroblastosis fetalis, the hemolytic disease of the newborn. This disease occurs frequently in newborns of Rh-negative mothers and Rh-positive fathers and more frequently in babies whose mothers have had multiple pregnancies. The disease occurs because Rh-negative women can become immunized to their babies' Rh-positive antigen during or following pregnancy. The IgG antibody to the Rh-positive antigen can cross the placenta and cause the lysis of the babies' red blood cells, causing a potentially fatal anemia. Passive immunization of the mother with anti-Rh antibodies before and after the birth of a Rh-positive child prevents her from developing a primary antibody response to the Rh factor since these anti-Rh antibodies destroy Rh-positive red blood cells in the mother. This forestalls the development of an Rh-incompatibility problem in a subsequent pregnancy.

Specifically labeled antibodies, especially monoclonal antibodies, are being tested as in vivo tracers for diagnosing and treating certain cancers. The possibility that therapeutic drugs can be coupled with specific antitumor antibodies may finally provide medicine with a "magic bullet" that has been searched for since the days of Paul Ehrlich, who discovered a chemical compound for the treatment of syphilis.

QUESTIONS

1 What is the significance of a rising antibody titer in a patient?
2 Describe the lattice hypothesis for the combination of antigen and antibody.

3 What is meant by the equivalence zone in a quantitative precipitin test?

4 List several ways in which precipitin tests can be useful.

5 Describe the Ouchterlony method of precipitin testing and explain how the results are interpreted.

6 Describe the technique of radial immunodiffusion and explain why it has advantage(s) over other methods of precipitin testing.

7 What is immunoelectrophoresis? Why is one type of immunoelectrophoretic test called rocket immunoelectrophoresis?

8 How is the precipitin test performed for the differentiation of streptococci?

9 How do the Widal test and the Weil-Felix test differ in purpose and procedure?

10 What are agglutinin-adsorption tests? How do they aid in the identification of bacteria?

11 Describe the hemagglutination test. What is its importance?

12 Describe the principle of the complement-fixation test.

13 Explain the following tests briefly: (*a*) RIA, (*b*) ELISA.

14 Compare the direct fluorescent-antibody technique with the indirect one.

15 What constitutes a positive test in the Neufeld quellung reaction? What is the basis for the positive reaction?

16 Describe some in vitro tests for cell-mediated immunity.

17 Indicate the purpose of the Schick test and the tuberculin test. What kind of immunity is involved in each case?

18 What causes erythroblastosis fetalis? How can this disease be prevented?

REFERENCES

Note that the references listed in Chaps. 32 and 33 are also useful for this chapter.

Hudson, L., and F. C. Hay: *Practical Immunology*, 2d ed., Blackwell Scientific Publications, Oxford, 1980. *A very useful and practical book giving an introduction to the techniques of immunology.*

Rose, N. R., and H. Friedman (ed.): *Manual of Clinical Immunology*, 2d ed., American Society for Microbiology, Washington, D.C., 1980. *A superb volume providing up-to-date laboratory methods for clinical immunology. Contains authoritative information on the best methods available for conducting specific immunologic tests. More than 130 authors.*

Thompson, R. A. (ed.): *Techniques in Clinical Immunology*, 2d ed., Blackwell Scientific Publications, Oxford, 1981. *A book that brings together the most useful laboratory tests in clinical immunology with detailed descriptions of the methods. Written by many practicing immunologists.*

Chapter 35   Epidemiology of Infectious Diseases

Humankind has long been beset by diseases that spread rapidly among a population, with devastating effects. How such pestilences were spread was unknown until the nineteenth century and was attributed to various natural and supernatural forces. The use of the techniques of **epidemiology** (the study of disease occurrence and distribution) helped to solve these mysteries. Moreover, epidemiological studies have had practical benefits by leading to an understanding of how to control the spread of infectious diseases.

Some pathogenic microorganisms are transmitted by one major mode, whereas others may be transmitted by several modes. Pathogenic microorganisms may be disseminated by food, water, aerosols (microorganisms that are dispersed in air), or the bites of various infected arthropods or other animals. They may also be transmitted by direct contact with an infected person or animal or by contact with inanimate objects (**fomites,** singular **fomes** or **fomite**) contaminated by an infected host. Epidemiological studies have not only concentrated on the transmission of causative agents of disease but have also revealed that the host population itself can play an important role in determining whether an infectious disease will be **pandemic** (having a large number of cases that occur on a global scale within a short time period); **epidemic** (having many cases in a particular geographic region within a short time period); **endemic** (having a low incidence but constantly present in a particular geographic region); or **sporadic** (having only an occasional occurrence).

How a pathogenic microorganism is transmitted does not necessarily depend on a knowledge of the nature of the microorganism, although this is very helpful. For instance, in 1855, nearly 30 years before Robert Koch discovered *Vibrio cholerae*, John Snow demonstated clearly that the causative agent of cholera was transmitted by drinking water contaminated with human feces; in a classic epidemiological study he traced the source of a London cholera epidemic to the Broad Street Pump in Golden Square. A more recent example is AIDS (acquired immunodeficiency syndrome): the modes of transmission of the causative agent of this disease were known since 1981, but identification of the probable causative agent of AIDS as being a retrovirus (HTLV-III) was not accomplished until 1984.

EPIDEMIOLOGICAL TECHNIQUES

The accumulation and organization of data about the occurrence and distribution of an infectious disease can provide clues as to the manner of transmission of the causative agent and the factors involved in acquiring a disease. The number of cases of a particular disease may be plotted versus time, geographic region, age, sex, race, occupation, or other parameters in order to determine what correlations may be present. Analyses of descriptive data are often helpful in determining the mode of spread of a disease or the factors contributing to outbreaks of the disease, as indicated by the following examples.

The finding of a *seasonal incidence* may provide important clues as to the mode of transmission of the causative agent. For example, an epidemic disease that occurs mainly during the colder months suggests an airborne mode of transmission, as in pneumonia, influenza, or chicken pox (Fig. 35-1). This is because during the colder months people are more likely to occupy crowded quarters and therefore are more likely to transmit microorganisms via aerosols generated through coughing and sneezing. On the other hand, the agent of a disease that occurs mainly in the warmer months would probably be transmitted by other means, e.g., by an arthropod that is prevalent during these months (as in tick-borne Rocky Mountain spotted fever) or by contaminated food (where warm temperatures may allow bacteria in improperly refrigerated food to multiply to high numbers; see salmonellosis, Fig. 35-2).

Geographic correlations may help to determine the mode of transmission. An epidemic occurring in a particular town or city suggests that a common factor

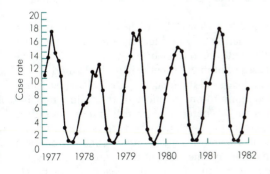

Figure 35-1. Reported case rates of varicella (chicken pox) per month per 100,000 population, United States, 1977–1981. The highest incidence occurs in the colder months, which is characteristic of an airborne infection. The peak incidence for 1981 was reached between March and May. (*Centers for Disease Control: Morbidity and Mortality Weekly Report, Annual Supplement Summary 1981, issued October 1982.*)

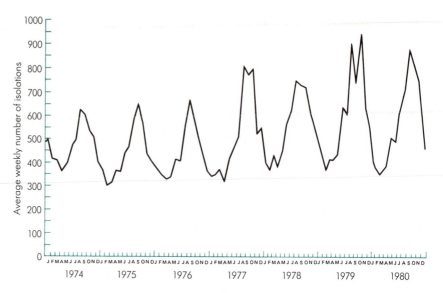

Figure 35-2. Reported isolations of salmonellas from humans, United States, 1974–1980. Each point represents the weekly average number of isolates for the month. The marked increase in the number of reported isolates during the warmer months suggests that conditions present during this time of year contribute to the spread of salmonellas. *(Centers for Disease Control: Morbidity and Mortality Weekly Report, Annual Supplement Summary 1980, issued September 1981.)*

Table 35-1. Reported Cases of Plague in Humans by State, United States, 1960–1981

Area	Total
United States	165
Arizona	22
California	17
Colorado	12
Idaho	1
Nevada	3
New Mexico	99
Oregon	7
Texas	1
Utah	2
Wyoming	1
Other states	0

SOURCE: Centers for Disease Control: Morbidity and Mortality Weekly Report, Annual Supplement Summary 1981, issued October 1982.

such as a particular water or food source may be involved, as in typhoid fever or bacillary dysentery. A sporadic disease restricted to a particular geographic area suggests that other factors may be involved; for example, human bubonic plague in the United States is limited to the western and southwestern regions and is acquired through occasional contact with the infected wild rodent population of those regions (Fig. 35-1).

Correlation of a disease with *age groups* affected by a disease may indicate factors of epidemiological importance. A disease that occurs mainly in the age group over 65 may be suggestive of a breakdown in immunity. For instance, most of the new active cases of tuberculosis in the United States occur in elderly persons (see Fig. 35-3), particularly males whose resistance has been lowered by factors such as malnutrition. Diseases that affect mainly children also suggest that a lack of active immunity may be a major factor. For example, whooping cough (pertussis) is most common in infants (Table 35-2). In contrast to many other infections, passive natural immunity acquired from the mother is relatively ineffective in protecting an infant against this particular disease, and thus vaccination of infants against whooping cough should be begun at about 2 months of age.

A correlation with *occupation* or *life style* can be made with some diseases. For instance, a disease that occurs mainly in veterinarians and slaughterhouse workers suggests that direct contact with the tissues of infected animals may be the mode of transmission, as in cases of human brucellosis in the United States. In another example, 91 percent of all cases of the disease called AIDS (see Chap. 33) have occurred in persons belonging to one or more of the following categories: homosexual or bisexual men (71 percent of cases); intravenous drug users such as heroin addicts (17 percent of cases); persons with hemophilia who receive injections of clotting factor prepared from the blood of many donors (1 percent of cases); heterosexual partners of members of these groups (1 percent of cases); and recipients of blood transfusions (1 percent of cases). This strongly

Table 35-2. Reported Cases of Pertussis (Whooping Cough) by Age, United States, 1982

Age Group	Cases
Under 1	960
1–4	474
5–9	181
10–14	88
15–19	35
20–24	24
25–29	22
30–39	29
40–49	9
50–59	2
60 +	2
Age unknown	60
Total for all age groups	1,886

SOURCE: Centers for Disease Control: Morbidity and Mortality Weekly Report, Annual Supplement Summary 1982, issued December, 1983.

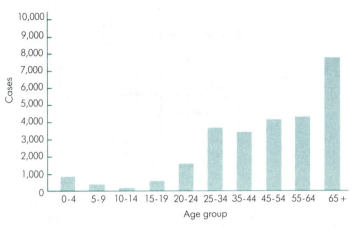

Figure 35-3. Reported cases of tuberculosis by age group, United States, 1980. *(Centers for Disease Control: Morbidity and Mortality Weekly Report, Annual Supplement Summary 1980, issued September 1981.)*

indicates that the causative agent (probably a retrovirus called HTLV-III—a member of the human T-cell leukemia virus group) can be transmitted from person to person by sexual contact or, less commonly, via blood or contaminated hypodermic needles.

Epidemiological Markers

To trace the origin and manner of spread of an outbreak of disease, it is often useful to determine whether the *same strain* of a microbial species is responsible for all the cases. For example, let us suppose that an epidemic of streptococcal sore throat occurs in one part of a city, and a few days later another outbreak occurs in a different area of the city. Are the two outbreaks caused by the same strain of *Streptococcus pyogenes* which has been transported across the city, or are the outbreaks completely unrelated to one another, each having a completely different origin?

Such questions could be answered easily if each microbial strain were to carry some sort of identification, like the numbers emblazoned on football uniforms that allow an observer to keep track of the players. The cells of microbial strains obviously do not have identifying numbers etched on their cell walls; however, they often possess other properties which can serve the same purpose. That is, they have properties that can be used to specifically characterize or **type** the strains within a species.

Subdivision by Antigenic Composition. The M proteins that occur in the cell walls of S. pyogenes can be used to classify the strains of this species into **serovars** (different serological types), since 63 types of M proteins occur and each strain possesses only one type. In the two outbreaks of streptococcal sore throat mentioned previously, if both were found to be caused by, say, type 45, this would strongly indicate that the same strain was involved. Many other pathogenic species can be subdivided into serovars, such as *Streptococcus pneumoniae* (80 serovars, on the basis of capsular antigens); *Haemophilus influenzae* (serovars a to f, on the basis of capsular antigens); and *Leptospira interrogans*

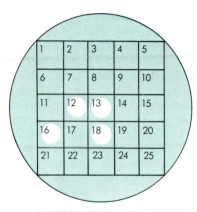

Figure 35-4. Phage-typing. The back of a Petri dish of agar medium was marked off into 16 squares. The surface of the medium was then swabbed with a strain of *Escherichia coli*. Each of 16 *E. coli* bacteriophages was inoculated onto the agar surface (phage 1 onto area 1, phage 2 onto area 2, etc.) After incubation for 6 h, zones of lysis developed where phages 12, 13, 16, and 18 had been inoculated. This phage-lysis pattern identified the *E. coli* strain as belonging to phage type A4B4, thus differentiating it from other strains of the species.

(180 serovars, on the basis of cell wall antigens). Influenza viruses are divided into three main types, A, B, and C, on the basis of their ribonucleoprotein (RNP) antigen (the S antigen). Thus all strains of type A share common RNP antigens; strains of types B and C have distinctly different ones. In addition, each type can be further divided into subtypes based on hemagglutinin (H) and neuraminidase (N) antigens.

Phage-typing. By applying a battery of different bacteriophages to a plate seeded with a particular strain, we can determine which phages will cause lysis of the bacteria and which will not. The pattern of susceptibility and resistance to phage lysis differs from one strain to another and can be used to characterize a strain as belonging to one or another phage type, or **phagovar** (see Fig. 35-4). For example, although the strains of *Salmonella typhi* cannot be distinguished from each other antigenically (i.e., they all belong to the same serovar), they can be subdivided into 33 distinct phagovars.

Resistance or susceptibility to various bacteriocins can also be used to type the strains of a species (see Chap. 12).

Biotyping. Differences in physiological properties can be used to classify strains into types **(biovars)** for epidemiological purposes. For instance, *Brucella abortus* can be subdivided into 8 biovars based on properties such as a growth requirement for CO_2, production of H_2S, and ability to grow in the presence of certain dilute dyes added to the culture medium.

Antibiotic Susceptibilty. Strains of *Staphylococcus aureus* are often differentiated on the basis of their susceptibility or resistance to a spectrum of antibiotics. The pattern of responses obtained constitutes an **antibiogram.** *S. aureus* can also be subdivided on the basis of either antigenic composition or phage lysis patterns.

ROLE OF THE HOST IN INFECTIOUS DISEASES

Carriers

Pathogenic microorganisms can be transmitted to healthy persons by **carriers**—persons who harbor the organisms. Such carriers can be of three types: (1) persons who have a clinical case of infectious disease; (2) persons who have had a disease and recovered from it, but still harbor the pathogenic organisms for some period of time **(convalescent carriers);** and (3) persons who harbor pathogens in their bodies, yet are not ill **(healthy carriers),** i.e., they have an **asymptomatic** infection.

The occurrence of healthy carriers indicates that merely acquiring a pathogenic microorganism may not automatically ensure that disease will result. Whether disease develops depends to a great extent on the natural resistance and immune state of the person. For example, 20 to 40 percent of the population harbors virulent S. *pneumoniae* in the upper respiratory tract, yet it is only when resistance is lowered that pneumococcal pneumonia develops. Factors that can decrease resistance toward pneumococcal pneumonia include alcoholic intoxication, anesthesia, or the occurrence of a mild primary respiratory infection such as the common cold.

Convalescent and healthy carriers may harbor pathogenic microorganisms for only a few days or weeks **(casual carriers);** but in some instances they may harbor them for months, years, or even for life **(chronic carriers).** For example, about 3 percent of persons recovering from typhoid fever become chronic carriers. In these carriers the typhoid bacilli have established a persistent, harmless infection of the gall bladder or bile ducts, from which they pass to the intestine and are excreted in the feces. Some typhoid carriers harbor the organisms in the urinary bladder. The stories of several persistent carriers have been recorded, but the most notorious is that of Mary Mallon, better known as Typhoid Mary, who was responsible for at least 10 outbreaks of typhoid fever, involving 51 cases and 3 deaths.

Chronic or casual healthy carriers can be treated with drugs to remove the focus of infection, or, in some cases, the organ harboring the microorganisms can be removed by surgery. For instance, the carrier state of a chronic typhoid carrier can usually be eliminated by the use of antibiotics (ampicillin) or by surgical removal of the gall bladder.

Herd Immunity

In order for an infectious disease to occur in epidemic form, the causative microrganism must be transmitted easily from one susceptible host to another within the population. Unless a sufficient proportion of the population is susceptible, the disease can occur only in an endemic or sporadic form. For instance, if 70 percent of schoolchildren in a population are immunized against poliomyelitis, epidemics of this disease are unlikely to occur even among the remaining 30 percent who were not immunized. These latter children enjoy what has been termed **herd immunity.** This is not true immunity but merely an expression of the unlikelihood that a susceptible individual will encounter the causative agent of the disease.

AIRBORNE TRANSMISSION

Many microbial pathogens have an airborne mode of transmission and cause infections of the respiratory tract. The infections caused by such airborne organisms tend to occur in epidemic form, appearing explosively and attacking large numbers of people within a short time. Their incidence usually increases during the fall and winter when people are more likely to occupy crowded quarters. The causative microorganisms occur in secretions from the nose and throat of infected individuals and can be transmitted directly to healthy indi-

Figure 35-5. High-speed photograph of an aerosol generated by sneezing. *(Courtesy of Marshall W. Jennison and the American Society for Microbiology.)*

viduals by **aerosols** (fine sprays producing droplets that remain suspended in air for a time) generated by coughs, sneezes (see Fig. 35-5), or even talking. Microorganisms that cause respiratory infections can also be transmitted indirectly via fomites such as drinking glasses, eating utensils, and handkerchiefs that have recently been used by an infected person.

Droplets and Droplet Nuclei

The size of bacteria-containing droplets expelled into the air by coughing and sneezing determines the time period during which they can remain suspended and also determines their ability to be trapped on the moist surfaces of the respiratory tract during inhalation. In general, larger droplets (10 μm or more in diameter) tend to settle out after traveling only a few feet. If inhaled, most become trapped in the nasal baffle and nasopharynx, from where they can reach the oropharynx by the downward flow of mucus. Smaller droplets (1 to 4 μm tend to evaporate rapidly, leaving **droplet nuclei** (the residue of solid material left after drying, e.g., bacterial cells). Such nuclei can remain suspended in air for hours or days, travel long distances, and serve as a continuing source of infection if the bacteria remain viable when dry. Viability is governed by a complex set of circumstances including (1) the atmospheric conditions, e.g., humidity, sunlight, and temperature; (2) the size of the particles bearing the organisms; and (3) the degree of susceptibility or resistance of the particular microbial species to the new physical environment. If inhaled, small droplets or droplet nuclei tend to escape being trapped in the nasopharynx; instead, they can reach the lungs and be retained there.

Infectious Dust

As indicated above, large aerosol droplets settle out rapidly from the air on various surfaces, such as textiles used by a patient, where they dry. Nasal and throat discharges from a patient also can contaminate surfaces and become dry. Disturbance of this dried material by bedmaking, handling a handkerchief having dried secretions on it, or sweeping a floor in the patient's room can generate dust particles which add microorganisms to the circulating air. The survival of microorganisms for relatively long periods in dust creates a significant hazard, particularly in hospital areas, and can contribute to **nosocomial diseases** (diseases acquired in a hospital). Tubercle bacilli have been isolated from the dust of sanitoria; diphtheria bacilli and hemolytic streptococci have been found in floor dust near patients or carriers harboring these organisms. The bacterial content of room air under various conditions is shown in Fig. 35-6. An illustration of the effect that occupants of a room have upon the numbers of airborne microorganisms is shown in Fig. 35-7.

Epidemiology of Influenza

One of the most familiar examples of airborne infection is epidemic influenza, or "flu" as it is commonly known. Epidemics of influenza occur in cycles; those caused by type A strains of the virus commonly follow a 2- to 3-year cycle and those caused by type B strains have a 4- to 6-year cycle. Type C strains rarely, if ever, give rise to epidemics; they cause subclinical infections or small outbreaks of the disease among children.

In addition to epidemics, influenza can occur in the form of pandemics having rapid global spread. Such pandemics illustrate the enormous geographic range and the rapid spread that airborne diseases can achieve. Great pandemics of influenza have occurred at intervals of 10 to 30 years or even longer and are caused by type A influenza virus strains. These strains have new subtype antigens, i.e., new varieties of the H (hemagglutinin) and N (neuraminidase) antigens that are different from those of the strains that preceded them. Antibodies against

Figure 35-6. Bacterial content of air in civilian and military establishments, as measured with a slit sampler. [From F. P. Ellis and E. F. Raymond, "Studies in Air Hygiene," Med Res Council (GB), Spec Rep Ser 262, 1948. By permission of the Controller of H. M. Stationery Office, London.]

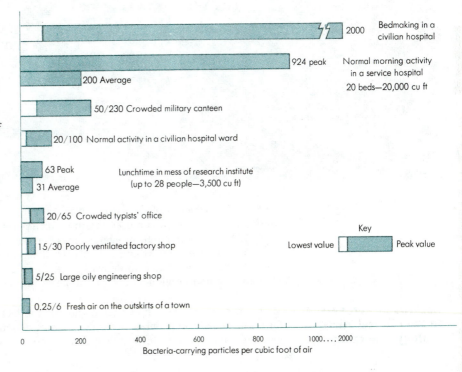

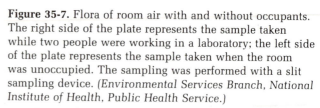

Figure 35-7. Flora of room air with and without occupants. The right side of the plate represents the sample taken while two people were working in a laboratory; the left side of the plate represents the sample taken when the room was unoccupied. The sampling was performed with a slit sampling device. *(Environmental Services Branch, National Institute of Health, Public Health Service.)*

Figure 35-8. Example of one type of safety cabinet used to protect personnel from infectious agents. (A) Open-front bacteriologic safety cabinet. (B) Pipetting operation in open-front safety cabinet. The technician is protected by the glass shield, rubber gloves, and gown. Room air is drawn into the cabinet at the front and ultimately passes through a HEPA (high-efficiency particulate air) filter before leaving the cabinet, thereby preventing the escape of infectious aerosols generated within the cabinet. [*Courtesy of the U.S. Naval Biological Laboratory and W. W. Umbreit (ed.), Advances in Applied Microbiology, vol. 3, Academic, New York, 1961.*]

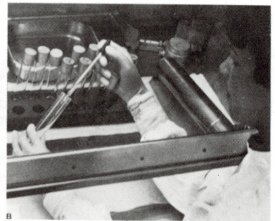

the H and N antigens of the virus are protective, and recovery from influenza confers a degree of active natural immunity on a population; however, antibodies formed by a host population against previous subtypes of influenza virus do not protect against new H and N subtypes. Therefore, when alteration of their antigenic makeup occurs, influenza viruses can cause new pandemics in populations that had previously been immune to influenza.

Hazards of Aerosols in the Clinical Microbiology Laboratory

Microbiologists are very much concerned by the fact that many routine laboratory procedures can generate aerosols of infectious microorganisms. Laboratory workers have acquired infections during microbiological investigations; more-

Influenza Pandemics—A Recurring Problem

The most disastrous influenza outbreak was the 1918–1919 pandemic, which was worldwide in its distribution and took more than 20 million lives. Although techniques for isolating and characterizing influenza virus were not available at the time, the strain is believed to be similar to one isolated from swine in 1930 which possessed the following variant H and N antigens: Hsw1N1. This was deduced from the fact that the blood serum of persons born between 1918 and 1929 contained antibodies to the swine virus H and N antigens, whereas the serum of persons born after that period did not. In other words, these antibodies provided an **immunological record** of the influenza virus strains that had been prevalent during the 1918–1929 period.

Marked changes in the H and N antigens occurred in 1947, 1957, and 1968, giving rise to strains having the antigenic variations H1N1, H2N2, and H3N2, respectively. Each change resulted in a pandemic. For example, after the emergence of the H2N2 variant, a pandemic began in the central part of mainland China in February 1957. From there it spread widely in China, then to Hong Kong, and then to other parts of the world. Shortly after September 1957, epidemics swept the United States. Although not as severe as the pandemic of 1918, it is estimated that nearly half the population of the United States became ill and more than 8,000 deaths were caused directly or indirectly by this new antigenic variant. Before this pandemic occurred, it had been found that the blood serum of people 70 to 90 years old already contained antibodies against the new strain. This was of interest because it implied that a similar strain may have caused an epidemic reported to have occurred in 1890. In 1968 the H3N2 variant appeared in Hong Kong and again a pandemic occurred. Still another variant was isolated in 1976 at Fort Dix, New Jersey; its antigenic type was HSw1N1, resembling that of the highly virulent strain thought to be responsible for the 1918 pandemic. This subtype was predicted to give rise to a new and serious pandemic; however, it failed to gain predominance over the H3N2 subtypes that were still prevalent. In 1977 still another subtype emerged, H1N1, similar to the subtype that had been prevalent in 1947, and spread throughout the world. The sequence in which these variants have emerged suggests that antigenic variation in type A virus may occur in a *repeating* or *cyclic fashion* rather than as an endless progression of new subtypes. Future comparisons of new variants with past variants will help to confirm or negate this hypothesis.

over, many infections that are not normally transmitted by aerosols in the general population, such as typhoid fever, can be acquired via aerosols in the laboratory. Certain technical manipulations create larger amounts of aerosols than others: among these are inserting a hot loop into a culture, streaking an inoculum on a rough agar surface, and blowing out the last drop of culture from a pipette. Aerosol formation from these sources can be avoided or greatly reduced by using good laboratory techniques. With highly contagious organisms such as *Francisella tularensis*, even a slight amount of aerosol can be dangerous since only one or a few cells can cause infection. Consequently, many safety devices have been designed to protect laboratory workers from this hazard. Bacteriological safety cabinets like that in Fig. 35-8 are available for working with particularly dangerous organisms.

WATERBORNE TRANSMISSION

Waterborne pathogens usually cause intestinal infections, such as typhoid fever, shigellosis, or cholera. Such infections are usually acquired by the consumption of polluted water containing human fecal matter from patients or healthy carriers. When human feces pollute a municipal water supply or other common source of drinking water, the outbreaks of intestinal disease tend to be of the epidemic type, and the source of drinking water is the common factor linking the various cases. The following example illustrates this principle.

Case History

A wedding reception was held in August at a country club in Pennsylvania. Over the next three days 90 of 119 guests developed gastrointestinal illness. Surveys revealed that only those who drank water at the reception developed the illness. A survey of club members who had played golf at the country club but who had not attended the wedding reception indicated that 60 of 113 golfers had also experienced gastrointestinal illness. Moreover, in the previous three months, 73 percent of the golfers surveyed had become ill. The drinking of water from fountains on the golf course was significantly associated with illness ($p = <0.01$). *Shigella sonnei* was isolated from stool samples from some of the wedding guests and golfers. Water for the country club and the golf course fountains came from an old private drilled well. The water was routinely chlorinated, but early in July the automatic chlorinator broke down and it was not functioning at the time of the outbreak. Coliforms, indicative of fecal contamination, could be cultured from the water supply of the club. Based on sample survey results, over 1,000 persons may have acquired shigellosis from the contaminated well.

The drinking of water is not always required for transmission of some waterborne infections. For instance, leptospirosis—a nonintestinal disease characterized by bacteremia and kidney damage—can be acquired merely by coming into contact with water contaminated with the urine from infected domestic or wild animals (e.g., by swimming in a farm pond frequented by infected cattle, or by working in a rat-infested sewer). The leptospires can penetrate the conjunctiva of the eye, abrasions in the skin, or mucous membranes of the nose and mouth.

FOODBORNE TRANSMISSION

Foodborne diseases may be intestinal diseases but can be other types as well. With regard to their epidemiology, they can be divided into two major categories. (1) In **foodborne intoxications,** such as botulism or staphylococcal food poisoning, the causative microorganisms produce an exotoxin in food; when a person consumes the food, the toxin is ingested and gives rise to the disease. (2) In **foodborne infections,** the causative organisms are ingested; these subsequently grow within the body and cause damage.

Epidemiology of Foodborne Intoxications

In staphylococcal food poisoning, human carriers are responsible for contaminating a food with enterotoxigenic strains of *Staphylococcus aureus*. These carriers usually harbor the organisms in the nose **(nasal carriers).** The sequence of events is usually as follows:

1 The hands of a carrier become contaminated with nasal secretions.

2 The carrier's hands inoculate food during its preparation.

3 The food is stored for several hours without being properly refrigerated, and during this period the staphylococci multiply and produce enterotoxin.

4 The food is consumed, raw or cooked. Cooking does not destroy the toxin; the crude form of the toxin is heat-stable and resists boiling for 30 min or more.

The foods involved are usually such items as milk products, custards, processed meat spreads, cream puff fillings, sandwich spreads, turkey stuffing, and potato salad. Prevention of staphylococcal food poisoning depends mainly on refrigeration of these food items so that multiplication of the staphylococci will not occur. The following example illustrates many of the principles involved.

Case History

Two scheduled airplane flights originating in Rome, Italy, and one charter flight originating in Lisbon, Portugal, landed in United States airports carrying a total of 246 passengers acutely ill with severe nausea and vomiting. Passengers on the first two flights were served identical lunches which consisted of salad, chicken, vegetables, rolls, and custard dessert. A survey revealed a significant association between illness and eating the custard dessert ($p = <0.005$). The lunch served to passengers aboard the third flight had a different entrée and vegetables but included the same custard dessert. On all three flights, first-class passengers and crew members were served different lunches without the custard dessert and none experienced illness. Results from several laboratories indicated the presence of high numbers of S. aureus in the custard. Epidemiologic investigation revealed that a catering facility located in Lisbon had provided the lunches for the three flights. The custard was produced from egg yolk, sugar, milk, gelatin, chocolate, gooseberry juice, and strawberry jelly; its preparation each morning required several pouring and chilling steps during a 4-hour period. It was then packed into individual passenger trays and stored in a holding area for 2 hours until placed aboard the plane. The holding area temperature was 62°F (rather than the usual refrigerator temperature) and apparently had been so for several weeks; therefore, the total time the custard was held at a temperature greater than 60°F was over 4 hours, sufficient to allow multiplication of enterotoxigenic staphylococci.

In botulism, soil is usually the source of endospores of the causative organism, *Clostridium botulinum*. Food is easily contaminated by soil and thus by the spores. If the food is an uncooked, processed food such as smoked fish or a cured ham, subsequent storage may allow the spores to germinate, the bacteria to grow, and the neurotoxin to be produced, if appropriate anaerobic conditions exist within the food. Alternatively, if the spore-contaminated food is canned, improper canning procedures (insufficient temperature or time) may allow some of the spores to survive. During a subsequent storage period the spores may germinate, and toxin may then be produced within the anaerobic interior of the can or jar. Botulism occurs sporadically in home-canned foods and only rarely in commercially canned foods. Home-canned foods are less subject to quality control, as illustrated by the following example.

Case History

Four women in Nevada ate a lunch which included a freshly prepared beet salad made from home-canned beets. In the next three days, three of the women had onset of botulism, and the severity of their symptoms corresponded to the amount of salad consumed. A sample of blood serum from one of the women taken prior to her treatment with botulinal antitoxin was positive for type A botulinum toxin by the mouse toxin-neutralization test. Type A botulinum toxin was similarly demonstrated in samples of leftover salad and leftover beets. Pressure cooking had not been employed during the canning of the beets. The beet salad reportedly tasted normal when eaten.

Epidemiology of Food-borne Infections

Foods such as meat, milk,, and eggs may come from an animal infected during its life by a pathogen. For example, infection of chickens, turkeys, swine, and cattle by certain *Salmonella* serovars is common. If such food is stored at a warm temperature, the salmonellas may multiply sufficiently to cause infection of persons who consume the food. This is illustrated by the following example.

Case History

During a one-week period, the Wisconsin Division of Health received reports that 7 individuals residing in three contiguous rural counties had become infected with *Salmonella* serovar *saintpaul*. All of the 7 persons had attended the same church dinner. A questionnaire returned by 188 of the 352 persons who had attended the dinner indicated that 19 persons had experienced recent gastrointestinal illness. Everyone attending the dinner ate chicken, but all 19 ill persons and only 53 of the 169 non-ill persons ate before 12:30 P.M. A food preparer, fearing a shortage of chicken, had hurriedly panfried a single batch of chicken for 10 minutes per side, and then placed the batch in a warming roaster at 175°F until served. These chickens were served until 12:30 P.M. but were removed when several persons complained of undercooked chicken.

A second way in which food may become contaminated is by means of human or animal carriers who have access to the food during its storage or preparation. The following example illustrates this kind of transmission.

Case History

During a two-month period, 18 cases of typhoid fever were diagnosed in Michigan among 310 people who had consumed a luncheon served at a community banquet hall. Of the 18 cases, 16 were confirmed by culturing *Salmonella typhi* from blood or stools. All isolates were phagovar E_1 and had the same antibiogram. No specific food could be incriminated. A chronic carrier of *S. typhi* was identified among the food handlers. This individual had participated in the preparation of all or most of the foods served, and S. *typhi* of the same phagovar and antibiogram as that obtained from the patients was isolated from a rectal swab.

The control of foodborne or waterborne infections depends primarily upon preventing the contamination of food and water supplies. This can be done effectively by means of such sanitary measures as proper disposal of human wastes, purification of water supplies, and use of sanitary methods in production and handling of food (including milk). Carriers of enteric organisms such as salmonellas and shigellas must not be permitted to participate in the handling and preparation of food. Detection of carriers, inspection of foods and food processing plants for occurrence of enteric pathogens, and identification of improperly treated water sources are important functions of local and state public health agencies.

TRANSMISSION BY DIRECT CONTACT

Transmission by Person-to-Person Contact

Of outstanding interest among the human direct-contact diseases are the venereal or sexually transmitted diseases (STDs) such as gonorrhea, syphilis, genital herpes, chancroid, nongonococcal urethritis, lymphogranuloma venereum, and granuloma inguinale.

Epidemiology of Gonorrhea. With the discovery during World War II that penicillin was effective in treating gonorrhea, it was predicted that this important STD would eventually be eliminated. Unfortunately, its incidence, which had hit an all-time low in the United States about 1958, took an upward turn, with a steady increase in the annual number of reported cases until 1978. Some of the factors involved were (1) the introduction of oral contraceptives and contraceptive intrauterine devices, which contributed to an increase in sexual freedom and to a decrease in use of spermicidal preparations and condoms (both of which afford some protection against gonorrhea); (2) the inability of public health departments and physicians to trace all the contacts of the many carriers and clinically infected persons; and (3) the emergence of certain strains of gonococci that produce penicillinase and, therefore, are resistant to penicillin (however, the number of isolates of this sort, although on the increase, is still relatively low). In the United States the annual number of reported cases of gonorrhea reached an all-time high of 1,013,436 in 1978, and the rate has decreased only slightly in subsequent years. A high incidence of gonorrhea has not been limited to the United States but has been reported in all parts of the world. The exposure and incidence is highest in the 15- to 29-year age group (Table 35-3).

Table 35-3. Reported Cases of Gonorrhea by Age in the United States, 1982

Age Group	Cases
0–14	10,453
15–19	235,086
20–24	363,135
25–29	195,037
30–39	121,208
40–49	25,972
50+	9,742
Total for all age groups	960,633

SOURCE: Centers for Disease Control: Morbidity and Mortality Weekly Report, Annual Supplement Summary 1982, issued December, 1983.

Gonorrhea can also be acquired by a newborn infant from an infected mother during passage through the birth canal, resulting in a blinding conjunctivitis (gonococcal ophthalmia neonatorum). It has been estimated that 10 percent of all cases of blindness may be due to this type of transmission of the gonococcus. For this reason it is a standard practice to place drops of silver nitrate solution or an antibiotic into the eyes of the newborn to prevent this conjunctivitis.

A rare third type of gonorrhea, vulvovaginitis, may occur in little girls, who may acquire the infection from fomites such as bedclothes, towels, and common bathtubs.

Control of sexually transmitted gonorrhea is accomplished by treatment of patients or carriers with penicillin (or spectinomycin in the instance of penicillin-resistant strains) and also by identification and treatment of other persons who have had contact with these patients or carriers.

Epidemiology of Syphilis. The syphilis spirochete is transmitted mainly by sexual intercourse; however, the organism can also be acquired by an infant in utero from an infected mother (congenital syphilis). The spirochete can also be transmitted by blood transfusion with contaminated blood. The incidence of syphilis, whose long-term effects are serious, is fortunately about 30 times less frequent than gonorrhea, but its distribution by age groups follows the same pattern.

In both syphilis and gonorrhea, the causative microorganisms are easily killed by drying and other environmental influences. This may help to explain why these organisms are restricted mainly to transmission by intimate contact.

Epidemiology of Genital Herpes. In recent years a marked increase in the number of cases of genital herpes has occurred. The disease is caused by the herpes simplex virus type 2. In the United States it has been estimated that there are 400,000 new cases of genital herpes each year and that between 5 and 20 million individuals have the infection. The disease cannot be cured by antibiotics, although the drug acyclovir shows promise as an antiherpes agent. The only method presently available for prevention of this and other viral diseases of the genital tract is the identification of patients and their contacts and avoidance of sexual intercourse with infected persons.

Transmission by Blood or Blood Products

Some microbial pathogens can be transmitted from person to person by transfusion of blood or blood products or by contact with traces of blood or blood serum on hypodermic syringes, syringe needles, tattoo needles, contaminated razors, and similar items that have been used by more than one person without being properly sterilized after each use. A classic example is viral hepatitis type B, a disease that is common among drug addicts. It is also associated with blood bank workers, dentists, and surgeons, who, because of their occupations, can come into contact with infected blood or blood products. There is some evidence that the hepatitis type B virus may also be transmitted by saliva and by sexual contact.

Transmission by Direct Contact with Animals

Some pathogens can be transmitted via contact with the tissues of infected animals. Diseases caused by such pathogens often have an occupational incidence, being contracted mainly by hunters, veterinarians, butchers, slaughterhouse workers, and the like. Examples are brucellosis, tularemia, and anthrax.

Epidemiology of Tularemia. Tularemia is mainly a disease of wild animals. About 1 percent of wild rabbits are infected, and over 90 percent of human cases are contracted from contact with these animals. The causative agent, *F. tularensis*, is an exceptionally small bacterium and has the ability to penetrate small abrasions on human skin and perhaps even unbroken skin. In the initial stage of the disease, a papule often appears on the skin of the fingers or hands and eventually becomes an open sore; this lesion indicates the site of entry of the pathogen into the body.

Case History

In November a 19-year-old man went deer hunting with friends and relatives in central Washington state. Four days later festering sores appeared on his hands, legs, and knees. Spiking fevers followed and eventually the disease was diagnosed as tularemia on the basis of development of a high antibody titer against *F. tularensis*. Despite repeated attempts to elicit a history of exposure to wild rabbits, none was obtained until after Christmas, when a relative recalled that the hunter had found a partially dismembered dead rabbit. The hunter had amputated the front paws for good luck charms, which he gave to another hunter in the party. The rabbit had been handled with bare hands that were bruised and scratched from the hunter's occupation as an automobile mechanic. The recipient of the "good luck charms" remained well; however, he had discarded the paws and they were therefore unavailable for culture.

Wound Infections

Infections may occur whenever a laceration or other type of wound is contaminated with pathogenic microorganisms. Wounds can be contaminated by organisms from the skin; from the object that inflicted the wound; or from various other sources such as soil, clothing, feces, or aerosols. For instance, the major causative agent of gas gangrene, *Clostridium perfringens*, is a normal inhabitant of the intestinal tract of humans and animals. Similarly, the spores of *Clostridium tetani*, the causative agent of tetanus, are common in animal feces and soil. Contamination from these sources frequently occurs in traumatic injuries, e.g., automobile accidents, gunshot wounds, battle injuries, and the like. Wounds most conducive to infection by anaerobic bacteria such as *Clostridium* or *Bacteroides* species are generally deep and ragged, with devitalized tissue. Consequently, the most important step in preventing such infections is to surgically remove the dead tissue and blood clots to prevent development of anaerobic conditions.

It should be recognized that surgical incisions are also wounds and can become contaminated, giving rise to postoperative infections. For instance, staphylococci are commonly present on the skin and, unless the skin is disinfected, they may gain entrance to the body via an incision. Surgical wounds may also become infected from environmental sources such as aerosols and infectious dust in the operating room or in the hospital ward or from contact with articles such as bedpans and bedclothes.

ARTHROPOD-BORNE TRANSMISSION

Arthropod-borne infections have an extensive distribution over the face of the globe. Through the centuries, such diseases have produced much suffering, economic loss, and death in the human population. On innumerable occasions, these diseases have reached pandemic proportions.

Arthropods associated with human infections serve as vectors for pathogenic microorganisms. A **vector** is an organism, such as an insect, that transports a pathogen. Some arthropods serve merely as mechanical vectors of pathogens. The common housefly, *Musca domestica*, is the classic example. The diseases transmitted by it includes salmonellosis, poliomyelitis, infectious hepatitis, amoe-

Table 35-4. Representative Protozoan Diseases of Humans Transmitted by Arthropods as Biological Vectors

Disease	Etiologic Agent (Geographic Distribution)	Biological Vector	Interrelationship of Arthropod-Pathogen-Human
Chagas' disease	*Trypanosoma cruzi* (Continental Latin America)	Reduviid bugs (*Triatoma* spp., *Panstrongylus* spp.)	Pathogen multiplies in midgut of insect; inoculated in humans by rubbing onto skin or into conjunctiva
African trypanosomiasis (sleeping sickness)	*T. gambiense* (West and Central Africa) *T. rhodesiense* (East and Central Africa)	Tsetse flies (*Glossina* spp.)	Pathogen multiplies in midgut and salivary glands of fly; humans inoculated by bite
Malaria	*Plasmodium vivax* *P. malariae* *P. falciparum* *P. ovale* (Regions with warm climates)	Mosquitoes (*Anopheles* spp.)	Pathogen completes sexual cycle, then multiplies by sporogony in mosquito; humans inoculated by bite
Leishmaniasis	*Leishmania donovani* (China, India, Africa, Mediterranean area, continental Latin America) *L. tropica* (Mediterranean area to western India) *L. brasiliensis* (Mexico to northern Argentina)	Sandflies (*Phlebotomus* spp.)	Pathogen multiples in midgut of fly; humans inoculated by bite

Table 35-5. Representative Bacterial Diseases of Humans Transmitted by Arthropods as Biological Vectors

Disease	Etiologic Agent (Geographic Distribution)	Biological Vector	Interrelationship of Arthropod-Pathogen-Human
Plague	*Yersinia pestis* (Africa, Asia, South America, western United States)	Rodent fleas (*Xenopsylla cheopis*); human fleas (*Pulex irritans*); body louse (*Pediculus humanus*)	Pathogen multiplies in gut of flea; humans inoculated by bite of flea
Tularemia	*Francisella tularensis* (North America, Asia, Europe)	Ticks (*Dermacentor* spp., *Amblyomma* spp., etc.); deerflies (*Chrysops discalis*)	Pathogen multiplies in gut and hemocoel (body cavity through which blood circulates); congenitally transmitted in some ticks; humans inoculated through bite or crushing of tick
Epidemic relapsing fever	*Borrelia recurrentis* (Asia, Africa, Latin America)	Body louse (*Pediculus humanus*)	Pathogen multiplies in tissues of louse outside gut; humans inoculated by crushing louse on skin
Rocky Mountain spotted fever	*Rickettsia rickettsii* (North America, Mexico, Colombia, Brazil)	Ticks (*Dermacentor* spp., *Amblyomma* spp., *Ornithodoros* spp., etc.)	Pathogen multiplies in wall of tick's midgut; congenitally transferred in tick; humans inoculated through bite

Table 35-5. (*continued*)

Disease	Etiologic Agent (Geographic Distribution)	Biological Vector	Interrelationship of Arthropod-Pathogen-Human
Scrub typhus	*Rickettsia tsutsugamushi* (Asia, Australia, Pacific Islands)	Red mites (*Trombicula* spp.)	Pathogen multiplies in gut of mite; congenitally transmitted in mite; humans infected through bite of larval mite
Rickettsialpox	*Rickettsia akari* (United States, Russia, Korea, Africa)	Mouse mite (*Allodermanyssus sanguineus*)	Pathogen multiplies in gut of mite; humans infected through bite
Classical typhus fever	*Rickettsia prowazekii* (Worldwide)	Body louse (*Pediculus humanus*)	Pathogen multiplies in epithelium of louse's midgut; humans inoculated through bite, feces, or crushing of louse on skin
Trench fever	*Rochalimaea quintana* (Europe, Africa, North America)	Body louse (*Pediculus humanus*)	Pathogen multiplies in midgut of louse; humans inoculated by feces or crushing of louse on skin
Murine typhus	*Rickettsia typhi* (Worldwide)	Fleas (*Xenopsylla cheopis* and others)	Pathogen multiplies in epithelium of midgut of flea; humans infected through bite

Table 35-6. Representative Viral Diseases of Humans Transmitted by Arthropods as Biological Vectors

Disease	Etiologic Agent (Geographic Distribution)	Biological Vector	Interrelationship of Arthropod-Pathogen-Human
Yellow fever	Yellow fever virus (a togavirus) (Africa, South America)	Mosquitoes (*Aedes aegypti, Haemagogus* spp.)	Pathogen multiplies in tissues of mosquitoes; humans inoculated through bite
Dengue fever	Dengue fever virus (a togavirus) (Southern and Southeast Asia, Pacific Islands, Northern Australia, Greece, Caribbean Islands, Nigeria, Latin America)	Mosquitoes (*Aedes* spp., *Armigeres oturbans*)	Pathogen multiplies in tissues of mosquitoes; humans inoculated through bite
Eastern, western, and Venezuelan equine encephalitis	Encephalitis viruses (belong to the togaviruses) (Western hemisphere)	Mosquitoes (*Aedes* spp., *Culex* spp., *Mansonia* spp.)	Pathogen multiplies in tissues of mosquitoes; humans inoculated through bite
Colorado tick fever	Colorado tick fever virus (an orbivirus) (Western United States)	Wood ticks (*Dermacentor andersoni*)	Pathogen multiplies in tissues of tick; humans inoculated through bite

biasis, and other enteric diseases. However, in most arthropod-borne diseases, the arthropod serves as a **biological vector,** i.e., one in which the pathogen undergoes a period of incubation or development. Various diseases in which biological vectors are involved are shown in the accompanying tables. Table 35-4 shows representative diseases of humans caused by protozoa and transmitted by arthropods. Table 35-5 shows representative human diseases of bacterial origin transmitted by arthropods. The majority of these bacterial diseases, as shown in the table, are caused by rickettsias—obligate intracellular parasites that are transmitted by lice, ticks, mites, or fleas. Table 35-6 shows representative human infections caused by viruses and transmitted by arthropods. Note that there are many kinds of encephalitis diseases, with different viral agents and varied geographical distributions. Also notice that mosquitoes and ticks are the main vectors involved in the transmission of the arthropod-borne viruses.

Examples of Arthropod-Borne Infections

Epidemiology of Plague. Plague pandemics ravaged Asia and Europe for centuries. The Great Plague, which started in 542, was reputedly responsible for over 100,000,000 deaths in 50 years. The Black Death, the great plague pandemic of the fourteenth century, was considered the worst catastrophe to strike Europe, and perhaps even the world. It resulted in the death of an estimated one-third of the world's population. The last pandemic of the nineteenth century began in central Asia in 1871 and spread to other parts of the world. In India alone during the years 1898–1918 it was responsible for 10,000,000 deaths.

The epidemiology of plague is complex. Plague is mainly a disease of wild rodents and is spread from animal to animal by fleas. In the United States, plague made its first appearance in 1900 at the port of San Francisco, being imported by rats on ships. The domestic rats of the city were soon infected and from them the disease spread to various wild rodents such as ground squirrels, prairie dogs, and wood rats. Thus plague spread eastward from the Pacific coast, and today a rodent reservoir of infection exists in the southwestern United States. A few sporadic human infections occur every year as the result of contact with infected wild rodents; however, epidemics of human plague have not yet occurred in this country.

Human plague epidemics follow a sequence of events. Domestic rats, i.e., rats that share human habitations, contract plague via fleas from infected wild rodents. The blood of infected rats contains large numbers of plague bacilli. When a rat flea ingests this contaminated blood, the action of a bacterial coagulase may cause a clot to form in the insect's proventriculus. This blockage prevents the access of food to the midgut (stomach) of the insect, and a flea in this condition becomes very hungry. As the rat hosts die, the fleas leave the corpses and begin to attack human hosts. Because of the blockage of the proventriculus, the human blood that is ingested by a flea becomes mixed with plague bacilli and is regurgitated into the flea bite. Fleas whose proventriculus is not blocked may excrete feces containing plague bacilli onto the skin during feeding; when a human scratches the flea bites, the flea feces can be rubbed into the bites, thereby transmitting the organism.

Once a human contracts bubonic plague, the disease can be transmitted from person to person via human fleas or human lice. This type of transmission requires heavily louse- or flea-infested populations, which were common in temperate regions such as Europe before the advent of sanitation. The wearing of thick clothing inhabited by lice or fleas was highly conducive to contracting not only plague but other diseases as well, such as louse-borne epidemic typhus. In some bubonic plague patients, the bacilli may reach the lungs and establish a pneumonia (pneumonic plague). This form of plague is highly contagious because it can be transmitted from person to person via aerosols and is no longer dependent on arthropods.

Control of plague depends first upon control and elimination of domestic rats and rat fleas; it is impractical to eliminate the wild rodent reservoirs of infection. Other preventive measures include immunization programs and elimination of human fleas and lice.

Epidemiology of Rocky Mountain Spotted Fever. Despite its name, Rocky Mountain spotted fever (RMSF) occurs throughout the temperate zones of the western hemisphere, and in the United States it is far more prevalent in the east than the west (Fig. 35-9). The causative agent, *Rickettsia rickettsii*, is transmitted in the east mainly by dog ticks and in the west by wood ticks. These arthropods serve as a reservoir of infection as well as vectors. The rickettsias are not pathogenic for the ticks; indeed, they become hereditary and are transmitted to the tick offspring by transovarian passage. The rickettsias are widely distributed throughout the body of a tick, and transmission to humans or animals occurs via infected saliva during biting. Although ticks probably constitute the primary reservoir of infection, there is also an animal reservoir of infection (wild rabbits, dogs, sheep, and rodents) in which the disease is perpetuated by associated ticks.

RMSF occurs mainly in the summer months and in persons engaged in outdoor pursuits, since such persons are most likely to be bitten by ticks. A characteristic rash usually appears first on the ankles and wrists and later spreads to the rest of the body. In many cases an initial diagnosis of measles, meningococcemia, scarlet fever, or other diseases involving a rash is made before RMSF is suspected; however, a history of tick bites is suggestive of RMSF. Prompt treatment with a tetracycline is essential; penicillin is not effective in treatment.

Figure 35-9. Reported cases of Rocky Mountain spotted fever by county, United States, 1981. *(Centers for Disease Control: Morbidity and Mortality Weekly Report, Annual Supplement Summary 1981, issued October 1982.)*

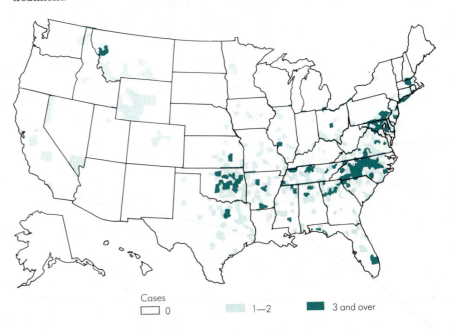

Case History

A 9-year-old boy from Nebraska became ill with a rash and a temperature of 102°F. Over the next three days these symptoms persisted and were accompanied by conjunctivitis, muscle aches, and joint pains. The rash became generalized and extended to the child's palms and soles, and lymph nodes were enlarged in the cervical area. The patient gave a history of having received multiple tick bites within two weeks prior to onset of the illness. A presumptive diagnosis of Rocky Mountain spotted fever was made and treatment was initiated with oxytetracycline. The child made a full recovery. The presumptive diagnosis was later confirmed by serological analysis of blood serum specimens collected at 4 and 34 days after onset of the illness: complement fixation tests indicated that the child had produced an active antibody response against RMSF antigen.

Epidemiology of Yellow Fever. Yellow fever is the most serious arthropod-borne viral disease of the tropics. For more than 200 years after the first identifiable outbreak in the Yucatan in 1648, it was one of the great scourges of the world. Although yellow fever now occurs in South and Central America and in Africa, during the nineteenth century it was also prevalent in the eastern United States, causing at least half a million cases. As late as 1905, New Orleans and other southern American ports had an epidemic that involved at least 5,000 cases and many deaths. The last indigenous case in the United States occurred in 1911; the last imported case occurred in 1923.

Yellow fever results from two basically different cycles of virus transmission, urban and sylvatic (jungle), as shown in Fig. 35-10. In the urban cycle, the virus is transmitted from person to person by bites of the *Aedes aegypti* mosquito. The blood meal taken from a patient contains the virus, which multiplies for 10 to 15 days in the mosquito's intestinal tissue. After the virus appears in the salivary glands, the mosquito can transmit the infection. Once infected, the mosquito remains infectious for the rest of its life.

Sylvatic yellow fever is caused by the same yellow fever virus but occurs in wild animals, mainly monkeys. The virus is transmitted among them, and sometimes to humans, by mosquitoes other than *A. aegypti*. In the rain forests of South and Central America, species of treetop *Haemagogus* or *Sabethes*

Figure 35-10. Relationship between enzootic (sylvatic) and epidemic (urban) transmission cycles of yellow fever.

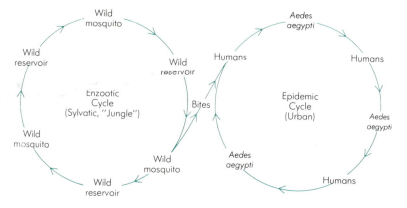

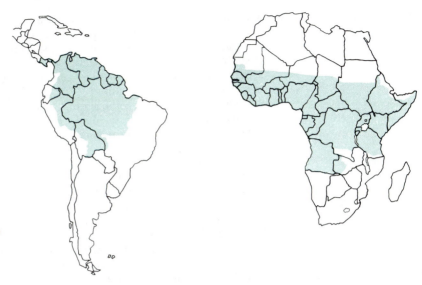

Figure 35-11. Regions where yellow fever is endemic (shaded areas) in South America and Africa.

mosquitoes maintain transmission in the wild animal reservoir. When humans enter the jungle, sporadic cases or local outbreaks may occur because of mosquito bites. In Africa, the mosquito-primate cycle is maintained by *Aedes africanus*, a species which seldom feeds on humans. However, the mosquito *Aedes simpsoni* feeds upon the primates encroaching on village gardens and it can then transmit the virus to humans. The threat of yellow fever always exists in urban areas of the tropical and semitropical regions because of the existence of the sylvatic cycles. Once yellow fever is reintroduced into an urban area, the urban human-mosquito-human cycle can be reinitiated, with the possibility of developing epidemics. The current endemic zones of yellow fever in South America and Africa are shown in Fig. 35-11. Why yellow fever has never invaded Asia despite widespread distribution of human-biting A. *aegypti* mosquitoes is an enigma of medical epidemiology.

Urban yellow fever can be prevented by eradicating A. *aegypti* mosquitoes or by suppressing their numbers to the extent that they no longer perpetuate infection. Control of the sylvatic form is impractical, however, because of the widespread animal reservoir of infection. However, jungle yellow fever can be effectively prevented in humans by immunization.

QUESTIONS

1 Distinguish between the terms *sporadic, endemic, epidemic,* and *pandemic,* and give examples of diseases to which these terms apply.

2 How does the seasonal incidence of influenza differ from that of (a) Rocky Mountain spotted fever and (b) salmonellosis? Indicate the basis for the differences.

3 Certain diseases, such as tuberculosis and pneumococcal pneumonia, have their highest incidence in elderly or aged persons. What might account for this?

4 Assume that two outbreaks of a bacterial disease occur in different areas of a city. How might you go about obtaining evidence that would confirm or negate the hypothesis that both outbreaks are caused by the same bacterial strain?

5 What evidence suggests that antigenic changes in type A influenza viruses may occur in a cyclic fashion?

6 Give an example of a pathogen that has a high rate of healthy carriers but causes a relatively low incidence of clinical disease. What other factors besides acquiring the organism may be important in occurrence of clinical disease?

7 *Streptococcus pyogenes* is the causative agent of streptococcal sore throat and is transmitted mainly by aerosols.
(a) How might it be transmitted by fomites?
(b) How might it be transmitted by infectious dust?
(c) How could it cause a wound infection?

8 To what hazards can aerosols contribute in the clinical microbiology laboratory, and how can these hazards be prevented?

9 Give an example of a waterborne disease in which ingestion of contaminated water is not required in order to contract the infection.

10 Give the sequence of events leading to (a) an outbreak of staphylococcal food poisoning and (b) an outbreak of botulism. How may such outbreaks be prevented?

11 Give the sequence of events leading to an outbreak of foodborne salmonellosis. How might such outbreaks be prevented?

12 Indicate the epidemiological differences between gonorrhea, vulvovaginitis, and gonococcal opthalmia neonatorum, all caused by *Neisseria gonorrhoeae*.

13 What is meant by the term *animal reservoir of infection?*
(a) Why might veterinarians and slaughterhouse workers be more likely to contract brucellosis than persons having other occupations?
(b) Why might hunters be more likely to contract tularemia than other persons?
(c) Why might a person in the southwestern United States be more likely to contract bubonic plague than a person in the eastern United States?

14 Describe the sequence of events leading to an epidemic of human bubonic plague. What are the differences between the person-to-person transmission of bubonic plague versus pneumonic plague?

15 Indicate the epidemiological differences between sylvatic yellow fever and urban yellow fever. Why would it be easier to eradicate the latter rather than the former?

REFERENCES

Baron, Samuel (ed.): *Medical Microbiology,* Addison-Wesley, Menlo Park, Calif., 1982. *The chapters of this text were individually prepared by authorities in the various fields of medical microbiology. Chapter 21 deals with the general topic of epidemiology of infectious disease.*

Davis, Bernard D., Renato Dulbecco, Herman N. Eisen, and Harold S. Ginsberg: *Microbiology,* 3d ed., Harper & Row, New York, 1980. *A comprehensive textbook on medical microbiology. The epidemiology of numerous infectious diseases is* covered throughout the various chapters dealing with pathogenic microorganisms.

Freeman, Bob A.: *Burrows Textbook of Microbiology*, 22d ed., Saunders, Philadelphia, 1985. *A classic medical microbiology text which deals with the biology of microorganisms. Chapter 13 addresses the general aspects of the epidemiology of infectious disease.*

Joklik, Wolfgang K., Hilda P. Willett, and D. Bernard Amos: *Zinsser Microbiology*, 18th ed., Appleton Century Crofts, Norwalk, Conn., 1984. *Also a comprehensive reference on medical microbiology, with detailed discussions of the pathogens and their transmission.*

Slack, John M., and Irvin S. Snyder: *Bacteria and Human Disease*, Year Book, Chicago, 1978. *A short text oriented toward a comprehensive treatment of the properties and activities of bacterial pathogens.*

Chapter 36   Microbial Agents of Disease: Bacteria

Although great attention has been paid to bacteria that cause disease in humans and animals, in fact relatively few of the thousands of bacterial species that occur in nature are pathogenic. Consequently, it is important for the clinical microbiologist to be able to distinguish a pathogenic species from the many harmless species that may also occur in clinical specimens. We usually cannot do this by merely observing a Gram stain from a lesion or streaking an agar plate: it is necessary to choose the proper culture media for isolation of the pathogen and to apply appropriate characterization tests in a systematic fashion. We must also determine the antimicrobial agents to which the pathogen is susceptible. All of these things need to be done *as quickly as possible*. Speed coupled with accuracy may make the difference between life and death, and the clinical laboratory is quite different in its aims and perspectives from the microbiology laboratory in an academic or research environment. Clinical microbiology is a demanding profession that requires a maximum of skill, responsibility, specialized knowledge of pathogenic microorganisms, and attention to a myriad of details attendant on the isolation and identification of those organisms.

In addition to bacteria which are pathogenic for humans and animals, some bacteria cause diseases of plants and insects and are consequently of great economic importance to agriculture.

This chapter serves only as an introduction to the subject of pathogenic bacteriology. It highlights the important features of some selected bacterial pathogens and the diseases which they cause. The major characteristics and taxonomic features of the genera discussed in this chapter have been described in Chaps. 13 and 14.

SPIROCHETES

The genera *Treponema, Borrelia,* and *Leptospira* contain species that are pathogenic for humans. Diseases caused by these species are listed in Table 36-1.

Treponema

The most important pathogen of this genus is *Treponema pallidum* subsp. *pallidum*, the causative agent of syphilis. Syphilis is a well-known and much dreaded sexually transmitted disease. Of untreated patients, about 25 percent exhibit a spontaneous cure, about 40 percent develop signs and symptoms of tertiary syphilis but do not die from the disease, and about 35 percent die from tertiary syphilis. Syphilis occurs only in humans and is transmitted by direct sexual contact (**venereal** syphilis) or by placental transfer from an infected mother to the fetus during the first 4 months of pregnancy (**congenital** syphilis). In 80 percent of the cases of congenital syphilis, the child is born asymptomatic but develops signs of syphilis several weeks or months later.

Venereal syphilis progresses in three stages. Only in the first two stages is the patient able to transmit the infection to others. The **primary stage** develops after an incubation period of 10 to 90 days during which the treponemes attach to host cells, multiply locally, invade the lymphatic system and blood, and become distributed throughout the body. The first sign of the disease is the **chancre,** a painless ulcer with a hard margin which develops on the genitals or on other areas of the body. Darkfield microscopy usually reveals treponemes in this lesion (Fig. 36-1). During the primary stage the patient does not feel ill, and the chancre heals within 25 to 40 days.

Table 36-1. Spirochetes Pathogenic for Humans

Genus and Species	Natural Host	Disease
Treponema		
T. pallidum		
subsp. *pallidum*	Humans	Syphilis, venereal (sexually transmitted) or congenital
subsp. *pertenue*	Humans	Yaws, a skin disease common in tropical countries in both hemispheres; transmitted by direct contact
subsp. *endemicum*	Humans	Nonvenereal endemic syphilis (bejel); occurs in the Middle East, Africa, Southeast Asia, and Yugoslavia; transmitted by direct contact
T. carateum	Humans	Pinta, a skin disease found in tropical countries in the Western hemisphere; transmitted by direct contact
Borrelia		
B. recurrentis	Humans	Epidemic relapsing fever; occurs in Asia, Africa, and Latin America; transmitted from person to person by body louse (*Pediculus humanus*)
B. hermsii B. parkeri B. turicatae B. hispanica and others	Wild rodents	Endemic relapsing fever; usually contracted by those engaged in outdoor activities; has a worldwide distribution; transmitted to humans by *Ornithodoros* ticks; a recently described disease called Lyme disease, characterized by a skin eruption, is caused by a *Borrelia*-like spirochete that is transmitted by *Ixodes* ticks
Leptospira		
L. interrogans	Various wild and domestic animals	Leptospirosis, characterized by a bacteremia, fever, muscular pain, and nephritis; transmitted by contact with or ingestion of water containing infected animal urine

The **secondary stage** develops after 2 to 6 months. A generalized eruption appears on the skin and mucous membranes of the body (Fig. 36-2), and the patient may have lymphadenopathy (swollen lymph nodes), malaise (a vague feeling of bodily discomfort), and a slight fever or headache. The treponemes

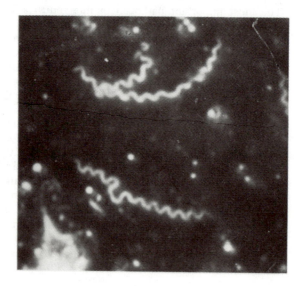

Figure 36-1. *Treponema pallidum* (X3000) in exudate, as seen by dark-field microscopy. *(Courtesy of General Biological Supply House, Inc.)*

Figure 36-2. Left: Secondary syphilis is ushered in by a widespread rash. Right: In secondary syphilis, lesions swarming with spirochetes may occur on the mucous membranes. Note the lesion on the lip. *(Courtesy of Armed Forces Institute of Pathology.)*

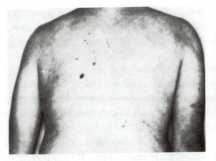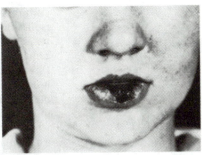

Figure 36-3. In tertiary, or late, syphilis, lesions called gummas rupture and result in ulcers. *(Courtesy of Armed Forces Institute of Pathology.)*

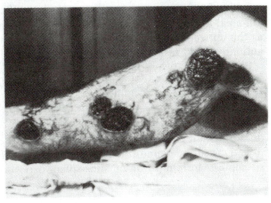

may be present in the lesions but are often hard to demonstrate. The skin and mucous membrane lesions disappear after 3 weeks to 3 months. A latent stage, which may last from 3 to 30 years, may intervene between the secondary and tertiary stages.

Approximately 75 percent of untreated patients will progress to the **late** or **tertiary stage** and may exhibit signs and symptoms ranging from mild to severe. Even when tertiary syphilis is mild, patients usually have 5 to 10 years taken off their life span and may also exhibit blindness, difficulty in hearing, and other symptoms of "early senility," i.e., being old before one's time. In more severe forms of tertiary syphilis, brain damage may occur, giving rise to **paresis** (general paralysis of the insane); the symptoms include failing memory, personality changes, insomnia, headaches, and delusions, and paresis may progress to the point where the patient must be institutionalized. **Tabes dorsalis** results from damage to the spinal cord: the patient becomes uncoordinated and has loss of reflexes, uncontrollable urinations, loss of perception and sensation, impotency, lancinating ("lightning") pains, blindness, and loss of hearing. In addition to these symptoms, damage to blood vessels such as the aorta or those nourishing the central nervous system can occur. Disfiguring granulomatous lesions called **gummas** may appear on or within various parts of the body (Fig. 36-3).

T. pallidum cannot be cultivated on laboratory media. It can only be grown, with difficulty, in the testes of rabbits or under special conditions in tissue culture. Consequently, the laboratory diagnosis of syphilis does not depend on isolation of *T. pallidum* but rather on the demonstration of antibodies in the patient's serum which have formed in response to the infection. Simple serologic screening tests are used initially for this purpose, such as the **VDRL** (Venereal Disease Research Laboratory) test or the **RPR** (rapid plasma reagin) test. These tests are simple and inexpensive, but they are **nonspecific;** i.e., they use a nontreponemal antigen, cardiolipin (usually prepared from beef heart muscle), to detect the presence of **Wassermann antibodies** (anticardiolipin antibodies), which occur in a syphilitic patient's serum. However, certain other diseases, e.g., malaria, lupus erythematosus, rheumatoid arthritis, and infectious mononucleosis, may also stimulate the formation of Wassermann antibodies. Thus a positive reaction must be confirmed by a **specific** test which can detect antitreponemal antibodies, i.e., a test in which a preparation of *T. pallidum* is used as the antigen. Two widely used specific tests are the fluorescent treponemal antibody **(FTA)** test and the *T. pallidum* hemagglutination **(TPH)** test.

For treatment of syphilis, penicillin is the drug of choice; however, the longer the disease progresses, the more difficult it is to cure. Immunization against syphilis is not yet possible because of the inability to culture *T. pallidum* for use in a vaccine. Recently, genes from *T. pallidum* have been cloned in *Escherichia coli*; this suggests a way to produce *T. pallidum* antigens for use in immunization without actually having to culture *T. pallidum* itself. With regard to active natural immunity, persons who have recovered from the disease are susceptible to reinfection on subsequent exposure.

MICROAEROPHILIC VIBRIOID GRAM-NEGATIVE BACTERIA

Campylobacter

The small, motile, vibrio-shaped organisms belonging to the genus *Campylobacter* (Fig. 36-4) are mainly parasites and pathogens of cattle, sheep, and other domestic animals. Some can also cause gastroenteritis and blood infections in humans; of these, *C. jejuni* has received the greatest attention. This species causes as many cases of diarrhea as *Salmonella* and *Shigella*. Transmission to humans occurs mainly by ingestion of food or water containing fecal matter from infected animals, and the disease has its highest incidence during the summer months. Person-to-person transmission rarely occurs. *C. jejuni* infects people of all ages, although it is isolated most frequently from persons 10 to 29 years old. The organism produces acute exudative and hemorrhagic inflammation of the wall of the small and large intestine, although the exact mechanism

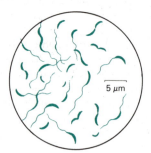

5 µm

Figure 36-4. Drawing of cells of *Campylobacter*. The flagella are not seen by ordinary staining. *(Erwin F. Lessel, illustrator.)*

by which illness is caused is not known. Some strains have been reported to produce an enterotoxin that is similar to the cholera enterotoxin. *C. jejuni* infections can be treated with erythromycin or tetracycline.

Bacteriological diagnosis depends on isolation of *C. jejuni* from stool samples. A medium containing various antibiotics is used to suppress other intestinal bacteria while allowing growth of *C. jejuni*. Because campylobacters are microaerophilic, cultures must be incubated under a gas atmosphere containing approximately 6 percent O_2. Growth occurs best at 42°C.

AEROBIC GRAM-NEGATIVE RODS

Pseudomonas

Several species of *Pseudomonas* are pathogenic, but most are opportunistic pathogens. The most serious animal pathogens are *P. mallei* and *P. pseudomallei*. *P. mallei* is a true parasite, occurring only in animal hosts; it is the causative agent of glanders, a disease to which nearly all warm-blooded animals are susceptible except cattle, pigs, and pigeons. *P. pseudomallei* is a tropical and subtropical soil organism that can cause a disease known as melioidosis in a large variety of animals and occasionally in humans.

Of the many species of *Pseudomonas*, the species most frequently encountered in human clinical specimens is *P. aeruginosa*. In addition to an endotoxin, many strains produce a lethal exotoxin (toxin A), which has a mechanism of action identical to that of diphtheria toxin (see Chap. 31). Other virulence factors include a hemolysin, a leukocidin, proteases which cause tissue necrosis, and an exocellular polysaccharide that inhibits phagocytosis. *P. aeruginosa* is commonly found in soil and water but can occur in many other places as well, including hospital environments; e.g., it has been isolated from disinfectant solutions, bedside water containers, hand creams, flower-vase water, sinks, respiratory equipment, and many other hospital items. It is the causative agent of many nosocomial infections (infections acquired in a hospital), including infections following surgery, burn infections, and urinary tract infections. With regard to the latter, catheterization (i.e., the insertion of a rubber or plastic tube called a *catheter* into the urethra in order to obtain urine specimens) can introduce *P. aeruginosa* into the body. Other factors that may predispose toward *P. aeruginosa* infections in hospital patients are immunosuppressive agents, antibiotics, and irradiation. Cystic fibrosis patients are especially prone to develop fatal infections. *P. aeruginosa* is resistant to many common antibiotics, although gentamicin and carbenicillin often are effective.

Legionella

These small Gram-negative rods (Fig. 36-5) cause legionellosis—a type of bronchopneumonia—in humans, and all *Legionella* species are potentially pathogenic. Humans probably acquire the organisms from environmental sources (see Chap. 13), and person-to-person transmission does not seem to occur.

Legionellosis was first recognized as a new form of pneumonia in 1976 as the result of an outbreak of illness at an American Legion convention in Philadelphia in which 182 persons were affected and 29 died. At first the causative agent could not be isolated, but the Centers for Disease Control in Atlanta were eventually able to culture the organism in guinea pigs and embryonated chicken eggs and on special laboratory media. Once the organism was isolated, it became clear that legionellosis was not a new disease and that it had existed at least as

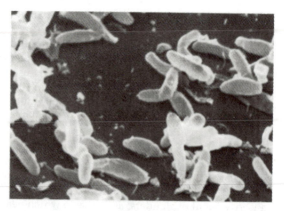

Figure 36-5. Scanning electron micrograh of the Legionnaires' disease bacterium, *Legionella pneumophila* (X16,200). (*Courtesy of D. D. Ourth, D. L. Smalley, and C. G. Hollis, Memphis State University.*)

far back as 1947. This was discovered by using the newly isolated bacterium as the antigen in immunofluorescence tests in order to detect the presence of serum antibodies against the organisms. The tests indicated that such antibodies were present in sera which had been preserved from various patients who, years earlier, had died of pneumonias of unknown origin.

Several species of *Legionella* have now been described, of which the best-known is *L. pneumophila*. This species may cause two kinds of disease syndromes. The first, known as Legionnaires' disease, is a severe bronchopneumonia characterized by fever, chills, and a nonproductive (dry) cough and sometimes by chest pain, abdominal pain, vomiting, diarrhea, and mental confusion. The fatality rate is 15 percent or higher. The disease tends to occur most frequently in men above the age of 50. Since 1976, more than 26 outbreaks and nearly 1,000 cases have been detected throughout the United States and in more than 15 other countries. The second type of legionellosis, termed Pontiac fever, is a less severe disease characterized by fever, chills, headache, dry cough, and muscle pain and is self-limiting and nonfatal.

Control of legionellosis is complicated by the wide environmental occurrence of the organisms, but it has been suggested that such measures as incorporation of germicides into the water of central air-conditioning cooling towers may be effective, especially during the warmer months when the incidence of legionellosis is highest. The drug of choice for treatment of *L. pneumophila* infections is erythromycin. No vaccine is presently available for prevention of legionellosis.

Brucella

Brucella species are mainly parasites and pathogens of domestic animals. However, three species—*B. melitensis* (occurring in sheep and goats), *B. abortus* (occurring in cattle), and *B. suis* (occurring in swine)—can cause brucellosis in humans, who acquire the disease from the animal hosts. Before the widespread use of pasteurization, unpasteurized milk from cows and goats was an important vehicle for many outbreaks of human brucellosis, and *B. abortus* and *B. melitensis* were the species most frequently encountered. At present, however, *B. suis* is the species most frequently isolated from human infection, being acquired through direct contact with infected swine or by inhalation of aerosols generated during processing of the meat from infected swine. Consequently, brucellosis is an occupational disease for veterinarians, butchers, and slaughterhouse workers.

In humans the disease is characterized by generalized aches and pains of the muscles and joints, headaches, chills and night sweats, and a prolonged, irregular (undulating) fever which continues into a chronic stage. Brucellas initially multiply within lymph nodes and later pass to the bloodstream. Here, antibodies formed by the patient act in conjunction with complement to cause bacteriolysis and liberation of **endotoxin,** which causes the fever response and other generalized symptoms. Many of the bloodborne bacteria are removed by macrophages of the liver, spleen, lymph nodes, and bone marrow. The bacteria continue to survive and grow within the macrophages; thus brucellas are mainly intracellular pathogens. Antibodies are ineffective in curing the infection because they cannot easily penetrate the cells within which the bacteria are growing. The only effective immunity in brucellosis is cell-mediated immunity, and the lesions within the body are **granulomas,** i.e., nodules which are composed of macrophages and T lymphocytes and which are similar to those occurring in tuberculosis (see later in this chapter).

Although spontaneous abortion is not a formal feature of human brucellosis, it is a very important feature of brucellosis in cattle and other domestic animals. Brucellas localize in the reproductive organs of these animals because of the presence of the 4-carbon sugar alcohol erythritol, a compound that is highly stimulatory to the growth of the bacteria. Erythritol does not occur in the reproductive organs of humans.

In the laboratory diagnosis of brucellosis, the organisms are difficult to isolate and usually diagnosis is based on demonstration of a rising level of anti-*Brucella* antibodies in the patient's serum.

Tetracycline is the drug of choice for treatment of brucellosis in humans. No vaccine is available for human immunization, but live attenuated vaccines are available for immunization of cattle, sheep, and goats.

Bordetella

The main human pathogen of this genus is *Bordetella pertussis* (Fig. 36-6), the causative agent of pertussis (whooping cough). This species attaches preferentially to ciliated bronchial epithelial cells and produces four virulence factors: (1) an **endotoxin;** (2) **pertussigen,** which induces lymphocytosis (excessive numbers of lymphocytes in the blood), causes increased sensitivity to histamine, and induces hypoglycemia; (3) the **HLT** toxin (heat-labile toxin), which may cause damage to the epithelium of the respiratory tract; and (4) an antiphagocytic **capsule.**

Whooping cough is mainly a childhood disease. About 15 percent of the cases and 85 percent of the deaths occur in children under 2 years of age. The disease is characterized by a paroxysmal cough that ends with an inspiratory crowing sound, or "whoop." During the paroxysms, cyanosis, vomiting, and hemorrhages of the nose, eyes, and even brain may occur. Inability to eat or retain food may result in malnutrition, making the patient susceptible to secondary infections. Bronchopneumonia is a common complication, and deafness and other permanent damage may result.

Diagnosis can be confirmed by isolation of the organism from nasopharyngeal swabs. The organism will not grow on ordinary culture media because such media contain toxic substances such as unsaturated fatty acids. Starch must be

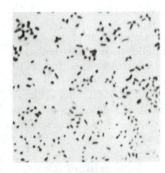

Figure 36-6. *Bordetella pertussis* (X1,300). Cells of this species range from 0.2 to 0.3 by 1.0 μm. *(Courtesy of J. R. Porter.)*

added to adsorb these toxic substances (detoxify the medium). A suitable starch-containing medium is Bordet-Gengou agar, to which penicillin is added to suppress the growth of other respiratory tract flora.

Erythromycin is the drug of choice for treating cases of whooping cough.

Vaccination of infants and children has reduced the incidence of whooping cough dramatically. Most pediatricians now give pertussis vaccine (a suspension of killed *B. pertussis*) along with tetanus and diphtheria toxoids to infants at 2, 4, 6, and 18 months of age. Vaccination is repeated at 4 or 5 years of age.

AEROBIC GRAM-NEGATIVE COCCI

Neisseria

The two most important pathogens of this genus are *Neisseria meningitidis* (Fig. 36-7), the causative agent of meningococcal meningitis, and *N. gonorrhoeae* (Fig. 36-8), which causes gonorrhea.

Neisseria meningitidis. Meningococci possess a capsule that helps to inhibit phagocytosis. Several serovars (antigenic types) occur and are differentiated by the polysaccharide composition of the capsule; serovars A, B, and C are the most common. Meningococci also possess a powerful endotoxin.

Humans are the only natural hosts for meningococci. The bacteria are transmitted by airborne means from patients with active infections and from convalescent or healthy carriers. Epidemics of meningococcal meningitis occur infrequently, but isolated cases may appear at any time. The disease is endemic in many large groups of "people in the herd," e.g., in the military and in college dormitories, where the majority of the population are young adults. Such groups experience rather frequent outbreaks due in part to introduction of new carriers or susceptibles. During an epidemic the carrier rate may be very high, even 80 to 90 percent, whereas the incidence of clinical disease is very low, 0.01 to 0.3 percent. Those who do develop the disease are very seriously ill and the mortality rate is high, 85 percent without treatment.

Meningococcal meningitis is characterized by excessive nasal secretions, sore throat, headache, fever, pain in the neck and back, loss of mental alertness, and sometimes a skin rash. Death can occur within 24 h after the symptoms begin, and therefore prompt diagnosis of meningococcal meningitis and treatment with penicillin is essential.

Presumptive diagnosis is made by demonstration of Gram-negative diplococci in stained smears of spinal fluid (Fig. 36-7). Confirmation is made by isolating the organism on prewarmed plates of a rich, blood-containing medium, with incubation under an atmosphere enriched in CO_2.

Penicillin is the drug of choice for treatment of meningococcal meningitis. A vaccine is available for epidemic control of meningococcal meningitis and consists of the purified capsular polysaccharides of serovars A and C; the serovar B polysaccharide appears not to be highly immunogenic.

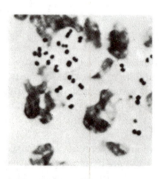

Figure 36-7. *Neisseria meningitidis* in cerebrospinal fluid, showing the characteristic diplococcus arrangement. The large objects are pus cells (neutrophils) (X1,400). *(U.S. Army photograph.)*

Neisseria gonorrhoeae. Virulence factors of gonococci include pili, which allow adherence of gonococci to host cells; an antiphagocytic capsule; and a potent endotoxin.

Gonorrhea is transmitted mainly by sexual contact, except for the forms called vulvovaginitis and gonococcal opthalmia neonatorum (see Chap. 35). In the

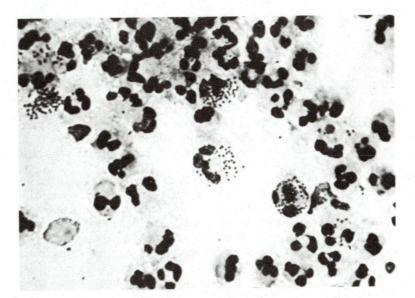

Figure 36-8. *Neisseria gonorrhoeae* in urethral exudate. Many of the bacteria occur within the cytoplasm of pus cells (neutrophils). These bacteria range from 0.6 to 1.0 μm in diameter. *(Courtesy of C. Phillip Miller.)*

male the primary site of infection is the urethra; in the female it is the cervix. The disease is usually more obvious in males because of pain during urination and a yellowish discharge from the urethra. The infection may extend to the prostate and epididymis. Between 20 and 80 percent of females are asymptomatic; those having symptoms may have painful urination, vaginal discharge, fever, and abdominal pain. In females the disease may ascend the genital tract and in 10 percent of cases the fallopian tubes become infected, leading to pelvic inflammatory disease (PID), a major cause of sterility. Other complications may include arthritis, endocarditis, and meningitis.

In the laboratory diagnosis of gonorrhea, smears from exudates often show the organisms inside neutrophils (Fig. 36-8). Isolation is done on **Thayer-Martin medium,** a rich blood-containing agar to which are added antibiotics to suppress the growth of other bacteria. *Neisseria* colonies are oxidase-positive, as indicated by a reddish-purple color that develops when the colonies are flooded with 1% tetramethyl-p-phenylenediamine reagent. Species identification is made on the basis of biochemical and serological tests.

Penicillin is usually used for treatment; spectinomycin is used for cases caused by penicillin-resistant strains. A vaccine is presently being developed, based on the ability of secretory IgA antibodies to prevent piliary attachment of the bacteria to host tissue.

FACULTATIVELY ANAEROBIC GRAM-NEGATIVE RODS

Escherichia

Although *E. coli* is part of the normal flora of the intestinal tract, certain strains can cause a moderate to severe gastroenteritis in humans and animals. **Enteropathogenic** strains colonize the jejunum and upper ileum of the small intestine and cause acute gastroenteritis in newborns and in infants up to 2 years of age. **Enteroinvasive** strains invade the epithelial cells of the large intestine and cause diarrhea in older children and adults. **Enterotoxigenic** (enterotoxin-producing) strains produce one or both of two different toxins: a **heat-stable toxin (ST)** and a **heat-labile toxin (LT).** Both toxins cause diarrhea in adults and infants. The LT stimulates adenylate cyclase activity in a manner similar to that of cholera toxin (see Chap. 31), whereas the ST stimulates guanylate cyclase activity. Enterotoxigenic strains of *E. coli* are often associated with traveler's diarrhea, a common disease contracted by tourists when visiting developing countries.

Other strains of *E. coli* which are usually harmless in their normal habitat (the intestine) can cause disease when they gain access to other sites or tissues. These diseases include urinary tract infections, septic infections, bacteremia, meningitis, pulmonary infections, abscesses, and skin and wound infections.

Salmonella

Over 2000 serovars of *Salmonella* exist, all of which are pathogenic for humans and often for animals. They are characterized by different combinations of **O antigens** (heat-stable outer membrane polysaccharides; see Fig. 5-21) and **H antigens** (heat-labile flagellar protein antigens). O antigens are designated by numbers 1 to 67. There are two major categories of H antigens, called phase 1 and phase 2. Each phase 1 antigen occurs in only a few serovars and is designated by a letter, a to z (z_1 to z_{59}). Fewer kinds of phase 2 antigens occur, but they are widely distributed among serovars and are usually designated by numbers. In addition to O and H antigens, a **Vi** antigen (capsular antigen, which can be destroyed at 100°C) occurs in *S. typhi* and a few other serovars. *Salmonella* serovars often have been assigned names as if they were species, but some are designated only by their antigenic formulas. Antigenic formulas are written as in this example: 6,7:r:1,7 (representing O antigens 6 and 7: phase 1 antigen r: phase 2 antigens 1 and 7). In the **Kauffmann-White scheme,** those serovars with particular O antigens in common are collected into O groups and arranged alphabetically by H antigens within the group. Table 36-2 gives the names of a few serovars commonly encountered in human infections, together with their antigenic formulas.

Salmonellas may cause three kinds of infections: enteric fever (typhoid or paratyphoid fever), gastroenteritis, and septicemia. Each of these has certain characteristic features.

Typhoid fever occurs only in humans and is caused by *S. typhi*, which is transmitted via food or water (see Chap. 35). The disease is characterized by a continued fever, inflammation of the intestine, formation of intestinal ulcers, and enlargement of the spleen. The disease begins in the small intestine, where the bacteria attach to the epithelium of the intestinal wall, penetrate this layer, multiply in the mesenteric lymph nodes, and eventually reach the bloodstream.

Table 36-2. Some *Salmonella* Serovars Commonly Encountered in Human Infections

| Serovar Name | Kauffmann-White | | Human Disease |
	O Group	Formula	
S. paratyphi A	A	1,2,12:a:–	Enteric (paratyphoid) fever or gastroenteritis
S. typhimurium	B	1,4,5,12:i:1,2	Gastroenteritis, septicemia, or focal infection
S. choleraesuis	C	6,7:c:1,5	Enteric fever or gastroenteritis
S. typhi	D	9,12 [Vi]:d:–	Enteric (typhoid) fever
S. enteritidis	D	1,9,12:g,m:–	Gastroenteritis

Here, lysis of some of the bacteria by the combined action of antibodies and complement results in the liberation of endotoxin, which causes various generalized disease symptoms such as fever. Some of the bacteria are excreted in the urine; others are removed from the blood by macrophages of the liver, spleen, lymph nodes, and bone marrow, in which they survive and multiply. Some pass from the liver to the gall bladder and bile ducts and are secreted into the intestine, where they establish a secondary infection and may cause diarrhea. Bacterial excretion in the feces may occur for weeks, and some persons become chronic carriers (see Chap. 35). Laboratory diagnosis of typhoid fever is based on isolation of the organism from blood samples during the first and second weeks or from stool samples during the second and third weeks. The most effective antibiotics for treatment of typhoid fever are ampicillin and chloramphenicol.

Recovery from an attack of typhoid fever does confer a lasting immunity, probably of the cell-mediated type, and second attacks in the same individual are rare. Immunization can be achieved by vaccination with acetone-killed whole cells of S. typhi and can confer protection for 3 years. Live attenuated vaccines for oral administration are presently being tested; these may stimulate the formation of secretory antibodies in intestinal fluid and thereby prevent initial-attachment of S. typhi to the intestinal wall.

Foodborne gastroenteritis is caused by many serovars of Salmonella, some of which are associated with animal as well as human infection. The epidemiology of the disease is described in Chap. 35. After an incubation period of 12 to 24 h, the illness commences with a severe headache followed by nausea, moderate vomiting, severe diarrhea, abdominal pain, and fever. The disease usually lasts for 2 to 5 days. Unlike typhoid fever, there is seldom a bacteremia, and the disease is mainly an intestinal infection. The mechanism by which the diarrhea is caused is not clear. It may be due merely to invasion of the intestinal wall, which occurs in both the small and large intestine and results in appearance of blood, mucus, and leukocytes in the stools, or it may be due to an enterotoxin. An enterotoxin has been demonstrated in certain Salmonella strains.

Laboratory diagnosis of Salmonella gastroenteritis depends on isolation of the causative Salmonella strain from stool specimens. Treatment of Salmonella gastroenteritis with antibiotics is not recommended except in severe cases (which usually occur in infants or elderly persons) because the disease is usually self-limiting and the drugs do not decrease the duration of the disease. To prevent dehydration of the body from the diarrhea, intravenous replacement of fluid and electrolytes may be required.

A vaccine against Salmonella gastroenteritis is impractical because of the large number of Salmonella serovars.

Septicemia can be caused by any Salmonella serovar, although S. choleraesuis is the most frequent. From the intestinal tract the bacteria presumably reach the bloodstream in a manner similar to that described above for S. typhi. In the bloodstream the bacteria multiply, causing a recurring high fever, chills, loss of appetite, and weight loss. Gastroenteritis seldom occurs, and although the organisms can be isolated from blood samples they cannot be isolated from stools. The disease may persist for a long time in chronic form. The bacteria in the blood are distributed to various parts of the body, where they may cause meningitis, pneumonia, abscesses, nephritis, osteomyelitis, or endocarditis. Ampicillin or chloramphenicol are effective for treatment of septicemia caused by *Salmonella*.

Shigella

Shigellas are restricted to humans as hosts. Four species exist, which correspond to four antigenic groups: *S. dysenteriae* (group A), *S. flexneri* (group B), *S. boydii* (group C), and *S. sonnei* (group D). These groups are based on O antigens only (shigellas have no flagellar H antigens and are nonmotile). In the United States, *S. sonnei* is common in the north and *S. flexneri* in the south; *S. dysenteriae* and *S. boydii* seldom occur.

The main virulence factor of shigellae is their endotoxin; however, some shigellas, such as *S. dysenteriae* type 1, can produce an enterotoxin (shiga toxin) as well, which causes fluid accumulation in the intestine. The mechanisms by which the shiga toxin causes this fluid accumulation is unknown. In addition to its action as an enterotoxin, the shiga toxin is also a cytotoxin that kills various kinds of tissue cells by inactivating the 60S subunit of eucaryotic ribosomes, thus inhibiting protein synthesis. Originally the shiga toxin was thought to be a neurotoxin because it caused paralysis when administered to rabbits. However, such paralysis could be rapidly reversed by treatment with antitoxin, which was not characteristic of the type of paralysis caused by a true neurotoxin (such as botulinum or tetanus toxins). Other evidence indicates that the shiga toxin probably causes paralysis by damaging the walls of blood vessels, resulting in fluid accumulation and pressure buildup in the central nervous system.

Shigellosis, also known as bacillary dysentery, is characterized by inflammation of the wall of the large intestine, with consequent diarrhea and stools containing blood, mucus, and pus. Fever is sometimes present, presumably due to absorption of endotoxin from the intestine. Unlike salmonellas, shigellas never penetrate beyond the intestinal wall, and blood cultures are invariably negative. Laboratory diagnosis depends on isolating shigellas from diarrheic stools or rectal swabs.

In severe cases of shigellosis, dehydration of the body may necessitate intravenous replacement of fluid and electrolytes. Ampicillin or a combination of trimethoprim and sulfamethoxazole can decrease the duration of the disease.

Yersinia

Three species of this genus have been studied extensively: *Yersinia pestis*, the causative agent of plague, and *Y. pseudotuberculosis* and *Y. enterocolitica*, which cause gastroenteritis.

Y. pestis is a plump, nonmotile Gram-negative rod which, when stained in smears of animal tissue by special methods, such as Wayson's method, exhibits bipolar staining (deeply stained area at both poles; Fig. 36-9). Unlike most pathogenic bacteria, *Y. pestis* grows best at 25°C rather than 37°C. Virulence factors produced by *Y. pestis* include (1) an antiphagocytic capsule termed fraction 1, composed of carbohydrate and protein; (2) the VW complex, a complex of a cell-wall protein (V) and a lipoprotein (W), and which is also antiphagocytic; (3) an endotoxin; and (4) the murine toxin, a protein toxin that is lethal for rats, mice, and other animals susceptible to plague. The role of the murine toxin in plague is not yet certain.

The epidemiology of bubonic and pneumonic plague is described in Chap.

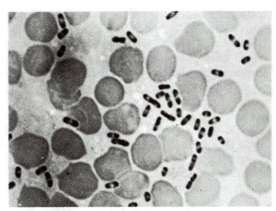

Figure 36-9. *Yersinia pestis* organisms show bipolar staining in this smear of mouse blood. The average size of this organism is 1.0 by 2.0 μm. The large round objects are erythrocytes. *(Courtesy of U.S. Naval Biological Laboratory.)*

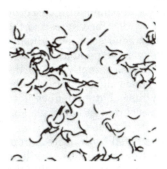

Figure 36-10. *Vibrio cholerae* (X1,500). *(Courtesy of J. Nowak, Documenta Microbiologica, part 1, "Bakterien," Gustav Fischer Verlag, Jena, Germany, 1927.)*

35. The bubonic form is characterized by chills, fever, nausea, vomiting, general weakness, and enlarged, inflamed lymph glands (bubos). The exudates from bubos are filled with plague bacilli. In untreated cases of bubonic plague the mortality rate is 50 percent. Pneumonic plague is a pneumonia characterized by a thin, watery sputum with bright red streaks of blood, and it has nearly a 100 percent mortality rate if untreated.

Streptomycin is effective for treatment of plague cases. Prevention of plague can be achieved by the control measures described in Chap. 35. Vaccines consisting of killed whole cells or live attenuated strains are available. The vaccines stimulate development of antibodies against fraction 1 and the V/W antigens, thereby enhancing phagocytosis.

Y. pseudotuberculosis is closely related genetically to *Y. pestis* and occurs in nearly all animal species, especially in rodents. It can cause acute mesenteric lymphadenitis in humans, a disease that is sometimes mistaken for appendicitis. In outbreaks, the initial cases are probably contracted by ingestion of food or water contaminated by feces from an infected animal, but secondary cases may result from the primary case via human fecal contamination. The virulence factors possessed by the organism are not known.

Y. enterocolitica also occurs in a wide range of animals and has much the same kind of transmission to humans as *Y. pseudotuberculosis*. It causes gastroenteritis, mainly in children, and in some parts of the world it is as common as *Shigella*. How widespread the infections are in the United States has not yet been evaluated. An enterotoxin similar to the ST toxin of *E. coli* has been demonstrated in laboratory cultures. Both *Y. enterocolitica* and *Y. pseudotuberculosis* grow best at 28 to 29°C, and isolation is often difficult on plates incubated at 37°C.

Vibrio

The most important species of this genus is *Vibrio cholerae* (Fig. 36-10), the causative agent of cholera. Other pathogenic species include two marine species: *V. parahaemolyticus*, which causes an acute gastroenteritis contracted by consumption of raw or inadequately cooked seafood, and *V. vulnificus*, which causes wound infections as well as fatal septicemias.

V. cholerae strains are assigned to various serovars based on their somatic O antigens. The strains that cause epidemic and pandemic cholera belong to serovar O1. For epidemiological purposes, this serovar is further divided into two biovars: "el tor," which is hemolytic on sheep blood agar, and "classical," which is nonhemolytic. The major virulence factor produced by serovar O1 strains is the **cholera enterotoxin.** The mechanism by which this toxin causes severe diarrhea is discussed in detail in Chap. 31. Also important for virulence is the ability of the vibrios to adhere to the epithelium of the small intestine; however, the mechanism of this adherence is not yet known.

Some strains of V. *cholerae* do not belong to serovar O1 and are called nonagglutinable (NAG) strains (because O1 antiserum does not agglutinate them) or noncholera vibrios (NCV). Such strains are widespread in freshwater and estuarine environments. They produce an enterotoxin similar to cholera enterotoxin and can cause a mild choleralike disease.

Cholera is a disease of antiquity and has been the cause of untold suffering and death. It was endemic in parts of India for centuries but began to invade the rest of the world in the early part of the nineteenth century, causing a series of six great pandemics between 1817 to 1923. A seventh pandemic, caused by the el tor biovar, began in 1960 in Hong Kong and by 1971 had spread to the Philippines, Indonesia, the Middle East, Russia, Africa, and even to Spain and Portugal. The Pakistani-Indian War of 1971, with its massive population shifts, resulted in thousands of deaths due to cholera. Although presently rare in the United States, cholera was prevalent during the nineteenth century. In the twentieth century cholera has been imported many times but epidemics have not occurred; however, in 1978 about a dozen cases occurred in Louisiana and a few other cases occurred in Texas in 1981.

Cholera is transmitted in water and food contaminated with excreta from patients and convalescent carriers. In the small intestine the organisms adhere to the epithelium, multiply, and produce the enterotoxin. No penetration of the intestinal wall or invasion of the body occurs. The symptoms include vomiting and profuse diarrheal (rice-water) stools that result in severe dehydration, loss of minerals, increased blood acidity, and hemoconcentration. Replacement of fluids, salt, and bicarbonates is essential to treatment, and supplemental tetracycline therapy can help to eliminate the vibrios from the intestinal tract.

Bacteriological diagnosis can best be made by isolating V. *cholerae* from the diarrheic stools. Selective media have a pH of 8.5, which inhibits most other intestinal bacteria. Identification is based on biochemical tests and agglutination of the cells by O1 antiserum.

A vaccine is available, consisting of killed whole cells; however, it only confers about 50 percent protection for about 3 to 6 months. In general, control by improved sanitation is preferable to immunization for prevention of outbreaks.

Haemophilus

The members of this genus require the **X factor** (heme) and/or the **V factor** (nicotinamide adenine dinucleotide, NAD) for growth. A medium called **chocolate agar** provides these factors and is prepared by adding blood to melted agar at 45°C and then heating the medium to 80°C to rupture the blood cells.

Of the *Haemophilus* species pathogenic for humans, one of the most important is *H. influenzae*, the leading cause of meningitis in children between 6 weeks and 2 years of age. Despite its name, *H. influenzae* does not cause influenza (which is caused by a virus); however, it can cause a severe secondary pneumonia in influenza patients. The major virulence factors produced by this species are (1) an antiphagocytic **capsule,** (2) an **endotoxin,** and (3) a heat-stable factor that causes **loss of cilia** from epithelial cells lining the respiratory tract. There are 6 capsular serovars, a to f; of these, serovar b is most frequently found in severe infections. Nonencapsulated strains commonly occur in the human respiratory tract, but these are much less virulent than encapsulated strains and act only as opportunistic pathogens.

H. influenzae has an airborne person-to-person mode of transmission. The most severe disease caused by this species is meningitis in children. The mortality rate in untreated cases is 90 to 100 percent, and even those who survive may be afflicted by deafness, speech impediments, behavioral problems, or other manifestations of damage to the central nervous system. The limits of the age group that is most susceptible to *Haemophilus* meningitis are determined by the fact that children 6 to 8 weeks to 2 to 3 years old no longer have the maternal bactericidal antibodies (acquired by passive natural immunization) that previously protected them, and they do not begin to make bactericidal antibodies of their own until about 3 years of age. No vaccine is presently available; however, an experimental vaccine consisting of purified b capsular polysaccharide may prove to be effective.

Another infection caused by *H. influenzae* is acute bacterial epiglottitis, which occurs mainly in older children. In this disease the epiglottis is inflamed and may become swollen to the extent that blockage of the trachea occurs and breathing is obstructed. Death can occur within 24 h of onset of the infection.

Immediate antibiotic therapy is necessary for cases of *Haemophilus* meningitis and epiglottitis because of the rapid course and high mortality rate of these infections. Initial treatment is accomplished by a combination of ampicillin and chloramphenicol. Ampicillin alone is often effective; however, nearly 10 percent of *H. influenzae* strains are now resistant to this antibiotic because of a β-lactamase that is coded for by a transmissible plasmid. Resistance to chloramphenicol occurs only rarely; however, this antibiotic may cause toxic effects in the body. Treatment with chloramphenicol may be discontinued if subsequent tests for β-lactamase production prove to be negative.

ANAEROBIC GRAM-NEGATIVE NONSPORE-FORMING RODS

Of the various genera of this group, *Bacteroides* and *Fusobacterium* have the greatest significance for clinical bacteriology, and *B. fragilis* is the most commonly encountered species. Members of *Bacteroides* and *Fusobacterium* occur as part of the normal flora of the intestine, oral cavity, nasopharynx, oropharynx, vagina, and urethra, and in these locations they are relatively harmless. However, if they gain access to other areas of the body, as by wounds, bowel surgery and other kinds of surgery, human or animal bites, dental extractions, uterine infection after abortion, or similar means, then they may establish severe infections that are usually characterized by abscess formation and tissue destruction. Abscesses may be formed in any part of the body, but most often near some mucosal surface that provided that portal of entry. Gas is frequently present in anaerobic infections, and if the lesions discharge fluid, this usually has a foul odor. Factors leading to such infections are those that cause depletion of oxygen in tissues and lower the oxidation-reduction potential, such as traumatic tissue injury (wounds), lack of local blood circulation, presence of facultative bacteria that can use up oxygen, or substances that are toxic to tissue. Anaerobic infections are often refractory to some ordinary antibiotics such as aminoglycosides. Identification of the bacterial species involved is helpful because many species have predictable patterns of susceptibility to various antibiotics; however, anaerobic infections are often polymicrobial (more than one species present), which may complicate treatment.

RICKETTSIAS AND CHLAMYDIAS

Rickettsia

These arthropod-borne intracellular parasites cause a variety of infections in humans. The epidemiology of Rocky Mountain spotted fever has been discussed in Chap. 35, and Table 35-5 lists other rickettsial diseases, the species which cause them, and their arthropod vectors.

One species, *Rickettsia prowazekii*, the causative agent of classical typhus fever (epidemic typhus), has been responsible for much human suffering and death. It is transmitted from human to human by means of body lice, which themselves die of the infection. Typhus fever begins with chills, fever, headache, generalized aches and pains, and exhaustion. The rickettsias multiply within the cells that form the walls of blood capillaries: damage to capillaries of the skin leads to a characteristic red or purple skin rash, and damage to brain capillaries results in neurological symptoms such as hallucinations, delirium, stupor, deafness, hand tremors, and interference with speech. The mortality rate for untreated cases is usually 10 to 40 percent but has been as high as 70 percent in some epidemics. The disease can be effectively treated with tetracycline or chloramphenicol. Laboratory diagnosis of typhus (and other rickettsial infections) is usually done by detecting the development of antirickettsial antibodies in the patient's serum.

Some who recover from typhus fever continue to harbor the rickettsias in a latent form in their lymph nodes. These persons serve as a reservoir to maintain the rickettsias between epidemics, and years after the primary infection these carriers may themselves experience a mild recurrence of typhus fever known as Brill-Zinsser disease.

Until recently it was believed that the human and the louse were the only hosts for *R. prowazekii*; however, the rickettsias have been found in flying squirrels in the southern United States, and it seems likely that occasional human cases of typhus which have appeared in this region may be attributable to these squirrels and their arthropod parasites.

Epidemic typhus is a disease that accompanies disaster, war, and famine. After it was introduced into Spain, possibly by soldiers who fought the Turks in Cyprus, typhus caused the death of 17,000 Spanish soldiers during the siege of Granada in 1489, where fewer than 3,000 were killed in combat. From Spain the disease spread to Italy. In 1528, when the French army was on the verge of victory in the siege of Naples, typhus struck down 30,000 soldiers, and those who were spared were forced to withdraw. Typhus epidemics had decisive effects on the war in the Balkans in the sixteenth century, the Thirty Years' War, the Napoleonic campaigns, and the Serbian campaigns during World War I. Epidemics occurred in Italy and Yugoslavia in World War II, and typhus spread from the German concentration camps to the civilian populations of Germany,

France, and England. It did not reach epidemic proportions in those countries, but in Japan and Korea, 30,000 cases were reported in 1946 and 1947. At present the disease is confined to a few endemic foci in Africa, Central America, and South America, and possibly in the southern United States.

Prevention of typhus epidemics can be achieved by applying insecticides to louse-infested humans and their clothing to destroy the arthropod vector. In addition, a vaccine consisting of killed whole rickettsias is available, although the immunity lasts only about a year.

Chlamydia

The chlamydias are obligate intracellular parasites that have a complex life cycle (Chap. 13). Unlike rickettsias, they are not transmitted by arthropod vectors.

Chlamydia trachomatis causes several types of infections. Some serovars cause trachoma, a chronic keratoconjunctivitis that is the single greatest cause of blindness in the world and is highly endemic in developing countries. The chlamydias are acquired by direct contact with a patient, fomites, or via flies.

In industrialized countries, other serovars of *C. trachomatis* are among the most common sexually transmitted agents of disease. Some infections are relatively mild, such as nongonococcal urethritis (NGU). (NGU may also be caused by a mycoplasma, *Ureaplasma urealyticum*.) Another relatively mild infection is inclusion conjunctivitis, an acute, purulent disease of the eyes that is acquired by a newborn from the mother's genital tract. The disease usually disappears spontaneously within a few months, even without treatment. Other *C. trachomatis* infections, more serious and invasive, may occur, such as lymphogranuloma venereum, which is characterized by enlarged lymph nodes in the analgenito region, inflammation of the rectum, and constitutional symptoms such as chills, fever, and headache.

Chlamydia psittaci causes psittacosis, an infection of birds which can be transmitted to humans, usually by inhalation of infectious dust derived from infected avian feces. For example, poultry handlers or workers in poultry processing plants may acquire the infection. Psittacosis has economic importance for agriculture: the mortality rate in flocks of domestic fowl can be as high as 30 percent. In humans the disease may range from a relatively mild respiratory infection to a severe pneumonia that may eventually involve the central nervous system, with encephalitis, coma, convulsions, and death.

Isolation of chlamydias is done by inoculation of tissue cultures or the yolk sac of embryonated chicken eggs; when stained by Giemsa's stain, the infected cells exhibit characteristic intracellular inclusions.

Among the antimicrobial agents used for treatment of chlamydial infections are tetracyclines, sulfonamides, and erythromycin. Effective vaccines for prevention of chlamydial infections are not available.

MYCOPLASMAS

Mycoplasma

The cell wall–less bacteria belonging to the genus *Mycoplasma* are parasites of mucous membranes and joints and are mainly pathogens of animals, but some species are pathogenic for humans. The excretion of **hydrogen peroxide** by mycoplasmas adherent to the surface of tissue cells appears to be an important factor in causing tissue damage.

Mycoplasma pneumoniae occurs only in humans and is the only *Mycoplasma* species that is found in the human respiratory tract. It has an airborne mode of transmission and is the causative agent of primary atypical pneumonia (also known as walking pneumonia). The disease most frequently affects persons 5 to 25 years old and is characterized by a gradual onset with headache, sore throat, fever, and cough. The mortality rate is less than 1 percent. Tetracycline is effective for treatment. On the basis of suitable isolation techniques, it has been estimated that *M. pneumoniae* is the causative agent in 25 percent of all pneumonias in young adults and 9 percent of the pneumonias in children. For isolation, high levels of penicillin are used in the culture media to inhibit other bacteria; these levels do not inhibit mycoplasmas because of their lack of a cell wall.

FACULTATIVELY ANAEROBIC GRAM-POSITIVE COCCI

Staphylococcus

The major pathogen of this genus is *Staphylococcus aureus* (Fig. 36-11), the causative agent of many suppurative processes ranging from localized abscesses which can occur anywhere in the body to fatal septicemias and pneumonias. *S. aureus* occurs in the nasopharynx, on normal skin, and in the intestines. Infections occur when staphylococci enter the body through breaks, cuts, and abrasions in the skin or mucous membranes.

Some of the virulence factors produced by human strains of *S. aureus* are described in Chap. 31: the **α toxin**, the **Panton-Valentine factor, protein A,** and **coagulase.** Other factors include the **δ toxin,** which damages tissue cells by its action as a phospholipase; and **lipase,** which catalyzes the hydrolysis of fats and oils on skin, in sebaceous secretions, and in blood plasma and may aid penetration of staphylococci into the body.

S. aureus causes localized infections in which the characteristic lesion is a walled-off "fort," the **abscess**—a cavity filled with pus cells (neutrophils), dead tissue, and bacteria. Fibrin is deposited at the periphery of the abscess; this may be aided by the staphylococcal coagulase. Neutrophils can penetrate the abscess

Figure 36-11. *Staphylococcus aureus* showing characteristic grapelike clusters of cocci (X1,300). *(Courtesy of U.S. Naval Biological Laboratory.)*

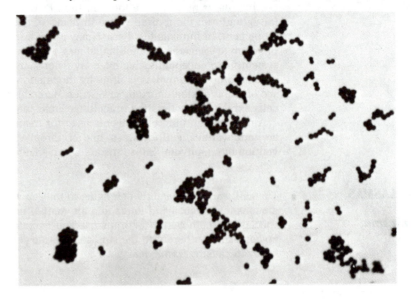

and, if not killed by the staphylococcal leukocidins, can phagocytoze the staphylococci. Phagocytosis can be greatly aided by antibodies; however, there is no circulation within an abscess and antibodies can penetrate the abscess only by a slow diffusion process. Antibiotics have a similar difficulty in reaching the staphylococci and may have to be administered to a patient for a long period in order to be effective.

Although S. aureus does not readily spread through tissues, the bacteria can sometimes be carried by the blood or within neutrophils to sites far removed from the original abscess. Indeed, nearly any organ or tissue may serve as a site for a secondary infection, leading to severe complications such as chronic osteomyelitis (bone infection that is difficult to treat because of poor penetration of the affected area by antibiotics), pneumonia, meningitis, endocarditis, and many more.

Another disease caused by S. aureus is toxic shock syndrome (TSS). Although TSS was first described in 1978, it may have occurred long before then as a complication arising from S. aureus infections. The disease occurs mainly in young women during a menstrual period, but males also may occasionally develop TSS from S. aureus infections. The symptoms, believed to be caused by **exotoxin C,** include fever, diarrhea, vomiting, shock, and a skin rash. Of 941 cases of TSS reported in 1980, 73 were fatal. TSS has been associated with the use of tampons, which may provide an environment for staphylococcal growth and toxin formation.

The treatment of staphylococcal infections is complicated not only by poor penetration by antibiotics into abscesses but also by the **multiple drug resistance** (resistance to several antibiotics, usually plasmid-mediated) that is exhibited by many clinical isolates of S. aureus. For instance, the use of penicillins for treating S. aureus infections would be desirable because of the bactericidal action of these antibiotics, but most clinical isolates of S. aureus are penicillin-resistant because they produce a plasmid-mediated β-lactamase. Certain semisynthetic penicillins, such as methicillin and oxacillin, are not destroyed by β-lactamase and thus can be effective in treating penicillin-resistant staphylococcal infections.

Certain strains of S. aureus that produce an **enterotoxin** can cause staphylococcal food poisoning. The epidemiology of this disease has been discussed in detail in Chap. 35. After consumption of an intoxicated food, symptoms begin within 1 to 6 hours and include severe nausea and vomiting and moderate diarrhea, but no fever. The disease lasts for about a day and, although extremely unpleasant, is not fatal.

AEROTOLERANT FERMENTATIVE GRAM-POSITIVE COCCI

Streptococcus

Certain Streptococcus species can cause infections in humans (Table 36-3). Identification of an isolate as belonging to a particular species depends on determination of the Lancefield group-specific antigens, on the type of hemolysis produced on blood agar, and often on a variety of additional biochemical tests. Of the pathogenic streptococci of human origin, S. pyogenes and S. pneumoniae are the most important.

Streptococcus pyogenes. This species comprises Lancefield group A. The cocci are arranged in long chains (Fig. 36-12), and colonies on blood agar are β-

Table 36-3. Some *Streptococcus* Species Pathogenic for Humans

Group and Species	Lancefield Group	Main Diseases Caused
Group A, B, and C streptococci		
S. pyogenes	A	Streptococcal sore throat (acute pharyngitis), scarlet fever, erysipelas, impetigo, acute glomerulonephritis, rheumatic fever, puerperal fever, septicemia
S. agalactiae	B	Meningitis and septicemia in newborn infants; occasionally other infections in adults; mastitis in cattle
S. equisimilis	C	Mild upper respiratory tract infections, erysipelas, puerperal fever; also various infections of animals
Group D enterococci		
S. faecalis	D	Urinary tract infections, endocarditis
S. faecium		
S. durans		
Group D nonenterococci		
S. bovis	D	Endocarditis
Viridans group		
S. sanguis	H	Endocarditis, dental caries
S, mutans	None	Endocarditis, dental caries
S. salivarius	K or none	Endocarditis
S. mitis	None	Endocarditis
Pneumococci		
S. pneumoniae	None	Lobar pneumonia

hemolytic (produce zones of clearing around the colonies). Major toxins or virulence factors produced by S. *pyogenes* are described in Chap. 31 and include the **M protein, streptolysin O** (SLO), **streptolysin S** (SLS), **erythrogenic toxin, streptokinase,** and **deoxyribonuclease** (DNase).

S. *pyogenes* is transmitted mainly by aerosols generated by carriers and clinical patients. Among the infections caused by S. *pyogenes* are streptococcal pharyngitis (streptococcal sore throat) and scarlet fever. These infections may give rise to complications, including inflammation of the middle ear (otitis media), mastoid bone (mastoiditis), sinuses (sinusitis), lungs (streptococcal pneumonia), heart valves (rheumatic fever), kidney (acute glomerulonephritis), and many others.

Streptococcal Pharyngitis. At least 300,000 cases of this disease occur each year in the United States. The disease is characterized by fever, enlargement of the lymph nodes of the neck, and a red, raw, and often bleeding throat surface. Scarlet fever is similar to the pharyngitis except for the skin rash that appears because of the erythrogenic toxin. The rash may extend to all parts of the body, appearing on the first or second day after onset of the disease. Clinical diagnosis of streptococcal pharyngitis and scarlet fever can be confirmed by isolating S. *pyogenes* from the pharynx. Specimens for cultural confirmation should be taken before antibiotic therapy is started. Penicillin and erythromycin are effective for treatment, which should be initiated as soon as possible because the

Figure 36-12. *Streptococcus pyogenes,* showing arrangement of the cells in chains (X1,600). *(Courtesy of J. Nowak, Documenta Microbiologica, part 1, "Bakterien," Gustav Fischer Verlag, Jena, Germany, 1927.)*

longer the steptococci continue to infect the body, the greater the chance that serious complications such as rheumatic fever may occur. If the organism cannot be isolated, circumstantial evidence for streptococcal infection can be obtained by detection of a rising level of antibodies in the patient's serum against certain streptococcal antigens, particularly SLO and DNase.

The only type of immunity that is protective against infection by S. pyogenes is humoral immunity: antibodies against M proteins act as opsonins, enhancing phagocytosis of the streptococci. Although it is theoretically possible to vaccinate against streptococcal infection by using M proteins as the immunizing antigens, as a matter of practicality it has not been possible to do so because of the great number of M proteins (over 60 types) that occur among the various streptococcal strains.

Rheumatic Fever. Inflammation and degeneration of the heart valves may follow approximately 3 percent of untreated cases of streptococcal pharyngitis or scarlet fever. Further damage to the heart increases with each subsequent streptococcal infection. The disease occurs most frequently in the preadolescent age group (3 to 10 years) and is a cause of death in the 5- to 20-year-old age group. Antibiotics are ineffective in curing the disease, and streptococci are usually not present in the heart or on the valves. The explanation of how the inflammation occurs in the absence of the bacteria is still not well understood, but it is likely to have an immunological basis. For instance, a similarity exists between certain S. pyogenes antigens and cardiac tissue antigens. In persons who have developed antibodies against the streptococcal antigens, the antibodies might cross-react with cardiac tissue and cause an inflammatory reaction.

Other Diseases. Acute glomerulonephritis (AGN) also appears to have an immunological basis. It is an inflammation, but not an infection, of the kidney that may occur following a S. pyogenes infection of the throat or skin. The disease can become chronic or even fatal; it can lead to kidney failure and to the necessity for the patient to be placed on a dialysis machine that carries out kidney functions.

Erysipelas is an acute skin infection caused by S. pyogenes. The lesions usually occur on the face and legs, and the skin becomes bright red, edematous, and covered with vesicles (fluid-filled blisters). The source of the streptococci may be the patient's respiratory tract, a carrier, or soiled linens, dressings, and other fomites. Control is best accomplished by good personal hygiene and by prophylaxis (preventive treatment) with antibiotics following known exposure.

Impetigo contagiosum is described by clinicians as a purulent dermatitis. Vesicular lesions appear most commonly on the face and hands but may cover the body and become crusted and ulcerated (form open sores). Impetigo is caused by S. pyogenes together with staphylococci and is transmitted by direct contact with a patient or contaminated objects. Because children are usually susceptible, epidemics are likely to spread rapidly through schools and camps unless controlled by good sanitary and hygienic practices.

Puerperal fever, sometimes called puerperal sepsis, is a streptococcal infection of the uterus of a mother and is acquired at childbirth. It is a serious disease but can be controlled by proper obstetrical practices. Sources of infection may

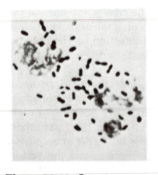

Figure 36-13. *Streptococcus pneumoniae, showing diplococcus arrangement (X1,500). (Courtesy of J. Nowak, Documenta Microbiologica, part 1, "Bakterien," Gustav Fischer Verlag, Jena, Germany, 1927.)*

be streptococci in the patient's genital, intestinal, or respiratory tract or on the skin of the patient or her attendants. Use of prophylactic antibiotics minimizes the danger of infection.

Streptococcus pneumoniae. Unlike S. pyogenes, this species is α-hemolytic and occurs mainly in the form of diplococci (Fig. 36-13). The major virulence factor is the **polysaccharide capsule,** which effectively inhibits phagocytosis; other virulence factors include an oxygen-labile hemolysin, **pneumolysin O,** which is similar in many respects to the streptolysin O of S. pyogenes.

Although pneumonia may be caused by many different organisms, at least 70 percent of bacterial pneumonias are caused by S. pneumoniae. Since 20 to 40 percent of the population harbors virulent strains of S. pneumoniae, contracting pneumococcal pneumonia is not merely a matter of acquiring the organism; the resistance of the host must also be lowered, as by a preliminary respiratory infection such as a common cold. In the lungs, pneumococci initiate an inflammatory response; fluid from nearby blood capillaries begins to fill the air sacs, and eventually the area of affected lung is no longer soft and spongy but rather is solid. Neutrophils also accumulate and attempt to phagocytoze the pneumococci; however, encapsulated pneumococci are highly resistant to phagocytosis unless the patient makes antibodies against the capsular polysaccharide. Even if the patient recovers, complications may subsequently occur, such as otitis media, mastoiditis, pneumococcal meningitis, septic arthritis, and endocarditis. Penicillin is the drug of choice for treatment of pneumococcal pneumonia; however, some strains of S. pneumoniae that are resistant to penicillin and other antibiotics have appeared in recent years.

Bacteriological diagnosis is based on isolating the organisms from the patient's sputum. A key test for identification is the ability of bile to induce lysis of pneumococci **(bile solubility test);** other α-hemolytic streptococci are not affected. In another test, diagnostic antisera are used to cause the pneumococcal capsules to become easily visible under a microscope **(quellung reaction).**

Only antibodies against the capsule of the pneumococcus are protective against pneumococcal pneumonia, and a vaccine has long been sought that would stimulate production of such antibodies. At first this seemed impractical because of the occurrence of over 80 kinds of capsular serovars. However, 14 of the serovars have been found to be responsible for at least 80 percent of the cases of pneumococcal pneumonia. Thus a vaccine consisting of the capsular polysaccharides from these 14 serovars was developed and is now licensed for use in the United States. Since pneumococcal pneumonia is a major killer of elderly persons, the vaccine is particularly recommended for this segment of the population.

AEROBIC/FACULTATIVELY ANAEROBIC GRAM-POSITIVE SPOREFORMING RODS

Bacillus

Although *Bacillus* cereus can cause a type of food poisoning in humans, only one species of the genus *Bacillus* is highly pathogenic for humans: *B. anthracis* (Fig. 36-14), the causative agent of anthrax. Two virulence factors are produced by this species: (1) the **anthrax toxin** (see Table 31-3) and (2) an antiphagocytic **polypeptide capsule.**

Anthrax is mainly a disease of herbivorous animals, particularly cattle and

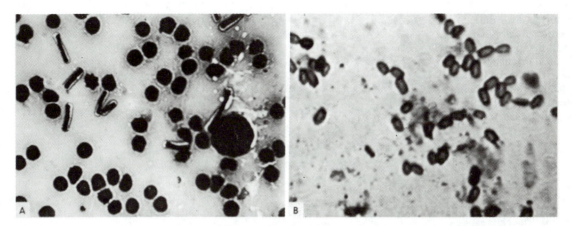

Figure 36-14. (A) Blood smear taken from a sheep that died of anthrax, showing vegetative capsulated B. *anthracis* (rod forms surrounded by a clear area). The large round objects are blood cells. (B) A group of B. *anthracis* spores. *(Courtesy of USDA, Bureau of Animal Industry.)*

sheep. Three forms of the disease occur, as defined on the basis of their mode of transmission. (1) **Intestinal** anthrax is frequently contracted by cattle and sheep that ingest anthrax spores by grazing on pastures, or on the hay produced from pastures, where anthrax-infected animals have died or been buried. Such pastures may remain infectious for many years. (2) **Cutaneous** anthrax is often contracted by cattle and sheep when anthrax spores get into a scratch or abrasion. Humans can also contract the disease by handling infected animals or products made from infected animals, such as hides, wool, and horn. (3) **Pulmonary** anthrax is contracted only by humans, who acquire the organism by inhaling dust from contaminated animal products. For instance, infectious dust may arise during the process of sorting or combing raw wool contaminated with anthrax spores. Regardless of the mode of transmission, anthrax begins as a localized infection of the skin, lungs, or intestine which, in fatal cases, develops into a septicemia.

Penicillin is the drug of choice for treatment of anthrax. A live attenuated vaccine is available for prevention of anthrax in cattle and sheep.

ANAEROBIC SPOREFORMING RODS

Several species of clostridia can cause disease in humans. Four of these species are discussed below.

Clostridium

Clostridium tetani. This species is the causative agent of tetanus. A characteristic feature of these bacteria is the formation of endospores at one end of the cell, giving the organisms a drumstick appearance (Fig. 36-15). The only virulence factor that is produced is a powerful **neurotoxin**. C. *tetani* occurs in the intestinal tracts of herbivorous animals and is widely distributed in soil. In wounds providing conditions favorable for its growth, C. *tetani* may grow and elaborate toxin. Such wounds are usually deep and ragged, with devitalized tissue in which aerobic or facultative organisms are also growing. The toxin becomes bound to nearby peripheral motor nerves and travels along these nerves to the central nervous system, where it exerts its effects (see Chap. 31). Symptoms include painful and violent contractions of the muscles, usually of the neck and jaw (restricting opening of the mouth, giving rise to the term *lockjaw*). This is

Figure 36-15. *Clostridium tetani.* Spores are terminal and swell the rods, producing a typical drumstick appearance. Cells range from 0.3 to 0.8 by 2.0 to 5.0 μm. *(Courtesy of General Biological Supply House, Inc.)*

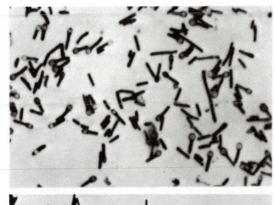

Figure 36-16. *Clostridium perfringens.* The cells range from 1.0 to 1.5 μm wide by 4.0 to 8.0 μm long. Spores are not present and are not usually formed in ordinary laboratory media. Spores may develop in special media; if formed, they are located in the center of the cell and do not appreciably distort the parallel sides of the rods. *(U.S. Army photograph from the Armed Forces Institute of Pathology.)*

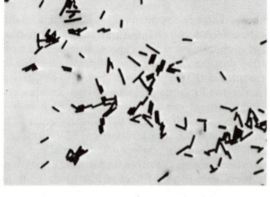

followed by paralysis of the thoracic muscles, frequently causing death from respiratory failure or cardiac failure. The mortality rate is 55 to 65 percent. Therapy involves the administration of muscle relaxants, and antitoxin is given to neutralize any toxin that has not yet become fixed to nerve tissue.

Immunization against tetanus is accomplished by a vaccine containing tetanus toxoid, which stimulates the body to produce antitoxin. Booster immunizations should be given every 10 years to maintain an adequate level of antitoxin. If a wound conducive to development of tetanus occurs in a nonimmunized person, tetanus antitoxin should be administered as soon as possible to neutralize any toxin that may be produced.

Clostridium perfringens. Several clostridia may cause gas gangrene in humans, but the majority of cases are caused by *C. perfringens* type A (Fig. 36-16), which occurs as part of the normal flora of the human intestine. The virulence factors produced by *C. perfringens* include the **α toxin** (a phospholipase that damages the membranes of erythrocytes and tissue cells), the **θ toxin** (an oxygen-labile hemolysin), the **κ toxin** (a collagenase that destroys connective tissue), the **μ toxin** (a hyaluronidase that may aid invasiveness), and the **ν toxin** (a deoxyribonuclease).

As in tetanus, gas gangrene results from contamination of wounds. The disease

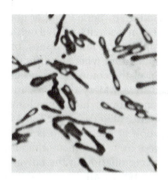

Figure 36-17. *Clostridium botulinum* (X1,500), showing many cells with oval, terminal, and subterminal spores. *(Courtesy of J. R. Porter.)*

is characterized by a toxemia, preceding and concurrent with the development of gas in the tissues. *C. perfringens* is highly invasive and spreads rapidly through tissues (see Chap. 31). Prevention of the disease rests on three measures: (1) surgical removal of dead tissue and blood clots that serve as breeding grounds for anaerobic bacteria, (2) administration of antitoxin, and (3) administration of antibiotics to stop growth of the bacteria. Once a case of gas gangrene has developed, the organisms are so rapidly invasive that it is difficult to halt their spread through tissues, and it may be necessary to amputate the affected limb to save the patient's life. An alternative treatment is the use of **hyperbaric oxygen:** the patient is placed in a chamber containing 3 atmospheres pressure of pure oxygen several times a day for half-hour periods. This treatment causes a high degree of tissue oxygenation that may halt the growth of the clostridia. However, none of these procedures may save the life of the patient if the clostridial toxins have become widely distributed throughout the body.

C. *perfringens* also causes a very common type of food poisoning. Various forms of raw meat or poultry may contain spores of *C. perfringens*. If the meat or poultry is cooked at a temperature which does not kill the spores and is subsequently kept at a temperature of 109 to 116°F for at least 2 hours, the clostridia may multiply. The food still appears and tastes normal, but after it is consumed the clostridia in the particles of ingested meat continue to grow and reach enormous numbers in the intestine. There they begin to sporulate, and during this process an enterotoxin is formed. The toxin causes abdominal cramps, diarrhea, and sometimes nausea and vomiting. The symptoms begin 8 to 24 hours after consumption of the food and last for 12 to 18 h.

Clostridium botulinum. This species (see Fig. 36-17) causes botulism, a paralytic disease of humans or animals. The disease is contracted by consuming food that contains the **neurotoxin** produced by these clostridia (see Chap. 35). The manner by which the toxin causes respiratory failure has been described in Chap. 31. The species is divided into 7 serovars, A to G, each of which produces an antigenically distinct kind of toxin. Botulism in humans is caused by *C. botulinum* serovars A, B, E, and (rarely) F. The spores of serovars A and B often occur in soil. Serovar E occurs mainly in the sediments of the Great Lakes and along the coasts of western North America and northern Japan, and the spores can contaminate fish or fish products. The epidemiology of foodborne botulism has been described in Chap. 35. Patients with botulism are treated with polyvalent antitoxin (i.e., antitoxin against toxin types A, B, and E) to neutralize any toxin that has not yet become fixed to nerve tissue. Patients experiencing respiratory difficulty may have to be placed on a mechanical respirator to prevent death from respiratory failure.

A type of botulism that is not acquired by ingestion of toxin can occur in infants 2 weeks to 6 months old. *C. botulinum* is not a normal part of the intestinal flora of humans and ordinarily cannot compete with normal flora organisms; however, in some infants whose normal intestinal flora is not yet well developed, ingested spores may be able to germinate and grow to some extent, producing enough toxin in the intestine to cause symptoms such as lethargy, excessive sleeping, poor head control, sluggish reflexes, and generalized weakness. In severe cases respiratory failure and death can occur.

Clostridium difficile. In the intestine of an adult person, *C. difficile* is either not present or is present in very low numbers. However, in patients treated with antibiotics such as ampicillin, clindamycin, chloramphenicol, tetracyclines, or cephalosporins, the normal flora may be inhibited to the point where *C. difficile* can compete effectively, producing a severe infection of the colon called pseudomembranous colitis. The disease is characterized by a watery diarrhea, inflammation of the colon wall, and a pseudomembrane (gray, white, or yellow patches on the intestinal wall). The mortality rate of untreated cases is 27 to 44 percent. Treatment of the disease by oral administration of vancomycin is usually effective. Two exotoxins appear to be involved in pseudomembranous colitis: an **enterotoxin,** toxin A, which causes fluid accumulation in the bowel, and a **cytotoxin,** toxin B, which kills tissue in the intestinal wall.

NONSPOREFORMING GRAM-POSITIVE RODS OF REGULAR SHAPE

Listeria

The cells of *Listeria monocytogenes* are short rods that may occur in chains. Four serovars can be distinguished based on surface antigens: 1, 2, 3, and 4; of these, 1 and 4 occur most often in human listeriosis. A **monocytosis-producing factor** can be extracted from the cells with chloroform and induces a hypernormal number of monocytes in the blood. An oxygen-labile **cytotoxic hemolysin** is also produced.

Listeriosis occurs in many species of mammals and birds. Symptoms are primarily neurological and are characterized by an acute encephalitis. Lesions (granulomas similar to those occurring in brucellosis, tularemia, or tuberculosis) are produced in many of the internal organs of infected animals. The organisms grow mainly within macrophages, and, as in brucellosis and tuberculosis, cell-mediated immunity is the only effective type of immunity. Humans can acquire *L. monocytogenes* from animals, although the mechanism of transmission is not known. In pregnant women, mild flulike symptoms may develop or the infection may be asymptomatic; however, the organisms may cross the placenta into the fetus and cause stillbirth or abortion. If the child is born apparently normal, meningitis or septicemia may develop 1 to 4 weeks later. Listeriosis mainly affects young children and aged persons; it also tends to occur in people whose immune mechanisms have been debilitated by some other type of illness such as alcoholism, diabetes, or cancer. Listeriosis can be treated effectively with penicillin G or ampicillin. No vaccine is presently available for prophylactic immunization.

NONSPOREFORMING GRAM-POSITIVE RODS OF IRREGULAR SHAPE

Corynebacterium

Three groups of corynebacteria occur: the animal or human parasites and pathogens, the plant pathogens, and the saprophytes found in soil and water. Human species include *C. diphtheriae*, the causative agent of diphtheria; *C. ulcerans*, which causes a diphtherialike disease; and *C. minutissimum*, which causes a skin disease called erythrasma.

C. diphtheriae (Fig. 36-18) produces several virulence factors. **Cord factor,** or **trehalose dimycolate** inactivates the mitochondrial membranes of phagocytes and other mammalian cells. **Diphthin** is a protease which inactivates IgA antibodies. **Neuraminidase** helps the bacteria to attach to mucous membranes of the throat by dissolving the mucus layer. **K antigens** are cell-wall proteins that

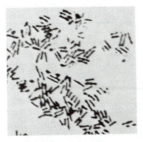

Figure 36-18. *Corynebacterium diphtheriae* (X1,300). *(Courtesy of J. Nowak, Documenta Microbiologica, part 1, "Bakterien," Gustav Fischer Verlag, Jena, Germany, 1927.)*

aid in attachment of the bacteria to host cells. The major virulence factor of C. diphtheriae is the diphtheria exotoxin. The gene that codes for this toxin (gene tox$^+$) is carried in the genome of a particular temperate bacteriophage called β, and a strain of C. diphtheriae produces toxin only when it harbors the β phage. Nontoxigenic strains of C. diphtheriae can be made toxigenic by infecting them with the β phage (lysogenic or phage conversion). The mechanism of action of the toxin as an inhibitor of protein synthesis is described in Chap. 31.

Diphtheria is mainly a disease of children. C. diphtheriae has an airborne mode of transmission and localizes in the tonsils, throat, and nose. The organisms are not invasive; however, the toxin produced by the organisms can circulate throughout the body, producing a general toxemia. Because of local necrosis of the throat cells, an inflammatory exudate may occur and develop into a tough pseudomembrane that can extend into the trachea and cause the patient to suffocate. Even when this does not occur, the systemic effects of the toxin are serious: the toxin is particularly damaging to heart muscle, nerve tissue, and kidneys. More than half the fatalities in diphtheria are due to cardiac damage. Treatment is by administration of diphtheria antitoxin; this should be done as early as possible because once the toxin has bound to mammalian cell membranes it can no longer be neutralized. Antibiotics are not effective in treatment because, unlike antitoxin, they cannot neutralize the toxin that has been produced. However, antibiotics can eliminate the carrier state in convalescent patients and thus help prevent further transmission of the disease. Laboratory confirmation of a clinical diagnosis is achieved by isolating the bacteria and demonstrating that they are toxigenic. Toxigenicity can be demonstrated by either an in vivo test using guinea pigs or an in vitro immunodiffusion test.

Once a very important disease and cause of death, diphtheria has become a clinical rarity in the United States because of two factors: (1) the detection of persons susceptible to diphtheria by means of the Schick test and (2) immunization of susceptible persons with diphtheria toxoid. The Schick test is performed by infection of a very small amount of diphtheria toxin into the skin. If the person is immune, the toxin is neutralized by antitoxin in the person's body and no skin reaction occurs; but if the person is susceptible (has no antitoxin), a local inflammatory reaction results.

MYCOBACTERIA

Mycobacterium

Both parasitic and saprophytic species of *Mycobacterium* occur; those that are pathogenic for humans are indicated in Table 36-4. Of particular importance is M. tuberculosis, which is responsible for over 90 percent of all cases of tuberculosis. It is almost exclusively a parasite of humans and has mainly an airborne mode of transmission.

M. tuberculosis can survive and multiply within phagocytic cells such as macrophages, and cord factor (trehalose dimycolate) in the bacterial walls can disrupt the respiration of mitochondria in phagocytes and tissue cells. Strains of M. tuberculosis possessing cord factor are virulent; strains lacking it are not. In cultures, the presence of cord factor is indicated when rough-looking colonies are formed due to the growth of the bacteria in cablelike arrangements (cords).

Pulmonary tuberculosis is a chronic, slowly advancing disease against which

Table 36-4. Some *Mycobacterium* Species Pathogenic for Humans

Group and Species	Clinical Significance
Tuberculosis group M. tuberculosis M. bovis	Both species are pathogenic and cause tuberculosis; they regularly exhibit susceptibility to antituberculosis drugs
Leprosy group M. leprae	Causes leprosy; has never been cultivated on laboratory media; can be grown in mouse footpads or in armadillos, where the temperature is favorable for growth (2 to 5°C below that of most mammals)

Runyon groups*
I Photochromogens. Nonpigmented when grown in the dark; yellow pigment formed when grown in the light; slow-growing
    M. kansasii
    M. marinum
	
	Causes a tuberculosis-like disease.
	Skin papules and ulcers; contracted from swimming in fresh or salt water; prefers temperature of 31°C for growth

II Scotochromogens. Red-orange pigment formed when grown in the dark or light; slow-growing
    M. scrofulaceum
    M. szulgai
	
	Causes cervical adenitis in young children
	Pulmonary disease, adenitis, bursitis

III Nonchromogens. Nonpigmented in dark or light; slow-growing
    M. avium/M. intracellulare group
 	
    M. xenopi
    M. ulcerans
	
	Tuberculosis-like disease in adults; lymphadenitis in children; usually resistant to ordinary antituberculosis drugs
	Tuberculosis-like disease
	Ulceration and necrosis of skin

IV Rapid-growing; may be pigmented or nonpigmented
    M. fortuitum
    M. chelonei
	
	Local abscess at the site of a trauma; occasionally causes a tuberculosis-like disease
	Has been isolated from patients with chronic respiratory disease

* The species in these four groups are environmental bacteria and do not appear to be directly transmissible from person to person.

cell-mediated immunity, rather than humoral immunity, is the major defense mechanism. The disease begins when the bacteria lodge within an air sac in the lungs. There they are rapidly ingested by macrophages, in which they multiply. The first evidence of infection is development of a hypersensitivity to the bacteria after about a month. This is determined by the **tuberculin test,** in which an extract of harmless proteins **(purified protein derivative,** or **PPD)** from *M. tuberculosis* is injected into the skin. A positive test is indicated by a red, swollen zone at the site of the injection within 48 h.

In untreated cases, as cell-mediated immunity develops, specific T lymphocytes migrate toward the bacilli in the lung and, upon contact with the bacilli, liberate lymphokines which attract macrophages to the area and cause them to accumulate. Eventually a small, pearl-gray nodule forms, consisting of bacilli, several concentric layers of macrophages, and an outer mantle of lymphocytes. This nodule, a type of granuloma, is called a **tubercle.**

In many instances, cell-mediated immunity develops sufficiently to be able to halt further advance of the infection. A lymphokine called **macrophage activating factor (MAF)** converts normal macrophages to **angry** or **activated macrophages,** which can arrest the growth of the tubercle bacilli. The tubercle becomes dormant and the infection remains subclinical. Tuberculin hypersensitivity will continue to exist as long as the dormant bacilli exist in the body—usually for a lifetime, unless the infection is eliminated by drug therapy.

In about 10 percent of untreated infections, cell-mediated immunity is not strong enough to halt the growth of the bacilli. The initial tubercle becomes larger and more of them develop. Within the enlarged tubercles, macrophages and tissue cells begin to die and fuse together to form an amorphous cheeselike mass **(caseation necrosis).** Even now the lesions may undergo a healing process called **calcification,** but if the infection progresses, eventually several tubercles coalesce to form an area of dead tissue that is large enough to be detected by a chest x-ray. As the area of dead tissue expands it may erode the wall of a bronchus, so that acid-fast bacilli begin to appear in the sputum coughed up by the patient (Fig. 36-19). The detection of acid-fast bacilli in sputum constitutes a presumptive laboratory test for tuberculosis, but isolation and identification of the mycobacteria are required for confirmation.

As the disease progresses the patient begins to exhibit loss of appetite, fatigue, weight loss, night sweats, and a persistent, worsening cough. If a blood vessel is eroded in the lungs, the sputum coughed up by the patient may become streaked with blood; also the tubercle bacilli may gain access to the blood and be transported to various parts of the body, establishing numerous secondary foci of infection. Death ultimately results when sufficient damage has occurred in the lungs or other vital organs.

The type of tuberculosis most prevalent in the United States today is *reactivation* tuberculosis, in which dormant bacilli from an old subclinical primary infection are no longer held in check and begin to proliferate rapidly. This form of the disease occurs most often in elderly persons whose resistance has been lowered by various factors such as malnutrition, alcoholism, and other stresses.

Chemotherapy is the most effective method for treatment of tuberculosis. If a person who has previously been tuberculin-negative becomes tuberculin-posi-

Figure 36-19. *Mycobacterium tuberculosis in sputum, acid-fast stain (X800). (U.S. Army photograph.)*

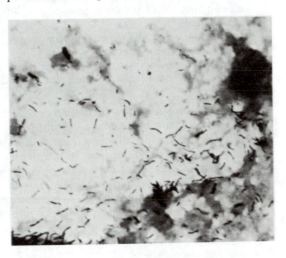

tive, antibacterial drugs are administered for several months to ensure that the infection will not progress to clinical tuberculosis. Treatment of a patient with clinical disease must be extended over a year or more because of the chronic nature of the disease and the walling off of the bacteria by tubercle formation. Among the effective drugs are isoniazid (INH), streptomycin, p-aminosalicylic acid (PAS), and rifampin.

A live attenuated vaccine(BCG vaccine) that induces cell-mediated immunity against tuberculosis is used in European countries and is about 80 percent effective. However, immunized individuals give a positive tuberculin test and thus this test is no longer of diagnostic value; for this reason the vaccine is not used in countries where the incidence of tuberculosis is relatively low, as in the United States.

BACTERIAL PATHOGENS OF PLANTS

Bacteria cause a wide variety of plant diseases characterized by such host reactions as galls (tumors composed of undifferentiated cells), wilts (loss of turgor), cankers (localized wounds or lesions resulting from necrosis of stems or bark), rots, deformed fruits, leaf spots, change in the color of plant parts, dwarfing, and retarded ripening of fruit. Bacterial plant pathogens may be disseminated in many ways—by wind, water, and soil movements; infected seeds and nursery stock; insect vectors; and infected farm tools, especially those used for pruning in orchards. Once inside the plant tissue, bacteria usually grow intercellularly. The following genera of bacteria contain plant pathogens; however, not all species in these genera are pathogenic.

Certain *Pseudomonas* species such as *P. syringae* cause leaf spot, leaf stripe, wilt, and similar diseases.

Xanthomonas spp., especially *X. campestris*, cause necrosis.

Erwinia spp. invade the tissues of living plants and produce dry necrosis, galls, wilts, and soft rots.

Agrobacterium spp. live in the soil or in the roots or stems of plants, where they produce galls (tumorlike growths).

The genus *Corynebacterium* contains plant pathogens that may be found in the soil as well as in diseased plants. They cause a vascular disease of alfalfa; ring rot of potatoes, grasses, and tomatoes; and diseases of many other plants.

Streptomyces spp. are responsible for potato scab, and a disease of sweet-potato roots and rootlets.

Spiroplasma spp. cause arthropod-borne disease in a wide range of crops and wild plants. The plants are often stunted and malformed.

Agrobacterium species are particularly noted for the ability to cause galls in plants. Galls are disorganized masses of plant cells, some large and swollen, others small and rapidly dividing (Fig. 36-20). As they develop, they may cut off the flow of water and nutrients in the plant, resulting in death. Crown gall, or plant cancer, may occur on fruit trees, sugar beets, and other broad-leafed plants where the stem comes out of the ground.

The causative agent of crown gall, *Agrobacterium tumefaciens*, has the unique ability of transforming normal plant cells into tumor cells. When the tumor

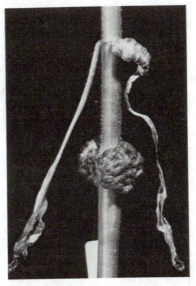

Figure 36-20. Crown gall, showing primary and secondary tumors and tumefaction in the petiole of one of the leaves of a sunflower plant (*Helianthus annus*). The tumor is the result of inoculation with *Agrobacterium tumefaciens*. (*Courtesy of Dr. Armin C. Braun, The Rockefeller Institute for Medical Research.*)

Figure 36-21. Bacterial wilt. Branches and stems of this tomato plant have wilted because of an infection with *Pseudomonas solanacearum*. (*Courtesy of USDA, Bureau of Plant Industry.*)

cells have become established, they continue to produce abnormal cells even in the absence of the bacterial pathogen; i.e., they multiply autonomously. The ability of *A. tumefaciens* to transform plant cells into cancer cells is due to a plasmid known as **Ti**; strains that do not possess this plasmid are not tumorigenic. During the course of infection of a wounded plant, the Ti plasmid is somehow transferred to the plant cells, where part of it is integrated into the nuclear material of the transformed plant cells. One unusual attribute of the transformed plant cells that is induced by the Ti plasmid is the ability to produce **opines** (unusual amino-acid derivatives such as octopine and nopaline); this creates a unique ecological niche for the cancer-inducing agrobacteria, since almost all strains of the bacteria can use the opines made by the cancer cells as a carbon and nitrogen source—an ability that is conferred on the bacteria by the Ti plasmid.

Agrobacteria are the only bacteria known to be involved in any form of cancerlike growth. As such they have served as a model system for mammalian cancer research.

Bacterial wilts are caused by the slime-producing bacteria that plug the passages in the plant through which water passes (Fig. 36-21). Wilt of sweet corn and some other varieties of corn is caused by *Erwinia stewartii*, bacterial wilt of cucumbers and muskmelons by *Erwinia tracheiphila*, that of tobacco by *Pseudomonas solanacearum*, and that of alfalfa by *Corynebacterium insidiosum*.

Cankers start in the water-conducting tissue of the plant, but they spread into the surrounding tissue. Bacterial canker of stone fruits decreases the yield of fruit or kills the trees. The most characteristic symptom of the disease is formation of canker accompanied by exudation of gummy substances, a symptom called **gummosis.** The bacterium that causes canker and gummosis is *Pseudomonas syringae*, which is not killed during the winter. Bacterial canker of

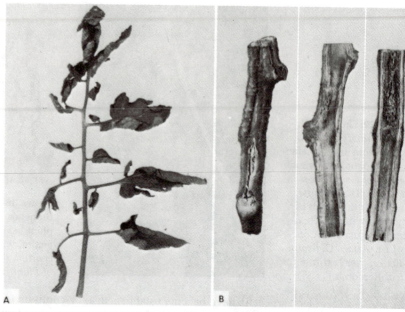

Figure 36-22. Bacterial canker. Leaves and stem of a tomato plant affected with bacterial canker, caused by *Corynebacterium michiganense.* (A) The curling and withering of leaflets that characterize this disease. (B) At the left is a portion of stem showing an open canker, and at the right are two stems cut lengthwise to show decay of the inner tissues. *(Courtesy of USDA, Bureau of Plant Industry.)*

Figure 36-23. Bacterial spot. These tomatoes show characteristic lesions of bacterial spot caused by *Xanthomonas campestris* pathovar *vesicatoria.* *(Courtesy of USDA, Bureau of Plant Industry.)*

tomatoes is due to *Corynebacterium michiganense* (Fig. 36-22). *Xanthomonas campestris* pathovar *citri* causes bacterial canker of citrus; in 1984 a variety of this organism was responsible for a serious outbreak of the disease in Florida.

Bacterial spot diseases usually occur on the leaves, but fruits and stems may be affected (Fig. 36-23). On peaches, bacterial spot is caused by *Xanthomonas campestris* pathovar *pruni,* and *P. syringae* pathovar *striafaciens* causes bacterial stripe of oats. Halo blight of oats is caused by *P. syringae* pathovar *coronafaciens,* which elaborates a toxic substance that produces yellowish areas surrounding the dead spots.

BACTERIAL PATHOGENS OF INSECTS

Some bacteria can cause insect diseases. Many of these diseases reach epidemic proportions, and if the insect species is beneficial, the diseases can cause great

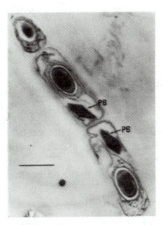

Figure 36-24. Electron micrograph of *Bacillus thuringiensis*, showing parasporal bodies (PB) and oval spores. The parasporal bodies are composed of a crystallized protein that is toxic for *Lepidoptera* larvae. The bar represents 1 μm. *(Courtesy of David J. Vitale and George B. Chapman.)*

harm. On the other hand, if the insects are of a harmful species, limitation of their population by disease may be desirable. Indeed, diseases of insects appear to be nature's control mechanism for the prevention of mass destruction of plants and animals by arthropods. Diseases of honeybees, known as foulbroods, exemplify the type of insect disease that must be prevented. Foulbroods are caused by sporeforming bacilli such as *Bacillus larvae*. One of the first bacterial diseases of harmful insects to receive great attention is characterized by dysentery and septicemia in locusts and grasshoppers. It is caused by a variety of *Enterobacter aerogenes*.

The idea that microorganisms may be used to control arthropod pests is not new, having been successfully used on a laboratory scale by Metchnikoff in 1879. Practical difficulties have deterred its development and application; however, highly successful results have been achieved in the control of the Japanese beetle by inducing milky disease in their grubs. The disease causes the blood of the sick grubs to become filled with bacteria and spores, giving it a milky appearance. The causative organism of type A milky disease is a sporeforming bacterium, *Bacillus popilliae*, and the causative organism of type B milky disease is *Bacillus lentimorbus*. The spores of these bacteria are highly resistant to desiccation, heat, and cold, and they survive in the soil for years. When these bacteria are introduced into soil where the grubs of the beetles develop, some of the grubs become infected; as the grubs die, more bacterial spores are introduced into the soil. This method has resulted in the virtual elimination of Japanese beetles in areas formerly heavily infested.

Bacillus thuringiensis is a sporeforming bacillus which causes a disease in lepidopterous larvae, such as the Mediterranean flour moth and other undesirable insects. The bacilli produce a toxin, which occurs in the form of a protein crystal **(parasporal body)** (Fig. 36-24). When the bacteria are ingested by larvae, the crystals dissolve and cause erosion of the gut epithelium. The toxin is not harmful to higher animals.

Although the use of insect pathogens has not been widely practiced, further study will surely lead to greater application because of the following obvious advantages:

1 *Permanency.* Once applied, bacterial spores such as those causing milky disease of Japanese beetles persist in the soil for a long time.
2 *Safety.* Most insect pathogens are harmless to plants and animals, and the danger of poisonous chemical residues is eliminated.

QUESTIONS

1 Distinguish between nonspecific and specific serologic tests for syphilis.
2 What special conditions are required for isolation of *Campylobacter jejuni*? What other organisms described in this chapter require special physical or chemical cultural conditions?
3 What kind of evidence is there that legionellosis occurred prior to the first recognition of the disease in 1976?
4 Zoonoses are diseases of animals that can be transmitted to humans. List the various zoonoses discussed in this chapter.
5 Why is humoral immunity developed by a patient relatively ineffective in contributing to recovery from brucellosis, listeriosis, and tuberculosis?

6 What general type of vaccine is most effective in preventing (a) brucellosis or tuberculosis, (b) pneumococcal pneumonia or meningococcal meningitis, and (c) diphtheria or tetanus?

7 List the bacterial pathogens that produce an enterotoxin and those that produce an antiphagocytic capsule.

8 What is the basis for antigenic subdivision of (a) *Salmonella*, (b) *Shigella*, (c) *Clostridium botulinum*, (d) *Haemophilus influenzae*, (e) *Streptococcus pneumoniae*, and (f) *Neisseria meningitidis*?

9 What are the characteristic features of an anaerobic infection? What general factors help to initiate an anaerobic infection? What bacterial species is the one most frequently isolated from anaerobic infections?

10 How and where might *Rickettsia prowazekii* maintain itself between typhus epidemics?

11 Name three organisms that can cause a pneumonia and indicate characteristics that can differentiate them.

12 Describe the characteristic kind of lesion produced by *Staphylococcus aureus* in tissues. Why might it be difficult for antibiotics to inhibit the staphylococci located in such a lesion? Why might it also be difficult for antibodies produced by the patient to inhibit the staphylococci?

13 In what fundamental way does rheumatic fever differ from most other kinds of diseases caused by pathogenic bacteria?

14 *Clostridium perfringens* food poisoning is not a true food poisoning like that caused by *Staphylococcus aureus* or *Clostridium botulinum*. Why? Is infant botulism a type of food poisoning? Explain.

15 What are the measures used for prevention of gas gangrene? What is the basis for the use of hyperbaric oxygen for treatment of a patient with gas gangrene?

16 In what fundamental way does the Schick test differ from the tuberculin test?

17 What is the basis for crown-gall formation in plants and why is this plant disease of interest to those doing research on animal cancer?

18 What advantages does biological control of insect pests have over control by chemical insecticides?

REFERENCES

See also references for Chaps. 31 and 35.

Agrios, G. N.: *Plant Pathology*, 2d ed., Academic, New York, 1978. *Deals with diseases of plants, disease cycles, parasitism, and pathogenicity; the mechanisms of infection and resistance are considered. An excellent reference for plant diseases in introductory microbiology.*

Finegold, S. M., and W. J. Martin: *Diagnostic Microbiology*, 6th ed., Mosby, St. Louis, 1982. *A reference text for medical microbiology laboratories and provides details on how to isolate, characterize, and identify pathogenic microorganisms.*

Krieg, A.: "The Genus *Bacillus*: Insect Pathogens," in M. P. Starr, H. Stolp, H. G. Trüper, A. Balows, and H. G. Schlegel (eds): *The Prokaryotes: A Handbook on Habitats, Isolation, and Identification of Bacteria*, vol. II, Springer-Verlag, New York, 1981, pp. 1743–1755. *This article describes the properties of bacterial species that are pathogenic for insects and indicates the practical importance of such bacteria.*

Lennette, E. H., A. Balows, W. J. Hausler, Jr., and H. J. Shadomy (eds.): *Manual of Clinical Microbiology*, 4th ed., American Society for Microbiology, 1985. *A comprehensive treatment on the isolation and identification of pathogenic microorganisms as described by various authorities in the field.*

Sun, M.: "The Mystery of Florida's Citrus Canker," Science **226**:322–323, 1984. *A summary of the epidemiology of the disease that led to the destruction of millions of seedlings in 1984 and threatened the entire Florida citrus industry.*

Chapter 37 Microbial Agents of Disease: Viruses

The fact that viruses are obligate parasites does not mean that they necessarily cause overt disease. Some viruses may cause latent or inapparent infections that never result in clinical signs or symptoms. Others can cause inapparent infections which may later give rise to clinical disease; here, the link between the virus and the disease has often been difficult to establish. Still other viruses are frank pathogens, causing disease in humans, animals, insects, or plants; these viruses have naturally received the most attention. Viruses that can cause human disease may be strictly limited to human hosts, or they may be mainly pathogens of animals, with humans serving only as accidental hosts. Viral pathogens that are restricted mainly or solely to animals or plants may cause great economic losses in agriculture. Laboratory identification of viral pathogens is more difficult than identification of bacterial pathogens, because we cannot ordinarily observe viruses directly with a light microscope or culture them on nonliving laboratory media. The use of tissue cultures and embryonated eggs does provide a means by which many viral pathogens can be cultivated in the laboratory; in addition, the antibody response of the patient can serve as circumstantial evidence for a particular infection. Viral diseases cannot be treated with the chemotherapeutic agents that are effective against bacteria (although a few antiviral drugs are being developed); consequently, great emphasis has been given to epidemiological and immunological methods by which viral infections can be prevented. For instance, such measures have resulted in total eradication of smallpox, and many other viral diseases have been greatly reduced in incidence. Moreover, recombinant DNA techniques are beginning to provide safer and less expensive vaccines than have been available in the past for prevention of viral infections, and they are also beginning to provide biochemical agents such as interferon, which have previously been difficult and expensive to obtain, for potential use in treatment of viral infections.

This chapter provides an introduction to the various viral pathogens and some selected diseases which they cause, as well as some of the ways in which these diseases can be diagnosed and prevented.

VIRUSES CONTAINING SINGLE-STRANDED (+) RNA

Picornaviridae (Picornaviruses)

Picornaviruses consist of small (22 to 30 nm in diameter) viruses composed of single-stranded RNA of the (+) type (which can function directly as mRNA in a host cell) contained within an icosahedral capsid (see Fig. 21-15). No lipid-containing envelope surrounds the nucleocapsid. The family includes **polioviruses, coxsackieviruses A and B,** and **echoviruses;** these picornaviruses are collectively referred to as **enteroviruses** because they are found in the intestines and excreted in the feces. They cause mainly intestinal infections and sometimes infections of the respiratory tract; neurological disease may also be produced. Other picornaviruses called **rhinoviruses** cause only the respiratory infections known collectively as the common cold. Still other picornaviruses, the **foot-and-mouth disease (FMD) viruses,** cause disease mainly in domestic animals.

Polioviruses. The morphology of polioviruses is depicted in Fig. 21-1A. Two outstanding characteristics of polioviruses are their affinity for nervous tissue and their narrow host range. Besides humans, most strains will infect only monkeys and chimpanzees. Three immunological types have been recognized. Type 1 is the common epidemic type, type 2 is associated with endemic infections, and type 3 is an occasional cause of epidemics.

Polioviruses are transmitted in nose and throat discharges and in the feces of infected individuals. Entry into the body, therefore, may be via the respiratory route or by the oral-intestinal route.

Polioviruses multiply in the oropharynx or intestinal mucosa, pass to the lymphatic system (tonsils or mesenteric lymph nodes), and eventually reach the bloodstream. In most instances the disease is *subclinical* (no symptoms). In 4 to 6 percent of cases, *nonparalytic* poliomyelitis occurs; this is characterized by fever and stiffness or pain in the neck and back muscles. In about 0.1 percent of cases, *paralytic* poliomyelitis occurs; the bloodborne virus infects the central

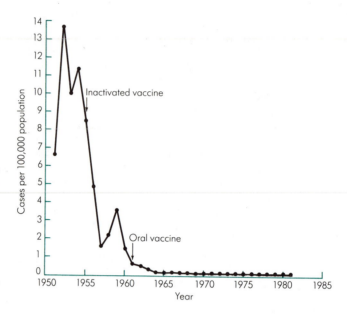

Figure 37-1. Reported case rates of paralytic poliomyelitis by year per 100,000 population, United States, 1951–1981. Arrows indicate when vaccination programs using the Salk (inactivated) vaccine and Sabin (oral) vaccine were begun. *(Centers for Disease Control: Morbidity and Mortality Weekly Report, Annual Supplement Summary 1981, issued October 1982.)*

nervous system and causes an inflammation of the gray matter of the spinal cord, especially the anterior horn, in which the cell bodies of motor neurons are located. When these cells are damaged, motor responses to the affected parts are weakened or destroyed. Paralytic poliomyelitis can result in death, although fatality rates have never exceeded 10 percent of the cases. The seriousness of the disease is also emphasized by the crippling effects on the survivors: one well-known example is that of Franklin D. Roosevelt, who acquired paralytic poliomyelitis as an adult and could walk only with the greatest difficulty for the rest of his life.

Two types of vaccines are available for immunization against poliomyelitis. The **Sabin vaccine** is an easily administered oral vaccine that is nearly 100 percent effective. It consists of live attenuated strains of the three immunological types. These strains infect the intestinal tract but, unlike virulent strains, they do not cause paralytic disease. Rather, they produce a long-lasting immunity by stimulating the formation of secretory IgA antibodies in the intestine and serum antibodies in the bloodstream. The secretory antibodies can neutralize the infectivity of virulent strains that may be subsequently encountered; thus these antibodies can prevent primary intestinal infection. The **Salk vaccine** is administered in a series of three intramuscular injections and is 70 to 90 percent effective. It consists of formalin-inactivated strains. This vaccine stimulates production of serum antibodies but not intestinal secretory IgA antibodies. Although the serum antibodies cannot prevent intestinal infection by a virulent poliovirus, they can prevent poliovirus in the bloodstream from reaching the spinal cord and causing paralysis.

In nations that have undertaken wide-scale immunization programs, paralytic poliomyelitis has decreased dramatically (Fig. 37-1). The vaccine presently used in the United States is the Sabin vaccine. Its use is not entirely without risk:

occasionally the type 3 mutant strain reverts to virulence, resulting in vaccine-associated paralytic poliomyelitis. The incidence is low (0.03 persons per 100,000 vaccinated persons) in comparison to disease rates of 7 to 13 per 100,000 population prior to vaccination programs (Fig. 37-1).

Rhinoviruses. Many kinds of viruses have been isolated from secretions from the respiratory tracts of persons with respiratory illness. Of these, rhinoviruses (Fig. 37-2) have been found to be major causes of the common cold, the most frequent of all human infections. Rhinoviruses differ from enteroviruses by being destroyed at pH values of less than 5 (such as would occur in the stomach) and by having an optimal growth temperature of 33°C rather than 37°C. These characteristics may help to explain why rhinoviruses are restricted to the respiratory tract.

Rhinoviruses are spread by droplets, by discharges from the nose and throat of an infected person, or by freshly contaminated fomites. After an incubation period of 12 to 72 h, the cold develops as an acute catarrhal infection of the nose, throat, sinuses, trachea, and bronchi, lasting approximately 2 to 7 days.

Antibodies appear in response to a rhinovirus infection, including secretory IgA antibodies in respiratory tract secretions. Such antibodies should protect against subsequent infections, yet a person may have several colds during a year. The reason is that there are at least 113 immunologically distinct types of rhinoviruses, and immunity against one type does not prevent infection by another type. No vaccine exists for prevention of the common cold because so many antigenic types would have to be included in the vaccine. Rhinoviruses are susceptible to human interferon, which may account for the fact that a person recovering from a cold often resists being reinfected for about 4 weeks.

Foot-and-Mouth Disease (FMD) Viruses. The FMD viruses are similar to the rhinoviruses and are the causative agents of foot-and-mouth disease, an acute communicable disease of cloven-hoofed animals, mainly cattle, sheep, goats, and swine. Humans are susceptible to infection and they may transmit the viruses to susceptible animals on their person or clothing. Animals which recover from the infection may act as asymptomatic carriers.

Entrance into a host is usually through abrasions of the skin and mucous membranes. In humans the disease is relatively mild and is characterized by

Figure 37-2. Electron micrograph of purified rhinovirus (X125,000). *(Courtesy of Frances Doane, University of Toronto.)*

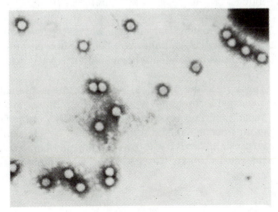

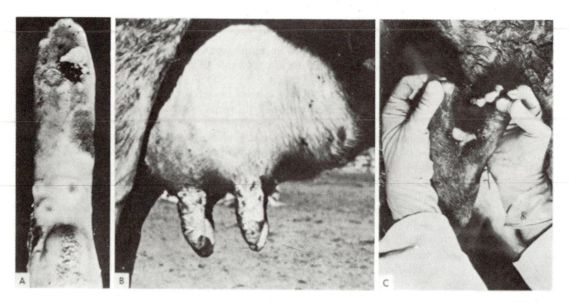

Figure 37-3. Foot-and-mouth disease in animals causes lesions of (A) tongue, (B) teats, and (C) foot. *(Courtesy of USDA.)*

fever, salivation, and a vesicular (blisterlike) eruption on the mucous membranes of the oropharynx and on the skin of the palms, soles, fingers, and toes. In animals, vesicles occur on the mucosa of the mouth, between the claws of feet, and on other parts of the body (Fig. 37-3). The mortality rate in animals is low, but FMD causes great economic losses due to reduced milk and meat production and to abortion. Laboratory diagnosis of FMD is done by complement-fixation, virus-neutralization, or mouse-inoculation tests. FMD has been eliminated from the United States by enforcement of quarantine, inspection, and regulations prohibiting importation of susceptible animals and animal products from countries where the disease exists.

Recovery from infection confers immunity to the particular viral serotype involved. Vaccines are available and are prepared from inactivated virus grown in tissue cultures; however, outbreaks of disease are frequently linked to incompletely inactivated vaccines or escape of live virus from research and production facilities. Recently a novel vaccine was developed by recombinant DNA techniques. A gene for an FMD viral coat protein was spliced into an *Escherichia coli* plasmid, and bacteria containing this plasmid then produced the viral protein antigen. The antigen was extracted from the bacteria and used as a completely safe immunizing agent. Such an approach avoids the dangers and difficulties associated with whole virus vaccines.

Togaviridae (Togaviruses)

Togaviruses are spherical (40 to 90 nm in diameter), have an icosahedral nucleocapsid contained within a lipoprotein envelope (the word *toga* means "coat"), and possess single-stranded RNA of the (+) type (see Fig. 21-15). Four groups occur which are differentiated by their antigenic properties: the **alphaviruses** and **flaviviruses** cause arthropod-borne infections of humans, **rubellavirus** causes rubella (German measles), and **pestivirus** causes several animal diseases.

Alphaviruses and Flaviviruses. Alphaviruses infect mainly animals and are transmitted to humans by mosquitoes. Some of the diseases caused by alphaviruses are indicated in Table 37-1. The diseases are characterized by a viremia which gives rise to fever, joint pains, and sometimes a skin rash. In some instances the bloodborne virus may invade the central nervous system and cause an encephalitis characterized by high fever, delirium, convulsions, paralysis, and coma. Mortality rates vary with the kinds of virus; for example, eastern equine encephalitis has a high mortality rate (50 to 70 percent), whereas western equine encephalitis virus infections are seldom fatal (3 percent).

Flaviviruses also have an animal reservoir and are transmitted to humans by either mosquitoes or ticks, depending on the virus species (Table 37-1). Like alphaviruses, most of the flaviviruses cause fevers or encephalitis.

The yellow fever virus differs from other flaviviruses by causing a disease in which the liver, kidney, and other organs are affected. In addition to generalized symptoms (headache, fever, muscle pains, backache), yellow fever is characterized by **albuminuria** (albumin in the urine, indicative of kidney damage); **hematemesis** (vomiting of blood); **hepatomegaly** (enlargement of the liver); and **jaundice** (yellowing of skin and mucous membranes, indicative of liver damage). The fatality rate for yellow fever is 25 to 30 percent. The epidemiology of urban and sylvatic yellow fever and the measures for control of the disease have been described in Chap. 35.

Rubivirus **(Rubellavirus).** Rubellavirus occurs only in humans and has an airborne transmission via respiratory secretions. The virus is the causative agent

Table 37-1. Some Examples of Alphaviruses and Flaviviruses

Virus	Reservoir	Vector	Occurrence	Disease in Humans
Alphaviruses				
Eastern equine encephalitis (EEE)	Birds	Mosquitoes	Western hemisphere	Encephalitis
Western equine encephalitis (WEE)	Birds	Mosquitoes	Western hemisphere	Encephalitis
Venezuelan equine encephalitis (VEE)	Rodents, horses	Mosquitoes	Western hemisphere	Acute febrile illness; sometimes encephalitis
Ross River	Mammals, humans	Mosquitoes	Australia, Pacific	Fever, rash, joint pains
Chikungunya	Monkeys, humans	Mosquitoes	Africa, India, Southeast Asia	Fever, rash, joint pains
Flaviviruses				
Yellow fever	Monkeys	Mosquitoes	Africa, South and Central America	Hemorrhagic fever, hepatitis, nephritis
Dengue	Humans	Mosquitoes	Tropics, worldwide	Fever, rash, joint pains
St. Louis encephalitis	Birds	Mosquitoes	Western hemisphere	Encephalitis
West Nile	Birds	Mosquitoes	Africa, Middle East, Europe	Fever, rash, joint pains
Japanese B encephalitis	Birds, swine	Mosquitoes	India, Japan, Far East	Encephalitis
Tickborne encephalitis	Rodents	Ticks	Europe, Asia	Encephalitis
Powassan	Rodents	Ticks	North America	Encephalitis

of rubella (German measles, or three-day measles)—a highly communicable disease that is unrelated to common measles (rubeola). Although rubella is communicable and usually occurs in epidemic form (every 7 to 10 years in the United States), the disease in general, as compared with rubeola, is of relatively short duration and of mild form.

After initial multiplication in the upper respiratory tract, the rubellavirus becomes distributed via the blood to the skin, lymph nodes, and joints. Respiratory symptoms do not usually occur. The disease is characterized by fever, malaise, lassitude, and other symptoms. A rash, which probably results from an immune response to viral antigens in infected skin cells, sometimes appears on the face and later on other parts of the body. A characteristic feature of the disease is swelling of lymph glands below the ear and at the nape of the neck.

The characteristic mild course of the disease can give a misleading impression of its importance. When rubella occurs in early pregnancy, it can cause serious congenital abnormalities in the fetus. During the 1964–1965 epidemic in the United States, it is estimated that there were 30,000 fetal deaths and that more than 20,000 children were born with defects of vision and hearing, heart disease, and mental retardation.

A vaccine prepared from live attenuated virus is available. Immunization with this vaccine in combination with the mumps and measles vaccine is recommended for all children at 15 months of age. Since the introduction of the rubella vaccine in 1969 there has been a 96 percent decrease in the incidence of rubella (Fig. 37-4).

Coronaviridae (Coronaviruses)

Coronaviruses have a helical nucleocapsid and contain single-stranded RNA of the (+) type. Their name is based on the occurrence of distinctive club-shaped particles of glycoprotein that project from the surface of the envelope and which give the effect of a crown or corona (see Fig. 21-15).

Figure 37-4. Reported cases of rubella (German measles) by year per 100,000 population, United States, 1966–1981. (*Centers for Disease Control: Morbidity and Mortality Weekly Report, Annual Supplement Summary 1981, issued October 1982.*)

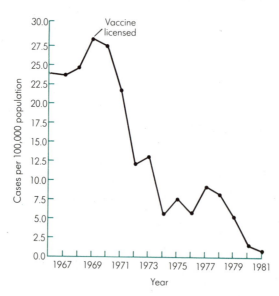

Like rhinoviruses, human coronaviruses represent causative agents of the common cold. At least three immunological groups exist. Other coronaviruses are important animal pathogens, causing diseases such as avian infectious bronchitis of chickens, mouse hepatitis, and gastroenteritis of piglets.

VIRUSES CONTAINING SINGLE-STRANDED (−) RNA

Orthomyxoviridae (Orthomyxoviruses)

The only members of the family *Orthomyxoviridae* are the influenza viruses. They are usually spherical, 80 to 120 nm in diameter (see Fig. 21-15); or they may be filamentous, up to several μm in length (see Fig. 21-11). They have a helical nucleocapsid with a core of single-stranded RNA of the (−) type (which serves as a template for synthesis of a complementary strand that acts as the messenger RNA). The RNA occurs in eight separate pieces to which protein subunits are attached. This ribonucleoprotein is wound up to form a rounded mass and is covered by a lipid-containing envelope. Protein spikes project from the envelope; these are hemagglutinins. Between the spikes are mushroom-shaped protrusions composed of neuraminidase. The role of the hemagglutinins and neuraminidases in attachment to host cells is discussed in Chap. 31.

The epidemiology of influenza and the role of antigenic variation in the periodic occurrence of pandemics are discussed in Chap. 35. Influenza is characterized by nasal discharge, headache, muscle pains, sore throat, a marked weakness and exhaustion, and a tendency to develop secondary bacterial pneumonias. During the disease the virus remains localized in the respiratory tract, where it kills ciliated epithelial cells. This killing effect may actually be due to cytotoxic T lymphocytes that respond to the viral antigens on infected cells.

A synthetic drug, amantadine, can prevent and even help to cure influenza caused by type A strains; it acts by preventing the penetration and uncoating of the virus after it attaches to a host cell (see Chap. 21). Another drug, ribavirin, inhibits not only influenza viruses but also other RNA viruses and DNA viruses by preventing the synthesis of guanosine monophosphate used for nucleic acid synthesis. The drug also inhibits host cells; however, the inhibitory effect is greater on the viruses because of their greater demand for nucleic acid precursors.

Vaccination against influenza is accomplished by use of formalin-inactivated virus. Immunity is based on the development of antibodies against the hemagglutinin antigens, thereby preventing viral attachment to host cells. Since immunity is subtype-specific, a mixture of several of the hemagglutinin subtypes most likely to cause infection is used in the vaccine. The antibodies most effective in preventing viral attachment are secretory antibodies present in mucus; however, inactivated vaccines mainly stimulate development of serum antibodies rather than secretory antibodies. The latter would be more efficiently produced in response to live attenuated vaccines, and such vaccines are currently under development.

Paramyxoviridae (Paramyxoviruses)

The paramyxoviruses have a structure similar to that of the orthomyxoviruses; however, they are larger (125 to 250 nm; see Fig. 21-15), the RNA is nonsegmented, and the hemagglutinin and neuraminidase activities occur together in a single kind of surface glycoprotein (HN). A second kind of surface glycoprotein (F) has hemolytic activity. The family includes **parainfluenza viruses** (see Fig.

21-1D), which cause 30 to 40 percent of all acute respiratory infections in infants and children. The infections range from mild coldlike disease to severe croup, bronchiolitis, and pneumonia. Other examples of paramyxoviruses include **respiratory syncytial virus,** the major cause of bronchiolitis and pneumonia in infants less than 1 year old; **Newcastle disease virus,** which causes avian pneumoencephalitis, an important disease of chickens; **measles virus;** and **mumps virus.**

Morbillivirus (Measles Virus). The measles virus occurs in respiratory tract secretions of patients who are in the early stages of measles and is transmitted by droplet infection. The portal of entry into the body is the upper respiratory tract or the conjunctiva.

Measles (rubeola) is one of the most common acute communicable human diseases, mainly affecting children but sometimes occurring in adults who have escaped previous infection. After initial multiplication, the virus becomes disseminated via the blood to the mucous membranes of the intestinal tract and urinary tract, to the skin, and to the central nervous system. Symptoms include fever; coldlike symptoms; cough; conjunctivitis; the occurrence of Koplik's spots (small bluish-white spots surrounded by a reddish area which occur on the mucous membranes of the cheeks and lips); and a characteristic red, blotchy skin rash. The rash probably results from an immune reaction with viral antigens on the surface of infected blood capillary cells, which in turn causes dilatation of the capillaries and leakage of blood into the tissues.

After exposure of a person to a known infection, prompt administration of gamma globulin (pooled antibodies from human blood) can prevent the disease or greatly modify its course. After measles develops, there is no effective treatment for the infection, although antibiotics may serve to check secondary bacterial invaders which may otherwise cause internal ear infection or pneumonia. Recovery from measles confers a high degree of immunity.

A rare, progressive, fatal disease of the central nervous system called subacute sclerosing panencephalitis (SSPE) may follow about 6 years after recovery from a case of measles. Although patients exhibit a high level of antibodies against the measles virus, it is not yet clear whether the SSPE virus is actually the measles virus or a variant of it.

The measles virus can be grown on tissue cultures of human or monkey cells and on chick embryos. Live vaccines prepared from such cultures by using the attenuated Edmonston strain are effective antibody stimulators, and immunization with the measles vaccine, in combination with the live rubella and mumps vaccines, is recommended for all healthy 15-month-old children. Vaccination programs begun in 1965 have resulted in a great decrease in the incidence of measles (Fig. 37-5).

Mumps Virus. Mumps virus is spread by droplets or fomites contaminated with infected saliva. Mumps (epidemic parotitis) is a common communicable disease that is endemic in most heavily populated areas. Epidemics are prevalent during the winter months, often occurring in schools and among military personnel. Most cases occur in children in the 5- to 15-year age group. Although mumps is a common childhood disease, it also attacks adults, in whom it may be a more serious infection.

Figure 37-5. Reported cases of measles by year per 100,000 population, United States, 1955–1981. For 1981 the incidence of reported measles reached a record low of 1.4 cases per 100,000 population—a 99.5 percent reduction from the prevaccine period of 1955–1962, when the average annual incidence was 299.5 cases per 100,000 population. *(Centers for Disease Control: Morbidity and Mortality Weekly Report, Annual Supplement Summary 1981, issued October 1982.)*

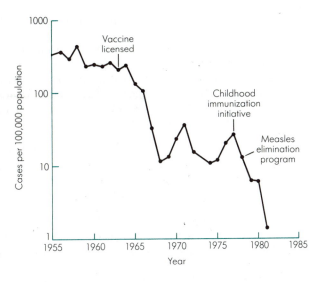

Figure 37-6. Reported cases of mumps per 100,000 population, United States, 1922–1982. *(Centers for Disease Control: Morbidity and Mortality Weekly Report, October 28, 1983.)*

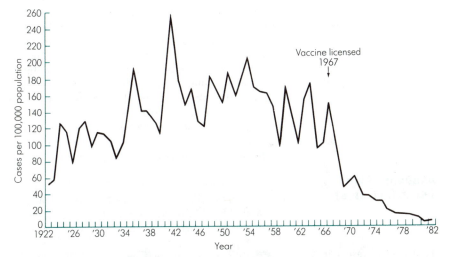

In the body a viremia is produced that distributes the virus to various glands, in which viral multiplication occurs. Mumps is characterized by painful swelling, particularly of the parotid glands, although the salivary glands, the testes, the ovaries, the pancreas, and other glands may be involved. A common manifestation of mumps in adults is orchitis (inflammation of the testes) in males or inflammation of the ovaries in females.

Immunization is achieved by means of a live attenuated vaccine. Active immunization with the combined mumps-measles-rubella vaccine is recommended for all children at 15 months of age. Since the introduction of the vaccine in 1967, a 97 percent decrease in the number of reported cases of mumps has occurred (Fig. 37-6).

Bunyaviridae (Bunyaviruses)

Bunyaviruses are enveloped viruses having a helical nucleocapsid. They contain single-stranded RNA which is composed of three segments and is of the ($-$) type. The viruses range from 90 to 100 nm in diameter. Bunyaviruses cause fevers or encephalitis in humans and/or abortion and hepatitis in domestic animals. They are transmitted by mosquitoes, biting flies, or ticks. Some examples are La Crosse encephalitis (North America, mosquito-borne); Bunyamwera fever (Africa, mosquito-borne); and Crimean-Congo hemorrhagic fever (Africa and Asia, spread by ticks).

Arenaviridae (Arenaviruses)

Arenaviruses are similar in structure to bunyaviruses; however, they have a large average diameter (110 to 130 nm) and their RNA is composed of two segments. The prefix *arena-*, meaning "sand," is based on the occurrence of electron-dense granules (probably host ribosomes) that occur within the virions (see Fig. 21-15). Arenaviruses cause asymptomatic infections in rodents, which constitute the natural reservoir of infection. Transmission to humans probably occurs by contact with infected rodent feces and urine. Human arenavirus infections include (1) a *meningoencephalitis,* a debilitating but rarely fatal disease that is caused by lymphocytic choriomeningitis virus (LCV) and (2) *hemorrhagic fevers* such as Lassa fever, in which there is a high fatality rate. Lassa fever was first discovered in Nigeria in 1969 and at least four other outbreaks have since occurred in Africa. Death rates have varied from 20 to 66 percent. The Lassa virus has been isolated from a species of rodent that is widely distributed in the region, and it is likely that this species constitutes a reservoir for the virus. It is possible that contact with the rodent, its secretions, or articles contaminated by it may be responsible for the initial human infections in an outbreak. Unlike most arenavirus infections, Lassa fever is highly contagious and can be spread from person to person. The mode of transmission is uncertain but may be via aerosols or direct contact, and strict isolation techniques must be used for patients. Fatal infections have been contracted from laboratory procedures involving clinical specimens.

Rhabdoviridae (Rhabdoviruses)

These viruses are 70 to 80 by 130 to 240 nm in size and have a characteristic bullet shape (Fig. 37-7). They have a helical nucleocapsid and contain single-stranded RNA of the ($-$) type. Surrounding the nucleocapsid is a lipid envelope with surface glycoprotein antigens.

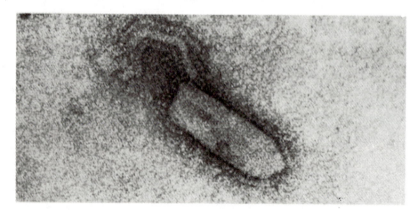

Figure 37-7. Vesicular stomatitis virus, showing the bullet shape that is characteristic of rhabdoviruses. Shown also is an irregular "handle" which may represent remains of a disrupted end (X200,000). *Courtesy of Margaret Gomersall, McGill University.)*

Discovery of a Killer Virus

The events surrounding the discovery of the Lassa fever virus were frightening. They began in 1969 at the Lassa mission hospital, located in a remote region of Nigeria, when a nurse contracted a strange and unidentifiable disease. The symptoms began with a severe sore throat, muscular aches, fever, and lassitude. During the next few days the patient showed an increase in temperature and development of ulcers in the back of the throat. Hemorrhaging of blood capillaries beneath the skin caused the skin to discolor. Despite intensive treatment, there was continued deterioration, with development of kidney failure, extreme difficulty in breathing, cyanosis (lack of oxygen in the blood), and cardiovascular collapse. The nurse died within a week after the illness began.

One of the nurses who had attended the patient also developed symptoms of the strange disease, became desperately ill, and died. About a week later, another nurse who had assisted at the autopsy began to develop symptoms of the illness. By the 12th day she was desperately ill. Arrangements were made for her to be flown to New York for treatment. There, in the isolation ward at Columbia Presbyterian Hospital, she slowly recovered from the illness. However, neither the causative agent of the disease nor an effective treatment for it were yet known.

Meanwhile, tissue and blood samples from all three victims had been sent to the Yale Arbovirus Research Unit in New Haven, Connecticut, where they had been inoculated into suckling mice and into tissue cultures. The tissue cultures showed cytopathic effects that were characteristic of a virus. Although the infant mice failed to become ill, the virus later proved to be highly fatal for adult mice. While information about the virus was accumulating, a Yale virologist who was studying the virus contracted Lassa fever. Although he was successfully treated with immune serum from the nurse who had recovered from the infection, another shock was to follow: a second Yale virologist, who had never worked with the Lassa virus and who had no known contact with it, inexplicably contracted the disease and died.

It was now clear that the usual safety precautions for dealing with infectious agents in a laboratory evidently were not sufficient for the Lassa virus. As a result, all research at Yale on the live virus was immediately halted; all live infectious material was transported to the new maximum security facility that had been constructed at the Centers for Disease Control (CDC) in Atlanta for containment and study of highly contagious pathogens, and where the Lassa virus was already under investigation.

At present, all studies of live Lassa virus, including the isolation and identification of the virus from clinical specimens, are restricted to laboratories such as those at the CDC which are specially equipped to deal with exceedingly hazardous microorganisms.

Rhabdoviruses cause disease in humans, animals, (vertebrates and invertebrates), and plants. One well known group of animal rhabdoviruses is the vesicular stomatitis virus (VSV) group, which causes vesicular (blisterlike) lesions in the mouth and other areas of the body of cattle, pigs, and horses. The mode of transmission is uncertain but may be by means of arthropods. Humans can contract a benign form of vesicular stomatitis characterized by malaise, fever, aches and pains, headache, nausea, and vomiting. Laboratory infections with VSV are common.

Rabies Virus. The most important member of the rhabdovirus family is the rabies virus, a parasite of domestic and wild mammals. Only one immunological type of this virus is known to exist. Transmission to humans occurs through the bite of an infected animal. Dogs, cats, bats, and skunks are most frequently the source of the virus infecting humans.

Rabies is essentially an overwhelming encephalomyelitis. In humans the incubation period from time of infection varies from 6 days to 1 year but is usually about 3 to 8 weeks. The development of symptoms and the length of the incubation period (in untreated cases) depend largely on the severity and location of the bite. It has been estimated that only 5 to 15 percent of all persons bitten by a rabid animal contract rabies. Symptoms in humans include severe headache and high fever, with alternating stages of excitement and depression. Patients have difficulty in swallowing, and slight stimuli incite muscular spasms in the throat and chest. Death usually follows paralysis or convulsive seizures. The mortality rate from rabies is nearly 100 percent.

If a person has been bitten by a rabid animal, the long incubation period for rabies allows time for measures to be taken to prevent the virus from reaching the central nervous system. These measures include a combination of passive immunization (by administration of immune human or horse globulin, which provides an immediate source of antibodies against the virus but which lasts for only about 14 days) and active immunization (administration of a rabies vaccine, to stimulate a longer-lasting production of antibodies by the patient).

Laboratory confirmation of rabies in the animal which has bitten the patient is done by any of several means: (1) detection of rabies virus antigen in clinical specimens by use of fluorescent antibodies; (2) isolation of the virus from saliva, urine, spinal fluid, or tissues by inoculation into the brains of mice; and (3) demonstration of inclusion bodies (**Negri bodies,** shown in Fig. 37-8) in the nerve cells of the brain. In all these tests, negative results do not exclude rabies, and for this reason the animal should not be destroyed prematurely but should be kept under observation for at least 2 weeks. If the animal develops symptoms of rabies, the human victim must be vaccinated.

In 1885, Pasteur produced an effective rabies vaccine by using brain tissue from rabbits in which rabies virus had been propagated. By drying the infected nerve tissue for increasing periods of time, the infectivity of the virus could be progressively diminished to the point where the preparation could be used to immunize persons who had been bitten by rabid animals. In 1919 Semple modified the Pasteur vaccine by adding phenol to completely inactivate the

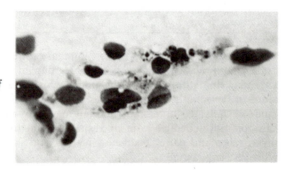

Figure 37-8. Negri bodies, inclusion bodies (dark, round structures) found in Purkinje cells and cells of the hippocampus, are diagnostic of rabies. *(Courtesy of J. Nowak, Documenta Microbiologica, part 2, "Pilze und Protozoen," Gustav Fischer Verlag, Jena, Germany, 1930.)*

virus. One difficulty with the Pasteur and Semple vaccines is that they sometimes cause an allergic encephalitis in the recipient, because of the presence of nerve tissue in the vaccines. In 1949 a different type of vaccine was developed. It contained a live attenuated strain of rabies virus—the Flury strain—which could be grown in embryonated duck eggs. This vaccine has been widely used; however, it has relatively low immunogenicity, and although it lacks large amounts of nerve tissue antigens, it still can occasionally cause allergic encephalitis. In 1980 an inactivated vaccine prepared from virus propagated in cultures of diploid human cells devoid of nerve tissue antigens was licensed for use. This new vaccine appears to be both safe and highly immunogenic.

VIRUSES CONTAINING DOUBLE-STRANDED RNA

Reoviridae

This family consists of icosahedral viruses, 70 to 80 nm in diameter, having a double capsid, no envelope, and containing double-stranded RNA with 10 to 11 segments. Among the groups included in the family which affect humans and animals are the orbiviruses, reoviruses, and rotaviruses

Orbiviruses are arthropod-borne viruses, transmitted by mosquitoes, ticks, or sandflies. They cause serious diseases in humans and animals; for example, Colorado tick fever virus causes an acute, generalized infection of humans and occurs in the western regions of the United States and Canada (see Table 35-6).

Reoviruses are widely distributed among mammals and are inhabitants of the intestinal tract. Although they have been isolated from humans with mild respiratory tract and gastrointestinal disease, they commonly occur in healthy individuals as well, and their relationship to disease is not yet clear.

Rotaviruses (Fig. 37-9) are intestinal viruses that occur in humans and animals. Their name comes from their wheellike shape. In recent years rotaviruses have been increasingly recognized as frequent and important contagious agents of diarrhea in infants and children (6 months to 2 years old). Indeed, the World Health Organization (WHO) has estimated that, on a worldwide basis, rotaviruses are responsible for one-half of the cases of infantile diarrhea requiring hospitalization. Like polioviruses, rotaviruses have a fecal-oral mode of transmission, infect the intestine, and are shed in stools. Rotavirus diarrhea is characterized by fever, severe diarrhea, vomiting, and dehydration. Diagnosis is based either on demonstrating the virus in stools (usually by means of an ELISA test; see Chap. 34) or on finding an increasing level of serum antibodies against the virus.

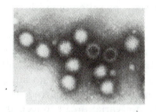

Figure 37-9. Electron micrograph of rotavirus (X80,000). *(Courtesy of M. Petric and Maria T. Szymanski, The Hospital for Sick Children, Toronto.)*

VIRUSES CONTAINING DOUBLE-STRANDED DNA

Poxviridae (Poxviruses)

Poxviruses are the largest of all viruses and are brick-shaped (200 to 260 by 250 to 390 nm) or ovoid. They contain double-stranded DNA, protein, and lipid and have a dumbbell-shaped nucleoid (see Fig. 21-15) surrounded by two membrane layers. The outer surface is covered with threads or tubules. Six genera and 27 species of poxviruses exist, most of which are animal pathogens. Of the human pathogens, the variola (smallpox) virus unquestionably has been the most important.

Variola (Smallpox) Virus. This virus is transmitted by droplet infection, either directly from an infected person to another person or by handling of articles infected by the smallpox patient. The virus is believed to lodge in the nasopharynx and to invade the regional lymphatic system. This is followed by dissemination of the virus via the bloodstream to all tissues and especially the skin. An initial fever occurs, followed by a rash consisting of small papules that appear first along the hairline and later on the face and other parts of the body. These papules or pustules become larger and are filled with fluid. The fever recurs and the patient becomes severely ill with generalized symptoms. As the pustules regress, scabs are formed, which leave the craterlike scars characteristic of smallpox. There are two kinds of variola virus: (1) *variola major*, which causes severe symptoms and has a fatality rate of 10 to 30 percent; and (2) the less virulent *variola minor (alastrim)*, with a fatality rate of only 0.1 to 0.3 percent. Except for the difference in virulence, the two viruses cannot be distinguished.

Since mild cases may be difficult to diagnose and may be confused with chicken pox (varicella, caused by a herpesvirus), the laboratory uses several methods of diagnosis to aid the clinician: (1) electron microscopy to visualize the brick-shaped virions in skin exudates, or, alternatively, the use of light microscopy to demonstrate inclusion bodies (**Guarnieri bodies,** probably the virus itself) in stained cells (see Fig. 21-21); (2) cultivation of the virus on the chorioallantoic membrane of embryonated chicken eggs; or (3) serological detection of the viral antigen in smallpox lesions.

Although smallpox has been widespread for thousands of years, both endemically and epidemically, no cases of smallpox have appeared anywhere in the world since 1977. This total eradication of smallpox was achieved because of several factors: (1) the virus had no animal reservoir of infection; (2) no subclinical carrier state occurred; (3) no dormant (latent) infections occurred; (4) an effective vaccine was available; and (5) no antigenic variation in the virus occurred (in contrast, for example, to influenza viruses). These factors permitted development of a grand strategy for eradication which could be implemented on a worldwide basis and which ultimately proved successful (Table 37-2).

Protection against smallpox involves both humoral and cell-mediated immunity and is achieved by intradermal inoculation with a live vaccine. The vaccine consists of a virus known as **vaccinia** which is closely related antigenically to variola virus but which does not ordinarily produce viremia. It is important that a live vaccine be used; inactivated vaccines prepared from vaccinia virus extracted from infected cells are ineffective because they lack an important immunizing antigen. This antigen can be acquired only by a live virus during its natural release from host cells.

Although smallpox immunization will eventually cease because of the eradication of smallpox, another potential use for vaccinia virus is being developed. In 1983 the gene for the immunizing antigen of a herpesvirus was successfully incorporated into the vaccinia genome by recombinant DNA techniques. The altered vaccinia virus produced not only its own surface antigens but the herpes antigen as well. Similarly, vaccinia strains that can produce antigenic components of hepatitis or influenza viruses have been developed. It is possible that a single vaccinia strain could be developed that can make several different viral antigens simultaneously. Such genetically modified vaccinia strains are of great interest because of their potential use as safe, inexpensive vaccines for prevention of several diseases concomitantly.

Table 37-2. Grand Strategy for Eradicating Smallpox

	Phase 1: Attack	Phase 2: Consolidation	Phase 3: Maintenance
Areas	Areas where over 5 cases of smallpox per 100,000 people occur per year and where less than 80% of all segments of the population shows scars of primary vaccination	Areas with less than 5 cases per 100,000 people and where over 80% of all segments of the population show scars of primary vaccination	Areas free of endemic smallpox for more than 2 years but geographically situated in endemic continental areas.
Vaccination	Systematic mass vaccination	Continuing maintenance vaccination	Continuing maintenance vaccination
Intelligence	Establish prompt and regular reporting of smallpox by all existing health facilities	Extension of case-detection system to assure that all suspected smallpox cases are reported	Continuing of case-detection system to assure that all suspected smallpox cases are reported.
Field investigations	Epidemiological investigation of major outbreaks throughout the country and of all cases in areas where systematic mass vaccination has been done	Prompt epidemiological investigation of all cases to establish sources of infection and to exclude the possibility of unreported cases	Each case to be investigated as an emergency by an epidemiologist
Laboratory	Establish techniques and methods for the submission and examination of specimens for confirmation of diagnosis	Specimens studied from all isolated cases and representative samples from each outbreak	Specimens studied from every suspected case
Containment	Localized, intensive vaccination in communities where cases or outbreaks occur; isolation of cases if feasible and disinfection	Vaccination and observation of case contacts; isolation of cases and appropriate disinfection; localized, intensive vaccination in community	Vaccination and observation of case contacts; isolation of cases and appropriate disinfection; localized, intensive vaccination in community

SOURCE: *World Health,* The Magazine of the World Health Organization, January–February 1968.

Adenoviridae (Adenoviruses)

The adenoviruses are icosahedral viruses, 60 to 90 nm in diameter, which contain double-stranded DNA and do not possess a lipid envelope. At each of the 12 vertices of the icosahedron there is a fiberlike projection (see Fig. 21-15). These fibers have hemagglutinin activity and also probably are responsible for viral attachment to host cells. Two major kinds of adenoviruses occur: those which are isolated from humans and other mammals and those which are isolated from birds. The human adenoviruses are divided into four subgroups based on their ability to agglutinate monkey or rat red blood cells. Each subgroup in turn contains various antigenic types.

Adenoviruses cause acute, self-limiting respiratory and eye infections and have an airborne type of transmission. Diseases caused by adenoviruses include acute febrile pharyngitis, which occurs most often in infants and children; pharyngeal-conjunctival fever in children; acute respiratory disease (ARD), which occurs mainly in military recruits; adenovirus pneumonia, a complication of ARD; and acute follicular conjunctivitis in adults.

Adenoviruses may also cause inapparent or latent infections, particularly of lymphoid tissue. Indeed, adenoviruses were initially discovered by accident:

Table 37-3. Oncogenicity of Subgroups and Serotypes of Human Adenoviruses

Hemagglutination Subgroup	Serotypes	Oncogenicity in Newborn Hamsters
I	3, 7, 11, 14, 16, 21	Weak
II	8–10, 13, 15, 17, 19, 20, 22–30	None†
III	1, 2, 4–6	None†
IV	12, 18, 31	High.

† But may cause in vitro transformation in tissue cultures.

tissue cultures had been prepared from apparently normal human tonsils and adenoids, yet these cultures exhibited degenerative changes (cytopathic effects) that were indicative of the presence of a virus.

Although adenoviruses are not known to cause cancer in humans, some strains, particularly those of subgroup IV, can cause cancer (are **oncogenic**) when inoculated into immunodeficient animals (newborn hamsters) (Table 37-3). These strains, as well as certain other strains that do not produce oncogenic effects in animals, can cause neoplastic transformation of cells grown in tissue culture (see Chap. 21). Such transformation occurs mainly in cells which, when infected, yield few or no viral progeny (**nonpermissive** cells). In the transformed cells, it can be shown that a portion of the viral DNA has become integrated into the genome of the host.

Herpesviridae *(Herpesviruses)*

Like adenoviruses, herpesviruses have an icosahedral nucleocapsid containing double-stranded DNA; however, the nucleocapsid is enclosed within a bilayered envelope from which extend numerous short projections (see Fig. 21-15). Moreover, herpesviruses are much larger than adenoviruses, having a diameter of 180 to 200 nm.

Herpesviruses cause disease in a wide variety of animals and in humans. One of the animal herpesviruses is Marek's disease virus (MDV), which affects the epithelium of feather follicles in chickens. The virus has special significance because it can also cause neoplastic transformation of chicken T lymphocytes, resulting in lymphoma—a type of cancer. Some human herpesviruses may also be related to cancer, although whether the relationship is a causal one is uncertain. A discussion of some human herpesviruses follows.

Herpes Simplex Virus Type 1 (HSV-1). Primary infections with HSV-1 are common and may be contracted by children over 6 months old. By adulthood the proportion of persons who have been infected is 50 percent or higher. On initial contact with HSV-1, more than 90 percent of persons develop a subclinical form of *primary* herpes. Clinical infections are usually self-limiting and include acute gingivostomatitis (vesicular eruption in the mouth), pharyngitis, cold sores, keratoconjunctivitis, or skin lesions (on nongenital areas). Occasionally, more severe and even fatal infections occur, such as encephalitis.

Following the primary infection (clinical or subclinical), people develop neutralizing antibodies and maintain them for the rest of their lives. Despite these antibodies, the virus continues to occur in a *latent* form. The latent virus may sometimes be reactivated by environmental factors such as heat and cold, by hormonal or emotional disturbances, or by other stimuli. This is *recurrent* herpes and is usually characterized by superficial vesicles (cold sores, fever blisters).

HSV-1 has been related to head and neck cancers, but whether it actually causes these cancers is uncertain. Under certain conditions, HSV-1 (and also HSV-2, discussed below) can cause neoplastic transformation in tissue cultures of hamster embryo cells and human fibroblasts.

Herpes Simplex Virus Type 2 (HSV-2). HSV-2 shares some antigens with HSV-1 but it also has some unique antigens and other distinctive properties. In recent years HSV-2 has been recognized as the causative agent of primary and recurrent disease involving the genital tract.

Primary genital herpes is most commonly transmitted by sexual contact. Infection is associated with small, painful blisters on the cervix, vagina, urethra, and anus in women; in men, the lesions are on the penis, in the urethra, or around the anus. The lesions become crusted and heal without leaving scars. The virus is present in the lesions, and the disease is most contagious when lesions are present; however, the disease may sometimes be contagious even when lesions are absent. Other aspects of the clinical syndrome include fever, painful urination, inflammation of the inguinal lymph glands, and genital soreness. Infection in pregnant women may result in serious neonatal disease with dissemination of the virus to the skin, eyes, central nervous system, and visceral organs of the newborn.

After the lesions disappear, the virus remains latent for periods from a few weeks to a year or longer, after which the symptoms may recur. Recurrent genital herpes is frequent. In women, the cervix is usually involved, but the infection is often subclinical; in men, herpetic vesicles are commonly present on the penis as in primary genital herpes. Several studies have shown a definite link between venereal herpes and cancer of the cervix or prostate, although it has not been proved that HSV-2 actually causes the cancer.

Genital herpes, like all virus diseases, cannot be cured by antibiotics. Topical application of the drug acyclovir can shorten the healing time in primary genital herpes but has little effect in treatment of the recurrent form; however, recent evidence suggests that oral or intravenous administration of acyclovir may help to suppress the recurrent form. Vaccines to prevent genital herpes are under development: recombinant DNA techniques are being used to insert genes for individual herpesvirus surface antigens into harmless bacteria. The antigens can then be produced by the bacteria and used for immunization, thereby avoiding the difficulties involved in propagating and inactivating infectious virus in order to make a vaccine. (See also vaccinia virus as a carrier of herpes antigens, described earlier in this chapter.)

Other Human Herpesviruses. Varicella-zoster virus (VZV) causes two diseases. Varicella (chicken pox) is a mild infection of children characterized by a vesicular skin rash. Herpes zoster (shingles) is a disease caused by activation of latent virus from a previous varicella infection. It occurs mainly in adults and is characterized by a vesicular eruption and a very painful inflammation of sensory nerves. Epstein-Barr virus (EBV) is the causative agent of infectious mononucleosis, a self-limiting disease which occurs in up to 80 percent of the population

Table 37-4. Some Polyomaviruses and Papillomaviruses

Virus	Diseases Caused
Polyomaviruses	
Mouse polyoma	Causes cancer when injected into newborn mice and hamsters; in tissue culture, transforms nonpermissive cells of secondary cultures of rat and hamster embryos; occurs widely in wild and laboratory mice, being transmitted to offspring by excretions and secretions
SV 40	Occurs in monkeys; first discovered as a latent virus in cultures of rhesus monkey kidney cells; causes sarcomas in newborn hamsters; transforms secondary cultures of hamster, rat, or mouse embryo cells
Human SV 40–like: BK, JC	Widespread in human populations; acquired during childhood; produce tumors in hamsters and transform hamster cells in vitro; BK is not known to cause any human disease; JC probably causes a rare, fatal disease (not a cancer) of the nervous system in adults (progressive multifocal leukoencephalopathy)
Papillomaviruses	
Infectious wart viruses	Cause benign tumors (warts) in humans; nearly all individuals have been infected by some wart viruses by the age of 20; cell-mediated immunity is probably effective in suppression or regression of disease
Rabbit papilloma	Causes warts in wild cottontail rabbits—usually benign but may become malignant; domestic rabbits are highly susceptible and tumors eventually become malignant

(peak incidence in the 15- to 20-year-old group) but which is usually subclinical. Patients exhibit fatigue, malaise, sore throat, fever, swollen cervical lymph nodes, and an increased level of blood monocytes. EBV is also thought to be the causative agent of Burkitt's lymphoma, a cancer of B lymphocytes; however, the causal relationship is not yet proved.

Papovaviridae (Papovaviruses)

The name *papovavirus* is derived from the names of three of the viruses included in the family: rabbit papilloma virus, mouse polyoma virus, and vacuolating agent SV 40. Papovaviruses are small, nonenveloped, icosahedral, double-stranded DNA viruses that induce tumors in animals. They are divided into two major groups, polyomaviruses (45 nm in diameter) and papillomaviruses (55 nm in diameter). The oncogenic potential of these viruses is indicated in Table 37-4. Except for the human wart viruses, none appear to be tumorigenic in humans.

VIRUSES CONTAINING SINGLE-STRANDED DNA

Parvoviridae (Parvoviruses)

Parvoviruses are small (about 20 nm in diameter), icosahedral, nonenveloped viruses that contain single-stranded DNA. Some parvoviruses are defective, i.e., they cannot replicate autonomously but only when the host cell is also infected by an adenovirus (helper virus); consequently, such parvoviruses are termed adenovirus-associated viruses (AAV). Humans are commonly affected with AAV, but the AAV apparently do not cause any overt disease and do not contribute to the symptoms caused by the adenovirus.

Other parvoviruses do not require a helper virus for replication; however, they do require that the host cell be in the S period of its growth cycle (i.e., the period during which DNA is being synthesized). Consequently, these viruses grow best in rapidly dividing cells. Parvoviruses can infect a variety of animals; for example, bovine parvovirus causes diarrhea and abortion in cattle. With regard to human infections, a parvovirus known as B19 has been implicated in aplastic crisis, a disease occurring in persons with sickle cell anemia: when rapid cell division occurs to produce precursors of new blood cells, the virus can multiply in these precursor cells and kill them. B19 has also been implicated in erythema infectiosum, a highly contagious rash occurring in children.

RNA TUMOR VIRUSES REQUIRING A DNA INTERMEDIATE FOR REPLICATION

Retroviridae (Retroviruses)

This heterogeneous family consists of enveloped viruses that contain single-stranded (+) RNA and cause tumors in chickens and mammals. An outstanding characteristic of retroviruses is the occurrence of an RNA-dependent DNA polymerase known as reverse transcriptase; indeed, the prefix retro- is derived from the activity of this enzyme. The action of reverse transcriptase is to synthesize a DNA strand complementary to the viral RNA strand (see Chap. 21). The single-stranded DNA subsequently serves as a template for synthesis of a complementary DNA strand, thereby resulting in double-stranded DNA. This double-stranded DNA is required for the tumorigenic properties of the viruses and becomes integrated into the host-cell genome. In this integrated form (provirus) it is transcribed to produce new viral RNA.

The relation between retroviruses and oncogenes is discussed in Chap. 21.

Human T-cell leukemia viruses (HTLVs) comprise a group of retroviruses that

---

Identification of the Probable Causative Agent of AIDS

Although the modes of transmission of the causative agent of AIDS had been known since 1981 (see Chap. 35), identification of HTLV-III as the probable causative agent of the disease was not accomplished until 1984. That a virus was the cause of AIDS had been suspected; however, the virus could not be isolated from the T cells of AIDS patients in quantities sufficient for study. This was because the T cells, which might have allowed the virus to proliferate when grown in tissue culture, failed to grow. Robert Gallo of the National Institutes of Health and his colleagues believed that if the virus was present in these T cells, it might merely be killing the cells before much viral proliferation could occur. A killing action on T cells would be consistent with the severe decrease in the level of T cells that occurs in AIDS patients.

The breakthrough came when a suitable permissive T-cell tissue-culture system was developed by Gallo and his colleagues. They found a T-cell line which allowed propagation of the AIDS virus but was not killed by the virus. With the new system, the virus was obtained for the first time in amounts sufficient to allow its antigenic properties to be determined. The virus was related to HTLV-I and HTLV-II but belonged to a new subgroup. The new virus was isolated from 18 of 21 individuals with "pre-AIDS" (mild, early symptoms of AIDS) and from 26 of 72 patients with clinical cases of AIDS. Further evidence for the role of HTLV-III in AIDS was provided when antibodies against the virus were detected in 43 of 49 AIDS patients. The antibodies were not detected in 185 of 186 control individuals, even in individuals having various other disorders of the immune system. Subsequent improved detection techniques have indicated the presence of antibodies against HTLV-III in 100 percent of AIDS and pre-AIDS patients tested but none in control individuals. These data provide strong circumstantial evidence that HTLV-III is the causative agent of AIDS.

are **lymphotrophic,**i.e., that either inhibit T-cell (T-lymphocyte) function, transform T cells, or kill T cells. There are presently three subgroups: HTLV-I, HTLV-II, and HTLV-III. HTLV-I is the subgroup most commonly isolated and causes a type of T-cell leukemia in adults which is endemic in certain areas of Japan, the Caribbean, and Africa. HTLV-II has been isolated from a patient with a different type of leukemia—hairy-cell leukemia. HTLV-III is the probable causative agent of acquired immunodeficiency syndrome (AIDS), a disease characterized by severely decreased levels of helper T lymphocytes (T cells that recognize an antigen and then facilitate a B cell response to it; see Chap. 33).

Examples of animal retroviruses include mouse mammary tumor viruses; avian sarcoma viruses; avian leukemia viruses; and murine, feline, porcine, bovine, and monkey leukemia viruses. It should be noted that transmission of some retroviruses is **exogenous,** or **horizontal;** i.e., a healthy animal contracts the infectious virus from an animal harboring the virus. For other retroviruses the transmission is **endogenous,** or **vertical;** i.e., the virus is carried in the form of noninfectious provirus (the viral genome is integrated into the host cell genome) and thus is transmitted hereditarily to offspring. Infectious virus occurs only upon spontaneous induction of the provirus or by treatment with mutagens.

MISCELLANEOUS VIRUSES

Hepatitis Viruses

Hepatitis A virus (HAV) is a small icosahedral virus 27 nm in diameter and contains single-stranded RNA; it is probably related to the picornaviruses. In contrast, hepatitis B virus (HBV) is an enveloped virus 42 nm in diameter and contains double-stranded DNA. HBV is also distinguished by a specific surface antigen designated **HBsAg,** or **Australia antigen.** The diseases caused by these two viruses are acute systemic infections primarily affecting the liver, but they differ in their epidemiology and other characteristics. HAV can infect chimpanzees and marmosets and can be propagated, with difficulty, in tissue cultures. HBV can be grown only in chimpanzees.

Viral Hepatitis Type A. This disease has also been called acute epidemic hepatitis, infective hepatitis, and short-incubation hepatitis. The major mode of transmission of HAV is similar to that of poliovirus, that is, by the fecal-oral route. Infections with HAV are endemic in nursery schools, mental institutions, and all establishments or societies where there is a high risk of fecal contamination. The disease is generally limited to humans.

Following ingestion, the acid resistance of HAV allows it to pass through the stomach into the small intestine. The virus infects the mucosal epithelial cells, replicates, and spreads to adjacent cells and, via the blood circulation, to the liver. During the preicteric (prejaundice) stage of the disease there is loss of appetite, fatigue, malaise, abdominal discomfort, and fever. The virus is shed in the feces until the onset of jaundice or slightly longer. With the appearance of jaundice the patient feels better, but the liver remains tender and palpable. Jaundice persists for 1 to 3 weeks. The disease is rarely fatal and recovery usually occurs gradually over a period of 2 to 6 weeks.

Diagnosis of type A hepatitis is based on (1) detection of the virions in the patient's feces by immune electron microscopy (a technique by which virions which have reacted with specific antibody become clearly distinguishable from other particles when observed with an electron microscope), (2) serologic tests for viral antigens, or (3) demonstration of a rising titer of antibodies in the patient's serum. No vaccine is presently available for active immunization against HAV, but passive immunization using pooled gamma globulin can often be protective.

Viral Hepatitis Type B. This disease is also known as serum hepatitis, homologous serum jaundice, and long-incubation hepatitis. The virus is transmitted by transfusion of blood or blood products, by contaminated hypodermic syringes, and possibly by saliva and by sexual contact. HBsAg appears in the blood during the incubation period, 2 to 8 weeks before the onset of jaundice, but disappears during convalescence.

Tumorigenic activity of HBV has been suggested by a correlation between hepatitis B and primary hepatocellular carcinoma, a type of cancer which is rare in the United States but common in certain parts of Africa, Mozambique, and Southeast Asia and in certain regions bordering the Mediterranean.

A hepatitis B vaccine was licensed in 1981 for immunization of persons who are at high risk of exposure. The vaccine consists of highly purified HBsAg obtained from the blood serum of apparently healthy carriers.

Viral Hepatitis, Non-A, Non-B. A form of viral hepatitis not caused by type A or type B viruses (i.e., non-A, non-B hepatitis, or NANB) was first described in 1974 and is now recognized as the most frequent form of post–blood-transfusion hepatitis in the United States. However, little is known of the fundamental nature of the NANB viruses. Moreover, there are no satisfactory tests for identifying NANB antigens or antibodies, and diagnosis of NANB hepatitis is based on exclusion of other types of viral hepatitis.

Slow Viruses

The classic slow virus diseases (kuru and Creutzfeldt-Jakob disease in humans, scrapie and transmissible mink encephalopathy in animals), and the unusual nature of the agents that cause them are described in Chap. 21.

VIRAL PATHOGENS OF PLANTS

Most plant viruses consist of single-stranded RNA surrounded by a protein capsid. Some, such as the wound tumor virus, have double-stranded RNA, and others, such as the cauliflower mosaic virus, have double-stranded DNA. In general, the shapes of plant viruses fall into two categories: rod-shaped and spherical. The rod-shaped viruses are commonly referred to as helical, and they vary from the rigid type to long, flexuous rods. The small spherical viruses are usually icosahedral.

Viroids constitute another group of agents that are pathogenic for plants. They are unusual in that they are composed of infectious RNA without any protein coat (see Chap. 21).

Pathogenic plant viruses are spread by insects, by infected vegetative parts of plants used for propagation, occasionally by infected seed, and by various other means. Many plant viruses are transmitted only by insects, including aphids, leafhoppers, whiteflies, and mealybugs, which carry infected plant juices from plant to plant. As an example, curly top of sugar beets is a highly destructive

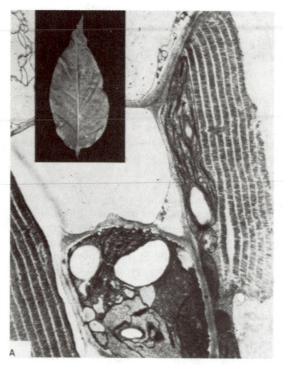

Figure 37-10. (A) Electron micrograph of an ultrathin section of tobacco tissue systemically infected with tobacco mosaic virus (TMV), showing tiers of virus particles in the cytoplasm. Virus particles are rigid rods 300 nm long (X6,000). Inset: Leaf of tobacco *Nicotiana tabacum)* exhibiting mosaic symptoms resulting from systemic infection by TMV. *(Courtesy of Prof. M. Kenneth Corbett, University of Maryland.)* (B) TMV particles photographed under an electron microscope. Individual particles are seen at the edge of the aggregation. *(From W. M. Stanley and T. F. Anderson, "A Study of Purified Viruses with the Electron Microscope," J Biol Chem, **139**:339–344, 1941; courtesy of the Journal of Biological Chemistry.)*

virus disease which is transmitted naturally by a leafhopper (or experimentally by inoculation with infected sap). When plants are propagated by grafting or budding or by tubers, roots, shoots, or other vegetative means, any virus present in the parent plant is carried over to the new plant. This important factor in the spread of virus diseases of plants can be controlled by the use of virus-free nursery stock for all new plantings. About 10 percent of plant viruses are seed-borne, and this mode of transmission can be important in certain diseases such as bean mosaic.

There are two major types of virus diseases of plants; these are differentiated according to the type of injury produced. The first group, the so-called mosaic diseases, causing mottling or spotting of leaves, is the largest and most important. These diseases are characterized by the production of yellowish spots or blotches and necrotic spots on the leaves and sometimes on the blossoms of plants. Cucumber and tobacco mosaic diseases and tomato spotted wilt are economically important examples of such infections (Fig. 37-10). The effects of some infections have been exploited in ornamental horticulture to produce unusual variegations on foliage or blossoms, examples of which are seen in certain tulips, in which the variegation is called **breaking.**

The second group of virus diseases of plants causes curling of leaves, yellowing, dwarfing, and sometimes excessive branching as well. This is the leaf curl and yellows group and includes such diseases as sugar-beet curly top, peach yellows, and strawberry stunt. Formation of tumors on roots and stems can also occur, as in wound tumor disease.

Many plant diseases caused by viruses result in considerable economic losses to the agricultural economy. High on the list of economic importance are virus diseases of sugarcane, sugar beets, stone fruits, tomatoes, potatoes, and other fruits and vegetables.

VIRAL PATHOGENS OF INSECTS

More than 250 viruses that are pathogenic for some insects have been isolated. Viral pathogens of insects affect the levels of insect populations in nature; moreover, some have economic importance, such as those which cause disease in silkworms or honeybees. Insect viruses usually cause latent infections and probably are transmitted vertically, to the offspring of the insect host, as well as horizontally, by ingestion of infectious virus. The latent infections seem to be activated to cause overt disease by certain environmental factors and also by various chemical and mutagenic agents. Some insect viruses are **occluded** (the virions are embedded within a protein matrix); others are **nonoccluded** (the virions are free within the infected cell).

Occluded Viruses

Polyhedrosis Viruses. Many insect diseases are **polyhedroses,** which are characterized by the formation of polyhedral inclusions in the infected cells. These inclusions, which can be seen with the light microscope, consist of a matrix of protein molecules that occur in a lattice arrangement and numerous virions which are embedded within the protein matrix. In **nuclear polyhedroses** the polyhedra occur within the nucleus of the infected host cell (Fig. 37-11). The virions within the polyhedra are rod-shaped, contain double-stranded DNA surrounded by a protein coat, and are enclosed within an envelope. The polyhedra enlarge to the point where they rupture the nucleus of the host cells and cause the cell to disintegrate. Nuclear polyhedroses occur mainly in the larvae of *Lepidoptera, Hymenoptera,* and *Diptera.* In **cytoplasmic polyhedroses** the polyhedra occur in the cytoplasm. The virions within the polyhedra are icosahedral and contain double-stranded RNA. Cytoplasmic polyhedroses occur mainly in *Lepidoptera* larvae.

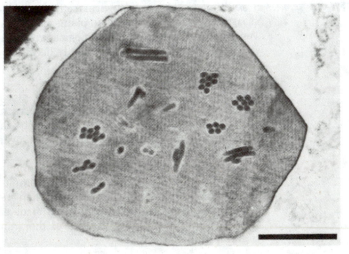

Figure 37-11. Electron micrograph of a thin section through a polyhedral inclusion body occurring in the nucleus of a cell of the southern armyworm, *Spodoptera eridonia,* infected with a nuclear polyhedrosis virus. The enveloped virions can be seen embedded within the polyhedral protein matrix. The bar represents 0.5 μm. *(Courtesy of Jean R. Adams, Insect Pathology Laboratory, USDA, Beltsville, Maryland.)*

Some nuclear polyhedrosis viruses have been used for biological control of insects. A virus preparation called Elcar (Sandoz, Inc., Crop Protection, San Diego, Calif.) has been used commercially in California for control of the cotton bollworm. Other preparations have been used experimentally for control of the gypsy moth in the northeastern United States and the spruce budworm in Canada.

Granulosis Viruses. Granuloses caused by DNA-containing viruses occur in *Lepidoptera* larvae and are characterized by small oval inclusions, each consisting of a protein matrix in which are embedded one or two rod-shaped virions. These inclusions occur mainly in the nucleus of the host cells.

Entomopox Viruses. Infected host cells contain spherical inclusions composed of a protein matrix and numerous virions that have a shape and structure resembling that of the classical poxviruses such as vaccinia. Entomopox viruses have been isolated from *Coleoptera*, *Lepidoptera*, *Orthoptera*, and *Diptera*.

Nonoccluded Viruses

Iridescent viruses are icosahedral viruses containing double-stranded DNA. Infected tissues acquire an iridescent appearance due to occurrence of microcrystals of the virus within the cytoplasm of the infected cells. Iridescent viruses occur in *Coleoptera*, *Diptera*, and *Lepidoptera*. One example is tipula iridescent virus (TIV), a DNA-containing virus that causes a disease of crane-fly (*Tipula paludosa*) larvae.

Another example of a nonoccluded virus is the *densonucleosis virus*, which is icosahedral, contains single-stranded DNA, and causes an infection of the greater wax moth, *Galleria mellonella*. Some examples of nonoccluded RNA-containing viruses are the *sigma virus* of fruit flies, the *bee paralysis viruses*, the *sacbrood virus* of honeybee larvae, the *Wassersucht virus* of *Coleoptera*, and several viruses that infect *Lepidoptera*, such as the virus of Flacherie—a fatal disease of silkworm larvae.

Other Viruses

In addition to these insect viruses, it must be remembered that many viral diseases of humans, animals, and plants are transmitted by arthropods. However, even when the arthropod is a biological vector (i.e., when the virus multiplies within the arthropod), it is not usually harmed by its viral infection.

QUESTIONS

1 For immunization against poliomyelitis, give advantages and disadvantages of (a) the Salk vaccine, and (b) the Sabin vaccine.
2 Why is it impractical to prepare a vaccine against the common cold?
3 Explain how secretory IgA can act to prevent infection by some specific viruses.
4 What role might recombinant DNA techniques play in the development of vaccines against viral diseases? Give examples.
5 List some antiviral drugs and explain their mechanisms of action.
6 Indicate the features of smallpox that made eradication of this disease possible. Give the essential features of the grand strategy that was used for eradication of this disease.

7 Rhinoviruses and coronaviruses can cause the common cold. How do these two kinds of viruses differ?

8 Could the following viruses easily be eradicated: (a) togaviruses, (b) orthomyxoviruses, (c) rabies viruses, and (d) adenoviruses? Explain your answer in each case.

9 What recent developments in rabies immunization have led to a safer yet effective vaccine?

10 Why is a live vaccine rather than an inactivated vaccine required for immunization against smallpox?

11 Contrast infections caused by herpes simplex virus type 1 (HSV-1) with those caused by HSV-2.

12 Contrast hepatitis virus type A with hepatitis virus type B.

13 Briefly define the following items:

Negri bodies	Guarnieri bodies
Variola minor	HBsAg
Adenovirus-associated virus	Reverse transcriptase
Polyhedroses	Latent infection
Horizontal transmission	Vertical transmission

REFERENCES

See also references for Chaps. 31, 35, and 36.

Cantwell, G. E.: *Insect Diseases*, vol. I, Marcel Dekker, New York, 1974. *This volume provides a comprehensive coverage of insect diseases caused by various kinds of microorganisms. Chapter 2 deals with the characteristics of viral pathogens and of the diseases they cause.*

Evans, A. S. (ed): *Viral Infections of Humans*, 2d ed., Plenum, New York, 1982. *A comprehensive treatment of the epidemiology, prevention, and nature of human viral diseases as written by many authorities.*

Fuller, J. G.: *Fever! The Hunt for a New Killer Virus*, Reader's Digest, New York, 1974. *This book presents a fascinating account of the discovery of the Lassa fever virus and the disastrous events which followed.*

Marx, J. L.: "Strong New Candidate for AIDS Agent," *Science*, **224**:475–477, 1984. *A succinct summary of the evidence linking HTLV-III to AIDS. The four original papers to which the summary refers are included in the same issue.*

Smith, K. M.: *Plant Viruses*, 5th ed., Chapman and Hall, London, 1974. *A textbook which covers the entire field of plant viruses in a concise and readable manner.*

Chapter 38

Microbial Agents of Disease:
Fungi and Protozoa

In addition to the bacteria and viruses, there are two groups of eucaryotic protists with members that are agents of disease. These are the fungi and the protozoa. We have discussed the biology of these microorganisms in Chaps. 17 and 19.

Fortunately, even though there exist over 100,000 species of fungi, only about 50 of these are known to be pathogenic for humans. Of these, only a few are pathogens; most of them are **opportunists** (i.e., able to cause infection in a compromised or weakened host). Fungal or mycotic diseases are assuming new importance because of the use or abuse of antibiotics in the treatment of bacterial infections. These antibiotics eliminate the natural microbiota which would otherwise suppress the growth of opportunistic fungi. The latter can then grow unrestrained and become pathogenic. Upon infection, initially there is acute inflammation with the accumulation of many polymorphonuclear leukocytes. Damage to the tissues is not caused by toxins but by development of allergic necrosis due to hypersensitivity of the host immune system to the fungi (see Chap. 33 on type IV hypersensitivity reactions). A compromised state in patients can also result from chemotherapy for cancer as well as from immunosuppressive therapy for organ transplantation, and opportunistic fungi can cause systemic infections in these circumstances. A few mycotic diseases are caused by fungi of **endogenous origin** (i.e., are part of the normal host microflora), such as candidiasis. Others are due to fungi of **exogenous origin** (i.e., coming from outside the body, such as the soil or bird droppings); an example of such disease is histoplasmosis.

Even though more than 65,000 species of protozoa have been described, only a few cause disease in humans. But these few protozoa have engendered untold misery for millions of people, especially in the rural areas of tropical countries (Table 38-1). The reason is that the control of many of these infections is

Table 38-1. Some Epidemiological Data on Protozoan Diseases

Disease	Thousands infected/yr	Thousands of cases/yr (Symptoms)	Death in thousands/yr
Malaria	800,000	300,000 (Fever, coma)	3,000
South American trypanosomiasis	12,000	1,200 (Heart disease)	60
Amoebiasis	400,000	1,500 (Dysentery, liver abscess)	30
Leishmaniasis	12,000	12,000 (Sores)	5
African trypanosomaisis	1,000	10 (Sleeping sickness)	5
Giardiasis	200,000	500 (Diarrhea)	Very few

dependent on an increased standard of living, as manifested by improved sanitation and better education, nutrition, and medical care. Furthermore, no vaccine exists for any of these diseases; preventive measures exist only for one, malaria. For most of these protozoan diseases, even diagnostic procedures are quite rudimentary and laborious. However, within the last few years, many commercial reagents and tests have become available for the **serodiagnosis** of major parasitic diseases.

DISEASES CAUSED BY FUNGI

Fungal diseases may be conveniently grouped into two types: (1) the **superficial mycoses** (diseases caused by fungi) or **dermatomycoses** and (2) the **systemic mycoses.**

Early concepts of the dermatophytic diseases attributed such infections to insects. Thus the Romans called these infections *tinea,* meaning "small insect larva." This term is still used as part of the clinical terminology for some dermatophytic diseases.

The fungi that cause superficial mycoses frequently are spread from animals to humans, a notable exception being athlete's foot, or ringworm of the feet, which is spread from person to person in locker rooms, swimming pool areas, and other locations. Fungi that cause systemic infections generally come from soil, vegetation, or bird droppings and are transmitted by air movements. Thus infection often starts in the lungs and then spreads to other organs.

Dermatomycoses

Fungus diseases that occur on the nails, skin, hair, and mucous membranes are referred to as superficial mycoses (see Fig. 38-1). Many of these fungi cause various forms of ringworm, or tinea, and the organisms that cause them are commonly called the **dermatophytes,** or ringworm fungi (Figs. 38-2 and 38-3). These fungi spread radially in the dead keratinized layer of the skin by means of branching hyphae and occasional arthrospores. Inflammation of the living tissue below is very mild and only a little dry scaling is seen. Usually there is irritation, erythema, edema, and inflammation at the spreading edge; this pinkish circle gave rise to the name *ringworm* (Figs. 38-4 and 38-5). These diseases are widespread and difficult to control, but fortunately they are often more

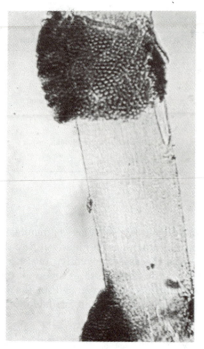

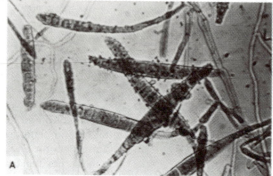

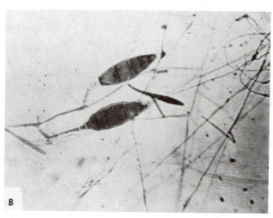

Figure 38-1. Black piedra is a fungus infection of hair, characterized by dark brown or black nodules on the hair shaft. It is caused by *Piedraia hortai*. *(Courtesy of Everett S. Beneke and the Upjohn Company.)*

Figure 38-2. Some ringworm of the scap is caused by various species of *Microsporum*, such as (A) *M. canis* and (B) *M. gypseum*. The large spindle-shaped conidia (spores) attached to the hypae are characteristic of species of this fungus. Such conidia are formed in artificial culture and are used to differentiate between the genera of dermatophytes. *(Courtesy of Everett S. Beneke, Michigan State University.)*

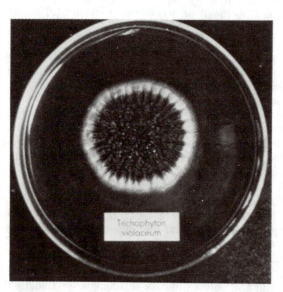

Figure 38-3. A colony of *Trichophyton violaceum*, a dermatophyte fungus. *(Courtesy of Centers for Disease Control, Atlanta, Ga.)*

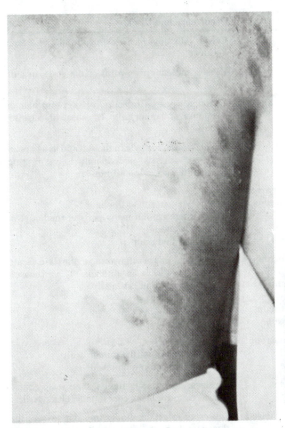

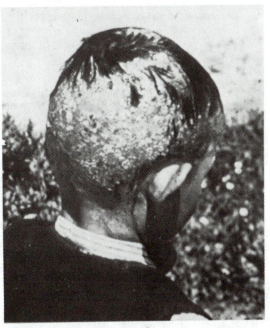

Figure 38-4. Ringworm lesions on the back of a patient caused by *Trichophyton verrucosum.* *(Courtesy of Centers for Disease Control, Atlanta, Ga.)*

Figure 38-5. Ringworm of the scalp caused by a species of *Trichophyton. (Courtesy of Centers for Disease Control, Atlanta, Ga.)*

annoying than serious. The causative microorganisms are sometimes present in the epidermal tissues without producing symptoms. They rarely, if ever, cause fatal infections. Transmission is commonly by direct contact with infected people or animals and by fomites. Dry skin is a fairly effective barrier against such diseases, but a "waterlogged" skin is vulnerable. This is why the sweat-laden, moist feet of athletes get infected with tinea, giving rise to the term *athlete's foot.* The most common dermatophytes are listed in Table 38-2.

The correlation of a dermatophyte species with a characteristic disease entity has been difficult because a single species can cause a variety of clinical symptoms in different parts of the body. Further, the same clinical manifestation can be caused by different species of dermatophytes. Thus dermatologists frequently have used terminology based on the part of the body involved. For example, tinea capitis is ringworm of the scalp (Fig. 38-5); tinea unguium, or onycho-

Table 38-2. The Dermatophytes

Group	Organisms	Occurrence and Disease
Epidermophyton	E. floccosum	Causes infections of the skin and nails on fingers and toes
Microsporum	M. audouinii	Causes epidemic ringworm of the scalp in children
	M. canis	Common cause of infection of skin and hair on cats, dogs, and other animals; causes tinea capitis of children
	M. gypseum	Occurs as a saprophyte in the soil and as a parasite on lower animals; occasionally found in ringworm of the scalp in children
Trichophyton	Gypseum subgroup	
	T. mentagrophytes	Primarily a parasite of the hair
	T. rubrum	Causes ringworm on many parts of the body; infects hair and scalp
	T. tonsurans	Infects hair and scalp
	Faviform subgroup	These fungi cause ringworm of the skin, scalp, and glabrous skin in humans; T. verrucosum causes ringworm in cattle also
	T. schoenleinii	
	T. violaceum	
	T. ferrugineum	
	T. concentricum	
	T. verrucosum	
	Rosaceum subgroup	
	T. megnini	Causes ringworm of the human scalp
	T. gallinae	Causes an infection in chickens
Miscellaneous	Piedraia hortae	Causes an infection of the hair and scalp characterized by hard, black concretions; black piedra
	Trichosporon beigelii	Causes an infection similar to above, except that the concretions are white; white piedra
	Malassezia furfur	Causes tinea versicolor, a generalized fungus infection of the skin covering trunk and sometimes other areas of the body
	Candida albicans	Causes candidiasis of skin, mucous membranes, and nails

mycosis is ringworm of the nails (Figs. 38-6 and 38-7); and tinea pedis is ringworm of the feet (Fig. 38-8).

Systemic Mycoses

The systemic, or deep, mycoses are mainly fungus diseases that often are serious or fatal. The organisms invade subcutaneous tissues or the lungs, from which they may spread to other organs of the body where they become established and produce disease. Many of them are airborne and enter the body through the respiratory tract; they may, however, enter by other portals.

Systemic mycoses appear to be increasing in importance, but their greater apparent incidence may actually be due to improved diagnostic methods and a greater appreciation of their importance. As a result of the great mobility of

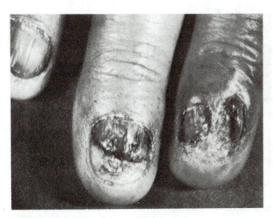

Figure 38-6. Onychomycosis is a disease of the nails caused by fungi. The infection pictured here was caused by *Trichophyton rubrum*. *(Courtesy of J. D. Schneidau, Jr.)*

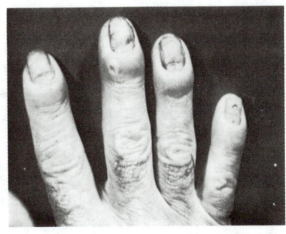

Figure 38-7. Onychomycosis caused by *Candida albicans*. *(Courtesy of J. D. Schneidau, Jr.)*

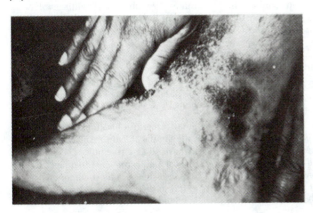

Figure 38-8. Tinea pedis, also known as athlete's foot, is caused by *Trichophyton rubrum*, *T. mentagrophytes*, and *Epidermophyton flocossum*. The chronic infection shown here is also manifested by the appearance of vesicles and is caused by *T. rubrum*. *(Courtesy of J. D. Schneidau, Jr.)*

Americans, more of them are exposed to fungi that are only prevalent in and endemic to specific, limited areas of the world.

Factors that predispose an individual to become more susceptible to systemic fungus infections include the following:

1 The presence of chronic debilitating diseases such as cancer, diabetes, leukemia, and tuberculosis
2 The use of newer types of drugs, such as antibiotics and hormones, which cause changes in the metabolism of the body or upset the normal relationships among the microorganisms on or in the body

Table 38-3. The Systemic Mycoses

Disease	Causative Organism	Characteristics of the Organisms	Characteristics of the Infection
Cryptococcosis	*Filobasidiella (Cryptococcus) neoformans*	These are yeastlike capsule-producing organisms that reproduce by budding; no hyphae or spores are formed; the cells grow well on ordinary culture media	This organism may infect any part of the body but usually starts in the lungs and spreads through the bloodstream; infections of the brain and meninges usually cause death; mode of transmission is not known, and spread from known cases in humans or other animals has not been established
Moniliasis	*Candida albicans* (order *Moniliales*)	*C. albicans* cells are yeastlike with pseudohyphae; they produce large, thick-walled, spherical chlamydospores; on ordinary culture media, pasty, smooth colonies having a yeasty odor develop	Monilia may infect any body tissue; it is found on mucous membranes of intestinal tracts of many healthy persons; infection with *C. albicans* in the mouth is called thrush; may also cause a mycotic endocarditis, pulmonary moniliasis, and vaginitis; may be spread by contact
North American blastomycosis	*Blastomyces dermatitidis*	These are large round cells with a single bud; intercalary and terminal chlamydospores appear in old cultures; optimum growth temperature is 37°C; on infusion-blood agar the colonies resemble *M. tuberculosis*; typical cells are found in body exudates and in cultures	Infection with *B. dermatitidis* resembles pulmonary tuberculosis with involvement of the lungs and pleura; it is characterized by chronic, granulomatous, suppurative lesions of any body tissue; it occurs only in the United States and Canada, most commonly among rural males aged 30 to 50; systemic infections are often fatal; it is not transmitted from humans or other animals to humans
South American blastomycosis	*Paracoccidioides (Blastomyces) brasiliensis*	These are yeastlike cells that are larger than *B. dermatitidis*; cells range from 6 to 30 μm in cultures and to 60 μm in exudates; parent cells give rise to many buds; smooth, waxy, yeastlike colonies appear on blood or meat medium after several days' incubation at 37°C	South American blastomycosis is clinically similar to North American blastomycosis; it also resembles coccidioidomycosis; it occurs most frequently in Brazil; lesions are most commonly found in the mouth and gastrointestinal tract and in the lymph nodes of the neck
Histoplasmosis	*Histoplasma capsulatum*	Small, oval cells found intracellularly in tissues; colonies on blood agar resemble *Staphylococcus aureus*; cells have single buds; at room temperature on Sabouraud's glucose agar, delicate branching, septate hyphae appear, with chlamydospores present in old cultures	This may occur as an acute or chronic, localized or disseminated infection of the reticuloendothelial system; clinically it may be confused with carcinoma of the nose, tongue, or pharynx, or with tuberculosis, Hodgkin's disease, or aplastic anemia; most infections regress spontaneously, but fulminating cases are usually fatal
Coccidioidomycosis	*Coccidioides immitis*	In cultures on Sabouraud's glucose agar these organisms develop as typical white- to buff-colored mold colonies that sporulate by arthrospore formation; in body exudates they are single-celled, thick-walled spherical organisms filled with endospores; chlamydospores are present in old cultures	This disease goes by many other names, such as valley fever, San Joaquin fever, and desert rheumatism; the fungus is highly infectious and widely distributed in the soil of certain areas of the United States; most cases are mild and transitory, but a few terminate fatally

Table 38-3. (continued)

Disease	Causative Organism	Characteristics of the Organisms	Characteristics of the Infection
Sporotrichosis	*Sporothrix schen-ckii*	Organisms are rarely found in tissues, but in experimentally infected rats the organisms are Gram-positive, resembling fusiform bacilli, in polymorpho-nuclear leukocytes; on Sabouraud's glucose agar incubated at room temperature, delicate, branching, septate hyphae with spherical or pyriform microconidia in clusters on lateral branches are found; on brain-heart-infusion agar, colonies are soft and composed of budding, yeastlike, cigar-shaped cells with a few mycelial elements	This is a chronic infection that usually begins as a subcutaneous nodule at the site of an injury; initial lesions resemble warts, boils or chancres; the organisms are disseminated in the body through the lymph channels to various lymph nodes; pulmonary involvement is infrequent; in rare cases involving spread of the organisms throughout the body, the patient may die

3 Local lesions caused by vitamin deficiency, irradiation, peptic ulcers, or other factors, which allow the fungi to get into the deep tissues

The fungi that cause the deep mycoses are important not only because the diseases they cause can be very serious, but also because the symptoms produced by some of them resemble tuberculosis or other diseases. It is essential that accurate diagnostic procedures be used in order that the most suitable treatment be used.

The principal systemic mycoses, their causative agents, the characteristics of these organisms, and the characteristics of the infections they cause are given in Table 38-3. Figure 38-9 shows the morphology in line drawings of some of these causative organisms. Note that many of them have a dimorphic growth habit, for example, a parasitic yeastlike phase and a saprophytic filamentous phase. Figure 38-10 shows the clinical symptoms of some of these diseases. Figures 38-11 and 38-12 illustrate the morphology of some of these etiologic organisms.

DISEASES CAUSED BY PROTOZOA

In adapting to their hosts, protozoa, like other animal parasites, have evolved many life-cycle patterns. While some species are parasitic during only one phase of their life cycle, others have adapted to more than one host during the different phases of their life cycle.

The host in which a parasite reaches sexual maturity and reproduction is termed the **definitive host.** If no sexual reproduction occurs in the life cycle of a protozoan, such as a trypanosome or an amoeba, the host that is believed to be the most important is arbitrarily identified as the definitive host. An **intermediate host** is one in which the other stages of the life cycle occur. For example, for the malaria protozoa, the mosquito is the definitive host, and humans or other vertebrates are the intermediate hosts. An animal (or human) that is

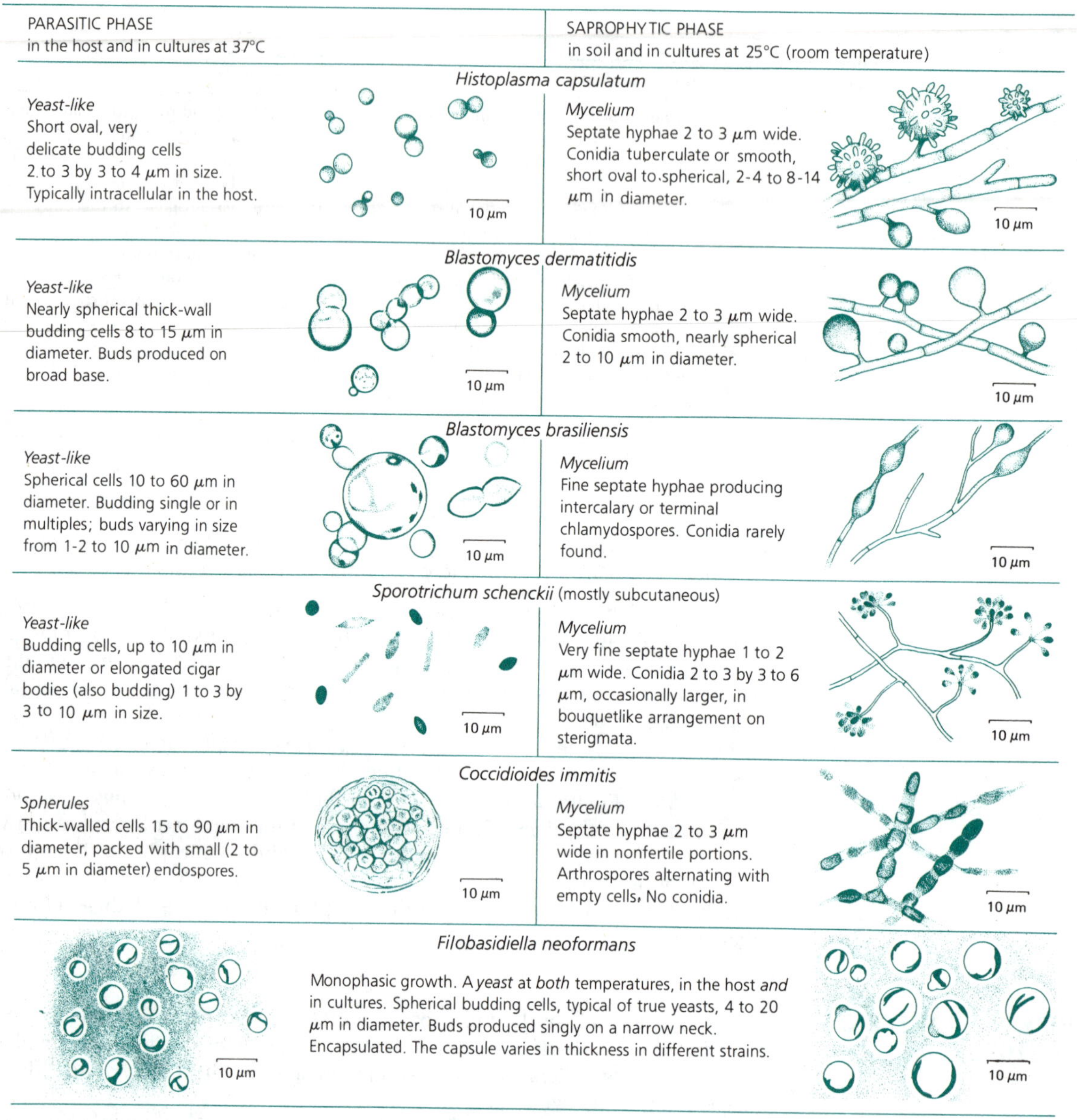
PARASITIC PHASE
in the host and in cultures at 37°C
SAPROPHYTIC PHASE
in soil and in cultures at 25°C (room temperature)
Histoplasma capsulatum
Yeast-like
Short oval, very delicate budding cells 2 to 3 by 3 to 4 μm in size. Typically intracellular in the host.
10 μm
Mycelium
Septate hyphae 2 to 3 μm wide. Conidia tuberculate or smooth, short oval to spherical, 2-4 to 8-14 μm in diameter.
10 μm
Blastomyces dermatitidis
Yeast-like
Nearly spherical thick-wall budding cells 8 to 15 μm in diameter. Buds produced on broad base.
10 μm
Mycelium
Septate hyphae 2 to 3 μm wide. Conidia smooth, nearly spherical 2 to 10 μm in diameter.
10 μm
Blastomyces brasiliensis
Yeast-like
Spherical cells 10 to 60 μm in diameter. Budding single or in multiples; buds varying in size from 1-2 to 10 μm in diameter.
10 μm
Mycelium
Fine septate hyphae producing intercalary or terminal chlamydospores. Conidia rarely found.
10 μm
Sporotrichum schenckii (mostly subcutaneous)
Yeast-like
Budding cells, up to 10 μm in diameter or elongated cigar bodies (also budding) 1 to 3 by 3 to 10 μm in size.
10 μm
Mycelium
Very fine septate hyphae 1 to 2 μm wide. Conidia 2 to 3 by 3 to 6 μm, occasionally larger, in bouquetlike arrangement on sterigmata.
10 μm
Coccidioides immitis
Spherules
Thick-walled cells 15 to 90 μm in diameter, packed with small (2 to 5 μm in diameter) endospores.
10 μm
Mycelium
Septate hyphae 2 to 3 μm wide in nonfertile portions. Arthrospores alternating with empty cells. No conidia.
10 μm
Filobasidiella neoformans
Monophasic growth. A yeast at both temperatures, in the host and in cultures. Spherical budding cells, typical of true yeasts, 4 to 20 μm in diameter. Buds produced singly on a narrow neck. Encapsulated. The capsule varies in thickness in different strains.
10 μm
10 μm

Figure 38-9. Some pathogenic fungal species causing systemic mycoses. *(Erwin F. Lessel, illustrator.)*

routinely infected with a protozoan or parasite which can also infect humans is termed a **reservoir host.** The major human diseases caused by protozoa are summarized in Table 38-4. Some of the important protozoan diseases are discussed below.

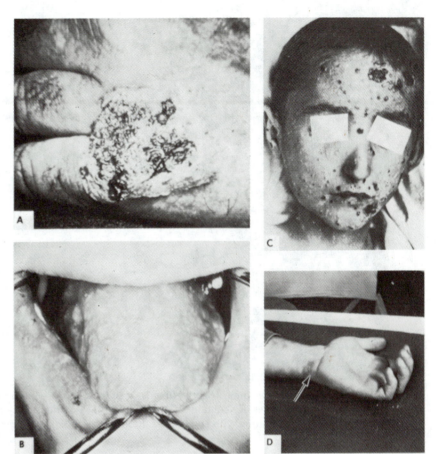

Figure 38-10. Clinical manifestations of some fungal diseases. (A) Blastomycosis of the hand. (B) Candidiasis on tongue and lips of a 19-year-old patient. *(A and B Courtesy of Centers for Disease Control, Atlanta, Ga.)* (C) Cryptococcosis on the skin of the face. (D) Coccidioidomycosis in a lesion of the arm. *(C and D Courtesy of Armed Forces Institute of Pathology.)*

Amoebiasis

Several protozoa cause intestinal diseases and are transmitted from person to person in infected food and water, by flies, and by direct contact. The most important of these, because of the great incidence of the disease and the fact that some fatalities occur, is *Entamoeba histolytica*. The most serious outbreak of amebiasis in many years occurred in 1933, when an epidemic originating in Chicago during the Century of Progress Exposition spread across the nation, and 1,400 cases and 4 deaths were reported.

Amebiasis occurs more commonly than is generally supposed. Authorities estimate that 10 million persons in the United States harbor *E. histolytica* and 2 million have symptomatic cases of the disease. People with amoebiasis may have few clinical indications of the infection, or they may have symptoms ranging from abdominal discomfort with slight diarrhea alternating with constipation to severe dysentery with blood and mucus in the stools. Abscesses may be formed in the liver or lungs or even in the brain.

Laboratory diagnosis depends on identification of *E. histolytica* in the stools. Search for the amoeba should be made as soon as possible after the stool is

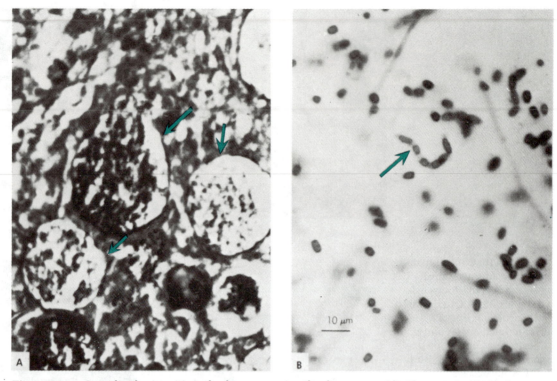

Figure 38-11. *Coccidioides immitis* is the fungus causing the disease coccidioidomycosis. (A) In the tissue phase it forms spherules containing endospores. (B) In culture the saprophytic filamentous phase forms arthrospores alternating with empty cells. *(Courtesy of L. Kapica and E. C. S. Chan, McGill University.)*

Figure 38-12. (Opposite page, top). Common airborne fungi can cause systemic mycoses. (A) *Candida albicans* in slide culture. (B) *Histoplasma capsulatum* in slide culture. (C) *Sporotrichum schenckii* from a culture (X1,200). *(Courtesy of Everett S. Beneke and the Upjohn Company.)*

passed to be sure of the presence of living, motile forms. Diagnosis requires differentiation from the nonpathogenic types of amoebas often found in the human intestinal tract.

Amebiasis can be controlled by the recognition of chronically infected persons, from whom infective cysts of the amoeba may be transmitted, and by proper sanitation and personal hygiene. The life cycle of the etiologic agent is shown in Fig. 38-13.

Malaria

The disease malaria has been known from antiquity and is aptly described as the single greatest killer of the human race. On a global scale, malaria is one of the most common infectious diseases of humans, causing much morbidity and significant mortality. As indicated earlier, each year more than 300 million people are gravely ill with malaria; about 3 million of these victims die of it.

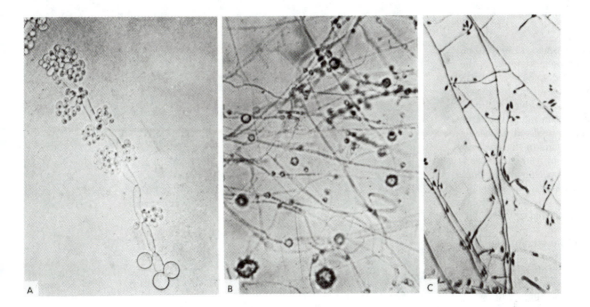

Table 38-4. Major Human Diseases Caused by Protozoa

Disease	Protozoan	Mode of Transmission to Humans	Definitive Hosts	Intermediate Hosts	Reservoir Hosts
Amoebiasis	Entamoeba histolytica	Ingestion (mature cyst)	Humans	None	Humans
Balantidiasis	Balantidium coli	Ingestion (mature cyst)	Hogs, humans	None	Hogs, humans
Giardiasis	Giardia lamblia	Ingestion (mature cyst)	Humans	None	Humans
Vaginitis	Trichomonas vaginalis	Contact (flagellate)	Humans	None	Humans
African sleeping sickness	Trypanosoma gambiense T. rhodesiense	Fly bite	Humans, animals	Tsetse flies (Glossina spp.)	Humans, animals
Chagas' disease	T. cruzi	Feces of bug	Animals, humans	Reduviid bugs	Armadillos, opossums, humans
Kala azar	Leishmania donovani	Fly bite	Humans, dogs	Sandflies (Phlebotomus spp.)	Dogs, humans
Oriental sore	L. tropica	Fly bite	Humans, dogs	Sandflies	Dogs, humans
Espundia	L. brasiliensis	Fly bite	Humans	Sandflies	Humans
Malaria	Plasmodium vivax P. falciparum P. ovale P. malariae	Mosquito bite	Anopheline mosquitoes	Humans	Humans

Figure 38-13. Life cycle of *Entamoeba histolytica,* the parasitic amoeba of humans that causes amoebic dysentery. Food or water contaminated with infective mature cysts is ingested. Excystation occurs in the host, releasing progeny amoebas, which become active amoebas called trophozoites in the intestine. Penetration of the intestinal mucosa by the trophozoites, with subsequent invasion of the portal circulation, may result in infection of the liver and other organs. Continued multiplication by binary fission and tissue destruction combine to result in abscesses. Thus the accompanying diarrhea is often tinged with blood. Cysts are passed in the feces and can infect other humans.

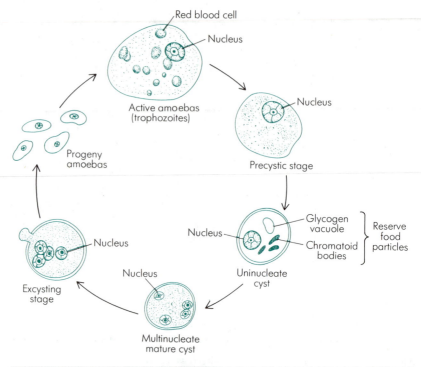

Figure 38-14. Malaria-infected youngster with enlarged spleen (area outlined on belly). *(Courtesy of Centers for Disease Control, Atlanta, Ga.)*

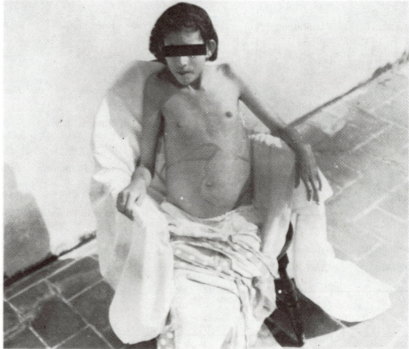

The disease has been virtually eliminated from the United States by control of its insect vector, the anopheline mosquito. But in these times of global travel, there is always the threat of contracting the disease in some country where it is prevalent and having it expressed only after returning to the United States.

Of the four species of *Plasmodium* protozoa that cause malaria in humans, *P. falciparum* and *P. vivax* are the two that most commonly cause infections. Symptoms usually occur 10 to 16 days after infection by mosquitoes. Paroxysms frequently begin with bed-shaking chills that are followed by high fevers, sweating, headache, and muscular pain. Fever cycles vary according to the species causing the infection. The symptomatic periods usually last less than 6 h. The spleen becomes enlarged and tender (Fig. 38-14); eventually the patient becomes weak and exhausted, and an anemia develops. The pattern of paroxysmal (periodic) illness interspersed with periods of well-being is characteristic of benign malaria, such as is caused by *P. vivax*, *P. ovale*, and *P. malariae*. In malignant falciparum malaria, the fever and symptoms are usually more persistent and also include edema of the brain and lungs and blockage of kidney activity.

Figure 38-15. Life cycle of *Plasmodium* species that cause malaria. Sexual reproduction takes place in the mosquito. Asexual reproduction takes place in the human host in liver cells as well as in the red blood cells. Sporozoites injected by the mosquito's bite enter liver cells via the blood stream and multiply asexually (schizogony). The resulting merozoites enter the red blood cells. Gametocytes formed from the merozoites are drawn up from the blood by the biting mosquito and are transformed into gametes in the insect's stomach. The zygotes encyst externally on the stomach wall and form oocysts; by asexual multiplication (sporogony), many sporozoites are formed which then invade the mosquito's salivary glands. From there they are injected by the mosquito's bite into another human victim.

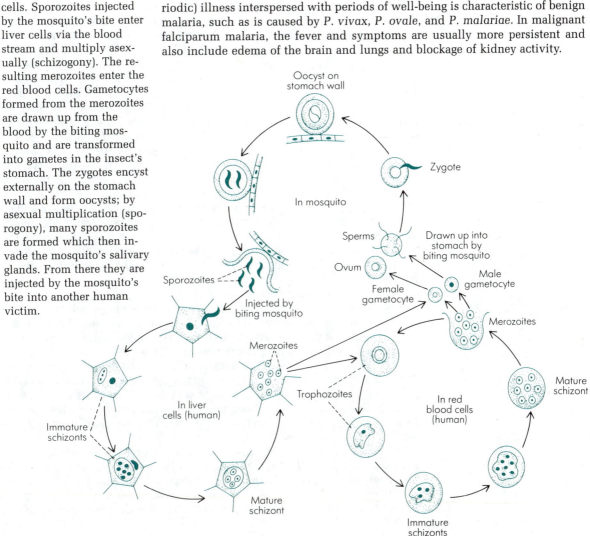

If not treated, benign malaria usually subsides spontaneously and recurs at a later date. Malignant malaria caused by P. falciparum has a high fatality rate if not treated promptly.

The malaria parasites are protozoa belonging to the group known as Sporozoa; their genus name is Plasmodium. Malaria is caused by more than fifty different species of Plasmodium; only four of them attack humans. The rest attack several hundred other animal hosts. The four species that cause human disease are P. falciparum, P. vivax, P. malariae, and P. ovale. P. falciparum causes the most serious form of the disease.

Plasmodium species have a complex life cycle, as shown in Fig. 38-15. When an Anopheles mosquito bites, its saliva, which contains the protozoan at the sporozoite stage of its life cycle, is injected into the bloodstream of the victim. The sporozoites quickly enter the liver, where they divide and develop into multinucleated forms known as **schizonts.** Within 6 to 12 days, the schizonts rupture and release the form known as **merozoites** into the bloodstream. These merozoites invade the host red blood cells, where they grow and divide to form more schizonts. These schizonts also rupture, destroying the erythrocytes and releasing more merozoites into the bloodstream to invade more red blood cells. The major symptoms of malaria are associated with rupture of the schizonts.

Some of the asexual merozoites in the patient's bloodstream develop into male and female gametocytes. When a mosquito bites, the gametocytes enter the mosquito's stomach where they become free male and female gametes. After fertilization occurs, the zygote passes to the outside of the stomach lining, where it develops into an oocyst containing many sporozoites. The mature sporozoites migrate to the salivary glands, from which they can be injected into the bloodstream of another victim to begin the cycle all over again.

Disease in malaria is caused specifically by the asexual erythrogenic cycle. The rupture of infected erythrocytes at the completion of schizogony occurs every 48 h with P. vivax and P. ovale and every 72 h with P. malariae, producing coincident chills and other symptoms. Synchronized, or coincident, schizogony and paroxysms of fever and chills are not, however, common with P. falciparum infection. The release of an endogenous pyrogen from injured cells may be the cause of the paroxysms of fever.

The high mortality rate of falciparum malaria is due in part to the high rate of reproduction of the asexual erythrocytic form of the parasite. The small veins and capillaries of the heart are clogged with parasitized erythrocytes; effective coronary blood flow and cardiac function are diminished.

The typical symptoms of malaria mimic a large variety of other human infections. Therefore the definitive diagnosis of the disease is made in the laboratory by the demonstration of the parasite in blood smears from patients (Fig. 38-16).

The indirect fluorescent-antibody and indirect hemagglutination tests are used in serologic diagnosis of malaria; however, antibodies are usually not detectable until after the second week of infection.

In North America, Europe, and probably northern Asia, malaria is largely a disease of the past. The cases that occur in North America and Europe have been contracted by persons visiting Asia, Africa, or Latin America. The influence of military involvement and of control measures on the number of cases reported in the United States is shown in Fig. 38-17. There were 1,103 cases of malaria

Figure 38-16. Malaria worker taking a blood sample from a child in a malaria eradication program in Thailand. The blood sample will be examined microscopically for the presence of parasites. *(Courtesy of Centers for Disease Control, Alanta, Ga.)*

in the United States in 1981. In the tropical and subtropical areas of the world, however, malaria is still the single most severe health problem today (Fig. 38-18).

Many African and American blacks are resistant to infection by *P. vivax*. It is suggested that this is due to the high frequency of a lack of a specific binding factor (Duffy determinants) on their erythrocytes, to which the protozoan must bind to infect. Also, the presence of the sickle cell trait (a genetic anomaly in the formation of the hemoglobin molecule), which is largely confined to blacks, confers some resistance to malaria.

The control of malaria depends on the elimination of the insect vector which transmits the disease. Eradication of the mosquito requires destruction of its breeding areas and killing of the larval stages and adults. These are not easy tasks because some mosquitoes develop resistance to insecticides, the behavioral patterns of others prevent their contact with insecticides, and in certain areas it is physically impossible to eliminate the breeding of mosquitoes.

At the individual level of control, netting can be used around sleeping areas; houses can be screened; and mosquito boots, insecticides, and mosquito repellents can be used. This type of control, of course, can be used for the prevention of all arthropod-borne diseases.

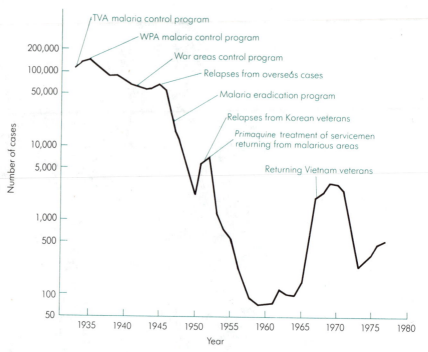

Figure 38-17. Reported cases of malaria by year in the United States, 1933–1977. The total number of cases each year has fluctuated with the application of control measures and the return of military personnel from areas of world where malaria is common. *(Centers for Disease Control: "Reported Morbidity and Mortality in the United States," Annual Summary 1977, issued September 1978.)*

Drug prophylaxis for the prevention of malaria can also be employed. For over 100 years, quinine was the only drug available. In World War II it was replaced by quinacrine, which in turn was supplanted by chloroquine and primaquine, the current drugs of choice. Chloroquine destroys merozoites in the blood, while primaquine destroys schizonts located in the liver. The combination of these two drugs is very effective against susceptible malaria parasite strains found in Africa, India, and Central America. Travelers to endemic areas are advised to take chloroquine phosphate.

Beginning in 1959, chloroquine-resistant strains of *P. falciparum* have appeared in many countries. An effective chemoprophylactic and therapeutic regimen for the chloroquine-resistant falciparum malaria is a combination of antagonists of folate metabolism in the parasite: sulfonamides and pyrimethamine.

There is no commercial vaccine available at the present time against malaria. A vaccine against the mosquito-borne infective form (the sporozoites) is currently in the advanced stages of development. But since an immune response to sporozoites apparently occurs naturally only after many exposures to infected mosquitoes, this vaccine must artificially immunize the host with antigens that the immune system may not readily respond to under natural conditions. It appears that the antibody formed prevents sporozoites from attaching to host cells.

Cultures of the asexual form of *Plasmodium* (merozoites) have been grown in vitro and serve as good sources of protozoa for vaccine development. Vaccina-

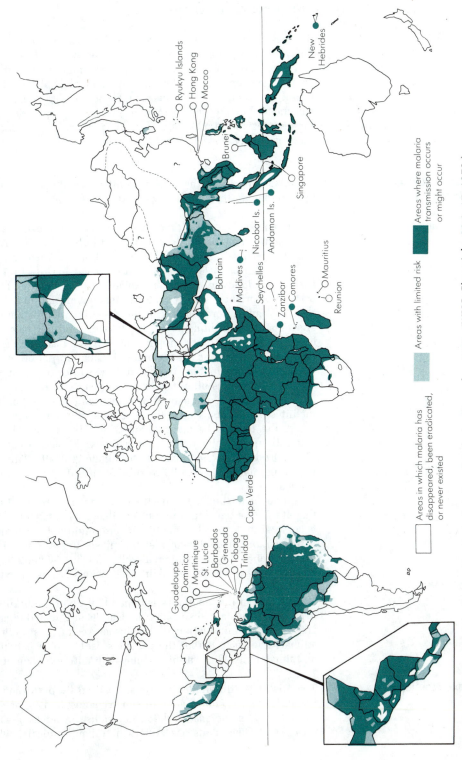

Figure 38-18. Epidemiological assessment of status of malaria, December 1976. (WHO Chronicle, **32**:9–17, 1978.)

Ryukyu Islands
Hong Kong
Macao

Brunei

Singapore

New Hebrides

Nicobar Is.
Andaman Is.

Seychelles
Zanzibar
Comores
Mauritius
Reunion

Bahrain
Maldives

Cape Verde

Guadeloupe
Dominica
Martinique
St. Lucia
Barbados
Grenada
Tobago
Trinidad

Areas in which malaria has disappeared, been eradicated, or never existed

Areas with limited risk

Areas where malaria transmission occurs or might occur

tion has been successful in simian malaria caused by isolated merozoites of *P. knowlesi*. The protection in rhesus monkeys was species-specific, stage-specific, and long-lasting. Owl monkeys that were immunized with *P. falciparum* merozoites were similarly protected. Unfortunately, optimal protection required the use of adjuvants that are unacceptable for human use. In the future, a malaria vaccine for human use may include a variety of stage-specific antigens. Although the prospects for a malaria vaccine have never been as good as in recent years, there are many obstacles yet to overcome.

Hemoflagellate Infections

Flagellated protozoa that are transmitted to humans by the bites of infected bloodsucking arthropods are referred to as hemoflagellates.

Leishmaniasis

Leishmaniasis is a parasitic disease caused by protozoa belonging to the genus *Leishmania*. The disease is encountered in India, China, Africa, the Mediterranean basin, Brazil, and other Central and South American countries.

Clinical forms of leishmaniasis are caused by infection with three different species: *L. donovani* in visceral leishmaniasis, commonly called kala azar; *L. tropica* in oriental sore (cutaneous leishmaniasis) and mucocutaneous leishmaniasis; and *L. brasiliensis* in the disease called espundia. These organisms are transmitted to humans by the bites of sandflies (genus *Phlebotomus*) harbored by dogs and other animals that serve as reservoirs for the parasites.

Clinically, kala azar has a variable incubation period, usually 2 to 4 months. Onset may be gradual or sudden, and the source of the disease may be acute or chronic. Symptoms often resemble malaria, with irregular, recurrent fever and leukopenia, with enlargement of spleen and liver. When the parasites invade the intestine, ulcers, secondary infections, and dysentery ensue. Relapses often occur, the disease becomes chronic, and the patient has a persistent fever, emaciation, and pigmented skin. Untreated cases become complicated with secondary infections and usually terminate in death. Patients who recover may have a permanent immunity to subsequent infections with *L. donovani* and *L. tropica*.

Cutaneous leishmaniasis begins with one or more small papules at the site of the bite. These develop into crusted ulcers. Spontaneous healing with scarring occurs within a year, leaving the patient immune to subsequent infections with *L. tropica*. In *mucocutaneous* leishmaniasis, the lesions are more extensive than in the cutaneous type and may involve mucous membranes of the mouth, nose, and throat, as well as the skin. The lesions result in extensive scarring and mutilation.

Diagnosis in the laboratory is dependent upon the finding of Leishman-Donovan bodies in stained smears from lesions or infected organs. Leishmaniasis is controlled by elimination of the insect vector, destruction of dogs or other animals known to harbor the parasites, isolation of patients, and treatment of all human cases with various pentavalent antimony compounds.

Trypanosomiasis

Two distinctly different diseases are caused by protozoan parasites belonging to the genus *Trypanosoma*. African trypanosomiasis, or sleeping sickness, is caused by *T. gambiense* and *T. rhodesiense*, whereas *T. cruzi* is the causative organism of Chagas' disease (American trypanosomiasis), which occurs in South

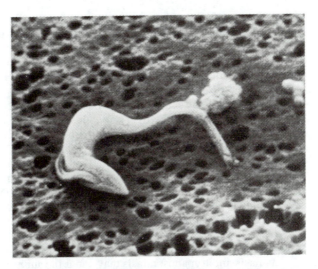

Figure 38-19. Scanning electron micrograph of an infective form of *Trypanosoma brucei* (X13,500.) Many parasitologists consider *T. gambiense* and *T. rhodesiense* to be subspecies of *T. brucei*. *(Courtesy of Peter R. Gardiner, International Laboratory for Research on Animal Diseases, Nairobi, Kenya, and the Journal of Protozoology.)*

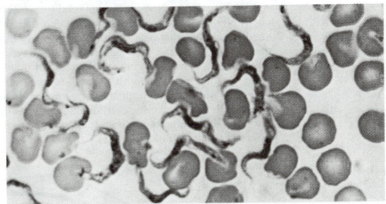

Figure 38-20. *Trypanosoma cruzi,* the causative agent of Chagas' disease (X960). *(Courtesy of A. Packchanian.)*

America and Mexico. Other species produce diseases in domestic and wild animals (see Fig. 38-19).

African sleeping sickness is transmitted to humans by bites of the tsetse fly (genus *Glossina*), and control of the disease depends upon control of these vectors and prompt treatment of human cases with tryparsamide (an organic arsenical preparation) or other drugs. Diagnosis may be made in the laboratory by demonstrating the flagellates in smears from blood, spinal fluid, or lymph nodes. In the early stages, African trypanosomiasis is characterized clinically by fever, headache, insomnia, lymphadenitis, anemia, and rash. Later, the central nervous system becomes involved, and symptoms include tremors, delusions, emaciation, lethargy, and somnolence. Untreated cases usually terminate in death.

T. cruzi (Fig. 38-20), the causative agent of Chagas' disease, is transmitted to humans by reduviid bugs (*Triatoma, Rhodinas,* and *Panstrongylus*), which are intermediate hosts as well as vectors for the parasites and in which the trypanosomes pass through a stage of their life cycle. As the insect feeds on an infected human or on animal hosts such as an armadillo or opossum, it ingests the trypanosomal forms, which multiply in the gut of the insect and pass into an infective stage of their life cycle. These parasites are excreted by the reduviid bugs on the human skin as the insect feeds. They enter the body through scratches and abrasions rather than by bites as the malaria plasmodia do.

Chagas' disease is an acute febrile disease more common in children than adults. During the acute stage it is characterized by fever and general discomfort, with enlargement of the liver and spleen. The face and eyelids become swollen and inflamed. Fatality is due to a meningoencephalitis or myocardial failure. *T. cruzi* is present in blood during the acute stage of Chagas' disease. Motile forms can be seen in coverslip preparations and smears stained with Wright's or Giemsa's stain. Animal-inoculation and complement-fixation tests also are used.

There is no known means of immunization. Prevention depends upon the avoidance of contact with the insect vector. Transmission from person to person has not been demonstrated, and there is no specific treatment for Chagas' disease.

The animallike flagellates, *Leishmania* and *Trypanosoma*, which cause leishmaniasis and trypanosomiasis may occur in four morphologically distinct stages representing consecutive developmental forms, as shown in Figs. 38-21, 38-22, and Table 38-5. The **amastigote** stage is an intracellular form that appears as a small round cell that has a distinct basal body but no flagellum. The **promastigote** is an elongated cell with a terminal flagellum. The **epimastigote** has a flagellum originating from the midportion of the cell. The **trypomastigote** form is the most mature stage, with a flagellum that begins in the caudal (tail) portion of the cell and forms an undulating membrane. In addition to the morphological forms, both genera manifest striking similarities in their life cycles. Both have two hosts, one being the human (or some other vertebrate) and the other being an insect vector (an invertebrate).

Figure 38-21. Morphological forms of the animallike flagellates. (*Erwin F. Lessel, illustrator.*)

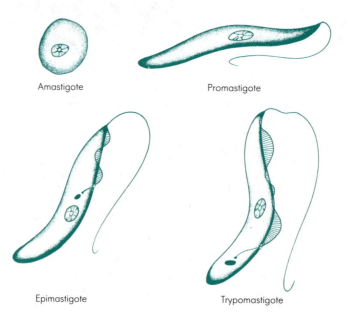

Amastigote

Promastigote

Epimastigote

Trypomastigote

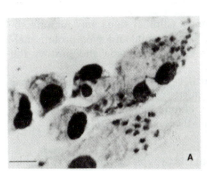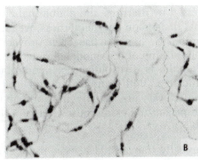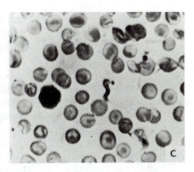

Figure 38-22. Some morphological forms of *Trypanosoma cruzi.* (A) Amastigote or intercellular form (bar represents 10μm.) (B) Epimastigote form. (C) Trypomastigote form, the most mature stage of the flagellate. Note the presence of red blood cells. *(Courtesy of Madeleine Faucher, McGill University.)*

Other Protozoan Infections

Balantidiasis

Balantidiasis is a disease caused by *Balantidium coli*, the only species of the ciliated protozoa known to cause disease in humans. *B. coli* is the largest of the parasitic protozoa that may be found in the human intestine; it can be as long as 200 μm. It resides primarily in the lumen of the intestine and obtains food by the ingestion of bacteria. Occasionally, however, *B. coli* may cause a bloody diarrhea similar to that present in amoebic dysentery.

B. coli is routinely present in hogs. Humans become infected by consumption of water or food contaminated by cysts present in swine feces. Most infections manifest no demonstrable symptoms.

Giardiasis

The etiologic agent of the disease giardiasis is *Giardia lamblia*. This is the only flagellated protozoan reported to date that causes intestinal disease. Infection in adults may be asymptomatic. The majority of overt cases of giardiasis are

Table 38-5. Host Sites of the Various Morphological Forms of Species of *Trypanosoma* and *Leishmania*

Species	Forms			
	Amastigote	Promastigote	Epimastigote	Trypomastigote
L. tropica	Human (skin)	Sandfly (midgut)	—	—
L. brasiliensis	Human (skin, mucosa)	Sandfly (midgut)	—	—
L. donovani	Human (liver spleen, lymph nodes)	Sandfly (midgut)	—	—
T. cruzi	Reduviid bug (gut); human (skin, liver, spleen, myocardium, CNS)	Reduviid bug (hindgut)	Reduviid bug (midgut)	Reduviid bug (feces); human (blood)
T. rhodesiense	—	Tsetse fly (salivary glands)	Tsetse fly (salivary glands)	Tsetse fly (proboscis); human (blood, lymph nodes, CNS)
T. gambiense	—	Tsetse fly (salivary glands)	Tsetse fly (salivary glands)	Tsetse fly (proboscis); human (blood, lymph nodes)

manifested by diarrhea and abdominal cramps. Clinical signs may include abdominal distension and tenderness, weight loss, anemia, and even protein malabsorption. *G. lamblia* has been proven recently to invade and destroy the epithelial lining of the duodenal mucosa.

G. lamblia is distributed worldwide, but it causes outbreaks in small local endemic foci. Both trophozoite and cyst stages can be found in feces. The infectious cysts are usually transmitted in contaminated water supplies, although person-to-person transmission of the disease has been reported.

Trichomoniasis

The sexually transmitted disease trichomoniasis, a type of vaginitis, is caused by *Trichomonas vaginalis*. The organism is found mainly in vaginal secretions and infects both men and women. Only the trophozoite stage of the organism is known. Thus its life cycle is presumably direct. The organisms reproduce within the vagina (in females) or the urethra (in males) by longitudinal binary fission. Infection in males is usually asymptomatic except in cases involving the seminal vesicles and prostate. In women, a thin, watery vaginal discharge is the most prominent symptom, usually accompanied by itching and burning. Both male and female partners should receive treatment with an effective antimicrobial agent such as metronidazole. The prevalence of trichomoniasis is estimated to be about 5 to 30 percent in the United States. The parasite is cosmopolitan in its distribution.

Toxoplasmosis

Toxoplasmosis is a disease caused by the protozoan *Toxoplasma gondii*. The disease in adult humans is frequently mild or asymptomatic. Mild symptoms include fatigue, fever, sore throat, malaise, rash, and headache. However, the disease in the immunocompromised patient is usually a severe, life-threatening one; the symptoms manifested are encephalitis (inflammation of the brain), myocarditis (inflammation of the myocardium or muscular tissue of the heart), and pneumonia. The disease can also be acquired congenitally by the human embryo or by newborn infants infected late in fetal development. The consequences of these infections are very severe. There is central nervous system involvement resulting in mental retardation and severe visual impairment or blindness; convulsions, fever, and an enlarged liver also may be present.

Toxoplasmosis is worldwide in distribution and is one of the most common infections of humans. Even in the U.S., serologic tests have indicated that over 50 percent of adults have been infected. The organism also infects all orders of mammals as well as many birds and some species of reptiles.

Toxoplasma gondii is a protozoan of the *Sporozoa* group. It exists in three forms: trophozoite, cyst, and oocyst. The cyst and oocyst are the principal forms in which the protozoan is transmitted. The crescent-shaped trophozoite form invades mammalian cells (except erythrocytes) and is the form found in acute human infections. The organism is spread by the ingestion of oocysts that are excreted in the feces of cats and by ingestion of undercooked pork and mutton containing cysts.

Diagnosis may be made by isolation of the parasite, by histologic demonstration of the organism in tissues, or by serology. A number of serologic tests are available. In the Sabin-Feldman dye exclusion test, living *Toxoplasma* cells are stained with methylene blue in the presence of the patient's serum. In the absence of specific antibodies, the cells are stained with the dye. In the presence of antibodies, the dye is excluded from the cells.

Pneumocystosis

As discussed in Chap. 33, *Pneumocystis carinii* is the infectious agent that causes the most common life-threatening secondary infection in AIDS (acquired immunodeficiency syndrome) victims. This microorganism is usually regarded as a protozoan belonging to the *Sporozoa* group although some workers consider it a fungus. The disease is called pneumocystosis and is an infection of alveolar spaces of the lung (pneumonia). The organism appears in vivo as aggregates of thick-walled cysts. The cysts are 6 to 9 μm in diameter, and each contains two to eight nucleated pear-shaped cells produced by binary fission. The organism has not been propagated in vitro. The source of human infections is unknown, although the microbe is suspected to be transmitted by fomites or respiratory droplets. Individuals susceptible to the disease are compromised persons, e.g., undernourished infants, immunosuppressed patients, and of course AIDS victims.

IMMUNE RESPONSE TO FUNGAL AND PROTO-ZOAN DISEASES

Humoral antibodies seem to play little or no role in host resistance to fungal infections. Cell-mediated immunity is the primary defense mechanism. Thus in patients with T-cell defects, opportunistic fungal infections are common.

Immediate or delayed-type hypersensitivities to fungi can be expressed in various ways. Inhalation of fungal spores may produce asthma. Other allergic reactions are manifested by eruptions of the skin which are associated with cutaneous infections due to dermatophytes (fungi causing skin infections) or species of *Candida*. In other cases an allergic reaction such as erythema nodosum (an eruption of pink to blue, tender nodules) may be part of the symptoms with systemic fungal disease, as in coccidioidomycosis. Hypersensitivity reactions may also be demonstrated by an intradermal test with a specific fungal extract, e.g., coccidioidin or histoplasmin. (Such tests are described in Chap. 34.)

In the case of protozoan infections, both cellular and humoral immunity to various protozoal antigens often can be demonstrated. Both sensitized T cells and antibodies arise during infection, but the exact role of each of them in combating and resolving the infection is unclear.

IgE antibodies are produced commonly, if not always, during protozoan infections. IgG and IgM antibodies are found frequently as well. Such antibodies are useful in the serodiagnosis of infections. Secretory IgA is also produced to prevent parasites from attaching to mucosal cells in the intestines. Not much is known at the present time about the exact role of cell-mediated immunity in recovery from protozoan infections, but it is thought to be an important mechanism for the elimination of these parasites. But unlike many viral and bacterial infections, parasitic infections rarely confer lifelong or complete immunity after primary infection.

THERAPEUTIC DRUGS FOR TREATMENT OF FUNGAL AND PROTOZOAN DISEASES

Most antibacterial antibiotics are ineffective in the treatment of fungal infections because the eucaryotic fungal cells lack the target sites of the procaryotic cells. Likewise, the few antifungal agents available are not effective against bacteria. One group of effective antifungal agents is the polyene antibiotics; the two most useful of these are nystatin and amphotericin B. Both act on the plasma membrane of the fungus, combining with membrane sterols, and cause leakage of intracellular potassium and other metabolites. Nystatin is effective in topical *Candida* infections but not in deep mycoses or even dermatophyte infections. Amphotericin B is effective against the deep mycoses, such as cryptococcosis and histoplasmosis.

Unfortunately, polyene antibiotics also have an affinity for human cells, lysing erythrocytes and causing leakage of potassium and other low-molecular-weight intracellular substances. Side effects from these antibiotics vary from a mild headache, chills, and fever to severe hemolytic anemia and acute nephritis (kidney inflammation). Another effective group of antifungal agents is the pyrimidine antimetabolites; the best example is 5-fluorocytosine, which is especially effective in the treatment of yeast infections. Because it is an analogue of pyrimidine, it interferes with pyrimidine metabolism or DNA synthesis.

The imidazole compounds are another group of antifungal agents. They have a very broad spectrum, high activity, and mild side effects. Miconazole reacts with cytoplasmic membranes and causes them to leak. It is very active against coccidioidomycosis and paracoccidioidomycosis. Ketoconazole is an oral imidazole with few side effects. It is active against several fungi, including *C. albicans*. Its mechanism of action is inhibition of ergosterol biosynthesis, an important step in synthesis of fungal cell walls and membranes.

The antibiotic griseofulvin is active against the dermatophyte infections (superficial skin infections) but not against the deep mycoses (systemic diseases caused by fungi). It interferes with protein and nucleic acid synthesis.

The drug treatment of parasitic (or protozoan) diseases varies widely and depends on the kind of disease organism implicated. Often all that is required is prevention of reinfection. In such cases the healthy host is capable of expelling the parasite. The use of drugs in therapy is complicated by the fact that most of the drugs used to eradicate protozoa are toxic to the host also. Thus the

Table 38-6. Some Drugs Used in Treating Protozoan Diseases

Disease	Drugs
Leishmaniasis	Sodium stibogluconate

Trypanosomiasis, early
    West African
    East African
	
	Pentamidine isethionate
	Suramin

Amoebiasis, dysenteric, intestinal	Metronidazole + iodoquinol

Falciparum malaria, acute
    Chloroquine-sensitive
 	
    Chloroquine-resistant
	
	Chloroquine phosphate
	    + primaquine phosphate
	Quinine sulfate
	    + pyrimethamine
	    + sulfadiazine

SOURCE: *Guide to Antimicrobial Therapy*, J. P. Sanford, West Bethesda, MD, 1984.

effectiveness of the drug must be weighed against its side effects and the necessity of treating a given patient. Unfortunately, for some parasitic diseases no effective drugs are available. Table 38-6 shows some drugs used in the treatment of protozoan diseases. It is apparent that the antibiotics effective against bacterial infections are not active against protozoa. This is because protozoa, like fungi, being eucaryotic cells, do not have the target structures for antibiotic activity, such as a peptidoglycan cell wall, 70S ribosomes, or a nucleus without a nuclear membrane, all characteristics typical of procaryotic cells.

QUESTIONS

1 Define the following terms: (a) opportunists, (b) microbes of endogenous and exogenous origin, (c) histoplasmin, (d) nephritis, (e) mycosis, and (f) dermatophytes.
2 Describe the factors that contribute to a rising incidence of mycotic diseases.
3 Why do protozoan diseases remain such a scourge to humans?
4 What is known about the role of cell-mediated immunity in fungal and protozoan diseases?
5 Describe the mode of action of these fungal antibiotics: (a) griseofulvin, (b) ketoconazole, (c) miconazole, (d) nystatin, and (e) amphotericin B.
6 What complication is encountered in using drugs in the therapy of protozoan infections?
7 Why are antibiotics that are effective against bacteria inactive against protozoa?
8 Are fungal infections transmitted by air or by contact? Explain.
9 What diseases are referred to as superficial mycoses?
10 Outline the factors that predispose a person to systemic fungus infections. Name three of these infections and identify their causative agents.
11 Describe the differences between a definitive host, an intermediate host, and a reservoir host in parasitic or protozoan diseases.
12 Describe the life cycle of *Entamoeba histolytica*.
13 Why is malaria such an important disease? How was the disease virtually eliminated in the United States?
14 Name the four species of protozoa that cause malaria in humans and identify the one that causes the most serious form of the disease.
15 Write an account of the life cycle of a *Plasmodium* species that causes malaria.
16 Chloroquine phosphate has been used as a prophylaxis for travellers to areas with endemic malaria. With the development of chloroquine-resistant strains of *Plasmodium falciparum* in many countries, what alternative drugs are recommended?
17 What difficulties have been encountered in the development of a vaccine against malaria?
18 Identify the specific etiologic agents of leishmaniasis and the clinical forms caused by each of them. Comment on their transmission and reservoir hosts.
19 What is Chagas' disease? Comment on its transmission, reservoir and intermediate hosts, and vector.
20 Discuss the similarities in the life cycles and morphological forms of *Leishmania* and *Trypanosoma* species.

21 Give a brief description of the diseases caused by the following protozoa: (a) *Balantidium coli*, (b) *Giardia lamblia*, and (c) *Pneumocystis carinii*.

REFERENCES

The references cited in Chaps. 17 and 19 are also applicable to this chapter.

Schmidt, G. D., and L. R. Roberts: *Foundations of Parasitology*, 2d ed., Mosby, St. Louis, 1981. *A clearly written and well-organized text intended for use in introductory courses in parasitology. In addition to the host-parasite relationship, pathogenesis, and epidemiology of the parasites, their biology, morphology, ecology, biochemistry, and immunology are also described. The first 10 chapters are of direct relevance to protozoology.*

Glossary

PRONUNCIATION GUIDE FOR GLOSSARY

Many of the terms in the glossary are followed by informal phonetic spellings in parentheses. A precise rendering of the pronunciation of each term from scratch is impossible without the use of a full-scale phonetic system, such as can be found in an unabridged dictionary. All that we intend is that a speaker (and speller) of English be able to form a workable idea of a term's pronunciation. Only words or word elements whose proper pronunciation may not be obvious are provided with these informal phonetic spellings.

In these informal phonetic spellings, stressed syllables are set in boldface type. The intended pronunciations of some of the letters and letter combinations we have used are:

Letter(s)	Pronunciation
ah	Broad **a** sound, as in **father, rotten**
ay	Long **a** sound, as in **ray, fade**
g	Hard **g** sound, as in **good, rug**
i	Short **i** sound, as in **sit, pacific,** except when used with final **e** (e.g., **tide**) or in **igh**
igh	as in **sight**
j	as in **jump** (used for soft **g**, as in **rage**)
o alone or preceded by consonant	long **o** sound as in **no, boat**
o followed by consonant	short **o** sound, as in **fox** (also rendered by **ah**)
oe (e.g., **doe**)	long **o** sound as in **foe**
oy	as in **toy, soy**
uh	unaccented syllable, as in **biology** (bye·**ahl**·uh·jee), **telephone** (tel·uh·fone), **arrest** (uh·**rest**)
ye	long **i** sound, as in **dye, rye, side**
zh	as in **revision, pleasure**

PREFIXES AND COMBINING FORMS USED AS PREFIXES

anti-, ant-. Against, inhibiting.
auto-. Self, independent.
bio-. Life.
cardi-, cardio-. Heart.
cary-, caryo-. Nucleus, nuclear.
centi-. Hundred.
cyt-, cyto-. Cell.
de-. From, removal.
dermato-. Skin.
di-. Two, double.
ecto-. Outside.
endo-. Within.
exo-, ex-. Outside, without.
hem-, hema-, hemo-. Blood.
hetero-. Different.
hist-, histo-. Tissue.
holo-. Complete, homogeneous.
hom-, homo-. Like, similar.
hyp-, hypo-. Deficiency, below.
hyper-. Excessive, above normal.
inter-. Between, among.
intra-. Within, into.
iso-. Equality, similarity.
kary-, karyo-. Nucleus, nuclear.
leuk-, leuko-. White.
macro-. Large.
meso-. Middle.
micro-. Small.

mono-. One.
multi-. Many.
myco-. Fungus.
necro-. Dead.
neo-. New.
noso-. Disease.
nucle-, nucleo-. Nucleus, nuclear.
olig-, oligo-. Few, deficiency.
oxy-. Oxygen in a compound.
pan-. All, many.
path-, patho-. Disease, pathologic.
peri-. About, around.
pneumo-. Pulmonary, respiration.
poly-. Many, diverse.
post-. After, behind.
pre-. Before.
pseudo-. False.
pyo-. Pus.
sacchar-, sacchari-, saccharo-. Sugar.
syn-, sym-. With, together.
tax-, taxi-, taxo-. Arrangement.
therm-, thermo-. Heat, temperature.
thi-, thio-. Sulfur present.
tox-, toxi-, toxo-. Poisonous, toxin, poison.
trans-. Through, across.
trich-, tricho-. Hair, filament.

SUFFIXES AND COMBINING FORMS USED AS SUFFIXES

-algia. Pain, suffering.
-ase. Enzyme.
-cide. Killer, killing.
-cyte. Cell.
-emia. Condition of blood.
-ia. Condition; abnormal or pathologic condition.
-iasis. Diseased condition.
-ism. Condition or disease.
-itis. Inflammation of a part.
-logy. Field of study.
-lysis. Dissolution or disintegration.
-oma. Tumor, neoplasm.
-osis. Process, disease, cause of disease.
-otic. Related to causing a process or condition.

-otomy. A cutting into.
-ous. Having or pertaining to.
-pathia. Disease.
-penia. Deficiency.
-phage, -phag. Ingesting, breaking down.
-rrhage, -rrhagia. Abnormal or excessive discharge.
-rrhea. Discharge.
-scope. Instrument for seeing or examining.
-taxia, -taxis. Arrangement or order.
-trophy. Nutrition, growth.
-tropic. Turning toward, having an affinity for.

abiogenesis (ay·bye·o·**jen**·uh·sis, ab·ee·o·**jen**·uh·sis). *See* **spontaneous generation.** (*Abio-* means nonliving; *genesis* means origin.)

abiotic (ab·bye·**ot**·ik, ab·ee·**ot**·ik). Pertaining to or characterized by the absence of living organisms.

abscess (**ab**·sess). A localized collection of pus in a cavity formed by tissue disintegration.

acervulus. An asexual fruiting body or reproductive structure in a fungus.

acid curd. Milk protein coagulated by acid.

acid dye. A dye consisting of an acidic organic grouping of atoms (anion), which is the actively staining part, combined with a metal; the dye has affinity for cytoplasm.

acid-fast. Retaining the initial stain and difficult to decolorize with acid alcohol. A property of certain bacteria.

actinomycetes (ak·tin·o·**mye**·seets). Gram-positive bacteria that are characterized by the formation of branching filaments.

activated-sludge process. The use of biologically active sewage sludge to hasten the breakdown of organic matter in raw sewage during secondary treatment.

active immunity. Specific resistance to disease acquired by individuals as a result of their own reactions to pathogenic microorganisms or to the products of such organisms.

active transport. The energy-requiring pumping of ions or other solutes across a cell membrane from a lower to a higher concentration.

adaptive enzyme. An enzyme produced by an organism in response to the presence of the enzyme's substrate or a related substance. Also called *induced enzyme.*

adenine (**ad**·un·neen). A purine component of nucleosides, nucleotides, and nucleic acids.

adenosine (uh·**den**·o·seen). A mononucleoside consisting of adenine and D-ribose, produced by the hydrolysis of adenosine monophosphate.

adenosine triphosphate (trye·**fos**·fate). A compound of one molecule each of adenine and D-ribose and three molecules of phosphoric acid; it plays an important role in energy transformations in metabolism. Abbreviation: ATP.

adenoviruses. A group of icosahedral double-stranded DNA viruses.

adjuvant (**aj**·oo·vunt). A substance that when injected together with antigen increases antibody production.

aerobe (**air**·obe). An organism that requires oxygen for growth and can grow under an air atmosphere (21% oxygen). *Compare* **anaerobe.**

aerosol. Atomized particles suspended in air.

aflatoxin (aff·luh·**tahk**·sin). The toxin produced by some strains of the fungus *Aspergillus flavus;* a carcinogen.

agar (**ah**·gar). A dried polysaccharide extract of red algae *(Rhodophyceae)* used as a solidifying agent in microbiological media.

agglutination (uh·gloo·tin·**ay**·shun). The clumping of cells.

agglutinin (uh·**gloo**·tin·in). An antibody capable of causing the clumping or agglutination of bacteria or other cells.

akinetes. Thick-walled single-celled nonmotile asexual resting spores formed by the thickening of the parent cell wall; formed by some cyanobacteria.

alga, pl. **algae** (**al**·guh, **al**·jee). Any member of a heterogeneous group of eucaryotic, photosynthetic, and unicellular or multicellular organisms.

alleles (uh·**leelz**). Two genes that are alternative occupants of the same chromosomal locus on a pair of homologous chromosomes.

allergy. A type of antigen-antibody reaction marked by an exaggerated physiological response to a substance in sensitive individuals.

allosteric enzymes (al·o·**stehr**·ik, al·o·**steer**·ik). Regulatory enzymes with a binding or catalytic site for the substrate and a different site (the *allosteric site*) where a modulator acts.

allosteric site. *See* **allosteric enzymes.**

alveoli. Air sacs of the lung.

amino acid (uh·**meen**·o). An organic compound containing both amino (—NH_2) and carboxyl (—COOH) groups.

aminoglycoside antibiotics. A class of antibiotics which disrupt the normal synthetic sequence of protein synthesis.

ammonification (uh·mon·i·fi·**kay**·shun, uh·mo·nif·i·**kay**·shun). The decomposition of organic nitrogen compounds, e.g., proteins, by microorganisms with the release of ammonia.

amphitrichous (am·**fit**·rik·us). Having a single flagellum at each end of a cell.

amylase (**am**·i·lase, **am**·i·laze). An enzyme that hydrolyzes starch.

anabolism (uh·**nab**·o·lizm). The synthesis of cell constituents from simpler molecules, usually requiring energy. *Compare* **catabolism.**

anaerobe (**an**·uh·robe). An organism that does not use O_2 to obtain energy, cannot grow under an air atmosphere, and for which O_2 is toxic. *Compare* **aerobe.**

anaerobic respiration. Respiration under anaerobic conditions in which a terminal electron acceptor other than oxygen is involved.

anamnestic response (an·am·**ness**·tik). The heightened immunologic reaction to a second exposure to an antigen.

anaphylatoxin. A complement-derived peptide, C5a, that causes the release of histamine from mast cells.

anaphylaxis (an·uh·fi·**lak**·sis). Hypersensitivity in an animal following the parenteral injection of an antigen.

anaplasia (an·uh·**play**·zhuh). Structural abnormality in a cell or cells.

angstrom (Å). A unit of length equal to 10^{-8} cm (1/100,000,000 cm). Conventionally used for measuring wavelengths. Also used to express dimensions of some intracellular structures of microbes.

antagonism. The killing, injury, or inhibition of growth of one species of microorganism by another when one organism adversely affects the environment of the other.

antheridium, pl. **antheridia.** A male gametangium.

anthramycin. An antitumor antibiotic.

antibiosis (an·tee·bye·**o**·sis). An antagonistic association between two organisms in which one is adversely affected.

antibiotic (an·tee·bye·**ot**·ik). A substance of microbial origin that has antimicrobial activity in very small amounts.

antibody (**an**·ti·bod·ee). Any of a class of substances (proteins) produced by an animal in response to the introduction of an antigen.

anticodon (an·tee·**ko**·don). A sequence of three nucleotides (in a tRNA) complementary to a codon triplet in mRNA.

antigen (**an**·ti·jen). A substance that when introduced into an animal body stimulates the production of specific substances (antibodies) that react or unite with the substance introduced (antigen).

antigenic determinant (an·ti·**jen**·ik). The part of an antigen molecule that, as the structural complement of certain chemical groupings on certain antibody molecules, determines the specificity of the antigen-antibody reaction.

antimicrobial agent (an·tee·migh·**kro**·bee·ul). Any chemical or biologic agent that either destroys or inhibits the growth of microorganisms.

antiseptic. Acting against or opposing sepsis, putrefaction, or decay by either preventing or arresting the growth of microorganisms.

antiserum (**an**·tee·seer·um). Blood serum that contains antibodies.

antitoxin (an·tee·**tahk**·sin). An antibody capable of uniting with and neutralizing a specific toxin.

aperture. The magnitude of the angle subtended by the optical axis and the outermost rays still covered by the objective.

aplanospore (ay·**plan**·o·spore). A nonmotile spore; an abortive zoospore.

apoenzyme (ap·o·**en**·zime). The protein moiety (portion) of an enzyme.

apophysis. The base of the sporangium.

apothecium. A sexual fruiting body in a fungus.

archaeobacteria. A major group of bacteria that includes the meth-anogens, the red extreme halophiles, and the thermoacidophiles, and which diverged from other bacteria at a very early stage in evolution. Also called *archaebacteria.*

arthropod (**ahr**·thro·pod). An invertebrate with jointed legs, such as an insect or a crustacean.

arthrospore (**ahr**·thro·spore). An asexual spore formed by the fragmentation of the mycelium.

ascitic fluid (uh·**sit**·ik). Serous fluid that accumulates abnormally in the peritoneal cavity.

Ascomycetes (ass·ko·migh·**see**·teez). A class of fungi distinguished by the ascus.

ascospore (**ass**·ko·spore). A sexual spore, characteristic of the *Ascomycetes,* produced in a saclike structure (an ascus) after the union of the two nuclei.

ascus (**ass**·kus). A saclike structure, characteristic of the *Ascomycetes,* in which ascospores are produced.

asepsis (ay·**sep**·sis). A condition in which harmful microorganisms are absent. Adjective: **aseptic** (ay·**sep**·tik).

aseptic technique. Precautionary measures taken to prevent contamination.

assay (**ass**·ay). The qualitative or quantitative determination of the components of a material, such as a drug.

assimilation (uh·sim·i·**lay**·shun). The conversion of nutritive material into protoplasm.

asymptomatic (ay·sim·tuh·**mat**·ik). Exhibiting no symptoms.

ATP. *See* **adenosine triphosphate.**

attenuation (un·ten·yoo·**ay**·shun). A weakening; a reduction in virulence.

autoclave (**aw**·toe·klave). An apparatus using pressurized steam for sterilization.

autogenous vaccine (aw·**toj**·uh·nus). A vaccine prepared from bacteria isolated from the patient to be treated.

autoimmune disease (aw·toe·im·**yoon**). A condition in which the body develops an immunological reaction against its own tissues.

autolysis (aw·**tol**·i·sis). The disintegration of cells by the action of their own enzymes.

autotroph (**aw**·toe·trofe). A microorganism that uses inorganic materials as a source of nutrients; carbon dioxide is the sole source of carbon. Compare **heterotroph.**

auxotrophic mutant (awk·so·**troe**·fik). An organism having a growth requirement of specific nutrients not necessary in the parental strain.

axenic culture (ay·**zen**·ik, ay·**zee**·nik). A microorganism of a single species, e.g., a bacterium, fungus, alga, or protozoan, growing in a medium free of other living organisms.

bacillus (buh·**sil**·us). Any rod-shaped bacterium.

bacteremia (bak·tuh·**ree**·mee·uh). A condition in which bacteria are present in the bloodstream.

bacterial filter. A special type of filter through which bacterial cells cannot pass.

bactericide (bak·**teer**·i·side). An agent that destroys bacteria.

bacterin (**bak**·tuh·rin). A suspension of killed or attenuated bacteria used for artificial immunization.

bacteriochlorophyll (bak·**teer**·ee·o·**klor**·uh·fil). A chlorophyll-like pigment possessed by anoxygenic photosynthetic bacteria.

bacteriocin (bak·**teer**·ee·o·sin). See **bacteriocinogenic factor.**

bacteriocinogenic factor (bak·**teer**·ee·o·sin·o·**jen**·ik). A plasmid in some bacteria that determines the formation of bacteriocins, which are proteins that kill the same or closely related species of bacteria.

bacteriolysin (bak·**teer**·ee·o·**lye**·sin). A substance that causes the disintegration of bacteria.

bacteriophage (bak·**teer**·ee·o·fayj). A virus that infects bacteria and causes the lysis of bacterial cells.

bacteriorhodopsin. A purple pigment that occurs in the cytoplasmic membrane of the group of archaeobacteria called the red extreme halophiles; similar to the rhodopsin that occurs in the retinal rods of higher vertebrates.

bacteriostasis (bak·**teer**·ee·o·**stay**·sis). The inhibition of the growth and reproduction of bacteria without killing them.

bacteriostatic. Inhibiting the growth of bacteria without killing them.

bacterium, pl. **bacteria** (bak·**teer**·ee·um, bak·**teer**·ee·uh). Any of a group of diverse and ubiquitous procaryotic single-celled microorganisms.

bacteroids. The morphological form of *Rhizobium* cells within the root nodules of legumes.

barophile. An organism that grows under conditions of high hydrostatic pressure.

basic dye. A dye consisting of a basic organic grouping of atoms (cation), which is the actively staining part, combined with an acid, usually inorganic; the dye has affinity for nucleic acids.

Basidiomycetes (buh·**sid**·ee·o·mye·**see**·teez). A class of fungi that form basidiospores.

basidiospore (buh·**sid**·ee·o·spore). A sexual spore produced following the union of two nuclei on a specialized clublike structure known as a *basidium.*

basidium (buh·**sid**·ee·um). A club-shaped specialized structure of the *Basidiomycetes* on which are borne the exogenous basidiospores.

BCG vaccine. Bacillus Calmette-Guérin vaccine; an attenuated strain of *Mycobacterium bovis* used to immunize against tuberculosis.

benthos (**ben**·thahss). A collective term for the organisms living along the bottom of oceans and lakes.

Bergey's Manual. An international reference work which classifies and describes bacteria.

beta hemolysis. A colorless, clear, sharply defined zone of hemolysis surrounding certain bacterial colonies growing on blood agar.

binomial nomenclature (bye·**no**·mee·ul). The scientific method of naming plants, animals, and microorganisms, so-called because species names are binomial, i.e., consist of two terms.

biochemical oxygen demand. A measure of the amount of oxygen consumed in biological processes that break down organic matter in water; a measure of the organic pollutant load. Abbreviation: BOD.

biodegradable (bye·o·dee·**grade**·uh·bul). Capable of being broken down by microorganisms.

biogenesis (bye·o·**jen**·uh·sis). The production of living organisms only from other living organisms. Compare **spontaneous generation.**

biogeochemical agents (bye·o·jee·o·**kem**·i·kul). Microorganisms that mineralize organic carbon, nitrogen, sulfur, phosphorus, and other compounds.

bioluminescence (bye·o·loo·min·

ess·unce). The emission of light by living organisms.

biomass (**bye**·o·mass). The mass of living matter present in a specified area.

biosphere (**bye**·o·sfeer). The zone of the earth that includes the lower atmosphere and the upper layers of soil and water.

biota (bye·**o**·tuh). The animal, plant, and microbial life characterizing a given region.

biovar. A subdivision of a species based upon physiological characteristics. Also called *biotype*.

blastospore (**blass**·toe·spore). A spore produced by a budding process along the hypha or by a single cell.

blood plasma. The fluid portion of blood. Also called *plasma*.

blood serum. The fluid expressed from clotted blood or clotted blood plasma.

bloom. A colored area on the surface of a body of water caused by heavy growth of plankton.

BOD. *See* **biochemical oxygen demand.**

botulism (**bot**·choo·lizm). Food poisoning due to the toxin of *Clostridium botulinum*.

Braun's lipoprotein. A cell-wall lipoprotein which anchors the outer membrane of enteric Gram-negative bacteria to the peptidoglycan layer.

bronchi. Branches of the trachea.

Breed count. A microscopic method of counting bacteria in a dried, stained film of milk.

Brownian motion (**brown**·ee·un). A peculiar dancing motion exhibited by finely divided particles and bacteria in suspension, due to bombardment by the molecules of the fluid.

bubo. A swollen infected lymph node.

budding. A form of asexual reproduction typical of yeast, in which a new cell is formed as an outgrowth from the parent cell.

buffer. Any substance in a fluid that tends to resist the change in pH when acid or alkali is added.

calorie. A unit of heat; the amount of heat required to raise the temperature of 1 g of water by 1°C.

capsid. The protein coat of a virus.

capsomere (**kap**·so·meer). A morphologic subunit of a capsid as seen by electron microscopy.

capsule. An envelope or slime layer surrounding the cell wall of certain microorganisms.

carotenoid. A water-insoluble pigment, usually yellow, orange, or red, which consists of a long aliphatic polyene chain composed of isoprene units.

carrier. A person in apparently good health who harbors a pathogenic microorganism.

catabolism (kuh·**tab**·o·lizm). The dissimilation, or breakdown, of complex organic molecules, releasing energy. A part of the total process of metabolism. *Compare* **anabolism.**

catalase (**kat**·uh·lase, **kat**·uh·laze). An enzyme that converts hydrogen peroxide to water and oxygen.

catalyst (**kat**·uh·list). Any substance that accelerates a chemical reaction but remains unaltered in form and amount.

cavitation (kav·i·**tay**·shun). The use of high-frequency sound waves in liquid to produce small bubbles that collapse violently, disintegrating microbial cells.

cecum. The distended intestinal pouch into which open the ileum, the colon, and the appendix.

cell. The microscopic, functionally and structurally basic unit of all living organisms.

cellulase (**sel**·yoo·lase). An extra-cellular enzyme that hydrolyzes cellulose to produce cellobiose.

cellulose. A complex polysaccharide consisting of many glucose molecules; the characteristic structural material of plant cell walls.

cell wall. A rigid external covering of the cytoplasmic membrane.

centrifuge. An apparatus that uses centrifugal force to separate or remove particulate matter suspended in a liquid.

chelating agent. An organic compound in which atoms form more than one coordinate bond with metals, keeping them in solution.

chemiosmotic theory. A theory which states that the energy liberated by the oxidation-reduction reactions of a respiratory chain can be conserved in the form of an electrochemical gradient of protons across the membrane; this gradient then is used to drive the synthesis of ATP.

chemoautotroph (kee·mo·**aw**·toe·trofe). An organism that obtains energy by oxidizing inorganic compounds. Carbon dioxide is the sole source of carbon.

chemolithotroph (kee·mo·**lith**·o·trofe). An organism that uses inorganic compounds as electron donors and relies on chemical compounds for energy.

chemoreceptor. Proteins located on the cytoplasmic membrane which sense gradients and are specific for various attractants and repellents.

chemostat (**kee**·mo·stat, **kem**·o·stat). A device for maintaining organisms in continuous culture; it regulates the growth rate of the organisms by regulating the concentration of an essential nutrient.

chemotaxis (kee·mo·**tak**·sis). The

movement of an organism in response to a chemical stimulus.

chemotherapy (kee·mo·**thehr**·uh·pee). The treatment of a disease by the use of chemicals.

chemotroph (**kee**·mo·trofe). An organism that uses chemical compounds for energy. *Compare* **phototroph.**

chitin. A polymer of N-acetylglucosamine present in the covering layer of arthropods and in the cell walls of many fungi.

chlamydospore (**klam**·id·o·spore). A thick-walled, resistant spore formed by the direct differentiation of the cells of the mycelium.

chlorophyll (**klor**·uh·fil). A light-trapping green pigment essential as an electron donor in photosynthesis.

chloroplast (**klor**·o·plast). A cell plastid (specialized organelle) in plants and algae that contains chlorophyll pigments and functions in photosynthesis.

chromatin body. Bacterial nuclear material.

chromatophore. A pigment-containing body; specifically applied to chlorophyll-bearing granules in bacteria.

chromogenesis. The production of pigments by microorganisms.

chromosome (**kro**·muh·sohm). A gene-containing filamentous structure in a cell nucleus; the number of chromosomes per cell nucleus is constant for each species.

cilium, *pl.* **cilia** (**sil**·ee·um, **sil**·ee·uh). On certain eucaryotic cells a relatively short hairlike appendage which is capable of a vibratory beating or lashing movement.

cis-trans. A genetic analysis in which the action of two genes on the same chromosome is compared with their action when on different chromosomes.

cistron (**siss**·trahn). The genetic unit that carries information for the synthesis of a single enzyme or protein molecule; determined by the cis-trans complementation test.

citric acid cycle. *See* **Krebs cycle.**

classification. The systematic arrangement of units (e.g., organisms) into groups, and often further arrangement of those groups into larger groups.

clone (**klohn**). A population of cells descended from a single cell.

coagulase (ko·**ag**·yoo·lase, ko·**ag**·yoo·laze). An enzyme, produced by pathogenic staphylococci, that causes the coagulation of blood plasma.

coccus (**kock**·us). A spherical bacterium.

codon (**ko**·dahn). A sequence of three nucleotide bases (in mRNA) that codes for an amino acid or the initiation or termination of a polypeptide chain.

coenocytic (see·no·**sit**·ik). A term applied to a cell or an aseptate hypha containing numerous nuclei.

coenzyme (ko·**en**·zime). The nonprotein portion of an enzyme.

coenzyme F_{420}. A flavinlike coenzyme unique to methanogenic bacteria and which fluoresces under ultraviolet light.

coenzyme M. A coenzyme unique to methanogenic bacteria and which is involved in methyl transfer reactions.

cofactors. Metal ions which function in combination with the enzyme protein and are regarded as coenzymes.

coliphage (**ko**·li·fayj). A virus that infects *Escherichia coli.*

colony A macroscopically visible growth of microorganisms on a solid culture medium.

colony-forming unit. The cell or aggregate of cells which gives rise to a single colony in the plate-count technique. Abbreviation: cfu.

colostrum. The first milk secreted by a mother after giving birth.

columella. The dome-shaped apex of the sporangiophore in some phycomycetes.

commensalism (kuh·**men**·suh·lizm). A relationship between members of different species living in proximity (the same cultural environment) in which one organism benefits from the association but the other is not affected.

communicable. Pertaining to a disease whose causative agent is readily transferred from one person to another.

competitive inhibition. The inhibition of the action of an enzyme by a nonsubstrate molecule occupying the site on the enzyme that would otherwise be occupied by the substrate.

complement (**kom**·pluh·ment). A normal thermolabile protein constituent of blood serum that participates in antigen-antibody reactions.

complement fixation. The binding of complement to an antigen-antibody complex so that the complement is unavailable for a subsequent reaction.

compromised host. A person already weakened with debilitating disease.

conidiophore (ko·**nid**·ee·o·fore). A hypha which bears conidiospores.

conidiospores (ko·**nid**·ee·o·spore). Any asexual spores which are formed at the tip of a hypha and which are not enclosed within a sac (as distinguished from sporangiospores). Also called *conidia.*

conidium (ko·**nid**·ee·um). An asexual spore that may be one-celled or many-celled and may be of many sizes and shapes. Also called *conidiospore*.

conjugation. A mating process characterized by the temporary fusion of the mating partners and the transfer of genes. Conjugation occurs particularly in unicellular organisms.

conjunctiva. The membrane covering the eyeball and lining the eyelids.

contamination. The entry of undesirable organisms into some material or object.

cord factor. A toxic mycolic acid derivative, trehalose dimycolate, which occurs in the cell walls of corynebacteria and mycobacteria.

crustose. A flat crustlike growth of lichens.

culture. A population of microorganisms cultivated in a medium.

cyst. A thick-walled dormant form of an organism which is resistant to desiccation, e.g., the cysts formed by certain bacteria such as *Azotobacter* or by various protozoa.

cytochrome (sigh·toe·krome). One of a group of iron porphyrins that serve as reversible oxidation-reduction carriers in respiration.

cytokinesis. The division of the cytoplasm following nuclear division.

cytolysis. The dissolution or disintegration of a cell.

cytopharynx. The region through which nutrients must pass to be enclosed in a food vacuole.

cytoplasm (sigh·toe·plazm). The living matter of a cell between the cell membrane and the nucleus.

cytoplasmic membrane. A thin layer under the cell wall consisting mainly of phospholipids and proteins; it is responsible for the selective permeability properties of the cell. Also called *plasma membrane*.

cytostome. The opening through which food is ingested in ciliates.

DAP. Diaminopimelic acid. A seven-carbon diamino acid that occurs as a component of cellwall peptidoglycan in some bacteria.

dark-field microscopy. A type of microscopic examination in which the microscopic field is dark and any objects, such as organisms, are brightly illuminated.

deaminase. An enzyme involved in the removal of an amino group from a molecule; ammonia is liberated.

deamination (dee·am·i·**nay**·shun). The removal of an amino group, especially from an amino acid.

decarboxylase. An enzyme that liberates carbon dioxide from the carboxyl group of a molecule, e.g., an amino acid.

decarboxylation (dee·kar·bock·si·**lay**·shun). The removal of a carboxyl group (—COOH).

decimal reduction time. The amount of time at a particular temperature sufficient to reduce a viable microbial population by 90 percent.

dehydrogenase. An enzyme which oxidizes a substrate by removing hydrogen atoms from it.

demineralization (dee·min·ur·ul·i·**zay**·shun). The process by which acid produced by bacteria dissolves the calcium salts of tooth enamel.

denature (dee·**nay**·chur). To modify, by physical or chemical action, the structure of an organic substance, especially a protein.

denitrification (dee·nigh·tri·fi·**kay**·shun). The reduction of nitrates to nitrogen gas.

dental plaque (plak). An aggregation of bacteria and organic material on the tooth surface.

deoxyribonucleic acid (dee·**ahk**·see·rye·bo·new·**klee**·ik). The carrier of genetic information; a type of nucleic acid occurring in cells, containing phosphoric acid, D-2-deoxyribose, adenine, guanine, cytosine, and thymine. Abbreviation: DNA.

deoxyribose (dee·ahk·see·**rye**·bose, dee·ahk·see·**rye**·boze). A five-carbon sugar having one oxygen atom fewer than the parent sugar, ribose; a component of DNA.

dermato-. A prefix meaning "skin."

dermatotropic (dur·muh·toe·**trope**·ik). Having a selective affinity for the skin.

dermis. The connective tissue layer under the epidermis.

desmid. Any of several freshwater algae.

detergent. A synthetic cleaning material containing surface-active agents which do not precipitate in hard water.

dextran (**deks**·tran). A polysaccharide (glucose polymer) produced by a wide range of microorganisms, sometimes in large amounts.

dialysis (dye·**al**·i·sis). The separation of soluble substances from colloids by diffusion through a semipermeable membrane.

dialyze. To pass through a semipermeable membrane.

diatomite. Silica-containing shells (cell walls) resulting from centuries of growth of diatoms.

diauxic growth (dye·**awk**·sik). Growth in two separate phases due to the preferential use of one carbon source over another; between the phases a temporary lag occurs.

differential stain. A procedure using a series of dye solutions or staining reagents to bring out differences in microbial cells.

dilution, serial. Successive dilution of a specimen; e.g., a 1:10 dilution equals 1 ml of specimen plus 9 ml of diluent; a 1:100 dilution equals 1 ml of a 1:10 dilution plus 9 ml of diluent.

dimorphic (dye·**more**·fik). Occurring in two forms.

dipicolinic acid. A compound found in large amounts in endospores. Abbreviation: DPA.

diplobacilli (dip·lo·buh·**sil**·eye). Bacilli occurring in pairs.

diplococci (dip·lo·**kahk**·sigh). Cocci occurring in pairs.

diploid (**dip**·loyd). Having chromosomes in pairs the members of which are homologous; having twice the haploid number.

disaccharide (dye·**sak**·uh·ride). A sugar composed of two monosaccharides.

disinfectant. An agent that frees from infection by killing the vegetative cells of microorganisms.

dissimilation (dis·sim·i·**lay**·shun). Chemical reactions that release energy by the breakdown of nutrients.

DNA. See **deoxyribonucleic acid.**

droplet nucleus. Airborne particle containing viable microbes.

ecology. The study of the interrelationships that exist between organisms and their environment.

ecosystem. A functional system which includes the organisms of a natural community together with their environment.

edema (eh·**dee**·muh). The excessive accumulation of fluid in body tissue.

effluent (**eff**·loo·unt). The liquid waste of sewage and industrial processing.

electrophoresis. An electrochemical process in which suspended particles with an electric charge migrate in a solution under the influence of an electric current.

endemic (en·**dem**·ik). With reference to a disease, one that has a low incidence but which is constantly present in a particular geographic region.

endergonic (en·dur·**gahn**·ik). Describing or pertaining to a chemical reaction which requires the input of free energy in order to proceed.

endoenzyme (en·doe·**en**·zime). An enzyme formed within the cell and not excreted into the medium. Also called *intracellular enzyme.*

endoflagella. See **periplasmic flagella.**

endogenous (en·**dahj**·uh·nus). Produced or originating from within.

endonuclease (en·doe·**new**·klee·ase, en·doe·**new**·klee·aze). An enzyme that excises a damaged segment of DNA.

endophytic (en·doe·**fit**·ik). Describing or pertaining to algae that are not free-living but live in other organisms.

endoplasmic reticulum (en·doe·**plaz**·mik ree·**tik**·yoo·lum). An extensive array of internal membranes in a eucaryotic cell.

endospore (**en**·doe·spore). A thick-walled spore formed in the bacterial cell. Very resistant to physical and chemical agents.

endosymbiont. An organism that lives within the body of the host without a deleterious effect on the host.

endothermic (en·doe·**thur**·mik). Describing or pertaining to a chemical reaction in which energy is consumed overall.

endotoxin (en·doe·**tahk**·sin). A heat-stable toxin which consists of lipopolysaccharide; it is located in the outer membrane of Gram-negative bacteria and is liberated only when the bacteria disintegrate.

enteric (en·**tehr**·ik). Pertaining to the intestines.

enterotoxin (en·tur·o·**tahk**·sin). A toxin specific for cells of the intestine.

enzyme (**en**·zime). An organic catalyst produced by an organism. See also **adaptive enzyme, endoenzyme, exoenzyme,** and **enzyme, constitutive.**

enzyme, adaptive. See **adaptive enzyme.**

enzyme, constitutive. An enzyme whose formation is not dependent upon the presence of a specific substrate.

enzyme, inducible. See **adaptive enzyme.**

enzyme, intracellular. See **endoenzyme.**

epicellular. On the surface of host cells.

epidemic (ep·i·**dem**·ik). With reference to a disease, one that displays a sudden increase in incidence in a particular geographic region.

epidemiology (ep·i·dee·mee·**ahl**·uh·jee, ep·i·dem·ee·**ahl**·uh·jee). The study of the occurrence and distribution of disease.

epidermis. The outer surface of skin.

episome (**ep**·i·sohm). A plasmid which can integrate reversibly with the chromosome of its bacterial host; in the integrated state it behaves as part of the chromosome, but it is also able to multiply independently of the chromosome.

esterase. One of a group of enzymes that hydrolyze esters.

estuary. A semienclosed coastal body of water which opens to the sea.

etiology (ee·tee·**ahl**·uh·jee). The study of the cause of a disease.

eubacteria. One of the two major groups of bacteria (the other being the archaeobacteria). Eubacteria have fundamental features that are considered to be typical of most bacteria.

eucaryote (yoo·**care**·ee·ote). A cell that possesses a definitive or true nucleus. *Compare* **procaryote.**

eutrophication. An aging process in lakes, during which the water becomes overly rich in dissolved nutrients; this results in the excessive development of algae and other microscopic plants, causing a decline in the levels of dissolved oxygen.

evapotranspiration. Evaporation from soil surfaces, lakes, and streams and by transpiration from plants into the atmosphere.

exergonic (ek·sur·**gahn**·ik). Energy-yielding, as in a chemical reaction.

exoenzyme (ek·so·**en**·zime). An enzyme excreted by a microorganism into the environment. Also called *extracellular enzyme.*

exogenous (ek·**sahj**·uh·nus). Produced or originating from without.

exons. Code-bearing sequences of a gene.

exonuclease (ek·so·**new**·klee·ase, ek·so·**new**·klee·aze). An enzyme that hydrolyzes a nucleic acid, starting at one end.

exospore. A heat- and desiccation-resistant spore formed external to the vegetative cell by a budding process; e.g., the exospores of the methane-oxidizing genus *Methylosinus.*

exothermic (ek·so·**thur**·mik). Describing a chemical reaction that gives off energy.

exotoxin (ek·so·**tahk**·sin). A toxic protein excreted by a microorganism into the surrounding medium.

exponential phase (ek·spuh·**nen**·chul). The period of culture growth when cells divide steadily at a constant rate. Also called *logarithmic phase* (commonly, *log phase*).

extrachromosomal genetic element (ek·struh·kro·muh·**so**·mul). A genetic element, called a *plasmid,* that is capable of autonomous replication in the cytoplasm of the bacterial cell.

exudate (**eks**·yoo·date). The more or less fluid material found in a lesion or inflamed tissue.

facultative anaerobe (fak·ul·**tay**·tiv). An organism which does not require O_2 for growth (but may use it if available), which grows well under both aerobic and anaerobic conditions, and for which oxygen is not toxic.

fastidious organism (fass·**tid**·ee·us). An organism that is difficult to isolate or cultivate on ordinary culture media because of its need for special nutritional factors.

fauna. The animal life characteristic of a region or environment.

feedback inhibition. A cellular control mechanism by which the end product of a series of metabolic reactions inhibits the further activity of an earlier enzyme of the sequence.

fermentation (fur·men·**tay**·shun). The anaerobic oxidation of compounds by the enzyme action of microorganisms; neither gaseous oxygen nor a respiratory chain is involved in this energy-yielding process. An organic compound is the electron acceptor.

F factor. The fertility or sex factor in the cytoplasm of male bacterial cells.

fibril, axial. A flagellalike structure located just beneath the outer membrane of spirochetes.

fibrinolysin (figh·bri·**nahl**·i·sin). A substance, produced by hemolytic streptococci, that can liquefy clotted blood plasma or fibrin clots. Also called *streptokinase.*

filamentous (fil·uh·**men**·tus). Characterized by threadlike structures.

filter, bacteriological. A special type of filter through which bacterial cells cannot pass.

fimbriae, sing. **fimbria** (fim·**bree**·ee, **fim**·bree·uh). Surface appendages of certain Gram-negative bacteria, composed of protein subunits. They are shorter and thinner than flagella. Also called *pili.*

fission (**fish**·un). An asexual process by which some microorganisms reproduce; transverse cell division in bacteria.

fission, binary. A single nuclear division followed by the division of the cytoplasm to form two daughter cells of equal size.

flagellates (**flaj**·uh·luts, **flaj**·uh·laits). Members of one of the subphyla of the phylum *Protozoa.*

flagellin. The protein monomer of bacterial flagella.

flagellum, pl. **flagella** (fluh·**jel**·um, fluh·**jel**·uh). A thin, filamentous appendage on cells, responsible for swimming motility.

floc. An aggregate of the finely suspended and colloidal matter of sewage.

floccule. An adherent aggregate of

microorganisms or other materials floating in or on a liquid.

flora (**flore**·uh). In microbiology, the microorganisms present in a given situation, e.g., intestinal flora, the normal flora of soil. *See* also **biota.**

fluorescence (floo·uh·**ress**·unce). The emission of light of a particular wavelength by a substance which has absorbed light of a shorter wavelength (for example, the emission of green light by molecules of fluorescein dye which have absorbed blue light).

fluorescence microscopy (migh·**krahss**·kuh·pee). Microscopy in which cells or their components are stained with a fluorescent dye and thus appear as glowing objects against a dark background.

foliose. Leaflike.

fomites (**fo**·mites). Inanimate objects that carry viable pathogenic organisms.

food poisoning. Stomach or intestinal disturbances due to food contaminated with certain microbial toxins.

formalin (**for**·muh·lin). A 37 to 40% aqueous solution of formaldehyde.

Forssman antigens (**force**·mun). Heterophile antigens widely distributed in nature.

fractional sterilization. The sterilization of material by heating it to 100°C (212°F) on 3 successive days with incubation periods in between.

Frei test (fry). A skin test to determine sensitivity to the agent that causes lymphogranuloma venereum.

fruticose. Shrublike.

fruiting body. A specialized, spore-producing organ.

fulminating infection. A sudden severe and rapidly progressing infectious disease.

fungicide (**fun**·ji·side). An agent that kills or destroys fungi.

fungus, pl. **fungi** (**fung**·gus, **fun**·jye). A microorganism that lacks chlorophyll and is usually filamentous; a mold or yeast.

fusiform (**fyoo**·zi·form). Spindle-shaped, tapered at the ends.

gametangium, pl. **gametangia.** Sex organelle of fungi.

gamete (**gam**·eet, guh·**meet**). A reproductive cell that fuses with another reproductive cell to form a zygote, which then develops into a new individual; a sex cell.

gamma globulin. A fraction of serum globulin that is rich in antibodies.

gas chromatograph. An instrument which allows the separation and identification of various volatile chemical compounds in a gaseous mixture by means of selective adsorption.

gastroenteritis (gass·tro·en·tur·**eye**·tis). Inflammation of the mucosa of the stomach or intestine.

gelatin. A protein obtained from skin, hair, bones, tendons, etc., used in culture media for the determination of a specific proteolytic activity of microorganisms or for the preparation of a peptone.

gelatinase (juh·**lat**·i·nase, juh·**lat**·i·naze). An exoenzyme that degrades gelatin.

gene (jeen). A segment of a chromosome, definable in operational terms as the repository of a unit of genetic information.

generation time. The time interval necessary for a cell to divide.

genome (**jee**·nohm). A complete set of genetic material; i.e., a complete set of genes.

genotype (**jee**·nuh·tipe, **jen**·uh·tipe). The particular set of genes present in an organism's cells; an organism's genetic constitution. *Compare* **phenotype.**

genus, pl. **genera** (**jee**·jus, **jen**·ur·uh). A group of very closely related species.

germ. A microbe, usually a pathogenic one.

germicide (**jurm**·i·side). An agent capable of killing germs, usually pathogenic microorganisms.

giardiasis. The presence of the protozoan *Giardia lamblia* in the human small intestine.

gingiva (**jin**·ji·vuh). The mucous membrane and soft tissue surrounding a tooth. Adjective: **gingival** (jin·ji·vul).

gliding motility. A type of movement across surfaces that is exhibited by some bacteria (e.g., myxobacters) which are devoid of flagella.

globulin (**glahb**·yoo·lin). A protein soluble in dilute solutions of neutral salts but insoluble in water. Antibodies are globulins.

glucan. A polymer of glucose.

glucose (**gloo**·kose). A carbohydrate classified as a monosaccharide and hexose, used as an energy source by many microorganisms. Also called *dextrose* or *grape sugar.*

glycogen (**glye**·kuh·jen). A carbohydrate of the polysaccharide group stored by animals. It yields glucose on hydrolysis.

glycolysis (glye·**kahl**·i·sis). Anaerobic dissimilation of glucose to pyruvic acid by a sequence of enzyme-catalyzed reactions. Also called the *Embden-Meyerhoff pathway.*

glyoxylate cycle. A sequence of biochemical reactions by which acetate is converted to succinic acid (a bypass of the Krebs cycle).

gnotobiotic (no·toe·bye·**aht**·ik). Pertaining to higher organisms living in the absence of all demonstrable viable organisms other than those known to be present.

Golgi apparatus (**gol**·jee). A membranous organelle in the endoplasmic reticulum of the cell.

gonidium. An asexual reproductive cell arising in a special organ in eucaryotes.

Gram-negative bacteria. Bacteria that do not retain the crystal violet–iodine complex when subjected to the Gram technique and thus acquire the color of the dye (usually a red dye) that is used to counterstain the cells.

Gram-positive bacteria. Bacteria that retain the crystal violet–iodine complex when stained by the Gram technique and thus appear dark blue or violet.

Gram stain. A differential stain by which bacteria are classed as Gram-positive or Gram-negative depending upon whether they retain or lose the primary stain (crystal violet) when subjected to treatment with a decolorizing agent.

granules, metachromatic. Intracellular granules of polyphosphate found in certain microorganisms; such granules stain a deep violet color when the cells are stained with dilute methylene blue.

groundwater. All subsurface water, especially that occurring in the zone of saturation.

growth curve. A graphic representation of the growth (population changes) of bacteria in a culture medium.

guanine (**gwah**·neen) A purine

base, occurring naturally as a fundamental component of nucleic acids.

Guarnieri bodies. Cytoplasmic inclusion bodies found in the epidermal cells of smallpox and chickenpox patients.

habitat. The natural environment of an organism.

halophile (**hal**·o·file). A microorganism whose growth is accelerated by or dependent on high salt concentrations.

hanging-drop technique. A technique in which microorganisms are observed suspended in a drop of fluid.

H antigen. A type of heat-labile protein antigen found in the flagella of certain bacteria.

haploid (**hap**·loyd). Having a single set of unpaired chromosomes in each nucleus; having the chromosome number characteristic of a mature gamete of the species. *Compare* **diploid.**

hapten. A simple substance that reacts like an antigen in vitro by combining with antibody but cannot induce the formation of antibodies by itself.

haustoria. Rootlike projections through which a fungus obtains nourishment from the alga in lichens.

HeLa cells (**hee**·luh). A pure cell line of human cancer cells used for the cultivation of viruses.

helical. Shaped like a corkscrew, with one or more turns or twists.

helix (**hee**·liks). A coiled spiral form.

hemagglutination (hee·muh·gloo·ti·**nay**·shun). The agglutination (clumping) of red blood cells.

hemoglobin (**hee**·mo·glo·bin). The constituent of red blood cells that gives them their color and carries oxygen.

hemolysin (hee·**mahl**·i·sin). A substance produced by microorganisms that lyses red blood cells, liberating hemoglobin. Also can refer to a type of antibody that acts in conjunction with complement to cause the lysis of red blood cells.

hemolysis (hee·**mahl**·i·sis). The process of dissolving red blood cells.

hemorrhagic. Showing evidence of hemorrhage (bleeding). The tissue becomes reddened by the accumulation of blood that has escaped from capillaries into the tissue.

hetero-. A prefix meaning "different."

heterocysts. Thick-walled cells formed by certain cyanobacteria. Heterocysts lack photosystem II but can fix molecular nitrogen, unlike the vegetative cells.

heterogamy (het·ur·**og**·uh·mee). The conjugation of unlike gametes.

heterokaryon (het·ur·o·**care**·ee·on). A cell having two nuclei that differ genetically.

heterologous (het·ur·**ahl**·uh·gus). Different with respect to type or species.

heterophile antibody (**het**·ur·o·file). An antibody that reacts with microorganisms or cells that are unrelated to the antigen that stimulated production of the antibody. The agglutination of *Proteus* spp. cells by serum from typhus fever patients is an example.

heterophile antigen. An antigen that reacts with antibodies stimulated by unrelated species.

heterothallic. Organisms in which one individual produces male gametes and another produces ova.

heterotroph (**het**·ur·o·trofe). A microorganism that is unable to use carbon dioxide as its sole source of carbon and requires one or more organic compounds. *Compare* **autotroph.**

histiocyte. A large phagocyte of the reticuloendothelial system. Also known as a *macrophage.*

holdfast. An adhesive base that attaches the thallus of certain microorganisms to a surface.

holoenzyme (ho·lo·**en**·zime, halh·o·**en**·zime). A fully active enzyme, containing an apoenzyme and a coenzyme.

homograft. A tissue graft with tissue from one species of animal to a recipient of the same species.

homologous. The same with respect to type or species.

homothallic. Plants which produce both male and female sex cells and can self-fertilize.

host. An organism harboring another organism as a parasite (or as an infectious agent).

humoral immunity (**hyoo**·mur·ul). Immunity arising from the formation of specific antibodies that circulate in the bloodstream in response to the introduction of an antigen.

humus. Highly complex organic residual matter in soil; not readily degraded by microorganisms.

hyaluronidase (high·ul·yoo·**ron**·i·dase, high·ul·yoo·**ron**·i·daze). An enzyme that catalyzes the breakdown of hyaluronic acid. Also called *spreading factor.*

hybridization. The act of producing hybrids, i.e., offspring from genetically dissimilar strains.

hybridoma. A hybrid cell resulting from the fusion of a myeloma cell with an antibody-producing B lymphocyte.

hydrologic cycle (high·dro·**lahj**·ik). The complete cycle through which water passes, from oceans, through the atmosphere, to the land, and back to the oceans.

hydrolysis (high·**drahl**·i·sis). The process by which a substrate is split to form products through the intervention of a molecule of water.

hypersensitivity. Extreme sensitivity to foreign antigens, e.g., allergens.

hypha (**high**·fuh). One filament or thread of a mycelium.

icosahedron (eye·kah·suh·**hee**·drun, eye·ko·suh·**hee**·drun). A solid formed of 20 triangular faces and 12 corners; the geometrical shape of many virions.

IDU. An antiviral agent; 5-iodo-2'-deoxyuridine.

immune serum. Blood serum that contains one or more specific antibodies.

immunity. A natural or acquired resistance to a specific disease.

immunization. Any process that develops resistance (immunity) to a specific disease in a host.

immunoelectrophoresis. A technique which employs a combination of immunodiffusion and electrophoresis to identify various antigens.

immunogenicity (im·yoo·no·jeh·**niss**·i·tee). The capacity to stimulate the formation of specific antibodies.

immunoglobulin (im·yoo·no·**glob**·yoo·lin). Any of the serum proteins, such as gamma globulin, that possess antibody activity.

imperfect fungi. Fungi that do not have a sexual cycle.

impingement. The entrapment of aerosol particles on a solid surface in a sampling device.

IMViC. A group of tests used to differentiate *Escherichia coli* from *Enterobacter aerogenes.*

inactivate. To destroy the activity of a substance; e.g., to heat blood serum to 56°C for 30 min to destroy complement.

inclusion bodies. Discrete assemblies of virions and/or viral components that develop within virus-infected cells.

incubation. In microbiology, the subjecting of cultures of microorganisms to conditions (especially temperatures) favorable to their growth.

incubation period. The elapsed time between the exposure to an infection and the appearance of disease symptoms, or the time period during which microorganisms inoculated into a medium are allowed to grow.

indicator. A substance that changes color as conditions change; e.g., pH indicators reflect changes in acidity or alkalinity.

induced enzyme. *See* **adaptive enzyme.**

induction. The production of an increase in the rate of synthesis of an enzyme, generally by the enzyme's substrate or a closely related compound.

infection. A pathological condition due to the growth of microorganisms in a host.

infectious. Capable of producing disease in a susceptible host.

inflammation. A tissue reaction resulting from irritation by a foreign material and causing a migration of leukocytes and an increased flow of blood to the area, producing swelling, reddening, heat, pain, and tenderness.

inhibition. In microbiology, the

prevention of the growth or multiplication of microorganisms.

inoculation (in·ahk·yoo·**lay**·shun). The artificial introduction of microorganisms or other substances into the body or into a culture medium.

inoculum (in·**ahk**·yoo·lum). The substance, containing microorganisms or other material, that is introduced in inoculation.

in situ (in **sigh**·too, in **sigh**·tyoo). In the original or natural location.

intercellular. Between cells.

interferon (in·tur·**feer**·ahn). An antiviral substance produced by animal tissue.

intracellular (in·truh·**sel**·yoo·lur). Within a cell.

introns. Extraneous pieces of the DNA of a gene.

invertase (in·**vur**·tase, in·vur·tase). An enzyme that hydrolyzes sucrose to glucose and fructose.

in vitro (in **vee**·troe). Literally, "in glass." Pertaining to biologic experiments performed in test tubes or other laboratory vessels. Compare **in vivo**.

in vivo (in **vee**·voe). Within the living organism; pertaining to the laboratory testing of agents within living organisms. Compare **in vitro**.

iodophores. Organic compounds of iodine.

isoantibody (eye·so·**an**·ti·bod·ee). An antibody, found only in some members of a species, that acts upon cells or cell components of other members of the same species.

isoantigen. A tissue-specific antigen present in one individual of a species but not in another. Also called *alloantigen.*

isoenzyme (eye·so·**en**·zime). Any one of a group of enzymes of different structural forms that possess identical (or nearly identi-cal) catalytic properties. Also called *isozyme.*

isograft (**eye**·so·graft). A graft of tissue from a donor of the same species as the recipient. Also called *homograft.*

isoniazid. A structural analog of pyridoxine used for the treatment of tuberculosis. Also called isonicotinic acid hydrazide, or INH.

isotonic. Pertaining to a solution that has the same osmotic pressure as that within a cell.

isotope (**eye**·suh·tope). Any of several possible forms of a chemical element, differing from other forms in atomic weight but not in chemical properties.

Kahn test. A flocculation test for the diagnosis of syphilis.

karyogamy. The fusion of gametic nuclei, as in fertilization.

karyotype. The diploid appearance of the set of chromosomes.

Kline test. A microscopic flocculation test for the diagnosis of syphilis.

Koch's postulates. Guidelines to prove that a disease is caused by a specific microorganism.

Krebs cycle. An enzyme system that converts pyruvic acid to carbon dioxide in the presence of oxygen, with the concomitant release of energy that is captured in the form of ATP molecules. Also called *citric acid cycle, tricarboxylic acid cycle.*

krill. A name applied to planktonic crustaceans.

lachrymal fluid. Tears.

lactose. A carbohydrate (disaccharide) that is split into glucose and galactose on hydrolysis. Also called *milk sugar.* Abbreviation: lac.

lag phase. The period of slow, orderly growth when a medium is first inoculated with a culture.

laminar airflow (**lam**·i·nur). The flow of air currents in which streams do not intermingle; the air moves along parallel flow lines.

Lancefield groups. Groups of streptococci based on various kinds of cell-wall polysaccharides.

latent. Not obvious or manifest; a disease carrier who shows no symptoms has a *latent* infection. Noun: **latency.**

LD$_{50}$. The dose (number of microorganisms or amount of toxin) that will kill 50 percent of the animals in a test series.

leghemoglobin. A hemoglobin-like oxygen-binding red pigment in the root nodules of legumes which protects the nitrogenase enzyme complex from being destroyed by excess oxygen.

leukocidin (loo·ko·**sigh**·din). A substance that destroys leukocytes.

leukocyte (**loo**·ko·sight). A type of white blood cell which is characterized by a beaded, elongated nucleus; a polymorphonuclear leukocyte.

leukocytosis (loo·ko·sigh·**toe**·sis). An increase in the number of leukocytes that is caused by the body's response to an injury or infection.

leukopenia (loo·ko·**pee**·nee·uh). A decrease in the number of leukocytes.

lichen (**lye**·kun). A symbiotic, or mutalistic, association of an alga and a fungus.

ligand (**lig**·und, **lye**·gand). A molecule that binds to a protein; e.g., one that binds to an enzyme and, through the role it plays in other processes, directly controls enzyme activity.

limnology. The study of the physical, chemical, geological, and biological aspects of lakes and streams.

lipase (**lye**·pase, **lye**·paze). An enzyme that catalyzes the hydrolysis of fats into glycerol and fatty acids.

lipid. A fat or fatlike substance.

lipolytic enzyme (lip·o·**lit**·ik). An enzyme that hydrolyzes lipids.

lipopolysaccharide. A complex molecular structure composed of sugars and fatty acids; occurs in the outer membrane of Gram-negative bacteria. Abbreviation: LPS.

liquefaction (lik·wi·**fak**·shun). The transformation of a gas or solid (e.g., a gel) to a liquid.

liter (**lee**·tur). A metric unit of volume containing 1,000 ml, or 1,000 cm³.

lithotroph (**lith**·o·trofe). An organism which uses reduced inorganic compounds as electron donors.

litmus. A lichen extract used as an indicator for pH and oxidation or reduction.

locus. In genetics, the site on a chromosome occupied by a gene, operon, mutation, etc.; in some cases, identifiable by reference to a marker.

logarithmic phase (log·uh·**rith**·mik). Commonly called *log phase. See* **exponential phase.**

lophotrichous (lo·**faht**·ri·kus). Having a polar tuft of flagella.

lumen. A hypha cavity filled by protoplasm. Also, a channel within a tubular organ, as in the lumen of the intestine.

lymph. The fluid within lymphatic vessels.

lymph nodes. Ovoid structures of the lymphatic system which range in size from 1 to several mm. and are widely distributed throughout the body. Also called *lymph glands.*

lymphocytosis (lim·fo·cye·**to**·sis). An abnormally high lymphocyte count in the blood.

lyophilization (lye·off·il·i·**zay**·shun). The preservation of biological specimens by rapid freezing and rapid dehydration in a high vacuum.

lysin (**lye**·sin). An enzyme, antibody, or other substance capable of disrupting or disintegrating cells (lysis).

lysis (**lye**·sis). The disruption or disintegration of such cells as bacteria or erythrocytes, e.g., by the action of specific antibodies plus complement.

lysogenic bacteria (lye·so·**jen**·ik). Bacteria that carry prophages.

lysogeny (lye·**sah**·juh·nee). The state of a bacterium that is carrying a bacteriophage (often as a prophage) to which it is not itself susceptible.

lysosomes (**lye**·so·sohms). Membrane-enclosed granules which occur in the cytoplasm of animal cells and contain hydrolytic enzymes.

lysozyme (**lye**·so·zime). An enzyme capable of digesting the cell walls of certain bacteria.

lytic phage (**lit**·ik). A virulent bacterial virus.

macrophage. *See* **histiocyte.**

macroscopic (mak·ro·**skahp**·ik). Visible without the aid of a microscope.

magnetosomes. Magnetite inclusions within a cell that allow the cell to become oriented as a magnetic dipole.

magnetotaxis. A movement of organisms in response to a magnetic field.

maltase (**mawl**·tase, **mawl**·taze). An enzyme that hydrolyzes maltose, yielding glucose.

maltose (**mawl**·tose). A carbohydrate (disaccharide) produced by the enzymatic hydrolysis of starch by diastase.

marine. Of or relating to oceanic and estuarine environments.

medium, pl. **media.** A substance used to provide nutrients for the growth and multiplication of microorganisms.

meiosis (mye·**o**·sis). A process occurring during cell division at different points in the life cycles of different organisms, in which the chromosome number is reduced by half, thus compensating for the chromosome-doubling effect of fertilization. *Compare* **mitosis.**

membrane filter. A filter made from such polymeric materials as cellulose, polyethylene, or tetrafluoroethylene.

meninges. The membranes that cover the brain and the spinal cord.

mesophile (**mez**·o·file). A bacterium growing best at the moderate temperature range 25 to 40°C.

mesosomes (**mez**·o·sohms). Membrane invaginations in the form of convoluted tubules and vesicles.

messenger RNA. The intermediary substance that passes information from the DNA in the nuclear region to the ribosomes in the cytoplasm. Abbreviation: mRNA.

metabolic pathway (met·uh·**bol**·ik). A series of steps in the chemical transformation of organic molecules.

metabolism (meh·**tab**·o·lizm). The system of chemical changes by which the nutritional and functional activities of an organism are maintained.

metabolite (meh·**tab**·uh·light). Any chemical participating in metabolism; a nutrient.

metachromatic granule (met·uh·kro·**mat**·ik). See **granules, meta-chromatic.**

metastasis (meh·**tass**·tuh·sis). The process of a malignant cell's detaching itself from a tumor and establishing a new tumor at another site within the host.

metazoa (met·uh·**zoe**·uh). Animals whose bodies consist of many cells.

methanogenic bacteria (meth·uh·no·**jen**·ik). Bacteria that produce methane gas under anaerobic conditions.

microaerophile (might·kro·**air**·o·file). An organism that requires low levels of oxygen for growth but cannot tolerate the level of oxygen (21 percent) present in an air atmosphere.

microbe (**migh**·krobe). Any microscopic organism; a microorganism. Adjective: **microbial** (migh·**kro**·bee·ul).

microbiology (migh·kro·bye·**ahl**·uh·jee). The study of organisms of microscopic size (microorganisms), including their culture, economic importance, pathogenicity, etc.

microbiota (migh·kro·bye·**o**·tuh). The microscopic flora and fauna.

microcysts (**migh**·kro·sists). Desiccation-resistant resting cells of myxobacteria. Also called *myxospores.*

micromanipulator (migh·kro·muh·**nip**·yoo·lay·tur). A device for the manipulation of microscopic specimens under a microscope.

micrometer (**migh**·kro·mee·tur). A unit of measurement: one-thousandth of a millimeter. Abbreviation: μm.

microorganism (migh·kro·**or**·guh·nizm). Any organism of microscopic dimensions.

microtome (**migh**·kro·tohm). An instrument for making thin sections of tissues or cells.

microtubules (migh·kro·**tyoo**·byoolz). Very thin rods that occur within all types of eucaryotic microbial cells.

mineralize. To convert into mineral substances.

minicell. A small daughter cell that arises from asymmetric septum formation during binary fission and which lacks DNA.

mitochondrion (migh·toe·**kahn**·dree·un). A cytoplasmic organelle in eucaryotic cells; the site of cell respiration.

mitosis (mye·**toe**·sis). A form of nuclear division characterized by complex chromosome movement and exact chromosome duplication. *Compare* **meiosis.**

mixotrophs. Chemolithotrophic heterotrophs which obtain energy by utilizing inorganic electron donors but obtain most of their carbon from organic compounds.

MLD. Minimum lethal dose. The smallest number of microorganisms or the smallest amount of a toxin that kills 100 percent of the animals in a test series; equivalent to LD_{100} dose.

modification. A temporary change or variation in the characteristics of an organism.

modulator. The regulatory metabolite that binds to the allosteric site of an enzyme and alters the maximum velocity. Also called *effector, modifier.*

moiety. A part of a molecule having a characteristic chemical property.

mold. A fungus characterized by a filamentous structure.

monoclonal antibody. A specific antibody produced by hybridoma cells.

mononucleotide (mah·no·**new**·klee·o·tide). The basic building block of nucleic acids (DNA and RNA); consists of a purine or pyrimidine base, ribose or deoxyribose, and phosphate.

monosaccharide (mah·no·**sak**·uh·ride). A simple sugar, such as a five-carbon or six-carbon sugar.

monotrichous (mo·**not**·ri·kus). Having a single polar flagellum.

mordant (**more**·dunt). A substance that fixes dyes.

morphogenesis (more·fo·**jen**·uh·sis). The process by which cells are organized into tissue structures.

morphology (more·**fahl**·uh·jee). The branch of biological science that deals with the study of the structure and form of living organisms.

MPN. Most probable number, a statistical expression of estimating cell number in a culture.

M proteins. Antiphagocytic protein antigens located on the surface of the cell walls of streptococci.

mRNA. See **messenger RNA.**

murein. See **peptidoglycan.**

mutagen (**myoo**·tuh·jen). A substance that causes mutation.

mutant (**myoo**·tunt). An organism with a changed or new gene.

mutation (myoo·**tay**·shun). A stable change of a gene such that the changed condition is inherited by offspring cells.

mutualism. A symbiosis in which two or more organisms living together benefit each other.

mycelium (mye·**see**·lee·um). A mass of threadlike filaments, branched or composing a network, that constitutes the vegetative structure of a fungus.

mycolic acids. High-molecular-weight α-branched β-hydroxy fatty acids which occur in the cell walls of corynebacteria, mycobacteria, and some nocardioform bacteria.

mycology (mye·**kahl**·uh·jee). The study of fungi.

mycophage (**mye**·ko·fayj). A fungal virus.

mycoplasma (mye·ko·**plaz**·muh). A member of a group of bacteria lacking cell walls.

mycorrhiza. A symbiotic association of a fungus with the roots of a higher plant.

mycosis (mye·**ko**·sis). A disease caused by fungi.

mycotoxin (mye·ko·**tok**·sin). Any toxic substance produced by fungi.

myxamoeba (mik·suh·**mee**·buh). A nonflagellated amoeboid cell that occurs in the life cycle of acellular slime molds.

myxospore (**mik**·so·spore). A desiccation-resistant resting cell of myxobacteria. Also called a *microcyst.*

NAD. Nicotinamide adenine dinucleotide, a coenzyme that functions in enzymatic systems concerned with oxidation-reduction reactions.

naked virion. A nonenveloped virus.

nanometer (**nan**·o·mee·tur, **nay**·no·mee·tur). A unit of length equal to one-billionth of a meter, or 10^{-9} m; 1 millimicrometer. Abbreviation: nm.

nasopharynx. The region of the respiratory tract above the soft palate.

Negri bodies (**neg**·ree). Minute pathological structures (inclusion bodies) found in certain brain cells of animals infected with rabies virus.

nematodes. Roundworms, many of which are pathogenic for plants.

Some are animal pathogens, and some are saprophytes.

neoplasm (**nee**·o·plazm). An aberrant new growth of abnormal cells or tissue; a tumor.

neurotoxin (nyoo·ro·**tahk**·sin). Any nerve poison, such as those produced by certain marine algae.

neutralism. A neutral interaction between two species in which there is no evident effect on either species.

nitrate reduction. The reduction of nitrates to nitrites or ammonia.

nitrification (nigh·trif·i·**kay**·shun). The transformation of ammonia nitrogen to nitrates.

nitrofurans. Synthetic antimicrobial agents derived from furfural.

nitrogen fixation. The formation of ammonia from free atmospheric nitrogen.

nitrogenous (nigh·**trahj**·i·nus). Relating to or containing nitrogen.

nomenclature (**no**·men·klay·chur). Any system of scientific names, such as those employed in biological classification.

nonseptate (non·**sep**·tait). Having no dividing walls in a filament.

nosocomial disease (no·so·**ko**·mee·ul). Describing or pertaining to disease acquired in the hospital.

nucleic acid (new·**klee**·ik). One of a class of molecules composed of joined nucleotide complexes; the types are deoxyribonucleic acid (DNA) and ribonucleic acid (RNA).

nucleoid. An indistinct area within a bacterial cell where the DNA is located. Also termed *chromatin body, nuclear equivalent,* or *bacterial chromosome.*

nucleolus, pl. **nucleoli** (new·**klee**·o·lus, new·**klee**·o·lye). A small body in a cell nucleus.

nucleoprotein (ne·klee·o·**pro**·teen). A molecular complex composed

of nucleic acid and protein.

nucleoside. A pentose sugar linked to a purine or pyrimidine base.

nucleotide (**new**·klee·o·tide). The basic building block of nucleic acids (DNA and RNA); consists of a purine or pyrimidine base, ribose or deoxyribose, and phosphate.

nucleus. The structure in a cell that contains the chromosomes.

numerical taxonomy. A method used in taxonomy to determine and numerically express the degree of similarity of every strain to every other strain in a particular group.

O antigens. Polysaccharide antigens which extend from the outer membrane of Gram-negative bacterial cells into the surrounding medium.

objective. In a compound microscope, the system of lenses nearest the object being observed.

ocular micrometer (migh·**krom**·uh·tur). A glass disk etched with equidistant lines that fits into the eyepiece of a microscope.

oidium, pl. **oidia** (oh·**id**·ee·um, oh·**id**·ee·uh). A single-celled spore formed by the disjointing of hyphal cells.

Okazaki fragments (oh·kuh·**zah**·kee). DNA strands replicated in small pieces.

oligodynamic action (ahl·i·go·dye·**nam**·ik). The lethal effect exerted on bacteria by small amounts of certain metals.

oncology (on·**kahl**·uh·jee). The study of the causes, development, characteristics, and treatment of tumors.

oogamy. The union of an egg cell and a sperm cell.

oogonium. In certain algae and fungi, the female sex organ that contains one or more eggs.

ookinete. The elongated mobile zygote of certain sporozoa, as that of a malarial parasite.

oospores. Spores formed after the fertilization of the eggs within the oogonium.

operator. A specific region of DNA at the initial end of the gene, where the synthesis of mRNA is initiated.

operon (**ahp**·ur·on). A cluster of genes whose expression is controlled by a single operator.

opportunistic microorganism. A microorganism that exists as part of the normal flora but becomes pathogenic when transferred from the normal habitat into other areas of the host or when host resistance is lowered.

opsonin (**op**·suh·nin). An antibody that renders microorganisms susceptible to ingestion by phagocytes.

order. In systematic biologic classification, a group of families.

organelle (or·guh·**nel**). A structure or body in a cell.

organotroph (or·**gan**·uh·trofe). An organism that uses organic compounds as a source of electrons.

osmosis (oz·**mo**·sis). The passage, due to osmotic pressure, of a fluid through a semipermeable membrane.

osmotic pressure (oz·**mot**·ik). The force or tension built up when water diffuses through a membrane.

ovum. An egg cell.

oxidase (ok·si·dase, ok·si·daze). An enzyme that brings about oxidation.

oxidase test. A test for the presence of cytochrome c; in a positive test, colonies become purple when treated with tetramethyl-p-phenylenediamine.

oxidation. 1. The process of combining with oxygen. 2. The loss of electrons or hydrogen atoms.

oxidation-reduction potential. A measure of the ability of an O/R system to absorb electrons, compared with that of the standard hydrogen O/R system. Symbol: E_h.

oxidative phosphorylation. The utilization of energy liberated by oxidation reactions in a respiratory chain to drive the synthesis of ATP.

palisade arrangement. Cells lined side by side, as in the genus *Corynebacterium*.

palmelloid. A stage in which masses of daughter cells without flagella develop.

pandemic (pan·**dem**·ik). A worldwide epidemic.

papain. A proteolytic enzyme found in the juice of the fruit and in the leaves of the papaya plant.

paramecium (pehr·uh·**mee**·see·um). A protozoan ciliate having cilia over the entire cell.

parasite. An organism that derives its nourishment from a living plant or animal host. A parasite does not necessarily cause disease.

parasitism (pehr·uh·sit·izm). The relationship of a parasite to its host.

parenteral (pur·**ren**·tur·ul). By some route other than via the intestinal tract.

Paschen bodies. Aggregates or colonies of virions growing in the cytoplasm of the host cell.

passive immunity. Immunity produced by injecting blood or serum containing antibodies.

pasteurization. The process of heating a liquid food or beverage to a controlled temperature to enhance the keeping quality and destroy harmful microorganisms.

pathogen (**path**·uh·jen). An organism capable of producing disease.

pathogenic. Capable of producing disease.

pebrine (pay·**breen**). A silkworm disease caused by a protozoan.

pellicle. A compound covering of membranes in protozoa; also, a film on the surface of fluid culture media due to the growth of microorganisms.

penicillin. The generic name for a large group of antibiotic substances derived from several species of the mold *Penicillium*.

pentose. A sugar with five carbon atoms; e.g., ribose.

pentose cycle. A pathway by which pentoses are metabolized or transformed in plants and microorganisms.

pepsin. A proteolytic enzyme from hog stomach tissues.

peptidase. An enzyme catalyzing the liberation of individual amino acids from a peptide.

peptide. A compound consisting of two or several amino acids.

peptidoglycan (pep·ti·doe·**glye**·kan). A large polymer that provides the rigid structure of the procaryotic cell wall, composed of three kinds of building blocks: (1) acetylglucosamine, (2) acetylmuramic acid, and (3) a peptide consisting of four or five amino acids.

peptone. A partially hydrolyzed protein.

peptonization (pep·tun·i·**zay**·shun). The conversion of proteins into peptones; the solubilization of casein in milk curd by proteolytic enzymes.

perfect fungi. Fungi with both an asexual and a sexual life cycle.

periphytes (**pehr**·i·fites). Microorganisms that become attached to surfaces, grow and form microcolonies, and produce a film to which other organisms become attached and grow.

periplasmic flagella. Flagella of the type which is possessed by spirochetes; they are located between the protoplasmic cylinder and the outer sheath; also called *axial fibrils* or *endoflagella.*

periplasmic space. The space between the cytoplasmic membrane and the outer membrane of Gram-negative bacteria.

periplast (**pehr**·i·plast). A surface membrane or pellicle of certain algae and bacteria.

peristalsis. Progressive and rhythmic contractions of the intestines.

perithecium (pehr·i·**theece**·ee·um). A spherical, cylindrical, or oval ascocarp that usually opens by a slit or pore at the top.

peritrichous (puh·**rit**·ri·kus). Having flagella on the entire surface of the cell.

permeability. The extent to which molecules of various kinds can pass through cell membranes.

permease (**pur**·me·ase). Any of a group of enzymelike proteins which are located in the cytoplasmic membrane and mediate the passage of nutrients across the membrane.

per os (pur ose, pur oss). Through the mouth.

peroxidase. An enzyme which catalyzes the reaction of hydrogen peroxide with a reduced substrate, resulting in the formation of H_2O and oxidized substrate.

pH. A symbol for the degree of acidity or alkalinity of a solution; pH = $\log$ (1/[H$^+$]), where [H$^+$] represents the hydrogen ion concentration.

phage (fayj). See **bacteriophage.**

phagocyte (**fag**·o·site). A cell capable of ingesting microorgan-

isms or other foreign particles.

phagovar. A subdivision of a bacterial species based on a particular pattern of susceptibility and resistance to various bacteriophages.

phenetic. In taxonomy, the relation based not on ancestry but on overall affinity or similarity.

phenol coefficient. The ratio between the greatest dilution of a test germicide capable of killing a test organism in 10 min but not in 5 min and the greatest dilution of phenol giving the same result.

phenotype (**fee**·no·tipe). That portion of the genetic potential of an organism which is actually expressed.

phosphatase (**fahss**·fuh·tase, **fahss**·fuh·taze). An enzyme that splits phosphate from its organic compound.

phosphatase test. A test to determine the efficiency of the pasteurization of milk. The test is based on the thermolability of the enzyme phosphatase.

phosphoglycerides. Straight-chain fatty acids ester-linked to glycerol phosphate.

phosphorylation (fahss·fo·ri·**lay**·shun). The addition of a phosphate group to a compound.

photoautotroph (fo·toe·**aw**·toe·trofe). An organism that derives energy from light and uses carbon dioxide as its sole carbon source.

photolithotroph. An organism which obtains energy from light and uses inorganic compounds as a source of electrons.

photolysis. Light-generated breakdown of water.

photoorganotroph. An organism which obtains energy from light and uses organic compounds as a source of electrons.

photophosphorylation. The utilization of light energy to drive the synthesis of ATP.

photoreactivation. The restoration to full viability by the immediate exposure to visible light of cells damaged by an exposure to lethal doses of ultraviolet light.

photosynthesis (fo·toe·**sin**·thuh·sis). The process in which chlorophyll and the energy of light are used by plants and some microorganisms to synthesize carbohydrates from carbon dioxide.

phototaxis. The movement of organisms in response to a change in light intensity.

phototroph (**fo**·toe·trofe). A bacterium capable of utilizing light energy for metabolism.

phycobilins. Water-soluble pigments, such as phycocyanin and phycoerythrin, which can transmit the energy of absorbed light to chlorophyll.

phycobilisomes. Granules on the surface of thylakoids, which contain phycobilin pigments.

phycology (fye·**kahl**·uh·jee). The study of algae.

phylogeny (fye·**lahj**·uh·nee). The evolutionary or ancestral history of organisms.

phylum, pl. **phyla** (**fye**·lum, **fye**·luh). A taxon consisting of a group of related classes.

physiology. The study of the life processes of living things.

phytoflagellate (fye·toe·**flaj**·uh·lut, fye·toe·**flaj**·uh·lait). A plantlike form of flagellate. Compare **zooflagellate.**

phytoplankton (fye·toe·**plank**·tun). A collective term for plants and plantlike organisms present in plankton. Compare **zooplankton.**

pilus. Any filamentous appendage other than flagella on certain Gram-negative bacteria.

pinocytosis. The uptake of fluids and soluble nutrients through small invaginations in the cell membrane that form intracellular vesicles.

plankton (**plank**·tun). A collective term for the passively floating or drifting flora and fauna of a body of water, consisting largely of microscopic organisms.

plasma. *See* **blood plasma.**

plasmalemma. The double-layered membrane surrounding the protoplasm in a hypha.

plasma membrane. *See* **cytoplasmic membrane.**

plasmin. An enzyme that dissolves the fibrin of blood clots.

plasmogamy. The joining of two cells and the fusion of their protoplasts in the process of sexual reproduction.

plasmolysis (plaz·**mahl**·i·sis). The shrinkage of cell contents as a result of the withdrawal of water by osmosis.

plastid. A pigmented inclusion body found in algae.

pleomorphism (plee·o·**more**·fizm). The existence of different morphological forms in the same species or strain of microorganism. Also called *polymorphism.*

polar. Located at one end or at both ends.

poly-β-hydroxybutyrate. A chloroform-soluble polymer of β-hydroxybutyric acid; occurs in the form of intracellular granules within certain bacteria and can be stained by fat-soluble dyes. Abbreviation: PHB.

polyene antibiotics. A chemical class of antibiotics that have large ring structures and increase cell permeability.

polyhedrosis. Any of several virus diseases of insects.

polypeptide (pahl·ee·**pep**·tide). A molecule consisting of many joined amino acids.

polysaccharide. A carbohydrate formed by the combination of many molecules of monosaccharides; examples of polysaccharides are starch, cellulose, and glycogen.

polysome (**pahl**·ee·sohm). A complex of ribosomes bound together by a single mRNA molecule. Also called *polyribosome.*

porins. Channel-containing proteins that span the outer membrane of Gram-negative bacteria.

potable (**po**·tuh·bul). Suitable for drinking.

PPLO. Pleuropneumonia-like organisms. These organisms belong to the order *Mycoplasmatales.*

Prausnitz-Küstner antibodies. The principal agents responsible for such allergies as hay fever, food and drug sensitivities, and anaphylaxes.

precipitin (pree·**sip**·i·tin). An antibody causing the precipitation of a soluble homologous antigen.

predation. The killing and eating of an individual of one species by an individual of another species.

primary treatment. The first stage in wastewater treatment, in which floating or settleable solids are mechanically removed by screening and sedimentation.

procaryote (pro·**care**·ee·ote). A type of cell in which the nuclear substance is not enclosed within a membrane; e.g., a bacterium or cyanobacterium. *Compare* **eucaryote.**

promoter. The binding site for RNA polymerase; it is near the operator.

prophage. The viral DNA of a temperate phage which becomes incorporated into the host DNA.

prophylaxis (pro·fi·**lak**·sis). Preventive treatment for protection against disease.

prostheca (pros·**theek**·uh). A semirigid extension of the cell wall and cytoplasmic membrane; it has a diameter that is always less than that of the cell.

protein. One of a class of complex organic nitrogenous compounds composed of an extremely large number of amino acids joined by peptide bonds.

proteinase (**pro**·teen·ase, **pro**·teen·aze). An enzyme that hydrolyzes proteins to polypeptides.

proteolytic (pro·tee·o·**lit**·ik). Capable of splitting or digesting proteins into simpler compounds.

protist (**pro**·tist). A microorganism in the kingdom *Protista.*

protonmotive force. The force which results from an electrochemical gradient of protons across a membrane and which can be used to drive ATP synthesis and certain other energy-requiring processes of a living cell.

protoplasm (**pro**·toe·plazm). The living substance of a cell. The term usually refers to the substance enclosed by the cytoplasmic membrane.

protoplast (**pro**·toe·plast). That portion of a bacterial cell consisting of the cytoplasmic membrane and the cell material bounded by it.

prototroph (**pro**·toe·trofe). An organism that is nutritionally independent and able to synthesize all required growth factors from simple substances.

protozoa (pro·tuh·**zo**·uh). Eucaryotic microorganisms with animal characteristics.

pseudopodium (syoo·doe·**po**·dee·um). A temporary projection of the protoplast of an amoeboid cell in which cytoplasm flows during extension and withdrawal.

psychrophile (**sigh**·kro·file). A "cold-loving" microorganism, capable of growing at 0°C or lower. Sometimes restricted to organisms that cannot grow above 20°C.

psychrotroph. An organism which is able to grow at 0°C but which grows best at a temperature above 20°C; also termed *facultative psychrophile.*

ptomaine (**toe**·main). Substances that are produced during the putrefaction of animal or plant protein and cause food poisoning.

pure culture. A culture containing only one species of organism.

pus. The fluid product of inflammation, containing serum, bacteria, dead cells, and leukocytes.

putrefaction (pyoo·truh·**fak**·shun). The decomposition of proteins by microorganisms, producing disagreeable odors.

pycnidium. An asexual fruiting body in fungi.

pyemia (pye·**ee**·mee·uh). A form of septicemia in which pyogenic organisms in the bloodstream set up secondary foci in organs and tissues.

pyocins. Bacteriocins produced by *Pseudomonas aeruginosa.*

pyogenic (pye·o·**jen**·ik). Forming pus.

pyrenoids. In chloroplasts, dense regions on which surface starch granules form.

pyrogen. A chemical which affects the hypothalamus, which regulates body temperature.

Quellung reaction. The increase in the visibility of a bacterial capsule that results from the reaction between the capsular antigens and specific anticapsular serum.

rabies (**ray**·beez). An acute disease of humans and other animals, caused by a virus. It is transmitted from infected animals to humans through bites.

radioisotope (ray·dee·o·**eye**·so· tope). An isotope that exhibits radioactivity.

RDE. Receptor-destroying enzyme. An enzyme that destroys the specific receptor by which a virus can attach to a susceptible cell.

reagin. Antibodies against cardiolipin, such as occur in syphilitic patients. Also termed *Wassermann antibodies.* The term is sometimes used to refer to the IgE antibodies which are involved in immediate hypersensitivities.

recombinant (ree·**kahm**·bi·nunt). A cell or clone of cells resulting from recombination.

recombination. The formation in daughter cells of gene combinations not present in either parent.

reduction. A chemical process involving the removal of oxygen, the addition of hydrogen atoms, or the gain of electrons.

regulator gene. A gene that controls the rate of enzyme synthesis in an operon.

rennet curd. The result of the coagulation of milk by the action of the enzyme rennin. Referred to as *sweet curd.*

rennin. An enzyme that transforms the soluble casein of milk into insoluble paracasein. The enzyme is obtained from the gastric juice of a calf.

replica plating. The replication of a pattern of colonies from one plate to another; a disk of sterile material (often velveteen) is pressed on the surface of the first plate, and the adhering bacteria are printed on the second.

replication. (1) In molecular biology, the production of a strand of DNA from the original. (2) In virology, the multiplication of a virus in a cell.

repression. A decrease in the rate of synthesis of an enzyme specifically caused by a small molecule which is usually the end product of a biosynthetic pathway, e.g., an amino acid or nucleotide.

resolution. In microscopy, the smallest distance by which two objects can be separated and still be visualized as separate objects.

resolving power. The ability of a microscope to distinguish fine details in a microscopic specimen.

respiration. An energy-yielding process in which electrons from an oxidizable substrate are transferred via a series of oxidation-reduction reactions to an exogenous terminal electron acceptor.

respiratory chain. A sequence of oxidation-reduction reactions by which electrons are transferred from an oxidizable substrate to an exogenous terminal electron acceptor.

reticulate body. In the developmental cycle of chlamydias, a noninfectious intracellular form that develops from an elementary body, is larger than the elementary body, and has a less dense arrangement of nuclear material. Also termed *initial body.*

reticuloendothelial system (ree·**tik**·yoo·lo·en·doe·**theel**·ee·ul). A system of cells in various organs and tissues, such as the spleen, liver, and bone marrow, that are important in resistance and immunity.

reverse transcriptase (tran·**skrip**· tase). An enzyme for the synthesis of a DNA molecule using RNA as a template.

rhizines. Short twisted strands of hyphae which serve as anchors in an alga-fungus lichenlike relationship.

rhizoid (**rye**·zoyd). A single-celled

or multicellular hairlike structure having the appearance of a root.

rhizosphere (rye·zo·sfeer). The soil region subject to the influence of plant roots and characterized by a zone of increased microbiological activity.

ribonucleic acid (rye·bo·new·**klee**·ik). A nucleic acid occurring in the cytoplasm and the nucleolus, containing phosphoric acid, D-ribose, adenine, guanine, cytosine, and uracil. Abbreviation: RNA.

ribosomal RNA (rye·bo·**so**·mul). The RNA of the ribosomes constituting about 90 percent of the total cellular RNA. Abbreviation: rRNA.

ribosome (**rye**·bo·sohm). A cytoplasmic structural unit, made up of RNA and protein, that is the site of protein synthesis.

rickettsias (ri·**ket**·see·uhs). Obligately parasitic bacteria of arthropods; many are pathogenic for humans and other mammals.

RNA. See **ribonucleic acid.**

RNA polymerase (**pahl**·im·ur·ase, pahl·**im**·ur·ase). An enzyme that synthesizes mRNA on a DNA template.

root-nodule bacteria. Bacteria belonging to the genus *Rhizobium*, family *Rhizobiaceae*, that live symbiotically in the nodules of roots of leguminous plants and fix atmospheric nitrogen.

*r*RNA. See **ribosomal RNA.**

RPR test. Rapid plasma reagin test; a macroscopic agglutination screening test for the detection of Wassermann antibodies in the serum of syphilitic patients.

rumen. The first chamber of the ruminant stomach.

saccharolytic (sak·uh·ro·**lit**·ik). Capable of splitting sugar compounds.

salmonellosis (sal·m··· ··nel·o·sis). An infection ·· *Salmonella* spp.

that affects the gastrointestinal tract.

sanitizer. An agent that reduces to levels judged safe by public health authorities the microbial flora in materials or on such articles as eating utensils.

saprophyte (**sap**·ro·fight). An organism living on dead organic matter.

sarcoma. A tumor made up of cells derived principally from connective tissue.

Schick test. A skin test used to determine a person's susceptibility to diphtheria.

schizogony (skiz·**og**·uh·nee). Asexual reproduction by multiple fission of a trophozoite (a vegetative protozoan).

schizont (**skiz**·ont). A stage in the asexual life cycle of the malaria parasites.

Schultz-Charlton reaction. A skin test used in the diagnosis of scarlet fever.

semiconservative replication. The replication of a complete DNA molecule in such a way that both the resultant double-stranded molecules contain one original and one new strand.

sepsis (**sep**·sis). Poisoning by the products of putrefaction; a severe toxic state resulting from an infection with pyogenic microorganisms.

septate. Possessing crosswalls.

septicemia (sep·ti·**see**·mee·uh). A systemic disease caused by the invasion and multiplication of pathogenic microorganisms in the bloodstream.

septic tank. A unit using an anaerobic system for the treatment of a limited volume of sewage.

septum. A crosswall.

serial dilution. The dilution of a specimen in successive stages. Thus a 1:100 dilution is achieved by combining one part

of a 1:10 dilution (one part of specimen plus nine parts of diluent, such as sterile water) with nine parts of diluent.

serology (suh·**rahl**·uh·jee, seer·**ahl**·uh·jee). The branch of science that is the study of serum.

serotype. See **serovar.**

serovar. A subdivision of a species based on its antigenic composition. Also called *serotype*.

serum. See **blood serum.**

sewage. Liquid or solid refuse (domestic and industrial wastes) carried off in sewers.

sewerage system. The system that collects and carries sewage from the source to the point of treatment and disposal.

sexual reproduction. Reproduction in which two cells (gametes) fuse into one fertilized cell.

sheath. A hollow tubular structure surrounding a chain of cells or a trichome; also refers to the covering that surrounds the flagella of certain Gram-negative bacteria.

shunt. An alternate pathway, a bypass.

siderophores. Iron-binding compounds formed by microorganisms.

simple stain. The coloration of bacteria or other organisms by applying a single solution of a stain to a fixed film or smear.

single-cell proteins. Microorganisms cultivated on industrial wastes or by-products as nutrients to yield a large cell crop rich in protein.

slime layer. A gelatinous covering of the cell wall. The term is sometimes used as a synonym of *capsule*.

sludge. The semisolid part of sewage that has been sedimented or acted upon by bacteria.

smear. A thin layer of material, e.g., a bacterial culture spread on a glass slide for microscopic examination. Also called a *film*.

soredia. In lichens, reproductive bodies which are knots of hyphae containing a few alga cells.

species. The basic taxonomic group; in bacteriology, a species consists of a type strain together with all the other strains that are considered sufficiently similar to the type strain to warrant inclusion in the species.

spectrophotometer. An instrument that measures the transmission of light, permitting the accurate analysis of color or the accurate comparison of the luminous intensities of two sources of specific wavelengths.

spermatium, pl. **spermatia.** A special male structure in fungi.

spheroplast (**sfeer**·o·plast). A Gram-negative bacterial cell with the peptidoglycan removed, leaving it devoid of rigidity.

spirillum (spye·**ril**·um). A rigid helical bacterium. A genus of helical bacteria is *Aquaspirillum*.

spirochete (**spye**·ro·keet). A helical bacterium which is flexible and has periplasmic flagella.

spontaneous generation. The origination of life from nonliving material. Also called *abiogenesis*. Compare **biogenesis**.

sporangioles. Walled vessels containing myxospores; occur in the fruiting bodies formed by certain myxobacters. Also termed *cysts*.

sporangiophore (spo·**ran**·jee·o·fore). A specialized mycelial branch bearing a sporangium.

sporangiospores. Asexual spores that develop within a sac (sporangium).

sporangium (spo·**ran**·jee·um). A saclike structure within which asexual spores are produced.

spore. A resistant body formed by certain microorganisms.

sporicide (**spore**·i·side). An agent that kills spores.

sporogenesis (spore·o·**jen**·uh·sis). 1. Reproduction by means of spores. 2. The formation of spores.

sporophore (**spore**·o·fore). A specialized mycelial branch upon which spores are produced.

sporozoite (spore·uh·**zo**·ite). A motile infective stage of certain sporozoans; it results from sexual reproduction and gives rise to an asexual cycle in a new host.

sporulation (spore·yoo·**lay**·shun). The process of spore formation.

stage micrometer (migh·**krom**·uh·tur). An instrument that functions as a ruler for the measurement of microorganisms under the microscope.

stalk. A nonliving ribbonlike or tubular appendage excreted by a bacterial cell.

staphylococci (staff·i·lo·**kahk**·sigh). Spherical bacteria (cocci) occurring in irregular, grapelike clusters.

starter culture. A known culture of microorganisms used to inoculate milk, pickles, and other food to produce the desired fermentation.

stationary phase. The interval directly following a growth phase when the number of viable bacteria remains constant.

stem cells. Formative cells in the bone marrow from which specialized cells, such as lymphocytes, arise.

stenothermophile. An organism which grows at 60°C or higher and which cannot grow at mesophilic temperatures; a true or obligate thermophile.

sterile. Free of living organisms.

sterilization. The process of making sterile; the killing of all forms of life.

steroid (**steer**·oyd). A complex chemical substance containing the tetracyclic carbon ring system of the sterols; steroids are often used as therapeutic agents.

sterol. Any of the natural products derived from the steroid nucleus.

stock cultures. Known species of microorganisms maintained in the laboratory for various tests and studies.

strain. All the descendents of a pure culture; usually a succession of cultures derived from an initial colony.

streaked-plate method. A procedure for separating cells on a sterile agar surface so that individual cells will grow into distinct, separate colonies.

streptobacilli (strep·toe·buh·**sil**·eye). Bacilli in chains.

streptococci (strep·toe·**kahk**·sigh). Cocci that divide in such a way that chains of cells are formed.

streptokinase (strep·toe·**kigh**·nase). See **fibrinolysin**.

structural gene. A gene coding for the structure of a protein.

subclinical. Pertaining to an infection so minor that there are no detectable clinical signs or symptoms of the infection.

subcutaneous (sub·kyoo·**tay**·nee·us). Beneath the skin.

substrate. The substance acted upon by an enzyme.

subterminal. Situated near the end, but not at the extreme end, of a cell.

sulfonamide. A synthetic chemotherapeutic agent characterized by the chemical group $-SO_2N<$.

supernate. The liquid over a precipitate or sediment; the fluid remaining after the removal of suspended matter.

superoxide dismutase. An enzyme which catalyzes the dismutation of superoxide radicals to form O_2 and H_2O_2.

suppuration. The formation of pus.

surface tension. The force acting on the surface of a liquid, tending to minimize the area of the surface.

susceptibility (suh·sep·ti·**bil**·i·tee). The state of being open to disease; specifically, the capability of being infected; a lack of immunity.

symbiosis (sim·bee·**o**·sis). The living together of two or more organisms; microbial association.

synchronous growth (**sing**·kruh·nus). Growth in a cell population in which all cells divide at the same time.

synergism. The ability of two or more organisms to bring about changes (usually chemical) that neither can accomplish alone.

synnemata. Compacted groups of erect hyphae which have a stem-like appearance.

synthetic medium. A medium composed of pure chemical compounds.

syntrophism. A type of mutualism involving an exchange of nutrients between two species.

syphilis. A venereal disease caused by *Treponema pallidum*.

systematics. The science of animal, plant, and microbial classification.

systemic (sis·**tem**·ik). Relating to the entire organism instead of to a part.

taxis (**tack**·sis). The movement away from or toward a chemical or physical stimulus.

taxon, pl. **taxa** (**tack**·sahn, **tack**·suh). A taxonomic group, such as a species, genus, or family.

taxonomy (tack·**sahn**·uh·mee). The classification (arrangement), nomenclature (naming), and identification of organisms.

teichoic aci (tel·**ko**·ik). Polymers

of ribitol phosphate or glycerol phosphate which occur in the walls of certain Gram-positive bacteria.

temperate bacteriophage. A bacteriophage capable of integrating with the host genome, being thus transmitted through cell divisions without causing host lysis.

temporal gradient. A change in the concentration of a chemical substance, or in the intensity of a physical agent such as light, with time.

terminal disinfection. The disinfection of a room or building after it has been vacated by a patient who had a contagious disease.

terminal infection. An infection with pathogenic microorganisms that terminates in the death of the host.

test. The outer shell of certain protozoa.

tetanus (**tet**·uh·nus, **tet**·nus). Lockjaw, a disease caused by *Clostridium tetani*.

tetracyclines. A chemical class of broad-spectrum antibiotics which inhibit protein synthesis.

thallophyte (**thal**·o·fite). A plant having no true stem, roots, or leaves; the group includes the algae and fungi.

thallospore (**thal**·o·spore). A spore that develops by the budding of hyphal or vegetative cells.

thallus (**thal**·us). A plant or microbial body lacking special tissue systems or organs; thalli may vary from a single cell to a complex, branching, multicellular structure.

therapeutic. Pertaining to the treating or curing of a disease.

thermal death point. The lowest temperature at which microorganisms are killed in a given time.

thermoduric (thur·mo·**dyoo**·rik). Capable of surviving an exposure to a high temperature.

thermolabile (thur·mo·**lay**·bile, thur·mo·**lay**·bil). Destroyed by heat at temperatures below 100°C (212°F).

thermophile (**thur**·mo·file). An organism that grows best at temperatures above 45°C.

thermostable (thur·mo·**stay**·bul). Relatively resistant to heat; resistant to temperatures of 100°C (212°F).

thylakoids. Flattened membranous sacs which contain the photosynthetic pigments of the cell. In cyanobacteria they occur within the cytoplasm; in photosynthetic eucaryotes they occur within the choloroplasts.

thymine (**thigh**·meen). One kind of pyrimidine.

tinea (**tin**·ee·uh). Ringworm, which is caused by fungi.

tissue. A collection of cells forming a structure.

tissue culture. A growth of tissue cells in a laboratory medium.

toxemia (tahk·**see**·mee·uh). The presence of toxins in the blood.

toxigenicity. The ability to produce a toxin.

toxin (**tahk**·sin). A poisonous substance, such as a bacterial toxin, elaborated by an organism.

toxin-antitoxin. A mixture of toxin and antitoxin containing slightly more toxin than antitoxin. This was formerly used to produce an active immunity.

toxoid (**tahk**·soyd). A toxin that has been treated to destroy its toxic properties without affecting its antigenic properties.

transcription. The process in which a complementary single-stranded mRNA is synthesized from one of the DNA strands.

transduction. The transfer of genetic material from one bacterium to another through the agency of a virus.

transfection. The introduction of phage hybrid DNA into the host cell.

transfer RNA. A specific RNA for each amino acid that becomes esterified to the terminal adenosine. Each of the 60 or so tRNAs has a specific trinucleotide sequence that interacts with a complementary sequence in mRNA. Abbreviation: tRNA. Also called *soluble RNA (sRNA)*.

transformation. The phenomenon by which certain bacteria incorporate DNA from related strains into their genetic makeup.

translation. The process in which genetic information in mRNA directs the order of assembly of the specific amino acids during protein synthesis.

transposons. Units of DNA which can move from one DNA molecule to another.

transverse binary fission (bye·nuh·ree). An asexual reproductive process in which a single cell divides transversely into two cells.

tribe. A division of the plant, animal, or microbial kingdoms containing a number of related genera within a family.

tricarboxylic acid cycle (trye·kar·bock·sil·ik). See **Krebs cycle.**

trichome. A single row of distinct cells in which there is a large area of contact between the adjacent cells (in contrast to a chain of cells).

trickling filter. A secondary treatment process in which sewage is trickled over a bed of rocks so that bacteria can break down organic wastes.

tRNA. See **transfer RNA.**

trophic stage (tro·f k). The vegetative stage of free-living protozoa.

trophozoite (tro·fo·zo·ite). The vegetative form of a protozoan.

trypsin (trip·sin). A proteolytic enzyme in pancreatic juice.

tubercle (tyoo·bur·kul). A nodule, the specific lesion of tuberculosis.

tuberculin (tyoo·bur·kyoo·lin). An extract of the tuberculosis bacillus capable of eliciting an inflammatory reaction in an animal that has been sensitized by the presence of living or dead tubercle bacilli. Used in a skin test for tuberculosis.

turbidimetry. A method of estimating bacterial growth or populations by the measurement of the degree of opacity (or turbidity) of the suspension.

turbidostat. A device for maintaining microorganisms in continuous culture; it monitors cell density and regulates the dilution rate to maintain the cell density at a constant value.

tyndallization. A process of fractional sterilization with flowing steam.

type species. The species that is the permanent reference example of the genus.

type strain. The strain that is the permanent reference strain of the species; it is the strain to which all other strains must be compared in order to be included in the species.

ultracentrifuge. A high-speed centrifuge used for the determination of the particle size of viruses and proteins.

ultrafiltration. A method for the removal of all but the very smallest particles, e.g., viruses, from a fluid medium.

ultrasonic waves. Sound waves of high intensity (beyond the audible range), used for the destruction of microbes or the cleaning of materials.

ultraviolet rays. Radiations from about 3900 to about 2000 Å.

undulant fever. Brucellosis, a disease caused by bacterial pathogens of the genus *Brucella*.

undulating. Exhibiting a wavelike motion.

upwelling. The rise of water from a deeper to a shallower depth in an ocean; the rise brings nutrients to the surface region.

urea. A soluble nitrogenous compound, $H_2N—CO—NH_2$, found in the urine of humans and other mammals.

urease. An enzyme that catalyzes the hydrolysis of urea.

vaccination. Inoculation with a biologic preparation (a vaccine) to produce immunity.

vaccine. A preparation of killed or attenuated microorganisms, or their components, or their products, that is used to induce active immunity against a disease.

vacuole (vak·yoo·ole). A clear space in the cytoplasm of a cell.

variant. An organism showing some variation from the parent culture.

vascular. Containing specialized vessels for the conduction of fluids: blood and lymph in animals, sap and water in plants.

VDRL test. Venereal Disease Research Laboratory test; a microscopic agglutination screening test for the detection of Wassermann antibodies in the serum of syphilitic patients.

vector. An agent, such as an insect, capable of mechanically or biologically transferring a pathogen from one organism to another.

vegetative stage. The stage of active growth, as opposed to the resting or spore stages.

venereal. Sexually transmitted.

V factor. Nicotinamide adenine dinucleotide, required for the growth of certain *Haemophilus* spp.

viable (vye·uh·bul). Capable of living, growing, and developing; alive.

vibrio (**vib**·ree·o). A bacterium that is curved with a twist; has less than one complete turn or twist (in contrast to a helical bacterium). *Vibrio* is a genus of Gram-negative bacteria.

Vincent's angina. An ulcerative condition of the tonsils and gums caused by a spirochete and a fusiform bacillus.

viremia (vye·**ree**·mee·uh). The presence of virus in the bloodstream.

virion (**vye**·ree·on). The complete mature virus particle.

viropexis. The engulfment of whole virions by cells in a phagocytic process.

virucide (**vye**·ru·side). An agent that kills viruses.

virulence (**vir**·yoo·lunce). The degree of pathogenicity exhibited by a strain of microorganisms.

virus (**vye**·rus). An obligate intracellular parasitic microorganism that is smaller than bacteria. Most viruses can pass through filters that retain bacteria.

viscidity. Stickiness; a characteristic produced by a bacterial sediment in broth that rises in a coherent swirl upon shaking.

Voges-Proskauer reaction. A test for the presence of acetylmethyl-carbinol to assist in distinguishing between species of the coliform group. Abbreviation: VP test.

volutin. *See* **granule, metachromatic.**

Warburg respirometer. An apparatus used to study enzyme reactions in which an exchange of gases takes place.

Wassermann antibodies. Antibodies against cardiolipin, such as occur in syphilitic patients; also called *reagins.*

Wassermann test. A complement-fixation test for syphilis.

Weil-Felix test. An agglutination test for typhus using *Proteus* spp. as antigens.

white corpuscle. A leukocyte of the blood.

Widal test. A slide agglutination test for typhoid or paratyphoid fever.

X factor. Heme; required for the growth of certain *Haemophilus* spp.

yeast. A kind of fungus that is unicellular and lacks typical mycelia.

yolk sac. The membrane covering the yolk of an egg.

zonation (zo·**nay**·shun). The distribution of organisms in zones; specifically, a stratification of certain kinds of algae at certain depths and locations in the ocean.

zooflagellate (zo·o·**flaj**·uh·lut, zo·o·**flaj**·uh·lait). An animallike form of flagellate. *Compare* **phytoflagellate.**

zoogloeal masses (zo·o·**glee**·ul). Masses composed of microorganisms which are embedded in a common matrix of slime.

zoonosis (zo·**ahn**·uh·sis, zo·o·**no**·sis). An animal disease transmissible to human beings.

zooplankton (zo·o·**plank**·tun). A collective term for the nonphotosynthetic organisms present in plankton. *Compare* **phytoplankton.**

zoospore (**zo**·o·spore). A motile, flagellate spore.

zygospore (**zye**·go·spore). A kind of spore resulting from the fusion of two similar gametes in some fungi.

zygote (**zye**·gote). An organism produced by the union of two gametes.

zymosan. An extract of yeast cell walls.

Name Index

Organism Index

Subject Index

Diffusion:
facilitated, *198, 199*
passive, *197, 198, 199*
Dilution, serial, 885, 898
Dimers of pyrimidine bases, 235
Dimorphic, 885
Dinoflagellates, *373, 384, 575*
Dioecious (unisexual) plants, 375
Diphtheria, 815
Diphtheria toxin, 695–696
Dipicolinic acid, 95, 885
Diplobacilli, 885
Diplococci, 885
Diploid, *351*, 885
Diplomonadida, 406
Disaccharide, 885
Disease(s):
germ theory of, 24–25
(*See also specific types of diseases, for example:*
Algae, and diseases; Bacteria, diseases
caused by; Fungi, diseases caused by;
Infectious diseases; Protozoa, diseases
caused by; Viruses, diseases caused by;
and specific names of diseases)
Disinfectant, 490, 885
Disk-plate technique for determining susceptibility
of microorganisms to antibiotics, *536*
Dissimilation, 885
DNA (deoxyribonucleic acid), 884–885
B form of, *213*
biosynthesis of, 209–218
excision repair of, *237*
replication of, 212, 215–218
linear mode of, 216–217
sigma ("rolling circle") mode of, 215, *216*
theta mode of, 215, *216*
semiconservative replication of strands of, 214–
215
structure of, 209–214, *211*
viral, 444, 447
Z form of, 213–214
(*See also Recombinant DNA technology;*
Transduction; Transformation, bacterial)
DNA base composition, *40*
DNA-dependent RNA polymerase, *220*
DNA homology experiments, 44
DNA ligase, 646
DNase (deoxyribonuclease), 700
Double-antibody-sandwich procedure, 754, *755*
Droplet nucleus, 770, 885
Dry heat, 478, *486*
Duffy determinants, 865
Dulse, 369
Dyes, 65–66, 498

EBV (Epstein-Barr virus), 461–462, 841–842
Echoviruses, 677
Ecology, 885
Ecosystem, 885
Ectocommensalism, 391
Ectodesmata, plant viruses penetrating through,
446
Ectoplasm, 394
Edema, 885
Effector molecules and enzyme regulation, 163
Effluent, 885
Eggs, microbial flora of, 619
Ehrlichieae, 278
Electromagnetic radiation, 480–483
Electromotive potential, 177
Electron(s), requirement for source of, 99
Electron flow, reversed, *207*
Electron microscope, 50, *58*
Electron microscopy, 64
limitations of, 63
scanning, 62–63
transmission, 57–62
Electron-transport chains, 177
Electrophoresis, 746, *747*, 885
Elementary body, 280–281, 459

ELISA (enzyme-linked immunosorbant assays),
734, *735*, 736
Elongation factor, 2, 695
Embden-Meyerhof pathway, 180, *181*
Encephalitis, 829, 834
Encystment of amoebas, 410
End-product (feedback) inhibition, 164, 886
End-product (feedback) repression, 168
Endemic, 885
Endergonic reaction, 172, 885
Endocommensalism, 391
Endoenzyme, 152, 885
Endoflagella, 80, 885
Endogenous, 885
Endonuclease, 236, 885
Endophytic, 885
Endoplasm, 394
Endoplasmic reticulum, 394, 885
Endospore, 94–96, 885
Endosymbiont, 282–283, 885
Endosymbiotic theory, 150
Endothermic, 885
Endotoxins, 87, 698–699, 885
characteristics of, 694
Energy, 171
production of: aerobic processes and, 185–191
anaerobic processes and, 179–185
photosynthesis and, 191–193
requirement for, 171
use of: in biosynthetic processes, 201–208
in nonbiosynthetic processes, 196–201
Energy-link control of activity of enzymes, 164
Energy-rich compound, 174
Enrichment medium, 134
Enteric, 885
Enterobacteriaceae, 270–272, 273
Enterochelin, 700
Enterotoxin, 695, 885
Entner-Doudoroff pathway, 183
Entomopox viruses, 848
Envelope, viral, 440
Enzyme(s), 39, 151–169, 885
action of: inhibition of, 160–161
nature and mechanism of, 156–158
activity of: conditions affecting, 158–159
determination of, 161–162
regulation of, 164, 165, 166
adaptive, 879
allosteric, 163–164, 879
anaplerotic, 191
characteristics of, 152–153
chemical and physical properties of, 153–155
classes of, *156*
constitutive, 166, 885
covalent modification of, 165
energy-link control of activity of, 164
feedback inhibition regulating activity of,
164
formation of, conditions affecting, 161
inducible, 166, 885
intracellular, 885
localization of, 62
nomenclature of, 155–156
oxygen inactivation of, 109
precursor activation and regulation of activity
of, 164
production of, 661
regulation of, 163
mechanisms of, 163–164
restriction, 426
synthesis of, 157
regulation of, 166–167, 168–169
repression of, 166–167
techniques for preparation of, 162–163
Enzyme-linked immunosorbant assays (ELISA),
734, *735*, 736
Enzyme-substrate reaction, 156–157
Enzyme system, 155, 164
Epibacteria, 573
Epicellular, 885
Epidemic, 885
Epidemiological markers, 767–768

Epidemiological techniques, 765–768
Epidemiology, 885
of infectious diseases, 764–785
Epidermis, 885
Episome, 247, 885
Epithelial cell layers, penetration of, 690–691
Epstein-Barr virus (EBV), 461–462, 841–842
Erysipelas, 809
Erythema nodosum, 873
Erythromycin, 524
Espundia, 868
Esterase, 886
Estuary, 582–583, 587, 588, 886
Ethylene oxide, 502–504, 5u7
Etiology, 886
Eubacteria, 45, 886
characteristics of, *316*
Gram-negative, cell walls of, 87–88
Gram-positive, cell walls of, 85, *86, 87*
Eucaryota, 332
Eucaryotes, 332, 886
enzyme regulation in, 169
Eucaryotic cells, features of, *9–10*
Euglenoids, *373, 379*
Euglenophycophyta, *373, 379*
Eustimatophycophyta, 373n.
Eutrophication, 582, 886
Evapotranspiration, 886
Excision repair, 237
Exergonic reaction, 172, 886
Exoenzyme, 152, 886
Exogenous, 886
Exons, 256, 886
Exonucleus, 236, 886
Exospore, 96, 886
Exothermic, 886
Exotoxins, 694–698, 723, 886
action of, 695
characteristics of, *694*
classification of, 695
examples of, *698*
potency of, 694
Exponential phase, 886
Extrachromosomal genetic element, 247,
886
Exudate, 886
Eye, 679–680
Eyespot of algae, 374

F factor (fertility factor; sex factor), 243–244, 886
F (sex) pilus, 82
Facilitated diffusion, *198, 199*
Facultative anaerobe, 109, 886
Facultative psychrophiles, 107
FAD (flavin adenine dinucleotide), 178
Fastidious heterotrophs, 103
Fastidious organism, 886
Fauna, 886
Feedback (end-product) inhibition, 164, 886
Feedback (end-product) repression, 168
Fermentation, 23, 183–184, 886
Fermented foods, 635–636, 638–639
Ferredoxin, 192
Ferritin-labeled antibody, 62
Fertility factor (F factor; sex factor), 243–244, 886
Fibril, 886
Fibrillar bundles in protozoan cytoplasm, 394
Fibrinolysin, 759, 886
Filamentous, 886
Filter:
bacterial, 881
bacteriological, 484–485, *487*, 886
Filterability, tobacco mosaic virus and, 438
Filterable viruses, 437
Filtration, 484–485
Fimbriae, *82*, 886
Finfish, microbial flora of, 620
Firmicutes, 12
Fission, 886
binary, 886
Five-kingdom concept, 11

910

Flagellates (*Mastigophora*), 403, 405–409, 886
  morphological forms of animallike, 870, 871
Flagellin, 78, 197, 886
Flagellum, 886
  of bacteria, 78–81, 197
    arrangements of, 78
    attachment of, 79
    hydrodynamics of, 79–80
    periplasmic, 80, 895
  of protozoa, 398–399
    rotation of, 72
Flavin adenine dinucleotide (FAD), 178
Flavin mononucleotide (FMN), 178
Flaviviruses, 829
Flavoproteins, 178
Fleas and plague, 782–783
Floc, 608, 886
Floccule, 608, 886–887
Flora, 887
Fluctuation test, 231
Fluorescence, 887
  of colonies of bacteria, 145
Fluorescence microscopy, 54, 55, 56, 64, 887
Fluorescent antibody technique, 55–56, 756, 757
Fluoride and enzyme inhibition, 161
FMN (flavin mononucleotide), 178
Foliose, 887
Fomites, 764, 887
Food(s):
  algae as, 368–369
  fermented, 635–636, 638–639
  microbial spoilage of, 624, 625, 626
  microbiological examination of, 626, 627, 628
  microbiology of, 618–641
  preservation of, 628–635
Food chain, 392
Food poisoning, 632, 633, 887
Food web in shallow estuary, 588
Food yeasts, 656
Foodborne infections, 776–777
Foodborne intoxications, 774–776
Foot-and-mouth disease vaccine, 648
Foot-and-mouth disease virus, 827–828
Foraminiferans, 408, 410
Formaldehyde, 501, 507
Formalin, 887
Forssman antigens, 758, 887
Forssman reaction, 723
Fractional sterilization (tyndallization), 477–478, 887, 901
Fragmentation in bacteria, 116
Free-energy charge, 172–174
Freeze-drying (lyophilization), 479–480, 891
  preservation by, 141–142
Freeze-etching, 61
Frei test, 887
Fruiting bodies, 307, 887
Fruits, microbial flora of, 619–620
Frustules, 377
Fruticose, 887
Fulminating infection, 887
Fungi, 11–13, 333–363, 887
  classification of, 345–346, 347
    cultivation of, 344–345
  diseases caused by, 851–857
    immune response to, 873
    therapeutic drugs for, 874–875
  distinguishing characteristics of, 334–335
  fruiting bodies of, 343
  imperfect, 889
  importance of, 334
  as insect parasites, 362–363
  morphology of, 335–337
  and nematodes, 362
  physiology of, 344
  reproduction in, 337–343
    asexual, 337, 338, 339
    sexual, 339–340, 341–343
  in soil, 546–547
  spores of: asexual, 337, 338, 339
    sexual, 340–341, 342, 343
  (See also Molds; Yeasts)

Fungicide, 887
Furazolidone, 530
Furfural, 530
Fusiform, 887

Galls, 818–819
Gametangia, 339, 356, 887
Gametes, 339, 887
  heterogamous, 375
  isogamous, 375
Gametogamy, 402
Gamma globulin, 887
Gamma rays, 482–483, 487
Gas chromatograph, 887
Gas gangrene, 812–813
Gas vacuoles, 93
Gaseous antimicrobial agents, 502–504
Gastroenteritis, 887
  Salmonella, 799
Gelatin, 887
Gelatinase, 887
Gene(s), 230, 887
  regulation and expression of activity of, 253–254, 255–256
  regulator, 167, 897
  structural, 167
Generation time, 887
  of bacteria, 120, 121
Genetic code, 221
Genetic engineering, 34, 256–257, 646–647
  benefits from, 648–649
  and nitrogen fixation, 561–562
  potential problems of, 649
Genetic relatedness, 43–45
Genital herpes, 841
  epidemiology of, 778
Genitourinary tract, normal microbial flora of, 684–685
Genome, 887
Genotype, 228–229, 887
Genotypic changes, 230–240
Genus, 42, 887
Geothermal vents, 577
Germ, 887
Germ sporangium, 356
Germ theory of disease, 24–25
Germfree animals, 674–676
Germfree equipment, 675
Germicide (microbicide), 490, 887
Giardiasis, 871–872, 887
Gingiva, 887
Gliding motility, 80, 887
Globulin, 887
Glomerulonephritis, acute, 809
Glucan, 887
Glucose, 887
  and energy production, 179–185
Glucose effect, 168
L-Glutamic acid production, 653–654
Glutaraldehyde, 501–502, 507
Glycogen, 887
Glycolysis, 180, 181, 887
Glyoxylate cycle, 190–191, 887–888
Gnotobiotic life, 675, 888
Golgi apparatus, 394, 888
Gonidium, 888
Gonorrhea, 796–797
  epidemiology of, 777–778
Gracilicutes, 12
Gram-negative bacteria, 261–263, 264–283, 888
Gram-negative eubacteria, cell walls of, 87–88
Gram-positive bacteria, 285–286, 287–298, 888
Gram-positive eubacteria, cell walls of, 85, 86, 87
Gram-positive filamentous bacteria of complex morphology, 320–321, 322–329
Gram stain, 66–67, 888
Gram-variable reaction, 67
Gramicidin, 514
Gramicidin C, 521
Granules, metachromatic, 93, 888, 902

Granulosis viruses, 848
Green olives, 639
Gregarinia, 404
Griseofulvin, 527
Groundwater, 571, 888
Group translocation, 198, 199–200
Growth curve, 888
Guanine, 210, 888
Guarnieri bodies, 459, 838, 888
Gullet, 411
Gummas, 791
Gummosis, 819
Gymnomycota, 346
Gyres, 577, 578

H antigen, 888
Habitat, 888
Halogens, 494–496
Halophiles, 888
  extreme, 317–318
Hanging-drop technique, 64–65, 888
Haploid, 351, 888
Haptens, 720, 888
Haptocysts, 397
Haustoria, 362, 387, 888
Heavy metals and their compounds, 497–498
HeLa cells, 456, 888
Helical, 888
Heliozoans, 408
Helix, 888
Hemagglutination, 888
Hemagglutination inhibition test, 752
Hemagglutination test, 751–752
Hemagglutinin, 689
Heme, 179
Hemoglobin, 888
Hemolysin, 695, 752, 888
Hemolysis, 888
α-Hemolysis, 288
β-Hemolysis, 288, 881
Hemorrhage, 699, 888
HEPA (high-efficiency particulate air) filters, 485, 486, 487
Hepatitis, viral:
  non-A, non-B, 845
  type A, 844–845
  type B, 845
Hepatitis viruses, 844–845
Herbicides, degradation of, 566–567
Herd immunity, 769
Herpes simplex virus:
  replication of, 448–449
  type 1 (HSV-1), 840–841
  type 2 (HSV-2), 841
Herpes zoster (shingles), 841
Herpesviridae, 451, 452, 461, 840–842
Herpesvirus, 436
Heterocysts, 305, 307, 888
Heteroduplex formation, 44
Heterofermentative bacteria, 185
Heterogamy, 888
Heterokaryon, 888
Heterologous, 888
Heteropolysaccharides, 83
Heterothallic, 888
Heterotrophs, 100, 103, 888–889
  cultivation of, 103
  fastidious, 103
Hfr (high-frequency recombination) strains, 244–247
High-efficiency particulate air (HEPA) filters, 485, 486, 487
High-energy-transfer compounds, 174
High-frequency recombination (Hfr) strains, 244–247
Histamine, 700
Histiocyte, 889
Histoplasmosis, 856
Holdfast, 312, 889
Holoenzyme, 153, 889
Homofermentative bacteria, 185
Homograft, 889

Mycology, 334, 893
Mycophage, 893
Mycoplasma(s), 91, 281–282, 446, 893
Mycoplasmataceae, 282
Mycoplasmatales, 282
Mycorrhizas, 363, 893
Mycoses, systemic, 853–855, 856–858, 893
Mycotoxin, 893
Myonemes, 394
Myoviridae, 422
Myxamoeba, 893
Myxobacterales, 307
Myxomycetes, 346
    life cycle of, 349
Myxospore, 893
Myxosporidia, 406
Myxozoa, 404

NAD (nicotinamide adenine dinucleotide), 178, 893
NADP (nicotinamide adenine dinucleotide phosphate), 178
Naked virion, 893
Nalidixic acid, 530
Nanometer, 893
Nasopharynx, 893
Natural killer (NK) cells, 714
Natural resistance, 703–706
Negative staining, 61
Negri bodies, 459, 836, 893
Neisseriaceae, 270
Nematodes, 362, 893
Neomycin, 523
Neoplasm, 893
Neothramycin, 529
Neuraminidase, 690
Neurotoxin, 695, 893
Neutralism, 549, 893
Nicotinamide adenine dinucleotide (NAD), 178, 893
Nicotinamide adenine dinucleotide phosphate (NADP), 178
Nitrate reduction, 557, 893
Nitrification, 555–557, 893
Nitrobacteraceae, 313–314
Nitrofurans, 529–530, 893
Nitrofurantoin, 530
Nitrofurazone, 530
Nitrogen, requirement for, 100
Nitrogen cycle, 553–554, 555–562
Nitrogen fixation, 557–561, 893
    nonsymbiotic, 559
    recombinant DNA and, 561–562
    symbiotic, 559, 560, 561
Nitrogenase, 558–559
Nitrogenase, 558–559
Nitrogenous, 893
NK (natural killer) cells, 714
Nocardioforms, 296–298
Nomenclature, 8, 45–47, 893
Nonionic detergents, 499
Nonseptate, 893
Nosocomial diseases, 771, 793, 893
Nuclear equivalent (nucleoid), 94, 893
Nuclear material of bacteria, 93–94
Nucleic acid:
    of bacteriophages, 419, 420, 421
    of viruses, 443, 444, 445
Nucleocapsid, 440
Nucleoid (nuclear equivalent), 94, 893
Nucleolus, 893
Nucleoprotein, 893
Nucleoside, 213, 893
Nucleotide, 209, 210, 893
    biosynthesis of, 214
Nucleus, 893
Null cells (lymphocytes), 714, 732
Numerical aperture, 52–53
Numerical taxonomy, 42–43, 893
Nystatin, 521, 527

O antigens, 87–88, 893
Objective, 893
914

Ocean:
    fertility of, 588–589
    (See also Aquatic environment)
Ocular micrometer, 893
Oidium (arthrospore), 337, 339, 880, 893
Okazaki fragments, 217, 893
Oligodynamic action, 893
Oligohymenophorea, 405
Oncogenes, 463
Oncology, 893
Oncovirinae, 461
Onychomycosis, 855
Oogamy, 375, 893
Oogonium, 339, 343, 893
Ookinete, 402, 894
Oomycetes, 346
Oosperes, 343
Oospores, 342, 343, 894
Opalinata, 403
Operator, 167, 894
Operon, 167, 894
    immunity, 431
    lac, 254, 255–256
Opines, 819
Opportunistic microorganism, 850, 894
Opsonins, 711, 894
Oral groove of ciliates, 396
Orbiviruses, 837
Order, 894
Organelle, 894
Organic carbon compound degradation, 562–563
Organotroph, 99, 894
Orthomyxoviridae, 451, 452, 831
Osmosis, 893
Osmotic pressure, 480, 894
    and food preservation, 633–634
Ovum, 375, 894
Oxidant (oxidizing agent), 176
Oxidase, 894
Oxidase test, 894
Oxidation, 176, 894
    sequence of, 179
β-oxidation, 188
Oxidation ponds, 610
Oxidation-reduction potential, 894
Oxidation-reduction reactions, 176–177
Oxidative phosphorylation, 177, 894
Oxidizing agent (oxidant), 176
Oxidoreductases, 156
Oxygen:
    requirement for, 100
    singlet, 110
Oxygen toxicity, 109–110

PABA (p-aminobenzoic acid), 524
Painted surfaces, deterioration of, 665–666
Palisade arrangement, 894
Palmelloid, 380, 381, 894
Pandemic, 894
Papain, 894
Paper, deterioration of, 665
Papillomaviruses, 842
Papovaviridae, 451, 452, 461, 842
Parainfluenza virus, 436, 831–832
Paramecium, 894
Paramyxoviridae, 451, 452, 831–833
Parasite, 262, 334, 894
    obligate, 103, 392
Parasitism, 552–553, 894
Parasporal body, 821
Parenteral, 894
Parvoviridae, 451, 452, 842–843
Paschen bodies, 459, 894
Passive diffusion, 197, 198, 199
Passive immunity, 894
Pasteurellaceae, 274–275
Pasteurization, 23, 478, 630–631, 894
    radiation, 634–635
Pathogen(s), 262, 894
    airborne transmission of, 769–770, 771–773
    arthropod-borne transmission of, 779, 780–781, 782–785
    blood transmission of, 778
    direct contact transmission of, 777–779

Pathogen(s) (Cont.):
    foodborne transmission of, 774–777
    waterborne transmission of, 774
Pathogenic, 894
Pathogenicity, 40–41, 688
Pathways, chemical, 152
PCP (pentachlorophenol), degradation of, 566
Pea enation mosaic virus group, 453
Pebrine, 24, 25, 894
Pedoviridae, 422
Pellicle, 396, 894
Penicillin, 517–519, 894
    discovery and development of, 468, 514
    production of, 658–659, 660
    structure of, 518
Pentachlorophenol (PCP), degradation of, 566
Penton, 419
Pentose, 894
Pentose phosphate pathway, 180–182
Pepsin, 894
Peptidase, 894
Peptide, 894
Peptidoglycan (murein), 85, 892, 894
    in Gram-positive filamentous bacteria, 321
    structure of, 203–204, 205
    synthesis of, 206
Peptidoglycan precursor, activation of, 204, 205
Peptone, 894
Peptonization, 894
Per os, 895
Perfect fungi, 894
Periphytes, 573, 894–895
Periplasmic flagella, 80, 895
Periplasmic space, 87, 895
Periplast, 371, 895
Peristalsis, 895
Peristome, 396
Perithecium (ascocarp). 343, 895
Peritrichous, 895
Permeability, 895
Permease, 199, 895
Peroxidase, 110, 895
Pertussis (whooping cough), 795–796
    reported cases of, by age, 767
Pesticides, degradation of, 566–567
Petroff-Hausser counting chamber, 125
Petroleum microbiology, 663–664
pH, 895
    of aquatic environment, 576
    and bacterial growth, 112–113
    and enzyme activity, 159
    and enzyme formation, 161
    and protozoa, 390
Phaeophycophyta, 373, 377
Phage(s) [see Bacteriophage(s)]
Phage-typing, 768
Phagocyte, 895
Phagocytosis, 28, 29, 709–714
    mechanism of, 711–712, 713–714
Phagolysosome, 712–713
Phagosome, 712
Phagovar, 768, 895
Pharyngitis, streptococcal, 808–809
Phase-contrast microscopy, 56–57, 64
PHB (poly-β-hydroxybutyrate) granules, 93, 896
Phenethicillin, 518, 519
Phenetic, 895
Phenol, 491, 492, 493, 507
    microbicidal action of, 492
Phenol-coefficient technique, 491, 505, 507, 895
Phenotype, 228–229, 895
Phenotypic changes due to environmental alterations, 229–230
Phosphatase, 895
Phosphatase test, 631, 895
Phosphodiester linkages in DNA, 210
Phosphoenolpyruvate-dependent sugar-phosphotransferase system, 199
Phosphoglycerides, 895
Phospholipids in cytoplasmic membranes, 89, 90
Phosphoric acid, 210
Phosphorus, requirement for, 100